2nd International Scientific and Practical Conference on Energy-Optimal Technologies, Logistics and Safety on Transport (EOT-2019)

MATEC Web of Conferences Volume 294 (2019)

Lviv, Ukraine
19 – 20 September 2019

Editors:

Oleh Voznyak
Yaroslav Bolzhelarskyi

ISBN: 978-1-5108-9775-5

Printed from e-media with permission by:

Curran Associates, Inc.
57 Morehouse Lane
Red Hook, NY 12571

CURRAN ASSOCIATES INC.
proceedings
.com

Some format issues inherent in the e-media version may also appear in this print version.

For additional information, please contact EDP Sciences – Web of Conferences at the address below.

EDP Sciences – Web of Conferences
17, Avenue du Hoggar
Parc d'Activité de Courtabœuf
BP 112
F-91944 Les Ulis Cedex A
France

Phone: +33 (0) 1 69 18 75 75
Fax: +33 (0) 1 69 28 84 91

contact-edps@webofconferences.org

Additional copies of this publication are available from:

Curran Associates, Inc.
57 Morehouse Lane
Red Hook, NY 12571 USA
Phone: 845-758-0400
Fax: 845-758-2633
Email: curran@proceedings.com
Web: www.proceedings.com

TABLE OF CONTENTS

DNIPRO NATIONAL UNIVERSITY
OF RAILWAY TRANSPORT NAMED AFTER ACADEMICIAN V. LAZARYAN

RAILWAY RESEARCH INSTITUTE (WARSAW)

FACULTY OF TRANSPORT OF WARSAW UNIVERSITY OF TECHNOLOGY

NON-GOVERNMENTAL ORGANIZATION "LINIA102.UA"

WESTERN CENTER OF THE UKRAINIAN BRANCH OF THE «INTERNATIONAL CENTER FOR SCIENTIFIC CULTURE - WORLD LABORATORY»

2-nd International Scientific and Practical Conference

"ENERGY-OPTIMAL TECHNOLOGIES, LOGISTIC AND SAFETY ON TRANSPORT" (EOT-2019)

September 19-20, 2019

Lviv, Ukraine

SCIENTIFIC COMMITTEE OF THE CONFERENCE

ORGANIZING COMMITTEE OF THE CONFERENCE

Conference languages: English, Ukrainian and Polish.

Proceedings editors:

Oleh Voznyak, Chairmen of NGO «Linia102.ua» board, PhD., asoc. prof., Lviv branch of Dnipro National University of Railway Transport named after Academician V. Lazaryan

Yaroslav Bolzhelarskyi, Deputy Chairman of the Conference Organizing Committee, PhD., asoc. prof., Director of Lviv branch of Dnipro National University of Railway Transport named after Academician V. Lazaryan

Coference website: http://eotconf.diit.edu.ua/

RAMADA

TÜVRheinland
Precisely Right.

RAMADA
Хто ми

PATRONAGE

УЗ

REFORM SUPPORT TEAM | MINISTRY OF INFRASTRUCTURE OF UKRAINE

GENERAL PARTNER

TÜVRheinland®

Precisely Right.

MAIN PARTNERS

amadeusmarine

Київ-Дніпровське МППЗТ

LSW

Petrosoft.pl®

Technologie Informatyczne Sp. z o.o.

PARTNERS

TABOR DĘBICA

МТРЗ

ЛЛРЗ

Консорціум "Трансмагістраль"

ДНІПРО ЦЕНТРАНС

ПРОЗОРІСТЬ ТА ЕФЕКТИВНІСТЬ У ЗАЛІЗНИЧНІЙ ГАЛУЗІ

ТМ «MAXI GROUP»!

ВО ГАБІОНИ ЗАХІД УКРАЇНА

ЦЕНТР BSC Безпеки Бізнесу

karpaty travel

HOTEL PARTNER

RAMADA

BY WYNDHAM

LVIV

Resolution of the 2-nd International Scientific and Practical Conference «Energy-optimal technologies, logistics and safety on transport»

Venue: Lviv, Hotel "Ramada-Lviv"
Date: 19-20 September 2019

Conference co-organizers were:
- Dnipro National University of Railway Transport named after academician V. Lazarian;
- Faculty of Transport of the Warsaw University of Technology;
- Railway Institute (Warsaw);
- Non-Governmental organization "Linia102.Ua";
- Western Center of the Ukrainian Branch of the «International Center for Scientific Culture - World Laboratory» (Lviv).

The conference took place under the patronage of the Reform Support Team of the Ministry of Infrastructure of Ukraine, Lviv Regional State Administration and JSC «Ukrainian Railway».

General partner of the conference was TÜV Rheinland Poland – the leading certification body acting on the market of certification and research in Poland.

Main partners of the conference included:
- Amadeus Marine company;
- PJSC «Kyiv-Dniprovsk Inter-Branch Enterprise of Industrial Railway Transport»;
- Petrosoft.pl Information Technology sp. z o.o.;
- Warsaw School of Applied Linguistics;

Partners of the conference included: TABOR Dębica Sp. z o.o., «Nikolaev Locomotive Repair Plant Ltd» , Lviv Locomotive Repair Plant, SPA «Dniprotechtrans», Consortium «Transmagistral», PA «GabionyZahid Ukraina», «Technoekotreiding Ltd», travel company «Karpaty Travel», PE «Business Security Center».

The following participants were involved in the work of International Scientific and Practical Conference «Energy-optimal technologies, logistics and safety on transport»:
- delegation of Lviv Regional State Administration headed by the First Vice-Chairman Vasyl Lozynskyi;
- delegation of Lviv City Council headed by the Deputy Director of the Department of Housing and Infrastructure Andrii Bilyi;
- delegation of the Ministry of Infrastructure of Ukraine;
- delegation of JSC «Ukrainian Railway»;
- delegation of ADIF company – the operator of railway infrastructure of Spain;
- Mr. John Orrison, infrastructure advisor, U.S. Treasury;
- Prof. Dr. Andrea Guiricin, CEO, TRA consulting.

The scientific committee of the conference received altogether 79 scientific works, 75 of which, on the basis of the results of an independent anonymous peer review by two reviewers, are recommended for publication in the scholarly edition Matec Web of Conference and 4 scientific works are recommended for publication in the Conference digest. Additionally, 9 abstracts were published in the Conference digest.

The results of the research in the following scientific directions were presented at the conference:
- energy-optimized technologies, energy efficiency and energy management on transport;
- ecology of rolling stock and objects of transport infrastructure;
- interoperability, safety and certification on transport;
- modern technologies of organization of international transportations and logistics;
- transport design;
- interaction of railways and industrial enterprises.

During the work of the conference 89 reports were heard. Scientific achievements in the work of the conference were covered by the representatives of the following scientific organizations and educational establishments:

from Belarus:
- Belarusian State University of Transport;

from Spain:
- ADIF company ;

from Italy:
- TRA consulting;

from the Republic of Poland:
- Warsaw University of Technology;
- Railway Institute;
- Rail Vehicles Institute „TABOR";
- Poznan University of Technology;
- WSB University in Gdansk;
- University of Economics in Katowice;

from the Russian federation:
- Ural State University of Railway Transport;

from the Slovak Republic:
- University of Žilina;

from the USA:
- U.S. Treasury;
- Geo Lab Drilling;
- Self-employed scientists.

from Ukraine:
- Dnipro National University of Railway Transport named after academician V. Lazarian;
- State Entreprise "State Road Transport Research Institute" ;
- Oles Honchar Dnipro National University;
- Institute of Geotechnical Mechanics named by N. Poljakov of National Academy of Sciences of Ukraine;
- Lviv State University of Life Safety;
- Lviv Research Institute of Forensic Science;
- National Aviation University;
- National Technical University of Ukraine "Igor Sikorsky Kyiv Polytechnic Institute";
- National Technical University «Kharkiv Polytechnic Institute»;
- National Transport University;
- Lviv Polytechnic National University;
- Odessa National Maritime University
- Pryazovskyi State Technical University;
- Prydniprovska State Academy of Civil Engineering and Architecture;
- Volodymyr Dahl East Ukrainian National University;
- Ukrainian State University of Railway Transport;
- Kharkiv National Automobile and Road University;
- Kharkiv Petro Vasylenko National Technical University of Agriculture;

from the Federal Republic of Germany:
- Technical University of Dresden;
- AEbt Applied Railway Technology;

from the Czech Republic:
- Leo Express Global a.s.;
- "SID International" company.

During the conference, the first Vice-Chairman of the Lviv Regional State Administration Vasyl Lozynskyi noted that it is extremely important for the country to have partners who are ready to work together to develop transport infrastructure. Further important issues include the provision of the scientific personnel of the branch which is successfully implemented by Dnipro National University of Railway Transport named after academician V. Lazarian, the issues of passenger transportation safety, their efficiency and speed. He also expressed the hope that the conference will become a powerful communication platform for discussing urgent issues of the transport industry and developing important contacts.

Deputy Executive Director of the regional branch office «Lviv Railway» for rolling stock and logistics Ihor Batiushyn: "The topic of this conference is directly related to the main tasks that are solved by our regional branch office and other regional branch offices of JSC «Ukrainian Railway». The tasks are to guarantee the necessary level of traffic safety, the speed of transportation of passengers and cargo, saving energy, creating a model of effective investigation of transport cases. The reports presented at the conference have many positive points regarding improving the investigation of transport events and avoiding them in the future. Such conferences are useful because experts have the opportunity to exchange information, obtain data on studies conducted not only in our country, but also in the EU countries and the USA, which creates the preconditions for achieving positive dynamics in transport.

Project manager of the railway sector reform of the Reforms Support Team Larisa Nazarenko outlined the main innovations of the draft law "On the railway transport of Ukraine" (registration number 1196-1), paying attention to the new tariff system, administration bodies and passenger transportations. Participants of the conference paid special attention to the discussion of implementation of Regulation No. 1370/2007 of the European Parliament and Council from 23 October, 2007 on the provision of services of passenger transportation by railway and road transport which is implemented by the draft law.

The draft law 1196-1 is the result of long-term work of European and Ukrainian experts. The rail industry needs immediate reform and the draft law should be the impetus that will create a solid legal framework for development.

It concerns such a pressing issue for the country as financing, including preferential transportation. Exactly due to the introduction of a new model of compensation for the provision of socially important transportation services on the basis of contract, the organization and financing of services for socially important passenger transportation by rail will be streamlined, in particular the quality of transportation will be controlled, the level of service and mobility of citizens will be raised, the break-even level of this transportation will be achieved. This in turn will contribute to further economic and social development of Ukraine, as well as fulfillment of international obligations of Ukraine.

Colleagues from the EU countries and from the USA shared their experience in implementing the latest and most energy-efficient technologies on transport, as well as product certification. The need to establish and maintain close links between organizations and institutions providing scientific support for transport projects is declared.

Railway infrastructure advisor of the Office of technical support of the US Department of the Treasury John Orrison presented the safety management system of the USA railways. Senior Advisor to the Transportation Industry of the World Bank Professor Andrea Guiricin presented the benefits of using intelligent transportation systems on the railways of the world, the experience of implementing high-speed rail by a private carrier, and the positive dynamics of market liberalization.

The conference participants provided recommendations on improving the level of transport safety through the use of intelligent systems of diagnostics of the technical state of transport infrastructure and information-measuring and diagnostic systems, raised pressing issues of environmental safety on transport, proposed new resource-saving technologies for implementation in the transport industry.

By results of the reports and discussions of the conference participants, the following conclusions and recommendations were obtained:

1. To recommend for print by topic in the electronic edition MATEC WEB of Conference the scientific articles which were reported by the conference participants and received two positive reviews if there is a signed agreement on publication. Also to submit for publication an article about the conference and its Decisions.
2. Due to the small number of works submitted in the scientific direction "Interaction of railways and industrial enterprises" and tendencies of development of railway transport of Europe, to combine the scientific directions "Interaction of railways and industrial enterprises" and "Modern technologies of the organization of international transportations and logistics" under one direction with the title "Modern technologies of transportation and logistics."
3. To recommend the scientific units of the conference organizers to develop areas of cooperation on the topic of the conference.
4. To recommend all participants of the conference to continue scientific and technical research in the field of transport infrastructure, international transportations, logistics, certification, transport design, increase of safety on transport taking into account world tendencies and needs of development of the transport industry.
5. To disseminate information about the conference and its digest to the relevant ministries of the EU countries.
6. To hold the next International Scientific and Practical Conference «Energy-optimal technologies, logistics and safety on transport» in October 2020 and to consider organizing it at Faculty of Transport of the Warsaw University of Technology and (or) Railway Institute.

Co-chairmans of the conference:

Rector of the Dnipro National University of Railway Transport named after academician V. Lazarian

prof. dr hab.
Oleksandr Pshinko

Dean of the Faculty of Transport of the Warsaw University of Technology

prof. dr hab. inż.
Marianna Jacyna

Director of the Railway Institute

dr. inż.
Andrzej Żurkowski

20 September, 2019
Lviv, Ukraine

edp sciences

Web of Conferences

Statement of Peer review

In submitting conference proceedings to *Web of Conferences*, the editors of the proceedings certify to the Publisher that

1. They adhere to its **Policy on Publishing Integrity** in order to safeguard good scientific practice in publishing.
2. All articles have been subjected to peer review administered by the proceedings editors.
3. Reviews have been conducted by expert referees, who have been requested to provide unbiased and constructive comments aimed, whenever possible, at improving the work.
4. Proceedings editors have taken all reasonable steps to ensure the quality of the materials they publish and their decision to accept or reject a paper for publication has been based only on the merits of the work and the relevance to the journal.

Title, date and place of the conference

2-nd International Scientific and Practical Conference
"Energy-optimal technologies, logistic and safety on transport"
(EOT-2019)
September 19-20, 2019
Lviv, Ukraine

Proceedings editor(s):

Oleh Voznyak

Yaroslav Bolzhelarskyi

Date and editor's signature

20.09.2019

2-nd International Scientific and Practical Conference "ENERGY-OPTIMAL TECHNOLOGIES, LOGISTIC AND SAFETY ON TRANSPORT" (EOT-2019) September 19-20, 2019

Preface

The International Scientific and Practical Conference "Energy-optimal technologies, logistic and safety on transport" was held in Lviv on September 19-20, 2019.

The conference is initiated by Dnipro National University of Railway Transport named after academician V. Lazaryan (DNURT), Non-Governmental Organization "Linia102.Ua" and the Western Center of the Ukrainian Branch of the «International Center for Scientific Culture - World Laboratory» (Lviv). Co-organizers of the conference also are the Railway Research Institute (Warsaw) and the Faculty of Transport of Warsaw University of Technology.

Directors and representatives of manufacturing enterprises, scientific institutions and educational institutions of transport industry of Ukraine and European Union, leading European experts in the transport industry, were invited to participate in the conference. New results of theoretical and experimental research and actual problems of science and technic development were heard and discussed during the conference in the following scientific directions:
– energy-optimized technologies, energy efficiency and energy management on transport;
– ecology of rolling stock and objects of transport infrastructure;
– interoperability, safety and certification on transport;
– modern technologies of organization of international transportations and logistics;
– transport design;
– interaction of railways and industrial enterprises.

The indicated directions are especially actual and relevant for the transport complex for both Lviv region and all EU countries.

The need for this conference arose in connection with the reform of the railway transport of Ukraine in accordance with the requirements of EU legislation. This implies a new model of the railroad market, the expected entry into the market of competitive, efficient, mobile participants - both of private and public ownership.

The format of the conference provided for the expansion of the site for establishing close links between scientists, manufacturers of products for rail transport, logistics operators and other transport companies, which had been created during the previous conference held on June 18-19, 2018.

Participiants of the conference were discussed important topics for improving the efficiency of the companies operating in the transport industry. Leading European speakers, representatives of the scientific organizations and transport companies and other participants were shared their best practices to the audience.

We hope that this scientific issue will be useful for both scientists and product manufacturers and will help to develop the transport industry.

See you at the following conferences.

Best regards,

Oleh Voznyak,
Chairmen of NGO «Linia102.ua» Board, PhD., asoc. prof., Lviv branch of DNURT

Yaroslav Bolzhelarskyi
Deputy Chairman of the Conference Organizing Committee, PhD., asoc. prof., Director of Lviv branch of DNURT

Analysis of the electric energy consumption components for mine train traction

Serhii Arpul[1], *Viktor* Artemchuk[1], *Mykola* Babyak[2], *Viacheslav* Vasilyev[1,*], *Hennadii* Hetman[1], and *Serhii* Marikutsa[1]

[1] DNURT, Department of Electric Rolling Stock of Railways, Lazariana St. 2, Dnipro, 49010, Ukraine,
[2] DNURT, Lviv branch, Department of Transport Technologies, I. Blazhkevich St. 12-a, Lviv, 79052,Ukraine

Abstract. The paper considers the issues of reducing the energy intensity of transportation at opencast mining enterprises, the relevance of which has now increased due to the rise in the cost of fuel and energy resources. It presents the study results concerning the cost structure of the electricity consumed by electric mine transport, which form the basis for the development of technical and operational measures to reduce the energy intensity of the transportation process. It is shown that the work to reduce the electricity consumption for mined rock transportation should be aimed at:
- Reduction of losses in the power circuits of the traction rolling stock due to the use of more advanced electric rolling stock and regulation of the degree of utilization of the installed traction power;
- Introduction of new contact materials for electrical circuits with the lowest possible resistivity, including for current collector plates;
- Introduction of measures to reduce energy consumption for power supply of auxiliary circuits;
- Development and implementation of rational train control techniques.
The introduction of energy-saving measures should include the development and application of effective methods for calculating individual norms of energy consumption and incentives for energy saving of the employees involved in the organization of the transportation process.

1 Introduction

Opencast mining enterprises are large consumers of electrical energy. Thus, for iron-ore mining plant, the total mine (i.e., taking into account all types of work) specific energy consumption in some cases reaches 50 kWh per 1 ton of mined ore. It is determined by the energy consumption of the entire complex of equipment that provides the technological process of production, and of all auxiliary workshops.

However, the analysis of the mine energy consumption structure shows that the most energy-intensive process is the transportation of mined rock. When used in public works, electric traction can account for 85% of the total cost of electrical energy in a mine.

The development of an open-cast mining method is accompanied by increased concentration of production, growth in the depth and spatial dimensions of the mines, transportation distances and the complexity of delivery of mined rock to the surface.

With the increase in mine depth, the distance of transportation and the corresponding costs of electricity grow.

Raising the lift height by 1 ton of mined rock using rail for every 100 m increases energy consumption by approximately 2.5 kWh. The same indicator for the rest of the mine equipment in other technological processes (excavation, drilling, drainage, lighting, etc.) is only 1.03-1.05 kW h.

It follows from the above that the development of mining enterprises that use electric traction in open-cast work will be accompanied by the increase in the transport component in transportation energy intensity. Therefore, the development of measures to save electricity in mines should include the implementation of reserves to save electricity costs for the transportation of rock with the aim of introducing energy-saving accounting technologies and reducing electricity consumption by rail, including using the advanced contact materials.

The purpose of the article is to analyse the reserves for reducing the energy intensity of rail transportation at mining enterprises that use electric traction during open-pit mining.

2 Literature review

The electricity cost recovery for train traction make up a sizeable share of the operating costs of railways, so the entire history of their electrification is inseparably linked with the search and implementation of reserves to reduce the energy intensity of transportation.

The issues of energy saving have acquired particular relevance in the last decade due to rising energy prices. The number of works devoted to saving electricity for

* Corresponding author: wasiljew@ukr.net

train traction has increased. In the field of the researchers' view there are both main links that determine the overall energy balance of electric traction - power supply devices and traction rolling stock [1 - 6]. Some researches are aimed at improving the interaction of rolling stock with infrastructure, including through the organization of rolling stock repair systems and the introduction of resource-saving technologies and modern electrical circuit materials [7 - 15]. A number of new studies are devoted to the improvement of methods for rationing electricity consumption for train traction [16, 17, 18].

The problem of saving train traction electricity in relation to main railway lines has been studied for almost the entire history of the railway electrification. As a result of the introduction of various energy-saving measures, the expansion of the electric traction range was accompanied by a significant reduction in the energy intensity of transportation. So, during the period 1950-1975, at the former USSR railways, the specific electric power consumption for train traction in all types of traffic decreased from 218 to 125 kWh/10^4t/km gross, i.e., by about 42% (see Fig. 1).

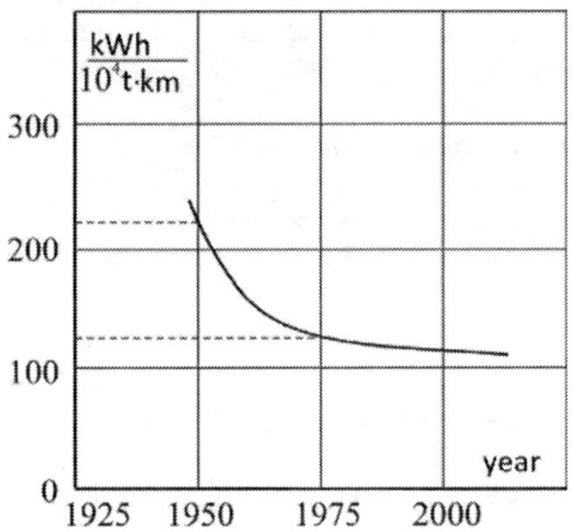

Fig.1. Specific energy consumption for train traction at the USSR railways

The energy intensity of mine electric transport is determined by a significant number of factors [19, 20, 21] (characteristics of the traction power supply system and the operating rolling stock, track gradient and layout, travel routes, traffic volume, train driving modes used, train crew qualifications, condition of the track superstructure, seasonality, organization of work of the traffic control service, etc.). The development of energy-saving measures and an effective method of rationing the energy consumption for traction in relation to the conditions of a particular enterprise should be based on the results of an impact analysis of these factors. First of all, it is advisable to present the total energy consumption in the form of components corresponding to the work of traction motors in overcoming the motion resistance forces differing in the nature of their origin. In addition, it is necessary to obtain a comparative assessment of the energy intensity of individual phases of the transportation process (loading, weighing, movement with cargo, etc.).

The main purpose of this analysis is to determine:
– Importance of the components (share) in the total consumption of electricity for traction;
– Degree of influence of individual factors on the energy intensity of transportation.

As the analysis of literary sources shows, the issues of reducing the energy intensity of transportation by open-cast railway transport have not been studied fully to date. In particular, they lack studies of the structure of the component costs for the transportation of the mined rock and the factors that determine it, which are necessary for the definition and realization of energy saving reserves. This paper attempts to fill this gap.

3 Subject matter

From a practical point of view, it is of interest to analyze the components of electricity consumption for:
– Traction (overcoming the motion resistance forces, covering losses in the brakes and in the electric circuits of the locomotive, power supply of auxiliary circuits);
– Performance of individual operations of the transportation process (loading, unloading, movement with cargo, movement into the pit, weighing);
– Power supply of auxiliary circuits (fan and compressor engines, heating and air conditioning devices, etc.).

The energy balance equation for a journey can be represented as

$$A = A_T + A_{AU} \tag{1}$$

and

$$A_T = A_r + A_i + A_{w_M} \pm A_P + A_{BE} + A_{BR} + \Delta A, \tag{2}$$

where A_T – electricity consumed for traction (work of traction engines and losses in power circuits); A_{AU} – electric power consumption for auxiliary power supply; A_r, A_i, and A_{w_M} – work of the motion resistance forces, respectively, for the curves and track gradients and the main one; A_P – work of forces to change the potential energy; A_{BE} – energy consumption during independent braking on excessive gradients; A_{BR} – energy loss in braking devices (excluding A_{BE}); ΔA – energy losses in electric traction circuits of an electric locomotive in traction mode.

To determine the electric energy consumption for traction and auxiliary services supply, we used the electronic kilowatt-hour meters of the type SKVT-F604, designed for electricity metering on the electric rolling stock of railways and public transport. The remaining components of the energy consumption were determined by calculation according to the method described in [22].

Based on the analysis of experimental and calculated data, it was established that the share of A_T in the total energy consumption is 88–92%. When the elevation difference, where the loading and unloading points are located, is 100 m or more, the largest share in the electricity consumption for the traction A_T is attributed to that associated with overcoming the track gradient motion resistance. Further, in descending proportion, there follow the consumption for overcoming the forces of resistance to train movement (main, from curves and additional motion), energy loss on excessive gradients and in power circuits, and auxiliary consumption. The relative share of the electricity consumption components for mine trains can be viewed from the data of Table 1.

Table 1. Components of energy consumption for mine train traction (numerator - in kWh, denominator – in % of total consumption)

Energy consumption	Travel direction		
	Loaded run	Empty run	Per Cycle
Electricity consumed for traction A_T	1730 / 100	190 / 100	1920 / 100
Electric power consumption for auxiliary power supply A_{AU}	113 / 6.53	7 / 1.1	120 / 6.25
Work of forces to change the potential energy A_P	1076 / 62.2	–442 / –70.0	634 / 33.02
Work the main one of the motion resistance forces A_{W_M}	276 / 15.95	136 / 21.5	412 / 21.46
Energy to overcome the additional resistance when the train is moving with cars forward A_{W_A}	32 / 1.85	5.5 / 0.9	37.5 / 1.95
Work of the motion resistance forces track gradients A_r	60 / 3.47	25 / 4	85 / 4.43
Energy consumption during independent braking on excessive gradients A_{BE}	22 / 1.27	25 / 4	47 / 2.45
Energy loss in braking devices (excluding A_{BE}) A_{BR}	– / –	370 / 58.5	370 / 19.27
Energy losses in electric traction circuits of an electric locomotive in traction mode ΔA	151 / 8.73	63,5 / 10	214.5 / 11.17

They correspond to the average values determined using the above expressions based on the results of three experimental journeys on a full cycle of traffic on the 12.35 km route with the end point height difference of 224.4 m. The experiments were conducted with a PE2[M] traction unit and a train of 9 cars 2VS-105 (train unladen weight M = 730 t; estimated weight of loaded train m = 1.730 t). In Table 1 A_{W_A} – energy to overcome the additional resistance when the train is moving with cars forward.

The energy consumption values averaged by a number of experimental journeys by travel direction in the form of histograms are shown in Fig. 2-4.

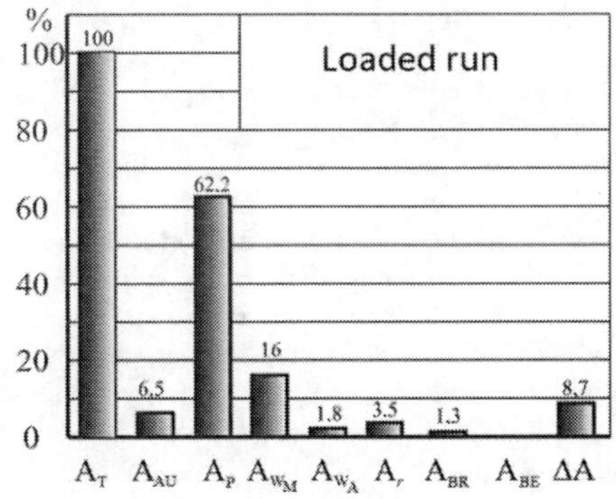

Fig. 2. Distribution of loaded run energy costs by the types of work

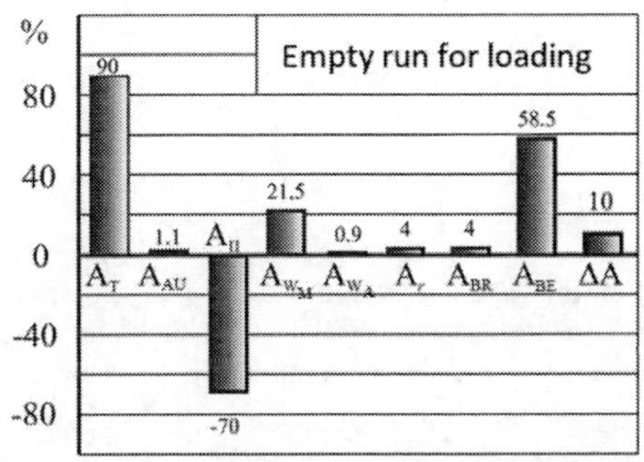

Fig. 3. Distribution of empty run energy costs by the types of work

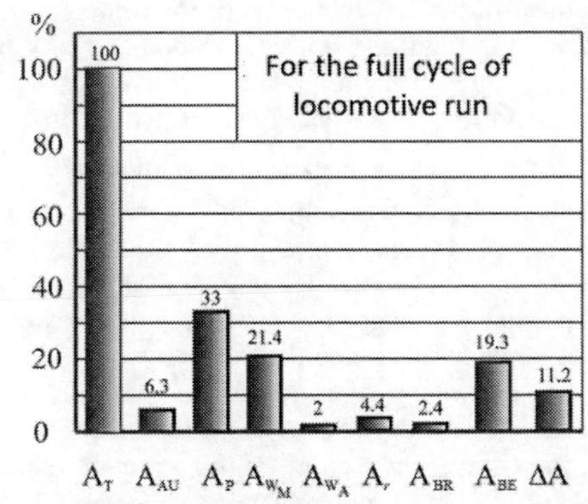

Fig. 4. Distribution of energy costs by the types of work for the full cycle of the locomotive run

Analysis of the data and calculated expressions shows that the total energy consumption and the proportion of its components depends mainly on the characteristics of the route (elevation of the end points; characteristics of the tracks in the places of the train stops and the number of the latter; length of sections with the train moving head forward and cars forward; the number and parameters of the track curves; the share of the length of fixed and movable tracks; track gradient in the places of loading and unloading). In the case under consideration, the largest share in the power consumption falls on the consumption A_P, associated with overcoming the track gradient motion resistance. Further, in descending proportion, there follow the consumption for overcoming the forces of resistance to train movement (main, from curves and additional motion), - 30.84%, energy losses on excessive gradients - 19.27%, in start circuits - 11.17%, and auxiliary consumption - 6.25%.

The main factors determining the structure of electricity consumption in terms of components and energy intensity of transportation should include primarily the rectified gradient of the route. The increase in the height difference at which the loading and unloading points are located, under otherwise equal conditions, leads to higher energy intensity of transportation and growth in the proportion of work for changing the potential energy of a train A_P.

Reducing the track gradient on the route is accompanied by a decrease in the proportion of the component A_P and an increase in the share of other components.

The level of the average motion speed has a certain effect. With its reduction, the component A_{w_M} decreases slightly and the auxiliary consumption increases significantly (both in absolute value and in relative share).

For this reason, in some cases (for example, when using traction units PE2^M), the absolute value and relative share of the components A_P, ΔA and A_{AU} depends on the level of voltage used in the power supply system of the traction rolling stock.

From the above data, it can be concluded that with current schemes for transporting the mined rock, regardless of the characteristics of the route gradient, the reserves for reducing power consumption for traction should be sought in decreased auxiliary consumption A_{AU} and energy losses in electric rolling stock ΔA, as well as on excessive gradients A_{BE}.

The structure of the considered energy consumption depends on the characteristics of the route: with an increase in the route length, the share of energy consumption for loaded run increases, and the proportion of the remaining components decreases. With current schemes for transporting the mined rock, regardless of the characteristics of the route gradient, the reserves for reducing power consumption for traction should be sought in decreased power for train movement A_T and auxiliary circuit consumption A_{AU}, reduced energy loss in electric rolling stock ΔA, while stopping and slowing down on excessive gradients.

The validity of the conclusions made above is confirmed by the results of the analysis of the structure of the operational power consumption and the energy components for auxiliary circuit power supply.

The peculiarity of work of electric transport of the open-cast mining enterprises is its cyclical nature. A typical transportation cycle includes:

- Movement to the open cut;
- Loading (of mineral product or overburden rock);
- Weighing (usually only of trains carrying mineral products);
- Movement with load (to the crushing plant or to dumps);
- Unloading of mined rock.

Electricity consumption for traction per cycle is as follows

$$A_T = A_{ROC} + A_{LD} + A_{LR} + A_{UNL} + A_{TW}, \qquad (3)$$

where the components of the right-hand side represent respectively the electricity consumption for:
A_{ROC} – train run to open-cut; A_{LD} – loading of mined rock; A_{LR} – loaded run; A_{UNL} – unloading of mined rock; A_{TW} – train weighting.

According to the results of the experimental journeys in the open-cast mines of UGOK and SSGOK by traction units PE2M, we obtained data on the relative share of the above-mentioned electricity consumption by operation.

Their values averaged by a number of experimental journeys in the histogram form are shown in Fig. 5.

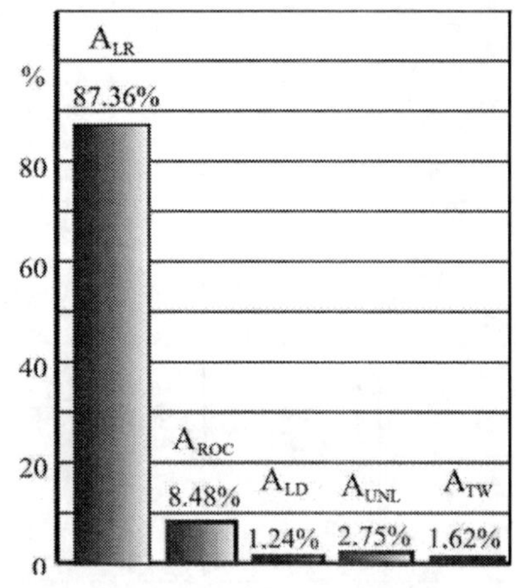

Fig. 5. Electricity consumption by operation for mine trains

It is easy to see that most energy is spent for the loaded run and to open-cut, so when collecting the statistical material necessary for rationing electricity consumption and developing measures to reduce the energy intensity of transportation, particular attention should be paid to choosing rational train control methods during the period of movement, since in this case, even a small percentage reduction in energy consumption can give a significant change in its total consumption.

The design engine power on the ruling gradient corresponds to the conditions of engine overload. The overload duration varies widely and depends on the specific gradient of the railway tracks. In the practice of operating electric locomotives, the inertia of the traction motor is always used in order to briefly generate power on the most difficult sections of the track when starting from a standstill. This is possible only if the temperature of the engine at the time of the overload is lower than allowed by the norms. At the end of the power boost period, the temperature and its overheating should not exceed the values allowed for this insulation class.

It was noted above that in the total electricity consumption, a significant share is attributed to the work of auxiliary equipment: compressor engines and fans of traction engines, control devices, lighting, heating, etc.

As applied to the main transport, the issues of saving energy consumption for powering the auxiliary circuits of electric rolling stock were investigated by a number of authors [22-25]. For mine electric transport this issue was not covered in the technical literature known to the authors.

At the same time, the search for savings reserves requires the study of the energy intensity of individual elements of auxiliary circuits, as well as an assessment of the amount of auxiliary energy expenditures during all phases of the transportation cycle. To solve this issue during the experimental journeys the necessary measurements were taken.

At the PE2M traction unit, the total power of auxiliary machines that are powered from the contact network is 390 kW:

1. Compressor drive engine KCT-6 - two engines of type TL110 (53.1 kW; 20.6 A; 3 kV).
2. Engines of type NB101 for driving centrifugal cooling fans of traction motors of electric locomotive and motor dump cars (six electric motors with a capacity of 41 kW each; 36 A; 1.5 kV) or DT-51 engines, with parameters and design identical to those of NB-101.
3. DK-604V type dynamotor-generator for supplying control circuits with 50 V voltage and recharging the battery (2x6 kW; 5.3 A; 2x1.5 kV).
4. Electric motors of type ETV-20 for driving axial cooling fans of the brake-start resistors (six motors with a power of 13 kW each; 200 V; 80 A).

Based on the foregoing, the auxiliary power consumption can be represented as

$$A_{AU} = A_{FE} + A_{CE} + A_{GE} + A_{CH}, \quad (4)$$

where the components of the right-hand side represent the electricity consumption for the power supply of the respective circuits: A_{FE} – fan engines; A_{CE} – compressor engines; A_{GE} – generator engines; A_{CH} – driver's cab heating.

The values of the components of auxiliary energy consumption, obtained experimentally, are given in the form of diagrams in Fig. 6...10, where the energy consumption is indicated respectively: FE - fan engines, CE - compressor engines, GE - motor-generators, CH - for driver's cab heating. It can be seen that during the loading (fig. 6) and unloading (fig. 7) period, the main part of the power consumption A_{AU} is due to the operation of the compressor engines (62-85%).

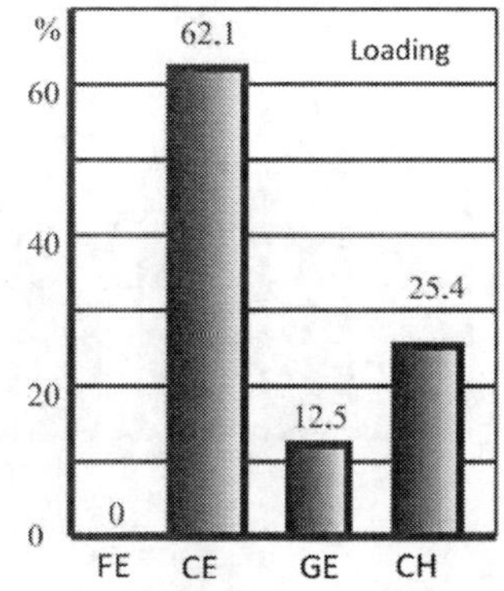

Fig. 6. Auxiliary electricity consumption during loading

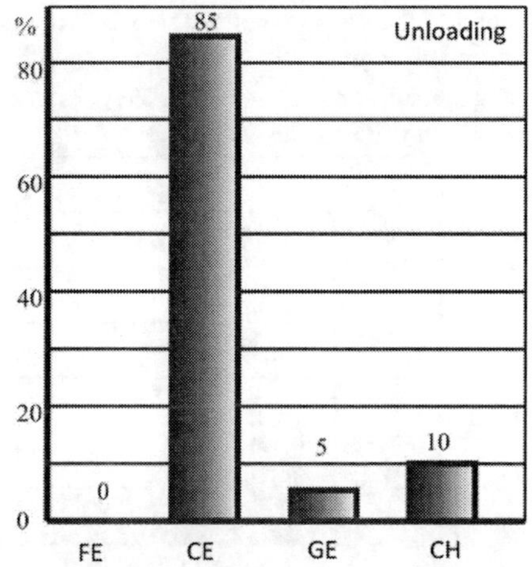

Fig. 7. Auxiliary electricity consumption during unloading

When running with a load (Fig. 8), the auxiliary consumption is mainly due to the operation of the fan engines of the traction motors (83.6%).

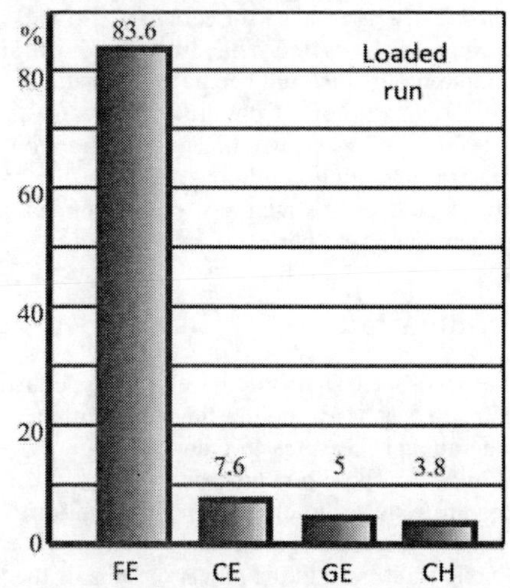

Fig. 8. Auxiliary electricity consumption during loaded run

When running to open-cut (Fig. 9), 28.5% of A_{AU} are due to the operation of the generator engines of control circuit, while the share of the compressor engines accounts for 44.3% of the energy consumption.

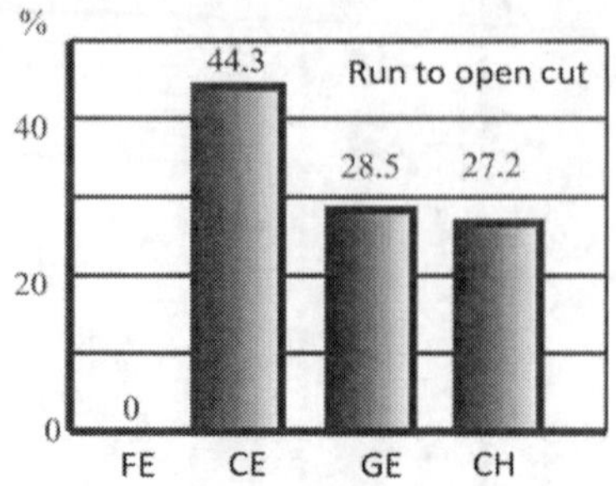

Fig. 9. Auxiliary electricity consumption during run to open-cut

For the full cycle of running (Fig. 10), more than half of the auxiliary energy consumption is due to the ventilation of the traction motors (59.3%). The operation of compressor engines takes 25.4% of A_{AU}, the power of the generator engine - 7.4%. The heating of the driver's cab in winter is 7.9% of A_{AU}.

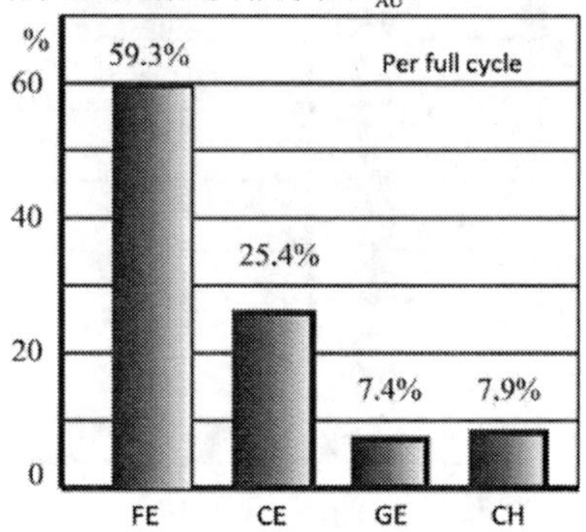

Fig. 10. Auxiliary electricity consumption structure per cycle

The above data correspond to the usual mode of operation of auxiliary machines in mine transport, which involves the switching on of traction motor fans only for the loaded run period. Consequently, in the case of longer operation of the fans, the share of auxiliary consumption in the total energy consumption will increase. It also follows from the above that if there is a reserve for heating the windings of the traction motors for specific operating conditions, then in this case it is possible to reduce the auxiliary power consumption by using the reduced fan performance mode.

4 Conclusions

1. The performed study of the electricity consumption structure is the basis for the development of technical and operational measures to reduce the energy intensity of the mine transportation process.
2. Work to reduce the electricity consumption for mine transportation should be aimed at:
- Reduction of losses in the power circuits of the traction rolling stock due to the use of more advanced electric rolling stock and regulation of the degree of utilization of the installed traction power;
- Introduction of measures to reduce energy consumption for power supply of auxiliary circuits;
- Development and implementation of rational train control techniques.
3. The introduction of energy-saving measures should include the development and application of effective methods for calculating individual norms of energy consumption and incentives for energy saving of the employees involved in the organization of the transportation process (first of all, train crews);
4. It is advisable to perform a comparative analysis of the energy consumption of the transportation process depending on the type of current and voltage level in the traction power supply system.
5. The results of the analysis of the components of electrical energy consumption for mine train traction given in this article can be used in the development of measures to reduce the energy intensity of transportation at open-cast mining enterprises.

References

1. S. M. Boyko Electrification of Transport. *On the development of the power supply system of mining enterprises using renewable sources of electricity.* **14**, 7-14 (2017) [in Russian]
2. S. P. Denisyuk Electrification of Transport. *Increasing the energy efficiency of rail transport enterprises based on the technological approach.* **14**. 78-85 (2017) [in Russian].
3. V. G. Sychenko Quality of electric energy in traction networks of electrified railways. (2015) [in Ukrainian]
4. O. N. Sinchuk Electric Transport. *Small commentary on the tactics of creating traction electromechanical complexes of combined mine electric locomotives.* **14**, 42-55 (2017) [in Russian]
5. I. B. Malyshko Bulletin of Dnipro Nat. Un-ty of Railway Transport n. a. Acad. V. Lazaryan. *Main directions of energy saving on the railway transport of Ukraine.* **13**. 36-38 (2006). [in Ukrainian]
6. O. I. Sablin, D. O. Bosyi, V. G. Kuznetsov, M. O. Babyak, Ye. M. Kosarev, P. V. Gubsky. Visnyk of Vinnytsia Polytechnical Institute. *Efficiency of electric energy recuperation in the electric transport system with inverter DC traction stations.* **2**, 72-78. (2016) [in Ukrainian]
7. M. O. Babyak. Visnik of the East Ukr. Nat. Un-ty. *Resource-saving technology for the use of current collector plates, taking into account their interaction with the contact wire.* **2**. 32-37 (2018) [in Ukrainian]
8. Babyak M, Horobets V, Sychenko V, Horobets Y. Eastern-European Journal of Enterprise Technologies. *Comparative tests of contact elements at current collectors in order to comprehensively*

assess their operational performance **12,** 13-21 (2018).

9. M. O. Babyak, V. L. Horobets, V. V. Artemchuk. Electrical Contacts and Electrodes. *Study of physical mechanical properties of pantograph plates used as current-moving elements of electric rolling stock.* - K.: IPM NAS of Ukraine. 89-100 (2016). [in Ukrainian]
10. Kalivoda, J., & Neduzha, L. *Simulation of Safety Against Derailment Tests of an Electric Locomotive.* Proc. of 25th Intern. Conf. "Engineering Mechanics 2019", 177-180 (2019). doi: 10.21495/71-0-177
11. Kyryl'chuk, O., Kalivoda, J., Neduzha, L. *High Speed Stability of a Railway Vehicle Equipped with Independently Rotating Wheels*. Proc. of 24th Intern. Conf. "Engineering Mechanics 2018", 473-476. (2018). doi: 10.21495/91-8-473.
12. Shydlovskui R. M., Babyak M. O., Artemchuk V. V. Visnik of the East Ukr. Nat. Un-ty. *Analysis of the possibilities of increasing the operational characteristics of the mechanical part of electric locomotives.* **1**, 240-244 (2016). [in Ukrainian]
13. Shydlovskui R. M., Babyak M. O., Artemchuk V. V. Electrification of Transport. *Modern state of reliability of the elements of the mechanical part of freight electric locomotives*. **12,** 92-97 (2016) [in Ukrainian]
14. L. Bondarenko, M. Babyak, S. Yakovlev, S. Istin, G. Moskalev. Science and Transport Progress. *Relationship Between Rolling and Slip Resistance in Rolling Bearings.* **3.** 161-170 (2016) - doi: 10.15802/stp2016/74760
15. S. V. Myamlin, V. L. Horobets, O. L. Yangulova, M. O. Babyak. Bulletin of Certification of Railway Transport. *About rational organization of the rolling stock repair system.* **10**, 58-60 (2016) [in Russian]
16. V. I. Mozgovoi. The World of Technics and Technology. *Electricity - consumption and accounting. Modern methods: Rational use and accounting of energy resources. Application of automated electric energy accounting system.* **11**, 56-58 (2005) [in Russian]
17. M. P. Silich. Industrial power engineering. *Energy Saving: Assessment of energy saving level in the region. Saving energy resources.* **3**, 2-6 (2006) [in Russian]
18. V. S Dziuban. Energy Saving. *Energy resource saving is a priority task of Ukrainian enterprises. Reduction of energy consumption by mining enterprises.* **11**, 10-11 (2006) [in Russian]
19. Yu. I. Lel Innovative transport. *Power engineering of mine transport.* **1.** 34-39 (2011) [in Russian]
20. S. M. Bro, A. S. Prygunov, S. A. Shypunov. Coll. of Res. Papers of the Nat. Mining Un-ty. *Comparative analysis of energy consumption by various types of mine transport.* **38**. 84-89 (2012). [in Russian]
21. Yu. I. Anistratov. *Method of determining the energy efficiency of technologies and mechanization of open-pit mining operations* (2011). [in Russian]
22. H. K. Hetman, V. E. Vasiliev. Bulletin of Dnipro Nat. Un-ty of Railway Transport n. a. Acad. V. Lazaryan. *Analysis of electric power consumption for mine train traction.* **36**. 70-74 (2011) [in Russian]
23. V. G. Kuznetsov, M. O. Ivanov, O. O. Matusevich. Bulletin of Dnipro Nat. Un-ty of Railway Transport n. a. Acad. V. Lazaryan. *Rationing of auxiliary electric power consumption for sectionalizing and parallel dc connection points.* **25**. 30-33 (2008) [in Ukrainian]
24. V. A. Gapanovich. *Energy saving on rail transport.* (2012). [in Russian]
25. H. K. Hetman, V. E. Vasilyev. Electrification of Transport. *Once more about determination of the power consumption for traction by means of partial disconnection of traction motors of the electric rolling stock* **5**, 33-40 (2013) [in Russian]

Comparing hydraulic and electromechanical drives by electric motor's power

Otto Grigorov[1*], *Galyna* Anishchenko[1], *Evhen* Druzhynin[1], *Vsevolod* Strizhak[1], *Mariana* Strizhak[1], and *Volodymyr* Malashchenko[2]

[1] National Technical University "Kharkiv Polytechnic Institute", 61000 Kharkiv, Ukraine
[2] National University "Lviv Polytechnic ", 79000 Lviv, Ukraine

The authors have made the analysis of energy consumption by positive-displacement hydraulic drive and asynchronous motor with phase rotor in the course of crane mechanisms operation at all stages of motion: acceleration, movement at nominal or intermediate speed, and deceleration. There have been determined and compared equivalent capacities per work cycle for electric and hydraulic drives of crane mechanisms depending on operating conditions group and operation time. There has been justified the decrease of rated capacity of electric motor of crane mechanisms' hydraulic drive as compared with the case of applying asynchronous motor with phased rotor for the same mechanisms.

1. Introduction

The price of a hydraulic drive's electric motor, with other conditions considered, is determined by its capacity, which in the long run affects capital expenditure on hoisting machine [1, 3–6]. That is why it is reasonable not also to determine electric drive's capacity, but also to compare it with that of electromechanical drive [2].

2. Scientific novelty

The scientific novelty of this work lies in the fact that for the first time an analytical expression was obtained to determine the equivalent power of the electric motor drive, taking into account the linear change in time of the power that is consumed by the engine during acceleration and generator braking. The validity of the obtained expressions was confirmed by tests on real bridge cranes of the Kharkov "Turboatom" factory with a lifting capacity of 30/5 tons, as well as on a bridge crane of the "Uzlovsky" machine-building factory with a lifting capacity of 15/3 tons and on the "Kirovets" frame-crane with a lifting capacity of 10 tons installed in the Kiev port.

3. Practical significance

The practical significance of this study lies in the fact that during field experiments on the bench (fig. 1), as well as on the above-mentioned real cranes, it was possible to reduce the equivalent power of the drive motor by 1.4 ÷ 1.6 times for any groups of operating modes of the movement (rotation) mechanism of the crane. So, for example, it was possible to reduce the power of the mechanism for moving the bridge crane with a lifting capacity of 15/3 tons at the "Uzlovsky" machine-building factory from 11 kW to 7 kW, and the turning mechanism of the hydrostatic drive of the "Kirovets" frame-crane with a lifting capacity of 10 tons from 55 kW to 40 kW. Note that in addition to reducing the equivalent power, there is a decrease in the cost of electricity for the completion of the crane operation cycle [7]. From the graphs given in [7] it is clearly seen that the ratio of energy costs in an electromechanical drive in relation to energy costs in an adjustable hydraulic actuator lies in the range from 1.5 to 3.1 in the absence of uniform crane movement. This ratio reaches 1 at such values of the time of uniform movement of the crane (from 28 to 75 seconds), which are not found in real operating conditions of the crane. The above data follows from the diagram of changes in power consumption of two compared drives (7, fig. 2). A decrease in dynamic loads on the metal structures of the crane by about 40% was also noted [8], [9].

4. Aim and problem statement

The aim of this work is to substantiate the decrease of rating capacity of drive electric motor in crane mechanisms' hydraulic drives as compared with the case of using asynchronous motor with phase rotor for the same mechanisms. To achieve the set aim it is necessary to solve a number of tasks: to analyze the character of power consumption by hydrostatic power drive and asynchronous motor with phase rotor in the course of operation of crane mechanisms at all stages of motion: acceleration, movement at nominal or intermediate speed, deceleration; to determine equivalent capacities of both types of power drives and to compare them.

* Corresponding author: ottow@kpi.kharkov.ua

5. Results

The comparative analysis has been performed on the results of industrial testing of two types of drives of motion mechanisms for overhead cranes with hoisting capacity of 30/5, 20/5, 15/3 tons, swing-out mechanism of frame-crane with hoisting capacity of 10 tons and benchmark tests (fig. 1).

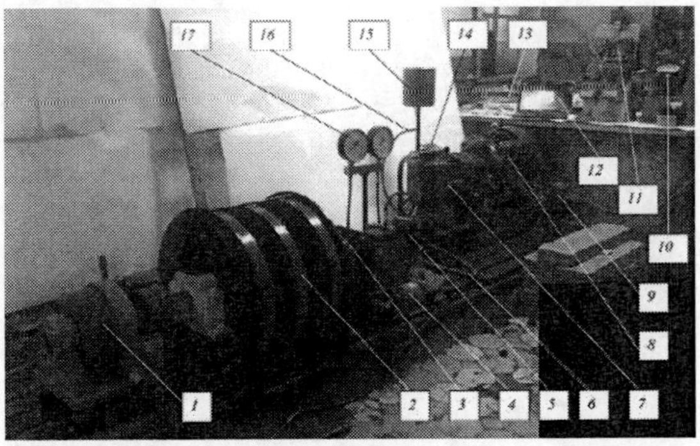

Fig. 1. The bench for testing movement and swing mechanisms of cranes with positive displacement regulated hydraulic drive: 1 – Weight cooled with running water to imitate static momentum of resistance to movement or swing; 2 – Flywheel weights to imitate the momentum of rotating masses energy (parameters of a wide range of hoisting capacity cranes are imitated); 3 – Radio strain sensors to transmit the twisting momentum to the amplifier and on to ADC; 4 – TMG-30P tacho-generator for recording hydromotor's rotation speed; 5 – IIM#5 hydromotor; 6 – High pressure pipes; 7 – IID#5 axial-piston pump; 8 – AO 62-4 electric motor; 9 – TMG-30P tacho-generator for recording rotation speed of electric motor as well as its operation in generation mode; 10 – Dial wattmeter; 11 – P004 power converter; 12 – Laptop; 13 – Amplifiers; 14 – Executive part of the control system; 15 – Refilling tank; 16 – Outlet pipe; 17 – Pressure gauges

The analysis shows that at uniform acceleration and deceleration there appear a number of discrepancies in the manner of drives parameters changes (fig. 2).

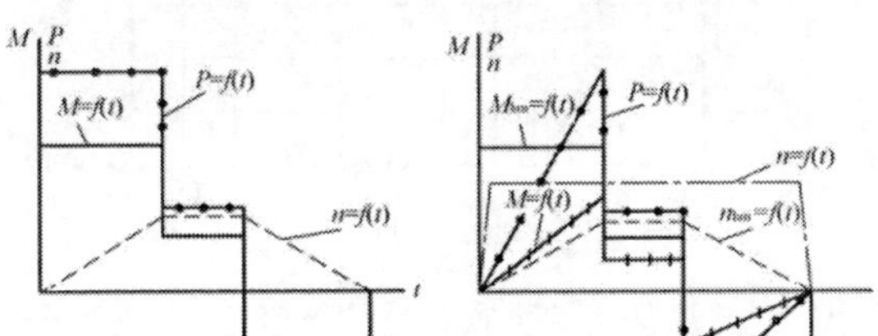

Fig. 2. Diagrams of changes in drives' parameters. a – electrical; b – hydraulic; M, P – rotation torque and power on the shaft of electric motor; n, n_{hm} – the rates of electric motor and hydromotor's rotation; M_{hm} – the momentum on the hydromotor's shaft; t – time.

Rotating torque on the rotor of hydraulic drive's electric motor, and thus the current spent on the motor's heating-up change in accordance with the linear law (proportionally to the speed of the crane's movement). Concerning a common electric drive (asynchronous electric motor with phase rotor) this value remains practically unchanged. Due to this fact, the thermal conditions within the hydraulic drive's motor are less intense, which permits in a number of cases to decrease the rated capacity of driving electric motors. Thus, it became possible to use a 7 kilowatt motor for hydraulic drive of a bridge-crane with hoisting capacity of 15/3 tons instead of an 11 kW motor, for the hydraulic swing drive of "Kirovets" frame-crane with hoisting capacity of 10 tons a 40 kW motor was used instead of 55 kW motor with the same time on starting-and-braking processes.

It is reasonable to compare the powers of electric motors of two different types of drives when used in various operation conditions. The numerical values are given for the bridge-crane with hoisting capacity of 30/5 tons (the span is 22.5 m; the total weight without cargo is 43.3 tons; the carriage weight is 10.632 tons; the rated motion speed is 76.5 m per minute; the rollers' diameter is 800 mm; the crane's motion mechanism's drive is separate).

The hydraulic drive for such a crane consists of asynchronous motors of A0 or A02 series, IID regulated pumps, IIM hydromotors, electronic tracking system to control pumps' effectiveness; electric drive includes metallurgical crane motors of MTV series. Hydraulic drive's electric motor is cut off from power during the breaks between the cycles of the motion mechanism's operation, there being no need in its constant operation.

The research were conducted in the operation conditions of both types of drives with operation periods (OP) of 15, 25, 40, 60% (groups of operation conditions of M1÷M8 correspondingly) for operation conditions of high and low intensity of transition processes, for two options of electric drive motor's deceleration – by means of opposition circuiting and mechanical brake for different cases of using drives at low speeds of movement ($e = V / V_{\text{rated}}$).

By transition process of high intensity are meant the operation conditions ensuring crane acceleration of $a_{cc} = 0{,}21\ \text{m/s}^2$, crane deceleration $a_{dec} = 0{,}41\ \text{m/s}^2$; at low intensity dimensions, and some other parameters.

By analyzing the curves in Fig. 4, the following conclusions can $a_{acc} = a_{dec} = 0{,}1\ \text{m/s}^2$. The high intensity operation conditions are ensured at the operation of the hydromotor and the pump under pressure of 160 MPa which is limited by safety valves adjustment. The low intensity operation conditions are stipulated by electric drive at controller's smooth switching from the zero position to the limit and back.

Considering the operation of a crane's motion and swinging mechanisms in the modes of acceleration, even motion, and deceleration, it is reasonable to select the drive electric motor's capacity by equivalent capacity P_e per cycle:

$$P_e = \sqrt{\frac{\int_0^t P^2 dt}{t}} \tag{1}$$

where P_e – current power on the shaft of electric motor; t – time.

The research has shown that in transition processes at the linear change in the pump adjustment parameter the power and torque momentum on the electric motor shaft, as well as consumed and recuperated power and the current in stator windings change linearly. That is why, as to an electrical hydraulic drive with taking into account its operation by trapeziform tachogram the expression (1) can be presented as follows:

$$P_e = \sqrt{\frac{\frac{1}{3}P^2_{p.max}\frac{1}{\eta^2_{red}\eta^2_{h.drive}}t_p + P^2_{p.x}\frac{1}{\eta^2_{red}\eta^2_{h.drive}}t_{p.x} + \frac{1}{3}P^2_{g.max}\eta^2_{red}\eta^2_{h.drive}\eta^4_{el.en.g}t_{h.t}}{t_p + t_{p.x} + t_{h.t} + \frac{1}{3}\theta}} \quad (2)$$

where $P_{p.max}$ – maximum positive power on the flywheel shaft at the end of the acceleration period; $P_{p.x}$ – positive power on the flywheel shaft at even motion; $P_{g.max}$ – maximum negative power on the flywheel shaft at the initial period of generative deceleration; η_{red} – reducer's efficiency factor at the end of acceleration period, at the beginning of generative deceleration, at even motion; $\eta_{h.drive}$ – total efficiency factor of the hydrostatic drive at the end of acceleration period, at the beginning of generative deceleration, at even motion; $\eta_{el.en.g}$ – electric motor's efficiency factor in generative mode of operation.

Time t_p of acceleration and $t_{h.t}$ of deceleration is written without 0.75 coefficient which is introduced for taking into account the worsening of the motor's cooling conditions in common electric drives in transition periods, because in hydraulic drives electric motor rotates in transition periods at about rated speed with no worsening of cooling conditions.

The 1/3 coefficient in front of θ shows that the motor's cooling conditions worsen when it is cut off from power supply.

The $\eta^4_{el.en.g}$ multiplier shows that the mechanical power of motor working as generator at the same electrical power conditioning its heating is $1/\eta^2_{el.en.g}$ times more than the motor's mechanical power at its operation in the motion mode [3]. For electric drive at accelerating and decelerating with constant rotation torques on the motor's shaft the expression (1) can be written as follows:

$$P_e = \sqrt{\frac{P^2_{p.max}\frac{1}{\eta^2_{red}}t_p + P^2_{p.x}\frac{1}{\eta^2_{red}} + P^2_{g.max}\eta^2_{red}\frac{\eta^2_n}{\eta^2_\sigma}t_{t.d}}{0,75\left(t_p + t_m\right) + t_{p.x} + \frac{1}{3}\theta}} \quad (3)$$

where $P_{p.max}$ – maximum negative power on the flywheel shaft at the initial period of its operation on countercurrent; $t_{t.d}$ – time of drive's deceleration while operating on countercurrent; η_n – a drive's efficiency factor while operating on countercurrent; η_σ – efficiency factor of motor's stator while operating on countercurrent.

Expression (3) was obtained bearing in mind that during the motor's accelerating and its operating on countercurrent under the integral sign in formula (1) there is indicated the power attributed to the crane's rated speed [3].

Thus, concerning a crane's electric drive, the power on the electric motor's shaft is taken as constant for convenience not only on the even movement stage, but also remains unchanged on the whole stage of accelerating ($P_{p.max}$) and decelerating by opposition circuiting ($P_{g.max}$).

The motors were selected by equivalent capacity P_e per cycle, tested for overload capacity and tolerable number of activations per hour, after which the rated power was determined by the largest of the needed capacities. The results of calculations for acceleration and decelerations of various intensities, as well as for different types of electric drive's deceleration are shown in fig. 3.

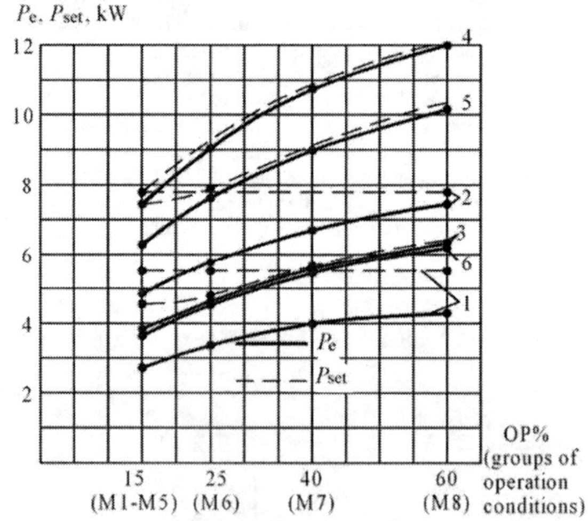

Fig. 3. Comparative graphs of rated and equivalent capacities and their relations for electric and hydraulic drives depending on the mechanism's operation time OP%:
P_e , P_e=*f*(OP%): 1, 2 – low and high (correspondingly) intensity of hydraulic drives' acceleration and deceleration; 3, 4 – electric drive, low and high (correspondently) intensity of acceleration and deceleration, decelerating by countercurrent; 5, 6 – electric drive, low and high (correspondently) intensity of acceleration and deceleration, decelerating by a mechanical brake.

Analyzing the graphs, the following conclusions can be made. In the same operation conditions equivalent capacity of hydraulic drive per working cycle of electric motor is less than that of electric drive which can be explained by favourable conditions of electric drive's load (electric motor initially accelerates idly, becoming loaded later) and by the absence of countercurrent deceleration. For electric drive the motor's rated capacity in the studied cases is determined by heating conditions (curves P_e and P_{set} coincide), i.e. by the value of P. This capacity grows with the intensity of drive's operation in terms of operation period OP%.

The hydraulic drive motor's rated capacity in the studied cases is determined, unlike that for electric drive, by overload capacity of motors of A0 and A02 series (lines P_{set} for all OP% are located above P_e curves) and are not determined by OP% (lines 1 and 2 run parallel to OP% axis [marked as OP% in fig. 3].

In the studied cases at low intensity of acceleration and deceleration the presence or absence of countercurrent decelerating does not affect the value of the electric drive's rated capacity (curves 3 and 6 coincide); at high intensity of acceleration and deceleration the use of only mechanical brake leads to the decrease in rated capacity by 15%.

In Fig. 4 there are shown the graphs built on the data of Fig. 3 with some addenda. Here, the adopted symbols $P_{set.el}$, OP% – rated capacity of electric drive's motor taken at OP% the drive is working at; $P_{set.el}$ =25% – the rated capacity of electric drive's motor is given for the OP=25%. The ratio $P_{set.el} / P_{set.h}$, OP=25% gives the idea of not only of the capacities ratio, but also to some extent of mass, be made. For a hydraulic drive the best ratio of rated capacities is obtained at high intensity of acceleration and deceleration, and intensive exploiting by OP% (curves 3, 4, 8, 10). In this case the ratio of $P_{set.el} / P_{set.h}$ may reach as much as 2.5. The worst ratio of rated capacities for a hydraulic drive is noted at low intensity of acceleration and deceleration with decelerating by mechanical brake of electric drive and light intensity of operation by OP% (curves 1, 2, 9).

This can be explained by the fact that in these operation conditions the choice of the hydraulic drive's rated capacity is to a great extent stipulated by overload capacity of industrial motors which is 1.3 to 1.5 times less than those for metallurgy cranage.

Correlation by equivalent capacity demonstrates that the ratio of $P_{set.el} / P_{set.h}$ practically doesn't depend on OP%, and is always to the advantage of hydraulic drive, varying between 1.3 …1.65 (curves 5, 6, 7) depending on the transition processes' intensity and the type of electric drive's decelerating.

In the practice of lifting-and-shifting machines exploitation there are cases when a drive designated for operation at a predetermined rated speed V_{rated} is used at the speed other than the rated speed ($V < V_{rated}$), for instance at an insufficient distance between bridge-cranes within the same span, or a small turning angle of a portal crane.

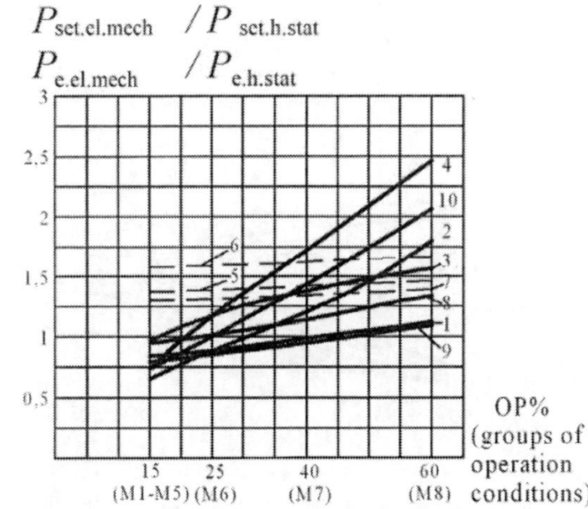

Fig. 4. Graphs of rated and equivalent capacities and their ratios for electric and hydraulic drives depending on the mechanism's operation period OP% (groups of operation modes):
1, 2 – $P_{set.el}$, OP% [marked as OP% in the Figure 4] ($P_{set.h}$; $P_{set.el}$ OP=25% $P_{set.h}$ – at low intensity of acceleration and deceleration; 3, 4 – $P_{set.el}$, OP% ($P_{set.h}$; $P_{set.el}$ OP=25% $P_{set.h}$ – at high intensity of acceleration and deceleration; 5, 6 – $P_{e.el}$ / $P_{e.h}$ – at low intensity of acceleration and deceleration; 7, 8 – ($P_{e.el}$ / $P_{e.h}$; $P_{set.el}$, OP%) $P_{set.h}$ – at high intensity of acceleration and deceleration, electric drive's decelerating by mechanical brake; 9, 10 – $P_{set.el}$, OP% ($P_{set.h}$; $P_{set.el}$, OP%) $P_{set.h}$ – at low and high (correspondingly) intensity of acceleration and deceleration, electric drive's decelerating by mechanical brake.

In Fig.5 there is the dependency $P_{e.el}, P_{e.h} = f(e)$. The analysis of the curves demonstrates that the electric drive's equivalent capacity is the more the less its velocity capacity is used (the less e is). The hydraulic drive's equivalent capacity is the more the more its velocity capacity is used.

The analysis of the curves demonstrated in Fig.6 makes it possible to come to the following conclusions. For a hydraulic drive the best relation of equivalent capacities is achieved at insignificant use of the drives' speed capacities ($e \to 0$). This relation is the higher, the higher intensity of transition processes is, and the higher the intensity of operation conditions by OP%. Thus, for example, at $e = 0.0333$ the ratio $P_{e.el} / P_{e.h} \approx 22$. The worse ratio of equivalent capacities for a hydraulic drive is viewed at the maximum use of drives' speed capacities ($e = 1$). In such a case at various operation conditions the ratio $P_{e.el} / P_{e.h}$ reaches 1.3 …1.65 which was seen earlier at the studying of curves 5, 6, 7 in Fig.4 assuming $e = 1$.

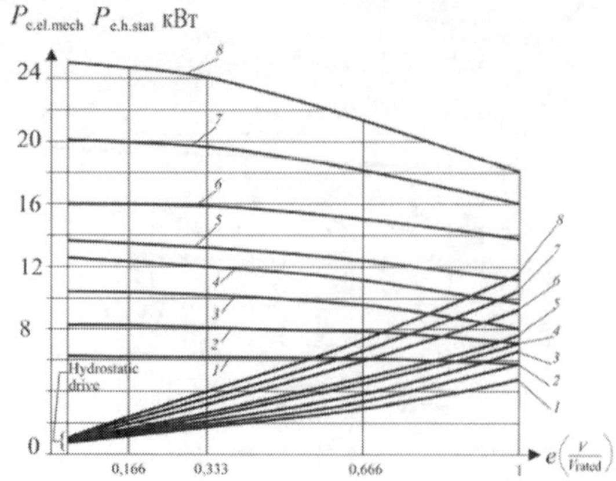

Fig. 5. The graphs of equivalent capacities $P_{\text{e.el}}, P_{\text{e.h}} = f(e)$ for electromechanical and hydraulic drives depending on their use by speed: 1 – small intensity of acceleration and deceleration, operation at OP=15%; 2 – the same at OP=25%; 3 – the same at OP=40%; 4 – the same at OP=60%; 5 – high intensity of acceleration and deceleration, operation at OP=15%; 6 – the same at OP=25%; 7 – the same at OP=40%; 8 – the same at OP=60%

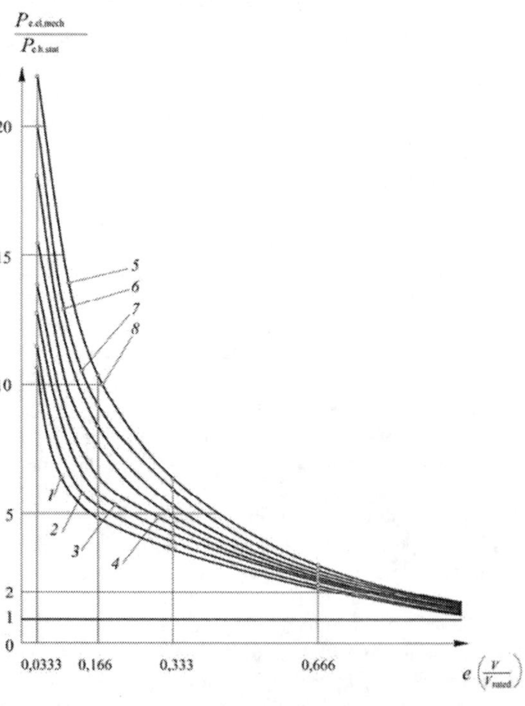

Fig. 6. The graphs of equivalent capacities relations $P_{\text{e.el}} / P_{\text{e.h}} = f(e)$ for electric and hydraulic drives.

On the grounds of the conducted research it can be stated that the rated capacity of hydraulic drive's electric motor when exploited in conditions of intensive acceleration, deceleration, and OP% is less than that of electric drive. Similar conclusions can be reached by performing calculations by equivalent momentum.

4. Conclusions

There has been performed the comparison of rated and equivalent capacities and their ratios for electric and hydraulic drives of crane mechanisms depending on groups of operation conditions and operation period. The advantages of hydraulic drive's electric motor by equivalent capacity as compared with electric drive increase in the case when operation conditions are characterized by insignificant use of the drives' speed capacities. At small intensities of acceleration and deceleration and light operation conditions hydraulic drive by the rated capacity of electric motor is less effective than electric drive on account of low overload capacity of electric motors. The expounded conclusions determine advisable areas of using drives from the perspective of getting maximum advantages in terms of the equivalent and rated capacity.

References

1. Bratslavsky H.L. *Hydrodynamic transmissions in construction and road machines.* – Moscow: "Mashinostroyenie" (Machine-building) PH (1976).
2. Grigorov O.V., Krauze F., Macwardt G.G. *Comparing efficiency factors and energy consumption in regulated hydrostatic and electromechanical drives* Collected scientific articles of KhDPU. High technologies in machine-building, – Pp. 72 – 78 (2000).
3. Petrov V.A. *Hydrovolumetric transmissions of self-propelled machines.* – Moscow: "Mashinostroyenie" (Machine-building) PH (1988).
4. Kordak R. *Der sekundärgeregelte hydrauliche Antrieb in mobilen Arbeitsgeräten* O+P „Ölhydraulik und Pneumatik", – № 11-12 (1995).
5. Ruckganer N. *Entwicklungstendenzen bei hydrostatischen Antrieben in mobilen Arbeitsmaschinen* O+P „Ölhydraulik und Pneumatik", – **№ 11-12** (1993).
6. W. Backe. *Technische Treds der Fluidec* O+P "Ölhydraulik und Pneumatik" **№ 11–12** (1995).
7. Grigorov O. *KKD, vytraty enerhii u rehulovanomu hidrostatychnomu ta elektromekhanichnomu pryvodakh* [Efficiency, energy consumption in controlled hydrostatic and electromechanical actuators]. Scientific, Technical and Production Journal „Lifting Structures. Special equipment". Odessa. **№1 (184)**, – Pp. 14-17 (2018). [in Ukrainian]
8. Grigorov, O., Druzhynin, E., Anishchenko, G., Strizhak, M., Strizhak, V. *Analysis of Various Approaches to Modeling of Dynamics of Lifting-Transport Vehicles* International Journal of Engineering & Technology. **№ 7 (4.3)** – Pp. 64-70 (2018)
9. Grigorov O., Anishchenko G., Petrenko N., Strizhak V., Turchin O., Radchenko V., Okun A. *Polipshennia roboty kranovykh metalevykh*

konstruktsii shliakhom zastosuvannia hidropryvodu ta deiakykh inshykh rishen v mekhanizmakh peresuvannia ta povorotu [Improving the performance of crane metal structures by applying hydraulic actuators and some other solutions in mechanisms of movement and rotation]. Lifting and Transport Engineering Scientific and Technical Journal. Odessa National Polytechnic University, **№ 2** – Pp. 4-14 (2019). [in Ukrainian]

10. Blanusa, B. New *Trends in Efficiency Optimization of Induction Motor Drives* New Trends in Technologies: Devices, Computer, Communication and Industrial Systems (2010).
11. Müller E., Engelmann J., Löffler T., Strauch J. Energieeffiziente *Fabriken planen und betreiben* [Plan and operate energy efficient factories] Berlin Heidelberg: Springer Verlag, (2009).

MATEC Web of Conferences 294, 01003 (2019) https://doi.org/10.1051/matecconf/201929401003
EOT-2019

Conceptual-innovative and energy-resource management on motor transport

Ramazan Khabutdinov[1], and *Arseniy* Khabutdinov[2]

[1]National Transport University, 1, Omelyanovich-Pavlenko, Kiev, 01010, Ukraine
[2]State Enterprise "State Road Transport Research Institute", 57 Peremohy Ave, Kyiv, 03113, Ukraine

Abstract. This paper discusses elements of the theory of conceptual-innovation and energy-resource management based on the system concept of technologically innovative providing of resource synergistic transport integration of a discontinuous terminal infrastructure of a transport system (TS) in accordance with transport demand and the systemic requirement of an innovative increase in the energy-resource quality of transportation services (ERQTS). It has been established that for the system enhancement of ERQTS are actual the development of innovation management theory and the implementation of conceptual improvement strategies for a number of subsystem and system factors: transport resources and technical properties, transport technologies, as well as innovative reproduction functions of the motor transport subsystem (MT) and integrative-equifinal functions of the motor transport system (MTS). The existing theory of service commercial management in MT is based on a simplified non-production and empirically conservative paradigm of knowledge about MT as a sphere of transport services (without product description) , therefore it does not fundamentally correspond to the above-mentioned concept of development of MTS and MT. This theory uses computational schemes for virtual transportation and service past costs, so it is non-innovative and technologically degenerate. A theory of conceptual and technologically innovative improvement of ERQTS, as well as the functions of MT and MTS, has been developed. A comprehensive model of the target function of the system-conceptual management of ERQTS is proposed, taking into account the improvement of the MTS and MT functions according to the principle of energy-resource synergy in the transport services life cycle. It takes into account: ways to eliminate the paradoxical fragmentation of types of knowledge about transport operations and processes, improving the properties of a car as a resource- technical means of transport production and reproduction, the essence of transport operations and technologies, as well as dysfunctions of subsystems of MTS.

1 Introduction

Distinctive features of the motor transport (MT) are as follows: a) a huge variety of transport processes factors (constructive-technical, road, traffic, operation, terminal and technology); b) high levels of trajectorial conflicts and resource expenses in transport operations; c) simplicity, lack of innovation knowledges in theoretical matrices "techno-logos" and "system-logos" in mathematical models of transport analysis. In the early stages of planning the development of technologically innovative transport processes (TP) and motor transport systems (MTS) is necessary to ensure the phasing of implementation of the concept and methodology of their resources- synergistic improvement. This concept is based on a new epistemological paradigm of technological consideration of transport as a sphere of material production and reproduction of transportation services (TS). At the same time MTS is considered as an integrity (on the basis of ensuring its fixihg - transport integration of discontinuous macroeconomics space of countries) of interacting transport and infrastructures elements [1]. The above methodology is based on the conceptual idea of justifying the cumulative effects of two ways of resource saving: a) by transport - using energy- resource quality management methods of transport services in their life-cycle projects taking into account the distinctive features of the motor transport; b) by infrastructure-based projects of minimization of dysfunctions of the road and terminal infrastructures.

It is accepted that life-cycle of transport services (LCTS) consists of three stages: a) justification of the conceptual quality of technical resources for projects of technologically innovative reproduction of transport services; b) operational production of transport services and consumption of the transport product by the subjects of the macrosystem; c) post operational justification of the conceptual requirements for new technical and technological support for future transport supply. In the resource-synergistic concept, terminal resource saving is ensured by the establishment of a rational logistical time for the implementation of specified chains of terminal operations in projects for the delivery of cargo or passengers lot [2].

* Corresponding author: habut20@gmail.com

In the road network infrastructure (RNI), resource saving is achieved by minimizing the indices of the trajectory conflicts and of the transport pro-anergic dysfunction of the road sections (TPADRS) [3]. The phenomenon TPADRS occurs when the microsystem "Driver-Car" is functioning under the conditions of the trajectory-conflict street-road network. It is characterized by a significant decrease (by 1.5..4 times) in the degree of use of the constructive-technical potential of the performance and energy efficiency of cars. Any traffic dysfunctions of road sections (low quality of road surfaces, high intensity of automobile traffic, increased trajectory conflicts, irrational traffic light regulation and traffic organization scheme) lead to an increase in the value of the indicator of TPADRS – ($K_{tpd} \rightarrow$ max, indicator of transport pro-anergic dysfunction of the road). The logical consequence of this growth is the increase in resources consumption in transport operations [3,4]. A necessary condition for the systemic and conceptual improvement of Road network infrastructure (RNI) is the minimization of the indicators of trajectory conflict and of TPADRS ($K_{tpd} \rightarrow$ min) of the roads along specified trajectories of automobile traffic.

The existing methodologies of management on Motor transport (MT) and the improvement of Motor transport systems (MTS) at the epistemological level do not correspond to the concept of resource synergistic improvement of MT and MTS. They are based on a very simplified epistemological paradigm: a) considering MT as a sphere of quasi-material transport services (without taking into account the properties of technical resources of transport and RNI elements) with unchanged parameters of cars and transport technologies; b) agglomerate consideration of MTS without regard to its transport and integrative purpose, as well as its structural and functional integrity mechanisms [2,5,6,7,11]. Due to the simplified epistemological paradigm, the existing management methodology has conceptual flaws: a) is: non-conceptual, non-productive, non-technological, non-technical, non-energy, non-innovative, non-systemic; b) the design scheme of the transpositional operation is used, which describes the virtual movement of the body with cargo (passengers) without taking into account the constructive-technical and resource-technical properties of the new cars, as well as the distinctive features of MT [6]; c) models of past operating expenses for transportation services are considered without taking into account the resource-cost and energy-cost production of the tangible product of transport [6,7]; d) intangible and technologically unjustified measures of the product of the transport are taken - ton-kilometers and passenger-kilometers, which have no physical meaning; e) at the level of axioms and basic design schemes, technical and technological innovations in MT are denied.

Purpose of the article: presentation of the main provisions of the conception and criterion apparatus for conceptual-innovative and energy -resource management on motor transport. Wherein the requirements of ensuring a system approach and the formation of a common strategic management methodology are taken into account, which includes models of important target functions of current management as particular cases.

2 Main part

The methods of conceptual innovation and energy resource management on MT must provide technological innovation of the Transport processes (TP), as well as the functions of the MT subsystem and of the MTS. Formed the following multiple model structure of MTS:

$$M(MTS) \in (MT, TI, RNI, TSI, MIIC, DSTRM, SSTSM, PS, GITDS, SFS, SF, STC, SSF) \quad (1)$$

where: *MT* – is the active subsystem "Motor transport"; *TI, RNI, TSI* – terminal, road network and technical service infrastructures (infrastructure subsystems); *MIIC* – mechanisms of internal interface connections; *DSTRM* – demand segments of the transport resources markets (automobiles, fuel, drivers and workers); *SSTSM* – supply segments of transport services markets; *PS* - is the purpose of the system; *GITDS* – the goal of innovation and technological development of the system; *SFS* – the subject of the functioning of the system; *SF* - many system functions; *STC* – system technology complex; *SSF* – sets of subsystem functions (in subsystems *MT, TI, RNI, TSI*).

The purpose PS of the MTS operation is individual (in the MT and TI subsystems) and mass (in the RNI and TI subsystems) transport integration of the discontinuous geometric space of the economic-social macrosystem (ESMS) in accordance with the transport demand, and also taking into account the operational and production requirements (adaptive performance, trajectory safety, transport energy efficiency) and terminal service requirements (timeliness of delivery, safety of goods and passengers, low cost) for motor transport processes [1]. The goal of managing the innovation and technological development of MTS (GITDS) is the long-term implementation of the concept of resource-synergistic transport integration of the macrosystem in a given transport and infrastructure space. Taking into account [8], a set of MTS system functions is formed: SF1 – provision of subsystem resources for given traffic trajectories; SF2 – transport-interface compositions of subsystem resources; SF3 – transport emergence (mass transport-technological processes) and traffic regulation; SF4 – transport integration; SF5 neutralize dysfunctions; SF6-conceptual modernization of subsystems and system functions; SF7-modernization equifinality (resource synergistic integration of a given macrosystem space and innovative improvement of the energy- resource quality of the transport services (ERQTS).

The system technology complex (STC) consists of the following technologies: TT – transport, TMT – terminal; TIT – transport - information; TMT – terminal; TMIT – terminal-informational; RCT – road construction; RMT – road maintenance; TFIT – traffic information; TST – technical – service. These technologies are included in the structures of the respective subsystems and together provide the implementation of two system functions SF2 and SF3.

At the same time, the theory of TT and the properties of its material and production components – TTC2 are of decisive importance when solving problems of strategic management [4]. In addition, the implementation of high TT (technically progressive, energy efficient and resource saving) will ensure resource synergistic improvement of a variety of MTS system functions SF and functions of the MT subsystem. The MT subsystem "Motor transport" is designed to meet the transport demand of macrosystem entities based on high-quality production and technologically innovative reproduction of transportation services using tactical (service–commercial) and strategic (conceptual–innovative and energy – resource) management methods.

The purpose of managing the innovative development of the MT subsystem is to increase the energy resource quality of of Transport services (ERQTS) taking into account the limitations of the TI and RNI subsystems based on maximizing the value of the objective function of technologically innovative projects and intensively resource-saving reproduction of Transportation services (TIIRSRTS) taking into account car markets. The functional structure of MT subsystem S_{MT} is as follows:

$$S_{MT} \in (MS, RTB, TTP) \qquad (2)$$

where *MS* – is the *MT* management superstructure (includes owners of capital, two types of R_{TM} management resources for providing: commercial tactics of MT-R_{TM1} and strategic management of R_{TM2}); *RTB* – resource and technological base of *MT*; *TTP* – transport - technological processes on MT.

As a functioning mode of transport, the MTS subsystem provides six of its functions (FMT): 1) FMT1 – transport production, in which the technological resources of MT – R_{TL} are involved and energetically transformed into a physical product W_p, the FMT1 function is based on transport technologies and human-machine labor and at the same time formed the emergent function MTS-SF3; 2) FMT2 – transport service, which is recorded on the basis of monitoring pairs of terminal events (sending - arrival of the joint venture); 3) FMT3 – transport capital turnover as a scheme of cyclical movement of working and advanced capital; 4) FMT4 – market supply of a technological resources of FMT1; 5) FMT5 – long-term and innovative-technological reproduction of transportation services on the principle of ensuring the technical and technological competitiveness of future transport supply; 6) FMT6 – car maintenance and repair (additional and non-essential function for most motor transport enterprises). The subject of functioning of the MT subsystem is the implementation of the first five functions of FMT1 .... FMT5 based on the chain of life cycle of transportation services (LCTS) projects and projects of conceptual and innovative improvement of transport technologies (TT), components of MT: RTB and TTP.

The transport technology (TT) is understood as a set of scientifically-described methods (man-machine and energy- resources) of high-quality production of an MT product. The structure of the S_{TT} transport technology is as follows [9]:

$$S_{TT} \in (TTC1, TTC2, TTC3.1, TTC3.2, TTC3.3\,) \qquad (3)$$

where *TTC1* - organizational forms of transport service on the basis of commercial projects of automobile delivery of goods or passengers; *TTC2* - resource-technical and production-process component of the TT; *TTC3.1* and *TTC3.2* - the gnoseological and mathematical support of the components of TT - TTC1 and TTC2); *TTC3.3* – rules and instructions for transport and terminal operations.
The structure of TTC2 includes the following elements:

$$TTC2 \in (Mp, R_T, R_E, R_L, P_{RT}, P_L, P_M, ET, ETRP) \qquad (4)$$

where *Mp* – party masses of cargo or passengers; R_T, R_E, R_L – carriers of technical (automobiles), energy (fuel) and labor resources (drivers) of MT, when they are connected in the structure of the car, the carrier of technological resources - R_{TL} of the MT subsystem is formed; P_{RT} – a set of properties of a car as a carrier of technical transport resources (as: a complex transport vehicle - P_{RT1}; a dangerous object of traffic control - P_{RT2}; a transportation means - P_{RT3}, potential maintenance facility - P_{RT4}; instruments of machine influences on the environment of motion - P_{RT5}; constructive and technical basis of the process of transforming transport resources - P_{RT6}; the element of the standard range of the rolling stock - P_{RT7}; MT technological capital and a source of transportation profit - P_{RT9}; P_L – many TT labor procedures; P_M – a set of machine procedures TT; *ET* – the process of vehicle energy conversion; *ETRP* – is the process of energy transformation of transport technological resources into a physical product MT - Wp (phenomenological basis of transport production).

Many machine procedures PM consists of the following elements:

$$P_M \in (P_{M1}, P_{M2}, P_{M3}, P_{M4}, P_{M5}, P_{M6}, P_{M7}) \qquad (5)$$

where is P_{M1} ... P_{M7} - TT machine procedures: energy conversion - P_{M1}; transmission - P_{M2}; traction coupling P_{M3}; trajectory-kinetic - P_{M4}; trajectory orientational - P_{M5}; trajectory-transgressive - P_{M6} (ensuring safe free space on the road in front of a moving car); aeration-toxication - P_{M7}.

Innovative development of TT is provided on the basis of design identification and substantiation of the complex of technical and technological innovations of NVTT:

$$NVTT \in (NVKT, NVGT, NVTS, NVTL, NVET) \qquad (6)$$

where *NVKT* – are design and technical innovations in the structural-parametric organization Kjr of a new car; *NVGT* – new vehicle parameters as a scientific and technical product; *NVTS* – innovations of transport service and TTC1; *NVTL* – technological innovations in the structure of TTC2; *NVET* – economic and technological innovations (resource-saving economy and TT).

The evolutionary model of the modular structural-parametric organization of the construction - Kjr of a new car (SPOCNC) for the analysis of NVKT and NVGT has the following form:

$$K_{jr}^{N} \in \left(x_{imp}^{N}, y_{iml}^{N}\right)_{jr}, \qquad (7)$$

where x_{imp}^{N} – is the updated subset of the p-th parameters of the m-th constructive-functional module in the i-th SPOCC device; N – is the upper index of innovation; y_{iml}^{N} – updated subset of l-th characteristics of the structure of the m-th structural and functional module and the i-th device SPOCC, (i = 1, 2 ... 4), (m = 1, 2 ... 4), (l = 11, 12, 13 ...); j – is the class index of the rolling stock (j = 1, n_c), where n_c– is the number of vehicle classes); r – is the index of the type of car in the j-th class (r = 1, n_v), where n_v– is the number of variants of cars with the implemented technical innovations).

The subject of the MTS operation (SOMTS) is: subsystem, compositional, interface, and path-trajectory support for implementing a variety of system functions (SF) for the complex conversion of technological resources R_{TL} of the transport into its physical product Wp and potential resources R_{mp} of the macrocystem ESMS transform into real resources R_{mr} according to logic scheme SOMTS:

$$SOMTS \in \left[\left(SF \rightarrow real\right) for: \left(R_{TL}(\theta_{rn}) \rightarrow W_p(\theta_{rn})\right) \rightarrow \left(R_{mp} \rightarrow R_{mr}(\theta_{TM})\right)\right] \qquad (8)$$

where W_p – is the physical product of transport [4]; θ_{rn} and θ_{mM} - road network and terminal trajectories of movement of carriers of technological resources R_{TL} of transport; → - operator of change of a condition of the carrier of resources.

The carriers of potential resources - R_{mp} of ESMS macrosystem are cargos (R_{mp1}) and passengers (R_{mp2}) before the transportation process begins. After the completion of this process, ESMS subjects, as consumers of transportation services, receive useful real resources R_{mp1} and R_{mp2}. Transport need of macrosystem is expressed in the necessity for transport transformation of potential consumer resources R_{mp} into real resources - R_{mr} in accordance with (8).

The macrosystem result of the transformations (8) is the transport integration of discontinuous geometric spaces of ESMS. At the same time, in transport operations, not only technological resources of transport are used, but also the material resources of two MTS infrastructures: terminal - R_{TI} and road - R_{RI}. According to the material and production paradigm of MT (motor transport), its physical product Wp (this is the route set of impulses of the amount of movement of the vehicle's operating mass) is based on the controlled processes of the ETRP-energy transformation of technological resources (R_{TL} → Wp) in adaptive transport operations. This takes into account three operational and production requirements: adaptive performance, trajectory safety and transport energy efficiency. In the production processes of a product in transport operations material production losses of technological transport resources ΔR_{TL} are formed, the value of which is determined by the amount of physical product W_p. However, in the theory of commercial management, only the accounting product of transport W_a is analyzed, as a part (40 ... 65%) from physical product W_p. It is known that an accounting product W_a is determined on the basis of a model of a virtual transport operation in which it is assumed that an "independent" movement of the net mass of cargo or passengers without taking into account the loss of technological resources ΔR_{TL} [6,7]. In addition, the state of the road infrastructure resources - RRI [9] has a significant impact on the production losses of the technological resources of the ΔR_{TL}.

In transport processes and in the system (MTS), two types of complex mechanisms of interaction and compositions are formed: a) technological resources of transport R_{TL} - with resources of road R_{RI} and terminal R_{TI} of MTS infrastructures; b) productive FMT1 and reproductive FMT5 transport functions - with SF system functions. Based on the technologically composite connection of these mechanisms, an integral structural - functional organization of the system (SFO MTS) is formed, which allows you to realize the purpose of MTS - transport integrate ESMS macrosystem, which is an existential condition for the functioning of this macrosystem.

The purpose of the technologically innovative improvement of the energy- resource quality of the transportation services (ERQTS) is a comprehensive reduction of energy costs and transport resources in the production and reproduction of this transportation services. It follows that the presence of a compositional relationship between the two transport functions (FMT1 and FMT2) and the many functions of the SF system allows the concept of resource synergy in SFO MTS based on the objective ERQTS control function in the life cycle of the transportation services. Based on the above, logical-procedural schemes of structural and functional organizations of the SFO MTS system and transport SFO MT are formed:

$$\begin{gathered} [SF1(R_{TL}, R_{TI}, R_{RI}) \rightarrow SF2(TT, STC) \rightarrow \\ \mathbf{FMT1}(TT, SF2, W_p) \rightarrow SF3(TT, \mathbf{FMT1}) \rightarrow \\ SF4(TT, \mathbf{FMT1}, \theta_{rn}, \theta_{mM}) \rightarrow (SF5 \rightarrow real) \rightarrow \\ SF6(TT, STC, FMT5, ERQ_{TS}) \rightarrow \\ SF7(TT_H, STC_M, SF3_{RS}, SF4_{RS}, SF6_{RS})] \rightarrow FESMS \end{gathered} \qquad (9)$$

$$\begin{gathered} [\mathbf{FMT1}(TT, SF2, W_p) \rightarrow \mathbf{FMT2}(W_p) \rightarrow \\ \mathbf{FMT3}(W_p, TPR)) \rightarrow \mathbf{FMT4}(R_{TL}, TT) \rightarrow \\ \mathbf{FMT5}(ERQTS, TT_{in})] \end{gathered} \qquad (10)$$

where *FMT1* and *FMT5* – are the functions of production and reproduction of the transportation services in transport; *ERQTS* – energy-resource quality of the transportation services; *TPR* – three production requirements for transport operations; TT_{in} – innovative TT; STC_M– the complex of the modernized technologies of the MTS system; SF_{RS} – system functions that correspond to the concept of resource synergy in the

MTS system; *FESMS* – functioning economic - socium system.

Based on the SFO MTS (9) and SFO MT (10) schemes, a systematic methodology of strategic management on MT has been developed, taking into account the following requirements [10]:

• structuring of MT and MTS functions, and components of transport technologies;

• the essential and technological analysis of transport production taking into account the properties of technical resources and dysfunctions of the road infrastructure (trajectory conflicts and transport pro-anergic phenomenon);

• structural and parametric analysis of vehicles with structural and technical innovations;

• physical, technical and technological causality of road transport operations, as well as the processes of transformation of the technological resources of transport and potential resources of the macrosystem;

• test and standard - comparative analysis of the transport energy efficiency of cars;

• a gradual increase in the values of indexes of energy-technological and energy-resource efficiency of the transport processes, taking into account road and operating conditions of transport operations;

• justification of a set of technical and technological innovations in transportation service life cycle projects;

• ensuring energy and technologically innovative resource-synergies in Motor transport system (MTS) based on the cumulative effect of transport and infrastructure resource saving;

• provision of innovative-technological and resource-intensive reproduction of transportation services based on innovative models of intensive surplus product of transport and future transport profits.

In accordance with these requirements and with the concept of resource-synergistic improvement of transport functions and the transport system, a conceptual-criterial apparatus has been developed for techno-innovative and systemic management of the energy- resource quality of transportation services. It is designed to solve innovative problems of strategic management in motor transport using a set of mathematical models and methods. Some methods are presented below:

• the theory of operational kinematics and energy by adaptive-discrete movement of the car as a carrier of technical resources of transport (with variable structural and parametric organization of the structure);

• calculation schemes and models of transport and technological operations, taking into account the processes of energy transformation of technological resources into a physical product of transport;

• testing and standard-comparative method of simulative analysis of the transport energy efficiency of vehicles with variable parameters for different projects of the life cycle of transportation services;

• the method of simulative analysis of energy technology efficiency of transport operations for different types of the road transportations, taking into account the changes in structural, technical, road and operational factors;

• the method of simulative analysis of technical and technological innovations of NVTT and the standard - comparative formation of projects for innovative and intensive resource-saving reproduction of transportation services.

The following are mathematical models for a simulative and standard-comparative analysis of some indexes of energy efficiency of automobiles and technological projects of transportation, as well as energy- resource quality of transportation services:

a) the dimensionless indexes of the transport energy efficiency of an automobile $P_E(K_{jr})$ with a given variant of the characteristics of its structural-parametric organization K_{jr} is determined as follows:

$$P_E(K_{jr}) = \frac{K_{VP}(K_{jr}, PR1, PR2)}{K_{EP}(K_{jr}, PR1, PR2)} \cdot \frac{\gamma_s}{(\gamma_s + \eta_e)} \to max \qquad (11)$$
$$K_{jr} \to Var, \; P_{R1} \to Var, P_{R2} \to Var$$

where K_{VP} and K_{EP} – are the energy coefficients of the vehicle speed and its mileage for the calculated test route (determined by the method of mathematical modeling of the processes of energy conversion of technological resources of transport R_{TL} into physical product W_p); P_{R1} and P_{R2} – properties of the road as rolling surface and as a communicative channel; γ_s and η_e – are the load capacity utilization factors (or bus capacity) of the vehicle and of its equipped weight.

An example of the results of the parametric analysis of the influence of the maximum power Nm (kW) of the engine of a VOLVO FH-12 FRUEHAUF DSND-32 auto train on a set of indicators of its energy efficiency in the test operation is shown in Figure 1.

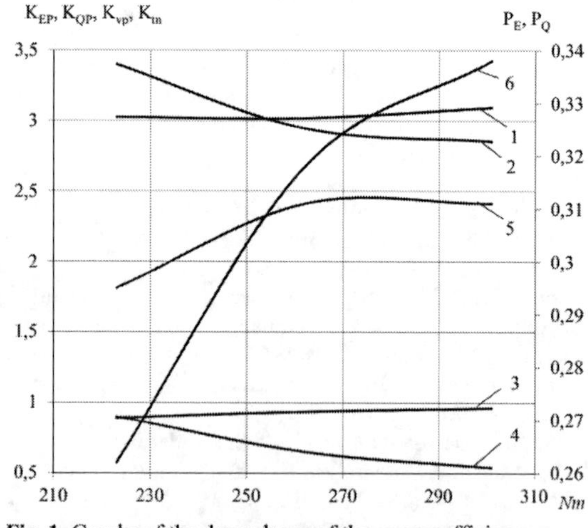

Fig. 1. Graphs of the dependence of the energy efficiency indicators of the VOLVO FH-12 + FRUEHAUF DSND-32 highway train on the maximum engine power Nm (kW) for the calculated test route:
1 - K_{EP}, 2 - K_{QP}, 3 - K_{vp}, 4 - K_{tn}, 5 - P_E, 6 - P_Q.
The following designations are taken in the figure: 1 – K_{EP}- coefficient of energy mileage; 2– K_{QP} - coefficient of fuel consumption; 3– K_{vp} - average velocity coefficient; 4– K_{tn}- indicator of the share of unsteady modes of movement of the road train in the test operation; 5– P_E - indicator of transport energy efficiency; 6 - P_Q - indicator of fuel efficiency.

Such test characteristics were obtained for modifications of different types of cars (small, medium and heavy load capacity, road trains and buses). Based on the test and multicriteria analysis of possible modifications of car types, two groups of conditions for their conceptual and innovative justification are determined taking into account their resource-technical properties of P_{RT} and machine procedures of P_M transport technologies according to expression (4): a) conditions for increasing transport energy efficiency and compliance with energy-saving transport technologies, which are achieved by maximizing the values of four criteria (K_{vp}, P_E, P_Q, TE) → max, where TE– is the indicator of the transport return of machine procedures PM transport technologies; b) the condition for an indirect assessment of the toxicity of a vehicle in a transport operation based on the requirement to minimize the values of two criteria (K_{QP} , K_{tn})→ min.

b) The index of transport efficiency TE of machine procedures P_M of transport technologies in test operations is determined as follows:

$$TE = \frac{M_p * V}{P_\alpha * t} \rightarrow max \quad (12)$$

where M_p –party mass of cargo or passengers; V and t – the speed and time of movement of the car in a given phase of the test operation; P_a – average thrust force of the car in a given phase of the test operation;

c) criteria for improving the energy-technology efficiency of road transportation:

• energy coefficient of the technological performance of the vehicle as a carrier of technical resources of transport - Kw:

$$K_w(K_{jr}, F_o, NVTT, t) \rightarrow max \quad (13)$$
$$K_{jr} \rightarrow Var, F_o \rightarrow Var, t \rightarrow Var$$

where $NVTT$ – is a set of technical and technological innovations in the life cycle of road transport services (constructive-technical, goods-technical, organizational, technological and economic-technological); F_o – many operational factors of the transport cycle [6]; t – variable time;

• energy coefficient of the cost of transportation taking into account the technologically innovative production of the Motor transport - Ks:

$$K_s\left(K_{jr}, F_o, Kep, Kvp, NVTT, \frac{C_f}{C_v}, t\right) \rightarrow min, \quad (14)$$
$$K_{jr} \rightarrow Var, F_o \rightarrow Var, t \rightarrow Var$$

where C_f / Cv – the ratio of specific fixed and variable costs of the transportation in the estimated car ride; K_{jr} – a set of characteristics of the structural and parametric organization of the design of a new car;

d) energy-equivalent criteria of the energy-resource efficiency of the road transportation and the quality of the transportation services – (ERQTS):

$$K_w(K_{jr}, F_o, NVTT, t) * W_{r0}(F_o) \rightarrow max$$
$$K_s(K_{jr}, F_o, NVTT, t) * S_{w0}(F_o) \rightarrow min \quad (15)$$

where W_{r0} and S_{w0} – are the indexes of the transportation productivity and transportation cost price) for the current time t_0, which are determined according to the scheme of the transposition process [6];

e) the dimensionless objective function COF_{RS} of the system-conceptual management of innovative functional resource- synergy projects at Motor transport (MT) and MTS - system is formed taking into account (9) .... (15) and has the following form:

$$COF_{RS} = \frac{PTh\,(NVTT, t)}{W_{ro} * T_{wo}} = \left(1 + \beta_R(NVTT, t)\right) \cdot \quad (16)$$
$$\left[\frac{m_T(t)}{m_a(NVTT, t)} - m_s(NVTT, t) * F_{TS}\right] \rightarrow max$$

where PTh – is the hourly transport profit; β_R – the index of the innovative increase in the energy-resource efficiency of the technological transportation project, with $\beta_R > 0$ the effect of the obtaining a technologically intensive surplus product of transport is provided, which is a necessary condition of transport resource synergy in technological projects; $m_T(t)$ – is the coefficient of the forecast growth of the tariff T_{w0} per 1 (t*km), $m_T(t) > 1$; m_a – is the index of the modernization reduction of the value of coefficient of the transport pro-anergic disfunction of the road - K_{tpd} on a given transportation route, $m_a = K_{tpd}(t) / K_{tpd}(t_0)$ $m_a<1$; m_s – is an index of innovative reduction in the value of the energy coefficient K_s of the cost of transportation; $m_s \leq 1$; F_{TS} – is the economic factor of the current organization of transportation services (the ratio of the cost of transportation and tariff T_{w0} for transportation) $F_{TS}=0.55...0.8$.

Model (16) with the system-conceptual management of innovative projects of functional resource synergy at MT and MTS provides the solution of three new development tasks: a) improving the energy- resource quality of the transportation services - ERQTS; b) the formation of projects of technologically innovative and resource-saving reproduction transportation services of TIRSRTS; c) analysis of the compliance of the modernization projects of the elements of the road and terminal infrastructures of the MTS with the resource-synergistic development concept.

The objective function $COF_{sm}(i)$ of strategic management at motor transport for the i-th development project (model of future profit in technologically innovative projects of transport processes) has the following form:

$$COF_{sm(i)}(t) = (COF_{RS(i)}(NVTT, t) \cdot W_{z0(i)} \cdot T_{w0(i)}) \rightarrow max,$$
$$i \in (1, N_{pr}),\ COF_{RS(i)}(t) > 0 \quad (17)$$

where Npr – is the number of versions for technologically innovative strategic management projects at motor transport.

3 Conclusions

1. Established the principal (from the point of view of ensuring the innovative development of transport technologies and processes) shortcomings of the existing theory of transport analysis and management; they are due to the previously adopted simplified paradigm of knowledge about transport as a sphere of commercial transport services, as well as the axiom of the immutability of the parameters of automobiles and transport technologies. Due to these shortcomings, these theories do not correspond to the goals of strategic management.
2. Technologically innovative concept of resource-synergistic transport integration of the country's economic and social macrosystem and a methodology for its implementation have been formed.
3. Established the structures and functions of both the transport system and the subsystem "Motor transport"; The ways of innovative resource saving on motor transport and in infrastructure subsystems are revealed.
4. A set of methods is proposed for the technologically innovative management of the energy resource quality of transport services and for resource-synergistic improvement of the functions of the transport system with its subsystems.
5. An analytical model of the objective function of strategic management in motor transport is proposed, which provides the concept of resource-synergistic transport integration of the economic and social macrosystem.
6. On the basis of this system methodology, it is recommended that: a) in motor transport implement technological and innovative strategies for the staged improvement of the energy resource quality of motor transportation services; b) in the road infrastructure -to ensure the minimization of the indicators of trajectory conflicts and transport pro-anergic disfunctions of road sections; c) in the terminal infrastructure - minimize logistical costs.

References

1. R. Khabutdinov, A. Khabutdinov, Visnyk NTU, **17**, 134 (2008)
2. A. Gadjinsky, Logistics (2008)
3. O. Petrashevskiy, A. Khabutdinov, Transport problems, **6**, 60 (2009)
4. R. Khabutdinov, O. Kotsyк, *Energy-resourse efficiency of the car* (1997)
5. K. Macconel, S. Brue, *Economics: Principles, problems and politics* (1992)
6. A. Vorkut, Motor transportations of goods (1986)
7. M. Khodosh, B. Daskovsky, *Organization, economics and management of cargo transportation by motor transport* (1989)
8. M. Setrov, *General principles of organization of systems and their methodological significance* (1971)
9. R. Khabutdinov, Transport problems, **3**, 164 (2006)
10. R. Khabutdinov, Visnyk NTU, **24**, 237 (2011)
11. G. Prokudin, O. Chupaylenko, O. Dudnik, A. Omarov, Eastern-European Journal of Enterprise Technologies Vol. **6**, Issue 3 (84). P. 54–61 (2016). doi: 10.15587/1729-4061.2016.85211

Energy-efficient and cost-effective methods of driving trains under the conditions of development of the electricity market

Alexander Pshinko[1], *Andrii* Afanasov[1], *Valeriy* Kuznetsov[2,*], *Vladislav* Skalozub[1], *Sergiy* Arpul[1], and *Alexander* Ivanov[1]

[1] DNURT, Dnipro, Ukraine
[2] Railway Institute, Poland

Abstract. The article for the first time considers the possibilities for the organization of cost-effective rail transportation on electrified section in terms of the functioning of the Ukrainian balancing electricity market. It is formulated the complex task of choosing effective mechanisms of purchase and sale on the electricity market on the basis of contracts for the implementation of markets "for the day ahead", as well as for the appointment of parameters and tariffs of the intraday market. The authors presented the models and results of calculations of optimum modes of trains' driving, which are calculated for various criteria of efficiency taking into account the cost of the consumed active and reactive electric energy. The technique of tariff formation for purchase of electricity on the balancing market was proposed.

Introduction

The report summarizes and discusses the results of research on the solution of a new and urgent task for railway transport - the formation of energy efficient and cost-effective methods for driving trains in the conditions of the development of the electric energy market in Ukraine. The hierarchical structures of complex tasks to ensure energy-optimal and cost-efficient electricity consumption processes of rail transportation are presented. A comparative analysis of their components, models and methods of implementation is conducted. Modern market forms and methods for providing traction power supply systems are considered, which require a revision of the methods for selecting the optimal train driving modes. An approach is proposed to solve the problem of reactive power compensation in power supply systems, which takes into account the charge for the consumption and generation of reactive power due to the formation of an appropriate train control criterion. There were given the optimal control modes in terms of consumption and cost management modes of trains on electrified sections, as well as rational control modes, taking into account the consumption of reactive energy, or by the criterion of the minimum cost of the consumed reactive power. For the calculation of driving modes, a model of stochastic optimal train motion control was used, as well as methods of fuzzy expert systems that provide rational modes of thrust based on simulation and processing control trips data.

Relevance

Railway Transport (RT) of Ukraine is one of the country's largest consumers of fuel and energy resources. Reducing operating costs is an important direction in improving the competitiveness of railways. One of the most effective ways to save resources is the introduction of energy-optimal train schedules in the main directions of transportation [1, 2]. In recent years, in connection with the formation and development in Ukraine of the Wholesale Electricity Market (WEM) [7], a new and relevant research area for railway has emerged - the formation of energy efficient and cost-effective train driving methods [3-5, 13]. In 1997 - 1999 railways of Ukraine were licensed for the right to transfer and energy supply on the basis of the regular tariff. The goal was to access cheap electricity on the Wholesale Market of Ukraine and reduce its own costs for its purchase. In 2002, the transition to WEM was first performed by the Pivdenno-Zahidna Railway, and in 2011, the last such transition was carried out by Odesskaya, Pivdenna and Lvivskaya railways.

The complexity of the task of optimal train control is determined by the completeness of taking into account the totality of factors characterizing the estimated section, model of the train, various conditions of the process of a train moving along a variable path profile, taking into account voltage changes at the current collector of a moving electric locomotive, arising forces, etc. [1, 3]. Taking into account the conditions of the wholesale electricity market has significantly changed and expanded the factors of the tasks of managing train traction modes [2, 5]. Differences in electricity tariffs for the periods of the day, the need to develop a package of mode maps (MM) and a corresponding adjustment of train control modes for the periods of the day, as well as taking into account differences in the cost of active and reactive energy, etc., have made relevant the requirements for more accurate calculation of electricity

* Corresponding author: vkuznetsov@ikolej.pl

consumption for train traction. For the first time, a significant impact of the conditions for the application of differentiated tariffs and tariffs on WEM for passenger train operation has been established in [2, 3]. When analyzing rational MM with variable electricity tariffs, a new important spatial-temporal relationship of the cost-energy indicators with the tasks of organizing passenger and freight rail traffic in general is emerged.

Further development and improvement of methods for calculating the MM of freight trains in terms of the wholesale electricity market and with variable electricity tariffs was obtained in [3, 4]. Here, for the first time, were studied the optimal modes (in terms of the cost of the electricity consumed) of train operation under the conditions of stochastic tariffs typical of the wholesale electricity market. At the same time, an adaptive method was developed for calculating the train driving statistics based on a fuzzy production model[6]. The method allows to use the results of experimental trips for the formation of traction modes, and provides sufficient for practice the accuracy of controlling the movement of the train without specifying a detailed description of the parameters of mathematical models. In [4], the piecewise-linear approximation of the tariff for electricity in the WEM was constructed, containing four intervals and allowing one to calculate the optimal MM in terms of the cost of the electricity consumed. For the WEM conditions, the criteria for evaluating train schedules were improved, taking into account the different costs of active and reactive energy. At the same time, on the basis of experimental data, the dependence of the power factor was obtained, which makes it possible to calculate the adjusted modes of train traction taking into account the different cost of reactive and active electric power.

In 2005 was established a single tariff for electricity. Following it, the railways had to sell electricity to their customers at a single tariff. At the same time, the railways of Ukraine lost the opportunity to receive certain advantages from electricity purchases at a much lower price of WEM, and also completely lost the incentive to adjust their own consumption schedule and the opportunity to obtain a certain economic effect from it.

In 2017, the Verkhovna Rada of Ukraine adopted the Law of Ukraine "On the electricity market" [7]. The purpose of the new electricity market is to introduce competitive mechanisms for the functioning of the market, free choice of contractors and ensure the right of the consumer to freely choose the electricity supplier. The law provides for various mechanisms for the purchase and sale of electricity bilateral agreements, the market "for the day ahead" and the domestic market. Tariffs at which the electricity supplier sells electricity at an unregulated rate to its consumers are determined and established exclusively through negotiations, and are not regulated by the state. Since July 1, 2019, a new model of the balancing market of WEM has been functioning in accordance with the provisions of this law.

In connection with the functioning in Ukraine of the model of the balancing electricity market, the content of the tasks of the scope of the operations of the RT on the WEM has changed, as well as the significance and role of the above mathematical models used to organize rational transportation in the electrified sections. Namely, in the models of calculating the optimal MM of train operation in the conditions of the WEM, the train flow and electricity tariffs were considered as given. In the modern conditions of work of the RT on the balancing market, the parameters of train schedules and variations in the volumes and tariffs for electricity by day should be jointly justified and calculated by simulating transportation processes. The previously mentioned and other similar models should be used to select the most efficient mechanisms for buying and selling electrical energy, when concluding bilateral agreements, implementing the day-ahead markets, and also for setting parameters and tariffs for the intra-day market. The given article is devoted to the formation and justification of ways to solve these new tasks of the operation of the Ukrainian railways in the WEM.

1. Setting targets for the calculation of energy efficient and cost-effective modes of driving trains in a balancing electricity market of Ukraine

Let us consider the issues of a comparative analysis of the structure of complex tasks to ensure energy-optimal and cost-efficient power consumption of rail transport processes, as well as differences in mathematical models and implementation methods. The content of the main technological task of choosing energetically optimal train control modes (depending on the profile of the track, speed limits, a given time of movement on a certain section, taking into account the stochastic properties of voltages on current collector, monitoring the temperature of the traction engines, as well as other characteristics) means to calculate such values of control parameters, for example, positions of the ERS controller, which provide the best values of the vector of control quality indicators when performing and the system of requirements for the given characteristics of the train movement [1, 3]. Different components of the vector function of the target are possible ($\overline{F}(F_k)_n = (F_1, F_2, ...)$; $F_1 = Y$ - traffic performance, $F_3 = C$ - cost of consumed electricity, $F_4 = T$ - requirements for the time of fright delivery, $F_5 = A$ - electricity consumption, etc.). They reflect various aspects of the operations of the RT and the requirements of the systems of various levels that organize and ensure the transportation process. The values of all components $\overline{F}(F_k)_n$ are essential in solving the problem of calculating the MM values $u(*)$ - the parameters for controlling the movement of a train at a point of section x at given time t point: $u(x,t,U_C(E,\mu(t,s),\sigma(t,s))) \in u_D$, where u_D – is the set of admissible controls. Further, the $U_C(*)$ voltage on the current collector of a moving electric locomotive at a point x, which has stochastic properties, is indicated, a E the EMF of the traction substations; $(\mu(t,s),\sigma(t,s))$ - respectively, the mathematical expectation and the standard deviation of the inter-train interval, which characterize the train flow associated with the movement of the calculated train on the section s. For brevity, "*" denotes sets of function arguments. Thus, already through the values $U_C(*)$, $(\mu(t,s),\sigma(t,s))$ as well as through some of the vector indicators $\overline{F}(F_k)_n$, the task of choosing the optimal train hauling modes at the level of the choice of $u(x,t,U_C(E,\mu(t,s),\sigma(t,s)))$ controls is included in the hierarchy of the railway systems. These systems are related to the implementation of the rail transport process as a whole. Figure 1 shows schematically the form of switching on a train, for which an energy-optimal (or at the minimum cost for the conditions of variable electricity tariffs as in [2, 4, 5, 10-11] is calculated for the section s, the multi-level structure of target indicators and other functional characteristics of the traffic .

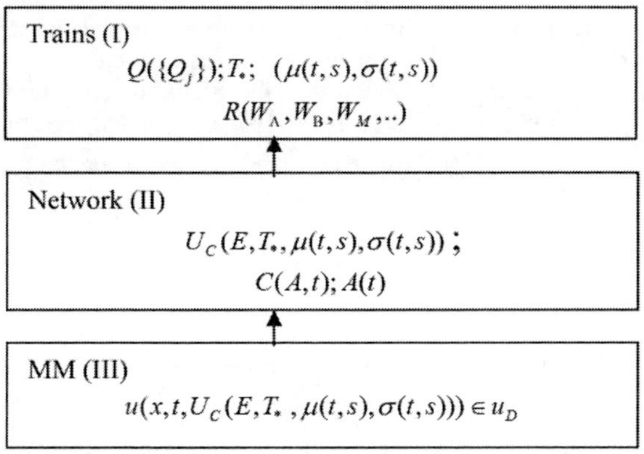

Fig. 1. The structure of the indicators for calculating the optimal MM

On fig 1, the following characteristics are indicated: $Q(\{Q_j\})$ – the volume of transportation by different categories of trains $Q_j, j \in \{g,n,e,v\}$, having different electricity consumption and driving modes, where $j=g$ it corresponds to freight trains, $j=n$ - passenger trains, $j=v$ - speed trains, $j=e$ - electric trains, etc .; T_* - requirements for the time of movement; $(\mu(t,s),\sigma(t,s))$ - the parameters of the accompanying train; $R(W_A,W_B,W_M,..)$ - vector of transportation resources, where W_A – locomotives, W_B – wagons, W_M – financial resources, etc.; $A(t)$ - indicators of electricity consumption by periods of the day t; $C(A,t)$ - the cost of electrical energy; the remaining characteristics are defined above.

The articles [2, 5] discussed an approach to solve a multi-level complex problem of MM optimizing. 1, from the standpoint of stochastic programming. At the same time, fig. 1, three levels of the model were distinguished: –1) train flows, - 2) electric traction network, - 3) train operation modes. Given the parameters of the train flow, the characteristics of the transportation processes for levels (II) - (III) are: electricity consumption for train traction, cost of electricity consumed for traction, losses in the traction network, average voltage level at the current collector, travel time on section, engine temperature, etc.

The qualitative difference and the complexity of the tasks of simulating of RT on the balancing electricity market, from the problems of choosing MM that are optimal in terms of cost or energy-efficient, which were previously used in railway transport of Ukraine, consists in changes in the models' structure of the systems fig. 1. In the conditions of a balancing WEM, the levels of model (I) and (II) should be swapped, or be realized together. At the same time, the planning of train flows should be carried out taking into account the tariff for electricity established on the basis of bilateral agreements, day-ahead markets, and the assignment of parameters for the intra-day market. Previously developed models of railway transportation processes should be used to ensure the completeness and accuracy of calculations of the optimal consumption and cost of electrical energy, as well as to obtain reliable MM for trains.

In this paper, further in the calculations, one of the known and complete models of the equations of train motion was used. It takes into account the parameters of the train length, temperature changes of the traction engine and other factors [1, 3]. This model turned out to be sufficient for the development of cost-effective MM for driving passenger and freight trains. A moving train seemed to be a flexible thread of length L_n with constant mass $p_n(z)$, $0 \le z \le L_n$. The dynamics of train motion is described by a system of differential equations (1) with initial conditions (2) [1, 2]

$$\frac{d(v^2)}{dx} = \frac{2\zeta}{(q+Q)(1+\gamma)}[F_e(x_e) - W(x_w) - B_T(x_B)]; \qquad (1)$$

$$\frac{d\tau}{dx} = \frac{\tau_*(I_D(v(u(x,t),U_C(x,t,\theta))) - \tau(x,t))}{v(x,t)T_D(I_D(v(u(x,t),U_C(x,t,\theta))))};$$

$$x(0) = x_0;\ v(0) = v_0;\ \tau(0) = \tau_0, \qquad (2)$$

where: x,t – coordinates of the way and time; - $u(x,t)$ – control (position number of the controller); - v – the speed of the center of mass of the train; - masses q,Q – of locomotive and train; - F_e – the power of the locomotive; - $U_C(*)$ – voltage in contact network; - $W(*)$ – resistance to the forward movement of the train; - B_T braking force acting on the train; - γ – the coefficient of inertia of the rotating masses; - $\zeta = g/1000, g$ – acceleration of gravity; - τ – heating temperature of the traction engines; - $\tau_* = \tau\infty$, T_D, I_D – thermal characteristics and currents of the traction engines; - $x_e = (v,u,U_C)$; $x_w = (x,v,u)$; $x_B = (v,k'(t),u,U_C)$, - $k'(t)$ – total force of pressing of brake blocks; - θ – random and uncertain factors of the task.

In (1) - (2), the time coordinate is t introduced to enable the choice of controls depending on the period of the day. The motion model also takes into account the random nature of the alternating voltage on the current collector, which occurs when the locomotive moves $U_C(*)$.

The resistance to the forward movement of the train in (1) - (a) took into account: the resistance to movement of the train from the longitudinal profile of the track, i - the slope value; resistance to movement from curves in terms of the path; i_r - resistivity from curves; area of permissible phase trajectories (starting and ending coordinates of the track; $\tau_{доп}$ - maximum permissible engine overheating temperature; $v^{max}(x)$ – permissible train speed). The restrictions imposed on controls $u(x)$ in accordance with the rules of technical operation included: $\tilde{U}$ - the scope of admissible controls; $I_д^{max}$ - maximum permissible current of engine; q_0 - vehicle axle load; - the $\psi_к$ coefficient of adhesion of the wheel with the rail; K'' - force of pressing of wheel pair brake block; $\varphi_к$ - the coefficient of friction of the brake block on the wheel. The train moving time on the section was represented by an isoperimetric constraint.

In the tasks of selecting energy-optimal train traction modes, the criterion for assessing the quality of the control law on the section $u(x)$ was the minimum of the functional representing electricity consumption.

$$A[u] = \int_{x_н}^{x_к} \frac{U_c I_{da}(v, u)dx}{v(x, u)}, \quad (3)$$

where is $I_{da}(v, u)$ the active current of the locomotive.

The task of choosing the optimal MM is following: it is necessary to calculate such a control law $u^*(x) \in \tilde{U}$ so that the corresponding solution of the system of equations (1) - (2) belongs to the admissible region, and the objective functional optimality (3) takes the minimum value.

The calculations of the optimal MM also took into account the alternating voltage at the current collector of a moving electric locomotive. It was estimated on the basis of the accompanying train traffic (ATT) model, as well as estimates of the voltage drop from the own current of the locomotive [2]. To do this in [3] were developed the models for the statistical estimation of the parameters of ATT for several types of trains and models for estimating the parameters of ATT based on the theory of fuzzy sets [6].

2. Optimal modes of conducting freight trains, taking into account the cost of active and reactive power

Along with the active power of the electricity supplier, the reactive power is also taken into account, which is payable at the established tariffs. The main part of the power in RT is consumed by locomotive electric engines, their asynchronous electric motors are consumers of reactive power. In the absence of compensating devices in the networks, the reactive current is supplied to the engines from the power supply networks, and is payable. In Ukraine, the fee for the consumption of reactive power is determined according to:

$$\Pi = \Pi 1 + \Pi 2 - \Pi 3 \quad (4)$$

where $\Pi 1$ - the main fee for the consumption and generation of reactive electricity; $\Pi 2$ - surcharge (for> $tg\phi$ 0.25) for insufficient equipment of the consumer's electrical network; $\Pi 3$ - payment discount in the billing period. The amount $\Pi 2$ of the allowance has a nonlinear dependence $tg\varphi$, and $tg\phi > 0,6$ can take values comparable to the main payment $\Pi 1$ for the consumption and generation of reactive electricity [4].

Reducing the value $tg\phi$ is a rather difficult task that can be solved by introducing rational MM using the criterion of minimizing the consumption of reactive energy or reducing the charge for the consumed reactive energy. In [4, 5] was used a functional to estimate the control indicator at the section $u(x)$, which took into account the costs of active and reactive energy of the form:

$$A[u] = \int_{x_н}^{x_к} \frac{U_c I_{da}(v,u) \cdot c(t) dx}{v(x,u)} + \int_{x_н}^{x_к} \frac{U_c I_{da}(v,u) \cdot \sin\phi \cdot (I_{da}(*)) \cdot D \cdot c(t) dx}{v(x,u)} + \Pi 2 \quad (5)$$

where $I_{da}(v,u)$ - the active current of the locomotive, the first member reflects the cost of consumption of only active energy, and the second - the cost of the reactive energy, D - coefficient of influence of the reactive power flow. $\Pi 2$ - surcharge (5), c *(t)* - the price coefficient of the tariff taking into account the time of day t;

$$\Pi 2 = \Pi 1 * C_{bas} * (K_f - 1) \quad (6)$$

In (6) $\Pi 1$ is the main cost for the consumed and generated reactive energy, $C_{баз} = 1$ the standard value of the coefficient of capital investments; $K_ф$ - the estimated coefficient for the value $tg\varphi$ of the consumer. For AC RT $\Pi 2$ there are no surcharges.

To construct the characteristic $\cos\phi$ (5) were used data from experimental trips on the locomotive VL-80t [4] and Skoda [9]. The approximating dependencies of the values on $\cos\phi(I_{da})$ the total current of the locomotive had the form:

$$f(x) = \frac{A}{B - e^{-Cx}} \quad (7)$$

Fig. 2 presents the results of tests and approximating their dependence.

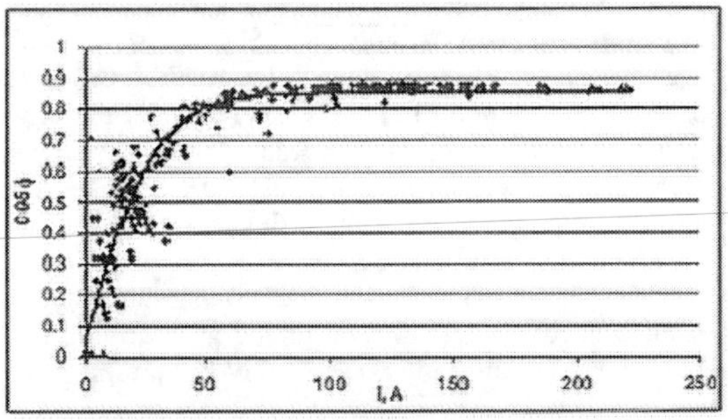

Fig. 2. Function approximation $\cos\phi(I_{da})$

The optimal modes of train control taking into account the values of active and reactive power, are shown at fig. 3 – fig. 5. Here for the same travel time over one track section are given the results of calculations of the train route section speeds, taking into account the flow rate and payment for the active and reactive power components. Horizontal lines limit speeds on track sections. The figures show these parameters – the time of movement on the section, electric energy consumption, as well as the number of time zones, if there were more than 1.

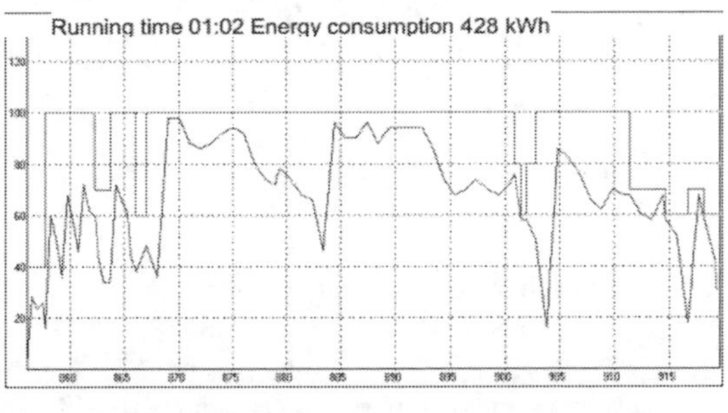

Fig. 3. Optimization of traction modes on the basis of total energy consumption

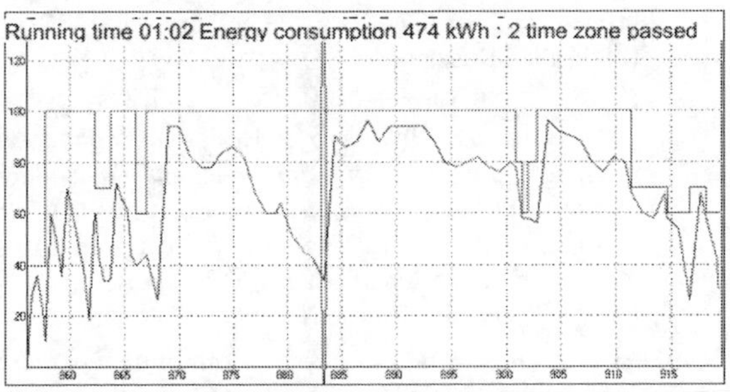

Fig. 4. Optimization of traction modes with regard to reactive energy cost

For the first case on fig. 3 the power consumption was minimal (the cost of energy is not taking into account). For the case on fig. 4 electricity consumption is bigger, but here more electric energy was used in the zone with a lower hourly rate to reduce the cost. For modes on fig. 5 the cost of energy is calculated more accurately taking into account the cost of reactive energy. We note the qualitative difference in the nature of train' control under various optimization criteria. In general, the calculations showed that the criteria developed and given by us, as well as models for optimizing train traction modes with the required accuracy provide both electricity metering and the expected estimate of its value in the framework of the new electricity market of Ukraine.

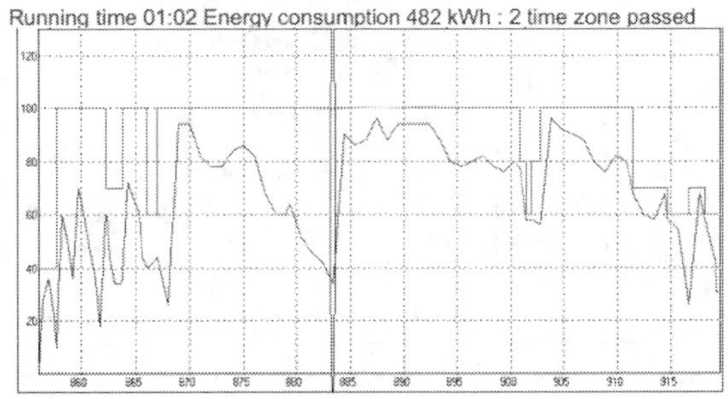

Fig. 5. Optimization of traction modes on the basis of the cost of full energy consumption

3. Methods of research and the formation of tariff models for the purchase of electricity on a balancing WEM

Previously, we pointed out a new task for simulating the power supply processes associated with the operation of RT on the balancing market — the selection and designation of rational parameters of the intra-day market. A reasonable choice of intraday tariffs of the WEM will then make it possible n for RT to plan the purchase of electricity and the transportation process in an advantageous way , fig. 1. The assignment of rational intraday tariffs of the WEM can be represented as a choice of relative hourly coefficients for a given base, single, tariff. As established in [2, 4, 8], variable tariffs require the development and implementation of MM packages used during different periods of the day. Since the change in intraday tariffs of the WEM cannot occur every hour, the problem of constructing an approximation of the tariff for electricity less than 24 in the number of relative price coefficients must be solved. To solve it can be used the methodology and procedures by which the four-zone approximation of the price of electricity in the WEM of Ukraine was previously obtained, presented in [4]. In the conditions of the new WEM of Ukraine, this technique should be part of a more general model for automating the choice for RT the intraday tariffs.

In order to clarify the content of the new tariff formation method, we will schematically show the stages of obtaining four zone approximation. On the basis of observational data was obtained an area of random estimates of the cost of electricity in the WEM over the periods of the day, were determined the averaged price coefficients. Further, calculations were made of the cost-effective MM when departing trains of a given weight every hour. Then, an approximation of the WEM hourly tariff was introduced using three or four zone tariffs, when the tariffs were set by relative coefficients. In table

1 are shown the calculated errors for approximate estimates of the cost of electricity consumption obtained using three or four zone tariffs for freight trains of different weights. On fig. 6 are presented comparative estimates graphically. The figure shows various models of variable tariffs for electricity: - DT, differentiated tariffs; - OPE, wholesale market rates; - OPE3, OPE4 tariffs for three and four zone approximation of cost in the WEM.

Table 1. Comparative analysis of the accuracy of WEM tariff approximation

Ratio tariffs	Train1 3000 t.	Train2 4050 t.	Train3 5025 t.	Train4 6000 t.	percentage of error
OPE/ OPE4	-0,3%	-1,5%	-1,5%	-1,9%	-1,3%
OPE/ OPE3	-3,2%	-2,5%	-1,6%	-2,9%	-2,5%

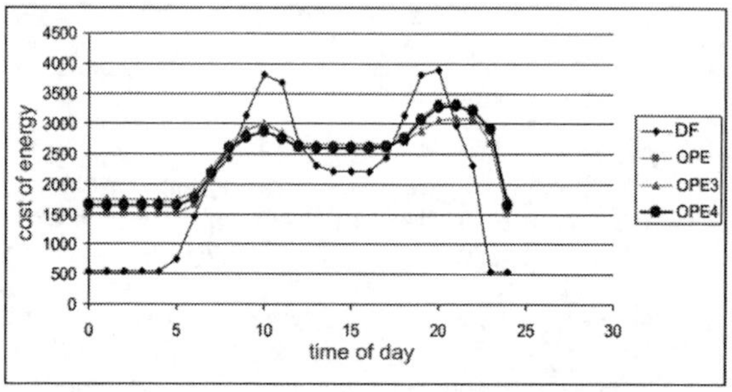

Fig. 6. Models of comparative estimates of the electric energy tariff

Thus, the four-zone model of approximation of tariffs made it possible with sufficient accuracy to replace the stochastic prices of the Ukrainian wholesale electricity market. An important feature of this method is that a rational model of approximation of tariff coefficients was chosen based on the analysis of the specified areas of their values by periods of the day, regardless of the train schedule. This circumstance serves as the basis for the decomposition of the tasks of modelling the power supply of the RT in the conditions of the balancing market of WEM. At the first stage of simulating of the electricity procurement, one should select three or four zone tariff models, and then set tariff coefficients using a simulation method and form a train schedule on the given section. Thus, the structure of the procurement model is as follows: Network - Train - MM.

The task of selecting zone tariff coefficients can be represented as an optimization model as follows: it is required to calculate four values of zone tariffs and build a timetable for a given set of trains of different categories (assign departure and arrival times taking into account the established system of restrictions) in order to minimize the total cost of electrical energy of the WME consumed for train traction, based on the optimal MM traffic with variable tariffs. The mathematical models and methods for selecting the optimal traction modes indicated earlier in the article should be used in solving this complex problem.

Conclusions

The problems of reducing operating costs through the introduction of energy and optimal at the cost of train schedules are among the highest priority for the railway transport of Ukraine. The introduction of the balancing electricity market in Ukraine in 2019 opens up new opportunities for organizing efficient transportation on the electrified sections of railways.

The article reviewed some new possibilities for using mathematical models for optimizing train driving modes according to the criteria for the minimum cost of consumed active and reactive electric energy. At the same time, a complex task was formulated for choosing the most efficient mechanisms for buying and selling electric energy in the wholesale electricity market of Ukraine, which are used in concluding bilateral agreements on the day-ahead markets, as well as setting parameters and tariffs for the intraday market.

The paper shows a significant difference in the optimal driving conditions of trains, which are calculated for different performance criteria, taking into account the charges for the consumed active and reactive energy. It has been established that the models used to optimize train traction modes are sufficient for WEM conditions. The method of forming zone tariffs presented in the article can be used for models in order to purchase of electricity in the balancing market.

References

1. E. P. Blokhin, A. N. Pshinko, G. V. Evdomaha, V. V. Skalozub, V.B. Zemlyanov *Selection of energetically optimal train operation regimes* Railway Transport of Ukraine, **No. 6**, P. 19 –22 (2001). [in Russian]
2. E. P. Blokhin, V. V. Kornienko, V. V. Skalozub *Resource-saving methods of train control with variable tariffs for electricity*. Sb. "Electrification and scientific and technical progress in railway transport". Russia, Saint-Petersburg, P. 112-118 (2004). [in Russian]
3. G. V. Evdomomaha, V. V. Skalozub, K. G. Zheleznov, A. P. Ivanov. *Modes of conducting passenger trains, taking into account the random voltage in the traction network and differentiated tariffs for electricity*. Bulletin of Dnipropetrovsk National University of Ralway Transport, **Iss. 3**, P. 65 -73 (2004). [in Russian]
4. A. P. Ivanov. *Study of train operation modes taking into account the cost of active and reactive energy*. Bulletin of Dnipropetrovsk National University of Transport and Transport, **Iss. 21**, P. 239–242 (2008). [in Russian]
5. V. V. Skalozub, V. G. Kuznetsov, D. A. Bosy, A. P. Ivanov. *Optimization of train operation modes on the basis of the criterion of minimum cost of consumed and reactive power*. Information-Control Systems for Railway Transport, **4 (78),** P. 111 - 115 (2008). [in Russian]

6. V. V. Skalozub, A. P. Ivanov. *Selection of train control modes based on fuzzy production model.* Bulletin of the Belarusian State University of Transport, Belarus, Science and Transport, **1-2**, P. 41 - 46. (2007) [in Russian]
7. Law of Ukraine «On the electricity market» . Access mode https://zakon.rada.gov.ua/laws/show/2019-19 (access date 09.07.2019). - Title from the screen.
8. V. G. Sychenko,V. G. Kuznetsov, D. O. Bosyi, O. I. Sablin. *Power engineering of traction networks.* Dnipro: Publ. PF "Standard-Service", 210 p. (2017). [in Ukrainian]
9. P. Hubskyi, V. Kuznetsov, A. Drubetskyi, A. Afanasov, M. Pulin. *Studying of the power modes in the traction line for ensuring the high-speed traffic.* Technology audit and production reserves. **Vol 5**, no **1(43)**. Industrial and technology systems. P. 42-51 (2018).
10. S. V. Arpul, G. K. Getman, A. I. Kyiko, Yu. V. Mykhailenko *The choice of rational parameters of the nominal mode of passenger electric locomotivesю.* Dnepropetrovsk: Makovetskyi, (2012) – 188 с. [in Russian]
11. A. M. Afanasov, A. Ye. Drubetskiy, S. V. Arpul, A. P. Khvorostyankina *Energy efficiency determination of loading-back system of electric traction machines.* Science and transport progress. Bulletin of Dnipropetrovsk National University of Transport and Transport – **Iss. 2(50).** – P. 67-73 (2014) [in Russian]
12. P. Daszkiewicz, M. Andrzejewski, MATEC Web of Conferences, *Preliminary analyzes in terms of the possibility of reducing energy consumption by the SM42 locomotive used in track works* **Volume: 118**, DOI: 10.1051/matecconf/201711800014, (2017).

Experimental investigation of actual situation of using and accounting of recovered energy of regenerative braking mode at the DC traction system

Oleh Bondar*, *Dmytro* Ustymenko, *Olena* Kurylenko, *Mykhailo* Kedria, *Oksana* Karzova, *Roman* Krasnov, and *Oleksii* Baliichuk

Dnipro National University of Railway Transport named after Academician V. Lazaryan, Department of Electrical Engineering and Electromechanics, Lazaryana Str. 2, room 125, 49000, Dnipro, Ukraine

Abstract. This paper deals with the problems of enhancement of registration and distribution of electric energy which can be obtained from DC electric locomotives by the regenerative braking. According to theoretical researches, it is possible to return approximately 8 - 12% of electric energy which is consumed for electric traction by using of a regenerative braking mode. However in recent years for electric traction networks of JSC "Ukrainian Railways", this index doesn't exceed 2.6%. So we consider that the regenerative braking is one of the most prospective ways for energy saving in electric railway transport. Results of the experimental researches which were executed on DC traction substations and VL11M6 electric locomotives are described in this paper. Instantaneous values of voltage and current in different modes have been obtained. The electric energy balance for experimental zone has been performed. Also the ways of distribution and consumption of energy of regenerative braking mode have been analyzed. On this basis of the executed investigations, recommendations about providing of supercapacitor energy storage system for Ukrainian railways are proposed.

1 Introduction

Railway transport is one of the largest and stable consumers of electric energy in Ukraine. Annual consumption of electric energy for traction by the Ukrainian railways exceeds 3800 MWh. At the same time the actual average tariff for the active electric power for traction annually grows so the problems of electric power efficiency at the railway transport are relevant for Ukraine. We consider that the regenerative braking is one of the most prospective ways for energy saving in electric railway transport. According to theoretical researches, it is possible to return approximately 8 - 12% of electric energy which is consumed for electric traction by using of a regenerative braking mode. However in recent years for electric traction networks of JSC "Ukrainian Railways" (UZ), this index doesn't exceed 2.6% [1].

Improvement of this situation can be reached by use of energy storage system (ESS) with supercapacitors (SC) [2-4]. Therefore in 2014 - 2015 years Cisdnieper Railway (PZ, a regional operator of the UZ) financed scientific researches for the purpose of improvement of an accounting system, distribution and a power consumption of a regenerated energy. Also propositions on implementation of ESS elements in 3.3 kV DC power supply system and into electric locomotives were developed. The experimental part of the specified scientific researches is described in this paper.

2 Choosing of experimental zone

For the reasonable choice of the railway section for experimental researches the analysis of turnover on sections of Cisdnieper Railway (PZ) is carried out by our researcher's group (head of group - D. Sc., Professor Andrii Mukha). Results for 2013 (orange bars) and 2014 (grey bars) is presented in Fig. 1. It is easy to see that the maximum turnover is registered on the sections that situated between Nyzhn'odniprovs'k-Vuzol station and Chaplyne station. Thus the experimental zone has to be located in borders of these sections. Also experimental zone must have high track gradients for effective electric energy obtaining by using of regenerative braking mode.

In such a way we chose a zone for the experimental researches between traction substation "Rozdory" (№. 19) and traction substation "Pys'menna" (№. 20). Thus, the experiment area consists of four block sections with a total length of 17.8 km. Chosen experimental zone has two tracks, which electrified by 3.3 kV DC power supply system with two ways feeding and parallel connection of overhead catenary systems of both tracks.

Also during the experiments it was decided to use VL11M6 electric locomotives which have the modern integrated system of diagnostics and monitoring.

* Corresponding author: bondardiit@gmail.com

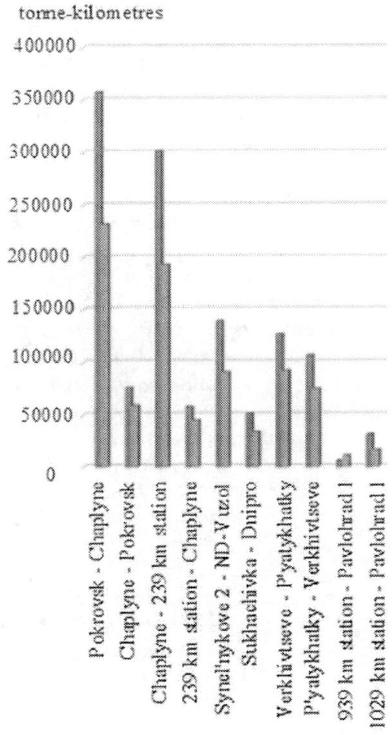

Fig. 1. Turnover at the most important sections of Cisdnieper Railway (PZ) in 2013 and 2014 years.

3 Experimental researches: stage 1

The purpose of the first stage of researches is experimental obtaining of time dependences of currents and voltages of traction substations and electric locomotives in the following modes: 1) "No-load mode"; 2) "Regenerative braking mode of electric locomotive", 3) "Regenerative braking mode of electric locomotive №1 and traction mode of electric locomotive №2".

Measurements into the traction substations were executed with using of kits of measurement equipment. Two kits of measuring equipment were installed at the traction substations. Thus, four such kits were used in experimental studies. Each kit consists of the insulated platform, an autonomous power source which consists of a rechargeable battery and the inverter, additional resistance (for voltage reducing), digital USB oscilloscope and the laptop with the specialized software.

That equipment was placed into location of 3.3 kV distributive devices of traction substations № 19 (see fig.2) and № 20 and connected to the electric traction network by the diagram which is presented in fig. 3.

The connection of the measuring equipment was carried out after a voltage disconnection. According to the diagram in Figure 3, a digital oscilloscope was connected through channel X to the secondary terminals of the RS shunt using a high-voltage cable. At the same time, the Y channel of this oscilloscope is connected to the measuring terminals of the voltage divider. Primary terminals of this additional resistance are connected to the "-'"bus of 3.3 kV distributive device. Thus, the common point X and Y channels of the oscilloscope have a potential of 3.3 kV relative to the earth. This fact necessitates the placement of equipment on insulated platform. To prevent the possibility of electric shock to researchers, the location of the equipment was fenced and the control of measuring equipment was provided by remote control using a wireless mouse-type manipulator.

Registration of time dependences of voltage on the pantograph and current of each electric locomotive is executed with using of its onboard integrated systems of diagnostics and monitoring.

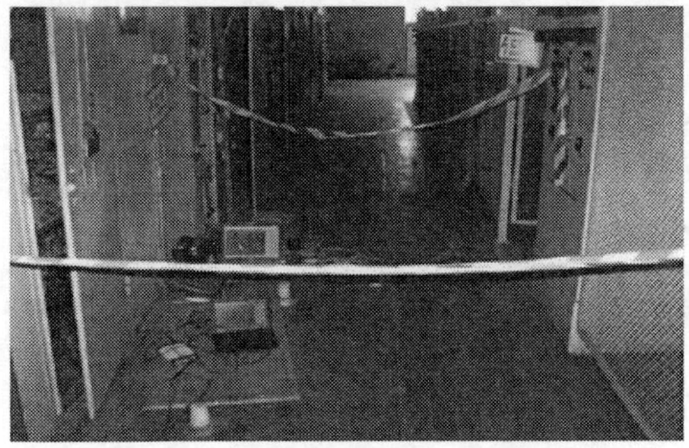

Fig. 2. Location of measurement equipment into the 3.3 kV distributive device of traction substation № 19.

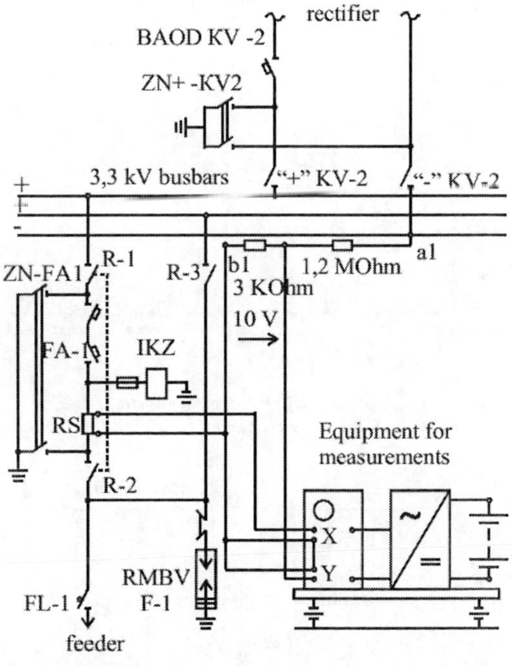

Fig. 3. Attaching diagram of measurement equipment for voltage and current registration.

In fig. 4 and fig. 5 an experimental time dependences of output voltage and FA 3 feeder current (traction substation №19) are presented.

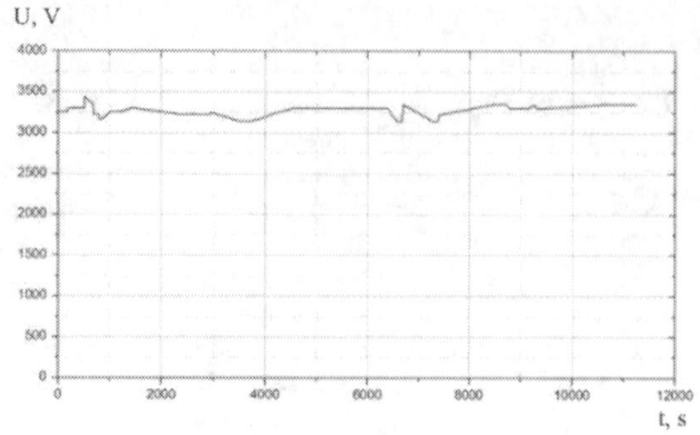

Fig. 4. Time dependence of output voltage of traction substation № 19 (after development).

From the theoretical point of view currents and voltages into the elements of the traction network are stochastic time functions. By statistical interpretation of the data, the laws of distribution of these random values are determined. In particular, it is defined, that voltage on the pantograph of an electric locomotive and its current in the mode of a regenerative braking can be described by the Gauss law. However the stochastic energy distribution of a regenerative braking W_R is described by the log-normal law according to expression (1).

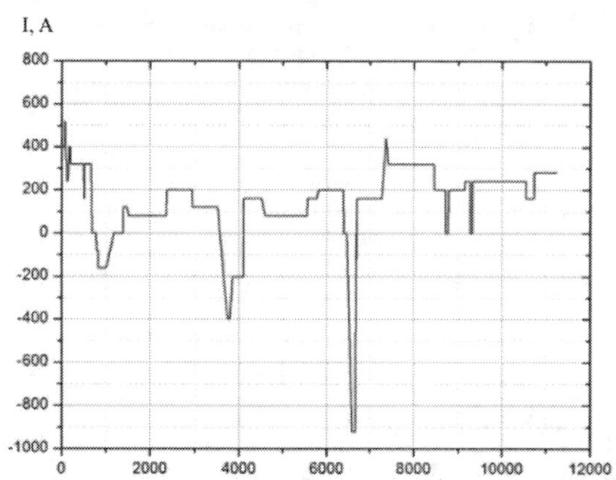

Fig. 5. Time dependence of FA 3 feeder current, traction substation № 19 (after development).

$$f(W_R)=\frac{1}{W_R\cdot\sigma_{W_R}\cdot\sqrt{2\cdot\pi}}\exp\left[-\frac{\left(\ln W_R-m_{W_R}\right)^2}{2\cdot\sigma_{W_R}^2}\right], \quad (1)$$

m_{W_R}, σ_{W_R} - are the mean and standard deviation.

4 Experimental researches: stage 2

The first stage of an experiment lasted 3 hours and 15 minutes. As it is possible to see from fig. 5, during the experiment, we registered time intervals with negative values of the feeder current. This fact indicates the presence of regenerative braking energy flows outside the experimental zone to other parts of the traction network along which electric locomotives move in the power consumption mode.

It is necessary to avoid such phenomena during the second stage of an experiment for the purpose of obtaining correct balance of electric energy for the experimental section. Therefore the following changes were made to conditions of fulfilment of the second stage of the experimental researches.

1. The zone of an experiment is expanded from traction substation "Rozdory" (№. 19) to the sectioning post 3.3 kV near Krasnoshchokove station. Total length of extended experimental zone - 47.6 km.

2. During experiment time on the section movement only of electric locomotive No. 1 and No. 2 is alowed.

3. Track №1 in operation only. Electrical feeding of track №2 is switched off.

4. Experimental zone electrically disconnected from another feeding zones.

5. Registrated electirc values: consumed and regenerated electric energy; voltages and currents of electric locomotives (instantaneous values) ; electric locomotive speed.

6. Type of electric locomotives is VL11M6. Mass of train №1 4458 tons, mass of train №2 5022 tons.

Modes of movement of trains along the experimental zone are shown in Fig. 6 and Fig. 7.

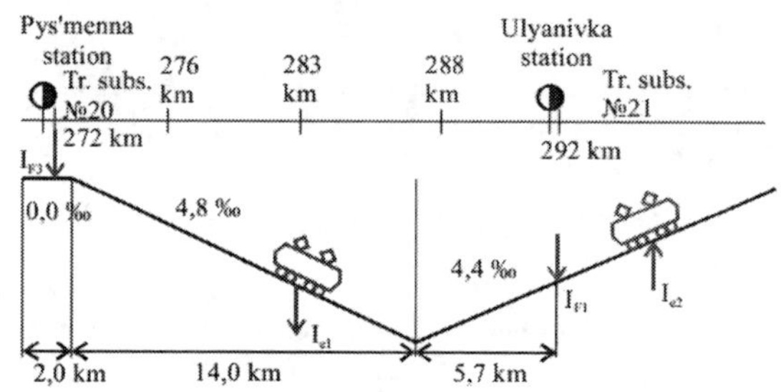

Fig. 6. Movement of trains along the experimental zone: traction mode of electric locomotive №1 and regenerative braking mode of electric locomotive №2.

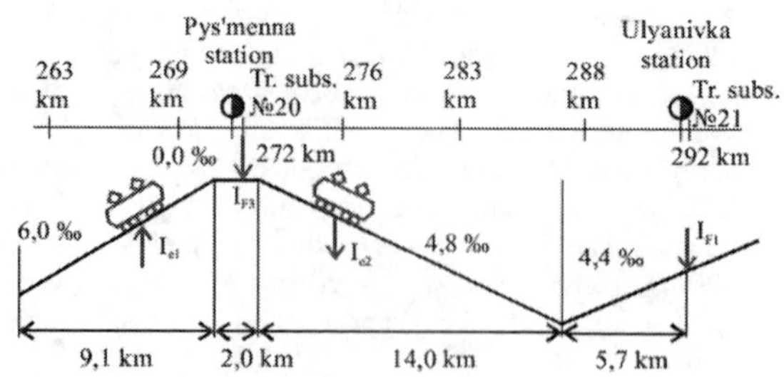

Fig. 7. Movement of trains along the experimental zone: regenerative braking mode of electric locomotive №1 and traction mode of electric locomotive №2.

As with the first stage of the experiment, the time dependences of the speeds, voltages, and currents of electric locomotives (Fig. 8-10) were obtained from the data of their built-in diagnostic and monitoring systems. These data made it possible to calculate the values of electrical energy that was consumed for train traction as well as the energy obtained by regenerative braking of electric locomotives №.1 and №.2 during the time of the experiment.

The results are presented in Table 1.

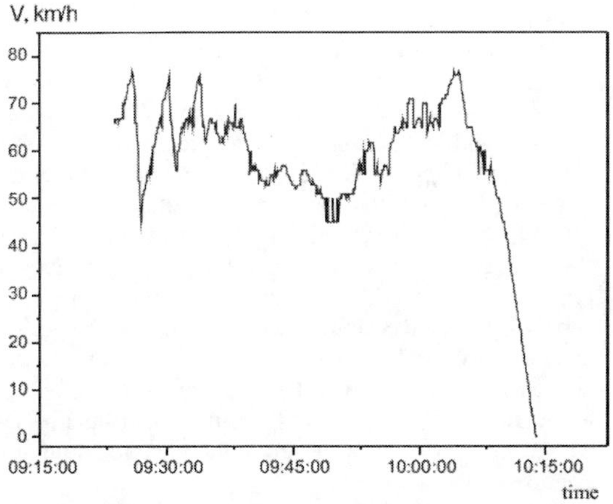

Fig. 8. Time dependence of train speed recorded from locomotive № 1 (experimental researches, stage 2).

Table 1. Consumed and regenerated energy of electric locomotives № 1 and №2.

Indicator	Electric locomotive № 1	Electric locomotive № 2
Energy for traction (according to the electricity meters data), kWh	1790	2190
Calculated electric energy for traction, kWh	1646	1558
Difference between experimental and calculated values of energy for traction, %	+8.75	+40.56
Energy of regenerative braking mode ((according to the electricity meters data), kWh	350	290
Calculated energy of regenerative braking mode, kWh	392	335
Difference between experimental and calculated values of recovered energy, %	-10.71	-13.43

According to the data from active electric energy meters of traction substations "Rozdory" (№19), "Pys'menna" (№20), "Ulyanivka" (№21), a temporary dependence of electric energy consumption was obtained during the experiment. The results are presented in Fig.11.

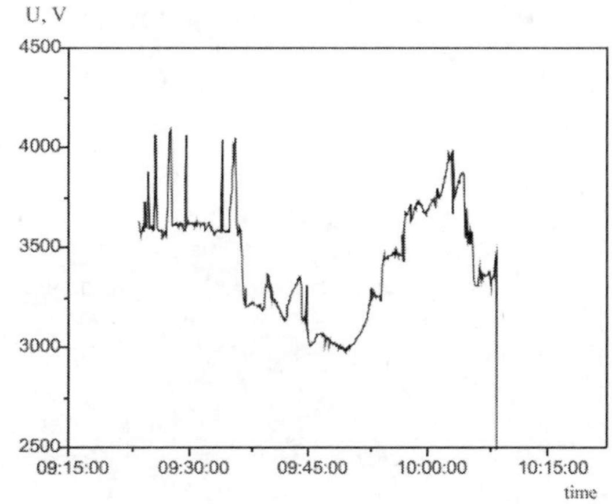

Fig. 9. Time dependence of pantograph voltage recorded from locomotive № 1 (experimental researches, stage 2).

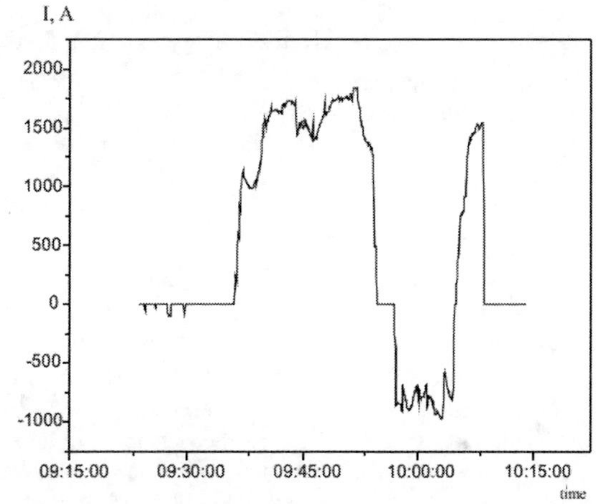

Fig. 10. Time dependence of current recorded from locomotive № 1 (experimental researches, stage 2).

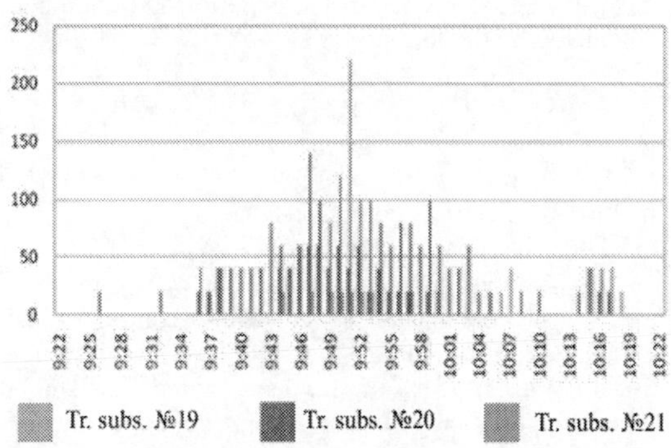

Fig. 11. Time dependence of electric energy consumption at the experimental zone (according to the electricity meters data).

This information makes it possible to determine the electrical energy transmitted by the traction substations to the electrical network of the experimental zone: W_S=2784 kWh.

Electric energy from regenerative braking mode of electric locomotives is $W_R = 392 + 335 = 727$ kWh (see Table 1). Thus, the total energy of all electric power sources at the experimental zone during the experiment is

$$W_{PS} = W_S + W_R, \quad (2)$$

$W_{PS} = 2784 + 727 = 3511$ kWh.

Electric energy losses in the traction network ΔW_T are calculated on the basis of experimental data based on the classical single-track model with two-way power supply [5] according to expression (2)

$$\Delta W_T = \frac{r \cdot 10^3}{T \cdot U^2} \cdot \sum_{k=1}^{n} \ell_{0k} \left[1{,}08T \cdot \left(1 - \frac{\ell_{0k}}{\ell}\right) \cdot \sum_{h=1}^{N} \frac{W_{kh}^2}{t_{kh\text{д}}} + \right.$$

$$\left. + 2W_{kt} \cdot \sum_{j=k+1}^{n} W_{jT} \left(1 - \frac{\ell_{0j}}{\ell}\right) \right], \quad (3)$$

where W_{kt}, W_{jT} – electric energy consumption by electric locomotives ;
T – time for calculations;
ℓ – length between traction substations;
ℓ_{0k}, ℓ_{0j} – distances from the traction substation to the middle of the sections k and j;
U – voltage of the traction network;
r – resistance of 1 km of traction network.

The results of the calculations show that the electric power losses in the traction network from the currents of the traction substations ΔW_T amounted to 149.1 kWh (approximately 5% of the energy supplied by the traction substations). At the same time, the regenerative braking energy losses in the traction network between electric locomotives ΔW_R reached 98 kWh (13.4% of the recovered energy).

Based on the obtained values, the divergence of the electric energy balance of the experimental zone can be calculated by the formula

$$\delta W = W_{PS} - W_{L1} - W_{L2} - \Delta W_T - \Delta W_R, \quad (4)$$

where W_{L1}, W_{L2} – electric energy consumed for traction by electric locomotives.

This divergence is $\delta W = 59.9$ kWh or 1.7% of the energy of power sources. However, if we use the data of electric power meters of electric locomotives when calculating this value, we will get the value of 803.1 kWh (23.45 % of the energy of power sources), which significantly exceeds the generally accepted norms and assumptions.

5 Location of ESS equipment: variants and proposals

There are two strategies for placing of ESS equipment with supercapacitors that is designed to store regenerative braking energy [2]. The first strategy (stationary ESS) is to arrange the ESS equipment at the traction substations (substation-inside ESS) or other elements of the traction power supply system (wayside ESS). Several literatures deal with the problems of implementations of the stationary ESS systems into DC traction power supply system of metro lines [2-3] and railway lines [6-9].

We suppose that in the system of traction power supply of Ukrainian railways (3.3 kV) for sections with heavy traffic can be used stationary ESS containing wayside and substation-inside elements. However, the development and implementation of this system requires large investments as well as a significant amount of scientific research. In this case, implementation of the ESS system is only an integral part of measures for improvement the energy efficiency of the traction power supply system. Therefore, when choosing the location and capacity of ESS system elements, it is necessary to take into account the possibility of implementation other innovations, for example, using distributed solar electricity generators to improve the voltage mode in the traction network. [10].

The second strategy involves placing the ESS equipment directly on the locomotive (on-board ESS) [4]. This option is advisable to use if the purpose of implementation an ESS is primarily the efficient use of regenerative braking energy. It is rational to take the capacity of the on-board ESS based on the most probable value of electric power (probability 0.94... 0.95) obtained in the process of regenerative braking. In turn, this value can be obtained by the method of stochastic analysis [11-13] using the data obtained during the experimental studies described in this paper.

A detailed comparison of both strategies for ESS storage by energy efficiency criteria is described in [14]. The results of calculation for various varieties of ESS are presented in tables 2 and 3.

Table 2. Calculation results for using substation-inside ESS (according to [14]).

Indicator	Traction substation without stationary ESS	Traction substation with stationary ESS	Difference, %
Electric energy consumption, kWh	100	93.6	-6.4
Electric energy loses into the traction network, %	3.5	3.6	+0.1
Minimum voltage, V	3017	3052	+1.2
Load factor of the main step-down transformer	0.25	0.23	-8.0
Load factor of the traction transformer	0.16	0.15	-6.2
Section bandwidth (even / odd direction), %	100	100	-
Minimum interval, min	12	12	-

Table 3. Calculation results for using on-board ESS (according to [14]).

Indicator	Electric locomotive without on-board ESS	Electric locomotive with on-board ESS	Difference, %
Electric energy consumption, kWh	100	81.1	-12.9
Electric energy loses into the traction network, %	6.5	1.9	-4.6
Minimum voltage, V	2888	3300	+14.3
Load factor of the main step-down transformer	1.39	0.49	-64.7
Load factor of the traction transformer	0.9	0.28	-68.9
Section bandwidth (even / odd direction), %	100	150.5	+50.5
Minimum interval, min	12	8	-33.3

Our own further calculations also showed that the use of the onboard ESS can significantly improve the efficiency of the traction power supply system. Electricity consumption for electric traction can be reduced by 18.9%. At the same time, the total losses in the transmission of electric energy are reduced by 12%. Also, the load of transformers and rectifiers at traction substations is significantly reduced.

6 Conclusions

As a result of the research we can draw the following conclusions.

1. It was experimentally confirmed that in the normal mode of operation of the experimental section, regenerative braking of an electric locomotive can occur when it is significantly removed from the electric locomotive in traction mode (distance of 20 km or even more). In this case, a substantial part of the regenerated energy is expended on heating the elements of the traction network.

2. Despite the fact that VL11M electric locomotive traction motors are DC electric machines, they are able to generate alternating low frequency voltages and currents (f = 0.01…9 Hz) into the traction network during regenerative braking of the electric locomotive, which are also stochastic values. In this case, in the experimental section, the voltage on the pantograph of an electric locomotive varies from 3102 to 3965 V and the regenerative braking current varies from 50 to 1734 A.

3. The value of electrical energy calculated by the results of monitoring (not according to the indications of electric power meters of the electric locomotives), which is recovered by an electric locomotive VL11M6, during one trip ranges from 807 to 1118.8 kWh. In this case, during one trip we observed 11…16 phases of regenerative braking with different duration. At each of these phases, the value of the recovered energy varies from 0.5 to 172 kWh.

4. Accounting for electric energy, which is consumed for traction of electric locomotives used in the experiment is unsatisfactory. The divergence between the calculated values and registered values by electric energy meters SKVT-D261 type can reach 40%. At the same time, the adequacy of the calculated values is verified by comparing with the data of high-precision electronic electric energy meters ABB EA05RL-P2C-3 type that are installed at the traction substations (divergence less than 1.7%).

5. Accounting of regenerated energy on electric locomotives used in the experiment is also unsatisfactory. The divergence between the data of the SKVT-D261 type electric energy meters and the experimentally determined values of the recovered energy can reach 13%. It is advisable to use meters with a higher accuracy class on electric locomotives.

6. We suppose that it is most expedient to accumulate the recovered energy directly on the electric locomotive using the electric energy storage system based on supercapacitors. Define of the rational parameters of ESS can be performed using stochastic analysis based on the time dependencies of the currents and voltages of electric locomotives that we obtained in the regenerative braking mode.

Finally we consider that implementation of using of ESS equipment is only s part of an integrated approach to the modernization of Ukrainian electric locomotives. Our research team also fulfils studies in areas of improvement the characteristics of pantograph contact inserts, the development of galvanic restoration technologies worn-out large-size details of the rolling stock, as well as the development and modelling of traction transformers for modern Ukrainian electric locomotives for the 27.5 kV AC electric power supply system.

References

1. A. Mukha, M. Kostin, O. Bondar, A. Nikitenko, O. Kurylenko, *Prerequsities of execution the experimental studies for definition of current state of using and accounting of the electrical rolling stock's regenerative energy on section of the "Pridniprovska" railway*, Electrification of Transport, Dnipropetrovsk, Ukraine, **№ 13**, pp. 54-58, (2017) [in Ukrainian]
2. B. Wang, Zh. Yang, F. Lin, W. Zhao, *An Improved Genetic Algorithm for Optimal Stationary Energy Storage System Locating and Sizing* , Energies, Basel, Switzerland, **7(10)**, pp. 6434-6458, (2014)
3. H. Xia, H. Chen, Zh. Yang, F. Lin, B. Wang, *Optimal Energy Management, Location and Size for Stationary Energy Storage System in a Metro Line Based on Genetic Algorithm*, Energies, Basel, Switzerland, **8(10)**, pp. 11618-11640, (2015)

4. Y. Geng, B. Li, Zh Yang, F. Lin, H. Sun, *A High Efficiency Charging Strategy for a Supercapacitor Using a Wireless Power Transfer System Based on Inductor/Capacitor/Capacitor (LCC) Compensation Topology*, Energies, Basel, Switzerland, **10(1)**, 135 pp. 1-17 (2017)
5. K. G. Markvardt, *Power supply of electrified railways*, Pub. "Transport", Moscow, (1982) [in Russian]
6. T. Ratniyomchai, S. Hillmansen, P. Tricoli, *Optimal Capacity and Positioning of Stationary Supercapacitors for Light Rail Vehicle Systems, International Symposium on Power Electronics*, Electrical Drives, Automation and Motion, IEEE, 807-812 (2014).
7. P. V. Radu, and Z. Drabek, "Analysis of wayside energy storage devices for DC heavy rail transport", *Proceedings of 13th International Conference "Modern Electrified Transport"* MET'2017, Warsaw, Poland, **Vol. 180**, 04001 (2018)
8. A. Szelag, T. Maciolek, Z. Drazek, M. Lewandowski, *Conditions for the effective use of traction substations with possible return of trains regenerative braking energy into a power network in a 3 kV DC supply system*, Logistyka 3, pp. 4683-4692 (2015)
9. W. Jefimowski, and A. Nikitenko, "Case study of stationary energy storage device in a 3 kV DC traction system", *Proceedings of 13th International Conference "Modern Electrified Transport"* MET'2017, Warsaw, Poland, **Vol. 180**, 02005 (2018)
10. V. G. Sychenko, D. O. Bosiy, E. M. Kosarev , *Improving the quality of voltage in the system of traction power supply of direct current*, Archives of Transport, Warsaw, Poland, **Vol. 35**, Iss. 3, pp. 63-70 (2015)
11. M. Kostin, *The methods for powers determination in the systems withstochastic electric processes*, Technical Electrodynamics, Kyiv, Ukraine, , **vol. 6**, pp. 6-8 (2006)
12. M. Kostin, A. Nikitenko, *Statistics and Probability Analysis of Voltage on the Pantograph of DC Electric Locomotive in the Recuperation Mode*, Przegląd Electrotechniczny, Warsaw, Poland, **No 2a**, pp. 273-275 (2013)
13. A. Nikitenko, M. Kostin, *The method of the correlation and dispersion defining of the total power components in the electric transport devices*, Science and Transport Progress. Bulletin of Dnipropetrovsk National University of Railway Transport, Dnipropetrovsk, Ukraine, **vol. 1(44)**, pp. 64-75 (2013)
14. V.T. Cheremisin, M. M. Nikiforov, V . L. Nevezak, *The choice of placement of equipment for energy storage at the DC experimental section according to the criterion of energy efficiency*, Science and transport, St. Petersburg, Russia, **2(6)**, pp. 48-52 (2013)

Fryze reactive power of trams in effective stochastic recuperation processes

Oleh Bondar[1], *Mikola* Kostin[1], *Andrei* Mukha[1], *Olha* Sheikina[1], and *Svitlana* Levytska[2]

[1] Dnipro National University of Railway Transport named after Academician V. Lazaryan, 2, Lazaryan Str., Dnipro, 49010, Ukraine
[2] Prydneprovsk state Academy of Civil Engineering and Architecture, 24-a, Chernyshevskoho Str., Dnipro, 49600, Ukraine

Abstract. Urban electric transport system, particularly tram systems, is not a direct current system not only in traction mode but in regenerative modes as both voltage on a collector and regenerative current are stochastic abruptly variable processes. The above- mentioned facts determine availability of Fryze's reactive power in this system that flows from a railway substation to trams, leads to incidental losses of energy and significantly reduces its quality. So evaluation of power effectiveness of the system in electrical trams operation is impossible without determining the level of reactive power in this system. We have analytical expression of reactive power by Fryze. Numerical calculations for trams type T3D and T4D in regenerative braking modes are done. Probabilistic statistical data processing operation of reactive power expressions is done. It is determined that reactive power changes in the limit of 10…100 kilo-volt ampere reactive with mathematical expectation – 37,0 kilo-volt ampere reactive. Statistical allocation of random power values are different. Numerical calculations of incidental losses, energy of recuperation are done and they range supplementary – 20% from total losses. It is stated that coefficient of reactive power of system route of trams is exceeding permissible value 0,25.

1 Research Objective

Municipal trams are the most widespread kind of electric rolling stock (ERS) of electric transport where there are often stops and starts and also regenerative braking (RB). The use of energy recovery in electric rolling stock is devoted to a significant number of scientific papers [1,2] which contain both theoretical aspects and the results of experiments, including on the rolling stock of railways. The methodology of the experiments can be applied to urban electric transport, this aspect and sanctifies the presented robot.

The further usage of recuperative energy allows to reduce energy consumption by 30…40 % at tram movement. However, electric power effectiveness of the tram system service is to be estimated not only according to the volume and the usage state of recuperative energy but also taking into consideration such energy indicators as active power losses ΔP and reactive power factor tgφ, that significantly depend on reactive power Q_F as to well- known expressions:

$$\Delta P = \frac{P^2 + Q_F^2}{U^2} R , \qquad (1)$$

$$tg\varphi = \frac{Q_F}{P} . \qquad (2)$$

In contrast these indexes are not nearly investigated in spite of the fact that the processes of RB have been applied for a long time, almost from the beginning of introduction of track return system. Probably, it is connected with the thought that as the traction motors of trams are DC machines, so generated in the RB modes voltage and current are also constant so then the reactive power is missing. Actually, recuperated voltage $U(t)$, and current $I(t)$ are variable ,even abruptly variable values that are typical for trams during the movement of ERS. In addition, they have stochastic nature (Fig. 1) due to technological probabilistic factors: weight and speed of wagons; rail profile; the mode of tram control (driver's qualification) etc. Therefore, the accuracy of estimation of electric power effectiveness of trams movement system in the RB modes is also defined by their reactive power level.

The object of the paper is to assess the impact of stochastic behavior of trams' traction voltages and currents on their Fryze's reactive power, energy indexes and technological losses of energy in regenerative modes.

* Corresponding author: sheikina.diit@gmail.com

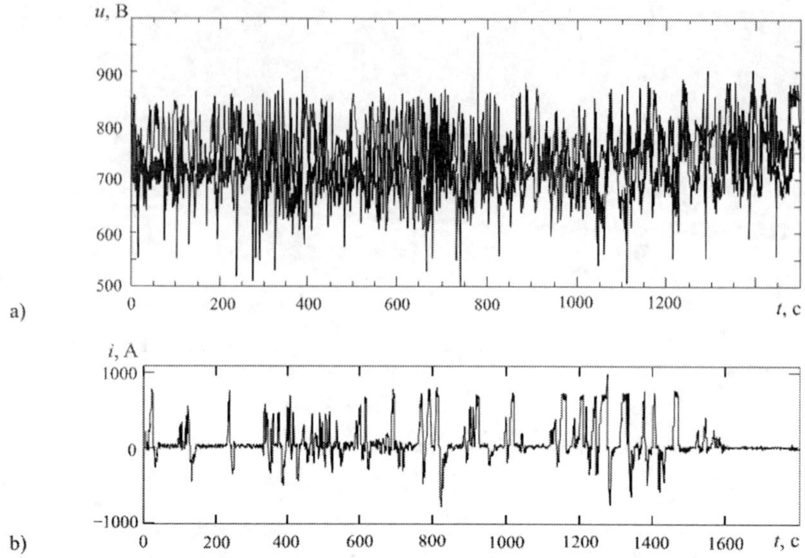

Fig. 1. Power tracing (a) and current (b) of tram in track regenerative mode: negative current values – recuperative current

2 Theoretical Assumptions in Determination of Powers.

The ambiguity of concepts and formula determining reactive power in electric circuits with non-alternating voltage and current is well-known [3]. The main evidence of reactive power in the mentioned circuits which are specific also for track return system are: inequality of active and total powers of loading (P≠S); distinction from zero angle of phase displacement between input voltage and current ($\varphi^{(k)}=0$); the evidence of energy return from loading into source; variation in time of input instantaneous load resistance. Specified evidence corresponds most likely (in term of quantity) to the concepts and expressions of reactive power as to K. Budeany and S. Fryze and also differential, integral and generalized reactive powers [3]. However, all scholars have the same idea as to the point of view in determining energy-efficiency as in fixed and so in transient state of system operation more correct and effective is the Fryze concept [4], according to which Fryze's reactive power is defined by well- known expression:

$$Q_F = \sqrt{S^2 - P^2}\ . \tag{3}$$

Stochastic nature of generated $U(t)$, $I(t)$ does not allow to use well-known in theoretical electrical engineering classical expressions for determining the reactive power, it is necessary to develop some specific methods that are performed particularly in [5]. Then in this paper we use the improved correlation method, the idea of which is given in [3] that is based on correlation theory of random stationary processes [6,7].

According to this theory, correlation function of voltage $K_U(\tau)$ and current $K_I(\tau)$ can be represented as mathematical expectations of scalar product of centered process ($U(t)$ or $I(t)$) and shifted to correlation window (time shift) τ copy of this process. Then, taking into account that unbiased estimation of mathematical expectation of stationary random process is its mean value for the time term of duration T of member function can be written as:

$$K_U(\tau) = \frac{1}{T}\int_0^T [u(t) - m_U]\cdot[u(t+\tau) - m_U]dt\ , \tag{4}$$

$$K_I(\tau) = \frac{1}{T}\int_0^T [i(t) - m_I]\cdot[i(t+\tau) - m_I]dt\ , \tag{5}$$

where m_U, m_I are mathematical expectations of time-independent processes as to the voltage and current.

In case of time shift τ=0, after simple transformations and integration of relations (4) and (5) we have expressions of active values of voltage U on a tram's current collector and its current I:

$$U = \sqrt{D_U + m_U^2}\ , \tag{6}$$

$$I = \sqrt{D_I + m_I^2}\ , \tag{7}$$

where D_U, D_I are dispersions as to the voltage and current that are equal to: $D_U = K_U(\tau = 0)$; $D_I = K_I(\tau = 0)$.

Taking account of (6) and (7), total power S recuperated by a tram will be written by well-known formula as:

$$S = UI = \sqrt{\left(D_U + m_U^2\right)\cdot\left(D_I + m_I^2\right)}\ . \tag{8}$$

Active power of recuperation P is defined using mathematical property of relative correlation function (in this case)of the voltage and current as mathematical expectations of scalar product of centered stochastic process of voltage $U(t)$ at shifted copy of centered process of current $I(t)$:

$$K_{UI}(\tau)=\frac{1}{T}\int_0^T[u(t)-m_U]\cdot[i(t+\tau)-m_I]dt=$$
$$=\frac{1}{T}\int_0^T u(t)i(t+\tau)dt-\frac{m_I}{T}\int_0^T u(t)dt- \quad (9)$$
$$-\frac{m_U}{T}\int_0^T i(t+\tau)dt+\frac{1}{T}\int_0^T m_U m_I dt.$$

During time shift equal to zero (τ=0), after integration and in case $m_{U\text{cp}}=m_U$, $m_{I\text{cp}}=m_I$, we have:

$$K_{UI}(\tau=0)=\frac{1}{T}\int_0^T u(t)i(t)dt-m_I m_U-m_U m_I+m_U m_I, \quad (10)$$

where $m_{U\text{cp}}$, $m_{I\text{cp}}$ are mean arithmetic values as to voltage and current for time interval T.

In (10) first term is an active power P recuperated by a tram, then:

$$P=m_U m_I+K_{UI}(\tau=0). \quad (11)$$

Then the required Fryze's reactive power is determined as:

$$Q_F=\sqrt{\left[\left(D_U+m_U^2\right)\cdot\left(D_I+m_I^2\right)\right]-\left[m_U m_I+K_{UI}(\tau=0)\right]^2}. \quad (12)$$

The flowing of reactive power QF in the system "traction substation - tram" is related to incidental technological losses of electric energy recuperation. These losses are connected with defective restored energy and that is why should be the main effectiveness criterion of energy processes in the RB tram modes.

Nowadays there are no devices and methods for experimental observation of incidental losses and they can be determined only by calculation according to the expression:

$$\Delta W_{\text{д}}=R\frac{Q_F^2}{U^2}\tau_{\text{р}}, \quad (13)$$

where R is active resistance of service area of electric traction network of tram power supply; U is effective value of voltage on a tram's current collector in the regenerative mode; τ - recuperation time.

3 Results and analysis of numerical calculations

Using the obtained expressions and monitoring of the material from experimental record 30 voltage $U(t)$ and current $I(t)$ implementations (Fig. 2) that are carried out according to the theory [9], numerical calculations of reactive power of trams of T3D and T4D types in regenerative braking modes were done during their operation on the routes 11 and 17 of Dnipro city. Correlation functions of voltage on current collectors of trams and their regenerative current that are shown in Fig.3.

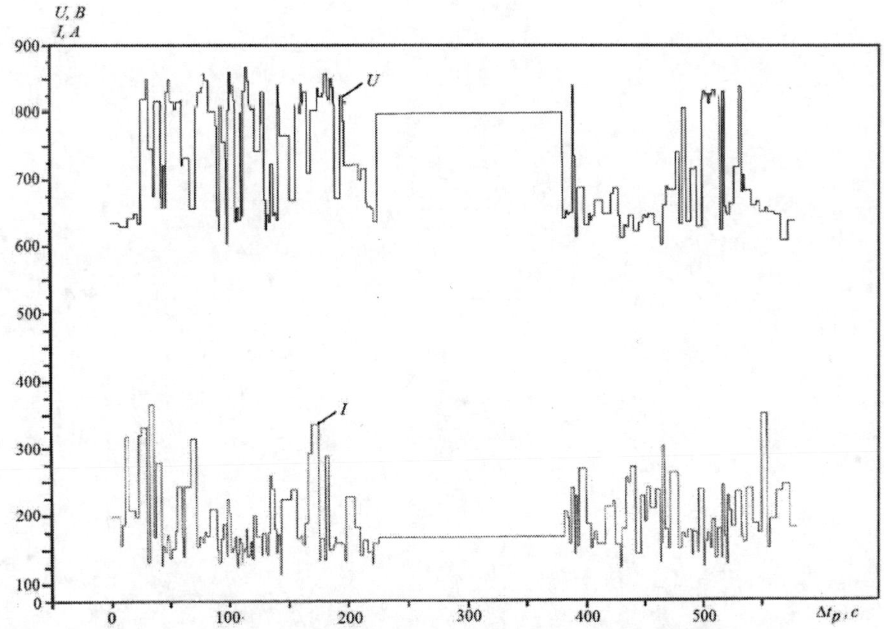

Fig. 2. Continuous voltage and current diagram of T4D tram in the recuperation mode

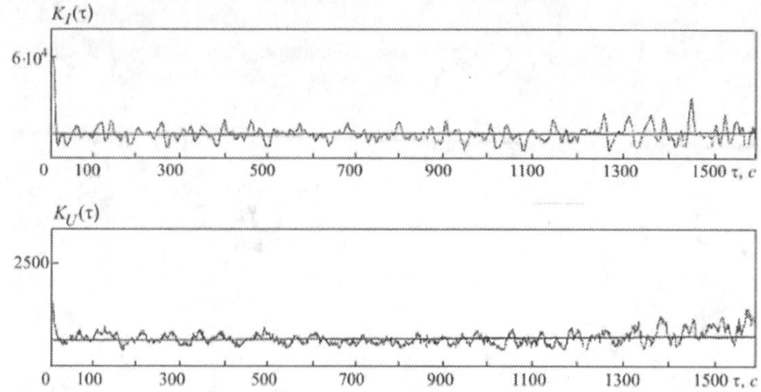

Fig. 3. Correlation functions of current and voltage of T4D tram in the recuperation mode

Abrupt random changes of voltage (Fig. 2, a) within 600…900 volt (at mathematical expectation 720 volt) and current (Fig. 2, b) at interval 100…350 ampere (at mathematical expectation 183.4A) lead to (as it is shown in Fig. 4) Fryze's reactive power Q_F of trams is a random value that changes within 10…100 kilovolt-ampere reactive for one rail ride on the pointed route. In this way, mathematical expectations irrespective of a tram type and tram number equals to ~ 37.0 kilovolt-ampere reactive. Statistical distributions of the value Q_F are different in relation to the routes: particularly, at the route 17 it is the law of Gauss with asymmetry coefficient 0.72 and excess 0.54, on the route 11- the value Q_F follows asymmetric normal law with dispersive function [10].

The above-mentioned availability of reactive power leads to sufficiently high values of reactive power factor tg in the limits of 0.1…0,486 (at acceptable value 0.25) and incidental losses $\Delta W_{д}$ of restored energy (Fig. 5), that range to 20% of total losses ΔW.

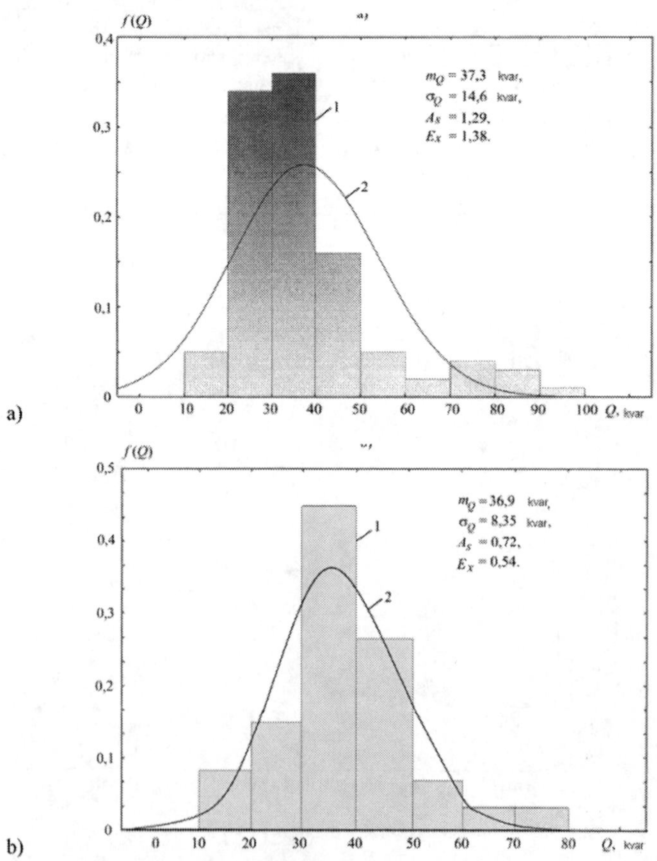

Fig. 4. Reactive power of T4D tram in the recuperation mode during the route № 11 (a) and route № 17 (b): statistical (1) and theoretical (2) distribution

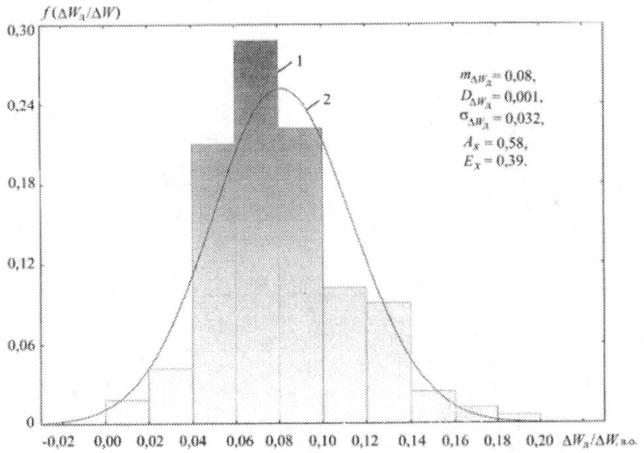

Fig. 5. Relative values of incidental losses of energy of recuperation in the system of electrically driven trams of type T3D and T4D: statistical (1) and theoretical (2) distribution

4 Conclusions

1. Abrupt random behavior of voltage change on a tram's current collector in regenerative mode leads to availability of Fryze's reactive power in the system.

2. In the phase of regenerative braking tram's voltage and current are stationary random process that gives the opportunity to define the reactive power and use correlation theory of random processes and particularly correlation–dispersion method.

3. Statistical distributions of reactive power are different according to the tram type and its route.

4. Availability of reactive power causes the occurring of incidental losses of restored energy in electric tram system and large (up to 0.486) coefficient values of reactive power tgφ.

References

1. A. Mukha, M. Kostin, O. Bondar, A. Nikitenko, O. Kurylenko, *Prerequsities of execution the experimental studies for definition of current state of using and accounting of the electrical rolling stock's regenerative energy on section of the "Pridniprovska" railway*, Electrification of Transport, Dnipropetrovsk, Ukraine, **№ 13**, pp. 54-58, (2017) [in Ukrainian]
2. A. Nikitenko, M. Kostin, *The method of the correlation and dispersion defining of the total power components in the electric transport devices*, Science and Transport Progress. Bulletin of Dnipropetrovsk National University of Railway Transport, Dnipropetrovsk, Ukraine, **vol. 1(44)**, pp. 64-75, (2013)
3. Czuchra W. *Ocena energochłonności tramwajów z napędem asynchronicznym.* 7tn International Conference «Modern Electric Transport in Integrated XXI st Century Europe», Poland, Warsaw, – pp. 160-164, (2005)
4. Kostin M., Nikitenko A. *Statistics and Probality Analysis of Voltage on the Pantograph of DC Electric Locomotive in the Recuperation Mode* Przeglad Electrotechniczny. Warsaw, Poland. **No 2a**. – pp. 273-275, (2013)
5. Sajenko Ju. L. *Reaktyvna potuzhnist' v systemah elektropostachannja z nelinijnymy navantazhennjamy.* Avtoref. dyss. na zdobuttja nauk. stupenja doktora tehn. nauk [Reactive power in power supply systems with nonlinear loads. Author's abstract. dis for obtaining sciences. doctor of technical sciences.]. L'viv, 36 p (2003) [in Ukrainian]
6. Domnin I. F., Zhemerov G. G., Krilov D. S. [i dr.] *Sovremennye teorii moshhnosti i ih ispol'zovanie v preobrazovatel'nyh sistemah jelektroniki Modern Power Theories and Their Use in Transforming Electronic Systems.* Tehnichna elektrodynamika – Technical electrodynamics, **no 1**, pp. 80-91 (2004) [in Russian]
7. Czarnecki L.: *On some misinterpretations of the instantaneous reactive power p-q theory.* IEEE Trans. Power Electron, **Vol. 19, Iss. 3**, pp.828-836 (2004)
8. Herrera R. et al.: *Present point of view about the instantaneous reactive power theory.* IET Power Electron., **Vol. 2, Iss. 5**, pp. 484-495 (2009)
9. Fryze S. *Wirk-, build- und scheinleistung in elektrischen stromkreisen min nicht sinsformigen verfaf von strom und spanning* / S. Fryze // *ETZ*, Germany, **No 25**, pp. 596-599; **No 26**, pp. 625-627; **No 29**, pp. 700-702 (1932)
10. Kostin M. O., Petrov A. V. *Metody vyznachennja skladovyh povnoi' potuzhnosti v systemah elektrychnoi' tjagy* [Methods of definition of components of total power in systems of electric

propulsion]. Tehnichna elektrodynamika – Technical electrodynamics, **no 3**, pp. 53-59 (2011) [in Ukrainian]

11. Pugachev V.S. *Teoriya sluchaynykh funktsiy i eye primeneniye k zadacham avtomaticheskogo upravleniya* [The theory of the random functions and its using in the problems of the self-operated control systems]. Moscow : Gostekhizdat Publ., – 659 p (1957) [in Russian]
12. Kostin N. A. *Nekanonicheskoye spektralnoye razlozheniye sluchaynykh funktsiy tyagovykh napryazheniya i toka v sistemakh elektricheskogo transporta* [Noncanonical spectral decomposition of random functions of traction voltage and current in electric transport systems.]. Elektrotekhnika i Elektromekhanika – Electrical engineering and electromechanics, **no. 1**, pp. 68-71 (2015) [in Russian]

Improvement of energy indicators of conventional electric driving asynchronous motors of non-standard technological equipment at the enterprises for repair of railway rolling stock through the use of energy-saving motors

Oleh Bondar, *Oleksii* Baliichuk, *Oksana* Karzova, *Roman* Krasnov, *Oksana* Marenych, and *Dmytro* Ustimenko

Dnipro National University of Railway Transport, Department of electrical engineering and electromechanics, 49010, Dnipro, Lazarian Str., 2, Ukraine

Abstract. The purpose of this work is to study the utilization efficiency of electric driving energy-saving motors of non-standard technological equipment meant for repair of railway rolling stock, in terms of improvement of energy performance, i.e. improvement of efficiency factors and power factors, and decrease of power losses as compared with the most commonly used "conventional" asynchronous motors. There are two options for improvement of the energy performance of electric drives of non-standard technological equipment: replacement of a conventional motor with an energy-saving one similar in terms of nominal voltage, rotational speed, power, etc.; changeover of stator winding from "delta" to "star". A special technique has been proposed to determine the effectiveness of this changeover in terms of improvement of energy indicators. Comparison of the real values of the energy indicators of conventional motors and energy-saving motors in the case under consideration is required to perform the technical and economic calculations when solving the problem of the expediency of replacement of conventional asynchronous motors with energy-saving motors in the abovementioned equipment.

1. Introduction

Nowadays, conventional asynchronous motors (AI, AIR, AIS, etc.) are most commonly used in electric drives of various units of equipment at the enterprises of Ukraine. The total installed power capacity of the motors in Ukraine is approximately 40 ... 50 million kilowatts [1].

A considerable amount of constant speed motors is used in various repair facilities, in particular, in the rolling stock repair providers, i.e. non-standard technological equipment meant for repair of locomotives, railway cars and multiple-unit rolling stock [2].

Technological processes at these enterprises permit the use of an electric drive without speed control, which is the most easy-to-operate, reliable, energy-efficient and cheap.

2. Analysis of literature data and problem statement

The improvement of energy performance of electric driving asynchronous motors (improvement of efficiency factor and power factor, decrease of power losses) comes at an opportune time, when energy resource saving has become a priority task [10, 11, 12, 13]. The question of energy saving is acute across the globe, but this topic is especially relevant for Ukraine, where the electricity costs per product unit are significantly higher than in the developed countries [1].

In this paper we developed the methodology and conducted research to increase the energy indicators of the electric drives of the abovementioned non-standard technological equipment through the use of energy-saving electric motors, which in recent years have found significant application in some countries (USA, Germany, Japan, etc.), but have very limited applicability in Ukraine. Energy-saving motors reduce the electricity costs while in operation due to higher efficiency and power factors ($\cos\varphi$) [1, 4, 14, 15, 17, 18, 19 and 20].

A complete set of non-standard technological equipment meant for repair of locomotives, railway cars and multiple-unit rolling stock is taken as a basic hardware [2].

Practice shows that operation of the abovementioned equipment is characterized by the fact that, depending on the technological process and type of rolling stock under repair, the load factor $k_н$ of electric driving motors of a significant number of devices (washing machines, conveyors, lifting mechanisms, etc.) can be equal to 0,2 ... 0.3 depending on the weight of the rolling stock unit under repair.

* Corresponding author: lejikbaliychuk@gmail.com

$$k_н = \frac{P_т}{P_н}, \qquad (1)$$

where $P_т$ - true (actual) power developed by electric motor;

$P_н$ - nominal (installed) power capacity of electric motor.

At the same time, the nominal rating power of electric driving motor must be such as to make the equipment work properly at maximum load ($k_н \approx 1$).

For example, an electric driving asynchronous motor of conventional cleaning machines for cleaning of traction electric motors works when $k_н \approx 1$ in case of cleaning of traction electric motor of the powerful electric freight locomotive, and when $k_н \approx 0{,}3$ in case of cleaning of traction motor of motor car unit.

The basic hardware includes more than fifty units of equipment fitted with conventional electric driving asynchronous motors. For example, manipulator for bogie frames, manipulator for automatic coupling devices, cleaning machines for cleaning of diesel motors, traction motors, etc.

The electric motor power falls within the range from 0.37 kW (manipulator for repair of draw clamps) to 150 kW (automated traction electric motor testing station AIS-TED).

The comparison of actual efficiency factors and power factors ($\cos\varphi$) of conventional motors and energy-saving motors in terms of different $k_н$ values is required for performing the technical and economic calculations in order to determine the advisability of the use of energy-saving motors. Note the fact that for performing the technical and economic calculations it will suffice it to know the weighted average values of efficiency factors and power factors, depending on the intended load.

3. Purpose and objectives of research

The purpose of this work is to assess the feasibility of electric driving energy-saving motors of non-standard technological equipment meant for repair of railway rolling stock in terms of improvement of energy indicators, i.e. improvement of efficiency factor (EF) and power factor ($\cos\varphi$), and decrease of power losses (ΔP) as compared with the most commonly used "conventional" asynchronous motors.

4. Materials and methods of research

The quantitative improvement of efficiency factor (η) and power factor ($\cos\varphi$) through the replacement of conventional asynchronous motors with energy-saving motors (at $k_н \approx 1$) can be estimated using the characteristics shown in Figure 1 [1, 4].

As can be seen from Figure 1, if the nominal rating power of the electric driving motor of any device is $P_н < 3$ kW, the replacement of a conventional electric motor with an energy-saving one will produce a gain in the value of $\cos\varphi$ only, whereas there will be a loss in the value of efficiency. For example, this is about a bearing shield assembly and disassembly press RSP-15 ($P_н = 2$ kW, 380 V), manipulator for automatic coupling devices KRS-AS ($P_н = 0{,}8$ kW, 380 V) and other equipment. This fact must be taken into account when performing the technical and economic calculations in order to assess the feasibility of replacement of conventional motors with energy-saving ones.

The range of power levels $P_н = 3...220$ kW includes all values of nominal rating power of the basic set of non-standard equipment ($P_{н\,max} = 150$ kW – engine of the automated traction electric motor testing station AIS-TED). Replacement of conventional asynchronous motors with energy-saving motors similar in terms of power in the specified range of power levels improves both efficiency (EF) and performance ($\cos\varphi$).

Indeed, the power of conventional motors of all devices included in the basic hardware is selected properly and it is unnecessary to change it. In other words, the technical characteristics of energy-saving asynchronous motors (nominal voltage, rotational speed, power, etc.) should be identical (or close) to the technical characteristics of conventional asynchronous motors. Thus, we can use the characteristics shown in Figure 1 [4] to conduct a comparative estimation of conventional asynchronous motors and energy-saving asynchronous motors in terms of energy indicators, which is required for performing the technical and economic calculations in deciding the question of the expediency of replacement of convenient asynchronous motors with energy-saving asynchronous motors in the basic hardware ($P_н = 3...150$ kW).

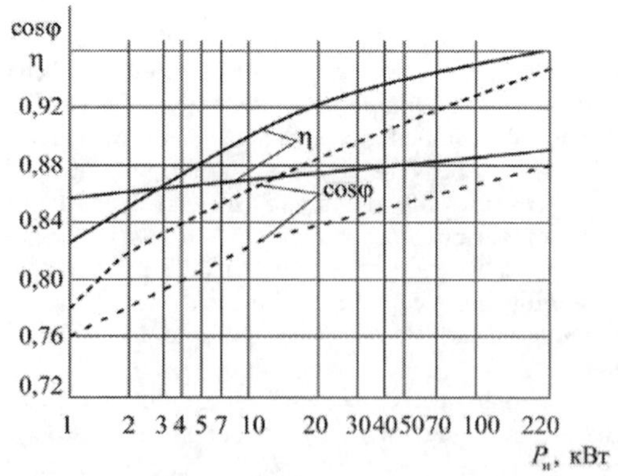

Fig. 1. Comparative estimation of the parameters of conventional asynchronous motors and energy-saving asynchronous motors.

When producing new devices in order to increase the energy indicators of the drive, it is expedient to immediately use the energy-saving motors with the required technical characteristics, if it is possible to acquire them [8, 16].

As noted above, according to the technological conditions, the load factor $k_н$ of the electric driving motors in a significant number of units included in the basic set of non-standard equipment may decrease to 0.2 ... 0.3 depending on the type of rolling stock under repair.

It is known that the energy indicators of an asynchronous motor decrease with the decrease of $k_н$ at the asynchronous motor shaft. In our case, the replacement of electric motors every time in case of underloading to improve the performance is difficult if not impossible according to the technological conditions. Therefore, the expediency of voltage reduction in underloaded asynchronous motors should be checked.

The following voltage reduction methods shall be used for constant speed motors:

- changeover of stator winding from "delta" to "star";
- segmentation of stator winding;
- voltage reduction in the power grids of enterprises by changeover of reducing transformer taps;

Changeover of stator winding from "delta" to "star" is the most easy-to-operate, reliable and cheap method recommended for the asynchronous motors systematically loaded less than 30% [3] (i.e. when $k_н \leq 0{,}3$), that is corresponding to the operating conditions of a significant number of units included in the basic set of non-standard equipment. Therefore, in our case, in order to increase the energy indicators of asynchronous motors of the equipment operating with $k_н = 0{,}2...0{,}3$, we recommend to changeover the stator winding from "delta" to "star" [9].

We offer the following methodology for calculation of energy indicators of asynchronous motors at a load factor $k_н < 1$ (in our case $k_н = 0{,}2...0{,}3$) [5, 6]. The technique is based on the equivalent circuit of an asynchronous motor (Figure 2) [7].

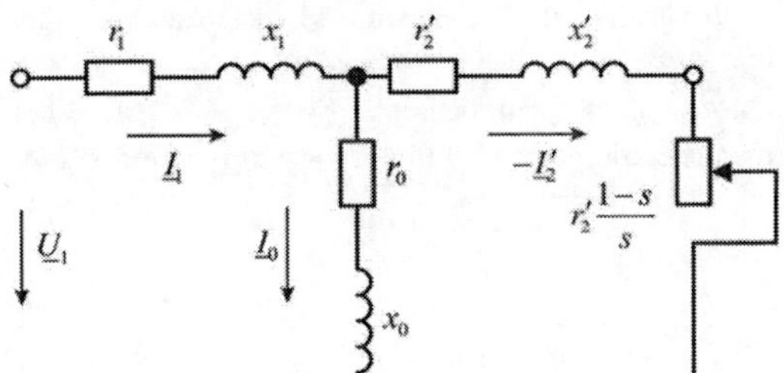

Fig. 2. Equivalent circuit of an asynchronous motor

Proposed calculation algorithm for Delta connection

1. Nominal value of phase excitation current (no-load current) $I_{ф0н\Delta}$

$$I_{ф0н\Delta} \approx I_{фн\Delta}\left(\sin\varphi_н - \frac{\cos\varphi_н}{b_н + \sqrt{b_н^{\,2} - 1}}\right), \qquad (2)$$

where $I_{фн\Delta}$ - nominal phase current for "delta" connection;

$\cos\varphi_н$ - power factor of an asynchronous motor in nominal mode;

$b_н$ - maximum torque ratio of an asynchronous motor under nominal load, voltage and frequency.

$$I_{фн\Delta} = \frac{P_н}{3U_{фн\Delta}\cos\varphi_н}, \qquad (3)$$

Where $P_н$ - nominal rating power of an asynchronous motor;

$U_{фн\Delta} = U_c$, where U_c - supply voltage.

2. No-load current ratio $\frac{I_{ф0н\Delta}}{I_{фн\Delta}}$.

3. According to Figure 3 we can determine the ratio $\frac{I'_{2н}}{I_{фн\Delta}}$ in view of found ratio $\frac{I_{ф0н\Delta}}{I_{фн\Delta}}$, where $I'_{2н}$ - adjusted current of a rotor for nominal ratings.

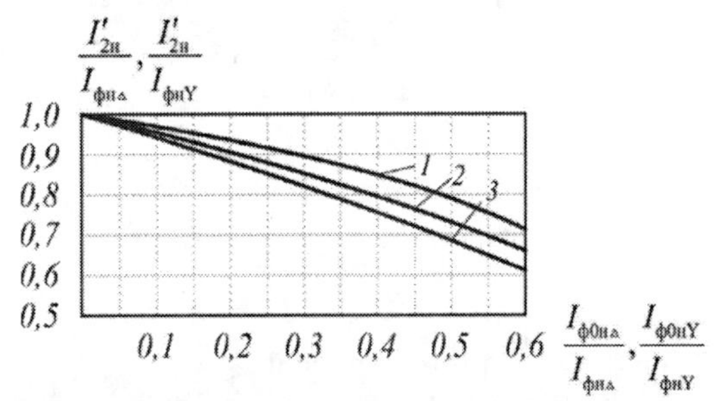

Fig. 3. Correlation between the ratio of adjusted current of an asynchronous motor rotor and the no-load current ratio under different maximum torque:
1 - $b_н = 3{,}0$; 2 - $b_н = 2{,}0$; 3 - $b_н = 1{,}6$.

4. Ratio of adjusted current of a rotor $\frac{I'_2}{I'_{2н\Delta}}$.

$$\frac{I'_2}{I'_{2н\Delta}} = \sqrt{k_н \frac{b_н + \sqrt{b_н^2 - 1}}{b_{c\Delta} + \sqrt{b_{c\Delta}^2 - 1}}}, \qquad (4)$$

where I'_2 - adjusted current of a rotor, when $k_н < 1$ and "delta" connection;

b_c - maximum torque ratio under off-nominal load, voltage and frequency.

$$b_{c\Delta} = \frac{b_н k_u^2}{k_н k_f^2}, \qquad (5)$$

where k_u - coefficient of voltage variation. For "delta" connection $k_u = 1$.

k_f - coefficient that is set to unity for the 50 Hz mains frequency, i.e. in our case $k_f = 1$.

5. Power factor $\cos\varphi'_{\Delta}$ (φ' - phase of adjusted current of a rotor relative to voltage U_1).

$$\cos\varphi'_{\Delta} = \sqrt{\frac{b_{c\Delta} + \sqrt{b_{c\Delta}^{\,2} - 1}}{2b_{c\Delta}}} \tag{6}$$

Therefore:

$$\sin\varphi'_{\Delta} = \sqrt{1 - \cos\varphi'^{2}_{\Delta}} \tag{7}$$

6. Reactive power $Q_{\text{н}\Delta}$ at a nominal voltage of mains and load factor $k_{\text{н}} < 1$.

$$Q_{\text{н}\Delta} = 3U_{\text{фн}\Delta} I_{\text{фн}\Delta} \left(\frac{I'_{2\text{н}}}{I_{\text{фн}\Delta}} \cdot \frac{I'_2}{I'_{2\text{н}\Delta}} \sin\varphi'_{\Delta} + \frac{I_{\text{ф0н}\Delta}}{I_{\text{фн}\Delta}} \right). \tag{8}$$

7. Stator phase current, when $k_{\text{н}} < 1$

$$I_{1\text{ф}\Delta} = I_{\text{фн}\Delta} \sqrt{ \left(\frac{I'_{2\text{н}}}{I_{\text{фн}\Delta}} \cdot \frac{I'_2}{I'_{2\text{н}\Delta}} \sin\varphi'_{\Delta} + \frac{I_{\text{ф0н}\Delta}}{I_{\text{фн}\Delta}} \right)^2 + \left(\frac{I'_{2\text{н}}}{I_{\text{фн}\Delta}} \right)^2 \left(\frac{I'_2}{I'_{2\text{н}\Delta}} \right)^2 \cos\varphi'^{2}_{\Delta} } . \tag{9}$$

8. Rated slip $s_{\text{н}}$

$$s_{\text{н}} = \frac{n_{\text{c}} - n_{\text{н}}}{n_{\text{c}}} \cdot 100\%, \tag{10}$$

Where n_{c} - synchronous speed.

9. Total losses $\sum\Delta P_{\Delta}$ in an asynchronous motor, when $k_{\text{н}} < 1$.

$$\sum\Delta P_{\Delta} = \Delta P_{\text{мех}} + \Delta P_{\text{стали ст.}\Delta} + \Delta P_{\text{об., доб}\Delta} \left(\frac{I_{1\text{ф}\Delta}}{I_{\text{фн}\Delta}} \right)^2 + \Delta P_{\text{рот}} \left(\frac{I'_2}{I'_{2\text{н}\Delta}} \right)^2, \tag{11}$$

Mechanical losses $\Delta P_{\text{мех}} = 0{,}01 P_{\text{н}}$.

Losses in the phase windings of the stator of an asynchronous motor, including additional losses

$$\Delta P_{\text{об.,доб}\Delta} = 3I^2_{\text{фн}\Delta} r_1 + 0{,}005 P_{\text{н}},$$

where r_1 - active resistance of one phase winding of the stator of an asynchronous motor.

With a reasonable degree of accuracy

$$r_1 = r'_{20}$$

r'_{20} - adjusted active resistance of a rotor, when $s = 0$.

$$r'_{20} = \frac{U^2_{\text{н}\Delta}(1 - s_{\text{н}})}{2c_1 (P_{\text{н}} + \Delta P_{\text{мех}}) b_{\text{н}} \left(1 + \frac{c_1}{s_{\text{н}} \left(b_{\text{н}} + \sqrt{b_{\text{н}}^{\,2} - 1} \right)} \right)},$$

where $U_{\text{н}\Delta}$ - nominal linear voltage of phase windings of the stator of an asynchronous motor.

Based on the experience of medium-powered asynchronous motor calculations, $c_1 \approx 1{,}03$.

Rotor losses $\Delta P_{\text{рот}} = \frac{1{,}01 P_{\text{н}} s_{\text{н}}}{1 - s_{\text{н}}}$.

Losses in stator steel

$$\Delta P_{\text{стали ст.}\Delta} = \sum\Delta P_{\text{н}} - \left(\Delta P_{\text{мех}} + \Delta P_{\text{об.,доб}\Delta} + \Delta P_{\text{рот}} \right).$$

Total loss in nominal mode

$$\sum\Delta P_{\text{н}} = \frac{P_{\text{н}}(1 - \eta_{\text{н}})}{\eta_{\text{н}}},$$

where $\eta_{\text{н}}$ - efficiency factor in nominal mode.

Iron losses do not depend on the load. Therefore, for all $k_{\text{н}}$ the losses will be the identical.

10. Power factor $\cos\varphi_{k_{\text{н}}\Delta}$, when $k_{\text{н}} < 1$

$$\cos\varphi_{k_{\text{н}}\Delta} = \frac{k_{\text{н}} P_{\text{н}} + \sum\Delta P_{\Delta}}{\sqrt{Q^2_{\text{н}\Delta} + \left(k_{\text{н}} P_{\text{н}} + \sum\Delta P_{\Delta} \right)^2}}. \tag{12}$$

11. Efficiency factor $\eta_{k_{\text{н}}\Delta}$, when $k_{\text{н}} < 1$

$$\eta_{k_{\text{н}}\Delta} = \frac{P_{\text{н}} \cdot k_{\text{н}}}{P_{\text{н}} \cdot k_{\text{н}} + \sum\Delta P_{\Delta}}. \tag{13}$$

Proposed calculation algorithm for Star connection

1. When the "star" connection of the phase windings of the stator of an asynchronous motor is performed, the phase voltage will decrease by a factor of $\sqrt{3}$. The magnetization current will decrease by a factor of 2 [5, 6].

$$I_{\text{ф0н}Y} = \frac{I_{\text{ф0н}\Delta}}{2}. \tag{14}$$

2. The total power of an asynchronous motor does not depend on the connection pattern of the phase windings of the stator.

Therefore, $3U_{\text{фн}\Delta} I_{\text{фн}\Delta} = 3U_{\text{фн}Y} I_{\text{фн}Y}$, where $U_{\text{фн}\Delta}$, $U_{\text{фн}Y}$, $I_{\text{фн}\Delta}$, $I_{\text{фн}Y}$ - phase nominal voltages and currents in the stator windings for "delta" and "star" connections, respectively. From here it follows that

$$I_{\text{фн}Y} = \frac{U_{\text{фн}\Delta} I_{\text{фн}\Delta}}{U_{\text{фн}Y}}. \tag{15}$$

No-load current ratio $\frac{I_{\text{ф0н}Y}}{I_{\text{фн}Y}}$.

3. According to Figure 3 we can determine the ratio $\frac{I'_{2н}}{I_{фнY}}$ in view of found ratio $\frac{I_{ф0нY}}{I_{фнY}}$.

4. Ratio of adjusted current of a rotor

$$\frac{I'_2}{I'_{2нY}} = \sqrt{k_н \frac{b_н + \sqrt{b_н^{\ 2} - 1}}{b_{cY} + \sqrt{b_{cY}^{\ 2} - 1}}}, \qquad (16)$$

where I'_2 - adjusted current of a rotor, when $k_н < 1$ and "star" connection;

$$b_{cY} = \frac{b_н k_u^2}{k_н k_f^2},$$

where $k_u = \frac{1}{\sqrt{3}}$, for "star" connection. $k_f = 1$ for the 50 Hz mains frequency.

5. $$\cos\varphi'_Y = \sqrt{\frac{b_{cY} + \sqrt{b_{cY}^{\ 2} - 1}}{2b_{cY}}}. \qquad (17)$$

Therefore

$$\sin\varphi'_Y = \sqrt{1 - \cos\varphi_Y'^{\ 2}}. \qquad (18)$$

6. Reacting power $Q_{нY}$ for nominal voltage of mains and load factor $k_н < 1$.

$$Q_{нY} = 3U_{фнY} I_{фнY} \left(\frac{I'_{2н}}{I_{фнY}} \cdot \frac{I'_2}{I'_{2нY}} \sin\varphi'_Y + \frac{I_{ф0нY}}{I_{фнY}} \right). \qquad (19)$$

7. Stator phase current, when $k_н < 1$

$$I_{1фY} = I_{фнY} \sqrt{\left(\frac{I'_{2н}}{I_{фнY}} \cdot \frac{I'_2}{I'_{2нY}} \sin\varphi'_Y + \frac{I_{ф0нY}}{I_{фнY}} \right)^2 + \left(\frac{I'_{2н}}{I_{фнY}} \right)^2 \left(\frac{I'_2}{I'_{2нY}} \right)^2 \cos\varphi_Y'^{\ 2}}. \qquad (20)$$

8. Maximum torque ratio in a winding changeover will decrease by a factor of 3, because the voltage applied to a motor decreases by a factor of $\sqrt{3}$. In order to maintain the operational stability, the load should be three times less than the nominal load. The last condition is met, when $k_н \approx 0,3$. Therefore, when $k_н \approx 0,3$, a slip remains equal to nominal $s_н = \frac{n_c - n_н}{n_c} \cdot 100\%$.

9. Total losses $\Sigma \Delta P_Y$ in an asynchronous motor, when $k_н < 1$

$$\Sigma \Delta P_Y = \Delta P_{мех} + \Delta P_{стали\ ст.Y} + \\ + \Delta P_{об.,\ добY} \left(\frac{I_{1фY}}{I_{фнY}} \right)^2 + \Delta P_{рот} \left(\frac{I'_2}{I'_{2нY}} \right)^2, \qquad (21)$$

When $s = s_н$, mechanical losses $\Delta P_{мех}$ will be identical to losses for "delta" connection. Iron losses $\Delta P_{стали\ ст.Y}$ for "star" connection will be three time less than for "delta" connection, because these losses are proportional to the square of induction (voltage). $\frac{U_{фн\Delta}^2}{U_{фнY}^2} = \frac{380^2}{220^2} = 3$.

Thus, $\Delta P_{стали\ ст.Y} = \frac{\Delta P_{стали\ ст.\Delta}}{3}$. Rotor losses for "star" connection will be identical to losses for "delta" connection.

$$\Delta P_{об.,добY} = 3I_{фнY}^2 r_1 + 0,005P_н.$$

10. Power factor $\cos\varphi_{k_нY}$, when $k_н < 1$

$$\cos\varphi_{k_нY} = \frac{k_н P_н + \Sigma \Delta P_Y}{\sqrt{Q_{нY}^2 + \left(k_н P_н + \Sigma \Delta P_Y\right)^2}}. \qquad (22)$$

11. Efficiency factor $\eta_{k_нY}$, when $k_н < 1$

$$\eta_{k_нY} = \frac{P_н \cdot k_н}{P_н \cdot k_н + \Sigma \Delta P_Y}. \qquad (23)$$

After performing the calculations for "delta" and "star" connections (when $k_н < 1$), we shall compare the energy performance of conventional motors and energy-saving motors.

Numerical example

As per abovementioned procedure, calculations are performed for motors with the following conditional technical characteristics.

Conventional asynchronous motor: $P_н = 20$ kW, $n_н = 1460$ rpm, $\cos\varphi_н = 0,84$ (Figure 1), $\eta_н = 0,875$ (Figure 1), $U_{н\Delta} = U_{фн\Delta} = U_c = 380$ V, $b_н = 2$, $k_н = 0,3$.

Energy-saving motor: $P_н = 20$ kW, $n_н = 1460$ rpm, $\cos\varphi_н = 0,884$ (Figure 1), $\eta_н = 0,92$ (Figure 1), $U_{н\Delta} = U_{фн\Delta} = U_c = 380$ V, $b_н = 2$, $k_н = 0,3$.

"Delta" winding arrangement of the stator of an energy-saving motor

$$I_{фн\Delta} = \frac{20 \cdot 10^3}{3 \cdot 380 \cdot 0,884} = 19,85 \text{ A}$$

$$\sin\varphi_н = \sqrt{1 - (0,884)^2} = 0,469$$

1. $I_{ф0н\Delta} \approx 19,85 \left(0,496 - \frac{0,884}{2 + \sqrt{2^2 - 1}} \right) = 4,6$ A

2. $\frac{I_{ф0н\Delta}}{I_{фн\Delta}} = \frac{4,6}{19,85} = 0,23$

3. As can be seen from Figure 3: $\frac{I'_{2н}}{I_{фн\Delta}} = 0,91$

4. $\frac{I_2'}{I_{2н\Delta}'}=\sqrt{0,3\frac{2+\sqrt{2^2-1}}{6,67+\sqrt{(6,67)^2-1}}}=0,29$,

where $b_c=\frac{2}{0,3}=6,67$

5. Power factor

$$\cos\varphi_\Delta'=\sqrt{\frac{6,67+\sqrt{(6,67)^2-1}}{2\cdot 6,67}}=0,997$$

$$\sin\varphi_\Delta'=\sqrt{1-(0,997)^2}=0,077$$

6. $Q_{н\Delta}=3\cdot 380\cdot 19,85(0,91\cdot 0,29\cdot 0,077+0,23)=$
$=56645\ \text{var}\approx 5,66\ \text{kVAr}$

7. $I_{1ф\Delta}=19,85\sqrt{\begin{array}{l}(0,91\cdot 0,29\cdot 0,077+0,23)^2+\\+(0,91)^2\cdot(0,29)^2\cdot(0,997)^2\end{array}}=7,18\ \text{A}$

8. $s_{н}=\frac{1500-1460}{1500}\cdot 100=2,67\%$

9. $\Delta P_{мех}=0,01\cdot 20000=200\ \text{W}$

$$\Delta P_{об.доб\Delta}=3\cdot(19,85)^2\cdot 0,149+0,005\cdot 20000=276,13\ \text{W}$$

$$r_1=r_{20}'=$$

$$=\frac{380(1-0,0267)}{2\cdot 1,03\cdot(20000+200)\cdot 2\cdot\left(1+\frac{1,03}{0,0267\left(2+\sqrt{2^2-1}\right)}\right)}=$$

$$=0,149\ \text{Ом}$$

$$\Sigma\Delta P_{н}=\frac{20000(1-0,92)}{0,92}=1739,1\ \text{W}$$

$$\Delta P_{рот}=\frac{1,01\cdot 20000\cdot 0,0267}{1-0,0267}=554,13\ \text{W}$$

$\Delta P_{стали\ ст.\Delta}=1739,1-(200+276,13+554,13)=$
$=708,84\ \text{W}$

$\Sigma\Delta P_\Delta=200+708,84+276,13\cdot\left(\frac{7,18}{19,85}\right)^2+554,13\cdot 0,29^2=$
$=991\ \text{W}=0,991\ \text{kW}$

10. $\cos\varphi_{k_{н\Delta}}=\frac{0,3\cdot 20+0,991}{\sqrt{(5,66)^2+(0,3\cdot 20+0,991)^2}}=0,778$

11. $\eta_{k_{н\Delta}}=\frac{20\cdot 0,3}{20\cdot 0,3+0,991}=0,868$.

Similar calculations are performed for an energy-saving motor connected in the Star, as well as for a conventional motor connected in the Star or in the Delta.

5. Research results

After having performed a similar calculations for an energy-saving motor and conventional motor, the authors obtained the following results:

- energy-saving motor connected in the Star: $\cos\varphi_{k_{нY}}=0,851$, $\eta_{k_{нY}}=0,888$;
- conventional motor connected in the Delta: $\cos\varphi_{k_{н\Delta}}=0,712$, $\eta_{k_{н\Delta}}=0,74$;
- conventional motor connected in the Star: $\cos\varphi_{k_{нY}}=0,824$, $\eta_{k_{нY}}=0,844$.

Thus, application of energy-saving motors instead of conventional motors increases the power factor (when $k_н\approx 0,3$) as follows:

$$\frac{\cos\varphi_{k_{н\Delta эн}}-\cos\varphi_{k_{н\Delta об}}}{\cos\varphi_{k_{н\Delta об}}}\cdot 100\%$$

(where $\cos\varphi_{k_{н\Delta эн}}$ - power factor for an energy-saving motor connected in the Delta, $\cos\varphi_{k_{н\Delta об}}$ - power factor for a conventional motor connected in the Delta).

Numerically: $\frac{0,778-0,712}{0,712}\cdot 100\%=9,3\%$.

Therefore: $\frac{\cos\varphi_{k_{н}Yэн}-\cos\varphi_{k_{н}Yоб}}{\cos\varphi_{k_{н}Yоб}}\cdot 100\%=$

$=\frac{0,851-0,824}{0,824}\cdot 100\%=3,28\%$.

$$\frac{\eta_{k_{н\Delta эн}}-\eta_{k_{н\Delta об}}}{\eta_{k_{н\Delta об}}}\cdot 100\%=\frac{0,868-0,74}{0,74}\cdot 100\%=17,3\%.$$

$$\frac{\eta_{k_{н}Yэн}-\eta_{k_{н}Yоб}}{\eta_{k_{н}Yоб}}\cdot 100\%=\frac{0,888-0,844}{0,844}\cdot 100\%=5,21\%.$$

For nominal ratings: $\frac{\cos\varphi_{k_{н\Delta эн}}-\cos\varphi_{k_{н\Delta об}}}{\cos\varphi_{k_{н\Delta об}}}\cdot 100\%=$

$=\frac{0,884-0,84}{0,84}\cdot 100\%=5,24\%$.

$$\frac{\eta_{k_{н\Delta эн}}-\eta_{k_{н\Delta об}}}{\eta_{k_{н\Delta об}}}\cdot 100\%=\frac{0,92-0,875}{0,875}\cdot 100\%=5,14\%.$$

Thus, application of energy-saving motors significantly improves the energy performance of an electrical drive of the equipment for repair of railway rolling stock, considerable part of which works with a load factor $k_н=0,2...0,3$.

6. Summary

1. The proposed methods and algorithm for energy performance calculation applied during replacement of conventional asynchronous motors with energy-saving ones and changeover of stator winding from Delta to Star with a load factor $k_н \approx 0,2...0,3$ can be applied when performing the technical and economic calculations to decide the question of feasibility of the energy-saving motors at the enterprises, which carry out repairs of various equipment.
2. Application of energy-saving asynchronous motors instead of conventional ones allows to improve the energy performance of electric drives of non-standard technological equipment meant for repair of railway rolling stock from about 3% to 17% for medium-powered motors.
3. The reported numerical values of power factors and efficiency factors, their ratios for energy-saving and conventional asynchronous motors are logically sound in the context of electrical machine theory, which confirms the accuracy of the proposed methodology for determining these energy indicators when the electric drive load of various devices is about 30% compared with the nominal ratings.
4. With due regard for the continuous rise in the electricity prices, energy-saving motors can be an excellent reserve for the improvement of energy performance of various electric drives at the repair facilities, especially if under technological conditions the load factor periodically decreases to 0.2 ... 0.3 over extended periods.

References

1. Zakladnyj O. M., Prakhovnyk A. V., Solovej O. I. *Energy saving by means of an industrial electric drive: a manual.* Kyiv: Kondar, 408 p. (2005)/
2. *Non-standard technological equipment for the repair of locomotives, cars and motor-car rolling stock.* "Trans-Atom" LLC Available online at: https://trans-atom.com/page/about.html
3. DSTU 3886 – 99. Energy Saving. Electric drive systems. Method of analysis and selection. Kyiv: Derzhstandart Ukrainy, – 54 p. (2000).
4. *Ways of energy saving* Available online at: https://lektsii.com/1-51568.html
5. Syromjatnikov I. A. *Modes of operation of asynchronous motors.* Moscow: Gosenergoizdat, 528 p. (1963).
6. Syromjatnikov I. A. *Modes of operation of asynchronous and synchronous motors.* Moscow: Energoatomizdat, 240 p. (1984).
7. Dubynec L. V, Momot O.I., Marenych O.L. *Electric machines. Transformers Asynchronous Machines: Basic Guide* Dnipropetrovsk: Dnipropetrovsk National University of Railway Transport Publishing, 208 p. (2004).
8. Dubynets L. V, Marenych O. L, Baliichuk O. Yu., Kortohus A. S. *Energy saving during operation of equipment with non-controlled electric drive in locomotive depot* Science and transport progress. Bulletin of the Dnepropetrovsk National University of Railway Transport. № 5 (65), P. 54 – 63.
9. Baliichuk O. Yu., Dubynets L. V, Marenych O. L, Lysenko O. O *Increasing of power factor of electric drives of non-standard equipment of enterprises for repair of rolling stock of railways.* Electrification of transport № 14, P. 31 – 36.
10. .Afanasov A. M, Shapovalov O. S. *Increasing of power efficiency of traction electric machines tests* Electrification of transport № 14, P. 28 – 30.
11. Lokarev V. Y. *On the system of evaluating the energy efficiency of a modern electric drive.* Bulletin of the National University of Shipbuilding № 6, P. 75 – 81
12. V. V. Dzhedzhula, I. Yu. Yepifanova. *Innovations in the system of management of energy saving of industrial.* Available online at: http://economyandsociety.in.ua
13. Lei Li, Haihong Huang, Zhifeng Liu Xinyu Li, Matthew J. Triebe, Fu Zhao. *An energy-saving method to solve the mismatch between installed and demanded power in hydraulic press.* Journal of Cleaner Production. – Vol. **139**. – P. 636-645 (2016). doi: 10.1016/j.jclepro.2016.08.063.
14. M. M. Shaikh, A. J. Memon, M. Hussain. *Data on electrical energy conservation using high efficiency motors for the confidence bounds using statistical techniques* Available online at: https://www.ncbi.nlm.nih.gov/pmc/articles/PMC4925456/. doi:10.1016/j.dib.2016.06.004
15. A. J. Memon, M. M. Shaikh. *Confidence bounds for energy conservation in electric motors: An economical solution using statistical techniques.* Energy. – Vol. **109**. P. 592-601. (2016). doi:10.1016/j.energy.2016.05.014
16. *Energy Saving Motors.* Available online at: http://europromtech.ru/energosberegaushie-dvigateli.html
17. Helduser, S. *Electric-hydrostatic drive an innovative energy-saving power and motion control system.* Proceedings of the Institution of Mechanical Engineers, Part I: Journal of Systems and Control Engineering, **213(5)**, 427–437. (1999). https://doi.org/10.1243/0959651991540250
18. L. Zarri, M. Mengoni, A. Tani, G. Serra, D. Casadei, J. O. Ojo, *Control schemes for field weakening of induction machines: A review*, Electrical Machines Design Control and Diagnosis (WEMDCD) 2015 IEEE Workshop on, pp. 146-155 (2015).
19. Bogdan Pryymak, Manuel Moreno-Eguilaz, *Characteristics of induction motor drives with torque maximization in field weakening region,* Electrical and Computer Engineering (UKRCON) 2017 IEEE First Ukraine Conference on, pp. 508-513 (2017).
20. Shuying Yang, Dawei Ding, Xi Li, Zhen Xie, Xing Zhang, Liuchen Chang, *A Novel Online Parameter Estimation Method for Indirect Field Oriented Induction Motor Drives*, Energy Conversion IEEE Transactions on, vol. **32**, no. 4, pp. 1562-1573 (2017).

Improving the energy efficiency of operation of elements of the structures of the rolling stock by means of surface engineering.

Vyacheslav Kopylov[1], *Oleg* Kuzin[2], and *Mykola* Kuzin[3,4,*]

[1] National Technical University of Ukraine "Igor Sikorsky Kyiv Polytechnic Institute", 03056, Kyiv, Prosp. Peremohy 37, Ukraine
[2] Lviv Polytechnic National University, 79013, Lviv, Bandera Str. 12, Ukraine
[3] DNURT, Lviv Branch, 79052, Lviv, I.Blazhkevych Str., 12-a, Ukraine
[4] Lviv Research Institute of Forensic Science, 79000 Lviv, Ukraine

Abstract. The manufacture of rolling stock components with a given life cycle based on the computerization of all design and production stages is the most promising way to increase the energy efficiency of its operation. The implementation of this approach requires the expansion of the use of technologies that provide parts with a high level of mechanical and operational properties. The state of the surface layers is important in the formation of the reliability parameters of high-loaded parts. The principles of using system approaches for the physical justification of the choice of the optimal modes of engineering grain boundaries depending on the operating conditions of products are presented. New approaches and algorithms have been created that allow quantitative and qualitative studies of the effect of technological treatments and the chemical composition of polycrystals on the strength of grain boundaries and the processes of softening of the boundary zones. Using the developed techniques, the ways to control the energy state of the internal interfaces by doping, microalloying and heat treatment were determined to increase the resistance to brittle fracture and wear of the steels used in the manufacture and repair of rolling stock.

1. Introduction

The operation of rolling stock with a given life cycle can significantly reduce the energy consumption of railways during transportation. The creation of such vehicles is possible based on the use of multifunctional computer systems at all stages of design, production and operation, consistently perform volumetric product design, design justification for its reliability and efficiency, the preparation of technological processes of manufacturing and engineering project management.

Computer engineering analysis programs make it possible to formulate new approaches to the choice of materials and optimize their processing technologies to improve product durability on the basis of a design basis.

Preparation of technological processes for obtaining products should include information on the parameters of the properties and behavior of materials during treatments and operating loads, making the creation of databases of computer modeling of the structure and properties of materials important. Solving these problems allows using the methods of computational materials science in digital engineering for the simultaneous design of materials and products with high reliability parameters. It should be noted that the development and use of such programs is limited by the lack of simultaneously established relationships between the damage resulting from technological processing, its development during the operation of products and the stages of the life cycle of parts and machines. Therefore, research in this direction are relevant.

The reliability indicators of the rolling stock largely depend on the nature of the contact of the mating parts or their interaction with liquid, gas and other media. The quality of the outer layers of parts significantly affects their weariness, strength, and corrosion resistance, which makes it necessary to form them with the given parameters of properties by the methods of surface engineering. When operating loads on the surfaces there are flows of defects that contribute to the development of damage.

Sources of defects are the boundaries and triple joints of the grains facing the surface of the parts [1, 2, 3]. With a further increase in voltage sources are activated away from the surface. Surface plastically deformed grains are stress concentrators that contribute to the deformation of neighboring grains [4, 5]. Near the boundaries and joints of the grains, a higher concentration of stresses occurs than for dislocations and their clusters [4].

A slight increase in local stresses in the zones near the grain boundaries contributes to the initiation of

* Corresponding author: kuzin.nick81@gmail.com

plastic shears due to the difference in the elastic moduli of the contacting grains. Landslides occur on the interfaces, where there are dangling interatomic bonds, which can spread deep into the material.

Strengthening a thin surface layer of the material by creating functional gradient structures by surface engineering methods delays the localization of the deformation and significantly improves the macromechanical characteristics of the parts. At the same time, durability growth is achieved due to changes in the stress-strain state in the local areas of parts, allows to effectively manage surface defects, the development of softening and destructive material processes, but requires an analysis of the role of grain boundary parameters and, in particular, their structural-energy state in the destruction processes .

In this regard, the use of new approaches to modeling and analyzing the mechanical behavior of internal boundary zones becomes important. The development of such approaches allows the scientific substantiation of technological solutions to increase the resistance to the formation of blight damages and cracks in parts of the rolling stock operating under conditions of intensive loads.

2. The object of study

The object of the study is the internal interface of the steel used in the manufacture of rolling stock, and their role in the formation of damage and intercrystalline destruction of alloys under the action of power loads, depending on the structural and energy state of the internal interfaces. The choice of the object of research is connected with the need to substantiate the methods of managing the structure to eliminate the intercrystallite destruction of low energy intensity of structural steels.

When exposed to alternating loads in the details of railway structures, scattering or non-localized damage is formed [6]. As a result of structural inhomogeneity of materials, plastic deformation zones are randomly distributed, in which microscopic cracks originate. In the case of the development of such cracks, one of them spreads faster, which leads to destruction. The formation and growth of the main fatigue crack is called the stage of localized fatigue damage.

Scattering damage significantly affect the rate of development of the main crack. In a material with a large number of microscopic cracks of the appropriate size, a trunk crack develops faster, absorbing small cracks during its development.

Conventionally, the stage of formation of damage when exposed to cyclic loads is divided into the formation of microstructure and physically short cracks and long cracks. Microstructures of short cracks are characterized by discontinuities, the dimensions of which are of the same order as the structural elements. For the transition from the microstructure of short cracks in physically short cracks take the transition from shear cracks to separation cracks. When a region of elastic-plastic deformations forms at the tip of a crack, a short crack transforms into a long crack [7, 8].

It should be noted that the structure of the material significantly affects the formation of scattered damage, but this relationship has not been fully detected, due to the complex multi-level structure of the alloys. To take into account the peculiarities of their behavior when exposed to external loads, the following hierarchical levels of material consideration are used:

1. Macro level. At this level, the material is considered as a continuous continual medium with properties whose parameters vary slightly in space. The typical size of a piece of material is commensurate with the entire product.
2. Meso level. The material is considered as a discrete-continuous medium. The level of consideration of this scale corresponds to the structural components and their boundaries.
3. Micro level. The level of consideration of lattice defects inside the crystallite.
4. Nano level. Consideration of the material corresponds to individual atoms (or their groups). At this level, a discrete model description is predominantly used.

Simultaneous model and computational consideration of all levels of material is a complex scientific problem, and therefore is currently used mainly in the theoretical direction [9].

In applied research, only one of the hierarchical levels is considered using the results obtained from analyzing the behavior of the material at other large-scale levels.

Of great importance is the need to determine the relative contribution of individual levels of the structural hierarchy in models of material behavior under external load [10, 11]. Such an analysis becomes possible on the basis of computer models of calculations, to ensure their reliability, they are supplemented by experimental data [12].

Computer simulation of materials and technologies for their processing includes the following steps: formulation of a physical model, mathematical modeling (development of a system of equations), generation of a computational grid, selection of programming algorithms, numerical solution, presentation of computational results.

At the present stage of development of computational materials science for computer modeling, the methods of molecular dynamics, finite element and hierarchical modeling are used. Molecular dynamics models the interaction of particles and examines phenomena at the nano level. The method of finite element modeling refers to the macro level and evaluates the behavior of the material by the methods of continuum mechanics. Between these scale levels there should be methods that have to bind nano-and macroscale models of material behavior. These are hierarchical models that are currently only at the beginning of their development [13]. To predict changes in properties in local volumes of materials, depending on the technological characteristics of their production, methods for parametric modeling of the microstructure begin to develop.

It should be noted that there are currently no databases on the properties of materials that determine their reliability with regard to technological processing processes, which is an unsolved problem in computer science.

3. The purpose and objectives of the study

The aim of the work is to create models and methods for controlling the reliability parameters of the steels used in the manufacture of rolling stock, by optimizing the structural-energy state of the interfaces between the grains of the structural components.

To achieve this goal, the following tasks were formulated:

1. To conduct a study of the influence of the structural-energy state of the interfaces between the grains of the structural components on the reliability parameters of the grinding steel.
2. To build a multilevel model of polycrystalline systems with variable parameters of the grain structure and the structural-energy state of the interfaces between the grains of the structural components and their influence on the performance properties of the alloys.
3. On the basis of the obtained experimental results and the constructed model, explore the main trends and the corresponding laws of the influence of the structural and energy state of internal interfaces on the properties of structural steels, conduct a scientific substantiation of technological solutions for the formation of a microstructure that provides for the output of polycrystalline systems from the area of operational instability.

4. Research existing solutions to the problem

In the materials of parts during the operation of railway structures in the conditions of the action of power loads there is an accumulation of damage. To assess the residual life of parts, it is important to establish the patterns of their formation, as well as the conditions for the transition from scattered to localized damage. In particular, the size of structurally and physically short cracks, as well as exploitation to the formation of a main crack.

The study of the formation of localized damage is carried out by methods of fracture mechanics. The structure of the material of the products has a significant impact on the stages of the formation of scattered damages. In this connection, for their research, they use the approaches of physical materials science, locally gradient mechanics of a deformable solid body, and also linear and nonlinear fracture mechanics. The multi-level structure of metals and alloys and the complexity of the processes influencing the formation of scattered damage require the use of new approaches associated with the development of computer materials science.

Structural features of metals and alloys have a significant impact on the value of their endurance limit. It is determined by the shape, size of the structural components, the presence of defects, and factors determining the intensity of this effect. Criteria for fracture mechanics allow describing the effect of defects on the fatigue fracture resistance of alloys, but they require determining their priority and size.

The use of fracture mechanics criteria is effective for those alloys for which the fundamental laws of the structure bonds and fracture resistance characteristics are established.

The heterogeneity of the structure of the alloys, the size, shape of the structural components, the difference in the modules of their elasticity, the particular stress state of the local metal volumes specify the conditions for the initiation of microscopic cracks and the transition from diffuse to localized damage at stresses exceeding the fatigue limits. Microscopically short cracks during the action of alternating loads arising at stresses lower than the fatigue limit, that is, the possibility of their formation under the operational loads of the railway structure.

Studies aimed at establishing the structural units that are responsible for the formation of scattered steel damages in similar chemical composition to those used in the manufacture of rolling stock, showed the correspondence of the sizes of microstructural short cracks to ferritic and perlitic grains [7]. It is shown that the size of the cracks in the transition from scattered to localized damage with bagatocycle fatigue is 0.01..0.02 mm. The influence of the material state of the boundary zones of the grains, the possibility of formation of damages and defects in them depending on the endurance limit on the grain size has been revealed [8]. At the same time, the correspondence of the size of such cracks in alloys to the size of the elements of their structure has not been unequivocally established.

The principles of determining structural parameters for various alloys that are responsible for the formation of short cracks are not fully substantiated.

5. Research methods

Modern methods of controlling the properties of polycrystalline alloys with minimal resources include technology based on the concepts of grain boundary engineering and the principle of grain-boundary structure design. Its main provisions include the selection of such technological modes of processing materials, ensuring the formation in them of the maximum proportion of grain boundaries (special boundaries), characterized by the minimum value of grain boundary energy [14]. Such boundaries are characterized by high fracture resistance due to the easy transfer of slip from one grain to another.

The widespread use of such approaches is hampered by the insufficient knowledge of the influence of the energy characteristics of grain boundaries of different levels of the hierarchy on the behavior of alloys under the influence of force loads. In this connection, the study of the structure of grain boundaries in alloys of various doping systems and the determination of the influence of their energy characteristics on the formation of grain-boundary damage and on the secondary grains of polycrystals becomes important.

In the details of railway transport, made from pre-steel grades, the grain size of austenite can vary widely, and in many cases grain is formed. The presence of crystals with different characteristics leads to the formation of grain boundaries that differ in structural and energetic state.

Studies were conducted on steel 40XHM (C – 0.40%, Si – 0.24%, Mn – 0.61%, Cr – 0.58%, Ni – 0.99%, Mo – 0.18%; international analog 8640 according to ASTM-SAE) industrial smelting. Billets were made of rolled products, which were annealed at a temperature of 840°C...870°C and subjected to quenching after heating in a salt bath from temperatures of 870°C... 1180°C. Then, tempering was done at a temperature of 600°C.

To study the state of grain boundaries with minimal changes in the structure, ion-plasma etching was used at the VUP-4 vacuum post [15]. During digestion, grooves were formed on the surface of the samples, which define the boundaries of the grains and phases. The depth and angle at the top of the digestive streams were determined experimentally using a Linnik interferometer (MII-4).

The analysis of the structural-energy state of the grain boundaries was performed using the metallographic estimation technique proposed in [16, 17]. To identify the relative change in the energy of the grain boundaries, the property of triple junctions was used, which consists in the formation of an equilibrium configuration of the boundaries at the place of their meeting. That is, the ratio of the surface energy of the tension of the grain boundaries λ and the angles α opposite to them is described by the analytical dependence of Hering – Young relationship [17].

6. Research results.

Analysis of the microstructure of steel 40XHM showed the formation of austenite grains when heated to a temperature of 860°C with an exposure time of 30 minutes, whose dimensions are in the range from 2 μm to 70 μm. The average grain diameter is 11.4 microns (Fig. 1, Fig. 2).

Heating to a temperature of 950°C leads to an increase in the average grain diameter of austenite, which is equal to 77.2 microns. The grain size is in the range from 20 μm to 220 μm, and the standard deviation is 44.2 μm (Fig. 1, Fig. 2).

After quenching from a temperature of 1050°C, the size of the former austenitic grains is in the range from 20 μm to 270 μm. Their average diameter is 98.7 microns, and the quadratic deviation is 52.0 microns (Fig. 1, Fig. 2).

When heated to a temperature of 1180 ° C, austenite grains with an average diameter of 328 μm are formed, and grains 800 μm in size also appear (Fig. 1, Fig. 2).

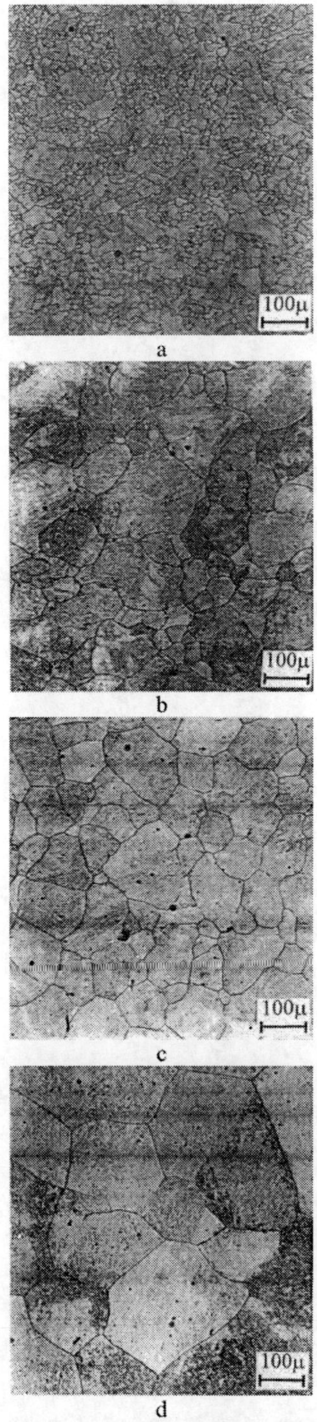

Fig. 1. Microstructures of steel 40XHM after quenching from temperatures of 860°C (a), 950°C (b), 1050°C (c) and 1180°C (d), quenching for 30 minutes.

The analysis of the grain roughness of steel after quenching from various temperatures showed that the quantitative indicator of grain roughness (D_{max} / D) after quenching from 860°C is equal to 4.72, 950°C - 2.91,

1050°C - 2.78, 1180°C – 4.57. That is, the grain-graininess is characteristic of steel after quenching from temperatures of 860°C and 1180°C.

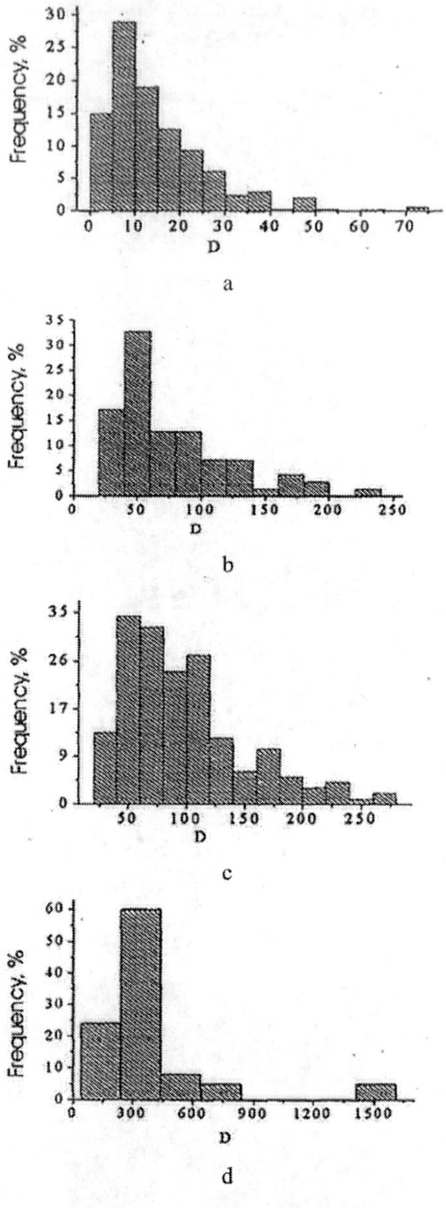

Fig. 2. The distribution of average diameters (D, μm) of austenite grains (a, b, c, d) in size in steel 40XHM after quenching from 860°C (a), 950°C (b), 1050°C (c), 1180°C (d), quench aging 30 min.

The results of determining the angles in the triple joints of the grains of steel 40XHM after quenching from 860°C, 950°C, 1050°C and 1180°C with an exposure time of 30 min are presented in Fig. 3.

An analysis of the relationship between the energy of the grain boundaries and the angles between them in the triple junctions showed their dependence on the grain size and grain size.

The proportion of grain boundaries, which are located opposite 80°...100° angles and are characterized by a higher relative energy value, is 10.0% for steel hardened from 860°C, 950°C - 16.67%, 1050°C – 10.0%, 1180°C - 16.67%. Most of the boundaries with a high value of grain-boundary energy after quenching from 860°C and 1180°C are in correlation with the results of a quantitative assessment of the roughness of steel.

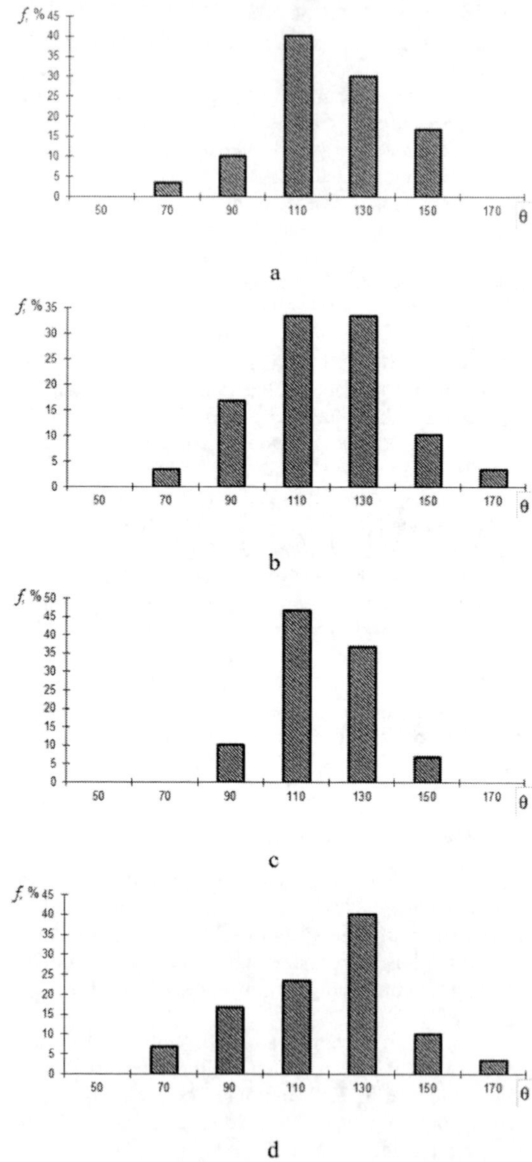

Fig. 3. Histograms of the distribution of flat angles between the boundaries of the former austenitic grains in steel 40XHM after quenching from temperatures of 860°C (a), 950°C (b), 1050°C (c), 1180°C (d), aging for hardening 30 min.

The results of the study of steel after quenching and tempering at 600 ° C are shown in fig. 4.

https://doi.org/10.1051/matecconf/201929401008

a

b

c

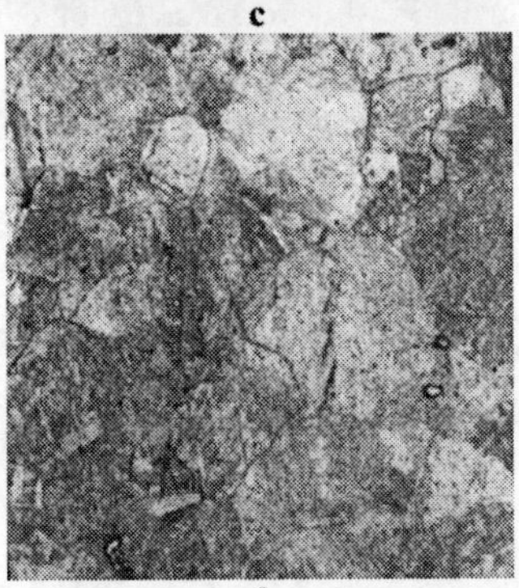

d

Fig. 4. Microstructure of steel 40XHM after quenching at temperatures of 860°C (a), 900°C (b), 950°C (c), 1050°C (d) and release at 600°C, shutter speed at tempering 30 min, x200.

The proportion of grain boundaries located opposite the corners 81° ... 100°, characterized by a high energy level of steel after quenching from 860°C is 13.33%, 950°C - 16.67%, 1050°C - 13.33%. The change in the number of grain boundaries with a higher energy level indicates the passage of the processes of their movement and the heterogeneity of phase precipitation during steel tempering (Tables 1, 2).

It should be noted that after quenching, most of the grains form boundaries, the angle between which is within 101°... 20°. After the holidays, the proportion of borders increases significantly, forming an angle of 121° ... 140°.

The presence of grain size after quenching from 950°C and 1180°C increases the proportion of triple grain joints, in which there are angles that are within 161°...180°, both after quenching and after improvement (Fig. 5).

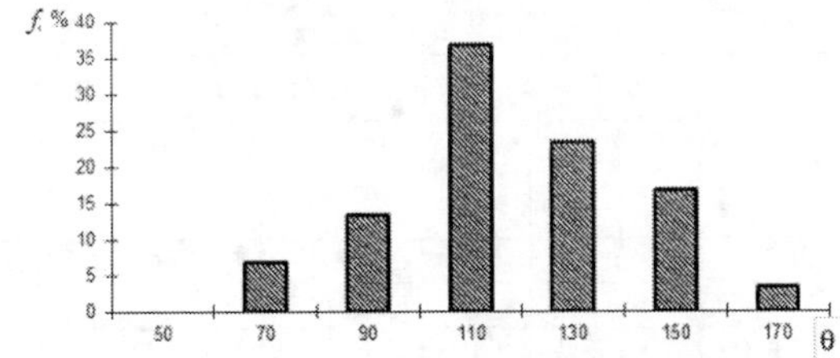

a

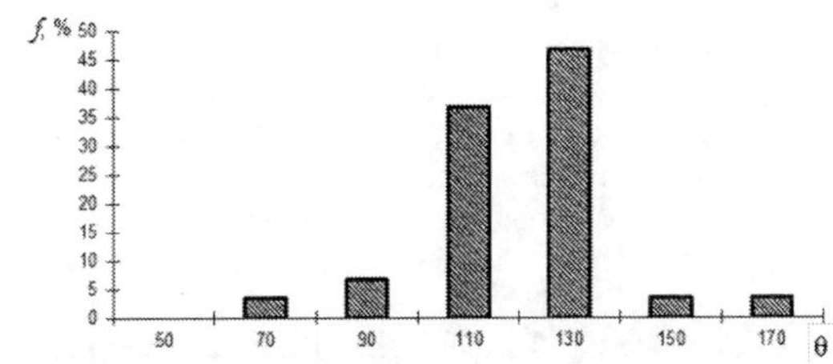

b

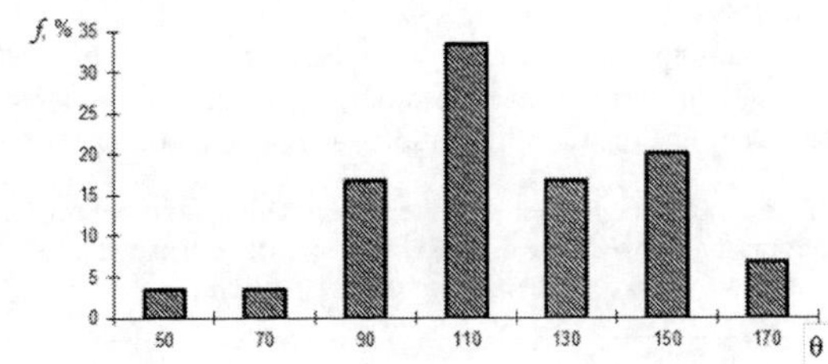

c

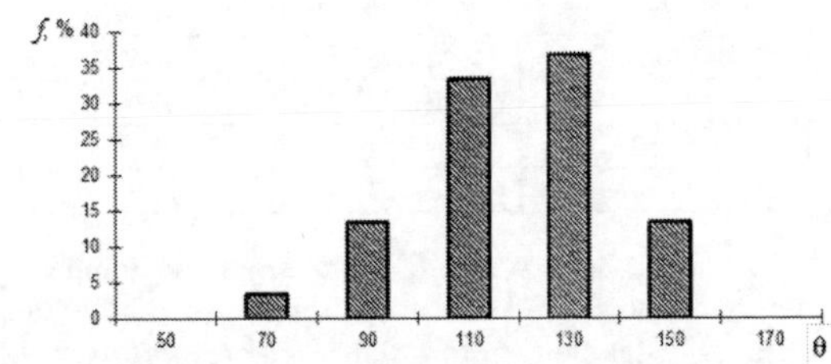

d

Fig. 5. Histogram distribution of flat angles between the boundaries of the remaining austenitic grains in 40XHM steel after quenching at temperatures of 860°C (a), 900°C (b), 950°(c), 1050°C (d) and release at 600°C .

Table 1. The fraction of the angles between the boundaries in the triple joints of 40XHM steel grains after quenching

Endurance for 30 minutes							
Tempe-ring tempe-rature	Share in the total number of measured angles,%						
	40 … 60	61 … 80	81 … 100	101 … 120	121 … 140	141 … 160	161 … 180
860°C	-	3,3	10,0	40,0	30,0	16,7	0,0
950°C	-	3,3	16,7	33,3	33,3	10,0	3,3
1050°C	-	-	10,0	46,7	36,7	6,7	-
1180°C	-	6,7	16,7	23,3	40,0	10,0	3,3

Table 2. The fraction of the angles between the boundaries in triple stacks of 40XHM steel grains after improvement (30-minute hardening)

Quenching, tempering 600°C							
Tempe-ring tempera-ture	Share in the total number of measured angles,%						
	40 … 60	61 … 80	81 … 100	101 … 120	121 … 140	141 … 160	161 … 180
860°C	-	6,7	13,3	36,7	23,3	16,7	3,3
900°C	-	3,3	6,7	36,7	4667	3,3	3,33
950°C	3,3	3,3	16,7	33,3	16,7	20,0	6,7
1050° C	-	3,3	13,3	33,3	36,7	13,3	-

The analysis of the influence of the structural and energy state of the grain boundaries on the processes of accumulation of damage and destruction was carried out on the basis of the results of the study of contact durability of steel under friction conditions. Testing for wear was carried out according to the scheme of "disk block" in the mode of dry sliding friction. The velocity of slipping the samples on the counterface v = 0.5 m / s, the load on the samples P = 300 N, the counterpart material - steel 45 (C - 0.45%; international analog 1045 according to ASTM-SAE), the hardness of the countertile - 42 ... 44 HRC [18]. After spin, each sample was worn for 10 hours. The values of mass intensity of wear were evaluated by the results of tests of six samples. The results of the studies are presented in table 3.

Table 3. The dependence of the mass intensity of wear on the tempering temperature of 40XHM steel after release at 600 °C (excerpt hardening under 30 minutes)

Tempering temperature t, °C	I_m, 10^{-8}	I_m/I_{mc}
860	2,599	100
900	2,698	104
950	3,228	124
1050	2,451	94
1180	2,821	109

As can be seen with the increase of the tempering temperature to 950°C, the wear resistance of the steel decreases. Standing from 1050°C increases its wear resistance. After quenching from 1180°C, the intensity of wear of steel increases.

Such a change in thickness is due to the significant influence of the structure on the accumulation of damage in terms of contact interaction. Reduced wear resistance after quenching from 950°C, 1180°C of improved steel is due to the growth of a fraction of triple grain boundary joints, in which the edges form corners 161° ... 180°.

According to modern notions, triple grains are one of the types of linear defects along which three grain boundary surfaces are contacted. In this regard, this type of joints is characterized by a more "loose" structure and increased specific energy parameters. Moreover, the contribution of such structural units in the processes of weakening and fracture of parts of structural elements is greater than the usual boundaries of grains of polycrystals.

We note that at the moment the issue of the structure on the nanowire area of the triple junction remains open.

At the energy level, the structure of this area of polycrystalline materials is taken into account with the aid of the Herring-Young ratios, which indicate the relative energy of grain boundaries, and its distribution, depending on the angles between them:

$$\left(\sum_{i=1}^{3}\sigma_i\vec{b}_i+\left(\frac{\partial\sigma_i}{\partial\varphi_i}\right)\vec{n}_i\right)=0, \quad (1)$$

$$\frac{\sigma_1}{\sin\lambda_1}=\frac{\sigma_2}{\sin\lambda_2}=\frac{\sigma_3}{\sin\lambda_3}, \quad (2)$$

where σ_i - energy of i boundary, $\vec{n}_i$ - unit vector which is normal to i boundary, $\vec{b}_i$ - unit vector along the junction of i boundary, φ_i - angle, that defines orientation of i boundary, λ_i - the angle opposite to i boundary ($i=1,2,3$).

As can be seen from equation (2), the energy equilibrium and the strongest is the junction in which $\lambda_1=\lambda_2=\lambda_3=120^0$. In the case of angle differences λ_i from 120°, it is possible energy inequalities of the boundaries σ_i, which leads to the appearance of a gradient of properties in the zone of "direct" triple junction, which is often the source of the destruction.

Its evaluation will be based on the ratio:

$$\Delta=|\sigma_1-\sigma_2|=\left|\sigma_1-\sigma_1\cdot\frac{\sin\lambda_2}{\sin\lambda_1}\right|, \quad (3)$$

where $|...|$ - module, Δ - increase in the energy of grain boundaries in the triple junction area.

To simplify the calculations we will assume that there is a leading angle λ_1, and λ_2, λ_3 have the condition:

$$\lambda_2=\lambda_3=\frac{1}{2}(360-\lambda_1). \quad (4)$$

In this case, the expression (3) will look:

$$\Delta=|\sigma_1-\sigma_2|=\left|\sigma_1-\sigma_1\cdot\frac{\sin(0.5\lambda_1)}{\sin\lambda_1}\right|. \quad (5)$$

In the dimensionless form (5) it transforms into the following equation:

$$\Delta = \left| 1 - \frac{\sin\left(\frac{\lambda_1}{2}\right)}{\sin \lambda_1} \right|. \quad (6)$$

The qualitative analysis of equation (6) shows that when the angle is changed λ_1 on the interval (0;180°) the Δ changes too. In this case, you can distinguish the characteristic three zones for the parameter Δ from λ_1: zone of minor parameter change Δ, $\lambda_1 \in (0;80^0) \cup (130^0;140^0)$, the zone of small Δ - $\lambda_1 \in (80^0;130^0)$, and the zone of sharp ("catastrophic") growth Δ - at $\lambda_1 \in (140^0;180^0)$. When λ_1 is in the limits $(140^0;180^0)$ this indicates a lack of a thermodynamic equilibrium at the junction, which allows us to assert about the significant difference in the energy of the boundaries forming the given joint, and its ability to form microdefects under load conditions.

The graphic representation of the obtained dependence (6) is shown in Fig. 6.

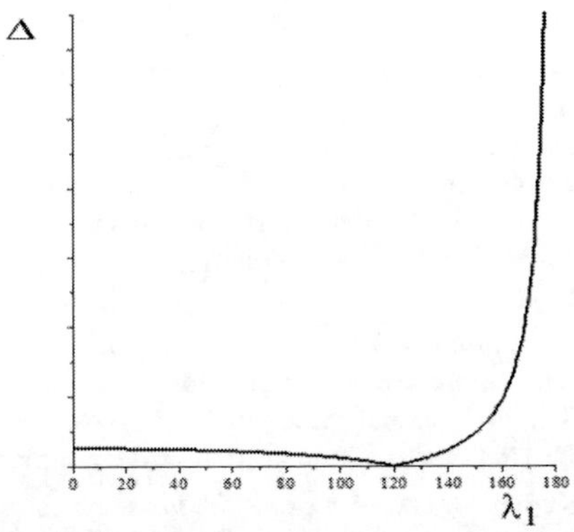

Fig. 6. Dependence of energy growth Δ of grain boundaries in the triple junction area in dimensionless form from angle λ_1.

The results obtained in the work point to the need to manage the structural and energy characteristics of triple joints to increase the reliability parameters of structural steels.

The study of improved steels showed a tendency to intergranular damage and cracks in external loads with an increase in the structure of the grain boundary fraction with a high energy level. It was established that the decrease of the proportion of such boundaries in the grains joints, which reach the surface of the parts, by changing the temperature and time of exposure to quenching, as well as the modes of release, allows us to achieve a significant resistance to the destruction of alloys in conditions of shock and contact loads.

Analysis of the research results showed that in the steel after the quenching from 860°C, the average values of the angles of triple grained joints are in the range from 70 ° to 150 °. Gantry from 950° extends the range of angles. Appear angles ranging from 70° to 170°. After quenching from 1050°C, the range of angles decreases from 90° to 150°. The largest share in the total number is 110° and 130°. Their total number is more than 82%.

After quenching from 1180°C, the range extends from 70° to 170°. The proportion of angles in the range from 100° to 140° while decreasing to 63%. The improvement leads to some change in the ratio between the corners of triple joints of grains, which is due to recrystallization processes. So the release after quenching at a temperature of 950°C contributes to the increase in the number of joints in which the angles between the borders are within the limits of 161° ... 180° (Table 2).

Determination of the average distance between the triple joints with high gradient energy showed that the distance between them along the boundaries of grains after quenching from 860 ° and release is up to 0.01 mm, and after quenching from 950 ° to 0.07 mm. The energy level of the grain boundaries between such joints is higher than other boundaries and approaches surface energy. Therefore, these boundaries are the places of origin of scattered damage (cracks).

The size of the cracks formed under the action of external loads in steel tempered from 950°C will be greater than after the quenching of 860°C. As a result, after the quenching from the higher temperature in the contact spots, the stage of development of localized damage and destruction is accelerated.

That is, multi-grain steel moves faster to the stage of formation of localized damage in the conditions of contact interaction, which leads to a decrease in its wear resistance. The established regularity is confirmed by tests of steel tempered from a temperature of 1180°C.

After quenching from 1050°C and improvement in steel, no triple joints with an elevated gradient of energy are observed. In addition to the formation of carbide heterogeneity, which corresponds to the mechanical friction pattern [20], this is one of the main factors in increasing the wear resistance of steel after the release.

Thus, the characteristics of the local volumes of grains significantly affect the stages of formation of scattered and localized damage in steels at contact loads. This ensures the possibility of introducing technologies for forming the structure of parts, which allows to significantly increase their reliability parameters - durability, resource, fault with minimum economic costs.

Conclusions

1. The structural and energetic state of the grain boundaries of the structural components significantly affects the shrinkage of steel 40XHM. Quenching from the temperatures at which the grain-graininess of steel is formed is accompanied by the formation of triple grain joints, the boundaries of which differ significantly in energy. The boundaries between the joints with a larger amount of energy correspond to the size of the scattered damage, which are formed when exposed to external loads.

2. The increase in wear resistance of steels being improved after quenching from 1050°C, besides the distribution of carbides, which corresponds to the mechanical scheme of friction, is associated with the formation of the great-cut boundaries of sorbitol grain separation with a more uniform value of grain boundary energy.
3. The propensity to increase energy in the field of triple joints with external loads depends on the proportion of boundaries that have a significant difference between the energies, and therefore reducing it helps to ensure the growth of resistance of the joints to the formation of microdefects due to the introduction of technologies for grain-boundary structure design to increase the parameters reliability of parts - durability, resource, reliability with minimal economic costs.

References

1. Shemiakin E. I. *Sinteticheskaia teoriia prochnocti* Fizicheskaia mezomekhanika **т. 2. - № 6.** – c. 63-69 (1999) [in Russian]
2. Panin A. V. *Zarozdenie i razvitie potokov defectov na poverchnosti deformiruemoho tverdoho tela* Fizicheskaia mezomekhanika. – **т.3. - № 1.** – c. 83-92 (2000). [in Russian]
3. Panin A. V. *Vliianiie sostoianiia poverchnostnoho sloia na mechanism plasticheskoho techeniia i soprotivlenie deformacii malouhlerodistoi stali* Fizicheskaia mezomekhanika **т.4. - № 4.** – c. 84-52. (2001) [in Russian]
4. Koneva N. A. *Istochniki polei napriazheniy v deformirovannych polikristalach* Fizicheskaia mezomekhanika **т.9. - № 3.** – c. 93-101 (2006) [in Russian]
5. Kelli A. *Vysokoprochnyie materialy* – Moskov.: Mir,. – 264 p (1976) [in Russian]
6. Troschenko V. T. *Ustalost metallov. Vliianiie sostoianiia poverchnosti i kontaktnoho vzaimodeistviia.* – K.: IPP NAN Ukrainy,. 661 p. (2009) [in Russian]
7. Kobayashi H. *A stress criterion for fatigue crack propagation in metals* Proc. of the Inter. Conf. on Mechan. Behavior of Mater. Kyoto, Japan Aug. 15-20, – **Vol. 2.** – P. 199-208 (1971).
8. Kinio T. *The role of prio austenite grains in fatigue crack initiation and propagation in low carbon martensite*/ T. Kinio, M. Shimizu, K. Ymada et all// Fatigue Fract. Eng. Mater. Stract. – **2, № 3.** – p. 231-299 (1979)
9. Makarov P. V. *Ob ierarkhicheskoi prirode deformacii i razrusheniia tvedych tel* Fizicheskaia mezomekhanika **т.7. - № 4.** – p. 25-34. (2004) [in Russian]
10. McDowell D.L. *Material design: a useful research focus for inelastic behavior of structural metals* Theoretical and Applied Fracture of mesomechanics in the 21th centure: current thinking on multyscale mechanics problems. – **v. 37.** – p. 245-259. (2001).
11. McDowell D.L. A *perspective on trends in multiscale plasticity* International Journal of Plasticity. – **v. 26.** – p. 1280-1309 (2010).
12. Curtarolo S. *The high through put highway to computational materials design.* Nature Materials. – **v. 12.** – p. 191-201 (2013).
13. Wargnier H. *Proposal for multi-material design procedure* Materials and Design. – **v. 56.** – p. 44-49 (2014).
14. Derhach T. A. *Nautchnoe obosnovanie vybora nizkolehirovannoi stali i technolohii izhotovleniia neftegazoprovodnych trub povyshennoi korroziinoi stoikosti* Sbornik nauchnych trudov «Stroitelstvo, materialovedenie, mashinostroenie». «Starodubovskie tshteniia - 12». **64.** – p. 202 – 210 (2012) [in Russian]
15. Kuzin O. A. *Vykorysnannia ionno-plazmovoho travlennia dlia analizu strukturno-enerhetychnoho stanu vnutrishnich poverchon rozdilu.* Visnyk Derzhavnoho universytetu «Lvivska politechnika». Optymizaciia vyrobnychych procesiv i tekhnichnyi control u mashynobuduvanni i pryladobuduvanni. – **№ 359** – C. 73-76 (1999). [in Ukrainian]
16. Derhach T. A. *Analiz kachestvennych kharakteristik korrozionno-stoikich aerito-austenitnych stalei s tceliu rasshirenia oblastey ich primeneniia.* Metaloznavstvo ta termichna obrobka metaliv. – **№ 3(66).** – c. 20 – 29 (2013) [in Russian]
17. Mak Lin D. *Granicy zeren v metalach* Moskov: GNTI. – 322 p. (1960) [in Russian]
18. Kuzin O. A. *Vplyv mikrostruktury na znoshuvannia stali 40Ch pislia polipshennia.* Visnyk Derzhavnoho universytetu «Lvivska politechnika». Optymizaciia vyrobnychych procesiv i tekhnichnyi control u mashynobuduvanni i pryladobuduvanni. **№ 371.** – P. 49-51 (1999). [in Ukrainian]
19. Kusyi Ya. M., Kuzin O. A., Kuzin M. O. *Vplyv tekhnolohichnoho marshrutu obroblennia na formuvannia miszzerennoi poshkoszhennosti* Schidno-Evropeiskii zhurnal peredovych tekhnolohiy. **№ 1/5 (79).** – C. 39-47 (2016). [in Ukrainian]
20. Kuzin O. A. *Struktura i protsesy znoshuvannia pokraschenykh stalei 40X i 40XHM.* Visnyk NU «Lvivska politechnika». Optymizaciia vyrobnychych procesiv i tekhnichnyi control u mashynobuduvanni i pryladobuduvanni. **№ 422.** – P. 104-113 (2001). [in Ukrainian]

Increasing the efficiency of highly concentrated coal-water fuel based on the simulation of non-Newtonian fluid flow

Nataliya Chernetskaya-Beletskaya[1], *Andriy* Rogovyi[2], *Igor* Baranov[1,*], *Alexander* Krut[3], *Maria* Miroshnikova[1], and *Nikita* Bragin[1]

[1] Volodymyr Dahl East Ukrainian National University, Department of logistics management and traffic safety vehicle, 93400 Severodonetsk, pr. Central 59-a, Ukraine
[2] Kharkiv National Automobile and Highway University, Department of theoretical mechanics and hydraulics, 61002 Kharkiv, Yaroslava Mudrogo Str, 25, Ukraine
[3] State scientific research, design and design institute of coal industry "Ukrndiproekt", 03142 Kyiv, Academician Palladin Avenue, 46/2, Ukraine

Abstract. The analysis of further prospects for increasing the efficiency of transportation of coal-water fuel in hydro-transport systems of industrial enterprises is carried out. The mathematical model of the spatial three-dimensional flow of coal-water fuel was developed on the basis of SST turbulence model based on the solution of Navier-Stokes equation. As a result of the calculation, the values of pressure loss, flow rate and velocity distribution over the cross section of the pipeline in the straight section and in the turn were determined, which allowed determining the energy consumption during coal-water fuel transportation in the industrial hydro-transport system. The performed studies allowed us to refine the mathematical model of water-coal suspension flow and, thus, improve the patterns of influence of hydro-transportation scheme and parameters of coal-water fuel on energy consumption for its supply to enterprise consumers. By means of mathematical model of non-Newtonian fluids flow, the patterns of influence of hydro-transport system parameters and transportation modes of coal-water fuel on its energy indicators in industrial hydro-transport systems are determined. The obtained results are related to reduction of energy consumption, optimization of enterprise transport network configuration and increase of efficiency of coal-water fuel transportation to enterprise energy facilities.

1 Introduction

The sustainable development of fuel and energy complex of Ukraine is associated with the use of alternative fuels, which are the most efficient and cost-effective for the energy sector of the country. The enterprises of thermal power and coal mining complex of Ukraine: mines and concentrating factories – have in their structure boiler-houses working on coal with the total annual demand of the enterprise up to 150 thousand tones. Meanwhile, further prospects for the use of coal fuels in Ukraine are inextricably connected with the use of coal-water fuel (CWF).

Coal-water fuels are widely used in such countries as China, USA, Russia, Italy, Sweden, Japan and Canada. Industrial hydro-transport systems (HTS) used for transportation of CWF have a wide application range: power industry, metallurgy, construction materials production, chemical industry. CWF consumers can be small, medium and large industrial enterprises of thermal energy and coal-mining complex: mines and concentrating factories. Industrial HTS (IHTS) can be transport arteries of fuel and energy or fuel and metallurgical complexes; can supply CWF to heating, heating and production and production boiler houses on the enterprise territory.

Increasing the efficiency of CWF transportation in IHTS means reducing the cost of transportation of 1 tkm of cargo at a constant cargo flow, at this the cost of transportation directly depends on transportation energy consumption. The disadvantage of pipeline hydro-transport is the increase in energy consumption , depending on the proportions of various fractions of granulometric composition (GMC) of the transportation product, as a result, in case of inappropriate composition, energy consumption for CWF transportation increases by up to 1.5-1.7 times [1].

1.1 Formulation of the problem

The problem of the functioning of IHTS is the absence of a specific criterion for assessing GMC of highly concentrated CWF (HCCWF) and that there are no recommendations how to prepare and adhere to appropriate proportions of its various fractions, resulting in increased energy consumption when transporting CWF.

* Corresponding author: baranov_90@ukr.net

The absence of CWF transportation scheme optimized according to technical and economic indicators and justified in terms of minimization of energy consumption of the parameters of the enterprise operation leads to an increase in the cost of CWF transportation by 19%. In addition, in order to estimate the energy consumption of HCCWF transportation and to increase the efficiency of HTS operation, it is necessary to clarify the flow patterns along the pipeline and to improve the mathematical models of process description of HCCWF transportation, which is a relevant scientific and practical task.

A significant number of theoretical and experimental studies of CWF transportation confirms the strong impact of HTS on energy efficiency of transportation process [2, 3].

The study of impact of various HTS options on CWF transportation parameters requires a large number of experimental studies providing for significant investment.

Presently, it is possible to perform the set task using the numerical methods of fluid flow problems, which are characterized by a minimum time consumption for calculation [4, 5].

This approach allows us to reduce the basic number of experimental studies by replacing some of the expensive experimental research with theoretical numerical ones.

To implement this approach it is necessary to validate mathematical models in order to obtain a minimum difference between the results of mathematical modelling and experimental research [6, 7].

Calculations on the mathematical models of CWF transportation processes in the pipeline and various elements of HTS configuration can reduce the time for improving the calculation methods of hydro-transportation parameters, obtain the most accurate characteristics of the systems, and define the limits of their use [8, 9]. Therefore, mathematical modelling is one of the most important and relevant tasks of HCCWF flow research [10].

1.2 Analysis of recent research and publications

In recent years, a significant number of works have been devoted to the simulation of Bingham fluids flow [2, 3, 5, 8, 9, 11, 12, 13, 14]. The works [11, 12] are devoted to the flow simulation of coal-water mixtures (CWM) in particular, but for the coal of Chinese and Indian regions. At the present time, transportation studies of HCCWF for the coal of fuel and energy complex of Ukraine, have not been conducted.

In [11, 14] the simulation of non-Newtonian fluids flow in the turns was performed, but regressive dependencies for all variants of rheological properties were not given. In article [15] the exact calculation of characteristics of the fluid flow are given, but only by a pipeline, which does not allow using the obtained correlation for various options of HTS configuration.

As the analysis of the research has shown, the greatest influence on transportation parameters of CWF in IHTS have the characteristics of raw coal, concentration and HTS , as well as the transportation velocity parameters , on which the rheological properties and energy consumption depend on the whole [2, 3, 5, 13].

2 Mathematical model of non-Newtonian fluids flow

In this paper, we used the SST turbulence model, which proved to be the best in the study of virtually any wall-bounded flow. When using the SST model, there appear significant errors in determining the velocity and pressure values at certain points, but the integral parameters are determined with its help with adequate accuracy and the error does not exceed 10% [16-19].

Calculation errors naturally arise in flow separation zones. However, given the computational appeal of this model in comparison with direct numerical simulation (DNS), the methods of large eddy simulation (LES) and detached eddy simulation (DES), the SST model has become virtually universal for uncomplicated flows. To solve this problem, Reyonolds averaged Navier-Stokes equations and continuity equations were used:

$$\frac{\partial u_i}{\partial t}+u_j\frac{\partial u_i}{\partial x_j}=F_i-\frac{1}{\rho}\frac{\partial p}{\partial x_i}+\frac{\partial}{\partial x_j}\left[\frac{\mu_{ef}}{\rho}\frac{\partial u_i}{\partial x_j}\right];\qquad \frac{\partial u_i}{\partial x_i}=0; \tag{1}$$

where x_j – Cartesian coordinates; u_j – velocity projections in Cartesian coordinate system; $\mu_{ef}=\mu+\mu_t$ – effective viscosity; μ_t – turbulent viscosity; μ – molecular viscosity.

System of equations of SST model [20]:

$$\frac{\partial(\rho k)}{\partial t}+\frac{\partial}{\partial x_j}(\rho u_j k)=\frac{\partial}{\partial x_j}\left(\mu_{ef}\frac{\partial k}{\partial x_j}\right)+P_k-\beta^*\rho k\omega;$$

$$\frac{\partial(\rho\omega)}{\partial t}+\frac{\partial}{\partial x_j}(\rho u_j\omega)=\frac{\partial}{\partial x_j}\left(\mu_{ef}\frac{\partial \omega}{\partial x_j}\right)-\rho\beta\omega^2+Cd_\omega+\alpha\frac{\rho}{\mu_t}P_k, \tag{2}$$

where k – kinetic energy of turbulent pulsations; ω – specific dissipation; P_k – generated member; Cd_ω – a cross member. Constants and descriptions of equations can be found in [20].

The modelling of the rheological parameters of Bingham fluid was carried out according to the formula [12, 21]:

$$\mu=\frac{\tau_0}{\gamma}+\eta, \tag{3}$$

where τ_0 – initial yield stress, γ – shear rate between fluid layers, η – structural viscosity.

Control volume approach and PISO algorithm (Pressure Implicit with Splitting of Operators) were used. Numerical schemes of the second order were used. The calculations were carried out until the residual in all the main parameters did not reach the value of 10^{-5}. In addition, the second condition for stopping the calculations was time invariance of costs. The constancy criterion of cost is the value of 1 %.

In order to form the flow in the boundary layer prismatic elements were used. To determine a minimal sufficient flow grid, calculations on three different grids were made and the grid with such a number of elements from which the integral parameters cease to change (the number of elements of 470 thousand) was selected. During the preliminary calculations, the effect of the parameter y^+ on the solution accuracy was also estimated. Thus, based on the calculations as well as recommendations given in [22], it was decided to use the grids with $y^+ < 4$.

2.1 Calculation results of coal-water fuel

At the moment, there is quite a number of software products that allow you to calculate the flow parameters of fluids and gases: OpenFoam, Ansys Fluent, Cosmos FlowWorks, Ansys CFX, FlowVision and many others. In this paper, the calculation was made using the ANSYS CFX (Academic Student Licence with noncommercial use only) software based on the Bingham model.

To verify the numerical calculations, experimental studies of CWF flow by direct pipeline were performed and pressure losses for realizing the given loss or CWF velocity were found [23]. Test bench for physical examination is equipped with measuring equipment with sufficient accuracy for calculations verification and then for adequacy verification. The maximum error of determining the pressure losses did not exceed 1 %.

As a result of the calculations, the values of pressure losses, flow rate and velocity distribution at the intersection of the pipeline in the straight part and in 90° turn (Fig. 1) were determined, which allowed us to determine energy costs when transporting CWF in IHTS. In stationary calculation, the trajectory of liquid particles flow has the form of flow lines, which coincide with the trajectories of particles.

As Fig. 1 shows, near the pipe axis there is a zone of approximately the same velocity, which qualitatively coincides with the results of experimental studies. This zone is a core flow, which is typical of Bingham fluids flow.

Thus, it may be affirmed that the numerical calculation results qualitatively coincide with the analytical description of Bingham flow and the results of experimental research. Because of the fact, that flow lines are straight lines, as well as when checking Reynolds numbers, it can be concluded that the flow is laminar. Therefore, calculations and comparison with experimental data of different models were performed: laminar, $k-\varepsilon$ and SST model (Fig. 2). The figure shows the error in calculating the pressure loss in percentages. As can be seen from the figure, the largest error is the flow calculation without taking into account the rheological law, which is the flow of the Newtonian fluid with the same viscosity. The laminar model and SST model, taking into account the rheological law, have the least error.

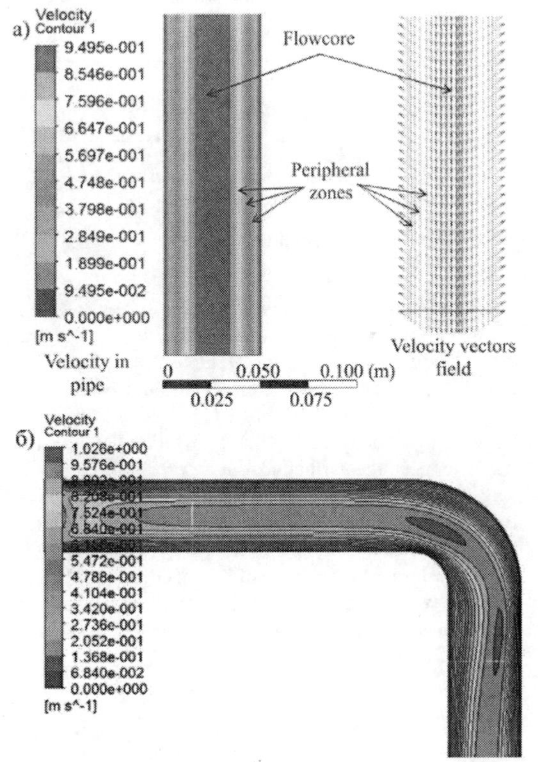

Fig. 1. Graphic representation of calculation results of CWF flow based on Menter SST model: a) – straight section (laminar flow); б) – shaped part of the pipeline (flow with elements of turbulence).

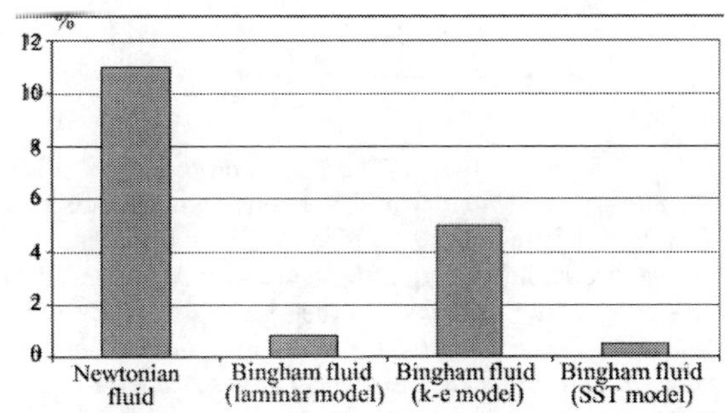

Fig. 2. Comparison of the accuracy of determining pressure loss according to different models.

According to the comparison of the calculation accuracy, it is of no particular importance, which model to select for calculation, laminar or Menter model, but regarding the versatility of model use , the SST model is more versatile and can be used for any flow type.

To verify the selected model, calculations of Bingham fluid flow of CWF with different parameters of rheological model are performed, the results of which are shown in Fig. 3. Here the calculation error of pressure losses in pipe at a different average velocity is

https://doi.org/10.1051/matecconf/201929401009

presented. The error is calculated by comparing the results of theoretical studies with the experimental data.

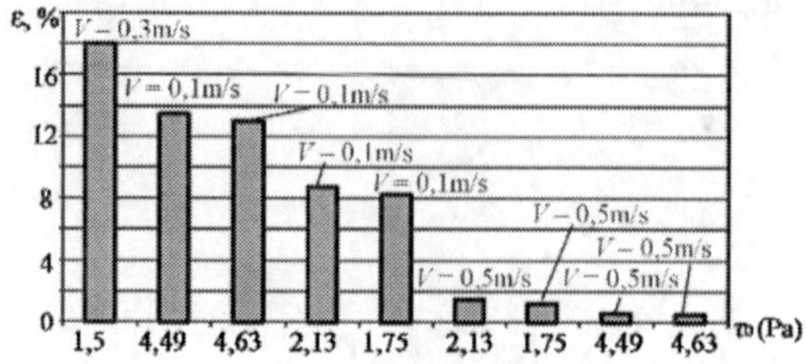

Fig. 3. Calculation error of pressure loss at Bingham fluid flow using SST-model.

As Fig. 4 shows, flow calculations with Reynolds numbers Re > 30, have the least error, that is, the greater the velocity of Bingham fluid is, the less is the calculation error according to SST-model of turbulence. In general, the error does not exceed 20 %, and with an increase in velocity makes 0,5 %, which is an acceptable result, and suggests that in the future, you can use the software package and SST model for calculation of CWF flow and use a mathematical model for optimization of fuel granulometric composition.

Granulometric composition (GMC) of CWF influences the viscosity of the rheological model, which further has an effect on the velocity profile and pressure losses of fuel flow (Fig. 4).

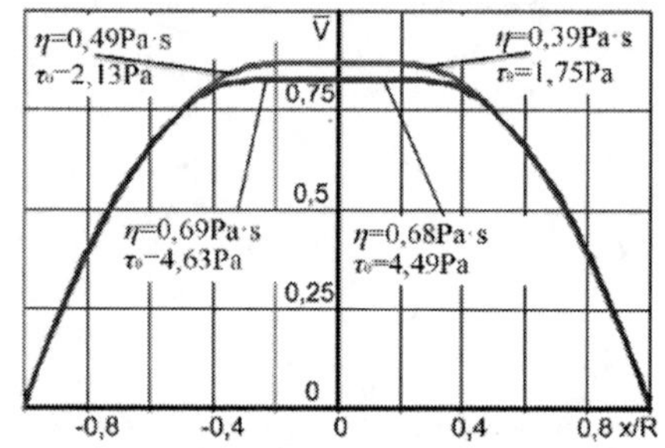

Fig. 4. Velocity profiles of CWF flow of different GMC.Fig. 4. Shows the velocity profiles obtained when calculating CWF of different GMC. With an increase in the initial shear stress, the maximum flow velocity decreases. In Fig. 4 velocity is assigned to the maximum velocity calculated according to Newtonian fluid model. To find the optimal values of initial shear stresses and structural viscosity, a numerical experiment on the basis of experiment planning methods was conducted in order to reduce pressure losses [21].

2.2 Calculation of coal-water fuel flow in bends

In the shaped elements of different configuration (elbows, bends) due to the change of flow structure, there appear centrifugal forces, directed from the center of the curvature to the outer wall of the pipeline. That is the reason for a pressure increase at the outer wall and its decrease at the inner wall during the transition of the flow from the straight pipeline section to the curved one.

When designing HTS and optimizing CWF flow, it is necessary to determine the pressure losses at bends with curvature radius of the standard values.

Fig. 5-7 show the velocity distribution of coal-water fuel in a bend with curvature radius $\bar{R}$·= 1,0 at different Reynolds numbers with rheological properties: η = 0,68 Pa·s, τ_0 = 4,49 Pa.

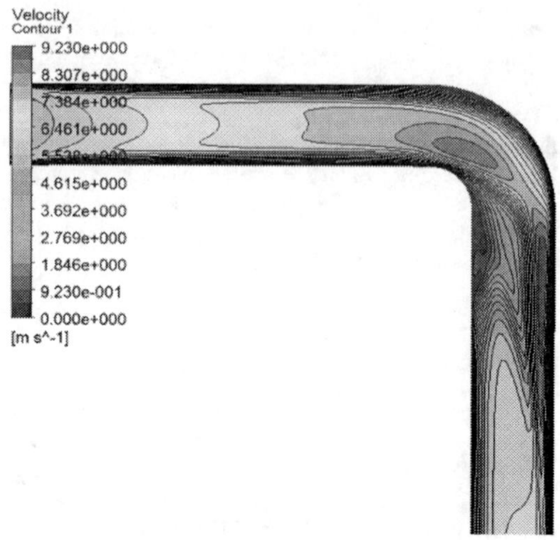

Fig. 5. Velocity distribution of CWF flow at 90° turn with curvature radius $\bar{R}$·= 1,0, Re = 440.

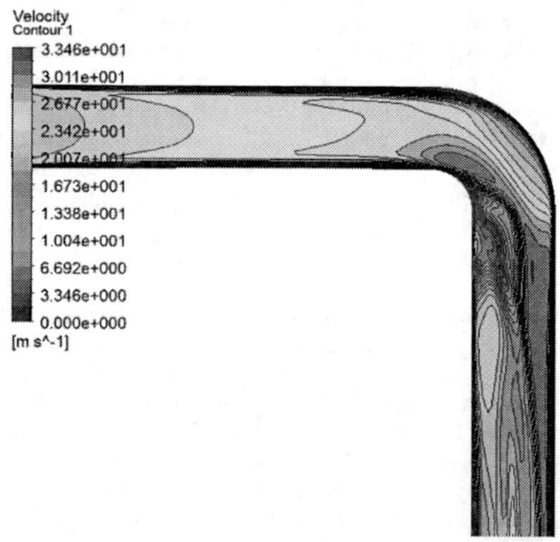

Fig. 6. Velocity distribution of CWF flow at 90° turn with curvature radius $\bar{R}$·= 1,0, Re = 1840.

With the increase of Reynolds numbers, there is a flow separation after the turn and an increase in hydraulic resistance coefficient [24]. Fig.8 shows the dependence of pressure losses in the bend with curvature radius $\bar{R}$·= 1,0 depending on the average velocity of CWF with rheological properties: η = 0,68 Pa·s, τ_0 = 4,49 Pa.

Fig. 9 shows a monotonous decrease in hydraulic resistance coefficient $\varsigma = 2\Delta p / \rho V^2$ of a bend with $\bar{R}$·= 1,0 with Reynolds number increase.

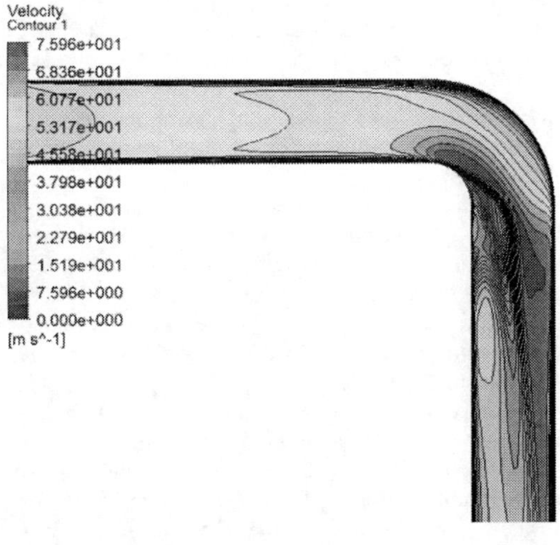

Fig. 7. Velocity distribution of CWF flow at 90° turn with curvature radius $\bar{R}$·= 1,0, Re = 4600.

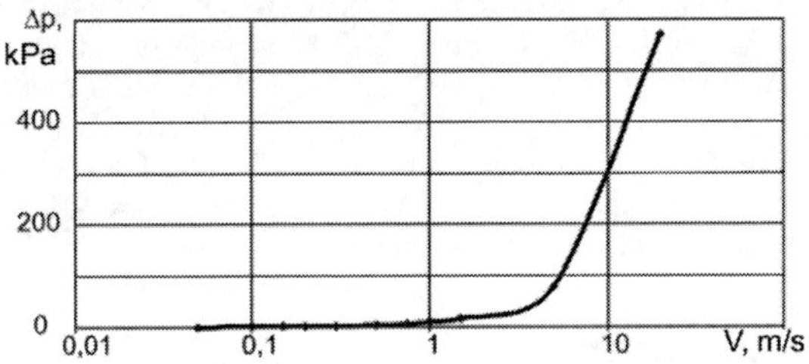

Fig. 8. Dependence of pressure losses in a bend with $\bar{R}$·= 1,0 on average flow velocity.

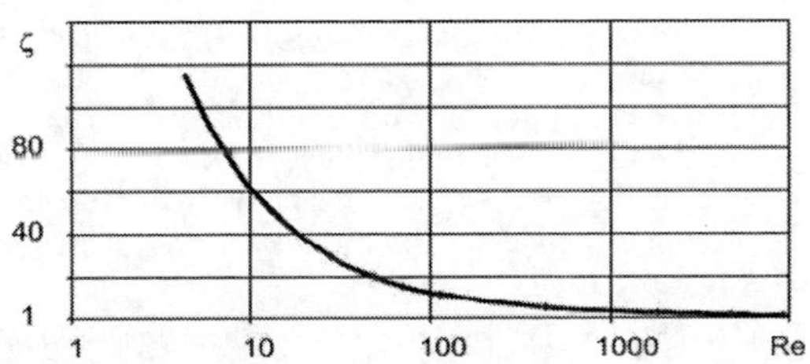

Fig. 9. Dependence of hydraulic resistance coefficient in a bend with $\bar{R}$·= 1,0 on Reynolds number.

Coefficient ς tends to a value of 1.0 with Reynolds number increase to value Re > 1000 [25]. For Newtonian fluid coefficient ς of such a bend is equal to 0.21 for smooth walls, which is almost five times less than for CWF with the stated rheological properties.

In order to take into account the hydraulic resistance of hydro-transport system sections, which include shaped parts, the following expression for specific pressure losses is used at the next stages of work

$$i = i_{л} + \frac{1}{2\ell_{д}}\sum_{j=1}^{n}\left[\kappa_j \zeta_j \rho V^2\right] \quad (4)$$

where $i_{л}$ – specific resistance of the straight part of section, Pa/m; $l_{д}$ – length of the section; j – index of the type of shaped part of the pipeline; n – number of types of shaped parts, which are part of the section; κ_j – number of shaped parts of j-type; ζ_j – coefficient of local resistance of shaped part of j-type.

3 Influence of granulometric composition of coal-water fuel on pressure losses

To obtain dependencies that have an effect on pressure loss during CWF transportation, a serious of calculations has been carried out with the helps of the proposed mathematical model for different GMC. It is established that the most important factors influencing the parameters of CWF hydro transportation are the concentration of the solid component (C) and the parameters characterizing the bimodality of GMC of CWF solid phase. This approach allows comparing the pressure loss for various options of bimodal composition and makes it possible to trace Γ_1, Γ_2 effect at different concentrations on CWF hydraulic parameters.

During the simulation, the following parameters are selected: D = 50 mm, V = 0.5 m/s.

For different Γ_1, Γ_2 on the basis of experimental research, rheological properties η and τ_0 were determined, and introduced into the calculation of the mathematical model.

Fig. 10 shows that pressure losses curves, depending on Γ_1/Γ_2 ratio (bimodality criterion (BC) have clearly defined minimum in the range of 1.7…2.0. In order to confirm the obtained conclusions regarding the minimum of hydraulic pressure losses during CWF transportation by pipeline, a comparison with experimental data was made, which is presented in Fig. 11.

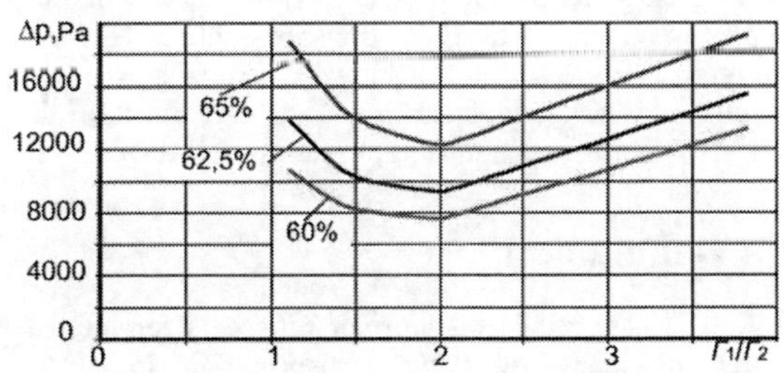

Fig. 10. Dependence of pressure losses on CWF granulometric composition.

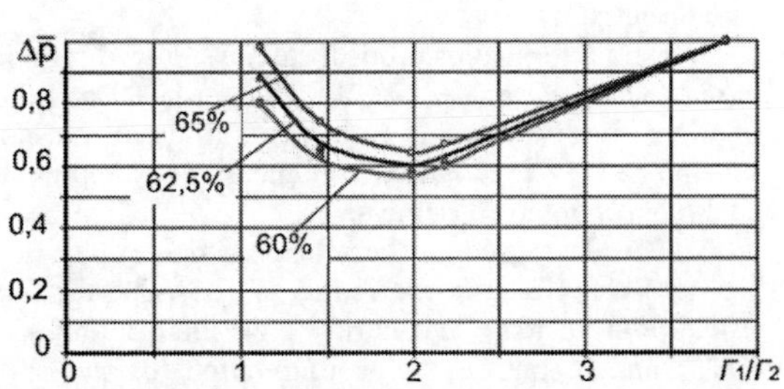

Fig. 11. Comparison of numerical calculation results with experimental results.

Pressure losses shown in Fig. 11 are attributed to the maximum pressure losses corresponding to losses during transportation of fluid with GMC having the ratio Γ_1/Γ_2=3.82. The lines correspond to the numerical calculations, the points correspond to the experimental studies. The difference between the experimental data and the numerical calculation does not exceed 1.6 %.

Thus, as it can be seen from Fig.11, with an increase in concentration, the relative pressure losses increase. Minimal losses correspond to the GMC with an index Γ_1/Γ_2=1.7...2.0. In addition, in [1] the division of bimodal variants of CWF into variants is proposed: those having a harmonious structure and those, which are inherently imperfect.

According to the estimated number of layers with a diameter, equivalent to particles of fine fraction, the number of such layers in the range of 2.2-2.6 should correspond to the harmonious structure. According to the results of the study [26] , the use of GMC of CWF (with Γ_1/Γ_2=2.33), allows reducing its transportation energy consumption to the HTS of the company "DTEK Dobropilvugillya LLC" by 86.13 thousand UAH/year compared with random granulometric composition, which was chosen as an output one.

As a result of using sequential-beam transportation of CWF, the total cost of pipeline laying has decreased by 893.89 thousand UAH in comparison with beam scheme. The effect of optimizing the topology of the transport network of distribution by CWF consumption points, optimizing the total pipeline costs, equipment and electric energy for transportation and selection of GMC of CWF amounted to 183.75 thousand UAH/year [26].

The total volume of capital investments in the proposed HTS project amounted to 7.47 million UAH, annual operating expenses amount to 199.89 thousand UAH. The overall economic effect of introducing IHTS instead of road transport is 4.08 million UAH/year, payback period of the project is 1.83 years. When using pipeline transport for CWF transportation, as HTS of the enterprise, the difference between capital costs is 0.34 million UAH, and operating costs are by 4.1 million UAH/year lower compared to road transport.

4 Conclusion

1. It is determined that further prospects for increasing the efficiency of CWF transportation in HTS of industrial enterprises are based on optimization of granulometric composition of solid fuel component and its concentration.

2. The performed theoretical studies allowed refining the mathematical model of CWS flow in IHTS and thus improve the patterns of influence of hydro-transport scheme and CWF parameters on energy consumption of coal fuel supply to consumers.

3. The developed mathematical model of the non-Newtonian fluids flow allowed us to determine the flow structure and energy loss during CWF transportation in IHTS and optimize the configuration of transport network, taking into account patterns of influence of GMC, rheological properties and transportation modes on transportation energy consumption.

4. It was established that according to the results of modelling, the specific pressure losses of transported by pipeline CWF have a local minimum in the studied range of change of the proposed criterion.

5. According to the results of theoretical studies of CWF flow taking into account the bimodality criterion and its rheological properties, the following conclusions can be made:

- According to flow numerical calculation results , a zone of approximately the same velocity is obtained near the pipe axis , which qualitatively coincides with the results of experimental studies. The results of numerical calculation qualitatively coincide with the analytical description of Bingham flow and the results of experimental studies.
- The curves of the pressure losses, depending on Γ_1/Γ_2 ratio, have a clearly expressed minimum in the range of 1.7...2.0. The difference between the experimental data and the numerical calculation does not exceed 1.6 % for formed dependencies of the loss on Γ_1/Γ_2. With an increase in concentration, relative loss of pressure increases.
- Hydraulic resistance coefficient ς of a bend with $\bar{R}$= 1,0 monotonously decreases with increasing Reynolds number and tends to values of 1.0 with Reynolds number increase to values Re > 1000. For Newtonian fluid coefficient ς of such a bend is equal to 0.21 for smooth walls, which is almost five times less than for CWF with stated rheological properties.

References

1. I.O. Baranov. *Improving the highly loaded coal-water slurry fuel transportation efficiency in the hydro-transport systems of industrial enterprises.* Abstract thesis for degree candidate of technical sciences. (Doctor of Philosophy), speciality 05.22.12 – Industrial Transport (275 – Transport Technology). – V. Lazarian Dnipropetrovsk National University of Railway Transport, Dnipro, 21 (2019).
2. T. Norton, D. Sun. *Computational fluid dynamics (CFD)–an effective and efficient design and analysis tool for the food industry: a review.* Trends in Food Science & Technology, **17**(11), 600-620 (2006).
3. M. Sorgun, M. Ozbayoglu. *Predicting frictional pressure loss during horizontal drilling for non-Newtonian fluids.* Energy Sources, Part A: Recovery, Utilization, and Environmental Effects, **33**(7), 631-640 (2011).
4. D. Syomin, A. Rogovyi. *Features of a working process and characteristics of irrotational centrifugal pumps.* Procedia Engineering, **39**, 231-237 (2012).
5. R. Peyret, T. Taylor. *Computational methods for fluid flow.* (New York: Springer Science & Business Media) (2012).
6. E. Fjodorov. *Metody izmerenija urovnja i granic razdela mnogofaznyh zhidkih sred.* Materialy VI

nauchno-prakticheskoj konferencii 27–30 maja 2015 g. – Tomsk, 65 (2015). [in Russian]

7. A. Thompson. *Basic hydrodynamics*. Elsevier Science. Library of Congress Cataloging in Publication Data, 177 (2013).
8. J. Wang, S. Wang, T. Zhang, F. Battaglia. *Mathematical and experimental investigation on pressure drop of heterogeneous ice slurry flow in horizontal pipes*. International Journal of Heat and Mass Transfer, **108**, 2381-2392 (2017).
9. F. Kabinejadian, D. Ghista. *Compliant model of a coupled sequential coronary arterial bypass graft: effects of vessel wall elasticity and non-Newtonian rheology on blood flow regime and hemodynamic parameters distribution*. Medical engineering and physics, **34**(7), 860-872 (2012).
10. C. Hervé. *The basics of plant hydraulics*. The Journal of plant hydraulics, **1**, 178-188 (2014).
11. M. Liu, Y. Duan. *Resistance properties of coal–water slurry flowing through local piping fittings*. Experimental Thermal and Fluid Science, **33**(5), 828-837 (2009).
12. A. Rawat, S. Singh, V. Seshadri. *Computational methodology for determination of head loss in both laminar and turbulent regimes for the flow of high concentration coal ash slurries through pipeline* Particulate Science and Technology. – T. 34. – №. **3**., 289-300 (2016).
13. K. Ekambara, R. Sanders, K. Nandakumar, J. Masliyah. *Hydrodynamic simulation of horizontal slurry pipeline flow using ANSYS-CFX*. Industrial & Engineering Chemistry Research, **48**(17), 8159-8171 (2009).
14. P. Csizmadia, C. Hős. *CFD-based estimation and experiments on the loss coefficient for Bingham and power-law fluids through diffusers and elbows*. Computers & Fluids, **99**, 116-123 (2014).
15. P. K. Swamee, N. Aggarwal. *Explicit equations for laminar flow of Bingham plastic fluids*. Journal of Petroleum Science and Engineering, **76** (3-4), 178-184 (2011).
16. J. Yin, L. Jiao, L. Wang. *Large eddy simulation of unsteady flow in vortex diode*, Nuclear Engineering and Design T. 240. – №. **5**. – 970-974 (2010).
17. A. Rogovyi. *Energy performances of the vortex chamber supercharger*. Energy, **163**, 52-60 (2018).
18. Y. H. Alahmadi, A. F. Nowakowski. *Modified shear stress transport model with curvature correction for the prediction of swirling flow in a cyclone separator* Chemical Engineering Science. Vol. **147**. – 150-165 (2016).
19. H. R. Thakare and A. D. Parekh. *Computational analysis of energy separation in counter–flow vortex tube*. Energy. Vol. **85**. – 62-77 (2015).
20. P. E. Smirnov, F. R. Menter. *Sensitization of the SST turbulence model to rotation and curvature by applying the Spalart–Shur correction term*. Journal of Turbomachinery. Vol. **131**. – №. 4. – 041010 (2009).
21. N. Chernetskaya-Beletskaya, A. Rogovyi, A. Shvornikova, I. Baranov, M. Miroshnikova, N. Bragin. *Study on the coal-water fuel pipeline transportation taking into account the granulometric composition parameters*. International Journal of Engineering & Technology. – Vol. **7**, Iss.4.3, 240-245 (2018).
22. G. Besagni, F. Inzoli. *Computational fluid-dynamics modeling of supersonic ejectors: Screening of turbulence modeling approaches*. Applied Thermal Engineering; **117**: 122-144 (2016).
23. N. Chernetskaya-Beletskaya, I. Baranov, M M. Miroshnikova. *Zabezpechennja stalogo rozvitku regionu: ekonomichni, upravlinski, pravovi ta informacijno-tehnichni aspekti: kolektivna monografija* ; za zag. red. Ju. I. Kljus., N. V. Shvec. – Severodoneck: vid-vo SNU im. V. Dalja ISBN 978-617-11-0118-0, 193-212 (2017). [in Ukrainian]
24. N. Chernetskaya-Beletskaya, A. Rogovij, I. Baranov, M. Miroshnikova. *Matematichna model prostorovoї trivimirnoї techiї vodovugilnogo paliva*. Visnik SNU im. V. Dalja, № **1** (242) – 159-164 (2018). [in Ukrainian]
25. N. Cherneckaya-Beleckaya, A. Rogovyi, I. Baranov, M. Miroshnykova. *Mathematical model flow coal-water fuel. Globalization of scientific and educational space. Innovations of transport*. Problems, experience, prospects: thesis, May 2018, Italy – Severodonetsk: Volodymyr Dahl East Ukrainian National University, 82-83 (2018).
26. I. Baranov. *Formuvannja optimizacijnoї modeli viboru parametriv transportuvannja vodovugilnogo paliva promislovim gidrotransportom*. Visnik SNU im. V. Dalja, № **3** (244), 7-13 (2018). [in Ukrainian]

Model for building traction information of suburban rolling stock on hydrogen fuel

Anatoly Falendysh[1*], *Volodymyr* Dzhus[2], *Olha* Kletska[1], *Oleg* Kosariev[1], and *Jan* Dizo[3]

[1] Ukrainian state university of railway transport, Heat Engineering and Heat Engines, Kharkov, Ukraine
[2] DNURT, Department Chair of rolling stock and railway, 79052 Lviv, I.Blazhkevych Str., 12-a, Ukraine
[3] University of Zilina, Department of Transport and Handling Machines, Zilina, Slovakia

Abstract. The materials consider the possibility of using hydrogen as a fuel for suburban rolling stock in Ukraine. The analysis of the use of hydrogen fuel in transport in various countries of the world is carried out. One of the main stages of calculating the rolling stock is the construction of its traction characteristic, which largely depends on the type of power plant. Therefore, a model has been developed for constructing traction characteristics of a suburban rolling stock when it is converted to hydrogen fuel. For this purpose, an analysis of power equipment with an energy source from a polymer of an exchange membrane fuel cell and methods of connecting their batteries to blocks is done. The scheme of power transmission for the rolling stock is proposed and its efficiency coefficient is preliminarily calculated. The main technical indicators of suburban rolling stock in Ukraine at the moment and using proton-exchange membrane fuel cells on it are considered. Preliminary consideration is given to the operation parameters of the electric circuit of the rolling stock under various modes of its motion. The traction characteristic of a suburban mobile suction on hydrogen fuel is constructed. The proposed model is implemented in the software product MathLab.

1. Introduction

The greater part of all transportations in Ukraine is provided by railway transport [1]. But at the same time, more than 90% of it is currently behind the world technical progress. In Ukraine, scientific works and practical research are mainly aimed at developing activities related to the extension of the life of the already existing rolling stock. At the moment, the new rolling stock in the country is being produced by Kryukovsky Carriage Works. He developed and produced an electric train ECP1 (Tarpan) and a diesel train DPKr-1 [2].

In the world there is a tendency to increase the speed of rolling stock movement and use of different types of thrust without harmful emissions into the environment. Therefore, the issue of developing our own high-speed and environmentally friendly rolling stock for Ukraine is topical. This is confirmed by the strategy for the development of "Ukrzaliznytsya" PAT until 2021, dated 12.09.2017 [3].

2. Research and discussion of the oresults

The use of hydrogen as a fuel in transport power plants is possible in two ways:

- as a fuel for internal combustion engines [4, 9];
- as a fuel cell, the principle of operation of which is based on electrolysis [5, 6, 9].

Work on the use of hydrogen fuel in transport along both tracks has been carried out in many countries around the world since the nineteenth century [4-10]. The use of hydrogen in railway transport has begun to receive attention only in the last 20 years. Therefore, further studies are continued for the use of hydrogen in power railroad transport usa- novs.

Scientists from the Tokyo Railway Research Institute (Japan) and the French company Alstom went on the second route of using hydrogen as an energy plant. For example, Alstom presented a prototype of the Coradia iLint rail car designed and built on the basis of the Lint 54 at the World Rail Transport Exhibition [10-13].

Also, the issues of the use of hydrogen in rail transport are given great attention in the United States, Britain and China. VNIIZhT scientists propose the use of hydrogen on the rolling stock of Russian railways [14]. Variants of the use of hybrid power plants on the modernized rolling stock in Ukraine are considered in [15-19].

Thus, the second version of the use of hydrogen as a fuel (the use of proton-exchange membrane fuel cells in hybrid power transmission) for a motor-car rolling stock was considered in [20]. A model for the technical and economic evaluation of the rolling stock on hydrogen fuel is proposed, the necessary equipment and its location for the modernization of the diesel train are

* Corresponding author: fap_hiit@ukr.net

https://doi.org/10.1051/matecconf/201929401010

preliminarily determined. But the issues of traction calculations for rolling stock on hydrogen fuel are not considered.

Therefore, consideration of carrying out traction calculations for rolling stock using hydrogen fuel is currently an urgent and timely task.

Based on preliminary calculations [20], for further work on the modernization of the rolling stock by power plants using hydrogen fuel, the diesel train DPKr2 (Fig.1).

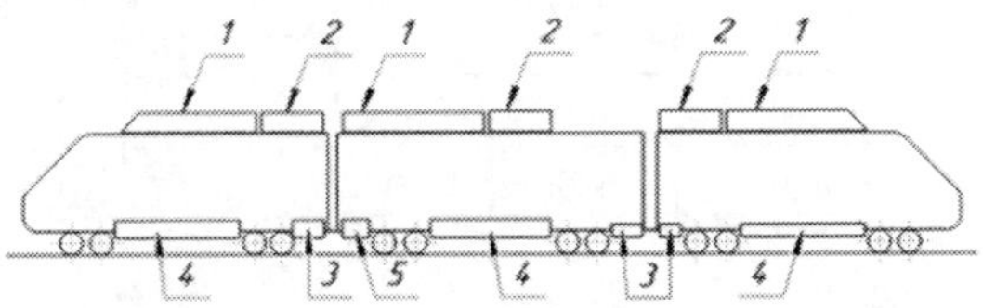

1 – Fuel tank, 2 – PEMFC, 3 – Invertor for traction motors, 4 – Li-Ion batteries, 5 – Invertor for indoor needs.

Fig.1. Location equipment on a multiple units rolling stock rolling stock with hydrogen fuel.

It is supposed to use the following equipment in advance: proton-exchange membrane fuel cells (PEMFC), fuel tanks, cryogenic for storage of hydrogen, storage batteries, drive trucks with traction electric motors.

For further calculations, the parameters of the main equipment operation and their technical characteristics were determined. The equipment was selected in the following order. First, the characteristics of traction electric motors were selected to provide the necessary tractive force. Then, taking into account the conversion of the electrical current in the inverter, the necessary voltage and current at the output of the battery pack was determined, the battery pack was formed and calculated, and the required PEMFC was subsequently determined which would provide the fastest and most economical charging of the battery pack. The main technical parameters for this equipment are presented in Table 1.

Table 1. The main technical characteristics of equipment for upgrading a suburban train.

Name	value
Traction asynchronous electric motor	
power, [kWt]	400
Maximum rotation frequency, [min^{-1}]	4126
Starting torque, [Nm]	3607
Accumulator batteries of the company "Panasonic" NCR18650PF	
Nominal capacity, [mAh]	2700
Rated voltage, [W]	3,7
Maximum discharge current, [A]	10
Maximum charging current, [A]	5,4
Weight, [kg]	0,047
amount NCR18650PF, [pieces].	12096
Cost, [USD]	$36288
Weight (without case), [kg]	568512
PEMFC	
Power rating, [kW]	240
Amperage, [A$_{dc}$]	0 – 1000
Voltage, [W$_{dc}$]	240–480
Maximum efficiency, [%]	60

The battery pack consists of "Panasonic" NCR18650PF batteries.

The connection of the batteries to the units occurs in stages in the same blocks, which are then connected together. This, first, is necessary because all connected batteries and units must have the same parameters, otherwise the parameter with the smallest value will be used in the connection.

Block A consists of 3 sub-blocks connected in parallel, the subblock consists of 4 batteries connected in series.

Block B consists of 3 sub-blocks connected in parallel, the subblock consists of 4 blocks A, connected in series.

Block C consists of 5 sub-blocks connected in parallel, the subblock consists of 4 blocks B, connected in series.

Block D consists of 2 sub-blocks connected in parallel, the subblock consists of 2 blocks C, connected in series.

The energy transfer scheme for a modern commuter train is as follows: POMTE → DC / DC converter → Rechargeable battery (120 kW) → DC / AC step-up inverter → Traction electric motor → Gear drive from the traction motor to the wheel pair.

Previously, the efficiency of this transmission was determined by the formula, %,

$$\eta_{dt} = \eta_{pemfc} \cdot \eta_{kdc/dc} \cdot \eta_{bat} \cdot \eta_{inv} \cdot \eta_{mot} \cdot \eta_{tr}, \quad (1)$$

where η_{pemfc} – efficiency of PEMFC,%, in nominal mode, we accept η_{pemfc}= 60%;

$\eta_{kdc/dc}$ – KDC/DC converter efficiency, %, $\eta_{kdc/dc}$ =97.5%;

η_{bat} – battery efficiency, %, η_{bat}=98,01%;

η_{inv} – efficiency of boosting inverter, %, η_{inv}=97,5%;

η_{mot} – efficiency of traction motor, η_{mot}=95%;

η_{tr} – efficiency of gear transmission from the traction motor to the wheel pair, %,η_{tr}=97%.

As a result of calculations, the overall efficiency of the power transmission scheme for the upgraded train was obtained, which is equal to 51.51%.

The next stage of the calculations was the determination of the parameters of the operation of the individual elements of the circuit under various modes of movement of a suburban train.

For the rolling stock in question, we determine the following modes of operation:

– traction mode (PT);

– idling speed (XX);

– Regenerative braking mode (TP);

– mechanical braking mode with the use of brake pads (TK).

In the following calculations we assume:

- when driving in traction mode when charging accumulator batteries more than 20%, the electric power from the storage batteries (SB) is transferred to the boosting inverter (BI), and from it to the TEM.

- when driving in the traction mode with a SB charge equal to 20% or less, and provided that the SB charge is not enough to continue driving to the nearest stop,

PEMFC is activated, which provides enough energy for the train to travel and power the auxiliary equipment.

- when driving idling, if the SB charge is less than 20% at the end of the rolling stock movement, the PEMFC turns on and continues to operate until the battery charge reaches the level at which the battery charge at the end of the train is at least 20 %. This condition may not be met only if after idle operation the regenerative braking mode is followed by further full stop, but in this case the battery charge can't be less than 5% at the end of the rolling stock in idle.

- braking mode with brake pads (BP) is intended only for complete stopping of rolling stock in 2 cases. 1 – with standard braking, the brake pads are pressed against the rim of the gearbox to stop the SS at a speed of 5 km / h; 2 – with emergency braking, brake pads are activated simultaneously with regenerative braking to ensure maximum braking force (controlled by a computer, PEMFC does not work).

- In the PT mode, the energy comes from the TEM through the inverter on the battery. PEMFC does not work.

- to increase the overall efficiency, the battery pack is fully charged from the network at the service points, before the rolling stock is sent to the flight.

When constructing the traction characteristics of a suburban train in accordance with [9-12], the following basic equations were used.

The torque on the wheel pair is determined by the formula, Nm,

$$M_k = M_{TM} \times \mu \times \eta_{gear} \tag{2}$$

where: M_{TM} – torque on the shaft of the traction motor, Nm;

μ - gear ratio;

η_{gear} - gear efficiency.

The driving force, which is accounted for by one traction motor, is determined by the formula, kN,

$$F_{kd} = \frac{2 \times M_k}{D_w} \tag{3}$$

where D_w – wheel diameter on rim (0,95м).

The tangential force of the traction of the entire rolling stock is determined by the formula, kN,

$$F_k = m \times F_{kd} \tag{4}$$

where m – number of traction electric motors in a suburban train.

The force of traction on the adhesion of wheels with rails is calculated by the formula, kN,

$$F_{сц} = \psi_к \times P_{сц} \tag{5}$$

where $P_{сц}$ - coupling weight of rolling stock, kN;

$\psi_к$ – coefficient of adhesion.

The results of calculations and construction of traction characteristics of the modernized suburban train are presented in Table 2 and in Fig. 2.

Table 2. Traction characteristics of a suburban train with an energy power plant on hydrogen fuel.

V, [km/h]	5	10	20	30	40
F_k, [N]	82688	82688	82688	82688	82688
V, [km/h]	50	60	70	80	90
F_k, [N]	82688	82688	75000	65000	60000
V, [km/h]	100	110	120	130	140
F_k, [N]	53000	47500	43500	40000	36500
V, [km/h]	150	160			
F_k, [N]	35000	34500			

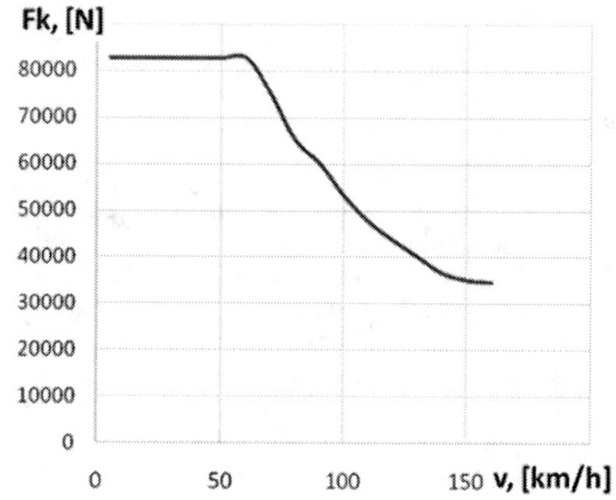

Fig. 2. Traction characteristics of a suburban train with an energy power plant on hydrogen fuel.

To verify the results obtained, traction calculations fulfilled The length of the route was chosen, the data for the site are shown in Table 3.

Table 3. Parameters of the study path.

Item number	Si, [m]	i, [°/oo]	Ri, [m]	Skri, [m]	alfa, deg	i used, [°/oo]
1	2	3	4	5	6	7
1	1400	0				0
2	1100	1,9	800	350		2,18
3	1800	2,3	900	550		2,54
4	1300	2	750	400		2,29
5	800	0	600	300		0,44
6	700	-3,1				-3,1
7	1100	-3,4				-3,4
8	700	0				0
9	1300	11				11
10	1400	5		650	20	5,38
11	1400	4,7		700	18	5.01
12	1300	1,6				1,6
13	900	3,8		700	15	4,06

Continuation of table 3.

1	2	3	4	5	6	7
14	800	2,8	900	300		3,09
15	400	3,9	800	150		4,23
16	500	0				0
17	1200	-8,1		600	25	-7,59
18	2300	-8,4		800	30	-7,94
19	2100	-8		650	24	-7,55
20	500	0				0
21	1600	-4.4	850	400		-4,19
22	1500	-4	900	500		-3,74
23	1200	-1				-1
24	1500	8		600	25	8,51
25	3300	8,5		650	22	8,91
26	1400	8,2		700	16	8,48
27	300	0				0
28	350	-3,4	800	150		-3,03
29	250	-3	900	150		-2,53
30	1900	-10,3				-10,3
31	500	0,6				0,6
32	1200	2,2				2,2
33	3500	-2,4				-2,4

The calculation of the retarding forces carried out according to known formulas [19].
The results of the calculations are presented in table 4.

Table 4. Calculation of the retarding forces acting on the train

V,[km/h]	5	10	20	30	40
F_k, [N]	82688	82688	82688	82688	82688
w'_o, [N/kN]	1,957	2,030	2,220	2,470	2,780
w''_o [N/kN]	0,141	0,168	0,237	0,326	0,434
W'_o, [kN]	4087	4238	4635	5157	5804
W''_o, [kN]	327	390	550	756	1009
W_o, [N]	4413	4628	5185	5913	6813
F_k-W_o, [N]	78275	78060	77503	76775	75875
f_k-w_o, [N/kN]	37,49	37,39	37,12	36,77	36,34
w_x , [N/kN]	2,454	2,525	2,720	2,985	3,320
W_x, [kN]	5123	5272	5679	6232	6931
W_x+W_o, [N]	5450	5661	6229	6988	7940
w_{ox}, [N/kN]	2,297	2,338	2,456	2,623	2,837
f_i k_p, [N/kN]	0,227	0,198	0,162	0,140	0,126
B_t, [N]	112,9	98,6	80,6	69,9	62,7
b_t+w_{ox}, [N/kN]	2,351	2,385	2,494	2,656	2,867
0.5b_t+w_{ox} [N/kN]	2,299	2,340	2,457	2,624	2,838

Continuation of table 4

V, [km/h]	50	60	70	80	90	100
1	2	3	4	5	6	7
F_k, [N]	82688	82688	75000	65000	60000	53000
w'_o, [N/kN]	3,150	3,580	4,070	4,620	5,230	5,900
w''_o [N/kN]	0,562	0,710	0,878	1,066	1,273	1,500
W'_o, [N]	6576	7474	8497	9645	10919	12318
W''_o, [N]	1307	1651	2040	2476	2957	3484
W_o, [N]	7883	9125	10537	12121	13876	15802
F_k-W_o, [N]	74805	73563	64463	52879	46124	37198
f_k-w_o [N/kN]	35,830	35,240	30,880	25,330	22,090	17,820
w_x , [N/kN]	3,725	4,200	4,745	5,360	6,045	6,800
W_x, [N]	7777	8769	9906	11190	12621	14197
W_x+W_o, [N]	9084	10419	11947	13666	15578	17681
w_{ox}, [N/kN]	3,099	3,409	3,768	4,174	4,629	5,131
f_i k_p, [N/kN]	0,116	0,108	0,102	0,097	0,093	0,090
B_t, [kN]	57,642	53,799	50,810	48,419	46,463	44,833
b_t+w_{ox}, [N/kN]	2,867	3,126	3,434	3,792	4,197	4,651
0.5b_t+w_{ox}, [N/kN]	2,838	3,100	3,410	3,769	4,175	4,630

Continuation of table 4.

V, [km/h]	110	120	130	140	150	160
F_k, [N]	47500	43500	40000	36500	35000	34500
w'_o, [N/kN]	6,630	7,420	8,270	9,180	10,150	11,180
w''_o [N/kN]	1,747	2,013	2,299	2,605	2,931	3,276
W'_o, [N]	13842	15491	17266	19166	21191	23341
W''_o, [N]	4058	4677	5341	6052	6808	7611
W_o, [N]	17899	20168	22607	25218	27999	30952
F_k-W_o, [N]	29601	23332	17393	11282	7001	3548
f_k-w_o [N/kN]	14,180	11,180	8,330	5,400	3,350	1,700
w_x , [N/kN]	7,625	8,520	9,485	10,520	11,625	12,800
W_x, [N]	15919	17788	19802	21963	24270	26723
W_x+W_o, [N]	19977	22464	25144	28015	31079	34334
w_{ox}, [N/kN]	5,681	6,280	6,927	7,621	8,364	9,155
f_i k_p, [N/kN]	0,087	0,085	0,083	0,081	0,079	0,078
B_t, [N]	43,453	42,271	41,246	40,349	39,558	38,855
b_t+w_{ox}, [N/kN]	5,701	6,300	6,946	7,640	8,382	9,173
0.5b_t+w_{ox} [N/kN]	5,682	6,281	6,927	7,621	8,364	9,156

Calculation of the operating parameters of the equipment of the suburban train, consumption and return of energy when moving along a given section of the path are made on the basis of the parameters obtained in the graphical construction of motion curves. A fragment of the results and of the calculations presents in table 5.

Table 5. Fragment of traction calculations of a suburban train on hydrogen fuel at the site.

Item number	S	V	t	Added	Capacit
1	2	3	4	5	6
-	[m]	[km/h]	[s]	[kWt]	[%]
1-2	27	5	19,44	0,162	98,28
2-3	86	15	20,64	0,172	96,41
3-4	135	25	19,44	0,162	95,64
4-5	164	35	16,87	0,140	94,76
5-6	196	35	20,16	0,168	94,62
6-7	65	31,69	7,38	0,061	94,26
7-8	224	38,38	21,01	0,175	93,01
8-9	287	48,38	21,36	0,178	91,43
9-10	351	58,38	21,64	0,180	89,47
10-11	236	61	13,93	0,116	88,18
11-12	573	73,62	28,02	0,233	85,55
12-13	741	83,81	31,83	0,265	82,48
13-14	486	90,99	19,23	0,160	80,57
14-15	1193	98	43,82	0,365	76,13
15-16	106	103,43	3,69	0,030	75,74
16-17	800	106,03	27,16	0,226	72,85
17-18	700	108,24	23,28	0,194	72,6
18-19	1100	107.35	36,89	0,307	72,30
19-20	700	108,94	23,13	0,192	69,75
20-21	1100	111,83	35,41	0,295	65,74
21-22	1400	109,075	46,21	0,385	65,29
22-23	229	104,95	7,86	0,065	65,22
23-24	344	100,29	12,35	0,102	66,97
24-25	339	91,21	13,38	0,111	68,67
25-26	276	81,21	12,23	0,102	70,03
26-27	211	71,21	10,67	0,088	71,14
27-28	222	60,60	13,19	0,109	72,21
28-29	148	50	10,66	0,088	72,91
29-30	123	40	11,07	0,092	73,46
30-31	74	30	8,88	0,074	73,76
31-32	49	20	8,82	0,073	73,93
...					
75-76	300	104,49	10,34	0,086	18,78
76-77	350	103,98	12,12	0,101	18,58
77-78	250	103,64	8,68	0,072	18,45
78-79	1900	105,54	64,81	0,540	17,41
79-80	500	106,88	16,84	0,140	17,14
80-81	1200	104,45	41,36	0,344	16,47
81-82	1289	101,87	45,55	0,379	15,72
82-83	252	97,64	9,29	0,077	17,94
83-84	308	89,27	12,42	0,103	20,57
84-85	251	79,27	11,40	0,095	22,66

Continuation of table 5

1	2	3	4	5	6
85-86	94	72,13	4,69	0,039	23,46
86-87	164	65	9,08	0,075	24,79
87-88	134	55	8,77	0,073	25,83
88-89	110	45	8,80	0,073	26,63
89-90	85	35	8,74	0,072	27,24
90-91	61	25	8,78	0,073	27,62
91-92	37	15	8,88	0,074	27,81
92-93	6	7,5	2,88	0,024	27,85
93-94	9	2,5	12,96	0,108	27,66
		0	120	1	36,50

The obtained traction characteristic and verification calculations show that the power of the power plant on hydrogen fuel will be sufficient for efficient operation of the suburban train.

3. Conclusions

1. In work the international experience is considered the use of hydrogen as a fuel on rolling stock. Based on the fuel and economic indicators for a suburban train, the use of PEMFC as a power plant was chosen.

2. Technical parameters of the main equipment for a suburban train with a hydrogen fuel power plant, a scheme for energy transfer and transmission efficiency in general have been determined.

3. For the suburban rolling stock, the main modes of operation in operation have been identified and assumptions have been made in calculations taking into account the features of proton-exchange membrane fuel cells.

4. An algorithm is proposed for constructing the traction characteristics of a suburban train on hydrogen fuel and a dependence of the traction force on the speed for the modernized rolling stock is constructed.

5. In the perspective of further research, the results could be extended by emission analysis and emission reduction of toxic compounds, for example [CO2].

6. In subsequent work, it is necessary to develop systems using proton-exchange membrane fuel cells on the train DPKr2, conduct their tests and specify the payback period of the train. Particular attention should be paid to solving the issue of logistics of production and supply of hydrogen to the points of refueling in the conditions of Ukraine.

References

1. Presentation: *Hydrogen Rail Test Projects in Germany*, Dr.-Ing. Ulrich Bünger (2017) URL: https://hydrail.appstate.edu/sites/hydrail.appstate.edu/files/12_bunger.pdf, 30p.; (data zvernennya: 16.11.2018).
2. Kryukovskiy Vagonostroitelnyiy Zavod [Elektronniy resurs]. –http://www.kvsz.com. – (16.11.2018).

3. Za strategiyu rozvitku UkrzalIznitsi do 2021 roku vklyuchno planuemo 5,9 mlrd dol. SShA kapitalnih investitsiy, z yakih blizko 4 mlrd dol. SShA, – na onovlennya ruhomogo skladu [Electronic resourse]: http://www.uz.gov.ua/press_center/up_to_date_topic/page-31/458113/. – (data zvernennya: 17.11.2018).
4. Article: Railpower GG20B [Electronic resource]: [2010] URL: https://en.wikipedia.org/wiki/Railpower_GG20B#Fuel_Cell_Testbed (date of the application 20.11.2018).
5. Hydrogenics corporate presentation & Renewable Hydrogen Activities [Electronic resource]: [May 2016] URL: https://www.slideshare.net/DenisThomas1/201605hydrogenicscompany-presentationcompressed (date of the application 16.11.2018).
6. Article: How a Fuel Cell Works [Electronic resource]: [2016] URL: http://www.proton-motor.com/how-a-fuel-cell-works/ (date of the application 17.11.2018).
7. http://tainy.net/13400-vodorodnoe-toplivo.html (date of the application 17.11.2018).
8. https://www.popmech.ru/technologies/371842-vodorodnyy-transport-tehnologiya-budushchego-ili-polnyy-proval/(date of the application 17.11.2018).
9. Vodorodne toplivo, ekonomika i ekologija http://www.newchemistry.ru/letter.php?n_id=1236 (date of the application 17.11.2018).
10. https://auto.today/bok/4339-kak-rabotaet-vodorodnyy-dvigatel.html (date of the application 18.11.2018).
11. V Germanii prezentovali poezd na vodorodnom toplive https://www.segodnya.ua/world/europe/v-germanii-prezentovali-poezd-na-vodorodnom-toplive-1071211.html (date of the application 16/11/2018).
12. Coradia iLint: V Germanii nachal kursirovat pervij v mire vodorodnij poezd. [Electronic resource]: [2016] URL: http:// iq.intel.ru/coradia-ilint-%D0%B2-%D0%B3%D0%/ (date of the application 25.07.2018).
13. Nemecki poezda poedut na vodorode. https://www.kommersant.ru/doc/3134887. (date of the application 16/11/2018).
14. Grigorovich D.N., Bulletin of the Joint Scientific Council, **6**, 37-50 (2013).
15. A. Falendysh, P. Kharlamov, O. Kletska, N. Volodarets, Transportation Research Procedia, **14**, 665-671 (2016).
16. Anatoliy Falendysh, Mykyta Volodarets, Olha Kletska, Viktoriia Hatchenko, *The impact of the type of operation on the parameters of a shuning diesel locomotive with hybrid power plant*, MATEC Web Conferences **133**.03003 (2017).
17. Merkisz J, Bajerlein M, Daszkiewicz *P 2012 The Influence of the Application of Photovoltaic Cells in City Buses to Reduce Fuel Consumption–(CO_2) and Exhaust Emissions (HC, PM, and NO_x)* Lecture Notes in Information Technology vol. 13.
18. A. Falendysh, TEKA. Commission of motorization and energrtics in agriculture, **12**, 58-63 (2012).
19. A. Falendysh, O. Chyhyryk, O. Kosariev, Kh: UkrDUZT, **175**, 12-23 (2018).
20. Osipov S.I., *Osnovy elektricheskoy i teplovoznoy tyagi*, M.: Transport, (1985).

Resource-Saving Method of Forming Information Infrastructure of Sorting Stations

Anatolii Kosolapov*

Dnipro National University of Railway Transport named after academician V. Lazaryan, Computer Department, Ukraine

Abstract. The paper proposes a resource-saving method of conceptual design of the information infrastructure of railway objects on the example of a distributed computer control system of a marshalling yard. This method allows determining the economically optimal degree of decentralization of the technical structure of the management system.

1 Introduction

The current stage of development of industrial and transport systems is characterized by a grace of technologies and automated functions based on the territorial distribution of resources and the wide use of artificial intelligence methods for solving problems of energy and technical resources, logistics and environmental safety. The integration of processes and systems leads to a synergistic effect and the formation of large intelligent transport systems [1]. In such systems, an important parameter affecting their efficiency is the multi-structure and its component - the technical structure [2]. For railway objects, information infrastructure, including servers, computers, controllers and information channels, are of particular importance. This is due to economic and environmental problems arising in case of failure of individual elements of the structure. For example, for a hump yard it could be a wagon battle, damage to environmentally hazardous goods, and so on. These circumstances increase the role of conceptual design of the information infrastructure of railway objects and use in the development process of resource-saving methods of system design [2].

The paper proposes such a method to determine the rational degree of decomposition of the computer system of the marshalling yard, minimizing production losses due to the unreliability of equipment.

The development of automation systems for marshalling yards in Ukraine and abroad was based on the creation of the first centralized computer systems based on mini-computers (SM-2) to build functionally distributed systems based on industrial microcomputers and microcontrollers (micro-DAT, SM-1800 / 1810, Advantech microcontrollers) [3, 4]. With the transition to decentralized (hierarchical) systems with a variety of options for distributed structures, the issue of the justification for the transition to decentralized management remains unresolved. Despite numerous attempts to find an answer to this question [5, 6, 7, 8], it remains relevant in the development of technical means of automation of hump yard. When considering the optimization of the structure of the hierarchical system, we will assume that the hump yard, as a technological object of control of the marshalling yard, has local automation devices. The centralized control system, optimizing the overall work of the hump yard, adjusts the work of local automation devices, forming adjustments for regulators (for example, retarder positions and switch operating apparatus). On this principle, the ASC was built for the Yasinovata station of the Donetsk railway [3].

With the growth of power of management computers (MC) there is an interest to consider the expediency of transferring the functions of local automation devices to higher-rank microcontrollers that provide local optimization of the subsystems [8].

The possibilities of automation systems were limited in terms of the implementation of complex control algorithms earlier, and MC did not differ greatly, then there are now technical possibilities to increase the power of both microcontrollers and MC. Therefore, the transfer of functions of local regulators on MC does not significantly affect its power (memory, speed, etc.).

However, such centralization of management significantly increases the responsibility of MC. Obviously, the failure of the MC violates the management of a whole hump yard, which leads almost always to great economic losses. Therefore, many authors consider [9, 10, 11], which determines the reliability of MC in the solution of this issue.

2 Problem statement and method of its solution

The hump yard consists of n controlled objects M_i, whose status O_i can be characterized by two parameters: y_i - adjustable parameter; x_i - control

* Corresponding author: kosolapof@i.ua

action (z_j - controlled, f_g - uncontrolled parameters, see Fig. 1).

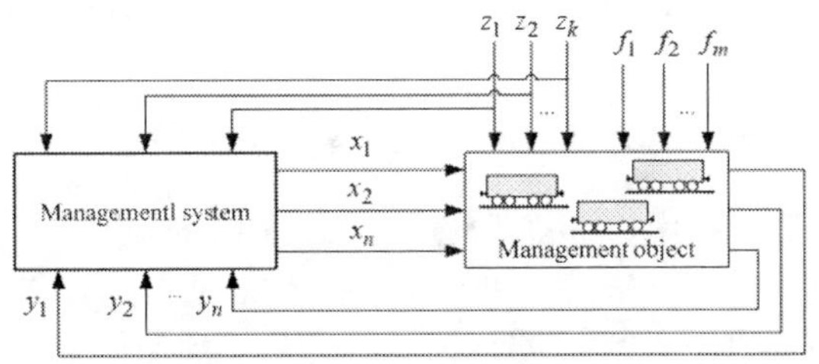

Fig. 1. A marshalling yard as a control object

Thus, the state of the system at each instant of time can be characterized by vectors: the state of the regulated parameter $Y = \{y_1,...,y_n\}$; of the position of the regulating body $X = \{x_1,...,x_n\}$ and the vector-setpoint $Y' = \{y_1',...,y_n'\}$. The vector-set Y' is a task in support of a regulated parameter and in general can differ from its true value - the vector Y. If the system does not have local regulators, then MC directly forms the vector X and issues it to the executive bodies. In this case, we have a centralized system (Fig. 2).

In the case of local regulators (LCs) implementing local control functions, MC's functions are simplified to formulate only the settings Y' for the LC. In this case, the hierarchical structure of the MS is discussed (Fig. 3).

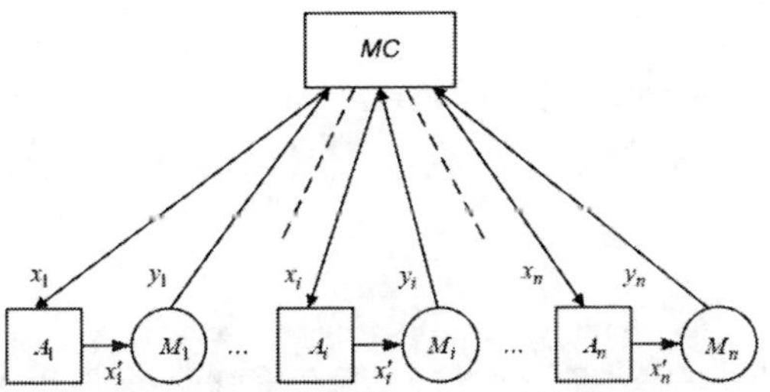

Fig. 2. The structure of the centralized control system

Let the system consist of MC and n local subsystems. In Fig. 2 and Fig. 3 the following symbols are acceptable: Managing System (MS), Actuator (A), Local Controller (LC), Executive Mechanism (M).

The implementation of the management algorithm is reduced to minimizing (maximizing) the function of the species

$$B = \varphi(Y, X) \tag{1}$$

(for example, minimizing the deviation of the speed of exit from the retarder position from the set by the control algorithm).

Optimization involves searching for such an optimal value of the control vector U*, so that the management function (1) at the given X takes the optimal value

$$B^* = \min(\max)\varphi(Y, X) \tag{2}$$

or

$$B^* = \varphi(Y, X^*). \tag{3}$$

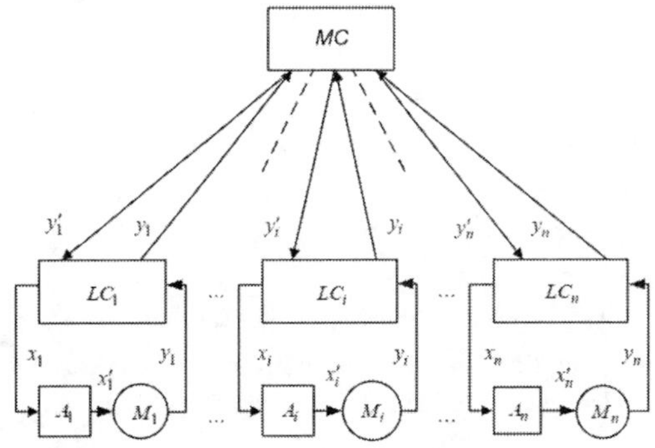

Fig. 3. Structure of the hierarchical system with local control subsystems

The evaluation of the variants of structures will be carried out on the criterion of the full cost of resources. To simplify the calculations we assume that all n control circuits are the same, the costs of operating systems are also the same and the unreliability of executive bodies can be neglected.

System management will be presented in the mode of time distribution as follows. According to the scheme in Fig. 2 (centralized control system) MC receives information about the state of the controlled object in the form of vectors Y in cycles in time T_c $(T_c, 2T_c,...,kT_c)$ intervals. As a result of information processing by algorithm (2), MC carries out the calculation of the new setpoint Y' and issuance of the control command to the executive bodies - a vector of the X position.

According to the scheme in Fig. 3 (hierarchical control system), which also receives information in the form of vectors Y. However, in this case, MC issues only a vector setting Y'. The calculation of the vector X is carried out by the local regulators LC; MCM only coordinates their work by adjusting the vector setting Y'.

Let's see what happens in the case of equipment failure in the kth time interval. In case of failure of MC in a centralized structure, the following situations are possible:

1) the work of the car classification is immediately terminated. Losses p_0 are determined by the downtime of the object and can be estimated as a complete loss of efficiency during the elimination of the malfunction (system recovery) T_{rec} :

$$p_0 = B^* T_{rec}$$

2) For some time after the failure of MC the system continues to function normally and then there is a violation of its work. In this case, the behaviour of the object may vary depending on the position in which the executive bodies will be. They can:

a) Remain in the position in which they were at the moment of the accident t_0, ie $X = X_0$;

b) Translated into a state of emergency $X = X_{em}$ characterized by values.
In the second situation, the damage p_1 will consist of two parts (Fig. 4), $y_i = v_i$ - speed of exit from the i-th braking position, $v_{i\min} < v_i(t) < v_{i\max}$).

At the moment time t_1, it is proportional to the difference between the optimal and actual values of efficiency:

$$p_{11} = (B^* - B_1)t_1 . \quad (4)$$

At the moment time t_2 there is an output parameter $v_i(t)$ at the permissible limits $v_i(t) > v_{i\max}$ and local protection disables the object, i.e., losses in this area p_{12} are determined by the total loss of efficiency:

$$p_{12} = B^* t_2 .$$

Complete losses for the second situation

$$p_1 = p_{11} + p_{12} = (B^* - B_1)t_1 + B^* t_2 ,$$

where

$$T_{rec} = t_1 + t_2 ; \; \lim_{t_1 \to \infty} p_1 = B^* T_{rec}$$

B_1 - Efficiency in case of not optimal control in the centralized system (manual control of sorting out of cars).

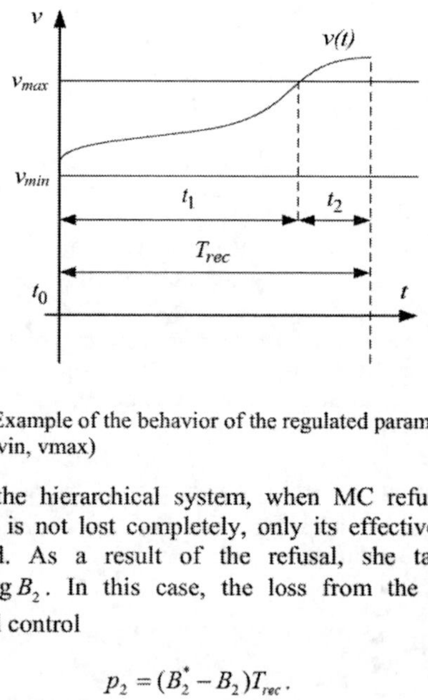

Fig. 4. Example of the behavior of the regulated parameter v (t) within (vin, vmax)

In the hierarchical system, when MC refuses, the control is not lost completely, only its effectiveness is reduced. As a result of the refusal, she takes on meaning B_2. In this case, the loss from the loss of optimal control

$$p_2 = (B_2^* - B_2)T_{rec} . \quad (5)$$

This damage, as a rule, is much less than the damage that is obtained when a CM system fails in a centralized system. Reducing the damage to a system with a hierarchical structure is achieved by increasing its value due to the presence of local regulators. In addition, the effect of loss reduction is reduced due to additional losses caused by the unreliability of local regulators. Let's take the following for further analysis

Assumption:

B^* - Loss per unit time when the car classification does not work;

B_1 - Loss per unit of time when the centralized system does not work MC (non-optimal control);

B_2 - Loss per unit of time when the hierarchical system does not work MC (non-optimal control);

b_i^* - Loss per unit of time when the i-th control circuit is not working

b_i - Loss per unit time with a sub-optimal value of the efficiency of the i-th circuit.

$$B^* > B_1 > B_2 \gg b_i^* > b_i .$$

Then the full costs for centralized P_1 and hierarchical P_2 systems can be estimated as follows:

$$P_1 = \left(B^* - B_1\right)\lambda_0 t_1 + B^* \lambda_0 t_2 + C_0^{centr} , \quad (6)$$

$$P_2 = (B^* - B_2)\lambda_0 T_{rec} + \\ + \sum_{i=1}^{n} \left[(b_i^* - b_i)\lambda_i t_{i1} + b_i^* \lambda_i t_{i2} \right] + C_0^{dec} + \sum_{i=1}^{n} c_i , \quad (7)$$

where λ_0 - the intensity of bugs CM in hierarchical and centralized systems; λ_i - intensity of failure of the i-th local regulator; $C_0^{cen}, C_0^{dcen}, c_i$ - the cost of CM respectively in the centralized system, in a decentralized (hierarchical) system and the cost of the microcontroller system of the i-th local regulator.

We will assume that

$$C_0^{cen} > C_0^{dcen} \gg c_i .$$

To simplify further analysis, we assume that the system is homogeneous in its composition, i.e.

$$b_i = b_n ; \; \lambda_i = \lambda_n ; \; c_i = c_n ; \; i = 1, 2, ..., n . \quad (8)$$

For the i-th control circuit, taking into account the possible failure of the regulator by analogy with the centralized control system (Fig. 4), we can write

$$t_{i2} = T_{rec\,i} - t_{i1} ,$$

where t_{i2} - the time during which the i-th parameter is within the permissible values; $T_{rec\,i}$ - time of repair of the i-th regulator; t_{i1} - the time during which the i-th parameter does not exceed the permissible limits.

In the first approximation we will accept that

$$t_{i1} = t_{n1} = t_1 , \; i = 1, 2, ..., n ; \\ T_{rec}^{cen}(\text{CM}) = T_{rec}^{dec}(\text{CM}) = T_{rec}(c_n) . \quad (9)$$

For further research, we introduce the coefficient

$$t_1 = rT_{rec},\ t_2 = (1-r)T_{rec}.$$

Then we have from (6) and (7)

$$P_1 = \left(B^* - B_1\right)\lambda_0 rT_{rec} + B^*\lambda_0(1-r)T_{rec} + C_0^{cen}, \quad (10)$$

$$P_2 = (B^* - B_2)\lambda_0 T_{rec} + (B^* - B_1)\lambda_n rT_{rec} + \\ + B^*\lambda_n(1-r)T_{rec} + C_0^{dec} + nc_n. \quad (11)$$

Let's turn to the integral index of reliability - the readiness factor

$$K_{avail}^{cen}(\mathrm{CM}) = K_{avail}^{dcen}(\mathrm{CM}) = K_0 = \\ = \frac{T_{mtf}}{T_{mtf} + T_{rec}} = \frac{1/\lambda_0}{1/\lambda_0 + T_{rec}}, \quad (12)$$

$$K_{avail}(c_n) = K_n = \frac{T_{\mathrm{mtf}(n)}}{T_{\mathrm{mtf}(n)} + T_{rec}} = \frac{1/\lambda_n}{1/\lambda_n + T_{rec}}, \quad (13)$$

where T_{mtf} - mean time to failure and

$$\lambda_0 = \frac{1-K_0}{K_0 T_{rec}};\ \lambda_n = \frac{1-K_n}{K_n T_{rec}}. \quad (14)$$

Then let's introduce the coefficient α :

$$K_n = \alpha K_0. \quad (15)$$

We substitute (14) and (15) in (10), (11), normalize full costs and introduce the notation

$$\delta_1 = \frac{B^* - B_1}{B^*};\ \delta_2 = \frac{B^* - B_2}{B^*}.$$

As a result, we get

$$\tilde{P}_1 = \frac{P_1}{B^*} = \frac{(1-K_0)(1 + r(\delta_1 - 1))}{K_0} + \frac{C_0^{cen}}{B^*}, \quad (16)$$

$$\tilde{P}_2 = \frac{P_2}{B^*} = \frac{\delta_2(1-K_0)}{K_0} + \\ + \frac{(1-\alpha K_0)(1 + r(\delta_1 - 1))}{\alpha K_0} + \frac{C_0^{dcen} + nc_n}{B^*}, \quad (17)$$

In Fig. 5, constructed graphs $\tilde{P}_1 = f_1(K_0)$ and $\tilde{P}_2 = f_2(K_0)$ under the following assumptions $r = 0,9$ and $\alpha = 1,1$.

From the graphs it follows that with the increase of the coefficient of readiness of loss in the hierarchical decentralized system, in comparison with the centralized decreases ($K_0^{(1)}, K_0^{(2)}, K_0^{(3)}$ - points on the graph where losses in both systems are the same). From (17) it follows that the losses in the hierarchical system are linearly dependent on the number and cost of local microcontroller regulators. Given the reliability indices (K_0, α) and the value of the subsystems in the technical structure of the control system (C_0^{cen}, C_0^{dcen} and c_n), the number of contours of local control n can be chosen based on the solution of equations (16) and (17).

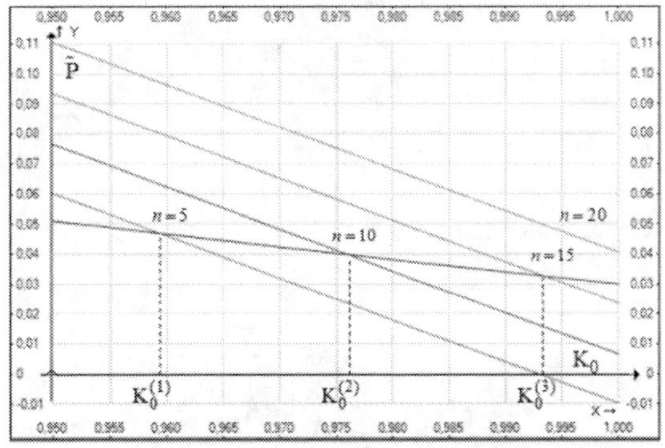

Fig. 5. Dependence of change of systems efficiency on reliability of MC and LC

If we compare (16) and (17), then introduce normalized coefficients

$$a_1 = 10^{-3}\delta_2 B^*;\ a_2 = 10^{-3}c_n;\ a_3 = 10^{-3}(C_0^{dec} - C_0^{cen}), (18)$$

we can obtain the necessary equations of two variables K_0 and n:

$$a_1(1-K_0)/K_0 + a_2 n + a_3 = 0.$$

Fig. 6, 7, 8 shows the planes that represent the geometric point of the points $K_0^{(i)}$ in which the losses in the centralized and hierarchical (decentralized) systems are identical and corresponds to the change of priorities in the construction of systems, depending on the readiness coefficients and the number of local regulators. Analysis of the results shows that rational solutions are located in the zone of high readiness and a small number of local subsystems. The zone of rational decisions increases if the value of the coefficient a_1 decreases. The number of local circuits does not exceed 15 subsystems.

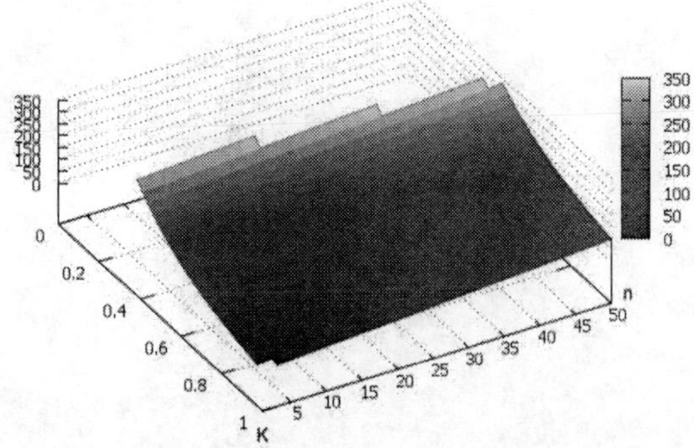

Fig. 6. Plans of equal losses of centralized and hierarchical systems ($a_1 = 200$)

https://doi.org/10.1051/matecconf/201929401011

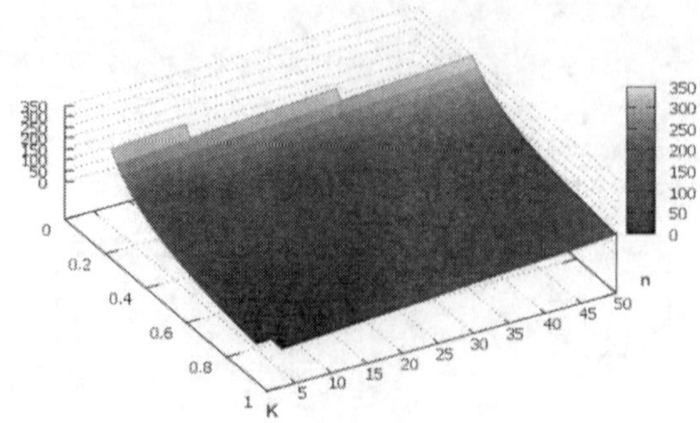

Fig. 7. Plans for $a_1 = 100$

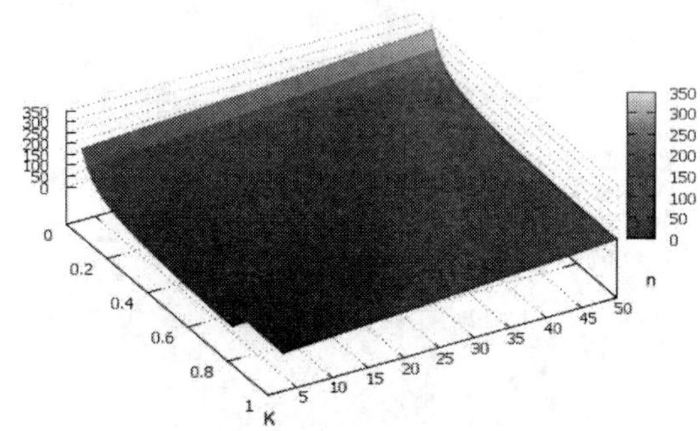

Fig. 8. Plans for $a_1 = 10$

3 Conclusions

It should be borne in mind that the results obtained are valid only for the case of finite and additive losses. For objects, the complete refusal of automatic devices which can lead to accidents such as catastrophes or associated with the possibility of human victims, the choice undoubtedly must be resolved in favor of the combined system.

References

1. Kosolapov A., Loboda D. *Semiotic-Agent-Ontological Approach To Design Intellectual Transport Systems.* American Journal Of Engineering Research (Ajer), Volume-**7**, Issue-7, pp.205-209 (2018)
2. Kosolapov A,, Loboda D. *Framework Conceptual Design Of Complex Real-Time Management System* (Codecs). European Journal of Advances in Engineering and Technology, **5(8)**, pp. 559-566 (2018)
3. A. A. Kosolapov, A. L. Blokhin, K. F. Boryak, et al. *The key role of transport in the modern world*: a monograph - Odessa: KUPRIENKO SV - 163 p. (2013)
4. *Automation Device and Computing 2012-2013* [Electronic resource] / Access mode: http://support.advantech.com.tw/ OnlineResources / SubIndex.aspx? Bu = 99B2E2BE-A7E3-4C58-9E35-B4F48A952AC7 & type = e Catalog. : Advantech (2013)
5. G.A. Krasovsky, L.V. Safris, A.A. Kosolapov, B.M. Filimonov, E.M. Shafit, E.V. Shcherbakov. Patent № 1073146, MKI, B61, L17 / 00. *A device for controlling the technological process of the dissolution of compositions on a sorting chamber.* Bul. №6. – USSR (1984)
6. V.I. Ivanchenko, N.N. Lyabakh, A.A. Sepety *A new approach to managing the process of dissolution of trains on a hump yard.* Works of RIIZhT. - Rostov-on-Don. pp. 34-41 (1984)
7. E. M. Shafit, I. V. Zhukovitsky, A. A. Kosolapov. *Development of a unified system of automated control of the process of disbanding trains at a sorting station.* Materials of temporary teams. St. Petersburg. pp. 81-93 (1996)
8. G. A. Shastova, A. I. Koykin. *Selection and optimization of the structure of information systems.* M.: Energy, 256 p. (1972)
9. Nechiporenko V.I. *Structural analysis of systems (efficiency and reliability*). M.: Soviet radio. 216 p. (1977)
10. A. M. Polovko, S. V. Gurov. *Fundamentals of the theory of reliability*. SPb. BHV-Petersburg.704 p. (2008)
11. Fritch V. *Application of microprocessors in control systems: Trans. with him.* M.: Mir,.464 p. (1984)

Risk-regulatory increase of trajectory safety, productivity and energy efficiency of motor transport operations in conditions of uncertainty

Arseniy Khabutdinov

State Enterprise "State Road Transport Research Institute", 57 Peremohy Ave, Kyiv, 03113, Ukraine

Abstract. The proposed method of complex increase of trajectory safety, productivity and energy efficiency of motor transport operations taking into account the laws of vehicle algorithmic control in difficult traffic conditions. Proceeding from the provisions of the theory of energy efficiency of a generalized type car and the laws of its adaptively-discrete energy of motion, the processes of functioning of the transport-ergatic systems "Driver-Vehicle" (TESDV) and control of the car using adaptive driving procedures (productive, safe, energy-efficient) in the conditions of information uncertainty of locally trajectory situations (LTS). In this work presents logic-procedural and mathematical models for the analysis of risk-regulatory algorithms for control of vehicle, taking into account the functioning of TESDV and quantitative assessment of operator-interface complexity of road sections. The basis of the method of estimating the information complexity of the operator-road interface in relation to any element of the road is the mathematical model of the entropy of the object (as a measure of disorder), which is used in technical cybernetics. For each section of the road, the total indicator of information complexity is determined, which takes into account three levels of uncertainty: low, medium and high. At medium and high levels of complexity of the road section, TESDV implements a risk-regulating automobile control algorithm. Proceeding from the essence of the regulatory labor procedures of driving, as well as their role in the formation of the modes of the TESDV, proposed results of operational simulation of algorithms regulation of locally-trajectory driving risks (LTR): counterproductive, sensory-tempo, incidental and anergic.

1 Introduction

Motor transport is a sphere of complex human-machine (ergatic) labor and energy-consuming transportation of goods and passengers in a conflict street-road network. Prior to motor transport operations (MTO) three requirements for their integrated performance are presented: "Productivity (P) - Trajectory safety (TS) - Energy Efficiency (EE)". These requirements in the MTO provide a productive element of the motor transport - transport ergatic system "driver-vehicle" (TESDV) based on labor and machine procedures for the control of the vehicle.

In the said process, the first two requirements (P-TS) are controversial in the procedures for adjusting the speed of the vehicle. When alternating dangerous and safe sections of the road, which belong to the length of the mileage vehicle in the given MTO, the driver alternately uses safe and productive motivation for the control of the vehicle. This leads to the motivated formation of uneven-intermittent TESDV modes.

In terms of the ergatic functioning of TESDV, this motion is adaptively discrete and trajectory. As a result of this movement a trouble-free MTO is performed. However, at the same time in the MTO is implemented a negative (in terms of technological cost-effectiveness of transportation) phenomenon of operational anergic of the vehicle (a phenomenon of simultaneous reduction of productivity and energy efficiency in the zones of locally-trajectory hazards (LTH) [1]. It is characterized by a significant and simultaneous decrease in the values of average speeds (1.5 ... 2.5 times) and indicators of its energy efficiency (in 2 ... 4 times). In this connection, ergatic bases of increase comprehensive success of motor transport operation are actual, which make it possible to improve the processes of TESDV's operation.

The methods of information analysis of the current speed of the car in the road traffic flow do not take into account the above three requirements for MTO, as well as the properties of the elements TEDSV and sections of the road [2].

2 Main part

At present, the following conditions are met for the fulfillment of certain requirements for the success of a MTO [3]. In particular, TESDV trajectory safety (TS) is carried out due to several factors:

* Corresponding author: habut20@gmail.com

a) timely detection of visual signs of LTH;
b) estimation of the level of complexity of the site of the vehicle trajectory for the driver-operator or for the on-board information and control system of the vehicle;
c) prevention of the development of the LTH into the trajectory incident (TI) is an event of a shock intersection of the vehicle trajectory with another object;
d) prevention of any possibility of realization of the mechanism of an emergency MTO (LTH → TI → Accident).

The demand for productivity (P) of MTO is also carried out on the basis of procedures of trajectory-adequate increase of the average speed of the vehicle in MTO in the absence of visual signs of the LTH. The MTO Energy Efficiency Requirement (EE) is met by rationally reducing the value of the car's anergic index with adaptively discrete TESDV movement in MTO, taking into account both TS and P requirements

These conditions of comprehensive performance of MOT must be carried out to the same extent as TESDV, as well as "unmanned" vehicle. It is established that taking into account the fuzzy level of information complexity of the site of the vehicle trajectory, TESDV can provide three levels of productivity of the MTO:
a) safe-successful MTO (SSTO, one requirement of TS);
b) functional-successful MTO (FSTO, two requirements- P and TS); c) technologically-successful MTO (TSTO, three requirements P, TS, EE).

In the implementation of these requirements, the driver can choose three types of adaptive vehicle control procedures: productive (P- procedures), safe (TS-procedures) and energy-efficient (EE- procedures) [3,5]. Based on the above three principles of ergative successful control of vehicle are formed:
a) principle of distance estimation of the fuzzy level of information complexity of the section of the trajectory of the car's motion;
b) principle of remotely- advance choice of a behavioral installation and the purpose of the control of the vehicle in the form of a variant of the success of MTO (SSTO, FSTO, TSTO);
c) principle of remotely- advance choice of local tactics and algorithm of adaptive driving procedures.
These principles should be implemented in the methods of modernization improvement of the processes of vehicle control or "unmanned" car as a transport robotic tool, taking into account the properties of elements TESDV.

Within the framework of TESDV, the driver-operator in controlling a vehicle implements a number of properties: psychophysiological [3], labor-motivational and labor-procedural [4,5]. By labor-motivational belong to the ability of the driver trajectory-adequately choose behavioral settings and management objectives - in the form of options for levels of MTO (SSTO, FSTO, TSTO). Labor-procedural properties enable the driver to select and implement six driving procedures P_{Li} (1) in accordance with the requirements of the complex success of the MTO [3,5].

As an element of TESDV in the control process, the vehicle implements four groups of operational and technical properties of a car [6]: complex energy transformer machine, dangerous traffic control object, a vehicle, the source of machine procedures P_{Mi} of transportation. The plurality of labor procedures of the driver (PL) and the machine procedures of the vehicle (PM) are as follows:

$$P_{Li} \in (P_{Li1}, P_{Li2}, P_{Li3}, P_{Li4}, P_{Li5}, P_{Li6}),$$

$$P_{Mi} \in (P_{Mi1}, P_{Mi2}, P_{Mi3}, P_{Mi4}, P_{Mi5}) \quad (1)$$

where P_{Li1}, P_{Li2}, P_{Li3}, P_{Li4}, P_{Li5}, P_{Li6} - labor procedures: sensory-perceptual, thought-analytical, sensory-tempo, predictive, solving, control, respectively; P_{Mi1}, P_{Mi2}, P_{Mi3}, P_{Mi4}, P_{Mi5} - machine procedures: energy-transforming, transmission, traction-coupling (or brake), trajectory-kinetic (formation of kinetic energy of the vehicle in accordance with the given trajectory of motion), trajectory orientation (the choice of direction of movement along the road line), respectively; i - the index of the type of labor and machine procedures in their motivational orientation (i = 1 - productive procedures, i = 2 - anti-incident, i = 3 - energy-efficient).

The process of driving a vehicle is divided into three stages: information-solving, motor and control. When operating TESDV in the first stage, the first four labor procedures of the driver P_{L1}, P_{L2}, P_{L3}, P_{L4} are used. In the second stage - the labor procedure of the driver P_{L5}, (with its four variants of actions on the management bodies of the vehicle). With the help of procedures P_{L5}, the driver adjusts the intensity levels of the chain of machine procedures of the vehicle P_M. In the third stage, the driver, with the help of the procedure P_{L6}, provides control of the parameters of the vehicle (speed, mileage, distance, acceleration, deceleration, etc.) in accordance with the objective of control.

An important initial task of the first stage of control a vehicle is the assessment of the fuzzy level of information complexity of the trajectory of the vehicle's for the driver. The method of solving this problem is presented below. The driver perceives information on trajectory elements of the road (with a traffic flow) in the section of the vehicle's traffic trajectory (SVTT), which are important for the qualitative performance of the first (information-solving) stage of traffic control. Two aspects of driver interaction with the road environment (operator-road interface) are important.

In the first aspect, the road is considered as an alternating rolling surface for a vehicle. This takes into account the state of the **four factors** of the road surface: the coefficient of adhesion of the wheels with the road, the angle of lifting the road, the roughness, the radius of the road curve. In the second aspect - as a conflict communicative channel for the adaptive movement of the vehicle. **Eleven factors** of this channel are considered: the number of lanes of the road; adjustable crossroads; unregulated crossroads; number of road signs; bus stop for public transport; public transport stops for drivers of other vehicles (not buses); SVTT in which the vehicle is moving at a constant speed; SVTT in which the vehicle is moving at an uneven velocity; unregulated ground pedestrian crossing; adjustable pedestrian crossing (ground); closed inspection area. The basis of the method for assessing the complexity of the information interaction of the driver with any element of the road (as the surface and the channel of movement) is a mathematical model of the entropy of the object (as a measure of disorder), which is used in technical cybernetics [7,8]. Entropy factor (disorderly measure) EN is defined as:

$$EN = \kappa \cdot ln(n) \tag{2}$$

where n - the number of possible (equally probable) states of the factor; k - a constant value, taken κ = 1,5. For $n = 1$ we have $EN = 0$, that is, at a single state of the factor, its entropy (disorderly measure) is zero. So, for n = 10 we have EN = 3,45. The greater the number of states n, the higher the operator-interface complexity of this factor of the road in relation to the chain of three driver's labor procedures (P_{L1}-P_{L2}-P_{L3}) at the first stage of the vehicle control. In the methodological plan, the value of EN_i for the i-th factor of the SVTT is equal to the magnitude of its operator-information complexity $P_{(OC)i}$, that is, $ENi = P_{(OC)i}$. These information models, which are based on the formalization of state sets of trajectory elements of the SVTT, are an important way of quantifying the level of complexity of the SVTT. The indicator of the total operator-interface complexity of a given section of the trajectory of the vehicle $P_{(OC)}$ (Δl) of length Δl, taking into account the foregoing, is defined as:

$$P_{oc}(\Delta l) = \sum_{i=1}^{4} P_{oci} \cdot n_{ei} + \sum_{j=1}^{11} P_{ocj} \cdot n_{ej} \tag{3}$$

where i - index of factors of the operator-interface complexity of interaction with the road as with the rolling surface; j - index of factors of operator-interface complexity of interaction with the road as a conflict communicative channel; n_{ei} and n_{ej} - the number of factors of the $_i$-th and $_j$-th species in the area of the trajectory of motion, respectively; Δl is the length of the specified area of the vehicle's trajectory.

Fifteen mathematical models were developed for estimating the complexity of each type of trajectory elements that take into account their operational status. For example, if the value of coefficient of adhesion of wheels to the road φ = 0,75, then the value of the operator-interface complexity parameter $P_{(OC)i} = 0$. For φ = 0,4, we have $P_{(OC)i}$ = 2,25. As an example, in table 1 shows the estimated values of operator-interface complexity parameters $P_{(OC)j}$ for various factors of the city bus route. It is assumed that each factor is detected once, that is, $n_{ej} = n_{ei} = 1$.

Based on formula (2), 15 calculation models were formed for the entropy assessment of the complexity of the driver's information interaction with 4 factors of the road as a vehicle's rolling surface and with 11 factors of the road as a communicative channel with conflict points. When calculating the entropy complexity indicators of the first group of road factors, their maximum, minimum and specified values are taken into account. Then calculating the entropy complexity indicators of the second group of road factors, the number of conflict points of various types and the traffic management scheme are taken into account. In addition, when determining the total indicator of the informational complexity of a road section, the entropy values of the kinematics of adaptive-discrete vehicle motion are taken into account: constant speed Vc, speed increment during acceleration of the vehicle + ΔV, decrease in speed when braking the vehicle -ΔV. Further, table 1 presents the calculated values of the information complexity indicators of some factors of the road as a communicative channel with conflicting points, as well as factors of the kinematics of vehicle movement Vc and + ΔV.

Table 1. Estimated values of parameters of operator-interface complexity $P_{(OC)j}$ for the given values of factors of interaction with the road on the city bus route.

n_L	n_{RI}	n_{UI}	n_{RS}	n_S	V	ΔV	n_{UPC}	n_{APC}	n_{ZCR}	Total $P_{(OC)i}$
3	1	1	3	1	60 km / h	30 km / h	1	1	1	
1,6	3,2	4,1	1,6	7,5	3,7	4,0	9,1	5,3	9,7	49,8

In table 1 the following notation is adopted: n_L - the number of lanes of the road; n_{RI}- the number of regulated intersections; n_{UI} - the number of unregulated intersections; n_{RS} - the number of road signs; n_S - the number of public transport stops; V - the constant speed of the vehicle in the road section; n_{UPC} - the number of unregulated pedestrian crossings; n_{APC} - the number of adjustable pedestrian crossings; n_{ZCR} - the number of zones of closed review. Conditions of uncertainty of the information state of the SVTT arise:

a) if the vehicle speed V becomes greater than the permitted Vt by Traffic Rules, ($V>Vt$);

b) if the total value of the operator-interface complexity indicator $P_{oc}\Delta l$ corresponds to the average and higher levels $P_{oc}(\Delta l) > 0,35P_{oc(max)}$;

c) if in the SVTT formed zones of closed review of different types, as well as the driver observes the implicit features of LTH, that is, at the value of the function of the f_{va} of the visual attribute, LTH f_{va} = 0,5.

In conditions of medium and high uncertainty, the MTO's comprehensive success is provided on the basis of the algorithm of adaptive regulation of locally-trajectory risks (LTR) of vehicle control [7]. Based on the essence of the processes of the TESDV, as well as the modes of adaptive-discrete motion of the vehicle, four models for evaluation and regulation of the LTR are proposed. These models are as follows:

$$r_{cp}=1\text{-}V/V_t\text{, } r_{cp}\rightarrow min\text{, } r_{cp} \in (0;1)\text{, } f_{vs} = 0 \tag{4}$$

$$r_{st}=a_1\cdot V/V_t\text{, } r_{st}\rightarrow min\text{, } r_{st} \in (0;1)\text{, } f_{vs} = 0,5 \tag{5}$$

$$r_i=\Delta V_a/V\text{, } r_i\rightarrow min\text{, } LTR\rightarrow invers\text{, } r_i \in (0;1)\text{, } f_{vs}=1 \tag{6}$$

$$r_a=1\text{-}K_{ac}/K_{ao}\text{, } r_a\rightarrow min\text{, } r_a \in (0;0,8)\text{, } f_{vs}=(0;\ 0,5) \tag{7}$$

where r_{cp}, r_{st}, r_i, r_a - the values of the parameters of LTR: counterproductive, sensory-tempo, incident and anergic, respectively; V - current speed of the vehicle; V_t - value of speed, regulated by Traffic Rules; f_{vs} - functional of the visual signs of the LTH; ΔV_a - anti-incident reduction of the speed of the vehicle, which ensures that the trajectory incident (TI) is prevented; K_{ac} and K_{ao} - constructive and operational indicators of car anergicity values; a_1 - constant dimensionless quantity; «*Invers*» - operator of anti-incident inversion of LTH in safety.

Two variants of algorithmic control of a vehicle are developed under the condition of procedural ensuring of the complex success of the MTO taking into account the levels of complexity (uncertainty) of the SVTT. First, at low levels of uncertainty of the information state of the SVTT ($P_{oc}(\Delta l) < 0{,}35P_{oc(max)}$), the principle of adapting the modes of the vehicle to four types of locally trajectory situations (LTS): SLTS (safe), PDLTS (pre-dangerous), DLTS (dangerous) and CLTS (critical).

Proceeding from this principle, four variants of situational algorithms of vehicle control and adaptive tactics vehicle speed control (productive, anti-incident and energy-efficient) have been formed. At medium and high levels of uncertainty of the information state of the SVTT ($P_{oc}(\Delta l) < 0{,}35P_{oc(max)}$), the principle of adapting the modes of vehicle movement to four types of LTR is used (see Expressions 4-7). Proceeding from this principle, three variants of risk-regulatory algorithms of vehicle control and adaptive tactics vehicle speed control (productive, anti-incident and energy-efficient) are formed. For the quantitative assessment of the quality of algorithmic control of the vehicle in multiphase MTO with anti-incidence phases, designed calculation schemes and mathematical models for analyzing the dimensionless indices of: average speed in MTO- K_V, the energy consumption of the vehicle K_E, the anergic of the vehicle K_A in the test MTO [5,6]:

$$K_p = V_A/V_{RS},\ K_p \in (0{,}5;\ 1{,}5, \quad (8)$$

$$K_E = E_F/E_{RS},\ K_E \in (1{,}15;\ 4), \quad (9)$$

$$K_A = K_E/K_{PV},\ K_A \in (1{,}0;\ 6), \quad (10)$$

where V_A and V_{RS} - average speed of the vehicle in the test operation and the value of the reference speed of the vehicle (V_{RS} = 11,1 m / s); E_F - energy consumption of the forces of resistance motion of the vehicle in the typical test operation (J); E_{RS} - energy consumption for moving the calculated reference vehicle at constant speed V_{RS}.

TESDV movement is carried out in multiphase MTO. In the general case, the structure of MTO includes four typical phases: a) VT - phase, V = const; b) ST-phase, "car acceleration " with $\Delta V > 0$; c) ZT- phase, " car braking " with $\Delta V < 0$; d) ZTn-phase, "car rolling", $\Delta V < 0$ in the absence of motor driver actions $P_{Li4} = P_{Li5} = 0$. As a starting expression for the formation of a techno-ergatic model of speed control of the car (according to the technical scheme) in multiphase MTO, the law of change of the kinetic energy of the car is known in Car Theory [9]:

$$\frac{dE_{ki}}{dt} = N_{1pi}(t) - N_{1i}(t), \quad (11)$$

where: E_{ki} cinetic energy of the car; N_{1pi}- power of the internal forces of the car, p = 1,2,3 (p = 1 - corresponds to the effective power of the engine of the car at the phases VT and ST in MTO; p = 2 - corresponds to the brake power of the engine at the compression modes of forced braking of the motor in phases ZT- "roll-up"; p = 3- corresponds to the power of friction forces in the brake mechanisms of the car in the phases of "braking" ZT); N_{2I} -the power of the resistance to the movement of the vehicle; i - index of phase number of MTO.

On the basis of (11) the general technical-ergatic model of regulation of speed of the car (as a complex machine) in the phases of transport operation is obtained [6]:

$$\begin{aligned} &V(P_{Li5},K_{jr},D_{rs},D_{cc},t) = A_1(P_{Li5},K_{jr},D_{rs},D_{cc},t) + \\ &+A_2(P_{Li5},K_{jr},D_{rs},D_{cc},t)\times\exp(B(P_{Li5},K_{jr})\times t) \end{aligned}, \quad (12)$$

where: A_1, A_2, B - are the parameters of the model of the generalized car velocity, which are functions of the sets of characteristics: labor motor procedures of the driver- P_{Li5}; of the set of Car Generalized Type Designs - K_{jr}; sets of road properties as rolling surface - D_{rs} and communication channel - D_{cc}; t- duration of the i-th phase of MTO; j - car class index by load capacity; r- is the index of the version of car designs in the j-th class.

Taking into account (11) and (12), the regularities of the kinematics of the controlled motion of the car at different phases of MTO (VT, ST, ZT) are obtained: velocities- $V_v(P_{Li5})$, $V_s(P_{Li5}, t)$, $V_z(P_{Li5}, t)$; the paths of motion are - $l_v(P_{Li5})$, $l_s(P_{Li5}, t)$, $l_z(P_{Li5}, t)$. Techno-ergatic models of energy of controlled motion of a car (according to the technological scheme of action of complex machine and source of machine procedures) were obtained, taking into account labor P_L and machine P_M procedures of TESDC in the analysis of energy consumption E_v (V_v) and fuel consumption Q_v (V_v, T_v) then:

$$\begin{aligned} &E_v(V_v) = l_v\cdot(A_3 + A_4\cdot V_v + A_5\cdot V_v^2 + \\ &+A_6\cdot V_v^3 + A_7\cdot V_v^4 + A_8\cdot V^5) \end{aligned}, \quad (13)$$

$$\begin{aligned} &Q_v(V_v, t_v) = t_v\cdot(B_4 + B_5\cdot V_v + B_6\cdot V_v^2 + \\ &+B_7\cdot V_v^3 + B_8\cdot V_v^4 + B_9\cdot V^5) \end{aligned}, \quad (14)$$

where: A_3, $A_4 \ldots A_8$, B_4, B_5, $B_6, \ldots$ B_9 - constant parameters, which depend on the technical parameters of the car, the characteristics of the road, the parameters of the h_{ev} engine control and the gearbox u_{gv} in the VT phases of MTO.

The energy of the forces of resistance to motion of the car in ST-phases (acceleration) is determined as follows:

$$\begin{aligned} &E_s(t,\Delta V_s) = P_{cs}(t)\cdot l_s(t_s) = B_{10}\Delta V_s + B_{11}(\Delta V_s)^2 + \\ &+B_{12}(\Delta V_s)^3 + B_{13}\cdot t_s + B_{14}\cdot t_s\cdot\Delta V_s + B_{15}\cdot t_s\cdot\Delta V_s^2 \end{aligned}, \quad (14)$$

where: B_{16}, B_{17}, B_{18}, B_{19}, B_{20}, B_{21}– constant parameters depending on the set of constructive technical parameters K_{jr} and control parameters h_{es} and u_{gsv} ; t_s – during the ST (acceleration) phase.

Also obtained the following model of fuel consumption in ST-phases:

$$Q_s(t,\Delta V_s)=B_{16}\cdot t_s+B_{17}\cdot\Delta V_s+B_{18}\cdot(\Delta V_s)^2+ \\ +B_{19}\cdot(\Delta V_s)^3+B_{20}\cdot t_s\cdot\Delta V_s+B_{21}\cdot t_s\cdot\Delta V_s^2 \quad (15)$$

where B_{16}, B_{17}, B_{18}, B_{19}, B_{20}, B_{21} – are constant parameters that depend on the set of structural and technical parameters K_{jr}, the control parameters of the car h_{es} and u_{gs}.

The magnitude of the coefficient of a car speed K_v in a typical test operation is defined as:

$$K_v=V_c/V_{cs},\ K_v\in(0,5;1,5) \quad (16)$$

where V_c – is the average velocities of the car in test transport operation; V_{cs} – is constant vehicle speed in a standart transport operation, in this case V_{cs}= 11.1 m/s.

The values of energy and fuel coefficients K_e and K_q in a test operation are defined as the ratio:

$$K_e=E_0/E_{s0},\ K_e\in(1,15;4), \quad (17)$$

$$K_q=Q_0/Q_{s0},\ K_q\in(1,1;3) \quad (18)$$

where E_o– is the total energy consumption of the resistance forces of the car motion in a test operation (joule); E_{so} energy consumption for moving a car in a standart transport operation at speed V = 11,1 m / s; Qo– is the total fuel consumption of the car in a test operation (grams or liters / km); Q_{so} – is the fuel consumption of the reference car at a speed of V = 11.1 m / s in a standart transport operation.

The energy-efficient tactics of control of the car in the test MTO is realized by minimizing the index of its operational anergicity (the phenomenon of operational decrease in the performance and energy efficiency of the car in adaptive transport operations in the presence of locally trajectory hazards):

$$K_a=\frac{K_e}{K_v}=\frac{K_e(K_v,K_{jr},Dr,P_M)}{K_v(K_{jr},P_{L5},f_{vs},d_{LTH},Dr,P_M)}\rightarrow\min \quad (19)$$

where P_m – machine driving procedures; P_{L5}– a subset of labor motor driving procedures; f_{vs} –visual sign of locally trajectory hazard (LTH); d_{LTH}–long distance to the LTH zone; ; D_r –set of road characteristics.

The mathematical construction of the model of the index of its operational fuel anergicity of the car K_{aq}, $K_{aq} = K_q / K_v$ has a similar form.

The dimensionless values of the energy efficiency coefficients -P_e and the fuel efficiency - P_q of the car are defined as the inverse values of the indicators of operational anergicity- K_a and K_{aq} energy indices:

$$P_e= 1/K_a,\ P_q =1/K_{aq}. \quad (20)$$

On the basis of formulas (13-18) and logical-procedural schemes of TESDV functioning, software and algorithms for multivariate modelling of the functioning processes of the ergatic 'Driver-Bus' system in test operations have been developed. They allow you to analyse a complex of dimensionless energy indicators TESDV taking into account changes in various factors of this system (structural, technical, road, transport and operational). As an example, table 2 presents data for analysing the effect of the distance between bus stops on the values of the set of energy efficiency indicators for a bus.

Table 2. Results of the analysis of the bus operation "Bogdan A092" between the stops at different lengths of the run

Dimensionless energy efficiency indicators of the bus	Length of run between stops ln, m					
	100	200	400	600	800	1000
K_E	4,69	4,1	3,79	3,35	3,08	2,9
K_A	9,44	6,48	4,66	3,63	3,11	2,8
K_{ns}	0,77	0,82	6,84	0,04	0,51	0,43
K_L	0,49	0,63	0,81	0,92	0,99	1,03
$П_E$	0,05	0,09	0,17	0,25	0,31	0,37
$П_{Eq}$	0,14	0,18	0,23	0,30	0,35	0,39

were K_E - energy consumption ratio, K_A - indicator of vehicle's energy, K_{ns} - the share of non-sustained modes in the operation, K_L - local productivity factor , $П_E$ - energy efficiency indicator, $П_{Eq}$ - fuel efficiency indicator.

Based on the theory of energy efficiency of cars of the generalized type [5], mathematical models were obtained for the determination and parametric analysis of the parameters of K_P, K_E and K_A. On their basis, the task of multivariate analysis of the influence of a number of factors (constructive-technical, labor procedures, road and operational) is solved. In table 2 and fig. 1 presents some of the results of the analysis.

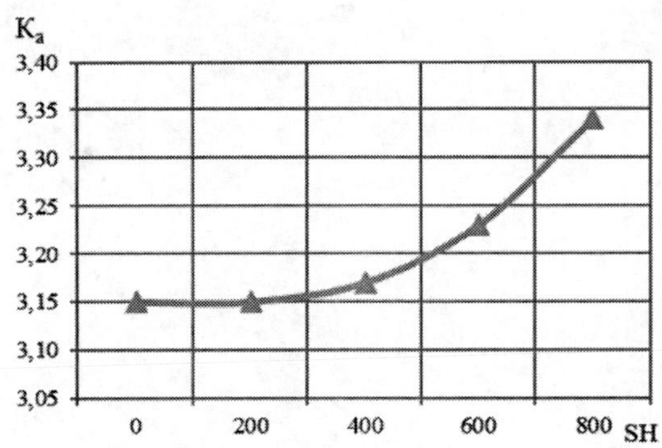

Fig. 1. Dependence of the indicator of the anergic K_A of the bus "Bogdan A092" on the value of the road's roughness index *SH* (cm/km)

3 Conclusions

1. On the basis of the entropy model of technical cybernetics, a method of quantitative estimation of the information complexity of a section of the road was

developed taking into account the diversity of states of the trajectory elements of the road and the traffic flow; it allows the driver-operator or on-board vehicle trajectory identification system to evaluate fuzzy (low, medium, high) levels of uncertainty of the trajectory section of the road to select situational (at low level) or risk- regulatory (at medium and high levels) car control algorithms.

2. It has been established that in order to comprehensively increase the productivity, trajectory safety and energy efficiency of transport operations in conditions of uncertainty, the driver should provide remote (according to visual signs of locally trajectory hazards) and preventive minimization of four types of locally trajectory driving risks: counterproductive, sensory-tempo (the possibility of deterioration in the perception of the driver of the road environment due to the high speed of the car), incident and anergic (the possibility of reducing the performance and energy efficiency of the car in difficult driving conditions).

3. Using design schemes of the theory of adaptive-discrete motion of a generalized type car, techno-ergatic models of controlled kinematics and energy of a car in multiphase transport operations have been developed; based on them, mathematical and software of procedural-simulation modeling of the functioning of the transport ergatic system "driver-vehicle" (TESDV) was developed.

4. Based on the results of a multivariate and test-simulation analysis of the functioning of the ergatic driver-bus system, a complex of dimensionless performance indicators and transport energy efficiency of cars in procedurally safe transport operations is proposed; these indicators ensure the development of two relevant methods in Motor transport: safe and energy-efficient driving, as well as the technologically successful car operation in accordance with the requirements of a comprehensive increase in adaptive performance, trajectory safety and energy efficiency of the road transportations.

References

1. A. Khabutdinov, Visnyk NTU, **1**, 519, (2015)
2. N. Kulbashnaya, N., Soroka, K. (2016). *Development of a model for choosing the driver's speed with regard to the road environment.* East European Journal of Advanced Technology, 3/2 (81), 22-29. DOI: 10.15587/1729-4061.2016.71489. (Scopus)
3. O. Petrashevskiy, A. Khabutdinov, Transport problems, **6**, 60, (2009)
4. V. Ilarionov, M. Koshelev, V. Mishurin, Driver and the car, (1985)
5. A. Khabutdinov, Project Management, Systems Analysis and Logistics, **8**, 204, (2011)
6. R. Khabutdinov, O. Kocjuk, Energy-resource efficiency of the car, (1997)
7. R. Sapozhnikov, Fundamentals of technical cybernetics, (1970)
8. A. Khabutdinov, Transport problems, **7**, 151, (2010)
9. A. Litvinov, Y. Farobin, Automobile: Theory of operational properties, (1989)

 https://doi.org/10.1051/matecconf/201929401013

Technological aspects of energy optimization during operating vehicles and increase their environmental safety

Mykola Karnaukh*, *Dmitriy* Muzylyov, and *Natalya* Shramenko

Kharkiv Petro Vasylenko National Technical University of Agriculture

Abstract. The paper discusses an actual scientific and practical problem of expanding the fuel base of the transport means by using biodiesel fuel in the form of ethyl esters made of rapeseed, sunflower and soybean oils. The choice of the optimal blend composition of diesel and biodiesel in appropriate operating conditions helps to provide the energy independence of transport companies on mineral hydrocarbons, reduce the anthropogenic influence on the environment and improve the environmental safety of transport. The research offers a new technological model for the production of biodiesel, which improves the quality and reduces the cost of biodiesel, reduces its negative impact on the elements of the vehicle fuel system. Reliability of fuel system elements is calculated. Mathematical expressions were obtained to determine the probability of failure-free operation of the fuel system and the probability of failure of its elements during operation at various fuel mixes. The assessment of the economic efficiency of the use of biodiesel as a fuel for vehicles was made.

1. Urgency of the problem

In anticipation of the expected shortage of mineral hydrocarbons, the development of alternative energy proceeds to a global scale. The tendency of searching for renewable resources is aimed at energetic and hence economic independence from the import of petroleum products, which is currently the primary task for our country.

An important direction in the implementation of unconventional energy resources is the environmental aspect, contributing to the reduction of anthropogenic impact on the atmosphere by reducing emissions of harmful substances.

According to the State Statistics Service of Ukraine, the main consumer of light petroleum products is the agricultural sector (27.8% of total consumption) and transport (26.9%) [1]. Based on these indicators, the use of alternative fuels in vehicles would help to solve the problems discussed above.

Since most of the transport work in the agriculture and transport industry is conducted by trucks, much of which are equipped with diesel engines [2], the issue of energy optimization of the operating vehicles and increasing their environmental safety when using biodiesel fuel is relevant.

2. Analysis of publications

Considering the relevance of the use of biodiesel as a fuel for vehicles, a lot of scientific works are focused on the study of this subject. The attention of scientists is aimed at distinguishing the physic-chemical properties of biofuels from mineral and the influence of these differences on the technical and operational characteristics of vehicles.

It has been stated that one of the main advantages of biofuel is its environmental aspect, which is manifested in the reduction of carbon dioxide emissions, products of incomplete combustion of hydrocarbons and solid particles [3]. At the same time, its use also has a negative impact, namely, it reduces the effective power of the engine with an increase in specific fuel consumption [4] and makes adjustments to the life of the parts of the fuel equipment [5]. The question arises of finding the optimal content of biodiesel in a binary mixture, which would allow to determine the total impact on the economic, environmental and operational component and to form a general positive effect.

Unfortunately, the studies of this issue has been insignificant, since they require costs of both material and time resources. In the work considering the adaptation of a diesel engine to bio-oil fuel compositions [6], the optimal biodiesel content limits are determined conditionally and range form 40 to 60%. A different approach to the solution of the problem depends on the gradation of the binary mixture with a certain step of the content of esters in fuel. In this case, there is no integrated approach and the indicator is taken into account that is closest to the maximum effect either by the environmental component [7] or by engine power characteristics [8].

The analysis of the literature does not give due any proper attention to the issue of seasonal operation of biofuel transport and its impact on the fuel system resource, which depends on a number of factors,

* Corresponding author: nikolay.karnauh@gmail.com

including the kinematic viscosity of the binary mixture. Therefore, the studies in this work are relevant and of a significant demand.

3. Research results

The observed trend in the development of biodiesel production is based on the methyl esters of vegetable oils, determines the use of various technologies of its production, which does not always determine its perfection and affects the quality of the final product. Non-compliance with the technological process along with the absence of a powerful filtration system from mechanical impurities and purification from saponification products, glycerol and residual methanol reduces the characteristics of biofuels, which leads to disruption of engine operation and shortening the time of its operation.

The methyl alcohol used in the esterification process is produced by a few chemical plants from natural gas, its reserves are no less limited than the reserves of petroleum products [9].

In view of the foregoing, it is reasonable to replace methyl alcohol with ethyl alcohol, which is a food product and does not have a detrimental aggressive effect. The use of ethyl alcohol, produced from renewable agricultural resources, will achieve independence from petroleum-based alcohols and ensure full utilization of the production capacities of distilleries. To implement this direction, a new technology for the production of biodiesel fuel has been patented, which allows to reduce the cost and improve the quality of the final product [10].

In the process of analyzing the work on expanding the fuel base of the means of transport, the issue of changing the reliability indicators of the diesel engine fuel system elements and the fuel system as a whole during operation is also relevant. A theoretical assessment of the probability of failure of each of the elements in the fuel system, as well as an assessment of the likelihood of system failure, will justify the timing of the maintenance or repair, which will increase the efficiency of using means of transport.

Using the mathematical apparatus of the theory of operations, we obtained an expression is obtained for assessing the reliability of the diesel engine fuel system at different fuel mixes.

$$P_0 = \frac{1}{1+\frac{\lambda_{0.1}}{\mu_{1.0}}+\frac{\lambda_{0.2}}{\mu_{2.0}}+\frac{\lambda_{0.3}}{\mu_{3.0}}+\frac{\lambda_{0.4}}{\mu_{4.0}}+\frac{\lambda_{0.5}}{\mu_{5.0}}+\frac{\lambda_{0.6}}{\mu_{6.0}}} \quad (1)$$

In the resulting formula, two parameters are used: the intensity of the failure of the fuel system's element $\lambda_{0.1}$-$\lambda_{0.6}$; recovery rate of a fuel cell element during maintenance (TO) - $\mu_{1.0}$-$\mu_{6.0}$.

Mathematical expressions were obtained to determine the probability of reliable operation of the fuel system during operation on ethyl ethers:

$$P_{0EEPO} = P_0 \exp(-c_E) \quad (2)$$

where P_{0EEPO} is probability of trouble-free operation of the fuel system when using mixed fuels (diesel fuel (DF) + ethyl esters of rapeseed oil (EEPO)); P_0 is probability of trouble-free operation of the fuel system on diesel fuel, the formula (1); c_E is the share of biofuels in diesel fuel DF.

For mixed fuel (95% DF + 5% EEPO), c_E = 0,05, for 100% EEPO, c_E = 1.

Dependence (2) is obtained from the simulation results and reflects the nature of changes in the reliability of the diesel fuel system when using different compositions of composite fuels, as it is shown in Fig. 1.

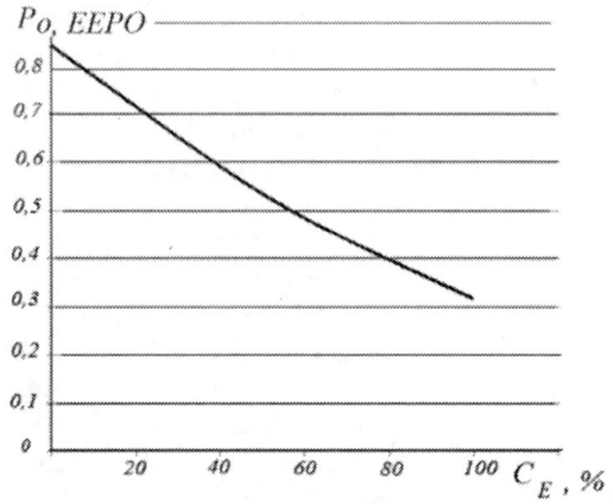

Fig. 1. Dependence of the change in the probability of trouble-free operation of the diesel fuel system on the content of the volume fraction of biofuel in diesel fuel

From the analysis of the dependence presented in Fig. 1, it follows that the probability of failure-free operation also decreases exponentially with an increase in the volume fraction of biodiesel in diesel fuel.

In fig. 2. dependencies are determined that establish the degree of influence of various types of biofuel and its volume fraction in petroleum diesel fuel on the reliability indicators of the fuel system. It is shown that mixed fuels based on ethyl esters of vegetable oils lead to a decrease in the coefficient of technical use of the fuel system. When the volume fraction (30% EEPO) in diesel fuel is exceeded, the technical utilization ratio K_{ma} decreases to 0.72.

Based on statistical data on fuel system failure during operation on various types of fuel (data for processing were obtained from agricultural enterprises where resource tests of vehicles on biodiesel were carried out) and mathematical modeling of the reliability of the fuel system, it has been stated that mixed fuels based on ethyl esters of vegetable oils are the most promising type of biofuel and expand the fuel base for vehicle operation.

The simulation results of the change in the probability of failure-free operation of the fuel system and the coefficient of technical use allow us to conclude that when using mixed fuels containing EEPO, it is possible to increase the volumetric content to containing methyl esters, where the concentration should not exceed (90% DF + 10% EEPO). However, such a suggestion requires the development of recommendations for the maintenance of fuel systems for summer and winter periods of operation, which will increase the likelihood of trouble-free operation of the fuel system to values when operating on petroleum diesel fuel.

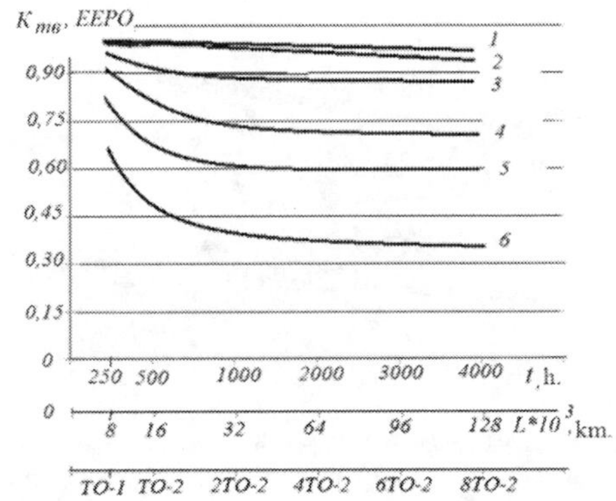

Fig. 2. Dependence on the change in the coefficient of technical use of the diesel fuel system from the use of various compositions of mixed fuels and the mileage of the truck: 1 - (DF) 2 - (95% DF + 5% EEPO) 3 - (90% DF + 10% EEPO) 4 - (70% DF + 30% EEPO) 5 - (50% DF + 50% EEPO) 6 - (100% EEPO)

During bench tests of a diesel engine on various blend compositions of diesel fuel and biological in the form of ethyl esters of vegetable oils of rapeseed, sunflower and soy, dependences of the effective power N_e and specific fuel consumption were obtained g_e. It has been found out that the use of ethyl ester of rapeseed oil reduces the effective power up to 9%, ethyl ester of sunflower oil up to 11%, ethyl ester of soybean oil up to 13%, while simultaneously increasing the specific fuel consumption by 10%, 12% and 14% respectively.

Also, we experimentally obtained dependences of the reduction of harmful emissions into the atmosphere of exhaust gases when using different types of composite fuels. It has been determined that the use of ethyl ester reduces carbon dioxide emissions CO by 37 ... 40%, the emission of products of incomplete combustion of hydrocarbons C_nH_m by 27 ... 29%, the emission of solid particles (soot) $TЧ$ by 16 ... 20% while simultaneously increasing the emission of nitrogen oxide NO_x by 2.1 ... 2.2%.

On the basis of experimental data, the following dependencies were obtained, characterizing the total emission of harmful substances into the atmosphere:

$$E = M_v \exp(-Q_{B\%}) \tag{3}$$

where E is the total emission of harmful substances into the atmosphere, g/m^3; M_v is the total emission of harmful substances into the atmosphere when the engine is running on diesel mineral fuel, g/m^3. For the diesel engine D-243 = 0.24; $Q_{B\%}$ is a dimensionless parameter that takes into account the reduction of harmful emissions, depending on the percentage of ethyl esters in diesel fuel.

The expression for calculating the exponent is as follows:

$$\begin{aligned} Q_{B\%} &= CO_{_{B\%}} \cdot m_{CO} + C_nH_{m\,B\%} \cdot m_{C_nH_m} + \\ &+ TЧ_{_{B\%}} \cdot m_{TЧ} - NO_{x_{B\%}} \cdot m_{NO_x} \end{aligned} \tag{4}$$

where $CO_{B\%}$ is the value of the reduction in emissions at a certain percentage of ethyl esters in the mixed composition; m_{CO} is the weight coefficient taking into account the "weight" (significance) in the total amount of harmful substances in the exhaust gases; $C_nH_{m_{B\%}}$ is the amount of emission reduction at a certain percentage of ethyl esters in the fuel; $m_{C_nH_m}$ is the weight coefficient taking into account the "weight" (significance) in the total amount of harmful substances in the exhaust gases; $TЧ_{B\%}$ is the value of reducing the emission of solid particles at a certain percentage of ethyl esters in the fuel; $m_{TЧ}$ is the weight coefficient taking into account the "weight" (significance) of solid particles in the total amount of harmful substances in the exhaust gases; $NO_{x_{B\%}}$ is the magnitude of the increase in emissions at a certain percentage of ethyl esters in the fuel; m_{NO_x} is the weight coefficient taking into account the "weight" (significance) in the total amount of harmful substances in the exhaust gases.

Based on the array of experimental data we obtained when testing a diesel engine on a bench using the least squares method, the following weights were obtained m_{CO} = 0.017; $m_{C_nH_m}$ = 0,0178; $m_{TЧ}$ = 0.0535; m_{NO_x} = 0.63.

Using the calculated values of the exponent, formula (4) and substituting it into expression (3) by modeling, we obtain the dependencies of the total emissions of harmful substances into the atmosphere when the diesel D-243 is operating on various mixed fuels containing ethyl esters. Dependencies are shown in Fig. 3.

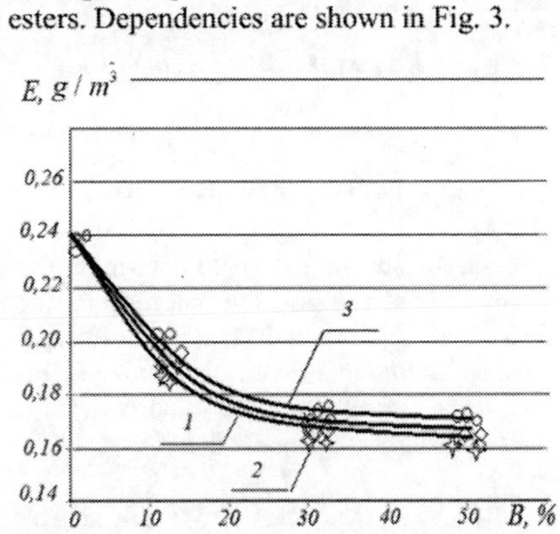

Fig. 3. Dependence on the change in the total emission of harmful substances: 1. ethyl esters of rapeseed oil; 2. ethyl esters of sunflower oil; 3. ethyl esters of soybean oil.

The analysis of the presented dependences allows to conclude that the intensive reduction of emissions of harmful substances into the atmosphere is typical for composite fuels $B10-B30$. Further increase in ethyl esters in diesel fuel does not create a greater effect. In our opinion, this is due to the contradiction of a simultaneous decrease CO, C_nH_m, $TЧ$ and increase NO_x. Due to the fact that the weighting factor of the indicator is NO_x quite high, there is a loss of the effect of the total reduction in the emission of harmful substances E, dimensionality g/m^3.

Further studies of the kinematic viscosity of the mixture of fuel compositions, the results of which are presented in Fig. 4. in the form of viscosity-temperature characteristics.

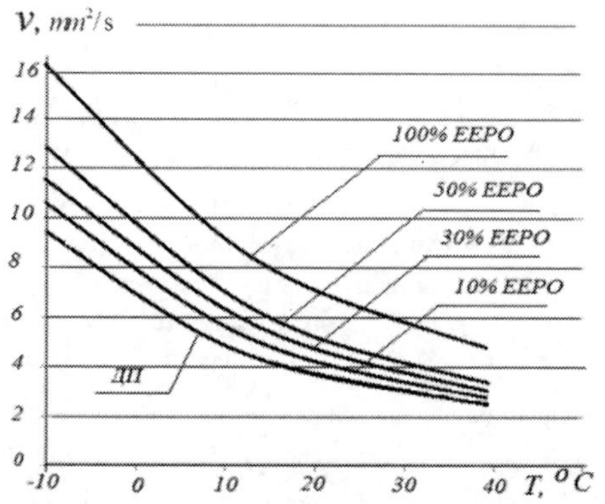

Fig. 4. Viscosity-temperature characteristics of diesel fuel in various volumetric concentrations of ethyl esters of rapeseed oil

These dependences are theoretical, since they were obtained by modeling using formulas (5) and (6) and the ASTM method D7152-11 and are intended to take into account the viscosity factor of mixed fuels when choosing the optimal volumetric concentrations of ethyl esters in diesel fuel for winter and summer periods of operation:

$$\nu_{EEPO} = 12{,}65\exp(-0{,}0245 \cdot T),\ mm^2/s \quad (5)$$

$$\nu_{DF} = 7{,}2 \cdot \exp(-0{,}028 \cdot T),\ mm^2/s \quad (6)$$

where ν_{EEPO}, ν_{DF} is the kinematic viscosity of ethyl esters of rapeseed oil and diesel fuel, mm^2/s; T is fuel temperature, $°C$.

Conducted studies allow us to determine the optimal ratio of the mixed composition, taking into account the technical and operational characteristics. For this purpose, in accordance with the second theorem of similarity and simulation (π-theorem), a complex optimization criterion is obtained K_{opt}:

$$K_{opt} = M_{kp} \cdot G_V^{-1} \cdot E^{-1} \cdot \nu_{EE}^{-1} = \frac{M_{kp}}{G_V \cdot E \cdot \nu_{EE}} \quad (7)$$

where $M_{кр}$ is the torque on the shaft of the diesel engine; G_V is volumetric fuel consumption of diesel, m^3/s; E is the amount of harmful substances emitted into the atmosphere, Fig. 5; ν_{EE} is kinematic fuel viscosity, m^2/s.

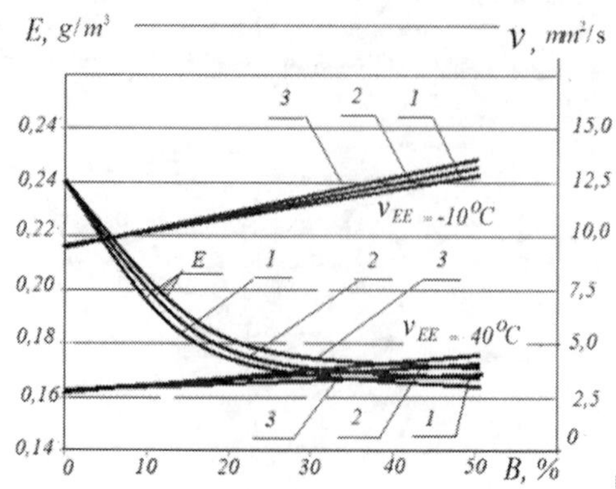

Fig. 5. Dependence on the change in the total emission of harmful substances and kinematic viscosity on the content of ethyl esters in diesel fuel:
1 - EEPO (ethyl esters of rapeseed oil);
2 - EECO (ethyl ethers of sunflower oil);
3 - EECoO (ethyl esters of soybean oil).

The criterion takes into account the volumetric fuel consumption of a diesel engine, the amount of harmful substances emitted into the atmosphere, the kinematic viscosity of the mixed fuel and the engine torque (Fig. 6)

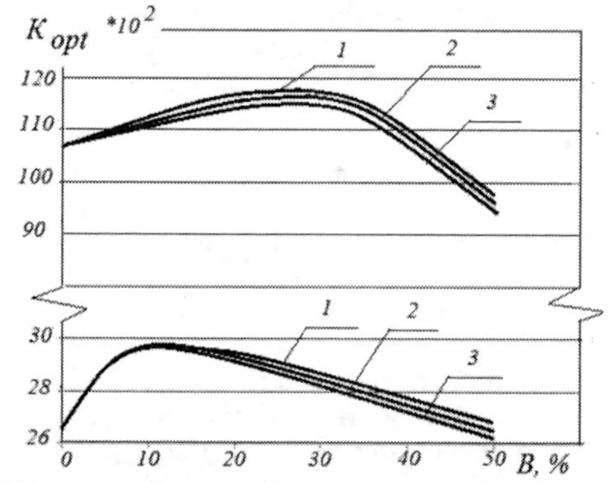

Fig. 6. Dependencies of the change in optimization criterion on the content of ethyl esters in diesel fuel:
1 - EEPO (ethyl esters of rapeseed oil);
2 - EECO (ethyl ethers of sunflower oil);
3 - EECoO (ethyl esters of soybean oil).

Analyzing the dependencies presented in Fig. 6, it can be stated: for the summer period of operation, it is necessary to use mixed fuels of not more than 30% of ethyl esters of vegetable oils; for the winter period of operation, the optimum value of the mixture composition of the fuel should contain no more than 10% of ethyl esters of vegetable oils.

To confirm the calculated data, experimental studies have been carried out to determine the fuel efficiency of the ZIL-5301 vehicle using optimal fuel mixes using main and city driving cycles, State Standard 20306-90.

From research it follows that during the operation of a vehicle on main roads with asphalt covering in the summer period of operation for mixed fuels containing 30% ethyl ethers, the fuel consumption increases compared to diesel fuel by 7.05 ... 8.67%. Smaller value refers to EEPO, and more to EECoO. For the winter period of operation, when mixed fuels contain 10% ethyl esters, fuel consumption increases by 3.31 ... 4.34%.

It is proved that during the operation of a car on the urban cycle in the summer period of operation on composite fuels containing 30% ethyl esters, fuel consumption increases by 7.1 ... 8.69% depending on the type of fuel. For the winter period of operation, when it is recommended to use no more than 10% of ethyl esters in diesel fuel, the consumption increases by 3.6 ... 4.22%.

According to the operational data, the dependences of the average fuel consumption of a vehicle on the mass of the cargo and the type of composite fuel for the winter and summer operating periods, which are approximated by the general regression equation, are obtained:

$$Q_{cp} = a + 9{,}35 \cdot 10^{-4} \cdot m,\ l/100\,km \tag{8}$$

where a is dimensionless coefficient depending on the type of fuel, period and operating conditions; m is weight of cargo, kg.

The use of the regression equation (8) allows us to make a forecast on the fuel consumption of the ZIL-5301 depending on various operating conditions. This calculation mechanism will allow planning and normalizing fuel consumption in automotive enterprises when using mixed fuels based on ethyl esters of vegetable oils.

An important factor affecting the feasibility of the use of biodiesel fuel in the operation of means of transport is the economic effect that is possible as a result of its use.

Carrying out the calculation of economic efficiency it is necessary to take into account two main components: the cost of production of 1 ton of fuel, which depends on the yield of the raw element (rapeseed, sunflower, soybean) and production technologies; reduction of anthropogenic impact on the environment by reducing the emission of harmful substances into the atmosphere together with the exhaust gases.

The impact of harmful substances can be estimated by the amount of damage in UAH, which must be spent on the restoration of the environment. Such a calculation is performed according to the methodology [11].

Analysis of the results of calculations of the cost of obtaining ethyl esters from various raw materials [12] shows that with an average yield, the cost of production of ethyl esters, especially from rapeseed and sunflower, is competitive compared to diesel fuel and can be used to expand the fuel base of vehicles. At the same time there is a reserve for reducing the cost of ethyl ethers, which is associated with an increase in the yield of the above crops. As a result, the use of ethyl esters of rapeseed oil as a promising fuel for vehicles, is economically feasible.

We performed an analysis of reducing the technogenic load on the environment due to the release of harmful substances into the atmosphere when using mixed fuel (90% DF + 10% EEPO) and (70% DF + 30% EEPO) in comparison with diesel fuel.

We performed an analysis for one Zil-5301 with a D-245 diesel engine and annual mileage of 64 thousand kilometers (2,000 hours of service).

According to the method, the amount of loss is calculated by the formula:

$$З = \sum_{i=1}^{n} m_i \cdot 1{,}1 \cdot П \cdot A_i \cdot k_m \cdot k_з \tag{9}$$

where $З$ is the amount of damage in UAH; $\sum_{i=1}^{n} m_i$ is total mass of pollutants emitted into the atmosphere, t; $П$ is the size of the minimum wage at the time of emission of harmful substances into the atmosphere, UAH; A_i is the dimensionless indicator of the relative hazard of the pollutant; k_m is the coefficient that takes into account the type of settlement; $k_з$ is the coefficient that takes into account air pollution in the locality.

The total mass of pollutants emitted into the atmosphere for the entire period of operation (1 year) is calculated by the formula:

$$\sum_{i=1}^{n} m_i = \frac{1}{1000} \cdot \left(\sum_{i=1}^{n} G_в + \sum_{i=1}^{n} G_m \right) \cdot T \tag{10}$$

where n is the number of engines, in this case, the calculation is carried out on a single engine; $\sum_{i=1}^{n} G_в$ is the total mass of air emitted into the atmosphere by the engine in one hour of operation, kg/h; $\sum_{i=1}^{n} G_m$ is hourly fuel consumption, kg/h; T is engine running time, h.

The total air mass emitted by diesel in one hour of operation:

$$G_в = 30 \cdot i \cdot V \cdot n \cdot \rho,\ \text{kg/h} \tag{11}$$

where i is the number of cylinders for D-245; V is volume cylinder, m^3. For D-245 $V = 1{,}18 \cdot 10^{-3}$, m^3; n - rated speed, rpm. for D-245 $n = 2400$ rpm; ρ - air density, kg/m^3. $\rho = 1$ kg/m^3.

The total mass of fuel for one hour of engine operation, that is, the hourly fuel consumption is determined by the formula:

$$G_m = \frac{g_e \cdot N_e}{1000},\ \text{kg/h}; \tag{12}$$

where g_e is specific effective fuel consumption, $g/kW \cdot h$. For D-245 $g_e = 245\,g/kW \cdot h$; N_e - diesel rated power, kW. For diesel engine D-245 $N_e = 73\,kW$.

To calculate the amount of losses caused to the environment by emissions of harmful substances by a

https://doi.org/10.1051/matecconf/201929401013

diesel engine according to the formula (9), it is necessary to determine the following initial data.

1. The minimum wage as of 01/01/2018 in Ukraine is 3200 UAH.

2. A dimensionless relative hazard indicator of the ith pollutant during operation on a diesel engine $A_{DF}=1$. When operating on mixed fuels, this figure will change. According to the measurements made in the third section of this work, the composition of the exhaust gases changes: the opacity of the exhaust gases decreases by 6.76 -10.6%; carbon dioxide emissions are reduced by 8.4 -18.94%; unburned hydrocarbons are reduced by 12 - 21.74%; nitrogen oxide emissions increase by 1.0 - 1.46%.

Taking into account formula, we calculated the average value of the reduction of harmful substances in exhaust gases during operation at (90% DF + 10% EEPO):

$A_{cp} = 1\text{-}Q_{10\%} = 1\text{-}0{,}172 = 0{,}828$, consequently, $A_{cp} = 0{,}828$.

When operating on (70% DF +30% EEPO):

$A_{cp} = 1\text{-}Q_{30\%} = 1\text{-}0{,}354 = 0{,}646$, consequently, $A_{cp} = 0{,}646$.

3. The coefficient k_m that takes into account the type of settlement is determined from the annex to the methodology, $k_m = 2{,}25$.

4. The coefficient that takes into account air pollution, according to the recommendations, we assume equal to 1.

The economic effect of the use of composite fuels is calculated as the difference between the size of the losses inflicted on the environment during the operation of one car during the year on the DF and (90% DF + 10% EEPO):

$$\Delta З = З_{DF} - З_{90\% DF+10\% EEPO} = 5692856 - 4713324 = \\ = 979572 UAH$$

on the DF and (70% DF +30% EEPO):

$$\Delta З = З_{DF} - З_{70\% DF+30\% EEPO} = 5692896 - 3677303 = \\ = 2015593 UAH$$

4. Findings

The concusions of our study are the following:

1. The source analysis showes that the operation vehicles on diesel oils based on veget methyl esters of vegetable oils and their mixtures makes possible a reduction in the smoke of exhaust gases and the content of products of incomplete combustion of fuel, which leads to a reduction of the environmental pollution. The physicochemical properties of vegetable oils and their fuels influence parameters of the fuel supply process and consequently change the characteristics of the injection and spraying of these fuels, the processes of blending and combustion, filtration through fuel filters of fine purification. High density and kinematic viscosity cause an increase in their cycle feed rate and hourly consumption of biodiesel compared to diesel. The source analysis figured out that methyl esters reduce the reliability of fuel system elements due to the negative impact of residual methanol on the sealing materials and filter elements, which requires the development of scientifically sound guidelines for the operation of biodiesel vehicles and biodiesel vehicles.

2. A structural analysis of the reliability of the fuel system of diesel vehicles was made by using the basic principles of the operations theory. Mathematical expressions were obtained to determine the probability of failure of fuel system operation and the probability of failure of its elements when operating on different types of fuel. The calculations were made to expose failures and restoration of working condition, the probability of failure of each element of the fuel system and the probability of failure of the fuel system as a whole on different types of fuel by using the developed mathematical apparatus and statistics. Dependencies obtained determine the degree of influence of different types of biofuels and its volume fraction in oil diesel on the reliability of the fuel system. The results of the simulation of the probability of failure-free operation of the fuel system and the coefficient of technical use suggest that the use of blended fuels EERO content can be increased to 30%, while the content of methyl esters in the mixed composition should not exceed 10%.

3. The dependence of the effective power and the specific fuel consumption applying mixed fuels based on ethyl esters of rapeseed, sunflower and soybean oils was obtained experimentally. The paper shows that the use of rapeseed oil ethyl esters reduces the effective power up to 9%, the sunflower oil ethyl esters up to 11%, the soybean oil ethyl esters up to 13%, while increasing the specific fuel consumption by 10%, 12% and 14% respectively. Dependences of reduction of emissions of harmful substances into the atmosphere in the exhaust gases were obtained by using different types of mixed fuels. It has been established that the use of ethyl esters reduces the emission by 37… 40%, the emission by 27… 29%, the emission of particulate matter (soot) by 16… 20% while increasing the emission by 5.4… 7.4%. The simulation obtained the dependence of the change in the total emission of harmful substances in the units of the exhaust gas volume for different mixtures of mixed fuels. It is calculated that the maximum effect achieve fuels containing 10 ... 30% of ethyl esters. A further increase in the percentage of ethyl esters in diesel does not have the effect of increasing the content of the exhaust gases.

4.The dimensionless criterion of optimization of the mixed composition of biofuels for operation of means of transport is obtained based on the theory of similarity and the method of dimensions. The criterion takes into account the volumetric fuel consumption by diesel, the amount of harmful substances emitted into the atmosphere, the kinematic viscosity of the mixed fuel and the torque on the crankshaft of the engine. The optimum content of ethyl esters of various oils in diesel fuel is established, which for the summer period of operation does not exceed 30% of ethyl esters, and for winter operation no more than 10%.

5. The fuel economy of the ZIL 5301 was evaluated when operating at the optimal mixed fuel depots according to the main and city driving cycles. It is

established that during the operation of the car on asphalt-paved main roads in the summer on mixed fuel containing 30% of ethyl esters, the fuel consumption increases in comparison with diesel fuel by 7.05… 8.67%. Less importance relates to EERO and more to EESO. For winter operation, when the blended fuel contains up to 10% of ethyl esters, the fuel consumption increases by 3.31 ... 4.34%. The dependence of the average fuel consumption by the car on the mass of the transported cargo and the type of mixed fuels for winter and summer operating periods, which are approximated by the general regression equation, were obtained. The equation makes it possible to predict the fuel consumption of a car when using blended fuels based on ethyl esters, as well as to develop normative indicators for the blended fuel consumption for the winter and summer operating periods, which is of practical importance for this work.

6. The evaluation shows economic efficiency of using ethyl esters of vegetable oils as fuel for vehicles. A reduction of damage caused to the environment by the release of harmful substances into the atmosphere by using a mixture of ethyl esters of rapeseed oil in comparison with the losses caused by the use of mineral diesel fuel has been carried out. The economic effect of reducing man-made environmental impact in the operation of one car with an annual mileage of 64 thousand kilometers on B10 fuel (90% DP + 10% EERO) will be UAH 979 572, on B30 fuel (70% DP + 30% EERO) economic the effect is equal to 2015593 UAH.

REFERENCES

1. The data of the State Statistics Committee of Ukraine, URL: www.ukrstat.gov.ua.
2. D.Muzylyov, A.Kravcov, N.Karnayh , N.Berezhnaja and O.Kutiya, *Development of a methodology for choosing conditions of interaction between harvesting and transport complexes*, Eastern European Journal of Enterprise Technologies, **2 (3)** (2016).
3. T. V. Dikun, P. I. Polyans'kij, *Analysis of the effectiveness of using biodiesel in internal combustion engines. Oil and gas,* Oil and gas energy, **1**, 86-93 (2015).
4. S. Chernenko, A. Atamas', *Economic and energy performance of the engine when using biofuel from rapeseed*, Bulletin of the Kremenchug State Polytechnic University named after Nikolai Ostrogradsky, **2**, 85–89 (2007).
5. V. A. Vojtov, M. V. Karnauh, O. B. Kalyuzhnij, M. S. Dacenko, *Studies of the features of biodiesel filtration through paper filters of fine cleaning of diesel engines*, Design, production and operation of agricultural machinery. **40 (1)**, 282-286 (2010).
6. A. M. Levterov, V. D. Savic'kij, N. Yu. Gladkova, *Development of methods for adapting diesel engines to bio-petroleum fuel compositions*, Mechanical Engineering Issues, **3**, 54-63 (2017).
7. Linus N. Okoro, Fadila I. Sambo, Lawal Mukhtar, Clifford Nwaeburu, *Thermodynamic and Viscometric Evaluation of Biodiesel and Blends from Olive Oil and Cashew Nut Oil*, Research Journal of Chemical Sciences, **1(4)**, 90-97 (2011).
8. J. S. Eloka-Eboka, A. C. Ibrahim, *Engine performance test of bio-diesel produced from african mesquite seed oil using pyrolysis.* African Journal of Physics, **3**, 241 -254 (2010).
9. M. V. Karnaukh, *Expanding the fuel base of automobile transport by using ethyl esters of fatty acids of vegetable oils*, Bulletin of the Kharkov National Automobile and Highway University, **49**, 47–51 (2010).
10. *The method of producing biodiesel in the form of ethyl esters of oils and fats:* pat. 65058 Ukraine: MPK S10L 1/19 (2006/1). №U201105331; zayavl. 26.04.2011; publ. 25.11.2011, bulletin, **22**. 4 (2011).
11. *Methodology for calculating the amount of compensation for losses caused as a result of regulatory emissions of pollutants into the atmosphere*, Ministry of Jusice of Ukraine, Jan. 21th 2009, **48** (2009).
12. V.A.Voytov, M. V. Karnaukh, M.S. Datsenko, *Estimation of the cost of manufacturing ethyl esters of vegetable fatty*, Bulletin of the Kharkov National University of Agriculture named after Peter Vasilenko, **108**, 105–115 (2010).

The concept of a hybrid traction power supply system

Viktor Sychenko[1,*], *Valeriy* Kuznetsov[2], *Yevheniy* Kosarev[1], *Petro.* Beh[1], *Yuriy* Sayenko[3], *Bohdan* Styslo[4], *Mihail* Pavlichenko[5], *Ihor* Vasilev[6], and *Mykola* Pulin[7]

[1] DNURT, [2] Railway Institute, [3] PGTU, [4] NTU "KhPI", [5] LLC DDK, [6] UrGUPS, [7] Lviv Railway

Abstract. The growing shortage of energy resources to provide heavy traffic and high-speed traffic necessitates the development of a new power supply scheme for a traction network using sources of distributed generation of electric energy. Application of asymmetric distributed traction power system with the use of distributed generation allows to increase electric energy efficiency of the feed system of rolling stock, reduce its material capacity and increase the reliability of the functioning of the railway system as a whole. The proposed structure of the hybrid electric traction system involves the creation of a new paradigm aimed at ensuring high demands on the voltage and specific power in the traction network, which will allow to improve the quality and safety of the transportation process when increasing the energy performance indicators. This paradigm will allow to form a new philosophy of building traction power systems. The key elements of this philosophy are decentralization of generation and consumption of electric energy, the use of intelligent electric equipment, distributed power supply management in a single information space of the energy process during transportation. The peculiarity of the proposed work is the optimal combination of technological processes of transmission, consumption of electric energy and transportation, taking into account the specific conditions of functioning of the electrified section: points of connection of power sources and their type, profile of the track, schedule of trains, types of rolling stock and equipment. Conceptually, the hybrid traction power system is an electric hub, in which the inputs are generating points, and the output is the power supply system of the rolling stock. The internal structure of the electric power objects of the hub is determined by the type of transformation and accumulation of electric energy and ways of managing these objects in real time.

1 Introduction

Nowadays were formed two systems of traction power supply: direct and alternating current. Existing experience of their operation proves that the speed of up to 250 km / h can be provided by existing systems of electric traction. It should be noted that the development and establishment of each of these systems was conditioned by the state of development of scientific and technological progress for a certain period of time [1]. The DC system from the moment of its appearance did not change structurally, but mainly changed the element base and used equipment. The AC system, on the contrary, from the very beginning of its application is constantly subject to various improvements. It can be stated that, not taking into account the declared advantages and prevailing on the implementation of the current system, the experience of its operation proves that, in general, the AC system no longer meets the requirements of the modern day. For all years of operation and modernization, the main drawbacks of this system were not eliminated: dangerous electromagnetic effects, asymmetric loading of phases and significant volumes of reactive energy transit [2]. Of course, scientific discussions about which electrification system is better, are conducted for a long time, but, as recent studies show, there is a gradual evolution of approaches from electrification of high-speed lines on alternating current up to the transition to a more thorough 24 kV DC system [3-5].

The methodological basis of this process is the development of modern power electronics, which allows implementation of the DC system with high voltage. Research of such systems is conducted long ago and it has already been proved that such systems in their technical and economic indicators far outweighed any advanced systems of AC electric traction [6]. Therefore, there are real prospects for the practical implementation of new DC systems taking into account the current state of power electronics. Getting high voltage in the contact network does not cause problems. For the rolling stock are known a number of possible variants of the converters structures. According to the authors, the ones that are currently used in power engineering for the two-stage DC power supply with the intermediate link of the alternating current are most suitable. It's known the new development of the company ABB on arcless chamber switches. The company offers high-voltage IGBT for inverters with a switching frequency of about 2 kHz, which is enough for use in electric locomotives.

Among other things, today is a conceptually grounded approach to the creation of distributed power

* Corresponding author: elpostz@i.ua

supply systems with high voltage and with the application of the principles of creating hybrid systems that use the following approaches: transmission of energy at high frequency, the use of amplitude-pulse-width modulation and active filtering, the use of power cables with characteristics adapted for the transmission of electric energy by single-phase voltage of high frequency [7].

At the same time, it should be noted that the existing DC system owith voltage in the traction network 3 kV is entitled to further application. Significantly strengthen traction power supply can not only the introduction of advanced technologies and modern equipment, but also the use of distributed power supply systems. The principles of constructing these systems were formulated by prof. Marquardt K. G. in the middle of the last century [8]. It is this ideology that should be entrusted to the development of new power supply systems and is already used in the construction AC systems (2x25 kV, 94 kV) and DC (6 kV). When it is used, the power of the contact network is carried out by means of converters connecting to the longitudinal power line of the elevated voltage. The economic effect is achieved by reducing the section of wires of the contact network, reducing energy losses, maintaining the necessary voltage level in the contact network and increasing the utilization factor of the power of the main power equipment at the decrease of its installed capacity.

Modern trends in the development of the requirements of society impose high requirements on traction power devices that have to provide the required amount of transportation work throughout the year, taking into account the unevenness of the movement and the difference in the weight of trains. In connection with the rise in energy prices has emerged the need for the development of a new energy policy, modernization of the equipment of traction power supply systems and the search for untraditional traction power supply systems. Energy efficiency in modern conditions is the most important factor in the competitiveness of railways in the domestic and foreign markets of transport services.

It should be taken into account that the choice of technical solutions for the creation or reconstruction of a traction power supply system (TPSS) for the provision of high-speed and high-speed passenger traffic is based on the main parameters of the projected line, which include: maximum permissible and operating speed; type of electric rolling stock (ERS) and its characteristics; principles of organization of transportation and schedules of trains [9]. From the listed parameters significantly depends on the ability of the load of individual power devices and the system as a whole. The parameters of the TPSS and the power circuit of the contact network are selected in accordance with the accepted standards for the permissible load of the power equipment of the traction substations, the voltage on the current collectors vers of the ERS and the heating temperature of the traction network wires. It should be noted that there is a steady tendency to change the requirements for TPSS - not only providing normalized voltage, but also providing the necessary energy intensity of the traction network [10]. In this context, it should be noted that in existing TPSS, with excess aggregate capacity at traction substations, it is necessary to increase the traction network to provide the necessary energy indices and stability of the system as a whole. Paradoxically, the current state of traction power supply of direct current is characterized by a growing shortage of electric energy, which raises the problem of using independent power sources to supply power traction network points (PP).

To date, one of the main vectors for the development of modern electricity is the evolutionary transition from centralized generation of electric energy to distributed smart generation systems that use renewable energy sources. Their application increases the maneuverability of energy flows, which leads to an increase in the quality of electricity supply under uneven schedule of electricity consumption. The application of distributed generation and accumulation of electric energy can reduce the energy content of the power systems, increase its reliability and vitality through the backup energy stored in the storage devices [11]. There are already examples in the world of successful use of electricity from renewable sources in the electric traction system. For example, in 2011, the train traffic on the Paris-Amsterdam route was unlocked using solar energy from photographic batteries installed on the train's route. Ukraine also implemented pilot projects for the parallel operation of the solar electric plants (SEP) and the external power supply system for power supply of 10 kV on Sambir traction supply station and wind enerrgy plants (WEP) for supplying 35 kV buses on the traction supply station StaryiSambir [12]. According to scientists, decentralization of electricity production is an important component of innovation and technological support for structural reform of the energy industry in order to increase its efficiency and, among other things, allows: to reduce electricity losses during transportation due to the maximum approximation of the source to the consumer, to reduce the number and length of main transmission lines, to reduce the consequences of accidents at central power plants and main transmission lines, to provide reciprocal multiple backup of power generating capacities.

The above stipulates the necessity of developing new approaches to the scheme of power supply of the traction network with the use of electric energy both from existing centralized power grids and from sources of distributed generation.

2 It is known circuit design decisions of distributed traction power supply systems

To date, there is no clear classification of distributed systems of traction power supply of direct current (DTPSS). It can be pointed out that three concepts of such systems have been formed [13, 14]:
- with increasing voltage in the traction network (6, 12, 24 kV);
- with a voltage of 3 kV in the traction network and AC power supply with an AC line (10, 35 kV);

- with a voltage of 3 kV in the traction network and the power supply of the direct current DC line (6, 12, 24 kV).

An option for implementing such a system is presented in Fig. 1) [13].

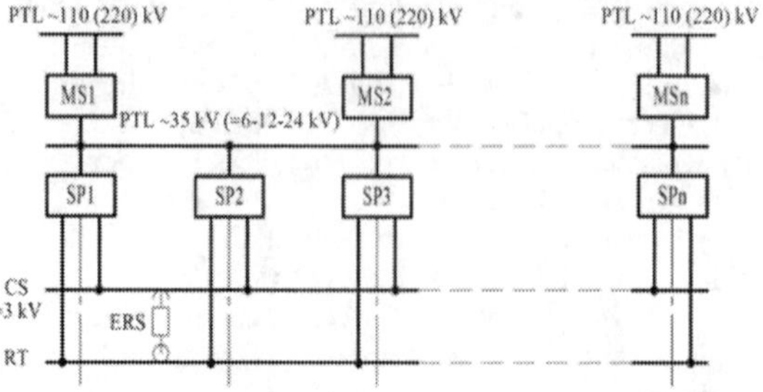

Fig. 1. Block diagram of distributed power system: MS - main substation, SP - supply point, RT – rail track; CS – catenary system; ERS – electric rolling stock.

Now in the Department of intellectual power supply systems in DNURT were formed two main approaches to circuit of DTPSS: system of distributed power with the use of alternative sources of electric energy (DTPSSA) [15] and a system of distributed power using single aggregate traction substations (DTPSSS) [16]. The implementation of the above-mentioned standards was carried out under the guidance of prof. Sychenko V.G. and Kuznetsov V.G. Options for the structural implementation of these systems are presented in Fig. 2 and 3, respectively, and the power supply parameters are given in Table 1.

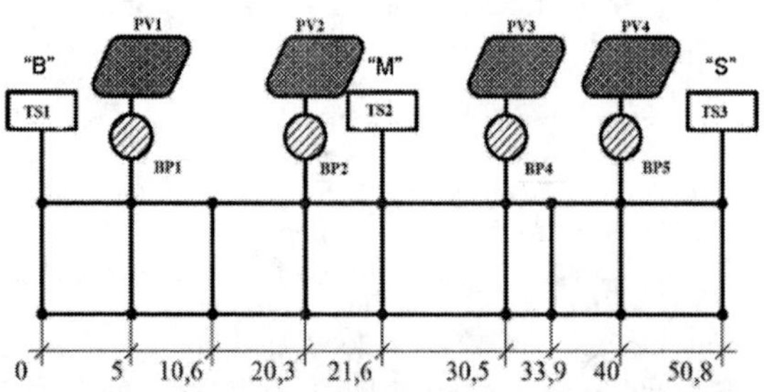

Fig. 2. Scheme of tested section of DTPSSA: TS – traction substation; BP – boost point; PV – photovoltaic power station.

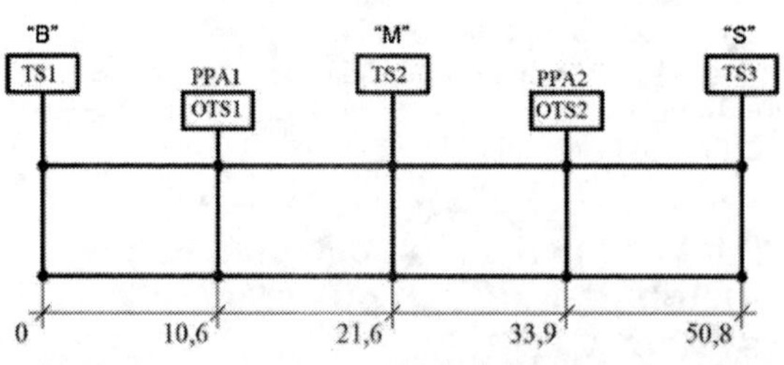

Fig. 3. Scheme of the section of DTPSSS with single-aggregate substations.

Each of these systems allows to provide the necessary requirements for the introduction of high-speed traffic and should be a priority direction for increasing the traction power supply of direct current in conducting the relevant feasibility study.

Table 1. Parameters of power supply systems.

<table>
<tr><th>Power, MW</th><th>TS1</th><th colspan="2">OTS1 (PPA1)</th><th>TS2</th><th colspan="2">(OTS2) PPA2</th><th>TS3</th></tr>
<tr><td>CTPSS</td><td>12,6</td><td colspan="2">-</td><td>25,2</td><td colspan="2">-</td><td>18,9</td></tr>
<tr><td>DTPSS</td><td>10,4</td><td colspan="2">5,2</td><td>10,4</td><td colspan="2">5,2</td><td>10,4</td></tr>
<tr><td>PSSA</td><td>12,6</td><td>6</td><td>2,5</td><td>12,6</td><td>2,5</td><td>6</td><td>12,6</td></tr>
<tr><td>No-load voltage,V</td><td colspan="7">3500</td></tr>
</table>

* Here CPSS - a centralized power supply system.

A comparative analysis of the energy performance of the proposed distributed systems and the existing centralized supply system is summarized in Table 2.

Table 2. Power indicators of traction power supply systems.

<table>
<tr><th colspan="2">Indicator</th><th>CTPSS</th><th>DTPSS</th><th>PSSA</th></tr>
<tr><td rowspan="4">Mode of voltage on the current collector of ERS, V</td><td>M(U)</td><td>3130</td><td>3291</td><td>3397</td></tr>
<tr><td>max(U)</td><td>3418</td><td>3500</td><td>3500</td></tr>
<tr><td>min(U)</td><td>2643</td><td>2901</td><td>3255</td></tr>
<tr><td>The value of the confidence interval</td><td>875</td><td>705</td><td>321</td></tr>
<tr><td colspan="2">Electricity consumption for traction, kWh</td><td>1495</td><td>1437</td><td>944.8 (from TS) +486.3 (from the PP)</td></tr>
<tr><td rowspan="2">Average losses power, kW</td><td>Excludinghigher harmonics</td><td>524</td><td>372</td><td>351</td></tr>
<tr><td>Taking into account the higher harmonics</td><td>631</td><td>428</td><td>409</td></tr>
<tr><td rowspan="2">Power losses, kWh</td><td>Excludinghigher harmonics</td><td>174,9</td><td>123,9</td><td>113,6</td></tr>
<tr><td>Taking into account the higher harmonics</td><td>210,6</td><td>142,7</td><td>126,1</td></tr>
<tr><td colspan="2">cosφ</td><td>0,77</td><td>0,79</td><td>0,81</td></tr>
<tr><td colspan="2">tgφ</td><td>0,8</td><td>0,76</td><td>0,74</td></tr>
<tr><td colspan="2">Efficiency</td><td>0,92</td><td>0,95</td><td>0,97</td></tr>
</table>

3 Hybrid system of traction power supply

3.1 Methodological principles of construction of hybrid TPSS

The energy infrastructure of the existing TPSS has a fairly simple structure: external power supply lines, traction substations with lowering and cconverter transformers (units), from which the electric rolling stock and non-traction consumers receive power. In this TPSS, a rather simple circuit for the transformation and transmission of electric energy is used. At the same time, in existing TPSS there is virtually no means of regulating real-time power consumption modes, which at low utilization of aggregate power leads to additional losses of electric energy, and, accordingly, financial losses of electricity supply stations.

The rapid development of distributed generation based on non-traditional and renewable sources of energy, both at the level of power supply systems and directly to consumers, and their integration into centralized systems require the implementation of new principles for the construction of these systems and the creation of intelligent management systems for them with developed information and communication support. The unification of disparate systems of different levels into a single technological complex will ensure the implementation of new functional capabilities, the use of more advanced technologies in operation, and the creation of integrated centralized distributed systems with coordinated control of their modes and active participation of consumers in the electricity process [17, 18].

The current stage of development of world energy is characterized by a change in technological paradigm, which involves reducing the priority of centralized control. At the same time, the market of affordable modern technologies for the intellectualization of equipment, control systems and energy accounting, telecommunications and information security, small generation and others are developing. [17]. The technological basis of providing these processes is the use of Smart-grid platforms. All this creates the preconditions for the construction of integrated intelligent energy systems that have a multidimensional structure of functional features and developmental properties, combining multicomponent, intelligence, efficiency, reliability, manageability, flexible use of conversion technologies, transport, energy storage and active participation in these processes of the consumer. Implementation of the task requires the transition to a new structure of the traction power system using the principles of multi-agent management of available sources of electricity generation and consumption.

It should be noted that multi-resource power supply systems today are one of the main directions of energetic development. Their application allows to achieve synergistic effect as a result of the specific advantages of each of the individual power supply systems (electric, heat, gas, water supply, etc.) [19].

The key technological elements of the multi-resource power supply system are: sources of centralized power supply, electrical networks and installations, small energy objects, intelligent switching devices and energy storage devices. The application of the multi-electric system allows improving the whole complex of performance indicators of the electrical system, the main of which are: optimization of consumer load schedules, optimization of the modes of generating stations and network infrastructure, improving the quality of electric energy, and increasing the reliability of electricity supply. Proceeding from this, we can formulate a new philosophy of building an TPSS, which is based on the concept of an energy hub [20, 21]. The peculiarity of the new philosophy is that the optimal set of technologies and devices for the electricity supply of a particular electrified area will be determined in each case separately, taking into account its properties, places of attachment of generation sources, profile of the track, schedule of trains, etc.

The power hub represents a model of a set of devices that implements the interaction between energy producers, transport infrastructure, on the one hand, and consumers - on the other hand. The main functions of the hub are the transfer, transformation and storage of energy resources. In a hub, power resources can be transmitted from input to output without changing the form (for example, electricity) or only with a significant change in quality (eg voltage transformation). In addition, energy resources can be transformed from one species to another, for example, a solar power plant converts solar radiation into electricity. The structure of the hub consists of four types of elements [20, 21]:

1. Inputs - flows of electric power of various origin;
2. Converters (transformation, conversion);
3. Storage facilities (storage devices, accumulators);
4. Outputs - flows of electricity coming from the hub to consumers (electric rolling stock, non-traction consumers).

The block diagram of an electric hub is presented in Fig. 4.

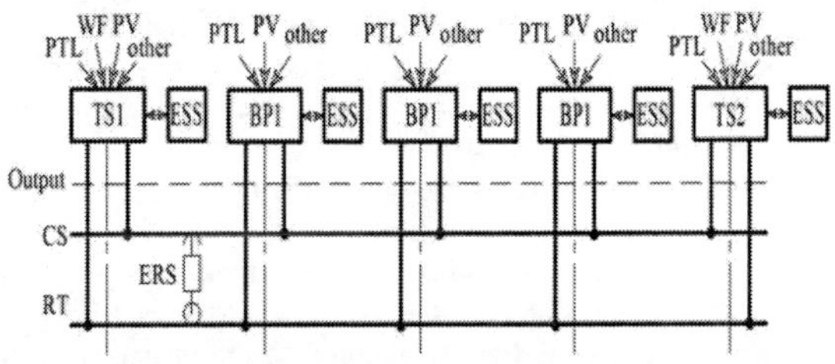

Fig. 4. Block diagram of an electric hub: TS – traction substation; BP – boost point; RT – rail track; CS – catenary system; ESS – energy storage system; PTL – power transmission line; WF – wind farm; PV – photovoltaic power station; ERS – electric rolling stock.

The proposed TPSS can be implemented with a high degree of intellectualization electricity accounting, distributed intelligent management of power consumption and intelligent forecasting. The peculiarity of this structure is the connection of wind power plants to traction substations, and the solar generation, both to traction substations, and for power supply of PP. This approach is caused by the circuitry of existing TPSS and is intended to minimize the material costs associated with replacement of equipment at the first stage of the implementation of the new system.

3.1 Energy storage devices

If earlier the electric storage devices (ESD) were considered as one of the means of energy saving, then with the development of eco-technologies, their application becomes a key task for the power industry. At present, the following main ESD are known: hydroaccumulating power plants, air-accumulating power stations, electrolytic capacitors, supercapacitors, electrochemical storage batteries, superconductive inductive energy storage units, fuel cells, hydrogen storage units, kinetic energy storage devices. In recent years, the greatest progress has been made in the development of hydrogen and kinetic ESD instead of earlier leading electrochemical storage devices [22].

The main areas of application of ESD in the electric power hub can be considered as:

- regulation of the load mode (alignment of day and night loading);
- regulation of the power flow of the consumer;
- providing more stable graphs of power output and loading schedule;
- increase of TPSS stability;
- improvement of the quality of electric energy with a sharp change in the traction load;
- reduction of the risk of systemic accidents.

Despite the wide range of electric energy storage devices, the specifics of their use in the energy sector imposes a rather serious constraint on the choice of a particular type, including energy, mass-size and economic indicators: energy intensity; service life; absence of "memory effect"; admissibility of fast charge modes; the ability to operate in the mode of large current overload; safety; low cost.An analysis of the characteristics of various types of electric power storage batteries [23] shows that each of the existing types of accumulators has advantages and disadvantages that should be taken into account when designing a system for the accumulation of electric energy in an electric power hub of TPSS.

According to experimental studies [24, 25], it is clear that the increase in the efficiency of the use of ESD can be achieved by expanding the frequency range of the efficient operation of the storage devices. For this purpose, it is expedient to include in the structure of the battery system different types of storage devices with different time characteristics of the charge-discharge cycle [26]. For example, although modern lithium battery batteries have sufficiently high frequency characteristics [24], their charging-discharging cycles are limited. In order to extend the lifespan of the amplifying point drives, it is expedient to allocate the reinforcement point tasks to two structurally coupled channels - low frequency (LFC) and high frequency (HFC). At fig. 5 shows a spectrograph of current consumed by an electric locomotive from a contact network, in which the intervals of frequencies are indicated Δf1 and Δf2 - respectively, the zone of action of the LFC and the HFC of the supply point.

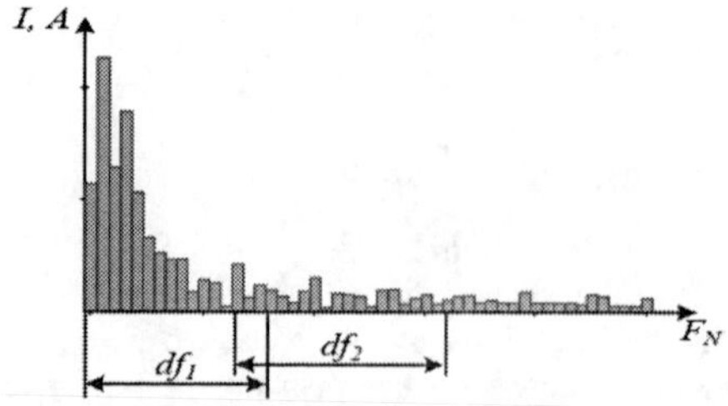

Fig. 5. To explain the zone of action of the LFC and the HFC of the power supply point.

The purpose of the LFC is to suppress all high-frequency current pulsation in the network. As an energy storage device in this channel it is expedient to use inertial flywheels in view of their high energy performance [27] compared with conventional electrolytic capacitors. As an energy storage device, the advantage of LFC should be givenLiFePO4 batteries that have improved power performance compared to traditional ones Li-Ion batteries [26].

To reconcile the energy characteristics of the ESD with the power supply (voltage level, charging level or discharge current) in the power supply point structure a matching converter is used. To effectively suppress pulsations in a wide frequency spectrum, it is advisable to carry out the APP according to the scheme of Fig. 6. Available in the system of matching converters in the English literature are called "cooperative", they are characterized by the existence of a common system of control of several separate converters working on a joint load.

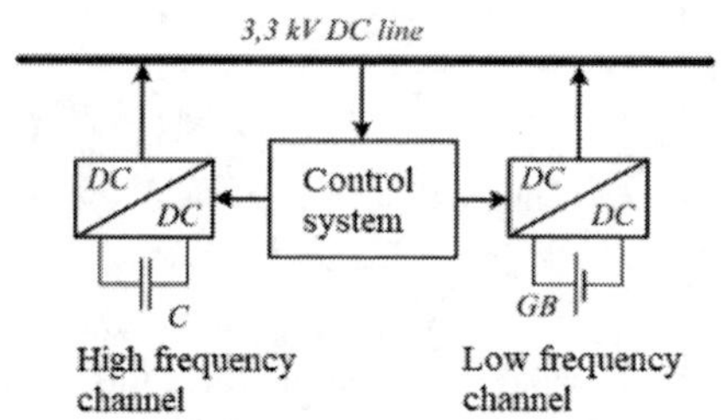

Fig. 6. The block diagram of a two-channel power supply point.

Since the boundaries of the typical charge / discharge time of the different types of drives vary in a sufficiently wide time range, frequency ranges Δf1 and Δf2 (Fig. 5) can be determined by superposition of energy and economic characteristics of storage devices of different types. Matching these storage devices should be done in real time, taking into account changes in load and generation (Fig. 7).

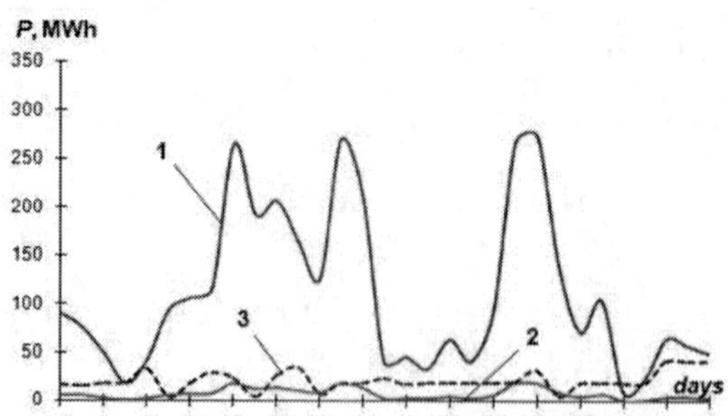

Fig. 7. Changes in load and generation: 1 – wind farm generation; 2 – photovoltaic generation; 3 – energy consumption.

In the development of algorithms for the functioning of intelligent devices in the hybrid TPSS it is necessary to take into account that the process of electricity consumption in the traction network occurs under the influence of heterogeneous perturbations: changes in the modes of operation of the electromotive force depending on the characteristics of the train timetable, the profile of the track and the available constraints, short-range separations of current collectors, the kind of transients, changes in the flows of generated energy, and so on. The mode of the voltage in the traction network under the influence of these perturbations has a non-stationary oscillatory nature with abrupt voltage changes that can lead to loss of voltage stability. As a result of the application of the quantitative approach to the calculation of the static stability of the traction power system, it was established that during the movement of the train in the real area there are zones of lack of voltage stability. The precise solution of the problem of stability evaluation is extremely complicated by the need to calculate nonlinear dependencies that determine the modes of operation of the traction power system and the electromotive composition [28]. The approach developed in [28] allows to calculate areas of stability of variant circuit designs of the traction network when introducing high-speed traffic and narrowing the range of voltage fluctuations. At the same time, it's not clear about the need for an asymmetric location of the PP, that means, depending on the specific conditions for ensuring voltage stability in the non-symmetric traction power system.

3.3 Simulation results

At fig. 8and, the block diagram of the BESD is presented, which includes a drive matching the DC / DC converter and charge controller. The DC / DC converter recharges or discharges the drive using the control strategy, according to the voltage in the contact network and the storage device charge (SoC). At fig. 8 b it's shown the equivalent diagram of battery modules, consisting of four elements [29].

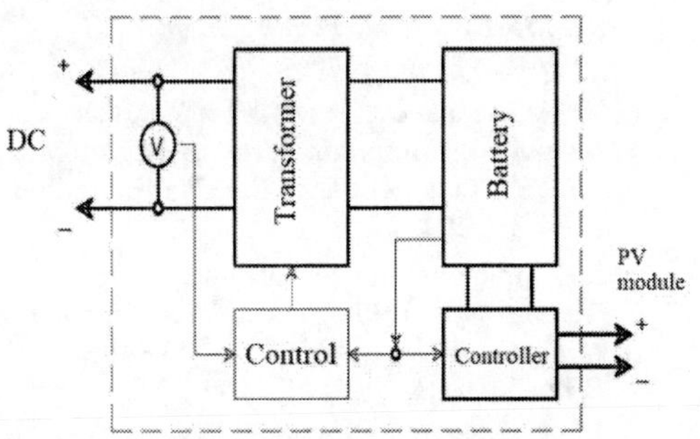

a.

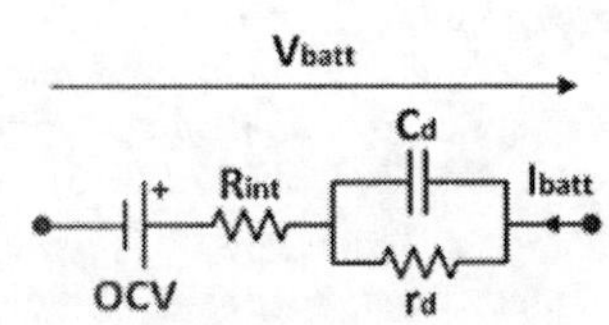

b.

Fig. 8. Battery energy storage system: a – structural scheme: b - equivalent.

The ideal voltage source is the non-load voltage (OCV), which is influenced by the charge level of the storage device (SoC); serial resistor Rvn represents the internal resistance of the storage device, while parallel RC link (rd and Cd) describes the transfer of charge and capacity respectively.

The electric model of the storage device can be described by Expressions 1 - 4, where the first expression represents the Kirchoff equation for the substitution circuit (Fig. 1 b), the second - a polynomial n degree of the relationship between non-load voltage and the charge level of the storage device. The third equation models the law of varying the level of charge, in accordance with the magnitude of the charge current, and the fourth expression is a differential equation describing the work of a parallel RC link.

$$U_{\text{H}}(t+dt)=OCV(t)-R_{\text{BH}}I_{\text{H}}(t)-u_d(t+dt) \quad (1)$$

$$OCV(SoC)=\alpha_n SoC^n+\alpha_{n-1}SoC^{n-1}+...+\alpha_0 \quad (2)$$

$$SoC(t+dt)=SoC(t)+\frac{U_{\text{H}}(t+dt)I_{\text{H}}(t)dt}{3600\cdot C_{\text{H}}} \quad (3)$$

$$u_d(t+dt)+r_dC_d\left(\frac{u_d(t+dt)-u_d(t)}{dt}\right)=\begin{cases} r_dI_{\text{H}}(t) \\ 0 \end{cases} \quad (4)$$

where, u_d - the voltage of parallel RC link;
dt - step sampling rate by time;
$\alpha_n ... \alpha_0$ - coefficients of interpolation;
C_{H} - capacity of the storage device.

Capacities of BESD for each PP are determined depending on the magnitude of the current of the amplifying current along the entire time of the passage of the ERS. By neglecting the amount of regeneration energy (assuming that energy was not supplied to the BESD, but only consumed throughout the entire range of motion), for discrete measurements of the current consumption, the capacity of the storage device is determined [30]:

$$C_{\text{H}}=\int_0^{T}\sum_0^{n-1}|i_n|dt \quad (3)$$

where, C - BESD capacity, A · h;
n - serial number of discrete measurement;
i - instantaneous value of the current consumption of the BESD, A;

Calculation of capacitance of storage devices of power points is realized on the example of the real electrified section of the Pridniprovskaya railway fig. 9.

The estimated section of the two-track, 128 km long, has 8 section between substations and receives power from 9 traction substations according to the given power schemes (Fig. 9). The non-load voltages and the internal resistances of the traction substations are determined based on their passport data. Traction network of the electrified area - M120 +2MF100 + A185 + R65. Places for PP installation: 9 km, 33 km, 51 km, 66 km, 76 km, 91 km, 109 km, 125 km. The train schedule is shown at fig. 10.

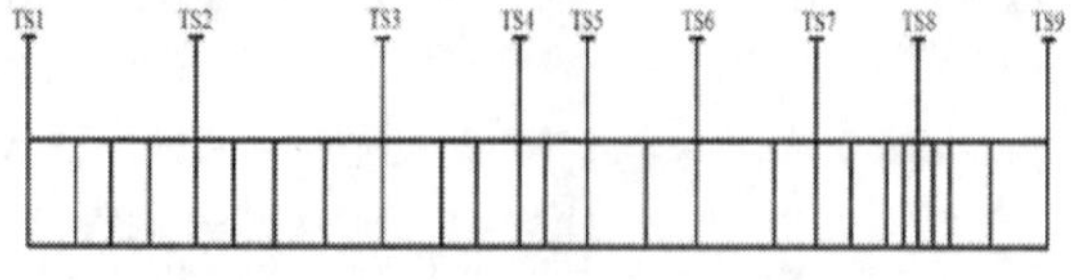

Fig. 9. Estimated section.

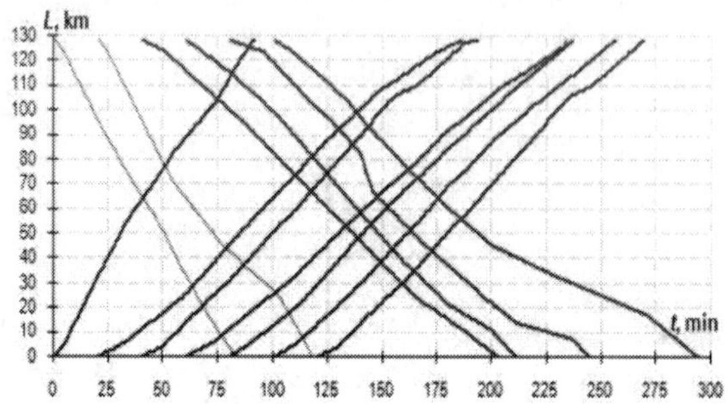

Fig. 10. Schedule of trains on section.

Based on the currents of amplification (Fig. 11) obtained as a result of the calculations carried out by means of a mathematical model, the capacities of the storage devices required for feeding the traction network and providing a predetermined level of voltage on current collectors of the ERS during the implementation of the train schedule are determined (Table. 3).

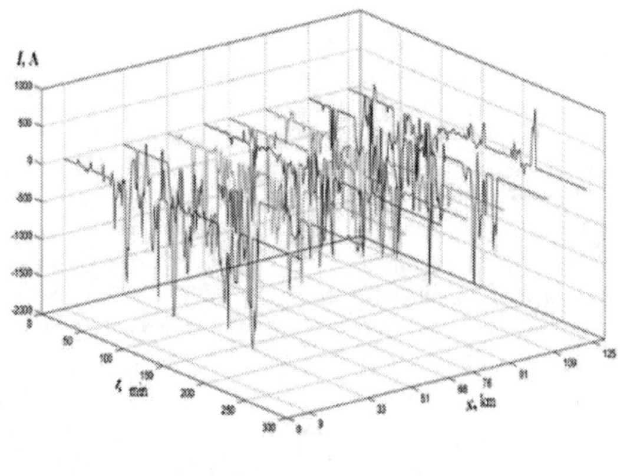

Fig. 11. Currents of BP.

Table 3. Capacities of the storage devices of BP.

№	BP 1	BP 2	BP 3	BP 4	BP 5	BP 6	BP 7	BP 8
Capacity A.·h	705	1210	565	390	630	455	990	425

4 Conclusions

1. The present state-of-the-art traction power supply systems with the introduction of high-speed traffic is due to the increasing requirements for voltage and power density, and, with increasing levels of traction load, the shortage of electric energy to meet these requirements.

2. The perspective direction of development of electric power systems is reduction of centralized supply of electric energy and centralized control of processes of its transfer and consumption with the use of Smart-technologies.

3. With the spread of the ideology of multi-resource generation, it is necessary to form a new ideology of building traction power supply systems to improve the technical and economic indicators of its functioning.

4. The proposed circuitry of the modern traction power supply system in the form of an electric hub allows to use the available resources for generating electric energy and optimally combine the technology of its transmission and consumption, taking into account the specific conditions of the electrified section.

5. It is shown that the system of accumulation of electric energy in the proposed system of traction power supply should have two channels to compensate for low-frequency and high-frequency oscillations of the load. The determination of the capacity of electric energy storage depends on the values of the consumed current, the nature of its change and the conditions for generating electric energy in the electric power hub of the traction power supply system.

References

1. V. G. Sychenko, *El. supply of high-speed railways.* Ukr. Railways. **No. 5-6 (23-24)**, P. 32-39 (2015)
2. V. G. Sychenko, D. A. Bosiy, *The quality of el. power use in el. traction syst.* Modern techn. Syst. analysis. Simulation. **№ 48,** P. 143-149 (2015)
3. B. A. Arzhanikov, A. T. Burkov, A. G. Galkin, V. A. Mansurov, I. O. Naboichenko, *Persp. of the devel. of the el. traction syst. of 24 kV high volt. for the Moscow-Yekaterinburg highway.* **No. 7**, P. 48-50 (2012)
4. V. V. Hananov, A. T. Burkov, D. V. Barch, *Innovations and innovative proc. in traction power supply.* Transport of the Russian Federation. **No. 6,** P. 34-39 (2012)
5. L. Abrahamsson, S. Ostlund, L. Soder, *HVDC feeding with OPF and unit commitment for el. railways. El. Syst. for Aircraft, Railway and Ship Propulsion* (ESARS) (2012)
6. V. G. Sychenko, D. O. Bosiy, V. V. Bozhko, E. M. Kosarev, Yu. A. Borisovskaya, Ya. P. Shcherbak, B. O. Borshch, *Comparative analysis of power supply syst. for high-speed highways.* El. of transport. **No. 10**, P. 29-37 (2015)
7. P. V. Hubskyi, *Power supply channels of distributed traction power supply syst.* El. of transport. **No. 15**, P. 23-30 (2018)
8. K.G. Marquard, *El. power supply of el. railways,* - 288 p. (1958)

9. V. E Marskiy, *Preparation of traction power supply for the org. of high-speed traffic on the St. Petersburg-Moscow line.* P. 15-19 (2010)
10. A. T. Burkov, S. A. Burkov, M. A. Sharpilova, *Modes of motion and features of traction calc. in determ. the loads on power supply systems of high-speed lines.* El. innovative techn., high-speed and high-speed traffic on railway transport. Materials of the VI Intern. Symposium Oct. 25 - 28, 2011. - St. Petersburg: PGUPS, P. 337 - 349 (2013)
11. V. G. Sychenko, A. A. Matusevich, A. V. Rogoza, M. E. Pavlichenko, I. L. Vasiliev, N. N. Pulin, *Enhancing en. efficiency in evolving power supply syst.* Bulletin of the National Technical University "Kharkiv Polytechnic Institute". - Kharkiv: NTU "KhPI", **No. 27** (1249), P. 182-186 (2017)
12. V. G. Sychenko, M. M. Pulin. *Energy-efficient syst. Of DC traction power supply with distributed gener.* Ukr. Railways, **No. 5-6 (47-48)**, P. 28-32 (2017)
13. A. N. Marikin, A. V. Myzintsev *New techn. in the constr. and reconstr. of traction subst.* - Moscow: Marshrut, - 220 p. (2008)
14. B. A. Arzhannikov, I. O. Naboichenko, *The concept of strengthening the syst. of traction power supply of a direct current of 3.0 kV: a monograph,* Yekaterinburg: UrGUPS (2015)
15. Ye. M. Kosarev, *Improvement of the volt. mode in the traction power syst. of el. DC railways.* Dissertation for the degree of candidate of techn. sc.. Dnipro, - 194 p. (2018)
16. P. V. Hubskyi *Increase of en. efficiency of the syst. of traction power supply of direct current with high-speed motion.* Dissertation for the degree of candidate of techn. sc. Dnipro, - 209 p. (2019)
17. N. I. Voropay, V. A. Stennikov, E. A. Barahtenko, *Integrated en. syst. as an innov. dir. of en. future* [Electronic resource]. Access mode: http://isem.irk.ru/energy21/papers/.pdf
18. A. V. Kirilenko, *Intelligent El. Power Syst.: Elements and Modes: Generally.* Institute of Electrodynamics of the National Academy of Sciences of Ukraine. - K.: IED NAN of Ukraine, - 408 p. (2014)
19. A. V. Mohilenko, D. A. Pavlyuchenko, *Multi-resource syst. of power supply: a new paradigm, intelligent techn. solutions and technologies.* Electro, **No. 1.** - P. 2-5 (2016)
20. M. Arnold, G. Andersson, Decomposed el. and natural gas, optimal power flow *//16th Power System Computation Conference,* Glasgow, Scottland, UK, July 26-30, 7 p. (2008)
21. M. Geidl, G. Andersson, *Optimal power flow of multiple en. carriers.* IEEE Transactions is Power Systems. - **Vol. 22. - No. 1** - P. 145-155 (2007)
22. V. Ya. Putilov, R. N. Shulga, *Some techn. and environmental aspects of the use of power supplies in the en. sector.* Electro, **No. 1.** - P. 6-11 (2016)
23. E. Sokol, V. Zamaruiev, S. Kryvosheev and oth., *Specificity of el. en. storage unit application.* IEEE First Ukraine Conference on El. and Computer Engineering (UKRCON). - Kyiv, - P. 432 – 435 (2017)
24. D. Cericola, PW Ruch, R. Ktz at al., *Simul. of a supercapacitor / Li-ion battery hybrid for pulsed applications.* Journal of Power Sources. - **N9.** - P. 2731-2736 (2010)
25. Pedram, Massoud, Naehyuck Ch. Et al, Hybrid el. en. storage syst. In Low-Power El. and Design (ISLPED) (2010)
26. B. O. Styslo, *Peculiarities of the use of el. en. storage devices in the railroad's el. supply syst.* Bulletin of the NTU "KhPI", Series: New solutions in modern technologies. - Kharkiv: NTU "KhPI". - **No. 9 (1285).** - P. 76-82 (2018)
27. M. A. Sokolov, V. S.Tomasov, R. P. Jastrzebski, Sc. and techn. bulletin of inf. techn., mech. and optics. # 4 (92). Comp. analysis of en. storage syst. Determination of optimal areas of app. of modern supermakhoviks. [Electronic resource]. Access mode: https://cyberleninka.ru/article/v/sravnitelnyy-analiz-sistem-zapasaniya-energii-i-opredelenie-optimalnyh-oblastey-primeneniya-sovremennyh-supermahovikov (2014)
28. V. Sychenko, V. Kuznetsov, Ye. Kosariev, P. Hubskyi, V. Belozyorov, V. Zaytsev, M. Pulin, *Devel. of the approach to ensuring the sustainability of the syst. of direct power supply syst.* East European Journal of Adv. Techn. **Iss. 5/2 (95),** - P. 47-57 (2018)
29. L. W. Yao, J. A. Aziz, *Modeling of a Lithium Ion Battery with Nonlinear Transmission Resistance,* Applied Power Electronics Colloquium (IAPEC), IEEE, - P. 104 - 109 (2011)
30. V. V. Zamaruev, B. O. Styslo, Ye. M. Kosarev, *Impr. the quality of el. en. in the syst. of power supply of rail transport by using el. en. storage devices.* Bulletin of the National Technical University "KhPI", **No. 27 (1249),** - P. 360 - 364 (2017)

The increasing of energy efficiency of international freight truck trains under the concept of energy innovative technologies development motor transport

Tatiana Kostiuk*, and *Ramasan* Habutdinov

National Transport University, Kiev, Ukraine

Abstract. At the motor transport (MT) the concept of technological and innovative energy saving is actual. However, the existing method of increasing the cost-effective efficiency of motor transport does not correspond to this concept because the basis of the theory of the motor transport process consists of the axioms, which neglected: a) the fragmenting effect of existing technological paradoxes on the integrity of knowledge about motor transport as a sphere of material production; b) the resource, production and technological essence of this process; c) innovative variability of parameters of new cars and transport technologies. In addition, the basis of the fundamental designed scheme of the transport operation is based on simplified mathematical description of the virtual transposition operations. These methodological shortcomings do not allow to analyze and substantiate innovative technical and technological projects of motor transport. For the possibility of realization of the actual concept of technological and innovative energy saving on MT, in NTU the theory and methods of a comprehensive increase of energy efficiency of a car (EREC) of a general type, as well as methods of simulation and test formation of energy saving of motor transport technologies were designed.

1 Formulation of the problem

The main concept of the development of motor transport and motor transport system (MTS) is to ensure the technological development of their components on the basis of the conceptual idea of innovative operational and technological energy- and resource-saving (IOTERS) in all transport enterprises. In connection to this, the system methodology IOTERS is actual, on the basis of which it would be possible to form mutually coherent and harmonized methods of ensuring IOTERS in three main components of motor transport: the controlling substructure of transport (CST), resource and technological base (RTB), transport and technological processes (TTP). Due to the fact that existing theories of economics of the organization of transport processes have significant drawbacks, and most importantly, it is non-technological, fragmentary, non-technical, non-technological, and does not take into account important processes energy transformation of resources into a tangible product of transport, so the theory of energy and energy type was developed. An epistemological obstacle to the coordination and harmonization of IOTERS methods for the mentioned components of motor transport is the methodological contradiction between the knowledge bases of specialists of the CST: organizers, transport managers and engineers [1]. This contradiction in many respects is stipulated due to the objective manifestations of six technological paradoxes (TP) of transport, as spheres of material The subset of TP paradoxes of vehicles is written as follows:

$$TP \in (TP1, TP2, TP3, TP4, TP5, TP6) \quad (1)$$

where TP1- a paradox of the form of transport product (product is material, but not real); TP2-paradox of the method of product realization (synchronous creation and consumption); TP3 is a paradox of the form of spatial-point dislocation of transport operations (transport operations are distributed in trajectory space of road network infrastructure, and terminal operations are carried out at certain points of the trajectory-in terminals); TP4- paradox of machine procedures of transport technologies (indirect and reactive character of machine actions on the subject of transportation; at the same intensity levels, these actions can be both productive-useful and trajectory-dangerous depending on the type of locally trajectory situations); TP5 is a substantive and valuable paradox of the processes of transformation of transport resources and transport production (the substance of resources is not transferred, and the cost of resources is transferred to the product). It should be noted that the fifth paradox violates the adequacy of the ratio of such product characteristics as "price-quantity", "price-quality." These ratios depend on the operational and technological conditions of transportation, which are not taken into account in the organizational theory of the transport process [1, 3].

* Corresponding author: tani4ka.kostyuk@gmail.com

https://doi.org/10.1051/matecconf/201929401015

TP6- a paradox of the form of the use of regime resources (the regime resources are converted into energy-driven time of the motor vehicle (MV)) and in this form is an important parameter of the transport operation and the physical product of transport.

The time of movement of MV on the local sections of the route of transportation is inversely proportional to the power function of the MV engine. It should be noted that this dependence is also not taken into account in the methods of organization of transportation [3]. Objective manifestations of paradoxes, on the one hand, lead to significant differences in the process of transport production from other spheres of material production.

On the other hand, the paradoxes lead to the discontinuity of the knowledge of different specialists of the CST. Firstly, organizers and economists use the simplified, techno-empirical and cost-effective methodology of transport services, which is based on the principle of unchangeability of technology and technology. Secondly, in the methodology the laws of the transposition process are considered but not transport. Thirdly, the properties of technical resources and the processes of their transformation into a transport product are ignored. In connection with the stated above, the existing methodology of transport services is: techno-empirical, non-technological, non-resourceful and anti-innovation (the principle of unchangeability of technology and technology). As a result the FUT (freezing undescribed technology) is used. It does not allow to realize the concept of innovation and technological development of transport processes and vehicles [3] and new DDT (described developed technology) principle [4], [5]. The indicated shortcomings of the existing methodology lead to the most negative consequences for the domestic transport companies which are operating in the field of international transportation. In the international road transport services markets the requirements for future energy efficiency and technical and technological competitiveness of transport proposals are more stringent.

2 Objectives and goals

The main goal and objective of the article is the analysis of transport energy efficiency of international freight truck trains with detailed consideration of the influence of structural, technical, road, operational factors of the transport process. The article presents the method of simulation analysis of energy efficiency of international road trains as resource - technical means of transport production and forecasting their suitability for energy saving technologies. This technique allows simulating the energy efficiency of the TTP and is an important component of the IOTERS methodology.

3 Main Part

The researched works concerning efficiency on transport are actual and taken into account [3], [6], [8] however in the theory of transport process only the cost-effective efficiency is considered, without taking into account technical and technological innovations. In addition, it should be pointed out that motor transport is fundamentally different from other modes of transport, due to suchfactors as high structural diversity, a wide range of changes in road and terminal conditions of transportation, resizing party masses and trajectory-conflict infrastructure. In works [1,2,7] it was established that the tasks of innovation and technological development of motor transport need to be solved on the basis of a new paradigm of the theory of management on transport - material, productive and innovative. The above papers present the main provisions of this new paradigm, as well as four corresponding conclusions. Firstly, the epistemological basis for the development of the IOTERS methodology, taking into account the subset of technological paradoxes (1) is the theory of energy-efficiency efficiency of the MT (EEMT) of the generalized (modular) type [4]. Secondly, the usefulness of the TTP for consumers of transport services is the result of the mass implementation of the processes of technological transportation of goods and passengers (TTGP). Each such process is a consequence of the driven process of ERW- the process of transformation technologic resources of transport into the physical transport product that is created under the given conditions. Thirdly, the tasks of innovative technological development of the TTP should be solved on the basis of the method of benchmarking - comparative analysis and forecasting of transport efficiency of new MV in test operations [4]. Fourthly, in order to create energy-saving transport technologies (TTs), it is necessary to solve the tasks of increasing energy efficiency of transportation: new MV, procedures of machine operations, processes of energy conversion MV and ERW.

The indicated tasks can not be solved using the known performance indicators of MV productivity and the cost of transportation due to the above mentioned methodological disadvantages [3]. Mathematical models of these indicators are based on the principle of invariance of technologies and on the calculation scheme of the transposition operation [3]. In regard to this, the indicated indicators are unsuitable for the improvement of transport technologies, and the model of the cost of transportation is anti-innovation. In work [4] it was established that the idea of improving transport technologies is a gradual increase in the value of the energy efficiency indicator (in a specific case - fuel consumption indicator) of transportation and this this the new approach in transport technology development of the motor transportation. To realize this idea, when substantiating the new rolling stock, the dimensionless indicator of the transport energy efficiency of the MV is used. Its value is determined by the formula

$$Ie = \frac{C_{ht}}{R_{Ie}} \tag{2}$$

where C_{ht} - heat transfer coefficient of the new MV in the calculation route, which takes into account the energy efficiency of the new MV in test operations (urban, trunk, mixed and benchmark), the estimated share of MV in urban conditions and distribution of MV runway according to road categories [4].

R_{le} - the reference value of the index of energy efficiency of a car.

On the basis of formula (2) the following calculation formula is received:

$$I_e = \frac{c_{ascr} \cdot \gamma_{st}}{c_{fm}(c_{rw} + \gamma_{st})} \quad (3)$$

where c_{ascr} - the dimensionless coefficient of average speed of MV on the calculated route:

$$c_{ascr} = \frac{S_a}{S_c} \quad (4)$$

where S_a – average speed of the MV on the calculated route;

S_c – reference (constant) speed;

γ_{st} – coefficient of static use of load-carrying capacity;

c_{fm} – dimensionless fuel coefficient of mileage of the MV;

c_{rw} – coefficient of running weight;

$$c_{fm} = \frac{f_{ce}}{f_{cr}} \quad (5)$$

where f_{ce} - the amount of estimated fuel consumption of the MV in the calculated route, f_{cr} - reference fuel consumption.

Calculated formulas for determining the coefficients speed of c_{sto} and mileage c_{fmto} in test operations are as follows:

$$c_{sto} = \frac{a_1(a_2 - a_3 \cdot \psi_a)}{a_4 \cdot t_d + a_5} \quad (6)$$

where $a_1, a_2, a_3, a_4 a_5$ – constant parameters, which are determined for test transport operations, taking into account the provided conditions of the MV movement and its constructive parameters;

ψ_a - average value of the resistance of the road,

t_d -time of dispersal of the MV to 60 km / h. (in urban cycle) and up to 80 km / h. (in the trunk cycle), s.

The value of the speed factor of the MV on the calculated route is determined by the formula:

$$c_{scr} = c_{suc} \cdot t_{uc} + c_{stc} \cdot (1 - t_{uc}) \quad (7)$$

where t_{uc} -share of time of work of MV in urban conditions;

c_{suc} and c_{stc} - the value of the coefficient for urban and trunk cycles, respectively.

The value of the coefficient of fuel mileage in test operations is determined as follows:

$$c_{fmto} = \frac{b_1 \cdot (b_2 \cdot \psi + b_3) \cdot c_{rt} \cdot F \cdot \rho}{b_4 \cdot C \cdot (\gamma_{st} + c_{rw}) + b_5 \cdot H_{mv} \cdot W_{mv}} \quad (8)$$

where b_1, b_2, b_3, b_4, b_5 - constant parameters, which are determined for test transport operations, taking into account the operating conditions of the MV and its constructive parameters.

ψ - value of the resistance of the road,

c_{rt} - coefficient of road traffic conditions;

F - the total fuel consumption rate per 100 km of dispersal;

ρ -is the volume mass of fuel (ρ = 0,76 for gasoline; ρ = 0,84 for diesel fuel);

C - load-carrying capacity of the MV, t;

γ_{st} - coefficient of static use of load-carrying capacity;

c_{rw} - coefficient of running weight;

H_{mv} and W_{mv} - height and width of the MV, m;

Fuel mileage factor on the calculated route:

$$c_{fmcr} = c_{fmuc} \cdot t_{uc} + c_{fmtc} \cdot (1 - t_{uc}) \quad (9)$$

where c_{fmuc} and c_{fmtc} – the value of the coefficient for urban and trunk cycles;

t_{uc} - share of time of ATZ in urban conditions.

We will analyze how the change of such constructive parameters of the road train as the wheel radius and the engine power will affect on the energy efficiency index . Our task is to choose from the existing range of cars models that ones, that will best meet our criteria and will be the most economical and this is the practical part of this theory.

Based on the calculated data, the diagrams of the dependences how the changes in wheel radius and engine power are influenced on the energy efficiency of the MT, such dependencies are presented in diagrams Fig. 1, 2. For existing freight truck trains, the power range changes from 360 to 600 kW, wheel radius from 0.4 to 0.6m according to the car's specifications.

On the basis of the multivariate calculation it was established that at a radius of 0.55 m, the energy efficiency indicator (I_e) reaches the maximum value, and when the engine power increases, the energy efficiency indicator decreases; the most optimal variants are with a power of 360 kW. The quantitative characteristics of these laws are obtained.

On the basis of comparing the values of the energy efficiency indicator in Figures 1 and 2 it can be concluded that freight truck trains with a radius of wheel of 0.55 m are more in line with the IOTERS concept, since the energy efficiency indicatior is higher by 19.5 -

21.8%, with the increase in engine power energy efficiency decreases by 18.1-25.3%.

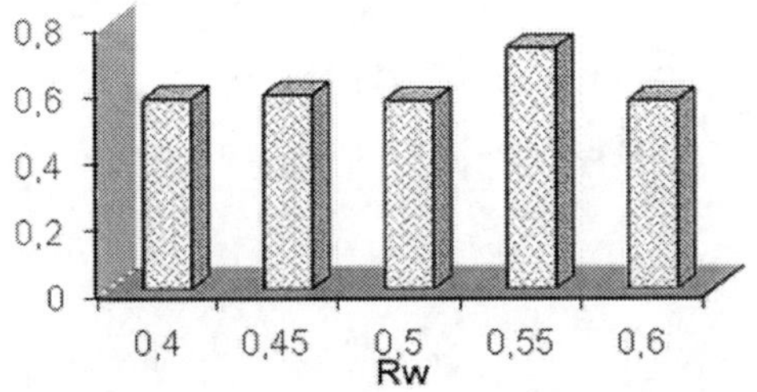

Fig. 1 - Diagram of the dependence of the energy efficiency index of the RT on the calculated route from the change in the radius of the wheel.

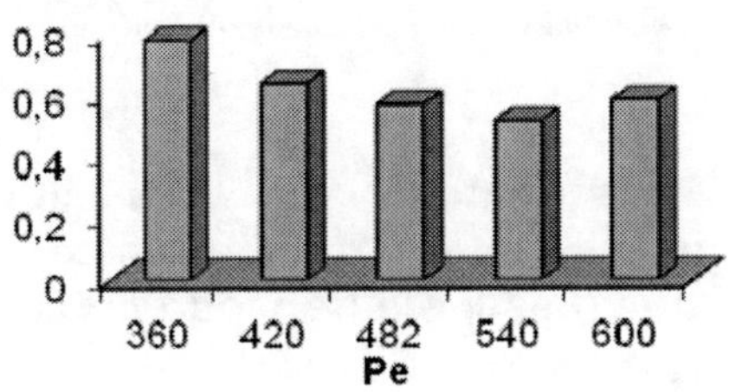

Fig. 2 - Diagram of the dependence of the energy efficiency index of the RT on the calculated route from the change in engine power.

Based on the conducted analysis, it is determined that, in contrast of calculations in the theory of the automobile, this technique takes into account several resource and technical properties (the car as a complex machine, the object of adaptive motion control in a multi-phase test operation, the car as a means of transport, etc., mathematical models adaptivelydiscrete kinematics and power engineering of the vehicle with variable parameters). Therefore, in practice, the desirability of operational choice and substantiation of the main parameters of freight truck trains taking into account the conditions of transportation is important.

3 Conclusions

1. It was discovered that the provision of technological development of motor transport is based on the conceptual idea of innovative operational and technological energy- and resource-saving (IOTERS) in all transport enterprises, because the existing methods do not allow to solve the problem of technological innovative development according to the concept of energy saving on transport

2. t was revealed that for the implementation of the IOTERS concept, which allows a detailed account of technical and technological innovations, it is necessary to use mathematical models of the theory of energy-resource efficiency of automobile.

3. The complex of mathematical models for the simulation and benchmarking analysis of energy efficiency of new MVs has been formed, which will enable the implementation of the concept of operational energy saving in all transport enterprises.

4. The technical characteristics of road truck trains are analyzed and quantitative characteristics of the influence of various technical factors on energy efficiency indicators are obtained.

5. The methodology of simulation and reference-comparative analysis of energy efficiency of new freight truck trains as resource and technical means of transport production is proposed. As a result, this methodology can be used at transport enterprises in order to improve energy efficiency of transportation and the introduction of new technologies, since it allows to select from the existing range of car models the ones that will best meet the theory of energy saving and will allow the introduction of innovations to improve the efficiency of freight transportation and the development of the transport enterprises at all.

References

1. R. Khabutdinov *Transport technological paradigm and methodology of novation management of motor-car transportations.* K.: Visnyk NTU, vol.**24**,part.2.,p. 237-240 (2011)
2. R. Khabutdinov *Methodological bases of transport-technological energology*, Scientific papers “Problem of transport”. K.: NTU, Vol. **3**. P 164–168 (2006)
3. A.Vorkut *Road transport of goods.* K.: High-school, 447 p. (1986)
4. R. Khabutdinov, A. Kotsyк *Energy-resourse efficiency of car.* K.: UTU (1997)
5. R. Khabutdinov *Systemic problem of innovative development of motor transport in conditions of theoretical denial of transport technologies and production.* K.: Visnyk NTU, vol. **34**, p. 491 -496 (2016)
6. H. Kravchenko *The essence of the economic category of efficiency and definition in transport.* D: The problems of transport economy, DNU, vol. **8**, (2014)
7. R. Khabutdinov *The methodology of conceptual and innovative management of technological development of vehicles.* K.: Visnyk NTU, vol. **29**, p. 409 -414 (2014)
8. P. Wang, R. Goverde *Multi-train trajectory optimization for energy efficiency and delay recovery on single-track railway lines*: Transportation Research Part B Methodological **105**, pp. 340-361 (2017) https://doi.org/10.1016/j.trb.2017.09.012

Analysis of emission tests of electromagnetic disturbances in diesel-electric locomotives

Andrzej Białoń, *Kamil* Białek*, and *Patryk* Wetoszka

Railway Research Institute, 50 Chlopickiego Jozefa Street, Poland

Abstract. This paper discussed the measurement of radiated emissions of electromagnetic disturbances generated by diesel - electric locomotive system. The measurement methodology, permissible values of a magnetic and electric field according to EN 50121-3-1 standard during driving and standstill for electric - diesel locomotive are presented. The article presents examples of test results and an assessment of the obtained measurement results in the electric and diesel traction according to the applicable requirements.

1 Introduction

In the railway area there are intentional and unintentional (stationary and non-stationary) sources of electromagnetic radiation that can interfere with the operation of electrical and electronic systems and devices. In the railway electromagnetic environment we distinguish the following sources of electromagnetic interference:

• stationary - derived from the traction network, power network supplying traction substations and the power supply line for railway traffic control systems;

• moving - disturbances generated by traction vehicles (electric traction units, electric and diesel locomotives).

Modern rail vehicles are equipped with devices, electrical and electronic systems that can be a potential source of interference, ie traction inverters, static converters, asynchronous motors, air conditioning systems, passenger information, safety (European Train Control System - ETCS), automatic braking of the train SHP, dead man's active equipment, communication devices (eg GSM-R, VHF radiotelephones), etc. The devices and systems used on rail vehicles are miniaturized, limited power consumption and high reliability of operation [3,4]. The introduction of these design constraints is associated with an increase in the probability of the level of disturbances generated by traction vehicles. Therefore, such an important problem is the continuous analysis of the state of the electromagnetic environment in the aspect of the intensity of the magnetic and electric field generated by rail vehicles moving along the railway route (electromagnetic environment diagnostics).

2 Measurement of emission of electromagnetic disturbances of radiated diesel-electric locomotive

2.1 Measurement methodology

The tests of electromagnetic compatibility of radiated radiation emissions are carried out in accordance with EN 50121-3-1 [1] standard in the frequency band of 150 kHz ÷ 1 GHz divided into two subband:

- measurement of the magnetic component of the field strength (the band 150 kHz ÷ 30 MHz),
- measurement of the electrical component of the field strength (the band 30 MHz ÷ 1 GHz).

To measure the magnetic field strength component, a frame (loop) antenna is used, while for measuring the electrical component a double-cone antenna and a log-and-periodic antenna. Figure 1 shows the permissible magnetic and electric field strengths for a traction vehicle (eg a diesel-electric locomotive) [4].

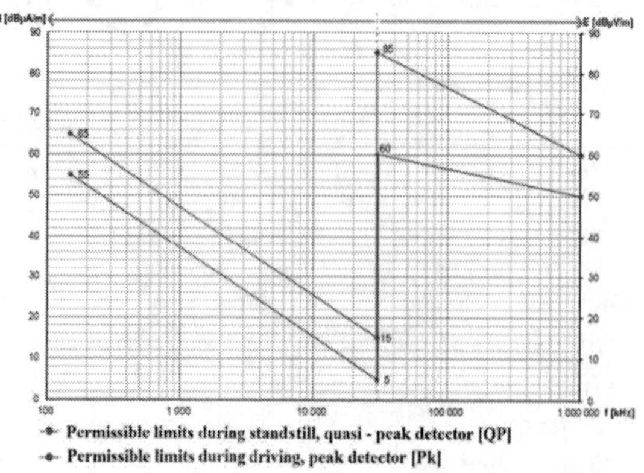

Fig. 1. Values for radiated disturbances for a traction vehicle (eg diesel and electric locomotive) [4].

* Corresponding author: kbialek@ikolej.pl

The measurements of the electromagnetic field intensity are carried out on the Experimental Track of the Railway Institute in Żmigród, on the railway track or standstill tracks, where the requirements of EN 55016-4-2 [2] (no trees, substations, transformers and railway traffic control equipment) within a radius of 30m are met. Other traction vehicle should not be on the same power section at a distance of at least 1km from the tested locomotive. Before testing radiation disturbances generated by rail vehicles, should be measured the background of the electromagnetic field environment in the frequency ranges given in the table below - Table. 1.

Table 1. Frequency range for background and driving mode [1]

Subrange [Hz]	Bandwidth [kHz]	Sweep time [ms]
150 k ÷ 1,15M	9 lub 10	37
1 M ÷ 11 M	9 lub 10	370
10 M ÷ 20 M	9 lub 10	370
20 M ÷ 30 M	9 lub 10	370
30 M ÷ 230 M	100 lub 120	42
200 M ÷500 M	100 lub 120	63
500 M ÷ 1 G	100 lub 120	100

The level of external disturbances - produced by other sources of the electromagnetic field should be lower by at least 6 dB from the permissible levels (Figure 1). If the measurement of the background electromagnetic field strength exceeds -6 dB for a frequency band, then this range should not be taken into account in the assessment of the results [1, 3, 6].

The diesel - electric locomotive is tested in two states: during standstill and while driving at a speed (50 ± 10) km / h. During this time all systems must be activated (air conditioning system, ETCS European Train Control System, passenger information system, communication devices, etc.), internal and external lighting on the vehicle. In the case of tests at a standstill, the measuring point should be determined by means of a test run as follows: the vehicle moves at a speed of about 10 km / h in front of the antennas in order to read the field strength value at the receiver. After the test run, the locomotive is placed in the place where the highest level of radiated disturbances is generated. The measurement is performed with a quasi-peak detector [3,4,8,9].

The tests must be carried out while driving a locomotive as electric and diesel. In the case of testing the locomotive while driving, the vehicle should accelerate or slow using 1/3 of the maximum traction in a given speed range. The measurement is performed with a quasi-peak detector. The measuring antennas for the above mentioned operating modes are set at a distance of 10 m from the axis of the track and are placed at different heights depending on the type of antenna used:
- frame - 1 m ÷ 2 m above the rail head,
- double conical and logarithmic - periodic 2.5 m ÷ 3.5 m above the rail head.

The height of the antenna is selected so as to obtain the maximum level of emitted electromagnetic field disturbances. Measurements should be made for the polarization of the vertical and horizontal antennas (V-vertical and H-horizontal). After completing the tests while the vehicle is stationary and while driving, the level of electromagnetic disturbances - BACKGROUND. Examples of measuring stations are shown in Figures 2, 3 and 4.

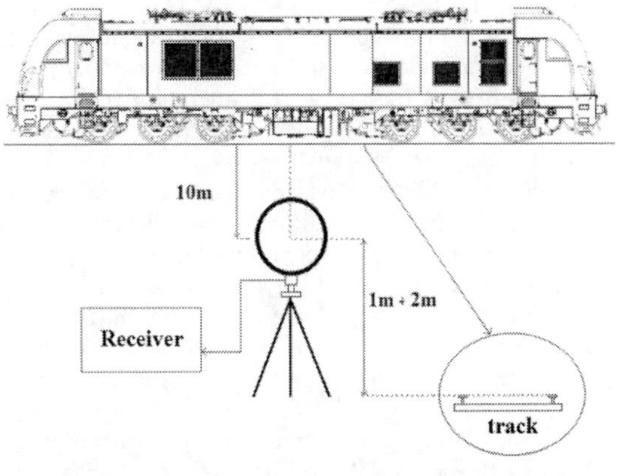

Fig. 2. Measurement of emission of radiated disturbances in the frequency band of 150 kHz ÷ 30 MHz with the use of a frame antenna.

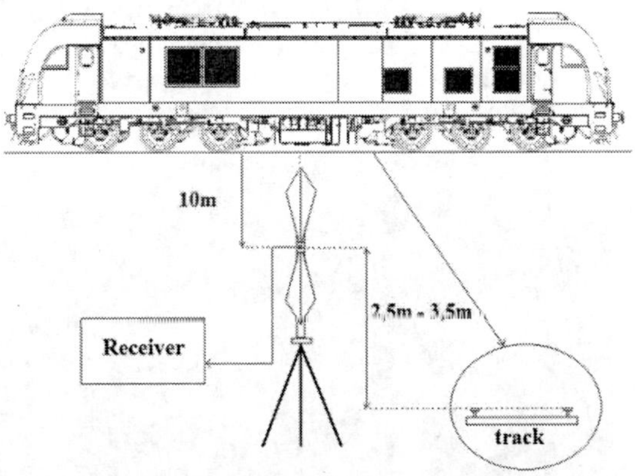

Fig. 3. Measurement of radiated emission in the 30MHz ÷ 230 MHz frequency band using a double-cone antenna.

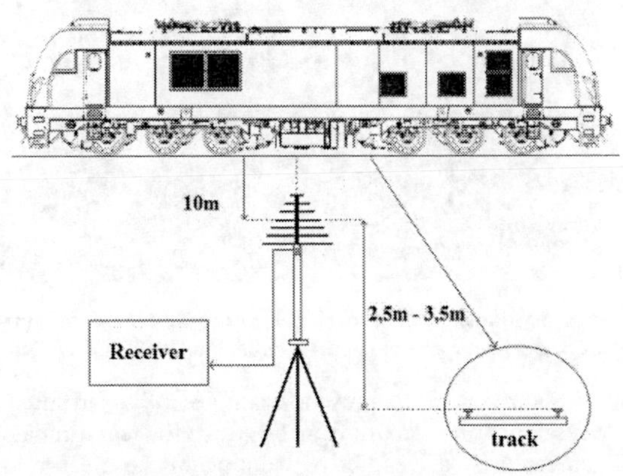

Fig. 4. Measurement of emission of radiated disturbances in the frequency band 230 MHz ÷ 1 GHz with the use of a logarithmic-periodic antenna.

2.2 Results of measurements of disturbances of radiated diesel and electric locomotive

The following are examples of the results of an exploration of an electric and diesel locomotive made on the Experimental District of the Railway Institute for a locomotive working in the electric and diesel traction. In the case of electric traction, the traction vehicle was powered with 3kV DC.

During the research on the locomotive all electrical, electronic and electrotechnical systems were included, which are used during normal operation (they take a certain current and generate conducted or radiated disturbances) [7]. The auxiliary systems were powered by means of a static converter. The measurements were made using a receiver and three measurement antennas: frame, double-cone and logarithmic - two-way for two modes of operation: standstill and driving. Tests using a double-cone and logarithmic antenna were made for both antenna polarization: vertical and horizontal. During the test standstill measuring antennas it was opposite the static converter and the locomotive inverter. In the case of testing during driving, the vehicle is traveling at a speed of 50 km / h at 1/3 of the maximum tractive force in a given speed range [1, 5, 9, 11].

Figure 6 shows a double-cone antenna with a mast during the measurement of emission of electromagnetic environment disturbances (BACKGROUND - also referred to in the literature on electromagnetic compatibility electrosmog). The background during the tests was measured before and after the measurements.

Fig. 5. Frame antenna during measurement of emission of electromagnetic environment disturbances (BACKGROUND).

Figures 6 to 14 present examples of the results of emission of electromagnetic disturbances radiated generated by a locomotive working in the electric and diesel traction in the frequency bands defined by the EN 50121-3-1 standard [1] and presented in Table 1.

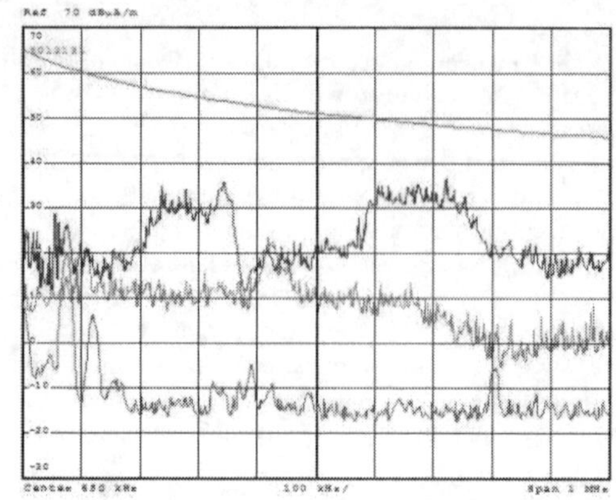

Background,
Driving 50 km/h - diesel traction,
Driving 50 km/h - electric traction

Fig. 6. The value of the emission of magnetic disturbances while driving, the band 150 kHz - 1.15 MHz.

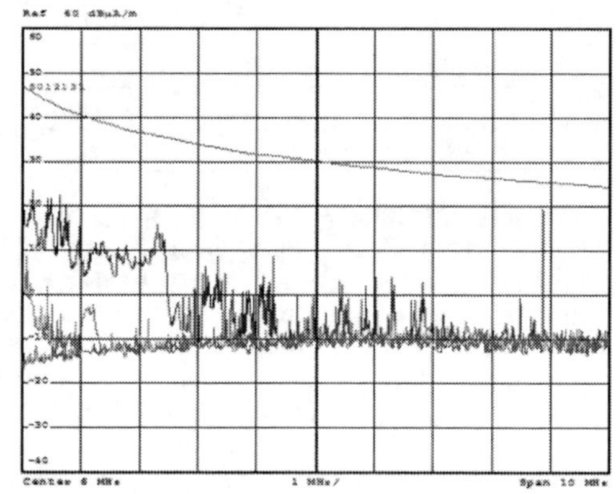

Background,
Driving 50 km/h - diesel traction,
Driving 50 km/h - electric traction

Fig. 7. The value of the emission of magnetic disturbances while driving, the band 1 MHz – 11 MHz.

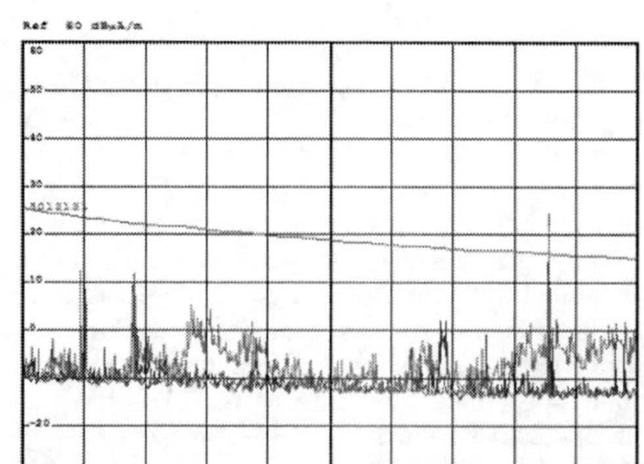

Background,
Driving 50 km/h - diesel traction,
Driving 50 km/h - electric traction

Fig. 8. The value of the emission of magnetic disturbances while driving, the band 10 MHz – 30 MHz.

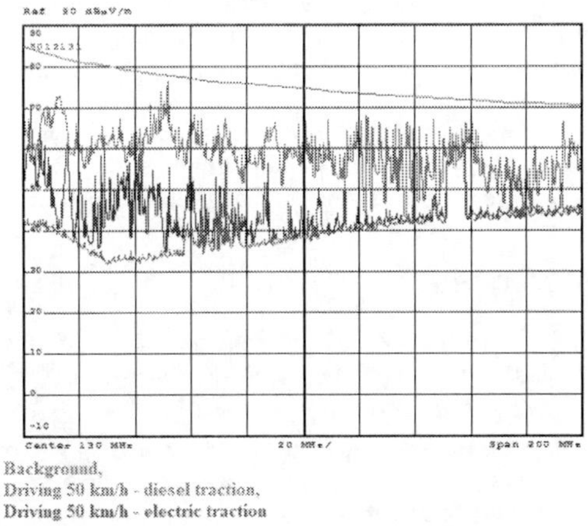

Fig. 9. The value of emission of electrical disturbances for vertical polarization of the antenna while driving, the band 30 MHz - 230 MHz.

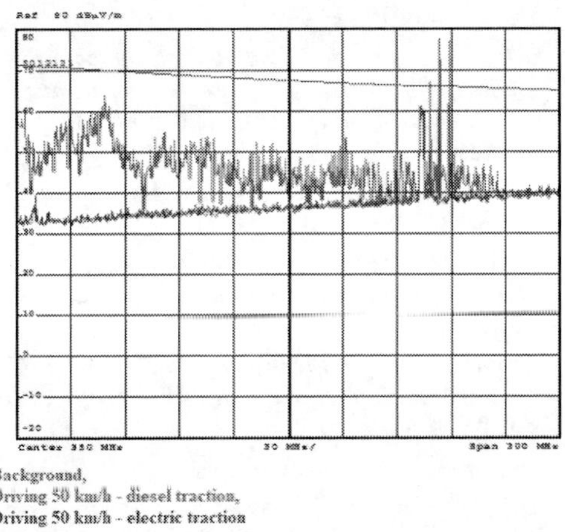

Fig. 10. The value of emission of electrical disturbances for vertical polarization of the antenna while driving, the band 200 MHz - 500 MHz.

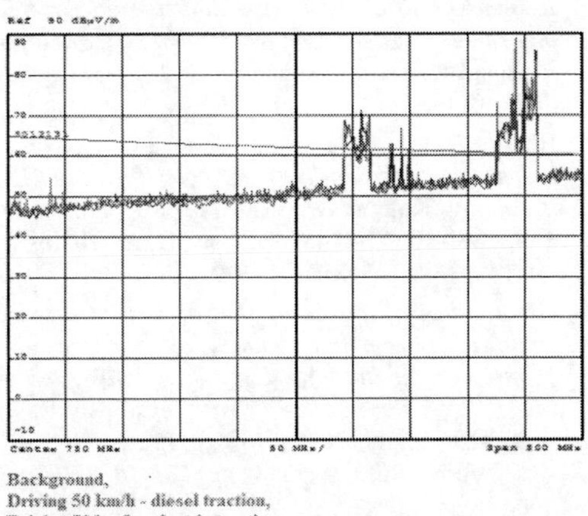

Fig. 11. The value of emission of electrical disturbances for vertical polarization of the antenna while driving, the band 500 MHz – 1 GHz.

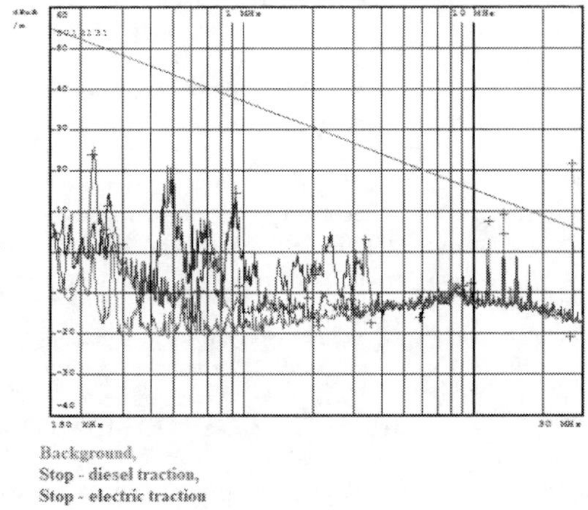

Fig. 12. The value of the emission of magnetic disturbances at a standstill, the band 150 kHz - 30 MHz.

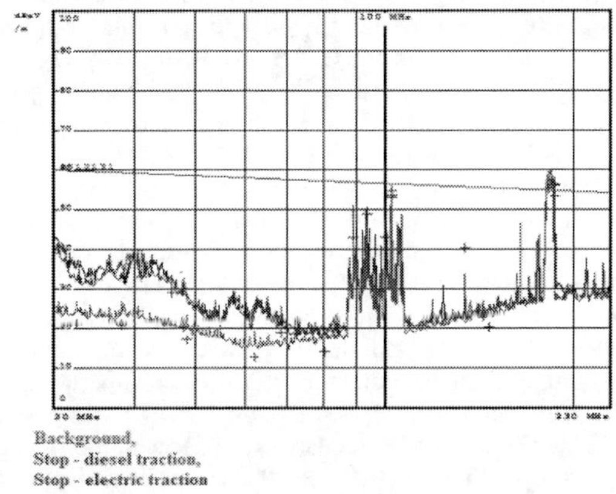

Fig. 13. The value of emission of electrical disturbances at a standstill for the vertical polarization of the antenna, the band 30 MHz - 230 MHz.

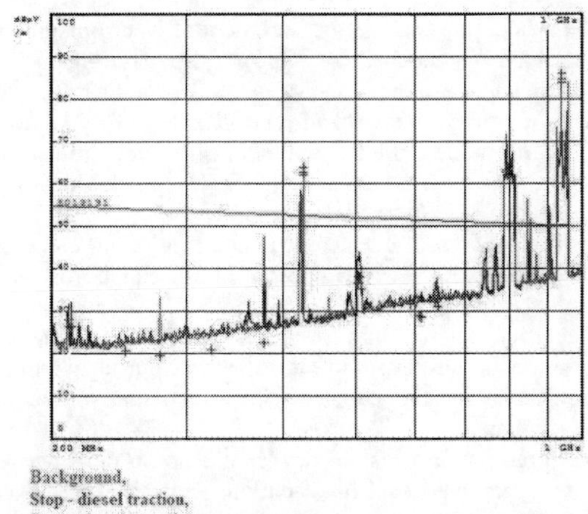

Fig. 14. The value of emission of electrical disturbances at a standstill for the vertical polarization of the antenna, the band 200 MHz – 1 GHz.

The presented measurement results take into account the applied measurement path in the electromagnetic field measurement system, the amplification / attenuation coefficients of the antennas and the attenuation of the cable as a function of frequency.

The presented results show that a locomotive working in the diesel traction system generated a higher level of electromagnetic disturbances. Measurements have shown that the source of elevated levels of disturbances in relation to the locomotive working in the electric traction is the generator set installed on the locomotive.

During the tests, values exceeding the permissible emission levels of radiated disturbances determined according to EN 50121-3-1 [1] were measured. The exceedances occurred on frequencies 427 MHz and in the bands respectively equal to: 790 MHz - 820 MHz and 925MHz - 970 MHz for two modes of operation: driving and standstill.

During the emission tests of radiated disturbances, the exceeding of the limit value on the 27 MHz frequency was also recorded in the driving and standstill. The source of exceedances is the balis antenna designed for communication in the ETCS vehicle-track system [9, 10, 11]. This overrun can be omitted as originating from a source intentionally placed in the railway system, which is in accordance with EN 50121-3-1 [1].

Pursuant to the recommendations of EN 50121-3-1 [1], the above ranges have not been evaluated because the values of external electromagnetic disturbances (Background) exceed the permissible level - 6 dB. The source of measured electromagnetic fields that reach high values are radiolocation and radio stations, navigation and telecommunications devices operating in the above frequency bands.

There were no exceedances of acceptable levels according to EN 50121-3-1 [1], the source of which is an electric-diesel locomotive.

3 Summary

The article presents the measurement of electromagnetic disturbances generated by a locomotive working in the electric and diesel traction. Measurements are carried out in the frequency range from 150 kHz ÷ 1 GHz for two operating modes: driving and standstill with the use of three measurement antennas: frame (loop), double-cone and logarithmic-periodic.

The sources of interference occurring on an electric-diesel locomotive with impulse control can be divided into two groups:

• external source that produces interference due to violation of uniform current collection during slipping pantograph on the contact wire - external source of interference,

• internal sources inside the vehicle body, i.e. traction motors, impulse systems, auxiliary equipment, power generator in the case of a locomotive operating in the traction system.

The locomotive working in the diesel traction generated a higher level of electromagnetic disturbances, because worked additional power generator on the vehicle. The measurements carried out allow to state that that traction vehicles have large an effect on the railway electromagnetic environment. The level of the intensity of the electromagnetic field generated by vehicles moving along the railway routes depends on many factors, e.g. vehicle type, pantograph structure, power system, energy conversion method, vehicle speed, etc.

References

1. EN 50121-3-1:2016 *Railway applications - Electromagnetic compatibility - Part 3-1: Rolling stock - Train and complete vehicle*
2. EN 55016-4-2:2011 *Specification for radio disturbance and immunity measuring apparatus and methods - Part 4-2: Uncertainties, statistics and limit modelling - Measurement instrumentation uncertainty*
3. Charoy A., *Kompatybilność elektromagnetyczna-zakłócenia w urządzeniach elektronicznych*, tom 1-3, Warszawa 1999-2000
4. Białek K., John Ł.: *Pomiary emisji zaburzeń od kolejowych pojazdów szynowych pracujących w trakcji wielokrotnej*, Przegląd elektrotechniczny; 03/2019 Str. 14; 00-950 Warszawa
5. Williams T., *EMC for product designer*, Nedition Newnes (1992)
6. Mark I. Montrose, Edward M. Nakauchi *Testing for EMC Compliance: Approaches and Technigues*, IEEE (2014).
7. Donald G. *Baker Electromagnetic Compatibility – Analysis and Case Studies in Transportation*, WILEY & Sons (2016).
8. Laskowski M., Dłużniewski A., John Ł.: *Problematyka generacji zaburzeń radioelektrycznych oraz ich pomiary w środowisku kolejowym*, Przegląd Telekomunikacyjny, nr **11** (2013).
9. Białoń A., Dłużniewski A., John Ł.: *Kompatybilność elektromagnetyczna taboru miejskiego i kolejowego – wymagania normatywne*, Przegląd Telekomunikacyjny, nr **10/2011**, str. 767-768 (2011).
10. Białoń A., Gradowski P., Toruń A.: *Nowoczesny system zarządzania ruchem kolejowym (ERTMS)*, Problemy Kolejnictwa Railway Reports, issue nr **148**, ISSN 0552-2145, Railway Research Institute, pp. 140, 144, 148-152 (2009).
11. Gradowski P.: *Upgrading the railway infrastructure technical parameters using the example of the Control-Command and Signalling subsystem with the EC verification certificate*, Railway Reports, issue **182** (March 2019), ISSN 0552-2145, Railway Research Institute, Warsaw, pp. 131-146 (2019).

Analysis of the exhaust emissions from hybrid vehicle during RDE test

Andrzej Ziolkowski[1,*], *Pawel* Daszkiewicz[1], *Lukasz* Rymaniak[1], *Paweł* Fuc[1], and *Pawel* Ukleja[2]

[1] Poznan University of Technology, Faculty of Transport Engineering , 60-965 Poznan, ul. Piotrowo 3, Poland
[2] Lexus Poznan, 61-961 Poznan, ul. Lechicka 7, Poland

Abstract. The introduction of the Euro 6c emission norm in 2017 resulted in a change of the type approval procedures. The most important of these was the replacement of the NEDC test with WLTC test with different procedures. In addition, the research was extended to include emission tests in real operating conditions (RDE). Such tests are enforced for heavy vehicles since 2014. PEMS apparatus was used for the measurements, which has been used by many research and development centers to carry out exhaust emissions measurements long before the applicable procedures were introduced. The article presents the methodology of conducting RDE measurements in accordance with the requirements defined by the Euro 6c norm. The focus in this case was primarily on the selection of the test route in the Poznań agglomeration. After determining its course, RDE measurements were made for a vehicle with a hybrid drive. The test route parameters have complied with the applicable requirements. All requirements were met and it was possible to analyze the exhaust emissions. The main focus being the determination of exhaust emissions from the entire test and in accordance with the EMROAD method. The CF (Comformity Factor) coefficients were also determined for both methods and compared with the applicable legal values. An analysis of the hybrid drive system operation was carried out, defining the degree of hybridization, which is the portion of the vehicle travel using only the electric motor of the drive system throughout the whole test. All analyzes were carried out in individual sections of the test: urban, rural and motorway.

1 Introduction

Increasing the efficiency and reducing the fuel consumption of drive systems is a complex multifaceted process that every car vehicle manufacturer performs according to an independently developed and adopted strategy [1-5]. This often requires major structural changes to be made both to the engine itself and to its accessories. Due to the cyclical changes in regulations regarding exhaust emissions, manufacturers are obliged to adapt their products to the limits as given. These regulations are the main factor determining the development of engines and propulsion systems. In terms of efficiency, the average individual CO_2 emissions are of key importance. It is the reduction in the consumption of conventional energy sources (primarily gasoline and diesel) by the road transport sector that is currently one of the main challenges awaiting manufacturers of motor vehicles. The implementation of the Euro 6c emissions norm for PC and LDV category vehicles and the introduction of the Euro VI standard for HDV (Heavy Duty Vehicle) vehicles required the use of an extensive exhaust aftertreatment system consisting of a series of catalytic converters and a diesel particulate filter. Each of these elements generates power losses resulting from the internal resistance of the exhaust gas flow, thereby increasing the share of energy losses in the exhaust system, which also increases the fuel consumption of the engine [6-10].

In the long-term, the constructors assume that it will be possible to replace the classic internal combustion engine with a fuel cell powered by hydrogen, which will become the main source of energy. It is also one of the main directions of drive systems development. Its large scale implementation now faces a number of difficulties for passenger cars, however, and there are only a few single vehicle models that are powered by hydrogen on the market. The hydrogen refueling infrastructure is also not developed. Electric vehicles using classic energy storage systems are now more popular. In their case, the most troublesome aspect is the time needed for recharging, which can last up to several hours. For this reason, among the driving alternatives, vehicles with hybrid propulsion systems are most often selected, characterized by cooperation of at least two energy sources, at least one of which must be a secondary source. The term "primary source of energy" should be understood as a source with a possibly constant expenditure, supplying energy to the system independently of the changes taking place in this system. It can be a thermal engine, a thermal motor coupled with

* Corresponding author: andrzej.j.ziolkowski@put.poznan.pl

a power generator or an electrical network. The original source of energy is characterized by irreversible work. The term "secondary energy source" (energy accumulator) is understood as a device that accumulates part of the primary energy as well as the kinetic energy of the vehicle when braking and supplies the power gathered at moments of peak external loads. It can be an electrochemical, hydraulic or mechanical battery. In work [11] was performed a preliminary study of the physicochemical properties of hybrid organic–inorganic gel electrolytes for Li-ion batteries. The secondary energy source is characterized by reversible operation. A hybrid vehicle is one in which propulsion energy can, in given operating conditions, be provided from two or more kinds or types of sources, or converters of energy, of which at least one source or transducer must be present in the vehicle [12]. The article presents the results of exhaust emission tests from a vehicle using hybrid drive performed in accordance with the RDE procedures, which are currently one of the main directions of development of emission testing procedures.

2 Test route selection in accordance with the RDE test requirements

In September 2017 the Euro 6c standard was introduced for the group of PC and LDV vehicles, which in addition to the change of exhaust emission limits introduced significant changes in how the exhaust measurement procedures should be performed. The applicable NEDC approval test was replaced by WLTC procedures and the obligation to perform emission RDE measurements under real operation conditions was introduced. This requires the use of equipment from the PEMS group. In order to ensure comparable measurement conditions, requirements for the test route, which will be used for testing, were also specified. The most important requirements include [13-14]:

a) the test route should consist of three sections:
- urban section, length min. 16 km, maximum vehicle speed of 60 km/h, lasting between 29-44% of the total test time with a tolerance of 10%,
- rural section, maximum vehicle speed between 60–90 km/h, length min. 16 km, lasting between 23–43% of the total test time with a tolerance of 10%,
- motorway section, vehicle speed over 90 km/h, length min. 16 km, lasting between 23–43% of the total test time with a tolerance of 10%,

b) the share of the time spent idle by the vehicle in the urban section should be in the range of 6-30%,
c) in the motorway section, the vehicle is expected to travel at a speed of at least 100 km/h for a minimum of 5 minutes,
d) the maximum speed of the vehicle should not exceed 145 km/h (it is permitted to travel at a higher speed, but its share must not exceed 3%),
e) the whole test should last between 90-120 minutes,
f) the test should start with the urban section, then it should move on to the rural and motorway sections (minor deviations from this order are acceptable),
g) the relative difference in the elevation between the starting and ending points of the test should not exceed 100 m,
h) the vehicle should be driven by a single driver in all sections,
i) pauses or stops in the measurements are not acceptable.

The test route selection in accordance with the presented guidelines requires experience in carrying out exhaust emissions measurements in real operating conditions. The authors of the publication have many years of experience performing these types of measurements and thanks to this a research route, which meets the requirements of the norm, has been developed along the routes of the Poznań agglomeration. It begins at the Poznan University of Technology and continues through the neighboring districts of the city (south-eastern part) and through its center (urban section). After driving through the city center, it leads towards the eastern city limits, then follows the main arterial roads towards the national road No. 92 (Fig. 1).

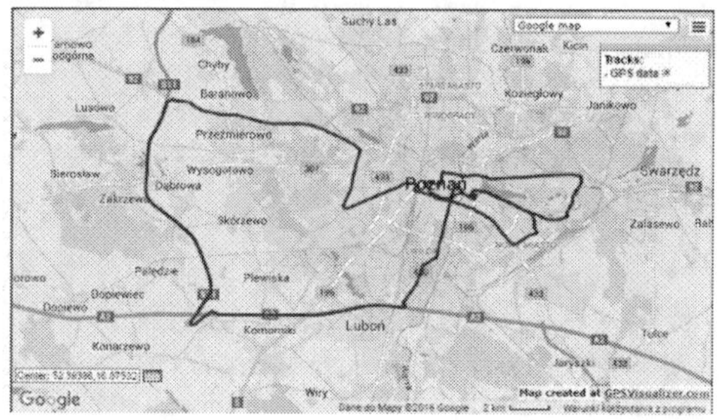

Fig. 1. Poznań-RDE road [15].

During this journey, there is a smooth transition from the urban to the rural section, which ends on the S11 expressway constituting the western city bypass route. The last stage of the test, which is the motorway section, begins there. This phase runs mainly through the southern bypass of the city, which is the A2 motorway. The motorway section ends after leaving the Poznań-Krzesiny junction. The last stage of the route is returning to the location where the test began. The total length of the developed research route is in the range of 75-80 km. Such a discrepancy in the length of the route is related to the required travel time, which should be at least 90 and a maximum of 120 minutes. Depending on the traffic, it is acceptable to finish the measurements before returning to the starting location, so as not to exceed the maximum RDE test duration.

3 Test vehicle characteristics

In RDE tests, a vehicle with a hybrid drive system with mixed configuration was used (Fig. 2). It is composed of a piston combustion engine constituting the primary source of energy and of the set of an electric motor with a Nickel-metal hydride battery acting as the secondary source of energy (tab 1.).

Fig. 2. View of tested vehicle.

Table 1. Characteristics of the hybrid vehicle used for the test.

Parameter	Value
Engine	
Type of engine	Spark ignition
Displacement	2.494 dm^3
Number of cylinders	4
Power output	133 kW @ 6000 rpm
Torque	221 Nm @ 4200–5400 rpm
Emissions standard	Euro 6c
Exhaust gas aftertreatment	TWC with lambda control
Electric motor/generator	
Type of motor	Permanent magnet, synchronous
Max. voltage	650 V
Power output	105 kW
Torque	300 Nm
Hybrid battery	
Battery type	Nickel-metal hydride
Nominal voltage	230.4 V
Number of cells	192
Hybrid drive system	
System type	Series/parallel full hybrid
Max. power	164 kW

The four-cylinder internal combustion engine has a displacement of 2.494 dm^3. It is equipped with a VVT-i timing DOHC type system with four valves per cylinder. The engine generates a maximum power of 133 kW at 6000 rpm and a maximum torque of 221 Nm in the range of 4200-5400 rpm. It also uses a direct gasoline injection system. The power of the electric motor is 105 kW and its maximum torque is 300 Nm. The motor produces a voltage not exceeding 650 V. The maximum power of the drive system of the tested vehicle is equal to 164 kW.

4 Equipment used for tests

The first instrument used for research was the Semtech-Ecostar. This is another device from the Setmtech® series – a successor to the Semtech Ds analyzer. The main difference between these devices is that Semtech-Ecostar is made up of separate modules (Fig. 3):

- FEM module for measuring the concentration of CO_2, CO, HC (NDIR analyzer),
- NO_x module for the measurement of NO_x (NO and NO_2) concentration using the NDUV analyzer,
- FID module to measure THC with a FID analyzer,
- PDM module to power supply system for each component.

Fig. 3. Semtech-Ecostar instrument used to measure emissions in accordance with the RDE methodology.

In the case when of all the modules are used (of the gas measurement section) they must be connected together by a CAN communication wire (green lines in Fig. 4). The FEM module is connected via a special cable with a flow meter for measuring thermodynamic parameters of exhaust gases (mass flow, temperature and pressure). This line is supplied a sample of exhaust gas that passed through the NDIR analyzer (measuring CO_2, CO and HC) is directed to the NO_x module with three connectors, wherein the measurement of NO and NO_2 takes place.

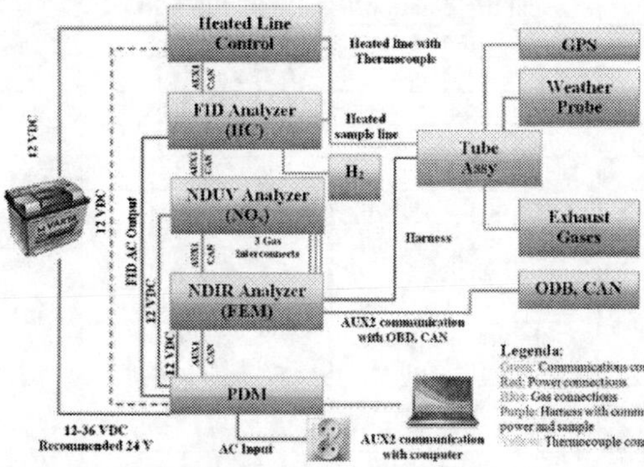

Fig. 4. Diagram of the Semtech-Ecostar system used for testing – section responsible for the measurement of gaseous compounds.

Apart from the exhaust gas sample a special conduit transmits the signal from the vehicle global positioning system (GPS) and weather sensor used to measure the atmospheric conditions (temperature, pressure, humidity), which are connected to the flow meter. A computer for system control and a signal converter from the vehicle's diagnostic system are connected using the AUX2 connector to the PDM, FEM and NO_x modules. Each module is equipped with one AUX1 and one AUX2 connection point. The signal from the power supply module for the heated wire is sent to the FID module. Semtech-Ecostar device can be powered by a PDM module, to which voltage is supplied from the grid or from the car battery. It is also possible to directly connect the modules to the mains. Due to the modular design of the instrument it can perform measurement of all gaseous compounds simultaneously, or separately using a single module.

5 Test results and analysis

Before analyzing the RDE test results, it is necessary to check whether the drive cycle met the requirements outlined in the norms. As described previously in the article the authors have selected a test route in the Poznań agglomeration dedicated to RDE research. Measurements for a hybrid vehicle were carried out on this route. The drive took place on a working day of the week in the afternoon hours at normal traffic levels. The test duration was 91 minutes and during this time the vehicle covered a distance of 70.3 km. The criterion regarding the duration of the journey has been met. The requirement for a minimum distance of 16 km for individual test sections (urban, rural, motorway) together with each section's overall share in the RDE test were also met. Other requirements that must be met are presented in Tab. 2. It was found that the measurements carried out met the requirements set out in the standard and further analysis of the obtained parameters is therefore possible.

Table 2. Characteristic of RDE road

No	Parameter	Unit [requerement]	Value
1.	Total trip distance	km [-]	70.3
2.	Total trip duration	min. [90–120]	91.0
3.	Cold start duration	min. [5]	5.0
4.	Urban distance	km [>16]	29.7
5.	Rural distance	km [>16]	17.7
6.	Motorway distance	km [>16]	22.9
7.	Urban distance share	% [29–44]	42.5
8.	Rural distance share	% [23–43]	24.9
9.	Motorway distance share	% [23–43]	32.6
10.	Urban average speed	km/h [15–40]	27.2
11.	Motorway speed above 145 km/h	km/h [<3% mot. time]	2.16
12.	Motorway speed above 100 km/h	Min. [>5]	9.28
13.	Urban stop time	5% [6–30]	21.9
14.	Start and end points elevation absolute difference	M [<100 m]	14.1
15.	Cumulative positive elevation gain	m/100km [<1200 m / 100 km]	301.1

The first test phase covered the urban section where the vehicle speed should not exceed 60 km/h. For the most part of this section, this rule was followed, however, there were single areas where the vehicle speed exceeded that limit and the test shifted into the rural section. The exhaust emission norms permit this to happen. The length of the urban section was 29.7 km, and the average speed was 27.2 km/h. In the rural section, which was the next phase of the RDE test, areas occurred where the vehicle only achieved speeds that qualified for the urban phase.

This resulted from the road congestion present during testing, where the traffic prevented a smooth drive through this section. The length of this section of the test was 17.7 km. In the case of motorway driving the vehicle reached a maximum speed of 156 km/h, but most of the time the speed did not exceeded 145 km/h. Therefore, the conditions stating that 3% of the time of the drive section with the speed exceeding 145 km/h and minimum 5 minutes travelling the speed above 100 km/h were both fulfilled. The smallest change in vehicle acceleration was observed for the motorway section (Fig. 5).

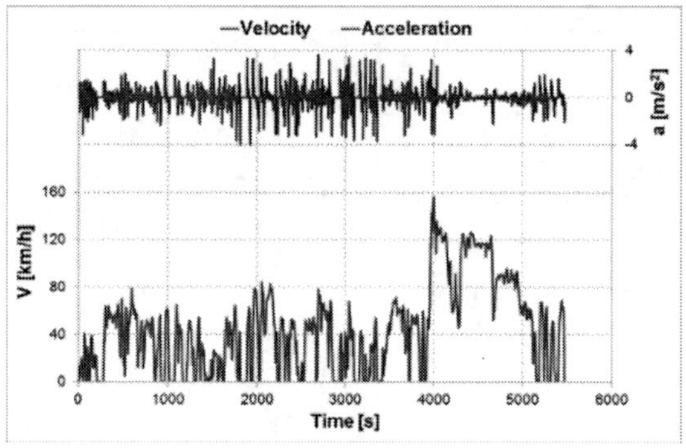

Fig. 5. The speed and acceleration of the vehicle during the RDE test.

In the case of vehicles with hybrid drives, it is also reasonable to analyze the share of drive time spent using only electric power for propulsion in the RDE test - this share was defined as hybridization or the degree of hybridization. To determine the degree of hybridization in the test drive, it was necessary to analyze the data recorded from the OBD system, and in particular the engine crankshaft speed. A basic 0–1 system was used, in which 0 means the use of the internal combustion engine and 1 means only the electric drive system is used. This article does not take into account the parallel operation mode of the drive system, where energy from the combustion engine and the electrical motor is used in combination. During the RDE test, the degree of hybridization was measured at 46%. This means that almost half of the travel time to the test drive was spent using only the electric motor for propulsion. This was evident especially in the urban and rural test sections (Fig. 6).

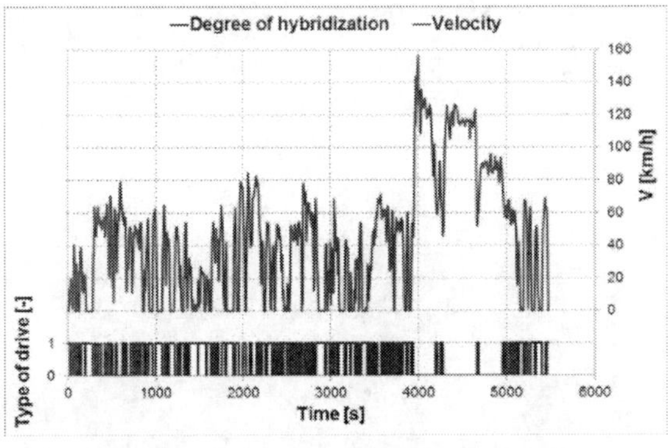

Fig. 6. Determination of the degree of hybridization during the RDE test.

In addition to determining the degree of hybridization for the entire RDE drive cycle, the operating conditions of the internal combustion engine for the individual test sections were also analyzed. The characteristics of the share of operating time in crankshaft speed and engine load intervals were used to represent the results. In the urban section, the share of the electrical power was observed to be dominant, because 58% of the whole operating time the internal combustion engine was not utilized at all (Fig. 7).

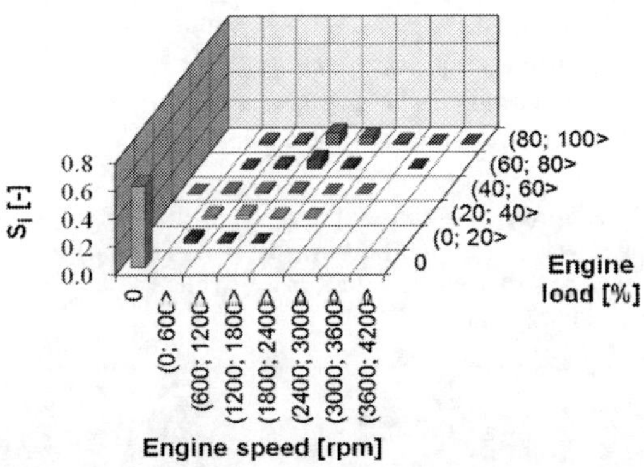

Fig. 7. Operating time density characteristics in the crankshaft rotational speed and engine load ranges for urban driving conditions in RDE test

In the remainder of this phase, it was registered that the engine was operating in the engine speed range 0-4800 rpm. In this range the largest share was for the engine load from two ranges (60; 80%> and (80; 100%>. They represented a total of 27% of total operating time.

In the rural section of the RDE test, the degree of hybridization of the propulsion system was much lower than in the urban section. It amounted to 22% (Fig. 8). However, the distribution of the operating points of the internal combustion engine was similar. The range of rotational speed of the crankshaft was increased to 5400 rpm. Driving with the highest load accounted for 52% of total engine operating time. The range of small and medium loads (0; 60%> was 23%.

The motorway phase was characterized by the throughout operation of the internal combustion engine (Fig. 9). No work point was registered in which only the electrical drive was used. The dominant feature here was the operation of an internal combustion engine with a maximum load in the range of (80, 100%>, which accounted for 80% of total operating time.

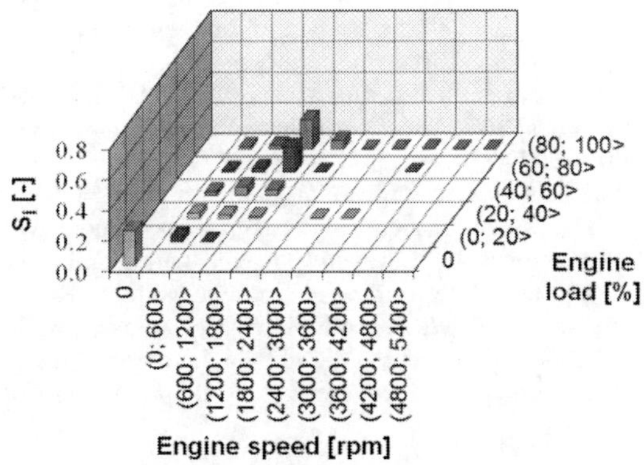

Fig. 8. Operating time density characteristics in the crankshaft rotational speed and engine load ranges for rural driving conditions in RDE test.

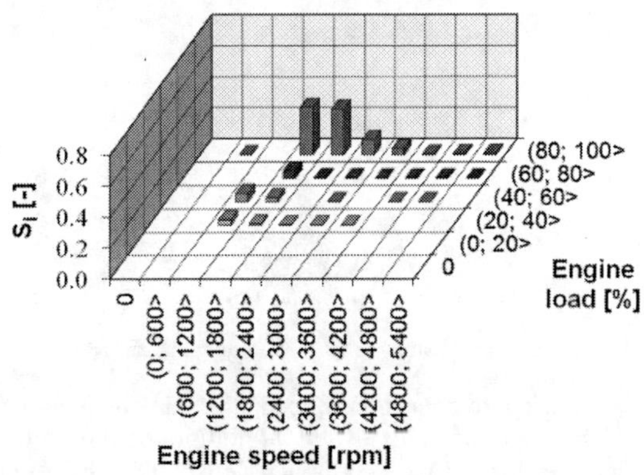

Fig. 9. Operating time density characteristics in the crankshaft rotational speed and engine load ranges for motorway driving conditions in RDE test.

This is a characteristic phenomenon for hybrid drive systems, because the motorway section travel has the highest energy demand, which can be covered exclusively by the combustion engine. There is also no recuperation of the braking energy, so in the motorway phase, the batteries are technically not replenished at all with electricity.

The next step in the analysis of the obtained test results is to determine the exhaust emissions, in particular road emissions of CO, NO_x, THC and CO_2. In this case, it was decided that the total exhaust emissions from the RDE route will be determined according to the MAW method (moving average windows), in the literature also referred to as the EMROAD method, which determines the measurement windows (on the basis of carbon dioxide emissions from the WLTC test) and on its basis determines the road emission in the RDE test. Only the CO and NO_x road emissions were determined using this method.

In the urban RDE test phase, the road emission was 0.37 g/km and was the smallest among the three drive cycle sections (Fig. 10). In the case of NO_x and THC, the highest values were achieved in the urban and motorway sections. It is worth noting that in the urban section the maximum value of NO_x road emissions was obtained, which could have been the result of the low conversion rate of the TWC catalytic converter, whose efficiency is closely related to the temperature of the exhaust gases. As mentioned earlier in the urban section, the degree of hybridization of the propulsion system was calculated to be almost 60%, which influenced the low temperature of the catalytic carrier in this test phase. Hence the dominant share of urban NO_x emissions in the RDE test. Such a situation would not occur for a vehicle with a conventional drive system, powered with an SI engine. In the case of performing measurements in the RDE test for such a vehicle type, the largest NO_x road emissions are expected to occur in the rural and motorway sections, as confirmed by other research results performed by the authors.

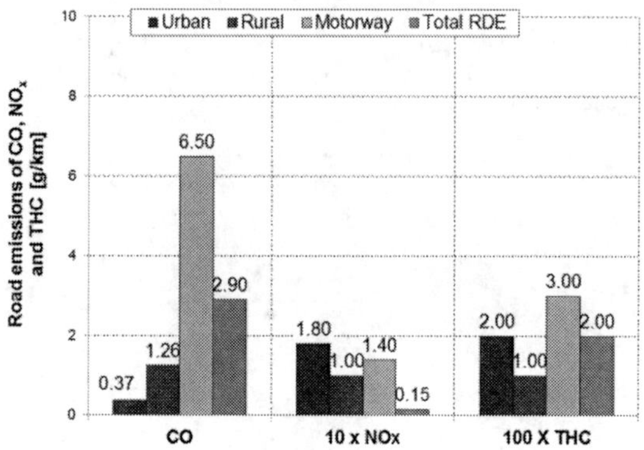

Fig. 10. Road emissions of CO, NO_x, and THC in RDE test.

In order to compare the road emission obtained in the RDE test to the values listed in the Euro 6c norm, it was decided that the CF coefficient, constituting the ratio of the total emission from the RDE test to the limit values specified in the norm, needs to be determined. The CF coefficient for CO was 2.9, for NO_x 2.5 and for THC 0.23 (Fig. 11). Therefore, since according to the norm that the CF coefficient for CO, NO_x should not exceed 2.1, this condition was not met in any of the compounds considered.

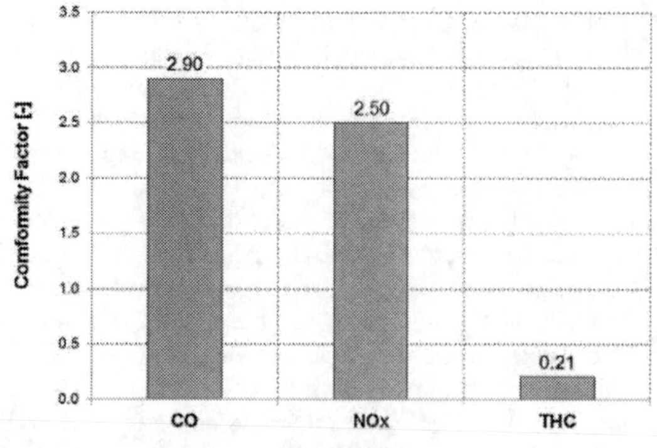

Fig. 11. Comformity factor (CF) of CO, NO_x, and THC determined during RDE tests (all results).

When determining the CF coefficient according to the EMROAD method for CO and NO_x road emissions, similar values were obtained as from the calculations made for the entire RDE test. Both compounds also did not meet the 2.1 limit value (Fig. 12).

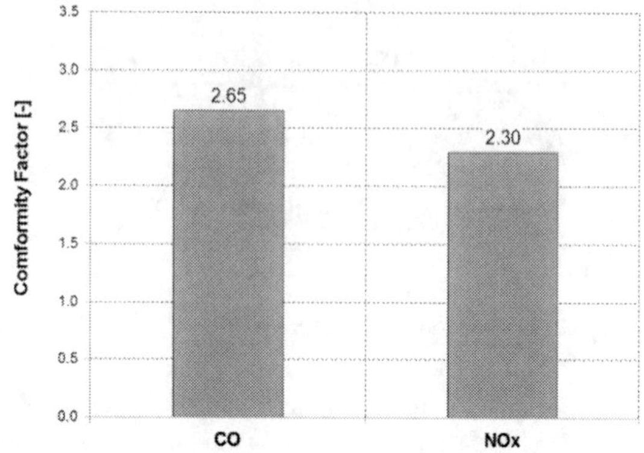

Fig. 12. Comformity factor (CF) of CO, NO_x, determined during RDE tests (EMROAD procedure).

The last stage of the analysis of the obtained test results in the RDE test was the evaluation of the vehicle fuel consumption. The carbon balance method was used to achieve this, in which the road emissions of CO, THC and CO_2 are used to assess the fuel use. The values of the first two compounds were presented earlier, while the CO_2 road emissions were as follows:

a) urban section – 174.6 g/km,
b) rural section – 114.7 g/km
c) motorway section – 180.9 g/km.
d) total – 161 g/km.

The mean fuel consumption value for the whole RDE test was 6.42 dm^3/100 km (Fig. 13). The smallest value was recorded in the rural phase, while the largest in the motorway phase.

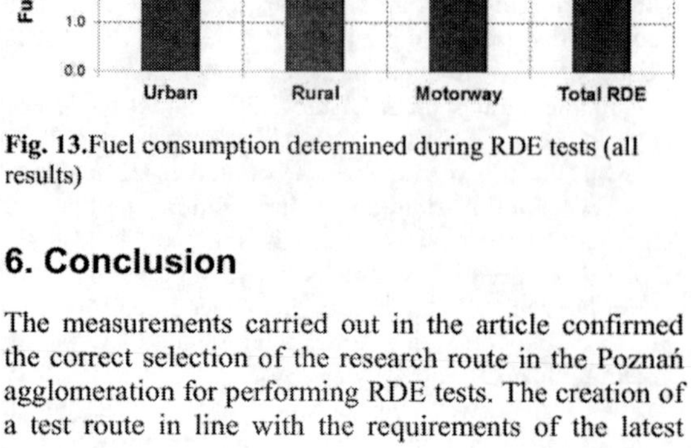

Fig. 13. Fuel consumption determined during RDE tests (all results)

6. Conclusion

The measurements carried out in the article confirmed the correct selection of the research route in the Poznań agglomeration for performing RDE tests. The creation of a test route in line with the requirements of the latest emission standards is currently one of the most difficult aspects of this type of research. For this reason, it was assumed that the route will be referred to as the Poznań-RDE test route and on it the authors will perform RDE tests in accordance with the applicable procedures for PC and LDV category vehicles. For the tested vehicle, the degree of hybridization during the whole test was 46% and its highest value occurred in the urban and rural sections. In the motorway section, the electric motor propulsion of the hybrid drive system was not used. The high share of electric motor use in the urban section ended up limiting the conversion rate of the TWC catalytic converter, which in turn resulted in a very large value of NO_x emission in this test section – it was higher than for the motorway section. The CF coefficients for CO and NO_x determined both from the entire RDE test and the EMROAD method were larger than currently defined by the emission norms. For this reason, it is necessary to perform further RDE measurements for motor vehicles in the PC and LDV category in order to check the correctness of the definition of the CF factor by the legislator.

References

1. K. Andrzejczak, J. Selech, Transport Problems **12**, 103-111 (2017), DOI: 10.20858/tp.2017.12.3.10.
2. J. Merkisz, M. Andrzejewski, A. Merkisz-Guranowska, I. Jacyna-Gołda, Journal of KONES Powertrain and Transport **21**, 219-226 (2014)
3. J. Merkisz, M. Idzior, J. Pielecha, W. Gis, *WIT Transactions on the Built Environment* **111**, 181-189 (2010), DOI: 10.2495/UT100171
4. J. Merkisz, P. Lijewski, P. Fuc, S. Weymann, Eksploatacja i Niezawodnosc – Maintenance and Reliability **15**, 364-368 (2013)
5. J. Zak, R. Jachimowski, P. Golebiowski, *Transport Means – Proceedings of the International Conference*, 772-777 (2016)
6. J. Merkisz, J. Pielecha, IOP Conference Series-Materials Science and Engineering **148**, UNSP 012078 (2016), DOI: 10.1088/1757-899X/148/1/012078
7. J. Pielecha, J. Merkisz, J. Markowski, R. Jasinski, E3S Web of Conferences 10, UNSP 00073 (2016), DOI: 10.1051/e3sconf/20161000073
8. J. Merkisz, J. Pielecha, W. Gis, Proceedings of the Ninth Asia-Pacific International Symposium on Combustion and Energy Utilization, 477-482 (2008)
9. J. Merkisz, M. Bajerlein, P. Daszkiewicz, *Lecture Notes in Information Technology* **13**, 106-113 (2012)
10. J. Merkisz, P. Lijewski, P. Fuc, S. Weymann, Applied Engineering in Agriculture **31**, 875-879 (2015), DOI: 10.13031/aea.31.11225
11. B. Kurc, Journal of Solid state electrochemistry **18**, 2035-2046 (2014), DOI: 10.1007/s10008-014-2451-x
12. D. Dolinski, E. Odachowska, Accident Analusis and Prevention **115**, 73-78 (2018), DOI: 10.1016/j.aap.2018.03.007
13. Commission Regulation (EC) 692/2008 of 18 July 2008 implementing and amending Regulation (EC) No 715/2007 of the European Parliament and of the Council on type-approval of motor vehicles with respect to emissions from light passenger and commercial vehicles (Euro 5 and Euro 6) and on access to vehicle repair and maintenance information, 2008
14. Commission Regulation (EC) 715/2007 of the European Parliament and of the Council of 20 June 2007 on type approval of motor vehicles with respect to emissions from light passenger and commercial vehicles (Euro 5 and Euro 6) and on access to vehicle repair and maintenance information, 2007
15. http://gpsvisualizer.com (access: 24.06. 2019)

The research was funded by European Union from European Regional Development Fund through the National Centre for Research and Development (Narodowe Centrum Badań i Rozwoju) – research project within the Smart Growth Programme (contract No. POIR.04.01.02-00-0002/18).

European Funds Smart Growth   Republic of Poland   European Union European Regional Development Fund

Atmosphere pollution modeling in the case of accident during rocket propellant transportation by trains

Mykola Biliaiev[1,*], *Viktoriia* Biliaieva[2], *Vitalii* Kozachyna[1], *Oleksandr* Berlov[3], *Olena* Gunko[1], and *Kateryna* Chernyatyeva[4]

[1] DNURT, Department hydraulics and water supply, 49010 Dnipro, Lazaryan Street, 2, Ukraine
[2] DNU, Department aerohydromechanics and energy and masstransfer, 49050 Dnipro, Kozakov Street, 18, Ukraine
[3] PSACEA, Assistant professor of health and safety, 49600 Dnipro, Chernyshevsky Street, 24a, Ukraine
[4] DNURT, Lviv branch, 79052 Lviv, I. Blazhkevich Street, 12a, Ukraine

Abstract. At present time, in Ukraine the intensive development of solid-propellant missiles takes place. These missiles are called «Grim», «Grim-2», etc. Transportation of rocket propellant very often is carried out by trains. In the case of accident during such transportation great amount of toxic chemicals may be emitted into atmosphere. It is very important to predict the atmosphere pollution level near railways transport corridors to obtain the realistic information about the size of possible zones of hitting. To forecast the atmosphere pollution in the case of rocket propellant burning in railway wagon numerical models have been developed. These models are developed to predict the atmosphere pollution in two scales. The first scale is the simulation of the atmosphere pollution near the railway tracks (so called "local scale"). The second scale is the simulation of the atmosphere pollution on the territory which is adjacent to the railway tracks («urban scale»). The forecast is based on the Lagrangian model of toxic chemical dispersion. The models allow also to predict acid rain formation in the case of solid propellant burning products dispersion into atmosphere. To solve the governing equations we used difference schemes of splitting. The results of numerical experiments are presented.

1. Introduction.

Pavlograd chemical plant (Ukraine) is developing and storing existing rocket propellants. Combustion products of these propellants are highly toxic substances. In case of this propellant burning huge amount of dangerous substances will be emitted in atmosohere. In this regard, an important task is to predict the formation of zones of the atmosphere chemical pollution [3].

It should be noted that for the transportation of solid rocket propellant rail wagons are mainly used (Figures 1, 2).

Fig. 1. Solid propellant missile RS-22 in the wagon

Of particular danger are emergencies arisen during the transportation of solid rocket propellant. Transport corridors are situated near small and big cities so the accidental missile solid propellant ignition in the wagon can cause death hitting of residents. So it is important to predict level of danger in case of accidental missile solid propellant ignition in wagon.

Fig.2. Solid propellant missile RS-22 in the wagon

In assessing the level of air pollution in this case, the following important circumstances should be considered:

1. The train in which solid rocket propellant is being transported cannot immediately stop in the case of a fire in the wagon.

* Corresponding author: diit.hydro.eco@gmail.com

2. In assessing the level of air pollution for the considered emergency, it is necessary to take into account that the emission of propellant burning products will occur at a certain time interval (10-15 minutes).

It should be noted that the models used in Ukraine for the prediction of accidental air pollution (for example, the regulatory methodology used in The State Emergency Service of Ukraine or the OND-86 model [6]) do not allow to take into account these circumstances.

To take into account these circumstances, it is necessary to develop specialized mathematical models that would allow to take into account the emission source movement and the time regime of chemically dangerous substances release.

This article describes the mathematical models that have been developed to assess the level of accidental air pollution caused by the solid rocket propellant burning. This work was carried out in cooperation with and commissioned by The State Emergency Service of Ukraine in the Dnipropetrovsk region. The aim of the research is to create mathematical models for assessing the level of air pollution, which take into account the main physical factors influencing the formation of contaminated areas and at the same time requiring little computer time.

2. Governing equations

To simulate the process of toxic chemical dispersion in atmosphere (local scale) 3D mass transport equation (equation of toxic chemical dispersion) is used [1, 4, 5, 7]:

$$\frac{\partial C}{\partial t}+\frac{\partial uC}{\partial x}+\frac{\partial vC}{\partial y}+\frac{\partial wC}{\partial z}+\sigma C=$$
$$=\frac{\partial}{\partial x}\left(\mu_x\frac{\partial C}{\partial x}\right)+\frac{\partial}{\partial y}\left(\mu_y\frac{\partial C}{\partial y}\right)+\frac{\partial}{\partial z}\left(\mu_z\frac{\partial C}{\partial z}\right)+ \quad (1)$$
$$+\sum Q_i(t)\delta(x-x_i(t))\delta(y-y_i(t))\delta(z-z_i)$$

where u, v, w are the velocity components in x, y and z direction respectively; C is the concentration of toxic chemical; σ is the parameter taking into account the process of toxic chemical decay or rain wash out; μ_x, μ_y, μ_z are the coefficients of turbulent diffusion in x, y and z direction respectively; x_i, y_i, z_i are the coordinates of point source of emission; $Q_i(t)$ is the intensity of toxic chemical emission; $\delta(x-x_i(t))\delta(y-y_i(t))\delta(z-z_i)$ is Dirac delta-function.

Setting the law of x, y coordinates changing in time we simulate the source of emission movement along the railway track. Position of railway track we model by markers in the numerical model.

In the developed numerical model, the following profile of velocity component u and coefficient of diffusion μ_z is used [5]:

$$u=u_1\left(\frac{z}{z_1}\right)^n, \mu_z=k_1\left(\frac{z}{z_1}\right)^m,$$

where u_1 is the velocity at height z_1; k_1=0,2; n=0,16; $m\approx1$. The following models to calculate the other diffusive coefficients are used [2]:

$$\mu_y=\mu_x,\ \mu_y=k_0u,$$

where k_0 is the empirical parameter.

The transport equation is used with the following boundary conditions [1, 5]:

- inlet boundary: $C|_{inlet}=C_E$, where C_E is the known concentration (very often C_E=0);
- outlet boundary: in numerical model the condition $C(i+1, j, k)=C(i, j, k)$ is used (this boundary condition means that we neglect the process of diffusion at this plane);
- top boundary and ground surface $\partial C/\partial n=0$.

To perform quick predictions of the atmosphere pollution in the case of accidental solid propellant burning in the railway wagon we also used 2D equation of mass transfer to simulate admixture dispersion in urban scale [7]:

$$\frac{\partial C}{\partial t}+\frac{\partial uC}{\partial x}+\frac{\partial vC}{\partial y}+\sigma C=$$
$$=\frac{\partial}{\partial x}\left(\mu_x\frac{\partial C}{\partial x}\right)+\frac{\partial}{\partial y}\left(\mu_y\frac{\partial C}{\partial y}\right)+ \quad (2)$$
$$+\sum Q_i(t)\delta(x-x_i(t))\delta(y-y_i(t))$$

where σ is the parameter taking into account the process of toxic chemical decay, rain wash out and interaction with earth surface.

We considered burning of propellant for missile "Grim", "Grim-2". According to open access literature, this propellant consists of: ammonium perchlorate, aluminium (Al) and astringent. The solid rocket propellant for missile has very important feature: it could not be exploded. It could only burn during some period of time.

For the case of atmosphere pollution prediction, it is important to know mass of different pollutants which is emitted during rocket propellant burning. To obtain this information, we must know interval of burning, mass of propellant in the wagon and reactions of chemical transformation.

In case of solid propellant burning for missile Grim or Grim - 2 we take into account chemical transformation of one important propellant component - NH_4ClO_4. In this case the reaction of NH_4ClO_4 chemical transformation is as following

$$2NH_4ClO_4=Cl_2+O_2+4H_2O+2NO$$

If we simulate dispersion in atmosphere products of chemical reaction we solve Eq. (1) or Eq.(2) for each component Cl_2, NO.

3. Numerical integration

To develop the numerical models the implicit difference schemes were used. Main features of the scheme for numerical integration of 3D equation of toxic chemical dispersion are shown below [7]. First, we split the basic equation of mass transport as following

$$\frac{\partial C}{\partial t}+\frac{\partial uC}{\partial x}+\frac{\partial vC}{\partial y}+\frac{\partial (w-w_S)C}{\partial z}=0$$
$$\frac{\partial C}{\partial t}=\frac{\partial}{\partial x}\left(\mu_x\frac{\partial C}{\partial x}\right)+\frac{\partial}{\partial y}\left(\mu_y\frac{\partial C}{\partial y}\right)+\frac{\partial}{\partial z}\left(\mu_z\frac{\partial C}{\partial z}\right) \quad (3)$$
$$\frac{\partial C}{\partial t}=\sum Q_i(t)\delta(x-x_i(t))\delta(y-y_i(t))\delta(z-z_i).$$

We represent the convective derivatives as following [1]:

$$\frac{\partial uC}{\partial x}=\frac{\partial u^+C}{\partial x}+\frac{\partial u^-C}{\partial x},\quad \frac{\partial vC}{\partial y}=\frac{\partial v^+C}{\partial y}+\frac{\partial v^-C}{\partial y},$$

$$\frac{\partial wC}{\partial z}=\frac{\partial w^+C}{\partial z}+\frac{\partial w^-C}{\partial z}$$

$$u^+=\frac{u+|u|}{2};\ u^-=\frac{u-|u|}{2};$$
$$v^+=\frac{v+|v|}{2};\ v^-=\frac{v-|v|}{2};$$
$$w^+=\frac{w+|w|}{2};\ w^-=\frac{w-|w|}{2}.$$

The approximation formulae were as following:

$$\frac{\partial}{\partial x}\left(\mu_x\frac{\partial C}{\partial x}\right)\approx\mu_x\frac{C^{n+1}_{i+1,j,k}-C^{n+1}_{i,j,k}}{\Delta x^2}-\mu_x\frac{C^{n+1}_{i,j,k}-C^{n+1}_{i-1,j,k}}{\Delta x^2}=$$
$$=M^-_{xx}C^{n+1}+M^+_{xx}C^{n+1},$$
$$\frac{\partial}{\partial y}\left(\mu_y\frac{\partial C}{\partial y}\right)\approx\mu_y\frac{C^{n+1}_{i,j+1,k}-C^{n+1}_{i,j,k}}{\Delta y^2}-\mu_y\frac{C^{n+1}_{i,j,k}-C^{n+1}_{i,j-1,k}}{\Delta y^2}=$$
$$=M^-_{yy}C^{n+1}+M^+_{yy}C^{n+1},$$
$$\frac{\partial}{\partial z}\left(\mu_z\frac{\partial C}{\partial z}\right)\approx\mu_z\frac{C^{n+1}_{i,j,k+1}-C^{n+1}_{i,j,k}}{\Delta z^2}-\mu_z\frac{C^{n+1}_{i,j,k}-C^{n+1}_{i,j,k-1}}{\Delta z^2}=$$
$$=M^-_{zz}C^{n+1}+M^+_{zz}C^{n+1},$$
$$\frac{\partial u^+C}{\partial x}\approx\frac{u^+_{i+1,j,k}C^{n+1}_{i,j,k}-u^+_{i,j,k}C^{n+1}_{i-1,j,k}}{\Delta x}=L^+_xC^{n+1}$$

$$\frac{\partial u^-C}{\partial x}\approx\frac{u^-_{i+1,j,k}C^{n+1}_{i+1,j,k}-u^-_{i,j,k}C^{n+1}_{i,j,k}}{\Delta x}=L^-_xC^{n+1}$$

$$\frac{\partial v^+C}{\partial y}\approx\frac{v^+_{i,j+1,k}C_{i,j,k}-v^+_{i,j,k}C_{i,j-1,k}}{\Delta y}=L^+_yC^{n+1}$$

$$\frac{\partial v^-C}{\partial y}\approx\frac{v^-_{i,j+1,k}C_{i,j+1,k}-v^-_{i,j,k}C_{i,j,k}}{\Delta y}=L^-_yC^{n+1}$$

$$\frac{\partial w^+C}{\partial z}\approx\frac{w^+_{i,j,k+1}C_{i,j,k}-w^+_{i,j,k}C_{i,j,k-1}}{\Delta z}=L^+_zC^{n+1}$$

$$\frac{\partial w^-C}{\partial z}\approx\frac{w^-_{i,j,k+1}C_{i,j,k+1}-w^-_{i,j,k}C_{i,j,k}}{\Delta z}=L^-_zC^{n+1}$$

Dependencies to compute concentration on the basis of the first equation from (3) are written as following:
- the first step of splitting ($k=n+1/4$):

$$\frac{C^k_{i,j,k}-C^n_{i,j,k}}{\Delta t}+\frac{1}{2}\left(L^+_xC^k+L^+_yC^k+L^+_zC^k\right)=0;$$

- the second step of splitting ($k=n+1/2$, $c=n+1/4$):

$$\frac{C^k_{i,j,k}-C^c_{i,j,k}}{\Delta t}+\frac{1}{2}\left(L^-_xC^k+L^-_yC^k+L^-_zC^k\right)=0;$$

- the third step of splitting ($k=n+3/4$, $c=n+1/2$):

$$\frac{C^k_{i,j,k}-C^c_{i,j,k}}{\Delta t}+\frac{1}{2}\left(L^-_xC^k+L^-_yC^k+L^-_zC^k\right)=0;$$

- the fourth step of splitting ($k=n+1$, $c=n+3/4$):

$$\frac{C^k_{i,j,k}-C^n_{i,j,k}}{\Delta t}+\frac{1}{2}\left(L^+_xC^k+L^+_yC^k+L^+_zC^k\right)=0.$$

The second equation from (3) was approximated using following formula:

$$\frac{C^{n+1}_{i,j,k}-C^n_{i,j,k}}{\Delta t}=\left[\mu_x\frac{C^n_{i+1,j,k}-C^n_{i,j,k}}{\Delta x^2}\right]+\left[\mu_x\frac{-C^n_{i,j,k}+C^n_{i-1,j,k}}{\Delta x^2}\right]+$$
$$+\left[\mu_y\frac{C^n_{i,j+1,k}-C^n_{i,j,k}}{\Delta y^2}\right]+\left[\mu_y\frac{-C^n_{i,j,k}-C^n_{i,j-1,k}}{\Delta y^2}\right]+$$
$$+\left[\mu_z\frac{C^n_{i,j,k+1}-C^n_{i,j,k}}{\Delta z^2}\right]+\left[\mu_z\frac{-C^n_{i,j,k}+C^n_{i,j,k-1}}{\Delta z^2}\right].$$

The approximation formula for the third equation from (3) is:

$$C^{n+1}_{ijk}=C^{n+A}_{ijk}+$$
$$+\Delta t\sum Q_{ijk}(t)\delta_l(x-x_i(t))\delta_l(y-y_i(t))\delta_l(z-z_i).$$

For the numerical integration of 2D equation, the following approach is used. First, we split the differential equation as follows:

$$\frac{\partial C}{\partial t}+\frac{\partial uC}{\partial x}=\frac{\partial}{\partial x}\left(\mu_x\frac{\partial C}{\partial x}\right),$$
$$\frac{\partial C}{\partial t}+\frac{\partial vC}{\partial y}=\frac{\partial}{\partial y}\left(\mu_y\frac{\partial C}{\partial y}\right), \quad (4)$$
$$\frac{\partial C}{\partial t}+\sigma C=\sum Q_i(t)\delta(x-x_i(t))\delta(y-y_i(t)).$$

Difference scheme of splitting for first equation from (4) is written as following:
- the first step of splitting:

$$\frac{C^{k}_{i,j,k} - C^{n}_{i,j,k}}{\Delta t} + L^{+}_{x}C^{k} = M^{+}_{xx}C^{k} + M^{-}_{xx}C^{n}; \quad (5)$$

- the second step of splitting:

$$\frac{C^{n+1}_{i,j,k} - C^{k}_{i,j,k}}{\Delta t} + L^{-}_{x}C^{n+1} = = M^{+}_{xx}C^{n} + M^{-}_{xx}C^{n+1}. \quad (6)$$

The difference schemes of splitting for the second equation from (4) are similar. To solve the third equation from (4) we used Euler method.

Worthy of note, that the unknown value of concentration C at every step of splitting in the described difference schemes is carried out using explicit formula of ' running calculation'.

4. Modeling acid rain formation

HCl emission during solid propellant burning results in acid rain formation. Interaction of HCl with water vapours in the atmosphere results in formation acid drops which fall on the earth. To simulate this process we use the following algorithm:

Step 1. We compute HCl vapours concentration spreading in the atmosphere at the particular moment of time.

Step 2. We compute of water vapours concentration spreading in the atmosphere at the paticular moment of time.

Step 3. We compute HCL acid amount (mass) with condition that in atmosphere maximum possible acid concentration in solution is 20%.

Step 4. We compute HCL acid migration in the atmosphere on the base of equation (1).

Step 5. We compute HCl concentration change taking into account that part of HCl goes to the formation of hydrochloric acid.

Step 6. We compute hydrochloric acid concentration changing in the atmosphere due to its gravity fall out on ground surface. Formula for new hydrochloric acid concentration value in each computation cell at new time moment (n+1) has the following view:

$$C^{n+1}_{ij} = C^{n}_{ij} - \frac{C^{n}_{ij} \cdot w_s}{H} dt,$$

where C^{n}_{ij} is hydrochloric acid concentration in the atmosphere at the moment of time n; C^{n+1}_{ij} is hydrochloric acid concentration in the atmosphere at the moment of time n+1; w_s is speed of hydrochloric acid drops precipitation; H is averaging height; dt is time step.

7. Calculation of hydrochloric acid amount settled on ground surface.

This algorithm was coded in additional subprogram.

5. Social risk assessment

Social risk assessment – the risk of toxic damage to people in the residential area – is the final stage in predicting the damage in case of hazardous substances emission.

By risk here we mean the number of people in the zone where the concentration of a toxic chemical exceeds the threshold concentration, which for HCl is equal or more than 4.5 mg/m^3. The developed code includes the subprogram «Hit-2», which calculates where in the region the concentration of the pollutant is equal to or exceeds this threshold value. Based on this data, an array is formed that determines those difference cells that correspond to this concentration. Based on this information, the affected area is built on the map, which allows quickly to assess the size of the area which is dangerous for life.

In addition, this subprogram calculates the number of people who possibly may die in the zone:

$$D = S^{'}_{D} \cdot P,$$

where D is number of died people; $S^{'}_{D}$ is space of the affected area; P is population density.

6. Results

We used the developed numerical models to predict HCl concentrations (HCl is the product of the solid propellant burning of RC-22 missile) in the case of the accident during its railway transportation. We assumed that at t=0 the solid propellant burning started in the wagon which was moving near Pavlohrad City. The main interest was to compute zones of hitting in the case of moving wagon with burning solid rocket propellant.

The duration of HCl emission was supposed to be 15min. The emission rate was 3500 g/s, the wind speed was u_1=6 m/s (we used data of meteorological base, this is typical wind for September, Dnipropetrovsk region, Ukraine). Train speed was 10 km/h.

The height of the plume that is formed during the propellant burning was calculated using the following empirical model [8]:

$$h = 4,71 \cdot \frac{Q^{0,444}}{u^{0,694}}$$

where u is wind speed, Q is the intensity of the heat emission.

This vertical plume is simulated as the set of vertical point sources using Delta function.

Figures 3-5 (Dimensions of the computational region are 1,6km x 1,3km) illustrate formation of contaminated zone near railway track.

 https://doi.org/10.1051/matecconf/201929402003

Fig. 3. Contours of HCl concentration, t=3min (level z=6 m)

Fig. 4. Contours of HCl concentration, t=9min (level z=6 m)

Fig. 5. Contours of HCl concentration, t=14 min (level z=6m)

We see that HCl plume covers quickly the inhabited territory near the railway track.

In Figures 6-9 (Dimensions of the computational region are 1,1km x 0,9km) we present results of numerical simulation of acid rain formation. In this scenario, the wagon with solid rocket propellant was at Pavlohrad Station. The train did not move.

Fig. 6. Contours of HCl (acid) concentration, t=0.7min (level z=4 m)

Fig. 7. Contours of HCl (acid) concentration, t=1.8min (level z=4 m)

Fig. 8. Contours of HCl (acid) concentration, t=2.2min (level z=4 m)

As one can see from Figures 6-9 zone of acid contamination is formed very quickly and has the form of «plume». Figures 10, 11 (Dimensions of the computational region are 1,1km x 0,9km) show hitting zones for different time steps after solid propellant ignition.

Fig. 9. Contours of HCl (acid) concentration, t=2.5min (level z=4 m)

Fig. 10. Zone of dearth hitting, t=2.1min

Fig. 11. Zone of dearth hitting, t=2.6min

Zones shown in Figures 10, 11 illustrate area where HCL concentration is equal or more than 4.5 mg/m^3. In these zones it will be danger for people life.

In case of modeling of another situation, user must make changes in input data block: he must input another wind speed, rate of emission, train speed, etc.

Worthy of note, that computational time was 10 s. It means that the developed numerical model can be used to compute hitting zones near the transport corridors to predict possible damage in case of accidents during transportation of rocket propellant.

The developed numerical model can be used in JSC "Ukrzaliznytsia" to assess risks of toxic hitting in case of accidents with toxic substances during railway transportation.

7. Conclusions

The paper shows the mathematical modeling for assessment of pollution level for the case of accident during solid rocket propellant transportation by railway. To compute contamination zones 3D and 2D equations of mass transfer are used. These equations take into account the main physical processes, which influence dispersion of toxic chemical in atmosphere. To solve governing equations difference scheme of splitting are used. The unknown value of toxic chemical concentration is calculated using explicit formulae.

Results of numerical simulation show that developed numerical models allow to obtain important information which is necessary in the case of PLAS development (regulatory document in Ukraine which is developed in the case of dangerous cargo transportation).

References

1. L.I. Antoshkina, *Ecological risk assessment in case of emergencies with chemically hazardous substances*, (Nauka i obrazovanie, Dnipro, 2008).
2. M.E. Berlyand, *Modern problems of atmosphere diffusion and atmosphere pollution*, (Gidrometeoizdat, Leninhrad, 1975).
3. M.M. Biliaiev, O.V. Berlov, Cal. of sc. works NGU, **42**., 160-167, 2013.
4. M.M. Biliaiev, NATO Science for Peace and Security Series C: Environmental Security, 87-91, (2012).
5. E.V. Bruyatskiy, Theory of atmosphere diffusion of radioactive releases, (Inst. hydromech. NAS of Ukraine, Kyiv, 2000).
6. Method of aftermath assessment of hazardous substances accidental release in case of emergencies at enterprises and transport, (Kyiv, 2001).
7. G.I. Marchuk. *Mathematical modeling in environmental problems*, (Nauka, Moscow, 1982).
8. K. Uork. *Air pollution. Sources and control*, (Mir, Moscow, 1980).

Comparison of motor coach exhaust emissions in real operating conditions for different testing procedures

Dawid Gallas[1,*], *Jerzy* Merkisz[2], *Maciej* Andrzejewski[1], *Sylwin* Tomaszewski[1], *Jan* Strzemkowski[1], and *Stepan* Kinter[3]

[1] Lukasiewicz Research Network - Rail Vehicles Institute "TABOR" , Poznan ul. Warszawska 181, Poland
[2] Poznan University of Technology, Institute of Combustion Engines and Transport, Poznan ul. Piotrowo 3, Poland
[3] DNURT, Lviv branch, I. Blazhkevich St. 12-a, Lviv, 79052,Ukraine

Abstract. The article discusses the results of applying real driving emissions (RDE) testing methods in measuring the exhaust emissions of a passenger rail vehicle. Portable Emissions Measuring Systems (PEMS) were used to measure the exhaust emissions from a motor coach in real operation, performed on a test track for selected drive cycle parameters. The road emission values of exhaust components have been calculated and compared to road vehicles and road exhaust emission limits based on the legal norms. The assessment was concluded with a comparison of the environmental cost, in terms of exhaust emissions, of transporting passengers using the tested rail vehicle relative to alternative road transport, based on measurements in real operating conditions.

1 Introduction

With the increasing awareness of the problems caused by CO_2 emissions produced by the transport sector, along with other toxic compounds emitted from spark and compression ignition combustion engines the focus placed on mitigating the most environmentally harmful aspects of the transport sector. Road transport has been the precursor to these measures, with increasingly demanding exhaust emission limits of harmful combustion products as well as testing and implementing new emission measuring devices and techniques into the legal framework of type approval. The most recent result of these efforts was the introduction of RDE (Real Driving Emissions) testing for road vehicles. This was the response of lawmakers to the problem of stationary dynamometer tests not being sufficiently representative of the actual exhaust emissions generated by vehicles. Spurred onwards by controversy caused by the defeat devices found in many brands of road vehicles meant to cheat the testing procedure, it culminated in the Euro 6 emission norm being a combination of both a WLTC (Worldwide harmonized Light vehicles Test Cycles) dynamometer test along with an RDE test in real driving conditions. Thus the new emission norms were defined as NTE (Not To Exceed) limit but allowed deviation for RDE test results based on the "conformity factor".

The emission norms for rail vehicles differ from the norms for road vehicles, but the combustion engines and exhaust aftertreatment technologies are similar for these two groups of vehicles [1, 2]. This, in turn, indicates a certain likelihood of the methods and limits pioneered for road vehicles to be eventually incorporated in similar legal provisions for rail vehicles and other NRMM (Non-Road Mobile Machinery). The current Stage V emission norm provides just two different sets of emission limit values based on two large categories of rail vehicles. This does not allow the norms to take into account the often widely different operating characteristics of various rail vehicles performing different types of work. In general, a specific exhaust emission values are used as limits, but this does not accurately represent the ecological costs of transporting passengers in passenger trains or motor coaches. Road vehicle limits, on the other hand, embrace the different natures and goals of different types of vehicles, resulting in specific emission values being used for measuring the ecological indicators of heavy and long-haul vehicles while for light and passenger vehicles road emission values are used. This allows for a more direct assessment of the mass of a given exhaust component being emitted per kilometer of travel, which can be further also divided by the number of passengers transported to obtain the emission per person per kilometer. These calculated values allow for a much easier emissions comparison, not only between different cars, but also between different modes of transport. In terms of mass passenger transport a direct comparison between rail passenger vehicles (motor coach or tram) and buses should be considered. Especially since the most modern urban bus solutions to improve their fuel economy and ecological indicators [3] have been the direct result of increasing the reliance of RDE testing for road vehicles belonging to the public transport sector [4, 5]. A direct comparison would serve as a guideline for municipalities in choosing between the investments in different public transport technologies.

* Corresponding author: d.gallas@tabor.com.pl

1.1 Test object and apparatus

The vehicle used for testing was a LINK motor coach (Fig. 1), powered by two diesel engines, each with a rated power of 390 kW and a swept volume of 13 dm^3. The tests were performed using only one of the engines in order to ensure results reliability.

The measuring apparatus was a PEMS type device AxionR/S+ produced by GLOBAL MRV (Fig. 2). This device was equipped with two exhaust gas analyzers working together to assess the exhaust emissions of CO_2, CO, HC, and NO. Additionally, the analyzer measured the O_2 content in the exhaust gas, but this data was used only for results validation. The PEMS device used an array of devices for measuring exhaust emissions such as Non-Dispersive Infrared (NDIR) and an electrochemical sensor. An airtight connection between the vehicle exhaust pipe and the flowmeter was created, and a probe was attached to a side channel to allow the flow of the exhaust gas into the AxionR/S+ analyzers (Fig. 3). The exhaust gas was transported with a heated line, to prevent condensation of volatile components on its walls. It should be noted that these devices and method are commonly applied to RDE exhaust emissions testing for road vehicles, as well as NRMM. The exhaust emission values presented further in the article are a result of using a road vehicle RDE testing and assessment methods for a rail vehicle in an effort to increase the comparability of the obtained data.

Fig. 1. The tested motor coach vehicle

The whole system was set up in such a way as to ensure there would be no risk of the added devices touching the above catenary, or extending outside of the permitted track gouge. The systems mounted on the roof were tightly secured to the vehicle with the use of special belts and suction pads.

1.2 Test method

The test drive was performed on a special track designated for testing rail vehicles in motion. The track was 7.7 km long and consisted of turns with radii of 600, 700, 800 and 900 meters. The maximum travel speed on the track was 160 km/h, but in the performed test drives only speeds of up to 100 km/h were used.

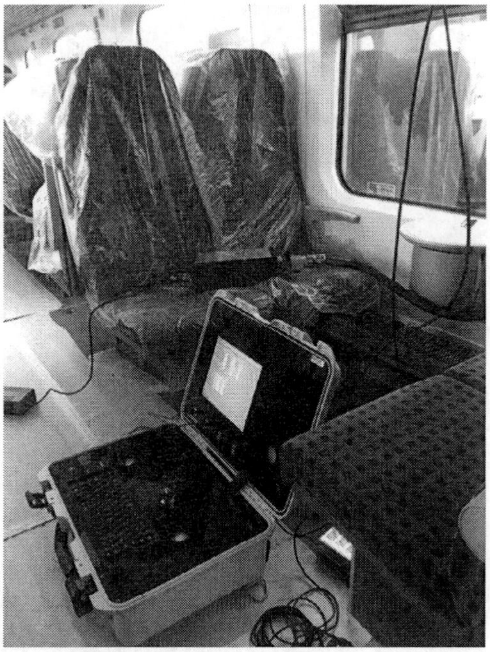

Fig. 2. AxionR/S+ device setup inside the motor coach, with a probe connected to the exhaust pipe

Fig. 3. Flowmeter mounted on the roof of the vehicle and the heated line for gas sampling

The exhaust emission data was gathered and processed in order to obtain emission values for CO_2, CO, HC, and NO. These results were compared to the road vehicles exhaust emission norms (Table 1), for further assessment of the motor coach ecological indicators. Do note that the particulate matter emissions are not considered in this research. Motion and speed of the vehicle were calculated based on the obtained GPS data.

Two drive cycles were performed to assess the impact of travel speed limit on the exhaust emissions. Both drives were done over the same route but the maximum travel speed limit for Drive 1 was 70km/h, while for Drive 2 it was 100 km/h. This allowed to numerically reflect the impact of travel speed on the vehicle ecological indicators.

https://doi.org/10.1051/matecconf/201929402004

2 Results

The exhaust emission results for the tested motor coach were presented as emitted mass of a compound per second and as road emissions (mass of compound emitted per km of travel). Figures 4 and 5 (previous page) show the measured exhaust emission results from Drives 1 and 2 respectively for carbon dioxide (CO_2), carbon monoxide (CO), hydrocarbons (HC) and nitrogen oxide (NO).

The total mass of each measured exhaust component emitted during each of the two drive cycles (1 and 2) was then used to determine the road emissions of the tested vehicle in real operation (Fig. 6). The change of the road emission value between Drive 1 and Drive 2 was also presented as a percentage of the value for Drive 1 (Fig. 7).

The road exhaust emission values were then divided by the number of passengers for a fully loaded vehicle, which in the case of a LINK motor coach is about 126 passenger seats in each vehicle set, excluding the driver. It should be noted, that in some cases the number of passengers may exceed the seat limit, however, allowing more passengers onboard to remain standing.

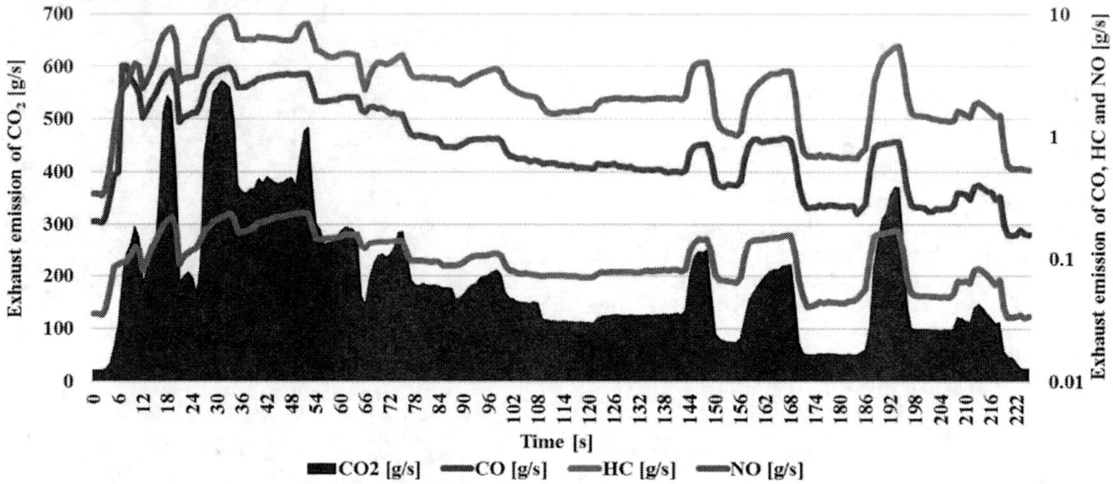

Fig. 4. Exhaust emissions represented in time for Drive 1

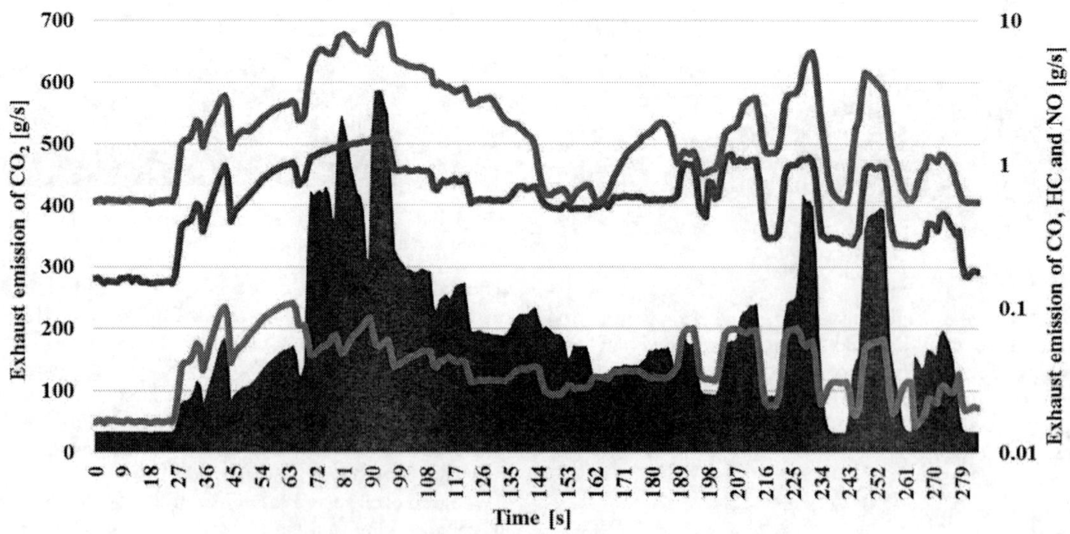

Fig. 5. Exhaust emissions represented in time for Drive 2

Table 1. Exhaust emission limits for light passenger cars

Name	In effect since	CO [g/km]	HC+NOx [g/km]	NOx [g/km]	PM [g/km]	PN [1/km]
Euro 3	Jan. 2001	0.66	0.56	0.50	0.050	-
Euro 4	Jan. 2006	0.50	0.30	0.25	0.025	-
Euro 5a	Sept. 2010	0.50	0.23	0.18	0.005	-
Euro 5b	Sept. 2011	0.50	0.23	0.18	0.0045	6×10^{11}
Euro 6b	Sept. 2015	0.50	0.17	0.08	0.0045	6×10^{11}
Euro 6d Temp	Sept. 2018	0.50	0.17	0.08	0.0045	6×10^{11}
Euro 6d	Jan. 2021	0.50	0.17	0.08	0.0045	6×10^{11}

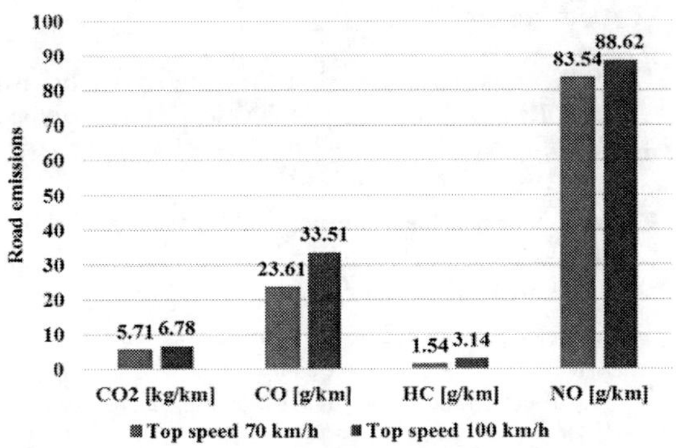

Fig. 6. Road emission values for test Drives 1 and 2.

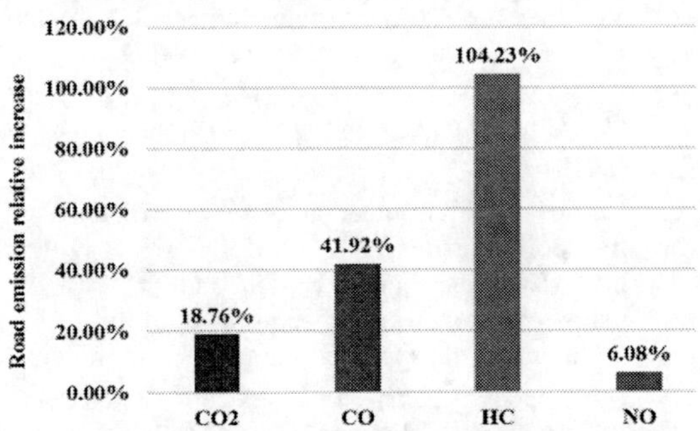

Fig. 7. Relative increase of road emission value for each of the measured exhaust components between Drive 1 and 2.

The exhaust emission value in grams per passenger per kilometer was shown to be higher than for passenger cars meeting the Euro norms (Fig. 8). It should be noted that the LINK motor coach did receive type approval as meeting the EURO III B exhaust emission norms for heavy-duty diesel engines, which are given in g/kWh. The obtained data can therefore be used to clearly show the differences in specific and road emission values, that stem from the different methods of calculation.

The obtained results indicate a high road emission values for nitrogen oxides with a relatively low fraction of hydrocarbons. Hydrocarbon emission was also most affected by the increase in top travel speed during the test. The increase in the share of vehicle acceleration in the test drive cycle did not result in a significant increase of nitrogen oxide emissions. The increase of carbon dioxide emission, however, was over 18%, which directly correlates with the vehicle fuel consumption. It was observed that the increase of top travel speed did not significantly reduce the total drive duration, with the mean travel speed for Drive 1 being equal to 25.67 km/h and for Drive 2 at 32.08 km/h.

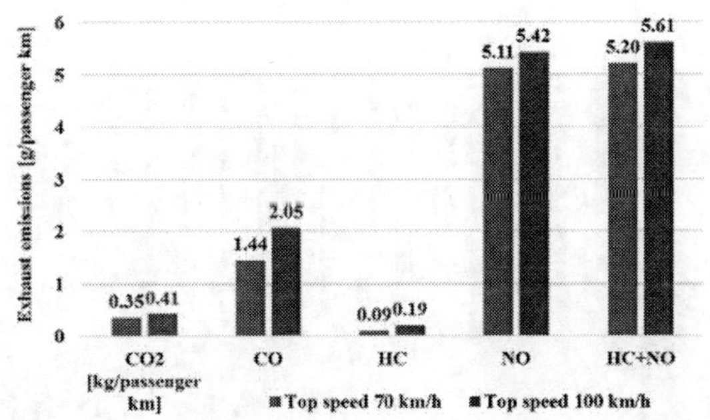

Fig. 8. Road emission values per passenger for test Drives 1 and 2.

3 Conclusions

The overall exhaust emission values were determined according to the measured emission data for the four main exhaust components (CO_2, CO, HC, NO). The European Environmental Agency claimed that the emission of carbon dioxide per passenger kilometer for rail transport is lower than for road transport or aviation (Fig. 9). This, however, does not take into account the toxic exhaust components contribution from each transport sector [6]. In order to properly represent the environmental costs of transport it becomes necessary to not only adjust the testing methods to be more representative (such as with the use of RDE tests), but also to provide a comprehensive assessment of real emission value for each toxic exhaust component per a single passenger being transported one kilometer. The use of specific emission values based on generated power allow an easier comparison of transport costs for goods and freight, but their accuracy in indicating the most environmentally friendly transport solution for passenger transport is questionable. Additional information could be used to improve the accuracy of the data comparison, such as the environmental and monetary costs of infrastructure, energy, maintenance, employment etc. It should be considered that the proper choice of maintenance procedures can have a significant impact on the lifelong environmental and economic impact of vehicles. There have been numerous attempts at quantitatively assessing the environmental and operating costs of public transport specifically [7], but the scope of assessment is limited by the comparability of the available data.

In terms of the method used it becomes clear that lowering the travel speed for rail vehicles results in lower exhaust emissions, mostly as a result of lower acceleration values. This allows the combustion engines to avoid operating in intermediate states with high load where their fuel efficiency and ecological indicators are not optimal. There is, however, a limit to how environmentally friendly a reduction in travel speed is, simply due to the vehicles still generating emissions even when at idle.

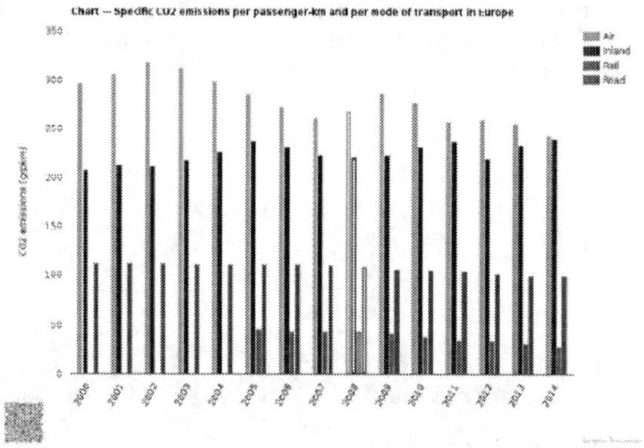

Fig. 9. Specific CO_2 emissions per passenger-km for different modes of transport [8]

The proportion of power used by the engines to move the cargo or passengers needs to be sufficiently higher from the proportion used just to overcome the internal resistances of the engine and generate the power to sustain auxiliary vehicle systems. Additionally, the economic limits of lowering travel speed are significant for rail operators. Since the single largest cost of operating a train line is the staff wages, reducing the number of manhours required to safely perform one full trip has a significant impact on the profitability of a given transport operation [9].

In the obtained results it was found that the overall transport cost for a motor coach is relatively high in the chosen setup and method. Reaching at minimum 350 g/pkm of CO_2, 1.44 g/pkm of CO, 0.09 g/pkm of HC, and 5.11 g/pkm of NO. This does not compare as favorably to road transport. It also indicates the importance of using an accurate method of establishing such values. Considering the data in Fig. 9 claimed by EEA it is not clear whether the CO_2 emission value was established with proper consideration of the share of electric rail vehicles in the total number of transported passengers. Nor is it clear whether the environmental costs of generating electricity used to power those vehicles was taken into account. This leads to data sets that can be used only in a narrow context and can be misleading when used out of its intended context.

From the presented results it can be concluded, that the environmental efficiency of rail transport might be overestimated in the official data, and that it may be more comparable with road transport depending on the circumstances. This, in turn, indicates the continued need for improving both the exhaust emissions measurement and assessment methods, as well as new exhaust aftertreatment technologies for rail transport.

References

1. P.W. Park, M. Downey, D. Youngren, C. Bruestle, ASME 2012, Internal Combustion Engine Division, Vancouver, Canada. ICEF2012-92198, pp. 591-600, doi:10.1115/ICEF2012-92198 (2012)
2. A. Merkisz-Guranowska, M. Andrzejewski, P. Daszkiewicz, D. Gallas, H. Stawecka, Pojazdy Szynowe **3** (2016).
3. M. Bajerlein, L. Rymaniak, P. Swiatek, A. Ziolkowski, P. Daszkiewicz, M. Dobrzynski, Applied Mechanics and Materials 2014, Volume: **518**, Pages: 108-113 (2014) DOI: 10.4028/www.scientific.net/AMM.518.108
4. L. Rymaniak, MATEC Web of Conferences, Volume: **118** (2017) Article Number: UNSP 00009, DOI: 10.1051/matecconf/201711800009
5. L. Rymaniak, P. Daszkiewicz, J. Merkisz, Y.V. Bolzhelarskyi, AIP Conference Proceedings, **2078** (2019)
6. European Environmental Agency (EEA) data – transport emissions of air pollutants https://www.eea.europa.eu/data-and-maps/indicators/transport-emissions-of-air-pollutants-8/transport-emissions-of-air-pollutants-6
7. K. Andrzejczak, J. Selech, TRANSPORT PROBLEMS, Volume: **12**, Issue: 3, Pages: 103-111, 2017. DOI: 10.20858/tp.2017.12.3.10
8. European Environmental Agency (EEA) data – specific CO_2 emissions per passenger chart https://www.eea.europa.eu/data-and-maps/daviz/specific-co2-emissions-per-passenger-3#tab-chart_1
9. European Commission Directorate General for Mobility and Transport – Study on the prices and quality of rail passenger services – Final Report 2016 by Steer Davies Gleave.

MATEC Web of Conferences **294**, 02005 (2019) https://doi.org/10.1051/matecconf/201929402005
EOT-2019

Diagnostics of the structural risk level in the sphere of waste management at Ukrainian railways

Zoriana Dvulit*, *Olga* Melnyk, *Yuliia* Chyrkova, *Nataliia* Petryshyn, and *Maryana* Bortnikova

Lviv Polytechnic National University, Department of Foreign Trade and Customs, 79000, Lviv, Ukraine

Abstract. The article focuses on the problem of the structural risk level diagnostics in the sphere of waste management at the railway transport objects. The specifics of the railway enterprises activity are the multi-vector business processes executed, which determines a significant nomenclature and volumes of formation, accumulation of wastes of different danger classes. A retrospective analysis of this problem on Ukrainian railways for 2007 – 2018 showed significant structural fluctuations both in terms of volume formation and in the costs and ecological payments for environmental protection in sphere of waste management. The method of sensitivity analysis, the peculiarity of which is to calculate the elasticity coefficients for determining the degree of environmental costs interaction in the field of waste management with the change of their components (capital investment, current costs, environmental tax and environmental services), is implemented. The corresponding elasticities for 2008 – 2018 showed a considerable variation of the most significant risk factors that were atypical for JSC "Ukrzaliznytsia" and 6 railways. The existing structure needs to be better diagnosed with a purpose of ensuring stability and predictability of environmental costs in order to reduce structural risk, rational environmental management and to ensure the ecology of rolling stock and objects of railway transport infrastructure.

Introduction

The problems of effective economic-ecological management of Ukrainian railways production activities and business operations received significant attention within the works of many scientists [1-14]. In particular, there have been investigated the key problems of railways management, mechanisms of environmental management (an organization-economical and economic-ecological one), of railway transport enterprises sustainable development; methodical approaches to assessing the environmental impact of railway transport enterprises activities; economic-ecological measures for upgrading environmental situation on the railways, etc.

JSC "Ukrzaliznytsia" and its 6 regional branches, when exercising their business operations, generate and accumulate vast quantities of wastes in different hazard categories. As a result, the company calculates and pays the environmental tax for waste disposal, allocates capital investments, provides for current expenditures and finances services at such direction of environmental protection as waste management. Retrospective study of the key financial and economic indicators in the area of waste management at Ukrainian railways and JSC "Ukrzaliznytsia" over the 12-years period, from 2007 to 2018, allows us to conclude about structural and dynamic fluctuations, both in natural and monetary equivalent. General costs for waste management consist of capital investments, current expenditures, environmental tax and environment-related services in the area of waste management. Regarding the structure of other waste management costs, they are characterized by the spasmodic oscillation: by an increase in some years (2008, 2011, 2012, 2016 and 2018) and decrease in the others (2009, 2010, 2013 – 2015, 2017). In particular, for 6 Ukrainian railways, these costs particularly halved in 2017 comparing with 2016, but increase in 2.25 times in 2018 in comparison with 2017.

In terms of the annual indicator "actually generated waste", a downward trend is observed in 2016 – 2018 (for reference: 2016 – 90.005 thousand tons, 2017 – 62.593 thousand tons, 2018 – 53.440 thousand tons). At the same time, in 2016, the share of the waste disposal was 33% in the structure of the environmental tax distribution; this share significantly increased and amounted to 58% in 2017, and 64.11% in 2018 [15] against the background of a general reduction in waste generation during these years.

Such a significant increase in the environmental tax for waste disposal is due, inter alia, to the ambiguity in the interpretation of the procedure for calculating and paying tax when placing scrap metal by some regional branches (by Ukrainian railways). This situation has developed because there is lack of a unified approach in JSC "Ukrzaliznytsia" on the issues of taxation for waste disposal. We should note that the conclusion of contracts for the provision of hazardous waste management services are subject to the Law of Ukraine "On Public Procurement" [16].

* Corresponding author: zdvulit@ukr.net

However, when competitive tendering by regional branches for the purchase of these services was announced, the auction was often canceled. This situation was due to either an insufficient number of participants, or non-compliance with the conditions of the tender documentation. In such cases, a negotiation procedure is envisaged, but due to the disagreement of prices for the provision of services, such agreements were not concluded.

Also, during 2017, regional branches submitted applications for holding auctions for the sale of ferrous and non-ferrous metal scrap. However, many of them have not been implemented due to the high starting price recommended by the commission of JSC "Ukrzaliznytsia".

In accordance with Article 14 of the Tax Code [17], the waste disposal is the permanent (final) placement or landfilling of waste in specially designated areas or facilities (waste disposal sites, storage facilities, landfills, complexes, building structures, subsurface sites, etc.), and their utilization is exercised on the basis of permission from authorized bodies. Consequently, business entities that carry out the temporary placement (storage) of waste before its destruction through export / transfer for disposal, placement or landfilling to specialized enterprises, are not payers of the environmental tax for waste disposal.

The inefficiency of management in this area leads not only to the erroneous environmental tax assessment, but also an increase in the activity riskiness. And this, in turn, leads to unpredictable and unplanned expenses. Generally, such a trend does not contribute to mission declared on the official portal on sustainable development of JSC "Ukrzaliznytsia", which is to ensure the sustainable development of Ukraine by providing quality and affordable transport and logistics services based on the principles of social and environmental responsibility, as well as by an effective model of company management that corresponds to the challenges of today and tomorrow [18]. In view of the foregoing, it became necessary to diagnose the degree of structural risk in the field of waste management on the Ukrainian railways.

Research methods

In order to fulfill the research, a sample of statistical data is used in accordance with the state statistical reporting form No.1 – environmental costs "Expenditures on environmental protection and environmental payments for the year of 20__" [19] for 6 regional railways of Ukraine. A method of quantitative risk analysis (sensitivity method) was used as an instrument for diagnosing and identifying the riskiest factors of influence on total environmental costs in the field of waste management.

Based on this, elasticity indicators are calculated to determine the degree of interaction of environmental costs in the field of waste management with changes in their components (capital investment, current expenditures, environmental tax and environmental services) for the period of 2008 – 2018. The scientific publications on the subject, domestic legislation, reports of JSC "Ukrzaliznytsya" regional branches became the information base of the research.

Results and discussion

The problems of risk management in the activities of enterprises have been and still remain the subject of study of many Ukrainian and foreign theorists and practitioners. In particular, M. Dyba in his work [20] identified the concept of risk and structural risk, developed a classification of financial risks; basic methods of risk assessment in risk management and the use of quantitative methods of risk analysis (sensitivity analysis) are highlighted in the work of D. Kozenkova and P. Nikitin [21]; the use of sensitivity analysis in the context of risk testing is presented in [22].

Other scientists substantiated the need to use mathematical models for risk management [23], methods for quantitative analysis of vulnerability to develop effective strategies for companies [24]. Application of environmental vulnerability analysis in risk assessment from a methodological standpoint is presented in [25]; advantages and disadvantages of using elasticity index in [26].

The principles and methods of system analysis, economic-mathematical methods and models, economic risk management of business entities for making effective management decisions are considered in [27, 28].

Questions of the nature and classification of risks in railway transport are given considerable attention in the work of A. Rachinska [29]. Other scientists [30, 31], who investigated the activities of specific railways, indicated the influence of factors that lead to the occurrence of corresponding risks.

When diagnosing and quantifying risk using simulation methods, the fluctuation of the initial value is estimated with random changes in input values, taking into account the degree of interdependence of random changes in input values. Sensitivity analysis refers to quantitative risk analysis.

This is a method of considering the factors of uncertainty that are specific for the assessment of an economic indicator (total costs of waste management).

Such a process provides for the implementation of 2 stages: the formation of a model, which determines the mathematical relationships between variables and the sensitivity analysis itself. It is the latter that makes it possible to identify the riskiest variables in the model associated with the assessment of the overall indicator.

This analysis involves the measurement of the sensitivity (vulnerability) of the main indicators, the effectiveness of a certain economic indicator (total costs of waste management) depending on a random change of factors. As a sensitivity indicator related to the change of a certain factor, it is proposed to apply an elasticity coefficient.

It should be mentioned that elasticity is a measure of the response of one variable, or function, to the change of the other one, that is, the argument. In turn, the elasticity indicator is a number that demonstrates the percentage change in a variable (function) as a result of a one-percent change in the argument. The total costs of waste management will be represented as a real-valued function of four factors (arguments):

$$y = f(x) = f(x_1, x_2, x_3, x_4), \tag{1}$$

where y – total expenditures on environment protection and providing environmental services in the area of waste management at regional branches (Ukrainian railways), UAH,

x_1 – capital investments in the area of waste management at regional branches (Ukrainian railways), UAH,

x_2 – current expenditures for waste management at regional branches (Ukrainian railways), UAH,

x_3 – an amount of environmental tax paid for waste disposal in specially designated places or facilities, except for the placement of certain types of waste as recyclable materials in regional branches (railways of Ukraine), UAH,

x_4 – payment for environment-related services in the area of waste management at regional branches (Ukrainian railways), UAH.

With the research purpose, we identify and systematize values of indicators x_1, x_2, x_3, x_4 – for each of the six regional branch (Ukrainian railways: Lvivska Railway, Donetska Railway, Odeska Railway, Prydniprovska Railway, Pivdenno-Zakhidna Railway, Pivdenna Railway) of JSC "Ukrzaliznytsia" and separately for JSC "Ukrzaliznytsia" on the basis of annual reports on the work of regional branches in the environmental protection area [15], as well as in aacordance with the state statistical reporting form No.1 – environmental costs "Expenditures on environmental protection and environmental payments for the year of 20__ " [19] for the period of 2007 – 2018. For JSC "Ukrzaliznytsia", it was taken the total value of each of the indicators of the six railways of Ukraine, other structural units that are part of this company was not taken into consideration.

According to the adopted system of hypotheses, the total costs of waste management (y) as a function of the main of the variables arguments (x_1, x_2, x_3, x_4) are proposed to be submitted using the following formula:

$$y = x_1 + x_2 + x_3 + x_4, \tag{2}$$

which will be considered an adequate model that is in compliance with the aim of the study and the adopted system of hypotheses. So, the I[st] stage of sensitivity analysis has been implemented. In the future, it is proposed to interpret the total costs of waste management (y) as a random variable, depending on random fluctuations of x_1, x_2, x_3, x_4 variables, which should be also treated as random quantities.

Stage II of the investigation is to calculate the elasticity indicators. Since the function is defined in a specific range of values of the 4 corresponding arguments, therefore, the elasticity index of this function is determined regarding the change of these 4 variables for each individual railway is defined as follows:

$$\varepsilon_{it} = \left(\frac{\Delta y_{it}}{\Delta x_{it}} \cdot \frac{x_{it}}{y_{it}} \right) \cdot 100, \tag{3}$$

where ε_{it} – the elasticity indicator of total expenditures related to environment protection in the area of waste management for the i-th railway at the t-th year, %,

i – the railways index ($i = \overline{1,6}$), 1 – "Lvivska Railway", 2 – "Donetska Railway", 3 – "Odeska Railway", 4 – "Prydniprovska Railway", 5 – "Pivdenno-Zakhidna Railway", 6 – "Pivdenna Railway",

t – the year index ($t = \overline{0,11}$), 0 – 2007, 1 – 2008, 2 – 2009, 3 – 2010, 4 – 2011, 5 – 2012, 6 – 2013, 7 – 2014, 8 – 2015, 9 – 2016, 10 – 2017, 11 – 2018.

The elasticity indicator for JSC "Ukrzaliznytsia" is determined by the following formula (4):

$$\varepsilon y3_t = \left(\frac{\Delta y y3_t}{\Delta x y3_t} \cdot \frac{x y3_t}{y y3_t} \right) \cdot 100, \tag{4}$$

where $\varepsilon y3_t$ – the elasticity indicator of total expenditures related to environment protection in the area of waste management for JSC "Ukrzaliznytsia" for the year t, %.

The main advantage of the elasticity indicator is that its value does not depend on the selection of units for measuring various factors. Consequently, the larger the absolute value of this economic indicator, the higher the degree of sensitivity will be. Therefore, the same is with a risk of changing a certain factor on which this indicator depends.

Note that the risks, which the overall costs of waste management are burdened with, are greater, the more the sensitivity (vulnerability), that is, the elasticity of this indicator before each factor changes, and the longer the range of possible variations of these factors (in the future) [27, 28].

This method gives an opportunity to carry out diagnostics based on the analysis of several possible options for the composition of the waste management total costs at individual regional branches.

As a result of the calculations, it is advisable to choose the option that is less susceptible to random changes in these 4 factors that affect the total amount of expenditure in the studied direction of environmental protection measures on the Ukrainian railways.

Only after it is confirmed that this indicator insignificantly depends on the main factors (the values of elasticity indicators are small in absolute value), you can refuse from further risk analysis.

On the basis of the calculation results for six Ukrainian railways and JSC "Ukrzaliznytsia" for the period of 2007-18, we have received the related values of 2008-2018 elasticity indicators represented in tables 1-7.

Table 1. Elasticity indicators, calculated for the regional branch "Lvivska Railway", %

Elasticity indicators	Years										
	2008	2009	2010	2011	2012	2013	2014	2015	2016	2017	2018
ε_1	53.18	-5.93	0.00	75.17	0.00	0.00	0.00	0.00	0.00	0.00	0.00
ε_2	-199.91	-50.44	-15364.38	754.55	33.98	-24.19	-0.67	13.87	20.44	775.57	246.42
ε_3	-223.78	0.00	-123.12	83.75	-11.06	2.86	-0.55	4.44	33.57	-1257.15	565.81
ε_4	53.18	10.72	0.00	75.17	41.86	18.84	5.91	185.21	-794.00	0.00	86.80

Table 2. Elasticity indicators, calculated for the regional branch "Donetska Railway", %

Elasticity indicators	Years										
	2008	2009	2010	2011	2012	2013	2014	2015	2016	2017	2018
ε_1	-7.74	20.33	-6.15	78.35	-51.74	0.00	0.00	0.00	0.00	0.00	7.47
ε_2	-32.96	-152.84	0.34	0.00	37.24	72.80	46.04	45.36	1376.42	202.49	-0.49
ε_3	97.89	94.00	9.70	160.47	152.63	-27.09	24.39	15.27	851.57	0.00	0.00
ε_4	20.08	168.80	-9.78	93.44	90.72	98.14	129.80	147.80	51.40	71.43	18.48

Table 3. Elasticity indicators, calculated for the regional branch "Odeska Railway", %

Elasticity indicators	Years										
	2008	2009	2010	2011	2012	2013	2014	2015	2016	2017	2018
ε_1	97.85	0.00	0.00	0.00	0.00	0.00	0.00	0.00	0.00	-11.12	129.06
ε_2	0.00	-1175.90	17.68	-13.00	12.95	157.83	-33.17	-0.04	25.27	3.10	67.70
ε_3	-16065.46	176.76	0.00	43.49	24.03	-3.47	0.00	0.00	0.00	-11.12	-114.80
ε_4	97.84	-1286.14	0.00	89.95	-468.51	61.17	72.17	0.51	113.31	5.05	78.46

Table 4. Elasticity indicators, calculated for the regional branch "Prydniprovska Railway", %

Elasticity indicators	Years										
	2008	2009	2010	2011	2012	2013	2014	2015	2016	2017	2018
ε_1	95.40	0.00	0.00	28.06	-66.89	49.14	0.22	27.34	17.54	-794.30	68.82
ε_2	226.23	396.79	-1222.10	-251.18	8.16	62.85	84.25	86.67	152.49	30.69	-796.16
ε_3	957.46	-1310.76	4068.17	32.23	25.41	-100.09	117.13	-2.26	-60.49	-34.74	-8.71
ε_4	82.34	20242.96	85.63	-329.76	-28.20	115.38	278.87	5794.28	257.52	35.90	228.23

Table 5. Elasticity indicators, calculated for the regional branch "Pivdenna Railway", %

Elasticity indicators	Years										
	2008	2009	2010	2011	2012	2013	2014	2015	2016	2017	2018
ε_1	88.25	89.98	13.86	0.00	92.01	3.41	0.00	0.00	0.00	-155.59	75.68
ε_2	274.20	5046.20	-462.98	35.69	-1831.75	-23309.54	66.32	53.41	12.60	72.57	2042.49
ε_3	190.23	-141.43	1353.72	8.83	-25.07	77.49	0.00	0.00	0.00	0.00	0.00
ε_4	78.88	1047.89	-1873.38	41.82	-1122.48	-7811.75	181.48	1829.44	-184.10	105.01	348.03

Table 6. Elasticity indicators, calculated for the regional branch "Pivdenno-Zakhidna Railway", %

Elasticity indicators	Years										
	2008	2009	2010	2011	2012	2013	2014	2015	2016	2017	2018
ε_1	97.16	51.48	0.87	455.41	60.88	0.00	0.00	9.92	42.69	0.00	0.00
ε_2	791.14	-516.04	-94.85	72.08	97.03	-124.51	128.48	61.45	-152.74	30.24	449.95
ε_3	171.64	3794.12	-60.27	76.50	46.36	0.00	0.00	0.00	42.63	-268.61	193.84
ε_4	97.16	-2167.15	-227.07	266.11	-201.51	-177.50	75.80	-854.84	214.53	137.08	51.64

Table 7. Elasticity indicators, calculated for JSC "Ukrzaliznytsia", %

Elasticity indicators	Years										
	2008	2009	2010	2011	2012	2013	2014	2015	2016	2017	2018
ε_1	89.01	62.88	9.04	118.84	74.81	3.15	0.13	12.20	-3.13	43.89	65.57
ε_2	1701.40	-5393.57	-425.64	129.78	232.45	-338.95	157.45	-278.43	5.00	43.53	-4265.48
ε_3	376.77	993.21	-244.70	50.65	332.88	39.09	54.52	-1.50	-2.61	-132.26	-365.99
ε_4	132.21	-283.22	-371.30	114.49	575.91	-375.20	91.13	68.33	11.35	79.37	102.13

The values of elasticity indicators for the period under review demonstrated dynamic fluctuations of risk factors both for individual railways of Ukraine, and for JSC "Ukrzaliznytsia" as a whole.

Thus, the Lvivska Railway in 2008 – 2018 was characterized by the various riskiest factors, as evidenced by the calculated elasticity indicators. In particular, in 2010 – 2011 the variable x_2 (current expenditures) was a significant risk factor, in 2008, 2017 and 2018 – the value of the environmental tax. And a factor "environmental protection services" was the most vulnerable in terms of risk in 2015-2016.

As for the Donetska Railway, the variable of the environmental tax x_3 was the highest in modulus during 2008, 2011 and 2012, in comparison with other variables. Consequently, this factor caused the greatest risk for the mentioned railway in the years indicated. This reflects a high degree of sensitivity, and, therefore, of risk. Such a factor as current expenditures acquired the largest absolute value of the elasticity index in 2016-2017, environmental protection services – in 2009, 2014, 2015.

It is worth noting that, based on the calculated values of elasticity indicators for variable x_1, we can conclude that for 11 years this risk factor was not sensitive, and it was not a risk factor at all.

The calculations for the regional branch "Odeska Railway" showed significant fluctuations of the main risk factors over the years. Thus, during the 4-year period (2011, 2012, 2014, 2016), the variable x_4 (environmental protection services) was a significant risk factor, as evidenced by the corresponding elasticity indicators. For 2009, 2010, 2013, the most sensitive in terms of the risk was the factor "current expenditures", in 2008 – the "environmental tax". The last two studied years were characterized by vulnerability to the "capital investments" factor.

The largest absolute values of the elasticity indicators of factor x_4 for the Prydniprovska Railway were demonstrated during 6 out of 11 years (in 2009, 2011, 2013-2016). The variable x_1 was a significant risk factor in 2012 and 2017, the variable x_2 – only in 2018, and variable x_3 – in 2010. Consequently, structural-dynamic risk fluctuations also occurred in the system of total waste management costs for this regional branch of JSC "Ukrzaliznytsia".

For the Pivdenno-Zakhidna Railway, structural and dynamic fluctuations in the risk of current expenditures and environmental protection services were most significant for 5 years for each of these factors (respectively: 2008, 2009, 2012, 2013, 2018; 2010, 2011, 2014 – 2016). Only in 2017, the "capital investment" factor caused the largest share of risk, while the environmental tax was not a risk factor during the researched period.

The environmental activities of the regional branch "Pivdenna Railway" were characterized by a high degree of risk due to the influence of all 4 factors. In particular, the factors "current expenses" and "environmental protection services" each caused the largest share of risk within 4 years, the variable x_3 – in 2009, x_1 – in 2011. It was concluded that environmental activity factors x_2 and x_4 for this railway also have a high degree of environmental sensitivity in the area of waste management.

Based on the systematization of the calculations, it was identified that in 2008 environmental tax was a risk factor both for each of the 6 regional branches and for JSC "Ukrzaliznytsya". This is due to the ambiguity of approaches to the collection of this tax on waste management.

The remaining 10 years demonstrated multi-vector sensitivity, that is, the sensitivity of the total waste management costs to changes in 4 factors (arguments), therefore, the risks were also unstable. When diagnosing JSC "Ukrzaliznytsia", which consists of 6 railways, as a whole, we arrive at the following findings. The magnitude of current waste management costs was the largest risk factor in 7 out of 11 years (2008 – 2010, 2014 – 2015, 2018).

The factor "environmental protection services" was sensitive or risky in 2012 – 2013, and the environmental tax only in 2017. Capital investments, according to calculations, were not a risk factor in any year, since the elasticity indicator was the lowest in absolute value. If we consider this factor for individual railways, it acquired the highest risk indicator values among others for the Odeska Railway in 2018, the Prydniprovska and Pivdenno-Zakhidna Railway in 2017 and for the Pivdenna Railway in 2011.

Summary

While summarizing the conducted research, we could conclude that under planning environmental protection expenditures in the field of waste management on Ukrainian railways, it is necessary to diagnose structural risk in order to identify possible vulnerable or most risky factors, primarily to determine, from many possible options, the optimal structure of waste management total costs that will provide minimal growth of the most significant risk factors. This approach further allows not only to plan efficiently the distribution of funds for environmental protection, but also to manage risks in the field of waste management.

The method of quantitative risk analysis applied in the work should be used for express diagnostics of risk factors, because it is somewhat limited. This method is based on the analysis of the impact of only certain factors on the total costs of waste management. But more significant for determining the riskiest factors is their integral influence.

Also, sensitivity analysis does not take into account interdependencies between certain factors. The lack of official data on the distribution of total costs in the field of waste management in the context of 4 hazard classes did not allow for the calculation of elasticity indicators to determine the degree of interaction of waste management environmental costs for 4 hazard classes with changes in their components.

Although the implemented method has certain limitations, however, it can be used to make express diagnostics of the most important risk factors in the retrospective review, which later allows you to see how common they are for all railways, to define whether they describe a certain general trend between the cost components inherent in both for a separate railway, and for JSC "Ukrzaliznytsia" as a whole. It can be also used to determine the riskiest factors for other types of environmental protection, not only for the railways of Ukraine, but also for other enterprises of various economy sectors.

The completed studies allowed us to diagnose the percentage change in waste management costs as a result of the interaction of the most influential risk factors, which is actually a method for analyzing structural risk. Such diagnostics contribute to efficient cost management in the field of waste management at railway enterprises.

The study showed that for individual business entities it is necessary to differentiate between the priority factors that determine the largest share of the risk in the field of waste management. The corresponding 2007–2018 sensitivity analysis for JSC "Ukrzaliznytsia" and its 6 regional branches (Ukrainian railways) showed a significant variation of the most important risk factors that were not typical for all railways. It is concluded about the instability of the waste management expenditures structure during the study period.

Therefore, such a structure requires a better diagnosis with a view to ensure the stability and predictability of such environmental costs in order to reduce structural risk, to provide environmental management and rational environmental activities, and, in general, environmental friendliness of rolling stock and railway transport facilities.

References

1. M. Chehovska, Rada po vyvchenniu produktyvnych syl Ukrainy NAN Ukrainy (2003) [in Ukrainian]
2. O. Chodakivsiy, Zaliznychnyi transport Ukrainy, **1**, 55-57 (2010) [in Ukrainian]
3. N. Shpak, Z. Dvulit, T. Luchnikova, W. Sroka, Engineering Management in Production and Services, **10**, 76-84 (2018)
4. N. Shpak, O. Sorochak, M. Hvozd, W. Sroka, Scientific Annals of Economics and Business, **65**, 215-226 (2018)
5. Z. Dvulit, O. Levchenko, Baltic Journal of Economic Studies, **3(5)**, 125–134 (2017)
6. I. Tokmakova, Visnyk ekonomiky transportu i promyslovosti, **29**, 227-229 (2010) [in Ukrainian]
7. O. Rybina, Ekonomichnyi prostir, **56/2**, 269-281 (2011) [in Ukrainian]
8. L. Pozdniakova, Ukrainska derzhavna akademiya zaliznychnoho transportu (2002) [in Ukrainian]
9. O. Martyniuk, S. Kharichkov, N. Andreeva, IPREED NANU (2011) [in Ukrainian]
10. V. Kopytko, Visnyk ekonomiky transport i promyslovosti, **49**, 58-64 (2015) [in Ukrainian]
11. N. Kolesnikova, KUETT (2007) [in Ukrainian]
12. Yu. Zelenko, S. Myamlin, Zbirnyk naukovych prac Dnipropetrovskoho natsionalnoho universytetu zaliznytchnoho transportu imeni akademika V. Lazariana. Problemy ekonomiky transportu, **7**, 47-53 (2014) [in Ukrainian]
13. H. Eitutis, Rada po vyvtchenniu produktyvnych syl Ukrainy Natsionalnoi akademii nauk Ukrainy (2010) [in Ukrainian]
14. O. Deineka, O. Mischenko, Visnyk Dnipropetrovskoho natsionalnoho universytetu zaliznytchnoho transportu imeni akademika V. Lazariana, **24**, 214-216 (2008) [in Ukrainian]
15. Results of the work of the rail transport on environmental protection for 2018, Ukrzaliznytsia (2019)
16. The Law of Ukraine «On Public Procurement», https://zakon.rada.gov.ua/laws/show/922-19 (2019)
17. Tax Code of Ukraine, https://zakon.rada.gov.ua/laws/show/2755-17 (2019)
18. Information portal on sustainable development of JSC "Ukrainian railways", http://sustainability.uz.gov.ua (2019)
19. Official site of State Statistics Service of Ukraine, www.ukrstat.gov.ua (2019)
20. M. Dyba, Visnyk natsionalnoho universytetu «Lvivska politechnika», **635**, 134–140 (2009) [in Ukrainian]
21. D. Kozenkov, P. Nikitin, Biznes inform, **10**, 248-253 (2012) [in Ukrainian]
22. N. Tarasevytch, A. Lytvynenko, Finansy, utchet, banki, **1**, 255-263 (2014) [in Russian]
23. J. Antucheviciene, G. Kou, V. Maliene, E.R. Vaidogas, Hindawi Publishing Corporation Mathematical Problems in Engineering, 3, (2016)
24. Åke J. Holmgren, Transport and Economics (2006)
25. H. J. De Lange, S. Sala, M. Vighi, J. H. Faber, Science of The Total Environment, **408**, 3871-3879 (2010)
26. V.G. Aschonitis, M. Gaglio, G. Castaldelli, E.A. Fano, Ecosystem Services, **20**, 66–68 (2016)
27. L. Maznyk, T. Berezianko, O. Bezpalko, A. Berher, Yu. Hryniuk, O. Drahan, O. Oliytchenko, 274 (2019) [in Ukrainian]
28. V. Vitlinckyi, P. Vertchenko, KNEU, 292 (2000) [in Ukrainian]
29. A. Ratchynska, Efektyvna ekonomika, **11** (2016) [in Ukrainian]
30. P. Kellermann, P. Bubeck , G. Kundela, A. Dosio, A. H. Thieken, Climate **4 (2)**, 25 (2016)
31. Shih-Tong Lu, Shih-Heng Yu, Dong-Shang Chang, The Scientific World Journal, 14 (2014)

Disposal of technological sludge of railway infrastructure enterprises

Yuliia Zelenko[1], *Maryna* Bezovska[1], *Anna* Leshchynska[1], and *Alexandr* Shnaiderman[1]

[1] Dnipro national university of railway transport named after academician v. Lazaryan, 49010 Dnipro, Ukraine

Abstract. The problem of improving environmental safety in the handling of oily waste and sludge is relevant for most industries. Significant amounts of waste oil adversely affect virtually all components of the environment. However, it is also a valuable hydrocarbon feedstock. Thus, a rational approach to the processing of sludge can have a significant environmental and economic effect. The shortcomings of traditional thermal technologies for the disposal of technological sludge and suggests using much cheaper mechanical methods were discussed in this article. For example, install decanters of various designs in a depot depending on the overall composition of the sludge and the composition of each of its main parts: the carbon-containing part, water and mechanical impurities. A modern scheme for the disposal of technological sludge from railway infrastructure enterprises has been proposed. Such a scheme allows to reduce the man-made load associated with their accumulation, circulation and minimization. The most promising example of the use of the proposed technology of utilization of oil sludge is its implementation at the local sewage treatment plants of locomotive and wagon depots.

1. Introduction

Process sludge (oil sludge) of the treatment facilities is visually thick, viscous pasty mass with separate inclusions of mechanical impurities, clusters of heavy oils and grease oils. Sludge is watered enough (water content from 20 to 70%). In most cases, it contains an average of 30% or more of petroleum products and has the combustion heat of 3000-5000 kcal/kg.

2. Materials and Methods

The chemical composition of the sludge is very complicated. In addition, not all of their fractions can be burned or recycled. The oil sludges are diverse in composition; they contain oil, water, oil emulsions, asphaltenes, tars, heavy metal ions, various mechanical impurities and radioactive elements. When stored for a long time, they separate into layers: the upper layer is an emulsion of oil with water, the middle layer is clarified water with petroleum products, the lower layer - solid phase (70%), impregnated with petroleum products. The lower layer (bottom sediment) contains 26-45% of resin-asphalt substances and 15-30% of paraffins, as well as combined oil, clay, sand, water, corrosion products.

Disposal of process sludge is somewhat complicated due to the significant difference in the composition and properties of the three fractions of process sludge, namely oil, water and solid. Disposal of oil sludge is possible in several ways:

- Mechanical;
- Physico-mechanical;
- Thermal;
- Using bio- and chemical technologies [1-4].

In accordance with the decisions made, spent petroleum products are subject to mandatory collection and disposal, and in some cases to destruction [5].

The most effective way of disposal is regeneration - the recycling of waste oils in order to completely restore their primary properties, which is usually carried out through complex multi-stage processing at specialized plants.

The constant emergence of new ways and means for solving the disposal problems, the creation of more advanced, waste-free methods of regeneration of waste lubricants, the issues of ecology and human health cause the need for constant attention to this problem, which is at the intersection of technology, ecology and economy.

The oil sludge of linear divisions of railways accumulates in the waste water of enterprises after the washing of cars, locomotives and their parts; therefore, the largest number thereof is formed in the railway passenger train and car-repair sheds, as it is here where the rolling stock is washed while being repaired.

For example, the sludge can be formed during the large periodic maintenance of rolling stock TP-2, which is carried out in special shops of the repair and maintenance depots. This type of repair presupposes body lifting, bogie rolling out, disassembling of individual units, removing of covers and caps, painting of diesel locomotives, painting of electric locomotives, as well as workpiece washing.

The oil sludge is also formed during the hoist repair of TP-3. It includes the removal of traction engines,

* Corresponding author: j.v.zelenko@gmail.com

rolling out of wheelsets, dismantling and disassembly of units for reliable inspection and repair, painting of interior and exterior surfaces with the previous mandatory washing [5-8].

For example, sludge formation is an accompanying process of the work done by the bogie, wheel, gearbox sections of the depot that carries out washing of cars, boxes, wheel sets, reduction gears.

The enterprises of different services annually produce between 6 and 9 tons of process sludge. If you look at the percentage ratio, you can see that a significant part (almost 40%) is occupied by them within the total amount of the passenger service waste.

Outside, industrial waste of process sludge is a mass of brown colour, which has a lump structure. More often, the process sludge falls into the third hazard class and is moderately hazardous.

Table 1 shows the results of chemical analysis of process sludge of the Dnipro passenger train depot.

Table 1. Analysis of process sludge

Metal content, mg/kg, hazard class	Form of presence				
	Total	Active	Water-soluble	Solubility, g/100 g	Content in waste, kg/t
Pb (1)	37	0	0	0	0.037
Cd (1)	absent				
Zn (2)	23.8	6.2	0.4	13	0.023
Ni (2)	13	3.5	1.3	1.2	0.013
Cu (2)	5	2	0.5	1	0.005
Cr (2)	10	0	0	0	0.01
Co (2)	absent				
Mn (3)	8.8	3.5	0	0	0.008
Oils and etroleum products	528046	-	-	0.002	500

During the researches for determination of heavy metals, the applied method is atomic absorption spectroscopy; for determination of petroleum products - chromatography. According to the analysis results, the sludge was classified as third-class toxicity, i.e. moderately hazardous [9, 10].

Today, process sludge is accumulated at railway enterprises and subsequently transferred under contracts to other organizations for disposal. The widely-used are physico-mechanical and thermal methods, which is explained by the specific composition of such sludge; various bio- and chemical technologies are also known.

The simplest methods are disposal of oil-containing waste at specialized landfills. But such methods do not solve the main problems and can only be used as auxiliary preliminary measures.

Among other physical and mechanical methods, the most technological, economic and simple method of recycling process sludge with a maximum moisture content up to 80% is a method of cube fractionation, which is a gradual evaporation, when the generated steam is carried-off continuously as it is released; only a liquid phase remains in the system and it is in equilibrium with the steam.

The process of gradual evaporation is usually carried out in cubic devices of periodic action. Often, the main product of distillation is fraction with the boiling temperature corresponding to that of paraffins. This method is quite energy-consuming, but the resulting product has a high price. The disadvantage of this method can be called the danger of creating and working with vacuum, which "accelerates" the fractionation.

One of the most common ways of oil sludge disposal is such thermal method as incineration. However, process sludges are limited combustible substances, so their incineration can be carried out only with the use of additional fuel. For example, for railway transport enterprises, it is proposed to burn them when mixed with solid fuels in furnaces of stationary locomotives located on the territory of the depot. It is recommended to check the ash remaining after incineration for the presence of heavy metals.

However, such technologies are low-efficiency, energy and financially expensive due to the high cost of boiler units, in addition, it is necessary to install additional expensive equipment for the purification of emissions into the air; they do not allow the full processing and disposal of oil sludge and do not provide environmental safety.

Biological methods are based on scientific developments in the field of accelerating natural processes of self-healing and self-purification of soils and reservoirs, that is, these methods are adapted for use on natural objects: earth mud, in case of oil spills on soil and water systems.

Chemical methods are based on the use of reagents that cause the oxidative decomposition of petroleum hydrocarbons. Chemical substances and decomposition products (carcinogenic and mutagenic substances - benzo(α)pyrene, phenols, polycyclic hydrocarbons and other hazardous compounds) are generally even more dangerous than primary petroleum contamination. Nevertheless, they found their application.

When selecting the process sludge disposal technology, it is necessary to take into account the frequency of generation of these wastes, their limited quantity, the considerable consumption of additional fuel that will support the incineration process. That is why, in order to take into account these features of the sludge of railway companies, we propose to use considerably cheaper mechanical methods. For example, to equip the depot with decanters of various designs depending on the general composition of the sludge and the composition of each of its main parts: hydrocarbon-containing part, water and mechanical impurities.

The decanter is a centrifuge with a digital conveyor. The rotor consists of conical and cylindrical drum parts, a screw conveyor and a feed mechanism pipe. Such a unit separates polluted oil, as well as oil sludge of various origins, into the hydrocarbon phase with a separation of water and solids [4].

The process is conducted due to the separation of the mixture into three phases based on the difference in their density using various ranges of high velocities and torques. from the remote control unit. Also, manufacturers note that in case of unauthorized shutdown of the The raw material (contaminated petroleum products) is fed through the feed mechanism pipe to the rotating part of the screw conveyor, where, under the action of centrifugal force, it is separated into the clarified product and the precipitate.

The clarified product is removed from the cylindrical part of the rotor, and the precipitate due to the difference in speeds of the screw and rotor enters the conical part, where it is dewatered. The dehydrated precipitate is unloaded at the narrow end of the conical part through special ports. The decanter is controlled fan the burning is maintained by natural draft.

Such installations provide the incineration of petroleum products in standard containers, the ability to dispose of waste directly at the place of their formation on a specially designated area of any depot. Separately we should note the higher quality of incineration compared with open burning, high level of explosion safety due to intensive blowing of the combustion chamber.

Since the oil sludges of the railway enterprises are sufficiently flooded, as they are the rolling stock washing waste, they do not need to be further diluted with water to reduce the percentage of mechanical impurities. It is recommended to use the decanters with a screw, which has a coating with special protection, for example, from ceramics or tungsten carbide.

Such measures are due to the presence in the process sludge of so-called "hard" mechanical impurities (sand, scales, metal shavings, etc.), which can cause the early wear of the working parts of equipment. Additionally, in order to prevent the penetration of large mechanical impurities into the decanter, it is expedient to install the mesh filters of different fineness of purification or vibration separators.

The decanter efficiency depends on such indicators of raw material as its uniformity and temperature. The first indicator will allow the decanter to operate in a continuous mode and avoid permanent human control over its operation parameters; because the settings of the decanter to the required parameters take the employee who serves it minimum 15 minutes.

The second indicator provides the required viscosity of raw materials and usually is 80-90ºC; the higher temperature can lead to overheating of the unit. As a rule, such heating is carried out directly in the sludge storage facility (due to heating of the intermediate containers) or "on the go" with the help of special heat exchangers when the oil sludge gets heated while passing through them [9, 11].

We can use regular mesh filters as a coarse filter. To obtain a product with a higher percentage of clarification, it is suggested to use the intermittent filters, for example, filter presses [4, 12, 13].

Filter presses have such advantages as significant filter surface, simple design, option to disconnect individual non-working plates.

Such a disadvantage as manual operation, which causes delays in the work, is not significant for the railway companies due to the small amounts of sludge generation. As a heat exchanger it is possible to offer, for example, shell-and-tube heat exchangers with U-tubes.

Practical studies have shown that for the optimal separation of the three phases of process sludge, different types of flocculants are needed for each case. Depending on the type of equipment used for processing oil sludge (centrifuges, filter presses, vacuum filters) the most suitable type of reagents is selected and used for further processing.

From the experience of using the cationic flocculants most effectively act on organic compounds, while anionic ones are more suitable for inorganic substances. Due to the diversity of sludge composition and properties, the selection of effective flocculants in each individual case should be carried out based on the preliminary laboratory and experimental-industrial tests.

3. Results

For example, there are widely known such flocculants as FT-410, PT-506, flocculants of Zetag-89 brand, technical polyacrylamide (PAA) and others. So the water-based cationic flocculant Zetag-89 manufactured by the Swiss company Ciba Specialty Chemicals is used in the oil refining industry in the sludge dehydration process in amount of 2 kg per ton of dry residue [14].

There are known research results of the sludge clarification in the presence of flocculants of the brands Zetag-89, Praestol 853 and polyacrylamide at a dosage of 10 g per ton of sludge. The obtained results showed that at low amounts of the flocculant it is possible to separate mechanical impurities from petroleum products rather accurately. The content of petroleum products in the lower precipitate layer using flocculants of the Zetag-89 and Praetol 853 brands was 6-10%, and in the precipitate without the use of flocculants - 18% [9, 14, 15].

As flocculant reagents, we tested surface-active substances (sulfonol, neonol AF 9-12, cocamidopropyl betaine) in conjunction with the classical reagent polyacrylamide (PAA), which is widely used for separating oil sludge, to accelerate the latter.

The test was as follows: the process sludge was heated to a temperature of 85°C; PAA was added in the amount of 1% by weight and stirred for 5 minutes; surfactant was added in the same amount as polyacrylamide, stirred for 5 minutes; the mixture was centrifuged for 30 minutes at a rate of 3000 rpm; then we observed the mixture separation (Fig. 1-4).

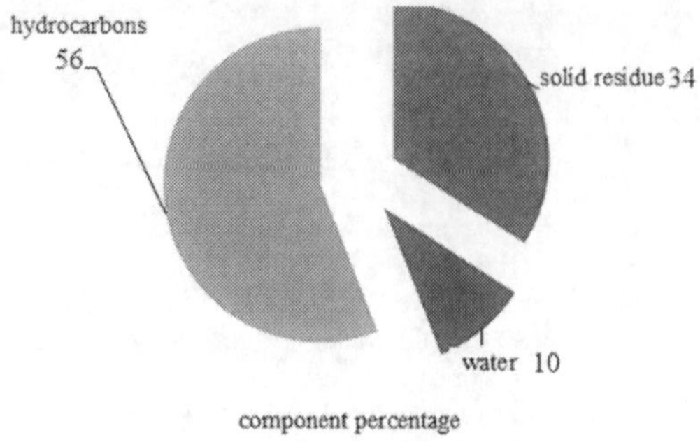

Fig. 1. Dependence of the degree of process sludge fractionation after the addition of polyacrylamide

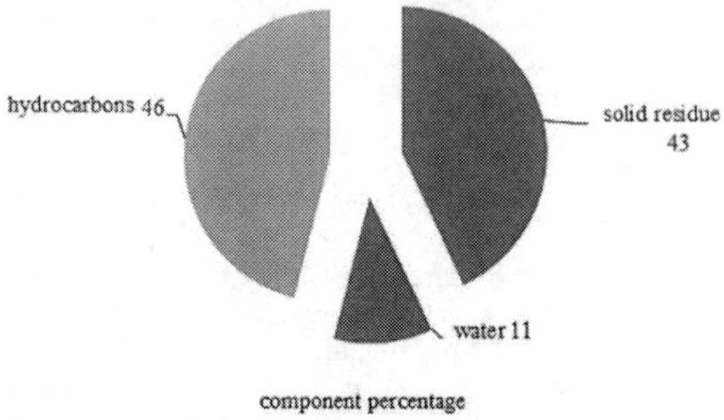

Fig. 2. Dependence of the degree of process sludge fractionation after the addition of polyacrylamide and sulfonol

As can be seen from the figures, the best result, that is, the effective separation of the mixture into three fractions - solid residue, water, hydrocarbons, was obtained by combining polyacrylamide with sulfonol. Therefore, we propose the use of polyacrylamide flocculants in combination with sulfonol ("technical" paste) [4].

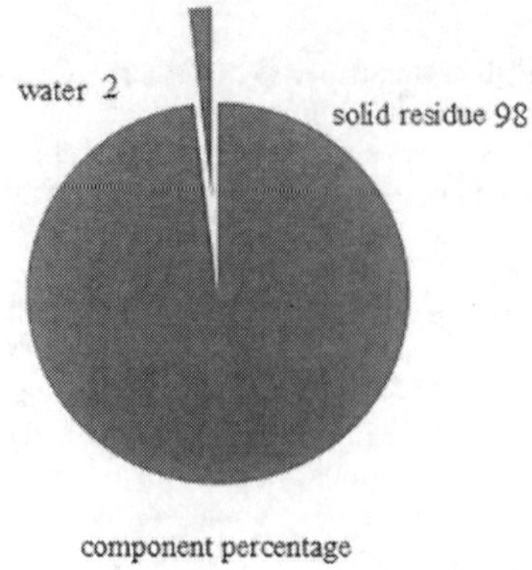

Fig. 3. Dependence of the degree of process sludge fractionation after the addition of polyacrylamide and neonol AF 9-12

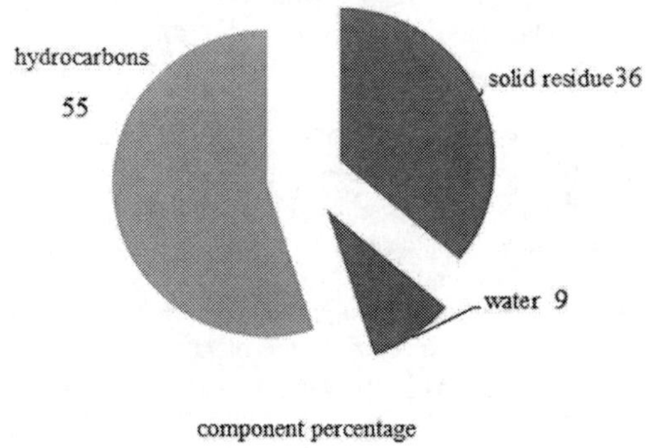

Fig. 4. Dependence of the degree of process sludge fractionation after the addition of polyacrylamide and cocamidopropyl betaine

Also we obtained kinetic dependences of the degree of process sludge fractionation, shown in Fig. 5-7 (except for the case of combination of neonol with PAA due to the worst result obtained - Fig. 3).

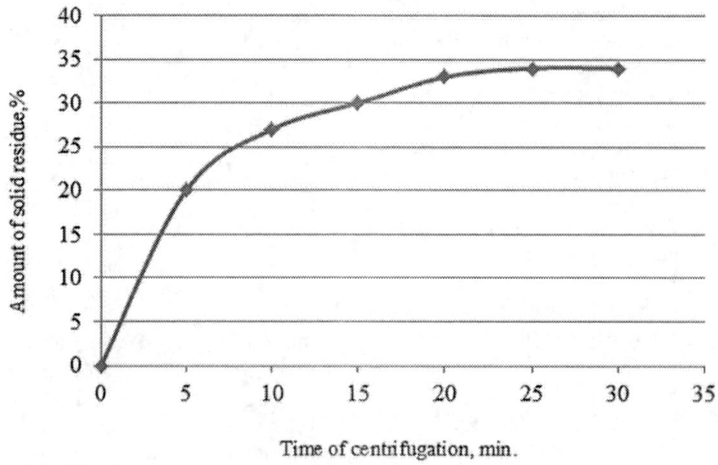

Fig. 5. Kinetic dependence of the degree of process sludge fractionation with the addition of polyacrylamide

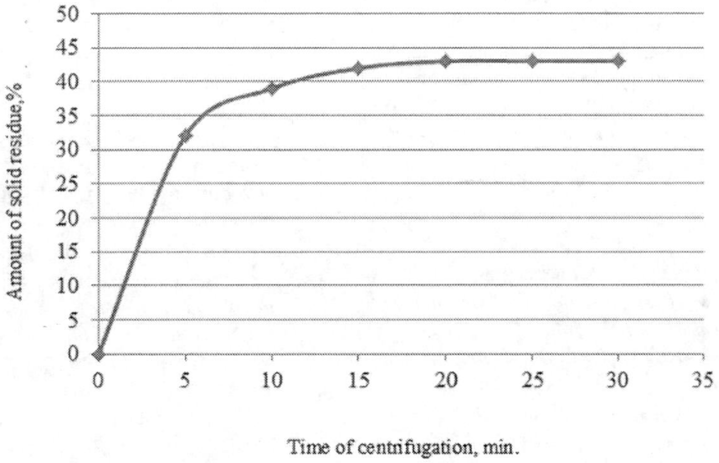

Fig. 6. Kinetic dependence on the degree of process sludge fractionation with the addition of polyacrylamide and sulfanol

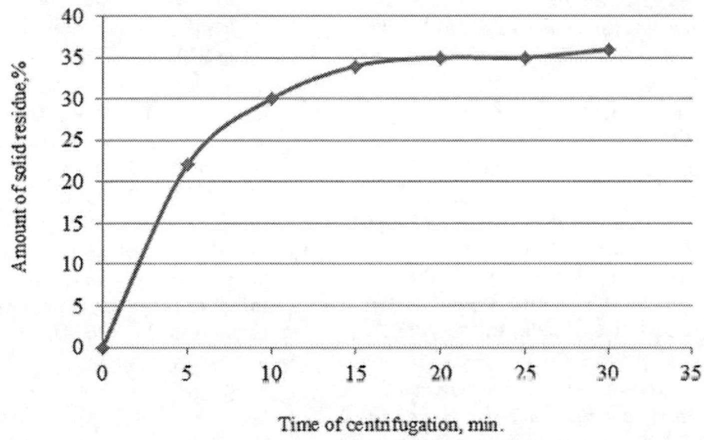

Fig. 7. Dependence of the degree of process sludge fractionation with the addition of polyacrylamide and cocamidopropyl betaine

4. Discussion

As can be seen from the figures, the amount of solid residue formed during the centrifugation process was selected as the tested parameter. Since in the case of PAA combination with sulfonol, the most qualitatively purified products (hydrocarbons and water) were obtained, as they contained the smallest amount of mineral and other impurities, we propose to use polyacrylamide flocculants in combination with sulfonol ("technical" paste).

The water remaining after sludge clarification is sent to the advanced treatment and can subsequently be used, as already mentioned above, for the preparation of a solution of reagents, as technical water in the process cycle of the enterprise or can be discharged into water bodies.

In order to remove the precipitate with a high content of mechanical impurities and heavy metals there are usually offered several options:

- Their disposal at the landfill;
- Incineration;
- Their gradual decomposition by means of biologically active sorbents at specially prepared sites.

In the conditions of depot it is possible to gradually accumulate such waste and dispose of them at the landfills.

5. Conclusion

Thus, in the course of work we investigated the industrial processes, resulting in process sludge waste. We analyzed their composition and properties and as a result it was found that they contain such dangerous components as heavy metals, which confirmed their rating as the third hazard class and the need for their detailed study.

We also proposed the principal scheme of reagent oil sludge processing with the use of the decanter and heat exchanger, which allows to obtain the purified products (hydrocarbons and water) that can be used in the technological processes of the enterprise; we carried out the experimental studies on the selection of flocculant reagents. The obtained kinetic dependences of the degree of process sludge fractionation confirmed the correctness of the choice of reagents and provided opportunities for optimization and prediction of the process of separation and clarification of oil sludge.

References

1. V.N. Plakhotnik, Ju.V. Onyshchenko, L.A. Yaryshkina L.A., *The environmental impacts of railway transportation in the Ukraine*, Transportation Research Part D: Transport and Environment, **10 (3)**, pp. 263-268 (2005).
2. Yu. V. Zelenko, *Naukovi osnovy ekolohichnoi bezpeky tekhnolohii transportuvannia ta vykorystannia naftoproduktiv na zaliznychnomu transporti monografiia*, Izdatelstvo Makovetskii, 242 p. (2010). [in Ukrainian]
3. Yu. V. Zelenko, M. S. Bezovska, A. L. Leschynska, *Rozrobka innovatsiinych resursozberihayuchykh tekhnolohii utylizuvannia naftovmischuyuchykh vidkhodiv*, Gurnal «Tekhnolohicheskii audit I rezervy proizvodstva», PP «Tekhnolohichnyi tsentr», p. 17–21, **(9)** (2013). [in Ukrainian]
4. M. S. Bezovska, Yu. V. Zelenko, *Rozrobka suchasnykh pryntsypiv utylizatsii tekhnolohichnykh shlamiv pidpryyemstv zaliznychnoi infrastruktury,* Zbirnyk naukovykh prats Donetskoho instytutu zaliznychnoho transportu Ukrainskoi derzhavnoi akademii zaliznychnoho transportu, Vudavnytstvo Donetskoho instytutu zaliznychnoho transportu, **27**, p. 159–162 (2011). [in Ukrainian]
5. *Rekomendacii z orhanizatsii resursozberihayuchykh zakhodiv na obyektakh lokomotyvnoho hospodarstva Ukrzaliznytsi. TST-0202*, Ministerstvo transport ta zviazku Ukrainy, Polohrafservis, 85 p. (2011). [in Ukrainian]
6. M. M. Uzdin, V. I. Kovalev, *Obschiy kurs zheleznykh doroh*, Akademiya, 252 p. (2012). [in Russian]
7. L. F. Khaskin, V. N. Matveev, *Ekonomika, orhanizatsiya i upravleniye lokomotivnym khoziaistvom*, Marshrut, 452 p. (2002). [in Russian]
8. S. Myamlin, L. Neduzha, Z.Urbutis, *Research of Innovations of Diesel Locomotives and Bogies*, Procedia Engineering, **134**, pp. 469-474, (2016).
9. M. S. Bezovska, Yu. V. Zelenko, V. M. Plakhotnyk, L. O. Yaryshkina, *Utylizatsiya naftovmisnykh tekhnolohichnykh shlamiv strukturnykh pidrozdiliv zaliznyts,* Zbirka tez dopovidey II Mizhnarodnoi Konferentsii z pytan povodzhennya z vidkhodamy vyrobnytstva ta spozhyvannya, p. 21–22 (2007). [in Ukrainian]
10. Yu. V. Zelenko, L. D. Tarasova, M. S. Bezovska *Pidvyschchennya rivnya ekolohichnoi bezpeky pry povodzhenni z vidpratsovanymy motornymy olyvamy zaliznychnoi infrastruktury: monographiya*, Vudavnytstvo «Litohraf», 150 p. (2016). [in Ukrainian]
11. Yu. V. Zelenko, S. V. Miamlin, *Ekoloho-ekonomichni zakhody stabilizatsii ta polipshennya stanu dovkillya na zaliznytshnomu transporti,* Zbirnyk naukovukh prats Dnipropetrovskoho natsionalnoho universytetu zaliznychnoho transport imeni akademika V. Lazariana «Problemy ekonomiky transportu», **7**, p. 47–53 (2014). [in Ukrainian]
12. M. S. Bezovska, Yu. V. Zelenko, L. O. Yaryshkina, O. B. Babenko, *Utylizatsiya naftoshlamiv ta mastyl, utvorenukh u strukturnykh pilrozdiliv zaliznyts*, Naukovo-praktychnyi zhurnal «Zaliznytchnyi transport Ukrainy», DNDTS UZ, **1**, p. 36–38 (2008). [in Ukrainian]
13. Zelenko Yu., Myamlin S., Sandovskiy M., *Scientific foundation of management of the environmental safety of oil product turnover in railway transport*, Litohraf, 332 p. (2014).
14. A. V. Hrechanikov, A. P. Platonov, S. H. Kovchur, A. S. Kovshur, *Novyie koahulyanty i flokulianty v protsessakh vodopodhotovki*, Vestnik Vitebskoho hosudarstvennoho universiteta, **23** (2012). [in Russian]
15. Zelenko Yu. V., Miamlin C, V., Neduzha L. O., *Parametrytshna ekolohiya na zaliznytshnomu transporti: pryntsypy, otsinka, kontrol, bezpeka: monohrafiya*, Litohraf, 240 p. (2014). [in Ukrainian]

Modeling of the atmosphere pollution from coal trains

Mykola Biliaiev[1,*], *Vitalii* Kozachyna[1], *Viktoriia* Biliaieva[2], *Mutiu Olatoye* Oladipo[3], and *Kateryna* Chernyatyeva[4]

[1] DNURT, Department hydraulics and water supply, 49010 Dnipro, Lazaryan Street, 2, Ukraine
[2] DNU, Assistant professor of aerohydromechanics and energy and masstransfer, 49050 Dnipro, Kozakov Street, 18, Ukraine
[3] GeoLab Drilling, Scientific researcher, 30680 Winder, Bill Rutledge Road, 800, USA
[4] DNURT, Lviv branch, 79052 Lviv, I. Blazhkevich Street, 12a, Ukraine

Abstract. Coal trains have an intensive impact on environment pollution. It is very important to predict adequately this impact during coal transportation. In Ukraine to predict atmosphere pollution from coal trains the regulatory model «OND-86» is used. This model does not take into account some important factors and coal train movement. We present numerical models to solve two problems:1) prediction of atmosphere pollution in the case of moving coal train; 2) prediction of atmosphere pollution in the case of additional special boards installation on the coal wagon. To solve these problems equation of coal duct convective- diffusive dispersion (Lagrange model of admixture dispersion) and equation of potential flow were used. Governing equations were numerically integrated using implicit difference schemes. For coding difference equations we used FORTRAN language. We present results of numerical experiments and laboratory experiments which illustrate the efficiency of the special additional boards installation on the coal wagon. These boards have «internal wing» and «external wing». The obtained results illustrate that installation of additional boards allows to minimize the atmosphere pollution near transport corridor.

1. Introduction

A problem of fugitive dust release from coal trains is of great importance in the problem of environmental protection and attracts scientist's attention in different countries [1-4].

For decreasing of the air pollution during coal transportation different measures are used. For example, one of the ways is wetting of a cargo with water or special solutions. Other way is to give special shape for cargo or to compact it. One of the most effective measures is using covers. Covers are placed on the wagon for full (Figure 1) or partial (Figure 2) covering of a cargo.

Fig. 1. Coal wagon with full covers (Australia) [5]

A literature search made by authors showed that research work in the field of dust release during coal transportation in wagons has not been conducted in Ukraine.

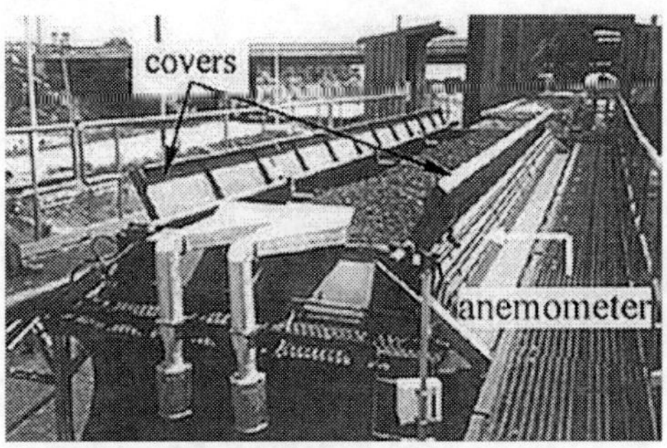

Fig. 2. Coal wagon with covers (Portugal) [1]

In addition, in Ukraine, to estimate the atmosphere pollution during coal transportation in open wagons the empirical model OND-86 is used. This model does not take into account some very important factors, for example: coal train movement, atmosphere turbulent diffusion, etc. Therefore, the prognostic results obtained with this model are not reliable. In addition, the problem of coal dust reduction from coal trains, practically, was not studied in Uktraine. Authors present some 'research directions" which are developed in Dnipro National University of Railway Transport named after academician V. Lazaryan (Ukraine, Project 0115U007227 "Environment protection at the elements of transport infrastructure").

* Corresponding author: diit.hydro.eco@gmail.com

2. Experimental section

For the experimental simulation, a 1:100 scale model of the open wagon 12-1592 type was used. This type of the open wagon is often used in Ukraine. The model of the open wagon is shown in Figure 3. We studied the case of "poor" loaded wagon.

Fig. 3. Model of the open wagon (poorly loaded).

Also a model of this open wagon with additional boards was used in the experiments. Additional board had an "internal" wing which was like a "shelf" on the board and an "external" wing (Figure 4). To show the form of these additional boards we present in Figure 4 the model of the wagon without front side additional board. It is supposed that additional boards can be made of wood or any other cheap material. Installation of these boards on the wagon does not imply special equipment.

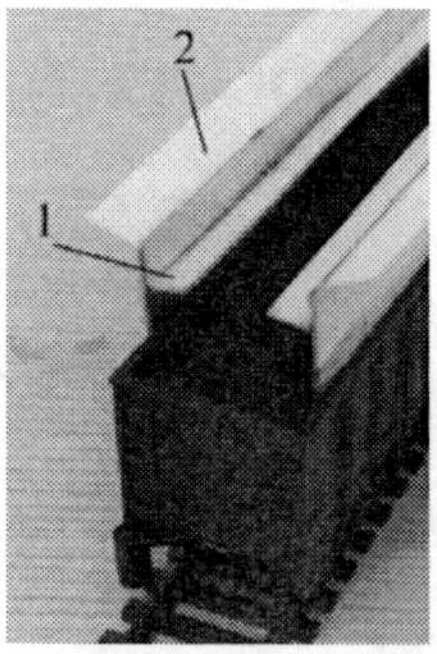

Fig. 4. Model of the open wagon with additional boards (model without front side additional board): 1 - "internal wing"; 2 – "external" board

Because of the problems with the wind tunnel equipment which is situated in Dnipro City, an industrial fan was used in the experiments to produce wind flow over the wagon. The Reynolds number in the experiments was $Re=10^5$. Coal from Mezhdurechenskaya Central Enriching Factory (Donetsk Region, Ukraine) was used in the experiments. Before each experimental run coal has been put into the drying stove for 3 hours. The temperature in the drying stove was +40^{0}C. After each experimental run coal dust which had been fallen on the surface of laboratory table was weighted. To weight coal dust we used balance KL168.

The small scale experiments showed that application of additional boards with "internal" and "external" wings reduces the amount of dust released approximately 22% of the quantity emitted from the open wagon without additional boards. The position of the "wings" on the additional board and their length influence the amount of dust released.

3. Mathematical model

Mathematical modeling of the atmosphere pollution during coal transportation was conducted in two scales: microscale and local scale.

To study atmosphere pollution near the wagon (microscale) we used fluid dynamics model of potential flow and equation of convective-diffuse dust dispersion. Model of potential flow allows to create a fast-calculating computer model. In this case the basic equation has the following form:

$$\frac{\partial^2 P}{\partial^2 x}+\frac{\partial^2 P}{\partial^2 y}=0 \qquad (1)$$

where P is velocity potential.

When applying this equation, it is assumed that the Y axis is directed vertically upwards.

Components of wind velocity are determined as

$$u=\frac{\partial P}{\partial x},\ v=\frac{\partial P}{\partial y}.$$

To solve (1) the following boundary conditions were used:

1) at the solid surfaces, upper boundary, the boundary condition was: $\partial P/\partial \mathbf{n}=0$, where $\mathbf{n}$ is unit vector of external normal to the boundary;

2) at the inlet boundary we set: $\partial P/\partial \mathbf{n}=V_n$, where V_n is known airflow velocity;

3) at the outlet boundary we set: $P=P_0$+const, P_0 is an arbitrary number (Dirichlet condition).

To simulate the coal dust dispersion from the wagon the mass transfer equation was used [6]:

$$\frac{\partial C}{\partial t}+\frac{\partial uC}{\partial x}+\frac{\partial (v-w_s)C}{\partial y}=$$
$$=\frac{\partial}{\partial x}\left(\mu_x\frac{\partial C}{\partial x}\right)+\frac{\partial}{\partial y}\left(\mu_y\frac{\partial C}{\partial y}\right)+ \qquad (2)$$
$$+\sum q_i\delta(x-x_i)\delta(y-y_i)$$

where C is dust concentration in the atmosphere; u, v are wind flow velocity components; w_s is dust fallout speed; $\mu=(\mu_x, \mu_y)$ are turbulent diffusion coefficients. q_i is pollutant emission rate at the point (x_i, y_i); $\delta(x-x_i)\delta(y-y_i)$ are Dirac's delta function, t is time.

Coal surface in the wagon is represented as a number of points (x_i, y_i). Pollutant emission rate q_i at any point (x_i, y_i) depends on the local wind speed. Local wind speed near the coal surface is computed using (1) and dependencies (2). To compute q_i we used empirical formulae [7] which include the local speed near the coal surface.

Initial and boundary conditions for (2) are discussed in [6].

For prediction of the atmosphere pollution in the transport corridor and in adjacent territory (local scale) in the case of coal train movement, we used the following equation of convective-diffusive dust dispersion [6]:

$$\frac{\partial C}{\partial t}+\frac{\partial uC}{\partial x}+\frac{\partial vC}{\partial y}+\frac{\partial (w-w_S)C}{\partial z}=$$
$$=\frac{\partial}{\partial x}\left(\mu_x\frac{\partial C}{\partial x}\right)+\frac{\partial}{\partial y}\left(\mu_y\frac{\partial C}{\partial y}\right)+\frac{\partial}{\partial z}\left(\mu_z\frac{\partial C}{\partial z}\right)+ \quad (3)$$
$$+\sum Q_i(t)\delta(x-x_i(t))\delta(y-y_i(t))\delta(z-z_i),$$

where C is dust concentration in the atmosphere; (x_i, y_i, z_i) are coordinates of dust emission point source on the surface of the cargo; $\delta(x-x_i(t))\delta(y-y_i(t))\delta(z-z_i)$ is Dirac's delta-function; $Q_i(t)$ is coal dust emission rate (intensity of dust emission from the wagon); t is time; μ_x, μ_y, μ_z are turbulent diffusion coefficients (computed using empirical formulae [8]); w_s is coal dust fallout speed; u, v, w are wind flow velocity components.

In the developed model each wagon is represented as a point source having emission rate $Q_i(t)$.This emission rate is calculated using empirical formula [7].

Worthy of note that coordinates x, y of each wagon depends on time t. These coordinates are changed in time and this change depends on the known train speed. This approach allows to simulate the movement of coal train with the speed we want to represent in the numerical experiment.

4. Numerical integration

Numerical integration of governing equations (1)-(3), was carried out using a rectangular difference grid. To create the form of the wagon, coal surface, geometrical form of the additional boards in 2D simulation (microscale level) we used "markers method" (porosity technique). In the case of 3D simulation "markers" show coal train route.

To solve (1) we wrote it in evolutionary form [7]:

$$\frac{\partial P}{\partial t}=\frac{\partial^2 P}{\partial^2 x}+\frac{\partial^2 P}{\partial^2 y}, \quad (4)$$

where t – fictitious time.

It is known, that when $t\to\infty$ the solution of this equation will approach to the solution of Laplace equation (1). To solve (4), it is necessary to set "initial' field of velocity potential in computational region at t=0. For example, before starting the calculation, we may set P=0 in the entire computational domain for t=0. Samarsky method [9] was used for numerical integration of (4). In this case difference equations of splitting are written as following:

1) first step of splitting:

$$\frac{P_{ij}^{n+\frac{1}{2}}-P_{ij}^{n}}{0.5\Delta t}=\frac{P_{i+1,j}^{n}-P_{i,j}^{n}}{\Delta x^2}+$$
$$+\frac{-P_{ij}^{n+\frac{1}{2}}+P_{i-1,j}^{n+\frac{1}{2}}}{\Delta x^2}+\frac{P_{i,j+1}^{n}-P_{i,j}^{n}}{\Delta y^2}+\frac{-P_{ij}^{n+\frac{1}{2}}+P_{i,j-1}^{n+\frac{1}{2}}}{\Delta y^2}; \quad (5)$$

2) second step of splitting:

$$\frac{P_{ij}^{n+1}-P_{ij}^{n+\frac{1}{2}}}{0.5\Delta t}=\frac{P_{i+1,j}^{n+1}-P_{ij}^{n+1}}{\Delta x^2}+$$
$$+\frac{-P_{ij}^{n+\frac{1}{2}}-P_{i-1,j}^{n+\frac{1}{2}}}{\Delta x^2}+\frac{P_{i,j+1}^{n+1}-P_{ij}^{n+1}}{\Delta y^2}+\frac{-P_{ij}^{n+\frac{1}{2}}-P_{i,j-1}^{n+\frac{1}{2}}}{\Delta y^2}. \quad (6)$$

At each splitting step, the unknown value of the velocity potential P is determined by the explicit formula of "running calculation" [6, 9]. The calculation is finished when the following condition is fulfilled:

$$\left|P_{ij}^{n+1}-P_{ij}^{n}\right|\le\varepsilon,$$

where ε is a small number (e.g., ε =0.001); n is an iteration number.

After determining the velocity potential field, the components of the air velocity vector are calculated at the sides of computational cell using dependencies:

$$u_{ij}=\frac{P_{ij}-P_{i-1,j}}{\Delta x}, v_{ij}=\frac{P_{ij}-P_{i,j-1}}{\Delta y}. \quad (7)$$

For the numerical integration of Eq. (2) we used the change – triangle difference scheme.

To integrate numerically equation (2) we represent the convective derivatives as following:

$$\frac{\partial uC}{\partial x}=\frac{\partial u^+C}{\partial x}+\frac{\partial u^-C}{\partial x},$$
$$\frac{\partial vC}{\partial y}=\frac{\partial v^+C}{\partial y}+\frac{\partial v^-C}{\partial y},$$

where

$$u^+=(u+|u|)/2, u^-=(u-|u|)/2,$$
$$v^+=(v+|v|)/2, v^-=(v-|v|)/2$$

For approximation of the derivatives we used the following expressions:

$$\frac{\partial u^+C}{\partial x}\approx\frac{u_{i+1,j}^{+}C_{ij}^{n+1}-u_{ij}^{+}C_{i-1,j}^{n+1}}{\Delta x}=L_x^+C^{n+1},$$

$$\frac{\partial u^-C}{\partial x}\approx\frac{u_{i+1,j}^{-}C_{i+1,j}^{n+1}-u_{ij}^{-}C_{ij}^{n+1}}{\Delta x}=L_x^-C^{n+1},$$

$$\frac{\partial v^+ C}{\partial y} \approx \frac{v^+_{i,j+1} C^{n+1}_{ij} - v^+_{ij} C^{n+1}_{i,j-1}}{\Delta y} = L^+_y C^{n+1} ,$$

$$\frac{\partial v^- C}{\partial y} \approx \frac{v^-_{i,j+1} C^{n+1}_{i,j+1} - v^-_{ij} C^{n+1}_{ij}}{\Delta y} = L^-_y C^{n+1} .$$

Time – depended derivative was approximated as following:

$$\frac{\partial C}{\partial t} = \frac{C^{n+1}_{ij} - C^n_{ij}}{\Delta t} .$$

For approximation of the second derivatives the following formulas were used:

$$\frac{\partial}{\partial x}(\mu_x \frac{\partial C}{\partial x}) \approx \tilde{\mu}_x \frac{C^{n+1}_{i+1,j} - C^{n+1}_{ij}}{\Delta x^2} - \tilde{\mu}_x \frac{C^{n+1}_{ij} - C^{n+1}_{i-1,j}}{\Delta x^2} = = M^-_{xx} C^{n+1} + M^+_{xx} C^{n+1},$$

$$\frac{\partial}{\partial y}(\mu_y \frac{\partial C}{\partial y}) \approx \tilde{\mu}_y \frac{C^{n+1}_{i,j+1} - C^{n+1}_{ij}}{\Delta y^2} - \tilde{\mu}_y \frac{C^{n+1}_{ij} - C^{n+1}_{i,j-1}}{\Delta x^2} = = M^-_{yy} C^{n+1} + M^+_{yy} C^{n+1}.$$

2D equation of the dust transfer can be written in the following difference form [6]:

$$\begin{aligned} &\frac{C^{n+1}_{ij} - C^n_{ij}}{\Delta t} + L^+_x C^{n+1} + L^-_x C^{n+1} + \\ &+ L^+_y C^{n+1} + L^-_y C^{n+1} + \sigma C^{n+1}_{ij} = \\ &= (M^+_{xx} C^{n+1} + M^-_{xx} C^{n+1} + \\ &+ M^+_{yy} C^{n+1} + M^-_{yy} C^{n+1}) + Q_{ij}\delta_{ij} . \end{aligned} \quad (8)$$

In this equation symbol δ_{ij} can be equal either "1" or "0", depending whether the source of dust emission is located in cell "ij" or not. Emission Q_{ij} *was* calculated as following:

$$Q_{ij} = Q_k / \Delta x / \Delta y ,$$

where Q_k is point source intensity in k-th point which is located in the difference cell "ij"

To solve the equation (8) we use the following 4 steps scheme of splitting [6]:

1) at the first step k=1/4 the difference equation is:

$$\begin{aligned} &\frac{C^{n+k}_{ij} - C^n_{ij}}{\Delta t} + \frac{1}{2}(L^+_x C^k + L^+_y C^k) + \frac{\sigma}{4} C^k_{ij} = \\ &= \frac{1}{4}(M^+_{xx} C^k + M^-_{xx} C^k + \\ &+ M^+_{yy} C^n + M^-_{yy} C^n) + \sum_{l=1}^{N} \frac{Q_l}{4} \delta_l ; \end{aligned} \quad (9)$$

2) at the second step k=n+1/2; c=n+1/4 the difference equation is:

$$\begin{aligned} &\frac{C^k_{ij} - C^c_{ij}}{\Delta t} + \frac{1}{2}(L^-_x C^k + L^-_y C^k) + \frac{\sigma}{4} C^k_{ij} = \\ &= \frac{1}{4}(M^-_{xx} C^k + M^+_{xx} C^c + \\ &+ M^-_{yy} C^k + M^+_{yy} C^c) + \sum_{l=1}^{N} \frac{Q_l}{4} \delta_l ; \end{aligned} \quad (10)$$

3) at the third step k=n+3/4; c=n+1/2 the difference equation is:

$$\begin{aligned} &\frac{C^k_{ij} - C^c_{ij}}{\Delta t} + \frac{1}{2}(L^+_x C^k + L^-_y C^k) + \frac{\sigma}{4} C^k_{ij} = \\ &= \frac{1}{4}(M^-_{xx} C^c + M^+_{xx} C^k + \\ &+ M^-_{yy} C^k + M^+_{yy} C^c) + \sum_{l=1}^{N} \frac{Q_l}{4} \delta_l ; \end{aligned} \quad (11)$$

4) at the fourth step k=n+1; c=n+3/4 the difference equation is:

$$\begin{aligned} &\frac{C^k_{ij} - C^c_{ij}}{\Delta t} + \frac{1}{2}(L^-_x C^k + L^+_y C^k) + \frac{\sigma}{4} C^k_{ij} = \\ &= \frac{1}{4}(M^-_{xx} C^k + M^+_{xx} C^c + \\ &+ M^-_{yy} C^c + M^+_{yy} C^k) + \sum_{l=1}^{N} \frac{Q}{4} \delta_l . \end{aligned} \quad (12)$$

The unknown value of the dust concentration from equations (9)-(12) on the new time layer is calculated with explicit formula of "running calculation". Initial condition for these equations is written as following:

$$C^1 \Big| = C(x, y, t^n), \; C^k \Big| = C^{k-1} \Big| .$$

For realizing of the boundary condition ∂C/∂n=0 on the solid walls, fictitious cells were used.

Main features of the difference scheme for integrating equation of dust convective-diffusive dispersion (3). Before numerical integration of this equation we split it as follows [6]:

$$\begin{aligned} &\frac{\partial C}{\partial t} + \frac{\partial uC}{\partial x} + \frac{\partial vC}{\partial y} + \frac{\partial (w - w_S) C}{\partial z} = 0, \\ &\frac{\partial C}{\partial t} = \frac{\partial}{\partial x}\left(\mu_x \frac{\partial C}{\partial x}\right) + \frac{\partial}{\partial y}\left(\mu_y \frac{\partial C}{\partial y}\right) + \frac{\partial}{\partial z}\left(\mu_z \frac{\partial C}{\partial z}\right), \\ &\frac{\partial C}{\partial t} = \sum Q_i(t) \delta(x - x_i(t)) \delta(y - y_i(t)) \delta(z - z_i). \end{aligned} \quad (13)$$

For numerical integration of the first equation from (13) we used the following formulae:

$$\frac{\partial uC}{\partial x} = \frac{\partial u^+ C}{\partial x} + \frac{\partial u^- C}{\partial x},$$

$$\frac{\partial vC}{\partial y} = \frac{\partial v^+ C}{\partial y} + \frac{\partial v^- C}{\partial y},$$

$$\frac{\partial wC}{\partial z} = \frac{\partial w^+ C}{\partial z} + \frac{\partial w^- C}{\partial z},$$

$$\frac{\partial u^+ C}{\partial x} \approx \frac{u^+_{i+1,jk} C^{n+1}_{ijk} - u^+_{ijk} C^{n+1}_{i-1,jk}}{\Delta x} = L^+_x C^{n+1},$$

$$\frac{\partial u^- C}{\partial x} \approx \frac{u^-_{i+1,jk} C^{n+1}_{i+1,jk} - u^-_{ijk} C^{n+1}_{ijk}}{\Delta x} = L^-_x C^{n+1},$$

$$\frac{\partial v^+ C}{\partial y} \approx \frac{v^+_{i,j+1,k} C^{n+1}_{ijk} - v^+_{ijk} C^{n+1}_{i,j-1,k}}{\Delta y} = L^+_y C^{n+1},$$

$$\frac{\partial v^- C}{\partial y} \approx \frac{v^-_{i,j+1,k} C^{n+1}_{i,j+1,k} - v^-_{ijk} C^{n+1}_{ijk}}{\Delta y} = L^-_y C^{n+1},$$

$$\frac{\partial w^+ C}{\partial y} \approx \frac{w^+_{ij,k+1} C^{n+1}_{ijk} - w^+_{ijk} C^{n+1}_{i,j,k-1}}{\Delta z} = L^+_z C^{n+1},$$

$$\frac{\partial w^- C}{\partial y} \approx \frac{w^-_{ij,k+1} C^{n+1}_{ij,k+1} - w^-_{ijk} C^{n+1}_{ijk}}{\Delta z} = L^-_z C^{n+1}.$$

where

$$u^+ = (u+|u|)/2, u^- = (u-|u|)/2, \quad v^+ = (v+|v|)/2, v^- = (v-|v|)/2 \text{, etc.}$$

Here we use designation w=w-w_s.

To solve the first equation from (13) we use the following 4 steps scheme of splitting:

1) at the first step the difference equation is:

$$\frac{C^k_{ijk} - C^n_{ijk}}{\Delta t} + \frac{1}{2}(L^+_x C^k + L^+_y C^k + L^+_z C^k) = 0 \qquad (14)$$

2) at the second step the difference equation is:

$$\frac{C^k_{ijk} - C^c_{ijk}}{\Delta t} + \frac{1}{2}(L_x C^k + L^-_y C^k + L^-_z C^k) = 0 \qquad (15)$$

3) at the third step we use difference expression (15);

4) at the fourth step we use difference expression (14).

The unknown value of dust concentration C is calculated from (14)-(15) using explicit formula of "running calculations". The second equation from (7) is similar to (4) that is why we can use Samarsky method [9] to solve it. To solve the third equation from (13) we used Eurler's method [9]. For coding difference equations we used FORTRAN language.

5. Results

Figure 5 shows coal dust concentration field over the open wagon (upper part of the wagon) with additional boards and in the case of cross wind.

From Figure 5 one can see that "internal" wings play role of a small "cover" which partially localize zone of dust emission in the wagon. But it is clear that for this case these "wings" are short. The "external" wings play the role of "flaps" and guide the wind flow with coal dust upward to the atmosphere layers with more high wind speed. Computing time to solve this problem was 5 s.

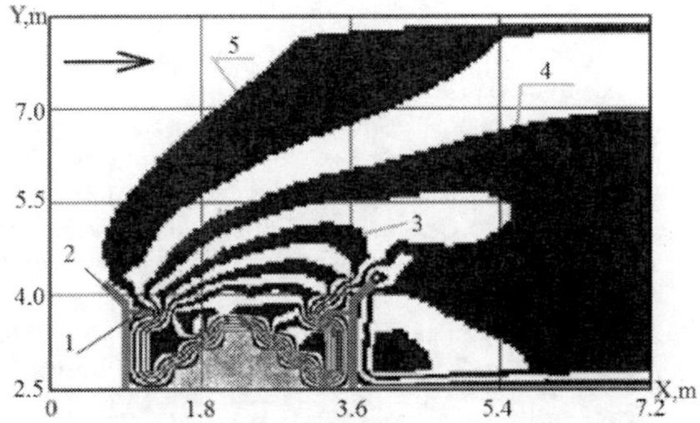

Fig. 5. Dust concentration field near the wagon with additional boards: 1 – "internal" wing, 2 – "external" wing, 3 – concentration C=0.36 mg/m^3, 4 – concentration C =0.22 mg/m^3, 5 – concentration C=0.04 mg/m^3

Figure 6 (scale 1:1000) shows predicted coal dust concentration at level z=5 m for coal train moving with speed 30 km/h and in the case of cross wind. In Figure 6 arrow indicates the direction of coal train movement. Numerical simulation has been carried out for site near Sinelnikovo Railway Station, Ukraine.

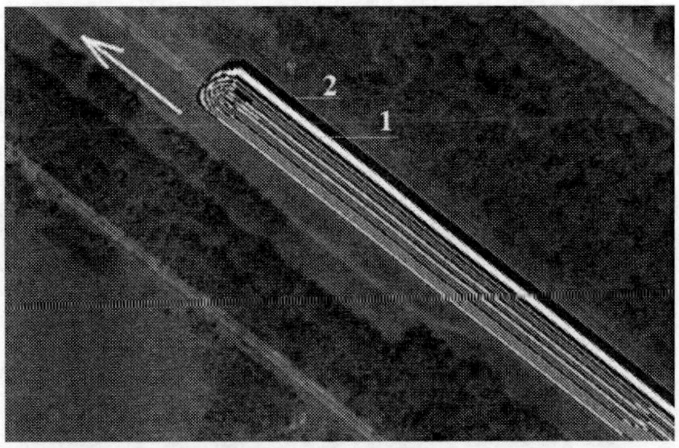

Fig. 6. Predicted coal dust concentration (level z=5 m, t=15 sec): 1 – C=0.31 mg/m^3; 2 – C=0.20 mg/m^3

Computing time to solve 3D problem was 3 sec.

6. Conclusions

In short, numerical models were developed to predict atmosphere pollution in case of fugitive dust emission from coal train. These models allow to take into account physical factors which influence pollutant dispersion in the atmosphere and which the Regulatory method OND-86 does not take into account. Small scale experiments showed that additional boards on the wagon can reduce coal dust emission.

Future research will focus on development quick computing 3D CFD model which allows to take into account the geometrical form of the wagon. This CFD model will be used do estimate effectiveness of additional boards installation on the wagon.

References

1. Ferreira, A.D., Viegas, D.X., Sousa, A.C.M. J. of Wind Eng. and Indust. Aerodyn. **91(10)**, 1271-1283 (2003).
2. Ferreira, A.D., Vaz, P.A. J. of Wind Eng. and Indust. Aerodyn. **92(7-8)**, 565-577 (2004).
3. Connell Hatch. Final Report Environmental Evaluation of Fugitive Coal Dust Emissions from Coal Trains Goonyella, Blackwater and Moura Coal Rail Systems Queensland Rail Limited. Queensland Government Environmental Protection Agency, Queensland (2008).
4. NSW Chief Scientist & Engineer. Final Report on the Independent Review of Rail Coal Dust Emissions Management Practices in the NSW Coal Chain. NSW Chief Scientist & Engineer, Sydney (2016).
5. Literature Review of Coal Train Dust Management Practices. NSW Environment Protection Authority (2014).
6. Marchuk, G. I. Mathematical modeling in the environmental problem. Nauka, Moscow (1982).
7. Vorobyov, Ye. O., Nikolenko, M.O., Khudyakova, I. O. J. News of AHI of DonNTU, **1**, 34-38 (2005).
8. Bruyatskiy, Ye. V. Atmosphere diffusion theory of radioactive emissions. Institufe of hydromechanics NAS of Ukraine, Kyiv (2000).
9. Samarskiy, A. A. Theory of differential schemes. Nauka, Moscow (1983).

Risk assessment in case of toxic chemical emission at railway transport

Viktoriia Biliaieva[1,*], *Polina* Mashykhina[2], *Ivan* Kalashnikov[3], *Oleksandr* Berlov[4], and *Ivan* Kravets[5]

[1] DNU, Department aerohydromechanics and energy and masstransfer, 49050 Dnipro, Kozakov Street, 18, Ukraine
[2] DNURT, Department hydraulics and water supply, 49010 Dnipro, Lazaryan Street, 2, Ukraine
[3] Kharkiv Branch Office «Design and Research Institute of Railway Transport» of the PJS «Ukrainian Railway», 61052 Kharkiv, Kotliar Street, 7, Ukraine
[4] PSACEA, 49600 Dnipro, Chernyshevsky Street, 24a, Ukraine
[5] DNURT, Lviv branch, 79052 Lviv, I. Blazhkevich Street, 12a, Ukraine

Abstract. Risk assessment during emission of toxic chemicals at railway transport is the problem of great scientific interest. To make such assessment we need special computer models. At present, in Ukraine,we have lack of such models. The authors present numerical models for territorial risk assessment in case of organized emissionsat railway transport (for example, emissions during locomotive movement) and in case of accident emissions (accident spills of dangerous cargo, emissions of NH_3 from railway tank, etc.). The basis of the developed numerical models is the system of fundamental equations of fluid dynamics. These equations are solved numerically using implicit schemes of splitting. The developed models allow to take into account some important factors which influence the territorial risk value: probability of atmosphere conditions, train route, transport infrastructure at railway stations, probability of emission site.Also the process of pollutant chemical transformation in the atmosphere is taken into account in the developed models. The developed models allow to predict territorial risk in case of moving source of emission (moving damaged railway tank).The results of numerical experiments are presented. These results illustrate territorial risk maps for different sites near Prydniprovska railway.

1. Relevance

During the operation of rail transport emissions of pollutants into the atmosphere can be divided into two classes. The first class is organized emissions to the atmosphere during the movement of diesel locomotives or other technical means. The second class of emissions is accidental emissions during transportation of dangerous cargo. In the event of an accident during the transportation of chemically hazardous substances, there is a risk of large-scale environmental pollution and damage to people. When considering both problems, a problem arises in assessing the territorial risk of toxic damage to humans [1 -4, 13, 15-20] . To carry out such assessment we need special mathematical models [5, 6, 7, 9, ,10, 12]. At present, such models are absent in Ukraine. Therefore, there is an important task of creating effective methods for assessing the territorial risk of various emissions of chemically hazardous substances in railway transport.

2. Governing equations

To simulate the process of toxic chemical dispersion in the atmosphere the transport equation is used [7, 8, 12, 14]:

$$\frac{\partial C}{\partial t}+\frac{\partial uC}{\partial x}+\frac{\partial vC}{\partial y}+\frac{\partial wC}{\partial z}+\sigma C=$$
$$\frac{\partial}{\partial x}\left(\mu_x\frac{\partial C}{\partial x}\right)+\frac{\partial}{\partial y}\left(\mu_y\frac{\partial C}{\partial y}\right)+\frac{\partial}{\partial z}\left(\mu_z\frac{\partial C}{\partial z}\right)+ \quad (1)$$
$$+\sum Q_i(t)\delta(x-x_i)\delta(y-y_i)\delta(z-z_i),$$

where u, v, w are the velocity components in x, y and z direction respectively; C is the concentration of toxic chemical; σ is the parameter taking into account the process of toxic gas decay or rain wash out; μ_x, μ_y, μ_z are the coefficients of turbulent diffusion in x, y and z direction respectively; x_i ,y_i ,z_i are the coordinates of point source of emission; $Q_i(t)$ is the intensity of pollutant emission; $\delta(x-x_i(t))\delta(y-y_i(t))\ \delta(z-z_i)$ is Dirac delta-function.

In the developed numerical model, the following profile of velocity component u and coefficient of diffusion μ_z is used [5]:

$$u=u_1\left(\frac{z}{z_1}\right)^n, \mu_z=k_1\left(\frac{z}{z_1}\right)^m, \quad (2)$$

* Corresponding author: diit.hydro.eco@gmail.com

where u_1 is the velocity at height z_1; k_1=0,2; n=0,16; m≈1. The following models to calculate the other diffusive coefficients are used:

$$\mu_y=\mu_x,\ \mu_y=k_0 u,$$

where k_0 is the empirical parameter.

The transport equation is used with the following boundary conditions [1, 5]:

- inlet boundary: $C|_{inlet}=C_E$, where C_E is the known concentration (very often C_E=0);
- outlet boundary: in numerical model the condition $C(i+1, j, k)=C(i, j, k)$ is used (this boundary condition means that we neglect the process of diffusion at this plane);
- top boundary and ground surface $\partial C/\partial n=0$.

To develop the numerical model the implicit difference scheme was used.

3. Modeling of land surface contamination

In case of accidents during transportation of toxic chemicals, an important problem is arisen: it is necessary justify economic losses, caused by land contamination. Worthy of note that is practically impossible to measure contamination area very quickly after an accident, because of lack of special equipment, big areas, which were affected by pollutant emission. So, the mathematical modeling is the main instrument to calculate the losses after the accident.

The following dependence is used to calculate mass of pollutant which is "captured" by land surface:

$$G=\int_0^T dt \iint_{S,z=0} (w_s+\mu\alpha)C(x,y,z=0,t)\,ds\,,$$

where S – specific area of the surface of the region (forest, crops, etc.), t – current time; α>0 – coefficient taking into account the "capture" of part of the impurity by the earth's surface [8]; T is the period of integration.

For calculation of this integral the formula of rectangles is used.

Worthy of note, that to calculate parameter G, it is necessary to know concentration field near the land surface. Besides, concentration near the land surface is changed during time. This concentration can be computed using equation (1).

4. Numerical integration

For numerical integration of the equation (1) finite-difference methods are used. First, equation (1) was split into a sequence of next equations:

$$\frac{\partial C}{\partial t}+\frac{\partial uC}{\partial x}=\frac{\partial}{\partial x}\left(\mu_x\frac{\partial C}{\partial x}\right),\ \frac{\partial C}{\partial t}+\frac{\partial vC}{\partial y}=\frac{\partial}{\partial y}\left(\mu_y\frac{\partial C}{\partial y}\right),$$

$$\frac{\partial C}{\partial t}+\frac{\partial (w-w_g)C}{\partial z}=\frac{\partial}{\partial z}\left(\mu_z\frac{\partial C}{\partial z}\right),$$

$$\frac{\partial C}{\partial t}+\sigma C=\sum Q_i(t)\delta(x-x_i(t))\delta(y-y_i(t)). \quad (3)$$

Each of first three equations from system (3) describes hazardous substances movement in one particular direction. This splitting is the bases for developing of locally-unidimensional difference scheme. For numerical integration of the first three equations, implicit alternately triangular difference scheme is used.

To build difference scheme, first of all the convective derivatives are represented as following [12]:

$$\frac{\partial uC}{\partial x}=\frac{\partial u^+C}{\partial x}+\frac{\partial u^-C}{\partial x},\ \frac{\partial vC}{\partial y}=\frac{\partial v^+C}{\partial y}+\frac{\partial v^-C}{\partial y},$$

$$\frac{\partial wC}{\partial z}=\frac{\partial w^+C}{\partial z}+\frac{\partial w^-C}{\partial z},$$

Where

$$u^+=\frac{u+|u|}{2};\ u^-=\frac{u-|u|}{2};\ v^+=\frac{v+|v|}{2};\ v^-=\frac{v-|v|}{2};$$

$$w^+=\frac{w+|w|}{2};\ w^-=\frac{w-|w|}{2}.$$

After that, approximation of the second derivatives is made by formulas from [12]:

$$\frac{\partial}{\partial x}\left(\mu_x\frac{\partial C}{\partial x}\right)\approx\mu_x\frac{C^{n+1}_{i+1,j,k}-C^{n+1}_{i,j,k}}{\Delta x^2}-\mu_x\frac{C^{n+1}_{i,j,k}-C^{n+1}_{i-1,j,k}}{\Delta x^2}=$$

$$=M^-_{xx}C^{n+1}+M^+_{xx}C^{n+1},$$

$$\frac{\partial}{\partial y}\left(\mu_y\frac{\partial C}{\partial y}\right)\approx\mu_y\frac{C^{n+1}_{i,j+1,k}-C^{n+1}_{i,j,k}}{\Delta y^2}-\mu_y\frac{C^{n+1}_{i,j,k}-C^{n+1}_{i,j-1,k}}{\Delta y^2}=$$

$$=M^-_{yy}C^{n+1}+M^+_{yy}C^{n+1},$$

$$\frac{\partial}{\partial z}\left(\mu_z\frac{\partial C}{\partial z}\right)\approx\mu_z\frac{C^{n+1}_{i,j,k+1}-C^{n+1}_{i,j,k}}{\Delta z^2}-\mu_z\frac{C^{n+1}_{i,j,k}-C^{n+1}_{i,j,k-1}}{\Delta z^2}=$$

$$=M^-_{zz}C^{n+1}+M^+_{zz}C^{n+1}.$$

For approximation of the convective derivatives formulas from [12] are used:

$$\frac{\partial u^+C}{\partial x}\approx\frac{u^+_{i+1,j,k}C^{n+1}_{i,j,k}-u^+_{i,j,k}C^{n+1}_{i-1,j,k}}{\Delta x}=L^+_xC^{n+1},$$

$$\frac{\partial u^-C}{\partial x}\approx\frac{u^-_{i+1,j,k}C^{n+1}_{i+1,j,k}-u^-_{i,j,k}C^{n+1}_{i,j,k}}{\Delta x}=L^-_xC^{n+1},$$

$$\frac{\partial v^+C}{\partial y}\approx\frac{v^+_{i,j+1,k}C_{i,j,k}-v^+_{i,j,k}C_{i,j-1,k}}{\Delta y}=L^+_yC^{n+1}$$

$$\frac{\partial v^-C}{\partial y}\approx\frac{v^-_{i,j+1,k}C_{i,j+1,k}-v^-_{i,j,k}C_{i,j,k}}{\Delta y}=L^-_yC^{n+1}$$

$$\frac{\partial w^+ C}{\partial z} \approx \frac{w^+_{i,j,k+1}C_{i,j,k} - w^+_{i,j,k}C_{i,j,k-1}}{\Delta z} = L^+_z C^{n+1}$$

$$\frac{\partial w^- C}{\partial z} \approx \frac{w^-_{i,j,k+1}C_{i,j,k+1} - w^-_{i,j,k}C_{i,j,k}}{\Delta z} = L^-_z C^{n+1}$$

Difference scheme for calculation chemically hazardous substance concentration is considered on the example of the first equation from system (3) because of similarity of the second and the third to the first one. Difference scheme is written as follow:
- on the first step of splitting:

$$\frac{C^k_{i,j,k} - C^n_{i,j,k}}{\Delta t} + L^+_x C^k = M^+_{xx}C^k + M^-_{xx}C^n; \quad (4)$$

- on the second step of splitting:

$$\frac{C^{n+1}_{i,j,k} - C^k_{i,j,k}}{\Delta t} + L^-_x C^{n+1} = M^+_{xx}C^n + M^-_{xx}C^{n+1}. \quad (5)$$

Difference scheme has the similar form for the second and the third equations from (3).

Unknown value of concentration C at each step of splitting (4), (5) is calculated with explicit formula of «running calculation». The last equation from (3) is solved with using of Euler method.

Further, the physical splitting is carried out on the differential level. When doing this splitting, the processes of convection, diffusion, «work» of sources and runoffs need to be separated. Splitting has following appearance [8]:

$$\frac{\partial C}{\partial t} + \frac{\partial uC}{\partial x} + \frac{\partial vC}{\partial y} + \frac{\partial (w - w_S)C}{\partial z} = 0$$

$$\frac{\partial C}{\partial t} = \frac{\partial}{\partial x}\left(\mu_x \frac{\partial C}{\partial x}\right) + \frac{\partial}{\partial y}\left(\mu_y \frac{\partial C}{\partial y}\right) + \frac{\partial}{\partial z}\left(\mu_z \frac{\partial C}{\partial z}\right) \quad (6)$$

$$\frac{\partial C}{\partial t} = \sum Q_i(t)\delta(x - x_i(t))\delta(y - y_i(t))\delta(z - z_i).$$

$$\frac{\partial C}{\partial t} + \sigma C = 0 \quad (7)$$

Need to be noted, that the third equation from (6) defines changing of hazardous substance concentration under the action of emission source with rate Q.

Equation (7) defines changing of hazardous substance concentration in the particular point under the chemical decay.

It is known, that to perform computational experiments other methods of numerical integration to control the correctness of numerical calculations need to be used. That is why developing of the other difference scheme for numerical integration of equation (1) is considered.

To describe this difference scheme, following dependencies are used [12]:

$$\frac{\partial C}{\partial t} \approx \frac{C^{n+1}_{ij} - C^n_{ij}}{\Delta t}, \quad \frac{\partial uC}{\partial x} = \frac{\partial u^+ C}{\partial x} + \frac{\partial u^- C}{\partial x},$$

$$\frac{\partial vC}{\partial y} = \frac{\partial v^+ C}{\partial y} + \frac{\partial v^- C}{\partial y}, \quad \frac{\partial wC}{\partial z} = \frac{\partial w^+ C}{\partial z} + \frac{\partial w^- C}{\partial z},$$

where

$$u^+ = \frac{u + |u|}{2}; \; u^- = \frac{u - |u|}{2}; \; v^+ = \frac{v + |v|}{2}; \; v^- = \frac{v - |v|}{2};$$

$$w^+ = \frac{w + |w|}{2}; \; w^- = \frac{w - |w|}{2}.$$

$$\frac{\partial}{\partial x}\left(\mu_x \frac{\partial C}{\partial x}\right) \approx \mu_x \frac{C^{n+1}_{i+1,j,k} - C^{n+1}_{i,j,k}}{\Delta x^2} - \mu_x \frac{C^{n+1}_{i,j,k} - C^{n+1}_{i-1,j,k}}{\Delta x^2} = $$
$$= M^-_{xx}C^{n+1} + M^+_{xx}C^{n+1},$$

$$\frac{\partial}{\partial y}\left(\mu_y \frac{\partial C}{\partial y}\right) \approx \mu_y \frac{C^{n+1}_{i,j+1,k} - C^{n+1}_{i,j,k}}{\Delta y^2} - \mu_y \frac{C^{n+1}_{i,j,k} - C^{n+1}_{i,j-1,k}}{\Delta y^2} = $$
$$= M^-_{yy}C^{n+1} + M^+_{yy}C^{n+1},$$

$$\frac{\partial}{\partial z}\left(\mu_z \frac{\partial C}{\partial z}\right) \approx \mu_z \frac{C^{n+1}_{i,j,k+1} - C^{n+1}_{i,j,k}}{\Delta z^2} - \mu_z \frac{C^{n+1}_{i,j,k} - C^{n+1}_{i,j,k-1}}{\Delta z^2} = $$
$$= M^-_{zz}C^{n+1} + M^+_{zz}C^{n+1},$$

$$\frac{\partial u^+ C}{\partial x} \approx \frac{u^+_{i+1,j,k}C^{n+1}_{i,j,k} - u^+_{i,j,k}C^{n+1}_{i-1,j,k}}{\Delta x} = L^+_x C^{n+1},$$

$$\frac{\partial u^- C}{\partial x} \approx \frac{u^-_{i+1,j,k}C^{n+1}_{i+1,j,k} - u^-_{i,j,k}C^{n+1}_{i,j,k}}{\Delta x} = L^-_x C^{n+1},$$

$$\frac{\partial v^+ C}{\partial y} \approx \frac{v^+_{i,j+1,k}C_{i,j,k} - v^+_{i,j,k}C_{i,j-1,k}}{\Delta y} = L^+_y C^{n+1},$$

$$\frac{\partial v^- C}{\partial y} \approx \frac{v^-_{i,j+1,k}C_{i,j+1,k} - v^-_{i,j,k}C_{i,j,k}}{\Delta y} = L^-_y C^{n+1},$$

$$\frac{\partial w^+ C}{\partial z} \approx \frac{w^+_{i,j,k+1}C_{i,j,k} - w^+_{i,j,k}C_{i,j,k-1}}{\Delta z} = L^+_z C^{n+1},$$

$$\frac{\partial w^- C}{\partial z} \approx \frac{w^-_{i,j,k+1}C_{i,j,k+1} - w^-_{i,j,k}C_{i,j,k}}{\Delta z} = L^-_z C^{n+1}.$$

The computational dependencies for the first equation from (6) are [12]:
- on the first step of splitting k=n+1/4:

$$\frac{C^k_{i,j,k} - C^n_{i,j,k}}{\Delta t} + \frac{1}{2}\left(L^+_x C^k + L^+_y C^k + L^+_z C^k\right) = 0; \quad (8)$$

- on the second step of splitting k=n+1/2, c=n+1/2:

$$\frac{C^k_{i,j,k} - C^c_{i,j,k}}{\Delta t} + \frac{1}{2}\left(L^-_x C^k + L^-_y C^k + L^-_z C^k\right) = 0; \quad (9)$$

- on the third step of splitting k=n+3/4, c=n+1/2:

$$\frac{C^k_{i,j,k} - C^c_{i,j,k}}{\Delta t} + \frac{1}{2}\left(L^-_x C^k + L^-_y C^k + L^-_z C^k\right) = 0$$

- on the fourth step of splitting k=n+1; c=n+3/4:

$$\frac{C^k_{i,j,k} - C^n_{i,j,k}}{\Delta t} + \frac{1}{2}\left(L^+_x C^k + L^+_y C^k + L^+_z C^k\right) = 0$$

Considered splitting scheme is called alternately triangular. On the bases of dependencies (8), (9) concentration of hazardous substances in the difference cells under the influence of air flow is computed using «running calculation».

Next calculation of concentration changing in cells under the influence of diffusion is made:
- on the first step of splitting [11]:

$$\frac{C^{n+\frac{1}{2}}_{i,j,k} - C^n_{i,j,k}}{\Delta t} = \left[\mu_x \frac{-C^{n+\frac{1}{2}}_{i,j,k} + C^{n+\frac{1}{2}}_{i-1,j,k}}{\Delta x^2}\right] + \left[\mu_y \frac{-C^{n+\frac{1}{2}}_{i,j,k} + C^{n+\frac{1}{2}}_{i,j-1,k}}{\Delta y^2}\right] + \left[\mu_z \frac{-C^{n+\frac{1}{2}}_{i,j,k} + C^{n+\frac{1}{2}}_{i,j,k-1}}{\Delta z^2}\right] \quad (10)$$

- on the second step of splitting [11]:

$$\frac{C^{n+1}_{i,j,k} - C^{n+\frac{1}{2}}_{i,j,k}}{\Delta t} = \left[\mu_x \frac{C^{n+1}_{i+1,j,k} - C^{n+1}_{i,j,k}}{\Delta x^2}\right] + \left[\mu_y \frac{C^{n+1}_{i,j+1,k} - C^{n+1}_{i,j,k}}{\Delta y^2}\right] + \left[\mu_z \frac{C^{n+1}_{i,j,k+1} - C^{n+1}_{i,j,k}}{\Delta z^2}\right] \quad (11)$$

On the bases of dependencies (10), (11) concentration of hazardous substances in the difference cells is computed using «running calculation».

The final stage is calculating of hazardous substance concentration under the action of emission sources and precipitations. So, for solving of the third equation from system (6) dependency (12) is used (obtained on the base of Euler method [11]):

$$C^{n+1}_{ijk} = C^{n+A}_{ijk} + \Delta t \sum Q_{ijk}(t)\delta_i(x - x_i(t))\delta_i(y - y_i(t))\delta_i(z - z_i). \quad (12)$$

Here the following designation of hazardous substance emission rate is used:

$$Q_{ijk} = Q_k / \Delta x / \Delta y / \Delta z .$$

where Q_k is known emission rate value of source.

To solve equation (7) dependency (13), obtained from Euler method, is used:

$$C_{ijk}{}^{n+1} = C_{ijk}{}^{n} - \Delta t \sigma C_{ijk}{}^{n} \quad (13)$$

Initial condition for each differential equation is written as following:

$$C^1\Big| = C(x, y, t^n);\ C^k\Big| = C^{k-1}\Big| .$$

5. Methodology of territorial risk assessment

As known, spreading of dangerous factor in case of toxic substances accidental emission has significantly probabilistic nature. This nature of spreading is depended from:
- the repetition of different velocity values, the direction of the wind;
- probability of atmospheric stability;
- accident scenario;
- place of emission, emission rate.

Therefore, in assessing the territorial risk, we will take into account these factors. Note that the zone of chemical contamination is different for different meteorological situations $P(W_i)$.

Indeed, at high air velocity the concentration of dangerous substance in the air decreases, and at low speed it increases, because there is no rapid dispersion. In addition, the process of dispersion of contaminations is influenced by the state of the atmosphere, which in the constructed models is taken into account by the vertical coefficient of atmospheric diffusion.

When constructing a method of risk assessment, by a specific meteorological situation the specific significance of the velocity and direction of the wind and the stability of the atmosphere is meant. As known, the probability of realization of a concrete meteorological situation for a given region is determined on the basis of statistical data (or on the basis of meteorological forecasting):

$$P(W_i) = N_\Pi / \mathrm{T}, \quad (14)$$

where N_Π is number of days (hours), corresponding to a particular meteorological situation; T is period of observing.

The probability of a person is in the zone of defeat will be determined as follows:

$$P(W)_\Sigma = \sum_{i=0}^{n} P(W_i), \quad (15)$$

where $P(W)_\Sigma$ is probability of all considered meteorological situations, in which a person is in a zone where the concentration of a dangerous substance exceeds a certain threshold, at which there is a damage of certain severity.

To calculate the probability of being in a zone of toxic damage when emitting a dangerous substance it is necessary, for a specific point of the calculated area, to perform calculations by the formula (15).

6. Algorithm of solving

The territorial risk assessment of chemical damage in the probable meteorological situation *P(W)* will be carried out in the following sequence:

1) at the first stage of the task solving, a data block is created to initiate the event (possible place of emission of a chemically hazardous substance, emission rate, emission mode, type of chemical agent);
2) at the second stage, a data block is formed about probable meteorological conditions $P(W_i)$, characteristic of the area where the railway is located;
3) at the third stage, the level of chemical contamination for probable meteorological conditions is calculated (at this stage the numerical integration of the masstransfer equation for a particular meteorological situation is carried out);
4) at the fourth stage, the zones where the concentration exceeds the threshold (for example, lethal concentration, Maximum permissible concentration etc.) in a specific meteorological situation are determined;
5) at the fifth stage, the construction of a matrix (risk matrix) for the given object is carried out.

To perform calculations in order to determine the value of the territorial risk of chemical damage in the case of an accident on the railway, the following parameters should be specified:

1. Dimensions of calculated zone.
2. Location of railway.
3. Location of inhabited area.
4. Meteorological parameters.
5. Hazardous substances emission rate.
6. Emission source coordinates.
7. Receptor coordinates.
8. Probability of each meteorological situation.

The result of the simulation is the matrix of the territorial risk or isoline of the concentration of the toxic substance. If necessary, the user can "include" in the calculation the probability of an accident location, and so on. Based on the developed algorithm, a computer program «PR-Prob» was created. FORTRAN was used for coding difference equations.

7. Results

The developed «PR-Prob» was used to solve the following problem. A train with liquid ammonia (NH_3) moves near Railway Station «Kamyanske» (Dnipropetrovsk Region, Ukraine) and at time t=0 the instant emission of ammonia takes place. It is assumed that at time t=0, the amount of emitted NH_3 was Q=6 ton (it means 10% of the total amount of liquid cargo). Because of this emission, toxic chemical cloud is formed near the railway (Figure 1, dimensions of the computational region: 1.75km x 1.3km).

The train keeps moving after the toxic chemical emission and long term emission of NH3 (Q=10kg/s) follows the instant emission. Therefore, there is following scenario of complex emission: «instant exhaust of toxic chemical + long term emission of it».

Formation of polluted zones in the atmosphere after the accident is shown in Figures 2, 3. Computation results were obtained for wind speed 7m/s, train speed 8m/s.

Fig. 1. Toxic chemical cloud after NH_3 emission, *t*=3.5min, level *z*=7m

Fig. 2. Toxic chemical cloud after NH_3 emission, *t*=12 min, level *z*=7m

Fig. 3. Toxic chemical cloud after NH_3 emission, *t*=18min, level *z*=7m

It is clear from Figures 2, 3 that in the case of accident large cloud of pollution is formed near the railway with further covering the inhabited area and creating the threat of toxic damage for people.

In Figure 4 the matrix of potential territorial risk for the meteorological situations, which are typical for this region in June, is formed.

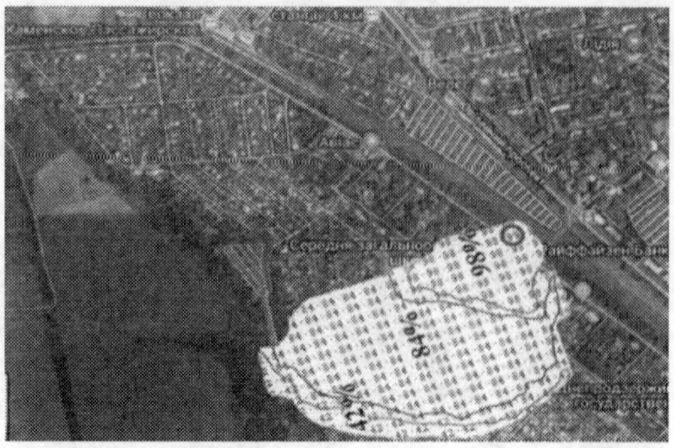

Fig. 4. Matrix of potential territorial risk, t=15min, level z=2m

Matrix of potential territorial risk shows the probability of people toxic damage at any time after the accident. This information allows us to separate zones, which are extremely dangerous for people in inhabited area for all possible weather conditions in the region.

In Table 1 we present computational data about the dynamics of land surface contamination. As we see from Table 1, the contaminated surface is increased during time.

Table 1. Dynamics of land surface contamination.

Time, s	112	169	239
Contaminated area, м^2	0,21×10^6	1,04×10^6	1,86×10^6

In Table 2 we present computational data about the dynamics pollutant mass settled on the land after the accident. This data allows to assess the damage after the accident and can be used to calculate economic losses, caused by the accident. So, in case of accidents during transportation, the developed model allows to make platform of scientific justification losses, caused by an accident.

Table 2. Pollutant mass settled on the land, g.

Time, min	Area
4	0,11×10^6
14	0,27×10^6
22	0,38×10^6

8. Conclusions

The paper shows the numerical approach to risk assessment in case of toxic chemicals emissions during railway transportations. This approach is based on numerical simulation of toxic chemical dispersion on the base of 3D equation of mass transport. For numerical integration of the governing equation, implicit difference schemes of splitting are used. The unknown value of toxic chemical concentration is calculated using explicit formulae on each step of splitting.

The proposed numerical models can be used to compute zones near railways which are influenced by the emissions from railway transport.

References

1. V.A. Kotlyarevskii, A.V. Zabegayeva, *Accidents and disasters: caution and aftermath liquidation*, (ASV, Moscow, 2001).
2. M.M. Biliaiev, E.Yu. Gunko, P.B. Mashykhina, *Mathematical modeling in tasks of ecological safety and emergencies monitoring*, (Akcent PP, Dnipro, 2013).
3. M.M. Biliaiev, O.V. Berlov, Cal. of sc. works NGU, **42**, 160-167, 2013.
4. O.V. Berlov, Sc. bul. of civ. eng., **1(75)**, 185-189, 2014.
5. M.E. Berlyand, *Modern problems of atmosphere diffusion and atmosphere pollution*, (Gidrometeoizdat, Leninhrad, 1975).
6. N.G. Gusiev, V.A. Belyaev, *Radioactive releases in biosphere*, (Energoatomizdat, Moscow, 1991).
7. Yu.A. Izrael, *Ecology and monitoring the environment state*, (Gidrometeoizdat, Moscow, 1984).
8. G.I. Marchuk. *Mathematical modeling in environmental problems*, (Nauka, Moscow, 1982).
9. P.B. Mashykhina, Bul. of Dn. nat. univ. of rai. tr., **27**, 138-142, 2009.
10. T.I. Rusakova, Bul. of Dn. nat. univ. of rai. tr., **6 (48)**, 32-45, 2013.
11. A.A. Samarsky, *Theory of difference schemes*, (Nauka, Moscow, 1983).
12. M.Z. Zgurovsky, V.V. Skopetsky, V.K. Khrusch, M.M. Biliaiev, *Numerical modeling of pollution spreading in environment*, (Nauk. dumka, Kyiv, 1997).
13. A.A. Shvyriaev, V.V. Menshykov, *Risk assessment of the impact of pollution in investigated region*, (MGU, Moscow, 2004).
14. M.M. Biliaiev, NATO Science for Peace and Security. Series C: Environmental Security, 87-91, 2012.
15. Michal Kisa, Ludovit Jelemensky and Jan Stopka, IChemE Symposium series № **153**, 1-8, 2007.
16. Ondrej Zavila, Pavel Dobes, Jakub Dlavka, Jan Bitta, The science for population protection, **2**, 1-8, 2015.
17. *Principles for environmental risk assessment of the sediment compartment: Proceedings of the topical scientific workshop*, (European Chemicals Agency, Helsinki, Finland, 2014).
18. *Risk Assessment Methodology for Hazardous Substances. How to assess the risk, cost and benefit of new hazardous substances for use in New Zealand*, (New Zealand, 2018).
19. *Risk Assessment for Toxic Air Pollutants: A Citizen's Guide*, (EPA, USA, 1991).
20. Werner K. Graber, Fritz Gassmann, Mathematics and Computers in Simulation (MATCOM), Paul Scherrer Institute, **52**, 413-426, 2000.

Test guidelines for evaluation real driving emission two-way vehicles

Michalina Kamińska[1,*], *Łukasz* Rymaniak[1], *Paweł* Daszkiewicz[1] and *Piotr* Lijewski[1]

[1] Faculty of Transport Engineering, Poznan University of Technology, Poland

Abstract. The article discusses the results of tests carried out under real operating conditions, based on which a custom test for road-rail vehicles was developed. The proposed test reflects the working conditions in which road-rail vehicles are used. The measurements were performed for a vehicle equipped with an internal combustion engine during five stages of operation. Measuring cycles included driving on a paved road, rolling on tracks and running on a track without load, with a load of 90 tons and with a load of 140 tons. The article compares the legislative guidelines with actual operation, thanks to which guidelines for the road-rail test for assessing the exhaust emission of harmful gases were developed. Based on the recorded data, a test proposal was described, taking into account the operating time share during the test, the average speed of travel and the duration of the test cycle. In addition, guidelines on the test procedure in the proposed test are discussed.

1 introduction

The type approval tests concerning the exhaust emission of harmful gases emitted from non-road vehicles are carried out on the engine dynamometer. Dynamometer measurements are to some extent designed to simulate the conditions of actual engine operation, but they are far from the real parameters of their operation. As proven in [1–7], the test cycles described do not reflect the real operating conditions. Thus, the test results do not make it possible to analyze the vehicle's real impact on the environment and human health. The qualitative and quantitative measurements of exhaust emissions in laboratory conditions often differ from the actual emission of exhaust gases. This is mainly due to the large variety of machines and vehicles that fall into the NRMM (Non-Road Mobile Machinery) group. In order to obtain reliable results, it is necessary to carry out tests in real operating conditions, similarly to road vehicles (cars and heavy vehicles). Currently, for passenger cars and heavy vehicles, tests are carried out in road conditions as described by the RDE (Real Driving Emissions) tests. These tests enable the verification of vehicles ecological indicators in a wide spectrum of their drive systems and allow to determine the conformity of their operation related to the relevant type approval standards. Such tests also make it possible to take into account the local ambient conditions.

The development of motor vehicles is primarily aimed at reducing their negative impact on the natural environment. This is achieved mainly through the introduction of new technologies in their construction. This applies to individual parts (improvement of the material quality) as well as entire propulsion systems, e.g. by their hybridization. In recent years, substituting the mechanical drive in functional systems (power steering, air conditioning, heating, etc.) with electric solutions is becoming increasingly popular in off-road and heavy vehicles, which requires the improvement of electrical systems parameters and the use of innovative solutions [8–10]. Road-rail vehicles are becoming more and more popular. Their functionality and operation economics allow for their wider use not only on the railway, but also in complex storage systems. Therefore, the proposed research activities become necessary to include them in various types of warehouse processes - for example by determining the economic balance. These results can be an extension to process analyzes, such as [11]. The implementation of the presented works can also be used in the analysis of the vehicle's life cycle. Measurements of emissions in various test drive cycles allow to determine the level of vehicle wear, as well as the wear of individual elements (mainly the engine construction solutions and exhaust aftertreatment systems). Such activities allow for detailed simulation and calculations of fleet operational costs, an example of such considerations is presented in [12]

2 Current legal regulations concerning the exhaust emissions measurement in road-rail vehicles

In the European Union, combustion engines used in NRMM (Non-Road Mobile Machinery) vehicles are tested using two test cycles. The total emission of pollutants measured in tests must not exceed the limit values defined in [97/68/EC].

The NRSC (Non-Road Stationary Cycle) test is a stationary cycle used to measure the emission of gaseous compounds and particulates. It is carried out for Stage I-

* Corresponding author: michalina.kaminska@put.poznan.pl

IV norms for CI engines with a net power of 19 kW ≤ P ≤ 560 kW operating at constant or variable speed. The test consists of a number of speed and load phases that cover the typical range of CI engines operating points used in off-road vehicles. For each of the test phases, the concentration of the tested exhaust component, the exhaust gas flow rate as well as the power and weighted averages of the measured values are determined. In a fixed sequence of operating conditions sets, the heated engine is tested continuously by taking a sample of the raw exhaust gas from the exhaust system. Particulate measurement is carried out by diluting the sample with conditioned air and passing such a prepared mixture through the filter (one sample for the entire test). The sample can also be collected on separate filters, one for each phase tested, and then the weighted average values for the entire test are calculated. The number of grams of each pollutant per kilowatt-hour is calculated in a strictly defined way. The test includes characteristics showing how to conduct a test cycle for the various devices defined in Directive [2004/26 / EC].

The NRTC (Non-Road Transient Cycle) test is a non-stationary cycle used to measure gaseous compounds and particulates emission. It is carried out as a part of Stage IIIB and IV norms for CI engines with net power in the range of 19 kW ≤ P ≤ 560 kW, working at variable speeds. It is important to take into account the variability of engine operation parameters. Two cycles are performed as a part of the test. The cold start cycle consists of conditioning the engine at 20-30°C. It is carried out after natural or forced cooling of the engine. The hot start cycle is performed in a warm state, 20 minutes after conditioning, immediately after the first cycle. During the test, the total exhaust emission of harmful compounds is obtained. The work output generated by the engine over the entire test cycle is determined by integrating power versus cycle time, using torque feedback and rotational speed brakes. The concentration of gaseous components for the entire test cycle can be determined using two methods. The first method is to measure the components concentration in the raw exhaust gas by integrating the analyzer signal in accordance with the procedure described in Directive [2010/26/EU]. It is also possible to measure the concentration in the raw exhaust gas in the CVS (Constant Volume Sample) system. Particulate measurement is carried out by taking a sample of exhaust gas through a suitable filter by means of a full or partial dilution method. The flow rate of diluted or non-diluted exhaust gas throughout the cycle is determined depending on the method used. To determine the mass of each pollutant measured per kilowatt-hour, the mass emission value is divided by the work output of the tested vehicle. The emission expressed in g/kWh is measured during each test drive cycle, for cold and hot start. The total exhaust emission is determined by applying 10% weight for cold start and 90% for hot start results.

Based on the operating parameters determined in the NRTC test, the time density characteristics were created as a function of engine rotational speed - load (rpm-Nm), as shown in Fig. 1. The distribution of operating parameters expressed in % is presented in half-closed intervals. The largest operating time share in single operating point (7.4%) occurs in the speed range of ⟨0%; 10%⟩ and load of ⟨0%; 10%⟩. The highest time density can be found for the range characterized by engine speed of ⟨50%; 110%⟩ and small and average torque values in the range ⟨0%; 60%⟩, in which the obtained values exceed 50%. Parameters exceeding 100% for the crankshaft speed are a result of the test methodology used, which requires temporarily overloadinf the engine during the test cycle.

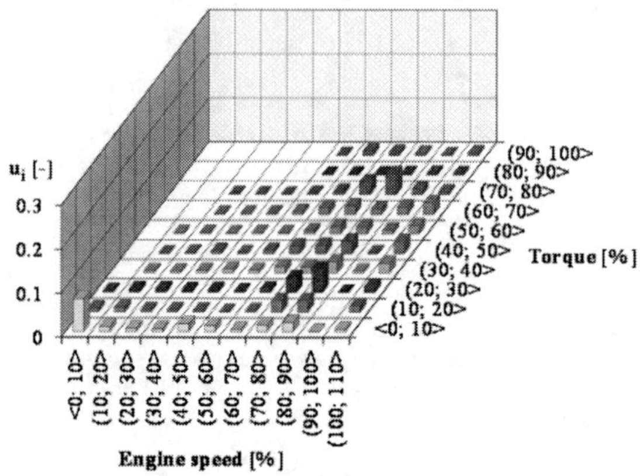

Fig. 1. Time densities for a standardized NRTC test in engine speed and load ranges

3 Test methodology

A road-rail tractor was used in the conducted tests (fig. 2, Table 1), it was equipped with an engine with a maximum power of 116 kW and a maximum torque of 640 Nm. According to the technical data, the vehicle complied with the Stage IV emission norms, thanks to, among others, exhaust aftertreatment systems, such as: DOC (Diesel Oxidation Catalyst) and DPF (Diesel Particulate Filter). The vehicle had a maximum traction power of 45 kN on the track and 55 kN on the road. The maximum speed was 25 km/h.

Table 1. Test vehicle parameters

Parametr	**Wartość**
Maximum power	116 kW
Maximum torque	640 Nm
Exhaust emission norm	Stage IV
Maximum pull on road	55 kN
Maximum pull on track	45 kN
Vehicle mass	8200 kg
Pull weight limit	140 tons
Maximum vehicle speed on road/track	25 km/h

The road-rail tractor was adapted to move both on the dirt road and railway tracks thanks to the use of circular guiding rollers to move along the tracks together with traction and buffer devices. Equipping the vehicle with additional devices made it possible to perform maneuver work necessary for the tests to be carried out.

Fig. 2. Tested vehicle – road-rail tractor

The PEMS (Portable Emissions Measurement System) measuring equipment was used to carry out the exhaust emissions measurements in real operating conditions of a road-rail vehicle. The concentration measurement of exhaust gases and its mass flow was made using the mobile SEMTECH DS apparatus. The exhaust emissions measurement was done by sending the gas sample into the device via a measuring probe maintaining the temperature of 191°C (in order to prevent the condensation of hydrocarbons before measurement). The device would then filter the solid particles from the exhaust gases that could affect the proper operation of the apparatus. The HC concentration is measured using the FID (Flame Ionization Detector) analyzer. The exhaust gases are further cooled down to 4°C, after which the measurement of the concentration of nitrogen oxides is carried out using the NDUV method (Non-Dispersive Ultraviolet). The concentration of CO and CO_2 is determined by using the NDIR method (Non-Dispersive Infra Red).

The measurement of particulate concentration was carried out using the AVL MSS 483 mobile analyzer (Micro Soot Sensor). This device operates based on the photoacoustic method, which enables high accuracy measurement (down to 1 μg/m³). The gas sample volume in the resonant measuring chamber is alternately heated and cooled. This affects the creation of vibrations of the elastic medium. Vibrations are recorded by sensitive microphones as a sound wave, and the concentration of solid particles is directly proportional to the generated sound characteristics.

The tests were supplemented with the ambient air measurement using the Micro PEMS Axion R/S+ mobile analyzer, which made it possible to also take into account the atmospheric pollution. Using the non-dispersive NDIR infrared analyzer and the electrochemical analyzer, the concentrations of gaseous compounds (CO_2, CO, HC, NO_x) were measured. A method based on the Laser Scatter effect (the measurement of particle velocity) was used to measure the particle concentration.

Data from the vehicle's diagnostic system was sent directly to the central measuring equipment unit. This allowed for continuous recording of the engine operating parameters (crankshaft speed, load generated). During the tests, a GPS position signal (Global Positioning System) was used to determine the position as well as the speed of the vehicle.

The research was divided into five stages. The first stage included the vehicle driving along an unpaved dirt road. Then the road-rail vehicle was rolled onto the railways. The third stage included a tractor equipped with a wheel set driving on the track without additional load. Then a load of 90 tons was added to the vehicle. The last stage consisted of driving with an extra load of 140 tons. Traffic parameters describing the performed test cycles are presented in Table 2.

Table 2. Driving parameters of the test vehicle

No.	Test drive cycle	Parameter			
		V_{max} [%]	$V_{śr}$ [%]	s [%]	t [%]
I	Dirt road driving	72	26	28	12
II	Moving onto the rail tracks	–	–	–	–
III	Track driving with no load	36	15	24	18
IV	Track driving with a load of 90 tons	36	8	38	48
V	Track driving with a load of 140 tons	18	2	10	22
I–V	Whole test	72	10	100	100

4 Operating parameters in real use

In order to develop the driving cycle, an analysis of road tests was carried out. For the measurements performed, the operating time shares were determined in the speed and load ranges of the test vehicle at individual stages of the test cycle and for the journey as a whole. All research stages were characterized by the highest values of time density, exceeding 60% in the low rotational speeds range of ⟨0%; 30%⟩ and at small loads (10%; 30%⟩. For road travel (Fig. 3) and track travel (Fig. 5) without load, the operating points were distributed mostly in the ranges of small and medium values of engine operating parameters ⟨0%; 50%⟩ and (10%; 40%⟩, where their share was more than 85% of the total drive time. Moving the test vehicle onto tracks (Figure 4) was characterized by a 98% share of operating time in a very narrow range of parameter values, defined by the speed range of ⟨0%; 10%⟩ and torque range of (10%; 30%⟩. The largest range of work parameters was found for a 90-ton load track drive (Fig. 6), for which the highest time-density values (85%) were obtained for almost the entire range of the rotational speed with a load in the range of (10%; 40%⟩. When driving with a load of 120 tons (Fig. 7), the largest work time (83%) was obtained for the smallest ranges of engine operation parameters ⟨0%; 10%⟩ and ⟨0%; 20%⟩. In Figure 8 a set of characteristics is presented which is a summary of all stages of research.

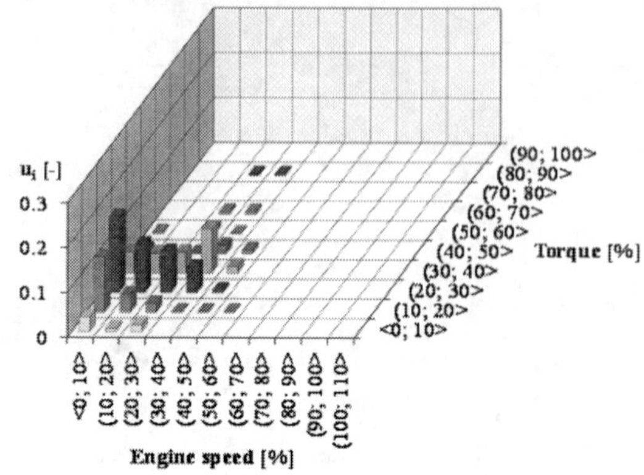

Fig. 3. Time density characteristic for the vehicle driving along the road without load presented in the engine speed and load intervals

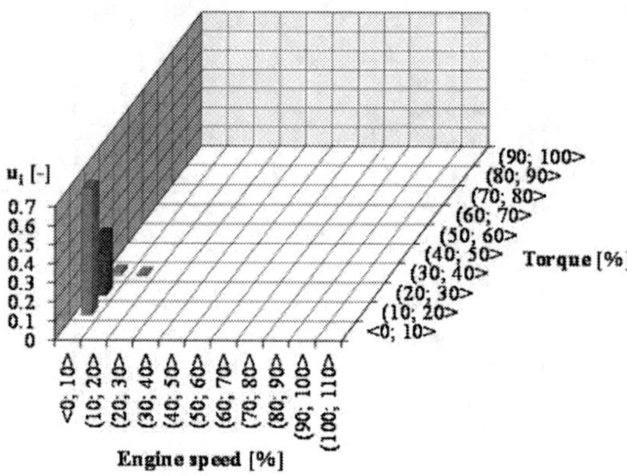

Fig. 4. Time density characteristic for the vehicle being moved onto rail tracks presented in the engine speed and load intervals

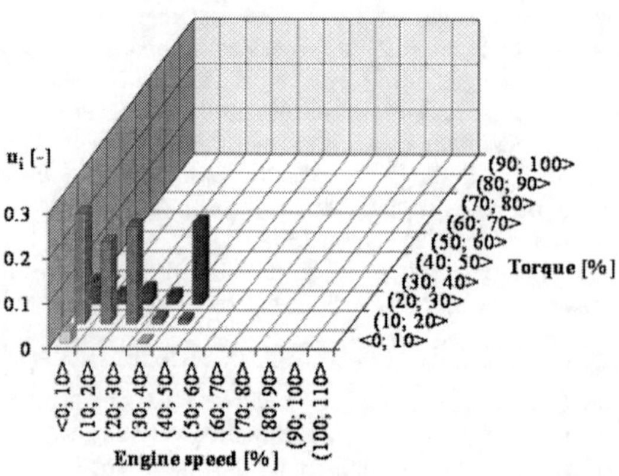

Fig. 5. Time density characteristic for the vehicle driving on the track without load presented in the engine speed and load intervals

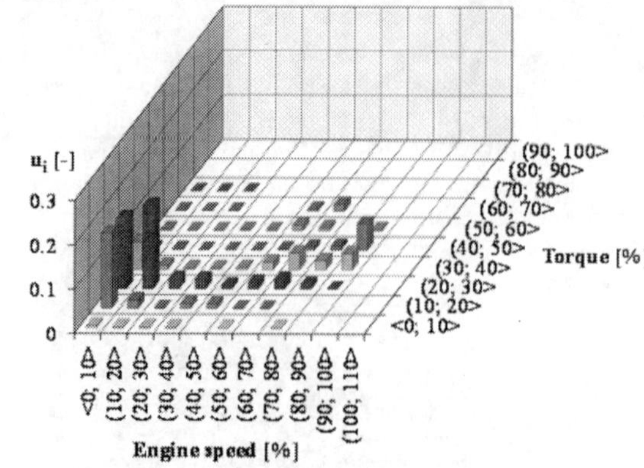

Fig. 6. Time density characteristic for the vehicle driving on the track with a load of 90 tons presented in the engine speed and load intervals

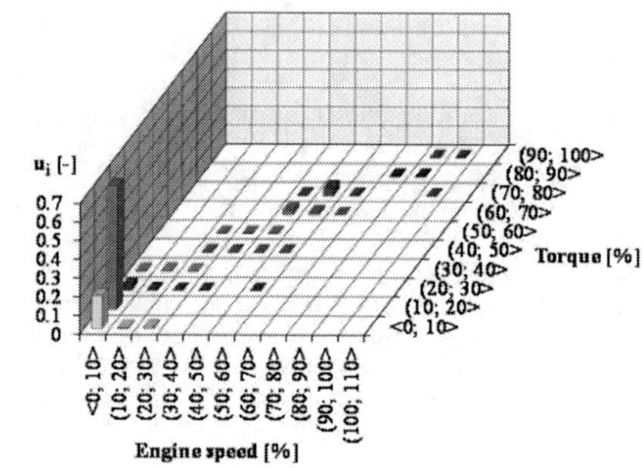

Fig. 7. Time density characteristic for the vehicle driving on the track with a load of 120 tons presented in the engine speed and load intervals

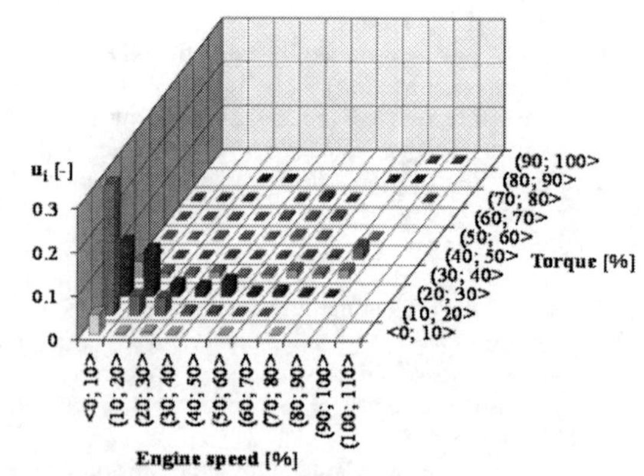

Fig. 8. Time density characteristic for the whole test drive cycle presented in the engine speed and load intervals

5 Comparison of legislative guidelines with real operation

Based on the previously obtained operating parameters, the characteristics of the absolute difference in operating time share between the NRTC test and the real drive of the road-rail tractor were created (Figure 9). The characteristics were limited to the values relevant to the tests (exceeding 1% share of operating time).

The largest difference in the share of operating time (29%) occurred in the parameters range within the engine speed range ⟨0%; 10%⟩ and torque range (10%; 20%⟩. This range of operating points with the largest difference in the operating time share is characterized by a low engine speed range of ⟨0%; 20%⟩ and low torque values in the range (10%; 30%⟩, in which the values obtained exceed the values determined in the NRTC test – marked in red in the figure. Significant differences (25%) occurred also in the range of high engine rotational speeds (70%; 110%⟩ and at medium and heavy loads (50%; 100%⟩. The presented range of operating points includes the points in which the values determined for the NRTC test exceeded the values obtained during the test - marked in blue in the figure.

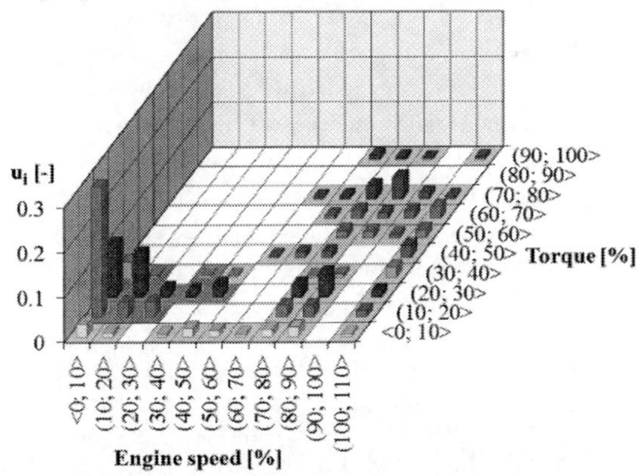

Fig. 9. The absolute difference in the operating time share between the NRTC test and the real driving test of the vehicle in the engine speed and load ranges: • – values greater than in the NRTC test; • – values smaller than in the NRTC test

The time density analysis was supplemented by the study of the specific exhaust emissions. The results obtained were the difference between the actual vehicle emission and the amount of pollutants found in the ambient air. They were divided into the five previously presented measuring cycles. In addition, overall ecological indicators for the entire test were presented.

The red line shown in Figure 10 marks the limit values for all toxic exhaust compounds as defined in the Stage IV norm. For carbon oxides, all cycles met the emission limit of 5 g/kWh, with the highest value measured for the first stage at 1.35 g/kWh. During the first cycle, the highest values of nitrogen oxide emissions (3.78 g/kWh) and hydrocarbons (0.8 g/kWh) were also observed. It should be noted that the NO_x emission limit of 0.4 g/kWh was met only for the cycle where the vehicle was switching to tracks. For hydrocarbons, none of the cycles were found to meet the HC emissions limit of 0.19 g/kWh. With respect to solid particles, the highest value of 0.13 mg/kWh was obtained during cycle IV. Stages I-IV were characterized by negative values which was caused by a significant concentration of PM in the ambient air. During all the drive cycles, the particulate emission limit of 0.025 g/kWh has been met. In terms of total emissions from the test, the vehicle met only the CO limit defined in the emission norms, as well as PM thanks to the effect of the particulate filter.

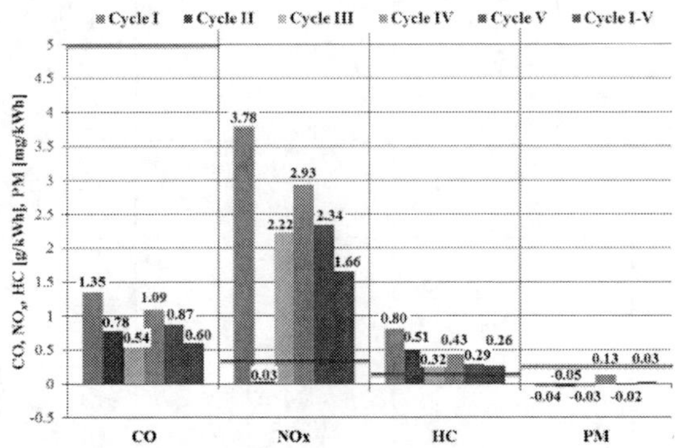

Fig. 8. 31 Specific emissions from the test vehicle

6 Guidelines for testing under real operating conditions - Real Road-Rail Emission (RRRE)

The performed tests should reflect the emissions from real road and rail traffic. The level of actual emissions often exceeds the emission values obtained when testing in laboratory conditions (in particular with regard to NO_x). Therefore, it is necessary to develop a new test procedure performed in real conditions for road-rail vehicles (RRRE - Real Road-Rail Emission), which will reflect the emission measured during its real operation. To determine the emission of harmful compounds in conditions of real operation (referring to Regulation (EC) No 715/2007), a number of requirements for the newly developed test are necessary. The RRRE test should cover the widest possible range of operating conditions of the propulsion system and vehicle movement to ensure compliance with the normal conditions in which the vehicles are to be used.

The tests should be carried out in dynamic conditions and in the ambient environment, for which requirements are specified in the Commission Regulation (EU) 2016/427 regarding the procedure for testing emissions in real driving conditions for light and commercial vehicles. In addition, for road-rail vehicles, it is necessary to set guidelines for the vehicle type and the equipment used.

The actual exhaust emissions should be determined in the five stages tests of vehicles reflecting the actual operation of road-rail vehicles (Table 3). The first of these is driving on a dirt road under normal driving conditions and at normal engine loads. The next stage should include rolling the object onto the railway tracks. The three remaining stages should relate to travel on rail tracks – with no load, with a load of 50% to 65%, and finally with a load greater than 85% of the maximum pull weight of the moved set.

Table 3. Driving parameters of the RRRE test

No.	Drive cycle	Share of cycle duration [%]	Mean vehicle speed [%]	Drive cycle duration [%]
I	Road driving	20	25	20
II	Moving onto the rail tracks	–	–	–
III	Track driving with no load	20	20	20
IV	Track driving with a load of 90 tons	40	10	40
V	Track driving with a load of 140 tons	20	5	20
Shares margin of tolerance ±10%				

Operational and transit requirements should include:

- The duration of the entire test should be no less than 60 minutes but should not exceed 90 minutes.
- Shares of time spent driving in individual stages. The cycles should be split into approximately 20% for driving on a dirt road, 20% for track travel without load and with a load of over 85% and 40% when travelling on track with a load of 50-65%. The accuracy of the obtained parameters should be within ± 10%, excluding the process of rolling onto tracks. The cycle of driving on rail track with a load from 50% to 65% must, however, reach a driving share of no less than 35% of the total test duration.
- Average drive speed, including idling, should not exceed 35% of the maximum vehicle speed. Idling should comprise at least 30% of the total driving time when moving on the track. The average road travel speed should be approximately 25% of the maximum vehicle speed. For the drive on the track, the vehicle should achieve an average speed of 20% for drive without load, 10% for the drive with a load of 50-65% and 5% for a drive with a load of not less than 85%.
- The route should be chosen in such a way that the test can be carried out without interruption, and the data is constantly being recorded by the test equipment. The road route travelled should be flat and paved, while for driving on the track it should include a section of a flat track with a maximum slope of 2‰.
- The test driver should have experience in working with this type of machine, perform cycles as standard, without matching the driving style, in the defined range of stress and focus (an example of evaluation can be found in [13]).

The requirements for apparatus should include:

- The test equipment must include analyzers belonging to the PEMS group to measure the concentration of exhaust gases and particulates.
- The tests should be supplemented with the exhaust emissions assessment measured by means of the mobile analyzer.
- Electricity to power the measuring devices should be supplied by means of an external power supply, energy should not be drawn directly or indirectly from the engine of the tested vehicle.
- The installation of the measuring devices should be carried out in such a way as to minimize its impact on exhaust emissions from the vehicle and its operation.
- Determining the exhaust mass flow should be performed with the use of measuring equipment functioning independently of the vehicle and not by relying on data from its diagnostic system.
- The location, elevation and speed of the vehicle should be determined using the global positioning system.
- The measurement of environmental conditions such as temperature, relative humidity, or atmospheric pressure should be carried out with the help of sensors or other devices that are not part of the vehicle equipment, it is acceptable to use special on-board systems [14].
- The tested parameters should be measured and recorded at a constant data acquisition frequency of 1 Hz or more.

7 Test proposal for the assessment of exhaust emissions of road-rail vehicles in real operating conditions

The test presented in this chapter is a proposal for testing road-rail vehicles. According to the guidelines presented, the authors of the study examined eight road-rail tractors. The obtained results confirm the possibility of achieving repeatability of the measurement tests and the implementation of a broader assessment of the ecological indicators of the vehicles of the tested group.

The conducted research indicates that the test for assessing pollutants for road-rail vehicles should include a larger share of operating time in the engine range characterized by low rotational speed ⟨0%; 20%⟩ and low torque values (10%; 30%⟩ in relation to the NRTC cycle. It would also be necessary to limit the share of work points in the range of high rotational speeds (70%; 110%⟩ and medium and high loads (50%; 100%⟩. It is also reasonable to extend the test duration from 1240 s to at least 3600 s to obtain a larger amount of measurement data. Dividing the test into the stages presented in this article allows for a thorough analysis of the data (Figure 11). The stage of driving on road should be characterized by a high variability of engine parameters for the range of small and medium parameters. At the same time, it should not exceed 30% of the total drive duration of the whole test. Rolling onto the rail track is to take place without interruption immediately after the end of the road drive cycle and immediately before the start of the track drive cycle. The track drive cycle should be characterized by lower variability of work parameters than is the case during the first stage, in the drive without load, the engine operation parameters range is to include medium and small values in the area of variability of the drive system operating parameters. For the drive with a load of 50-65%, the vehicle should operate at small and medium loads and achieve high values of crankshaft rotational speed. This stage is characterized by the largest share of operating time, not less than 35% of the entire test duration. The drive stage with a load of more than 85% is characterized by variability of parameters in a wide range of engine operation.

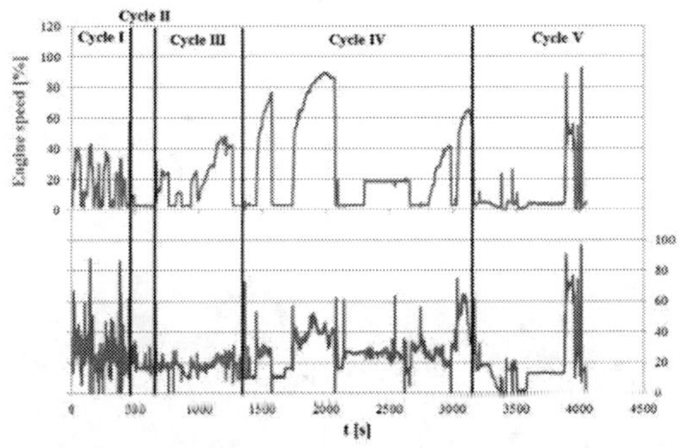

Fig. 11. Characteristics of engine rotational speed and torque expressed in the RRRE test

8 Conclusions

Due to the need to protect the environment, an increasing public awareness, as well as the development of more precise and capable emission measuring devices, various types of research are conducted to assess the exhaust emissions from various vehicle groups. Currently, normative guidelines for real driving tests for passenger cars and heavy vehicles are introduced. Leading research institutes deal with the issues of field testing of other vehicle and machine groups, including road-rail facilities. The article presents the tests of a road-rail tractor in accordance with the adopted original research method.

Characteristics of time density as a function of engine rotational speed and load created based on the operating parameters as determined in the NRTC test are characterized by the largest share of 50% operating time in the range of high engine rotational speeds of ⟨50%; 110%⟩ and small and medium torque values at ⟨0%; 60%⟩. Characteristics of time density as a function of engine rotational speed and load created based on the operating parameters as determined in the designed custom test, are characterized by the largest share of 60% operating time in the range of low engine rotational speeds ⟨0%; 30%⟩ and small loads (10%; 30%⟩.

Specific emission values obtained throughout the whole test exceeded the emission limit values for nitrogen oxides (1.66 g/kWh) and hydrocarbons (0.26 g/kWh). The exhaust components whose emission values met the limit values were the carbon oxides (0.6 g/kWh) and solid particles (0.03 mg/kWh).

Based on their research experience, the authors proposed a new test cycle designed specifically and exclusively for road-rail vehicles. It unifies the measurement method of road-rail vehicles and modifies the currently applicable type approval tests, which, as it as shown in the article, do not accurately reflect the real operation of the tested vehicles.

The guidelines for the test implementation were also described in the article. They refer mainly to the method of conducting measurements, which should be divided into five stages, including both driving on a paved road, rolling onto the tracks, as well as driving on tracks with various additional loads. Some focus was also placed on operational requirements and specific requirements for the tested vehicles, where the vehicle operating time shares, their average speed and duration of individual measurement cycles were all determined. The requirements for the measuring equipment that can be used to measure the concentration of exhaust gases and particulates and the exhaust emissions in ambient air were also described. Further research includes plans to define the adjustment coefficients - CF (Conformity Factor). They will describe the permitted level of exceeding the emission limit values given in the currently applicable regulations. However, to accomplish this, it is necessary to carry out research work on a larger group of vehicles.

References

1. M. Bajerlein, P. Daszkiewicz, M. Dobrzyński, Ł. Rymaniak, M. Siedlecki, Combustion Engines, **3**, 800-804 (2015)
2. P. Fuć, P. Lijewski, A. Ziółkowski, M. Siedlecki, Combustion Engines, **3**, 417-424 (2015)
3. J. Merkisz, P. Fuć, J. Pielecha, Publishing House of the Warsaw University of Technology (2014)
4. J. Merkisz, D. Michalak, M. Bajerlein, P. Molik, Ł. Rymaniak, A. Ziółkowski. Autobusy – Technika, Eksploatacja, Systemy Transportowe, **3**, 111-119 (2012)
5. J. Merkisz, D. Michalak, M. Bajerlein, M. Nowak, Ł. Rymaniak, A. Ziółkowski, Autobusy – Technika, Eksploatacja, Systemy Transportowe, **3**, 104-110 (2012).
6. J. Merkisz, J. Pielecha, P.Fuć. The Committee of Civil and Water Engineering of PAN (2013)

7. J. Merkisz, Ł. Rymaniak Combustion Engines, **3,** Pages: 21-26 (2017)
8. B. Kurc, Journal Of Solid State Electrochemistry, **18**, 2035-2046 (2014)
9. J. Merkisz, P. Fuć, 10th WFC10 Congres Proceedings, **3**, 479-483(2008)
10. K.T. Wojciechowski, J. Merkisz, P. Fuć, J. Tomankiewicz, R. Zybała, J. Leszczyński, P. Lijewski P. Nieroda, Combustion Engines, **3,** 60-71 (2013)
11. R. Jachimowski, P. Golebiowski, M. Izdebski, D. Pyza, Archives of Transport, **41**, 31-42 (2017)
12. K. Andrzejczak, J. Selech, Transport Problems, **12**, 103-111 (2017)
13. D. Dolinski, E. Odachowska, Accident Analysis And Prevention, **115**, 73-78, (2018)
14. J. Merkisz, R. Grzeszczyk, Advanced Microsystems For Automotive Applications 2010: Smart Systems For Green Cars And Safe Mobility, Edited by: G. Meyer, J. Valldorf, 347-356, (2010)

Artificial neural network based detection of neutral relay defects

Volodymyr Havryliuk[*]

Dnipro National University of Railway Transport named after Academician V. Lazaryan, dept. of automatic and telecommunication, 49010 Dnipro, Lazaryan Street 2, Ukraine

Abstract. The problem considered in the work is concerned to the automatic detecting and identifying defects in a neutral relay. The special design of electromechanical neutral relays is responsible for the strong asymmetry of its output signal for all possible safety-critical influences, and therefore neutral relays have negligible values of dangerous failures rate. To ensure the safe operation of relay-based train control systems, electromechanical relays should be periodically subjected to routine maintenance, during which their main operating parameters are measured, and the relays are set up in accordance with technical regulations. These measurements are mainly done manually, so they take a lot of time (up to four hours per relay), are expensive, and the results are subjective. In recent years, fault diagnosis methods based on artificial neural networks (ANN) have received considerable attention. The ANN-based classification of relay defects using the time dependence of the transient current in the relay coil during its switching is very promising for practical utilization, but for efficient use of ANN a lot of data is required to train the artificial neural network. To reduce the ANN training time, a pre-processing of the time dependence of relay transient current was proposed using wavelet transform and wavelet energy entropy, which makes it possible to reveal the features of the main defects of the relay armature, contact springs, and magnetic system. The effectiveness of the proposed approach for automatic detecting and identifying of the neutral relays defects was confirmed during testing of the relays with various artificially created defects.

1 Introduction

Most railway signalling systems that are currently in use were built over half century ago and are mainly based on the use of electromechanical contact components (transmitters, relays, etc.). Although the computer-based automation systems have appeared in the last decade, the part of electromechanical signalling systems on Ukrainian railways is still exceed 90%. Most of the existing relay-contact systems are already outdated. The functional safety of railway electromechanical signalling systems is based on the use of so-called neutral relays (NRs), which have specific design that provides strong asymmetry of their output when all possible adverse for the safe system's operation factors arise [1].

In order to ensure the safe operation of relay signalling systems, NRs should be subjected to periodic routine maintenance, during which their main operating parameters are measured and regulated in accordance with technical regulations [2]. Since these measurements are usually performed manually, they require a lot of time (up to four hours per relay), are expensive, and their results are exposed to the influence of subjective factors [2, 3]. Automation of measurements of most NRs' electrical parameters, such as the minimum pick-up voltage (operating voltage), must drop-out voltage (must release voltage), coil resistance, contacts resistance, etc., can be easily implemented based on modern measuring equipment. But the automation of the mechanical parameters measuring of NRs encounters serious difficulties. The mechanical and electrical forces that act on the armature of NR when it is switched determine the switching time of the NR, simultaneous or non-simultaneous switching of the contacts, contact pressure, etc. The technical condition of the relay armature, contact springs, magnetic system and design parameters of relays, such as the height of the antimagnetic pin, inter-contact air gap, paramagnetic gap between yoke and core, etc., strongly influence on mechanical parameters of NRs.

As was shown in [4, 5], the electrical transient current that flows in the relay coil when NR is turned on and off reflects the mechanical processes occurring during the armature movement, and, therefore, the features of the transient current can be used to monitor the technical condition of the armature and magnetic system of NRs. The use of wavelet transform for detecting defects of the armature of neutral relays investigated in [6]. Wavelet decomposition of signal current was used for detecting disturbances in it [7]. The multi-resolution wavelet analysis of voltage and current signals acquired by distance relays was used to identify various faults in signals [8]. The wavelet technique for fault detection in transmission line during power swing was proposed in [9].

[*] Corresponding author: vl.havryliuk@gmail.com

Discrete wavelet transform for improving the accuracy of an unbalance current protection relay due to transient fault and inrush current signals was investigated in [10]. The use of an artificial neural network (ANN) to detect and identify NRs' defects by analyzing of the time-dependences of their transient current is promising for effective and prompt monitoring of NR technical conditions [3]. However, training the ANN for detecting all possible defects of the armature and the magnetic system by using the time-dependences of the transient current of the NRs is a very difficult and inefficient procedure for practical application. Combination of neural network and wavelet transform for islanding detection of distributed generation in a small-scale network was proposed in [11].

The purpose of this work is to investigate the effectiveness of the proposed new method of monitoring the technical condition of the NRs, which consists of two stages. At the first stage, the NRs' time-dependences of the transient currents are measured when NRs are energized, and then these dependences are analyzed using the wavelet transform and wavelet energy Shannon entropy for revealing the features of the main defects of armature and magnetic system of neutral relays. At the second stage, NRs' defects are detected and identified using an ANN classifier, which has been preliminary trained using the revealed features of relay defects.

2 Mathematical descriptions of electromechanical processes in relay

NR consists of a soft iron magnetic system, which is composed of a magnetic core with a coil, a yoke, a moving armature, and springsets with contacts of change-over type, which linked to armature. The magnetic flux generated by the electric current in the NR coil varies along the core due to the leakage flux in the air between the core and the yoke of the relay. Therefore, the use of magnetic equivalent circuits (MEC) to simulate a magnetic field in a relay proposed in [12] is not enough correct.

Typical time dependence of the transient current in relay coil during voltage switching is shown in Fig. 1. The transient current plot can be divided into six segments. First three segments (from $t=0$ to $t=t_3$) correspond to relay energizing, and second three (from $t=t_3$ to $t>t_5$) correspond to its de-energizing.

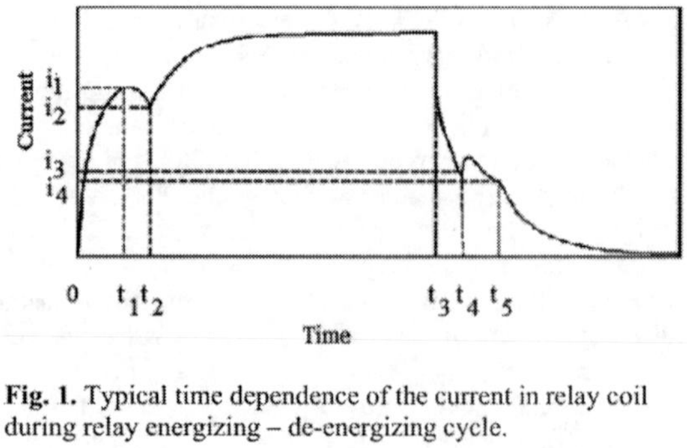

Fig. 1. Typical time dependence of the current in relay coil during relay energizing – de-energizing cycle.

In the first segment from moment $t=0$ to $t=t_1$, the relay armature does not move, since the magnetic force is not strong enough to begin movement, and relationship for the relay energy balance can be written in a form [4, 6]

$$W_E(t)=W_{EL}(t)+W_{FL}(t)+W_{FS}(t), \quad (t \in t..t_1), \qquad (1)$$

where $W_E=\int_0^t U\,i(t)\,dt$ is a total energy supplied by the electric source; $W_{EL}=\int_0^t i(t)^2 R\,dt$ is the energy that is dissipated as heat due to the active resistance of the relay coil; $W_{FS}=\int_0^{\Psi(t)} U\,d\Psi(t)=\int_0^t U\frac{d\Psi(t)}{dt}dt$ is the energy stored in the magnetic and electric fields; W_{FL} is the energy loss due to magnetic hysteresis, eddy currents in the magnetic system, and the dielectric losses; U is the voltage applied to relay coil; R is the active resistance of relay coil; $i(t)$ is the instantaneous current value; $\Psi(t)$ is the instantaneous interlinkage flux in relay core. The current $i(t)$ in the first segment exponentially increases in time.

The second segment from t_1 to t_2, which corresponds to the movement of the armature, is accompanied by the conversion of part of the total energy, supplied by the electric source W_E at the first stage and accumulated in the magnetic and electric fields W_{FS}, into the mechanical work W_M of the motion of the armature. The current in the second segment decreases.

In the third segment, the relay armature is completely attracted to the core at the moment t_2, and the relay current again exponentially increases with time.

Due to the fact that the core and the yoke of NRs are located parallel at a distance significantly smaller than their length, it allows to use the theory of lines with distributed parameters to describe the magnetic flux flowing in them [4]. Differential equations for magnetic flux Φ_M and magnetic potential U_M in the cross section of the core with the coordinate x can be written as

$$\frac{d\Phi_M(x)}{dx}=-g_M U_M(x), \qquad (2)$$

$$\frac{dU_M(x)}{dx}=\frac{e_M}{l}-R_M\Phi_M(x), \qquad (3)$$

where g_M is the permanence of leakage per unit length (p.u.l.) of the core, $e_M=i\,w$ is the magnetomotive force (p.u.l.), i is the current in the relay coil, w is the number of winds in the coil (p.u.l.), R_M is the magnetic reluctance of the core and yoke (p.u.l.) expressed by

$$R_M=\frac{1}{\mu_0\mu_C S_C}+\frac{1}{\mu_0\mu_Y S_Y}, \qquad (4)$$

where μ_0 is the permeability of free space, μ_C and S_C are the permeability and cross-section of the core, μ_Y and S_Y are the permeability and cross-section of the yoke. Equation (2), (3) have well-known solutions [4]

$$U_M(x) = C_1 sh(\gamma_M x) + C_2 ch(\gamma_M x), \quad (5)$$

$$\Phi_M(x) = \frac{e_M}{lR_M} - \frac{1}{R_{MO}}\left[C_1 ch(\gamma_M x) - C_2 sh(\gamma_M x)\right], \quad (6)$$

$$R_{MO} = \sqrt{\frac{R_M}{g_M}} \text{ and } \gamma_M = \sqrt{R_M g_M}, \quad (7)$$

where C_1 and C_2 are constants of integration, values of which are determined from boundary conditions for the start $x = 0$ and the end $x = l$ (near the armature) points of the magnetic circuit.

Magnetic interlinkage is expressed by

$$\Psi = w\int_0^l \Phi(x)dx. \quad (8)$$

Electromagnetic attraction force of the armature to the coil expressed as

$$F_M = -\frac{dW_M}{d\delta}, \quad (9)$$

where δ is instant value of the air gap between the core and the armature, and W_M is the magnetic energy stored in magnetic field

$$W_M = \int_0^{\Psi_0} id\Psi \quad (10)$$

Movement of the relay armature when relay is energized or de-energized can be described by equation [4]

$$M_A\ddot{\delta}_C + F_F\dot{\delta}_C + F_E\delta_C = (F_M - F_W); \quad (11)$$

where F_M is the equivalent magnetic attraction force reduced to equivalent mass centre (r.e.m.c.), F_W is the equivalent weight of the moving components of relay (r.e.m.c.); F_E is the equivalent elastic force due to deformation of contact springs (r.e.m.c.); F_F is the equivalent force of friction (r.e.m.c.), M_A is the equivalent mass of relay armature (r.e.m.c.), δ_C is the reduced distance between mass centre of the moving components and the magnetic core.

When a voltage V is applied to relay coil the electrical process in the system is described by equation

$$V = Ri(t) + \frac{\partial\Psi}{\partial t}. \quad (12)$$

Note, that (12) accounts only the energy losses in a form of heat and not eddy current and hysteresis losses. By substituting (10) in (9) with using (12) the expression for the electromagnetic attraction force can be written as

$$F_M = -\frac{1}{d\delta/dt}\left[Ui - Ri^2\right]. \quad (13)$$

Therefore, F_M can be calculated on the basis of the measured time dependence of the transient current, provided that the speed of the armature movement was simultaneously measured during the switching of the relay.

Thus, the electromagnetic force of the armature attraction to the core is determined by the magnetic flux in the magnetic circuit, and as a result, the defects in the magnetic system that can be in the form of increased gap between the core and the yoke, or shorted winds in the coils, lead to a decrease in the electromagnetic attraction force.

Changing of any force in (11) lead to change of mechanical parameters of NRs. The most common cause of changes in mechanical parameters is curved contact springs that occurs during the operation of the NRs.

3 A brief overview of wavelet transform and artificial neural network

3.1. Wavelet transform

Three main types of wavelet transform (WT) are usually considered, namely, continuous wavelet transform (CWT) [13], discrete wavelet transform [14], and discrete wavelet packet transformation (DWPT) [15].

Wavelet analysis is realized by projecting the original signal $f(t)$ onto a sequence of basic functions, known as wavelets. There are two basic wavelet functions: the father wavelet $\phi(t)$ called as the scale function and the mother wavelet $\psi(t)$ called as the wavelet function, which can be scaled and translated to form the basis for the Hilbert space of square-integrable functions $L^2(R)$. The father and the mother wavelets are defined as [15]

$$\phi_{j,k}(t) = 2^{-\frac{j}{2}}\phi\left(2^{-j}t - k\right), \quad (14)$$

$$\psi_{j,k}(t) = 2^{-\frac{j}{2}}\psi\left(2^{-j}t - k\right), \quad (15)$$

where $j = 1,...,J$ is the scaling parameter in a J-level decomposition, and k is a translation parameter $(j,k \in Z)$. The long-term trend of the function is fixed by the father wavelet, the integral of which is 1, while the mother wavelet, the integral of which is 0, describes the fluctuations from the trend. The continuous wavelet transform of the square-integrable time signal $f(t)$

consists of the scaling $a_{j,k}$ and wavelet coefficients, $b_{j,k}$ [15]

$$a_{j,k}(t)=\int\phi_{j,k}(t)f(t)dt\,, \tag{16}$$

$$b_{j,k}(t)=\int\psi_{j,k}(t)f(t)dt\,. \tag{17}$$

DWT can decompose the approximate part in each level, while the detail part remains unchanged. Consequently, the resolution of DWT is not high enough to separate the fault signals from the extra components. Differing from DWT, DWPT can decompose the detail part signals from the extra components. Therefore, wavelet packet method is a generalization of wavelet decomposition that offers a richer range of possibilities for signal analysis.

In wavelet analysis a signal is split into an approximation and a detail. The approximation is then itself split into a second-level approximation and detail, and the process is repeated. For n-level decomposition, there are $n+1$ possible ways to decompose signal.

$$f_{2n}(t)=\sqrt{2}\sum_k h(k)f_n(2t-k)\,, \tag{18}$$

$$f_{2n+1}(t)=\sqrt{2}\sum_k h(k)f_n(2t-k)\,, \tag{19}$$

where $f(t)$ is the original signal.

In recent years, wavelet entropy has become widespread tool used for the study of complex non-stationary processes. Wavelet packet energy Shannon entropy (WPESEE) is described by the expression [16]

$$E=-\sum_{k=1}^{N}P_{i,j,k}\log\left(P_{i,j,k}\right), \tag{20}$$

$$P_{i,j,k}=\frac{W_{i,j,k}}{W_{i,j}}, \tag{21}$$

where $P_{i,j,k}$ is the probability of the k-th coefficient, and $W_{i,j,k}$ is the energy of the k-th coefficient in the j-th node at i-th level of the wavelet decomposition; $W_{i,j}$ is the total energy in the j-th node at i-th level.

3.2. Artificial neural network

Methods based on artificial neural networks have recently attracted much attention due to their ability to analyze complex, non-linear processes. ANN has a variety of architectures, but the most widely used is the feed forward multilayer perceptron (MLP) network trained by back propagation algorithm. A MLP consists of at least three layers of nodes: an input layer, a hidden layer and an output layer [17].

In a single-layer network with R input elements and S neurons, the input vector $\mathbf{X}=\left[x_1,x_2,..,x_R\right]^T$ is transformed into the output vector $\mathbf{A}=\left[a_1,a_2,..,a_S\right]^T$ in accordance with the expression

$$\mathbf{A}=\mathbf{f}\left(\mathbf{Wp}+\mathbf{b}\right), \tag{22}$$

where $\mathbf{W}=\left[w_{ij}\right],\left(i=1...R,\ j=1...S\right)$ is the weight matrix, $\mathbf{b}$ is a bias vector, $\mathbf{f}$ is an activation function or transfer function. The most common activation function includes linear function, tangent hyperbolic function, threshold function and sigmoid function.

4 Measurement Techniques

The measurements were carried out using a special laboratory bench that consists of stable voltage source, 14-bit analog-to-digital converter, connected to PC, and the voltage switching device. The measuring equipment and technique are described in [6]. The appearance of the measuring equipment is shown in Fig. 2.

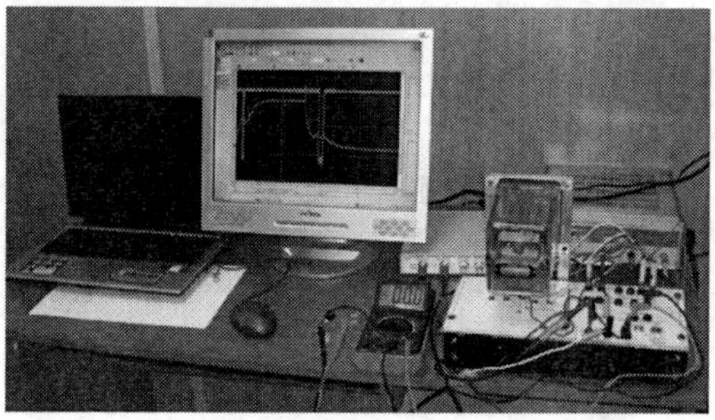

Fig. 2. The appearance of the measuring equipment

To reveal the features of relay armature defects, a transient current in relay coil as well as current in all contacts were recorded by PC when relay energized and de-energized. The contacts in NR have double-throw structure. Therefore, the back and front contacts of the same contact group were connected in parallel and included in the break of the measuring circuit on the one side, while the common contact was included in the same break on the other side. Thus the current in measurement circuit through contacts was interrupted in time when common contact moved from front to back contact or in return direction. For investigations ten groups of different types of NRs were chosen. Each group contained NRs with different artificially created defects of armature or magnetic system. These defects are listed in Table 1. Up to 100 experiments were carried out in each group with varying degrees of defects. Paper presents only results for one type of NR, namely NMSH 4-600, with nominal coil resistance of 600 Ohms, and the nominal switching voltage of 12 V. The results obtained for other types of NR were, in general, similarly and were used for training ANN, and for statistical confirmation of the conclusions made in the work.

5 Results and discussion

5.1. Revealing of relay defect features

The time dependences of the transient currents of NR strongly depend on the defects of their armature (Fig. 3, 4). The parameters of the transient process when NR switching were characterized by pick-up time t_a, the maximum value of the current in the time interval corresponding to the armature movement I_m, and the drop-out (release) time when the armature moves to the initial position t_r. Time t_a was determined as the time interval from the moment, when the voltage is applied to the relay coil, to the time corresponding to the local minimum on the transient current curve, which corresponds to the point between the first and second segments of the transient current (Fig. 1).

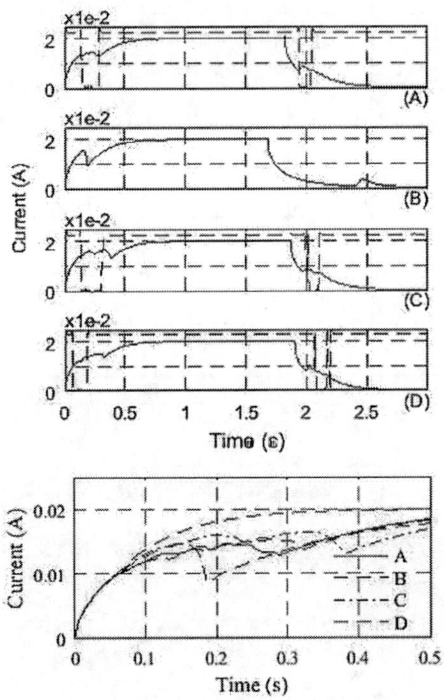

Fig. 3. The time dependences of the transient currents for relay of types (A), (B), (C). (D). Dotted line - current in contacts.

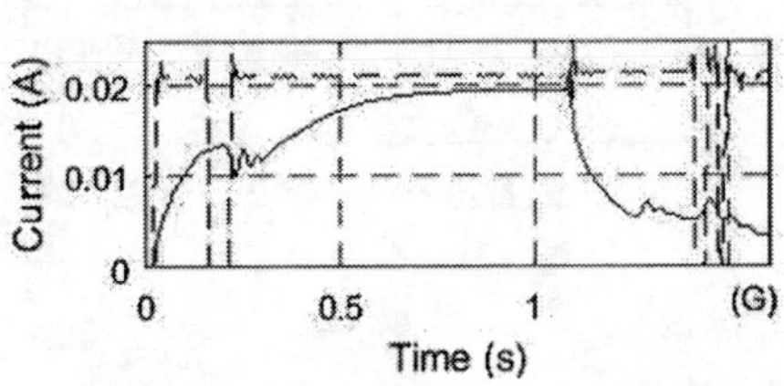

Fig. 4. The time dependence of the transient current for relay of types (G). Dotted line corresponds to current through contacts.

All mechanical defects of the relay (such as bending contact springs, or increase of the resistance force due to fixing various loads on the armature, as well as defects of the magnetic circuit, in the form of shorting turns of the relay coil, or an increase in the paramagnetic gap between the core and the yoke) lead to an increase in the t_a, I_m parameters, and to decrease t_r parameter (Fig. 3, 4). For the relay with deliberately increased friction force the t_a, t_r, I_m parameters have slightly larger values compared to the same parameters for a relay in good condition. Consequently, the observed effect of defects in relay armature and magnetic system on the parameters of the transient current allows the use of these parameters as features of defects. To compare the parameters of the transient current for different relays and for the convenience of training ANN, these parameters were converted into relative form as the ratio to similar parameters for relays in good condition, and the obtained values are presented in Table 1. For relays with armature defects the ssecond segments of the transient current curves, which corresponds to the armature movement, contain some features in a form of swells and sags (Fig. 3, 4), the value, position and number of which correlate with the technical conditions of the relay contact springs. An increase in the number and degree of deformation of the contact springs leads to an increase in the number, and values of amplitude and duration of sags and swells at the second segment of the transient curve of the relay. For relays with armature bounces (rebounds) after it striking to the relay core on the transient current curve after the time t_a, when the armature is fully raised, the damped oscillations are observed. For the relays with magnetic circuit defects the values of the current I_m and the switching time values t_a, t_r are higher than similar parameters for NR that is in good condition (type A), but wavy segments with excessive sags, swells or oscillations that are observed for relays with curved contact springs (type C, D), do not appear on the transient current curves.

As mentioned earlier (Fig. 1), when the relay is energized, its armature does not move in time intervals corresponded to the first and third segments of the transient current curve. If the magnetic flux in the core is not saturated, one can make a reasonable assumption that the transient current in the first and third segments increases exponentially in time, and the time constants of the current increase depend on the technical state of the magnetic system.

The values of the time constants were calculated from, and their normalized values T_{1n} and T_{3n}, calculated as a ratio to the corresponding values for the relays that is in good condition, are performed in Table 1. Obtained values are in satisfactory agreement with the values calculated from the directly measured coil resistance and inductance for two stable position of the armature.

The values of time constants for the first T_{1n} and the third T_{3n} segments are not significantly depended on the armature defects for the relay of type A, B, C. D, E, F, I, J (Table 1), but have a lower values for relays with magnetic system defects (NRs of type J and H).

Moreover, the ratio T_{1n}/T_{3n} for defects in the form of an increased paramagnetic gap between the core and the yoke has a slightly larger value compared to the similar ratio for a magnetic system with shorted winds in coil. These resultes makes it possible to use T_{1n} and T_{3n} as features of the magnetic system defects.

Thus, the features of the transient current curve are very sensitive to defects in the armature and the magnetic system due to their influence on the magnetic and mechanical forces acting in the system when the relay is switched. But these features are small compared to the amplitude values of the transient current during relay switching, Therefore, the direct use of transient relay currents to detect defects using the ANN classifier is inefficient for practical use because it requires the processing of redundant data of sampled transient current that is not related to the features of defects. To reduce the amount of data for the ANN classifier, it is reasonable to preliminary reveal the main features of relay defects using wavelet transform.

Table 1. Features of relay defects

Labels of relay defects	Type of defect	Features of defect							Output vector
		Pick-up time	Release time	I_0	T_1	T_3	WPESE 1	WPESE 2	
A	The relay in good conditions	1	1	1	1	1	1	1	[00000000]
B	With removed contact group	0.625	0.704	0.782	1.011	0.991	0.881	0.97	X
C	Common contact springs are bent to the back contacts	1.331	0.891	1.081	0.991	0.995	1.129	1.01	[00000001]
D	Common contact springs are bent to the front contacts	0.98	1.121	1.075	0.990	0.995	1.117	1.011	[00000010]
E	Contact springs are randomly bent	1.415	1. 450	1.155	1.021	1.012	1.21	1.042	[00000100]
F	Armature bounce	0.972	0.942	0.977	0.977	0.991	1.021	1.311	[00001000]
G	Shorted turns in NR coil	1.212	0.951	1.212	0.910	0.923	1.011	1.012	[00010000]
H	Increased gap between core and yoke	1.341	0.921	1.287	0.972	0.970	1.014	1.011	[00100000]
I	Increased mechanical counterforce to anchor	1.410	0.754	1.202	0.991	0.994	1.023	1.021	[01000000]
J	Increased friction force	1.081	1.011	1.06	0.988	0.998	1.025	1.024	[10000000]

5.2. Wavelet analysis of transient currents

To reveal the features of defects on transient current curve, the values of which exceed a certain acceptable level, the measured transient currents for relay with different defects were analyzed by using DWPT. Transient current decomposition was performed using a three level DWPT.

The distributions of the energy in the DWPT nodes for NRs with armature defects are shown in Fig. 5, and the time dependences of the wavelet coefficients at nodes (3.2), (3.3), and (3.6) for relay that is in good condition (of type A) and for relay with bent contact springs (of type D) are shown in Fig. 6. These nodes are selected since the features of the relay armature defects are most distinguishable in them.

As can be seen, defects of NRs are affected on distribution of the energy in the DWPT nodes (Fig, 5). Defects of the armature lead to a strong increase in the DWPT decomposition coefficients. In Fig. 6 results are presented for type (D) relays only. For other NRs with armature defects, the results are similar.

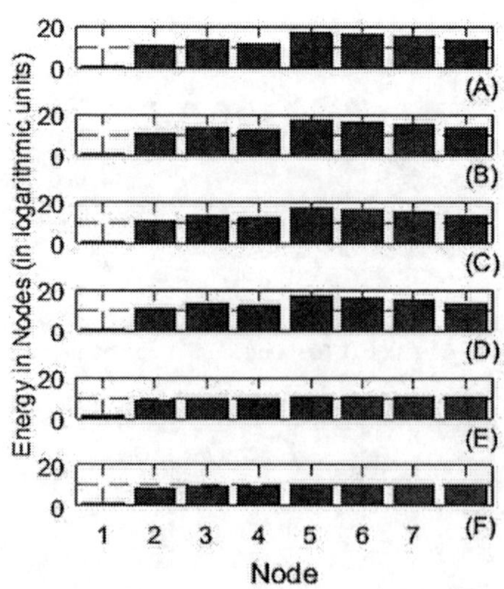

Fig. 5. The distribution of the energy in the DWPT nodes.

The DWPT decomposition coefficients increase with increasing number and degree of deformation of the bent contact springs, and these coefficients can be used for revealing defects of this type.

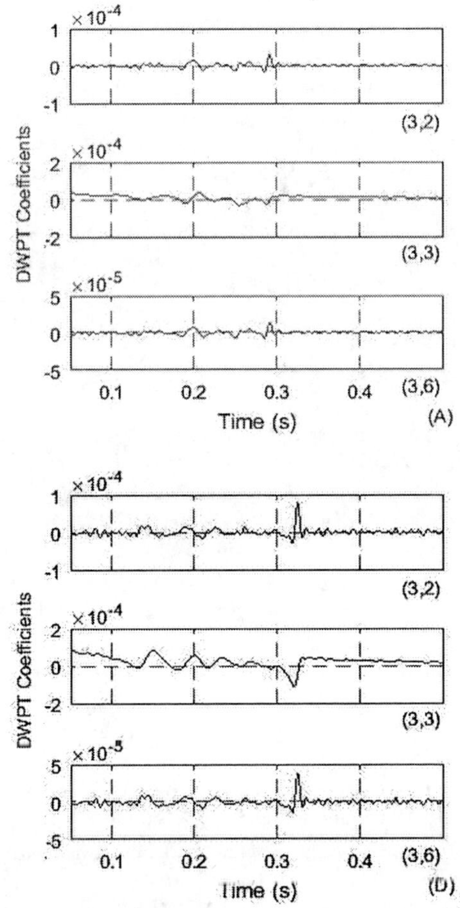

Fig. 6. The time dependence of the DWPT coefficients for relays of type (A) and (D).

For relays with a completely removed contacts (type A), the time dependences of the DWPT coefficients at nodes (3.2), (3.3), (3.6) have one narrow peak at time 0.19 s, which corresponds to the moment when the armature is touched to magnetic core. The bending of several contact springs in relay of E type leads to the appearance of several swells and sags on the time dependence of the DWPT coefficients, whose amplitudes increase with increasing degree of bending of the contact springs.

Therefore, WT-analysis of the transient currents of the relay allows to reveal the features of the main armature and magnetic system defects, but their automatic recognition and identification still require large amounts of the DWPT coefficients. For automatic prompt detecting of NR armature defects, the wavelet packet energy Shannon entropy (WPESE) was proposed to use. The WPESE values were calculated for transient currents of the NR in the time intervals $0..t_a$ and $t_a..0.75t_a$, and then obtained values were normalized by dividing on the WPESE values for the similar type of the NR, that is in good technical conditions. The relative values of entropy for two time intervals ($0..t_a$ and $t_a..0.75t_a$), labelled respectively as WPESE1 and WPESE2, are given in Table 1. Armature defects of NR in the form of bent contact springs (NR types C, D, E) cause a strong increase in WPESE1 values, while defects in the magnetic circuit of the relay have a much smaller effect on WPESE1 values. At the same time, the relay defect in the form of anchor rebound after striking to the NR core, leads to a noticeable increase in WPESE2. Thus, the wavelet entropy can be used as a feature for the effective detection of armature relay defects.

5.2. Neural network Classifier

Based on the features of the main defects of the NR (Table 1), the Artificial Neural Network architecture was designed as three layer feed-forward network with seven inputs, eight outputs, a hidden layer, and sigmoid function neurons. Each input node is connected to a hidden layer node, and each hidden node is connected to an output node in the same way.

The target output vector has all zero components if the NR is in good condition, and when a certain defect occurs, the corresponding component of the target vector takes a value equal to one. The total amount of data prepared for ANN training consisted of 900 input-output (target) vectors, which contain 100 vectors for NR in good technical conditions, and 800 vectors for NR with various defects belonging to the eight main defects group listed in Table 1. Part of these data (60%) was used for ANN training purposes, 20% of the data for validation and 20% of the data for ANN testing. The testing data comprises of both data for relay in good condition, as well as data for NR with defects and are totally different from the ANN training data.

The error back-propagation algorithm with Levenberg-Marquardt optimization technique was used to train ANN. The initial weights and biases of the network were generated automatically by the program. The mean square error achieved during ANN training is 0.0027. After training, the ANN performance is assessed using 180 test data. A trained ANN correctly classified 179 data from 180, which indicates a total detection rate of 99.4%. The testing and the validation curves have similar characteristics which is an indication of efficient training.

6 Conclusions

A new method for the automatic detection of defects in a neutral relays based on ANN is proposed. To reveal the features of the armature and magnetic system defects of NR, the time dependences of the transient current when relay is switched are measured and analyzed using wavelet packet transform and wavelet energy Shannon entropy.

The dependence of the transient current on time when the relay is turned on can be divided into three characteristic segments. In the first segment, the relay armature does not move, since the magnetic force is not strong enough to begin movement. The current in the first segment exponentially increases in time.

In the second segment, the armature moves, and part of the total energy stored in the first stage in the magnetic and electric fields of the magnetic circuit is converted into the mechanical work of the armature movement, and this process is accompanied by a certain decrease in the relay current,

In the third segment, the relay armature is completely attracted to the core, and the relay current again exponentially increases with time. The time constants of the exponential increase in the transient current of the relay in the first and third segments can be used to detect defects in the magnetic system of the relay. The features of the transient current in the second segment can be used to identify defects in relay armature.

Since the changes in the transient current curve caused by the armature defects are small, the discrete wavelet packet transform was used to reveal features of the defects. For automatic prompt detecting of armature defects, the wavelet energy Shannon entropy has been proposed to use.

To detect and classify defects of NR the three layers feed-forward artificial neural network with seven inputs, eight outputs, a hidden layer, and sigmoid function neurons was used. A trained ANN correctly classified 179 data from 180, which indicates a total detection rate of 99.4%.

The effectiveness of the proposed approach for automatic detecting and identifying of the neutral relays defects was confirmed during testing of the relays with various artificially created defects.

References

1. D. Efanov, A. Lykov, G. Osadchy. 2017 IEEE East-West Design & Test Symposium (EWDTS). IEEE (2017)
2. V. Gavrilyuk, V. Dub. Bulletin of Dnipropetrovsk National University of Railway Transport named after Academician V. Lazaryan. **12**,12 (2006)
3. V. Dub, V. Gavrilyuk. Archives of Transport System Telematics. **2**, 2 (2009)
4. M. V. Vitenberg. *Calculation of Electromagnetic Relays* (Moscow: Energija, 1975)
5. G. M. Korsunskij, A. F. Mironenko, P. S. Fedorets. Engineering Communications. Ser. Wired Communication Equipment. **38**, (1979) (in russian)
6. V. Havryliuk. Science and transport progress. Bulletin of Dnipropetrovsk National University of Railway Transport named after Academician V. Lazaryan. **66**, 6 (2016)
7. V. Havryliuk. 2019 IEEE 2nd Ukraine Conference on Electrical and Computer Engineering, (UKRCON-2019). IEEE (2019)
8. C. Pang, M. Kezunovic. MedPower (Thessaloniki, Greece. 2008)
9. S. K. Mohanty, P. K. Nayak, S. Banerjee. 2018 IEEE International Conference on Power Electronics, Drives and Energy Systems (PEDES). IEEE (2018)
10. T. Patcharoen, A. Ngaopitakkul. 2019 IEEE/IAS 55th Industrial and Commercial Power Systems Technical Conference (I&CPS). IEEE (2019)
11. E. Shahryari, M. Nooshyar, B. Sobhani. International J. of Ambient Energy. **40**, 3 (2019)
12. E. R. Laboreo, C. Sagues, and S. Llorente. IEEE Transactions on Industry Applications (2016)
13. S. G. Mallat. *A Wavelet Tour of Signal Processing representation* (San Diego. Academic Press: Elsevier, 1999).
14. S. G. Mallat. IEEE Transaction on Pattern Analysis and Machine Intelligence. **11**, 7 (1989)
15. N. Lu, F. Wang, F. Gao. Ind. Eng. Chem. Res. **42** (2003).
16. R. R. Coifman, M. V. Wickerhauser. IEEE Transactions on Information Theory. **38**, 2 (1992).
17. G. Dreyfus. *Neural networks: methodology and application* (Springer Science & Business Media, 2005)

Classification of quality conditions of a traction motor frame insulation of locomotives

Michail Kapitsa[1], *Vasiliy* Laguta[2,*], and *Yuriy* Kozik[3]

[1] Dnipro National University of Railway Transport named after Academician V. Lazaryan, Department of Locomotives, 49010 Dnipro, Lazaryan Street 2, Ukraine
[2] Dnipro National University of Railway Transport named after Academician V. Lazaryan, Department of automation and telecommunications, 49010 Dnipro, Lazaryan Street 2, Ukraine
[3] LLC "Applied Railway Technologies", Head of Technical Projects, Department of Inspectors, Adam-Klein-Str. 26, Nürnberg D-90429, Germany

Abstract. The objective of the article is to determine the quality conditions of a locomotive traction motor frame insulation based on observations over the recovery voltage and classification of a tested insulation wear degree as well as to determine estimates of operating time until the insulation restoration or replacement with a new one. To classify the insulation quality conditions and to forecast its residual life until the restoration, a cluster analysis and discriminant analysis are used based on indicators reflecting the properties of an insulation recovery voltage curve. The study defines an information content received from frame insulation recovery voltage readings as compared with its resistance from the perspective of testing the insulation condition and predicting its residual life. Statistical characteristics of clusters corresponding to the insulation condition stages allow determining the limit values of predicting parameters and estimating the residual insulation life prior the restoration. To classify the insulation condition, an integrated evaluation of the insulation condition quality is proposed. Integrated evaluation is performed according to the recovery voltage curve and is used as an indicator during clustering. An experimental study of the proposed method practically confirmed its effectiveness.

1 Introduction

The analysis of "Ukrzaliznytsia" (Ukrainian Railways) locomotive fleet condition shows that the amount of damage happened to the TM winding insulation per one mln km of run has recently averaged from three to five cases. Therefore, an increased attention is given to the problem of improving the operational reliability of TM insulation.

Systems of maintenance, repair and in-between diagnostics in the locomotive economy of Ukrainian railways are periodic and subject to regulation. To perform the majority of measurements the equipment should be withdrawn from the operation. However, such diagnostic system does not meet modern requirements. Continuous monitoring of the TM insulation technical condition becomes an optimal solution enabling reorganizing of the operational diagnostics. Automated continuous control along with predicting has several advantages over the periodic control, i.e. the reduction of checking periods, consideration of effects caused by operational factors, accumulation of diagnostic information, remote provision of information at various levels, etc. The introduction of automated monitoring will allow scheduling of equipment repairs and, if necessary, urgent TM withdrawing from operation. This approach will allow transferring from the outdated system of periodic maintenance to the maintenance matching the real technical condition of an object. Continuous monitoring systems need specific procedures, algorithms and models necessary for their effective functioning.

The condition of frame insulation of winding located at the main poles of traction electrical machines (hereinafter referred to as frame insulation) is a serious factor ensuring reliable TM operation. The implementation and development of works in respect of the frame insulation service life management was stimulated by the decision to extend the design service life of locomotives. Important elements of the service life management programme are non-destructive diagnostics of the insulation condition and prediction of its service life. Improving methods for predicting the insulation condition implies the clarification of allowed values of aging coefficients.

One of the methods that allows possible predicting of the frame insulation residual life may be the procedure of classification of its condition according to the values characterizing the recovery voltage curve. The article offers a procedure of classification, classification values, allowed values of parameters for predicting the residual life of TM NB-406 frame insulation.

* Corresponding author: vasilij.laguta@gmail.com

2 Literature review

Aging and degradation of the insulation placed in electrical machines determine operational service life of the latter. Aging processes depend on the nature and values of electrical stresses, mechanical forces applied, operating temperatures and environmental conditions affecting (permanently or temporarily) the insulation.

The dynamic assessment of insulation conditions is given and highlighted in. Authors indicate that during the insulation aging process such phenomena as aging, degradation and destruction can be observed. Aging and degradation of insulation directly affect the destruction of insulation, but there is no clear line between aging and degradation. Differences between the phenomena of degradation and breakdown (sudden failure) of insulation are also given in [1]: "The essential difference between the degradation and breakdown phenomena consists in the fact that the degradation is a process that takes place for a long period of time, while the failure is a process that occurs suddenly and it is catastrophic, insulation being unable to support the nominal voltage after its failure".

In [2, 3] changes in the insulation quality condition is associated with the process of aging and is considered from the molecular physics perspective. Molecular changes in the insulation contribute to the deterioration of electrical, mechanical and other properties thus leading to the development of mechanisms of insulation degradation.

The specific requirements in respect of the maintenance of locomotives running on the railways of Ukraine necessitate a slightly different approach to determining the stages of the electrical machines insulation quality in time than the approaches considered above. Regarding the frame insulation, to define its aging stages it is necessary to determine the type of restoration technological operation:

– to restore the frame insulation through soaking treatment;

– to replace the frame insulation with a new one.

The automatic diagnostics of the insulation of traction motors attracts more and more attention due to the accelerated insulation degradation observed in recent decades caused by the use of electronic pulse-width static converters in new types of rolling stock that generate powerful electromagnetic interference in a wide frequency range [4].

Several methods of dielectric diagnostics were developed to assess the insulation condition: polarization and depolarization current measurement (PDC – polarization and depolarization current) [5], frequency domain spectroscopy (FDS) [6], recovery voltage measurement (RVM), [7, 8]. Measuring of the recovery voltage is a good method for obtaining information on the insulation condition.

Insulation residual resource can be determined via mathematical modeling of the insulation aging process.

Models [9, 10, 11] are designed bearing in mind insulation structural changes occurred due to electromechanical stresses capable to destruct crystallites or damage interfaces in composite materials.

In model [12], authors proceed from the fact that a space charge formed in the insulation sooner or later becomes the cause of aging. The charge accumulated in different insulation zones locally increases the electric field, increases the mechanical stress, and then chemical bonds are destroyed.

The model presented in [13] proceeds from the assumption that a space charge is formed because of aging, and the aging process itself is due to high electric fields that destroy chemical bonds in the insulation.

The most favourable models for predicting the residual insulation resource still are statistical models.

Models that use neural network modelling of insulation condition diagnosing should be distinguished among the statistical models [14]. This work applies models based on feed-forward artificial neural networks. In [15] researchers propose exponential models and models based on Weibull's distribution law. To determine the residual resource the regression models are used as well. Cox's regression model [16] is the most common regression model since it is not related to assumptions on the tolerance time distribution. The model assumes that the resource wear process occurs independently of operational impacts.

The disadvantages of statistical models include the fact that all of them require a large amount of reliable statistical data.

3 Materials and Methods

3.1 Analysis of VL-8 locomotive TM failures (depot Nizhnedneprovsk- junction)

Damaged VL-8 locomotives were observed at depot Nizhnedneprovsk-junction railway yard (Ukraine). The prerequisite for selecting locomotives to be statistically relevant was the fact that locomotives underwent complete overhaul (CR) and TMs were fully restored. Locomotive traction motors are characterised with fairly high level of failure, i.e. 12% (provided that a locomotive has 6 elements). Almost a third of TM failures are failures of insulation of TM excitating windings, Fig. 1.

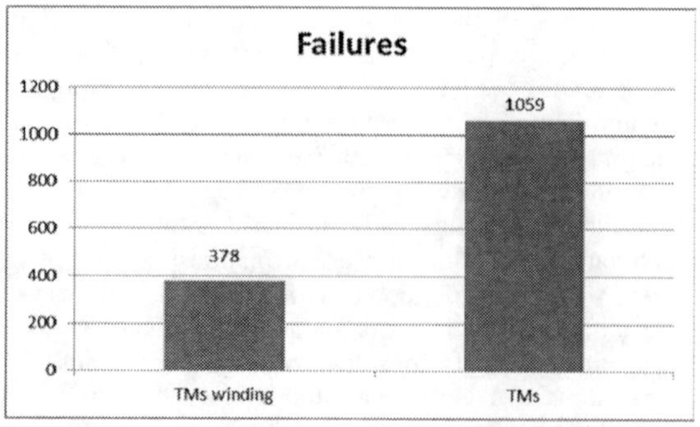

Fig. 1. Failures of winding insulation relative to the total TM failure rate.

Fig. 2 graphically represents the dependence of failures of TM excitating winding isolation and the TM itself.

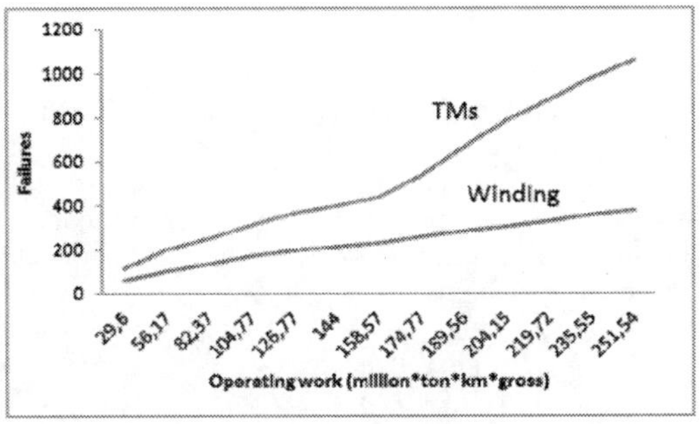

Fig. 2. Dependence of TM winding isolation failures and TM failures in relation to operating work (million*ton*km*gross).

Table 1 indicates dynamic failures of major structural elements of a TM. Special attention should be paid to the TM, where a tendency of failure increase is observed.

Table 1. Percentage-wise ratio of unplanned repairs of traction motors per their major structural elements

TM element	2014	2015	2016	2017
Armature	70,5	73,5	69,5	64,4
Pole winding	11,6	8,55	10,5	16,8
Cable connections	8,53	6,84	16,8	9,9
Motor armature bearings	9,3	11,1	3,16	8,91

Measurements were performed and research was carried out in respect of the insulation resistance change in TM main poles excitation winding depending on the run during summer and winter periods. It turned out, that the insulation resistance, measured by using a megohmmeter with a relative accuracy, does not demonstrate any established regularity and is not subject to any descriptive law. That is, this parameter is not informative in terms of diagnostics and prediction of the coil insulation condition. Several attempts were made to establish the dependence of insulation resistance on the run and seasonal "winter-summer" use (Fig. 3, 4), which also did not yield the expected results.

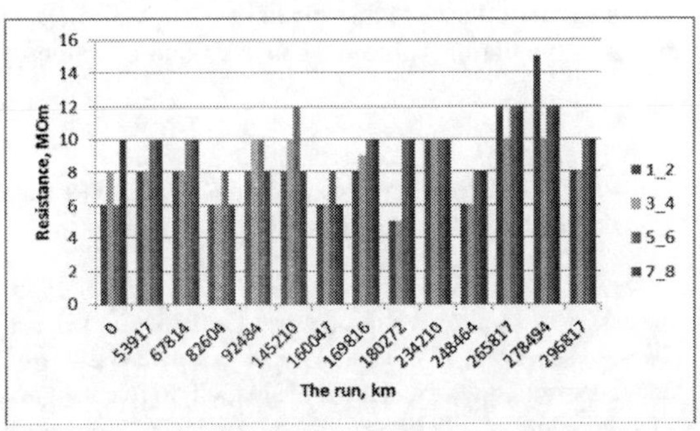

Fig. 3. Changes in TM main poles excitation winding insulation resistance depending on the run (summer). 1_2, 3_4, 5_6, 7_8 – poles.

Measurement of insulation resistance by using a procedure recommended in "Rules for Locomotive Repairing" is to only control this parameter so that it does not exceed the allowed limits. To obtain an informative diagnostic parameter, which could be used to make conclusions on the insulation condition at a given time and to predict changes in its future condition, it is necessary to develop new procedures and methods.

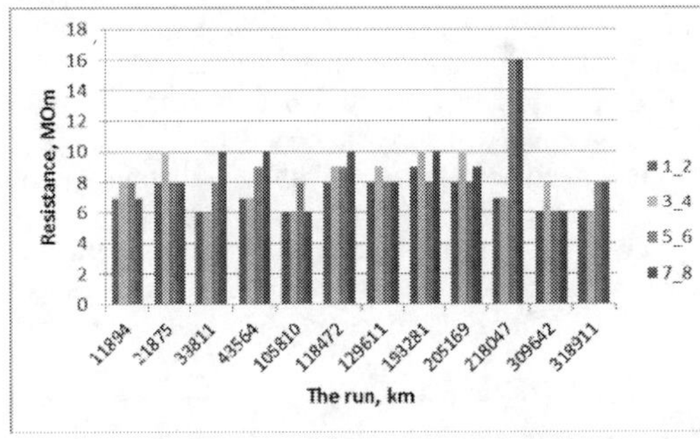

Fig. 4. Changes in TM main poles excitation winding insulation resistance depending on the run (winter). 1_2, 3_4, 5_6, 7_8 – poles.

Measures to eliminate TM winding failures will increase the reliability of TM and reduce penalties charged due to work failures, demanded by operational departments from the depot.

3.2 Problem statement

The vast majority of known insulation resistance measuring methods and devices are designed to be used in an auto-mode. In connection with this, methods for assessing the insulation condition are necessary, which allow for both autonomous and continuous monitoring of the insulation condition. Despite the considerable amount and variety of existing methods and devices, no classification of the insulation wear share is available, which makes it difficult to decide on the timing of its restoration.

The objective of this work is to determine quality conditions of the traction motor frame insulation according to the recovery return voltage data based on the classification of the wear level of the insulation being diagnosed, as well as to determine operating run estimates until the restoration or replacement of the insulation with a new one.

3.3 Integrated evaluation of the insulation quality condition

An integral evaluation can take into account not only the magnitude of some process, but also the rate of its change, etc. Integrated evaluation is intended to produce a general evaluation of the damping rate and a value of the process deviation from a given value without determining the former and the latter individually. The simplest integral evaluation can be an area between the values of the studied process and its steady state value.

The main characteristic used to determine indicators of the insulation condition is the recovery voltage. The recovery voltage yields the fullest information on the insulation aging process.

Characteristics of the insulation condition are conditionally divided into two types: direct and indirect. Direct characteristics include those, which are directly derived from the recovery voltage plot (measurements). Indirect characteristics include those, which are calculated based on the recovery voltage observational data. An integral evaluation of the insulation condition is attributed to the indirect characteristic. We will select and define the characteristics based on the information on recovery voltage observational data.

Integrated evaluation of the insulation condition allows characterizing the recovery voltage curve with a certain number, which comprehensively evaluates the curve as a whole over a given interval. As a rule, such evaluations are functionals and are expressed as integrals.

Let us have a look at some integral characteristics. Since the conditions of measurements can be different at different times (voltage supply instability, meteorological conditions, etc.), the sequence of measured recovery voltage values should be normalised. The range of recovery voltage values U_k, $k=\overline{1,n}$ should be reduced to a single interval $u_k \in [0,1]$ using dependence

$$u_k = \frac{U_k - U_{\min}}{U_{\max} - U_{\min}},\ k=\overline{1,n}, \quad (1)$$

where u_k is a normalised value of k-th measurement; U_k – is measured value of recovery voltage; $U_{\min}$ and $U_{\max}$ – minimum and maximum values of return voltage respectively; n – number of measurements. Considering the peculiarity that $U_{\min}=0$, then equation (1) will be simplified down to

$$u_k = \frac{U_k}{U_{\max}},\ k=\overline{1,n}. \quad (2)$$

Normalised values u_k should be approximated using dependence $u(t)$ in measurement interval $t \in [0, t_C]$. In the study, the approximation of measurement of the recovery voltage was performed using cubic splines.

Integral indicator of the insulation condition J_0 in based on the work evaluation, which is done by the insulation dipoles under the influence of a constant electric field. The magnitude characterizing the work of dipoles may be limited by a recovery voltage curve and the time axis. Formally, the recovery voltage drop time $U(t)$ goes to the infinity. The evaluation of work of dipoles at self-discharge, evaluation time interval t_C should be chosen in such a way as to allow $U(t_C)=U_{\max}/2$, where $U_{\max}$ – is the recovery voltage maximum value. Now, we can yield the target evaluation from the equation

$$J_0 = \int_0^{t_C} u(t)\,dt. \quad (3)$$

A similar way of evaluating di-electrical properties is considered when determining the reliability of capacitors.

Integrated evaluation (3) takes into account only the additive quality of the recovery voltage curve. An integral evaluation of the recovery voltage curve can be made via evaluating

$$J_{01} = \int_0^{t_C} \left[u(t) + \gamma \cdot u'(t)\right] dt,\ \gamma > 0. \quad (4)$$

Integrated evaluation (4) takes into account both additive quality of the recovery voltage curve and the curve change rate (recovery voltage growth rate and the rate of self-discharge). It should be noted that normalising of a scale for the recovery voltage derivative $u'(t)$ is performed in the same way as for the recovery voltage measurements.

The assessment of recovery voltage curve quality also can be performed using an integral quadratic criterion

$$J_{21} = \int_0^{t_C} \left[u^2(t) + \gamma \cdot \left(u'(t)\right)^2\right] dt,\ \gamma > 0, \quad (5)$$

or

$$J_{12} = \int_0^{t_C} \left[u(t) + \gamma \cdot \left(u'(t)\right)^2\right] dt,\ \gamma > 0, \quad (6)$$

parameter γ is a calibration parameter. It is chosen depending on the importance of a derivative (summand).

A type of equation used for the integral evaluation depends on the type of frame insulation and is chosen experimentally. In general, the insulation quality will be characterized by two indicators (J, t_C), where J – is an integral evaluation of the recovery voltage curve; t_C – time of curve evaluation (the upper limit of integration of the integral evaluation equation). As an integral evaluation value we can take one of values J_0, J_{01}, J_{21}, J_{12}.

3.4 Identification of classes of NB-406 TM frame insulation quality conditions

When predicting using the method of *classification*, it is necessary to solve two problems: first, to build a set of classes; secondly, it is necessary to evaluate the attributes and, according to the results of evaluation, to relate an object of predicting to this or that class. Solving of the first task requires processing a large amount of statistical data obtained during the operating period and using technical objects or performing special experiments. The possibility of forming classes largely depends on the successful selection of diagnostic attributes. These attributes should sufficiently enough characterize the course of processes leading to the loss of the object's workability, and their accurate evaluation should not present great difficulties. The success of solving the second task is largely determined by the accuracy of relating an object to a known class based on the evaluation results. The class is characterised by a certain tendency of change occurred to the object to over time.

In the study the insulation quality is characterised by a pair of indicators (J_{01}, t_C), where J_{01} - integral evaluation of the recovery voltage curve calculated in accordance with equation (4), t_C - time of recovery voltage curve evaluation. The distribution per clusters were performed using the method of balls. Isotonic and isomorphic distribution per clusters was used. Clustering was carried out for 38 curves and allowed defining three groups of curves (clusters), i.e. insulation conditions. Upon clustering 7 curves were removed from the received clusters.

The recovery voltage curves were constructed for the frame insulation showing different degrees of wear. In addition, the statistics included objects with the restored frame insulation of the main poles winding.

4 Results and Discussion

In accordance with the results obtained during the cluster breakdown procedure, we introduce the following stages of the frame insulation condition: 1st stage (cluster K_1) – "satisfactory condition"; 2nd stage (cluster K_2) – "restored condition"; 3rd stage (cluster K_3) – "high risk condition". The introduced stages are conditional and may be interpreted as follows. The first stage means new insulation or insulation being used for a negligible period or its condition may be referred to as "new". As a rule, such insulation does not need substantial maintenance measures during technical check-up. The second stage means insulation, which is characterised by the wear of almost a half of operating run until the first failure. Regular restoration of the insulation is recommended at this stage. The third stage is characterised by an increased risk of insulation failure occurrence (ranging from the increased possibility of partial discharge to the insulation breakdown). At this stage, the last restoration or the complete restoration of the insulation should be planned (replacement of frame insulation). This work does not consider the moment of a complete wear. We presume that when the recovery voltage reaches 40-50 V at 25th second of measurement, insulation should not be restored and is subject to be replaced with a new one. Further usage of such frame insulation can cause its failure while a locomotive is moving. In accordance to the introduced insulation condition stages, a set of recovery voltage curves, which can be referred to them, will be called as follows: the 1st curve group, the 2nd curve group, the 3rd curve group. Fig. 5 shows examples of recovery voltage curves: with maximum U_{max} voltage (the 1st curve group), with minimum U_{max} voltage (the 3rd curve group), a curve from the second curve group is in the middle of the Fig. 5.

Using other methods of clustering it is possible to obtain a different number of clusters (insulation conditions). The study used integral evaluations (5) and (6) as indicators that differ from the above (4), but in this case, the distribution of curves per clusters did not correspond to experimental observations.

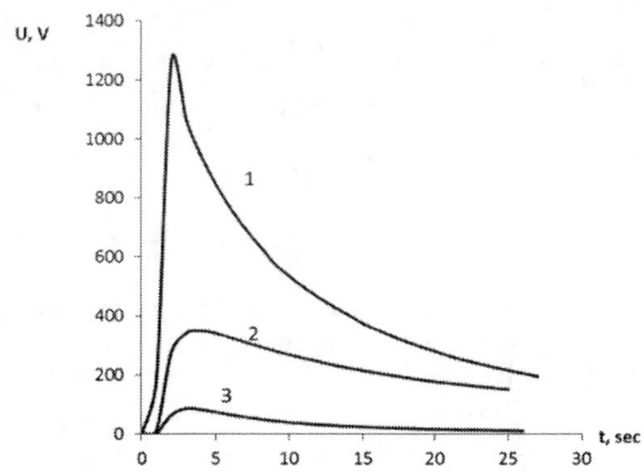

Fig. 5 – Example of the recovery voltage curves at various stages of insulation condition:
1st curve corresponds to the 1st stage, i.e. "satisfactory condition"; 2nd curve corresponds to the 2nd stage, i.e. "restored condition"; 3rd curve corresponds to the 3rd stage, i.e. "high risk condition".

Calculation of values of critical radii R_i to determine a cluster which a new object belongs to (recovery voltage curve of the diagnosed frame insulation) was performed according to the method of balls.

For received clusters the values of classification indicators should be averaged, $\overline{J}_i = \overline{\tilde{J}}_{01}, \tilde{J}_{01} \in K_i$ and $\overline{t}_i = \overline{\tilde{t}}_C, \tilde{t}_C \in K_i$. Values $\tilde{J}_{01}$ and $\tilde{t}_C$ are respective normalized values, similar (1)-(2), integral evaluation J_{01} and time t_C. Operating run $\overline{T}_i$ of frame insulation coils is averaged as well since it falls into the studied cluster of curves $K_i, i = \overline{1,3}$.

Verification of the normal law of distribution of values of operating run of frame insulation for the constructed clusters was carried out.

The verification of the distribution law has shown that operating run values demonstrate normal distribution characteristics with the level of significance $\alpha = 0,05$. For operating run averaged values $\overline{T}_i$ confidence intervals $(\overline{T}_i - \Delta T_i, \overline{T}_i + \Delta T_i)$, $i = \overline{1,3}$ were defined. The value $\overline{T}_i$ can be interpreted as an average operating run at which the frame insulation reaches i – th condition. Cluster parameters and confidence intervals of insulation operating run for each cluster are presented in Table 2.

Table 2. Characteristics of classification clusters used to determine isolation quality conditions

Condition	$\overline{J}_i = \overline{J}_{01}$	$\overline{t}_i = \overline{t}$	R_i	$(\overline{T}_i - \Delta T_i, \overline{T}_i + \Delta T_i)$ thnds km	Point of restoration according to the repair rules, thnds km
1st stage	0.87	0.43	0.15	(165, 405)	350
2nd stage	0.71	0.74	0.26	(580, 940)	700
3rd stage	0.13	0.54	0.17	(865, 1175)	1050

The quality condition of the diagnosed insulation (new diagnosed object) and values of the recovery voltage curve $(\tilde{J}, \tilde{t}_C)$ are calculated if the following precondition is met

$$(\tilde{J} - \overline{J}_i)^2 + (\tilde{t}_C - \overline{t}_i)^2 < R_i^2, \; i = \overline{1,3}. \quad (7)$$

If the fulfilment of inequality (7) is not unique, then the object's membership in a cluster is defined using a smaller evaluation

$$\Delta_i = \left| (\tilde{J} - \overline{J}_i)^2 + (\tilde{t}_C - \overline{t}_i)^2 - R_i^2 \right|, \; i = \overline{1,3}. \quad (8)$$

The normalisation of indicators (J, t_C) of a new object being tested is performed within the whole set of indicators $\{J_i\}_{i=\overline{1,N}}$ and $\{t_{Ci}\}_{i=\overline{1,N}}$ respectively, N - the number of indicators – the sum of elements (J, t_C) included in clusters $K_i, i = \overline{1,3}$. Normalisation (1)-(2) is fulfilled if only two inequalities are simultaneously true

$$J \le J_{\max} \text{ and } t_C \le t_{C\max}. \quad (9)$$

Upon testing a new object it is necessary to add new object (J, t_C) to the cluster, which the tested object belongs to and to re-calculate necessary statistical characteristics of the modified cluster.

If one of inequalities (9) is untrue, it is necessary to perform the classification once again taking into the new case isolation tested object with subsequent determining its resource.

Depending on the membership of a diagnosed object in some cluster, the insulation condition will be characterised by the operating run value until its current restoration (including the replacement of an old insulation for a new one). The operating run value until its first restoration τ_1 is defined using equation

$$\tau_1 = \overline{T}_1 + \Delta T_1 - T, \; T < \overline{T}_1 + \Delta T_1, \quad (10)$$

where T - is current operating run workout of the diagnosed insulation;

evaluation of the operating run value until the second restoration τ_2

$$\tau_2 = \overline{T}_2 + \Delta T_2 - T, \; \overline{T}_1 + \Delta T_1 < T < \overline{T}_2 + \Delta T_2; \quad (11)$$

evaluation of the operating run value until the replacement of insulation for a new one or it complete renovation τ_3

$$\tau_3 = \overline{T}_3 + \Delta T_3 - T, \; \overline{T}_2 + \Delta T_2 < T < \overline{T}_3 + \Delta T_3. \quad (12)$$

If inequality (7) is untrue for any cluster, then it is possible to evaluate the insulation quality condition by using its workout until the time of its restoration according to the repair rules defined with one of the following values

$$\tau_1 = 350 - T, \; T < 350; \quad (13)$$

$$\tau_2 = 700 - T, \; 350 < T < 700; \quad (14)$$

$$\tau_3 = 1050 - T, \; 700 < T. \quad (15)$$

To determine whether the studied recovery voltage curve belongs to this or that cluster is also possible via discriminant analysis. In this case, it is necessary to have available learning samples of insulation quality conditions (clusters are filled with test objects during the maintenance process).

The recognition whether the tested frame insulation belongs to a cluster in accordance with equation (7) - (8) should be applied at the initial stage. When clusters contain a sufficient number of elements, then it is more appropriate to perform discriminant analysis.

The above method of determining the quality condition of frame insulation is acceptable within one depot, where locomotives are usually operated in the same way.

Conclusion

The classification of recovery voltage curves allows determining frame insulation quality conditions.

Clustering attributes are as follows: J - integral evaluation of a recovery voltage curve determined from equation (4), t_C - time of recovery voltage curve evaluation.

The breakdown of recovery voltage curves of the frame insulation into clusters allowed distinguishing three stages of the insulation quality condition and mark them "satisfactory condition", "restored condition", "high condition".

Software for classification of insulation quality conditions and discriminant analysis for determining the insulation condition has been developed.

The research presented in this article made to order and funded by Pridniprovska Railway Company, Ukraine. The authors are grateful to the management of the Depo Nizhnedniprovsk-Uzel engineering department for their assistance in obtaining the necessary data for research.The authors also express their gratitude to V.M. Zubenko (the engineer of the testing center of Pridniprovska Railway Company) for test assistance of the TMs frame insulation. Authors would also like to thank O. Voznyak – PhD, associate professor, Lviv branch of Dnipro National University of Railway Transport, Ukraine, for informational support.

References

1. Notingher, PV & Plopeanu, M, "Accelerated Development of Electrical Trees. Part I: Initiation of Trees", *EEA - Electrical Engineering, Electronics, Automatics*, Vol. 57, No. **4**, pp. 11-19, (2009).
2. Crine, J-P & Vijh, AK, "Molecular Approach to the Physico-Chemical Factors in the Electric Breakdown of Polymers", *Applied Physics Communications*, Vol 5, No. **3**, pp. 139-63, (1985).
3. Lewis, TJ, "Ageing - A Perspective", *IEEE Electrical Insulation Magazine*, Vol. 17, No. **4**, pp. 6-16, (2001).
4. V. I. Havryliuk, "Modelling of the return traction current harmonics distribution in rails for AC electric railway system", 2018 *International Symposium on Electromagnetic Compatibility* (*EMC EUROPE*), IEEE, pp. 251-254, (2018).
5. Junhao Li, Yanming Li, "Influencing Factors Investigation of Polarization and Depolarization Current Measurements on Assessment Aging Condition of Power Transformer", *International Review of Electrical Engineering (IREE)*, Vol. 5 No. **6**, pp. 2637-2643, (December 2010).
6. A. A. Shayegani, E. Gockenbach, H. Borsi, H. Mohseni, "Investigation on the Transformation of Time Domain Spectroscopy data to frequency domain data for impregnated pressboard to reduce measurement time", *Electrical Engineering*, Vol. 89 No. **1**, pp. 11-20, (June 2005).
7. T. K. Saha, M. Darveniza, D. J. T. Hill, T. T. Ie, "Electrical and chemical diagnostics of transformer insulation part A: Aged transformer samples", *IEEE Transactions on Power Delivery*, Vol. 12, pp. 1547–1554, (October 1997).
8. Michail Kapitsa, Vasiliy Laguta, Yuriy Kozik, "Selecting the Parameters of The Diagnosis of Frame Insulation Condition in Electrical Machines of Locomotives", *International Journal of Engineering & Technology,* Vol. **7** (4.3), pp. 110-114. DOI: 10.14419/ijet.v7i4.3.19718, (2018).
9. Jones, JP, Llewellyn, JP & Lewis, TJ, "The contribution of field-induced morphological change to the electrical aging and breakdown of polyethylene", *IEEE Transactions on Dielectrics and Electrical Insulation*, Vol. 12, No. **5**, pp. 951- 966, (2005).
10. Lewis, TJ, "Polyethylene under Electrical Stress", *IEEE Transactions on Dielectrics and Electrical Insulation*, Vol. 9, No. **5**, pp. 717-29, (2002).
11. Rowe, SW, "Electrical Ageing of Composites: An Industrial Perspective", *Proceedings of International Conference on Solid Dielectrics*, Winchester, UK, July 8-13, pp. 401-406, (2007).
12. Dissado, LA, Mazzanti, G & Montanari, GC, "Elemental strain and trapped space charge in thermoelectrical aging of insulating materials. Life modelling", *IEEE Transactions on Dielectrics and Electrical Insulation*, Vol. 8, No. **6**, p 966- 971, (2001).
13. Mazzanti, G & Montanari, GC, "Electrical aging and life models: the role of space charge", *IEEE Transactions on Dielectrics and Electrical Insulation*, Vol. 12, No. **5**, pp. 876-890, (2005).
14. LS. Skrupskaya, A.O. Oliinyk, A.A. Sakhno, "Model building for current transformer paper-oil insulation state diagnostics", *ISSN 2074-272X. Elektrotexnika i elektromexanika*, No. **2**, pp. 48-51, (2014).
15. O.A. Sakhno, "The continuous monitoring and remaining life prediction system for high-voltage current transformers". – Manuscript. *The thesis for Ph.D. degree obtaining in the specialty 05.14.02 – Electric Power Stations, Grids and Systems*. National University "Lviv Polytechnics", Lviv, pp. 7-14, (2012).
16. Cox D.R., "RegressionModels and Life-Tables M", *Journal of the Royal Statistical Society, Series B (Methodological)*, Vol. 34, No. **2**, pp. 187-220, (1972).

Concept of the railway safety, security and cybersecurity functional integrity levels

Marek Pawlik[*]

Railway Research Institute, 50 Chlopickiego str., 04-275 Warsaw, Poland

Abstract. Observed growing changes in the character and severity of the risks in rail traffic safety and rail transport security are associated with present development of utilized technical solutions. New hazards are coming out, besides known ones, including hazards associated with cyber-crime. As a result it is fully justified to undertake works dedicated to collect and settle all risks associated with technical solutions using modern technologies for acquisition, computing and transfer of the data, which are vital from the rail traffic safety and rail transport security point of view. Article defines rail transport systems safety, security and cybersecurity functional integrity levels thanks to knock-out and differentiating questions regarding identified key safety related functionalities. Proposed methodology was used for safety and security verification of a chosen homogenous rail transport system separated from the overall Polish railway system. Results have shown discrepancies in utilized protection measures. Proposed methodology can be used for assessment of existing systems as well as for specifying scopes of investments both for infrastructure and rolling stock modernizations. Applicability range covers railway transport, light rail services, metro, urban rail transport systems as well as rail based transport systems using autonomous vehicles.

1 Introduction

1.1 Technical safety and personal security

It is unambiguous, that all technical solutions, which are utilized by railway transport, are being defined, constructed and maintained having safety in mind. It is not so easy to declare, that railway system with all its solutions is safe, as incidents and accidents happen since time to time even if the amount of them is relatively low. Severity of the railway accidents is frequently high and therefore it is a real challenge to define how safe the railway system has to be to be declared as a safe one.

From the very beginning of the railway history safety was based on technical solu-tions and procedures. Technical solutions were, and still are, constructed in a way ensuring safe operation in case of errors, faults, failures, which may cause system malfunctioning. As degraded operational circumstances do appear during long lifecy-cles of the railway solutions it was, and still is, required to apply fail-safe principle. It means, that neither faults or failures nor errors or extreme external conditions, e.g. temperatures or loads, can lead to dangerous situations understood mainly as giving the train permission to run too far or too quick. As a result faults, failures, errors and extreme external conditions thanks to own inherent characteristics of the solutions lead to shifting responsibility from technical systems to staff and procedures in de-graded operational circumstances.

Applying fail-safe principle was, and still is, not appropriate for some technical so-lutions – for ensuring appropriate endurance of the tracks, embankments, bridges, as well as vehicles' bodies, running gears and auxiliary systems. In that respect safety since the beginning was, and still is, based on widely accepted codes of practice e.g. UIC leaflets, EN standards, OTIF specifications.

Moreover applying fail-safe principle is not appropriate for technical solutions which are utilizing electronic, programmable systems and modules for which huge catalogue of possible faults, failures and errors forefend verifications of the appropri-ateness and completeness of the fail-safe principle. That mainly applies to railway control command and signalling equipment.

It was obvious from the beginning, that technical safety is a must, but insufficient requirement. Railways had, and still have, to ensure personal safety on stations and in trains. They had, and sill have, to ensure safety of the cargo as well as minimized undesirable influence on environment. Up to the beginning of the twenty first century that was ensured by procedures and dedicated staff – by railway police. Presently they are more and more supported by electronic equipment, and therefore relationships between technical safety and personal & goods security starts to be blotted.

[*] Corresponding author: mpawlik@ikolej.pl

1.2 Railway traffic safety and rail transport security

Presently technical experts responsible for railway infrastructure – for tracks, switches and crossings as well as underlying embankments, bridges and viaducts as well as accompanying platforms, pathways and station buildings and their accoutrements apply nearly only technical solutions, which are fully in accordance with abovemen-tioned codes of practice e.g. Vignole railway rails fully compliant with EN 13674 standard series. To large extend approach based on codes of practice apply also for traction power supply and rolling stock as adequate technical documents directly define requirements for materials and constructions as well as for verifications of the final solutions.

Codes of practice also exist for control command and signalling equipment installed both trackside and onboard. However in that respect technical documents define functionalities and safety verification rules generally omitting materials and the way how to construct technical solutions. This is largely linked to relatively quick development of technical solutions utilized for control command and signalling. As a result fail-safe principle and codes of practice are not sufficient for ensuring technical safety. Since twenty years in that respect railway signalling solutions respect so called Safety Integrity Levels defined by EN standards dedicated to reliability, availability, maintainability and safety – RAMS standards [3÷7]. Control command and signalling systems and devices inherent safety is assessed in case of electronic, digital and programmable solutions against Safety Integrity Level SIL-4 defined by tolerable hazard rates

10-8÷10-9 for hazardous events per hour, which may be caused by random failures and external factors and recommended technics for minimization of hazards associated with human mistakes. The SIL based approach is required for control command under interoperability directive [2], which is applicable to new technical solutions. At the same time risk based approach is required under railway safety directive [1]. Both however omit security challenges, which the railways are facing all the time.

Railways are using presently technical solutions based on electronic, digital, pro-grammable components not only for control command and signalling but also for supporting security staff – for generating emergency alerts, video monitoring, communication, supporting rescue and evacuation. In that respect SIL-4 based approach is not required. As a result it is reasonable to ask whether safety and security are fully ensured and whether they are ensured similarly well in relation to different risks which are presently common. Safety and security impact reference model was defined to answer those questions.

2 Safety and security impact reference model SSIRM

Safety and security impact reference model SSIRM is based on identification of functions which are supported by electronic, digital and programmable solutions. Safety is treated not only as an overall requirement which has to be ensured in normal operation but also as inherent characteristics of the technical solutions which ensure safety in degraded modes of operation. Key functionalities, understood as groups of individual functions, which influence railway traffic safety are shown at figure 1.

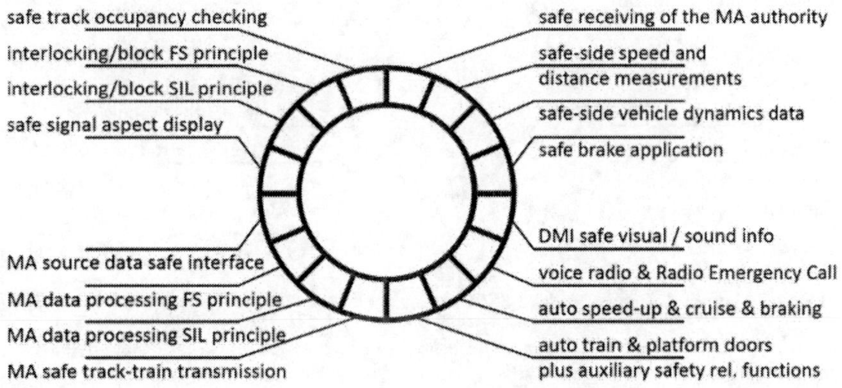

Fig. 1. Safety and Security Impact Reference Model SSIRM, safety related functionalities
FS – fail-safe principle, SIL – safety integrity level SIL-4 principle, MA – digital movement authority, DMI – driver machine interface /own elaboration/

Safety domain.

Sixteen functionalities can be subdivided into five groups. The first group (upper left) represents classic signalling equipment installed trackside. The interlocking and block systems are due to respect fail-safe principle and ensure Safety Integrity Level SIL-4. This would however not ensure safety if track occupancy checking or displaying signal aspects do not respect safety rules. Applying fail-safe and SIL-4 principles are required for individual solutions, however overall verification has to be performed taking into account dependences between elements working together.

The second group (lower left) represent control command equipment installed trackside. Its role is to prepare an electronic movement authority for the train on the basis of data taken from signalling equipment. The way how the data is taken cannot change the data in the wrong side neither on signalling nor on control command side. Data processing has to respect both fail-safe and SIL-4 principles. Prepared move-ment authority has to be sent in a safe way. Data acquisition, data processing and data transmission have to technically safe in all operational circumstances.

The third group (upper right) represents control command equipment installed onboard of the traction vehicles. Movement authorities have to be received, verified and respected. Obtaining digital movement authority is worth only if onboard equip-ment ensures reliable and safe information about relationship between location and speed of the train in relation to the braking curves imposed by authorities taking into account distance and speed measurements as well as considering vehicle dynamics. From safety point of view it is necessary to ensure also safe application of the brakes taking into account both full service brake and emergency brake.

Digital movement authority available onboard can be utilized by manually operated trains and by automatically operated vehicles. Therefore the four remaining functionalities are subdivided into two groups (lower right). Group fourth representing cab signalling by visual and audible information as well as radio communication between train driver and trackside staff together with emergency calls. The fifth group representing automatic driving – automatic speed-up, cruise and braking as well as functions which are necessary for safe access and egress including platform doors control and auxiliary auto train functionalities. The sixteen fields on the ring can be used to represent safety aspects by colors.

Security domain.

Control command and signalling functionalities have to be complemented by security related ones. Security domain is shown at figure 2. It is also composed by five groups of functionalities. The first group (upper left) represents solutions which are due to ensure basic passenger safety. Passenger information systems are very important both trackside and onboard in case of emergency, to prevent panic, to support evacuation etc. Railways are also due to ensure fire safety and electrical safety. More and more that is also supported digitally and therefore has to be considered.

The following groups (lower left) represent systems ensuring protection against crime and vandalism as well as enhanced protection for passenger health. The second group covers solutions enabling passing alarms to dedicated staff and video monitoring systems. The third group covers emergency call installations as well as medical equipment like e.g. automatic external defibrillators AED.

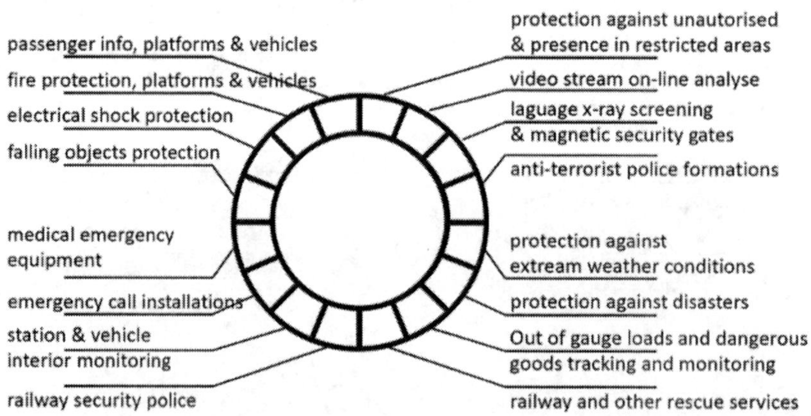

Fig. 2. Safety and Security Impact Reference Model e-SSIRM, security related functionalities /own elaboration/

The fourth group (upper right) is representing enhanced technical means against crime, vandalism and terrorism. It covers protection against unauthorized access and presence based on simple solutions as well as video stream analyzers, which are able to detect persons entering restricted areas e.g. passing from one platform to another over tracks, unattended luggage, fake crowd, running persons, etc. As a result appro-priate information is automatically identified and communicated to security staff immediately providing chance to react in due time and not only to document hazardous events for further investigations. Additionally stations and platforms can be protected by language screening and magnetic gates. That is already in use in some cases, however it is questionable especially in case of traffic in agglomerations.

The fifth group (lower right) represents technical protection means against natural and construction disasters. Also such technical solutions are already in use in some places e.g. in case of high speed lines going through seismic areas, in case of long railway tunnels and so on.

3 Enhanced impact reference model e-SSIRM and safety, security and cyber-security Functional Integrity Levels FIL

Safety of the data transmission.

It is visible for experts, that risk associated with data transmission systems has to be treated very seriously, especially as wireless communication technics are more and more used for train control [9, 10]. However simple SSIRM model does not fully take into account data transmissions. As a result of considerations regarding cybersecurity it was therefore enhanced by adding data transmissions in a form of connections between different fields on the safety ring and between different fields on the security ring. Adding connections representing data transmission systems on one side enables showing safety by colors similarly to the fields. On the other side adding connections provides flexibility which ensures easy way for showing different transmission media arrangements covering individual transmission systems and complex transmission systems serving different functionalities as well as wired and wireless transmission systems and their relationships with safety and security functionalities.

Such representation enables easy way to visualize systems which have to respect safety related requirements for transmission systems [7] together with their relationships with equipment components which have to respect safety related requirements for hardware modules [6] and for software modules [5]. The overall assessments are however supported only visually while it would be helpful to create an add on to Safety Integrity Levels SIL-4 requirements which are mandatory for individual technical systems supporting safety.

Functional integrity levels for safety, security and cybersecurity.

Ten groups of functionalities, five dedicated to data transmission based technical means supporting railway traffic safety and five dedicated to data transmission based technical means supporting railway transport security, were used to prepare sets of questions containing two types of questions – knock-out questions and differentiating questions. The knock-out question can receive value "0" and value "1". The differentiating question can receive values "1" and value "2". Additional set of questions was prepared for transmission systems to reflect resistance against internal malfunctioning, extreme external conditions and cyber-crime. Also in that respect knock-out and differentiating questions were defined.

Examples of the safety related knock-out questions.

1 Whether all tracks with their full length are covered by track occupancy systems?
2 Whether all track occupancy systems fully apply fail-safe rules?
3 Whether all track occupancy systems utilizing electronic, programmable or simply digital solutions have safety cases elaborated by producers and verified by independent safety assessors proving SIL-4?
4 Whether all station interlockings, line block systems and level crossing protection systems fully apply fail-safe rules?
5 Whether all station interlockings, line block systems and level crossing protection systems have safety cases elaborated by producers and verified by independent safety assessors proving SIL-4?
6 Whether all track sections are protected by visible track-side signals or visualised in the cabs on-board in all trains permitted to run?
7 Whether all signal types are included in safety cases elaborated by producers and verified by independent safety assessors proving SIL-4?
8 Whether all technical solutions for data acquisition from interlockings, line block systems and level crossing protection systems (used for digital movement authorities) do not influence, under any foreseeable circumstances, safety of the interlockings, block systems and level crossing protection?
9 Whether all technical solutions used for data acquisition from interlockings, line block systems and level crossing protection systems fully apply fail-safe rules?
10 Whether all technical solutions for data acquisition from interlockings, line block systems and level crossing protection systems have safety cases elaborated by producers and verified by independent safety assessors proving SIL-4?
11 Whether all technical solutions for data processing for electronic movement authorities have safety cases elaborated by producers and verified by independent safety assessors proving SIL-4?
12 Whether all messages, especially those containing movement authorities, contain data enabling sender authentication?
13 Whether all messages, especially those containing movement authorities, contain data enabling verification of validity?
14 Whether on-board control command equipment verifies sender authentications and message validities of all received messages, especially those containing movement authorities?
15 Whether on-board control command equipment properly estimates and takes into account errors of the distance and speed measurements?
16 Whether on-board control command equipment properly estimates and takes into account train braking dynamics?
17 Whether all on-board control command elements are fully taken into account in safety cases elaborated by producers and verified by independent safety assessors proving SIL-4?
18 Whether on-board control command equipment provides drivers with visualisation of the electronic movement authorities? or Whether on-board equipment automatically changes speed using on-board control command elements?

19 Whether voice radio communication is provided for drivers and signalmen ? or Whether on-board equipment automatically controls and commands on-board and trackside equipment (e.g. pantograph, main circuit-breaker, train doors and platform doors)?
20 Each knock-out question may have positive answer (YES = 1) or negative answer (NO = 0). The overall value is a product of all of them. Even a single negative answer is a knock-out for safety of a whole solution.

Differentiating questions for safety.

1 Whether control command messages contain data which are used by on-board control command equipment for verification of completeness and coherency of all received messages?
2 Whether on-board control command equipment verifies cryptographic protection of all received messages?
3 Whether drivers are informed by control command about latest places for starting braking and warned before equipment interventions?
4 Whether automatic braking interventions are using more than one braking mode (full service braking and emergencbraking)?
5 Whether receiving emergency signal automatically initiates braking which ensures stopping in a place appropriate for evacuation and for security and rescue staff interventions?

Differentiating questions for Security.

1 Whether emergency medical equipment, especially automated external defibrillators, are available in all stations in areas accessible for passengers and provides with appropriate signs and instructions?
2 Whether video-monitoring system used for providing security is equipped with video-stream analyser ensuring immediate generation of security warnings?
3 Whether luggage scanning is provided?
4 Whether protection against possible natural disasters is provided?
5 Whether tracking of dangerous goods is provided?

Differentiating questions for Cybersecurity.

1 Whether in case of detecting loss of communication for signalling automatic reconfiguration of communication system takes place or automatic switch on of the backup communication system takes place to ensure traffic control by technical means (and not only procedures)?
2 Whether safety related personnel is equipped with backup communication means?
3 Whether in case of detecting loss of communication for control command automatic reconfiguration of communication system takes place or automatic switch on of the backup communication system takes place to ensure train running supervision?
4 Whether in case when control command system is out of order trains can be driven on the basis of the signal aspects displayed on the track-side signals?
5 Whether technical systems and devices supporting security, especially video-monitoring systems are provided with backup power supply?

All questions are defined and described in a dedicated monograph [8].

Knock-out questions for safety are shown in table 1. as an example. Differentiating questions for safety, security and cybersecurity are shown in table 2. Each knock-out question may have positive answer (YES = 1) or negative answer (NO = 0). The overall value is a product of all of them. Even a single negative answer is a knock-out for safety of a whole solution. Each differentiating question also may have positive or negative answer, however in this case YES = 2 while NO = 1. Safety, security and cybersecurity are therefore represented by a vector.

$$\left[safety\ ,security\ ,cybersecurity \right] \quad (1)$$

in other notation

$$[\ \mathrm{SF, SC, CS}\ ] \quad (2)$$

where:
SF – product of all answers regarding safety,
SC – product of all answers regarding security,
CS – product of all answers regarding cybersecurity.
The Functional Integrity Level for safety, security and cybersecurity, FIL level, is defined as a sinus of an angle between vector and reference geometrical plane, for which maximum vector is perpendicular.

$$FIL_{SF,\,SC,\,CS} = sin\left(\begin{matrix} 32 & 0 & 0 \\ 0 & 32 & 0 \\ 0 & 0 & 32 \end{matrix} ,\left[SF,SC,CS \right] \right) \quad (3)$$

For $SF \neq 0$ and $SC \neq 0$. and $CS \neq 0$.
If $SF \neq 0$ or $SC \neq 0$ or $CS \neq 0$
then $FIL_{SF,\,SC,\,CS} = 0$.

where:
$FIL_{SF,\,SC,\,CS}$ is a safety, security and cybersecurity functional integrity level

An angle between vector and geometrical plane (represented by matrix) may only be right (=90°) or acute (<90°). Maximum FIL value equals "1" (as a sinus of 90°) when products of the answers regarding safety, security and cybersecurity are equal to each other. Growing discrepancies between products of the answers causes dropping of the FIL keeping it > zero for non-zero values of the SF, SC and CS.

4 Conclusion

Presently technical means based on data acquisition, processing, transmission and storage are widely used for supporting railway traffic safety as well as rail transport security. Therefore safety, security and cybersecurity should be seen as complementary topics, which have to be provided on similarly high level. It is not reasonable to provide high safety for some functionalities and no safety for the others. The cyber-attacks which have already happened did not affected control command or signalling equipment but passenger information systems and timetabling.

References

1. Directive (EU) 2016/798 of the European Parliament and of the Council of 11 May 2016 on railway safety (EU OJ L 138/102).
2. Directive (EU) 2016/797 of the European Parliament and of the Council of 11 May 2016 on the interoperability of the rail system within the European Union (EU OJ L 138/44).
3. European Standard EN 50126-1:2017, Railway applications —The specification and demonstration of reliability, availability, maintainability and safety (RAMS) – Part 1: Ge-neric RAMS Process.
4. European Standard EN 50126-2:2017, Railway Applications - The specification and demonstration of reliability, availability, maintainability and safety (RAMS) – Part 2: Sys-tems Approach to Safety.
5. European Standard EN 50128:2011, Railway applications — Communication, signalling and processing systems — Software for railway control and protection systems.
6. European Standard EN 50129:2003/AC:2010, Railway applications - Communication, signalling and processing systems - Safety related electronic systems for signalling.
7. European Standard EN 50159:2010, Railway applications - Communication, signalling and processing systems - Safety-related communication in transmission systems.
8. Pawlik M.: Railway safety and security functional reference model built on data transmission based systems (Referencyjny model funkcjonalny wspierania bezpieczeństwa i ochrony transportu kolejowego przez systemy z transmisją danych). ISBN 978-83-7814-908-8, Warsaw University of Technology Publishing House (Oficyna Wydawnicza Politechniki Warszawskiej), Warsaw 2019.
9. Lewiński A., Perzyński T., Ukleja P.: *Possibility of Use the Wireless Communication in Protection of Rail Traffic on the Regional Line*, Railway Reports, issue **175** (June 2017), ISSN 0552-2145, Railway Research Institute, Warsaw 2017, pp. 53-58
10. Gradowski P.: *Upgrading the railway infrastructure technical parameters using the example of the Control-Command and Signalling subsystem with the EC verification certificate*, Railway Reports, issue **182** (March 2019), ISSN 0552-2145, Railway Research Institute, Warsaw 2019, pp. 131-146

Determining the causes of rolling stock derailment from the track using modern research methods

Andriy Kuzyshyn[1,2,3*], *Andriy* Batig[3], *Julia* Sobolevska[2], *Sergei* Kostritsa[1], *Lyudmila* Ursulyak[1], and *Stepan* Dovhaniuk[1]

[1] Dnipro National University of Railway Transport named after Academician V. Lazaryan, 49000 Dnieper, Ukraine
[2] Lviv branch of Dnipro National University of Railway Transport named after Academician V. Lazaryan, 79052 Lviv, Ukraine
[3] Lviv Research Institute of Forensic Science, 79000 Lviv, Ukraine

Abstract. The analysis of cases of rolling stock derailment from the track at PJSC «Ukrzaliznytsia» over the past 5 years has been carried out and their main types are given. It is noted that the rapid development and improvement of computer technology allows the use of computer and mathematical modelling in the study of cases of rolling stock derailment from the rail track, which makes it possible to take into account a combination of factors that have the main influence on the behavior of rolling stock when moving along the track. It has been established that the first step is the construction of a mathematical model of an object using a selected type of mathematical description: algebraic, differential, integral equations, discrete mathematics, set theory, and others. And the second step is the choice of a method for solving these models. The simulation methods are given, which, depending on the programming style, are divided into: procedural-oriented, object-oriented, logical-oriented, rule-oriented, and oriented to constraints. An object-oriented programming of a freight car has been developed, which will make it possible to evaluate its main dynamic indicators, in particular, indicators of stability from derailment. This in turn will facilitate the establishment of the main causes of rolling stock derailment from the rail track.

1. Introduction

In accordance with the Regulations [1], transport events during trains traffic and maneuvers that threaten the safety of traffic are classified according to characteristics as disasters (accidents with serious consequences), accidents and incidents.

At present, one of the main and most significant categories of an accident is rolling stock derailment, which has been analyzed for the last 5 years at PJSC «Ukrzaliznytsya» (Fig. 1).

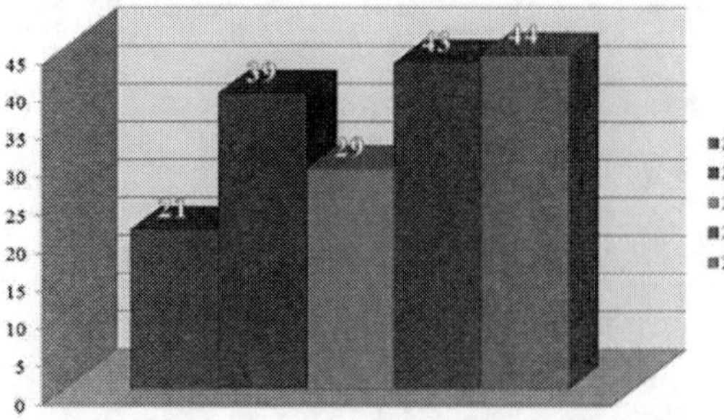

Fig. 1. The number of cases of rolling stock derailment on the railways of PJSC «Ukrzaliznytsya» for the last 5 years

In accordance with [2], all cases of rolling stock derailment can be divided into the following main types:

– derailment due to track disassembly (ousting by railhead flange of one rail because of its elastic inclination and, as a result, the second wheel falling into the middle of the track from another rail);
– derailment due to rolling of a wheel ridge on the railhead with subsequent falling of the second wheel of the same wheel-pair into the middle of the track);
– derailment due to unacceptable horizontal transverse curvature (shift) of a rail-sleeper grid by wheels of a defective bogie when the train is braked or because of its temperature release;
– derailment due to rolling of a cut ridge of a defective bogie at the turnout;
– derailment due to arrow cut;
– derailment due to rolling stock collision;
– derailment due to rail breakage.

It should be noted that the study of the process of rolling stock derailment and identification of causes that led to it is a rather complex, labor-intensive work, which requires high qualification of a scientist, considerable time consumption and in-depth research with the use of modern scientific tooling applied in the field of dynamics of mechanical systems.

The analysis of cases of rolling stock derailment shows that the causes of these derailments can be obvious (explicitly expressed), as well as non-obvious (not explicitly expressed).

In this case, «hidden» causes may be conditioned by the simultaneous coincidence of a number of factors that

* Corresponding author: kuzyshyn1993@gmail.com

form the most adverse (worst) combination of circumstances that have developed in a certain state of the rolling stock unit and the section of the track where derailment occurred.

In order to take into account the impact of a number of different parameters and to determine the most significant of them in relation to the circumstances of rolling stock derailment, it is necessary to use methods of mathematical and computer simulation.

The use of these methods will allow to assess the force impact of rolling stock and the track, to set limit values of a plurality of parameters for both rolling stock and the track for the possibility of preventing railway transport incidents and ensuring traffic safety.

2. Analysis of literary sources and problem statement

A considerable amount of scientific works is devoted to the research of rolling stock dynamics with the help of mathematical models.

Academician V.A. Lazaryan first applied methods of mathematical modeling to study the stability of undisturbed train traffic, forced oscillations of locomotives and cars, as well as stationary and transitional modes of trains movement [3].

Also, among well-known works, one should highlight the work of Romen Yu.S. [4], where the results of rolling stock dynamics with the use of computation were given. However, in this paper, the mathematical model has been simplified, which gave rise to doubts about full adequacy of the model as for the actual process of interaction of rolling stock with the track.

Particular attention deserves the work of Yong He [5], which proposed the method of object-oriented modeling that can be used to predict some aspects of dynamic behavior of rolling stock. In simulation of each object behavior, a virtual prototype of rolling stock was formed, and the result of the simulation gave the opportunity to determine the parameters of rolling stock interaction with the track. However, this method does not take into account accidental nature of track unevenness, which has a major impact on dynamic parameters of rolling stock.

In paper [6], the authors determined the impact of rail track unevenness on dynamic forces in the contact wheel-rail. Two two-dimensional and three-dimensional numerical models have been developed that describe the process of interaction of rolling stock with the railway track. The disadvantage of this research is that when the difference in unevenness between the right and left rails increases, the results obtained in simulation process differed significantly from experimental ones.

In paper [7], when modeling rolling stock dynamics, the model has been developed that describes three-dimensional dynamics of rolling stock. For simulating rolling stock dynamics, Matlab-Simulink and Adams Rail graphic environments of model simulation were used.

Also, it should be noted that in Lviv Research Institute of Forensic Expertise, cases of rolling stock derailment are analyzed with the approaches developed by D.Sc. Sokol E.M. [8-9]. The verification of these approaches in practice has confirmed the possibility of their application in cases of rolling stock derailment due to the following causes: the presence of significant difference in the diameter of wheels in car wheel-pairs, thin wheel ridges, not-standard distance between inner faces of wheels, significant longitudinal and transverse gaps between a side frame of a bogie and an axlebox.

To study the dynamic characteristics of freight cars, one can use the model that was developed at Moscow State University of Railways [10]. This model is spatial and takes into account a considerable part of rolling stock parameters, however, when simulating interaction of a wheel ridge with the inner face of the railhead, the origin of guiding force is not taken into account, which may lead to unreliable estimation of power interaction of rolling stock with the track.

Significant contribution to the development of the theory of rolling stock interaction with the track was posited by the scientists of the European Union, as it is depicted in the works [11-17].

The use of mathematical models when studying dynamics of modern rolling stock is depicted in works [18-19]. The authors' model of diesel train DPKr-2 allows to take into account a certain number of its parameters, but does not take into consideration the change in thermodynamic processes occurring in pneumatic springs of a car while running along the track, which has a random change in unevenness both in vertical and horizontal planes.

Consequently, for the present time, a lot of programs and software complexes have been developed for computer experiments on the study of rolling stock dynamics that allow perform realistic imitation of railway transport behavior while running on the railway track. Examples of foreign software packages can be: ADAMS/Rail, Medyna, Nucars, Simpack and Vampire. For studying with the help of the method of mathematical modeling of dynamic loading of cars, there are also software complexes «Dynamics of Rail Vehicles» («DYNRAIL») and «Universal Mechanism».

However, the use of existing mathematical models and software complexes does not allow to depict in full measure the behavior of power interaction of rolling stock and the track and take into account the simultaneous set of different parameters of rolling stock running gear and the track.

3. Purpose and objectives of the research

The purpose of this work is to improve the mathematical model of a freight car and to develop object-oriented programming, which will allow to assess the main dynamic parameters of a car, in particular, indices of stability against derailment and to determine the main causes of car derailment in the presence of non-obvious (not explicitly expressed) reasons.

Research objectives:

– analyze the calculation scheme of a freight car in *ZX*, *YX*, *ZY* planes;

– to improve the mathematical model of a freight car taking into account the guiding force;

– to develop object-oriented programming for the study of the process of power interaction of a freight car with the track.

4. Investigation of power interaction of a freight car with the rail track

The main method of studying complex processes and systems today is a computational experiment with computer simulation. This concept was introduced by Academician AA Samara as a method and technology of the research based on the construction and analysis of a computer model of an object being studied.

The first step of computer simulation is to construct a mathematical model of an object with the help of the chosen type of mathematical description: algebraic, differential, integral equations, discrete mathematics, set theory, and others.

The second step of computer simulation is the choice of the method of these models developing. This method should be the method of approximate or accurate computing, which, in turn, can be algorithmic and implemented on a computer. The implementation of a computational experiment can be either in the form of a finished, already known program or a program package, or a new program, compiled in the chosen programming language.

After calculations, the stage of processing, analysis and study of the results comes, as well as their use for correcting the computational experiment and obtaining conclusions about the behavior of the object of research.

System simulation technology includes the sequence of such actions (Fig. 2):

1. Setting the goal of modeling;
2. Analysis of the object of modeling and obtaining its known properties;
3. Analysis of obtained properties in terms of the goal of modeling, determination of essential features;
4. Formalization of the initial problem;
5. Construction of a mathematical model;
6. Model calculation;
7. Analysis of the obtained model for conformity with the purpose of simulation;
8. Checking the adequacy of the model;
9. Implementation of calculation.

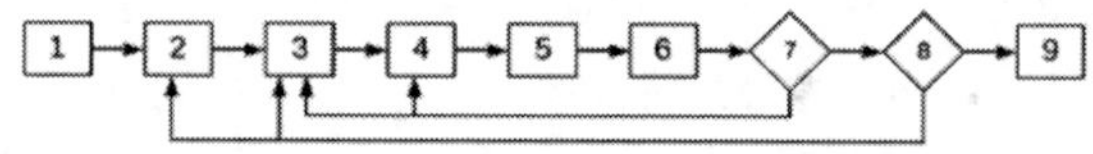

Fig. 2. Stages of constructing systems' models and their interconnections

In this case, the object of the study is a freight car, which spatial computation scheme is given at (Fig. 3-5).

The spatial computation scheme of a freight car consists of 11 solids (bodywork, four side frames, four wheel-pairs and two truck bolsters) [10].

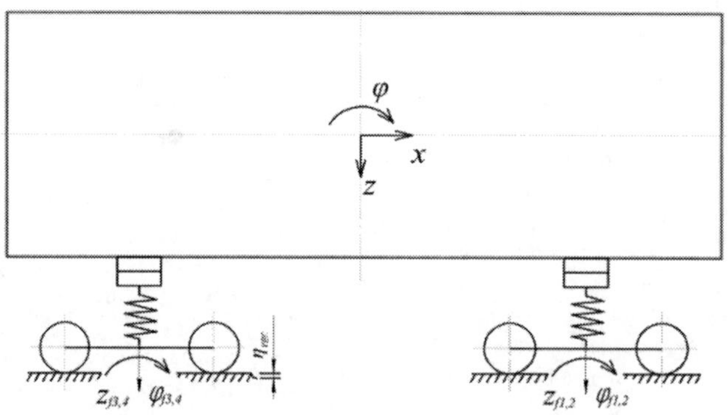

Fig. 3. Computation scheme of a freight car in plane *ZX*

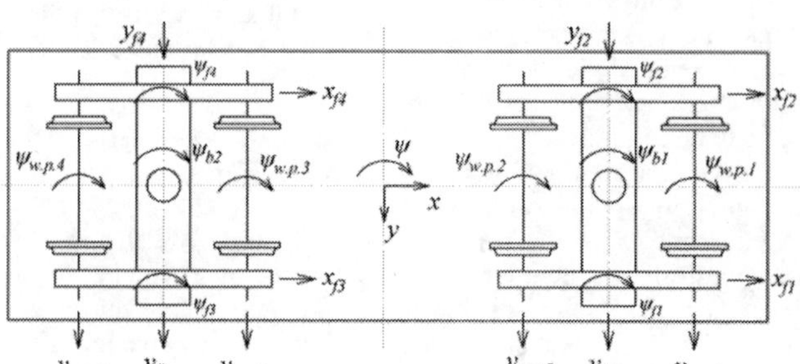

Fig. 4. Computation scheme of a freight car in plane *YX*

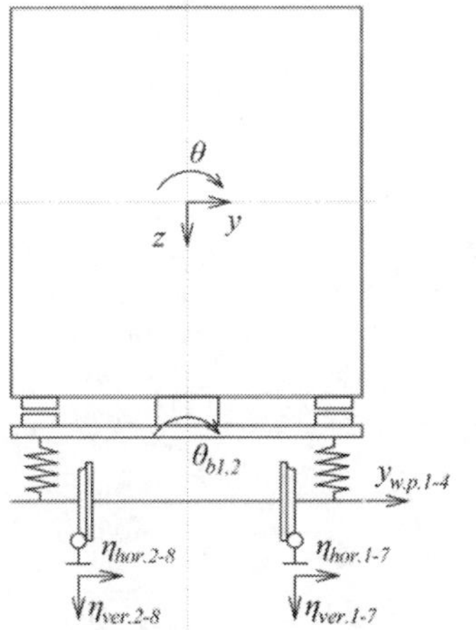

Fig. 5. Computation scheme of a freight car in plane *ZY*
$z, y, x, \varphi, \psi, \theta$ – generalized body coordinates; $z_{fi}, y_{fi}, x_{fi}, \varphi_{fi}, \psi_{fi}$ – generalized coordinates of a bogie frame ($i = 1-2$ frame number of the bogie); $y_{w.p.j}, \psi_{w.p.j}$ – generalized coordinates of wheel-pairs ($j = 1-4$ wheel-pair number), θ_{bk}, ψ_{bk} – coordinates of truck bolsters ($k = 1,2$ number of truck bolster); $\eta_{ver.l}, \eta_{hor.l}$ – vertical and horizontal unevenness under wheel-pairs of a freight car ($l = 1-8$ wheel-pair number)

On the basis of constructed spatial calculating scheme of a freight car, its mathematical model was compiled [10]. This mathematical model has been improved taking into account the guiding force and consists of 30 differential equations of the second order.

Some assumptions were made, namely:

– the body, truck bolsters, sidewalls and wheel-pairs were considered as absolutely solid bodies, since their stiffness significantly exceeds the rigidity of the elastic elements that connect them;
– fluctuations of bouncing, lateral displacement, twitching, galloping, lateral swinging, twisting are considered;
– the values of stiffness are considered to be the same for corresponding elements of spring suspension of different bogies and wheel-pairs;
– the elastic forces are considered to act on the axis of the corresponding elastic element;
– the stiffness in the contact point of a wheel and the rail is not taken into account;
– a wheel-pair and interacting with it mass of the track move jointly.

Here are some basic differential equations of the mathematical model of a freight car, which describe the process of its power interaction with the track.

Body

$$m\ddot{x} + \sum_{j=1}^{5} P'_{fxj} + \sum_{j=1}^{5} P''_{fxj} + \sum_{j=1}^{5} P'_{sxj} + \sum_{j=1}^{5} P''_{sxj} + N_{x1} + N_{x2} - N_{x3} - N_{x4} + N_{x5} + N_{x6} - N_{x7} - N_{x8} = 0$$

$$mH\ddot{\theta} + m\ddot{y} + \sum_{j=1}^{7} P'_{fyj} + \sum_{j=1}^{7} P''_{fyj} + \sum_{j=1}^{7} P'_{syj} + \sum_{j=1}^{7} P''_{syj} + \sum_{j=1}^{8} T_{yj} = 0$$

$$m\ddot{z} + \sum_{j=1}^{5} P'_{fzj} + \sum_{j=1}^{5} P''_{fzj} + \sum_{j=1}^{5} P'_{szj} + \sum_{j=1}^{5} P''_{szj} + \sum_{j=1}^{8} N_{zj} - mg = 0$$

$$mH\ddot{y} + \left(mH^2 + J_{\mathrm{sup}.x}\right)\ddot{\theta} + R_{\mathrm{b}1}e_{\mathrm{c.p.1}} + R_{\mathrm{b}2}e_{\mathrm{c.p.2}} + R_{\mathrm{bear.1}}B_1 - R_{\mathrm{bear.2}}B_2 + R_{\mathrm{bear.3}}B_3 - R_{\mathrm{bear.4}}B_4 = 0$$

$$\begin{aligned}
&J_{\mathrm{sup}.y}\ddot{\varphi} + P'_{\mathrm{fz}1}\left(L_1 + e'_{\mathrm{f}1}\right) + P'_{\mathrm{fz}2}\left(L_1 + e'_{\mathrm{f}2}\right) + P'_{\mathrm{fz}3}\left(L_1 - e'_{\mathrm{f}3}\right) + P'_{\mathrm{fz}4}\left(L_1 - e'_{\mathrm{f}4}\right) + P'_{\mathrm{fz}5}L_1 + P''_{\mathrm{fz}1}\left(L_1 + e''_{\mathrm{f}1}\right) + P''_{\mathrm{fz}2}\left(L_1 + e''_{\mathrm{f}2}\right) + \\
&+ P''_{\mathrm{fz}3}\left(L_1 - e''_{\mathrm{f}3}\right) + P''_{\mathrm{fz}4}\left(L_1 - e''_{\mathrm{f}4}\right) + P''_{\mathrm{fz}5}L_1 - P'_{\mathrm{sz}1}\left(L_2 - e'_{\mathrm{s}1}\right) - P'_{\mathrm{sz}2}\left(L_2 - e'_{\mathrm{s}2}\right) - P'_{\mathrm{sz}3}\left(L_2 + e'_{\mathrm{s}3}\right) - P'_{\mathrm{sz}4}\left(L_2 + e'_{\mathrm{s}4}\right) - \\
&- P'_{\mathrm{sz}5}L_2 - P''_{\mathrm{sz}1}\left(L_2 - e''_{\mathrm{s}1}\right) - P''_{\mathrm{sz}2}\left(L_2 - e''_{\mathrm{s}2}\right) - P''_{\mathrm{sz}3}\left(L_2 + e''_{\mathrm{s}3}\right) - P''_{\mathrm{sz}4}\left(L_2 + e''_{\mathrm{s}4}\right) - P''_{\mathrm{sz}5}L_2 + N_{z1}\left(L_1 + E_1\right) + N_{z2}\left(L_1 + E_2\right) + \\
&+ N_{z3}\left(L_1 - E_1\right) + N_{z4}\left(L_1 - E_4\right) - N_{z5}\left(L_2 - E_5\right) - N_{z6}\left(L_2 - E_6\right) - N_{z7}\left(L_2 + E_7\right) - N_{z8}\left(L_2 + E_8\right) - \sum_{j=1}^{5} P'_{fxj}H_1 - \\
&- \sum_{j=1}^{5} P''_{fxj}H_2 - \sum_{j=1}^{5} P'_{sxj}H_3 - \sum_{j=1}^{5} P''_{sxj}H_4 - N_{x1}H_{\mathrm{k}1} - N_{x2}H_{\mathrm{k}2} + N_{x3}H_{\mathrm{k}3} + N_{x4}H_{\mathrm{k}4} - N_{x5}H_{\mathrm{k}5} - N_{x6}H_{\mathrm{k}6} + N_{x7}H_{\mathrm{k}7} + N_{x8}H_{\mathrm{k}8} = 0
\end{aligned}$$

$$J_{\text{sup}.z}\ddot{\psi} + P'_{\text{fy}1}(L_1 + e'_{\text{f}1}) + P'_{\text{fy}2}(L_1 + e'_{\text{f}2}) + P'_{\text{fy}3}(L_1 - e'_{\text{f}3}) + P'_{\text{fy}4}(L_1 - e'_{\text{f}4}) + P'_{\text{fy}5}L_1 + P'_{\text{fy}6}(L_1 + e'_{\text{f}6}) + P'_{\text{fy}7}(L_1 - e'_{\text{f}7}) +$$
$$+ P''_{\text{fy}1}(L_1 + e''_{\text{f}1}) + P''_{\text{fy}2}(L_1 + e''_{\text{f}2}) + P''_{\text{fy}3}(L_1 - e''_{\text{f}3}) + P''_{\text{fy}4}(L_1 - e''_{\text{f}4}) + P''_{\text{fy}5}L_1 + P''_{\text{fy}6}(L_1 + e''_{\text{f}6}) + P''_{\text{fy}7}(L_1 - e''_{\text{f}7}) -$$
$$- P'_{\text{sy}1}(L_2 - e'_{\text{s}1}) - P'_{\text{sy}2}(L_2 - e'_{\text{s}2}) - P'_{\text{sy}3}(L_2 + e'_{\text{s}3}) - P'_{\text{sy}4}(L_2 + e'_{\text{s}4}) - P'_{\text{sy}5}L_2 - P'_{\text{sy}6}(L_2 - e'_{\text{s}6}) - P'_{\text{sy}7}(L_2 + e'_{\text{s}7}) -$$
$$- P''_{\text{sy}1}(L_2 - e''_{\text{s}1}) - P''_{\text{sy}2}(L_2 - e''_{\text{s}2}) - P''_{\text{sy}3}(L_2 + e''_{\text{s}3}) - P''_{\text{sy}4}(L_2 + e''_{\text{s}4}) - P''_{\text{sy}5}L_2 - P''_{\text{sy}6}(L_2 - e''_{\text{s}6}) - P''_{\text{sy}7}(L_2 + e''_{\text{s}7}) +$$
$$+ T_{y1}(L_1 + d_1) + T_{y2}(L_1 + d_2) + T_{y3}(L_1 - d_3) + T_{y4}(L_1 - d_4) - T_{y5}(L_2 - d_5) - T_{y6}(L_2 - d_6) - T_{y7}(L_2 + d_7) -$$
$$- T_{y8}(L_2 + d_8) - T_{\text{bear}.1}B_1 + T_{\text{bear}.2}B_2 - T_{\text{bear}.3}B_3 + T_{\text{bear}.4}B_4 - M_{\text{fric}.1} + M_{\text{fric}.2} = 0$$

Truck bolsters

$$J_{x\text{b}1}\ddot{\theta}_{\text{b}1} + N_{z1}a_1 - N_{z2}a_2 + N_{z3}a_3 - N_{z4}a_4 + \sum_{j=1}^{5} P'_{\text{f}zj}b'_{\text{f}j} - \sum_{j=1}^{5} P''_{\text{f}zj}b''_{\text{f}j} - R_{\text{bear}.1}B_1 + R_{\text{bear}.2}B_2 - \sum_{j=1}^{7} P'_{\text{f}yj}h_{\text{b}1} - \sum_{j=1}^{7} P''_{\text{f}yj}h_{\text{b}1} = 0$$

$$J_{z\text{b}1}\ddot{\psi}_{\text{b}1} - N_{x1}a_1 + N_{x2}a_2 + N_{x3}a_3 - N_{x4}a_4 - \sum_{j=1}^{5} P'_{\text{f}xj}b'_{\text{f}j} + \sum_{j=1}^{5} P''_{\text{f}xj}b''_{\text{f}j} + T_{y1}d_1 + T_{y2}d_2 - T_{y3}d_3 - T_{y4}d_4 + P'_{\text{fy}1}e'_{\text{f}1} +$$
$$+ P'_{\text{fy}2}e'_{\text{f}2} - P'_{\text{fy}3}e'_{\text{f}3} - P'_{\text{fy}4}e'_{\text{f}4} + P'_{\text{fy}6}e'_{\text{f}6} - P'_{\text{fy}7}e'_{\text{f}7} + P''_{\text{fy}1}e''_{\text{f}1} + P''_{\text{fy}2}e''_{\text{f}2} - P''_{\text{fy}3}e''_{\text{f}3} - P''_{\text{fy}4}e''_{\text{f}4} + P''_{\text{fy}6}e''_{\text{f}6} - P''_{\text{fy}7}e''_{\text{f}7} + M_{\text{fric}.1} +$$
$$+ T_{\text{bear}.1}B_1 - T_{\text{bear}.2}B_2 = 0$$

Sidewalls

$$m_{f1}\ddot{x}_{f1} - T_{x1} + T_{x3} + T_{fx1} + T_{fx3} - \sum_{j=1}^{5} P'_{\text{f}xj} + N_{\text{bx}11} - N_{\text{bx}12} - N_{\text{bx}31} + N_{\text{bx}32} = 0$$

$$m_{f1}\ddot{y}_{f1} - T_{y1} - T_{y3} + T_{fy1} + T_{fy3} - \sum_{j=1}^{7} P'_{\text{f}yj} - N_{\text{by}11} + N_{\text{by}12} - N_{\text{by}31} + N_{\text{by}32} = 0$$

$$J_{zf1}\ddot{\psi}_{f1} - T_{y1}d_1 + T_{y3}d_3 + T_{fy1}l_1 - T_{fy3}l_3 + P'_{\text{fx}1}f'_{\text{f}1} - P'_{\text{fx}2}f'_{\text{f}2} + P'_{\text{fx}3}f'_{\text{f}3} - P'_{\text{fx}4}f'_{\text{f}4} - P'_{\text{fy}1}e'_{\text{f}1} - P'_{\text{fy}2}e'_{\text{f}2} + P'_{\text{fy}3}e'_{\text{f}3} +$$
$$+ P'_{\text{fy}4}e'_{\text{f}4} - P'_{\text{fy}6}e'_{\text{f}6} + P'_{\text{fy}7}e'_{\text{f}7} - N_{\text{by}11}l_1 + N_{\text{by}12}l_1 + N_{\text{by}31}l_3 - N_{\text{by}32}l_3 = 0$$

Wheel-pairs

$$m_{w.p.1}\ddot{y}_{w.p.1} + F_{y1} + F_{y2} - T_{fy1} - T_{fy2} + R_1n_1 - R_2n_2 + N_{\text{by}11} - N_{\text{by}12} - N_{\text{by}21} + N_{\text{by}22} + N_{nap1}n_{flan.1} - N_{nap2}n_{flan.2} = 0$$

$$J_{z\,w.p.1}\ddot{\psi}_{w.p.1} + F_{x1}S_1 - F_{x2}S_2 + T_{fx1}D_1 - T_{fx2}D_2 + N_{\text{bx}11}D_1 - N_{\text{bx}12}D_1 - N_{\text{bx}21}D_2 + N_{\text{bx}22}D_2 = 0$$

where m – mass of the body and two truck bolsters; $J_{\text{sup}.x}, J_{\text{sup}.y}, J_{\text{sup}.z}$ – moments of body inertia relative to the axes x, y, z which pass through the center of its weight; P_{fx} and P_{sx} – longitudinal reactions of springs of suspension kits for the first bogie in the train (index «f») and the second (index «s») bogie, respectively; N_{x1-8} – projections on the axle x of normal forces acting between truck bolsters and friction wedges; H – height of the gravity center of the body; P_{fy} and P_{sy} – transverse reactions of springs of suspension kits for the first and second bogies, respectively; T_{y1-8} – horizontal transverse force of friction between friction plates and wedges; P_{fz} i P_{sz} – vertical reactions of springs of suspension kits for the first and second bogies; $R_{\text{b}1}$ and $R_{\text{b}2}$ – vertical reactions on axial bearings; $e_{\text{c.p}.1}$ and $e_{\text{c.p}.2}$ – transverse displacements relative to the center of application of vertical reactions on axial bearings; $R_{\text{bear}.1\text{-}4}$ – vertical reactions on sliders; $T_{\text{bear}.1\text{-}4}$ – friction forces on sliders; $M_{\text{fric}.1}$ and $M_{\text{fric}.2}$ – moments of friction forces on axial bearings when truck bolsters turning relative to the body; S_{1-8} – distance from the center of wheel-pairs' gravity to the points of a wheel and the rail contact; D_{1-8} – distance from the gravity centers of wheel-pairs to the middle of the necks of axes; R_{1-8} – vertical reactions between the wheel and the rail; $n_{1,2}$ – the taper of the wheel rim at points of contact; N_{nap} – guiding force, which arises when working face of the flange is interacting with the side working face of the railhead; $n_{flan.}$ – the taper of the flange.

To calculate the spatial mathematical model and study the process of power interaction of a freight car with a rail track, the authors of the article developed object-oriented programming using computer environment Maple.

Maple is a software package, a computer algebra system (more precisely, a system of computer mathematics), focused on complex mathematical calculations, data visualization and modeling.

To solve this problem, a block diagram of the algorithm was constructed, which is shown at (Fig. 6).

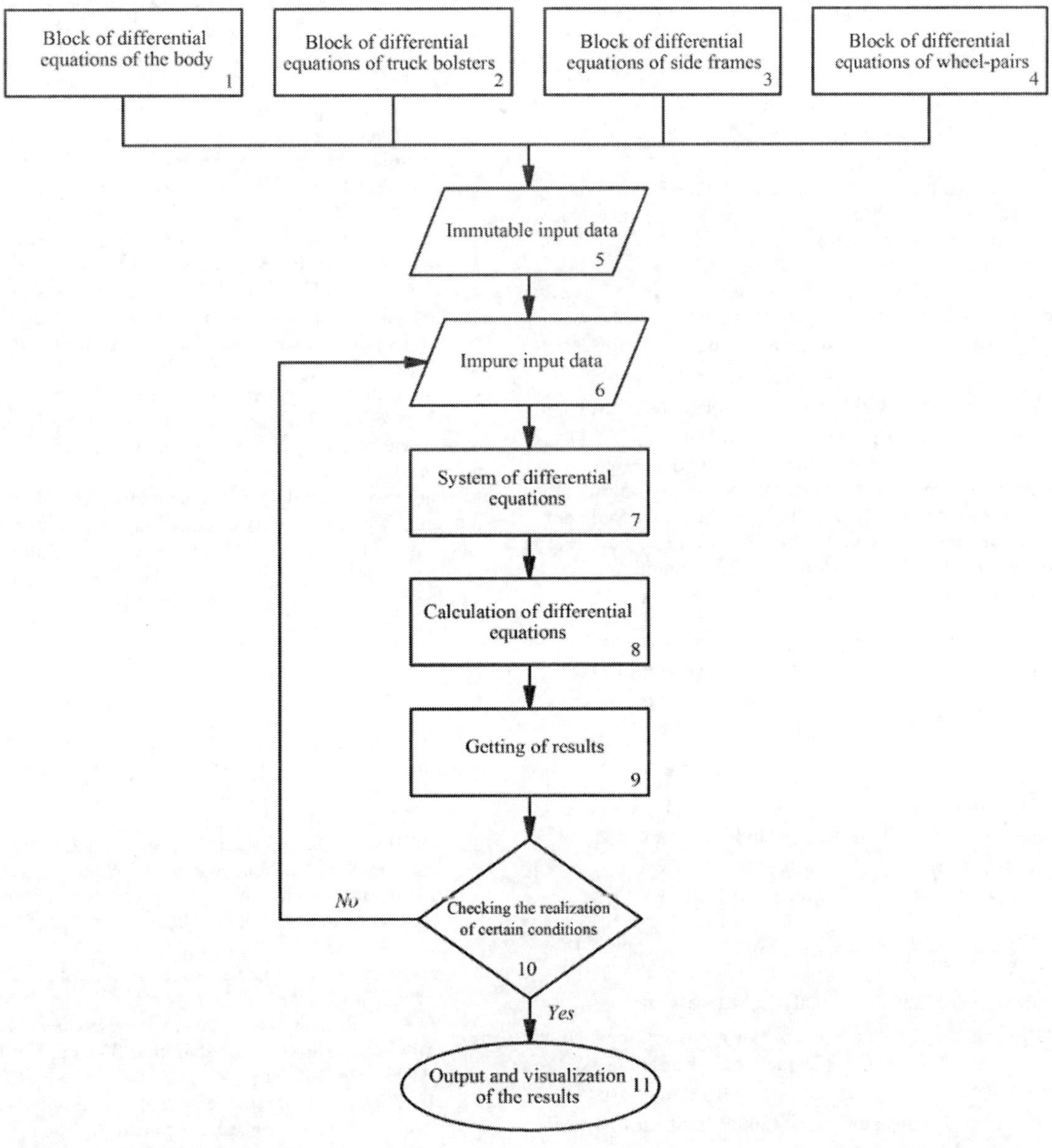

Fig. 6. Block diagram of the algorithm for studying the dynamic behavior of rolling stock in its interaction with the track

The block diagram of the algorithm depicts a sequence of blocks interconnected by arrows indicating the sequence of performance and the relationship between blocks. Inside the blocks their short content is written.

Input data are given by a separate block through an operator *table* (F). Where F is a table or an array.

Blocks of differential equations of a body, truck bolsters, side frames and wheel-pairs are described using the procedure. The procedure is called an application module that has an independent value and performs one or more operations, usually quite complex and different from the operations performed by the built-in operators and functions.

Sequence of the procedure call is as follows:

proc (parameterSequence) :: returnType; localSequence; globalSequence; optionSequence; descriptionSequence; usesSequence; statementSequence end proc;

where parameterSequence – formal parameter declarations;

returnType – (optional) assertion on the type of the returned value;

localSequence – (optional) names of local variables;

globalSequence – (optional) names of global variables used in the procedure;

optionSequence – (optional) names of procedure options;

descriptionSequence – (optional) sequence of strings describing the procedure;

usesSequence – (optional) names of modules or packages the procedure uses;

statementSequence – statements comprising the body of the procedure.

For analytical and numerical solution of differential equations in Maple software package, the dsolve command is used, which looks like:

$$\text{dsolve}\left(\text{eqns, vars, option}\right)$$

where eqns – is the system of equations with respect to non-invertible functions vars ;

option – additional conditions that allow to specify the method of solving the task.

Calculations of the mathematical model were carried out using numerical Runge-Kutta method of 4-5 orders.

Often, when modeling the dynamics of a rail train, there are cases of logic use, which is due to the closure of certain gaps or realization of certain conditions. To do this, an operator if is used which format looks like:

if < conditional expression > then < statement sequence >
elif < conditional expression > then < statement sequence >
else < statement sequence >
end if

To visualize obtained results Maple allows to use computer graphic capabilities. The main command is plot in the standard library and does not require a previous call.

The structure of the command is as follows:

$$\text{plot}\left(\{\text{func1, func2...}\}, x = \text{a...b}, y = \text{c...d, option}\right)$$

where func1, func2 – analytical expressions;

x = a...b – interval of a variable change x (segment on the abscissa axis);

y = c...d – interval on the ordinate axis.

Thus, the mathematical model built in software package Maple allows:

– to study the process of power interaction of rolling stock with the rail track;

– to evaluate the main dynamic parameters of rolling stock, in particular, indices of its stability against derailment;

– to determine safe operating conditions when running on the rail track, which has deviations from its maintenance standards;

– to evaluate the impact of a plurality of different parameters and to determine the most significant of them in relation to circumstances of rolling stock derailment.

5. Conclusions

1. Analysis of the calculation scheme of a freight car, and its running gear, in particular, has shown that it can be used for the study of dynamic parameters, but the study of derailment cases requires improvement.
2. The mathematical model of a freight car has been improved, namely, the process of interaction of a wheel and the rail, which was realized with respect of guiding force at the moment when the wheel's flange presses on the lateral working flange of the rail.
3. On the basis of the improved mathematical model of a freight car, the block diagram of the algorithm for constructing a software product was developed, with the help of which it is possible to automate the study of the process of power interaction of a freight car with the rail track.
4. Using the step-by-step sequence of processes, which are depicted in the block diagram of the algorithm, object-oriented programming was developed with the help of Maple computer environment.
5. The use methods of mathematical and computer modeling will allow to assess the force impact of rolling stock and the track, to set limit values of a plurality of parameters for both rolling stock and the track for the possibility of preventing railway transport incidents and ensuring traffic safety.

Bibliography

1. *Pologennya pro klasyfikatsiyu transportnykh podii na zaliznytshnomu transporti.* Zatverdzheno Nakazom Ministerstva infrastruktury Ukrainy 03.07.2017 № 235. (2017) [in Ukrainian]
2. TSRB-0036 Metodychni vkazivky schodo poryadku sluzhbovoho rozsliduvannya prychyn skhodzhen rukhomoho skladu z reyok na zaliznytsiakh Ukrainy. Zatverdzheno Nakazom Ukrzaliznytsi 21.06.2012 №194-TSZ. (2012) [in Ukrainian]
3. V. A. Lazaryan Dinamika vagonov, Transport, 256 p, (1964) [in Russian]
4. Yu. S. Romen *O dvizhenii zheleznodorozhnoho ekipazha v krivykh uchastkakh puti*, Vestnik VNIIGT, **6** (1964) [in Russian]
5. He Y *Object-Oriented Modeling and Simulation of Railway Vehicle Systems*, Applied Mechanics and Materials, **26-28** 900-904 c (2010)
6. J. Sadeghi, A. Khajehdezfuly, M. Esmaeili *Investigation of rail irregularity effects on wheel/rail dynamic force in slab track: Comparison of two and three dimensional models*. J Sound Vibration, **374** 228-244 c (2016)
7. J. Auciello *Determination of wheel/rail contact points in the simulation of a railway vehicle dynamics*, WIT Transactions on Engineering Sciences, **62** 261-270 c (2010)
8. E. N. Sokol *Skhody s relsov i stolknoveniya podvizhnoho sostava (Sudebnaya ekspertiza. Elementy teorii i praktiky)*, Transport Ukrainy, 368 p (2004). [in Russian]

9. E. N. Sokol *Krushenija zheleznodorozhnyh poezdov (Sudebnaja jekspertiza. Jelementy teorii i praktiki)*, Feniks, 355 s (2007) [in Russian]
10. Chan Fu Than. *Dinamika gruzovogo vagona pri nelinejnyh svjazjah kuzova s telezhkami: dis. ... d-ra teh. nauk: 05.22.07*, Moskva, MIIT. 287 s (2010) [in Russian]
11. W. Huimin, B. Kerchof *Management of wheel/rail interface to prevent rail rollover derailments* Proceedings of the Institution of Mechanical Engineers, Part F: Journal of Rail and Rapid Transit published online (2014)
12. M. Sebès, L. Chevalier, J. Ayasse, H. Chollet *A fast-simplified wheel–rail contact model consistent with perfect plastic materials*, Vehicle System Dynamics: International Journal of Vehicle Mechanics and Mobility, **50**, pp 1453-1471 (2012)
13. J. Piotrowski, W. Kik *A simplified model of wheel/rail contact mechanics for non-Hertzian problems and its application in rail vehicle dynamic simulations*, Vehicle System Dynamics, **1-2 (46)**, pp 27-48 (2008)
14. M. Sysyn, U. Gerber, D. Gruen, O. Nabochenko, V. Kovalchuk *Modeling and vehicle based measurements of ballast settlements under the common crossing*, European transport, **71** (2019)
15. M. Sysyn, O. Nabochenko, V. Kovalchuk, U. Gerber *Evaluation of railway ballast layer consolidation after maintenance works*, ACTA Polytechnics, **1 (59)**, pp 77-87 (2019)
16. J. Kalivoda, L. Neduzha *Simulation of Safety Against Derailment Tests of an Electric Locomotive*, Proc. of 25th Intern. Conf. «Engineering Mechanics 2019», pp 177-180 (2019)
17. O. Kyryl'chuk, J. Kalivoda, L. Neduzha *High Speed Stability of a Railway Vehicle Equipped with Independently Rotating Wheels*, Proc. of 24th Intern. Conf. «Engineering Mechanics 2018», pp 473-476 (2018)
18. A. Kuzyshyn, A. Batig, S. Kostritsa [et al.] *Research of safety indicators of diesel train movement with two-stage spring suspension*, BulTrans 2018: 10th Intern. Sci. Conf. on Aeronautics, Automotive and Railway Engineering and Technologies. – Sozopol: **234** (2018)
19. V. Kovalchuk, A. Kuzyshyn, S. Kostritsa [et al.] *Improving a methodology of theoretical determination of the frame and directing forces in modern diesel trains*, Eastern-European Journal of Enterprise Technologies, **6/7 (96)**, pp 19-25 (2018)

Development of an algorithm for investigation of technical state of wagons' running gears during their derailment

Andriy Batig[1*], *Petro* Hrytsyshyn[1,2,3], *Andriy* Kuzyshyn[1,3,4], *Andriy* Milyanych[3], *Oleh* Voznyak[1,3], and *Yuriy* Tereshchak[3]

[1] Lviv Research Institute of Forensic Science, 79000 Lviv, Ukraine
[2] Western center of the Ukrainian branch of the World laboratory, 79021 Lviv, Ukraine
[3] Lviv branch of Dnipro National University of Railway Transport named after Academician V. Lazaryan, 79052 Lviv, Ukraine
[4] Dnipro National University of Railway Transport named after Academician V. Lazaryan, 49000 Dnieper, Ukraine

Abstract. A freight wagon is a collection of a large number of parts and assemblies that form a single structure. During wagon operation, there are gradual changes in the characteristics and parameters of its structural elements. That is, the parameters of its constituent elements change – their numerical values increase or decrease that result in a change of the unit's technical state, and, in turn, affects the totality of its performance. Gradually accumulating, the changes in the parameters of the structural elements of a wagon reach such a level that radical, sometimes abrupt qualitative change occurs. A malfunction that has not been repaired timely may lead to a failure of a wagon structural element, which, in turn, may result in rolling stock derailment. It has been established that the loss of stability of freight wagons during their movement is most often due to their unsatisfactory dynamic properties, which can be explained by design features and technical state of running gears. In this regard, the authors of the article have developed an algorithm for investigation technical state of running gears of freight wagons and determined the effect of their parameters' deviation on rolling stock operation with possible further derailment.

1. Introduction

The ability of the railway transport to meet the needs for passengers and cargo transportation with unconditional security of traffic determines the development and achievement of its leading position in the transportation market. One of the most important problems on railway transport is the safety of trains traffic both at design stage and in the process of transport operation. At the same time prevention of transport incidents on railways is of primary importance [1].

Of the total number of transport incidents on the Ukrainian railways, a significant part of the cases is rolling stock derailment, and the task is a cause-and-effect analysis of factors that contribute to their development.

Such factors may be the characteristics and parameters of the technical state of the running gears of wagons that have derailed, the state of the railway track at the section of derailment, the formation scheme and mode of train operation, etc.

The investigation of cases of railway traffic accidents has shown for the recent years that there is an increasing number of cases of empty freight wagons derailment in trains. Such cases were most often provoked by unsatisfactory dynamic properties of wagons, which are primarily determined by the design features and technical state of their running gears and the condition of the railway track [2].

Therefore, the development and improvement of methods for assessing the impact of the rolling stock technical sate on its dynamic behavior is an important part in traffic accidents prevention on railways.

2. Analysis of literary data and problem statement

A considerable amount of scientific work is devoted to investigation of trains stability and effect their running gears technical state on traffic safety.

Among them, it is necessary to highlight the work [3], which outlines the main types of rolling stock derailment; their main causes have been considered and recommendations as for the maintenance of rolling stock and railway track technical state have been given to prevent railway traffic accidents.

The problem of derailment of empty wagons was investigated in [4]. The authors of this work propose a new scheme of a spring set of a freight bogie 18-100, which includes three-component wedges and spring-loaded sliders.

The calculations and tests on the use of such elements have shown the improvement of dynamic parameters of empty wagon movement and the reduction stress of side frames of bogies.

In the work [5] the main attention is paid to the investigation of the impact of various factors of the technical state of freight wagons running parts

* Correspond ing author: batigasha1992@gmail.com

(deviations in size both in sliders and in the wedge system) on their main dynamic indicators. This made it possible to assess objectively the impact of the technical state of bogie sliders on traffic safety indicators.

The authors of the paper [6] have considered the issues of stability against wheel flange of a freight wagon rollover on the head of the rail for different models of empty four-wheeled freight wagons with bogie models 18-100 (TSNII-H3). In this paper, the authors investigated the influence of technical characteristics of a wagon for the period of time during which a wheel flange rolling onto rails occurs in the process of wagon movement in curved sections of the track.

In the paper [7], the authors carried out a thorough investigation of train movement stability and the mechanism of its derailment; the impact of defects in the running parts of rolling stock and rail track on the occurrence of railway traffic accidents was established.

In the paper [8], the authors performed the statistical analysis of causes of the railway traffic accidents that occurred on the US railways between 2001 and 2010. It showed that the main causes of accidents and disasters on the US railways were rails breakage due to improper braking at speeds up to 10 mph. In turn, bearings malfunction, wheels breakage, defects of axes and necks were the main causes of railway traffic accidents during trains running at speeds greater than 25 mph.

In the paper [9], the author notes that one of the important tasks concerning rolling stock maintenance in operating condition is the diagnosis of axle-box unit of the running gear, which allows timely detect malfunctions of their elements and thus avoid the occurrence of railway traffic accidents. On the basis of this, the author of this work carried out theoretical and experimental researches of axle-box bearings by diagnosing their parameters and it was established that the rate of various types of defects development in the elements of their structure is not the same.

The methodical manual that describes many obvious and hidden causes of rolling stock derailment on Russian railways is the work [10]. It contains methods and technics of conducting official investigation of accidents with trains caused by the force influence of rolling stock units on the elements of the upper structure of the track. Along with the statement of recommendations on determining the causes of rolling stock derailment, some proven in practice recommendations in the field of maintenance of the technical state of rolling stock and rail track during operation are provided.

In the railway-transport expertise, approaches as for establishing the technical state of rolling stock were developed by Sokol E.M. In his monograph [11], the main attention was paid to the development of the theory of rolling stock derailment in the case of wheel rolling onto the head of the rail, in the event of a rail track breaking, as well as in the case of vehicles collision. The author of this work noted that in order to establish the causes of rolling stock derailment and the mechanism of a railway accident occurrence, the grounded research should be conducted concerning rolling stock technical state, the railway track, the devices of signaling, centralization and interlocking as well as communication, and actions of persons who are responsible for their maintenance.

In the work [12] the authors specified the objects of railway transport expertise and, based on the description of railway accident circumstances, the recommendations were given as for establishing technical state of elements of the rolling stock structure in accordance with DSTU 2860-94.

As can be seen from the analysis of the above-mentioned literary sources, the problems of ensuring stability of rolling stock movement, determining the causes and conditions of its derailment remain urgent at present. Identification of such causes is a rather difficult task with respect to a large number of factors contributing rolling stock derailment and uncomplete information as for various causes or their coincidence in a specific rail accident. In turn, the availability of an algorithm for investigating the impact of the technical state of wagons' running gears on the probability of their derailment would greatly simplify and deepen the detection of causes of railway traffic accidents.

3. Purpose and objectives of the research

The aim of the work is to investigate the influence of the technical state of wagons' running gears on the cases of their derailment.

Objectives of the investigation:

– to analyze the causes of the rolling stock derailment on the railways of Ukraine for the last five years;

– to investigate the influence of design features and technical state of wagon's running gears on railway traffic incidents;

– to develop an algorithm for assessing the impact of the technical state of running gears of freight wagons on their derailment.

4. Investigation of causes of freight wagons derailment on the railways of Ukraine for the past five years

There have been 2746 railway and transport incidents on Ukrainian railways for the last five years, where 639 cases are conditioned by the malfunctions in the construction of freight wagons [13].

All malfunctions of wagons that resulted in railway transport incidents can be distributed as follows: shock-traction equipment malfunctions – 8.29%, brake equipment malfunctions – 38.81%, malfunctions of bodies and frames – 6.1%, malfunctions of running gear – 43,97%, others – 2,82% (Fig. 1).

As can be seen from (Fig. 1), the greatest percentage of malfunctions of wagons falls on their running gear.

The malfunctions of rolling stock running gears that were the causes of railway transport incidents include:

– excessive wear of the supporting surfaces of axlebox slots of the lateral frames, sloping surfaces of over-sprung beams, thrust-bearings, friction wedges and slats;

https://doi.org/10.1051/matecconf/201929403005

– over standard gaps between the guiding rails of the axlebox bodies and side frames, the bogie sliders and wagon frames;
– weakening and breakage of the end mounting of axlebox, bearings;
– spalling, sliders, buildups on the rolling rail surface, fine-dimensional flanges of wheel-pairs, their vertical cutting and sharp-end rollover.

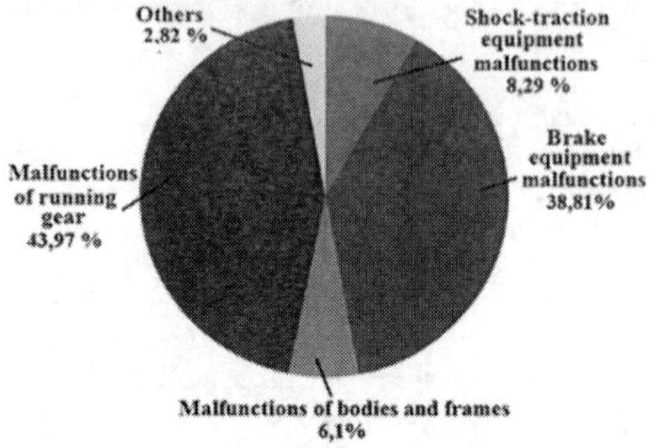

Fig. 1. Diagram of malfunctions of freight wagon distribution, which were the causes of railway traffic accidents for the last 5 years

The appearance of above-mentioned malfunctions of the running gears is due to the fact that they are one of the most important elements of the wagon structure, through which the transfer of all loads from the body to the track, as well as the adoption of a significant part of traction and braking effort takes place.

Therefore, the presence of wear, defects of design and operational nature in the elements of the running gear structure directly affects the operational reliability of freight wagons and it is one of the main reasons for stability loss during their movement and, consequently, derailment.

There have been about 176 derailments of freight wagons in trains on the railways of Ukraine for the last 5 years [13].

Depending on the reasons and in accordance with the existing classification, the reasons for rolling stock derailment were as follows: due to discontinuity of the track – 30.14%, train displacement – 3.98%, rail track shift – 1.13%, rollover of the cut wheel flange at the turnover – 2,27%, arrow cut – 11,36%, collision of rolling stock – 18,18%, rail breakage – 0,57%, others – 32,3% (Fig. 2).

Under other causes of rolling stock derailment, according to the authors of this work, which should be included in the general list, there are cases of rolling stock derailment due to its unsatisfactory interaction with the rail track, damage of rolling stock structure components, rolling stock rollover on outside objects.

Accidents of rolling stock derailment due to the damage of rolling stock structure components, rolling stock rollover on outside objects amounted to 5% of the total number of such types of railway accidents. In turn, rolling stock derailment due to its unsatisfactory interaction with the rail track occurred in 27.3% of cases.

Such cases of railway traffic incidents are possible under that are necessary and sufficient condition.

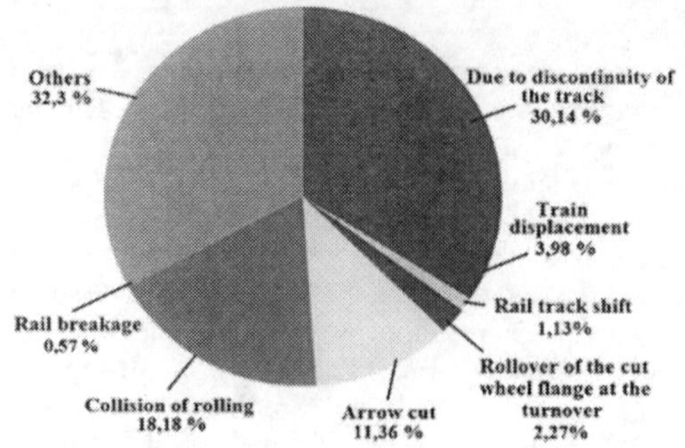

Fig. 2. Diagram of the distribution of reasons of freight wagons derailment for the last 5 years

The analytical expression of the required condition for rolling the wheel flange on the head of the rail has the form (Fig. 3) [11]:

$$\left(P_2 b_2 + Y_p r\right) - \left(P_1 b_1 + G_\kappa l\right) > 0\,, \qquad (1)$$

where P_1, P_2 – load on a wheel-pair, *kN*; b_1, b_2 – distance from the points of forces application P_1, P_2 to the middle circle of wheel rolling of the second wheel, *m*; Y_p – frame strength, *kN;* G_κ – weight of a wheel-pair, *kN*; r – radius of a wheel, *m*; $2l$ – distance between rolling circles of wheels, *m*.

The sufficient condition for rolling the wheel flange on the head of the rail is as follows:

$$l \sin\theta_{max} = f\,, \qquad (2)$$

where θ_{max} – the maximum value of the angle, at which the axis of the track should return in cross-section plane of the railway, so that the flange is to be on the rolling surface of the rail head, *rad*; f – height of the wheel flange, *m*; $2l$ – distance between middle circles of wheels rolling of a wheel-pair, *m*.

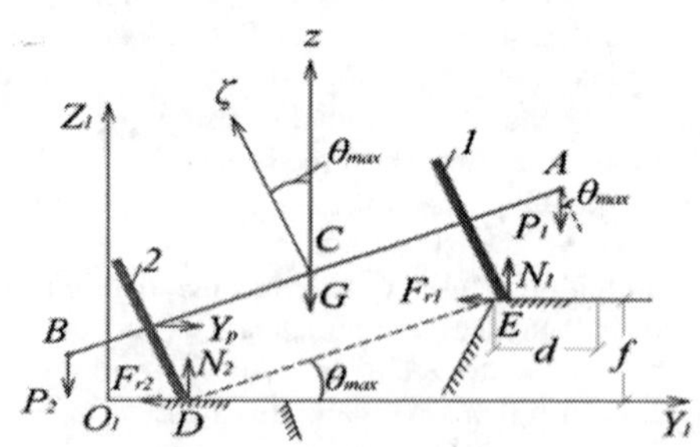

Fig. 3. Rolling wheel onto the rail head

After analyzing the cases of rolling stock derailment, which have occurred due to poor interaction of rolling stock and the rail track for the last five years, it is possible to distinguish the following characteristics:
– in the vast majority, empty wagons with a high center of mass (tank wagons, covered wagons), with a short base (hopper type) and with a rigid body (tank wagons) are liable to derailment;

– wagon derailment took place both on the straight sections of the track and in the curved sections of the track with a radius of up to 1000 m;
– wagon derailment took place both on the run and during braking;
– mainly, wagon derailment occurred at speeds above 50 km/h;
– technical state of rolling stock running gear and railway track had a significant impact on the development and occurrence of railway accidents.

In addition, there were occasions of freight wagons derailment, when parameters of their running gears and rail tracks approached the allowed limits.

Such cases of derailment are primarily due to structural disadvantages of bogie model 18-100, which contribute to safety indicators deterioration of freight trains. Among them there are such as [2]:
– large unsprung mass that interacts directly with the rails;
– intense wear of wedge oscillation dampers and non-compliance of damping characteristics in empty and loaded states of the wagon;
– insufficient constructive clearance and the level of friction forces between a truck bolster and a side bar of a bogie;
– intense wear of supporting surfaces of axial bearing units, axle-box slots;
– low critical speed in terms of movement stability.

This far-not complete list confirms that one of the main factors that leads to freight wagon derailment is structural features and the state of their running gears.

Therefore, investigation of the technical state of freight wagons running gears in the case of a railway accident is one of the most important identification and diagnostic tasks, the solution of which allows not only to determine the reasons for rolling stock derailment, but also to state limiting values of parameters of its technical condition, at which there is a threat to safety traffic.

5. Development of an algorithm for assessing the impact of the technical state of wagons' running gears on their derailment

Investigation of the technical state of wagons running gears is carried out both during the official investigation of railway traffic accidents and in the course of forensic railway-transport expertise. Such investigation is to compare the numerical values of actual parameters of wagons' running gears with the normative ones, establishing the conformity (non-conformity) of their numerical values, and, in case of non-conformity, referring to specific points of normative documents; the conclusion is issued whether the technical state of the running gear conforms to the requirements of normative documents or not.

In turn, actual (factual) parameters of the running gears of derailed wagons are determined within the framework of the official investigation on the basis of data obtained using special measuring devices (templates), and commission acts are issued. These acts are included into the materials of official investigation as well as the materials provided to the expert for investigation. Basic information of these documents is also put by the investigating judge into output data of the decision on the appointment of forensic railway-transport expertise [11].

When investigation the rail-transport incidents, there are some difficulties in investigation of the technical state of rolling stock.

One of the reasons is that its elements undergo significant force influence, which can lead to their deformation, destruction during rolling stock derailment. It results in the impossibility of establishing conformity of a large number of rolling stock parameters with the requirements of normative documents, whose portion of influence on the development of railway accident could be significant [14]. Also, when investigation the technical state, it is difficult to establish the origin of malfunctions of elements of the rolling stock structure, since the development of cracks in them, over-wearing can begin in periods, when their normal work is expected. Such malfunctions can occur both in operation and be the result of poorly done repairs.

One of the most difficult cases for investigation is rolling stock derailment, when technical state of its elements and the state of the rail track, according to the materials provided for research, have no deviations from permissible values. The causes of derailment in such cases are considered "hidden" and may be conditioned by simultaneous coincidence of a number of factors that form the most adverse (worst) combination of circumstances that developed by the state of the rolling stock and the section of the track on which the derailment took place. In addition, there are cases when there is a significant proportion of inaccurate information, which doesn't allow to identify the true cause of the rolling stock derailment [2].

In this case, it is possible to facilitate and speed up the implementation of forensic railway-transport expertise by constructing an algorithm for assessing the impact of the rolling stock technical state on the emergence and development of a railway accident.

This research algorithm should include the following steps (Fig. 4):
– collection of initial data on the state of the wagon running gears;
– verification of data on the state of the wagon running gears;
– choice of parameters for the wagon running gears and their evaluation criteria;
– comparison of numerical values of the parameters of wagon's running gears with the normative.

As a result of checking the parameters of the wagon's running gears, the conclusion is made about their technical state, which may be properly, faulty and non-operable at the moment of a railway accident. If the technical state of the wagon's running gears is not in a cause-and-effect relation with the railway-transport incident, then the corresponding conclusion about its study is issued.

Otherwise, if there is such a connection, then in order to be able to determine the influence of the technical condition of the wagon on the onset of a railway accident in the totality of circumstances accompanying the derailment, it is necessary to develop a computer model of the dynamics of a train with a detailed display of the rolling stock unit. Such model will cover a whole range of influential factors and identify the most significant of them with regard to the circumstances of derailment by means of a computer experiment.

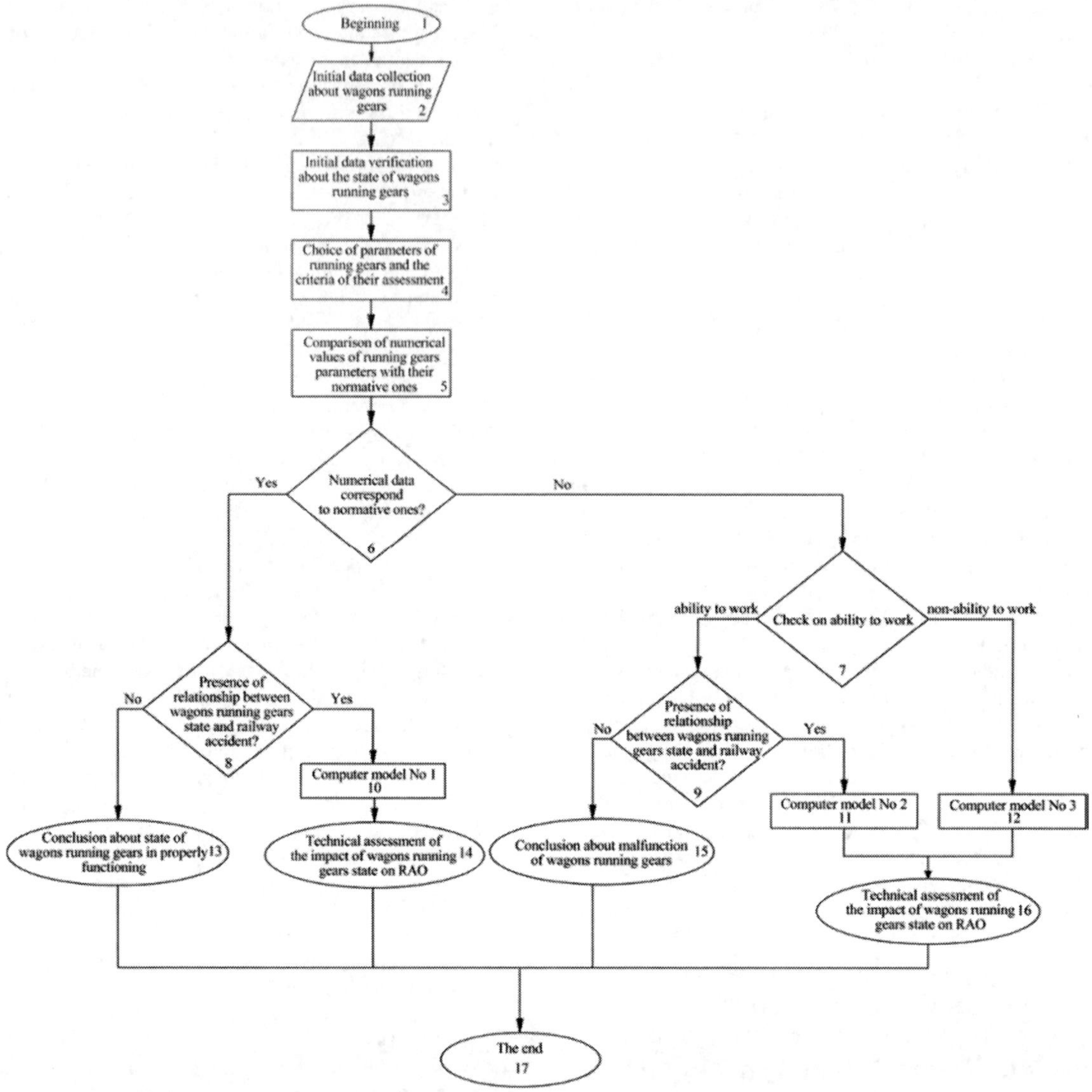

Fig. 4. Algorithm of investigation the technical state influence of freight wagons' running gears on railway accident occurrence

The process of constructing the model for determining possible causes of rail-transport incidents is in the following steps:

– analysis of information about the railway-transport accident;
– development of the spatial calculation scheme and the mathematical model of the wagon that derailed first;
– creation of the computer model of train dynamics with an acceptable degree of detail allowing to take into account a certain set of input parameters;
– verification of adequacy of the computer model;
– formation of a set of parameters that accompanied rolling stock derailment;
– conducting researches using a computer model;
– determination of the degree of influence of the parameters of wagon's running gears technical state on the railway-transport accident.

Of course, the choice and formation of parameters that accompanied rolling stock derailment depends on the type of a mathematical model. This can be a model of elements wear of rolling stock structure, their fatigue failure, a model of dynamic interaction of rolling stock and the rail track, etc.

For the mathematical model construction of rolling stock and the rail track interaction, the necessary parameters can be:

– speed of the train at the section of derailment;
– scheme of train formation regarding wagons loading and automatic coupling devices in the train;
– mode of train driving;
– characteristics of the first derailed wagon:
– type of a wagon, inertia and geometric parameters of the

body;
– parameters of wagon's running gears with the display of their technical state (drifting of axle-box slots, wedges and axial bearings as well as characteristics of friction units):
– actual dimensions of wheels and wheel-pairs;
– parameters of the railway track, etc.

It should be noted that the connection between the process of derailment and any event factor cannot be established on the basis of logical consequences of only one calculation option, since the circumstances of derailment are inherent uncertainty, which is amplified by the dynamic formulation of the problem. Information obtained on the basis of the official investigation of a transport incident allows us, at the best, to formulate the list of factors pertaining to the incident with only a certain degree of certainty. It is impossible to find out what caused derailment with complete confidence, but it is possible to determine which factors among all possible contributed to the development of an emergency. This confidence should be based on statistical data obtained by conducting and analyzing multivariate calculations based on the developed model of rolling stock dynamics at the section of derailment [2].

Such model provides an opportunity to expand significantly the limits of the virtual experiment, to accelerate the process of investigation the dynamic parameters of rolling stock, to facilitate the establishment of limit values of the parameters of its technical state and to assess the consequences change of these parameters when investigation railway traffic accidents.

Conclusions

1. The analysis of the causes of rolling stock derailment on the railways of Ukraine for the past five years has shown that more than half of the total number of derailments is due to rolling of the wheel flange of the wheel-pair onto the rail head in the presence of malfunctions of freight wagons' running gears.
2. To determine the influence of the technical state of wagons' running gears on the process of their derailment, the algorithm for research that includes a computer experiment was developed.
3. The developed algorithm for investigation the causes of rolling stock derailment allows to determine the limits of the parameters of its technical state, to evaluate the consequences of these parameters change, transferring the formation of requirements for the system of rolling stock maintenance into the plane of evidentiary dynamics.

Bibliography

1. I.V. Klimenko. *Development of theoretical bases and methods of assessment improving the movement safety of the railway rolling stock*: diss. ... dr. tech. sciences: 05.22.07, DNUZT, 284 p., (2015)
2. R.Yu. Demin. *Development of methods and means of research on maintenance of technical operation of railway rolling stock*: diss. ... dr. tech. sciences: 05.22.07, Volodymyr Dahl East Ukrainian National University, 475 p., (2018)
3. V.S. Lysyuk. *Causes and the mechanism of the vanishing wheel from the rail. The problem of wear of wheels and rails*, publ. Transport of Russia, 215 p., (2002)
4. A.N. Savchuk, A.V. Shatunov, A.G. Reidemeister, M.A Hrychaniy, V.A. Ryzhov, M.I. Lukhanin. *Improving the dynamics of empty wagons on running gears model 18-100*, Bulletin of the DNUZT, **№18**, p. 137-142, (2007)
5. S.V. Miamlin, L.O. Nedouzh, A.O., Shvets. *Technical condition of the side bearing as one of the factors influencing the dynamics of freight wagons*, Donetsk iof Railway Transport, **№35**. p. 65-72, (2013)
6. V.D. Danovich, V.V. Rybkin, A.G. Reidemeister, N.V. Khalipova, A.P. Tryakin. *On the issue of stability against wheel rolling in on the rail for empty freight wagons*, Bulletin of the DNUZT, **№3**, p. 90-96, (2004)
7. *The investigation of derailments*, Indian railways institute of civil engineering, Pune, 400 p., (2014)
8. X. Liu, M.R. Saat, C.P.L Barkan. *Rail Transportation and Engineering Center, Department of Civil and Environmental Engineering*, University of Illinois at Urbana–Champaign, 205 North Mathews Avenue ХангVolume: **2289**, issue: 1, p. 154-163. (2012)
9. V.G. Ravlyuk. *Determination of the technical condition of axial bearings of rolling stock by the method of vibrodiagnostics*, Eastern-European Journal of Enterprise Technologies, **№2 (7)**, p. 11-15, (2015)
10. L.O. Gracheva, A.D. Khamoev, L.N. Kosarev, A.V. Martynyuk. *Causes derailment of rolling stock in trains and measures to prevent accidents and crashes*, publ. Intext, 93 p., (1995)
11. E.N. Sokol. *Derailment and collisions of rolling stock (Forensic Expertise. Elements of the theory and practice)*, publ. Transport of Ukraine,. 386 p., (2004)
12. D.V. Bobyr, I.V. Bolzhelyarsky. *Establishment of the technical condition of rolling stock in the forensic railway transport expertise*, Theory and practice of forensic expertise and forensic science, **№10**, 484-491 p., (2010)
13. *Analysis of the railway traffic safety in the structure of Ukrzaliznytsia for 2014-2018 years*, Kyiv, Ukrzaliznytsia (2014-2018).
14. M. Sysyn, U. Gerber, O. Nabochenko, D. Gruen, F. Kluge *Prediction of Rail Contact Fatigue on Crossings Using Image Processing and Machine Learning/Methods*, Urban rail transit, **№5(2)**, pp. 123-132, (2019).

Improving the safety of railway subgrade when it is strengthened using soil-cement elements

Vitalii Krysan[1], *Volodymyr* Krysan[2], *Volodymyr* Petrenko[3], *Oleksii* Tiutkin[3], and *Volodymyr* Andrieiev[3,*]

[1] Head of REMBUD LLC Dnipro, Ukraine
[2] Head of PARITET LLC, Dnipro, Ukraine
[3] Dnipro National University of Railway Transport named after Academician V. Lazaryan, Lazaryan str., 2, Dnipro, Ukraine

Abstract. The article identifies the main parameters of the drilling-mixing technology, which is the most effective in fixing weak soil bases during the construction and restoration of transport, industrial and civil structures. The difference of the technology developed by the authors is that the strengthening process is carried out at low pressure (0.15 ... 0.25 MPa). The relevance of the research is that the proposed technology requires less cost with high rates of restoring the strength of soil foundations. To prove the high quality of the technology, laboratory studies were carried out to determine the optimal characteristics of the soil-cement element, as well as the proportions and composition of the fixing solution. The dependences of the strength of soil-cement elements in the air-dry condition with cement content from 7% to 23% with water-cement ratio in solution 1 / 0.3 and with cement content from 13% to 37.5% with water-cement ratio in solution 1/0,6. During the experimental-industrial studies of the author's technology, the embankment was constructed with the soil-cement-reinforced elements for the access road approaches at one of the facilities in Kirovograd region, which ensured safety in the operation of a complex transport structure.

Introduction

With the transition of railways to increased train speeds, forecasting of the stress-strain state of the subgrade becomes relevant.

The instability of the subgrade of railways may depend not only on operational factors, poor properties of soils, its components. Therefore, the transition to high-speed trains (200 km/h and more) requires a thorough study of the condition of the subgrade and its corresponding reconstruction.

The main task is to find engineering solutions aimed at improving the operational reliability of earthworks and the normal provision of a long service life. Scientific and engineering proposals must meet modern requirements for the operation of railways, since the transition of railways to increased speed without modernization and partial or complete restructuring of engineering structures is impossible. There are no references to solving this problem in the world.

The following basic requirements are imposed on the subgrade of railways:

- it must be strong, stable, reliable and durable;
- all surfaces of the subgrade, devices in the right of way must be planned and protected so that atmospheric water does not stagnate anywhere, and its maximum outflow to the side or special drainage structures are ensured, and the flowing water does not erode the slopes and base;
- the construction of the subgrade should provide minimal costs for its construction and operation with the maximum possible mechanization of work.

In addition, the subgrade must be interconnected with the highway, maintainable, and ensure long-term operation when skipping modern and promising types of rolling stock at maximum train speeds and rated load.

The main element of the subgrade, a roadbed, takes the greatest load from the rolling stock. The roadbed and its structures should ensure the safety of trains with given loads and speeds at the set load capacity. The roadbed, its elements and structures must comply with the standards of strength and stability during the entire period of operation.

The reliability of the roadbed is determined by the elastic characteristics, which in turn determine the stress-strain state of the elements of the track superstructure and its durability. As it is known, only a roadbed, which has residual deformations, is attributed to a deformed one, and elastic deformations, which have a sufficiently small value, are not taken into account. But residual deformations should not be allowed and the subgrade should be such that under no force or climatic actions such deformations would arise. This determines the reliability and stability of the roadbed.

The characteristic determining the quality of the subgrade is its elasticity. The distribution of stresses in the subgrade is rather uneven. Herewith, the unevenness of the stress state of the rails due to the non-uniform

* Corresponding author: avs_diit@ukr.net

elasticity of the track can reach up to 60-70%. The structure of the subgrade soil determines the nature of the stress distribution. The subgrade is considered sound if it does not have intense residual deformations. The stresses from the train load are unevenly distributed over the cross section, which indicates the presence of shear static and dynamic stresses. Due to the increased intensity of the subgrade loading, the condition of the track deteriorates. In some cases, there is a concentration of increased deteriorations of the track and the failure of its elements. The reason for this condition may be decreased subgrade elasticity. When overhauling the subgrade, it is necessary to comply not only with its strength and stability, but also with the elastic characteristics. Their value is set depending on the influence on the track elastic state both by absolute value and by the ranges of oscillations.

Deformations of the subgrade and its structures arise due to:

a) Insufficient bearing capacity of the soil from which it is built;

b) Non-conformity of the track superstructure strength and the load from the rolling stock;

c) Insufficient protection of the subgrade soil from the adverse effects of climatic and geotechnical factors.

As is known, oscillations arise in the railway roadbed under the influence of train load, the amplitude of which depends on many factors, including train speed and load on the rolling stock axle.

The experience gained during the construction and operation of high-speed rail lines shows that the forces transmitted by rolling stock to the subgrade at speeds exceeding 160 km/h are not very different from those transmitted at speeds of 140-150 km/h. But as the speed rises, the subgrade vibrations increase dramatically.

Improper use of weak soils for the construction and further operation of railway embankments from the same soil leads to a loss of bearing capacity of railway embankments. In turn, this may lead to non-compliance with the rail traffic safety requirements[3]. Therefore, an increase in the carrying capacity of the subgrade soil directly leads to an increase in safety in railway transport.

One of the most promising technologies for securing weak soil bases during the construction and restoration of transport, industrial and civil structures is the drilling-mixing technology. Its main product is soil-cement elements that strengthen and reinforce the soil foundation.

The soil-cement element technology consists in mechanical mixing and additional hydromonitor development of a certain soil volume. At the same time, there is an additional high-quality mixing of the soil-cement mix, which forms the body of the soil-cement element. Using the jet-mixing bit invented by us (declarative patent for invention No. 62868A), it is possible to perform hole deepening for the element and lift the tool with rotation without changing the direction of rotation of the working tool. As a result, a soil-cement body is formed in the soil thickness, while the moisture content of the soil massif does not affect its formation. Soil-cement elements can be created in both dry and water-saturated soils.

One of the most important technological factors in the drilling-mixing technology of rock consolidation is the selection and use of fixing solutions. This is due both to the specifics of the proposed method and the requirements for fixed rocks when solving various construction problems.

The fixing solutions should have:

1) Ability to form new soil-cement structures with desired physical, mechanical and filtration properties;
2) Ability not to exfoliate during jetting;
3) Ability to harden under water;
4) Setting time, providing the necessary conditions;
5) Ensuring the implementation of the technological process of preparation and jetting;
6) Satisfactory miscibility with the ground being fixed;
7) Necessary filtrate loss after the process completion;
8) Sufficient uniaxial compression strength.

Soil-cement elements, if necessary, can be reinforced in the upper part, and for their entire length. As reinforcement, reinforcing bars, space framework, I-beams, channel bars or other structural materials can be used depending on the purpose. Moreover, the reinforcement can be performed in the first hours before the soil-cement material sets, by ordinary immersion of reinforcing elements. The technology of soil consolidation by this method is aimed at increasing the carrying capacity of the subgrade body, which in turn leads to rail traffic safety.

Methodology

When checking the operated embankments for reliability and stability using existing standards, it was established that most of them meet the requirements and are aimed at improving traffic safety. At the same time, restoring activities only support the subgrade in a more or less stable state. It has not been possible yet to reduce the number of deformed sections and their share remains within 10–12% of the total length of lines operated, therefore large sums are spent annually on the subgrade repair. Obviously, the main cause of subgrade deformations is the growth of the dynamic load from trains, moving objects. Accounting for the dynamic impact of train load in the work of the subgrade is one of the most important areas of research in transport geomechanics and geotechnics [1, 2]. All researches can be divided into 5 directions.

The first direction is connected with the determination of the dynamic characteristics of soils and, first of all, the strength properties of soils.

Another direction is experimental simulation and first of all centrifugal testing of models.

The third direction may be field studies.

The fourth direction is vibration diagnostics, which can be complemented by test boring and drilling.

The fifth direction is the subgrade calculations taking into account the dynamic impact. Dynamic calculations are divided into 3 parts:

1. Subgrade stability calculations.
2. Elastic settlement calculations.
3. Roadbed residual deformation calculations.

On the basis of the analysis performed, we proposed a technique for studying the subgrade state by the mathematical modeling method using field and laboratory studies of soils that are available to the personnel of geotechnical stations of railways. Herewith, the subgrade state study can be carried out promptly, taking into account specific conditions, and with minimal costs. In addition, the presence of the calculation complex in the railway management system will allow changing the approach of geotechnical personnel to solving the problems of improving traffic safety and reliability of operation of the subgrade and track facilities.

Currently, a number of ways have been developed to strengthen weak soils with the creation of soil-cement elements, including injection of compressed air and water under high pressure (50...70 MPa) into the foundation. The essence of the technology developed by the authors is that the strengthening process is carried out at the sufficiently low pressure (0.15...0.25 MPa). In this case, the soil destructing is performed by a special device - a jet-mixing bit (declaration patent for invention No. 62868) [6]. Thus, the process of dispersing and mixing the soil is the simultaneous impact of the bit and the jet of water-cement mixture, which is fed through the hole in the monitor.

The setup of soil-cement elements can be made instead of footing excavation and arrangement of the groundbed on the subsiding soils, bedding for various purpose foundations, as well as during the embankment restoration. Portland cement, slag portland cement (DSTU B.V.2.7.46.96), sulphate-resistant cement (DSTU B.13.07-85-99) of grade 400 or higher are used for the construction of soil-cement elements. To ensure the mixture mobility and adjust the setting time, plasticizers, superplasticizers and retarders are introduced in an amount necessary to ensure the design parameters (1...4% of the cement volume). Water-cement ratio is selected depending on the engineering-geological conditions, porosity and soil moisture, its particle size distribution, as well as based on the results of work on the experimental plot[4].

Findings

In order to determine the optimal characteristics of the soil-cement element, we carried out the laboratory studies of the proportions and composition of the fixing solution, as well as the ratio of its quantity in the volume of the test soil (collapsing and loess loam) [5]. The soil characteristics determined by engineering and geological surveys are shown in Table 1.

Table 1. Soil characteristics

Natural condition of soil			
γ=14.48 kN/m³	φ=24°	C=23.0 kPa	E=13.0 MPa
Water saturated soil			
γ=17.94 kN/m³	φ=13°	C=6.0 kPa	E=3.0 MPa

where: γ - Soil density; φ - angle of internal friction; C - Specific adhesion; E - strain module.

To obtain optimal characteristics, we conducted two series of experiments to determine the strength characteristics of soil-cement at different ratios of water, cement and plasticizer [7]. For this purpose, there was prepared a fixing solution of water and cement with the addition of a plasticizer, preventing the mixture from splitting, in a special container. The volumetric method allowed measuring the required amount of fixing solution and soil, that was mixed in a steel cylindrical vessel until a homogeneous composition of the mixture was obtained. In the first series of experiments, the fixing solution was prepared at the rate of 300 kg of cement per one cubic meter of water. The plasticizer was added at the rate of 2.5 kg per 1000 kg of cement. The ratio of soil and fixing solution was 1/1, 1/2, 1/3, 1/4.

The results of determining the strength of soil-cement elements in the air-dry condition with a cement content of 7% to 23% with a water-cement ratio in the solution 1/0.3 and with a cement content of 13% to 37.5% with a water-cement ratio in the solution 1/0,6 are presented in Figure 1.

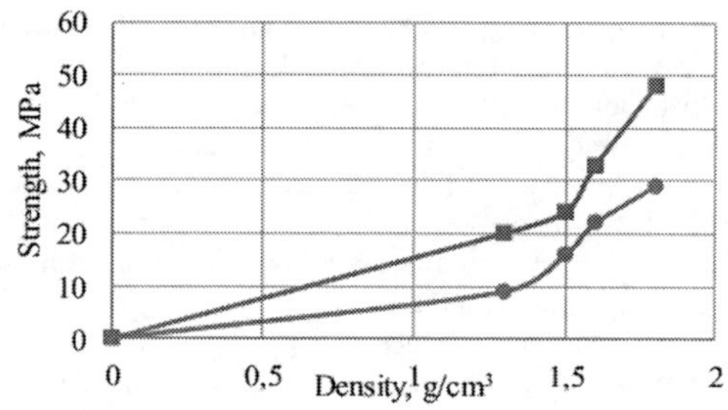

Fig. 1. Soil-cement element strength dependency graph:
Row 1 - with the cement content from 7% to 23%;
Row 2 - with the cement content from 13% to 37.5%

The experiments showed the increased strength of the samples with the rise in cement percentage and, accordingly, their density. Herewith, in the second series of experiments, the same soil was taken (collapsing and loess loam), and the fixing solution was prepared at the rate of 600 kg of cement per one cubic meter of water. As in the first experiment, the plasticizer was added at the rate of 2.5 kg per 1000 kg of cement. The ratio of the soil and the fixing solution was taken, as in the first experiment, 1/1, 1/2, 1/3, 1/4 (see Fig. 1, curve 2).

Changing the water-cement ratio and obtaining the characteristics of the soil-cement element while reducing the amount of water in the solution was carried out in order to obtain solutions that, when used, would not soak the soil outside the body of the soil-cement element, which would prevent the development of subsidence phenomena in loess soils. The experiment confirms the increase in the strength of the samples with increased percentage of cement. According to the graphs of the dependences obtained we can predict the strength of the soil-cement element when the percentage of cement changes for its other ratios.

A large class of structurally unstable soils is composed of loess subsidence soils, in which a structural disruption with the occurrence of significant subsidence occurs when they are soaked under load.

Drawdowns are called local fast-flowing vertical deformations of soils caused by a sharp radical violation of the structure and accompanied by a partial or complete loss of resistance to disturbed soil masses, and with excessive moisture - squeezing soils to the sides.

The practice of building on loess soils has shown that subsidence can reach significant values. The property of loess soils to lose the stability of their structure upon wetting determines the peculiar construction qualities of these soils that require special consideration.

Loess soils occur in a significant part of Ukraine. For the practice of construction, it is very important to be able to distinguish subsidence loess soils from ordinary soils, to know the peculiarities of the mechanical properties of subsidence soils and to foresee the influence of these properties on the structures under construction.

The mixing process cannot be considered only as mechanical mixing of components in a certain volume of the material. Mixing should ensure optimal conditions for the physicochemical processes of the formation of the technological properties of soil-cement elements.

Due to the fact that the fixing solution is supplied to the well bottom under pressure of 0.1...0.3 MPa, and flows out of a small diameter nozzle, turbulent motion of soil particles occurs in the flow, which move randomly not only in the direction of flow, but also perpendicular to it. This process is accompanied by interpenetration of the particles from one layer into another and the number of their collisions increases, which creates a more uniform structure of the soil-cement element. In all cases, when performing work, the fixing mixture jet flows out of the nozzle into the medium consisting of particles of the soil destroyed up to certain parameters and the fixing substance existing there.

To determine the zone of penetration of the fixing solution into the original soil during the formation of soil-cement elements, in the experimental plot of works, soil was sampled in the immediate vicinity of the soil-cement element. Then we determined the soil characteristics.

In addition, these laboratory studies were compared with the studies conducted for the construction project development. The results of these studies are presented in Table 2.

Table 2 - Results of field studies of soil properties

Soil parameters	Sampling from the mass wall at a distance of 50 mm from the soil-cement element				Control samples		Soil survey characteristics
	1	2	3	4	1	2	1
Natural moisture	0.19	0.17	0.18	0.19	0.09	0.18	0.14
Soil density, g/cm^3	1.53	1.56	1.63	1.62	1.46	1.61	1.72
Porosity	0.52	0.50	0.48	0.49	0.49	0.49	0.44
Relative subsidence with control soaking	0.046	0.046	0.0355	0.0463	0.0411	0.0308	0.0157
Elasticity modulus E, MPa	1.98	2.38	2.63	3.00	3.79	3.09	2.049
Angle of internal friction φ, degrees	27	–	25	–	29	27	24
Specific cohesion C, MPa	0.010	0.016	0.016	–	0.014	0.0093	0.0155

During the experimental-industrial studies, the embankment was constructed with the soil-cement-reinforced elements for subgrade approaches. The width of the embankment on top was 13 meters, and at the bottom it was variable, from 25 to 40 meters, height - 9 meters, length - more than 1000 meters. At present, the embankment settles, and its slopes, washed out by storm and melt waters, are sagging.

An engineering-geological element (EGE) should be considered an engineering-geological body, represented by one rock, statistically uniform in some indicator of properties, selected in each case based on the requirements of the calculation of a particular engineering-geological process that is performed during the design of the structure.

The volume and configuration of the IGE is established, taking into account geological data and data on the type, design and technical characteristics of the structure. When calculating the settlement of a structure (the compaction value of the geological environment at its base), the NE active zone should be divided into engineering-geological elements, i.e., engineering-geological bodies, statistically uniform in modulus of general deformation - compressibility factor. When calculating the filtration, the filtration zone should be divided into IGE, statistically uniform in value of the filtration coefficient.

To obtain the soil characteristics of the embankment, we carried out the geological engineering surveys. It was established that the embankment body consists of eight layers of soil, most of which are not compacted and according to their characteristics by DBN B.2.3-4-2000 cannot be the body of the embankment. Relatively homogeneous layers have a thickness of from 0.5 m to 3.5 m, not maintained in plan. In addition, at the base of the embankment there is topsoil with a thickness of up to 90 cm, represented by black loam, solid, with an admixture of organic matter. Furthermore, at the base of the embankment there is a topsoil and a layer of loess subsiding soils, which, when soaked, will give a subsidence of up to 40 mm.

From the survey materials it follows that the heaping of the embankment is performed without proper compaction with soils of different composition with a large amount of organic matter. It should also be noted that a drainage layer should be arranged at the embankment bed, which was absent.

It is known during backfill operation the soil moisture in it is higher than in the adjacent massif. For this reason, at the contact of the embankment with its bed, the soil moisture will rise and the embankment bed will subside. In addition, the physico-mechanical characteristics of loess deteriorate, which would entail the settlement of the embankment. Given that the height of the embankment reaches 9 meters and it varies along the length of the profile, the settlements will be extremely uneven. The soil skeleton creep is the decisive factor in the progress of the deformation of the embankment over time. Transport flow, which creates dynamic loads, is also a factor that increases the intensity and amplitude of the embankment settlements during vibration creep.

Based on the foregoing, guided by technical, economic and environmental considerations, it was proposed to reinforce the existing embankment and its foundation with a system of soil-cement elements ensuring the transfer of permanent and temporary loads to the non-subsiding layer (Fig. 2).

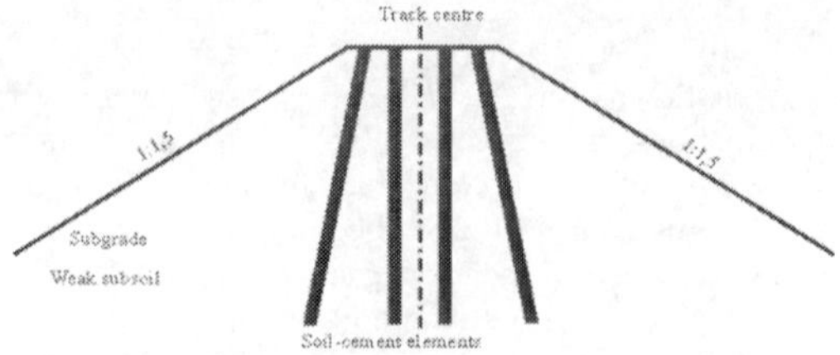

Fig. 2. Scheme of reinforcement of the existing railway embankment and its foundation with soil-cement element system

As a result of industrial experimental studies performed on the considered object, there were obtained the following technological parameters of soil-cement elements, created with the help of drilling-mixing technology of rock consolidation (Table 3).

Table 3. Technological parameters of element formation using the drilling-mixing technology

Engineering geological element (EGE)	Monitor immersion speed v, m/min	Monitor rotation speed *w*, rpm	Axle pressure *P*, MPa	Soil-cement element diameter *d*, m	Strength of soil-cement element for uniaxial compression σ^{av}_{comp}, MPa
EGE-2 sand	1…2	30	2.5…5.0	0.6…0.9	0.42
EGE-2a sandy loam	1…2	30	2.5…5.0	0.9…1.04	0.47
EGE-3 sandy loam	1…2	30	2.5…5.0	0.7…0.76	0.52

Conclusions

As a result of laboratory and experimental-industrial experiments, we established the basic laws of distribution of the physical mechanical properties of soil-cement elements. The values of the following characteristics were obtained for similar indicators of moisture and subsidence: increase in the density of soil-cement elements and a strain modulus, as well as decrease in their porosity with increased cement content in the experimental sample of the element.

The implementation of the developed method of reinforcing the high ground embankment made it possible to obtain the strengthening soil-cement elements in real volume with diameters increased to 0.6...1.0 m with their high strength.

In the course of experimental-industrial work, we established the possibility of arranging soil-cement elements in soils during railway construction, the foundation of which are subsiding loess soils. This allows to effectively perform work near existing facilities, as well as to carry out reconstruction of railway embankments with increased load on the roadbed, which will improve the railway transport safety.

References

1. Mitchell, J. K., Willem, Villet C. B., Dimillo, A. F. *Soil Reinforcement for stabilization of Earth Slopes and Embankments* Public Roads,. Vol. **48**. – No. 3. pp. 88-95 (1984)
2. Elliott, Robert P. Permanent *Deformation of Subgrade Soils* Fayetteville: Mack-Blackwell Transportation Center, 216 p. (1998)
3. Ginzburg, L. K. *Protivoopolznevye sooruzhenija: monografija* [Anti-landslide structures: monograph] - Dnepropetrovsk: Lira LTD., 188 p. (2007)
4. Zotsenko, M. L., Vinnikov, Yu. L., Lartseva, I. I., Shokarev B. C., Krysan V. I. *Kharakterystyky shtuchnykh tverdykh gruntiv, yaki polipsheni burozmishuval'nym metodom* [Characteristics of artificial solid soils, which are improved by the mixing method] Svit geotehniki. – No. 3. - P. 14-18 (2011)
5. Shapoval V.G., Sedin V.L., Shapoval A.V., Morklyanik B.V., Andrieiev V.S. *Mekhanika gruntov* [Soil mechanics] Textbook. Dnepropetrovsk: Porogi, - 168 p. (2010)
6. Romanenko V. I., Krysan V. I. *Mix-drilling bit* Declarative Patent for Innovation No. 62868 of 15.03.2007.
7. Petrenko V.D., Pshinko P.O., Tiutkin O.L., Marochka V.V., Kulazhenko E. Yu. *Method of subgrade strengthening* Patent for utility model No. 114248 of 10.03.2017.
8. V. D. Petrenko, O. L. Tiutkin, I. O. Sviatko *Estimation of subgrade strengthening influence using soilcement elements* Science and Transport Progress. No. **4 (64)**. – P. 161-168 (2016)
9. D. Ignatenko, O. L. Tiutkin, V. D. Petrenko, A. M. Alkhdour IJCIET, **10**(01), (2019)

Investigation of train shunt problem in track circuits of signalling and interlocking devices

Adam Razgonov, *Anton* Zhuravlyov, *Kateryna* Yashchuk*, and *Vadym* Shcheka

Dnipro National University of Railway Transport named after Academician V. Lazaryan, Department of automation and telecommunications, 49010 Dnipro, Lazaryana Str., 2, Ukraine

Abstract. The problems of electric track circuits calculating were investigated in the works of various authors, it is noted that there is no complete clarity on the definition of shunt resistance. The value of the calculated train shunt resistance in many respects affects the area of signaling systems rail channel stable operation and train traffic safety. It is relevant to study the possibilities of changing the standard value of and the conditions for the execution of track circuits shunt mode, especially for sections of railway with low resistance insulation of the rail line. The surface area of the "wheel-rail" skating, taking into account the different conditions of interaction dynamics and wear of rails and wheel elements were obtained in article. It is shown that the reduction of the normative value of the resistance of the train shunt is appropriate for signaling systems, especially for sections of railway with low resistance insulation of the rail line. An approximate estimate of the geometric parameters of the membrane that forms on the "wheel-rail" surface is obtained; proposed a more complete electrical replacing scheme for the train shunt circuit, which allows analyzing the influence of the circuit elements parameters values and the frequency of the signal current. A method for calculating the correlation dependence of the train shunt resistance from wheel-rail surface membrane parameters is proposed.

1 Introduction

1.1 Urgency of the problem and the results of studies of the film on the rail head surface

Track circuits are road transmitters that determine the location of a mobile unit, which is very important information for train situational awareness in real time. This makes it possible to quickly respond and take appropriate measures both in case of occurred unfavorable train situation and in the very likelihood of its occurrence. When designing new track circuits, as well as analyzing the operation of existing ones, the calculations are made. A prerequisite for the correctness of these calculations takes into account the numerous factors, including shunt resistance (R_{sh}). But in this direction some problematic issues concerning the definition of R_{sh} arise [4–7, 10]. This is due to the fact that this parameter depends on a large number of factors acting on it. A characteristic feature is that these factors are independent of each other and are different in their nature. It is difficult, since in order to obtain a complete picture of the physical processes taking place during train movement and to provide a shunt mode, by means of which the location of a mobile unit is determined and which is also subject to mandatory calculation, all factors should be taken into account. At the same time, the stable operation of the rail channel of signaling and interlocking, as well as traffic safety of trains largely depends on the values of the calculated resistance R_{sh}. Hence, it seems relevant to study the possibilities of changing the standard value of R_{sh} and the conditions for fulfilling the shunt mode of track circuits, especially for the track sections with a reduced resistance of rail line insulation. We will review the publications and some technical solutions on the problem of the train shunt resistance. Researchers from different countries (France, Japan, Germany, etc.) constantly return to this problem. It is noted [6, 7, 10] that there is no complete clarity on this issue, since the formation of the resistance of rail head contamination occurs under the influence of dynamic, electrochemical processes, temperature, humidity, speed of movement, etc. The shunt resistance also depends on the dynamics of changes in the "wheel-rail" contact area in the process of continuous movement of points – A-spots of film break-down at the contact point [11]. The experiments on the measurement of $R_{sh}(x)$ during the movement of electric train [5] show that in the absence of poor conductive films, the resistance amounts to a few μOhm, despite the fact that R_{sh} of heavy trains with a shiny rail surface is consistently measured in mOhm.

Also, the work [5] established that the resistance of the "wheel-rail" make-before-break contact nonlinearly depends on the applied voltage (semi-conductor property) and decreases abruptly to a certain critical value determined by the thickness and composition of

* Corresponding author: k.i.yaschuk@gmail.com

the film; the influence of the current frequency from zero to several kHz on R_{sh} is insignificant; the influence of the so-called coherent effect is manifested in the time lag in contact establishment, as a result of which the value of R_{sh} increases with increasing speed. At high speed, the duration of contact is so small that the film between the wheel and the rail does not have time to completely break.

In article [4] the resistance of the “wheel-rail” contact is investigated. The resistance in the frequency range 200-2000 Hz behaves as an inductive one – its impedance is an increasing function of frequency; R_{sh} depends on the modes of train movement and increases in the following sequence: stop – braking – thrust – coasting retardation. As well, in this paper, the simplified versions of the equivalent circuits for the “wheel-rail” contact are proposed. Below is a more complete version of the equivalent circuit of the train shunt`s circuit contact.

Train shunt observations show that non-conductive film on the rails is formed during the trains` movement and in winter conditions with frosts below -30..-40°C.

In article [10], the characteristics of a random stationary process (hereinafter SP) of changing R_{sh} of the underground cars are determined. It is established that when moving, R_{sh} may exceed the norm of 0.06 Ohm. Moreover, R_{sh} of a light mobile unit sometimes reaches several Ohms, which does not ensure proper operation of the track circuits. The R_{sh} density function well approximated by the α-distribution is constructed:

$$f(R_{sh}) = \frac{C_p}{R_{sh}^2\sqrt{2\pi}} \exp\left[-\frac{1}{2}\left(\frac{\beta}{R_{sh}} - \alpha\right)^2\right], \qquad (1)$$

where the parameters of α = 1.19, β = 0.184, $C_p = \int_{R_{sh1}}^{R_{sh2}} f(R_{sh}) \cdot \alpha R_{sh} = 1.085$; and the value ρ = $\mu\alpha$ = = 0.103 from the median $\mu = 0.912$.

The SP correlation function approximated by the expression of the form $k(\tau) = Ae^{-\alpha_0\tau}\cos(\omega_0\tau)$, where $A = 1$, $\alpha_0 = 0.88$, $\omega_0 = 2.61$ was calculated.

Let us turn to the problem of research [8] on measuring the coefficient of “wheel-rail” adhesion ($K_{sh} = 0{,}1..0{,}45$) from the degree of contamination, moistening of the track superstructure, properties and structure of colloidal contaminations of the contacting surfaces – composition, particle size and film thickness.

The results obtained are of interest for explaining some properties of a film on a contaminated rolling surface of “wheel-rail” from the point of view of the interaction physics of chemical, mechanical and electrical processes that affect the film structure and the electrical resistance of train shunt.

The contamination samples were taken in two ways:

1. The rails were coated with a 10% celluloid solution in acetone, then after 15-20 minutes a film of celluloid was formed on the rail head surface, which was easily removed with contaminations.

2. Contaminations were scraped from the rail head surface with a scalpel.

Scraped films were subjected to photometric measurement in order to determine the thickness of the rail contamination layer (Table 1). After that, the films were dissolved in acetone and the solid particles were separated [8].

Table 1. Film thickness distribution

Measurements	External part		Tread middle	
	h, µm	β, grad	h, µm	β, grad
Average value	33.7	83.7	7.4	47
Dispersion	4.1 – 57	16 – 15	4 – 11.2	21 – 93

Note: β is the angle between the horizontal and the tangent to the rolling surface.

The contaminations scraped by the second method were dried and the moisture content was determined.

After benzene washing followed by separation, the powdered contaminations were directed to a microscope survey and to X-ray, chemical and spectral analyses. Microscope survey showed that the contaminations are powders with particles of indeterminate forms. The statistical curve of particle size distribution [8] is shown in Fig. 1.

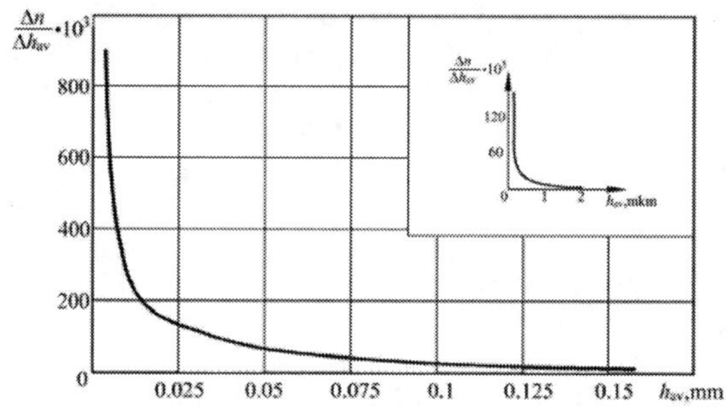

Fig. 1. The probability density distribution of the film particles on the "wheel-rail" surface

X-ray diffraction pattern showed iron particles with a highly deformed crystal lattice, a line of compounds of iron oxide hydrate, calcium carbonate, as well as Fe, Fe_3O_4 and SiO_2, herewith less than half of iron and its oxides were found.

The results of chemical analyses indicate that the main components of surface railheads` contaminations are oxides of iron and silicon. The percentage ratio between them depends on the state of ballast section and silica sand supply under the wheels of locomotives, as well as the sand brought in by the wind currents. Spectral analyses of the samples indicate the content of 6–10% Al, Si, Ca and Mn, C, S, Mg, Pb, Ni, Cu in insignificant quantities. The averaged thickness of the contamination film layer in the middle of the rail head is 1 µm for light and shiny surfaces, and up to 4..9 µm for grayer ones (Table 1). Thus, from the analysis of the source [8] we can conclude that:

1. Mostly the rails are covered with colloidal contaminations with a film thickness of 1 to 20 µm, and

their composition is particles of the wear of wheel and rails (30 – 70%) and impurities carried by the wind currents with sand from the subgrade.

2. Small-sized solid particles that make up the bulk of wheel and rail contaminations form structured systems and have structural and rheological properties. Further studies of the rheological properties of contaminations will help to reveal the physical nature and electrical properties of the films.

Therefore, it can be assumed that iron and silicon in contaminations and under conditions of relatively powerful specific pressure on the contact surface form film crystalline structures with nonlinear semiconductor properties that provide electrical resistance of the train shunt.

The article [7] presents the results of statistical studies of R_{sh} and the surface leakage coefficient ρ of the track circuits, as well as the influence of the values of these criteria on the sizes of the track circuits` stability regions. It is shown that a change in the above mentioned criteria in the direction $\rho > 0$ and $R_{sh} < 0.06$ Ohm allows significant expansion of stability region of the track circuits. For example, at $\rho = 0.5$, the limiting insulation resistance in the rail line 800 m DC track circuit drops 1.6 .. 1.8 times; if we take in the calculations $R_{sh} = 0.03$ Ohm, then this resistance in the code track circuit with a length of 1000 m will decrease from 0.37 to 0.13 Ohm·km.

The track circuit stability is determined by the following coefficient:

$$\varepsilon = \frac{r_{i1}(R_{sh} = 0,06)}{r_{i1}(R_{sh} < 0,06)}, \quad (2)$$

where r_{i1} – is the minimum limiting insulation resistance at $R_{sh} = var$ [6].

Fig. 2 presents the probability distributions of the shunt resistance obtained by measuring with a microhmmeter [7].

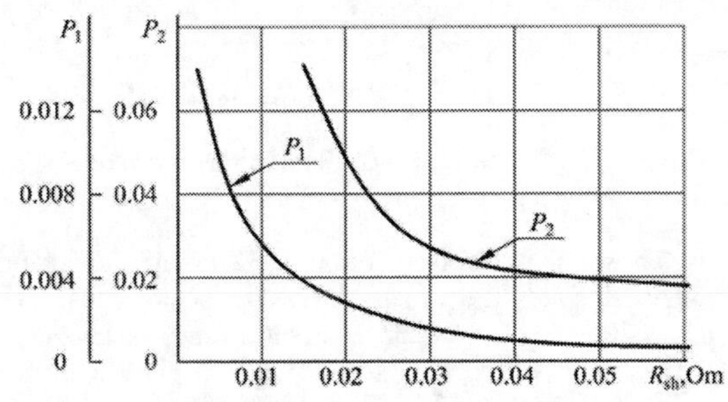

Fig. 2. The probability distribution function of shunt resistance

As one can see from Fig. 2 when reducing the shunt resistance to less than 0.03 Ohms, the probability $P(R_{shf} > R_{sh})$ sharply increases, which, apparently, from the viewpoint of signaling and interlocking systems` security prevents the adoption of absolute values of $R_{shf} < 0,03 - 0,04$ Ohm as the calculated ones. The authors [6, 7] admit the possibility of reducing the standard shunt (up to 0.03 Ohms) in automatic block systems, taking into account the resistance of two wheel sets, assuming that the location of such wheel sets at the wayside track circuit is possible only in the traffic light area.

Meanwhile, considering the calculated data of the track circuit with low insulation resistance, in our opinion, it is quite possible to calculate the track circuit using the existing normals with R_{sh} = 0.03..0.04 Ohms. At the same time, it is possible to estimate the real gain by reducing the limiting insulation resistance R_{i1}.

2 Influence of dynamic effects on the processes of film material formation on the "wheel-rail" contact

In the process of the train movement, a reliable electrical contact along the contact line of the wheel rim and the rail is required. Calculations of the train control schemes on the way take into account the static state of the transition resistance R_{sh} between the rails no more than the standard value of 0.06 Ohms. Experience has proven that it is the dynamic effects of axial loadings over the contact resistance area of "wheel-rail" have the most significant influence on the shunt. The load transmission from the wheels to the rail occurs at a small area, which is interesting to us, the material on it is experiencing a three-dimensional stress state.

The research of stress distribution in rails by the methods of the elasticity theory [1, 2] showed that their values exceed the yield strength of modern rail steel. It is shown that the design and metal of the rails in the contact area of the rail and wheel can withstand higher stresses without plastic deformations.

The contact areas and contact stresses essentially depend on the contact form of the bodies: the rolling surface of the wheels (new or machined ones) has a conical shape, but during calculations the surface is considered cylindrical (Fig. 3).

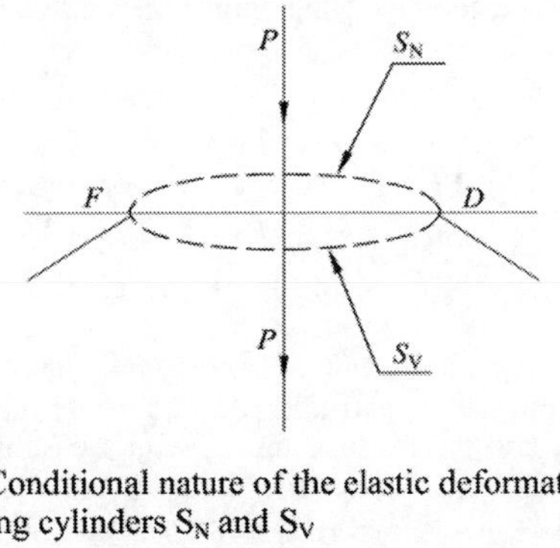

Fig. 3. Conditional nature of the elastic deformations of contacting cylinders S_N and S_V

In this case, the pressure intensity within the contact surface is approximately distributed according to the law of the semi-ellipsoid $Ax^2 + By^2 = C$, where A, B – depend on the radius of bodies` curvature, and $C$ is the value of bodies` convergence due to elastic deformation.

The dependence of the force on the stress σ_{max} is determined by the following formula:

$$P = \sigma_{max} \cdot \frac{2}{3} \cdot \pi \cdot a \cdot b, \tag{3}$$

where P is the load, kg; a, b – ellipse semi-axes, cm.

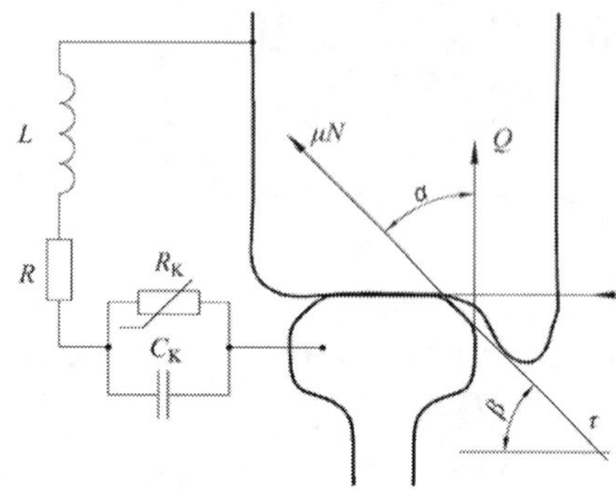

Fig. 4. The distribution of forces acting on the "wheel-rail" contact

The volume of the stress semi-ellipsoid [1]:

$$V = \frac{2}{3} \cdot \pi \cdot a \cdot b \cdot l, \tag{4}$$

where $\frac{A}{B} = \frac{R_1}{R_2}$ is the ratio proportional to the semi-axes a and b: $A = \frac{1}{2 \cdot R_2}$, $B = \frac{1}{2 \cdot R_1}$, l is the length of the contact surface area.

From this we can find that

$$\sigma_{max} = \alpha \cdot \sqrt[3]{\frac{P \cdot E^2}{R^2}}, \tag{5}$$

where α is a coefficient depending on the ratio $\frac{A}{B} = \frac{r}{R}$, E is the elasticity modulus of the rail steel.

The values of r and R represent a smaller and larger radius of curvature. It is considered that at σ_{max} – a dangerous point of stress, which lies at a depth below the area $\sigma_{calc} = 0,6\sigma_{max}$), in the center of the ellipse $\sigma_c = n_1\sigma_{max}$ along the major semi-axis $\sigma_s = n_2\sigma_{max}$, where the coefficients n_1 and n_2 are given in the works [1, 3].

The bogie in the process of hunting in the track curve rests upon the support line with the wheel flange of the first set and the rear one moves with a clearance 25..30 mm "rail track – flange". Herewith the friction force $F_{tr} = \mu \cdot \theta$, arises, where $\mu = 0,25$ is the coefficient of friction, and θ is the gross weight which puts the pressure on the axis. It follows that the action forces of the bogie weight P_b on the support and internal rails become different in value (up to 6-7% from 0.5 P_b). This changes the "wheel-rail" contact area of the bogie. The values of loadings and unloadings of the wheels are estimated by a dynamic factor:

$$K_d = \frac{P_b}{P_{st} - q}, \tag{6}$$

where P_{st} is a steady load on the axle, q is direct weight (pressure) of the wheel.

According to estimations of the works [1, 3], the factor K_d varies depending on the speed of movement in the range from 0.28 to 0.66 at the speeds of 60-120 km/h. It should be borne in mind that K_d is the result of various body oscillations during rolling, pitching and jumping. The change in the value of K_d under these effects may be in the range of 30–50%. It was found that the oscillation period of the wheel-rail pair reaches 35-45 ms.

As the rails wear out, the rolling surfaces of the rail head change, their radii increase from 300 to 320 mm. The actual increase in the radius of the rail head curvature leads to an increase in the rolling surface:

$$R_a = \frac{m}{m-1} \cdot R_c, \tag{7}$$

where $m = \frac{R_c}{R}$, and R_c is the radius of curvature forming the rolling surface.

Herewith, in the track curves, the steering wheel contact with the outer rail occurs along smaller radii of curvature, which causes an increase in stresses and deformations in the outer rails [2, 3] (Fig. 5).

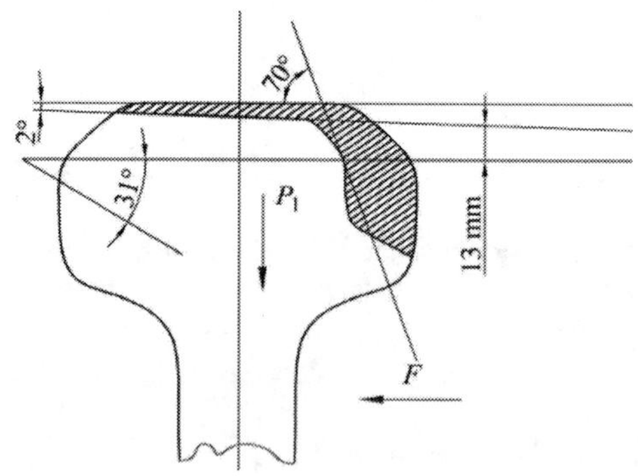

Fig. 5. Cross-section of the wheel in the curve of 368 m after passing 104 million tons of cargo

The wear intensity of rails [2, 3] mainly depends on the radius of the track curvature R_{cur}. Thus, with $R_{cur} \leq 400$ m, lateral wear occurs 2.5 times faster than in the sections where $R_{cur} \leq 600$ m. With significant side wear the wheels with their flanges start rolling on the galling foundation of the lateral edge of the rail. If the force P_1 pressing the wheel down, will be greater than the force $P_2(F)$, which impedes it, then the classic scheme of negotiating of the bogie`s first wheel will be observed when the flanges climb onto the outer rail and the wheels do not derail under the action of the horizontal force F. It can be seen that when the rail is worn, the area S_c of the "wheel-rail" contact increases (by 15–20%) and in the curved track sections it may exceed the values obtained below.

For example, we give the calculation of the contact stresses in the rails and the evaluation of the "wheel-rail" rolling surface. If we calculate the dynamic wheel loading on the rail [12] with the axial loading of a four-axle heavy car 10.2 tons, the movement speed 85 km/h, the rail type P50, we obtain the following:

$$\theta_{\text{dyn}} = \frac{P}{g} = 18669 \text{ kg} \tag{8}$$

The wheel radius is R = 47.5 cm, the radius of the rolling surface is r = 30 cm. Hence, $\frac{A}{B} = \frac{r}{R} = 0,632$ [2, Table. 33], $\alpha = 0.459$. Then from the formula (see above)

$$\sigma_{\max} = 0.459 \cdot \sqrt[3]{\frac{18669 \cdot (2.1 \cdot 10^6)^2}{47.5^2}} = 15200 \text{ kg/cm}^2 \tag{9}$$

The coefficients n_1=0.23; n_2=0.241, then the stresses in the directions of the rolling area axes will be: $\sigma_{\text{tc}} = 0.23 \cdot 15200 = 3500$ kg/cm^2, $\sigma_{\text{ext}} = 0.241 \cdot 15200 = 3620$ kg/cm^2, being at the yield strength limit.

Estimation of the rolling area: $S_{\text{era2}} = \frac{P}{\sigma_{\max}} = \frac{18700}{15200} = 1.23$ cm^2, in the danger area, with stress σ_{tc} obtained $S = \frac{18700}{3620} = 5.17$ cm^2.

To determine the fictitious radius of the rail head with the wear of rolling surface 3–4 mm, the axle loading 10.2 tons, the radius of curvature R_a = 170 mm, and the radius r = 80 mm, we will find:

$$m = \frac{170}{80} = 2.1; \; R_a = \frac{2.1}{1.1} \cdot 8 = 15.3 \text{ cm} \tag{10}$$

Then $\frac{A}{B} = \frac{15.3}{47.5} = 0.322$, $\alpha = 0.586$;

$$\sigma_{\max} = 0.586 \cdot \sqrt[3]{\frac{18700 \cdot (2.1 \cdot 10^6)^2}{47.5^2}} = 19400 \text{ kg/cm}^2.$$

From [3] n_1=0.22; n_2=0.23. Therefore: σ_{tc} =5240 kg/cm^2; σ_{ext} =4460 kg/cm^2, and $\sigma_{\text{calc}} = 0,6 \cdot \sigma_{\max} = 11640$ kg/cm^2 kg/cm^2.

Hence, $S_K = \frac{19400}{4460...5240} = 4,34...3,7$, and approximately it turns out that the surface area of the "wheel-rail" contact can be expected in the range 1.23..4.34 cm^2.

3 Redistribution of efforts on the contact areas "wheel-rail"

The rail head metal in the process of wheel hunting or when moving in the curved track sections works in the mode of compression or sliding; at the same time a side jolt is observed, in which the forces of the areas contact are established on the inner and opposite rail as $P_1 = P_0 + \Delta P$ and $P_2 = P_0 - \Delta P$, respectively. It is established that the pressure difference is ΔP = 2.25–3.75, and the difference of stresses is greater than $15 \cdot 10^3$ kg/cm^2.

As the rolling surface [1] wears out instead of the conicity generator line (Fig. 1) at the end of the junction plane to the flange "throat" the changes in the rolling – sliding surfaces appear, where $S_K = S_K^* + S_{sr}$, , where S_K^* – is the rolling surface, S_{sr} is the sliding surface of the flange along the lateral edge of the rail head.

The sliding condition of the wheel flange is assessed by the Nadal criterion [3]:

$$\frac{Y}{Q} < \frac{\tan\beta - \mu}{1 + \mu \cdot \tan\beta} \leq 1, \tag{11}$$

setting the ratio of the forces acting on the wheel: Y – force, pushing the wheel up, Q – force pressing on the wheel down, keeping it inside the track. Y and Q create possible flange wear, which threatens to strike along the switch point`s end, if the gap "switch point – stock rail" is more than 4 mm.

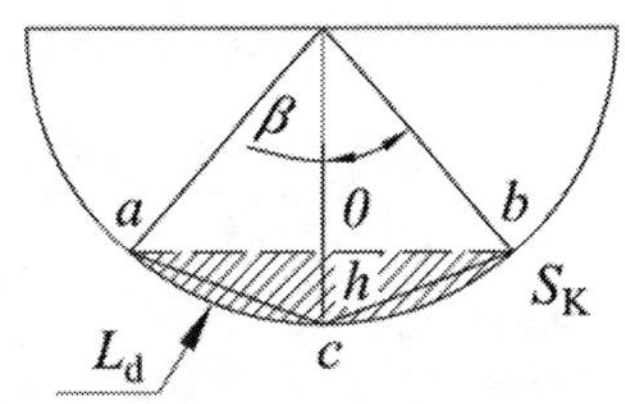

Fig. 6. Sliding area "wheel flange-lateral edge of the rail"

Suppose that the length of the generator line of sliding surface has a radius R=25 cm, and the slide angle is determined from the Nadal condition, then it is possible to determine the sliding area of a flange from *abc* sector triangle (Fig. 6), or by the ellipse (segment) formulas.

If we assume that due to the flange wear, the slide angle exceeds 100%, then the sliding area S_K can reach the values of 2 cm^2 and higher. However, these values of the area are approximate and require experimental verification. It should be noted that the train shunt resistance, on average, is relatively stable during the dynamic "wheel-rail" interaction when moving in a straight line, when the influence of the bogie wheel is observed, or in the curve and flange sliding along the lateral edge of one rail head.

4 Electrical equivalent circuit of film and estimation of parameters of its elements

Fig. 4, a variant of the equivalent circuit (EC) of the make-before-break contact is considered – the film material and the ferromagnetic "wheel - rail" material conjugated with it on both sides. The circuit contains linear elements: wheel inductance L, wheel resistance $R = R_K + 2 \cdot R_\mu$, presumably active resistance of ferromagnetic layers (substrate) – ΔR_μ and film capacitance C_K, as well as a nonlinear element – active resistance $R(P_K, S_K, U)$, which is a function mainly of pressure, film area and stress. Moreover, the film mainly consists of iron oxide Fe_2O_3, sand and red dust stuck to the rail.

Above is an estimation of the geometric parameters of the film, and below – the order of values of the electrical elements of the EC.

The impedance of the equivalent circuit is expressed as:

$$Z = j\omega L + \frac{R \cdot \frac{1}{j\omega C}}{R + \frac{1}{j\omega C}} = \frac{j\omega L - RCL\omega^2}{1 + j\omega RC} \quad (12)$$

Following [4] after transformations (4) we will obtain:

$$Z = \frac{j\omega L - RCL\omega^2 + R}{(1 + j\omega RC)} \cdot \frac{(1 - j\omega RC)}{(1 - j\omega RC)} \quad (13)$$

And finally, we get resistance in a complex form:

$$Z = \frac{R + j \cdot (\omega L - \omega^2 R^2 C + \omega^3 R^2 C^2 L)}{1 + \omega^2 R^2 C^2} \quad (14)$$

Which makes it possible to determine the resistance modulus of the train shunt:

$$|Z| = \frac{\sqrt{R + (\omega L - \omega^2 R^2 C + \omega^3 R^2 C^2 L)}}{1 + \omega^2 R^2 C^2} \quad (15)$$

The electrical resistance of the film formed during the motion, as a rule, is not destructible, but it can be broken down [5]. With a relatively long contact time, so-called coherent effect appears – with increasing speed of movement the film does not have time to be broken down, but the joint resistance remains relatively high. Such a state can more often occur when moving the cars of light-weight electric trains.

Capacitance C depends only on the form of the dielectrics material and can be approximately compared with the capacitance of the area at the transition of the tread-rail, whose area reaches 5 cm^2.

Capacitance is determined by the formula:

$$C = \frac{\varepsilon \cdot S}{l} \quad (16)$$

Insulation thickness of h = 7.4 μm, average dielectric constant for insulating material ε = 8, then the capacitance: $C = \frac{8 \cdot 5{,}1 \cdot 10^{-4}}{7{,}4 \cdot 10^{-3}} = 5{,}4 \cdot 10^{-4}$ μF.

If we assume that the film resistance (approximately, as in silicon diodes) is 0.5 Ohm, then the denominator of resistance according to the formula (15) at a frequency of $10 \cdot 10^{-3}$ Hz is approximately equal to one, and the modulus

$$|Z| = \sqrt{R^2 + (\omega L + \omega^2 R^2 C + \omega^3 R^2 C^2)} \approx$$

$$\approx R = 7 \cdot 10^{-4}..0.5 \quad (17)$$

i.e. the modulus is approximately equal to the active resistance of the "wheel-rail" contact in the range from 10^{-4} to 0.5 Ohms (in the presence of films).

Calculating the derivation (with respect to the argument ω) of the resistance Z, and equating it to zero, we obtain the biquadratic equation for ω:

$$3RLC\omega^4 + 4R^2LC^2\alpha\omega^2 + \alpha^2 = 0 \quad (18)$$

where $\alpha = L - CR^2$.

If $\alpha < 0$ or $L < CR^2$, then the real roots (18):

$$\omega^2 = \frac{-2\alpha \pm \alpha}{3R^2LC^2} > 0 \quad (19)$$

From equation (18), the value of EC inductance (Fig. 2) is obtained, $L = 0{,}12 \cdot 10^{-7}$ H or 1.2 μH

Equation (18) makes it possible to use the resistance law Z at different signal frequencies, resistance variations R and its effect on the shunt resistance, etc.

In conclusion, it can be noted that with low contact resistance of the "wheel-rail" contact and the corresponding values of R, C and $L > CR^2$ (19) is a practically increasing linear dependence on frequency. With bad contacts, when $L < CR^2$, the dependency $Z(\omega)$ decreases. The dependencies with large transition resistances of the contact are of the greatest interest. In such conditions at the stations it is more profitable to use overfrequencies of the tonal spectrum (up to 3 kHz). This will make it possible to improve the double rail plan of the station using shorter track circuits and achieve the reduction in the insulating joints.

An overview of the problem state of rail circuits` shunting and the obtained approximate values of the electrical parameters of the EC affecting the shunt mode make it possible to better understand the physical nature of the dynamics of electrical and mechanical interactions when solving the "wheel-rail" contact problem and recognize as appropriate more detailed studies of such an urgent problem of signaling and interlocking devices.

It is of interest (Fig. 1, 2) external qualitative similarities in the nature of the probability distributions of shunt resistances and the sizes of contamination particles. It can be argued that both physical random processes – the film formation on rails and the value of its electrical resistance are dependent and correlated with each other.

Let us turn to the analytical proof of such a statement.

5 The correlation dependency estimation of the parameters of the film and train shunt

Let us consider the research results of interrelated problems [7, 8]. In one of them [8], as noted above, the effects of contamination of the friction surfaces of rails and wheels and the estimation of their influence on the friction coefficient (0.1-0.45), as well as the tractive characteristics of locomotives, were investigated.

According to the results of processing statistical material [8], the following functions are presented: the function $f(h_{av})=\frac{\Delta n}{\Delta h_{av}}\cdot 10^3$ of probability density of the average film thickness on the rolling surface of rails and the function $F(h_{av})$ of this random variable distribution (Fig. 7).

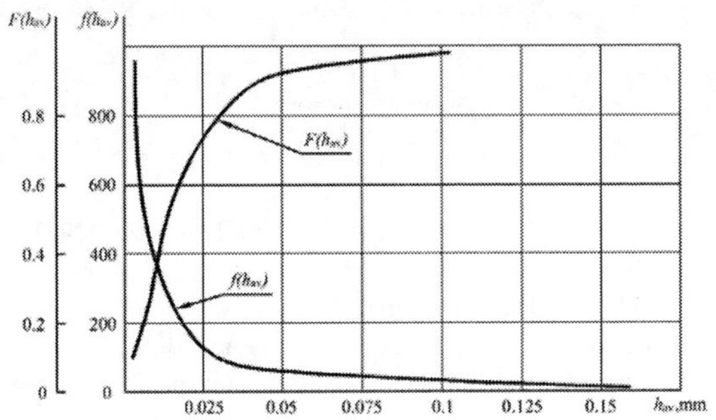

Fig. 7. Function graphs f(h_{av}) and F(h_{av})

The construction of the function $F(h_{av})$ is performed in the following sequence.

From the experimental density function, the number of contamination particles in each discharge was determined (from 0 to 4000). In case of equality $\frac{\Delta n_i}{\Delta h_{av}}\cdot 10^3=800$, where n_i is the number of particles hitting to discharge $\Delta h_{av}=4{,}5$ mm (Fig. 7), the value $n_1=800\cdot 4{,}5=3600$, is determined, similarly for $n_2=600\cdot 4{,}8=2880$, etc. The total sample size for the number of hits in the range from 0 to 800 was N = 18900. Hence, the frequencies of hits in each discharge were determined: $p_1=\frac{3600}{18900}=0{,}1905$, $p_2=\frac{2880}{18900}=0{,}1524$, etc.

According to the results of calculations, the probability distribution function of the particles from the value of the average film diameter was constructed [8] (Table 2).

Table 2. Frequencies and particle sizes by discharges

$F(h_{av})$	0.98	0.69	0.45	0.34	0.19
$h_{av}\cdot 10^{-3}$, mm	56.0	18.0	8.2	4.8	3.25

From Figure 7 it follows that for the approximation of the function $f(h_{av})$ the exponential distribution law can be applied.

Experimental studies of the train shunts resistance [7] were carried out in two specially equipped track circuits with a length of 12.5 m, located on the track sections with autonomous traction. According to the proposed method, the probability that a signal exceeds a given resistance level R_{sh} is measured by the formula $P_{R_{sh}}=\frac{t_i}{T}$ where t_i is the total duration of exceeding the i-th level over the observation period T = 8760 h. Based on the statistical materials, the distribution functions of the random variable $F(R_{sh})$ and the probability density $f(R_{sh})$ of resistance R_{sh} (Fig. 8) were additionally constructed with estimations of the random variable up to $5\cdot 10^{-3}$ Ohms (with data less than this estimation, the lines were broken).

When statistics were obtained to the left of the 0.0075 Ohm resistance, the density function could be approximated by the Rayleigh distribution law:

$$f(R_{sh})=\frac{R_{sh}}{\sigma^2}\cdot\exp(-\frac{R_{sh}^{\ 2}}{2\cdot\sigma^2}), R_{sh}\geq 0, \sigma>0 \quad (20)$$

where R_{sh} depends on the mathematical expectation $M(R_{sh})$.

The descending part of the probability density function more than $(5-10)\cdot 10^{-3}$ Ohm is reliably described by the exponential distribution

$$f(R_{sh})=\lambda\cdot\exp(-\lambda\cdot R_{sh}) \quad (21)$$

where the parameter $\lambda=\frac{1}{m_{R_{sh}}}=45{,}4$ 1/Ohms.

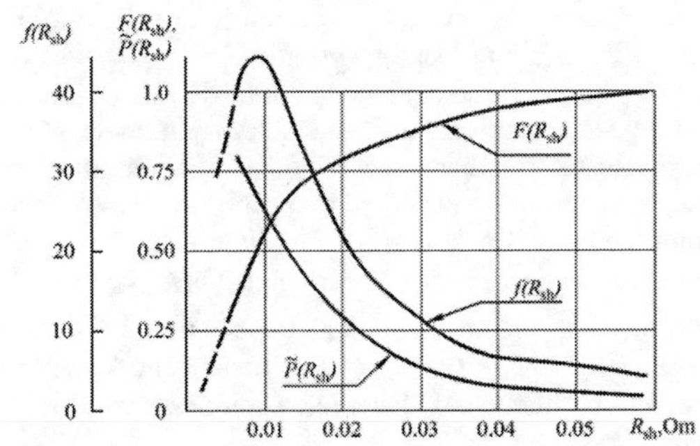

Fig. 8. Graphs of distribution functions $F(R_{sh})$, probability density $f(R_{sh})$ and dependencies $\tilde{P}(R_{sh})$

Table 3 shows the probabilities of the distribution function of the train shunt resistance and film thickness [7]

Table 3. Probabilities of the distribution function

$F(x)$	0.2	0.4	0.6	0.8	1.0
$R_{sh}\cdot 10^{-3}$, Ohm	5.12	7.9	16.5	32.15	-
$h_{av}\cdot 10^{-3}$, mm	3.0	8.2	18.1	30.0	-

The authors of work [7] claim that:

1. Due to the increase in the probabilities $P(R_{sh}> 0.06$ Ohm), according to the conditions for increasing the track circuit`s operational performance, a significant decrease in the standard value R_{sh} cannot be achieved.

2. The changes` dynamics of the function $\tilde{P}(R_{sh})$ (Fig. 8) shows that for automatic block systems operating with a reduced insulation resistance of the rail line, the train shunt resistance of two axles of wheel sets equal to 0.03 Ohm can be assumed as a normative value. On the one hand, the probability $P(R_{sh}=0.03$ Ohms) relative to 0.06 Ohms does not increase so much, on the other hand, the use of calculations with a shunt resistance of 0.03 Ohms at a critical point of the shunt location – near intermediate block signal – does not pose a safety hazard because the movement is carried out on the restrictive signal.

Fig. 8 shows that over 85% of the wheels pass along the rails at $R_{sh}< 0.02$ Ohm.

Note that by reducing the normative value of the train shunt resistance to 0.03..0.04 Ohms, the performance area of rail circuits` operating modes is expanded, mainly due to increase in the power supply voltage and, as a result, a reduction in the limiting insulation resistance R_{il} of the rail line.

The analysis of the physico-chemical nature of the film material and the obtained stochastic dependencies of random variables (Fig. 7, 8, [7, 8]) – the geometrical dimensions of the film covering the rolling surface "wheel-rail" and the electrical resistance of the train shunt – indicate that the film characteristics influence the value of R_{sh}.

The film parameters at the "wheel-rail" contact not only affect the friction coefficient and the locomotives' traction characteristics, but also make the main contribution to the formation of the electrical resistance modulus of the train shunt and, as a consequence, the performance criteria for rail circuits and the safety of signaling and interlocking systems.

To quantify the dependency intensity of y – the train shunt resistance on the x – the film parameters, let us turn to a mathematical model that studies such dependencies [9]. We consider in the first approximation that there is a linear correlation relationship between the random variables $x(h_{av},\text{mm})$ and $y(R_{sh},\text{Ohm})$. This means that the mathematical expectation y is linearly dependent on the value x. The dependency graph (the regression line y on x) has the equation $M(y)=\rho\cdot x+b$, where ρ and b are some unknown constants. The value of the correlation coefficient r_{xy} is also unknown.

The results of independent observations [7, 8] over the values of x and y, obtained in pairs on the same probabilistic space of their appearance, are presented in Figures 7 and 8, as well as in Table 3. From these sources, the necessary information to search for ρ, b and r_{xy} estimations was obtained. The values of these estimations correspond to the minimum of the function $F(\rho,b)=\sum_{k=1}^{n}(\rho\cdot x_k+b-y_k)$ and the equation systems for partial derivatives $\frac{\partial F}{\partial \rho}$ and $\frac{\partial F}{\partial b}$, are obtained, as required by the extremum conditions and the method of least squares:

$$\rho=\frac{n\cdot\sum_{k=1}^{n}x\cdot y_k-\sum_{k=1}^{n}x_k\cdot\sum_{k=1}^{n}y_k}{n\cdot\sum_{k=1}^{n}x_k^{\,2}-\left(\sum_{k=1}^{n}x_k\right)^2} \tag{22}$$

$$b=\frac{n\cdot\sum_{k=1}^{n}x_k^{\,2}\cdot\sum_{k=1}^{n}y_k-\sum_{k=1}^{n}x_k\cdot\sum_{k=1}^{n}x_k\cdot y_k}{n\cdot\sum_{k=1}^{n}x_k^{\,2}-\left(\sum_{k=1}^{n}x_k\right)^2} \tag{23}$$

To estimate the correlation coefficient, let us use the relation

$$r_{xy}=\rho\cdot\frac{\partial x}{\partial y}, \tag{24}$$

where σ_x and σ_y are the mean square deviation of x and y, respectively.

The values of the point estimations for the distribution parameters of the mean square deviations σ_x and σ_y the mathematical expectations m_x and m_y of the film`s geometric dimensions and the train shunt resistance, obtained from statistical observations, are presented in Table 4 [7, 8] and are used below to calculate the correlation coefficient and the regression line.

Table 4. The values of the point estimations

Parameter	Mathematical expectation	Mean square deviations
x	$11.6\cdot10^{-3}$ mm	$21.7\cdot10^{-3}$ mm
y	$22.0\cdot10^{-3}$ Ohm	$41.0\cdot10^{-3}$ Ohm

By substituting the variables in expressions (3-5), we obtained the values of the parameters $\rho=-1.61$, $b=0.025$ and the correlation coefficient $r_{xy}=0.82$. The obtained value of r_{xy} indicates a fairly close relationship between the train shunt resistance and the film thickness of contaminations formed on the rolling surface of the wheel on the rails.

The volume of statistical data obtained during the research: 18900 in [7] and 2050000 in [8] is sufficient to correctly calculate the maximum likelihood estimations for the parameters of the distributions taken here and to estimate the correlation coefficient.

The linear regression equation for y on x has the following form: $M(y)=-1{,}61\cdot x+0{,}025$. Calculations by this equation are performed on the average measured values of the parameter x.

6. Conclusions

1. Using the calculations of the dynamic interaction of wheels of rolling stock and rails we obtained the area of the "wheel-rail" rolling surface, taking into account the different conditions of the dynamics of interaction and wear of rails and wheel elements.
2. It is shown that reducing the normative value of the train shunt resistance to 0.03..0.04 Ohms is advisable for signaling and interlocking systems, especially for the railway sections with a reduced resistance of rail line insulation.
3. It has been established that train wheel sets moving on a track circuit, in more than 85% of cases, provide a train shunt resistance of less than 0.02 Ohms.
4. An approximate estimation of the geometrical parameters of the film formed on the "wheel-rail" rolling surface was obtained; more complete electrical equivalent circuit of the train shunt circuit has been proposed, which allows analyzing the influence of the parameter values of the circuit elements and the signal current frequency, in particular, on the train shunt resistance modulus.
5. It was proposed a method for calculating the correlation dependence of the train shunt resistance from the film parameters on the "wheel-rail" rolling surface, which made it possible to obtain a quantitative estimation of the correlation coefficient ($r_{xy} = 0.82$), indicating a close relationship between dynamic and physico-chemical formation processes of film material characteristics and train shunt resistance.

References

1. A.Ja. Kogan. *Dinamika puti i ego vzaimodejstvie s podvizhnym sostavom* – M.: Transport 326 s. (1997) [in Russian]
2. G.M. Shahunjac *Zheleznodorozhnyj put', 3-e izd., pererab. i dop.* – M.: Transport, 479 s. (1987) [in Russian]
3. M. A. Chernyshev. *Prakticheskie metody rascheta puti*. – M.: Transport, 226 s. (1967) [in Russian]
4. Walter J. *Au problème du choix de la fréquence de courant alimentant le circuit de voie* Revue Generale des Chemins defer. # **6**. – P. 22-25 (1986)
5. Hirosi Okumura. *The resistance of the train shunt in the track circuits* Quarterly Report of the Railway Technical Research Institute Japan National Railway. V. **3**, N. 3 (1962)
6. A.P. Razgonov. *O nadezhnosti rel'sovyh cepej* Avtomatika, telemehanika i svjaz'. # **3**. – S. 36-39 (1986) [in Russian]
7. V.A. Minin, V.S. Dmitriev, V.S. Luchinin. *Ocenka znachenij kojefficienta poverhnostnoj utechki rel'sovyh cepej i soprotivlenija poezdnogo shunta* Vestnik VNIIZhT. # **2**. – S. 52-54 (1986) [in Russian]
8. Ju. M. Luzhnov, M. V. Rusakova, R. G. Cherepashenec. *Zagrjaznenija poverhnostej katanija trenija rel'sov i koles podvizhnogo sostava* Vestnik VNIIZhT. # **4**. – S. 38-41 (1972) [in Russian]
9. I.M. Petrushko, V. I. Afanas'ev i dr. *Kurs vysshej matematiki* – M.: Izdatel'stvo MJeI. 304 s. (2004) [in Russian]
10. Ju.A. Kravcov, B.M. Stepenskij. *Verojatnostnaja model' poezdnogo shunta dlja legkih podvizhnyh edinic i otcepov* Trudy UJeMIIT. – Sverdlovsk.: UJeMIIT, vyp. **64**. – S. 58-60 (1980) [in Russian]
11. R. Hol'm. *Jelektricheskie kontakty*. – M.: Inostrannaja literatura. 410 s (1961) [in Russian]
12. DSTU 7571:2014: Ruhomij sklad zaliznic'. Normi dopustimogo vplivu na zaliznichnu koliju shirinoju 1520 mm (2014) [in Russian]

Justification of parameters of wheelset axle fatigue strength test-bench for railway rolling stock

Oleksii Kuropiatnyk[1,*], *Serhii* Raksha[1], and *Pavlo* Anofriev[1]

[1]Dnipro National University of Railway Transport named after Academician V. Lazaryan, Department of Applied Mechanic and Material Sciences, 2 Lazaryana St., 49010 Dnipro, Ukraine

Abstract. The operation safety of the railway rolling stock depends directly on the strength of the base part of the wheelset – its axle. Therefore, in recent years numerous studies of theoretical and experimental nature, both for means of rail transport, and for machines containing rail track equipment are carried out. This paper substantiates the main parameters of the fatigue test bench for axles for wheelsets of railway rolling stock. We performed an analysis of the load schema for bench tests, determined the dependence of the bench work force on the parameters of the wheelset axle. In addition, we substantiated the design and parameters of vibrator. For the research, we adopted the following criteria of rationality: work force, energy consumption and longitudinal dimension of the test bench. The design is considered the best, if these criteria acquire the smallest values. The results of the research indicate that the load scheme "cantilever beam" is rational for the fatigue bench tests of the axles for wheelsets of railway rolling stock. The best design of the vibrator have to provide the test bench work force by rotating the unbalanced mass around the stationary wheelset axle. The rational vibrator contains a sectoral eccentric mass rotating at an angular velocity of about 150 s^{-1}. In this case, the eccentric mass value is about 80 kg, and its eccentricity is 135 mm.

1 Introduction

In many industries, the widespread used was gained by railway transportation vehicles. They are used for transportation loads and people within the precincts of an enterprise, open-pit mine or in mines (intra-company industrial conveyance facilities) or ensure traffic between certain enterprises (external industrial conveyance facilities, including the railway transport).

Each of railway transportation vehicles includes a bogie that contains, in particular, a wheelset (WS). This component part containing an axle as a basic part, running wheels (or a single wheel for semi-axles) and a gear wheel (for driving axles).

The operational safety of a railway transportation vehicle is directly related to the strength of wheelset basic part, i.e. its axle. Therefore, latterly numerical studies of theoretic and experimental nature are conducted, both for railway transport vehicles [1, 2, 3], and for motor vehicles that contain railway running equipment [4, 5, 6].

Before assembly of a wheelset certain number of axles from a batch is subjected to testing. For this purpose, special equipment is used, in particular, test-benches that imitate application of forces equivalent to operational loads.

In accordance with interstate standard [7] the following types of bench tests were established for wheelset axles:

– Testing when loaded with static vertical and lateral horizontal forces;

– Testing when loaded with tangential forces of threshold adhesion values with consideration of wheelslip;

– Fatigue strength testing when loaded with cyclic rotational bending.

The latter of the types of testing is the most complex because it renders operation of the bench under conditions of significant dynamic loading. Quite a number of studies were conducted on the issues of establishment of a rational pattern or loading, determination of design and performance. Results of such studies got their reflection in a number of safety documents [8, 9, 10], and in the standards database [11, 12].

The analysis of information sources showed that the problem of a rational design development and justification of wheelset axles fatigue strength test-bench parameters for railway transport is a pressing one, but still remains unsolved.

The purpose of this work is to justify basic parameters of wheelset axles fatigue strength test-bench for railway transport.

In order to accomplish a determined objective the following problems require solving:

1) Justification of a rational pattern of loading under bench testing;

2) Identify dependence of test-bench operational loading from parameters of wheelset axle being tested;

* Corresponding author: kuropyatnick@gmail.com

3) Justify the design for vibrator, i.e. sources of test-bench operating force;

4) Determine parameters of the vibrator.

For the purpose of obtaining and analyzing the results, the methods of mathematical modelling were used involving special-purpose software complexes.

2 Justification of a rational pattern of loading

General requirements to conducting wheelset axle fatigue strength bench testing are set out in the interstate standard [11]. According to this standard the wheelset axle is stipulated to be set on two hinged supports, one of which is to be set as close as possible to the box journal face, and the other one – at the section between surfaces of running wheels installation (Fig. 1, a). Therewith, test-bench operating force (F_1) acts in radial direction on the surface of the running wheel installation located between supports. In such a case the pattern of loading can be presented as a beam on two hinged supports with a transverse concentrated force between them (Fig. 1, b). The standard does not regulate the magnitude of this force, and beam sizes are indicated in relation to design and type of the axle.

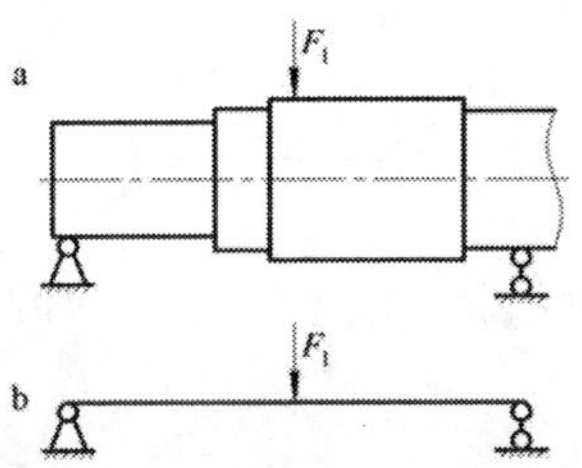

Fig. 1. Installation layout (a) and loading pattern (b) of wheelset axle as per standard [7]

Another interstate standard [12] specifies that cyclic rotational bending of wheelset axle can be implemented through rotation of the imbalance weight installed on the box journal (location of hinged stationary support in Fig. 1, a). Moreover, the axle can be press-fit into the running wheel (or into the temporary process hub installed instead of the wheel), fixed against shifting (Fig. 2, a). Under such conditions the pattern of loading represents a beam constrained from one end with its console under applied concentrated force F_2 (Fig. 2, b). The magnitude of this force, as in the previous case, is not regulated, and the dimensions of the pattern depend on wheelset axle sizes.

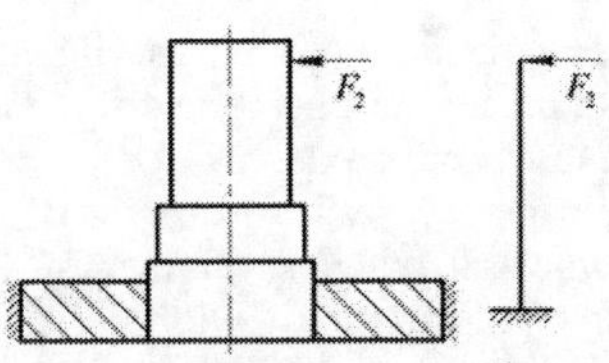

Fig. 2. Installation layout (a) and loading pattern (b) of wheelset axle as per standard [8]

It should be noted that in case of testing as per the pattern in Fig. 1 or Fig. 2 only a part of a wheelset axle is installed on the bench.

Another two patterns of axle loading should be noted that to a greater extent correspond to actual conditions of wheelset usage. Each of these is based on presenting the wheelset axle as a beam on two hinged supports. In one case the support points are located on surfaces of running wheels installation, and the test-bench operational force (F_3) applied to box journals imitates the loading from the bogie (Fig. 3, a). The other pattern provides a location of support points on box journals with application of test-bench operational force (F_4) to running wheels installation surfaces thus representing an imitation of rail track force response impact to bogie action (Fig. 3, b).

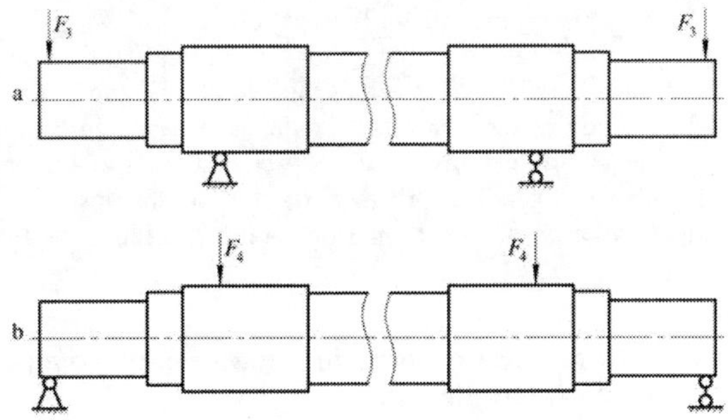

Fig. 3. Wheelset axle installation layouts approximated to service conditions

The common lack of both layouts as per Fig. 3 the requirement of testing axles as a whole that increases test-bench overall dimensions as compared with layouts as per Fig. 1, 2. Therefore, let's hereinafter consider the first two layouts only:

– "Free-ended beam" (Fig. 1, b) – installation of wheelset axle on two hinged supports with application of the operational force radially to running wheels installation surface located between supports;

– "Cantilever beam" (Fig. 2, b) – fixation of wheelset axle using an intermediate piece (running wheel or temporary process hub) with application of test-bench cantilever operational force to the box journal.

In order to justify a rational pattern of loading let's adopt the following criteria:

– Test-bench operational force (and, as an effect, power consumption);

– Overall dimension lengthwise.

The best one is the pattern, wherein both criteria gain lowest values.

2.1 Comparison by operational force magnitude

When conducting the fatigue strength tests the wheelset axle cyclic bending is implemented. Disregarding the loading pattern the bending stress in any cross-section is determined by the formula

$$\sigma = M / W , \quad (1)$$

where M – bending moment in the design cross-section; W – section modulus of the design cross-section.

Considering that the section modulus W depends only on wheelset axle dimensions, the selection of a rational pattern of loading is made by the magnitude of the bending moment.

The bending moment is determined as follows:

1) For the "free-ended beam" pattern

$$M = 0.5F_1x; \tag{2}$$

2) For the "cantilever beam" pattern

$$M = F_2x. \tag{3}$$

In these formulas F_1 and F_2 are the test-bench operational forces for the first and the second patterns of loading accordingly; x is the coordinate of the design cross-section (the same for both patterns).

The formulas presented show that in case of gaining the same bending stress values the operational force as per the loading pattern 1 has to be twice as great as compared to the operational force of test-bench with the loading pattern 2. Therefore, by the criterion of the test-bench operational force the rational one is the "cantilever beam" loading pattern.

2.2 Comparison by the test-bench overall dimension lengthwise

The test-bench overall dimension lengthwise directly depends on geometrics of the loading pattern (beam). The same portion of wheelset axle is being tested. One end of the beam is the box journal face; it has one of hinged supports located thereon (for the "free-ended beam" pattern) or the cantilever loading is applied thereon (for the "cantilever beam" pattern). However, the distance from this beam end to the opposite one for different loading patterns is not the same. The other end of the cantilever beam (as a point) corresponds to one of cross-sections of the running wheel installation surface on the wheelset axle, whereas for the free-ended beam this point is located further, between two running wheels installation surfaces. Thus, the length of such beam is greater, therefore by criterion of the test-bench overall dimension lengthwise the rational loading pattern is the "cantilever beam".

It follows from the above that by operational force criteria (and therefore, power consumption as well) and by test-bench overall dimension lengthwise the rational loading pattern is the "cantilever beam", for which the specified values will gain the lowest magnitude.

3 Determination of a dependence of the test-bench operational force from wheelset axle parameters

According to the formula (1) the bending stresses in the axle cross-section, described by the coordinate x, depend on two factors: bending moment that is determined by the formula (3) for the selected loading pattern ("cantilever beam"), and the section modulus of the design cross-section. Such a cross-section is the fillet from the box journal to the step between the journal and the running wheel installation surface (Fig. 4).

The section modulus of the design cross-section with the diameter of d_x is determined by the formula

$$W = \frac{\pi d_x^3}{32}. \tag{4}$$

Considering the nature of testing (fatigue strength testing) and particulars of cyclic loading implementation (by the symmetric cycle), let's adopt $\sigma = \sigma_{-1}$, where σ_{-1} – axle fatigue limit under symmetric loading.

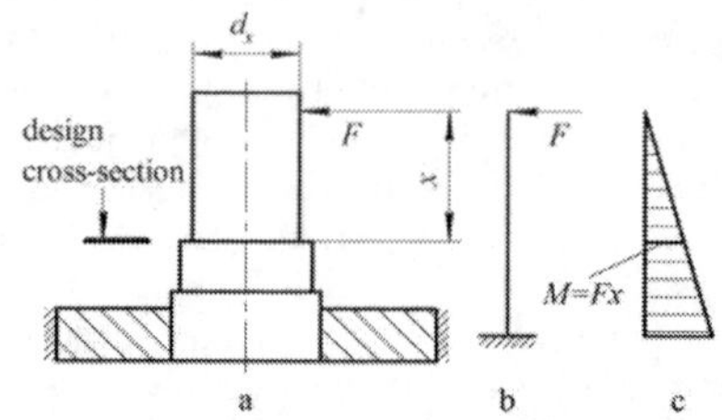

Fig. 4. The pattern for determination of the bending moment in the design cross-section: a – wheelset axle installation layout; b – loading pattern; c – distribution of bending moments

Through combination of the formulas (1), (3) and (4), we obtain:

$$\sigma = \frac{M}{W} = \frac{32Fx}{\pi d_x^3} = \sigma_{-1}. \tag{5}$$

Therefore, the test-bench operational force is

$$F = \frac{\pi d_x^3}{32x}\sigma_{-1}. \tag{6}$$

It should be noted that the design cross-section diameter d_x and the coordinate x are wheelset axle characteristics that are regulated by the relevant standard [11]. The fatigue strength σ_{-1} is determined as per the standard through testing [12], and therefore for each nominal size of wheelset axle the test-bench force F can be determined in advance. The expression (6) also shows that when designing the test-bench intended for testing axles of several nominal sizes it is required to determine not only the geometry variability of specimens under test, but various magnitudes of the operational force as well.

4 Justification of vibrator design

Within this work let's consider as vibrator a test-bench element that creates an operational force and ensures its application to wheelset axle.

The test-bench cyclic operational force can be applied to wheelset axle in several ways:

1) Using a pulsator (variability of the force as per the harmonic law);

2) By means of rotating the imbalance weight (debalance) around the rigid wheelset axle;

3) By means of rotating the wheelset axle with a debalance installed thereon;

4) By means of pressing with a test-bench working tool on the rotating wheelset axle while maintaining the constant magnitude of pressure corresponding to the amplitude of cyclic operational force.

Considering that the test-bench operational force has to ensure cyclic rotational bending of the wheelset axle as per symmetric cycle [12] we come to conclusion that the first way is the most complex one in the context of vibrator structural design.

Since the wheelset is much heavier than the debalance, the vibrator will consume more power in case of the test-bench operational force application in the third or the fourth way.

Hence, by the criteria of design complexity and power consumption the best one is the vibrator that ensures the test-bench operational force application by rotating the imbalance weight (debalance) around the rigid wheelset axle. In case of using it the criteria used gain the smallest magnitude. Let's assume it as a basis to carry out further studies.

The operating principle of the vibrator of chosen design is based on the fact that in the course of debalance rotation a force emerges being proportional to the weight (m), eccentricity (e) and angular speed (ω) of debalance rotation:

$$F = me\omega^2 . \quad (7)$$

The vector of this force is constant by magnitude, and the variability of its direction ensures the wheelset axle rotational bending.

The vibrator has the following components (Fig. 5): 1) motor; 2) elastic coupling; 3) multiplying gear; 4) gear-type coupling; 5) debalance.

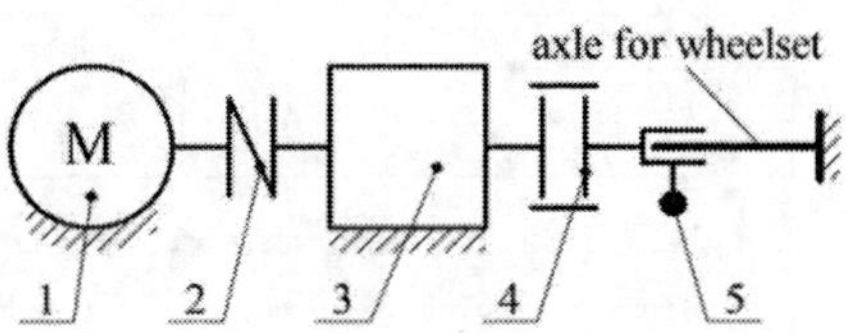

Fig. 5. Kinematic layout of the vibrator

Therewith, the multiplying gear may be optional; in such a case the duration of testing corresponding to the basis of testing [12] is significantly increased.

5 Determination of vibrator parameters

The basic element of the vibrator creating the test-bench operational force and determining its magnitude is debalance having the following parameters:

– Shape of debalance;
– Eccentricity of debalance (e);
– Weight of debalance (m);
– Angular speed of debalance rotation (ω).

The first and the last characteristics relate to the number of debalance parameters rather conditionally: the debalance shape is impossible to be measured (however, it can be assigned a sequence number), and the angular speed ω depends on the motor shaft rotational frequency and the multiplier gear ratio (considering the independence nature of these values their direct influence on the value ω is numerically ambiguous).

Let's determine rational debalance parameters subject to adherence to a certain test-bench operational force (F). The criterion of rationality is vibrator power consumption.

Weight (m) and eccentricity (e) of the debalance are linked by several dependencies:

1) Test-bench operational force; according to the formula (7)

$$F = me\omega^2 ; \quad (7^*)$$

2) Weight of debalance

$$m = \gamma V(e) , \quad (8)$$

where γ – density of the material, whereof the debalance is made; $V(e)$ – debalance volume against its eccentricity.

Upon the condition of (7*) the nature of debalance weight dependence from its eccentricity is hyperbolic. The debalance volume (and hence, the weight) are proportional to its eccentricity. Therefore, identification of dependencies $m(e)$ from expressions (7*) and (8) allows laying down parameters m and e for a debalance of any shape at a certain rotational frequency ω.

For clear presentation of the procedure of determination of debalance rational parameters let's consider the test example. Test item – axle for freight wagons. Axle dimensions required for calculation [11] (Fig. 6):

– Box journal diameter d_1 = 150 mm;
– Box journal length l_1 = 210 mm.

Thickness of debalance will be taken equal to thickness of the bearing, whereon it will be installed. Due to presence of wheelset axle scenes during the test we will use a double-row barrel-type radial roller bearing with the following parameters:

– Internal diameter $d = d_1$ =150 mm;
– External diameter D = 270 mm;
– Width B = 73 mm.

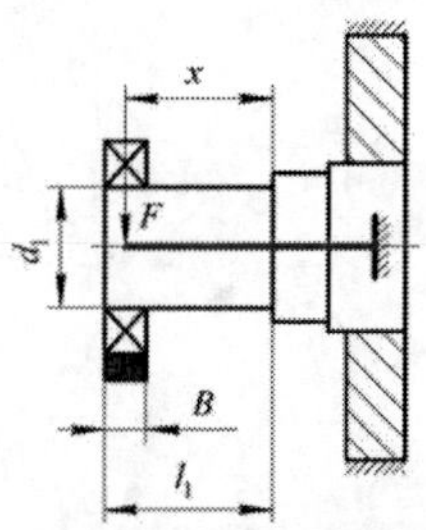

Fig. 6. Fragment of wheelset axle and loading pattern

According to the adopted designations the coordinate of the design cross-section in the loading pattern is:

$$x = l_1 - 0.5B . \quad (9)$$

According to the formula (6) test-bench operational force is:

$$F = \frac{\pi d_x^3}{32x}\sigma_{-1}, \quad (6^*)$$

where $d_x = d_1$ – design cross-section diameter (location of fillet on the box journal; see Fig. 4); σ_{-1} = 145 MPa – axle fatigue strength in the design cross-section [12].

Lumping together (6*) and (9), we will obtain:

$$F = \frac{\pi d_x^3}{32(l_1 - 0.5B)}\sigma_{-1}. \quad (10)$$

For the test example F = 274.4 kN.

From the formula (7) the debalance characteristic me (product of debalance weight and its eccentricity) is:

$$me = \frac{F}{\omega^2}. \quad (11)$$

The calculation is performed for three values of angular speed of debalance rotation provided the direct drive is used (without multiplying gear). For the test example we obtain the following characteristics of rebalances:

– n_1 = 3000 rpm; ω_1 = 314 s^{-1}; me = 2.78 kg·m;
– n_2 = 1500 rpm; ω_2 = 157 s^{-1}; me = 11.1 kg·m;
– n_3 = 1000 rpm; ω_3 = 105 s^{-1}; me_1 = 24.9 kg·m.

The use of lower values ω leads to excessive duration of testing.

Let's consider three types of debalances (Fig. 7):
a) Rectangular;
b) Circular;
c) Sectoral.

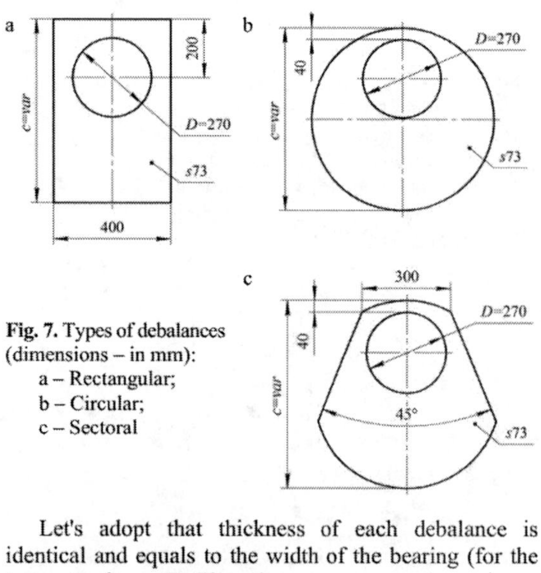

Fig. 7. Types of debalances (dimensions – in mm):
a – Rectangular;
b – Circular;
c – Sectoral

Let's adopt that thickness of each debalance is identical and equals to the width of the bearing (for the test example $s = B$ =73 mm).

Multiple-option 3D-simulation of debalances with varying of the value c ($c = var$, Fig. 7) allowed to obtain the following dependencies (m – in kg; e – in m):

– For rectangular debalance

$$m(e) = 379e + 54.77; \quad (15)$$

– For circular debalance

$$m(e) = 705e^2 + 242e + 21.6; \quad (16)$$

– For sectoral debalance

$$m(e) = 877e^2 + 338e + 20.8. \quad (17)$$

Graphic representation of functions (12)–(14) and (15)–(17) in the same coordinate plane allows to define the weight m and eccentricity e of each type of debalances that ensure the required value of the test-bench operational force. Values m and e are coordinates of the intersection points of relevant curves (Fig. 8). The results of the plot analysis are presented in Table 1–3.

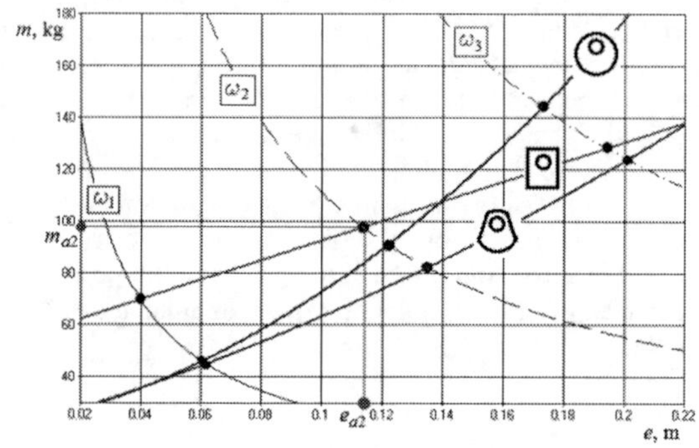

Fig. 8. Plot for determination of weight and eccentricity of the debalance

Table 1. Parameters of Rectangular debalance.

ω, s^{-1}	m, kg	e, m	I, kg·m²
ω_1 = 314	69.9	0.037	2.86
ω_2 = 157	97.2	0.114	5.89
ω_3 = 105	128.0	0.191	12.23

Table 2. Parameters of Circular debalance.

ω, s^{-1}	m, kg	e, m	I, kg·m²
ω_1 = 314	46.3	0.060	1.54
ω_2 = 157	90.7	0.119	4.89
ω_3 = 105	144.9	0.171	12.00

Table 3. Parameters of Sectoral debalance.

ω, s^{-1}	m, kg	e, m	I, kg·m²
ω_1 = 314	44.5	0.064	1.47
ω_2 = 157	81.6	0.133	4.42
ω_3 = 105	125.2	0.198	10.78

Let's determine the power consumed for debalance rotation using the formula

$$E = 0.5I\omega^2. \quad (18)$$

Results of calculations are summarized in Table 4.

Table 4. Power consumption for debalance rotation.

ω, s^{-1}	*E*, kJ, for debalance		
	rectangular	circular	sectoral
$\omega_1 = 314$	141.0	75.9	72.5
$\omega_2 = 157$	72.6	60.3	54.6
$\omega_3 = 105$	67.4	66.2	59.4

As it is obvious, the lowest are power consumption for sectoral debalance rotation at the angular speed of $\omega_2 = 157\ s^{-1}$.

The studies showed that further increasing of the debalance rotation frequency leads to a decrease of its weight, eccentricity and moment of inertia. However, due to high angular speeds of debalance rotation the magnitude of the test-bench operational force begins to be influenced by inaccuracies of fabrication and mounting of wheelset axle and vibrator elements. Thus, when using (in vibrator design) the motor with shaft rotation frequency of $n \approx 3000$ rpm and multiplying gear with the ratio $u \approx 1/15$ we will obtain $\omega = 4710\ s^{-1}$. Under such conditions the eccentricity of sectoral debalance $e = 5.9 \cdot 10^{-4}$ m, that by the order of magnitude concurs with the wheelset axle runout tolerance.

By analyzing the results obtained we come to the conclusion that by the criterion of power consumption the rational is the sectoral debalance rotating at the angular speed of around $150\ s^{-1}$. Therewith, the debalance weight is around 80 kg, and the eccentricity is 135 mm. Under such conditions the criterion specified gains the lowest value.

5 Conclusions

1. For the purpose of conducting the bench testing to determine the railway wheelset axle fatigue strength the rational is the "cantilever beam" loading pattern, which make the rationality criteria (operational forces, power consumption and test-bench overall dimension lengthwise) to gain the lowest values.

2. The test-bench operational force depends on wheelset axle dimensions (diameter and box journal length), therefore the universal test-bench has to provide the possibility of varying the operational force depending on axle nominal size.

3. By the criteria of design complexity and power consumption the best one is the vibrator that ensures the test-bench operational force application by rotating the imbalance weight (debalance) around the rigid wheelset axle. In case of using it the criteria used gain the smallest magnitude.

4. For testing the axles for wheelsets of the considered size, by criterion of power consumption the rational is the vibrator design with sectoral debalance rotating at the angular speed of around $150\ s^{-1}$. Therewith, the debalance weight is around 80 kg, and the eccentricity is 135 mm. Under such conditions the criterion specified gains the lowest value.

References

1. M. Novosad, R. Fajkoš, B. Řeha, R. Řezníček. Procedia Engineering **2**, 2259–2268 (2010). https://doi.org/10.1016/j.proeng.2010.03.242
2. A. Răduţă, C. Locovei, M. Nicoară, L. Cucuruz. WSEAS Transactions on Applied and Theoretical Mechanics **5**, 197–207 (2010).
3. M. Filippini, M. Luke, I. Varfolomeev, D. Regazzi, S. Beretta. Structural Integrity Procedia **4**, 11–18 (2017). https://doi.org/10.1016/j.prostr.2017.07.013
4. S. Raksha, P. Anofriev, V. Bohomaz, O. Kuropiatnyk Naukovyi Visnyk Natsionalnoho Hirnychoho Universytetu **2**, 108–115 (2019). https://doi.org/10.29202/nvngu/2019-2/16
5. S. Myamlin, J. Kalivoda, L. Neduzha. Procedia Engineering **187**, 688–695 (2010). https://doi.org/10.1016/j.proeng.2017.04.439
6. S. Raksha, O. Kuropiatnyk, P. Anofriev, D. Onopreychuk, I. Kovalov. MATEC Web of Conferences **230**, 01010 (2018). https://doi.org/10.1051/matecconf/201823001010
7. GOST 31373. *Wheelsets of locomotives and motor-powered units. Strength calculations and tests* (2008).
8. SU 1076816. *Ustanovka dlia ispytaniia osei kolesnykh par na ustalost pri krugovom isgibe*. A. Adiklis, et. al. (1984).
9. RU 2650327. *Stend dlia ispytaniia koles i osei kolesnykh par na soprotivlenie ustalosti*. E. Kochetkov, et. al. (2018).
10. RU 2651629. *Stend dlia ispytaniia koles i osei kolesnykh par na soprotivlenie ustalosti i sposob provedeniia ispytanii*. A. Bidulia, et. al. (2018).
11. GOST 33200. *Axles for wheelsets of railway rolling stock. General specifications*. (2014).
12. GOST 33783. *Wheelsets of the railway rolling stock. Methods of strength performance determination*. (2016).

Mathematical modelling of the communication channel between the rail circuit and the inputs devices of automatic locomotive signalization

Oksana Hololobova[1], *Serhii* Buriak[1,*], *Volodymyr* Havryliuk[1], Ihor Skovron[2], and Oleksii Nazarov[2]

[1] Dnipro National University of Railway Transport named after Academician V. Lazaryan, Department «Automatics and Telecommunications», 49010 Dnipro, Lazaryan street 2, Ukraine
[2] Dnipro National University of Railway Transport named after Academican V. Lazaryan, Department «Transport junctions», 49010 Dnipro, Lazaryan street 2, Ukraine

Abstract. In modern practice of operating under traffic safety conditions, the traffic light signal must be transmitted to the locomotive that moves to it, and duplicate in the driver's cab. However, this communication channel is not protected from external interference. In order to prevent the occurrence of code failure, it is necessary to create conditions under which the automatic locomotive signalling system will distinguish between signals with useful information, from signals with false information. The best way to solve this problem at the first stage is to model the devices. Using the simulation tools of graphical environment of simulation modelling Simulink from Matlab software environment, the software model of the communication channel between the railroad and the input devices of automatic locomotive signalling system was constructed. The created mathematical model with the actual parameters allows us to obtain diagnostic signs of a proper condition, on the basis of which the research is aimed at the identification, recognition and definition of various types of malfunctions, failure, damages and defects in the work of the constituent elements of the system and the signal transmission channel of the automatic locomotive signalling system.

1. Introduction

In modern practice of operating under traffic safety conditions, the traffic light signal must be transmitted to the locomotive that moves to it, and duplicate in the driver's cab. This problem is performed on most main lines by automatic locomotive signalling of a continuous type with numerical coding (ALSN). However, such a communication channel is not protected from external interference [1, 2], the most powerful among which are the components of the traction current, electromagnetic fields of power lines, the asymmetry of the rail circles, the zone of joints, and many others [3-6]. Also, the speed of the train, the vibration caused by the movement and the probability of failure of the outdoor units are very influential. The effect of these factors leads to crashes and failures in the system, which significantly affects the safety of trains [7, 8]. Methods of studying the principles of ALSN used in operation are not perfect. In accordance with the procedure of interaction, the locomotive brigade reports about detecting a violation in the ALSN system during traffic to employees of the traffic service, which record the case and transmit information for consideration by employees of the signalling and communication service. Employees of this service inspect the place where the codes failed, and check the voltage in the track circle. According to the operating experience, if along the track or between the rails it is the rail prepared to replace, the most probable reason for the failure of the codes is its residual magnetization, which mostly appears in the process of cargo operations using electric magnets. In most cases, the causes remain unclear. The highest reliability of the recording and analysis of various system failures is the use of a laboratory car and automated processing of the information received, but this method requires additional financial costs for the maintenance of the laboratory car and technical experts, as well as the need for a direct exit to the relevant site, which is not always convenient and an expedient way [9-11]. Taking into account the great importance of ensuring traffic safety, improving the work of ALSN in order to improve the reception of signals and reduce the impact of interferences of any origin remains relevant at present [12].

In order to prevent the occurrence of code crashes, it is necessary to create conditions under which the system will independently distinguish between signals with useful information, from signals with errors. At the first stage, it is necessary to determine the reference signals used by the system for functioning, and the signals with the most common failures in the transmission of information. The best way to solve this problem is to simulate the operation of the devices, since it allows to

* Corresponding author: ser.buryak@gmail.com

carry out the necessary number of experiments without leaving the line of rotation of the trains and to investigate the effect of all the constituent elements of the system separately, both in normal operation and under the influence of obstacles. The most convenient way to solve this problem is mathematical modelling. The mathematical model allows us to study the system in detail, since it allows us to analyze the internal processes of the constituent elements and their interaction with each other and with other objects. The advantages of simulation include repeatability, which allows conducting a sufficient number of experiments under the necessary conditions for the operation of the communication channel between the receiving devices and the rail line. Changes in the simulated signal can be investigated depending on the type and intensity of the effect on the object of the study, or on its individual components. Simulation also provides certainty that allows if necessary to take into account or not take into account their random processes. In addition, on the model example, it is possible to further develop and apply various circuitry solutions to reduce the impact of interferences on the functional capabilities of the ALSN system and the introduction of an automated complex to recognize the type of fault [13-15].

Let's conduct research of ALSN work by constructing a mathematical model of the communication channel of its input devices with a rail line.

2. Structure and system devices

All devices that are part of the ALSN can be divided into tracks (transmit the code signal) and locomotive (accept the code signal). Each pass-through traffic light (TL) in a relay locker installed with a track coding equipment consisting of a coding scheme (CS), a code track transmitter (CTT), a transmitter relay (kvT) having its own contact in the primary circuit of the code transformer (CT), the secondary winding of which is connected to the track circuit through the choke-transformer CT1 [16]. The transmitter converts the signal indications of the traffic light into the corresponding combination of the numerical pulse code, which consists of a certain number of pulses and pauses of different durations (fig. 1) [16].

In the event of a short circuit of the insulating joints, the coded signals from the neighbouring block section get into the track circuit, which threatens the safety of trains. By alternating the transmitters at each signal unit of the route, the necessary shift in the passage of code pulses in adjacent rail circuits and protection from the influence of code signals from adjacent block sections are achieved.

In the composition of locomotive devices ALSN (fig. 2) [16] includes receiving coils (RC), filter (F), locomotive amplifiers (LA) with pulsed relay (P), decoder block (DB), electropneumatic valves of autostop (EPV), locomotive traffic lights (LTL), locomotive speedometer (SM), arm (button) of vigilance (BV), button (BR) for ignition on a locomotive white traffic light instead of red, as well as a tumbler (switch) DZ to change the interval of time periodic vigilance of the machinists [16].

For the unimpeded passage of the traction current, by passing the insulating joints installed chokes-transformers (CT). They are used to transmit the track equipment ALSN code signals to the rail circuit, in additional CT preventing the flow of signal current between adjacent block sections [17, 18].

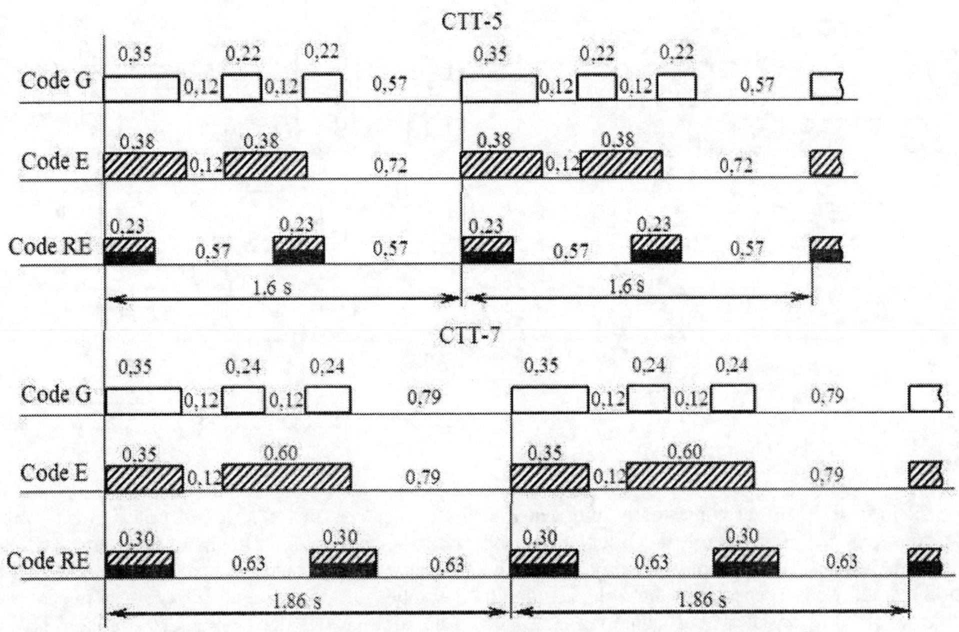

Fig. 1. Time characteristics of numeric codes produced by transmitters of the CCT-5 and CCT-7 types

The ALSN track devices code current I_c on one of the rail threads is directed toward the locomotive, locked through its first wheelset and returns to the power source on the second rail thread (fig. 3) [16]. The leakage of the alternating current pulses in the rails is accompanied by the formation around the rails of the alternating magnetic field F_c (F_{cl} – around the left rail, F_{cr} – around the right), which permeates the receiving coils and induces an EMF in them (e_{kl} – in the left coil, the e_{kr} – in the right coil), in which moves the locomotive receiving coils that are suspended in front of the first wheel pair on each side [16]. The direction of the code currents I_c in each of the rails is opposite (in the left rail the current is directed to "us" and is conventionally indicated by a point, and in the right one - "from us" and is shown by a cross), therefore the induced by these currents EMF (e_kl and e_{kr}) also have a counter direction. The height of installation of receiving coils over the level of the head of the rail is 150 – 220 mm. Power lines of the magnetic field, crossing the turns of the RC, give them a alternating EMF, the value of which depends on the value of the code current in the rails and the height of the installation of coils. So, at the height of the location of the RC above the level of the head of the rail 150 mm and the code current in the rails 10 A 50 Hz in the coil, respectively, is given by an EMF, the value of which is approximately 0,65 – 0,75 V.

To add EMF of both coils they are included sequentially. Two connected successively receiving coils at a frequency of 50 Hz should have a Q-factor of 3,5 – 4,0 and an inductance of 14,0 Gn EMF which is given in open and disconnected from the control panel coils, with a current in the line 10 A must be 1,3 V. The minimum code current that can be received by the receiving coils for different types of traction and kind of current is from 1,2 A to 2,0 A [17].

It is necessary to take into account that simultaneously with the signal I_C, the traction current I_T flows through the main threads, which, passing through the main winding of the CT, forks and flows along each rail line in one direction. In this case, the traction current creates a magnetic field around each rail, covering the receiving coils of the locomotive and directing EMF in them (e_{tl} – in the left coil, e_{tr} – in the right coil). The directions of these EMF are the same, since the directions of the traction currents that created them are the same. With respect to the traction current, the coils are inversely connected, therefore, e_{tl} and e_{tr} are mutually destroyed one other and not interfere with the reception of the signal current I_k. If there is a so-called longitudinal asymmetry of the traction current (the characteristics of the left rail thread do not coincide with the characteristics of the right one), the values of the e_{tl} and e_{tr} induced in the receiving coils are different and full compensation of these EMF does not occur, thus there is an interfering effect of the traction current on the ALSN devices which may cause the system to malfunction and endanger the safety of trains.

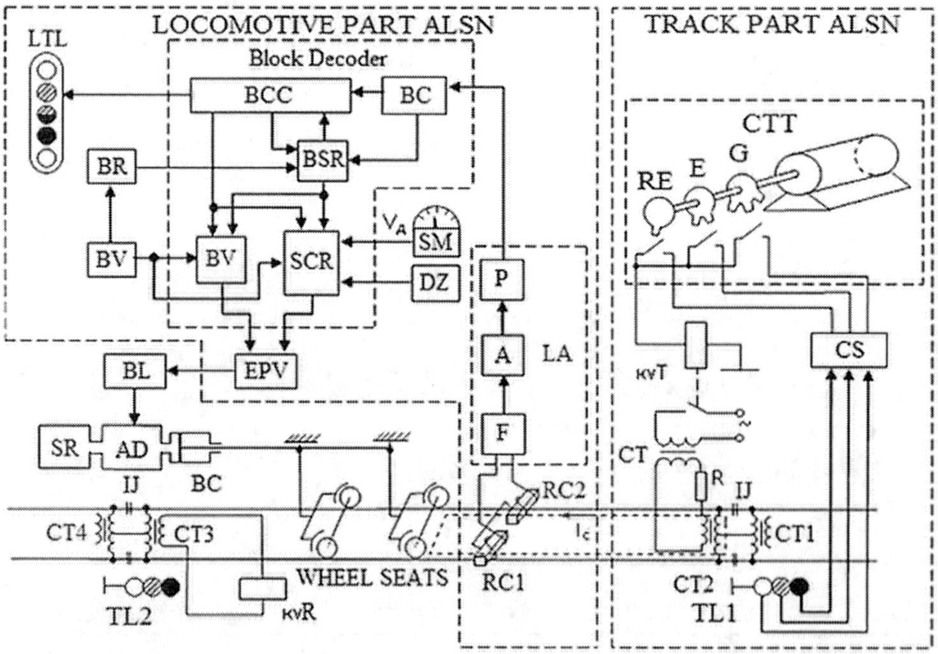

Fig. 2. Block diagram of ALSN

Listed in the RC EMF act consistently and generate the code current I_k passing through the filter (F) to the locomotive amplifier (LA). The pulsed relay (P) at the output of the amplifier is a repeater of the code, sending it to the decoder (BD) as an encrypted signal reading. The decoder contains a series of relays, which are combined into several blocks. Blocks of the counter (BC) – includes relay-counters, which provide the calculation of the number of pulses and intervals between them. Block signal relay (BSR) – includes signal relay "G" (green light), "E" (yellow light), "RE" (red-yellow light), which create corresponding circuits of power supply of signal lamps of locomotive traffic light. Controlling Conformity Block (BCC) – ensures control over the compliance of the accepted code and status of the BSR signal relays by comparing them. The Conformity Block periodically after 5 – 6 s connects the signal relays to the relay-counters, in order for the locomotive traffic light to turnthe desired fire. Thus, the change in the lights of the locomotive traffic light occurs with a delay of 5 – 6 seconds. This time corresponds to the reception of three series of code pulses. Speed control unit – contains a speed control relay (SCR) that interacts with a locomotive speedometer. Block of vigilance (BV) – carries out control of vigilance of the machinist. When changing the light of a locomotive traffic lights, for example from green to yellow, the electric circuit of the coil of the electropneumatic valve (EPV) ruptures and an acoustic signal sounds for 7 to 8 seconds. By the end of this time, the machinist must press the lever (button) of vigilance (BV) and thereby restore the supply circle of the EPV reel and stop the sound of the whistle [3].

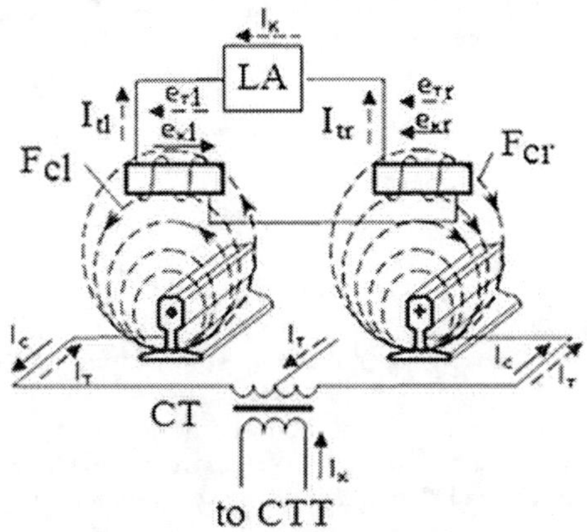

Fig. 3. Current direction in the rails and the receiver coils

If this does not happen, at the end of 7 – 8 s the EPV breaks and releases air from the brake line (BL), which leads to a redistribution of air pressure in the air distributor (AD) and the combination of spare reservoirs (SR) with brake cylinders (BC). In this situation, the train enters the emergency braking mode.

3. Mathematical model

To understand the processes taking place in the ALSN system and constructing a mathematical model of the communication channel between the input devices of the system in the form of an equivalent circuit for replacing the input filter and the track circle through which the signal current flows through the tools of the MATLAB software environment and the Simulink extension, let's construct a mathematical model [19, 20]. To do this we use the most visual and easy to perceive the version of the representation of the filter scheme at a frequency of 50 Hz (fig. 4) [17].

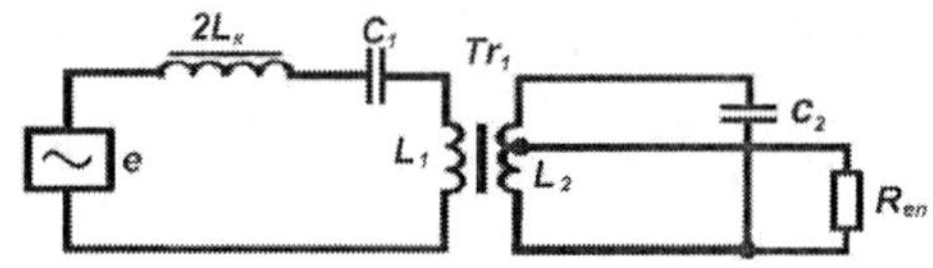

Fig. 4. Filter pattern 50 Hz

The first circuit includes the inductance of the two receiving coils, as well as the inductance of the coupling L_1 formed by the input circuit of the transformer Tr_1 and the C_1 capacitor. The Q-factor of this contour is 3-4. Also, this circuit includes a source e, which generates ALSN signals. The second contour is made by another winding of a transformer with inductance L_2 and a capacitor C_2 (these filter elements are located in the amplifier block). The quality of this contour is equal to 10. From the part of the winding of the transformer Tr_1, the received signal is fed into the first cascade of an amplifier with a load. Due to the inductive coupling and the setting of the contours in the resonance, they create a filter that passes with only a certain frequency band, which is a collection that has filtering properties. Such a filter is electrically equivalent to two consecutively connected circuits: a sequence of elements of the first circuit (receiving coils and capacitor C_1) and a parallel circuit with an inductive resistance communication and a C_2 capacitor. This system of contours has two frequencies of a successive and one of a parallel resonance.

The filter has a bandwidth of approximately 14 Hz and suppresses interference with a frequency below 10 Hz induced by the fluctuations of the receiving coils in the magnetic field of the traction current during the movement of the locomotive as well as by the influence of harmonics with a frequency of 100 Hz and above that appear under influence of contact network.

We created the mathematical model using the following blocks (fig. 5):

Scope – oscilloscope;

AC Voltage Source – Variable Voltage Source;

Ideal Switch – switch;

Current Measurement – ammeter;

2Lk – inductance of two receiving coils;

C1, C2 – capacitors;

Tr1 – transformer;

Rvx – load of the amplifier.

Signal Builder – block for generating the following codes:

G – generator of the green light code (G);

E – code generator for yellow light (E);

RE – red-yellow code generator (RE).

The parameters of these generators, create combinations of intervals and pulses of a certain duration that accurately reflect the combinations of the corresponding code signals are shown on the oscilloscope (fig. 6). Electrically, according to the current technical conditions, the receiving coil is characterized by the following parameters: active resistance to alternating current with frequency 50 Hz more than 650 Ohms, inductance 7,1 ± 0,35 Gn, Q-factor 3,5.

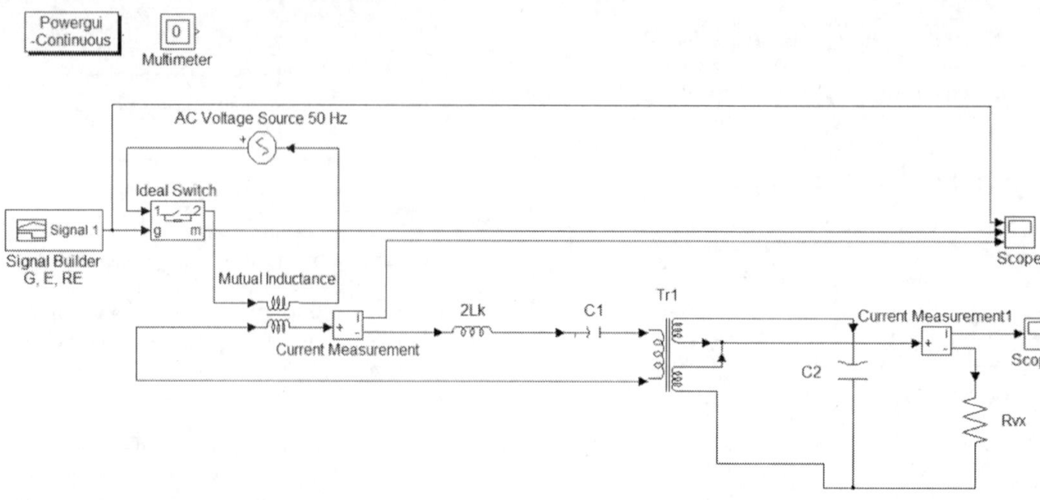

Fig. 5. Mathematical model of input devices ALSN

The filter transformer and its capacitors are included in the design of the amplifier itself (type УК25/50М). The primary winding I (820 turns) is included with the capacitors C1 (МБГП type, 0,75 μF ± 5%, 200 V) in the outlet of the receiving coils. The secondary winding II (1700 turns with taps from 60, 75, 90 and 105 turns) has an additional sectional winding III (161 turns with taps from the 92nd and 115th turns), with the help of which the second contour with the C2 capacitors (capacity 4 μF ± 5%, 200 V) is tuned to resonance at a frequency of 50 Hz. The magnet wire of the transformer has a fixed air gap of 0,9 mm to stabilize the inductance of its windings.

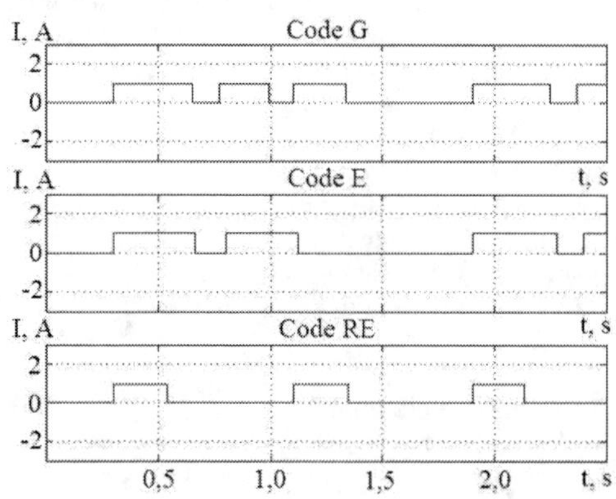

Fig. 6. Code signals G, E, RE

Unlike the specified parameters of the ALSN system, the mutual inductance between the rails and receiving coils must be calculated. In order to determine the value of the mutual inductance of the signal transmission channel to the locomotive, on first stage it is necessary to perform measurements of the voltage and current in the receiving coils using an ammeter [21]. The values obtained in normal conditions for the coil are I=0,00818 A and U= 19,25 V. Next, let's take a look a signal transmission channel which contains of a rail line, an ALSN receiving coil, and an ALSN code device. In order to make it possible, it is necessary to determine the active specific resistance of the rack thread to the alternating current by the formula (1)

$$R_a = \frac{2,8 \cdot \sqrt{\mu \cdot \rho \cdot \varphi}}{u} \quad (1)$$

where $\mu = 100$ – relative magnetic permeability of the track circle;

$\rho = 0.21 \cdot 10^{-6}$ Ohm · m – resistance of the metal of the track circle;

$\varphi = 50$ Hz – frequency of the current of the rail circle;

u – the perimeter of the cross section of the rails.

Then, substituting the values of the parameters for the formula (1) we obtain the necessary value:

$$R_a = \frac{2,8 \cdot \sqrt{100 \cdot 0.21 \cdot 10^{-6} \cdot 50}}{0,070} = 1,296 \text{ Ohm}.$$

The next step is to determine the internal specific inductance of the track circle by the formula:

$$L_i \approx 0.0955 \cdot \frac{R_a}{\varphi} = 0.0955 \cdot \frac{1,296}{50} = 4.947 \cdot 10^{-4} \text{ Hn}$$

Let's substitute the obtained values for the formula for determining the mutual inductance and calculate its value:

$$M = \frac{1}{2\iota\omega} \cdot \left[\frac{U}{I} - R_a - R_к \right] + \frac{1}{2} L_i + \frac{1}{2} L_к \quad (2)$$

where $L_к = 7,1$ Гн – inductance of the ALSN coil;

$R_к =$ 110 Ohm – resistance of the ALSN coil.

$$M = \frac{1}{2 \cdot 0,7 \cdot 2 \cdot 3,14 \cdot 50} \cdot \left[\frac{19,25}{0,00818} - 1,296 - 110 \right] +$$
$$+ \frac{1}{2} \cdot 4,947 \cdot 10^{-4} + \frac{1}{2} \cdot 7 = 5,11 \text{ Hn}$$

4. Results

The time dependencies of the current change at the filter output when the system input codes G, E, and RE are obtained using Matlab (fig. 7, fig. 8, fig. 9) [22, 23].

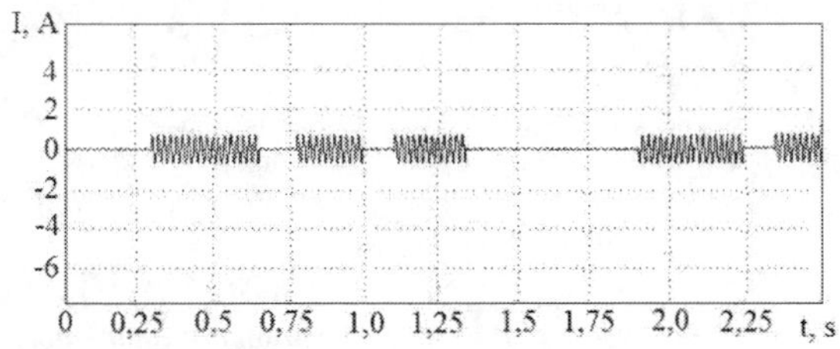

Fig. 7. Coded signal G at the filter output

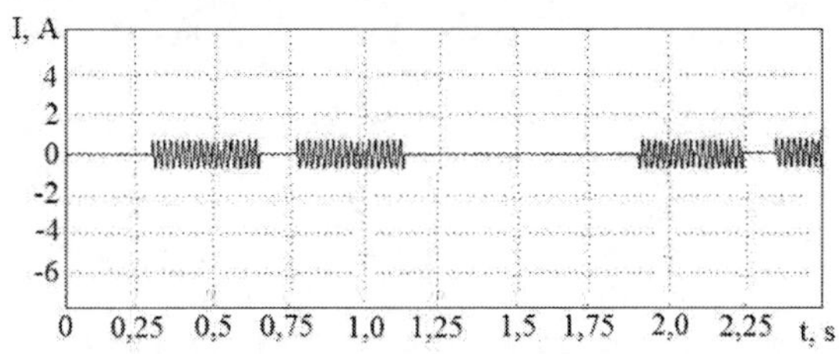

Fig. 8. Coded signal E at the filter output

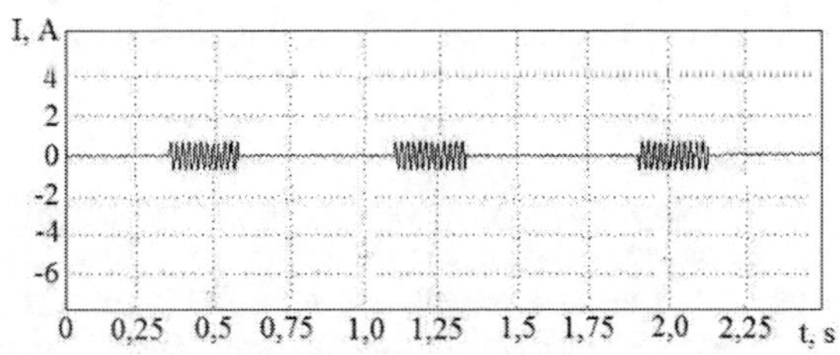

Fig. 9. Coded signal RE at the filter output

For comparison, we give an example of an image of ideal code signals measured by a laboratory car (Fig. 10, Fig. 11, Fig. 12).

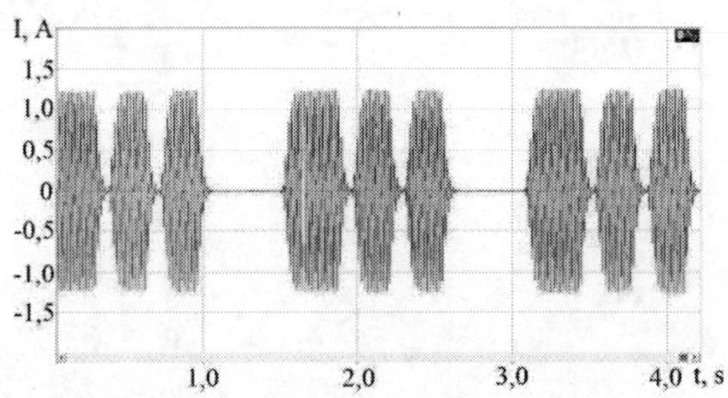

Fig.10. Coded signal G

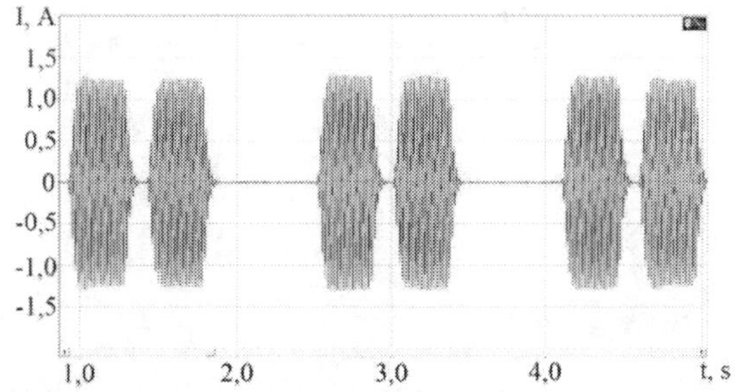

Fig. 11. Coded signal E

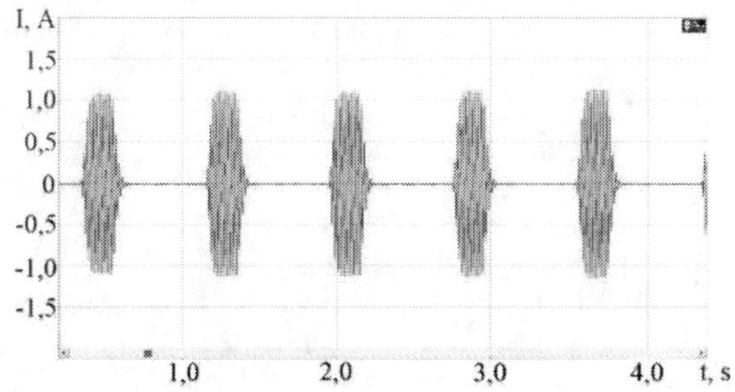

Fig. 12. Coded signal RE

From the results of the simulation, it can be seen that the developed model reproduces the ideal signal in real-world devices.

According to the work results of the mathematical model of the communication channel between the input devices of the system of automatic locomotive signalling of a continuous type with numerical coding in the form of an equivalent circuit for replacing a filter with a frequency of 50 Hz and a track circle, on which the signal current flows, it can be established that the signal at the output of the model satisfactorily coincides with real measurements, which testifies to the adequacy of the built system.

5. Conclusions

On the basis of the analysis of the structure and functioning of the devices of the ALSN system, a mathematical model of the communication channel of its input devices with the ALSN rail line was developed taking into account the mutual inductance between the rails and the locomotive receiving coils. The results of modeling at the filter output are presented at the input of all types of code combinations. Since the signals obtained from the model, and measured from the laboratory car, agree satisfactorily, then the developed model of input devices of the receiver ALSN corresponds to its real analog. This makes it possible to experimental study with the ALSN in the conditions of interference and distortion of the code signal, as well as research and determination of the behavior of the circuit during normal operation and in the event of interference without leaving the area of interference in the virtual environment of the mathematical package MATLAB and its extension Simulink.

Thus, the ability to simulate the emergence of probable complex situations that lead to code crashes and the development of ways to detect, prevent and protect the action of devices from them was created. There is also the possibility in MATLAB and Simulink to develop and apply various circuitry solutions to reduce the impact of interference on the functionality of ALSN.

References

1. V. Gavrilyuk, T. Serdyuk, Telemetric system for the control of signal parameters of rail circuits. Transport systems telematics. II International Conference. Katowice-Ustron (Poland), 185-190 (2002)
2. V. Gavrilyuk, A. Zavgorodnyj, Computer simulation of electromagnetic interference from railway electric power system harmonics. Archives of transport system telematics, Vol. **2**, № **1**, 33-37 (2009)
3. Vl. V. Sapozhnikov, I. M. Kokurin, V. A. Kononov, A. A. Lykov, A. B. Nikitin *Jekspluatacionnye osnovy avtomatiki i telemehaniki: Uchebnik dlja vuzov zh.-d. transporta.* M., Marshrut, 247 (2006) [in Russian]
4. T. E. Kijakina, *Prichiny sboev v rabote avtomaticheskoj lokomotivnoj signalizacii, metody reshenija problem* Tehnicheskie nauki v Rossii i za rubezhom: materialy II mezhdunar. nauch. konf., 47-49 (2012) [in Russian]
5. V. L. Grigor'ev, A. V. Kotel'nikov, *Optimizacija jelektroprovodnosti rel'sovogo styka pri propuske tjazhelovesnyh poezdov*. ASI **8**, 13-16 (2005) [in Russian]
6. V. I. Shamanov, Ju. A. Trofimov, *Assimetrija tjagovyh tokov pod katushkami.* ASI **11**, 37-39 (2008) [in Russian]
7. S. V. Lukojanov, *Sboev kodov ALSN na skorostnom uchastke stalo men'she.* ASI **9**, 22-25 (2011) [in Russian]
8. I. N. Sheverdin, V. I. Shamanov, *Vlijanie tjazhelovesnyh poezdov na rel'sovye cepi i ALS,* ASI **8**, 24-29 (2004) [in Russian]
9. E. V. Gorenbejn, *Sboi kodov ALSN i ih uchet.* ASI **7**, 8-22 (2012) [in Russian]
10. V. B. Leushin, *Analiz prichin sboev v sisteme ALSN.* ASI **4**, 20-25(2013) [in Russian]
11. S. V. Lukojanov, *Sboev kodov ALSN na skorostnom uchastke stalo men'she.* ASI **11**, 34-36 (2011) [in Russian]
12. N. N. Baluev, *Razvitie sredstv ZhAT. Strategija i taktika.* ASI **8**, 2 (2012) [in Russian]
13. S. Ju. Burjak, V. I. Gavriljuk, O. A. Gololobova, *Issledovanie vremennoj zavisimosti i spektral'nogo sostava signala v cepi strelochnyh jelektrodvigatelej peremennogo toka. Nauka i progress transporta.* Vestnik DNUZhT. D., DNUZhT, **54**, 7-22 (2014) [in Russian]
14. S. Ju. Burjak, V. I. Gavriljuk, O. A. Gololobova, M. A. Kovrygin, *Distancionnoe diagnostirovanie sostojanija strelochnyh perevodov po vremennoj harakteristike i spektral'nomu sostavu tokovoj krivoj.* Nauka ta progres transportu. Vestnik DNUZhT. D., DNUZhT, **56**, 7-24 (2015) [in Russian]
15. A. M. Safarbakov, A. V. Luk'janov, S. V. Pahomov, *Osnovy tehnicheskoj diagnostiki detalej i oborudovanija: uchebnoe posobie.* Irkutsk, IrGUPS, 128 (2007) [in Russian]
16. S. V. Shvecov, A. S. Talyzin, N. V. Esin *Metodicheskie ukazanija k vypolneniju laboratornyh rabot pri izuchenii discipliny «Bezopasnost' vozhdenija poezdov»* Omsk, 48 (2010) [in Russian]
17. N. F. Kotljarenko, A. V. Shishljakov, Ju. V. Sobolev, I. 3. Skripin, N. A. Shishljakin, *Putevaja blokirovka i avtoregulirovka: uchebnik dlja vuzov.* M., Transport, 408 (1983) [in Russian]
18. G. Theeg, S Vlasenko, *Railway Signalling and Interlocking. International Compendium. Hamburg, Eurail-press Publ.*, 448 (2009)
19. O. Beucher, *Introduction to MATLAB & Simulink: a project approach. Hingham, Massachusetts New Delhi, Infinity Science Press LLC*, 390 (2008)
20. I. V. Chernyh, *Modelirovanie jelektro-tehnicheskih ustrojstv v MATLAB SimPowerSystems i Simulink,* M., DMK Press, 288 (2008) [in Russian]
21. L. A. Bessonov, *Teoreticheskie osnovy jelektrotehniki.* M., Vysshaja shkola, 515 (1996) [in Russian]
22. L. F. Chaparro, *Signals and Systems Using MATLAB Dep. of Electrical and Computer Eng. Univ. of Pittsburgh. Amsterdam, Boston, Heidelberg, London, New York, Oxford, Paris, San Diego, San Francisco, Singapore, Sidney, Tokyo: Academic Press*, 752 (2011)
23. R. J. Tervo *Practical Signals Theory with MATLAB Applications, Hoboken, New Jersey, Wiley*, 486 (2014)

Method of railway line audit in terms of the condition of railway infrastructure

Magdalena Bartoś, and *Piotr* Gołębiowski*

Warsaw University of Technology, Faculty of Transport, 00662 Warsaw, Poland

Abstract. The most important problems concerning the organization of railway traffic is continuous provision and systematic improvement the level of safety of services provided by railway transport enterprises. This safety results from many factors – including condition of the infrastructure, state of superstructure or the way of conducting railway traffic. This article covers railway infrastructure within its scope. In this regard include, among others take appropriate actions to monitor its condition. One of the methods may be the audit of the railway line in terms of infrastructure condition. The purpose of the article is to present the method of railway line audit in the area of infrastructure. The concept of railway line audit from the point of view of infrastructure status was defined in the paper. The audit method was presented along with the identification of problems that may occur as a result of the audit and suggestions for solutions to these problems. The presented method was verified for a selected railway line in Poland - railway line No. 12 Skierniewice – Łuków. The possibilities of improving the condition of this line were proposed together with the estimation of the amount of expenditures necessary to make therefore and time gains achieved through improvement.

1 Introduction

Nowadays, rail transport is the safest type of land transport. The main factors influencing this fact should include, among others [1]:
- technical condition of the railway infrastructure,
- rolling stock technical condition [2, 3],
- technical condition of control command and signalling equipment [4],
- traffic organization [5],
- employees' competences,
- culture of public safety [6].

However, as in any transport mode, there may be a risk of a dangerous situation that can be caused by various factors. It can be classified into two groups:
- technical and organizational factors in which the railway system itself is the source of problems (e.g. bad condition of the track surface, inefficient rolling stock or control command and signalling equipment),
- external factors, caused by problems unrelated to the railway system (e.g. as a result of third-party activities or as a result of atmospheric phenomena).

In order to minimize the possibility of occurrence of any of the factors causing the emergence of danger, it is necessary to systematically carry out research of elements that may be exposed to their operation. The results of these research indicate the irregularities in the system. The main entities conducting research are infrastructure managers and railway undertakings. The mentioned units are responsible, among others, for safe operation of railway transport and railway traffic by exercising control over railway infrastructure. They have a huge impact on the risk in the railway system, which is why their representatives can quickly and effectively identify problems.

Research can be carried out in many ways. They include, among others, continuous control of individual units' data, analysis of the causes and effects of railway accidents [7-9], taking into account required safety standards and improving employee qualifications. Noteworthy is also the direct way of conducting research, i.e. inspection of individual elements of the railway transport system on the ground. A total of research methods to identify problems and where they occur can be described as an audit.

Audits can affect various areas of the railway transport system - including the organization of railway traffic [10, 11]. Therefore, in order to properly organize railway traffic and constantly improve the quality of services provided, it is necessary to systematically check the condition of the railway line, i.e. carry out its audit. This process allows an in-depth analysis of problems threatening the safety of railway traffic, as well as identifying factors that negatively affect, among others, for the time of travel. The results of the audit also allow estimating the amount of expenditures necessary to improve the condition of the line.

The purpose of the article is to present the method of railway line audit in the area of infrastructure. The concept of railway line audit from the point of view of infrastructure status was defined in the paper. The audit method was presented along with the identification of

* Corresponding author: pgolebiowski@wt.pw.edu.pl

problems that may occur as a result of the audit and suggestions for solutions to these problems. The presented method was verified for a selected railway line in Poland - railway line No. 12 Skierniewice – Łuków. The possibilities of improving the condition of this line were proposed together with the estimation of the amount of expenditures necessary to make therefore and time gains achieved through improvement.

2 Audit of the railway line

As mentioned in the introduction, the purpose of the article is to present the method of audit of the railway line from the point of view of the infrastructure condition. There is no explicit definition of railway line audit in the literature. One can only meet many definitions of the audit and the railway line.

In the literature, the concept of audit appears in a different context. The first of these concerns the verification of safety management systems (SMS), the implementation of which is required, among others, infrastructure managers [12-14]. The audits also concern activities related to technical maintenance of control command and signalling equipment [15]. Another context is the design of property security systems for travellers [16]. An important group of audits are also energy [14, 17, 18], organization [19], environmental [20-24] and financial audits [25]. Audit also plays a very important role in maintenance systems that are used on the railways [26-29]. It should be noted that systematic audits allow to maintain a safety culture in rail transport [7]. Interesting are tourist audits, whose area of interest is the functioning of railways in regions attractive for tourists [30].

Based on the considerations and the analysis of literature [31-40] the definition of railway line audit was developed. It reads as follows: **railway line audit is a professional, independent and objective examination of the railway network element together with the necessary infrastructure intended for railway traffic conduction in order to determine whether its condition is correct due to the quantitative and qualitative criteria.**

Quantitative criteria determining the condition of a railway line allow for a specific and actual characterization of a given quantity described in numbers, e.g. speed. On the other hand, the quality criteria are characterized by a lack of measurability and an independent character. These can include, among others, subjective assessment of passenger comfort.

Therefore, the audit of the railway line consists of characterizing the selected railway line in order to identify the problems encountered, and then on presenting proposals to solve them after the initial analysis of the time gains and the calculated amount of expenses related to the repair of identified problems. The introduced solutions are primarily aimed at improving railway traffic safety and railway infrastructure security.

Fig. 1 presents a general scheme of the audit of the railway line, which can be presented in the following form:

- **STEP 1** – adoption of the value of the variable defining the direction of the railway line k ($k \in \mathbf{K}$) equal to 0 (**K** – set of directions of the railway line),
- **STEP 2** – increasing the value of the variable determining the direction of the railway line k ($k \in \mathbf{K}$) by 1 in order to consider all directions,
- **STEP 3** – checking if the value k ($k \in \mathbf{K}$) determined in step 2 is less than or equal to the number of sets of directions; if so then go to step 4; otherwise the end of the algorithm's work is finished,
- **STEP 4** – adoption of the value of the variable defining the problem number n ($n \in \mathbf{N}$) equal to 0 (**N** – set of problems regarding the state of infrastructure),
- **STEP 5** – increasing the value of the problem number n ($n \in \mathbf{N}$) by 1 to consider all problems on the railway line,
- **STEP 6** – checking if the value of n ($n \in \mathbf{N}$) determined in step 5 is less than or equal to the number of problems occurring on the railway line; if so, go to step 7; otherwise, go to step 2,
- **STEP 7** – adoption of the value of the variable defining the kilometre value of the railway line km ($km \in \mathbf{KM}$) equal to 0 (**KM** – the set of kilometres constituting the railway line),
- **STEP 8** – increasing the value of the variable determining the kilometre value of the railway line km ($km \in \mathbf{KM}$) by 0.001 in order to consider all the railway lines in succession,
- **STEP 9** – checking if the value of km ($km \in \mathbf{KM}$) set in step 8 is less than or equal to the final kilometre value of the railway line; if so, you can go to step 10; otherwise, go to step 5,
- **STEP 10** – checking if the problem n ($n \in \mathbf{N}$) found in a given kilometre on the railway km ($km \in \mathbf{KM}$) was found; if so, go to step 11, otherwise go to step 8,
- **STEP 11** – record of a specific problem n ($n \in \mathbf{N}$) occurring at a specific kilometre on the railway km ($km \in \mathbf{KM}$); then go back to step 5.

In summary - the audit of the railway line consists in reviewing the railway line in each direction. For each problem identified in item 3 of the article, a railway line is viewed meter by meter if a given problem occurs.

3 Identification of problems possible to occur as part of the audit of the railway line

As mentioned in point 2 of the article, the audit of the railway line consists of a review of each meter of line for each direction of the railway line for the occurrence of individual problems. Therefore, in order to conduct an audit of the railway line, problems that may arise as a result of the audit should be identified.

Typical problems regarding the condition of the railway infrastructure that can be identified as part of the audit of the railway line are included in Table 1.

All the problems listed in Table 1 belong to the quality criteria for assessing the condition of railway infrastructure. The effect of most of them is to reduce the speed of trains, which is directly related to the lengthening their travel time.

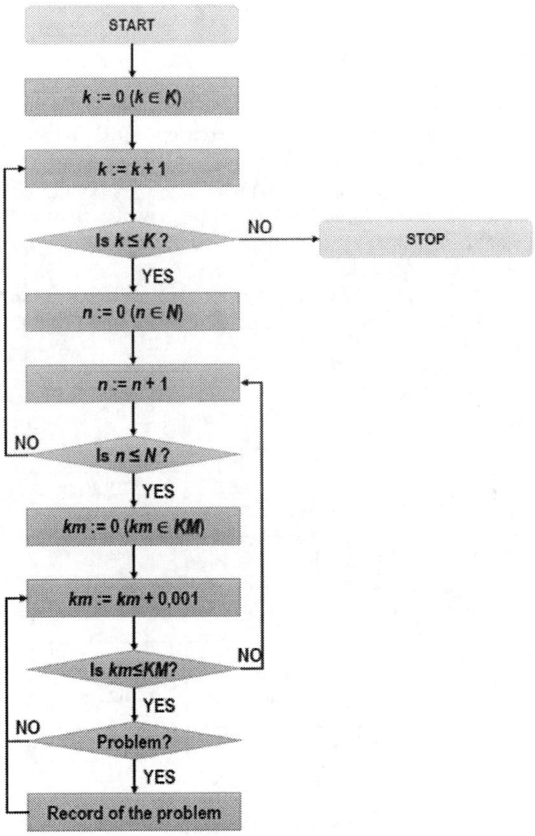

Fig. 1. The course of the railway line audit (source: own study)

Table 1. Typical problems regarding the condition of railway infrastructure (source: own study based on [35, 41, 42])

No	Problem	Nature of the problem	Impact on railway traffic
1	bad technical condition of the track		
a	cracked rail	damage of stretch of rails	reducing train speed or closing the track and stopping train traffic
b	crookedness and deformation of the track	deviations from the correct track construction system with the possibility of continuously increasing the clearances	reducing train speed or closing the track and stopping train traffic
c	damaged turnout	damage to any of the turnout elements, incorrectly mounted turnout on the switch sleepers, defective connection of turnouts	switching off the turnout from operation and immobilizing the switch in the position for authorized driving, closing the turnout for rolling stock driving, limiting the speed of trains
d	bad technical condition of the substructure of the track	settlement of the track, track bedding, damage to slopes or side ditches, obstruction of culverts	reducing train speed or closing the track and stopping train traffic
2	sudden track failure	unforeseen problem	closing the track and stopping train traffic
3	wrong geometric layout of the track	too short straight track section after the obtuse crossing of turnout, too small radius of rounding arch	reducing the speed of trains on a given section
4	inadequate ballast layer	fast abrasion of ballast, failure to ensure the stability of sleepers, inappropriate surface water drainage	reducing train speed or closing the track and stopping train traffic
5	turnout technical parameter	wrongly selected radius or slant of the turnout	reducing the speed of trains on a given turnout
6	small radius of the arch	radius of the arch not adjusted to the current speed	reducing the speed of trains on a given arch
7	damaged control command and signalling equipment and systems	damage of track side and linear equipment	reducing the frequency of trains on the open line, reducing train speeds
8	damaged traffic protection devices at road-crossings	damage of track side and linear equipment	reducing the speed of trains on a given section
9	damaged telecommunications devices		
a	faults in wired communication devices	damage of track side and linear equipment	introduction of non-interlocked block and speed limit on road-crossings up to 20 km/h, extension of dispatch of train times at operating control points
b	faults of radio communication devices	damage of track side and linear equipment	temporary trains announcing using general telecommunication devices or mobile telephony and speed limit on road-crossings up to 20 km/h, extension of dispatch of train times at operating control points
c	GSM-R faults	damage of track side and linear equipment	reducing the speed of trains, operating of trains traffic based on control command and signalling equipment and systems of the base layer
10	damaged energy devices		
a	damaged telecommunica-tions devices	damage of track side and linear equipment	communication break, impact on traffic as for case 9a or 9b
b	interruptions in the supply of electricity	lack of electricity in the catenary	total break in the train traffic

No	Problem	Nature of the problem	Impact on railway traffic
c	catenary faults	damage to the catenary	closing of the traffic for electric traction, running of trains with diesel traction, depending on the scale of damage - conducting one track two-way traffic on the second active track
11	carried out track or road works	passing the train on neighbouring track with limited speed	reducing train speed or closing the track and stopping train traffic
12	exceeded the exposure factor on the road-crossing	too high road traffic volume in relation to the volume of railway traffic	reduction of train speeds to 20 km/h at road-crossing
13	damage of complete long-arm gate	device failure	reduction of train speeds to 20 km/h at road-crossing
14	no service at road-crossings	no devices on the road-crossing or devices are working but they are not running	reduction of train speeds to 20 km/h at road-crossing
15	temporary crossing for pedestrians	pedestrian crossing in the level of rails created temporarily	reduction of train speeds to 20 km/h at pedestrian crossing
16	occurrence of crisis or emergency situations	unforeseen problems	closing the track and stopping the trains traffic or limiting the speed of trains
17	no service at the station	closing the station for technical and operational needs, no devices at the station or devices are working but are not activated	possible speed limit for trains
18	the impact of weather conditions	lowering the embankment, equipment failure due to lightning, flood, deformation etc.	speed reduction or temporary suspension of train movement until the obstacle is removed

They cause delays in the running of passenger trains and longer delivery times for goods in freight traffic. As we know, one of the most important factors in choosing the type of transport for passengers and customers is travel time. In order to shorten it significantly, the maximum speed should be increased and the restrictions on the railway line should be eliminated.

4 Determining ways to solve identified problems

In railway transport, the solution of problems in the area of railway infrastructure can be twofold: revitalization or modernization of damaged elements. In order to abolish conditions on railway lines, the exact definition of both processes should be familiarized in order to correctly assign them to the listed problems.

Revitalization is a comprehensive and in-depth set of activities carried out to improve infrastructure conditions. The purpose of the performed activities is to restore the value and significance of degraded areas and objects (in the event of losing their original functions) to improve the economic, social and environmental situation.

Modernization [43] includes the execution of construction works in order to modernize and improve the infrastructure, striving to increase its utility value.

Therefore, revitalization, in contrast to modernization, which assumes improvement of the railway line parameters and increasing the permissible speed of trains, is only a reconstruction of the original condition of the railway line.

Problems on the railway lines identified in point 3 can be solved thanks to the proposed solutions contained in Table 2.

Table 2. Proposals for solutions of the problems related to the condition of railway infrastructure (source: own study)

No	Problem	Proposals for solutions		
		Revitalization	Modernization	Others
1	bad technical condition of the track			
a	cracked rail	X	-	-
b	crookedness and deformation of the track	X	-	-
c	damaged turnout	X (repair or replacement for the same)	X (replacement on a different type)	-
d	bad technical condition of the substructure of the track	X	-	-
2	sudden track failure	X	-	-
3	wrong geometric layout of the track	-	X	-
4	inadequate ballast layer	X	-	-
5	turnout technical parameter	-	X	-
6	small radius of the arch	-	X	-
7	damaged control command and signalling equipment and systems	X (repair)	X (replacement)	-
8	damaged traffic protection devices at road-crossings	X (repair)	X (replacement or assembly of new ones)	-
9	damaged telecommunications devices			
a	faults in wired communication devices	X (repair)	X (replacement or assembly of new ones)	-
b	faults of radio communication devices	X (repair)	X (replacement or assembly of new ones)	-
c	GSM-R faults	X (repair)	X (replacement or assembly of new ones)	-

No	Problem	Proposals for solutions		
		Revitalization	Modernization	Others
10	damaged energy devices			
a	damaged telecommunications devices	X (repair)	X (replacement or assembly of new ones)	-
b	interruptions in the supply of electricity	-	-	X
c	catenary faults	X (repair)	X (replacement or assembly of new ones)	-
11	carried out track or road works	X	-	-
12	exceeded the exposure factor on the road-crossing	-	X	-
13	damage of complete long-arm gate	X	-	-
14	no service at road-crossings	X	-	-
15	temporary crossing for pedestrians	-	X	-
16	occurrence of crisis or emergency situations	X (repair)	X (replacement)	-
17	no service at the station	X	-	-
18	the impact of weather conditions	X (repair)	X (replacement)	-

Summing up the presented proposals for solutions of problems regarding the condition of infrastructure on railway lines, it can be noticed that for 25 problems, up to 19 can be eliminated by revitalization process. Comparing the cost of both processes, it can be seen that modernization consumes more costs. Thus, this means that the audit of the railway line leads to the identification of cheaper solutions of the constraints and difficulties. This factor is very important from the point of view of the infrastructure manager. Therefore, one of the main advantages of the method of solving defined problems proposed in the article is the need to make low financial outlays.

5 Audit of railway line No. 12 in terms of infrastructure condition

Based on the algorithm presented in point 2 of this article, an audit of the railway line No. 12 Skierniewice - Łuków was carried out in order to find all the problems occurring on the open lines and operating offices. The line review was started from the odd track (from the Skierniewice station to the Łuków station), subsequently searching for existing defects from the set of problems specified in point 3 in each kilometre.

As a result of the audit, identified problems were presented in Table 3.

Table 3. Audit of the railway line No. 12 (source: own study based on [41])

Problem no	Problem no from table 1	Problem	Place where problem occurred [km]
ODD DIRECTION			
1	12	exceeded the exposure factor on the road-crossing	4,545 – 4,555 5,485 – 5,495
2	16	occurrence of crisis or emergency situations (one-track two-way traffic due to closing of track No 2)	13,732 – 26,199
3	5	turnout technical parameter (occurrence of scissors crossing)	13,732 47,963 89,024 97,350 103,700
4	11	carried out track or road works	103,260 – 103,650
5	17	no service at the station	109,300 – 110,935 141,176 – 143,186
EVEN DIRECTION			
1	12	exceeded the exposure factor on the road-crossing	4,545 – 4,555 5,485 – 5,495
6	2	sudden track failure (closing the track due to landslide)	13,732 – 26,199
7	1d	bad technical condition of the substructure of the track	25,780 – 25,830
8	4	inadequate ballast layer	25,780 – 25,830
3	5	turnout technical parameter (occurrence of scissors crossing)	13,732 47,963 89,024 97,350 103,700
4	11	carried out track or road works	103,260 – 103,650
5	17	no service at the station	109,300 – 110,935 141,176 – 143,186

6 Proposals for solutions to identified problems for railway line No. 12

The railway line audit method, proposed in the article, assumes that for the identified problems related to the railway line, a time profit and the amount of expenditure necessary to make their repair will be presented.

The biggest problem on the railway line No. 12 is the numerous limitations on the speed of train traffic associated with the poor condition of railway infrastructure along the entire length of the line. Therefore, the line parameters should be improved (to revitalize and/or modernize the line fragments in order to restore the construction speed). Additionally, at km 4,550 and 5,490 road-crossings should be reconstructed in order to raise their category from D to B, because currently they are not equipped with appropriate train protection systems and devices. Changing the category means mounting on half-gate crossings and automatic traffic lights. Raising the category will allow for eliminating the problem related to the exposure factor.

Proposals for solutions of identified problems on the railway line no. 12 assume that all stations and passenger stops will be open for passenger trains. Therefore, the Parysów and Jedlanka stations located between Pilawa and Łuków should be opened for commercial and operational purposes. In addition, it is necessary to rebuild the platforms, which will be fully adapted to the needs of people with disabilities.

In addition, option 1 assumes the restoration of the construction speed over the entire line, the addition of necessary turnouts and the reconstruction of scissors crossings into single turnouts. In option 2 of the proposed solutions, all the changes introduced in the first option were taken into account and, additionally, the reconstruction of the closed track on the section Puszcza Mariańska - Mszczonów was assumed. Option 3 consists of the changes introduced in the second option and additionally the construction of the Góra Kalwaria - Kępa Gliniecka open line with the railway bridge on the Vistula river for track 2. Additionally, it is necessary to reconstruct the track system at both exploitation points. The creation of a new facility will allow to shorten the time of trains' travel as a result of preventing the narrowing of the railway line from two-track to one-track.

In order to accurately compare all options, calculations were made, thanks to which a time profit was obtained as well as the amount of necessary outlays for each option. Detailed results are provided in Table 4.

Table 4. Comparison of solution options (source: own study)

Option	Time profit from the current state [min]		Outlays [PLN]
	pass. trains	freight trains	
option 1	70,81	84,99	231 348 243
option 2	75,09	90,95	238 387 256
option 3	76,02	93,62	293 399 611

Summarizing the results of the calculation of the time profits and costs of all proposed variants relative to the current state of the railway line no. 12, it can be concluded that the most advantageous in terms of time profit is option 2. Rejection of option 3 is due to the fact that there are too high costs of the proposed solutions in relation to the obtained time profit, in which the difference between option 2 is: 0.93 min for passenger trains and 2.67 min for freight trains.

7 Summary and conclusions

The audit allows to detect irregularities affecting the safety of railway traffic. Therefore, it is necessary to systematically control the railway lines in terms of their infrastructure condition in order to improve the provided services. The presented audit process in this article is based on a professional, independent and objective research of all elements of the railway network together with the necessary infrastructure in terms of quantitative and qualitative criteria, in order to check whether its condition is correct.

The condition of the infrastructure on the railway line No. 12 is bad. Many sections of the track surface are suitable for exchange, platforms and buildings at stations and passenger stops are devastated and overgrown, do not meet the requirements related to the transport of people. As a result of the audit, eight main problems affecting the existing limits of maximum speed of the trains were found on the railway line. These include: poor technical condition of the track, poor technical condition of the substructure of the track, sudden track failure (landslide), inadequate ballast layer, technical parameter of the turnout, track or road works, exceeded exposure factor at the road-crossing, closing of the station for technical and traffic needs.

The solution of the problems is possible by carrying out works allowing for raising the maximum speed to the construction speed. Three options have been developed, including the following: revitalization of railway tracks, construction of new platforms at passenger stations and stops, assembly of new turnouts, reconstruction of road-crossings to upgrade their category and construction of a new single-track railway bridge on the Vistula River.

Currently, the route of the railway line no. 12 passenger trains overcome in 185,68 min, and freight in 194,73 min. Due to the changes introduced in individual options, the time of train travel has been significantly shortened by the following time:

- in option 1: passenger trains travel less by 70.81 min, freight trains by 84.99 min,
- in option 2: passenger trains travel less by 75.09 min, freight trains by 90.95 min,
- in option 3: passenger trains travel less by 76.02 min, freight trains by 93.62 min.

The basic costs at the level of works performed in the first option are estimated at PLN 231 million. This amount increases with the introduction of additional solutions: by PLN 7 million in option 2 and by PLN 62 million in option 3.

References

1. Urząd Transportu Kolejowego, *Sprawozdanie ze stanu bezpieczeństwa ruchu kolejowego w 2016 r.*, Warszawa (2017)
2. M. Jacyna, P Gołębiowski., E. Szczepański, Proceedings of 19th International Scientific Conference Transport Means, 160 – 168 (2015)
3. A. Szeląg, Archives of Transport, **42(2)**, 75-83 (2017)
4. M. Jacyna, J. Kukulski, P. Gołębiowski, S. Jasiński, W. Wychowański, Journal of KONES Powertrain and Transport, **25(4)**, 533-546 (2018)
5. M. Jacyna, P. Gołębiowski, D. Pyza, Archives of Transport System Telematics, **11(1)**, 27-33 (2018)
6. R. Burdzik, B. Nowak, J. Rozmus, P. Słowiński, J. Pankiewicz, Archives of Transport, **44(4)**, 15-24 (2017)

7. S. Clarke, Work & stress, **12(3)**, 285-292 (1998)
8. J. Santos-Reyes, A. N. Beard, Civil Engineering and Environmental Systems, **20(1)**, 1-21 (2003)
9. J. Santos-Reyes, A. N. Beard, R. A. Smith, Proceedings of the Institution of Mechanical Engineers, Part F: Journal of Rail and Rapid Transit, **219(2)**, 47-65 (2005).
10. P. Gołębiowski, Logistics and Transport, **41(1)**, 91-100 (2019)
11. M. Jacyna, P. Gołębiowski, M. Krześniak, Scientific Journal of Silesian University of Technology. Series Transport, **96**, 59-69 (2017)
12. R. Licow, F. Tomaszewski, Problemy Kolejnictwa, **171**, 51-56 (2016)
13. M. Sitarz, TTS Technika Transportu Szynowego, **19(5-6)**, 56-61 (2012)
14. Z. Tian, S. Hillmansen, C. Roberts, P. Weston, N. Zhao, L. Chen, M. Chen, IET Electrical Systems in Transportation, **6(2)**, 41-49 (2016)
15. I. Jasiński, R. Białas, Zeszyty Naukowo-Techniczne Stowarzyszenia Inżynierów i Techników Komunikacji w Krakowie. Seria: Materiały Konferencyjne, **2(113)**, 227-239 (2017)
16. P. Królak, M. Sergot, M. Butlewski, Logistyka, **6**, 14392-14399 (2014)
17. A. Grzegory, M. Kałas, Zeszyty Naukowo-Techniczne Stowarzyszenia Inżynierów i Techników Komunikacji w Krakowie. Seria: Materiały Konferencyjne, **2(116)**, 165-170 (2018)
18. M. Jacyna, P. Gołębiowski, M. Urbaniak, Communications in Computer and Information Science: Tools of Transport Telematics, 199-210 (2016)
19. P. C. Cacciabue, Applied Ergonomics, **36(6)**, 709-718 (2005)
20. G. He, L. Zhang, Y. Lu, Environmental management, **44(3)**, 579-589 (2009)
21. I. Jacyna-Gołda, P. Gołębiowski, M. Izdebski, M. Kłodawski, R. Jachimowski, E. Szczepański, Journal of Vibroengineering, **19(7)**, 5627-5638 (2017)
22. I. Jacyna-Gołda, J. Żak, P. Gołębiowski, Archives of Transport, **32(4)**, 17-28 (2014)
23. M. Jacyna, M. Wasiak, K. Lewczuk, N. Chamier-Gliszczyński, T. Dąbrowski, Annual Set The Environment Protection, **20**, 1007–1025 (2018)
24. J. Żak, P. Gołębiowski, K. Popiela, Transport Means 2018. Proceedings of 22nd International Scientific Conference. Part II, 950 – 955 (2018)
25. M. J. Broniewska, Zarządzanie i Finanse, **1(3)**, 75-90 (2013)
26. A. Boholm, Risk management, **12(4)**, 235-255 (2010)
27. J. R. Edwards, Accounting and Business Research, **16(63)**, 251-263 (1986)
28. M. Jabłoński, A. Jabłoński, Zeszyty Naukowo-Techniczne Stowarzyszenia Inżynierów i Techników Komunikacji w Krakowie. Seria: Materiały Konferencyjne, **2(104)**, 183-196 (2014)
29. J. Kwaśnikowski, G. Gramza, *Analiza wybranych zakłóceń w ruchu kolejowym*, Poznań (2007)
30. M. Suszczewicz, Turystyka Kulturowa, **8**, 20-40 (2015)
31. A. Herdan, *Audyt wewnętrzny jako narzędzie wspomagające efektywny nadzór korporacyjny w spółkach akcyjnych*, Kraków (2009)
32. M. Jacyna, K. Lewczuk: *Projektowanie systemów logistycznych*, Warszawa (2016)
33. A. Jezierski, Logistyka, **5**, 91-93 (2007)
34. W. Pająk, Zeszyty Naukowe Uniwersytetu Ekonomicznego w Krakowie, **774**, 69-83 (2008)
35. PKP Polskie Linie Kolejowe S.A., *Instrukcja o prowadzeniu ruchu pociągów Ir-1*, Warszawa (2017)
36. Polskie Wydawnictwa Naukowe, *Słownik języka polskiego*, http://www.sjp.pwn.pl/
37. A. Skoczylas-Tworek: *Audyt we współczesnej gospodarce rynkowej*, Łódź (2015)
38. J. Sysak (red.), *Drogi kolejowe*, Warszawa (1982)
39. K. Towpik, *Infrastruktura transportu kolejowego*, Warszawa (2009)
40. J. Twaróg, Gospodarka Materiałowa i Logistyka, **7-8**, 154-157 (1998)
41. PKP Polskie Linie Kolejowe S.A., *Dodatek 2 do wewnętrznego rozkładu jazdy pociągów zawierający wykaz ostrzeżeń stałych*, Warszawa (2018)
42. PKP Polskie Linie Kolejowe S.A., *Instrukcja o kontroli biegu pociągów pasażerskich i towarowych Ir-14*, Warszawa (2011)
43. J. Mikulski, K. Gorzelak, Archives of Transport, **44(4)**, 47-54 (2017)

Methodology of risk assessment and forms of environmental safety management for the transport of dangerous goods by railway transport

Yuliia Zelenko[1], *Oleh* Dzhus[2,*], *Volodymyr* Dzhus[3], and *Dmytro* Yanchenko[1]

[1] Dnipro National University of Railway Transport named after Academician V. Lazaryan, 49010 Dnipro, st. Lazarian, 2, Ukraine
[2] Lviv Research Institute of Forensic Science, 79024 Lviv, st. Lipinskogo, 54, Ukraine
[3] Dnipro National University of Railway Transport named after Academician V. Lazaryan, Lviv branch, 79052 Lviv, st. I. Blazhkevich, 12a, Ukraine

Abstract. The authors examined the scientific basis for managing environmental and economic risks in the transport of dangerous goods by rail. The importance of optimizing the monitoring of environmental safety and the organization of environmental risk management was noted. The paper presents the interrelation of the functional strategy of ensuring the reliability of the transportation process with other strategies. A new functional strategy and a schematic diagram of the environmental management system for making management decisions to optimize environmental management in railway transport have been proposed. A conceptual apparatus has been developed to identify and determine environmental risks and the principle of their management during the transport of dangerous goods by railway transport.

1 Introduction

Modern society in its activities often goes into risks and unstable states that are associated with emergencies of different levels. Accordingly, there is a need to determine and predict potentially hazardous conditions and develop adequate methods for their management.

One of the important directions of the presented work is the development of strategies for managing environmental risks during transportation and handling hazardous products during railway transportation with the purpose of risks minimizing, which is one of the priorities of environmental safety on the railways today.

Ensuring a high level of environmental safety and, accordingly, minimizing environmental risks is based not only on the availability of certain tools and skills, but also on the ability to simulate the situation, predict its development and make decisions that are adequate the situation that has arisen.

One of the most important aspects of maintaining positive environmental performance of the railway transport is: to optimize the monitoring of environmental safety level and the management of environmental and economic risks [1-3].

2 Development of methodology for the organization of environmental activities on the railway transport

The development of relations between the society and the nature requires the use of new means of information processing, which are grounds for the decision-making system ensuring environmental quality. In this case, information technology is considered as the most optimal tool. Under information technology we mean a set of methods, production, software and technological tools, integrated into the technological chain, which ensures collection, storage, processing, output and spread of information. At the same time environmentally sound solutions depend not only on the quality and quantity of information, but also on its prompt receipt.

Statistics confirm that in recent years, due to intensive development of fundamental sciences (mathematics, probability theory, statistics, etc.), there were objective prerequisites for creation of methods for risk measuring based on the introduction of appropriate permissible measure.

The full-fledged solution of environmental management tasks in areas of dangerous products transportation, in particular, depends on the systematic analysis of a large number of heterogeneous data, as well as the solving a large number geographical, technological and industrial tasks (Fig. 1). Reliable and timely receipt of necessary information often plays a decisive role in the optimization of nature management, which is becoming extremely important in the context of intensification of transport complex development.

The development of environmental management systems during nature management organization on the basis of modern information technology will not only reduce the frequency of emergencies associated with pollution of environmental elements, but also optimize the organization of nature use, because high concentration of

* Corresponding author: djusov22@gmail.com

resources can lead to mutually exclusive types of anthropogenic activity in adjacent areas. In this case, the environmental management system is designed to identify the best options for environmental management, taking into account environmental and economic factors.

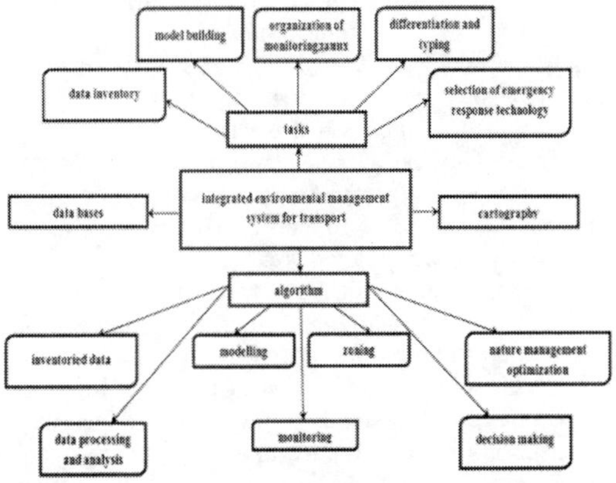

Fig. 1. Principal scheme of the integrated system for managing the environmental situation on transport.

In order to minimize the economic and environmental damage caused by traffic accidents during dangerous products transportation, special attention was paid to preventive measures, including simulation of emergency warning systems. The process of preventive system modeling for preventing railway accident occurrence and ensuring environmental safety during transportation includes the stage of a terrain model construction containing different layers of digital map information, including receptivity maps, as well as the model of dangerous goods behavior.

In addition, the analysis of the existing conceptual apparatus and the existing regulatory framework confirmed the need for harmonization of industry terminology in the gradation of occurrence of railway accidents with different degrees of severity and the national system of concepts. The system approach to this problem of assessment and management traffic accidents risks, in particular, during implementation of the technological process of dangerous goods transportation by the network of railways involves solving the two-stage task of stochastic programming:

– development of methods for quantification of risks of dangerous transport situations;

– development of mathematical models of emergency consequences on rail transport and the choice of rational measures to reduce the risks of their occurrence and the size of damage from emergencies that have already developed.

In this case, one of the priority tasks is to determine the degree of danger (degree of vulnerability) of objects and operations that are carried out during railway transport operation [5].

The degree of danger depends on the probability of danger occurrence and its intensity, which is determined by the degree of territory vulnerability.

The authors offer a term "the receptive status of the object (vulnerability)" which i The receptive status of the object (vulnerability) is characterized by the magnitude of losses caused during fixed time by the action of a dangerous process of certain intensity on the territory or object. Vulnerability is an important component of a risk that affects its decline. For today, management is mainly limited to engineering protection of objects, rather than extending to the most dangerous processes management. Thus, the problems of risk management are reduced to managing the vulnerability of territories (objects) under the influence of dangerous processes.

The assessment of recipients' vulnerability stems from the assumption of their possible partial damage as a result of a dangerous process. The receptive status of territories should be assessed, as a whole, in the territory and in the context of its separate regions. In existing techniques, the question of depicting territories specifics, in general, remains neglected.

The vulnerability of recipients (objects) can be evaluated according to recipients' types, which makes the method universal and easy to use. To do this, first of all, it is necessary to determine a complete list of recipients in their most fractional classification on each area of action. The main recipients are usually: agriculture, forestry, communal household. The general scheme of analysis, assessment and management of environmental risk is as follows:

1. detection and prognosis of hazardous processes, their intensity, repeatability, area of action;
2. zoning of the territory;
3. assessment of the receptive status of objects;
4. prediction of the development of secondary synergetic hazards;
5. mapping of risks;
6. establishment of permissible levels of risk and decision-making on management.

The assessment of risk degree of processes and vulnerability of territories and objects is carried out on the basis of experimental and theoretical studies regarding the database of territories and objects receptive status during the event [5].

Thus, the receptive status of the object ( ) equals the degree of potential vulnerability of the object with its anthropogenic component.

The risk, as the probability of a dangerous accident occurrence, can be classified according to various features: the sources of risks occurrence and the nature of risks' consequences. The sources of risks include geological processes, seismic phenomena, human factor, financial instability, technical innovations and even political events. According to these sources, the risk is called geological, seismic, financial, innovative, and political. Thus, the safety criterion is the permissible level of risk.

3 Conceptual principles of ensuring environmental safety in the developed methodology

Environmental safety in the organization of hazardous products transportation by railway depends on the reliability of engineering structures, compliance with established operating modes, the reliability of technological systems operation, organization and compliance with the established sequence of work and operations performance. Taking into consideration the performance of all works with the direct participation of specialists of different levels, the security of made decisions to a large extent depends on the skill and diligence of the staff.

The functional strategy of ensuring reliability of the transportation process should become the priority option for functioning all departments and units of transport infrastructure (Figure 2).

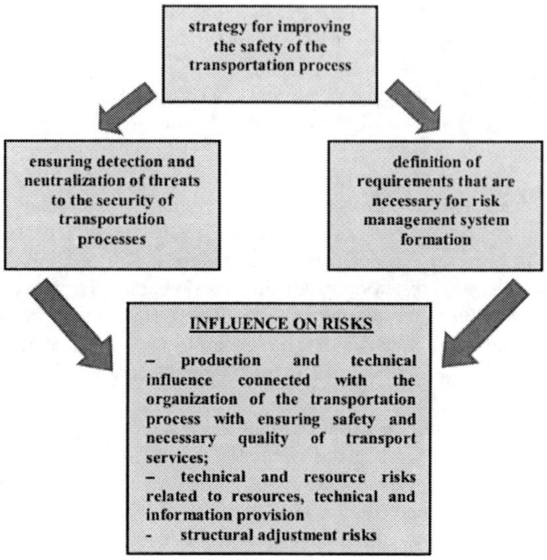

Fig. 2. Scheme to increase the level of safety of the transportation process by means of risk management [6].

To form a new functional strategy for improvement of transportation process safety, you need:

1. analysis of weak points of the existing functional strategy;
2. guarantee of the information reliability;
3. adaptation and processing of the regulatory base;
4. advanced training of the bodies of operational management and control;
5. constant control over regulatory requirements observance.

The analysis of the regulatory framework observation for ensuring safety processes on the railway revealed that the security was achieved so far, to a certain extent, by intensifying maintenance works, the growth of the number of unplanned repairs of rolling stock, the increase of the number of defectoscopists and other strategically incorrect steps. Maintaining stable position and transition to a guaranteed level of traffic safety should be based on the concept of "acceptable" risks.

The classification of hazard indicators on the railway transport should include two interrelated groups of danger indices characterizing the consequences of accidents (material and social damage) and probabilistic indices (the probability of an accident due to some risk factor, etc.).

It is established that according to available statistical information about dangerous transport situations and train crashes it is possible to evaluate only the compliance of indices of safety state of trains with the exponential distribution law with constant and variable intensity of accidents. During preparation of plans for railway lines and main railways for freight trains formation, which regulate trains traffic, and during dangerous goods transportation, in particular, the substantiated values of indicators that ensure traffic safety using the methodology of risk analysis and anthropogenic risks, in particular, are not taken into account. Thus, at present, there are no conditions for a reasonable comparative choice of routes that are acceptable according to their risk indicators and often are laid in densely populated areas.

Thus, the authors have set and implemented the tasks of simplifying and improving the methodology for assessing (taking into account) the risks of transport accidents and economic losses of railway lines, separate territories and specific routes for dangerous goods transportation, with regard existing dangerous crossings with pipelines and motor transport, the presence of adjacent industrial and urban structures (in the areas of mass transshipment of oil cargo, in particular), as well as various factors of natural hazards.

From the standpoint of the natural component of risks, the following environmental factors must be taken into account: seismo-geological; showers, floods, flooding, storms (for coastal areas); hurricanes, tornadoes.

The necessity to account these types of natural hazards is due to the specifics of railway transport, which infrastructure damage should be timely determined and traffic on dangerous sections should be stopped. These under-predicted transient processes bear the main danger.

From the position of the technogenic component of risks for building maps, industry-specific accident data should be used. Technogenic risks include:

– risks of accidents due to non-normative technical condition of infrastructure elements;
– risks from interaction with communication networks at crossings;
– risks from interaction with motor highways at crossings.

The lack of a single method for assessing the damage caused by accidental spills of dangerous products (oil, gasoline, etc.) into environmental components, and methodological approaches for assessing the environmental risk of railway lines operation complicate the development of environmentally sound solutions to minimize the consequences of emergency situations for the environment.

The strategy of minimizing the impact on the environment should be aimed at applying methods for restoration of contaminated areas and liquidation of waste and soils that contain them [7].

In connection with it, there is a need to improve the state of ecological safety at the phase of emergency situations elimination during dangerous goods transportation, by solving optimization problem of determining the optimal amount, location and the capacity of complexes, which technological support allows to process the entire volume of possible waste (petroleum products and substances of chemical industry, in particular) formed as a result of technological processes and emergency situations.

Taking into account all the above-mentioned aspects, the methodology for assessing ecological risk of accidents on railway transport during dangerous goods transportation on a heavy traffic route is proposed. Knowledge of the probability of emergency situations occurrence and the value of losses that they can inflict on the environment allows predict the magnitude of contaminated areas and the amount of waste generated during elimination of emergency consequences (the spill of dangerous goods).

One of the priority guidelines of existing and emerging theories of risks is the development of principles of managing them. For risk management we accept determination, development and implementation of measures to reduce the level of risks.

The scheme of developed risk management algorithm describes the procedure for finding an acceptable level of risks associated with dangerous natural and anthropogenic processes. In this scheme, the assessment of permissible risks is assumed in the decision-making block (Fig. 3).

The scientific basis for determining basic quantitative risk indicators is grounded on the statistical model of safety transportation by railway and statistical substantiation of the type of casual events stream. This allows to select a specific model that describes the state of safety traffic by obtaining quantitative estimation of risk indicators.

It should be noted that operational environmental damage includes losses from the economic activity L_{ea} and payments for pollution of the environment.

Since the railway is a complex multifactorial system that contains both long-distance linear objects (routes for dangerous goods transportation) and local objects with significant potential risk (oil stores, petroleum terminals, fuel-and-bulk discharge refueling units), it is necessary to analyze the peculiarities of the criteria application for comparing risks in various options for reducing the risk of the most dangerous situation, when dealing with oil products, and the expected loss of social components and infrastructure.

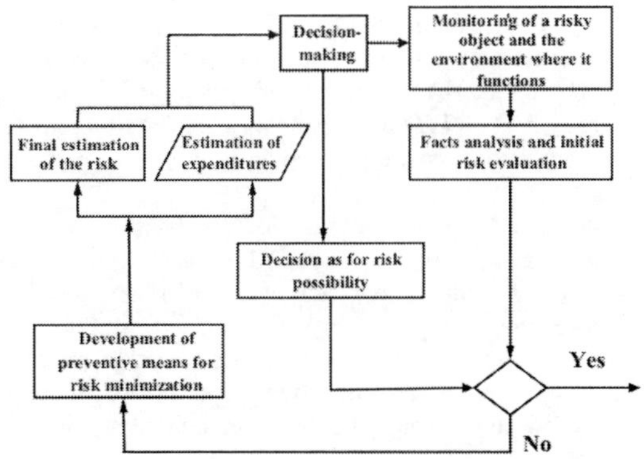

Fig. 3. Block diagram of the principle of risk management.

In the general case, for the analysis and quantification of risks, there are different methods, among which the most widely used are [3, 4, 5]: statistical method – Monte-Carlo method; analytical method; decision tree method and probabilistic approach, etc..

Each of the existing methods has its advantages and disadvantages and is used in particular situations; there is no universal method applicable to all cases. Under these conditions, to provide an objective choice of the most efficient methods of risk assessment 3 methods of parallel quantitative analysis were applied on the basis of screen-testing (figure 1) carried out by the authors.

Since each method is not without flaws, it is advisable to use several different methods in practice. The authors used the method of expert evaluation and the method of sensitivity analysis (analysis of the receptive status of objects). The results obtained by different methods vary, but the study of differences between them has revealed the factors contained in some methods and are absent in others, which affect the accuracy of evaluation and the reliability of results.

Table 1. Comparative characteristics of quantitative risk assessment methods

Characteristics	Methods of quantitative risk assessment						
	Statistical	Analytical	Tree of decision making	Expert evaluation	Normative	Sensitivity analysis	Analogues
Conditions of use	Known data of past periods of economic activity	Availability of detailed information about the project or activities	Known risk facts and the effects of their impact	Lack of information	Known key performance financial indicators	Availability of detailed information about the project or types of activity	Presence of analogues and the immutability of conditions of economic activity performance
Assessment	Absolute	Relative	Absolute	Relative	Relative	Relative	Absolute
Accuracy of evaluation	Not high	Medium	High	Not high	Not high	Medium	Not high
Expenses	Medium	Minor	Considerable	Considerable	Minor	Minor	Medium
Accounting of each of the risk factors impact	Minor	Impossible to determine	Highest	Practically impossible to determine	Impossible to determine	Practically impossible to determine	Minor

Based on the existing statistical data on main railway network operation in Ukraine, author's own research and expert assessments, the causes of failures were identified (delays, accidents and technological losses) and the likelihood of failure in different scenarios was identified (Table 2) [6, 7].

Analyzing the results of determining the likelihood of emission according to these scenarios, we see that the greatest danger of occurrence of railway traffic accidents (hereinafter – RTA) and technological losses of petroleum products is observed under violation of operation rules on the railways (Table 2).

Reducing risks that is the probability of these emissions' occurrence should be based on increased control over the timely (terms in accordance with the requirements of regulatory documents) repair works, control over the state of rolling stock, track and equipment, as well as improving the quality of training and knowledge testing of service personnel (operators).

Table 2. Risks of oil product emissions on railway transport under different scenarios.

№	Scenario (cause of emission)	Probability of emission, in shares	Probability of emission, in % of general quantity
1.	Operation rules violation	0,4701	47,01
2.	Wear	0,3213	32,13
3.	Defective construction and mounting works	0,1724	17,24
4.	Natural factors	0,0305	3,05
5.	Corrosion, external interference	0,0095	0,95

For the purpose of constant monitoring the authors have developed a new methodological approach according to which risk factors of both linear and local railway objects include receptive status and complex natural and technology-induced status of the geographical area of the object [6].

To create a database for electronic maps of receptive status and complex natural and technology-induced risks along the main routes of movement of trains with oil products the authors suggest a methodology of intensity recalculation of nature hazards. The analysis of natural and anthropogenic risks takes into account the degree of loading of railway lines (cargo tension).

Estimation of emergency situations probability at main railways, as well as the magnitude of damage to the components of the environment, allowed to calculate the ecological risk of accidents.

Taking into account the relatively small time to choose the rational solution and a large amount of related information (location of the source of an emergency situation at the station, rolling stock on adjacent tracks, adjacent production facilities and housing development, their population and fire characteristics, tactical and technical data of fire trains, ways of evacuation, etc.), it is expedient to automate the process of information collection and processing.

As the risk indicators are not similar for different main line sections the authors suggest using the total risk for the present moment which can be presented as a total of risks for railway sections with an equal probability of accidents and damage amount. The characteristic objects of negative impact during emergency emissions on railways are soils and underground and surface water bodies. Therefore the amount of ecological risk of a particular railway main line section was defined by the formula:

$$R_{з_l} = P_{авар} \cdot \frac{\left(\sum_{i=1}^{n} з_{зем_i} l_{зем_i} + \sum_{j=1}^{k} з_{вод_j} l_{вод_j}\right)}{l_{маг}} \qquad (1)$$

where: $P_{авар}$ is the probability of an accident; l is the length of a section; $з_{зем_i}$ is the ecological damage caused by pollution of the land by oil products of the *i-th* ecological vulnerability; $l_{зем_i}$ is the length of the section of the main railway passing on the territory of the land of the *i-th* receptive status (environmental vulnerability); $з_{вод_j}$ is the ecological damage caused by pollution (by oil products) of **j** water basin; $l_{вод_j}$ is the length of the main railway section crossing or passing along the *j-th* water basin.

The risk assessment of transportation of oil products along the specified routes of high intensity was carried out according to the formula:

$$R_{ЗМ} = R_m + R_{np} \qquad (2)$$

where: R_m are anthropogenic risks; R_{np} are natural risks.

In this case technology-induced risks were defined as follows:

$$R_m = R_{НС} + R_{КМ} + R_{аМ} \qquad (3)$$

where: $R_{нс}$ is the risk of damage from accidents due to non-normative state of infrastructure elements; $R_{км}$ is the risk of damage due to the failure of communication networks and product lines with possible formation of striking effect; $R_{ам}$ is the risk of the damage associated with the presence of highway crossings with possible traffic rules violation.

Risk of the damage associated with failure of communication networks and product lines with possible formation of a striking effect:

$$R_{КМ} = \sum_j r_{КМ} \cdot p_N \cdot \Omega_{ЗК}(\kappa) \qquad (4)$$

where: $r_{км}$ is the risk of an accident in communication networks and at product lines; p_N is the probability of a train with petroleum products movement in the zone of detonation or outbreak of products that have flown out of the product pipeline; $\Omega_{ЗК}(\kappa)$ is the distribution function of the loss associated with accidents in communication networks and at product lines at railway bed crossing points, which is determined by the volume of oil product that has leaked and/or burst into fire in the area of an accident.

Risk of the loss associated with passing of highway crossings:

$$R_{aM} = \sum_{j} p_{aM} \cdot \Omega_{3n}(n) \quad (5)$$

where: p_{aM} is the probability of an accident at the crossing with participation of a train with oil products; Ω_{3M} *(n)* is the distribution function of the loss during RTA at the crossing.

In course of processing the obtained results the prediction of development dynamics of a dangerous (emergency) situation will be of unquestionable interest.

The main component of risks within large railway stations is the risk of social damage (up to 80%), which can be lowered by means of the use of bypass ways around densely populated areas.

Thus for the purpose of constant monitoring and possible risk management on railways in general the authors suggest a methodology of identification and assessment of risks arising during the transportation, storage and usage of oil products on railway on the basis of a comprehensive systemic evaluation and taking into account the receptive status of objects and their resistance to negative factors.

According to the suggested methodology the risk assessment is carried out on the basis of three indicators - risk frequency or probability; risk significance and dimension according to the receptive status of the object. To ensure their comparability and to make the qualitative assessment easier a ball scale is offered. (Tables 3; 4) [6, 7].

Table 3. Frequency or probability of dangerous products emergency emissions.

Point	Meaning	Frequency or probability
1	Very seldom	Once per 7 years and more rarely (probability up to 5%)
2	Seldom	Once per 5 years (or probability 25%)
3	From time to time	Once per 3 years (or probability 40%)
4	Frequently	Once per 1 year (or probability 80%)
5	Very frequently	Once per half a year and more frequently (probability > 95%)

The map-matrix of risks allows to assess the relative significance (ponderability, severity) of each risk (as compared to other risks), and to determine the objects of risk exposure, which are critical and require the development of measures for their management.

Construction of a risk map allows:

1) determine the potential for risks retention within the limits that can be applied to all railway operations;
2) to develop a list of critical risks for the railways and ensure appropriate processes for managing them;
3) determine the priority of risks and develop the distribution of financial resources.

The risk map is divided into several areas, highlighted in different colors.

1) Green Zone – risks with low probability of occurrence and (or) no significant impact on the biosphere (safe operation zone).
2) Blue zone – risks that have an average probability of occurrence or average potential impact (zone of relative tolerance).
3) Purple zone – risks with a high probability of occurrence or average potential impact (zone of environmental alertness).
4) Red zone – risks that are critical for a railway object or due to the high probability of occurrence or due to serious damage potential that may affect the financial viability of the railway (critical area).

Risk identification numbers are plotted on the map (regarding the registry) in accordance with frequency (probability) of risk occurrence and its degree.

Developed risk map is a graphic representation of an object's proximity to critical risks and should be a mandatory addition to the report.

Table 4. Frequency or probability of dangerous products emergency emissions.

Points	Degree of impact	Receptive status of an object $\sum_{i=1}^{n} RSt_i$, points	Potential damage of the risk occurrence
1	Minor	< 8	Absence of any consequences in case of risk occurrence
2	Low	9-16	Consequences of risk occurrence are not significant
3	Medium	17-24	Consequences of risk occurrence are significant, but can be completely removed
4	Considerable	25-32	Consequences of risk occurrence are very significant, but can be corrected to a certain degree
5	Critical	> 33	In the case of risk occurrence, the railway will almost be unable to recover from consequences

Relative magnitude (significance) of risk is determined by the matrix map.

The priority of risks is determined according to the position of each of the risks on the risk map (Figure 4):

Group 1 – catastrophic risks – red area of the risk map – risks with the highest priority;

Group 2 – average risks – violet area on the risk map – risks of the second priority;

Group 3 – low risks – green area on the risk map – risks within the company's ability to maintain monitoring and control.

Inside each of the groups, the priority of risks is determined on the basis of the point of risk significance.

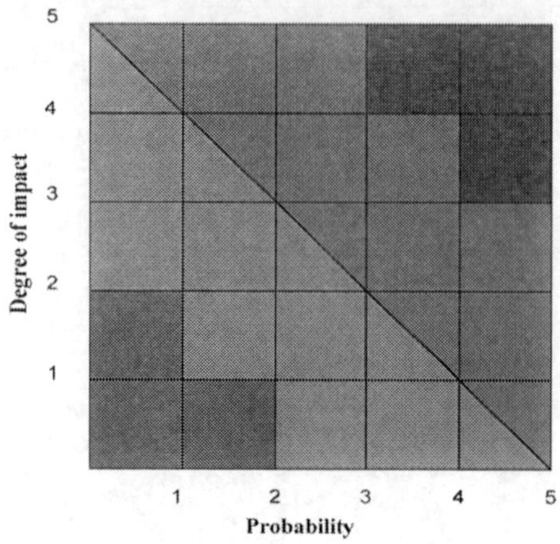

Fig. 4. Map-matrix for determining the importance (significance) of risks [2].

Each of the risks included in Priorities 1 and 2 is assessed on the basis of the following factors:
1) analysis of the causes of risks occurrence (scenarios of losses);
2) analysis of the potential impact of risks on the financial indicators of the Society – inherent risks (without taking into account risk management methods) and residual risks (residual risk after application of risk management methods);
3) analysis of the correlation of risks with other risks (repayment of risk negative effect in one unit with positive one in the other unit – the principle of compensation, or increase of negative effect in connection with occurrence of other risks – the principle of domino).

Thus, the methodology is a simple, efficient and affordable example of identification, assessment and further risk management capabilities on the railways. In this case, the final formulation of a permissible risk becomes necessary.

4 Models of management by ecological and information systems

For efficient management of transportation process, not only reliable data and their timely receipt, but also the possibility of their operative interpretation, which lead to the use of the geoinformation approach, are needed. It combines graphical visualization tools with spatial analysis and data sharing tools that facilitate the efficient organization and data management [6, 8, 9, 10].

From the analysis of environment condition in the areas of potential risk of oil and oil product pollution conducted by the authors, it appears that the variety of factors determining integral sensitivity of nature objects in the main line area stipulates the necessity of systemic approach to the problem of prevention and elimination of emergency situations connected with accidental oil spills. In connection with this an important task was to develop methods of design, calculation and comprehensive assessment of sensitivity of a chosen main line section which required the usage of modern geoinformational technologies. In accordance with this the integral sensitivity of the section of Pyatykhatky locomotive changing point area was considered as a total of sensitivity indices of biota existing here in a certain stage of life cycle. For the objects of environmental management the authors have accepted a relative indexation, that's why during the assessment of integral sensitivity only the presence of some kind of manufacturing is taken into account.

The geoinformation approach and GIS mapping allowed to develop the understanding of the information flow regulation system, ranging from cartographic support to the decision-making system, resulting in effective participation in regulatory mechanisms.

As the basis of the ecological security system, topographic maps and plans specially digitized have been used. The difference between cartographic projections raises the task of harmonizing the underlying basis for adequate perception of information. The geoinformation approach provides the opportunity to optimize the program of environmental monitoring in the area of transportation, storage and use of petroleum products, supporting various analytical operations, providing spatial data management and information transmission services, as well as their cartographic representation at the decision-making stage [8].

Integration of databases of different structures allows using the system for solving various application problems. This system is open, therefore, depending on the emerging tasks, it is possible to construct new modules, their algorithmic and software, and thus, their visualization.

For systematized analysis, as well as evaluation and ranking of objects of railway transport according to their potential environmental impact, the authors recommend using maps of receptive status of objects – sensitivity and stability of environmental objects to anthropogenic loading.

Receptive status maps are part of the scientific and methodological support of the ecological safety system, which allow developing a response strategy for elimination of RTA (accidents). This is one of the main resources that management and decision-making bodies need to take when making decisions, assessing the consequences of accidents (emissions), as well as calculating necessary response facilities and their locations, based on typical conditions for the given area.

Receptive status maps are not just a set of electronic maps with the ability to simulate different map configurations, but also provide access to the database of objects marked on this map. Availability of attributive information of cartographic objects and communication with databases provides qualitatively new possibilities for analysis of available information. The considerable amount of data contained in receptive status maps, the possibility of optimal organization, their processing and operation, determined the need for creating information system that allows operational access to necessary data for decision making in the face of pollution threat to natural ecosystems. The basic scheme of environmental safety management on the railways and mapping of receptive status is presented in Figure 5.

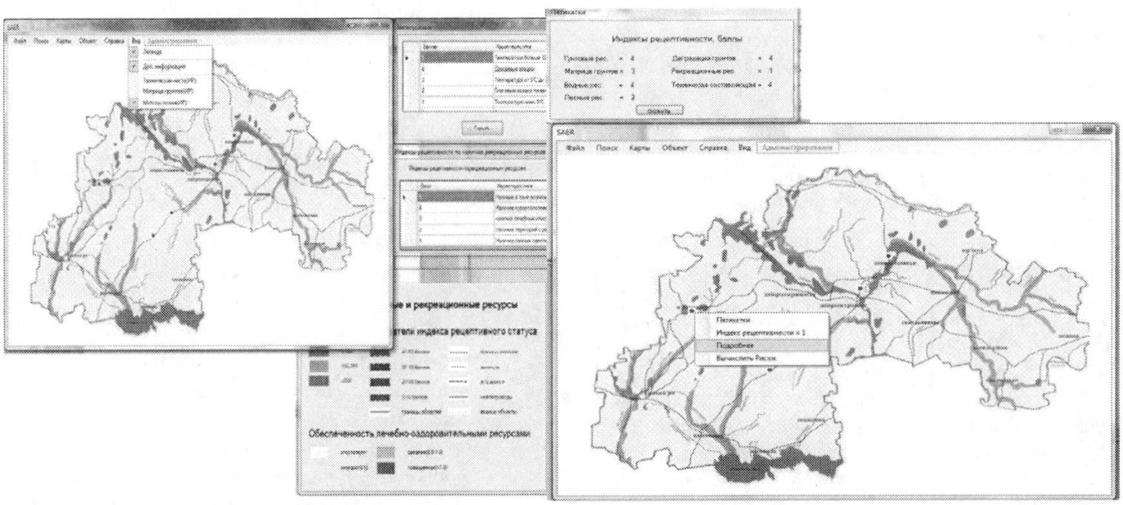

Fig. 5. Dialogue boxes of the software package.

In addition, the sensitivity maps make it possible to identify quickly priority areas at emission elimination, to simulate and predict the process associated with oil emissions, as well as to assess the damage caused by the discharge of petroleum products.

The information-analytical block of the decision support system provides:

- accumulation and provision of information that is necessary for decision-making in emergencies;
- determination of risks of environmentally valuable areas contamination and the choice of the best methods for pollution elimination;
- determination of environmentally vulnerable to oil pollution objects of the environment;
- calculation of losses incurred by the emission;
- representation of the most valuable objects located in the zones of main lines operation;
- the ability to integrate information for spatial operations realization.

The efficiency of various blocks of the information-analytical system increases due to the use of distributed databases and the creation of complex models. It includes five key components: hardware, software, data, methods and executors.

Thus, the developed map data bank that is connected to any spatially distributed data of operation zone of railway lines and units has become the basis of the whole system.

The functioning of a suggested software package (figure 5) is performed with the help of database management which contain specially designed blocks and spreadsheets.

The informational system makes it possible also to optimize and reflect in cartographic form the routes of advances of forces and funds into the area of spill, sites of storage and utilization as well as to create alert systems and communications for localization and elimination of accidental oil spill consequences (fig. 4).

The procedure of determination of receptive status of object [11]. The maps of sensitivity of main line areas to oil pollution are presented as a graphic model of ecological situation in the area of accidental spill risks. They show the distribution of environmental objects and provide a possibility to establish priorities while protecting the most vulnerable components in course of oil spill elimination.

The maps of sensitivity contain not only the information about the environmental state of the area of potential pollution and make it possible to measure the vulnerability of biota to oil influence but also give a possibility to make a prompt prognosis of the process connected with accidental spills and estimate the economic harm inflicted.

The solution of the task to choose the best technology of a railway section cleaning is presented as a sequence of operations that give a possibility to make the algorithm of the procedure. Every section presented on the map was referred to a certain index in order that the user has a possibility to choose an adequate technology. After filling in the form proposed by the module that includes data about the amount and type of accidental oil product emission the user consequently gets the justification of the choice of technology which is the most appropriate in these circumstances. Such job was done for a section of Prydniprovsk railway in the area of Pyatykhatky locomotive changing point.

The developed maps of sensitivity made it possible to identify the sites of potential pollution of the chosen area. This area comprises sites having different sensitivity to oil pollution.

The methodology of automated selection of technology of cleaning sections of railway main lines during elimination of accidental spills of oil and oil products developed by the authors is based on the defining the degree of ecological sensitivity of nature objects in the railway area and options of reacting that provide the least environmental impact. Thus it helps to optimize the ecological and economic result of elimination of accident consequences.

A particular complex of methods and technologies of elimination of ecological consequences of oil emissions is chosen with the help of specific matrices developed by the authors according to clearly differentiated scenarios of emission development (e.g. Scenario 1. A leakage from a tank or a container for storing heavy oil products onto the surface of the ground without the threat of geofiltration; Scenario 2. A leakage from a tank or a container for storing light oil products onto the surface of the ground with further geofiltration; etc.)

Conclusions

1. The principal scheme of the ecological situation management system on railway transport is proposed, which allows to carry out systematic analysis of heterogeneous data: monitoring of geographical, technological and production tasks, and provides the possibility of full, reliable and timely decision-making management as for the technical system of railway transport.
2. On the basis of analysis of statistic data about emergency situations arising during transportation of oil products, a hierarchical structure of malfunctions of socio-technical system of railway transport is developed.
3. A methodology of ecological and economic assessment of emergency risk during transportation, storage and usage of oil products on railway transport is developed.
4. With the usage of the main methods of emergency risk management during transportation of oil products along railways the authors suggested a complex of measures aimed at the optimization of the process of situation assessment and making adequate decisions on reducing the size of environmental damage.
5. A conceptual device for identifying and determining environmental risks is proposed.

The proposed scientific and methodical device gives opportunities for operative management of ecological and economic risks on the railway, in particular, when dealing with oil products, by continuous monitoring, using GIS-technologies, and system analysis on the basis of the special software program that is specifically developed and functioning.

Bibliography

1. Y. Zelenko, *Development of methodological foundations of resource-saving environmentally friendly technologies in the operation of railway transport:* dissertation, 456 p. (2015).
2. S. Boychenko, Y. Zelenko, *Optimization of risk management of emergency situations during the transportation of petroleum products by rail,* pp 112–117 (2010).
3. *Risk assessment for public health from air pollution: guidelines,* 40 p. (2007).
4. E. Henley, J. H. Kumamoto, *Reliability of technical systems and risk assessment,* 528 p. (1984).
5. Bridget M. Hutter, *Regulation and Risk,* Occupational Health and Safety on the Railways, 389 p. (2001).
6. Y. Zelenko, S. Myamlin, M. Sandovskiy. *Scientific foundation of management of the environmental safety of oil product turnover in railway transport: monograph,* 332 p. (2014).
7. Y. Zelenko, *Scientific basis for the environmental safety of transportation technologies and use of petroleum products in railway transport: monograph,* 242 p. (2010).
8. S. Trepak, Y. Zelenko, *Geographic information systems as a tool for regulating environmental safety in the process of transportation and use of petroleum products in railway transport*, pp. 205-206 (2011).
9. Mohd Rapik Saat, Charles J. Wertha, David Schaeffer, Hongkyu Yoonb, Christopher P.L. Barkan, *Environmental risk analysis of hazardous material rail transportation,* Journal of Hazardous Materials, 264, pp. 560-569 (2014).
10. S. Cafiso, A. Di Graziano, N. Di Blasi, *Risk assessment on railway transportation of hazardous materials*, Risk Analysis V: Simulation and Hazard Mitigation, pp. 97-106 (2006).
11. Y. Zelenko, V. Tarasov, S. Triepak, *Certificate of registration of copyright to the work, computer program "The system of analysis of ecological risks on railways (SAER)"*, (2012).

Non-destructive control data analysis of railroad foundation constructions

Alexander Trofimov[1,*], *Albina* Kuzmenko[1], *Halyna* Nesterenko[2], *Svitlana* Avramenko[3], *Mykhailo* Muzykin[3], *Nikolaj* Mormul[4], and *Anatoliy* Sokhatsky[5]

[1] UCF, Department of Transportation systems and technologies, 49000, Dnipro, Volodymyra Vernadskoho str., 2/4, Ukraine
[2] DNURT, Department of Operational Work Management, 49010, Dnipro, Lazaryan str., 2, Ukraine
[3] DNURT, Department of Life Activity Safety, 49010, Dnipro, Lazaryan str., 2, Ukraine
[4] UCF, Department of Applied Mathematics and Informatics, 49000, Dnipro, Volodymyra Vernadskoho str., 2/4, Ukraine
[5] Institute of transport system and technologies of National academy of sciences of Ukraine, Department 7, 49005, Dnipro, Pisargevskoho str., 5, Ukraine

Abstract. The evaluation of the parameters of multi-layered foundations (railroad basis, foundations of railway structures, etc.) plays an important role in ensuring the safe movement of trains. The method of estimating the mechanical and geometric parameters of such foundations based on the solutions of inverse problems for multi-layered elastic packets is proposed. As input data for such problems the measured displacements of certain points on the package surface are used. The method allows estimating the parameters of the a priori distribution of unknown variable parameters, identifying and excluding outliers of the measured data from the created model, and constructing a posteriori estimation of the unknown parameters probability density with acceptable resolution. Proposed method can be used to create a new generation of equipment intended for non-destructive monitoring and estimating of the condition of the railroad basis and the foundations of artificial structures. The appropriate software of such vehicles based on the developed methods of data processing can be developed. The use of such equipment allows to operatively analyzing the state of individual areas of the railroad to decide on the need of repairing or replacing the railroad base or foundation of other elements of railroad infrastructure.

Introduction

Problem statement. The issues of non-destructive monitoring and control of the mechanical and geometrical parameters of the roadbed are of great importance in the practice of safe railways operation. The uncertainty of the train movement process over long distances greatly complicates the transportation planning and ensuring traffic safety. This is due to factors associated with the risks in the process of organizing the movement of trains, when it is necessary to ensure high reliability of all railway subsystems, in particular infrastructure subsystems. Therefore, the evaluation of the parameters of multi-layered foundations (railways, foundations of railway structures, etc.) plays an important role in ensuring the safety of trains [1]. It is known that the consequences of train derailments caused by the collapse of railway embankments often exceed the consequences of accidents caused by the destruction of the rail track itself [2]. In addition, the cost of restoring the damaged subgrade and associated infrastructure is usually significantly higher. This raises the problem of creating methods that are more accurate for analyzing the diagnostic parameters of the roadbed structures to predict possible unallowable deformations and/or damage.

Review of resent researches and publications. There is a variety of works devoted to roadbed state diagnostics. For example, E.V. Nepomnyashchikh and K.A. Kirpichnikov in [3] set the tasks of diagnostics of the roadbed, namely: determining the parameters, that define the deformation properties of the roadbed; predicting the values of such parameters in time and their influence on the deformations occurrence of the roadbed construction; development of new principles, methods and technical means of diagnosing in order to detect "bottlenecks" in such structures. V.D. Petrenko, D.O. Yampolsky, I.O. Svyatko devoted his work [4] to the analysis of the existing modern and classical methods of numerical modeling of behavior of soil samples under the action of a static load. They proposed a new computational model, in which, in addition to the classical approach to the analysis of the stress state of the roadbed soil, the deformation state is taken into account. In the article [5] by scientists Luchko Y.Y. and Kravts, I. B., the condition of the roadbed on the railways of Ukraine is analyzed, the requirements for it and the methods of monitoring are considered. The attention is focused on the importance of monitoring the state of

* Corresponding author: atrof2222@gmail.com

railway infrastructure facilities in the context of introducing high-speed traffic on the Ukrainian railways. The ground penetrating radar method (GPR) as one of the most promising methods for monitoring of the roadbed is proposed. The world and domestic experience in the use of GPR for non-destructive monitoring of the condition of the roadbed, ballast layer and artificial structures is also given. Its advantages are efficiency of work and low labor intensity. However, the methods proposed in this work provide information only about the geometrical parameters of the railroad bed base and do not give any information of its mechanical properties.

The purpose of the article. The main goal of this work is to develop methods for analyzing data from scanning devices designed to determine the surface points displacement velocities of the loaded railway bed, which helps to recognize the weakness of the roadbed structures and prevent unallowable deformations and destruction of the it and, thus, increase the safety of train traffic.

1 Analysis of methods and technologies for determining the diagnostic parameters of the roadbed

The diagnostic parameters for assessing the condition of the roadbed can be divided into three groups: geometric (the shape and size of the soil base layers that make up the roadbed); mechanical (elastic deformation moduli, Poisson's ratios, elastic wave propagation velocities, plasticity, creep parameters, etc. of base layer materials); physical (indices of electrical and thermal conductivities, conductivity parameters of penetrating radiation, characteristics of moisture saturation, porosity, etc.). The most important in terms of assessing the deformation properties of the soil base are the first two groups of diagnostic parameters. Physical parameters mainly serve to estimate more accurately the geometric and mechanical characteristics.

For obtaining the diagnostic parameters one use the following methods.

1. *Engineering and geodetic.* They include the scanning of transverse embankment profiles, the binding of boreholes and geophysical profiles.

2. *GPR.* The method is based on determining the geological characteristics of the soil foundation using the parameters of short high-frequency electromagnetic pulses from the generator, transmitted and received through antennas located on the ground surface. Devices that allow such sounding are called ground penetrating radars. By ground penetrating radars it is possible to determine the geometrical characteristics of material layers that make up the soil foundation. One can also estimate some of the physical characteristics of the layers (moisture saturation, porosity, etc.). GPR techniques are also adjacent to the methods of electrodynamic sensing.

3. *Method of loading trains.* It consists in the determination of the elastic sediment of the railway track under the action of a passing loading train [3].

4. *Method of measuring the surface points elastic displacement of the roadbed under the passing trains.* The method is based on measuring the amplitude of the oscillation of the roadbed surface under the action of passing trains. Measurements are carried out using low-frequency sensors installed at various points on the subgrade surface.

5. *Seismic method.* In the seismic method, the velocities of elastic waves excited in the ground bed by a shock load are recorded. Taking into account this information, the characteristics of the velocities of propagation of shock waves and their amplitude-frequency parameters in the soil massif are investigated, which makes it possible to draw conclusions about the structure of the subgrade.

Technologically diagnostic examination is carried out in two stages. At the preliminary stage, "problematic" sections of the roadbed are identified, which are then subjected to more thorough examination at the main stage (detailed inspection stage).

Requirements for the preliminary stage are high survey speed, low cost, the possibility of collecting as much data as possible for preliminary analysis. The procedures of the main stage (seismic sounding, measurements using low-frequency sensors) are usually much more expensive and time-consuming.

The main tools of the preliminary stage of diagnostics are usually GPR sounding and the use of loading trains. The main disadvantage of such an organization is that georadars are generally not useful for assessing the mechanical properties of the investigated subgrade layer materials, and information from load trains (deformation of the track under load) is usually insufficient for adequately estimation of the entire layered foundation deformation properties. The procedures of the main stage of diagnosis are quite informative, but they are expensive and have a longer duration in time. At the same time, the existing level of development of measuring devices and data processing methods makes it possible to create diagnostic tools that combine high data acquisition rates with sufficient coverage of the most "informative" part of the deformable foundation.

Such diagnostic tools, that allows to obtain the small displacement (or velocity) of deformable surface points, call "deflectometers". Modern deflectometers are already widely used in the practice of roads building and operating for diagnosing the condition of the pavement [6,7]. Such devices are divided into two classes. Devices of the first class allow estimating the displacement of the test surface points. They can be called "displacement deflectometers". Such tools use, as a rule, optical sensors-rangefinders [11], mounted on the bottom of the body of a special vehicle that creates the necessary loading of the diagnosed pavement. The disadvantage of such tools is that they must be equipped with a complex system of precise positioning of rangefinder sensors, as well as a speed integrator to determine accurately the time intervals between fixing the results of two adjacent displacement sensors at the same point of the investigated surface.

Another significant problem is the vibration of the mounting base of optical sensors, the amplitude of which can significantly (sometimes by several orders of magnitude) exceed the values of the measured displacements. This problem may partly be solved by installing damping devices, as well as accelerometers-integrators, which take into account the vertical and horizontal movements of the sensor mounting base.

Deflectometers of the second type, called "velocity deflectometers", lack these deficiencies [8]. They use optical Doppler sensors to determine the velocities of certain points of the investigated surface, which serve as the initial data for calculating the parameters of the roadbed. As a rule, such instruments are also equipped with rangefinder sensors for simultaneous scanning of the surface shape of the foundation under study. Such combined tools can be installed on the locomotive (see Fig. 1) and used to scan the interesting parameters of the train roadbed construction in online mode with high speed. This approach will provide the minimum amount of time and money without changing the train schedule.

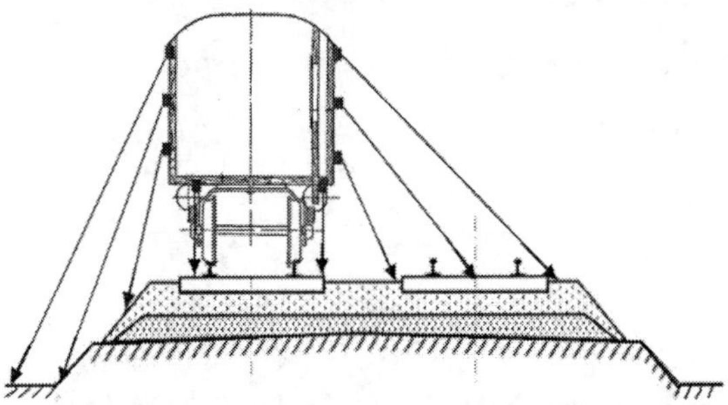

Fig. 1. The scanning of the roadbed by deflectometers installed on the locomotive.

It is possible to install deflectometers at different angles to the testing surface, which will increase the measurement flexibility, making it possible to focus the scan on the most informative part of the deformable foundation.

At the present stage of development of testing devices, dynamic deflectometers of the considered types are equipped with a number of scanning sensors of some units [6], which is clearly not enough to obtain any detailed information about the parameters of the testing multi-layered subgrade. As a rule, the use of such devices consists in finding a certain integral index (for example, SNP, Adjusted Structural Number, [9]) by means of certain empirical formulas, which characterizes the residual strength of the testing structure. At the same time, the existing development of hardware measurement tools, as well as progress in the development of effective data processing methods, allows using a larger number of input data (moreover, data from both deflectometer sensors and georadar and other devices), which, obviously, will make it possible to estimate more parameters of layered foundation, and perhaps with greater accuracy. This will allow to create a new generation high-speed tools for monitoring the state of the roadbed with online interpretation of obtained measurements.

Data processing supplied by the considered tools is based on certain models of layered foundations, discussed below.

2. Models of layered foundations

Parameters taking into account during the modeling of layered foundations can be divided into such main groups (Fig. 2).

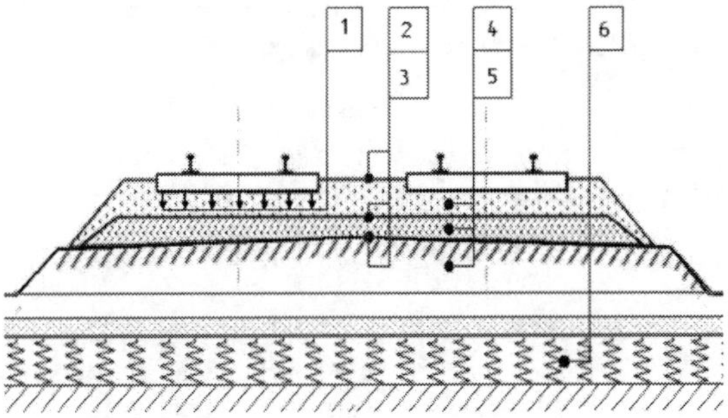

Fig 2. Layered foundation model.

1. **Parameters of the distributed load** that this layered packet is subjected to. Define the spline or other approximating function.
2. **Characteristics of layered packet geometry.** This group of parameters determines the number and thickness of the layers, the forms of the boundary between layers and top free surface defined by splines or other approximating functions.
3. **Properties of the interaction between layers.** These are type of the interaction between layers (hard grip or the possibility of slipping according to Amontons-Coulomb law etc.) and its parameters.
4. **Mechanical properties of the packet layers.** These, primarily, the elasticity constants (for example, shear modulus G and Poisson's ratio ν), parameters characterizing plastic, viscoelastic etc. properties of layers material and also fatigue development rate.
5. **Other physical parameters of layers material.** These include degree of compaction, porosity, permeability, viscosity etc.
6. **Properties of the ground subgrade** where the considered packet is located on. Fig. 2 shows the Pasternak's subgrade model that is described by two parameters: shear modulus G_S and "spring" stiffness rate k, but one can use models with less (Winkler foundation model) of more (Kerr subgrade model) parameters, however.

All or some of these properties (include discretized version of mentioned functions) form a model parameters vector $\mathbf{m}=(m_1,m_2,\ldots)^T$. Mathematical model of layered foundation formulated in displacement velocities can be expressed as an operator equation

$$\mathbf{A}[\mathbf{m}](\mathbf{v}) = \dot{\mathbf{f}}[\mathbf{m}]. \quad (1)$$

Here $\mathbf{v}=\mathbf{v}(\mathbf{x})=\dot{\mathbf{u}}$ is a vector-function of displacement velocities of packet points (point over symbol means as usual time derivative) defined in composite domain $\Omega=\Omega_1\cup\Omega_2\cup\ldots\cup\Omega_K$ (Ω_i, $i=\overline{1,K}$ – domain, holds by the i-th layer of the packet, K-th layer is in the bottom of packet), $\mathbf{f}[\mathbf{m}](\mathbf{x})$ – volumetric forces vector, $\mathbf{A}[\mathbf{m}](\cdot)$: $\mathrm{H_v}[\mathbf{m}]\to\mathrm{H_f}[\mathbf{m}]$ – vector operator, which components defined as

$$\mathrm{A}_i[\mathbf{m}](\mathbf{v})=\frac{\partial\dot{\sigma}_{ij}[\mathbf{m}](\varepsilon_{kl}(\mathbf{v}))}{\partial x_j},\quad i,j,k,l=1,2,3. \tag{2}$$

We use the summation over repeated indexes as is customary in tensor analysis. Components of stress velocities tensor $\dot{\sigma}$ relates with small deformation velocity tensor components $\dot{\varepsilon}$ by the defining relations (fundamental law of material)

$$\dot{\sigma}_{ij}=\dot{\sigma}_{ij}[\mathbf{m}](\dot{\varepsilon}_{kl}). \tag{3}$$

It can be linear elastic law, Prandtl-Reuss plastic flow equations or other more sophisticated plasticity (or viscoelasticity) theory equations. Small deformation velocity tensor components defines through the displacement velocities by Cauchy-Green formulas

$$\dot{\varepsilon}_{kl}(\mathbf{v})=\frac{1}{2}\left(\frac{\partial \mathrm{v}_k}{\partial x_l}+\frac{\partial \mathrm{v}_l}{\partial x_k}\right). \tag{4}$$

Structure of functional space $\mathrm{H_v}[\mathbf{m}]$ of displacement velocities $\mathbf{v}=\mathbf{v}(\mathbf{x})$ defines by the system of restrictions: boundary conditions

$$\dot{\sigma}_n(\mathbf{x})=\dot{\mathbf{T}}[\mathbf{m}](\mathbf{x})=(\dot{\mathrm{T}}_1(\mathbf{x}),\dot{\mathrm{T}}_2(\mathbf{x}),\dot{\mathrm{T}}_3(\mathbf{x}))^{\mathrm{T}},\quad \mathbf{x}\in\mathrm{S}_\sigma,\ i=1,2,3, \tag{5}$$

$$\mathbf{v}(\mathbf{x})=\mathbf{v}_\mathrm{u}[\mathbf{m}](\mathbf{x}),\ \mathbf{x}\in\mathrm{S_u},\ \mathrm{S}=\partial\Omega=\mathrm{S}_\sigma\cup\mathrm{S_u}, \tag{6}$$

($\mathbf{n}$ is outer normal unit vector to the surface of the packet, $\sigma_{ni}=\sigma_{ij}n_j$ – coordinates of normal component of the stress vector) and condition of interaction between packet layers and between the packet and the subgrade. In case of hard concatenation («gluing») of layers i and k on the contact surface $\mathrm{S}_{ik}[\mathbf{m}]=\partial\Omega_i\cap\partial\Omega_k$ we have

$$\mathbf{v}^{(i)}(\mathbf{x})=\mathbf{v}^{(k)}(\mathbf{x}),\ \mathbf{x}\in\mathrm{S}_{ik}; \tag{7}$$

in case of contact with slipping of scabrous layers according to Amontons-Coulomb friction model

$$\begin{cases}\mathrm{v}_n^{(i)}(\mathbf{x})=\mathrm{v}_n^{(k)}(\mathbf{x}),\\ \dot{\sigma}_\tau^{(i)}(\mathbf{x})=\mu_{ik}[\mathbf{m}](\mathbf{x})\dot{\sigma}_n^{(k)}(\mathbf{x}),\end{cases}\quad \mathbf{x}\in\mathrm{S}_{ik}, \tag{8}$$

where $\mathbf{n}$ is normal vector to the contact surface S_{ik}, $\tau\perp\mathbf{n}$ is unit tangent vector determined as

$$\tau(\mathbf{x})=\frac{\mathbf{v}_\tau^{(k)}(\mathbf{x})-\mathbf{v}_\tau^{(i)}(\mathbf{x})}{|\mathbf{v}_\tau^{(k)}(\mathbf{x})-\mathbf{v}_\tau^{(i)}(\mathbf{x})|},\ \mathbf{v}_\tau^{(i)}=\mathbf{v}^{(i)}-\mathrm{v}_n^{(i)},\ \mathbf{v}_\tau^{(k)}=\mathbf{v}^{(k)}-\mathrm{v}_n^{(k)},\ \mathbf{x}\in\mathrm{S}_{ik}, \tag{9}$$

$\mu_{ik}=\mu_{ik}[\mathbf{m}](\mathbf{x})$ is friction coefficient between layers i and k.

Winkler model of packet interaction with the subgrade defines by the relation (we neglect friction between K-th packet layer and the subgrade)

$$\begin{cases}\dot{\sigma}_n^{(K)}(\mathbf{x})=k_\mathrm{S}[\mathbf{m}](\mathbf{x})\mathrm{v}_n^{(k)}(\mathbf{x}),\\ \dot{\sigma}_\tau^{(K)}(\mathbf{x})=0,\end{cases}\quad \mathbf{x}\in\mathrm{S}_K, \tag{10}$$

where S_K is the contact surface of the K-th layer and the subgrade, $k_\mathrm{S}[\mathbf{m}](\mathbf{x})$ is the local stiffness coefficient of the subgrade.

Equation (1) forms so-called "forward problem": having certain parameters vector $\mathbf{m}$ one needs to obtain the vector of displacement velocity $\mathbf{v}(\mathbf{x})$

$$\mathbf{v}=\Re(\mathbf{m}), \tag{11}$$

where $\Re(\cdot)$: $\mathbf{H_M}\to\mathbf{H_v}[\mathbf{m}]$ is a forward operator, $\mathbf{H_M}$ is model parameters vector space. For an "inverse problem" formulation one should takes into account an operator $\aleph(\cdot)$: $\mathbf{H_v}[\mathbf{m}]\to\mathbf{H_D}$, that for each $\mathbf{v}(\mathbf{x})$ defines a "model data vector" of the problem

$$\mathbf{d}=\aleph(\mathbf{v})=\aleph(\Re(\mathbf{m}))=\mathbf{g}(\mathbf{m}), \tag{12}$$

where $\mathbf{H_D}$ is a model data space, $\mathbf{g}(\cdot)$: $\mathbf{H_M}\to\mathbf{H_D}$ is a "forward problem operator". As a model data, one should consider a radial velocity $\mathbf{v}_i$ of certain roadbed point A_i, scanned by each used deflectometer (Fig. 1):

$$\mathrm{d}_i=l_{ij}\mathrm{v}_{ij},\ j=1,2,3,\quad i=1,2,\ldots,K_d, \tag{13}$$

where l_{ij} are direction cosines of i-th deflectometer scanning ray, K_d is number of deflectometers used.

The essence of inverse problem is "to invert" the forward problem operator $\mathbf{g}(\cdot)$: for each (may be noisy) measured by the deflectometers vector $\mathbf{d}$ find the corresponding model parameters vector $\mathbf{m}$, that satisfies (12).

Method for solving the formulated inverse problem is proposed in [13]. This method is based on information-probabilistic (Bayesian) approach to solving the inverse problem [12]. According to this approach, solution of this problem is a posterior probability density of estimated model parameters vector $\mathbf{m}$. For this approach, the "prior" (or "a priori") information about estimated parameters $\mathbf{m}$, expressed as a correspondent prior probability density, plays a very important role. Without such information sometimes is impossible to obtain the qualitative inverse problem solution.

3. Multigrid solutions of the forward problem

To investigate the solutions of the considered inverse problems for a layered foundation, gradient methods to minimize a certain residual functional, and statistical modeling methods (Monte-Carlo) to study the behavior of this functional near extreme points are used. When using any of these methods, the key issue is the effective solution of the "direct problem" for layered packet with specified characteristics defined by a certain (fixed) model parameters vector **m**. For this purpose, one should solve two such main associated issues.

1. *Choose an appropriate discretization scheme.*
2. *Develop fast numerical algorithm for solving obtained system of grid equations.*

The choice of a successful discretization scheme determines the accuracy of the obtained solution of a differential problem (1). For both the most popular Finite Element and Finite Difference Methods the building of appropriate computational grid defines the quality of entire scheme of discretization. In [14] the problem of creating practical grid for an elastic layered packet was considered. The recommendations are: 1) use a regular grids as the most suitable for further calculations; 2) use interpolation formulas as grid construction procedures (Fig. 3).

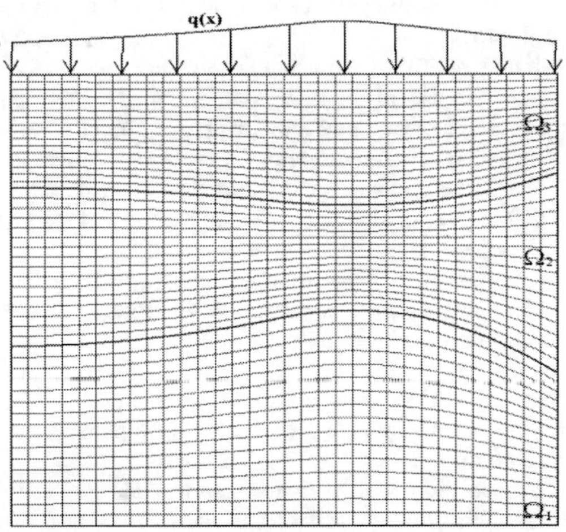

Fig. 3. Regular grid created for the layered packet by interpolation formulas technique.

As a result of the discretization process we obtain the system of algebraic equations

$$\mathbf{A}^{\mathbf{h}}\mathbf{v}^{\mathbf{h}} = \mathbf{f}^{\mathbf{h}} \qquad (14)$$

(**h** is a discretization parameter), that approximates the source differential problem (1). The main task is to effectively (numerically) solve the system (14).

Today one of the most advanced for such purpose (especially for elliptic problems like (1)) is a Multigrid Method [15], first proposed by R.P. Fedorenko in 1962. It's efficiency is defined by the fact, that it needs $O(N)$ arithmetic operations to approximately solve (14) "at the level of discretization error", where N is the number of unknowns in the system (14). This method is based (in its classical Fedorenko's formulation) on two main ideas. First, instead of solving the system (14) on only one grid, one considers the discretization of the problem (1) on the system of nested grids that forms a "hierarchy of grid levels". So the discretized problems

$$\mathbf{A}_k\mathbf{v}_k = \mathbf{f}_k, \quad k=1,2\ldots,K_{\max} \qquad (15)$$

are considered, where the index k=0 (the lowest grid level) corresponds the coarsest grid and the system (15) contains a little number of unknowns, that provides the fast solving of it by a direct method (Gauss or Cholesky decomposition). Index number $k=K_{\max}$ (the highest grid level) corresponds the finest grid (grid with the smallest mesh-size), that is our goal for system (14) to solve.

The solving procedure begins from the lowest grid level k=0. Then obtained solution interpolates to the higher level k=1, where it is refining, and then corrected solution interpolates to the next level k=2 and so on. This procedure makes up the essence of so-called Full Multigrid (FMG) Algorithm.

The method of précising the approximate solution on the level k is the most sophisticated issue in entire multigrid theory. The main idea of this method is to use the means of coarser grids for correction of the solution on the current level k. This procedure calls Coarse Grid Correction (CGC) Algorithm or CGC-Cycle.

Algorithm CGC_CYCLE(k, $\mathbf{f}_k$, $\mathbf{u}_k^{(0)}$, ν_1, ν_2, γ)

/ Algorithm for approximate solution of a system of linear equations* $\mathbf{A}_k\mathbf{v}_k=\mathbf{f}_k$*, where k is current grid level,* $\mathbf{v}_k^{(0)}$ *is initial approximation,* $\mathbf{A}_k$ *is system matrix,* $\mathbf{f}_k$ *is right-hand side,* ν_1 *is number of pre-iteration of relaxation scheme,* ν_2 *is number of post-iterations,* γ *is a cycle parameter,* $\mathbf{v}_k^{(1)}$ *is refined approximate solution. */*

Begin

1. If k=0, then the system (15) is solved by direct method. As a result, one obtains the value $\mathbf{v}_k^{(1)}$. Go to step 9.
2. (*Pre-relaxation*). Perform ν_1 smoothing iterations by the formula

$$\widetilde{\mathbf{v}}_k^{(0)} \leftarrow \mathrm{S}_k^{\nu_1}\mathbf{v}_k^{(0)}, \qquad (16)$$

where S_k is a *relaxation scheme operator*.

3. Calculate the residuals by

$$\mathbf{r}^k \leftarrow \mathbf{A}_k\widetilde{\mathbf{v}}_k^{(0)} - \mathbf{f}_k. \qquad (17)$$

4. *Projection* of the residuals to the coarser grid (level $k-1$)

$$\mathbf{r}^{k-1} \leftarrow \mathrm{R}_k^{k-1}\mathbf{r}^k, \qquad (18)$$

where R_k^{k-1} is a projection operator from finer to coarser grid.

5. Calculation and refinement of the correction vector $\mathbf{e}^{k-1}$ on the coarser grid, that is solving the coarse grid equation

$$\mathbf{A}_{k-1}\mathbf{e}^{k-1} = \mathbf{r}^{k-1} \qquad (19)$$

recursively by the same algorithm, beginning with null initial approximation

$\widetilde{\mathbf{e}}_k^{(0)} \leftarrow \mathbf{0}$
For i←1 **to** γ **do**
$\widetilde{\mathbf{e}}_{k-1}^{(i)} \leftarrow$ **CGC_CYCLE**($k-1$, $\mathbf{r}^{k-1}$, $\widetilde{\mathbf{e}}_{k-1}^{(i-1)}$, ν_1, ν_2, γ)

6. *Interpolation* of the obtained correction $\widetilde{\mathbf{e}}_{k-1}^{(\gamma)}$ to the finer grid (level k)

$$\mathbf{e}^k \leftarrow \mathrm{P}_{k-1}^{k}\widetilde{\mathbf{e}}_{k-1}^{(\gamma)}, \qquad (20)$$

where P_{k-1}^{k} is an interpolation operator from coarser to finer grid.
7. Correction

$$\hat{\mathbf{v}}_k^{(1)} \leftarrow \tilde{\mathbf{v}}_k^{(0)} + \mathbf{e}^k. \qquad (21)$$

8. (*Post-relaxation*). Perform ν_2 additional smoothing iterations by

$$\mathbf{v}_k^{(1)} \leftarrow S_k^{\nu_2}\hat{\mathbf{v}}_k^{(1)}. \qquad (22)$$

9. Return the result $\mathbf{v}_k^{(1)}$.
End

Full Multigrid Algorithm can be formulated as follows.
Algorithm **FMG_CYCLE**(ν_1,ν_2,γ,μ_k)
/ Algorithm for approximate solution of a system of linear equations* $\mathbf{A}_{Kmax}\mathbf{v}_{Kmax}=\mathbf{f}_{Kmax}$ *on the highest grid level* K_{max}. *Parameters* ν_1, ν_2, γ *are the same as in algorithm* **CGC_CYCLE**, μ_k *is a number of CGC-cycles performed on the grid level k. */*
Begin

1. Solve the system (15) by direct method on the grid level k=0. As a result we obtain the value $\mathbf{v}_0^{(1)}$.
For k=1 **to** $K_{\max}$ **do**
Begin
2. Interpolation of the obtained approximate solution $\mathbf{v}_{k-1}^{(1)}$ to the finer grid:

$$\mathbf{v}_k^{(0)} \leftarrow P_{k-1}^{k}\mathbf{v}_{k-1}^{(1)}. \qquad (23)$$

Interpolation operator may be differ from one used in CGC-cycle.
3. Refining of the solution on the grid level k.

$$\mathbf{v}_k^{(1)} \leftarrow \mathbf{v}_k^{(0)}.$$

For i←1 **to** μ_k **do** $\mathbf{v}_k^{(1)}$ ← **CGC_CYCLE**(k, $\mathbf{f}_k$, $\mathbf{v}_k^{(1)}$, ν_1, ν_2, γ)
End
4. Return the result $\mathbf{v}_{Kmax}^{(1)}$.
End

The main goal of the FMG-algorithm is to obtain an approximate solution $\mathbf{v}_{Kmax}^{(1)}$, that satisfied

$$\|\mathbf{v}_{Kmax}-\mathbf{v}_{Kmax}^{(1)}\| \le c h_{K_{max}}^2, \qquad (24)$$

where $\mathbf{v}_{Kmax}$ is the "exact" solution of (15) and h_{Kmax} is a characteristic mesh-size on the highest grid level (we assume bilinear finite elements used, that provide the second order of approximation), constant c is independent on the mesh-size.

Although the theory promises $O(N_{Kmax})$ complexity of the FMG-algorithm for achieving (24) for wide classes of problem, in practice it's not easy to get such efficiency for a specific problem. It may be related with incorrect choice of the parameters for CGC- and/or FMG algorithms in problems where physical or geometric anisotropy is presented. Therefore, the issues of a suitable tuning of the multigrid algorithm components play a key role in successful application of this method.

In [16] the question of appropriate choice of the multigrid parameters for the boundary-value problem (1) for the multi-layered foundation is investigated. This problem has not physical but geometric anisotropy: the length of the roadbed area considered is larger than its width and height. The determined parameters of the Multigrid Algorithm are: type of the relaxation operator (16); interpolation and projection operators (20), (18); method for creating the coarse-grid matrix $\mathbf{A}_k$ on the levels $k<K_{max}$, numbers ν_1, ν_2 of pre- and post-relaxation, number γ of correction refinement procedures, numbers μ_k of précising steps in FMG-algorithm.

It has been established, that the best choice, taking into account computational work expenses, is as follows.

Relaxation scheme is block-wise Gauss-Seidel operator on vertical lines, ν_1=1, ν_2=0, γ=1 (such variation of the CGC-cycle calls "V-Cycle"), μ_k=3. The rest of parameters is typical. This choice of parameters provides the standard $O(N_{Kmax})$ multigrid efficiency.

Prospects for further research. The following areas of improvement of methods for assessing the diagnostic parameters of the roadbed are promising.

1. Because of significant importance of the "prior" information for perfect inverse problem solution analysis, the main directions for further research may concern the methods of involvement of the information obtained by another considered methods of diagnostic parameters estimation (georadation, seismic, etc.), creation of the detailed roadbed maps with such data and precise binding technique development (including geolocation methods).
2. Improvement of inverse problem solving approaches (including optimal discretization procedures for unknown parameters vector **m**) for obtaining the best resolution.
3. Development of modern approximation technique (including neural networks etc.) for fast online interpretation of scanning data.

References

1. T. Butko, M. Muzykin, A. Prokhorchenko, H. Nesterenko, H. Prokhorchenko, *Transport and Telecommunication Journal*, **20 (3)**, 215 (2019)
2. A. Glikman, *NTF "GEOFIZPROGNOZ" po zakazu stroitelnogo zhurnala*, Electronic resource. Access mode: https://www.newgeophys.spb.ru/ru/article/railways_crash/
3. E. Nepomnyashih, K. Kirpichnikov, Diagnostika sostoyaniya zheleznodorozhnogo puti, ZabIZhT, (2012) 109 c.
4. V. Petrenko, D. Yampolskij, I. Svyatko, *Nauka ta progres transportu*, **4 (46)** (2013)
5. J. Luchko, I. Kravec, Visnik Odeskoyi derzhavnoyi akademiyi budivnictva ta arhitekturi, **71** (2018)
6. A. Nikolaides, *Highway Engineering Pavements, Materials and Control of Quality*, International Standard Book Number-13: 978-1-4665-7997-2 (eBook - PDF), (2014)
7. P. Ullidtz, Elsevier Science Publishers B.V., (1987)

8. G. Flintsch, S. Katicha, J. Bryce, B. Ferne, S. Nell, B. Diefenderfer. *Assessment of Continuous Pavement Deflection Measuring Technologies*, (2013)
9. G. Morosiuk, M. Riley, J.B. Odoki, *The Highway Development and Management Series*, **6**, (2004)
10. A. Das, *Analysis of Pavement Structures*, International Standard Book Number-13: 978-1-4665-5856-4 (eBook - PDF), (2015)
11. G. Berkovic, E. Shafi, *Advances in Optics and Photonics* **4**, 441–471 (2012)
12. A. Tarantola, *Inverse Problem Theory*, 0-89871-572-5 (pbk.), (2005)
13. A. Trofimov, Visnik AMSU, Series Technical Science, **2 (52)**, 48 (2014)
14. A. Trofimov, J. Petrova, Visnik AMSU, Series Technical Science, **2 (54)**, 48 (2015)
15. U. Trottenberg; C. W. Oosterlee; A. Schüller, *Multigrid*, Academic Press, 978-0-12-701070-0, (2001)
16. A. Trofimov, Visnik UMSF, Series Technical Science, **1 (55)**, 119 (2016)

Optimization of the design of relay valves for rail vehicles braking systems in the context of train traffic safety

Sylwin Tomaszewski[1], *Franciszek* Tomaszewski[2,*], *Włodzimierz* Stawecki[1], *Patryk* Urbański[2], and *Mariusz* Far[1], and *Yaroslav* Bolzhelarskyi[3]

[1] Łukasiewicz Research Network – Rail Vehicles Institute "TABOR", ul. Warszawska 181, 61-055 Poznań, Poland
[2] Poznan University of Technology, Pl. Marii Skłodowskiej-Curie 5, 60-965 Poznań, Poland
[3] Dnipro National University of Railway Transport named after Academician V. Lazaryan,, Lviv branch, 79052 Lviv, I. Blazhkevich Street, 12a, Ukraine

Abstract. The article presents the functions of the relay valve and its place in brake systems of rail vehicles. The relay valve were divided according to their type, the structure and principle of their operation were discussed, and examples of rail vehicles and modules in which relay valves are installed are presented. Because of their function in braking systems the relay valves should be characterized by high reliability and correct carrying out the tasks in the braking system of rail vehicles. The article describes selected structural factors affecting the operation of the relay valve and which are important for ensuring the safety of train traffic. Such factors are the lubricants used in the relay valves and the way of controlling it simultaneously with two signals with the function of an alternative that eliminates unreliable elements of the braking system. The analysis of the test results of the lubricant parameters and the testing of the relay valves on the test bench showed that the correctness of the relay valves depends essentially on the properties of the lubricant used, especially at low temperatures. On the basis of the tests carried out, a grease was selected that at low temperatures meets the criteria for correct and reliable operation of relay valves in rail vehicles. In terms of the control method, a comparison of two control signal delivery systems to the relay valve has been made, a new design solution is presented that ensures greater operational reliability and thus greater safety of the train traffic.

1 Introduction

The braking system is an indispensable element of every vehicle. It is used to reduce speed and stop the vehicle, especially in an emergency situation, in the shortest possible time and with the greatest certainty. It is also one of the basic vehicle systems and a key system in terms of ensuring safety for vehicle users as well as for people and objects in their surroundings. The failure of braking systems is one of the main causes of traffic accidents caused by motor vehicles. Therefore, ensuring the braking system's efficiency and reliability is the main goal at the stage of construction, manufacture and operation of all types of vehicles operating in the world. The basic transport branches, in which the aim is to ensure operational safety by maintaining the highest quality, efficiency and reliability of braking systems, are road transport and rail transport [1,2,3].

The article provides a partial analysis of the brake system of rail vehicles. For the proper functioning of entire braking system in every vehicle, which is a part of the train set, there must be the suitable apparatus on it that receives the signals about braking or releasing of the brakes and standardized actuators in order to obtain similar braking performance of all wagons on the train. This applies to the most responsible braking devices, such as the distributor valve and the relay valve or relay valves of brake cylinders, emergency braking accelerator, weighing valve and anti-slip system [4]. The required parameters of the rail vehicle braking system are defined in the relevant UIC leaflets and European standards.

The article presents the exemplary of tests results carried out on the relay valve and describes relations between selected structural factors affecting the parameters it achieves, which have a direct impact on the maintaining of train traffic safety. On the basis of the obtained test results, directions of introducing appropriate changes in the technical documentation of the relay valves were proposed, which will ensure the fulfillment of current requirements during the operation and the increase the reliability of its operation, ensuring a high level of safety of the operated vehicles.

2 Relay valve – types, construction and application

The relay valve is a device used to amplify the pneumatic signal with the required pressure (through the control system) in the cylinder (s). In the rail vehicle brake system the relay valve can be installed at a some distance from the distributor valve on the separate bracket (Fig. 1) or on a common bracket with the

* Corresponding author: franciszek.tomaszewski@put.poznan.pl

distributor valve (so-called the compact system) or on the pneumatic board of the vehicle (wagon) [5].

Figures 2 and 3 present examples of relay valves from the relay valve family for filling the vehicle brake cylinders of traction units.

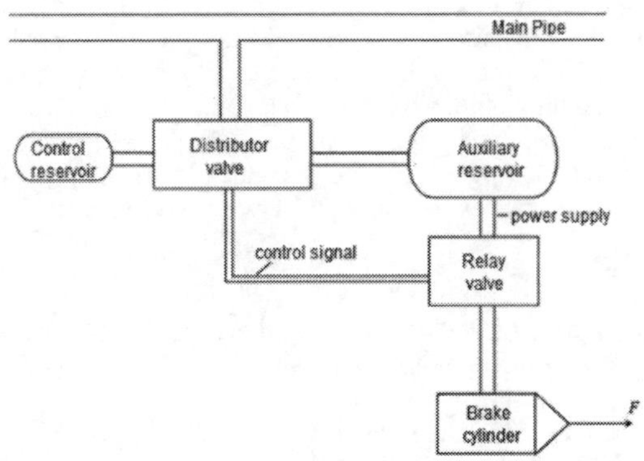

Fig. 1. Location of the relay valve in the rail vehicle brake system [5]

Fig. 2. Relay valve for filling the brake cylinders with regulated ratio

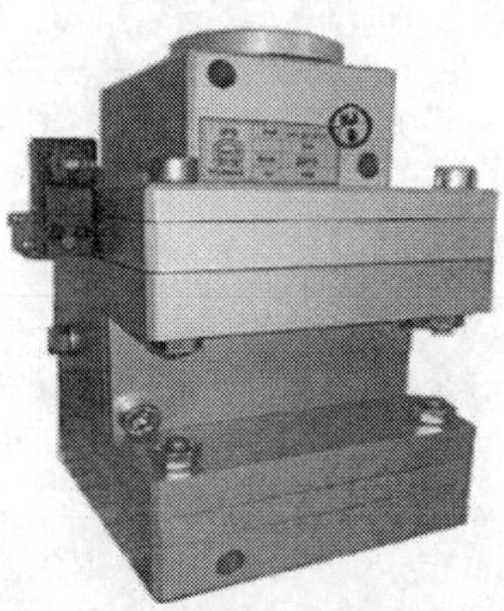

Fig. 3. Double stage relay valve for filling the brake cylinders with regulated ratio and with two control signals

The division of relay valves by types is defined by the European norm EN 15611+A1 [6]:

1. Single stage relay valve type A

Relay valve with one fixed ratio, where the ratio can be less, equal or greater than 1.

2. Multi stage relay valve type B

Relay valve with more than one fixed ratio, where the ratios can be less, equal or greater than 1.

3. Relay valve type B1

Multi stage relay valve that can change ratio during a brake application.

4. Relay valve type B2

Multi stage relay valve where a change of ratio cannot take place during a brake application.

5. Empty/load relay valve

Specific type of multi stage relay valve (type B1 or B2) with only two stages, giving an empty (tare) or a loaded output pressure proportional to input pressure dependant on the load signal input.

6. Variable load relay valve type C

Relay valve with a continuously changeable ratio, where a load signal is used to change the ratio.

7. Multi stage variable load relay valve type C1

Relay valve with a continuously changeable ratio, where a load signal is used to change the ratio and with a multi stage feature added.

8. Two (multi) input relay valve type D

Relay valve (type A, B or C) with two (or more) input pressure, controling a single output pressure.

9. Variable load relay valve with kinked characteristic type E

Variable load relay valve (type C) with a special non-linear characteristic, with automatically comes into operation when the vehicle has more than a certain load.

The relay valves presented in Figures 2 and 3 are mainly used in braking systems of modernized Electric Traction Units (Fig. 4) and modern locomotives. The relay valves developed and made at the Lukasiewicz Research Network – Rail Vehicles Institute "TABOR" are usually mounted as the modular apparatus on the pneumatic brake boards of rail vehicles (Fig. 5, 6).

Fig. 4. The Electric Traction Unit EN57 after modernization

In the relay valve construction (Fig. 7) a number of solutions based on actuators, such as pistons sealed with diaphragm transferring the appropriate forces, valves with a rubber O-ring and U-type sealing rings, springs and a precise piston-slide system with a rolling diaphragm the smooth change of pressure in the cylinders as a function of the load. In the construction of this relay valve it is provided for a very wide range of regulation of its parameters in order to adjust its characteristics to different needs and types of rail vehicles [5].

Fig. 5. The modular pneumatic brake board of rail vehicle

Fig. 6. The modular pneumatic brake board of rail vehicle

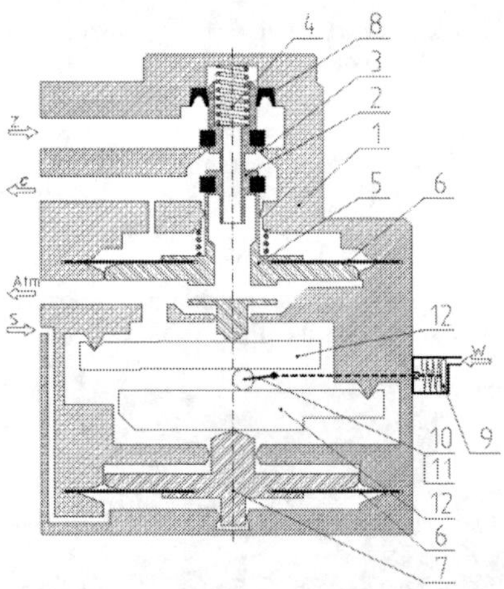

Fig. 7. An illustrative construction of the single stage pressure transformer developed by Łukasiewicz Research Network – Rail Vehicles Institute "TABOR" [7]
(Z – power supply, C – cylinder, S – control signal, W – pneumatic signal of the weight of the vehicle, 1 – relay valve body, 2 – filling valve, 3 – valve seat, 4 – spring, 5 – upper piston with a venting rod, 6 – membrane, 7 – lower piston, 8 – sealing ring, 9 – piston - slide system, 10 – roller, 11 – inclined swing lever, 12 – levers)

The main element of the relay valve consists of a venting rod connected to the piston (5) and the diaphragm (6). It moves in the body (1), up or down depending on the pressure difference occurring on both sides of the diaphragm (6). The supply air (Z) from the auxiliary reservoirs is constantly supplied to each relay valve. When the distributor valve generates a control signal (S), which reaches to the chamber under the lower piston (7) and diaphragm (6), the piston (7) moves upwards and the rod (5) opens the valve (2) via the lever (12), which allows the supply air to enter onto the output side (C) of the relay valve to the brake cylinder capacity. In the opposite situation, when the control signal (S) decreases, the valve closes the supply flow (Z), and the air on the output side (C) of relay valve escapes from the cylinder into the atmosphere through the hollow rod. The pneumatic signal of the weight of the vehicle (W) changes the lever ratio (12) between the pistons (7) and (5) [5].

3 Impact of the applied lubricants on relay valve operation

To meet the requirements demanded from the relay valves with a complicated structure, it is essential to make their subassemblies very precisely and to assemble and adjust them carefully. Because of observation of a certain scatter of the relay valve parameters, there was a need to carry out the detailed tests, allowing to determine the influence factors and optimize the structure in order to ensure the highest possible repeatability and reliability of operation of the relay valves. One of the selected factors of influence was the type of lubricants used in the relay valves [7].

During the analysis of the operation (functioning) of the relay valves in operation in winter, it was found that there was the increase of failure of these devices during this period. Therefore, it was decided to carry out the correctness tests of the relay valves operation at sub-zero temperatures.

During the tests of the relay valve in the climatic chamber it was observed that the type of the used grease had the influence on the operation of the relay valve at different temperatures due to the changes in the resistances of motion. Therefore, three selected types of lubricants were tested in a climatic chamber [7].

The lubricant tests were carried out in the Laboratory for Exploitation Materials of the Poznan University of Technology. The measured parameter was the consistency of the lubricant, specifically its hardness index. For the measurements of the hardness of lubricants, a PL-12DC laser penetrometer, manufactured by the Institute for Sustainable Technologies - National Research Institute in Radom, was used, shown in Figure 8.

The laser penetrometer allows to determine the consistency of the grease. The correct consistency must ensure that the grease does not generate too high resistances. The consistency is graded according to the scale developed by the NLGI (National Lubricating Grease Institute) presented in Table 1. The softer is the grease the lower number is on the scale. During the test, the depth of penetration of the standard cone into a sample of grease is measured in tenths of a millimeter (Fig. 9).

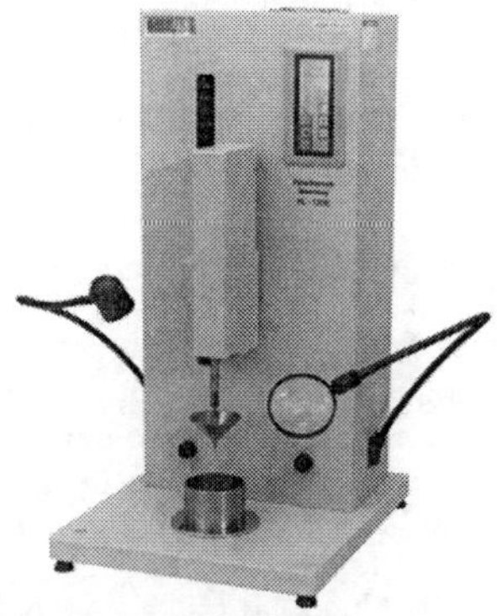

Fig. 8. Laser penetrometer PL-12DC [7]

Fig. 9. Laser penetrometer PL-12DC – view during the test [7]

It is an automatic device, in which it is used the modern laser technology to measure the penetration. The penetrometer is intended, among others, for testing the consistency of petrochemical products (e.g. lubricants, asphalts, paraffins), food products, cosmetic products, as well as mortars and plastic explosives [8,9].

Table 1. Classification of greases under NLGI consistency class [7]

Number NLGI	Penetration according to ASTM (10^{-1} mm)	Appearance at room temperature
000	445 - 475	very liquid
00	400 - 430	liquid
0	355 - 385	semi-liquid
1	310 - 340	very soft
2	265 - 295	soft
3	220 - 250	medium hard
4	175 - 205	hard
5	130 - 160	very hard
6	85 - 115	extremely hard

The following lubricants popularly used in Łukasiewicz Research Network – Rail Vehicles Institute „TABOR" were tested:

- "Vecolit EPX 00" – semi-liquid gear lubricant and central lubrication,
- "Elf MULTI" – lithium calcium grease for bearings,
- "TF silikon + teflon" – silicone grease with the addition of Teflon.

The above lubricants were tested at three temperatures: 24 ° C (room temperature), 0 ° C and (- 26 ° C), which were obtained using a laboratory freezer [7].

The results of the tests carried out in the laboratory are listed in Table 2. A graphical image of the change of the consistency of various greases as a function of temperature is shown in Figure 10.

Table 2. The test results of greases consistency as a function of temperature [7]

Name of lubricant	Vecolit EPX 00			Elf MULTI			TF silikon + teflon		
Temperature	-26 °C	0 °C	24 °C	-26 °C	0 °C	24 °C	-26 °C	0 °C	24 °C
Measurement I	157,4	359,8	443,2	126,0	242,1	273,1	284,3	290,0	308,8
Measurement II	174,2	358,7	437,5	117,0	234,5	266,0	274,3	280,2	305,4
Measurement III	160,7	369,7	446,0	136,1	229,4	263,8	290,4	291,4	306,1
Average	164,1	362,7	442,2	126,4	235,3	267,6	283,0	287,2	306,8

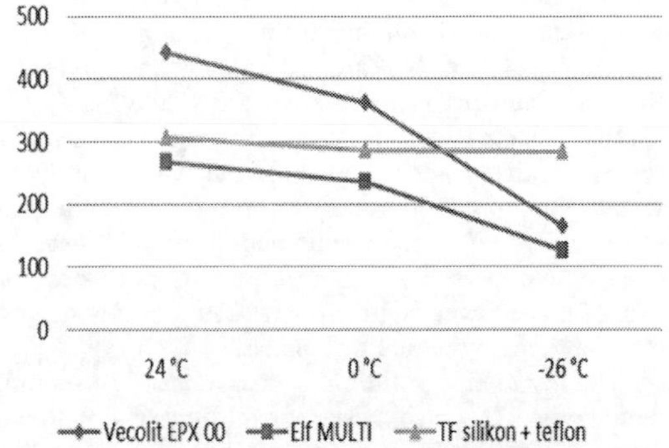

Fig. 10. Consistency of greases as a function of temperature [7]

Figure 11 shows the view of one of the relay valve's elements (valve) with the used Elf MULTI grease and Vecolit EPX 00 grease (used in the relay valve) at temperature -26°C. It is clearly seen that the used lubricants, under the influence of low sub-zero temperature, changed their consistency from the liquid into a solid consistency, which is disadvantageous for the correct operation of the relay valve [7].

The carried out tests have shown that this causes occurrence of large resistances of movement and leads to the suspension of the valve, and thus to the lack of full release of the cylinder (C) in response to the control signal (S), and the occurrence of large leak in the power supply system (Z) of the braking system of the rail vehicle [7]. The pressure characteristics in the relay valve system during operation with frozen grease and the suspended valve are presented in Figure 12.

Fig. 11. View of the used Elf MULTI grease (on the left) and Vecolit EPX 00 grease (on the right) in the relay valve operating at temperature of -26 ° [7]

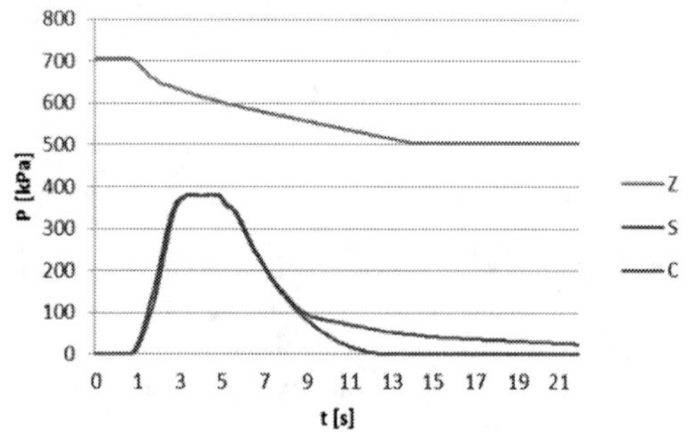

Fig. 12. Exemplary course of the test with a suspended valve of relay valve [7]

However, Figure 13 shows the valve with the used TF silicone + teflon grease, which during operation at -26 °C did not change its consistency and thus provided the correct operation of the relay valve at the sub-zero temperatures.

Fig. 13. View of the used TF silikon + teflon grease in the relay valve operating at temperature of -26 °C [7]

As a result of the above analysis of lubricants, the previously used Elf MULTI grease for TF silicone + teflon grease has been changed. However, further observations of pressure transformers during operation have shown that this grease adversely affects the operation of the relay valve systems having elements made of silicone rubber (piston diaphragms, rubber bellows). Therefore, it was selected the grease, which used in the specific elements of the relay valve construction allows it to function correctly and reliably.

4 Elimination of unreliable elements of the rail vehicles braking system

As already mentioned above, the basic element of pneumatic brake systems of railway vehicles is the relay valve supplied from the auxiliary tank with compressed air, filling brake cylinders during braking. In brake systems of traction units or pre-defined vehicles, the cylinder relay valve is controlled by pneumatic signals coming from the pneumatic brake control system (from the distributor valve) and from the electro-pneumatic brake system. In locomotives, the cylinder relay valve can be controlled by a signal from the distributor valve and from the direct auxiliary brake system.

Until now, the most commonly used system was a relay valve controlled by a signal from these two sources through a double-return valve realizing the alternative function (Fig. 14). This arrangement ensures the cylinder pressure on the vehicle due to a stronger control signal.

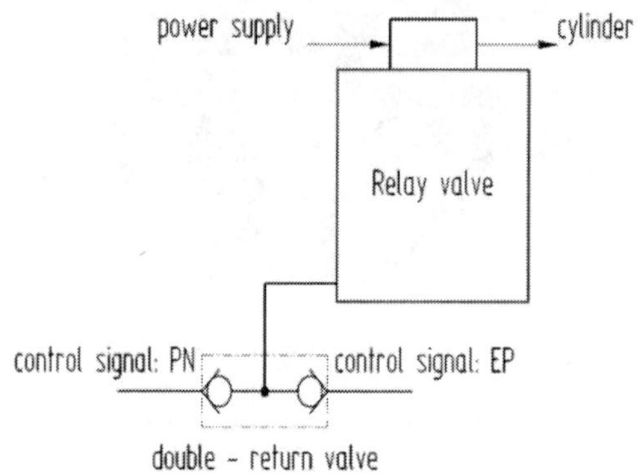

Fig. 14. Diagram of supplying a control signal to the relay valve with a double-return valve

Experience collected as a result of many years of observation, many double-return valves in operation has shown that they sometimes fail, reducing the reliability of the entire system. These failures occurred as a result of slight contamination of the valve or loss of lubricating grease after a long service life. As part of the ongoing development work on increasing the functionality and reliability of the braking equipment in Łukasiewicz Research Network – Rail Vehicles Institute "TABOR", a control system for a relay valve controlled from two sources with an alternative function was developed, in which the commonly known and used double-return valve was completely eliminated from this system. A relay valve with two control pistons was developed (Fig.15), the lower control piston being a moving piston relative to the upper control piston.

Thanks to this solution, it was possible to simplify the relay valve control system by eliminating the double-return valve (Fig.16) and to ensure greater reliability of the system, because the newly developed control system based on two control pistons is completely resistant to dirt that immobilized the double-return valve and does not require special lubrication, which ensures correct operation of the brake system in the long-term operation of the vehicle between inspection repairs.

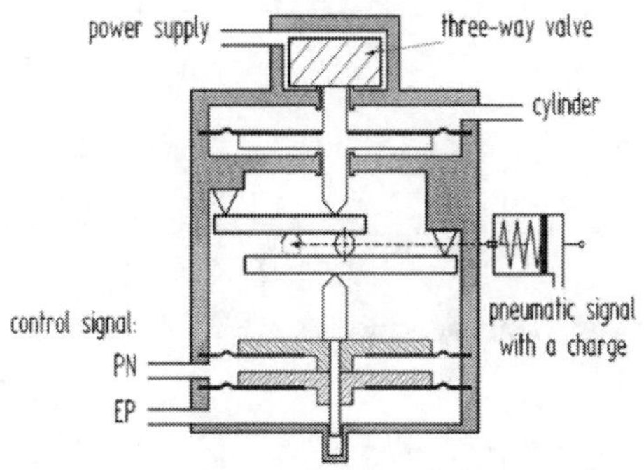

Fig. 15. Construction of the relay valve with a control system with an alternative function

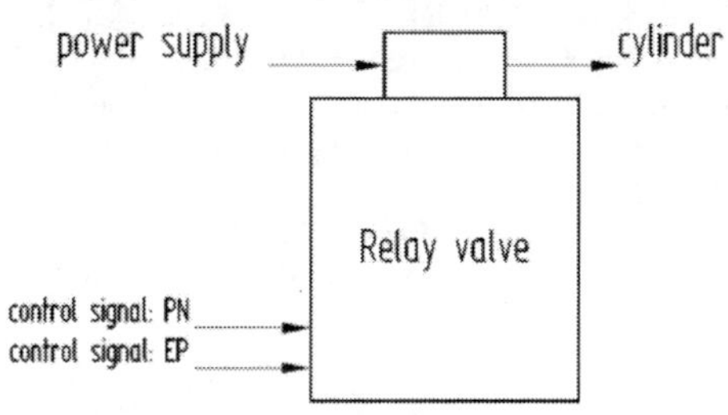

Fig. 16. Diagram of supplying a control signal to the relay valve with an alternative function

Any interruption in the delivery of the control signal causes that the rely valve will not work or will work incorrectly, which results in the lack or incorrect pressure in the brake cylinders and, consequently, no braking or lengthening its distance. This condition directly affects the reduction of train traffic safety.

5 Conclusions

Ensuring the safety of vehicle users as well as people and objects in their surroundings is the main and most important purpose of all vehicle builders. In particular for mass transport vehicles, such as rail vehicles, this is the topic with the highest priority. Therefore, every effort should be made to ensure that every element of the vehicle responsible for safety works with the greatest certainty and has a high reliability factor. In the case of a rail vehicle, such element is the cylinder relay valve located in the brake system.

The relay valves as one of the most responsible elements of the rail vehicle braking system are subjected to continuous analyzes in Łukasiewicz Research Network – Rail Vehicles Institute "TABOR" in order to improve the structure and improve the reliability of their operation. The article identifies two structural factors that are important for the optimization of the structure in the direction of improving the operational reliability and thus ensuring the safety of train traffic.

One of these factors was the applied lubricant. Therefore presented the test results of lubricants in order to select the lubricant with the best parameters as a function of temperature, in application to the relay valve. During the analysis of the test results it was found that the grease used in the relay valve must have the stable parameters in the wide range of temperatures. Especially at low temperatures, it must keep as liquid consistency as possible. Therefore, the silicone grease was introduced in the appropriate systems of relay valve, which has a practically constant consistency in the range of temperature from -30 °C to +40 °C. On the basis of the carried out tests the grease was selected, which used in the specific elements of the relay valve construction allows it to function correctly and reliably, and thus guarantees reliable operation of the rail vehicle brake system.

The second factor was the method of controlling the relay valve with two control signals from the alternative function, the use of which eliminated the unreliable system with a double-return valve. Such a solution not only ensures greater reliability of the braking system and increases train safety, but also simplifies the construction of this system.

References

1. Tomaszewski F., Tomaszewski S.: RAIL VEHICLES, Volume: **4**, Pages: 1-10, Published: (2015)
2. J. Merkisz, L. Rymaniak, EKSPLOATACJA I NIEZAWODNOSC-MAINTENANCE AND RELIABILITY, Volume: **19**, Issue: 4, Pages: 522-529, DOI: 10.17531/ein.2017.4.5, Published: (2017)
3. P. Daszkiewicz, M. Andrzejewski, MATEC Web of Conferences, Volume: **118**, Article Number: UNSP 00014, DOI: 10.1051/matecconf/201711800014, Published: (2017)
4. Kaluba M.: PROCEEDINGS OF THE INSTITUTE OF VEHICLES. Volume: **1(60)**, Published: (2006)
5. Tomaszewski F., Tomaszewski S.: RAIL VEHICLES, Volume: **4**, Pages: 14-19, Published: (2014)
6. EN 15611+A1: Railway applications – Braking – Relay valves, (2011)
7. Tomaszewski S., Tomaszewski F., Strzemkowski J.: RAIL VEHICLES, Volume: **3**, Pages: 18-24, Published: (2018)
8. M. Andrzejewski, D. Gallas, P. Daszkiewicz, A. Merkisz-Guranowska, H. Stawecka, MATEC Web of Conferences, Volume: **118**, Article Number: UNSP 00015, DOI: 10.1051/matecconf/201711800015, Published: (2017)
9. L. Rymaniak, MATEC Web of Conferences, Volume: **118**, Article Number: UNSP 00009, DOI: 10.1051/matecconf/201711800009, Published: (2017)

Railway transportation of dangerous goods: a bibliometric aspect

Tetiana Kolesnykova[1,*], *Olena* Matveyeva[1], *Lev* Manashkin[2], and *Maxym* Mishchenko[3]

[1] Dnipro National University of Railway Transport named after Academician V. Lazaryan, Scientific and Technical Library, Dnipro, Ukraine
[2] Self-Employed, West Hartford, Connecticut, USA
[3] Dnipro National University of Railway Transport named after Academician V. Lazaryan, Department of Economics and Management, Dnipro, Ukraine

Abstract. The purpose of this paper is to research and define the promising worldwide scientific trends in the field of railway transportation of various dangerous goods. To obtain relevant empirical data, the authors reviewed the world literature on paper topic using Scopus and Web of Science citation bases. We determined that this research was focused on several major thematic areas: 1) automation and telematics systems; 2) navigation systems; 3) logistics; 4) energy; 5) locomotives; 6) freight cars; 7) materials; 8) rails; 9) impact on the environment and people. The article used mapping, ensuring a visual perspective for researchers and helping to understand general situations in specific subject areas of the research. This study provides useful information concerning the development of the field of research for the railway transportation of dangerous goods, identifying those academics (authors, countries and institutions) that have made the greatest contribution to its development and defining the priority research directions

1 Introduction

The topic of dangerous goods refers primarily to the transport of the respective goods including the interim storage, if caused by the transport.

Transportation of dangerous goods (TDG) include the activities related to the movement of dangerous goods from their place of manufacture or storage to their destination with the preparation of cargo, packaging, vehicles and crew, reception of goods, carrying out cargo operations and short-term storage of goods at all stages of their transfer [1].

Dangerous goods (DG) are items or substances that when transported are a risk to health, safety, property or the environment [2]. There are the following types of hazard connected with dangerous goods: flammability, explosion hazard, harmfulness, toxicity, infectious hazard, corrosion, oxidative properties of goods, radioactivity [1, p. 21].

The rapidly growing needs of society bring with them, on the one hand, the development of industry, new techniques and technologies, and on the other, the use of new types of hazardous substances. Currently, a large number of hazardous substances are transported by rail, which increases the risk of incidents during transportation. Thus, the statistics of rail transportation in the USA (2015 - Aug. 2018) shows an increase in the volume of traffic, including of DG: oil and oil products – by 12,780 carloads, or by 28.8%; chemistry – by 11,911 carloads, or by 7.8% (https://www.advis.ru/php/view_news.php?id=9009E291-1A24-A640-A324-64F6255B9A92).

The measures to minimize risks in TDG to acceptable levels are discussed by M. Šolc and M. Hovanec [3]. The most important measure here is the rules for the transport of dangerous goods, the responsible implementation of which by each of the countries will significantly improve safety. For example, 46 countries in Europe, Asia and Africa (as of May 1, 2019), united by international traffic, agreed and signed the Convention concerning International Carriage by Rail (COTIF). Appendix C – Regulations concerning the International Carriage of Dangerous Goods by Rail (RID) [4]. The regulations relate to procedural and transportation requirements and exemptions to ensure the safety during carriage and entered into force on January 1, 2019. Changes in RID 2019 include the classification, packing and labelling of items containing dangerous substances, a revision of the classification codes for corrosive substances, new packing instructions for damaged or defective lithium batteries and destructive testing of gas cylinders.

Given the recognized urgent need in different countries to develop the topic of DG transportation by rail, it becomes important to clearly understand the current state of research activity in this area.

At the same time, thorough bibliometric studies do not exist so far that illustrate the overall research activity in the field of TDG over a long time span.

* Corresponding author: kolesnykova@diit.edu.ua, chief.library@gmail.com

This article aims to study the flow of publications of the global research environment in the field of transportation of dangerous goods by a particular mode of transport – rail, using bibliometric analysis.

The purpose of this paper is to research and define the promising worldwide scientific trends in the field of railway transportation of various dangerous goods.

2 Literature review

Railway transport plays a key role in the transportation of dangerous goods (DG). It is for this reason that improving the safety of rail transport is a priority for both industry and governments.

The consolidation of a body of knowledge (BOK) in the domain of formal methods for railway systems safety would not be possible if there had not been done much research work already in this field since many years.

Measuring the size of the global flow of publications on various aspects of cargo transportation and its quality characteristics are very important, including for allocation of funding and prioritization of resources.

So, scientists of United Kingdom, studying the current state of research in the field of Decision support systems (DSS) decision support systems for logistics, established the following. The bibliometric analysis showed that DSS for sustainable logistics is an emerging field; however, it is still evolving but at a slower pace [5]. The inherent key themes were decision models and frameworks to address sustainable logistics issues covering transport, distribution and third-party logistics. The most prominent sustainable logistics issue was carbon footprinting.

In the article [6], in relation to the necessary identification of established knowledge in the field of railways, the authors applied lattice-theoretical methods for analysing formal concepts in order to structure and systematize large volumes of relevant bibliometric data from the rail literature flow.

The influence of random factors and events that can lead to an accident and threaten not only the safety of rail transport, but also the life and health of people, as well as pose a threat to the environment, are always relevant for researchers [7, 8].

The results of bibliometric analysis in the field of road transport medicine have shown that researches are increasing annually, and the cooperation of scientists from different countries is growing [9]. But still, these rates are relatively low compared to other research areas.

Studies of rail transport show that they have a high level of penetration of connections in various interdisciplinary areas, especially technical ones [10]. The efforts of the inventors, enshrined in their technical solutions, are aimed at ensuring improved safety, reducing the negative impact on the environment, the use of renewable energy sources, energy efficiency.

Chinese scientists, using bibliometric analysis, systemic induction and deduction, explore the topic of urban rail transport safety in several developed countries (the USA, UK, Japan and South Korea) in its current state and trends, including the future vision of traffic development and preventative security measures [11].

3 Method

This research is characterized as theoretical in nature. The study was conducted using bibliometric analysis methods, with recognized scientometric analysis tools combined with visualization. To obtain relevant empirical data, the authors reviewed the world literature on paper topic using Scopus (Elsevier) and Web of Science (Clarivate Analytics) citation bases.

Bibliometric analysis is a method of displaying the main authors, journals and keywords by a specific topic. J.L.S. Santos, M. Uriona-Maldonado, and R.N.M.D. Santos [12] state that these methods are tools based on a scientifically recognized methodological theoretical basis that allows the use of statistical and mathematical methods to compare information from bibliographic records of documents stored in databases.

Our paper used mapping that provides a visual perspective for researchers and helps to understand general situations in specific subject areas of the study.

To manage and tabulate the collected data we used two software programs: (i) Bibexel and (ii) VosViewer.

In the work we used the Bibexel program (https://bibliometrie.univie.ac.at/bibexcel/), created by Olle Persson, a professor of library science and computer science at the Swedish Umeå University [13]. It allows analyzing the metadata, namely, the links between the co-authors and the analyzed abstracts (to identify the most "top" keywords in them) [14].

VosViewer, developed by N.J. van Eck, and L. Waltman [15] is a software for visualizing knowledge blocks of documents, especially suitable for analyzing large-scale data sets and building complex networks. In particular, it can be used to build bibliometric networks based on data downloaded from bibliographic databases, such as the Web of Science (CC) and Scopus. We used the VosViewer version 1.6.11.

In order to identify and evaluate changes relating to scientific activity and consider a more recent situation specifically, we analyzed two time intervals (t). The first time interval is from 1975 to June 2019 (t1), the second is from 2015 to June 2019 (t2).

Search strategies. Using the Scopus and WoS topic search, the following query was applied: dangerous goods, railway ('dangerous goods' AND railway).

To clarify the data, we used the subject areas: 1) automation and telematics systems; 2) navigation systems; 3) logistics; 4) energy; 5) locomotives; 6) freight cars; 7) materials; 8) rails; 9) impact on the environment and people.

Categorization of data. The categorization technique was as follows. Bibliometric data of all types of publications were extracted from each database for two time intervals – t1 and t2. The data were then classified according to different parameters (for example, year of publication, authors of publication, place of work of authors, country, type of document published, source, language of publication, areas of knowledge, funders).

For the collection and processing of references, we used EndNote software – a bibliographic manager that works in integration with the Web of Science.

Scientometric analysis. Transportation of dangerous goods is an interdisciplinary research area, which includes the integration of several sciences: engineering, chemical, medical, environmental sciences, etc. Unfortunately, world science is not sufficiently represented by research on this topic using bibliometrics, which is a statistical method for studying scientific publications, citations, patent documents and reports. But although bibliometric analysis is a process of quantitative analysis, its results are still qualitative [16].

One of the research methods of this article is based on scientometrics and involves combining software output information with manual interpretation to obtain final results or conclusions. As mentioned earlier, we used scientometric analysis software (VosViewer, BibExcel), which assists researchers in scientific statistics and network mapping of literature.

4 Findings

The analysis of two time intervals for identifying and evaluating changes related to scientific activity showed the following results. During the first time interval from 1975 to June 2019 (t1), there were identified 193 Scopus publications (p1) and 87 WoS (CC) publications (p2). In the subsequent period from 2015 to June 2019 (t2), the same search strategy identified: 38 Scopus publications (p1) and 40 WoS (CC) publications (p2) (Figure 1).

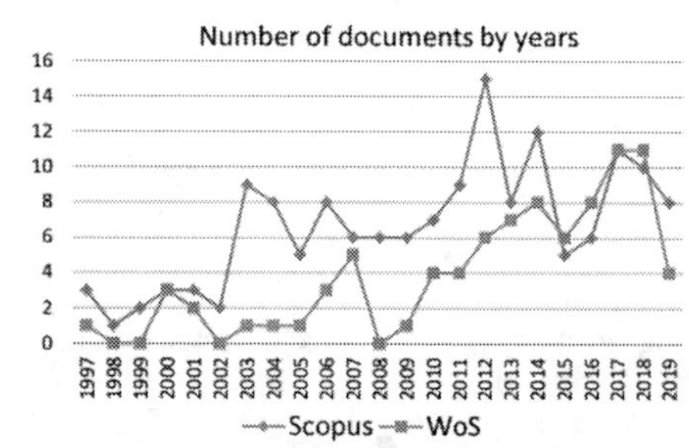

Fig. 1. Number of documents by years (Scopus, WoS)

When refining data using 9 subject areas, it was found that the same article may relate to different subject areas.

The trend shown in Table 1 and Table 2 demonstrates how research on the transport of dangerous goods attracts the attention of many scientists around the world.

The top subject areas (by the number of articles in each of the databases) for the entire observation period (t1) became the areas "Materials", "Rails", "Logistics", "Impact on the environment" (Table 1).

But although the data on top research areas for the last 5 years (t2) for Scopus and WoS differ significantly in percentage, they still show an increase in new areas: "Automation and telematics systems", "Navigation systems", "Logistics", "Energy". There is stated a decrease in interest in the "Locomotives" subject area (Table 2).

Table 1. Number of documents by thematic areas

Thematic areas	Number of documents			
	Scopus		Web of Science	
	For all the time (t1)	In 5 years (t2)	For all the time (t1)	In 5 years (t2)
Automation and telematics systems	6	3	1	1
Navigation systems	6	3	4	3
Logistics	34	14	5	4
Energy	15	8	3	2
Locomotives	7	2	2	0
Freight cars	17	6	3	3
Materials	114	23	27	13
Rails	101	27	37	17
Impact on the envi-ronment and people	33	10	6	3

Table 2. The percentage of new articles

Thematic areas	Scopus	WoS
Automation and telematics systems	50.00%	100.00%
Navigation systems	50.00%	75.00%
Logistics	41.18%	80.00%
Energy	53.33%	66.67%
Locomotives	28.57%	0.00%
Freight cars	35.29%	100.00%
Materials	20.18%	48.15%
Rails	26.73%	45.95%
Impact on the environment and people	30.30%	50.00%

Main journals. The documents obtained from Scopus (p1=193) and WoS (p2=87) were published in various sources (Table 3 and Table 4).

Table 3 shows that "Safety Science" (SNIP 2.251, CiteScore 4.49), "Accident Analysis and Prevention" (SNIP 2.047, CiteScore 3.82), "Tiedao Xuebao Journal of the China Railway Society" (SNIP 0.894, CiteScore 0.77) have been the journals of interest to authors for a long time (t1 and t2). The journals "Safety Science" and "Accident Analysis and Prevention" are also indexed in WoS. The journal "Safety Science" is in the first quartile by WoS categories "Engineering, industrial" and "Operations research & management science". The "Accident Analysis and Prevention" journal is in the first quartile by the categories "Ergonomics", "Public, environmental & occupational health (SSCI)" and "Social sciences, interdisciplinary", and in the second quartile by the category "Transportation".

The "Journal of Hazardous Materials" published 2 of the most highly cited to date articles on the topic of TDG. The first article by P. H. Bottelberghs "Risk analysis and safety policy developments in the Netherlands" (123 citations in Scopus) [17] is affiliated with the Ministry of Housing, Netherlands. The second article by B. Fabiano, F. Currò, E. Palazzi, R. Pastorino

"A framework for risk assessment and decision-making strategies in dangerous good transportation" (110 citations in Scopus) [18] is affiliated with the University of Genoa, Italy.

Table 3. Number of documents by sources (Scopus) (At least two documents)

For all the time		In 5 years	
Source name	**Number of documents**	**Source name**	**Number of documents**
Journal of Hazardous Materials	20	Accident Analysis and Prevention	3
Gigiena I Sanitariia	11	Safety Science	3
Safety Science	7	MATEC Web of Conferences	2
Tiedao Xuebao Journal of the China Railway Society	7	Tiedao Xuebao Journal of the China Railway Society	2
Accident Analysis and Prevention	6	Procedia Computer Science	2
Clinics in Occupational and Environmental Medicine	5	Transportation Research Procedia	2

Table 4. Number of documents by source (WoS) (At least two documents)

Source name	Number of documents (for all the time)	Two documents in 5 years
Accident Analysis and Prevention	2	
Advanced Materials Research	2	
Canadian Journal of Civil Engineering	2	
Case Studies on Transport Policy	2	2
Chemical Engineering Transactions	2	
Communications in Computer and Information Science	2	
Computers Operations Research	2	
Interdisciplinary Description of Complex systems	2	2
Journal of Loss Prevention in the Process industries	2	
Journal of Rail Transport Planning Management	2	2
Proceedings of Spie	2	
Safety Science	2	2
Transport	2	
Transport Means Proceedings of the International Conference	2	
Transportation Research Part C Emerging Technologies	2	
Transportation Research Part E Logistics and Transportation Review	2	

As indicated in Table 4, only for the last 5 years (t2) publications on the transportation of dangerous goods by rail have been presented in the journals "Case Studies on Transport Policy" "Interdisciplinary Description of Complex systems", "Journal of Rail Transport Planning Management", "Safety Science". The results show that journals do not have an increase in publications in the field of TDG in recent years.

The journals "Case Studies on Transport Policy" (SNIP 1.06, CiteScore 1.58), "Journal of Rail Transport Planning Management" (SNIP 1.123, CiteScore 1.78), "Safety Science" (see above) are also indexed in Scopus.

The most productive authors. As for the most productive authors, we identified 12 scientists who have 4 or more publications in Scopus for t1 (Figure 2).

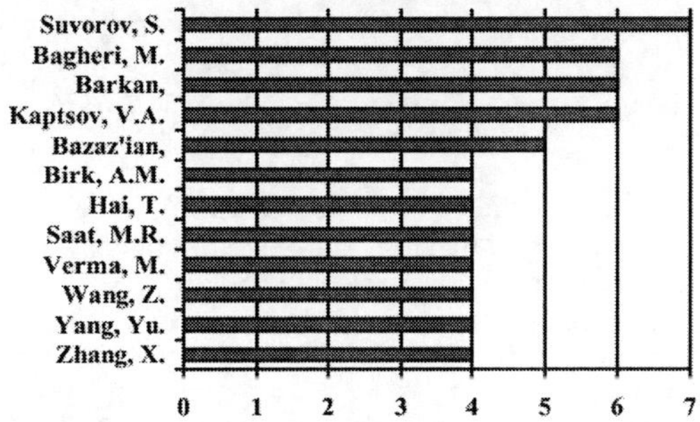

Fig. 2. Authors, who published at least 4 papers (1975-2019, Scopus)

These are the following researchers:

• Suvorov, S. V. – Inst Zheleznodorozhnoj Gigieny MPS, Moscow, Russian Federation, h-index 2, total documents in Scopus – 71.
• Bagheri, M. – Iran University of Science and Technology, Tehran, Iran, h-index 6, total documents in Scopus – 28.
• Barkan, Christopher P.L. – University of Illinois at Urbana-Champaign, Urbana, United States, h-index 22, total documents in Scopus – 104.
• Kaptsov, V. A. – Joint Stock Company "Russian Railways", Moscow, Russian Federation, h-index 2, total documents in Scopus – 121.
• Bazaz'ian, A. G. – non-affiliated, h-index 1, total documents in Scopus – 8.
• Birk, A. M. – Queen's University, Kingston, Kingston, Canada, h-index 18, total documents in Scopus – 167.
• Hai, Tao – Beijing Jiaotong University, Beijing, China, h-index 2, total documents in Scopus – 5.
• Saat, M. R. – Association of American Railroads, Washington, D.C., United States, h-index 11, total documents in Scopus – 28.

• Verma, M. – McMaster University, DeGroote School of Business, Hamilton, Canada, h-index 12, total documents in Scopus – 36.
• Wang, Zhe – Southwest Jiaotong University, Chengdu, China, h-index 1, total documents in Scopus – 4.
• Yang, Yuefang – Beijing Jiaotong University, Beijing, China, h-index 2, total documents in Scopus – 11.
• Zhang, Xingchen – Beijing Jiaotong University, Beijing, China, h-index 5, total documents in Scopus – 68.

Iranian author M. Bagheri has a joint publication with authors from Canada F. Saccomanno, S. Chenouri, L. Fu "Reducing the threat of in-transit derailments involving dangerous goods through effective placement along the train consist" (33 citations in Scopus) [19].

The number of authors having 4 or more publications in WoS is rather insignificant, therefore we identified authors who have 3 or more papers (Figure 3).

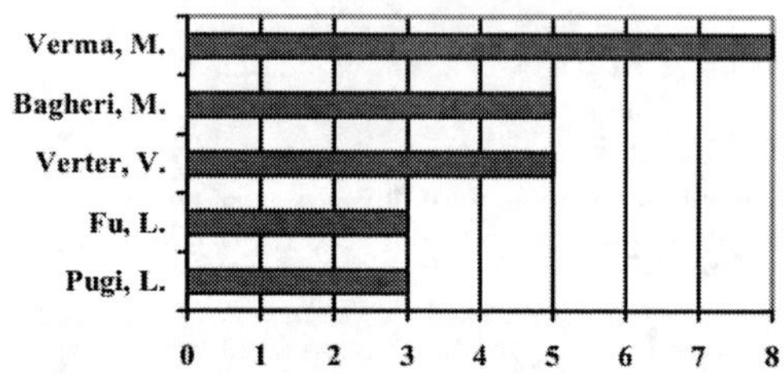

Fig. 3. Authors, who published at least 3 papers (1997-2019, WoS)

These are the following researchers:
• Verma, M. – McMaster University, Hamilton, Canada, h-index 14, total documents in WoS – 36.
• Bagheri, M. – Iran University of Science and Technology, Tehran, Iran, h-index 7, total documents in WoS – 24.
• Verter, V. – McGill Univ, Montreal, Canada, h-index 29, total documents in WoS – 79.
• Fu, Liping – Jiangsu Normal Univ, Xuzhou, Peoples R China, h-index 23, total documents in WoS – 181.
• Pugi, L. – Univ Florence, Florence, Italy, h-index 16, total documents in WoS – 126.

The authors M. Verma and V. Verter from Canada and M. Bagheri from Iran have a joint publication "Transport Mode Selection for Toxic Gases: Rail or Road?" (23 citations in WoS) [20].

The temporal evolution of keywords. Then we decided to trace the temporal evolution of keywords (Scopus) and categories (WoS) (Table 5 and Table 6).

Table 5. Most used keywords (Scopus)

For all the time (t1)		In 5 years (t1)	
Keywords	**Number of documents**	**Keywords**	**Number of documents**
Dangerous Goods	111	Dangerous Goods	18
Railway	88	Railroads	18
Railroads	80	Railroad Transportation	14
Article	56	Railroad Accidents	12
Risk Assessment	52	Risk Assessment	12
Hazardous Substances	49	Railway	11
Railroad Transportation	47	Freight Transportation	10
Human	45	Hazardous Substances	7
Hazardous Materials	35	Rails	7
Accidents	33	Transportation	7

The results show that among the most used keywords for t2 there are no words "Human" and "Hazardous Materials". But "Rails" and "Freight Transportation" began to be used more often.

To get the top keywords visualization, we used the VosViewer program. To work with it, the data from the Scopus database were saved in the format ris. With the help of Bibexel, which specializes in metadata processing, a file was created containing annotations of 193 documents.

Table 6. Most used Categories (WoS)

For all the time		In 5 years	
Categories	**Number of docu-ments**	**Categories**	**Number of docu-ments**
Transportation science technology	23	Transportation	10
Transportation	17	Transportation science technology	10
Operations research management science	14	Operations research management science	4
Engineering civil	8	Engineering civil	3
Engineering mechanical	8	Engineering electrical electronic	3
Engineering electrical electronic	7	Engineering industrial	3
Engineering industrial	7	Engineering mechanical	3
Computer science artificial intelligence	5	Social sciences interdisciplinary	3
Computer science interdisciplinary applications	5	Telecommunicat ions	3
Engineering chemical	5	Computer science information systems	2

The results were then imported into the VosViewer program for visualization. In cluster analysis, the link layer was taken at least 10.

It is clearly seen (Fig.4) that the TDG research topic has formed 4 clusters, and each cluster has its own color.

Conventional cluster names: "Research & Development" – green, "Train & Assessment" – red, "Impact & Article" – yellow,, "Railway & Security" – blue. There is a significant correlation between keywords in each cluster. Reading the map "The keywords co-occurrence network of TDG studies", we see that the most intense and extensive connections between the co-authors and the analyzed annotations formed keyword nodes in the red cluster. These are the subtopics associated with the research of TDG.

Continuing further to monitor the temporal evolution of keywords, we turned to Categories (WoS) for t2 and discovered the emergence of new terms: "Social sciences interdisciplinary", "Telecommunications", "Computer science information systems" (Table 6).

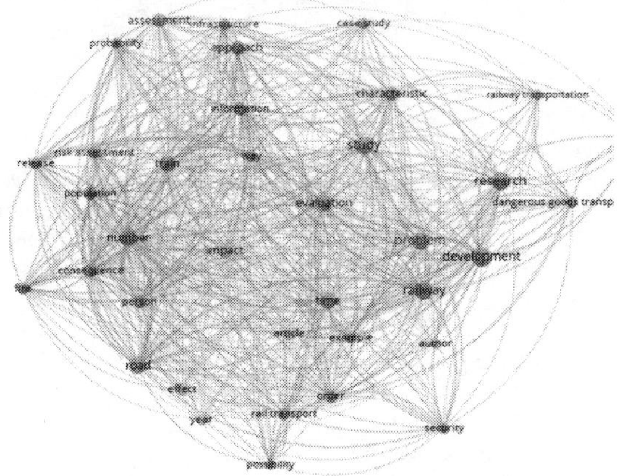

Fig. 4. The keywords co-occurrence network of TDG studies.

Table 7. Organizations, the authors of which have published at least 3 papers

Scopus, 1975 – 2019		WoS, 1997 - 2019	
Organization	**Number of documents**	**Organization**	**Number of documents**
Beijing Jiaotong University	15	Beijing Jiaotong University	8
Southwest Jiaotong Univ	10	Mcgill University	6
University of Illinois at Urbana-Champaign	6	Iran University Science Technology	5
University of Waterloo	5	Mcmaster University	4
Iran University of Science and Technology	4	Memorial University Newfoundland	4
Queen's University, Kingston	4	University of Genoa	4
McGill University	3	University of Florence	3
Lanzhou Jiaotong University	3	University of Waterloo	3
Università degli Studi di Genova	3		
Università degli Studi di Roma La Sapienza	3		
Vilniaus Gedimino Technikos Universitetas	3		
Desautels School of Management	3		

The organizations and countries. The categorization by organization whose authors published at least 3 papers for t1, shown in Table 9, shows the top 12 universities of the world in the Scopus database. The results for the WoS DB are more modest. The leader in the number of publications in each of the databases (Scopus and WoS) became Beijing Jiaotong University (China).

According to the search results, all TDG research works originate from 29 **countries** (or territories) of the world. 18 publications had no information about the origin of the authors. Top 10 countries whose scientists were more productive for t1, according to **Scopus** DB: China (40), United States (29), Canada (19), Italy (12), Germany (9), France (6), Netherlands (6), Iran (5), Switzerland (5), Lithuania (4). But for t2, the researchers of most of these countries have lost interest in this topic. Researchers from China (8), Canada (6), Italy (3), France (4) are still active.

The results of the analysis of data obtained from the WoS database for t1 with a similar search are as follows: Canada (16), Peoples r China (16), Italy (12), USA (8), Iran (7), Germany (6), France (4), England (3), Greece (3), Lithuania (3). But for t2, only researchers from 4 countries were interested in the topic of TDG: Canada (7), Peoples r China (6), Italy (4), USA (5).

Number of documents by type. The most popular types of documents in the Scopus and WoS databases have become Article, Conference (Proceedings) Paper, Review (Table 8).

Table 8. Number of documents by type (Scopus, WoS)

Document type	Number of documents	
	For all the time Scopus / WoS	**In 5 years Scopus / WoS**
Article	128 / 41	20 / 20
Conference (Proceedings) Paper	43 / 46	16 / 17
Review	9 / 3	0 / 3
Conference Review	8 / 0	2 / 0
Note	2 / 0	0 / 0
Editorial	1 / 0	0 / 0
Erratum	1 / 0	0 / 0
Letter	1 / 0	0 / 0

Documents by branches of knowledge. The multidisciplinarity of the TDG topic is confirmed by the presence of publications in 20 branches of knowledge. The distribution of the number of publications (at least 4) for t1 and t2 is shown in Table 9.

Categorization by language. In the general search, the overwhelming majority of all analyzed publications were written in English: Scopus – 77.72% (t1) and 97.37% (t2); WoS – 95.4% (t1) and 97.5% (t2). The second language in the number of publications was Chinese, but only in an insignificant number: Scopus – 8.29% (t1) and 5.26% (t2); WoS – 2.3% (t1) and 2.5% (t2). All other languages, although present in Scopus publications for t1, for example, Russian (12), German (9), French (3); but for t2 there were no such publications (Table 10).

Table 9. Documents by branches of knowledge

Branch of knowledge	Number of documents		% new documents
	For all the time	In 5 years	
Engineering	88	21	23,86%
Medicine	52	8	15,38%
Social Sciences	50	17	34,00%
Environmental Science	36	4	11,11%
Computer Science	22	5	22,73%
Chemical Engineering	16	3	18,75%
Business, Management and Accounting	10	3	30,00%
Mathematics	10	4	40,00%
Decision Sciences	9	2	22,22%
Chemistry	5	2	40,00%
Energy	5	3	60,00%
Agricultural and Biological Sciences	4	2	50,00%

Table 10. Number of documents by languages (Scopus, WoS)

Language	Number of documents	
	For all the time Scopus / WoS	In 5 years Scopus / WoS
English	150 / 83	37 / 39
Chinese	16 / 2	2 / 1
Russian	12 / 0	0 / 0
German	9 / 1	0 / 0
French	3 / 0	0 / 0
Dutch	1 / 0	0 / 0
Italian	1 / 0	0 / 0
Polish	1 / 0	0 / 0
Portuguese	0 / 1	0 / 0
Spanish	1 / 0	0 / 0

Categorization by Financial Sponsors. The results of the study fix a fairly small financial support for the TDG topic in different countries. According to the Scopus database the financial sponsors are 18 organizations: Natural Sciences and Engineering Research Council of Canada (n=5), Association of American Railroads (n=2), College of Natural Resources, University of California Berkeley (n=2), National Natural Science Foundation of China (n=2), National Research Council (n=2). Another 13 organizations funded one project each: Alberta Innovates; Centre National pour la Recherche Scientifique et Technique; City, University of London; Dow Chemical Company; Dwight D. Eisenhower Army Medical Center; Generalitat Valenciana; Institut d'Estudis Catalans; Korea National Railroad College University; Ministero dell'Istruzione, dell'Università e della Ricerca; Ministry of Infrastructure, Transport and Networks; National Research Council Canada; Northwestern University Transportation Center; Rutgers, The State University of New Jersey.

From the publications presented in the WoS, it follows that financial support for research was provided by 18 organizations. Herwith, 3 publications were supported by the Natural Sciences and Engineering Research Council of Canada (n = 3). Another 17 organizations supported one project each: Department of Civil and Environmental Engineering at Rutgers; Ansaldo Sts.; Association of American Railroads BNSF Railway; collaboratively Center for Advanced Infrastructure and Transportation an Office of the Assistant Secretary for Research and Technology US, Department of Transportation Designated National University, Transportation Center at Rutgers the State University of New Jersey; French Program of Research and Innovation in Terrestrial Transport Predit; Fundamental Research Funds for the Central Universities; Generalitat Valenciana Spain; Ministry of Education and Science of the Republic of Srbia; Mitacs Globalink Scholarship; National University Rail Center a US Dot Rita University Transportation Center; Natural Science Foundation of China NSFC; School of Engineering at Rutgers; Security Capacity Construction Foundation of Civil Aviation Authority of China; Social Sciences and Humanities Research Council of Canada; University of Genova. We note that according to the Scopus and WoS databases, the largest amount of researches was supported by the National Sciences and Engineering Research Council of Canada.

5. Conclusions

In the present study, we evaluated the flow of publications of the global research environment in the field of transportation of dangerous goods by a particular mode of transport – rail. We used a combination of bibliometric tools, scientometric analysis and new visualization methods.

The growth of international and domestic transport of dangerous goods leads to increased risk for transportation hubs and nearby settlements, as well as the emergence of environmental and safety hazards. Of course, these questions are in the focus of attention of researchers, as evidenced by the increase in the number of publications over the past 5 years in the Scopus database – 38 (t1 = 193) and Web of Science – 40 (t1 = 87). The Scopus database indexes more publications on the TDG topic than the WoS database. In general, from 1975 to June 2019 (t1), the oscillatory tendency remained with respect to the frequency of published works. For example, a decrease in the number of publications in 1998-1999 and 2007–2009 can be associated with global financial crises that began in 1998 and 2008. The likelihood of another global spiral of financial problems (possibly in 2020) increases the likelihood of reducing the amount of research on the TDG topic.

However, the increase in the number of publications can also be affected by the increase in the number of such type of document as Conference (Proceedings) Paper, affecting the search procedure. This was observed in our study. For t2 in each of the databases, the number of Article and Conference (Proceedings) Paper became almost equal. So, in Scopus – respectively 20/20, WoS – 16/17.

The multidisciplinarity of the TDG topic is confirmed by the presence of publications in 20 branches of knowledge.

Bibliometric visualization using VOSviewer (Fig. 2) allowed us to identify two research clusters with the most powerful interrelationships. This is first of all the red "Train & Assessment" cluster with sub-topics: approach, infrastructure, risk assessment, release, consequence, number, person, probability, population, way, information. In the green "Research & Development" cluster it was shown that the most intense and extensive links between co-authors and analyzed annotations were formed by the following keyword nodes - sub-topics: study, evaluations, railway transportation, characteristic, problem, case study, china, dangerous goods.

In addition to bibliometric visualization, analyzing statistical data by the number of documents, we have noted in the last 5 years (from 2015 to June 2019) for Scopus and WoS a decreased interest of researchers in the subject area "Locomotives" and increased interest in new subject areas: "Automation and telematics systems", "Navigation systems", "Logistics", "Energy". Categories (WoS) for t2 also demonstrate the activity of researchers in new areas "Social sciences interdisciplinary", "Telecommunications", "Computer science information systems". It is obvious that research in these areas are promising global scientific trends in the field of transportation of various dangerous goods by rail.

Both visualization and statistical analysis provide insights into the research environment and can help with the choice of the research topic and the outlook for TDG researchers.

The main journals of interest to the authors for a long time include "Safety Science", "Accident Analysis and Prevention" and "Tiedao Xuebao Journal of the China Railway Society". But 2 of the most highly cited to date articles on TDG were published in "Journal of Hazardous Materials". Their authors are P. H. Bottelberghs (Netherlands) and B. Fabiano, F. Currò, E. Palazzi, R. Pastorino (Italy).

According to the search results, all TDG research works originate from 29 countries (or territories) of the world. In recent years (t2), by Number of documents by countries and territories, the leading positions belong to the researchers from: by Scopus database - China, Canada, Italy, France; WoS DB - Canada, China, Italy, USA.

Analysis of the author publication activity showed that the most productive authors are from the following countries: Russian Federation, Iran, United States, Canada, China.

Analysis by organization identified the leader in the number of publications in each of the databases (Scopus and WoS) - Beijing Jiaotong University (China). But the analysis in the aspect of joint research by scientists from different countries shows a very low level of interaction.

Categorization by language made it possible to establish an unconditional increase in the number of publications in English. In each of the databases (Scopus and WoS) for t2, this number is more than 97%.

There is an obvious link between how the state invests in scientific research and its economic power. But the results of our study fix a small financial support for TDG topic in different countries. Financial sponsors are 18 organizations. According to the Scopus and WoS databases, the largest number of researches was supported by the National Sciences and Engineering Research Council of Canada.

6. Originality / Value

As far as the authors know, this is the first study focused on a bibliometric analysis of the global flow of scientific publications on the transport of dangerous goods by rail. The results of the study can be a guide, stimulating and directing interested researchers for further study.

References

1. Yu.V. Zelenko, I.L. Zhuravel, A.M. Okorokov, O.M. Patlasov, A.M Boichenko, and H.I. Nesterenko. *Posibnyk dlia kursu spetsialnoho navchannia z pytan perevezen nebezpechnykh vantazhiv na zaliznychnomu transporti. Chastyna 1. Bazovyi kurs*. Available at http://eadnurt.diit.edu.ua/bitstream/123456789/8821/1/Textbook.pdf (2015)
2. *BusinessDictionary.* Available at http://www.businessdictionary.com/definition/dangerous-goods.html (n. d.)
3. M. Šolc, and M. Hovanec. The Importance of Dangerous Goods Transport by Rail. *Naše more* **62**, 181-186. DOI: 10.17818/NM/2015/SI17 Available at http://www.nasemore.com/the-importance-of-dangerous-goods-transport-by-rail/ (2015)
4. *Convention concerning International Carriage by Rail (COTIF). Appendix C – Regulations concerning the International Carriage of Dangerous Goods by Rail (RID).* Available at https://otif.org/fileadmin/new/3-Reference-Text/3B-RID/RID%202019%20E.pdf (2019)
5. F.H. Qaiser, K. Ahmed, M. Sykora, A. Choudhary, and M. Simpson. Decision support systems for sustainable logistics: a review and bibliometric analysis. *Ind Manag Data Syst*, **117(7)**, 1376-1388. DOI: https://doi.org/10.1108/IMDS-09-2016-0410) (2017).

6. S. Gruner, A. Kumar, and T. Maibaum. Towards a Body of Knowledge in Formal Methods for the Railway Domain: Identification of Settled Knowledge. *CCIS*, **596**, 87-102. DOI: https://doi.org/10.1007/978-3-319-29510-7_5 (2016).
7. L. Yaryshkina, and A. Boychenko. Development of the Efficient Technology for Eliminating Environmental Aftermaths in Transport. *Transport Problems*. **5(10),** 113-118 (2010).
8. A. Samarska, and Y. Zelenko. Assessment of the Railway Influence on the Heavy Metal Accumulation in Soil. *Science and Transport Progress*, **4(75),** 25-35. DOI: http://dx.doi.org/10.15802/stp2018/140551 Available at http://stp.diit.edu.ua/article/view/140551/147604 (2018)
9. B. Groneberg-Kloft, D. Klingelhoefer, S. E. Zitnik, and C. Scutaru. Traffic medicine-related research: a scientometric analysis. *BMC Public Health*, **13**, 541. DOI: https://doi.org/10.1186/1471-2458-13-541 (2013)
10. S. Kolesnykov. Improvement of Life Support Systems of Passenger Rolling Stock: Patent Review. *Science and Transport Progress*, **1(73),** 44-55. DOI: http://dx.doi.org/10.15802/stp2018/123460 Available at http://stp.diit.edu.ua/article/view/123460/120313 (2018)
11. C. Wei, R. Ma, and X. Wang. An Analysis of Policy and Management Mode of Urban Rail Transit Safety in Developed Countries. *Proceedings of 2014 International Conference on Public Administration*, **II,** 961-967 (2014)
12. J.L.S. Santos, M. Uriona-Maldonado, and R.N.M.D. Santos. Inovação e conhecimento organizacional: um mapeamento bibliométrico das publicações científicas até 2009. *Revista Organizações em Contexto*, **7(13)**, 31-58 (2011)
13. F. Åström, R. Danell, B. Larsen, and J. Schneider (Eds.) *Celebrating Scholarly Communication Studies: A Festschrift for Olle Persson at his 60th Birthday*, 9-24 (ISSI, 2009)
14. T. Kolesnykova, and O. Matveyeva. Performance Assesment of the Research: Bibliometrics. *Science and Transport Progress*, **4(58),** 7-22. DOI: http://dx.doi.org/10.15802/stp2015/49190 Available at http://stp.diit.edu.ua/article/view/49190/49145 (2015)
15. N.J. van Eck, and L. Waltman. Software survey: VOSviewer, a computer program for bibliometric mapping. *Scientometrics*, **82(2)**, 523–538 Available at https://akademiai.com/doi/pdf/10.1007/s11192-009-0146-3 (2010)
16. J. Chen, Y. Su, H. Si, and J. Chen. Managerial Areas of Construction and Demolition Waste: A Scientometric Review. *IJERPH*, **15(11),** 2350. (2018)
17. P. H. Bottelberghs. Risk analysis and safety policy developments in the Netherlands. *J Hazard Mater*, **71(1-3)**, 59-84. DOI: 10.3390/ijerph15112350 (2000).
18. B. Fabiano, F. Currò, E. Palazzi, R. Pastorino. A framework for risk assessment and decision-making strategies in dangerous good transportation. *J Hazard Mater*, **93(1),** 1-15 (2002)
19. M. Bagheri, F. Saccomanno , S. Chenouri, and L. Fu. Reducing the threat of in-transit derailments involving dangerous goods through effective placement along the train consist. *Accid Anal Prev*, **43(3)**, 613-620. DOI: 10.1016/j.aap.2010.09.008 (2011)
20. M. Bagheri, M. Verma, V. Verter. Transport Mode Selection for Toxic Gases: Rail or Road? *Risk Analysis*, **34(1)**, 168-186. DOI: 10.1111/risa.12063 (2014)

Reducing noise influence on an audio frequency track circuit

Inna Saiapina[1,*], *Mykhailo* Babaiev[2], and *Olha* Ananieva[3]

[1] State University of Infrastructure and Technologies, Telecommunication Technologies and Automation Department, 9 Kyrylivska str., 04071, Kyiv, Ukraine
[2] Ukrainian State University of Railway Transport, Department of electroenergy, electrical equipment and electromecanics, 7 Feierbakh sq., 61050 Kharkiv, Ukraine
[3] Ukrainian State University of Railway Transport, Department of automation and computer telecontrol train traffic, 7 Feierbakh sq., 61050 Kharkiv, Ukraine

Abstract. In connection with electromagnetic interference influence on the track circuits, the purpose of the research is finding the means to increase the noise immunity of an audio frequency track circuit. The authors propose a new engineering solution, which enables reducing the effect of noises on the input of the track receiver in the intervals between signal current pulses. The proposed noise-immune audio frequency track circuit is based on inserting a delay line, an adjustable single-pulse generator and a controlled electronic switch into the existing audio frequency track circuit equipment. To analyze its efficiency, the operation of the audio frequency track circuit was simulated under conditions of traction current disturbances, impulse and fluctuation interferences with the known parameters. The results show that proposed device for railway transport allows to increase a signal-to-noise ratio on the track receiver input from 8% to 30%, depending on the interference parameters and the level of the useful signal.

1. Introduction

Track circuits belong to the most popular systems to provide information on position and movements of trains and ensure safety of the railway transportation process [1, 2]. They are the main track sensors at the Ukrainian railways, and the correct operation of the automatic blocking, dispatching centralization and automatic locomotive signalling systems, and, consequently, the safety of train traffic depends on their functioning.

Audio frequency track circuits are a perspective type of the track circuits due to a number of advantages: absence of block joints, which are one of the most unreliable elements of the track circuits; elimination of the need to install impedance bonds; the possibility of power supply to two adjacent track circuits from one common source, etc [3 - 5]. Railways of Ukraine are equipped with audio frequency (or also called tonal) track circuits of the following modifications: TRC 3 and TRC 4. TRC 3 has the following signal carrier frequencies: 420, 480, 580, 720 and 780 Hz. In the TRC 4 carrier frequencies are 4500, 5000 and 5500 Hz. The power to audio frequency track circuits is supplied by an AM tone signal. However, the operating conditions of audio frequency track circuits are characterized by the impact of a variety of different interferences (Figure 1) [6-12].

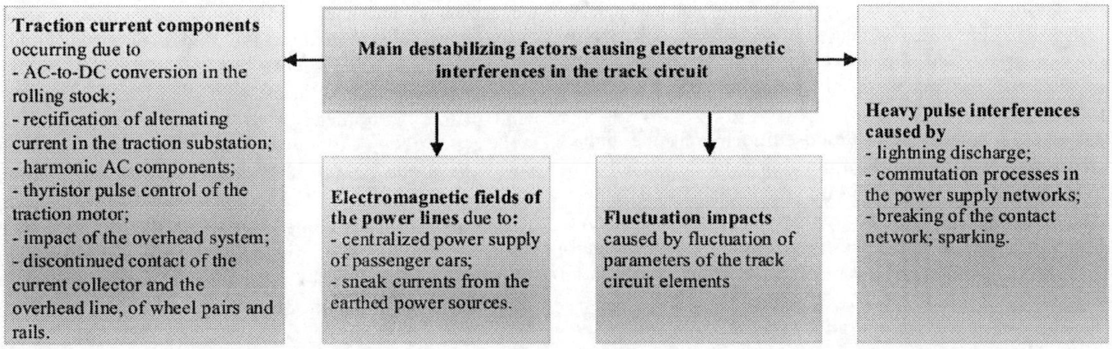

Fig. 1. Main factors causing interferences in track circuits

* Corresponding author: inna.saiapina@gmail.com

https://doi.org/10.1051/matecconf/201929403015

Materials [13, 14] show that the most industrial and atmospheric disturbances are characterized by a rapid change in the electric current causing occurrence of electric and magnetic fields that interfere with the reception of a useful signal in the track circuits. The highest levels of impulse noise occur during switching, accompanied by a short circuit or a circuit interruption. When the alternating current passes the circuit with a nonlinear resistance, interferences from harmonic and combinational components of the fundamental frequency appear.

In [15] it is mentioned that in the range of low values, fluctuations of the ballast resistance have significant impact on the input voltage of the track receiver, which can cause a malfunction of the tone track circuit due to fluctuation interferences.

Interference caused by the operation of asynchronous traction motors with impulse control, which most new types of electric rolling stock are equipped with, is particularly dangerous for the operation of audio frequency track circuits. Their operation causes broadband electrical noise in the reverse traction current circuit, which are within the operating frequency range of these track circuits [16-19].

Papers [20, 21] consider the methods of protection of railway automation systems against over-voltage, describe the effect of traction current on jointless track circuits and ways of protecting against it. In work [21], a study of increasing the noise immunity of a track receiver (TR) by using various methods of modulation and demodulation of the rail control signal was performed.

However, these existing methods of increasing noise immunity cannot completely prevent the effect of interference on the receiving equipment of the track circuits which require further improvement. Authors propose an engineering solution to improve noise immunity, which will reduce the time of interference effects on the track receiver of audio frequency track circuits [22, 23]. To study its efficiency, the operation of audio frequency track circuits when exposed to different interferences was simulated.

2. Solutions to improve the noise immunity of the track circuits

An engineering solution [23] to improve the noise immunity of audio frequency track circuits has been developed to reduce the effect of interference at the input of the track receiver in the intervals between the status monitoring signal pulses of the audio frequency track circuits. To construct the proposed noise-immune audio frequency track circuit the following new elements have been further introduced into the established audio frequency track circuit equipment: an adjustable delay line (DL), single pulse generator (SPG) and managed electronic switch (ES) (marked with yellow color on Figure 2).

The principle of operation involved opening the controlled electronic switch at the input of the track receiver in the absence of a useful signal. The effect of interference is reduced due to the implementation of guard intervals in the pauses between the useful signal pulses, during which the interference from the transmission channel of audio frequency track circuits cannot reach the input of the track receiver.

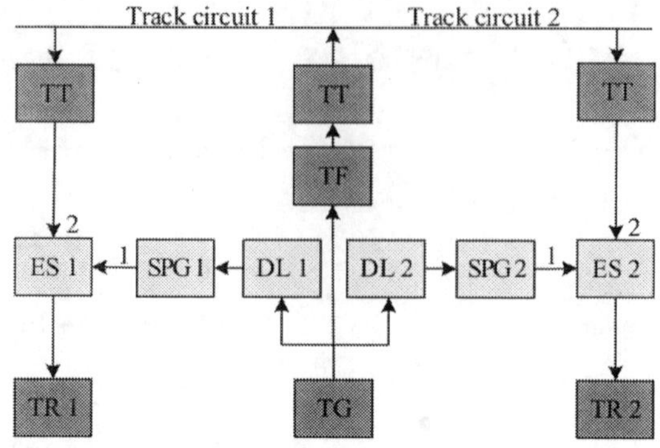

Fig. 2. Block diagram of noise-immune audio frequency track circuit

Moreover, the length of the pulse generated by the single-pulse generator and thus the length of the guard interval can be adjusted.

This device is intended for use in systems with centralized equipment arrangement. The block diagram (Figure 2) shows the feeding end of the track circuit, which includes a track generator (TG), a track filter (TF), a track transformer (TT) and two relay ends of adjacent rail circuits consisting of a track relay (TR) and a track transformer (TT). The model of the device was developed in SIMULINK® software (Figure 3)

The generator block consists of three functional blocks: a carrier frequency generator, a pulse generator and a modulator. The carrier frequency generator produces a sinusoidal signal of a specified frequency and amplitude, which is fed to the first input of the modulator. The second input of the modulator is supplied with a pulse signal with a modulation frequency. The modulator performs the operation of multiplying a sinusoidal signal of the carrier frequency with a pulse signal, which has a modulation frequency. The form of the signal on the output of the track generator block is shown on Figure 4a. The track filter block is modelled on the basis of capacitive, active and reactive resistance elements. The cable link line, as well as the rail line, is made in the model using the Simulink 'Transmission Line' block. The model type in the block options is chosen as 'Line with distributed parameters', its primary parameters and length are specified. The track transformer block consists of a transformer with a transform coefficient of 38, which is connected through a protective resistor. The input resistance of the adjacent sections of the rail line from both the supply and the relay end is taken into account by the inclusion of the corresponding complex resistances shown in Figure 3 as 'Adjacent rail line' blocks.

The experimental setup for the modelling was as follows: frequency from the carrier frequency generator – from 420 to 780 Hz; modulation frequency of the pulse generator – from 8 or 12 Hz; output signal voltage of the track generator block – from 1 to 8 V; specific active resistance of the rail line – from 0.935 to 1.24 Ohm/km; specific inductance of the rail line – from 1.59 to 1.82 mH; insulation resistance of the rail line – from 0.5 to 50 Ohm/km; length of the cable link line – from 0 to 6 km; length of the rail line – from 50 to 1100 m.

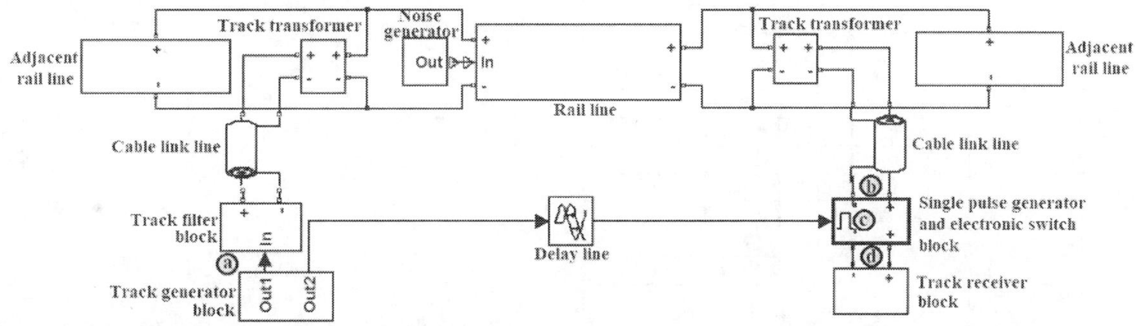

Fig. 3. Simulink model of the proposed device

To check the adequacy of the model, the sampling of voltage values on track circuit elements in the model was compared to the voltage values on the real track circuit using the following criteria of mathematical statistics: the Mann-Whitney criterion, the Fisher criterion, the Student's test. The results confirmed the adequacy of the model with the confidence probability of 95 % [22].

Since one generator encodes the track circuit and the delay line (Figure 3), the generated signal (Figure 4a) simultaneously arrives at 2 inputs: the track filter input and the delay line.

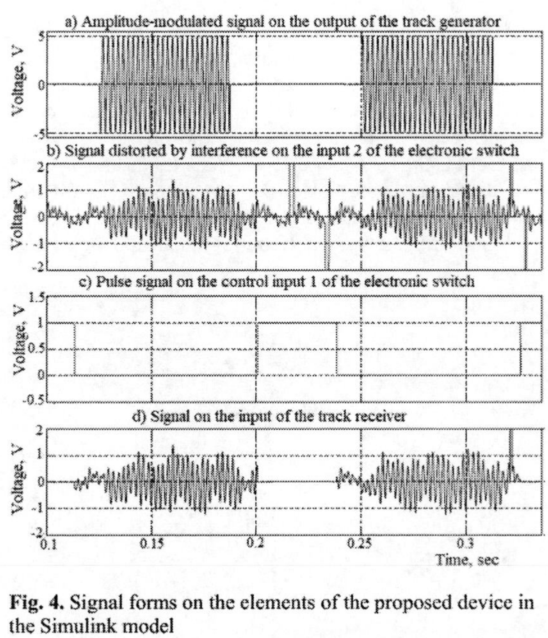

Fig. 4. Signal forms on the elements of the proposed device in the Simulink model

From the track filter through the track transformer, the signal enters the track circuit. If it is free, the signal together with the track circuit interference (Figure 4b) passes through the track transformer to the input 2 (Figure 2) or input, matched ‘b’ (Figure 3), of the electronic switch.

The delay line is designed for the time of passage of the useful signal from track generator to the input 2 of the electronic switch in the audio frequency track circuits. As the signal from the delay line arrives at the input of a single pulse generator, it generates a single pulse (Figure 4c). When the pulse appears on the control input 1, the electronic switch interrupts the circuit; as a result, interference from input 2 of the electronic switch cannot reach the input of the track receiver during the pulse signal (Figure 3d). Thus, the possibility of interference to the track receiver within the guard interval between the status monitoring signal pulses of the audio frequency track circuits is excluded.

3. Efficiency analysis of the proposed solution

To analyze the efficiency of the proposed engineering solution, the effect of the interference from traction current, as well as impulse and fluctuation interference on the operation of the audio frequency track circuit was simulated using SIMULINK® model (Figure 3).

For this purpose, the noise generator block of the model synthesizes interference, which includes impulse interference, fluctuation interference and a harmonic signal from traction current disturbances: interference from a direct traction current, interference from alternating traction current and interference from direct traction current and thyristor engine control. The nature of the effect of interference from the traction current depends on their harmonic composition, which is determined by the harmonic coefficient.

The interference from a direct traction current according to [5] consists of three main harmonics: the one with the frequency of 50 Hz has harmonic content factor 0.072; the one with the frequency of 200 Hz has harmonic content factor 0.013 and the one with the frequency of 300 Hz has harmonic content factor 0.017.

The interference from an alternating traction current according to [5] consists of odd harmonics, shown in Table 1.

Table 1. Harmonic components of alternating traction current in the rail line

The harmonic order	Frequency, Hz	The percentage of harmonic content, %	The harmonic order	Frequency, Hz	The percentage of harmonic content, %
3	150	30	19	950	0,11
4	200	0,3	21	1050	0,07
5	250	15	23	1150	0,10
6	300	0,1	25	1250	0,09
7	350	5	27	1350	0,09
8	400	0,05	29	1450	0,2
9	450	2	31	1550	0,04
11	550	1,5	33	1650	0,2
13	650	1,2	35	1750	0,2
15	750	0,8	37	1850	0,1
17	850	0,5	39	1950	0,1

The interference from direct traction current and thyristor engine control according to [5] consists of currents with certain frequencies and peak amplitudes shown in Table 2.

Table 2. Maximum currents of interference in the rail line during operation of thyristor converters

Frequency, Hz	25	50	75	125	175	225	275	325	375
Amplitude, A	4,7	1,1	6,2	3,8	2,6	2,0	3,4	2,6	2,2

Pulse noises in track circuits are usually caused by lightning discharges, emergency and switching processes in power supply systems. To simulate the effect of pulse interference on the track circuit, a random pulse sequence model is used which was obtained in [7] as a result of recording the interferences in the railway. According to it, the amplitude of pulses is within the range from -47 to 51 V. Mathematical expectation for the pulse duration is 3.52 ms, the dispersion is 17.01 ms^2, the mean square deviation is 4.12 ms. For inter-pulse intervals, the mathematical expectation is 171.7 ms, the dispersion is 0.2030 s^2 and the mean square deviation is 0.451 s. The duration of the interference is

$$\omega_{imp}(n) = \lambda e^{-\lambda n}, \tag{1}$$

where $\lambda = 284{,}057$, while duration of inter-pulse intervals according to the maximum likelihood criterion of the is subject to the gamma distribution

$$w_{\text{int}}(n) = \frac{b^a}{\Gamma(\text{a})} n^{a-1} e^{-bn}, \tag{2}$$

where $\Gamma(x) = \int_0^{\infty} z^{x-1} e^{-z} dz$ - gamma function with parameters a=0.0109 and b=1.216.

An example of implementation of the described pulse interference is shown in Fig. 5.

Fluctuation interference can be characterized as a continuous process, consisting of multiple pulses with random amplitude. The main cause of the fluctuation interferences in the track circuit is its own noises caused by the fluctuation of parameters of its elements, as well as other noise of the surrounding space.

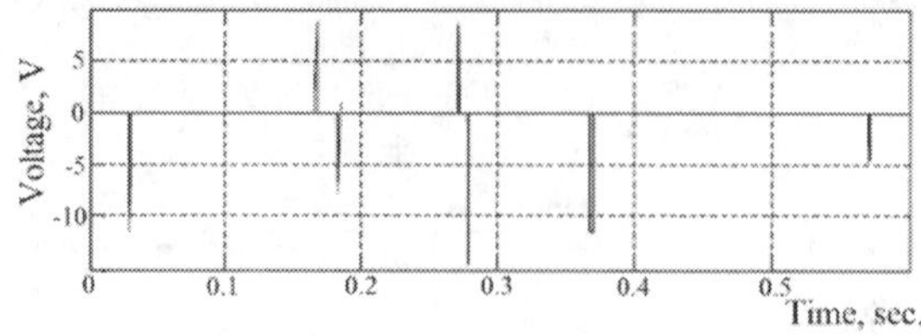

Fig. 5. An example of pulse interference on the track receiver input

The model of fluctuation interference affecting the track circuit is a white Gaussian noise with the indicated values of interference characteristics: zero mathematical expectation and unit dispersion. An example of implementation of fluctuation interference is shown in Fig. 6.

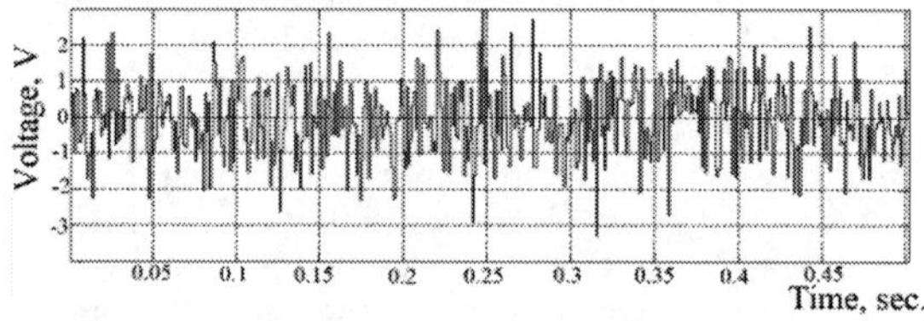

Fig. 6. An example of fluctuation interference on the track receiver input

To estimate the increase in noise immunity with the proposed engineering solution, the signal-to-noise ratio (SNR) (3) at the input of the track receiver has been calculated using the device when exposed to the described interference on the transmission path, and without using it. The voltage signals U_{signal} and $U_{interference}$ on the input of the track receiver (point 'd' in Figure 3) were measured using the Simulink model. Expressions (4) and (5) allow to determine signal and interference power.

$$SNR_{dB} = 10 \cdot \lg\left(\frac{P_{signal}}{P_{interference}}\right) \tag{3}$$

$$P_{signal} = \frac{1}{T} \int_{-\frac{T}{2}}^{\frac{T}{2}} U_{signal}{}^2(t)dt \tag{4}$$

$$P_{interference} = \frac{1}{\tau} \int_0^{\tau} U_{interference}{}^2(t)dt \tag{5}$$

where SNR_{dB} is a signal-to-noise ratio, calculated in dB; P_{signal} is the average power dissipated by the signal during the time interval; T is a signal period; $U_{signal}(t)$ is the useful signal; $P_{interference}$ is the average power dissipated by the interference signal; τ is a time interval; $U_{interference}$ is a voltage of an interference signal.

Based on expressions (3-5) using Mathcad® software, comparative signal-to-noise ratios have been obtained when exposed to interference from traction current in the sections with DC and AC electrification, as well as interference from thyristor control of the traction motor, impulse interference and fluctuation interference. The simulation results are shown in Tables 3 and 4.

Table 3. Signal/noise ratio when exposed to interference from traction current.

	SNR when the device for increasing noise immunity is not used, dB		*SNR* when the device for increasing noise immunity is used, dB		% increase in SNR due to the use of device for increasing noise immunity	
	U_{s1}=1,2V	U_{s2}=1,5V	U_{s1}=1,2V	U_{s2}=1,5V	U_{s1}=1,2V	U_{s2}=1,5V
When exposed to interference from a direct traction current	5,269	6,852	5,982	7,566	13,5 %	10,4 %
When exposed to interference from alternating traction current	2,117	3,701	2,788	4,372	30,6 %	18,1 %
When exposed to interference from direct traction current and thyristor engine control	5,084	6,668	5,965	7,549	17,3 %	13,2 %

Table 4. Signal/noise ratio when exposed to impulse and fluctuation interference.

	SNR when the device for increasing noise immunity is not used, dB	*SNR* when the device for increasing noise immunity is used, dB	% increase in SNR due to the use of device for increasing noise immunity
	U_{s1}=1,25V	U_{s1}=1,25V	U_{s1}=1,25V
When exposed to impulse interference	1,13	1,39	23 %
When exposed to fluctuation interference	7,97	8,64	8,4 %

The SNR increase percentage at the input of the track receiver is the ratio of SNR obtained when using the device for improving noise immunity to SNR obtained without using it, and is expressed as a percentage, at the corresponding useful signal voltage U_s on the input of the track receiver.

4. Conclusions

This article proposes a device for improving noise immunity of an audio frequency track circuit with the centralized equipment arrangement, which contributes to reduction of the effect of interference on the input of the track receiver in the intervals between the signal current pulses. The reliability of the railway automatics systems which are responsible for the safety of the transportation process depends on the noise immunity of the track circuit. That is, the proposed solution can improve the safety of trains. Noise immunity of the audio frequency track circuit is improved due to the introduction of a delay line, an adjustable single pulse generator and a controlled electronic switch. The device work is based on cutting the interference in the interval between the useful signal pulses. In order to evaluate how effective the proposed engineering solution increases the noise immunity, the operation of the audio frequency track circuit affected by interferences from traction current, thyristor control, pulse and fluctuation noise was simulated. A signal/noise ratio (SNR) is obtained at the input of the track receiver both during the normal operation of the audio frequency track circuit and when the proposed device to increase noise immunity is used. On the basis of their comparison, the percentage value (% increase in SNR) is obtained, which reflects the degree of noise immunity at the input of the track receiver of the audio frequency track circuit. According to the results of the analysis, this degree is from 8% to 30%, depending on the level of the useful signal and the parameters of the interference.

References

1. J. Wybo, Track circuit reliability assessment for preventing railway accidents. Safety Science, **118**, 268-275 (2018) DOI. https://doi.org/10.1016/j.ssci.2018.03.022
2. T. de Bruin, K. Verbert, R. Babuška, Railway Track Circuit Fault Diagnosis Using Recurrent Neural Networks. IEEE Transactions on Neural Networks and Learning Systems, **28**, 523-533 (2017) DOI: https://doi.org/10.1109/TNNLS.2016.2551940
3. V. Havryliuk. Audio Frequency Track Circuits Monitoring Based on Wavelet Transform and Artificial Neural Network Classifier. 2019 IEEE 2nd Ukraine Conference on Electrical and Computer Engineering (UKRCON-2019), 491-496 (2019)
4. Railway Group Standards, "TI.21 Audio Frequency Track Circuits", GKRC0761 Iss 1Use, Jun 1996.
5. V.S. Arkatov, Yu. V. Arkatov, S.V. Kazeev, Yu. V. Obodovskiy *Track circuits of mainline railways.* (Moscow, Transport Publ., 2006) (in Russian)
6. A.D. Manakov, V.A. Kudryavtsev, A.T. Os'minin, Methods of determining the interference current in rails of double subway track circuits. Russ. Electr. Engin., **87**, 282–285 (2016) (in Russian) DOI: https://doi.org/10.3103/S1068371216050102

7. K.E. Blachev, Imitating modeling of a digital signal receiver of system of an interval handling of traffic of trains. Vestnik of Samara State Technical University: Technical Sciences Series. **2 (22)**. 223-227 (2008) (in Russian)
8. K.V. Honcharov Investigation of transient processes in tonal track circuits. Science and Transport Progress. Bulletin of Dnipropetrovsk National University of Railway Transport named after Academician V. Lazaryan, **4 (46)**, 7-17 (2013) (in Russian) DOI: https://doi.org/10.15802/stp2013/16567
9. Serdyuk T.N., Zavgorodniy A.V., Gavrilyuk V.I. Measurement of electromagnetic interference in the return traction network. Bulletin of Dnipropetrovsk National University of Railway Transport named after Academician V. Lazaryan, **29**, 134-139 (2009). (in Russian) DOI: https://doi.org/10.15802/stp2015/55352
10. Standard EN 50121. *Railway applications. Electromagnetic compatibility*. (London : BSI, 2015) DOI: https://doi.org/10.3403/BSEN50121
11. K.V. Honcharov Methods of protection of tonal track circuits from influence of pulse interferences. Bulletin of DNURT named after Academician V. Lazaryan, **38**, 188-193 (2011).
12. I. O. Saiapina. Analysis of an electromagnetic interference impact on the tonal track circuit's receiving equipment. Information-operating system on railway transport, **3**, 24-31 (2013) (in Russian)
13. T. N. Serdiuk, V Gavrilyuk. Interaction of electrosupply traction system of direct current with the rail circuits. Information-operating system on railway transport, **4**, 108-112, (2007) (in Russian)
14. V. I. Shcheka. The organization of rail circuits protection from electromagnetic interference from the contact network's side. Electromagnetic compatibility and safety on railway transport, **1**, 84-90, (2011) (in Ukrainian) DOI: https://doi.org/10.15802/ecsrt2011/35202
15. K. V Honcharov. Improving the stability of tonal track circuits under fluctuations of ballast resistance. Science and Transport Progress. Bulletin of Dnipropetrovsk National University of Railway Transport, **6(48)**, 23-31 (2013) (in Russian) DOI: https://doi.org/10.15802/stp2013/19674
16. V.I. Havryliuk. Modelling of the return traction current harmonics distribution in rails for AC electric railway system. 2018 International Symposium on Electromagnetic Compatibility (EMC EUROPE). IEEE, 251-254 (2018) DOI: https://doi.org/10.1109/EMCEurope.2018.8485160
17. V. I. Shcheka. Mathematical modelling of track circuits' protection system from electromagnetic disturbances. Electromagnetic compatibility and safety on railway transport, **12**, 33-40 (2016) (in Ukrainian) DOI: https://doi.org/10.15802/ecsrt2016/105492
18. A. M. Beznaritnyy, V. I. Gavrilyuk, I. O. Romantsyev, V. I. Shcheka. Electromagnetic compatibility research of return traction network with signaling devices, centralization and blocking. Science and Transport Progress. Bulletin of Dnipropetrovsk National University of Railway Transport, **3(51)**, 7-14 (2014) (in Russian) DOI: https://doi.org/10.15802/stp2014/25778
19. V. I. Havryliuk, V. I. Shcheka, V. V. Meleshko. Tests of new types of rolling stock on electromagnetic compatibility with signaling and communication devices. Science and Transport Progress. Bulletin of Dnipropetrovsk National University of Railway Transport, **5 (59)**, 7-15 (2015) (in Russian) DOI: https://doi.org/10.15802/stp2015/55352
20. A. A. Antonov, P. E. Maschenko., A.S. Shapovalova. Effect of Tractive Current on Long-welded Track Circuits. *World of transport and transportation*, **1**, 46-51, (2010) (in Russian)
21. K. V. Honcharov, Comparative analysis of modulation and demodulation methods of the control signals of rail line. *Science and transport progress. Bulletin of DNURT named after Academician V. Lazaryan*, **42**, 12-19, (2012) (in Russian)
22. Saiapina, I.O. Improvement of methods and means to increase audio frequency track circuits noise immunity. *Thesis of PhD Kharkiv* (2017) (in Ukrainian)
23. Babaiev M.M., Koshevyy S.V., Sotnyk V.O., Romanchuk V.B., Ananieva O.M., Saiapina I.O. Track circuit. *Patent UA, no. 101093*, (2013) (in Ukrainian)

Research on measurement of electromagnetic fields generated by electric and combustion powered rolling stock.

Dominik Adamski*, *Krzysztof* Ortel, and *Juliusz* Furman

Railway Research Institute, 04-275 Chłopickiego 50, Poland

Abstract. Axle counters are more and more often applied in train detection systems. The wheel sensor is a main part of each axle counter system. In parallel, more and more complex railway vehicles, especially traction ones, are a potential source of interferences influencing the operation of these train detection systems. It is the reason to verify the electromagnetic compatibility (EMC) between the signalling equipment, particularly train detection systems and new vehicles in the process of obtaining the permission for their exploitation. The measurement of interfering magnetic fields generated by vehicles is one of tests to be carried out. For the simplification and unification purpose of the applied interference test methods the EN 50238 standard and TS 50238-3 technical specification were developed. The specification defines unified testing procedures. However, it is necessary to verify if it may replace different testing methods used in particular European states. It is the goal of the European research project financed from the TEN-T network resources.

1 Introduciton

The widespread use of metering systems for track vacancy detection by various railway managements has highlighted problems of these devices' resistance to interferences generated by rolling stock. Another related aspect is also equipping rail vehicles with electric and electronic equipment with extensive structure. The experience of many railway managers shows that modern vehicles can cause disturbances in the axle counters' operation. Therefore the standardization activities on the European level have been ongoing for many years in order to unify and simplify procedures related to particular rolling stock types' placing in service on the European railway network. These efforts are also largely focused on the so called process of frequencies management resulting in the determination of maximum levels of interference for particular frequency bands.

The legal conditions in the field of railway interoperability require rolling stock manufacturers to meet the requirements specified in the Technical Specifications for Interoperability (TSI). The main document concerning the subject discussed in this article which defines the requirements for rolling stock is the document ERA/ETRMS/033281, which presents, among others, limits of magnetic field strengths for AC traction and the method of measuring these fields. It is worth noting that currently there are no uniform European requirements regarding the limits of magnetic fields for the DC traction system, they are so-called open points whose requirements are defined at the level of the national law.

Maximum levels of electromagnetic interference within the scope of national requirements are defined in the project of the Railway Research Institute No. 4430/10 "Determination of maximum levels and interference parameters for railway signalling equipment", while in the European law these levels are included in the technical specification CLC/TS 50238-3:2013: "Railway applications. Compatibility between rolling stock and train detection system - Part 3: Compatibility with axle counters" [8] and PN-EN 50592: 2017-04 "Railway applications - Testing of rolling stock in terms of electromagnetic compatibility with axle counters" [6]. Issues generally related to the subject of compatibility regarding the interaction of rolling stock with train detection devices are also defined in the following standards: PN-EN 50238 and PN-EN 50617-2: 2015-12 "Railway applications - Technical parameters of train detection systems concerning the interoperability of the trans-European rail system - Part 2: Axle counters"[7].

2 Methodology of measurement

In compliance with the provisions contained in the ERA/ETRMS/033281 document, PN-EN 50592:2017-04 and PN-EN 50617-2:2015-12 standards, measurements of magnetic fields generated by rolling stock should be carried out with a standardized antenna (length 15 cm, width 5 cm and height 5 cm) (Fig. 1) independently for the three measuring planes: X, Y, Z (Fig. 2). The maximum values of magnetic field strength from traction vehicles are determined for three frequency bands: 27

* Corresponding author: dadamski@ikolej.pl

kHz – 52 kHz, 234 kHz – 363 kHz and 740 kHz – 1250 kHz.

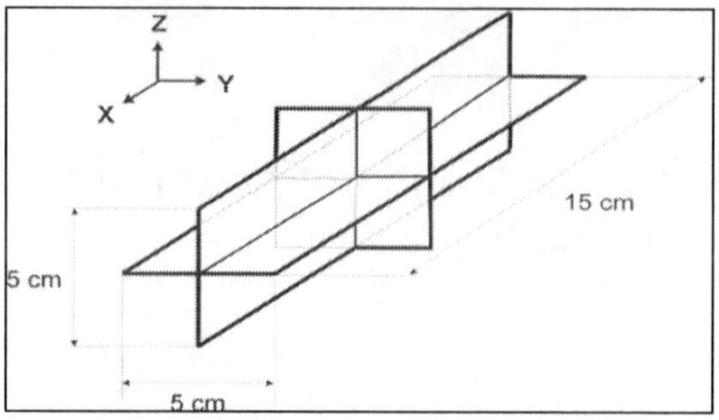

Fig. 1. Recommended dimensions of measurement antenna [8].

Fig. 2. Method of mounting the measurement antenna [4].

The Railway Research Institute has measurement equipment that meets the binding European requirements for this type of research. The measuring equipment includes (Fig. 3): portable laptop computer with data processing and archiving software, two measurement antennas, three oscilloscope cards, two TNB modules with built-in impedance transformers, USB hub with external power supply, measuring leads.

Fig. 3. View of the magnetic field test stand (laboratory) [2].

The tests of electromagnetic fields consist in passing a particular vehicle over measurement antennas installed in the track, as shown in figure 4, on which axle counters are installed or will be installed, and the measurement of the generated magnetic fields values. The tests are carried out in various vehicle operation configurations, such as: traveling at various speeds, starting, electrodynamic braking, travels with on-board devices activated (air conditioning, heating, etc.). Voltage values induced in antennas during the passage of the tested vehicle are recorded on the above-mentioned oscilloscope cards. The recorded results are then sent to the measuring computer, where the FFT analysis is successively carried out using specialized software. This analysis results in acquiring characteristics of the generated magnetic fields values as a function of frequency, which were compared with the limit values determined for each of the measured planes, in accordance with the previously mentioned normative requirements.

Fig. 4. Measurement antennas mounted in a railway track [3].

Two measuring antennas largely reduce the number of performed travels and allow a comprehensive assessment of the tested rolling stock, regardless of the possible asymmetry of the location of the interference source on the vehicle in relation to the track centre, which is particularly important during tests conducted on the railway route.

Examples of tests carried out for a locomotive operated in a 3 kV DC power supply system and in diesel traction are presented below. These tests concerned the assessment of meeting the requirements in the process of vehicle approval for the network managed by PKP PLK S.A.

3 Test results

The test stand was located on the Test Track Centre of the Railway Research Institute in Żmigród. The measurements were carried out for the basic operational states of the traction locomotive (starting, braking, driving at a constant speed) and for the cases of the absence of the vehicle near the measurement antennas (the so-called background measurement). Figure 5 presents examples of results of the background measurement recording, whereas figure 6 shows the magnetic field strength recorded during the start-up of the locomotive in the electric and diesel traction The red colour indicates the limit values of magnetic fields compliant with the requirements of the ERA/ETRMS/033281 document and the [6] standard, while the measured values are shown in blue.

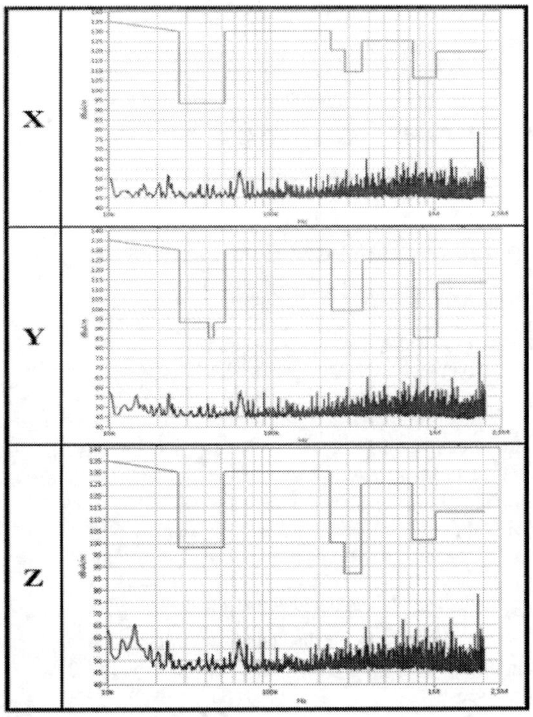

Fig.5. Magnetic field strength in the X, Y, Z planes recorded in the track without a locomotive (background measurement) [1].

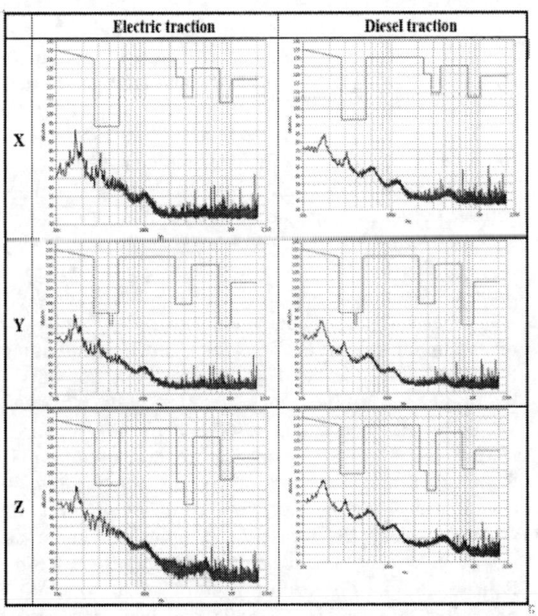

Fig.6. Magnetic field strength in the X, Y, Z planes recorded in the track during a start-up of a diesel-electric locomotive [2].

5 Conclusions

The examples of test results presented in the article confirm that they are different for the same traction vehicle for runs in electric and diesel traction. This proves the necessity to perform measurements of the impact of magnetic fields emitted by diesel-electric traction vehicles both in the electric and diesel traction, as they constitute an essential point of electromagnetic compatibility tests for rolling stock running on railway tracks equipped with train detection equipment. It is reasonable to carry out tests on the emission of magnetic field strengths in the case of introducing new and modernized rolling stock into service. This allows eliminating sources of interference from vehicles, exceeding the maximum values [9], already at the stage of approval tests.

The presented test method complies with the requirements in force and allows for a precise assessment of the measured values in relation to the binding maximum values. The Railway Research Institute has calibrated measurement equipment, a test stand on the test track circuit and a team of qualified employees with extensive experience in the mentioned above research.

References

1. Adamski D., Ortel K., Zawadka Ł.: *Unified verification method of electromagnetic compatibility between rolling stock and train detection systems.* Global Debate on Mobility Challenges for the Future Society, 15-16.11.2018, Warsaw.
2. Adamski D., Białoń A., Furman J., Ortel K., Zawadka Ł.: *Pola magnetyczne generowane przez pojazdy w systemie trakcji prądu przemiennego.* Technika Transportu Szynowego TTS 11/2016.
3. Adamski D., Białoń A., Furman J., Ortel K., Zawadka Ł.: *Magnetic fields generated by vehicles in alternating current traction system.* Wydawnictwo PiT Kraków, 2016.
4. Białoń A., Furman J., Ortel K., Zawadka Ł.: *Wpływ pojazdów różnej generacji na poziomy emisji pól magnetycznych.* Logistyka 4/2015.
5. EN 50238 Railway Applications Railway applications – Compatibility between rolling stock and train detection systems.
6. EN 50592:2017-04 Railway Applications Railway applications - Testing of rolling stock for electromagnetic compatibility with axle counters.
7. EN 50617-2:2015 Railway Applications - Technical parameters of train detection systems for the interoperability of the trans-European railway system - Part 2: Axle counters.
8. Technical Specification CLC/TS 50238-3:2013 Railway applications. Compatibility between rolling stock and train detection systems. Compatibility with axle counters.
9. Gradowski P.: *Upgrading the railway infrastructure technical parameters using the example of the Control-Command and Signalling subsystem with the EC verification certificate,* Railway Reports, issue **182** (March 2019), ISSN 0552-2145, Railway Research Institute, Warsaw 2019, pp. 131-146

Stability of the Railway Subgrade under Condition of Its Elements Damage and Severe Environment

Vitalii Kovalchuk[1,*], *Mykola* Sysyn[2], *Olga* Nabochenko[1], *Andrii* Pentsak[3], *Oleh* Voznyak[1] and *Stepan* Kinter[1]

[1] Dnipro National University of Railway Transport named after Academician V. Lazaryan, Lviv Branch, I. Blazhkevych str., 12-a, Lviv, 79052, Ukraine
[2] Institute of Railway Systems and Public Transport, Technical University of Dresden, Dresden, Germany
[3] Lviv Polytechnic National University, S. Bandery str., 12, Lviv, 79013, Ukraine

Abstract. Investigation of damages of the subgrade slope in combination with its overwetting and rolling stock loading, which significantly affects safety traffic of the railway transport, has been carried out. The complex method of slope stability estimation of the subgrade is developed, which includes calculation of loads and vibration action of rolling stock on the main site of the subgrade, as well as the dynamic model of vibrations propagation in the body of the subgrade embankment and the model of plastic deformations accumulation. Dynamic and nonlinear plastic models are based on a finite-element model of the cross section of the subgrade. The plastic model takes into account the characteristics of soil strength of the subgrade, depending on the area of the vibration load impact. The developed method allows to carry out the estimation of external and internal factors impact on occurrence of subgrade destruction, which is of practical value for safety state estimation of the railway transport.

INTRODUCTION

Railway subgrade embankment is one of the most critical parts of railway infrastructure. The failures in railway embankments require continual maintenance and repair works that leads to enormous maintenance costs, decrease of reliability and availability of railway transportation [1, 2]. Moreover, embankments failures, together with other unfavourable factors can cause the reduction of slope stability and risk of train derailment [3,4,5]. The stability of the railway subgrade is simultaneously influenced by a number of factors: maintenance works implementation the state of culverts, atmospheric precipitation, loading of rolling stock. In addition to these factors, the factor of subgrade state is also important due to the heterogeneous geological structure and weak soils of the embankment and the foundation, that is, the presence of the defect subgrade. In assessing the stability of the embankment, it is necessary to determine the contribution of each of the factors to the occurrence of subgrade landslide. A study of moisture values of subgrade construction materials depending on climatic factors is presented in [6].

The impact of the rolling stock on the stability of the subgrade is both static and dynamic loading of sleepers on the ballast prism and the subgrade. On the one hand, static and dynamic loading leads to the increase of landslide stresses, which, when exceeding the permissible value, causes plastic deformation, and on the other hand, dynamic vibrational loading results in the loss of friction and adhesion between soil particles. The loading of the rolling stock, in its turn, depends on many factors of rolling stock and the track, namely, the speed of movement and others. The influence of rolling stock dynamic loading on railway subgrade and ballast materials is presented in [7-10]

The critical issue is the quantitative assessment of interconnections of the effects of maintenance, state and characteristics of both the subgrade and the superstructure, taking into account rolling stock loading and the state of its running parts [11-12]. The basis for solving this problem is mathematical modeling of the processes of elastic and irreversible deformations of the subgrade in mutual connection with rolling stock loads on the upper structure of the track. Such formulation of the problem allows to take into account not only the possibility of rolling stock derailment as a result of subgrade subsidence, but also to depict the process of state transition of a railway from operable and working to marginal and non-working. At present, the model of elastic and plastic deformations of the subgrade is most widely used on the basis of the finite element method and the criteria of plasticity of Mohr-Coulomb or Drucker-Prager. Proceeding from practical possibility of obtaining information on soils characteristics and the characteristics of railways subgrade, as well as the necessary of calculations accuracy, the problem of elasticity and plasticity in flat formulation is optimal. In addition to model developing, it is important to take into account the impact of vibration loading of the rolling stock on the mechanical characteristics of soils in the transverse section of the subgrade body.

* Corresponding author: kovalchuk.diit@gmail.com

Along with considered advantages of the models of elastic and plastic deformations of the substructure of the track, there are also a number of shortcomings, that is mathematical complexity of models, the need for the development of specific software for practical application, the need for output for the calculation of the parameters of soil characteristics, which are determined experimentally, although they are not always available. In this regard, it is advisable to consider simpler methods for calculating the borderline state of the subgrade, which, under condition of some uncertainty of background data on soils, can give an answer about potential destruction of the subgrade. Due to the simplicity of these techniques and their possible implementation without the use of specialized software, they can be more widely used in assessing the safety of the railway transport.

The present paper deals with the study of slope of railway embankments under condition of its damage and severe environment and operational loading. For assessing the subgrade stability, the calculation of vertical loading of a sleeper on the ballast that arises during movement of corresponding components of a train with actual speeds in accordance with the existing standard methodology and instructions is carried on. To determine the effect of dynamical vibrational loads on the mechanical properties of soils, it is necessary to calculate the distribution of dynamical displacements of soil points of the embankment, which is performed using the principle of dynamical theory of elasticity. This takes into account vibrational dynamical loading caused by fluctuations of unsprung masses of a train as a result of movement along separate track irregularity, continuous wheel irregularity and the presence of wheel sliders. The solution of the dynamical problem is an iterative solution of a static task with regard inertial forces. As a result of calculation of the impact of dynamical vibrational loading on mechanical properties of soils, the field of distribution of specific adhesion and the angle of internal friction in the cross section of the embankment were obtained, which were taken into account in calculations of plastic deformations later.

Calculation of vertical force action of a sleeper on the ballast during trains traffic

Rolling stock, moving by the railway, transfers dynamic loading on the rails, sleepers, ballast and the subgrade. The acting forces of rolling stock depend on a number of factors, the most important of which are the weight of rolling stock, speed, mechanical characteristics of the suspension and the running gear of rolling stock, geometric state of the railway track and its mechanical characteristics. The vertical forces of rolling stock acting on the track and their distribution in the elements of the track are determined according to the "Rules of calculating railway track strength and stability" being valid for the railways of Ukraine [13]. The distribution of vertical forces from rolling stock to the rails is determined by the methodology of a number of authors [5–8].

Calculation of vertical force of a sleeper acting on the ballast consists of two parts:
1) calculation of vertical dynamic loads of rolling stock acting on rails;
2) calculation of vertical forces distribution acting on the elements of the track.

Dynamic loads from a single wheel acting on a rail and from a group of wheels are calculated in accordance with clauses 2.4.1–2.5.4 of the Rules TsP-0117 [13].

Calculation of impact-vibration loads of rolling stock

This calculation is carried out for determining the distribution of oscillation amplitude A in the subgrade and further evaluating shock-vibration loads impact on strength parameters C_{dyn} and φ_{dyn} for each type of soil of the subgrade.

For this purpose, a developed model of dynamic oscillations of the subgrade in the transverse plane is used. The starting data for calculation are:
1) mechanical properties of soils obtained from the results of geological survey and reference literature: deformation module U, specific gravity m and Poisson coefficient;
2) external vibration loads on the ballast layer at the given speed of rolling stock are taken as a sum of impulses of harmonics of the function:

$$F(t) = F_p(t) + F_{ti}(t) + F_{swi}(t) + F_{cwi}(t) \quad (1)$$

The resulting dynamic loads on the ballast layer is:

$$F(t) = p_s^{dyn} \begin{pmatrix} \frac{S_p^2}{S^2}\sin(\frac{2\pi}{T_p}t) + \frac{S_{ti}^2}{S^2}\sin(\frac{2\pi}{T_{ti}}t) + \\ +0,05\frac{S_{swi}^2}{S^2}\sin(\frac{2\pi}{T_{swi}}t) + 0,95\frac{S_{cwi}^2}{S^2}\sin(\frac{2\pi}{T_{cwi}}t) \end{pmatrix} \quad (2)$$

where p_s^{dyn} is the amplitude of the dynamic force of the sleepers on the ballast per 1 running meter of the track, which is calculated in accordance with the Rules [13]; S_p^2, S_{ti}^2, $0,05S_{swi}^2$, $0,95S_{cwi}^2$ are components of the composition of distribution laws of the mean deviation of dynamic vertical loads; T_p, T_{ti}, T_{swi}, T_{cwi} are periods of impact-vibration loads.

The basis for calculating dynamic loads action of rolling stock on the track is the Rules [13], where the rule of forces superposition is taken into account that includes the calculation of the total effect of all dynamic loads of different periods of oscillation: from 10-50 m with fluctuations of over-sprung mass of a carriage, 3–5 m from unevenness of the railway up to 15–20 cm with separate roughness of rolling stock wheels. For the given calculation, only those oscillation frequencies are important that are relevant to experimental studies of soil vibrational dislocation – 11–50 Hz [14]. The component of oscillations of over-sprung structure in formula (1) in the speed range up to 55 km/h corresponds to frequencies of 0.5–2 Hz, so it is not taken into account in the calculation.

Example of components of formula (1) calculation S_{ti}^2, $0,05S_{swi}^2$, $0,95S_{cwi}^2$ and their periods of oscillation T_{ti}, T_{swi}, T_{cwi} at the speed of 55 km/h for a four-axial freight car on the bogies TsNII-X3-0 and for electric locomotive VL-80s is given in Table 1.

The calculated values of the components of the dynamic forces and their periods are used to construct the impact vibration load function $F(t)$ (1), which is further used in simulation of oscillations in the subgrade.

The function $F(t)$ is presented by one load impulse and, in order to ensure its smooth detailing, such a step of time is chosen that within the period of oscillation there were at least 10 points of calculation. With a multiple reserve, the time discretization is taken to be 0.001 sec.

Table 1. Components of dynamic loads composition and their periods of oscillation

Velocity of movement, km/h	Components of loads composition, кN				Oscillation periods, sec		
	S_{ti}^2	$0,05S_{swi}^2$	$0,95S_{cwi}^2$	$\overline{S}^2$	T_{ti}	T_{swi}	T_{cwi}
for a four-axial freight car on the bogies TsNII-X3-0							
55	19,47	6,77	0,29	34,36	0,065	0,052	0,054
for electric locomotive VL-80 s							
55	47,01	3,33	0,30	52,53	0,065	0,087	0,071

The graph (Fig. 1) shows the change in time of total dynamic loads on the ballast layer at rolling stock speed of 55 km/h.

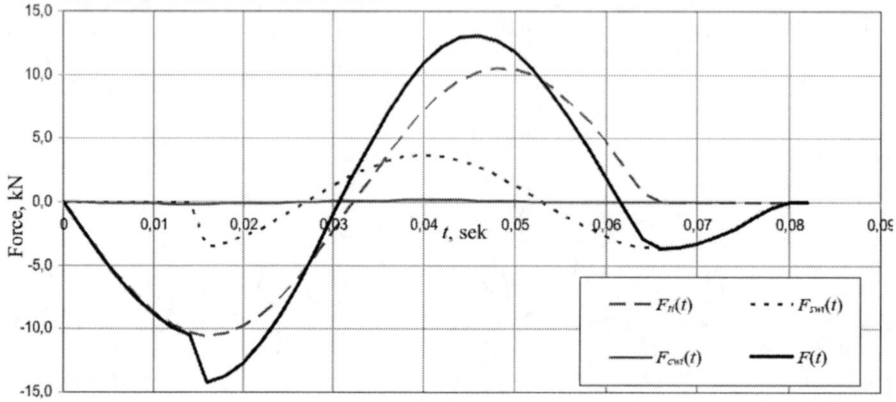

Fig. 1. Chart of time variation of the total dynamic loads on the ballast layer

Calculation of dynamic fluctuations distribution in the subgrade

For calculation of the impact of shock-vibration loads of rolling stock on the propagation of dynamic oscillations in the subgrade, a finite-element model is used. According to this model, the transverse section of the subgrade (Fig. 2) is divided into separate elements, whose properties correspond to one or another type of soil in accordance with the papers of geological research. The division into finite elements is carried out with the aim of detailed description of separate zones of the cross-section with different soil properties. To increase calculating accuracy the slope zone from the side of possible soil sliding, is divided into smaller finite elements.

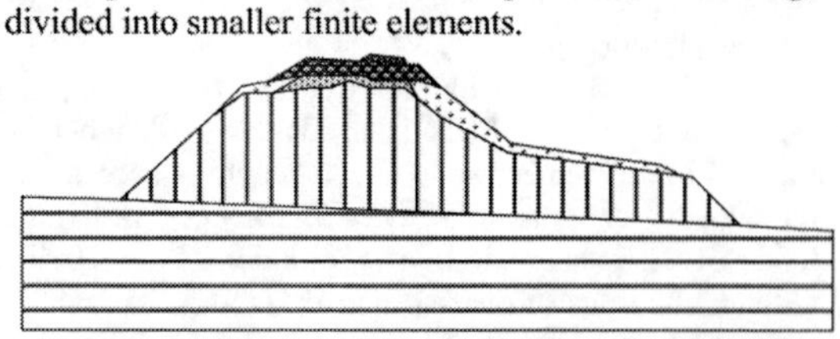

Fig. 2. The scheme of soil distribution in the cross section of the embankment

In order to provide an acceptable calculation time and, taking into account the available computer resources, the size of finite elements in the depth of the embankment is enlarged. The scheme of finite elements distribution in the cross section of the embankment and its constituent soils are shown in Fig. 3. The total number of quadrangular 8-node elements is 687, the total number of nodes of the elements is 2196. Estimated external dynamic loading from the sleeper is taken as distributed into nodes 12, 13, 18, 19, 24, 25, 3 0, 31, 36, 37, 42, 43, 48, 49 and 54.

The geometrical characteristics of the transverse profile and soil properties of the subgrade should be taken according to geological survey data. For this calculation it is assumed that the body of the embankment consists of the following soils:

Soil 1. Base of the embankment: clay of dense consistency and semi-solid consistency with the following characteristics: soil density – 17,8 kN/m³; deformation module U – 16 MPa; yield index (consistency) I_L – 0.08; plasticity number I_p – 0.27. The internal friction angle ϕ – 16 ° and the specific adhesion c – 36 kPa are taken in accordance with [14, 16] regarding yield index and plasticity number.

Soil 2. Body of the embankment: loam with clay layers, with consistency from semisolid to mild plasticity with the following characteristics: soil density m – 17,5 kN/m³; number of plasticity I_p – 0.14; yield index (consistency) I_L – 0.28. The internal friction angle ϕ is 22° and the specific adhesion c – 25 kPa taken with regard [15, 16] according to yield index and plasticity number. Deformation module is adopted according to [9] and is confirmed by a number of sources [17, 18] = 13 MPa.

Soil 3. Sub-ballast layer: ballast sand with inclusions of crushed stone, non-uniform, of medium density and dense. Data are taken according to reference literature and regulatory documents with regard to DBN V.2.1-10-2009 and DIN 18196 [15, 18-20] according to classification SE, SU; the angle of internal friction ϕ – 32–38°, accepted ϕ – 38° and density m –19,0 kN/m^3 (in the state of natural humidity); deformation module can vary within 48–100 MPa (from medium compacted state to compacted one), taken – 75 MPa; specific adhesion can vary within 2–6 kPa, taken – 6 kPa.

Soil 4. Ballast plume: quartz sand with gravel inclusion, siliceous gravel and loam, inhomogeneous, from large to medium size, loose. Characteristics are taken in accordance with reference literature and regulations with regard to DBN B.2.1-10-2009 and DIN 18196 [15, 18-20] classified SI, SU, SW; internal friction angle ϕ – 28–35°, accepted ϕ – 32° and density m – 18.0 kN/m^3 (in the state of natural humidity); deformation module can vary within 18–50 MPa (in the loose state), taken U – 30 MPa; specific adhesion – 2-4 kPa, accepted c – 2 kPa.

Soil 5. Ballast: crushed stone ballast. Characteristics are taken in accordance with reference literature and regulatory documents, taking into account the fact that crushed stone ballast is in compressed state and without contaminants with regard to [19, 20]. According to GE classification, the angle of internal friction ϕ = 37–65°, accepted ϕ=63° (for sub-ballast layer ϕ=58°) and density m =20,0 kN / m^3; deformation module can vary within 150–300 MPa (in the loose state), accepted U =150 MPa; the characteristic of specific clutch in crushed stone is absent, but this material has a similar property of another nature, that is spike-tooth adhesion, which is 0.8–0.92 kPa and accepted c =0.9 kPa (for sub-ballast layer c =0.5 kPa).

Possible distribution scheme of soils in the transverse section of the embankment is shown in Fig. 4.

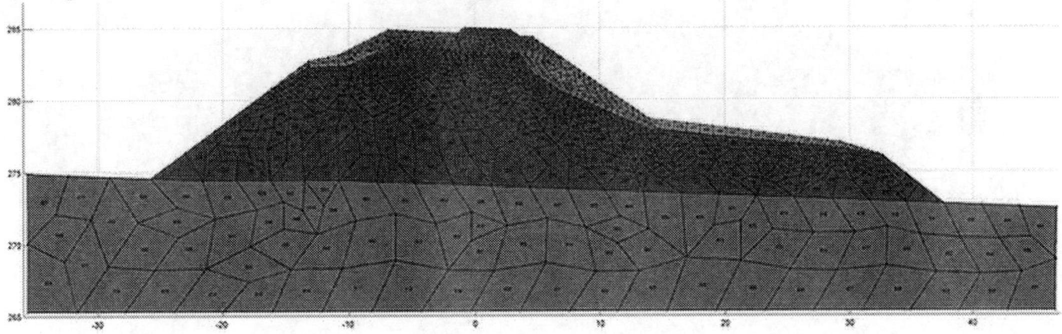

Fig. 3. Scheme of finite elements model in transverse section of the embankment

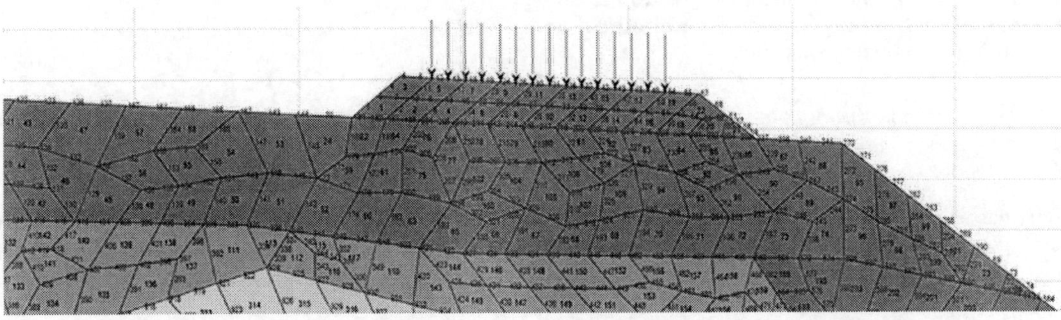

Fig. 4. Scheme of external dynamic loads distribution from the sleeper to the nodes of the ballast layer

To simulate the passage of dynamic oscillations in the transverse section of the embankment, a time interval of 0–0.2 seconds is chosen that corresponds to several periods of oscillation at each point of the cross section of the embankment. It is not expedient to perform calculations over a longer period of time, since the greatest interest in the study is the maximum values of amplitudes of oscillations, and after the first periods of oscillation, their attenuation occurs. Fig. 5 shows variations of dynamic displacement of the points of the subgrade at a different depth below the middle part of the sleeper in the interval of 0.2 seconds after the start of dynamic loads application. The calculation is based on the developed software.

Fig. 6 shows distribution of oscillations amplitudes of soil points dynamic movement in the cross section of the subgrade. The graph shows that the largest amplitudes of oscillations reach 110 mkm and are at the point of dynamic loads application from the sleeper. At the same time amplitudes of oscillations quickly fade in depth starting from the surface of the subgrade and along the slopes.

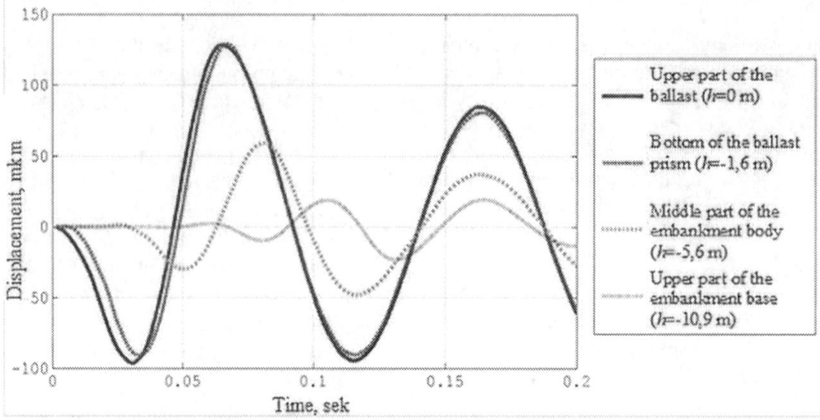

Fig. 5. Oscillation of subgrade points under the sleeper (ballast – subgrade – base of the embankment)

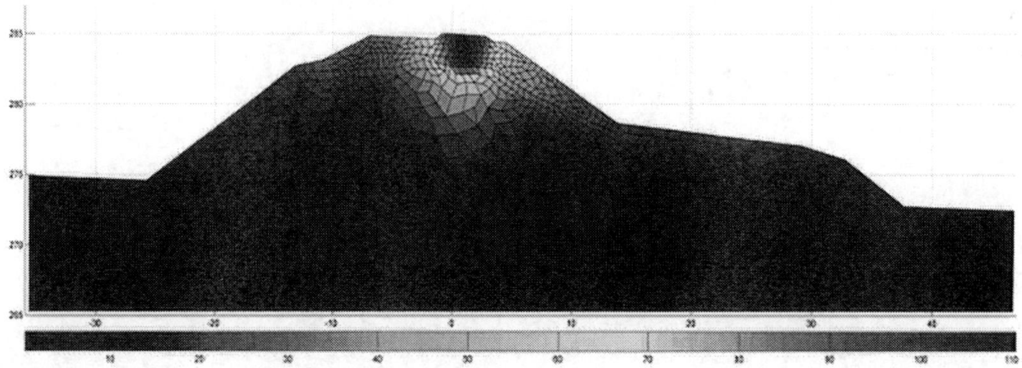

Fig. 6. Distribution of oscillations amplitude of dynamic displacements in the cross section of the embankment

Method of elastic and plastic deformations of the subgrade body, taking into account the impact of vibro-loading and moisture on mechanical properties of soils

Calculation of soil strength characteristics of the subgrade, taking into account vibro-dynamic impact of rolling stock.

Under the action of vibrational dynamic loads of rolling stock, there is a significant reduction in the value of subgrade soils bearing capacity. To evaluate the impact of vibrational dynamic action of rolling stock on the characteristics of soil strength, the technique developed by the scientists of PSURT Prokudin I.V., Kolos A.F., and others is used and approved by the Ministry of Railways of the Russian Federation for the assessment of the quality of the railway track base [20]. This technique is the basis of many modern studies of the subgrade stability [21, 22]. According to it, and in order to evaluate the effect of vibrational dynamic action on soil deformative and strength characteristics, the indices of relative reduction of specific adhesion and the angle of internal friction of the soil are used, which are determined in parts or percentages by the formulas:

$$K_c = \frac{C_{st} - C_{dyn}^{min}}{C_{st}} \quad K_\varphi = \frac{\varphi_{st} - \varphi_{dyn}^{min}}{\varphi_{st}}, \tag{3}$$

$$K'_c = \frac{C_{dyn}^{min}}{C_{st}} \quad K'_\varphi = \frac{\varphi_{dyn}^{min}}{\varphi_{st}}, \tag{4}$$

where C_{st}, φ_{st} – specific adhesion and angle of internal friction of the soil, which are determined under the action of static loads; C_{dyn}^{min}, φ_{dyn}^{min} – specific adhesion and angle of internal friction of the soil, which are determined under the action of maximum vibro-dynamic loads.

The most important factors affecting soils strength are soil moisture, its density, the magnitude of stresses pulsation or the amplitude of oscillations, soil deformation to vibro-dynamic effect and stressed state of the subgrade soil. Patterns of clutch change, angle of internal friction and deformation module with increasing vibro-dynamic impact are the main factors that determine the state and ability of soils to resist external loads.

Taking into account the proportional dependence of oscillations amplitudes on voltage pulsation, the law of strength change and deformation characteristics, the magnitude of soils oscillation amplitudes is described by the following formulas:

$$C_{dyn} = C_{st}\left[K_c' + K_c \exp\left(-K(A - A^{н})\right)\right], \quad (5)$$

$$\varphi_{dyn} = \varphi_{st}\left[K_\varphi' + K_\varphi \exp\left(-KA\right)\right], \quad (6)$$

where, K_c, K_c', K_φ, K_φ' are limiting values of relative decline of characteristics that determine soil deformation to vibro-dynamic loads; K, K' are coefficients of vibrational destruction and vibrational deformation; A is the resulting amplitude of oscillation, at which the characteristics are determined; $A^{н}$ is an initial amplitude of oscillations, at which the reduction of characteristics does not exceed 3–5%.

The study of soils of different composition and state with the largest volumetric mass obtained at standard soil consolidation with the humidity that is different from the optimum one, allowed to reveal the values of characteristics reduction, their correlation, coefficients of vibrational destruction and vibrational deformation, as well as initial amplitudes. The data about them are shown in Table 2.

To calculate the parameters C_{dyn} and φ_{dyn} for each soil of the subgrade and their distribution within its cross-section, the computation of the distribution of the resulting amplitude of oscillations A in the subgrade is made using the developed model of dynamic oscillations of the subgrade in the transverse plane. Dynamic loads on the ballast layer from rolling stock are determined according to the TsP-0117 [19] being valid for the railways of Ukraine taking into account the main factors of vibrational loads.

Table 2. Indices of soil characteristics decline [14]

Indices	Name of the soil and its consistency							
	Clays with I_L			Loam with I_L			Sandy loam	
	0–0,15	0,15–0,45	0,45	0–0,15	0,15–0,45	0,45	solid	plastic
K_c	0,15	0,55	0,13	0,2	0,5	0,15	0,1	0,60
K'_c	0,85	0,45	0,87	0,80	0,50	0,85	0,90	0,4
K_φ	0,09	0,45	0,10	0,10	0,40	0,08	0,07	0,4
K'_φ	0,81	0,55	0,90	0,90	0,60	0,92	0,93	0,60
K'	0,005	0,01	0,015	0,006	0,011	0,02	0,006	0,025
K	0,01	0,012	0,018	0,008	0,012	0,015	0,007	0,02
$A^{н}$	10	15	0	15	20	0	10	20

Fig. 7 and Fig. 8 show the results of calculations of distribution parameters C_{dyn} and φ_{dyn} taking into account vibrational impact.

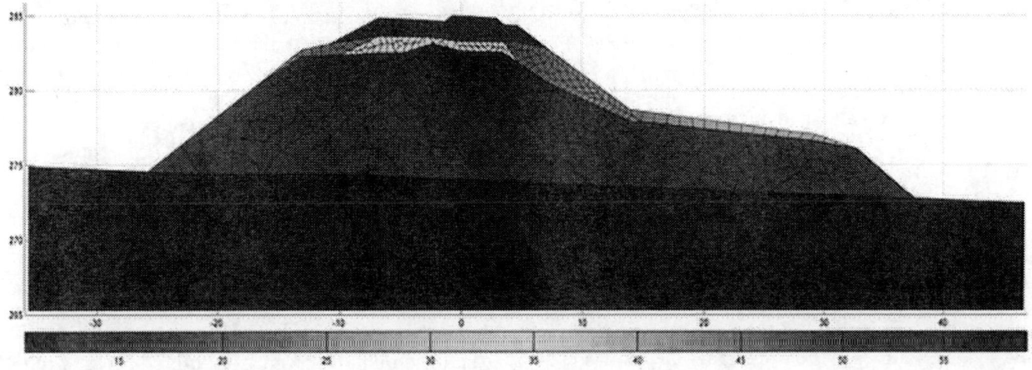

Fig. 7. Scheme of distribution of the angle of internal friction of the soil φ_{dyn} in the cross section of the subgrade, taking into account vibrational impact

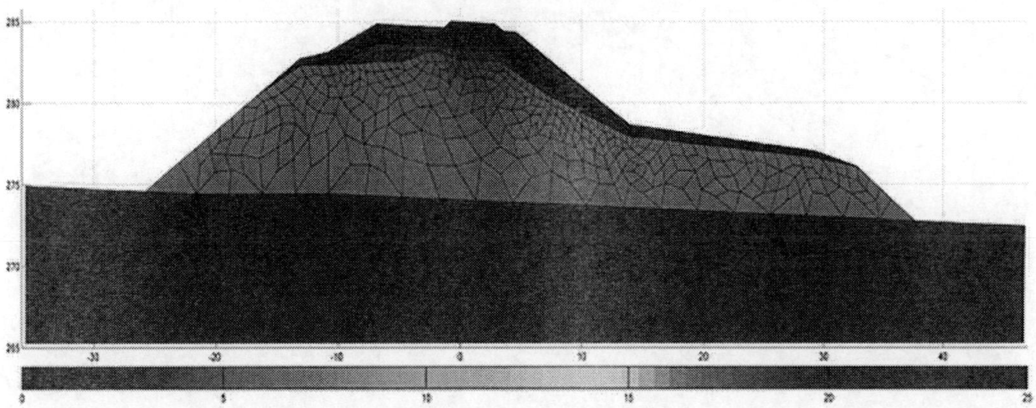

Fig. 8. Scheme of distribution of specific soil cohesion C_{dyn} in the cross section of the subgrade, taking into account vibrational impact

Charts in Fig. 7 and Fig. 8 show that a significant vibrational dissociation occurs in the distribution of the angle of internal friction, namely in the body of the embankment made of loam and ballast plume, which are close to the zone of vibration loads application. Vibrational loads have also an impact on the reduction of specific adhesion in the loam of the body of the embankment.

Example of calculation of the railway track stability under the impact of adverse operational, hydrometeorological conditions and its elements damage.

Calculations of plastic deformations of the embankment of the subgrade

The displacement of the slope of the subgrade can be influenced by a number of factors in their mutual combination:

Factor 1. Low safety margin of the existing subgrade, due to the presence of inhomogeneous and weak soils in its composition (gravity loads, constant factor).

Factor 2. Dynamic loading of rolling stock, which depends on the type and speed of trains, running gear of rolling stock, geometrical state of the track, etc.

Factor 3. Reduced soils strength due to their humidification by short intense atmospheric precipitation.

Factor 4. Dismantling of a part of holding elements of the subgrade and the embankment during replacement of a pipe culvert by a repair crew.

To determine the degree of impact of these factors separately and in a mutual combination, it is necessary to make calculations of the coefficient of stability of the slope according to the developed model of deformation of the subgrade in all possible combinations, as shown in Table 3.

Thus, there are only 8 cases of calculating the stability of the subgrade: 1. In the absence of loads of rolling stock, atmospheric precipitation and repair works; 2. Under the impact of rolling stock without atmospheric precipitation and repair works; 3. Under the impact of atmospheric precipitation without trains traffic and repair works; 4. Under repair works performance without atmospheric precipitation and trains traffic; 5. Under the impact of rolling stock and atmospheric precipitation without repair works; 6. Under the impact of rolling stock and repair works without atmospheric precipitation; 7. Under repair works performance without atmospheric precipitation and trains traffic; 8. Under the impact of rolling stock, repair works and atmospheric precipitation.

Table 3. Possible cases of factors combination

Calculation variants	Factor 1	Factor 2	Factor 3	Factor 4
1	x			
2	x	x		
3	x		x	
4	x			x
5	x	x	x	
6	x	x		x
7	x		x	x
8	x	x	x	x

It is known from the geological section of the subgrade that from the side, where the slope of the subgrade has occurred, there is a ballast plume (Soil 4) with a thickness up to 1 m (Fig. 9). Proceeding from this, we can assume that there was a significant moisture of upper drainage soils of the embankment, namely Soils 3-5. It is known from a number of sources that soil moisture significantly reduces their strength characteristics [23, 24, 25], namely, sliding characteristics: specific adhesion C_{st} and the angle of internal friction φ_{st} ($tg(\varphi_{st}) = f_{st}$) under condition of moisture decrease:

$$C_v = (0{,}5...0{,}6)C_{st}, \quad f_v = (0{,}75...0{,}85)f_{st} \tag{7}$$

We accept extreme lower limits for lowering landslide characteristics of soils: 0.5 for specific clutch and 0.75 for the angle of internal friction, respectively.

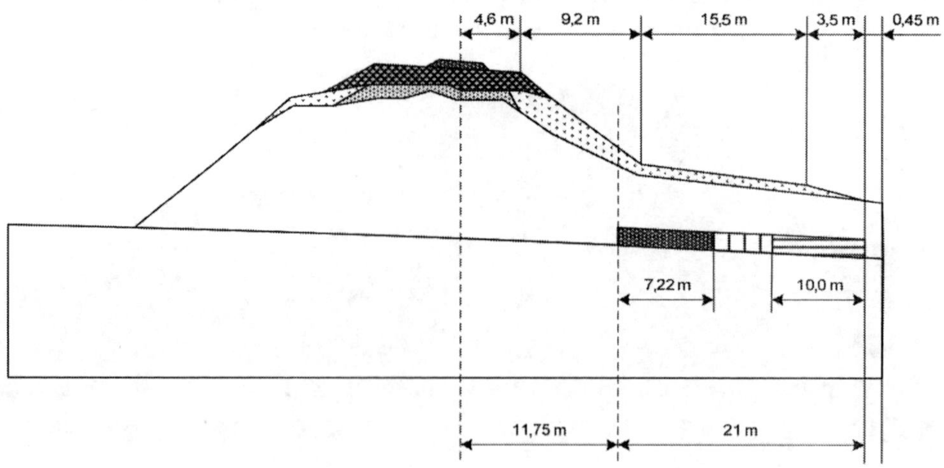

Fig. 9. Cross-section of the embankment and the scheme of pipe culvert dismantling

Calculation of stability when performing repair works and under the impact of adverse factors

When performing subgrade dismantling for the distance of 11.75 m from the axis of the unpaired track without atmospheric precipitation, without trains traffic and the coefficient of parameters change of sliding strength less than 1.18 is the attenuation of intensity of average subsidence (red line in Fig. 10) to 2 mm. At higher values of the coefficient there is a gradual increase in subsidence rate, that is, the destruction of the subgrade, in this case, will be gradual (Fig. 11, 12). Coefficient of stability of the embankment k_y=1.18 is adopted.

The results of determining stability coefficients for various combinations of factors impact are summarized in Comparative Table 4.

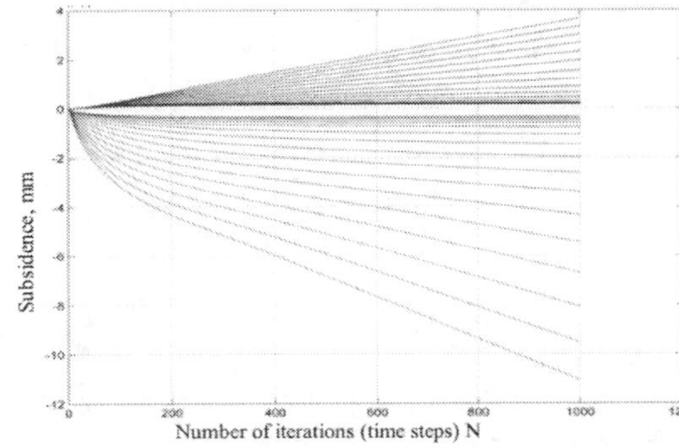

Fig. 10. Average and uneven subsidence of points of a ballast prism surface depending on the step of "pseudo-time" under the increase of the coefficient of parameters change of sliding strength of soils

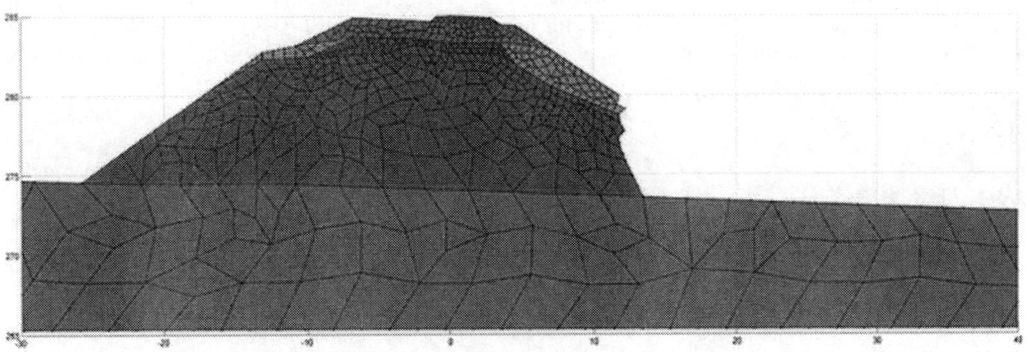

Fig. 11. Transverse profile of the embankment after dismantling of the subgrade to destruction

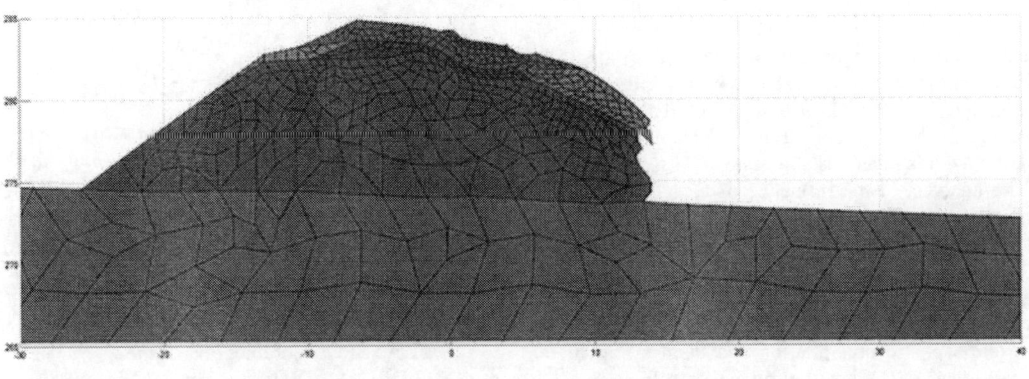

Fig. 12. Plastic deformations of the transverse profile of the embankment at the moment of displacement beginning (deformation increased by 500 times)

Table 4. Possible cases of influence of factors combination and safety margin coefficients

Calc. variants	Factor 1	Factor 2	Factor 3	Factor 4	k_y
1	x				2,1
2	x	x			1,45
3	x		x		2,05
4	x			x	1,18
5	x	x	x		1,36
6	x	x		x	0,98
7	x		x	x	1,1
8	x	x	x	x	0,92

Analysis of the influence of repair works and precipitation on the strength of the subgrade.

The results of the study show that the subgrade of the embankment under the impact of rolling stock at the speed of 55 km/h and without intensive moistening, has a safety margin against rapid loss of stability of 1.45.

When dampening the upper drainage layers of the subgrade with storm atmospheric waters, the safety margin against rapid loss of resistance decreases to 1.36, although, at the same time, there are slow deformations that do not threaten safety traffic. When disassembling the subgrade at a distance of 11.75 m from the axis of the track and absence of a heavy shower, the subgrade is in the near-crucial state without a stock of stability.

Calculations show that with significant moisture and disassembled subgrade to a distance of 11.75 m from the axis of the track, the embankment is in the stable state with a stock of stability of 1.1, but in case of atmospheric precipitation, the embankment has a coefficient of stability equal to 0.98, that is, it is in an unstable state.

Thus, considering the influence of each of the factors separately, it can be seen that the effect of factor 4 (disassembly of the subgrade) is the largest and reduces the resistance by 44 %, factor 2 (rolling stock) reduces the resistance by 31 %, and factor 3 (storm atmospheric precipitation) has the least impact – about 3 %.

Considering the effect of factor 3 (storm atmospheric precipitation) and factor 4 (disassembly of the subgrade) and excluding the impact of factor 2 (rolling stock), it can be stated that storm atmospheric precipitation caused 6 % decrease in stability, and disassembly of the subgrades caused 32 % of decrease in the stability of the subgrade.

CONCLUSIONS

The vibrational loading impacts the reduction of specific adhesion and the internal friction angle of the body of the embankment of the railway track.

When dampening the upper drainage layers of the subgrade with storm atmospheric precipitation, the safety margin against rapid loss of resistance decreases to 1.36, which does not endanger safety traffic.

Calculations show that in the disassembled subgrade to a distance of 11.75 m from the axis of the track the embankment is in the stable state with a stability reserve of 1.1, but in case of atmospheric precipitation the embankment has the coefficient of stability equal to 0.98, that is, it is in an unstable state.

The analysis of the influence of each of the factors separately, shows that the influence of subgrade dismantling is the largest, and it reduces the resistance by 44%, the impact of rolling stock reduces the stability of the subgrade by 31 %, and the impact of atmospheric precipitation is about 3 %. Considering the effect of storm atmospheric precipitation and dismantling of the subgrade, and excluding the impact of rolling stock, it can be stated that the storm atmospheric precipitation caused 6 % decrease in the stability of the subgrade, and the dismantling of subgrades caused 32 % resistance decrease.

REFERENCES

1. Briggs, KM, Loveridge, FA and Glendinning, S *Failures in transport infrastructure embankments.* Engineering Geology, 219. pp. 107-117 (2017)
2. Iżvolt, L., Šestáková, J., Šmalo, M. *Tendencies in the development of operational quality of ballasted and ballastless track superstructure and transition areas.* IOP Conference Series: Materials Science and Engineering, **236(1)**, 012038 (2017)
3. Zhang, Q., Gao, Y., Zhang, H.-X., Xu, F., Li, F. *Risk identification, assessment and monitoring design of high cutting loess slope in heavy haul railway.* Structural Monitoring and Maintenance, **5(1)**, pp. 67-78 (2018)
4. Qian Zhang, Jing Wang, Wenyu Wang, Songsong Bai & Peng Lin, *Study on slope stability due to the influence of excavation of the high-speed rail tunnel*, Geomatics, Natural Hazards and Risk, 10:1, 1193-1208 (2019)
5. M. P. Sysyn *Vplyv dsbratciinoho navantagennia ruchomoho skladu na robotu hruntiv zemlianoho polotna* Zaliznytchnyi transport Ukrainy. **№ 2.** – P.P. 51-54 (2012) [in Ukrainian]
6. Libor Ižvolt, Peter Dobeš, Alžbeta Pultznerová, *Monitoring of Moisture Changes in The Construction Layers of the Railway Substructure Body and Its Subgrade*, Procedia Engineering **161** 1049 – 1056 (2016)
7. Sysyn Mykola, Nabochenko Olga, Kovalchuk, Vitalii, Gerber Ulf *Evaluation of Railway Ballast Layer Consolidation After Maintenance Works* ACTA POLYTECHNICA, **59(1)**, Pages: 77–87 (2019), DOI: 10.14311/AP.2019.59.0077. Accession Number: WOS:000462321600009.
8. Sz. Fischer, f. Horvát, *Investigations of the reinforcement and stabilisation effect of geogrid layers under railway ballast*, Slovak journal of civil engineering, Vol. **XIX**, No. 3, 22 – 30 (2011) DOI: 10.2478/v10189-011-0015-y
9. Plasek, Otto; Hruzikova, Miroslava; Svoboda, Richard; INFLUENCE OF UNDER SLEEPER PADS ON TRACK QUALITY, AKUSTIKA, 23:28-33, 2015
10. Szabolcs Fischer, Breakage Test of Railway Ballast Materials with New Laboratory Method, Periodica Polytechnica, Civil Engineering, 61 (4), pp. 794–802, 2017, https://doi.org/10.3311/PPci.8549
11. Sysyn Mykola, Gerber Ulf, Nabochenko Olga, Gruen Dmitri, Kluge Franziska *Prediction of Rail Contact Fatigue on Crossings Using Image Processing and Machine Learning* URBAN RAIL TRANSIT, Volume: **5**, Issue2, Pages: 123-132 (2019) DOI: 10.1007/s40864-019-0105-0. Accession Number: WOS:000471653400004.
12. Sysyn Mykola, Gerber Ulf, Gruen Dimitri, Nabochenko Olga, Kovalchuk, Vitalii *Modelling and vehicle based measurements of ballast settlements under the common crossing* EUROPEAN TRANSPORT-TRASPORTI EUROPEI, Issue: **71**, Article Number: 5, (2019) Accession Number: WOS:000462060400005.
13. Danilenko E. I., Rybkin V. V. *Pravyla rozrakhunkiv zaliznychnoi kolii na micnist i stiikist* (CP/0117). Kyiv: Transport Ukrainy – 168 p. (2006) [in Ukrainian]

14. Shachuniants G. M. *Ustroistvo zheleznodorozhnogo puti* Moskva: Transzheldorizdat, (1944) [in Russian]
15. Kohan A. Ya. *Dinamika puti i eho vzaimodeistvie s podvizhnym sostavom* Moskva: Transport,. – 326 p. (1997) [in Russian]
16. Chernyshev M. A. *Prakticheskie metody rascheta puti.* / Чернышев М.А. – М.: Транспорт, 1967. – 236 с. [in Russian]
17. Prokudin I. V. Hrachev I. A., Kolos A. F. *Organizatsiia pereustroistva zheleznych doroh pod skorostnoe dvizhenie poezdov*, Moskva: Marshrut,. – 716 p. (2005) [in Russian]
18. *Osnovy ta fundamenty sporud. Osnovni polozhennia proektuvannia. DBN V.2.1-10-2009.* Kyiv: Minrehionbud Ukrainy. – 104 p. 2009 [in Ukrainian]
19. Danilenko E. I. *Zaliznychna koliia. ulashtuvannia, proektuvannia i rozrakhunki, vzaiemodiia z rukhomym skladom.* Kyiv, Inpres,. – Tom **2**. 456 p. (2010) [in Ukrainian]
20. *Rekomendacii po intensivnoi tekhnolohii i monitoring stroitelstva zemlianykh sooryzhenii na slabykh osnovaniiakh* Moskva: «Timr»,. 96 p. 2005 [in Russian]
21. Simmer K. *Grundbau: Bodenmechanik und erdstatische Berecnungen* Stuttgart: Teubner B.G. GmbH, – 430 p. (2009)
22. Lichtberger B. *Handbuch Gleis: Unterbau, Oberbau, Instandhaltung, Wirtschaftlichkeit.* Hamburg: Tetzlaff Verlag, 318 p. 2003
23. Vorlesungensammlung von Lehrstuhl für Grundbau, Bodenmechanik, Felsmechanik und Tunnelbau. – München: TUM Zentrum Geotechnik, pp.1-28 (2011)
24. Kornev D. A. *Ocenka naptiazhenno-deformirovannoho sostoianiia zheleznodorozhnykh nasypei s primeneniem obiemnykh geomodelei* Avtoreferat dissertacii na soiskaniie uchenoi stepeni kandidata tekhnicheskich nauk spec. 05.23.11 – Proektirovaniie i ctroitelstvo doroh, metropolitenov, aerodromov, mostov i transportnykh tonnelei. SHUPS, Novosibirsk, 28 p. (2009) [in Russian]
25. Daniliants E. S. *Prohnoz naprizhenno-deformirovannoho sostoianiia zheleznodorozhnoho zemlianoho polotna s utchetom vibrodinamicheskoho vozdeistviia poezdov* Avtoreferat dissertacii na soiskaniie uchenoi stepeni kandidata tekhnicheskich nauk spec. 05.22.06 – Zheleznodorozhnyi put, izyskaniie i proektirovanie zheleznykh doroh. DVHUPS, Khabarovsk, 24 p. (2009) [in Russian]

Stress-strain state analysis of the leading car body of DPKr-2 diesel train under action of design and operational loads

Sergei Kostritsa[1,*], *Alexander* Pshinko[1], *Lyudmila* Ursulyak[1], *Andriy* Kuzyshyn[1,2], *Maxim* Kramarenko[3], and *Alex* Grechkin[3]

[1] Dnipro National University of Railway Transport named after Academician V. Lazaryan, 49000 Dnipro, Ukraine
[2] Lviv branch of Dnipro National University of Railway Transport named after Academician V. Lazaryan, 79052 Lviv, Ukraine
[3] Public joint stock company «Kryukovsky Railway Car Building Works», 39621 Kremenchug, Ukraine

Abstract. Purpose. Provision of strength and durability of the main structural element of DPKr-2 diesel train - the leading car body. **Methodology.** A spatial solid-state 3-D model of the body is built and durability calculations are carried out concerning action of loads stipulated by regulatory documents operating in Ukraine. In particular, the following main estimated modes are considered: mode 1 – a notional safety mode which takes into account the possibility of considerable longitudinal forces arising during shunting movements, transportation and accidental collision; mode 2 – an operational mode which takes into account forces acting on a train during acceleration to constructional speed, coasting or braking from this speed while passing a curve. **Results.** Based on the results of theoretical and experimental studies a conclusion has been made that the leading car body construction of DPKr-2 diesel train meets the requirements of regulatory documents regarding strength and durability. **Practical relevance.** A complex of calculation and experimental work concerning assessment of stress-strain state of the leading car body of DPKr-2 diesel train under action of design and operational loads allowed the creation of construction which meets not only operational requirements but also strength and durability ones.

1. Introduction

Diesel trains are one of the main kinds of rolling stock which is successfully exploited in many countries of the world for suburban and interregional passenger transportation on unelectrified sections of railways. One of such latest developments of the national producer is DPKr-2 diesel train designed and manufactured at Kryukovsky Railway Car Building Works.

DPKr-2 diesel train consists of three railcars: 2 leading cars of model 63-7083 and one intermediate car of model 63-7084 (Fig. 1).

The main technical characteristics of DPKr-2 diesel train are given in Table 1.

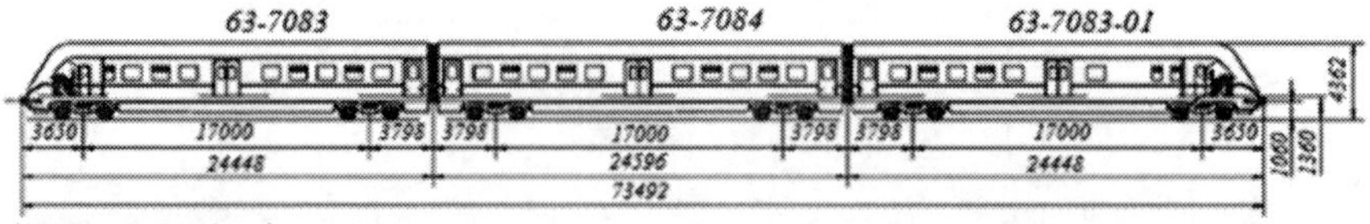

Fig. 1. Scheme of DPKr-2 diesel train

Table 1. Main characteristics of DPKr-2 diesel train.

Parameter, dimensionality	Standard for a car
Track width, mm	1520
Car gauge according to GOST 9238	T
Car framework, mm - leading car, model 63-7083 - intermediate car, model 63-7084	 17000 17000
Car width, mm	3500
Car height, mm	4400
Height of automatic hitch axis from the top of rails, mm	1060 ± 20
Car length along the hitch axes: - leading, mm - trailer, mm	 24596 24596
Passenger capacity/number of seats, pc:	637/289
Car tare-weight, t: car of model 63-7083 car of model 63-7084	 63,5 61,2
Maximum static pressure of a wheelset on the rails, tf: - leading car - intermediate car	 20 20
Engine unit capacity, kW	3x390
Maximum operating speed, km/h	140
Length of turnover section, km	up to 1000
Height of boarding platforms from the top of rail, mm millimeters	high – 1100 low - 200
Amount of fuel in every car, l	1000

To provide operation reliability and safety of DPKr-2 diesel train, specialists of Dnipro National University of

* Corresponding author: kossa571@gmail.com

Railway Transport and PJSC "Kryukovsky Railway Car Building Works" have conducted a complex of theoretical and experimental studies, in course of which the conformity of diesel train construction to the requirements of regulatory documents operating in Ukraine [1-2] and the requirements of statement of work has been checked. One of the main kinds of studies mentioned above was the assessment of strength, including fatigue strength, of the main load-bearing elements of DPKr-2 diesel train – frames of body and bogies.

The complex of calculation and experimental work of stress-strain assessment of the bogie frames of DPKr-2 diesel train under action of design and operational loads is outlined in work [3]. General issues related to design, manufacturing and putting into operation of new types of bogie frames of self-propelled rolling stock in the countries of the European Union are presented in works [4-7].

Quite a big number of works, e.g. [7-13], are devoted to the issues of theoretical and experimental research of strength and durability of welded car bodies, but none of them uses legal framework for multiple unit rolling stock operating in Ukraine.

2. Research methodology

To assess the strength of railway rolling stock, including self-propelled one, a legal framework based on fundamental research in the field of material fatigue is used in Ukraine and abroad. However, criteria for this assessment differ qualitatively. In particular, in Ukraine the fatigue strength assessment is carried out using fatigue safety factor [1-2] which should not exceed the normative value. In EU countries the fatigue strength assessment is carried out upon allowable stresses, which, in turn, are determined using ultimate stress charts [14-15]. In work [16] a comparison of the regulatory documents mentioned above is conducted which leads to the conclusion about good convergence of results obtained with regard to standards applicable in Ukraine and EU countries.

Strength assessment, including fatigue strength one, of DPKr-2 diesel train leading car was conducted according to the requirements [1] in two steps:

– theoretical studies aimed at choosing parameters of main load-bearing elements of the body and establishing installation positions of measuring devices during tests;

– experimental studies (static strength tests, navigation strength tests and collision tests) aimed at determining actual parameters characterizing the construction strength.

While dimensioning a diesel train car the ANSYS Workbench [17-18] package of applications (PA) is used which is characterized by a high degree of strength calculation automation at the stage of project-related research. This PA permits any combination of static pressures and allows for stress state at the elastic stage of both the whole car body construction and its separate nodes to be analyzed with a different degree of detailing.

An estimated 3-D model of the leading car body designed with the help of the ANSYS package of applications (PA) is presented in (Fig. 2). The model contains 1071131 elements and 1646383 nodes.

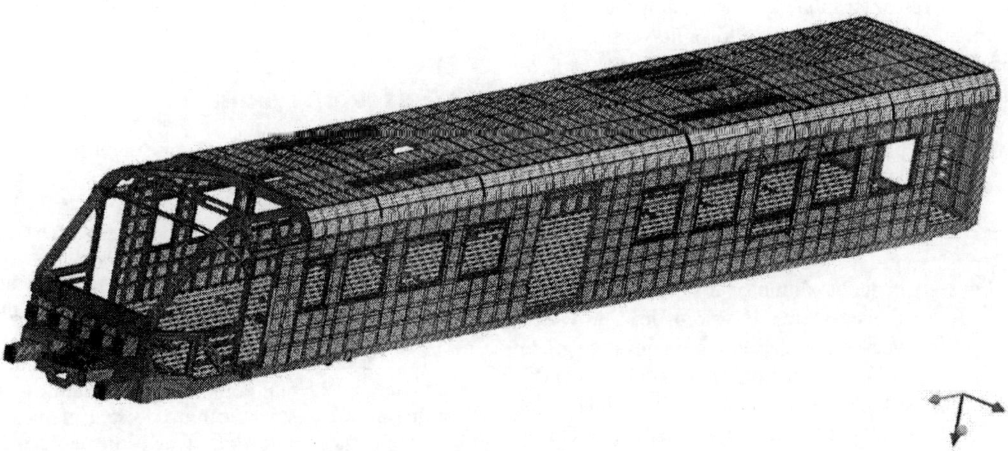

Fig. 2. Estimated 3-D model of the leading car body of DPKr-2 diesel train

Separately, the above given estimated scheme of the body is presented by a set of substructures integrated into a general estimated scheme. This approach allowed an independent preparation and adjustment of input data to be made for each substructure. After that system assembly and task solution can be performed. While designing the finite-element model of the body, flat quadrangular and triangular finite elements were used and an assumption was introduced that the construction material works in the elastic stage of deformation and has permanent characteristics – modulus of elasticity $E = 2,0 \cdot 10^5$ Mpa and Poisson's ratio $\mu = 0,3$ [19]. Car resting on bogie is presented in the estimated scheme by introduction of kinematic links into nodes which correspond to the support.

Car strength, according to [1], is calculated for 2 main estimated modes:

Mode 1 – a notional safety mode. This mode takes into account the possibility of considerable longitudinal forces arising during shunting movements, transportation and accidental collision;

Mode 2 – an operational mode which takes into account forces acting on a train during acceleration to constructional speed, coasting or braking from this speed while passing a curve.

During calculation for mode 1 the following main forces were applied to the estimated model:

– normative longitudinal compressive force of 1500 kN acting along automatic hitch axes;

– gross gravitational force of the body of 699 kN, in this case gravitational forces of equipment elements were applied at the points of their attachment to the body framework.

During calculation for mode 2:

– normative longitudinal tensile force of 800 kN;

– longitudinal inertia forces acting on separate nodes and elements which were identified by multiplying their mass by acceleration of the body;

– gross gravitational force of the body of 699 kN;

– vertical dynamic force which was identified by multiplying gross gravitational force of the body by the normative value of vertical dynamics ratio in the second stage of hanging which equals 0.2;

– centrifugal inertia force of the body proceeding from undamped acceleration while moving around a curve which equals 0,7 m/s^2.

Strength assessment for mode 1 was carried out upon allowable stresses, for mode 2 – upon allowable stresses and fatigue safety factor. Allowable stresses for estimated modes 1 and 2 are given in table 5.1 of standards [1].

To determine fatigue safety factor n the following proportion was used:

$$n = \frac{\sigma_{-1}}{k\sigma_v + \psi\sigma_m}, \quad (1)$$

where σ_{-1} – endurance limit of a standard sample in the symmetrical cycle of loading which was chosen depending on the steel grade using table 5.2 of standards [1];

σ_m – average stress of a cycle which was determined in every structural element under action of gross vertical static load;

σ_v – amplitude of dynamic stresses which was identified from the proportion $\sigma_v = k_d\sigma_m$, where k_d – vertical dynamics ratio in the second stage of suspension. On the basis of dynamic calculation results [20] of a diesel train $k_d < 0,1$. Normative (maximum permissible) value of vertical dynamics ratio is 0,2. Therefore during calculation the maximum value of $k_d = 0,2$ was accepted;

ψ – a coefficient reflecting the sensitivity of a metal to cycle asymmetry (if $\sigma_m > 0$ $\psi = 0,3$, if $\sigma_m < 0$ $\psi = 0$);

k – effective coefficient reflecting endurance reduction of a detail in relation to endurance limit of a standard sample.

Coefficient k was determined from the proportion:

$$k = \beta_k \frac{k_1 k_2}{\gamma m}, \quad (2)$$

where β_k – effective coefficient of stress concentration was accepted as equal to 1;

k_1 – coefficient reflecting material heterogeneity of the detail, for rolling $k_1 = 1,1$;

k_2 – coefficient reflecting the influence of inner stresses in the detail depending on detail size $k_2 = 1,0...1,2$;

γ – coefficient reflecting the detail sizes which was determined according to the graph given in figure 3,5 of standards [1];

m – coefficient reflecting the surface state of a detail, for rolling $m = 0,8$.

According to requirements [1], fatigue safety factor must be not less than normative value -2.

Based on conclusions from the analysis of calculation results for modes 1 and 2, installation positions of measuring devices during static strength tests, navigation strength tests and collision tests were established.

3. Research results

Figure 3 illustrates fields of equivalent (according to the Mises-Hencky theory) stresses in the structure of the leading car body of DPKr-2 diesel train as well as nodes in which there are maximum stresses under action of loads corresponding to estimated mode 1.

Table 2 presents allowable stress values and maximum stresses obtained as a result of calculation for estimated mode 1.

The presented results show that under loads simulating accidental collision and shunting movements maximum stresses in the most loaded elements of the body construction do not exceed allowable values, i.e. strength in estimated mode 1 is ensured.

Figure 4 illustrates fields of equivalent (according to the Mises-Hencky theory) stresses in the structure of the leading car body of DPKr-2 diesel train as well as nodes in which there are maximum stresses under action of loads corresponding to estimated mode 2.

Table 3 presents allowable stress values, maximum stresses and fatigue safety factors obtained as a result of calculation.

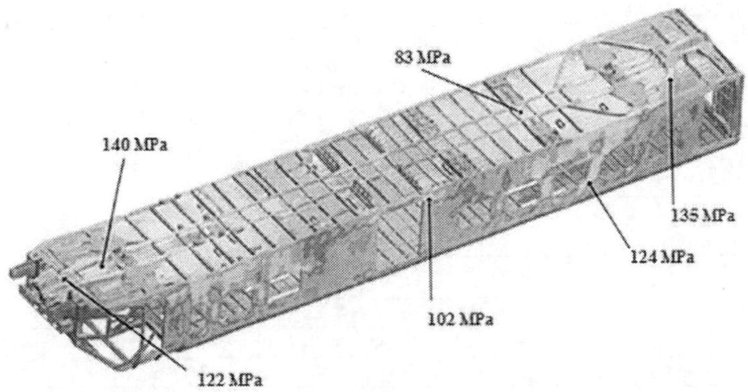

Fig. 3. Stress state of the structure of the leading car body of DPKr-2 diesel train under action of loads corresponding to estimated mode 1 (view from below)

Table 2. Maximum stresses in the main structural elements of the body under action of loads corresponding to estimated mode 1.

Structural element of the body	Steel grade	Allowable stresses, mPa	Equivalent stresses, mPa
Underframe			
Spine beam (head part)	325-09Г2С	292	122
Centre bearer	295-09Г2С	265	140
End girder	345-09Г2С	327	93
Cantilever unit diagonals	345-09Г2С	327	135
"Power pack" mount beams	345-09Г2С	327	85
Side wall			
Window opening vertical struts	08X18H10	194	127
Doorway vertical struts	08X18H10	194	102
Lower door beam	08X18H10	194	88
Cab framework			
Cab door step	345-09Г2С	327	95
Doorway pillar	345-09Г2С	327	83
Roof			
Longitudinal beam	08X18H10	194	39
Cross-beam	08X18H10	194	37
Air conditioner support	08X18H10	194	42

Note to table 2. Low alloy steel 09G2S is an analogue of European steel DIN 16Mn6, stainless steel 08X18H10 – European steel DIN 1.4301.

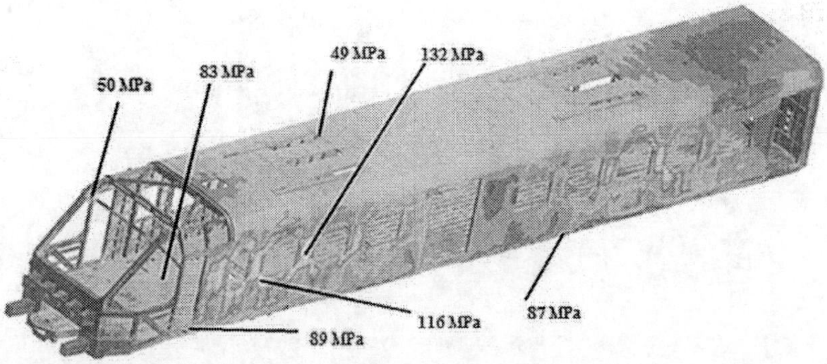

Fig. 4. Stress state of the structure of the leading car body of DPKr-2 diesel train under action of loads corresponding to estimated mode 2.

Table 3. Maximum stresses in the main structural elements of the body and fatigue safety factors under action of loads corresponding to estimated mode 2.

Structural element of the body	Steel grade	Allowable stresses, mPa	Equivalent stresses, mPa	*n*
Underframe				
Spine beam (head part)	325-09Г2С	205	103	3,1
Centre bearer	295-09Г2С	195	83	3,6
Side rail (the middle of the wagon)	325-09Г2С	205	62	5,1
"Power pack" mount beams	345-09Г2С	220	92	3,6
Stair mechanism cross-beam	345-09Г2С	220	80	4,1
Side wall				
Vertical strut	08Х18Н10	138	112	2,6
Strut under window	08Х18Н10	138	96	3,0
Upper window opening beam	08Х18Н10	138	76	3,6
Lower window opening beam	08Х18Н10	138	87	3,3
Cab framework				
Cab door step	345-09Г2С	220	89	3,7
Inclined beam of the cab	345-09Г2С	220	50	5,7
Roof				
Longitudinal beam	08Х18Н10	138	36	7,9
Air conditioner support	08Х18Н10	138	49	5,8

The analysis of stress state of the leading car body shows that maximum stresses in the most loaded elements of body construction do not exceed allowable values and fatigue safety factor is significantly higher than the normative value 2.

Based on the results of theoretical studies described above, a strain gauge layout chart has been developed for the purpose of conducting static strength tests, navigation strength tests and collision tests. Figures 5-7 illustrate examples of strain gauge layout during the testing mentioned above.

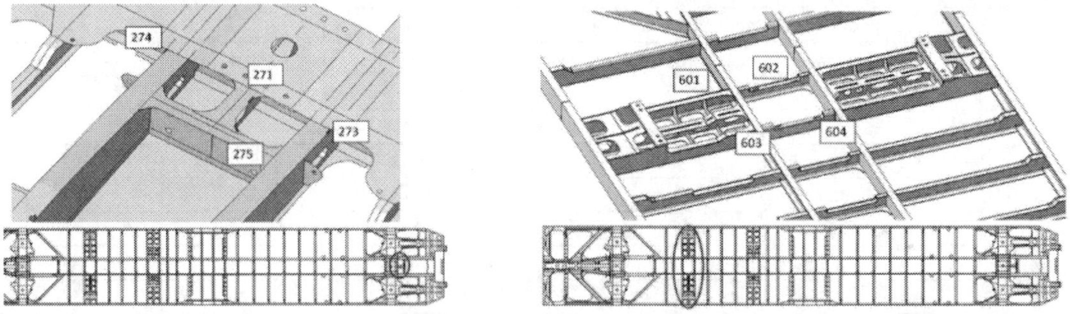

Fig. 5. Strain gauge layout on the underframe

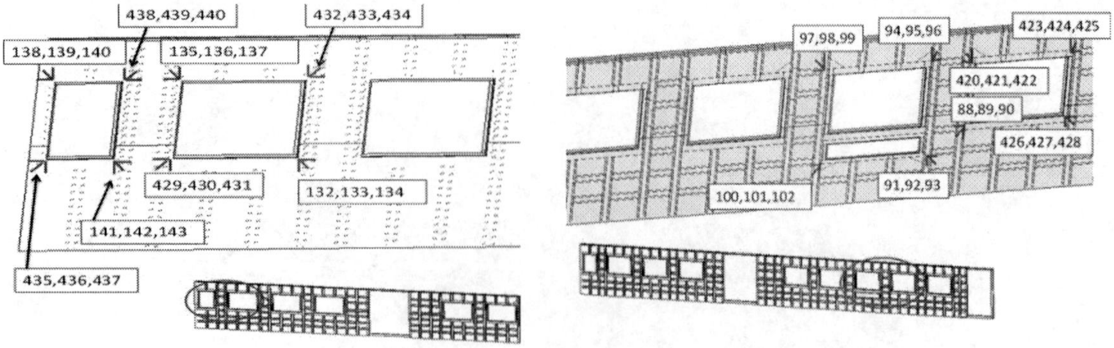

Fig. 6. Strain gauge layout on the side walls

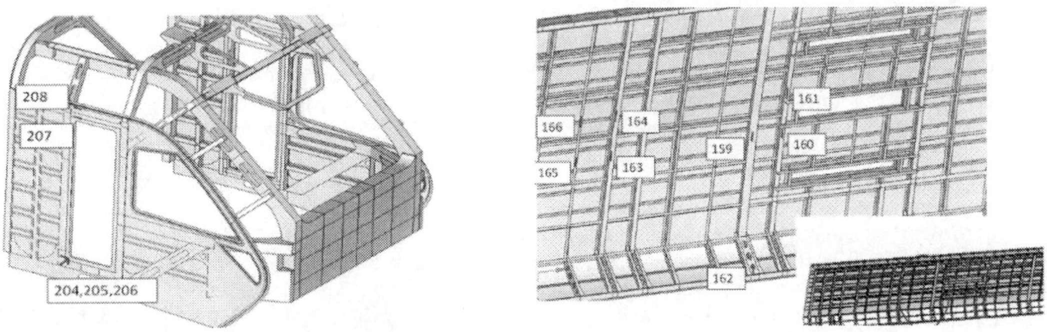

Fig. 7. Strain gauge layout on the cab framework and roof

The analysis of static strength testing and navigation strength testing results showed that the minimum value of fatigue safety factor occurs in the area of power unit fastening to the underframe and equals 3.7 (point 252, Fig. 5).

According to the collision testing results, the most loaded is the junction area of spine and longitudinal beams where stresses reached the values of 210 mPa (points 271, 275, Fig. 5) with the force of impact of 1750 kN.

On the basis of theoretical and experimental research results it was concluded that the construction of the leading car body of DPKr-2 diesel train meets the requirements of regulatory documents against the criteria of strength including fatigue one.

4. Scientific novelty and practical relevance

1. For the first time a spatial finite-element model of a diesel train body of a new generation is built and calculation of strength under action of standard loads is made.
2. An estimated scheme of the body is presented by a set of substructures integrated into a general estimated scheme. This modular approach made it possible to make an independent preparation and adjustment of input data for each substructure and after that perform the system assembly and task solution.
3. Modular approach to the estimated scheme design makes it possible to use it for the strength assessment of the body of diesel trains of new generations changing only the structural elements that are subject to upgrading.

Conclusions

The complex of calculation and experimental work on stress-strain state assessment of the leading car body frame of DPKr-2 diesel train under action of design and operational loads made it possible to create a construction that meets both operational standards and strength and durability requirements.

Bibliography

1. *Regulations on calculation and assessment of the strength of load-bearing elements and the dynamic qualities of multiple units*, Moscow, 146 p. (1997)
2. *DSTU 4493: 2005 Wagons of long-distance passenger diesel and electric trains. Security requirements*, Kiev, State standard of Ukraine, 16 p
3. S. Kostrytsia, S. Molchanov, M. Kramarenko, A. Hrechkin, D. Laktionov *Stress-strain state assessment of the bogie frame of DPKr-2 diesel train under action of design and operational loads*, Science and Transport Progress. Bulletin of Dnipropetrovsk National University of Railway Transport, **№.1(79)**, pp 128-138 (2019)
4. B. Li, Y. Zhao *Strength assessment for bogie frame based on UIC standard*, Machinery, **№10**, pp 9-12 (2012)
5. T-S. Jung-Won, H. Hyun-Moo, J. Hyun-Kyu *Fatigue Design Evaluation of Railway Bogie with Full-Scale Fatigue*, Advances in Materials Science and Engineering, pp. 1-11 (2017)
6. M. Kassner *Fatigue strength analysis of a welded railway vehicle structure by different methods*, International journal of fatigue, **№34**, pp. 103-111 (2012)
7. *EN 13749:2011 Railway applications – wheel sets and bogies–methods of specifying structural requirements of bogie frames*, European Committee for Standardization, (2011)
8. A. Shukri, D. Willem, Y. Anas *Modeling aspects of the design of railway vehicle structures and their crashworthiness*, Proceedings of the Institution of Mechanical Engineers, Part F: Journal of Rail and Rapid Transit, **№230 (6)**, pp. 1575-1589 (2016)
9. M. Płaczek, A. Wróbel, A Baier *Computer-aided strength analysis of the modernized freight wagon* Modern Technologies in Industrial Engineering (ModTech2015), IOP Conf. Series: Materials Science and Engineering 95, (2015)
10. D. Peng, R. Jones, T. Constable *Tools and methods for addressing the durability of rolling stock*, Engineering failure analysis, **№34**, pp. 278-289 (2013)

11. S. Slavchev, V. Stoilov, S. Purgic *Static strength analysis of the body of a wagon*, Journal of the Balkan Tribological Association, **№21 (1)**, pp. 38–57 (2015)
12. O. Fomin, J. Gerlici, A. Lovskaya, K. Kravchenko, P. Prokopenko, A. Fomina, V. Hauser *Research of the strength of the bearing structure of the flat wagon body from round pipes during transportation on the railway ferry*, Horizons of Railway Transport 2018, MATEC Web of Conferences 235, (2018)
13. P. Šťastniak, M. Moravčík, L. Smetanka *Investigation of strength conditions of the new wagon prototype type Zans*, MATEC Web of Conferences 254, (2019)
14. *ERRI B12/RP17 Programme of Stresses to be carried out on Wagons with Steel Underframe and Body Structure*, European Rail Research Institute, **№8** (1996)
15. *EN 12663:2010 Railway applications – Structural requirements of railway vehicle bodies – Part 1: Locomotives and passenger rolling stock (and alternative method for freight wagons)*, European Committee for Standardization, (2011)
16. S. Kostrytsya *Assessment of fatigue strength of a self-propelled rolling stock on the results of certification tests in Ukraine and the EU*, Railway problems. Railway reports, **№61**, pp. 29-36 (2017)
17. K. Bosov *Ansys for constructors*. DMK Press, 248 p. (2016)
18. A. Kaplun *Ansys in the hands of an engineer. Practical guide*, Librokom, 270 p. (2016)
19. G. Pisarenko, A. Yakovlev, V. Matveev *Handbook of resistance to materials*, Science. Dumka, Kiev, 165 p. (1988)
20. A. Kuzyshyn, A. Batig, S. Kostritsa [et al.] *Research of safety indicators of diesel train movement with two-stage spring suspension*, BulTrans 2018: 10th Intern. Sci. Conf. on Aeronautics, Automotive and Railway Engineering and Technologies. – Sozopol: **234** (2018)

Study of rail vehicles movement characteristics improvement in curves using fuzzy logic mechatronic systems

Mikhail Kapitsa[1], *Evgen* Mikhailov[2,*], *Sergii* Kliuiev[2], *Stanislav* Semenov[2], and *Maksim* Kovtanets[3]

[1]DNUZT, Department of locomotives, 49010 Dnipro, Lazaryan Street 2, Ukraine
[2]EUNU, Department of logistic management and traffic safety, 93405 Severodonetsk, Central Avenue, 59A, Ukraine
[3]EUNU, Department of railway and road transport, lift and care systems, 93405 Severodonetsk, Central Avenue, 59A, Ukraine

Abstract. The article deals with the effectiveness of reducing the level of force interaction of the rail vehicle wheels with rails in curved sections of the track through the use of mechatronic position control systems for wheel pairs in the rail gauge in the horizontal plane. The approaches to the creation of such a mechatronic system operating on the principles of fuzzy logic are described. To determine the angles of attack of wheels on the rails, it was proposed to use the acoustic emission indicators of the contact of the wheel with the rail. To determine the direction of curvature of the rail track, it is advisable to use data from navigation systems. The study of the dynamics of the rail vehicle during the passage of a curved section of the track in real time was carried out using the Matlab/Simulink software package. The proposed mechatronic control system for the position of the wheel sets in the horizontal plane allows to ensure their optimal installation under various driving conditions in the rail gauge. This makes it possible to minimize the angles of attack of the wheels and reduce the forces of the horizontal interaction of the wheels with the rails.

1 Introduction

Analysis of the scientific and technical information has shown that the dynamic properties of the rail carriage in the horizontal plane dependent on the speed of its movement, the state of the chassis and carriage track. Wear in the wheel-rail system depends on several factors: the level of forces in the points of contact of the wheels and rails, sliding velocity at these contacts, the values of the angle of attack, the coefficient of friction, hardness, wheel and rail material.

One way of improving the dynamic characteristics of a rail carriage is the rational choice chassis construction [1-4].

Reduction of interaction forces, improved dynamic performance in a curve and a decrease in the wear intensity of rails and flanges may be controlled by wheelset position on the rail track.

2 Analysis of recent research and publications

The problem of reducing the interaction forces of the wheels and the rails when driving railway vehicles in the curve explored by I. V. Biryukov, E. P. Blokhin, M. F. Verigo, A. L. Golubenko, N. I. Gorbunov, Y. Demin, S. V. Doronin, V. V. Evstratov, A. A. Kamaev, V. A. Kamaev, V. N. Kashnikov, A. N. Konyaev, M. L. Korotenko, V. S. Koss, S. M. Kutsenko, V. G. Masliev, G. S. Mihalchenko, I. I. Nikolaev, M. L. Radchenko A. M. Savoskin, M. I. Spiryagin, V. I.Spiryagin, V. P. Tkachenko and etc.

The basic requirements that apply to new railway vehicles are to increase traffic safety and decrease the level of dynamic interaction with the rail.

Passage curves passing is accompanied by an increase of wheels wear and rails intensity [5 - 10]. Important areas of reduction of wear in curves is the application of lubricant on the contacts of the wheels with the rails and reduction of forces acting on them, reducing efforts and guides the angle of attack (due to the choice of optimal circuit parameters and spring suspension of railway vehicles) [2, 13, 14].

One of the most promising ways to reduce the rate of wear in the "wheel-rail" is to use rational constructions of railway vehicles, supplemented with active position control system wheelset in curve [11 - 12].

To determine how to control the position of the wheelset on the rail track, it is necessary to know the characteristics of the controlled object and the noise coming into the control unit.

Railway vehicle control system can be divided into two groups (Fig. 1.)

1) The passive control system, allowing to change the kinematic scheme depending on how (straight or curved) it is moves;

2) The active control system, which curves have a force effect of adjusting the wheelsets angle of attack on the rails and minimizing lateral force.

Active control systems are the most advanced solutions for the problems of wear of wheel sets.

In the low speed range, the efficiency of mechatronic systems determined by the decrease of resistance to

* Corresponding author: sergistreet@gmail.com

motion and wear in the wheel-rail contact. In the range of high speeds, special attention should be focused on the stability of the motion of wheel pairs in the horizontal plane.

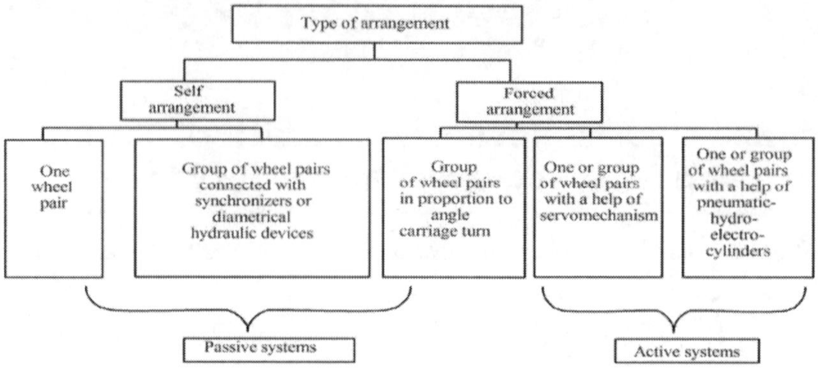

Fig. 1. Classification of device for radial wheel sets installation in curve

For the development of modern microprocessor technology, it becomes possible to use mechatronic control systems operating on principles of expert systems, fuzzy logic, neural networks, associative memory and others.

Table 1 shows the results of analysis techniques and algorithms of existing mechatronic control systems.

Table 1. Construction of mechatronic systems for forced rotation of the wheelset

Mechatronics system		
object of control	Control system	
	control principle	way of management
traditional pair of wheels	feedback, the principle of feedforward	Optimal control, optimal control with genetic algorithm, linear-quadratic Gauss control, proportional control, sliding mode control, fuzzy logic, proportional-integral controller
Wheel set with independent wheel rotation	the principle of feedback	proportional, integral and differentiating control, adaptive control, robust control, optimal control

Currently, the most important is the use of mechatronic solutions for traditional wheel sets, which are widely used in railway transport. Further development increases the need for technologies mechatronic systems for wheel sets with independently rotating wheels. Currently, the construction of railway vehicles with independently rotating wheels (IRW) is not yet widely available, although work in this direction is actively underway.

Studies show that when using wheel sets with IRW missing their centering in the rail track. In this regard, there is a long ridge overrunning of one of the wheels on the rails and an increased level of wear and motion resistance [8, 22, 27]. Therefore, while the design of the chassis with independently rotating wheels are used, mainly in urban rail carriages with low floors, where the wheel flange wear is small due to low speed and small load axis [22, 23].

In the case of such structures on the main rail transport is necessary to adopt additional measures to ensure the self-centering of wheel sets in rail track. Of the few examples of practical applications they can be called bogies Talgo system operated quite successfully for several decades because of the original concept direction wheel sets in a track [24].

Certain reduce slippage wheel flanges on the rails can be achieved by improving the design schemes of the wheels themselves, for example, by allowing the independent rotation of the supporting and guide surfaces [25, 26].

Also investigated a limited possibility of wheel rotation in a wheel pair relative to each other through the use of innovative designs axle and wheel mounting thereon units [8, 22, 23].

One of the most currently developed areas of railway vehicles with IRW dynamics improvement, reduce wear of wheels and rails and driving resistance is using of mechatronic systems for controlling the position of wheel pairs with IRW in the rail track in a horizontal plane. The results of numerous theoretical and experimental studies [28-30] confirm the effectiveness of this approach.

One way to reduce the horizontal transverse forces arising during the passage of railway vehicles of the curves is to provide a controlled motion wheel sets [15].

It was theoretically and experimentally proved that traditional solutions and passive methods have already reached their capabilities, and further reducing the solution to the problem of wear and the force interaction may be accomplished by using of active control systems. The algorithm of such a control system is that the wheel set turning in a horizontal plane on a railway line at an angle with the motion of the actuator curve to minimize the angle of attack of wheels on the rail [16].

Control action on wheel pairs opens up new possibilities to improve quality of the dynamics on curve, reduces dynamic influence on the rail and wear. However, existing control systems radial wheel set installation until providing the necessary wheel on the rail angle of attack control, due to the complexity of the measurement of this parameter [15].

3 The purpose of the study

The aim of the article is to examine the impact on the reduction of the force interaction of railway vehicles and the rail wheel set mechatronic position control system of the rail track, operating on the basis of fuzzy logic.

4 The results of the railway vehicles dynamics study in the curve

A disadvantage of existing devices wheel set turning in the horizontal plane is the absence of the automatic position control of the wheel set on the track. Therefore, it is advisable to create a high-speed active control system of the position of the wheel set on the track is not on the basis of operational parameters of movement, but with the control wheel angle of attack on the rail and the direction of movement in a curve.

The authors proposed to use the audio signal system setting for radial wheel pairs installation based on parameters of the contact wheel pair and operating on fuzzy logic principles.

Automatic control of the position of wheel set on the track should provide:

- stabilizing the motion of the wheels et in straight sections of the track (increase in wavelength yaw);
- minimizing the interaction forces between wheel set and rails;
- handling and uniqueness.

Additional requirements to the system of automatic control of the position of the wheel set on the track:

- the presence of anti-aliasing in the subsystem "state variables - active management";
- reliable operation in a predetermined mode for a predetermined period of time;
- reliable and correct functioning in an emergency such as a power failure and subsequent recovery, blown fuses, broken wires;
- the content of the minimum possible number of components and a minimum possible length of the connecting wiring lines.

Using the pulling device (drive: pneumatic, hydraulic, electromechanical) pair is transmitted to the wheel force control. The most important issue in the construction of an algorithm controlling the position of the wheel set on the track is to determine the time the driving force for changing and the establishment of control algorithm of this magnitude.

As shown in [17] for radial installation of the wheel set in a rail track necessary to ensure the equality of the longitudinal wheel traction forces on a single axis (or zero power in the case of rolling without traction or braking) and the equality of lateral cohesion forces for all wheel sets rail vehicle. This can be achieved if the angle of attack for the two wheel sets (for example, biaxial carriage) can be adjusted and is equal, and if the carriage moves along the central axis in the track curve. This can be achieved by setting equal angles of rotation of wheel sets relative to the bogie frame. But it is almost impossible to achieve in real-world conditions such equality of angles [18]. Therefore, the mechatronic control system was designed to force the wheel sets. In the proposed system the forced radial setting of the wheel set by means of additional actuators whose operation is controlled by a microprocessor system (see. Fig. 2) by a special algorithm. The proposed model takes into account the parameters of the acoustic emission from the wheels in contact with the rail, obtained experimentally [4].

In this algorithm analyzes the acoustic signal parameters of the wheels in contact with the rail for determining the value of the angle of attack. In this position of the vehicle on a track section recorded in accordance with the data of GPS system. Also takes into account the lateral position of the wheel set relative to the track and the rotation angle of the carriage with respect to the vehicle body.

Regulator regulating effect is formed with the curvature direction on the track section. The angle of attack is determined by the corresponding displacement of the actuator stem.

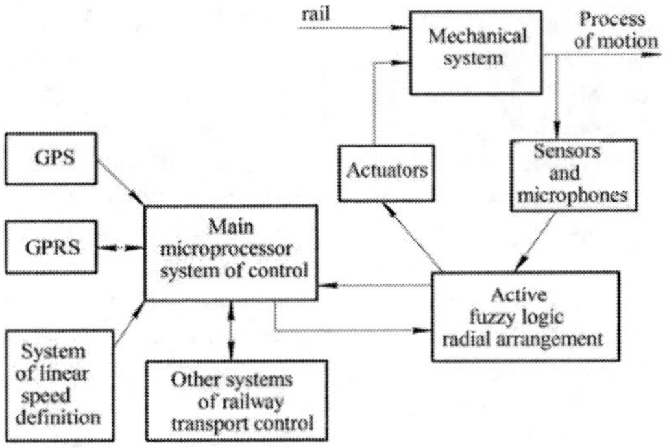

Fig. 2. The mechatronic system radial installation of wheel sets

To simulate the interaction of the rail carriage with a rail track has been selected Matlab/Simulink simulation environment. Mae T., Lee II assumptions were taken to initial prerequisites for creating the model. [3].

Basic mechatronic control system generates a control action for turning the wheel pair in the horizontal plane of the track at an angle which minimizes the angle of attack of wheels on the rail or shift to a non-contact motion with the rail head.

The control circuit with a feedback controller operating according to the principles of fuzzy logic, is shown in Fig. 3.

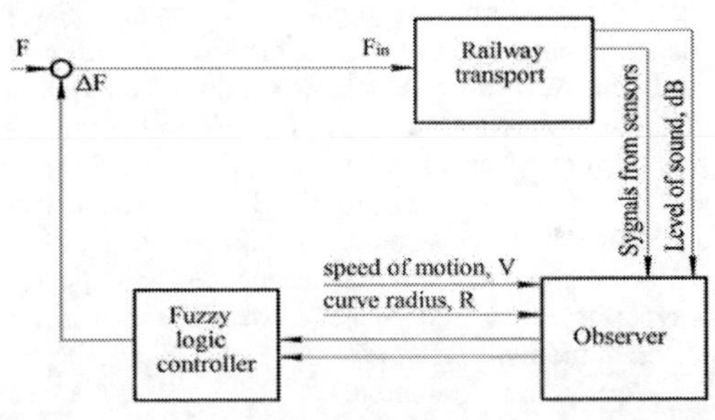

Fig. 3. The control circuit of the controller using fuzzy logic for the wheel set control active position

Where F - force value, which is realized in the drive certain initial moment. The value of this force can be obtained for a particular observer in a special database curve. The database may be generated based on theoretical studies. In our case, the value of the force F is set and depends on the curve radius R and the velocity V. The next value of the force depends on the time when F (t) should be equal to the previous value of the force.

The corrected value for ΔF drive force can be obtained in accordance with fuzzy logic rules. With the use of the mechanical system analysis and numerical methods for solving the problem of optimal choice made state variables - the difference between the ideal and the calculated angle of rotation of the wheelset relative to the bogie frame.

One of the problems in developing a fuzzy logic control system is to obtain information about the angle of attack of the wheel set on the rail. Existing methods for determining the angle of attack without installing a stationary sensor on the rail base is used for indirect measurement and calculation of the angle of attack do not allow to measure it in real time and automatically.

The possibility of determining the angle of attack of a wheel set on the rail on the basis of acoustic emission method is observed. In order to explore the use of acoustic radiation from the contact rail-wheel for determining the angle of attack on the rail wheels have been carried out experimental studies of the mechanism of generation of the acoustic signal in this contact [19, 20]. Experimental studies were carried out on the model of the roller station and locomotive 2TE116.

It proved experimentally that the acoustic emission spectrum is not identical. At certain frequencies of sound pressure level is significantly higher than the average equivalent sound pressure level.

According to the experimental studies, based on the parameters of the acoustic emission from the wheels in contact with the rail was installed informative feature and the experimental dependence of the sound pressure level of deviation has been obtained on the dominant frequency of the equivalent sound pressure level at different angles of attack on the rail wheels.

The functional dependence of the sound pressure level of the angle of attack on the acoustic emission dominant frequency was used to determine the angle of attack of wheels on the rails when forming the control cylinder impact to the wheel of a rail vehicle when traveling on curve.

Determination of the dominant acoustic emission frequency of the wheel-rail contact was made using a signal processing algorithm using wavelet transformation using a wavelet function for any moment of time t. According to statistical data of sound pressure level depending on the angle of attack on the rail wheels was performed neuro-fuzzy adaptation, based on which set of rules has been received and the surface to determine the wheels on the rail angle of attack.

To receive the digital sequence was performed wavelet transformation using a wavelet function for the different moments of time t.

Wavelet transformation:

$$W_f(t,a) = (1/\sqrt{a}) \cdot \int \Psi[(x-t)/a] \cdot f(x)dx \quad (1)$$

where $f(x)$ – is the research signal; $\Psi(x)$ – is a wavelet-function; $\Psi(x)=(\cos x+i\cdot\sin x)\cdot e^{-x2/50}$; t – is the time; and a – is a scale of wavelet-function.

The frequency spectrum obtained by processing acoustic emission signals in accordance with the algorithm of fast Fourier transform.

However, the acoustic method for determining the angle of attack can not determine the direction of movement (the angle mark). Determining the direction of the curve on the section of track seems reasonable if you use satellite navigation and inertial system [21].

Joint processing of the calculated coordinates carried by the synchronization signals from the satellite receiver and the inertial unit. Improving the accuracy was achieved through the suppression of errors in each system.

Scheme joint operation of the satellite receiver and the inertial unit shown in Fig. 4.

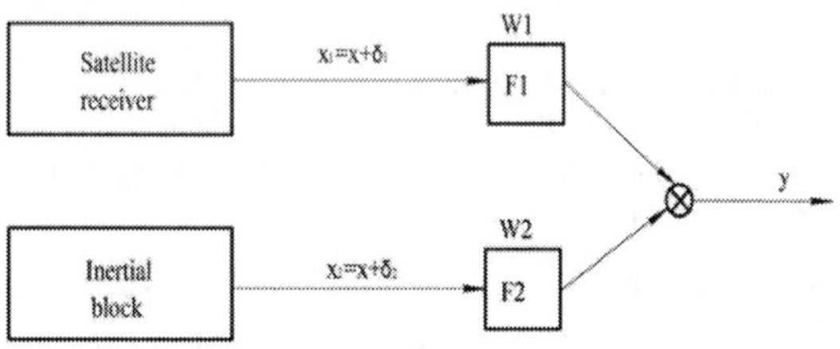

Fig. 4. Block diagram

Where y – coordinate of the crew location; W1, W2 – filter transfer functions; F1, F2 – filters; x1, x2 – representation of the desired signal x(t); δ1 and δ2 – view satellite receiver errors and inertial unit

$$y = (W1 + W2) \cdot x + W1 \cdot \delta_1 + W2 \cdot \delta_2 \quad (2)$$

$$\delta_1(T) = b + m \cdot \sin(n \cdot t) \quad (3)$$

where b – is a permanent part (constructive error); m – is an amplitude of dynamic error; m · sin($n\cdot t$) – is a high-frequency constituent of error (dynamic error), n – is angular frequency of vibrations, t – is the time.

$$\delta_2(T) = r \cdot t, \quad (4)$$

where r – is the average speed of inertial block drift.

As for the input and output variable defined triangular and trapezoidal functions properties because they are simple and common in use. Group of linguistic rules used to form the input and output variables. This group includes tags: N and P – negative and positive values, respectively; S, M and B – small, medium and large values, respectively; ZE – mean value equal to zero. Parameter ranges can be obtained either from experimental data, or from simulation results. The ranges of the input variables obtained from the results of the quasi-static forces analysis for basic carriage.

Following a set of rules (Fig. 5) was selected for the simulation of fuzzy logic system.

Actuator function control is required to turn the wheel pair is shown in Fig. 6.

A set of rules for the proposed fuzzy logic systems with fuzzy logic controller is shown in Table 2.

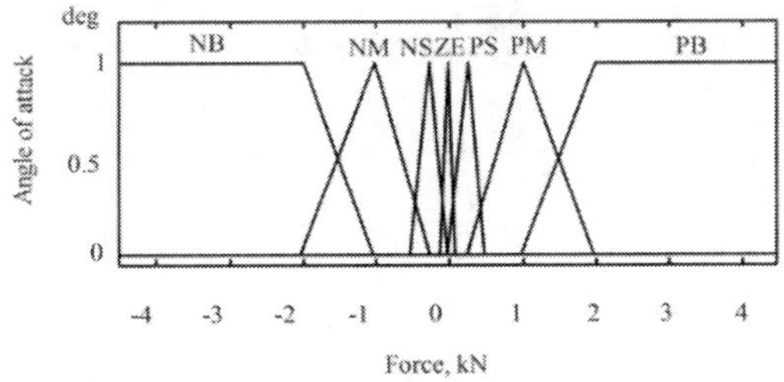

Fig. 5. Function of property variable difference between the ideal layout and the measured angles of the wheel set: NB – negative big, NM – negative medium, NS – negative small, ZE – zero, PS – positive small, PM – positive medium, PB – positive big

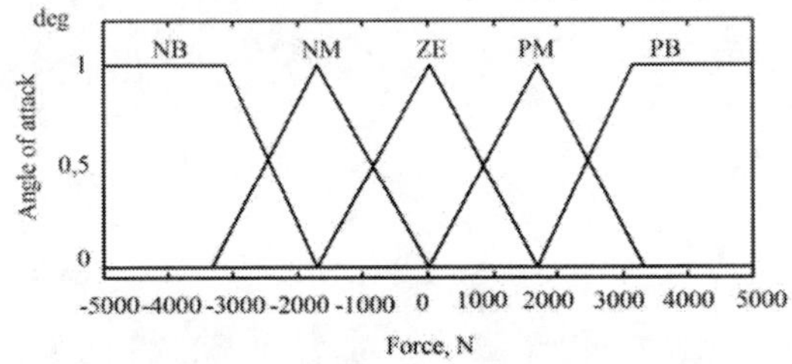

Fig. 6. The function variable values drive efforts properties: NB – negative large, NM – negative mean, ZE – zero; PM – positive average, PB – positive big.

Table 2. A set of rules for the fuzzy logic controller

	$\gamma_{IDEAL}-\gamma$						
	NB	NM	NS	ZE	PS	PM	PB
ΔF	NB	NM	NS	ZE	PS	PM	PB

5 Discussion of the research results of railway vehicle dynamics in a curve controlled on the basis of fuzzy logic

Theoretical study of movement of the locomotive type 2TE116 model track curve performed on the original data, which correspond to the parameters of the experiment shown in [20].

Result comparison of the modeled locomotive movement of serial 2TE116 and locomotive with a fuzzy logic control system of the wheel sets on a curve with a radius of 300 m at a speed of 70 km/h are shown in Fig. 7. For a system simulation used the same output data and the simulation technique, which are applied in Iwnicki [19].

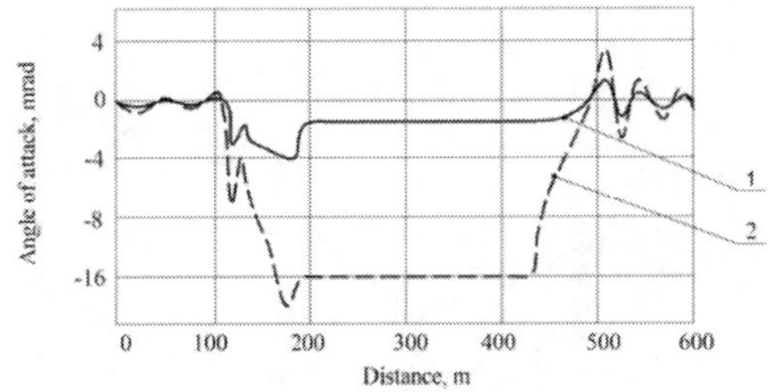

Fig. 7. The theoretical values of first wheel set angle of attack of the 2TE116 locomotive at a controlled and uncontrolled locomotive motion on curve (R = 300 m, V = 70 km/h): 1 – controlled movement; 2 – uncontrolled movements

Work proposed mechatronic system was also tested in the simulation carriage movement with the main parameters diesel train type DEL-01.

To create a non-linear simulation model of diesel train traffic software Simpack was used. The proposed method of controlling the forced installation of the wheel sets has been implemented in a software product Matlab/Simulink. A vehicle model created in Simpack, was associated with a control system designed in Matlab/Simulink, using SIMAT interface that is used for both special modeling software packages.

Fig. 8, 9 show the values obtained for wheel angles of attack and lateral movement of the front and rear wheel pairs as a function of distance traveled for standard design and construction with the mechatronic bogie wheel sets active setting.

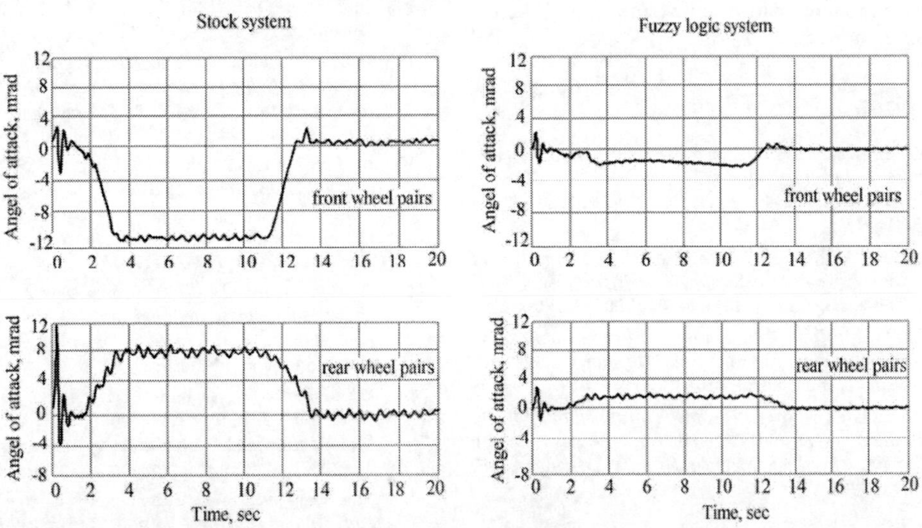

Fig. 8. Results diesel train movement simulation DEL-01 (R = 300 m, V = 90 km/h)

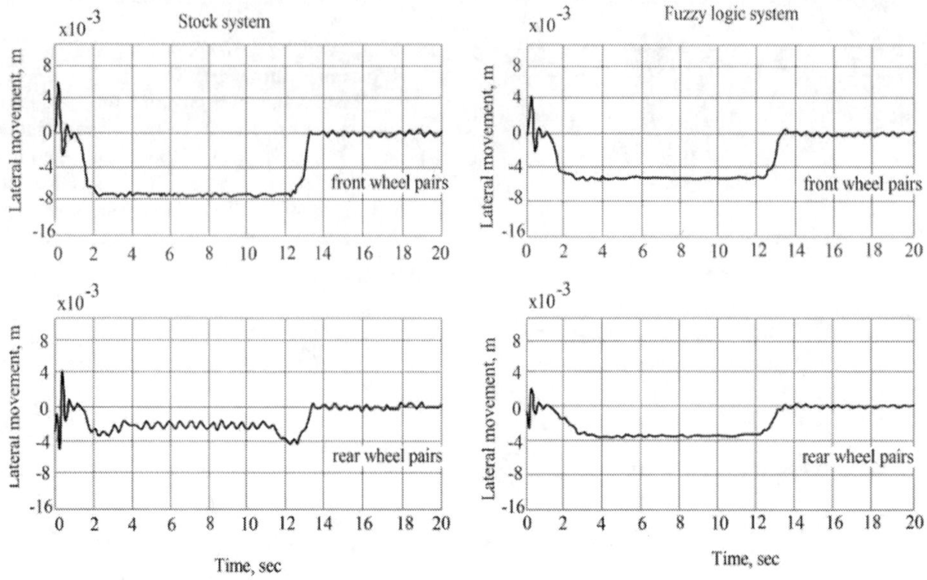

Fig. 9. Results of a diesel train DEL-01 motion simulation (R = 300 m, V = 90 km/h)

6 Conclusions

Presented mechatronic active position control system of a rail wheel pair in the track allows to obtain its optimum setting for different modes of a rail carriage motion, thereby reducing the forces on the guides for 35% and to minimize the angle of attack to zero. Fuzzy logic control allows to provide efficient management of complex mechatronic system and has advantages as compared with control by proportional control. Using of fuzzy logic in this case is allow to avoid the creation of a complex mathematical model for the simultaneous actuators using of each wheel set.

One of the methods for wheel set relative to the rail position evaluating is to use acoustic emission parameters from contact of wheels with the rails. For practical use of this method it is necessary to create an appropriate database, depending on the characteristics of a rail carriage motion, part of railway track and friction conditions in contact.

References

1. V. Ulshin, S. Klyuev, E. Zubar, *Metod avtomaticheskogo kontrolya ugla nabeganiya kolesa na rels*. Visnik SNU im. V.Dalya. Lugansk, **4 (193)**, 259-262 (2013) [in Russian]
2. S. Klyuev, A. Klyuev, *Metodiki aktivnogo upravleniya telezhkoy relsovogo transportnogo sredstva*. Visnik SNU Im. V.Dalya. Lugansk, **4 (134)**, 67-74 (2009) [in Russian]
3. T. Mei, H. Li, *Control design for the active stabilization of rail wheelsets*, Journal of Dynamics Systems, Measurements and Control, **130** (2008)
4. H. Scheffel, H. Tornay, *The mechanism of the rotable lemniscate suspension applied to bogies having selfsteering wheelsets*, Vehicle System Dynamics, **17**, 368-380 (1988)
5. A. Teplyakov, *Puti snizheniya intensivnosti iznosa grebney kolesnyih par lokomotivov:* Diss. kand. tehn. nauk, Habarovsk, 197 (2004) [in Russian]
6. B. Kossov *Snizhenie nagruzhennosti hodovyih chastey lokomotivov i puti*. Avtoreferat diss. dokt. tehn. Nauk, Kolomna, 339 (2001) [in Russian]
7. A. Golubenko, A. Kostyukevich, I. Tsyganovskiy, V. Nozhenko, *The influence of a rail lateral bending on the stress - strain state of a wheel - rail contact*, TEKA Commission of Motorization and Power Industry in Agriculture, **11a**, 78-84 (2011)
8. A. Golubenko, *Stseplenie kolesa s relsom: Monografiya*, 2-e, pererobl. ta dop. Lugansk, 476 (1999) [in Russian]
9. V. Vasin, V. Berezin, A. Kokorev, V. Nadelyuev, *Effektivnost radialnoy ustanovki kolesnyih par shetiosnogo lokomotiva*. Problemyi razvitiya lokomotivostroeniya. Lugansk, **26** (1990) [in Russian]
10. S. Bruni, R. Goodall, T. Mei, H. Tsunashima, *Control and monitoring for railway vehicle dynamics*, Vehicle System Dynamics, Ed. **45**, 733-779 (2007)
11. J. Choi, S. Park, J. Kim, *Dynamic adhesion model and adaptive sliding mode brake control system for the railway rolling stocks*, Proc. IMeche Part F: J. Rail and Rapid Transit, **221**, 313-320 (2007)
12. M. Spiryagin, V. Spiryagin, I. Kostenko, *Modelling of a controlled tractive wheelset for a bogie of a railway vehicle based on noise spectrum analysis*, TEKA Commission of Motorization and Power Industry in Agriculture, **11a**, 232-244 (2011)
13. D. Nyuland, *Upravlyaemost podatlivoy zheleznodorozhnoy telezhki na krivolineynom uchastke puti. Konstruirovanie i tehnologiya mashinostroeniya*, **3**, 365-375 (1969) [in Russian]
14. I. Kirichenko, A. Kashura, M. Kashura, *Design of process of friction of rolling with sliding*, TEKA Commission of Motorization and Power Industry in Agriculture, **8a**, 63-71 (2008)
15. M. Guskova, V. Ruban, *Parametryi konstruktsii podvizhnogo sostava i puti, vliyayuschie na podrez*

grebney bandazhey kolesnyih par lokomotivov // Vestnik RGUPS. Rostov, **2**, 32-35 (2000)

16. R. Goodall, S. Bruni, T. Mei, *Concepts and prospects for actively-controlled railway running gear*. Vehicle System Dynamics, **44** (suppl), 60-70 (2006)
17. S. Iwnicki, *Handbook of Railway Vehicle Dynamics*, CRC Press, UK (2006).
18. V. Masliev, *Nauchnyie osnovyi vyibora konstruktorsko-tehnologicheskih parametrov ustroystv dlya umensheniya iznosa bandazhey koles lokomotivov,* Diss. dokt. tehn. nauk, HPI, Harkov, 497 (2001) [in Russian]
19. V. Spiryagin, S. Klyuev *Eksperimentalnoe issledovanie zvukovogo signala, ispuskaemogo s kontakta koleso-rels pri razlichnyih usloviyah ih vzaimnoy ustanovki.* Zb. (MehTriboTrans-2011)", Rostov-na-Donu, 214-217 (2011) [in Russian]
20. M. Spiryagin , Soo L. Kwan, Hee Y. Hong, V. Spiryagin, S. Klyuyev, *Study on using noise for development of active steering control system of rail vehicle.* Proceedings of the 23nd National Conference and Exposition on Noise Control Engineering (Noise-Con 2008), 28-31 July 2008, Dearborn, Michigan. USA. Curran Associates, Inc., 499-506 (2009)
21. V. Spiryagin, S. Klyuev, E. Zubar, *Vyibor metoda opredeleniya koordinat mestopolozheniya lokomotiva pri upravlyaemom dvizhenii kolesnyih par v relsovoy kolee.* Visnik SNU im. V. Dalya. Lugansk, **5 (194) Ch. 2**, 144-146 (2013) [in Russian]
22. Marcel Ujfaluši. *How can design of wheelsets improve future rolling stock performance?* Global Railway Review. Issue 5, September 10 (2016)
23. A. Bracciali, *"Railway Wheelsets: History, Research and Developments"*, International Journal of Railway Technology, **5 (1)**, 23-52, doi: 10.4203 / ijrt.5.1.2 2016 (2016)
24. Talgo. [Electronic resource]. - Access Mode: https://www.talgo.com/
25. E. Mikhailov, S. Semenov, V. Tkachenko, S. Sapronova, *Reduction of kinematic resistance to motion of the railway vehicles*. MATEC Web of Conferences *235*, 000 Horizons of Railway Transport, 33-40 (2018)
26. A. Shiler, *Analysis and simulation of new wheel pair construction.* Proceedings of the 25th DAAAM International Symposium Procedia Engineering. Vienna. Austria, 1714-1723 (2014)
27. V. Tkachenko, S. Sapronova, I. Kulbovskii, O. Fomin, *Research into resistance to the motion of railroad undercarriages related to directing the wheelsets by a rail track.* Eastern-European journal of enterprise technologies. Kharkiv, PC «Technology center». **5/7 (89)**, 65-72. (2017)
28. J. Perez, J. Busturia, T. Mei, J. Vinolas, *Combined active steering and tract in for mechatronic bogies vehicles with independently rotating wheels.* IFAC Mechatronic Systems, 737-742 (2002)
29. D. Derbyshire, *Development of an intelligent sensing technique for active control of railway vehicles with independently rotating wheels.* Manchester Metropolitan University. A thesis for the degree of Master of Philosophy, 87 (2015)
30. N. Farhat, P. Ward Christopher, Roger M. Goodall, Roger Dixon. *The benefits of mechatronically-guided railway vehicles: A multi-body physics simulation study.* Mechatronics. **51**, May 2018, 115-126 (2018)

The intensity of rail failure flow

Oleksandr Patlasov, and *Yelyzaveta* Fedorenko*

Dnipro National University of Railway Transport named after academician V. Lazaryan, Department of Railway track and Track facilities, Ukraine

Abstract. The article reveals the quantitative changes in the intensity of failure accumulations in areas with gradual introduction of high axial loads. To obtain the actual values, the authors received three-dimensional array of intensity of rail failure flow. According to the results of the research, it was discovered that the main cause of failure of the rails are the defects in the main contact-fatigue origin. The main factor that impacts the intensity of the rail failure flow is the number of transmitted tonnage. The size of the axial load and curvature of the line pours weight on this process. The proposed rail failure model is a model that can be used to optimize the track maintenance system. The authors conducted a study on the identification of quantitative changes in the intensity of accumulation of track failures in areas with high axial loads and evaluated their influence on the planning of repair and road works. On the basis of the obtained results it is possible to estimate the effect of the axial load on the intensity of changes in the state of the flow rail failure flows.

Introduction

For rational management of the track facilities and the choice of the most appropriate upper structures of the track, it is necessary to determine the expected service life of the upper structure of the track as a whole and its individual elements. An analysis of the actual state of the rail facilities and repair planning on the actual state of the track is an urgent problem.

The most dangerous failures when operating the track are broken rails. To prevent this, there is a control system to detect cracks in the metal at an early stage of development and to ensure the removal of potentially hazardous rails prior to their destruction.

In order to predict the service life of the rails, it is necessary to evaluate the factors affecting them, to describe the work of rails with the help of a mathematical model, in which it is necessary to take into account the influence of the main factors of the operation of the railway track and determine the quantitative changes in the intensity of the accumulation of rail failures in areas with the gradual introduction of high axial loads.

Existing norms for determining the periodicity of carrying out track repair work depend on the screened tonnage, as well as additional criteria, including a single output of rails, which takes into account the actual state of the track. In turn, the conditions of the tracks depend on gages, radiuses and lengths of curves, their total length, ascents and descents of different lengths.

The intensity of the rail failure flow rates is established on the basis of massive observations of their work in different operating conditions.

Theoretical and experimental researches on intensity of development of defects of rails, influence on them axial loading, low temperatures, plan of the line, volumes of transportations are devoted to work of domestic and foreign scientists: IS. Baulin, S.A. Kolotushkin, E.A. Shur, NI Karpushchenko, VS Lysyuk, VO Pevzner, A. Yu. Abdurashitov, VV Ribkin, VV Kosarchuk, J. V. Ringsberg, V.A. Reichart, M. Taraf, E.H. Zachaf, O. Ususoadi, A. Zegul. [1-16].

In [17] experimental dependences of the intensity of the flow of failures were obtained.

Given that increasing the capacity of cars up to 25 tonnes / axle is considered as one of the priority areas for the development of railways, it is advisable to consider the effect of such cars on the condition of the track, namely on the condition of the rail economy in the conditions of heavy freight trains.

Taking into account that increasing the carrying capacity of carriages up to 25 t/axle is considered as one of the priority directions of development of railways, it is expedient to consider the influence of such carriages on the state of the track.

The main objective of this study is to identify quantitative changes in the intensity of accumulation of failure flows in areas with high axial loads.

Methodology and results

It is estimated that the productivity of carriages with a load of 25 tf/axle is higher by 7-10% due to an increase in inter-repair mileage and a reduction in the cost of traction in calculating the carriage of one ton of cargo. But a heavier carriage more intensively wears a track and leads to malfunctions [18].

* Corresponding author: rapunzeeelliza@gmail.com

With the development of tonnage in the process of operation, rails accumulate various damage, deformation, fatigue defects, resulting in a decrease in their reliability, more often failures occur and there is a need to reduce speed or stop the movement of trains.

The main reason for the track failure are mainly defects of contact-fatigue origin. The occurrence of contact-tired defects is a long process. Significantly influences its.formation of the load on the axle from the rolling stock.

Defects originate at the stage of smelting, bottling and metal processing, and then appear at the rolling and heat treatment. The inefficiency of control at all stages of production and release of rails leads to the supply of a track complex of rails with defects of production origin.

The main criterion for total rail replacement for new and old ones is the continuous wear of the rails above the standard value or the same number of rails, as with the appointment of the appropriate type of modernization or major repair of the track. In this case, the type of accompanying work, the number of elements of fastenings and sleepers, which must be replaced, is determined by the distance of the track according to the results of commission inspections.

The complete replacement of rails, the replacement in curved sections, as well as the replacement of rails from external threads of the curved parts of the track to the straight ones, are intended provided that the rails are drained to the limits set by the Ukrzaliznytsya [19].

The main types of damages, deformations and defects of the rails are cracks, exfoliation, deflection, erosion, influx, corrosion of the metal, mechanical damage to the rails in the form of bends, screw heads or soles, internal fatigue defects in rail metal, etc.

The intensity of the single output of rails depends on the tonnage (runtime), load intensity, dynamic loads loaded on the rolling stock of the structure of the track and its condition, plan and profile of the track and other factors.

The function of the failure rate depends on the average axial load on the axle, the screened tonnage, the radius of the curve:

$$\lambda = f\left(P_{av,} T, R\right) \quad (1)$$

where $P_{av,}$ – average axle load in tf/axle;

T – screened tonnage in million tons gross;

R – curves radius in meters.

However, the intensity of the rail failure flows depends not only on the magnitude of the axial load, but also on the following other factors:

- type and structure of the upper structure of the track,
- rail steel quality,
- line plan and profile,
- load capacities,
- screened tonnage,
- types and number of repairs,
- speed of movement,
- climatic conditions, etc.

By analyzing the function of the dependence of the intensity of the rail failure flow, we can predict that with increasing axial load, the intensity of the rail failure flow increases, as well as inter-repair terms reduce. This is due to the fact that with the increase of the axial load the life of the track decreases, which is the most expensive element of the upper structure of the rail, shortening the life of the transfer services and increasing labor costs.

To solve these issues, a study was conducted on double tracks of the Pridneprovska railway distillation on Pryshyb station - Burchatsk station running line (1174 km, PC 3). The photo of the experimental site is shown in Fig. 1

Fig. 1. Photo of the experimental site

As input data, a train No. 1762 was considered, which was moving according to the schedule, and consisted of the VL-11 locomotive, five carriages with an axial load of 25 tf/axle and 51 carriages with an axle load of up to 23.5 tf/axle.

The scheme of the experimental train is shown in Fig. 1. The train was moving at a speed of about 75 kmph.

Carriage dead weight and cargo weight are shown in Table 1. The experimental train was moving from the side of the Pryshyb station in the direction of Burchatsk station.

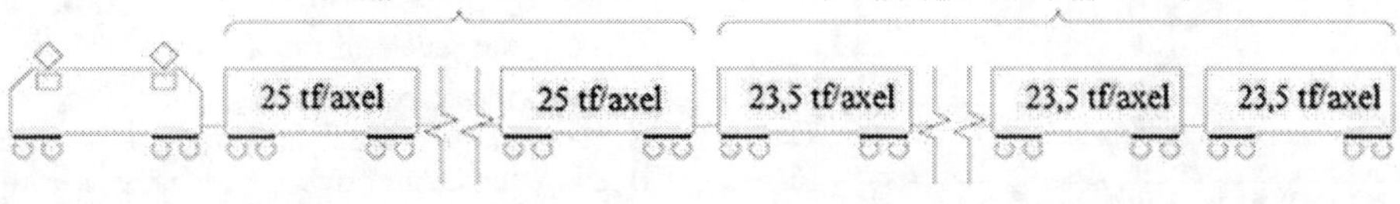

Fig. 2. Experimental train schem

Table 1. Load capacity of train carriages

No.	Carriage No.	Dead weight, tons	Tonnage, tons	Total, tons	Average axle load, tf/axle
0	VL-11	184	0	184	23,000
1	62139845	23,5	76	99,5	24,875
2	62139878	23,5	76	99,5	24,875
3	62139852	23,5	75	98,5	24,625
4	63584981	23,5	76	99,5	24,875
5	62139860	23,5	76	99,5	24,875
average 25 tf/axle		**23,5**	**75,8**	**99,3**	**24,825**
6	55081020	23,5	70	93,5	23,375
7	62783170	23,5	70	93,5	23,375
8	56732332	23,6	71	94,6	23,65
9	65576373	23,5	70	93,5	23,375
10	52183282	23,5	70	93,5	23,375
11	55210926	23,5	70	93,5	23,375
12	53007878	23,6	71	94,6	23,65
13	55531487	23,6	70	93,6	23,4
14	62073077	23,6	70	93,6	23,4
15	55267934	23,6	70	93,6	23,4
16	56653389	23,6	69	92,6	23,15
17	59787184	23,5	70	93,5	23,375
18	52283546	23,5	71	94,5	23,625
19	52244332	23,5	70	93,5	23,375
20	55732184	23,5	70	93,5	23,375
21	62782917	23,5	70	93,5	23,375
22	54008180	23,5	70	93,5	23,375
23	62783022	23,5	70	93,5	23,375
24	62783063	23,5	70	93,5	23,375
25	56264443	23,5	71	94,5	23,625
26	53148474	23,5	70	93,5	23,375
27	60736675	23,5	70	93,5	23,375
28	56936529	23,5	70	93,5	23,375
29	62073556	23,5	70	93,5	23,375
30	55483846	23,5	71	94,5	23,625
31	60808706	23,5	70	93,5	23,375
32	62649314	23,5	70	93,5	23,375
33	55440044	23,5	71	94,5	23,625
34	55531537	23,5	70	93,5	23,375
35	62783436	23,5	70	93,5	23,375
36	56462955	23,5	69	92,5	23,125
37	55371645	23,5	69	92,5	23,125
38	56843808	23,5	70	93,5	23,375
39	62479860	23,5	70	93,5	23,375
40	56592603	23,5	71	94,5	23,625
41	55459895	23,5	71	94,5	23,625
42	57656985	23,5	70	93,5	23,375
43	63679112	23,5	70	93,5	23,375
44	56367360	23,5	70	93,5	23,375
45	56846421	23,5	70	93,5	23,375
46	56479066	23,5	71	94,5	23,625
47	61236154	23,5	70	93,5	23,375
48	62309620	23,5	70	93,5	23,375
49	58140310	23,5	70	93,5	23,375
50	56002819	23,5	69	92,5	23,125
51	56850043	23,5	70	93,5	23,375
52	53587937	23,5	70	93,5	23,375
53	53586806	23,5	70	93,5	23,375
54	62932181	23,5	70	93,5	23,375
55	56421084	23,5	69	92,5	23,125
56	61102281	23,5	70	93,5	23,375
Average 23,5 tf/axle		**23,51**	**70,08**	**93,59**	**23,398**

Today, Ukraine already has a rolling stock with an axle load of 25 tf/axle and its implementation is considered, but the actual axle load is 23.5 tf/axel. To calculate, let us compare the value of the intensity of the rail failure flow, depending on the average of the actual and increased axial load.

The actual average axle load for the carriage of cargo shall be the following (2):

$$P_{\text{av cargo}} = \frac{T_{car}}{n_{axis}} \tag{2}$$

where T_{car} – dead weight of a loaded carriage in tons;

n_{axis} – number of axels in the carriage;

Actual average axle load for empty carriages (3):

$$P_{\text{empty cargo}} = \frac{T_{car}}{n_{axis}} \tag{3}$$

where T_{car} – dead weight of an empty carriage in tons;

n_{axis} – number of axels in the carriage;

The following initial documentation of the distances and the service of the track was used for obtaining the initial data: longitudinal profiles, rail-sleeper-ballastinc map, record of rails taken from the main tracks (form PU-4), annual reports of the track distance.

For the selected experimental sites and each element of the plan, the following parameters were determined:

- load capacity;
- average axle load;
- description of plan and profile;
- the term of the last major repair;
- screened tonnage after the last upgrade.

From each removed rail from a track, the following aspects were fixed:

- direction;
- track number;
- kilometer and number of the rail;
- day, month, and year of rail removal;
- defective classification number [20].

One of the main indicators that characterize the system reliability of the track is the parameter of the rail failure flow. The rate of failure of the rails is the number of defective rails, which are replaced by 1 km referenced to the unit of the passed tonnage:

$$\lambda_{ijk} = \frac{N_{ijk}}{L_{ijk} \cdot \Delta T_{ijk}} \tag{4}$$

where N_{ijk} – the number of rails replaced on the experimental plot, pcs;

L_{ijk} – length of the experimental plot, km;

ΔT_{ijk} – screened tonnage, million tons gross;

i, j, k – the levels of the axial load test factors, the line plan and the screened tonnage.

To obtain the actual values it is necessary to get three-dimensional array of intensity of rail failure flow. The size of the array is determined by the number of levels of each of the test factors (axial load, plan of the line, screened tonnage) [21].

The algorithm of calculation begins with the sorting of data on each replaced rail in relations with levels of influencing factors. As a result of sorting, the values of the indexes i, j, k are found. Thus, the number of replaced rails is determined at all levels of the analyzed factors.

The next step is to determine the length of the track segment, which corresponds to the characteristics of each element of the three-dimensional array L_{ijk}. Each elementary segment is understood as a segment of the track, where any chosen for the analysis change. Moreover, it was taken into account that one and the same real elementary segment can be entered into a cell of the array, which are different values of the screened tonnage. For example, if the factor "screened tonnage" is divided at the level of 50 million tons gross, then the length of one and the same elementary segment can be entered in the first, second, etc. levels, as the same that through this real segment of the track the tonnage of 0 to 50 million tons gross can be screened, from 50 to 100, etc. It is also taken into account that during a certain period of operation of the section of the track there are no data on the failure of rails in this area. In this case, the length of this section in the corresponding cell of the screened tonnage is not entered. If the level of the screened tonnage is lower than the level of the factor at the site, then the "length" changes in proportion to the actual tonnage missed.

After filling all the cells of the three-measuring array with the number of replaced rails and the "length" of the sections according to the formula (4), the intensity of the rail failure flow is determined.

Taking into account that the levels of all factors being analyzed are intervals, for each value of the intensity of the rail failure flow, the value of the screened tonnage, the axial load and the radius of the curve were determined.

For the implementation of the dispersion analysis, factor 3 was adopted: the line plan (straight and curves with a radius of 600 m) and the screened tonnage of 800 million tons gross, and variants of axial load, namely:
5% of carriages loaded 25 tf/axel, 95% - up to 23.5 tf/axel
10% of carriages loaded 25 tf/axel, 95% - up to 23.5 tf/axel
25% of carriages loaded 25 tf/axel, 95% - up to 23.5 tf/axel
50% of carriages loaded 25 tf/axel, 95% - up to 23.5 tf/axel
75% of carriages loaded 25 tf/axel, 95% - up to 23.5 tf/axel
100% of carriages loaded 25 tf/axel, 95% - up to 23.5 tf/axel

The results of calculating the intensity of the rail failure flow are shown in Table 2.

Table 2. Results of calculating the intensity of the rail failure flow

Average axial load, tf / axel	The curvature of the site.	Missed tonnage, million tons	The parameter of the flow of the rail, pcs
23,46935	0.0143478	800	0,118149899
23,5407	0.0143478	800	0,118160136
23,75475	0.0143478	800	0,118190848
24,1115	0.0143478	800	0,118242033
24,46825	0.0143478	800	0,118293219

Conclusion

According to the results of researches, it has been discovered that the main cause of track failure are mainly defects of contact-fatigue origin. The main factor that impacts the intensity of the rail failure flow is the number of transmitted tonnage. The size of the axial load and curvature of the line pours weight on this process. The track failure model is proposed. Which can be used to optimize the track maintenance system.

Literature

1. Baulin I. S., Dyakov V. N., Uskova O. N. et al. *Investigation of the mechanism of contact-fatigue damage* (defect 82 and 64) / Vesnik VNIIZhT. **No. 4**, 1962 .- p.34-39
2. Kolotyshkin S. A., Dyakov V. N., Poroshin V. L., Baulin I. S. *Detection of internal longitudinal-inclined cracks in the track heads* / Vesnik VNIIZhT. Edition 1, 1973, p. 41-43
3. Tsukanov P. P. *Operational stability of tracks of modern production* / Research on heavy type tracks. Collection of VNIIZhT scientific papers. M .: 1961. P. 4-31
4. Shur E. A., Konyukhov A. D. *Influence of residual stresses in hardened tracks on the emergence and propagation of fatigue cracks in cyclic bending* / Residual stresses and strength of railway tracks. VNIIZhT. Edition.491
5. Shur E. A. *On the choice of permissible stresses under the strength of track calculations* / Vesnik VNIIZhT. Edition 8, 1977, pp. 38-41
6. Shur E. A., Kiselev T. N., Poroshin V. L., Beyserov M. S. Durability of tracks with cracks / Vesnik VNIIZhT. Edition 5, 1984, p. 48-52
7. Karpushchenko N. I. *Planning of repairs of the railroad according to its tho stato* // G. K. Shchepotin Improving the reliability and efficiency of the railroad. / Under edition if N. I. Karpushchenko. Collection of scientific papers – Novosibirsk, 1991, - p. 122.
8. Lysyuk V. S. *Fundamentals of the methodology for calculating failures and inter-repair resources of the railway track for damages of tracks* // VNIIZhT. - M., 1983. - 57 p. CNIITEI MPS Dep 25.02.83, № 2120 ZHD-D83 // Railway Transport Magazine abstract, No. 6, 1983.
9. Pevzner V.O. *To improve planning of track works.* // Track and track facilities, 1993, № 9, - P. 10-12.
10. Abdurashitov A. Yu. *The regularities of the formation of contact-fatigue defects.* // Track and track facilities, 2002, No. 11, pp. 16-20.
11. Rybkin V. V. *Mathematical model of track failure due to fatigue defects* // A. M. Patlasov. Aspects of the interaction of the track and the rolling stock. Works of DIIT. - D., 1990. - P. 52-58.
12. Ringsberg J.W. *Cyclic ratcheting and failure of a pearlitic rail steel* // Fatigue & Fract. Engng. Mater.& Struct. – 2000. – 23. – P. 747 – 758

13. Reichart V.A. *Analysis of rail defects* // Path and track economy. - 2001. - №4. - P. 22 - 25
14. Kosarchuk V.V., Agarkov O.V. *Prediction of durability of rails by the criterion of occurrence of contact fatigue cracks* // Proc. Sciences. to wash DUIT. Transport Systems and Technologies Series. - 2012. - Vip. 20. - P. 77 - 90.
15. Taraf M., Zahaf E.H., Oussouaddi O., Zeghloul A. *Numerical analysis for predicting the rolling contact fatigue crack initiation in a railway wheel steel* // Tribology International. – 2010. – 43. – P. 585 – 593.
16. Kosarchuk V.V., Danilenko E.I., Agarkov O.V., Rafalsky O.Y. *Influence of the profile of car wheels on the level of contact stresses in rails* // Coll. Sciences. to wash DUIT. Transport Systems and Technologies Series. - 2018. - Vip. **31**. - P. 3
17. Rybkin V. V., Umanov M. I., Bal O. M. *Individual track output as an important factor for forecasting the period of repair of a track* // Bulletin of the Dnipropetrovsk National University of Railway Transport named after academician V. Lazaryan. - 2007. - Ed. **18**. - P. 62-70. - Access mode: http://nbuv.gov.ua/UJRN/vdnuzt_2007_18_16
18. Rybkin, V. V. *Stress-deformed state of the track in the interaction of wagons with increased axial load* / V. V. Rybkin, A. M. Patlasov, V. I. Klimov // Problems of mechanics of railway transport. Increase of reliability and perfection of structures of rolling stock: theses. doc. All Union. conf. (Dnipropetrovsk, May 1988). - Dnipropetrovsk, 1988. - P. 96.
19. Regulations on conducting scheduled maintenance and repair works on railways of Ukraine: CP-0287: Order of Ukrzaliznytsia dated 11/3/2014 № 470-CZ / A. Babenko, G. Linnik, K. Moiseenko [et al]. - Kyiv, 2015. - 45 p.
20. Classification and catalog of defects and damage of elements of switching transfers and tracks of Ukrainian railways: CP 0060 - 2000.
21. Patlasov A. M. Improvement of the system of planning of repairs of a railway tracks: diss. Cand. of Tech. Sciences: 05.22.06 / Alexander Mikhailovich Patlasov; St. Petersburg Institute of Ing. Railway Transport Engineering - St. Petersburg, 1991. – p. 167.

Actual utilisation of maximum line speed - Polish and Ukrainian experience

Andrzej Massel*

Railway Research Institute (Instytut Kolejnictwa), Warsaw, Poland

Abstract. The overall condition of the Polish railway infrastructure has been recently improved. The condition of railway infrastructure and the level of its technical equipment are reflected in the maximum train speeds allowed on particular sections of railway lines. The analysis covers the utilisation of the maximum line speed by the long-distance trains in Poland in the 2018/2019 timetable. The database has been prepared, covering start-to-stop runs of Express Intercity Premium (EIP) trains). The analysis shows that typical value of speed utilisation ratio for EIP trains operated with ED250 EMU is approximately 0.82-0.83. The correlation ratio between the average maximum line speed and the train commercial (start-to-stop) speed is rather good. The data from the Polish railways have been compared with Ukrainian ones. The results of the research show clearly, that the selection of passenger rolling stock for particular route can have significant impact on its day-to-day operation.

1 Introduction

The European railways have recently been significantly transformed through implementation of the EU legislation imposing "unbundling", i.e. strict separation of infrastructure management and operation of passenger and freight services.

The Directive on the interoperability of the rail system within the European Union states, that the quality of rail services depends, inter alia, on excellent compatibility between the characteristics of the network (in the broadest sense, i.e. the fixed parts of all the subsystems concerned) and those of the vehicles (including the on-board components of all the subsystems concerned) [1]. Performance levels, safety, quality of service and cost depend upon that compatibility. Performance of the rail system can be evaluated according to numerous criteria and described through various parameters, like capacity, speed, accessibility.

Overall performance of railway system in Poland was evaluated by Wróbel [2]. This assessment was focused on the integration of timetables for passenger trains co-financed by competent authorities in the framework of Public Service Contracts (PSC). The examples of parameters used for this particular assessment were: number of connections in the directions specified in the Transport Plan, average waiting time and average connection time for stations and for individual directions.

The issue of performance evaluation is valid in case of all transport systems. It is very important to develop methods for selection of optimum system suited to specified local conditions. For example there numerous works comparing light rail systems (LRT) with bus rapid transit systems (BRT). One of the key features of particular transit system is its commercial speed. Therefore it is necessary to investigate factors influencing that speed. According to works by Kühn [3], in case of LRT systems, the direct relation exists between percentage of separated right-of-way and commercial speed. For completely (or almost completely) segregated LRT routes, the typical commercial speeds of 29-32 km/h can be achieved. If only 50% or less right-of-way is separated, the commercial speed usually does not exceed 20 km/h (data from Cologne) [3]. The commercial speed influences the service that can be provided for customers, but also determines the number of vehicles and drivers that will be needed. The commercial speed is also related to the maximum speed of the vehicle, and to braking and acceleration characteristics. However, TRL/INRETS field surveys revealed little difference between electric trains and buses. Research using multiple regression analysis for such factors as station spacing (i.e. the distances between stations), suggests that the inherent difference between busways and LRT is not statistically significant. [3].

Various aspects of railway transport system have been assessed by the European Court of Audit (ECA). For example in 2016, ECA evaluated rail freight transport across Europe [4]. The poor performance of rail freight transport in terms of volume and mod-al share in the EU is not helped by the average commercial speed of freight trains. On some international routes freight trains run at an average speed of only around 18 km/h. This is due to weak cooperation between the national infrastructure managers. In Central and Eastern

* Corresponding author: amassel@ikolej.pl

European Member States, the average speed is between 20 and 30 km/h. The analysis o ECA showed, however, that the situation is significantly better in some freight corridors, where the average is speed is around 50 km/h. This values is closer to the average speed of trucks (around 60 km/h). [4].

A comprehensive performance audit on the long-term strategic planning of high-speed lines in the EU, on the cost-efficiency (assessing construction costs, delays, cost overruns and the use of high-speed lines which received EU co-funding), and on the sustainability and EU added value of EU co-funding was carried out in 2018 [5]. The analysis was focused on the speed on the audited high-speed lines. It shows, that, on average along the course of a line, trains run at only around 45 % of the line's design speed (in European conditions usually at the level of 300 km/h). Only two lines operate at average (commercial) speeds of more than 200 km/h, and no lines operate at an average speed above 250 km/h. The lowest speed yield on a completed high-speed line is on the Madrid-León high-speed line (39 % of design speed). The cross-border Figueres - Perpignan section also only operates at 36 % of its design speed, because it accommodates mixed traffic. Average speed so far below the design speed indicates that an upgraded conventional line would have been enough to achieve the objectives set, at a much lower cost [5].

Important aspect influencing performance of railway system is the quality of transport offer and its structure. The problem was investigated by Żurkowski, who proposed classification of passenger services. The following categories have been distinguished: intercity, interregional, regional and agglomeration. [6].

Charkina developed the proposal for new classification of passenger trains in Ukraine taking into account the working period, the territory of service, the maximum distance of movement, average journey speed, time of a trip, a type of a rolling stock and sources of financing [7].

Organisational aspects of implementation of accelerated passenger train services in Ukraine have been investigated by Matusevich [8]. Their introduction is treated as a mean of improvement of efficiency of passenger transport. The practical result of the research study was an algorithm for the introduction of new economic classification of passenger trains.

The aim of the present research is to verify method for evaluation of actual utilisation of the maximum line speed. Another goal is to identify factors having influence on utilisation of the maximum line speed in case of long-distance passenger trains on the Polish railway network.

2 The railway infrastructure in Poland

The vast majority of railway lines in Poland is managed by PKP Polskie Linie Kolejowe S.A. (PKP PLK). According to the Annual Report of the company, the total length of railway lines forming PKP PLK network is 18513 km (December 2017). The total length of tracks at that time was 35967 km, 27120 km of which were route tracks (i.e. tracks on open sections) and principal main tracks at stations and 8847 km were other station tracks.

The total length of electrified railway lines is 11816 km and total length of tracks equipped with overhead catenary lines is 24697 km [9]. This means, that approximately 63.8% of railway lines in Poland are electrified.

The technical condition of the railway network in Poland is not uniform. It is noteworthy, however, that as a result of the maintenance and repair works as well as performed investment tasks, the overall condition has been recently improved. The length of railway line tracks graded as good in terms of technical condition (as at 31 December 2017) represented 58.9% of the total track length, which is almost 23% increase in comparison to the status from 31 December 2010, when only 36% of tracks were in a good condition (tracks in good condition – tracks operated in line with the assumed parameters, only maintenance work is required).

Simultaneously the percentage of railway line tracks graded as unsatisfactory has been reduced from 29% in 2010 to 15.6% in 2017. Tracks in unsatisfactory condition are tracks characterised with significantly lower operation parameters (low timetable speed, large number of local speed limits, lower permissible loads), which qualify railway tracks for comprehensive replacement.

The condition of railway infrastructure as well as the level of its technical equipment are reflected in the maximum train speeds allowed on particular sections of railway lines. The maximum speeds on all operated railway lines managed by PKP PLK are presented in Fig. 1.

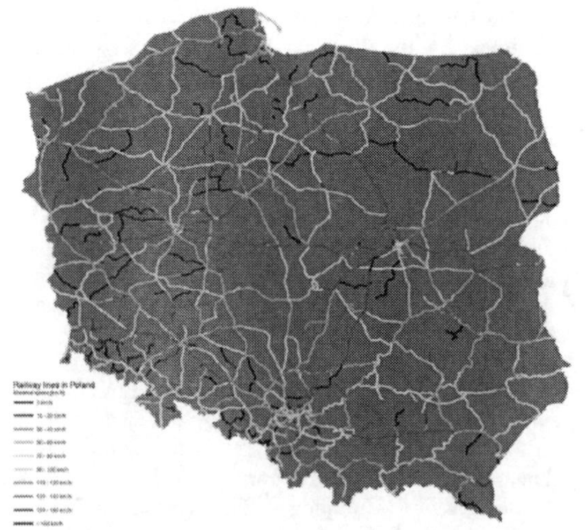

Fig. 1. Maximum speeds on railway lines in Poland (2017). Source: Own elaboration of IK

It should be noticed, that the structure of maximum speeds on the Polish railway network has been significantly changed recently. In December 2010 only 20.2% of main tracks were cleared for operation at the speed of 120 km/h or more. In 7 years, till December 2017, the share of main tracks with the maximum speed of 120 km/h or more has been increased to 37.8%. Moreover, since December 2014 the maximum speed of 200 km/h is in force on the part of line 4 Grodzisk Mazowiecki – Zawiercie (Central Trunk Line). In the 2014/2015 timetable, the total length of line allowed for 200 km/h was 87 km (i.e. the total track length was 174 km). Four years later, in the 2018/2019 timetable, two sections on the Central Trunk Line are operated at this speed and located between Grodzisk Mazowiecki and Zawiercie (76.7 km) and between Wloszczowa Polnoc and Zawiercie (58.3 km). The total length of tracks for 200 km/h is approximately 271 km.

Simultaneously, the percentage of main tracks with the maximum train speed lower than 80 km/h has been reduced from 43.8% in 2010 to 29.5% in 2017. The key issue is, however, to what extent the maximum line speeds are reflected with the train commercial speeds, i.e. the speeds which are perceived by the end-users of railway transport – the passengers.

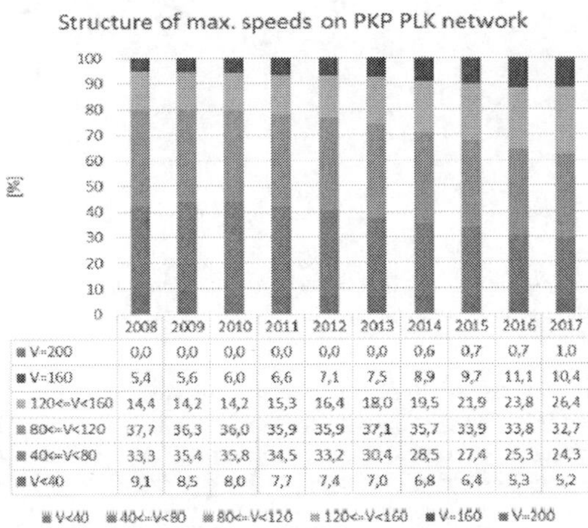

	2008	2009	2010	2011	2012	2013	2014	2015	2016	2017
V=200	0,0	0,0	0,0	0,0	0,0	0,0	0,6	0,7	0,7	1,0
V=160	5,4	5,6	6,0	6,6	7,1	7,5	8,9	9,7	11,1	10,4
120<=V<160	14,4	14,2	14,2	15,3	16,4	18,0	19,5	21,9	23,8	26,4
80<=V<120	37,7	36,3	36,0	35,9	35,9	37,1	35,7	33,9	33,8	32,7
40<=V<80	33,3	35,4	35,8	34,5	33,2	30,4	28,5	27,4	25,3	24,3
V<40	9,1	8,5	8,0	7,7	7,4	7,0	6,8	6,4	5,3	5,2

Fig. 2. Changes in structure of maximum speeds on railway lines in Poland from December 2008 till December 2017. Source: Author's own elaboration on the basis of PKP PLK data

3 Express train services in Poland

In Poland comprehensive network of express trains was implemented in early 1960s. The essential role of these daytime trains was to establish the fastest possible connections between Warsaw and regional administrative centres. In 1984 the operation of the express trains making use from newly built Central Trunk Line (CMK) was started. Initially the offer included one pair of non-stop trains between Warsaw and Cracow as well as one pair of trains from Warsaw to Gliwice via Katowice. For the first time in the history of the Polish railways the maximum speed of these trains was set as 140 km/h. In the following few years (until 1989) the network of express trains running non-stop (or at least with very limited number of stops) was expanded. Moreover, in 1988 the maximum speed of passenger trains on CMK line was increased to 160 km/h.

In December 2014 the new ED250 Pendolino electric motor units (EMUs) were introduced on the Polish railway network [10]. Generally they serve the main north-south axis from Gdynia and Gdansk through Warsaw to Cracow and Katowice. Moreover selected trains between Warsaw and Wroclaw through Opole are also operated with these EMUs. These trains make good use of the modernised and revitalised infrastructure.

Table 1 shows changes in journey times between Warsaw and the largest cities in the country. Three milestones have been selected: 1989 - start of political and economic transformation in Central-Eastern Europe, 2004 – accession of Poland into EU, 2019 – present year (journey times according to the timetable valid from December 2018 till March 2019).

Table 1. Evolution of journey times in Poland [hh:mm]

Section	Length [km]	Time 1989	Time 2004	Time 2019
Warsaw - Cracow	293	2:45	2:45	2:17
Warsaw - Lodz	133	1:30	1:50	1:18
Warsaw - Wroclaw	422	5:08	4:50	3:36
Warsaw - Poznan	302	3:01	2:47	3:26
Warsaw - Gdansk	328	3:34	4:00	2:51
Warsaw - Szczecin	515	5:27	5:32	6:28
Warsaw - Bydgoszcz	288	3:40	3:40	3:19
Warsaw - Bialystok	183	2:33	2:21	2:17
Warsaw - Katowice	298	2:42	2:27	2:19

It is clear the in majority of cases the journey times have been significantly reduced in comparison with the year 1989. The most notable shortening of journey time (more than 1 1/2 hours) is in the case of Warsaw – Wroclaw route. The journey time between Warsaw and Poznan, however, is much longer than in 1989. The reason is ongoing modernisation of the line, requiring deviation of all trains (they are operated through Gniezno).

4 Utilisation of maximum speed – the case of the Polish railways

The analysis covers the utilisation of the maximum line speed by the long-distance trains in Poland in the 2018/2019 timetable. The database has been prepared, covering 46 start-to-stop runs of Express Intercity Premium (EIP) trains). All involved trains are operated by PKP Intercity on commercial basis.

The official infrastructure data of PKP PLK were used to calculate exact distances and mean maximum speeds for all sections.

To characterise maximum train speeds it is necessary to adopt some statistical measures. The most important parameter seems to be the highest maximum speed on the section in question (V_{max}). This value defines the requirements for rolling stock making the full use from the infrastructure capabilities [11].

The differentiation of the maximum speed along the line has significant influence on journey time and, consequently, on line capacity. It is taken into account with the harmonic weighted mean $V_{0\ max}$, calculated according to the formula (1):

$$V_{0\ max} = \frac{\sum_{i=1}^{n} l_i}{\sum_{i=1}^{n} \frac{l_i}{V_{i\ max}}} \quad (1)$$

where $V_{i\ max}$ is maximum speed on section of track i, and l_i is the length of (sub)section i. Harmonic weighted mean is a quotient of the total length of the line (or the network) and the sum of theoretical journey times on particular (sub)sections with the constant speed. It is clear, that $V_{0\ max} \leq V_{max}$.

In the case of Central Trunk Line the maximum speed of 200 km/h is allowed on two sections (total length approximately 135 km), whereas the maximum speed of 160 km/h is in force on the remaining part of the line (see Fig. 3). The only exceptions are the station of Idzikowice (from km 80.2 to 82.4) and the final 2 kilometres at the end of the line close to Zawiercie station. As a result the weighted mean value of maximum speed for that line is $V_{0\ max}$=180.6 km/h.

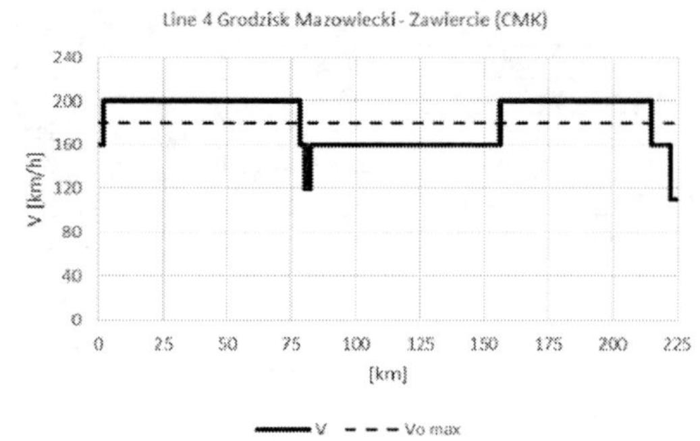

Fig. 3. Example of the speed profile with long sections passed at uniform maximum speed. Source: Author's elaboration on the basis of official data of PKP PLK S.A.

The journey times have been collected for all start-to-stop sections. They were extracted from official public timetables and from working timetables valid from 9 December 2018 [12].

Practical measure for assessment of railway offer in passenger traffic is the value of commercial speed V_c, which is calculated according to the formula (2):

$$V_c = \frac{l}{\sum t_r + \sum t_S} \quad (2)$$

where l is the length of train route, t_r are the journey times on consecutive sections and t_s are the stopping times. It should be noticed, however, that for the analysis of utilization of the maximum line speed, the start-to-stop average speeds V_s are more convenient measure. Start-to-stop average speed was commonly used for presentation of the fastest train services in professional journals before World War II. Good example was the series of articles "British Express Train Services" in the Railway Magazine [13]. Similar analyses were also compiled for express train services in other European countries. It is noteworthy that the start-to-stop train speeds are the basis for train classification in the World Speed Survey, published bi-annually in Railway Gazette International until now [14].

As the parameter characterising the utilisation of the maximum line speed, the speed utilisation ratio I_s has been defined (3):

$$I_S = \frac{V_S}{V_{0\ max}} \quad (3)$$

where V_s is the start-to-stop average speed of the train on particular section and $V_{0\ max}$ is the weighted average maximum line speed on the same section.

In general, the values of start-to-stop average speeds of EIP trains show very good correlation with (harmonic) mean maximum speeds (Fig.4).

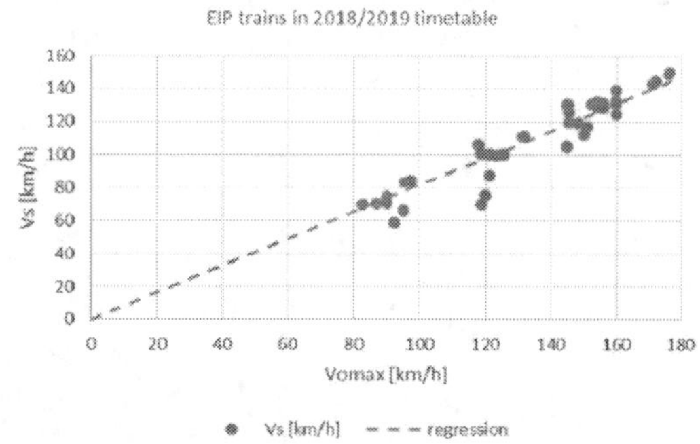

Fig. 4. Correlation between average maximum speeds and start-to-stop speeds for EIP trains in the timetable valid from 9 December 2018

In case of 46 runs of these trains (served with ED250 EMUs) the correlation coefficient r equals to 0.942. The regression equation for this group of train services can be written as follows (4):

$$V_S = 0.818\ V_{0\ max}\ . \quad (4)$$

It should be noted, that neglecting a few sections affected with track works results in even better correlation and in higher values of speed utilisation ratio.

There are several sections with particularly favourable utilisation of maximum line speed, for example: Warszawa Wschodnia – Ilawa Glowna (I_s=0.900) and Lebork – Slupsk (I_s =0.903).

For the start-to-stop runs of EIP trains, covering Central Trunk Line (at which 200 km/h is achieved) the speed utilisation ratio is also quite good: Warszawa Zachodnia – Krakow Glowny (I_s=0.858), Krakow Glowny - Warszawa Zachodnia (I_s=0.859).

5 Acceleration of passenger train services in Ukraine

Novelty on the Central-Eastern European railways is the gradual replacement of overnight trains (composed of couchete and sleeping cars) by EMUs operated as daytime trains. The best examples can be shown in Belarus and Ukraine.

In Belarus such restructuring of passenger offer was possible after electrification of selected lines of BC network and simultaneous raising the maximum speed. up to 140 km/h. The central location of the capital city of Minsk. at the intersection of main railway lines in Belarus. makes it very easy to start new daytime connections.

In the year 2002, the first fast daytime Intercity services were implemented on the Kyiv – Kharkiv route in Ukraine (branded as Stolichny Express). They were composed of the new passenger open-saloon cars and operated at the maximum speed of 140 km/h. In the following years similar intercity services were introduced also on other important routes.

Since 2012 several intercity trains in Ukraine are operated with modern EMUs manufactured by Hyundai and Skoda (Fig.5). They serve routes linking Kyiv with major regional centres as Lviv, Dnipro, Kharkiv.

Fig. 5. HRCS2 EMU of Ukrainian Railways serving the Lviv – Kyiv Intercity+ train (2012)

Table 2 shows the changes in journey times between Kyiv and major regional centres valid for 1989, 2004 and 2019 timetables. It is noteworthy that the numbers from the year 2004 reflect the significant improvement on Kyiv – Kharkiv and Kyiv – Dnepropetrovsk (now Dnipro) routes thanks to implementation of Stolichny Express.

Table 2. Evolution of journey times in Ukraine [hh:mm]

Section	Length [km]	Time 1989	Time 2004	Time 2019
Kyiv - Kharkiv	486	9:44	5:50	4:45
Kyiv – Dnipro	532	9:18	5:45	5:45
Kyiv – Kryvyi Rih	448	9:08	9:04	5:36
Kyiv – Lviv	572	9:19	8:30	5:09
Kyiv – Mariupol	1032	19:35	17:23	17:54
Kyiv – Mykolaiv	526	9:22	8:22	6:27
Kyiv – Odessa	654	9:29	9:35	7:15
Kyiv - Zaporizhia	657	13:32	10:16	7:18

Using the maximum speed and journey time data from the 2013/2014 timetable it was possible to calculate the average maximum speeds and start-to-stop speeds for a few sections:

1) Kyiv – Mirgorod, 246.5 km, V_{0max}=133.6 km/h, V_s=122.2 km/h, I_s=0.915;
2) Darnitsa – Mirgorod, 232.4 km, V_{0max}=138.5 km/h, V_s=127.9 km/h, I_s=0.924;
3) Mirgorod – Poltava K., 88.4 km, V_{0max}=121.9 km/h, V_s=110.5 km/h, I_s=0.906.

All above runs were made by trains operated with HRCS2 EMUs. It should be noticed, however, that the present journey times on respective sections are a few minutes longer than in the year 2013.

6 Conclusions

Quality of rail services depends, to large extent, on excellent compatibility between the characteristics of the infrastructure and the rolling stock. Good example of such compatibility is adjustment of speed characteristics of EMU or loco hauled train to the maximum speeds allowed on the route and to their differentiation.

Comprehensive data base including the runs of long-distance trains in Poland in the 2018/2019 timetable has been prepared. The analysis shows that typical value of speed utilisation ratio I_s for EIP trains operated with ED250 Pendolino EMU is approximately 0.82-0.83. The correlation ratio between the average maximum line speed V_{0max} and the train commercial (start-to-stop) speed V_s is rather good.

The differences between start-to-stop average speeds and average maximum speeds result from time losses due to train acceleration and braking at the beginning and the end of each run but also in all locations, where maximum speed changes. It should be remembered that timetable recovery margins are added to guarantee timekeeping.

The highest values of the speed utilisation ratio are observed for the longest sections passed without intermediate stops. On such sections the influence of acceleration and braking is relatively minor. I_s values at the level of 0.9 have been identified for some runs of EIP trains, operated with ED250 trainsets, for example between Warszawa Wschodnia and Ilawa Glowna stations (distance 204.7 km).

The utilisation of maximum speed is negatively influenced by significant differentiation of the speed profile (frequent and large changes of speed along the line). The most effective utilisation of line capabilities is in the case of Electric Motor Units with distributed power and high power-to-weight ratio.

Similarly good results have been achieved by the Ukrainian Railways (UZ) thanks to the implementation of new generation of rolling stock, in particular modern EMUs. Despite numerous changes in maximum speeds on route, these trains are able to reach start-to-stop speeds in the range of 120 km/h without exceeding the maximum speed of 160 km/h.

The results of the research show clearly, that the selection of passenger rolling stock for particular route can have significant impact on its day-to-day operation.

Therefore it is very important to take the infrastructure characteristics (gradients, maximum speed profile) into account at the stage of drafting specifications for the rolling stock to be used on particular route. Similarly it is important to carry out in-depth analysis of rolling stock performance as a part of feasibility study for construction of the new railway infrastructure or modernisation of existing one [11]. This is important field of applicability of the method and the results of the study as the tool to compare large number of various variants of the future infrastructure investment. The method allows to check, how the speed profile for designed alignment will be "consumed" in practice with the rolling stock to be used and with the planned stopping pattern.

References

1. Directive 2016/797 of the European Parliament and of the Council of 11 May 2016 on the interoperability of the rail system within the European Union. Official Journal of the European Union 26.05.2016, L**138**, p. 44-101 (2016)
2. I. Wróbel. *Evaluation of the Integration of Train Timetables for Public Service Trains.* Problemy Kolejnictwa (Railway Reports), Vol. 62, Issue **178**, 73-79 (2018)
3. F. Kühn. *Bus rapid or light rail transit for intermediate cities?* Urban Mobility for All. Proceedings of the 10th International CODATU conference, Lome, Togo, p.357-365 (2002)
4. European Court of Auditors. *Rail freight transport in the EU: still not on the right track.* Special Report no 08 (2016)
5. European Court of Auditors. *A European high-speed rail network: not a reality but an ineffective patchwork.* Special Report no 19 (2018)
6. A. Żurkowski (Editor). *High-Speed Rail in Poland. Advances and Perspectives.* Taylor & Francis Group, London (2018)
7. T. Charkina. *Improved classification of passenger trains in Ukraine.* Proceedings of the Dnipropetrovsk National University of Railway Transport named after Academician V. Lazaryan «The Problems of the Transport Economics», **7** (2014)
8. A. Matusevich. *Launching Fast Passenger Trains, and Improvement of the Current Passenger Trains Classification in Ukraine*, Problemy Kolenictwa (Railway Reports), Vol. 58, Issue **165**, 75-83 (2014) [in Russian]
9. PKP PLK, *Annual Report 2017*, (Warsaw, 2018)
10. A. Massel. *Acceleration of railway traffic in Poland.* Technika Transportu Szynowego **1-2**, 15-22 (2015) [in Polish]
11. A. Massel. *Infrastructure and operation - research on utilisation of the maximum train speed profile.* Proceedings of 7th Transport Research Arena TRA 2018, April 16-19, Vienna (2018).
12. PKP PLK. *Sieciowy Rozklad Jazdy Pociagow* (Network Train Timetable), valid from 9 December 2018, Warszawa (2018)
13. "Mercury". *British Express Train Services in 1935.* The Railway Magazine, **1**, 5-12 (1936)
14. J. Hartill. *Italy joins the premier speed league. World Speed Survey.* Railway Gazette International **7**, 28-31 (2017)

Algorithm of rescue units logistic support planning in the process of regional life safety systems development

Oleksandr Prydatko[1], *Vasyl* Popovych[1,*], *Igor* Malets[1], *Valentyn* Prydatko[2] and *Ivan* Solotvinskyi[1]

[1]Lviv State University of Life Safety, 79007, Lviv, Ukraine
[2]Higher vocational school of Lviv State University of Life Sciences, 79007, Vinnytsya, Ukraine

Abstract. The problem of resource support planning in the process of reforming and developing rescue units of regional life safety systems is highlighted in this research. The absence of methods that allow optimizing the required amount of rescue equipment of the region in accordance with the optimal safety criterion is emphasized. Formulated the task of optimizing the safety criterion with the parameter of the number of rescue equipment and limitation in the form of logistical support capability. According to the results of previous studies, the algorithm and its software were developed for implementation in relation to planning of logistic support of rescue units (in the context of providing rescue equipment), which allows to determine the index of the optimal number of rescue units. The developed algorithm was tested on the example of administrative-territorial units of Lviv region, which showed its effectiveness.

1. Introduction

The processes of change of the administrative-territorial structure in the country in accordance with the Strategy of reforming the state administration give rise to changes in the system of protection of the population from emergency situations. As it is known, the functions of prevention and response to emergencies of various nature are assigned to the State Emergency Service of Ukraine (SES). According to [1] system protecting the population from emergencies ZAZ Mr. aye changes by reforming existing and c t Rennes new pro with iynyh rescue units (central Subordinates) and volunteer rescue units (local Subordinates). We have defined the proper combination of professional rescue units of the Emergency Situations of Ukraine and voluntary fire brigades (teams), which is focused on a certain administrative-territorial unit (united territorial community) as a regional system of life safety.

As reform processes existing and creation of new rescue units require the number of parameters for their optimal functioning. One of the key indicators of the operation is the optimal number of rescue equipment, to be dispersed in the rescue 's formation 's designated area. Of course, there are a number of other indicators, for example, types of rescue equipment, staffing structure of rescue units, personnel and its qualifications, etc. However, in this research we will focus only on determining the optimal index of rescue equipment for the purpose of planning the logistics of rescue units. Experience has shown that during the reform processes of the SES of Ukraine, attention to identifying and analyzing material resource needs in accordance with the needs of the region (in the context of providing rescue equipment) is often neglected. As a rule, the number of rescue equipment is determined in accordance with the requirements of [2] , which boils down to the adoption of a stable structure of rescue formation (the number of rescue equipment) in accordance with a certain category . This approach does not allow for a differentiated approach to determining the required amount of rescue equipment depending on the needs of the studied region and contradicts the coordination functions of logistics. In our opinion, the amount of rescue equipment should be optimal to the indicator of its demand, expressed, for example, as a safety criterion. Adjusting this ratio should be done by optimization. However, before formulating such a task, a brief overview of the scientific achievements in the research field is given.

Scientific concepts on optimization of time and content of operations on the example of milk collection in logistic systems of agricultural production are reflected in work [3]. The authors presented a range of studies under different conditions of transportation operations. The article presents an algorithm for matching the content of work performed in logistic systems of milk collection with production conditions. The authors proved the feasibility of daily planning and coordination of the content of works and production conditions in the logistics and milk collection system.

The issue of cost optimization for finding flexible supplier networks in the process of supply chain management is reflected in [4], where the authors present a model of strategic and operational decision-making in the process of cargo delivery. The problem of prompt decision-making in integrated transport networks is considered in [5]. The importance of horizontal co-operation in the supply chain management process for

* Corresponding author: popovich2007@ukr.net

small and medium-sized freight carriers in the area of unloaded transportation is demonstrated in the article.

Scientific works [4, 5] have significant scientific results and on solving optimization problems in the field of transport logistics. However, in these works, all optimization methods are not focused on the problems of logistical support and the study of resource requirements at the planning stage. Planning issues are considered in [6, 7]. In particular, in scientific work [6], recommendations were formulated and the process of their implementation was described for joint planning of logistic connection of two separate business units. With regard to scientific work [7], the authors studied the variety of actions and features of decision-making in the process of planning activities. The complexity of the process of planning difficult work is highlighted in the paper, and based on the theory of management it is hypothesized that in order to solve more complex problems it is necessary to expand the range and variety of actions. Research findings show that increasing task complexity, with no time constraint, leads to a variety of actions in the planning process. However, this work does not consider the harmonization of parameters that change will affect the efficiency of the planning process.

The review of scientific advances in both transportation logistics and logistics planning can continue for a long time. However, most research is focused on solutions and applications that are not related to safety. It t th further review of scientific papers focus on issues of logistics nd ensure it rescue units.

Scientific papers [8, 9] focus on the study of restructuring of regional fire extinguishing systems. Much attention is paid to finding methods of determining the optimal locations of rescue groups, their staffing Second structures and and equipment. However, these studies did not address the issue of logistical support for rescue equipment, depending on the needs of the region being studied. The number of rescue equipment for newly formed rescue units was adopted by the authors in accordance with the existing standards, depending on the defined criterion of threats of the research region. Another scientific work [10] reveals the features of formation of zones of responsibility rescue units within the administrative unit, but a question of logistics support and planning, the required number of missing equipment is available. Other work by the author team [11] examines innovative mechanisms for the allocation of resources to ensure the effective functioning and development of spheres and civil protection. However, the aforementioned work outlines the mechanisms for allocating resources for the functioning of the socio-economic component, and there is no issue of resource provision for rescue equipment.

It is worth mentioning the scientific work [12], which highlights innovative approaches to the creation of rescue formations to protect forests from fires. The described mechanisms take into account the peculiarities of the functioning of the designed rescue units, but they do not take into account the features of the resource.

So, reviewed scientific papers that focus on the research processes of regional development life safety systems, largely covering the methods and mechanisms for planning logistics support rescue units in the condition 's reform and restructuring. And those works that are relevant to the issue of resource support is not the answer to the question: how to determine the optimal number of rescue equipment and to distribute it between the rescue units of the researched region at the stage of material resources planning. Part of this issue is encountered in previous scientific papers, in particular [13], which discusses a differentiated approach to determining the structure of the rescue team and its equipment depending on the features of the region. The result of this work is the development of an information system structure to support managerial decision-making. In addition, the system itself does not contain methods, models and mechanisms for determining the optimal number of rescue equipment, which causes the continuation of these studies.

Therefore, according to the results of the analysis of the investigated branch, no method was found that allowed to determine the optimal number of rescue equipment in the process of development of the regional life safety system. In order to determine the volume of resources Software search optimization method, which consists in the settlement value of "safety of the region – economic feasibility is used". As you know, the optimization process is to find the best option and most profitable relationships. It is proposed to determine the optimal amount of rescue equipment basised of standards for its demand. The output is the number of calls to rescue units per unit of time (the need to involve rescue equipment). In order to determine the normative number of rescue equipment in previous studies [14], it was proposed to use queuing methods. In [14] a detailed review of methods of queuing theory was carried out and the possibility of its application was established to determine the index of the optimal number of rescue equipment to support managerial decision making in the development of regional life safety systems.

Considering the stated problems, the existing achievements in the field of research and the results of previous research in the work set a goal: to implement a method of optimizing the amount of rescue equipment in the form of an algorithm for identifying and analyzing the needs of material resources (in the context of providing rescue equipment) at the planning stage, provided achieving the optimum value of the safety criterion. To accomplish this, you must complete the following tasks:

- to formulate the task of optimizing the amount of rescue equipment of the studied region;
- implement a method of optimizing the number of rescue equipment based on queuing theory in the form of an algorithm with subsequent software implementation using Java;
- to research the effectiveness of the developed algorithm on the example of determining the optimal number of rescue equipment for the united territorial communities of Lviv region (Ukraine).

2. Theoretical study

In the process of finding the best value for the safety criterion and the amount of rescue equipment in the region, it is necessary to formulate an optimization problem that includes the criteria and optimizations, parameters and their limitations. As for the criterion, as indicated, it is a safety criterion, presented in the form of a conditional meter, which varies in the range 0…1 and characterizes the probability of refusal to respond immediately to an emergency event due to the absence of free (not activated) rescue equipment. And on the possibility of timely response depends on the criterion of safety of citizens of the region. The parameter y for the optimization problem is the number of rescue units (technicians) in the studied region. Changing this setting can affect the efficiency of the process (increasing the safety criterion). And finally, the restrictions that are related to the economic component and affect the ability to logistically provide rescue units of the studied region. Of course, increasing the number of rescue equipment to the maximum will maximize the safety criterion, but this tendency entails an increase in staffing, an increase in the area of premises and maintenance costs, an increase in the cost of training of personnel. Therefore, in the outlined optimization problem, the main limitation is the ability to provide logistics. Accordingly, the optimization process will be to find the best (most favorable) ratio of safety criterion and the amount of rescue equipment, taking into account the indicator of logistical capacity.

The mathematical model of the optimization problem is described in detail in [14], but for a better understanding of the developed algorithm and its sequence, we will conduct its abstract review (review of the main stages). As stated in the previous article proposed method of optimizing presented two version. The first is the case when the number of calls to rescue units, and therefore the required rescue equipment, is considered as a multi-channel queuing system with an unlimited queue. Arrives simultaneously several challenges involved in rescue equipment to respond to them until the number of calls would not exceed the number of available engineer units in the investigated area. In this case, all other calls form a queue.

In the second embodiment, we consider the case when the analyzed system has only one channel of service. In this case, with the increased intensity of calls (orders to the system), denials of service may occur, which is not acceptable for the rescue sector.

Adapted optimization method involves 14 steps. Let's look at them in more detail.

1. In determining the actual number of service channels N_k. The number of service channels is assumed to be equal to the number of functioning rescue units in all existing units in the studied region.

2. To determine the density of the order flow for execution λ .

3. Determining intensity of orders within one for a guiding channel service (Rescue department) μ. Given the known values of the intensity of calls (execution orders) and the intensity of their execution, it becomes possible to build a graph of the states of operation of the system. Figure 1 shows, for example, a graph of the states of operation of the regional safety system, with those two service channels, ie two rescue units in all units of the researched region.

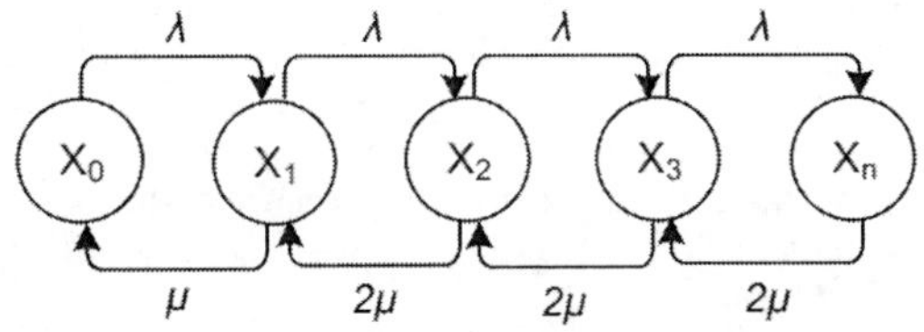

Fig.1. State graph of the functioning of a two-channel life safety system

4. In the formation of consolidated order execution density α.

5. Determining the probability that all channels of the system are free, that is, no rescue department is involved in performing tasks assigned to P_0.

6. Determination of the probability that the number of channels of the system k is involved in order fulfillment, which reflects the probabilistic indicator of the demand for rescue units in the studied region P_k.

7. Verification of the results of the probability calculation. If all the previous calculations are correct, the sum of the probabilities P_0 and all P_k should be as close as possible to 1.

8. To determine the downtime of the system per unit time t_{pr}.

9. In order to determine the likelihood of refusal to accept the order P_{vid}, which describes the fact of simultaneous involvement of all rescue departments of the researched region.

10. Calculation relative system capacity queuing q.

11. Determining actual Q and a nominal Q_H productivity (throughput) system queuing.

12. Determining expectation number of orders, which are both serviced $M(k)$. This point of the proposed method is crucial, since the calculation of the maximum number of simultaneous orders should make the decision to increase or decrease the number of rescue units, and therefore units of rescue equipment.

13. Determination of probability of formation of conditional queue for execution of orders P_{ch}. Of course, the concept of the queue for the case under study is an unacceptable indicator, so the probability of its formation must be determined for the overall characteristics of the system. In the case of a queue for the execution of orders in the real system, rescue departments adjacent to the study areas are involved.

14. In case of positive significance likelihood of queues is necessary to determine the total duration of existence queue for change t_{ch}.

Based on the calculations, a decision is made regarding the need to optimize the safety criterion. The value of the safety criterion is inversely proportional to the calculated value of P_{ch}. Accordingly, the higher the probability of queue formation, and therefore the likelihood of failure to respond immediately to an emergency, the safety criterion decreases. This stimulates the regulation of optimization processes by varying the parameters, and in our case , the service channels to meet all the needs of the system. If the likelihood of queues is close to zero, the criterion of increasing safety and necessity of varying parameters and zoom Annie channel s service is absent. If the parameters are varied (increase or decrease in the number of service channels), it is necessary to determine the probable value of the safety criterion of the upgraded system by performing a recalculation. In the case of optimal results of verification calculations, the number of maintenance channels should be increased by the value that was laid down for their carrying out.

Taking into account the abstract presentation of the optimization method, which is based on the statistical methods of calculating the parameters of queuing systems, the algorithm of logistical planning of an unusual amount of rescue equipment has been developed, which can be used in the development of regional life safety systems. Algorithm is the original version for writing application language Java with a view automated calculations.

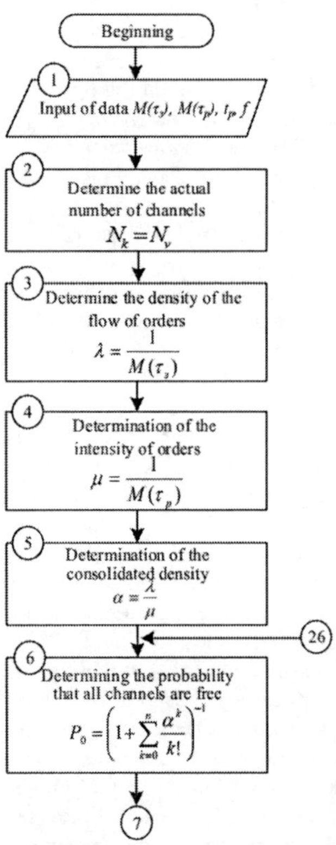

Fig. 2. Algorithm of planning of logistic support of rescue formations (determination of index of optimum number of rescue equipment of multichannel system)

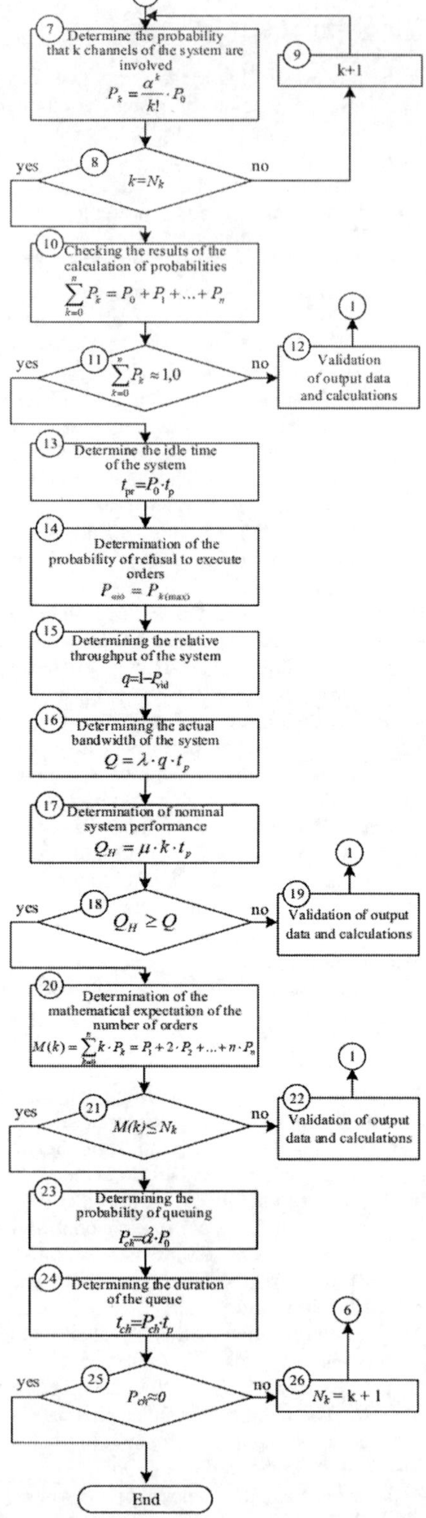

Fig. 3. Continuation of the algorithm of planning of logistic support of rescue formations (determination of the index of optimum number of rescue equipment of a multichannel system)

Next, consider the option of functioning of the regional life safety system as a single-channel queuing system, that is, in the studied region on the balance of rescue units is only one rescue department. The scheme of such a system has the form of a graph of states whose order execution rates do not depend on the vertex number (Fig. 4). In this case, due to the increased intensity of order receipt, refusals of service may occur, and thus the safety criterion will be invalid. In practice, such a situation encourages the involvement of subdivisions of neighboring regions.

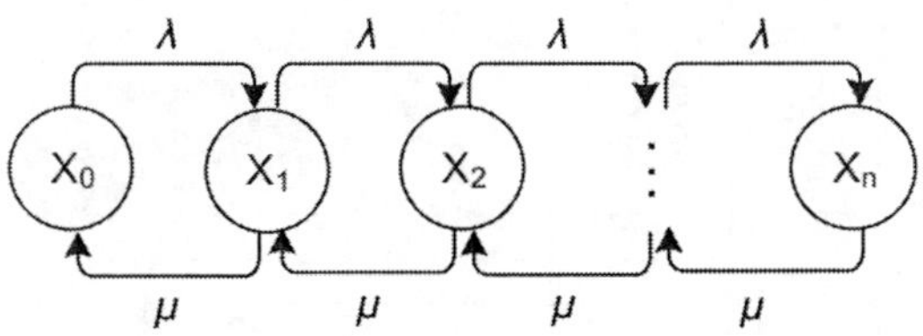

Fig. 4. State graph of a single-channel queued queuing system

In the course of solving the optimization problem for a single-channel system, the following steps of the proposed method should be performed [14]:

1. Similarly to the previous case, the calculation begins with determining the input stream density λ, the order execution rate μ and the summary order execution density α.
2. Determining the probability of the system being in the unloaded state P_0.
3. Determination of the duration spine work (use) system t_v.
4. Determining the idle time of the system without work t_{pr}.
5. Determination of the working time ratio of the system C_t.
6. Determination probably spine formation queue of orders (call) and the total duration of its existence during the shift of P_{ch}, t_{ch}.
7. Determining the probability of refusal to execute orders due to the presence of too large queue P_{vid}.
8. Determining the relative bandwidth the spine system capable q.
9. Determination of real and nominal Q, Q_H spine productive system.
10. Determination lengths and the queue of orders $M(r)$ and central duration rear waiting time $M(t_{ch})$.

Similarly, the previous version, based on the calculations determined values of safety criteria and decide on necessities rear change parameters (increase service channels). In the case of a decision to increase the service channels, the control calculations shoul d be made by the first method. Figures 5 and 6 are presented as algorithm logistical planning to ensure single-channel systems, life safety.

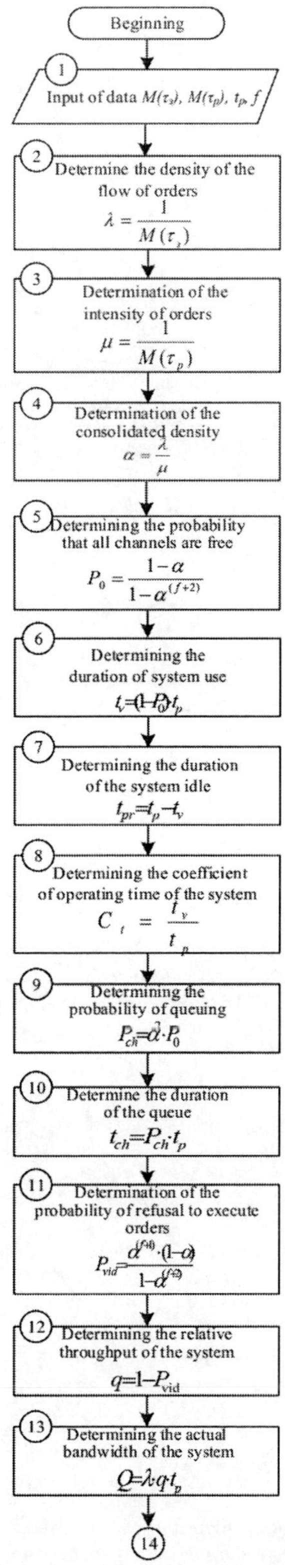

Fig. 5. Algorithm of planning of logistic support of rescue formations (determination of index of optimum number of rescue equipment of single-channel system)

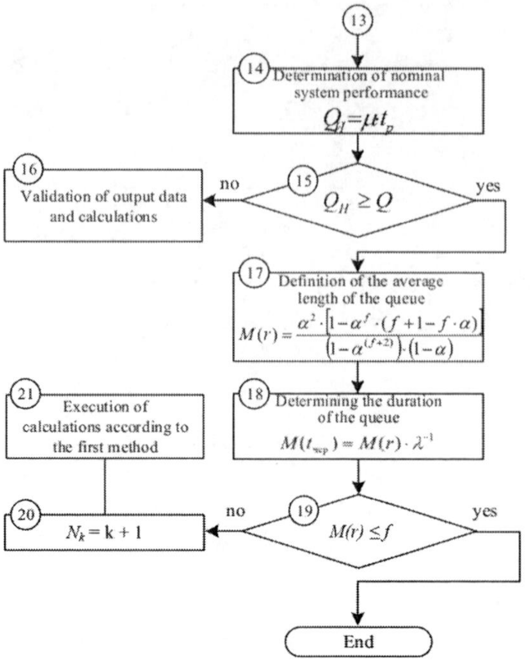

Fig.6. Continuation of the algorithm in planning logistics software rescue units (determination of the optimal number of rescue vehicles single-channel system)

3. Approbation of research results

In order to determine the effectiveness of the adapted optimization method and the algorithm based on it, it was tested by performing the calculations based on empirical studies.

As objects of observation selected three administrative units of Lviv region t and the analysis of the intensity of the involvement of existing rescue units. The list of objects of observation includes the Slavic United Territorial Community (# 1), the union of 13 settlements around the city of Chervonograd and the village of Sosnovka in the Lviv region (# 2) and the Sikhiv administrative district of the city of Lviv (# 3). The main initial parameters for the calculation include: the number of rescue units and equipment available in the research region; density of order flow; intensity of order fulfillment. The results of the empirical studies are summarized in Table 1.

Table 1. Observation results

The object of observation	Number of subdivisions of the studied region	Number of equipment units (rescue units)	Average order density, swap / day	Intensity of execution of orders, substitute / day
# 1	2	2	5	4
# 2	2	3	5,5	6,25
# 3	2	4	6,4	6,6

Carrying out the calculations in accordance with the proposed algorithm, it is established that the probability of queues and orders formation in # 1 is 52 %, # 2 – 32%, # 3 – 35%. According to the results obtained, the safety criterion for # 1 is 0.48, # 2 is 0.68, and # 3 is 0.65. Considering the results of the calculations in accordance with the above algorithm, it is necessary to make a decision on the planning of the resource provision of rescue units (in the context of providing rescue equipment). The results obtained indicate that the safety criterion for # 2 and # 3 is a valid value, unlike # 1. In the latter case, the decision should be made to increase the number of rescue equipment (service channels) and to recalculate with the changed parameter.

For clarity, we present a graphical dependence of the change of the safety criterion of the investigated regions on the parameter of the amount of rescue equipment concentrated for servicing these regions (Fig. 7).

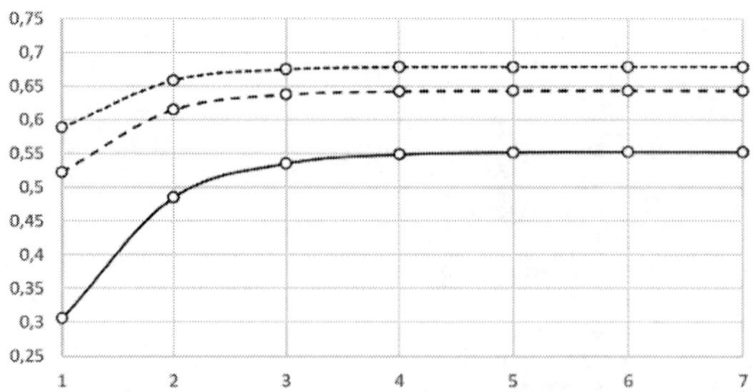

Fig.7. Graphical dependence of the safety criterion (Y-axis) on the number of rescue equipment of the investigated region (X-axis)

Approbation algorithm demonstrated by the existing units, but its main purpose is to implement the planning material s resource s in the creation, restructuring, reform and development of the rescue forces at the regional level. The calculations are inconclusive for making recommendations for restructuring the regional emergency services, it is not taking into consideratiòn Fire first and manmade first load and region. The economic and social component of the region also remains unaddressed. The presented results only demonstrate the effectiveness of the proposed method of optimizing the safety criterion and the number of rescue equipment of the investigated region. The developed algorithm is proposed for systematic use as one of the methods of decision support for the development of regional life safety systems.

Conclusions

1. On the basis of a detailed study of the issue of planning processes for logistical support of rescue units at the stage of their reforming, the lack of a differentiated approach to determining the optimal number of rescue equipment depending on the needs of the region, which led to the formulation of the task of optimizing the safety criterion and the number of rescue technology and the condition of rescue technology.
2. With the help of algorithmic representation of mathematical calculations , the method of optimization of the number of rescue equipment was implemented, based on the statistical methods of ting the parameters of queuing systems, to identify and analyze the needs of logistical support of rescue units of regional life safety systems.
3. By adopting the developed algorithm , its effectiveness was established on the example of optimization of the safety criterion and the indicator of the optimal number of rescue equipment for certain territorial communities of Lviv region (Ukraine).

References

1. The Resolution of the Cabinet of Ministers of Ukraine (2017) "*Strategiia reformuvannia systemy derzhavnoi sluzhbyUkrainy z nadzvychainykh sytuatsii*" [Strategy of reforming the system of the State Service of Ukraine for Emergency Situations], available at: www. zakon5. rada. gov. ua/ laws/ show/ 61-2017-%D1%80
2. The Resolution of the Cabinet of Ministers of Ukraine (2013). "*Poriadok funktsionuvannia dobrovilnoi pozhezhnoiokhorony*" [The order of functioning of voluntary fire protection], available at: www.zakon2.rada.gov.ua/laws/show/564-2013-%D0%BF
3. Tryhuba, A., Pavlikha, N., Rudynets, M., Tryhuba, I., Grabovets, V., Skalyga, M., Tsymbaliuk, I., Khomiuk, N., Fedorchuk-Moroz, Valentyna. *Studying the influence of production conditions on the content of operations in logistic systems of milk collection*. Eastern-European Journal of Enterprise Technologies, Vol. **3**, no. 3 (99), pp. 50-63 (2019)
4. Liotta, G., Stecca, G., Kaihara, T. *Optimisation of freight flows and sourcing in sustainable production and transportation networks*. International Journal of Production Economics, **164**, 351-365 (2015)
5. Buijs, P., ("Hans") Wortmann, J. C. *Joint operational decision-making in collaborative transportation networks: the role of IT. Supply Chain Management*: An International Journal, **19 (2)**, 200-210 (2014)
6. Buijs, P., Alvarez, J. A. L., Veenstra, M., Roodbergen, K. J. *Improved Collaborative Transport Planning at Dutch Logistics Service Provider Fritom*. Interfaces, **46 (2)**, 119-132 (2016)
7. Fransoo, J. C., Wiers, V. C. S. *Action variety of planners: Cognitive load and requisite variety*. Journal of Operations Management, **24 (6)**, 813-821 (2006)
8. Tryhuba, A., Ratushny, R., Bashynsky, O. & Shcherbachenko, O. *Identification of firefighting system configuration of rural settlements*. MATEC Web of Conferences, Vol.**247** (2018) doi: 10.1051/ matecconf / 201824700035
9. Shcherbachenko, O. *Establishing scenarios of fire extinguishing systems development in united territorial communities*. Bulletin of Lviv State University of Life Safety, Issue 17, pp. 14-23 (2018)
10. Sydorchuk, O., Ratushnyi, R., Bondarenko, V., Ratushnyi, A., & Tryhuba, A. *The formation of territorial coverage areas of firefighting units in administrative district*. Bulletin of Lviv State University of Life Safety, **9**, pp. 110-116 (2018)
11. Kvashuk, V., Rak, Yu., Bondarenko, V. *Mechanism of management of resources of resources at projects of development of folding social and economic systems*. Management of Development of Complex Systems, **15**, pp. 25-29 (2013)
12. Zaver, V. B. *Method of management by project configuration of the of fire protection system improvement at the mountain forest district*. Eastern-European Journal of Enterprise Technologies, No.**1/11(55)**. pp. 16-20 (2012)
13. Prydatko, O., Smotr, O., Borzov, Yu., Solotvinskyi, I., Didyk O. *Informational System of Project Management in the Areas of Regional Security Systems' Development*. 2018 IEEE Second Conference on Data Stream Mining & Processing, Issue 2, pp. 187-192. (2018)
14. Prydatko, O., Smotr, O., Martyn, Ye., Prydatko, V., Solotvinskyi, I. *Optimization of methods of the theory of mass service for the decision of applied tasks of development of regional life safety systems*. Information Processing Systems, **2 (157)**, pp. 146-153. (2019).

Capital investments in the context of sustainable development of railways

Olesia Kharchenko[1], *Andrii* Okorokov[1,*], *Roman* Vernigora[2], *Pavlo* Tsuprov[2], *Oleksandr* Papakhov[1], and *Nataliia* Logvinova[1]

[1]DNURT, Management of Operational work department, 49010, st. Lazariana 2, Dnipro, Ukraine
[2]DNURT, Transport hubs department, 49010, st. Lazariana 2, Dnipro, Ukraine

Abstract. Ukraine, due to its geographical position and developed transport infrastructure, has significant potential as a transit country, primarily in the supply chain of trade between Asia and Europe. According to the estimates of the British Institute for Transport Rendell Ukraine, the transit coefficient of Ukraine is 3,75 (with a maximum of 5); This is the best indicator among the countries of Europe (for comparison, in Poland, which occupies the second place, this indicator is 2,92).
Nowadays, Ukraine uses its transit potential extremely insufficiently and inefficiently; over the past 10 years, the volume of transit traffic through the territory of Ukraine has declined by more than 2,5 times, primarily due to the fall in transit traffic by rail. One of the reasons for this situation is, of course, the general political and economic situation in the country. However, there are a number of other negative factors that prevent Ukraine from fully and effectively using its transit potential. These problems lie both in the purely technical plane (deterioration of transport infrastructure and rolling stock, lack of capacity of the main transport routes, underdevelopment of logistics terminals in the country and at land borders, difference in width with the European railway system) and the legislato-thick (high level of port dues, bureaucratization and regulation of customs procedures, lack of flexible tariff policy for carriers, etc.).

1 Introduction

Analyzing the problems of transport complex of Ukraine one can certainly denote that at present there is a practical absence of methods for optimal distribution of capital investments directed at the development of fixed assets of transport system, in particular the railway transport. Significant wear-out of rolling stock and infrastructure objects, the absence of new generation means of transport, the low speed of movement as well as insufficient level of service lead to the reduction of efficiency of the country's transport complex functioning and appear a serious obstacle to its integration into the European and world transport system.

The solution of problems of the country's transport complex lies in the complex and systemic approach that should contribute to [1]:

- creation of favorable conditions for the attraction of additional freight traffic for the transport system of the country;
- increase of competitiveness of railway transport both on internal and external markets of transport services;
- increase of the level of economic and social welfare of the industry staff;
- support of environmental conservation.

The Ministry of Infrastructure of Ukraine together with the representatives of the EU have developed a document - "The transport strategy of Ukraine for the period up to 2020", in which the main directions of the development of transport sector of Ukraine are defined as follows [2]:

- upgrading of the transport system and the increase of its functioning efficiency;
- meeting the needs of national economy and population in transportation, improvement of quality and affordability of transport services;
- provision of timely cargo delivery;
- improvement of the system of administration of transport industry;
- increasing the capacity of transport network;
- increasing the level of safety on transport;
- acceleration of rate of integration of national transport system into the European and world transport system.

The fulfillment of key issues of this strategy is impossible without the construction of modern and efficient system of management of the country's transport system. At present the most promising paradigm of management of complex systems, which railway transport can be surely referred to, is the concept of sustainable development [3].

The concept of sustainable development is aimed at replacing the concept of economic growth. In the second half of the 20th century productivity and profit were seen as an indicator of successful economic performance of the country. Thus the state of environment and social

* Corresponding author: andrew.okorokoff@gmail.com

problems were relegated. Recently more and more experts have expressed doubts as to whether the principles of economic growth are fundamental to maintaining the pace of development of mankind. Therefore nowadays leading countries transition to the concept of sustainable development, according to which a number of economic, social, cultural, ecological and other factors are comprehensively taken into consideration.

Ukraine, due to its geographical position and developed transport infrastructure, has a significant potential as a transit country, primarily in the logical chain of trade exchange between Asia and Europe. It is estimated by the British Randall Institute for Transport Issues that the transit ratio of Ukraine is 3.75 (while 5 is maximal); this is the best figure among European countries (for comparison, Poland, ranking second, has the ratio of 2.92) [4]. In addition, the basis of transport system of Ukraine is formed by the railway transport which provides for more than 60% of total freight traffic and more than 35% of passenger traffic. Moreover, the railway industry is one the biggest taxpayers in the country and more than 240000 job opportunities provided by the industry determine its high social importance [5]. Major advantages of railway transport in the current circumstances are its cost-effectiveness and environmental friendliness. This means of transport is characterized by an extensive use of electricity for mass transportation which presents a critical factor for competitiveness of railway transport in the context of reducing its negative impact on the environment [6].

The territory of Ukraine is crossed by three international pan-European railway transport corridors (ITC) number 3, 5 and 9, international corridors number 5 and 8 of the Organization for Cooperation between Railways, and the international transport corridor TRACECA. The overall length of the network of international railway transport corridors going through the territory of Ukraine is 3162 km. ITC network mainly consists of double-track electrified sections equipped with an automatic blocking system and having high throughput and carrying capacity [7].

However, for the full utilization of ITC and its integration into common European transport network, the systemic upgrading of the respective infrastructure needs to be carried out. This involves the task of the most rational distribution of investment resources that can be allocated to modernize ITC in order to obtain the maximum effect. Accordingly, the research in the area of efficiency increase of the railway functioning from the point of view of sustainable development is nowadays a relevant scientific and practical undertaking.

2 Bibliography analysis

The concept of sustainable development consists in the achievement of meeting the vital needs of the present generation of people without depriving future generations of such possibility because of exhaustion of natural resources and environment degradation [3, 8].

In the context of this definition and main directions of transport sector development it can be said that sustainable development of railways must incorporate such rational distribution of investments that would eventually provide for the balanced solution of the issues of social and economic development, the reduction of negative impact on the environment and would also address the existing and future needs of different branches of the economy for transport services.

Analyzing theoretical advances in the field of efficiency increase of functioning of railways and their units one can note that most works [9-12] use complex economic indicators as the criterion of efficiency. However, there are also methods and approaches considering the processes of efficiency increase of railway unit functioning on the basis of technological indicators [13, 14] or quality indicators [15]. A number of works use for this purpose indicators that take into account the ecological dimension of technological processes as well [6, 16, 17]. At the same time the analysis of works mentioned above shows that in fact they lack mathematical models and methods for qualitative and quantitative assessing of activities addressing complex and systemic approach on the basis of the concept of sustainable development.

At the same time, the strategy of development of one of the biggest transport companies of Great Britain NetworkRail is oriented towards increasing the safety and reliability as well as improving the price/quality ratio for the clients; and the principles of sustainable development lie at the heart of this strategy [18].

In [19] it is pointed out that the construction and functioning of complex systems are based on the realization of the principle of systemic approach which is demonstrated, first of all, by integration and an unambiguous definition of parameters of interaction of all elements of the system. Consequently, considering the tasks of efficiency increase of railway functioning, it is necessary to take into account not only the many factors determining the main parameters of technological processes and the characteristics of their arrangement, but also factors connected with ecological and social aspects of transport organization, and the quality of provided services. Therefore addressing such challenges requires a complex approach stipulated in the concept of sustainable development.

3 Purpose and tasks of the study

The purpose of the study is to obtain the dependency of functions of elasticity of capital investments along the main directions of sustainable development on the parameters of demand for the railway transport services. In doing so, the most effective instrument to use for obtaining the dependencies is simulation. Analyzing the obtained dependencies one gets the solution of the problem of the most rational distribution of investment resources that can be allocated for the upgrading of the railway network or its separate section, e.g. an international transport corridor.

To reach the objective the following tasks have been solved by the authors:

- to make the mathematical model of the task: to form the target function and limitations;
- to draw up a plan of experimental studies that makes it possible to estimate the nature of dependencies of indicators of elasticity of capital investments along the directions of sustainable development on the parameters of demand for the transport services;
- to create an economic and mathematical model of forward-looking development of backbone transport networks;
- to analyze the results of the experiment which form the basis for determining functional dependencies of indicators of elasticity of capital investments along the directions of sustainable development on the parameters of demand for the transport services.

4 Research methodology

4.1 Formalization of the task

According to recommendations [3, 8] the main directions of sustainable development of a system, in particular one of transport, are improvement of technical and technological parameters, preservation of environment, social programmes and guarantees, ensuring the quality of provided services. It is suggested to estimate the sustainable development of railways on the basis of integral criterion of efficiency offered in the work [20]:

$$E_ц = E_{тех} + E_{эк} + E_{соц} + E_к \rightarrow max \quad (1)$$

where $E_{тех}$ - indicator characterizing the operating efficiency of railway units towards improving the technical and technological parameters (e.g. from the point of view of resource conservation);

$E_{эк}$ – ecological component reflecting change of emission of harmful substances by railway units during realization of a set of actions expressed in monetary equivalent;

$E_{соц}$ – social component of the process of railway functioning that can be assessed by the amount of contributions to social programmes, salary increase, staff education, etc.;

$E_к$ – indicator of quality of customer service that can be assessed by the change of wagon turnover value expressed in monetary equivalent.

If the sum of investments in the execution of actions connected with the areas of sustainable development Cyp, is unchanged during a definite period t, then the target function of task (1) of ensuring the most effective option of sustainable development of a railway or its separate section can be presented as

$$F_ц = \frac{E_ц}{C_{ур}} = \delta_{тех} \cdot \varepsilon_{тех} + \delta_{эк} \cdot \varepsilon_{эк} + \delta_{соц} \cdot \varepsilon_{соц} + \delta_к \cdot \varepsilon_к \rightarrow \max \quad (2)$$

where $\delta_{тех}$, $\delta_{эк}$, $\delta_{соц}$, $\delta_к$ - the share of money in the total amount along directions of development of resource-saving technologies, reduction of harmful impact on the environment, ensuring the social component of functioning and quality increase of client service respectively;

$\varepsilon_{тех}$, $\varepsilon_{эк}$, $\varepsilon_{соц}$, $\varepsilon_к$ - functions of elasticity of capital investments along directions of development of resource-saving technologies, reduction of harmful impact on the environment, ensuring the social component of functioning and quality increase of client service respectively.

Optimization of target function (2) with regard to variables $\delta_{тех}$, $\delta_{эк}$, $\delta_{соц}$, $\delta_к$ must be implemented with the following limitations

$$\begin{cases} \delta_{тех} + \delta_{эк} + \delta_{соц} + \delta_к = 1 \text{ or } \sum \delta_i = 1 \\ \delta_i \le \delta_i^{\min} \end{cases} \quad (3)$$

where $\delta_i^{\min}$ - set lower limit of the share of general investments made to ensure the i-th direction of sustainable development.

The function of elasticity of expenditure of resources ε_i, on the i-th direction of sustainable development is defined as the ratio of effect E_i, obtained from the realization of actions, to the value of resources, used in doing so, P_i:

$$\varepsilon_i = \frac{E_i}{P_i} \quad (4)$$

It should be noted that for the social component, the sum of allocated funds serves directly as the effect, therefore $\varepsilon_{соц}=1$.

At the same time, in order to determine the numerical value of efficiency indicator, a prior explanation of numerical values of a number of relevant technical and operational, and technical and economic parameters is required. Moreover, these indicators must characterize an option of functioning of a railway ground (in this case of an international transport corridor) which is rational from the perspective of sustainable development. A rational option of functioning for a ground is determined on the basis of known features of demand for transport services, taking into account the features of the current transport network. [21].

4.2 Formulation of an economic and mathematical model of a railway ground

To determine the numerical values of separate indicators of a railway ground (e.g. ITC), including the characteristics of demand for transport services (the intensity of transportation process) for further calculation of efficiency indicators E_i, is a difficult task on account of a non-determined character of the railway transportation process. In this regard the authors have designed an economic and mathematical model of a railway ground (ITC) based on the methods of graph theory and non-linear optimization.

A transport corridor is presented as a directed graph $G(V, L)$ [22], where V is a multitude of vertices, each of which corresponds to a certain junction (a station, a border crossing, a point of cargo flow emergence or release) of the corridor; L is the multitude of edges each of which corresponds to a railway section between the junctions of the corridor.

Main parameters of the model are: X – one-year volume of transportation along the transport corridor; d – the level of technical facilities of the corridor; i – a random element of the transport corridor; p – the sort of cargo transported along the corridor; K_i - the cost of upgrading of a transport corridor element; E – annual operational expenses on cargo transportation and maintenance servicing of transport devices in corridors; t – a period (stage) of capital investments (years, months) within the specified reference period T. Then X_{tid} - vector characterizing the amount of work at the stage t of network element i with it being equipped to the technical level d; K_{tid} - the cost of reconstruction (construction) at the stage t of element i to the level d; E_{tid} - operational expenses at the stage t on element i equipped to the level d. Thus every combination *{tid}* characterizes a certain option of development of the transport corridor: at the stage t the technical condition of element i is being changed to the level d, which corresponds to a certain set (vector) of technical limitations b_{tid} (e.g. throughput or carrying capacity).

It is accepted that for every point of departure and arrival of goods (passengers) $v_g \in V$ the amount of arrival (departure) of every specific cargo p_k at every stage of upgrading t is determined by the function $Q(vg, pk, t)$.

It is necessary that among the multitude M of possible options of upgrading *{tid}* such subset M^* is identified, that with specified standard ratio of capital investment efficiency F during the reference period T the following condition is fulfilled:

$$\sum_{M^*} \frac{K_{tid} + E_{tid}}{(1+F)^t} \rightarrow \min \qquad (5)$$

In (1) capital K_{tid} and operational E_{tid} expenses are added together for all elements *{tid}* of subset M^*; in addition to this the values of elements X_{tid} and *{tid}*, minimizing the target function (5), must satisfy the following conditions (limitations):

1. The specified amounts of arrival and departure of all kinds of cargo p_k for all points v_g of the transport network must be carried out at every stage of upgrading t:

$$\sum_{l} X_{tpvl} = Q(v_g, p_k, t) \qquad (6)$$

Here the summation in (6) is carried out for all edges l along which cargo p_k can be transported and which connect the junction v_g under consideration with adjacent junctions; in this case X_{tpvl} is accordingly the amount of cargo p_k, transported at the stage t from the junction v_g along the edge l.

2. Amounts of transportation work on all elements i, upgraded at the stage t to the level of technical equipment d, must meet corresponding technical limitations b_{tid}:

$$f(X_{tid}) \le b_{tid} \qquad (7)$$

If only maximum amount of cargo flow for element i of transport corridor (throughput or carrying capacity) is meant by limitations b_{tid}, then limitation (7) takes the form of:

$$\sum_{p} X_{tpj} \le b_{tid} \qquad (8)$$

Additional limiting conditions can be introduced into the model, e.g. concerning the amount of capital investments at every stage of investment project, the terms of transportation, the priority of upgrading of transport corridor elements, etc.

The solution of task (5) generally requires finding extreme value of non-convex functionality in the space of rather large dimensionality having many linear and non-linear limitations, which needs a considerable amount of computational work to be accomplished. One of the possible ways of solving this task is using methods of vector optimization, as a result of which a subset of Pareto-efficient solutions is obtained. It is appropriate to verify the obtained solutions using methods of simulation which is an effective instrument of evaluation of various options of investment level distribution in the process of upgrading the infrastructure in transport corridors [21, 23, 24].

4.3 Development of a plan of experiment

The developed model of ground functioning was used as a framework for carrying out experimental research in the field of efficiency increase of its operation due to achieving sustainable development. The developed programme model highlights the intensities of freight train traffic over a transport network and the number of cars in trains as the main numerical parameters allowing the description of a probabilistic demand for railway services. In the course of experimental research for the purpose of defining functional dependency of elasticity indicators of resource expenditure along the directions of sustainable development on the characteristics of demand for services, a mathematical expectation of train traffic intensity μ_ζ and a mathematical expectation of the number of cars μ_ψ in one freight train are considered. Therefore, the purpose of simulation exercise is to determine the following functional dependencies:

$$\begin{cases} \varepsilon_{\text{тех}} = f(\mu_\zeta, \mu_\psi), \\ \varepsilon_{\text{эк}} = f(\mu_\zeta, \mu_\psi), \\ \varepsilon_{\text{к}} = f(\mu_\zeta, \mu_\psi). \end{cases} \qquad (9)$$

To carry out a full factor experiment a plan of experimental research has been drawn up that makes it possible to estimate the non-linear nature of dependencies (9). The number of test series N_s in the full factor plan of the experiment is defined as the number of all possible combinations of levels of variation of incoming factors [25]:

$$N_s = \prod_{i=1}^{N_f} l_i \qquad (10)$$

where N_f - the number of incoming factors;
l_i - the number of levels of variation of the i-th factor.
According to (10), the number of test series in the experiment by the definition of the combination of functional dependencies (9) amounts to 12.

The numerical results of modelling are a set of four variables – functions of resource expenditure elasticity $\varepsilon_{\text{тех}}$, $\varepsilon_{\text{эк}}$, $\varepsilon_{\text{к}}$ and a feedback function $F_{\text{ц}}$ (2).

To determine the sufficient number of tests in series using the developed model for one of ITCs of Ukraine a trial experiment with 100 tests in every series has been conducted. The analysis of trial experiment outcome reveals that to achieve correct results of modelling in experiment series it is sufficient to carry out 52 tests.

4.4 Analysis of simulation experiment results

The analysis of samples obtained as a result of simulation experiments revealed that in all series random values of functions of capital investment elasticity in directions of reduction of harmful impact on the environment $\varepsilon_{\text{эк}}$ and quality increase of client service $\varepsilon_{\text{к}}$ have normal distribution; on the other hand, the function of elasticity of capital investments in development of resource-saving technologies $\varepsilon_{\text{тех}}$ has gamma-distribution in all test series.

Figures 1-3 illustrate distribution histograms of random values of elasticity functions $\varepsilon_{\text{тех}}$, $\varepsilon_{\text{эк}}$ and $\varepsilon_{\text{к}}$ in the first test series.

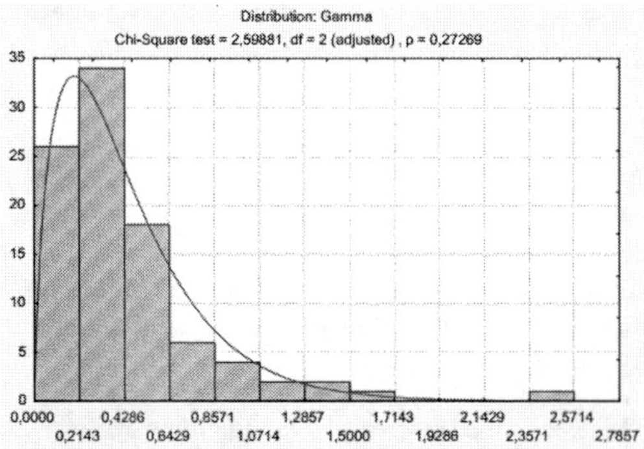

Fig. 1 Distribution histogram of elasticity indicator of capital investments in development of resource-saving technologies.

The value of feedback function $F_{\text{ц}}$ is determined on the basis of elasticity functions according to (2). In this case the value of resource shares δ_i in directions of sustainable development which are defined by the results of solving the task of linear programming (2) (e.g. by simplex method) act as normalizing values.

The correctness of experiment results, i.e. their reproducibility was confirmed using Cochran's test.

To process the results of experimental research the methodology of regression analysis [26] of samples obtained in consequence of the experiment is used. So about 40 possible regression models characterizing the kind of elasticity functions (9) were examined. Among the obtained regression models for the function of elasticity of capital investments in development of resource-saving technologies $\varepsilon_{\text{тех}}$, the power-law model is characterized by the highest adequacy (the lowest deviation of experimental values from values obtained with the help of the model). At the same time the models of present sum are the most adequate for functions of capital investment elasticity in directions of reduction of harmful impact on the environment $\varepsilon_{\text{эк}}$ and quality increase of client service $\varepsilon_{\text{к}}$.

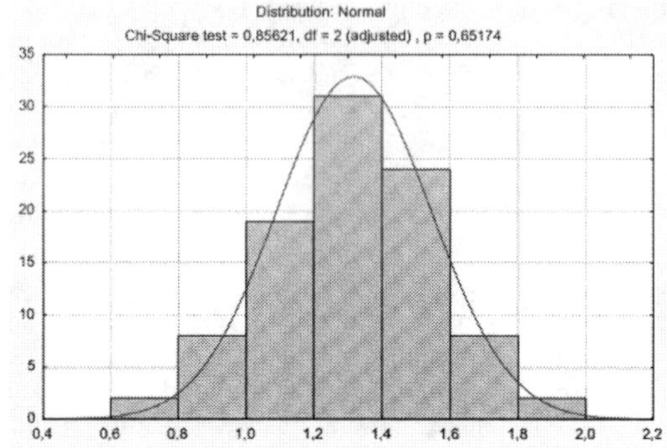

Fig. 2 Distribution histogram of elasticity indicator of capital investments directed at reduction of harmful impact on the environment.

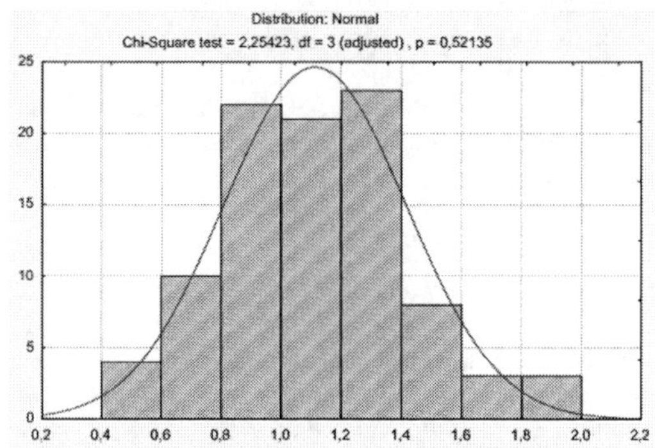

Fig. 3 Distribution histogram of elasticity indicator of capital investments directed at quality increase of client service.

Thus the functions of elasticity of capital investments along the directions of sustainable development of a railway ground depending on the parameters of demand for the transport services (9) is defined as the following set:

$$\begin{cases} \varepsilon_{\text{рес}} = \mu_{\zeta}^{0,477} \cdot \mu_{\psi}^{0,043}, \\ \varepsilon_{\text{эк}} = 0,370 \cdot \mu_{\zeta} + 0,206 \cdot \ln \mu_{\psi}, \\ \varepsilon_{\text{соц}} = 1, \\ \varepsilon_{\text{к}} = 0,054 \cdot \mu_{\zeta} + 0,268 \cdot \ln \mu_{\psi}. \end{cases} \qquad (11)$$

4.5 Research results

The obtained dependencies (11) in practice make it possible to take decisions concerning the choice of the most promising directions of resource expenditure (investments) on the basis of data about the parameters of demand for transportation (the intensity of wagon flow – parameters μ_ζ and μ_ψ) on a railway ground (e.g. within ITC). It is revealed that the highest value of resource expenditure share δ_i corresponds to the direction for which the value of elasticity function ε_i is maximal in relation to the values of elasticity functions for other directions. The values of elasticity functions ε_i lie in intersecting intervals. Therefore it is obvious that for the set of sustainable development directions of the highest priority there exists variability which is determined by the parameters of demand for a transport service within a railway ground (ITC).

As an example, figure 4 illustrates graphs of elasticity function values (11) depending on the train traffic intensity μ_ζ (with the fixed train composition μ_ψ =40 cars).

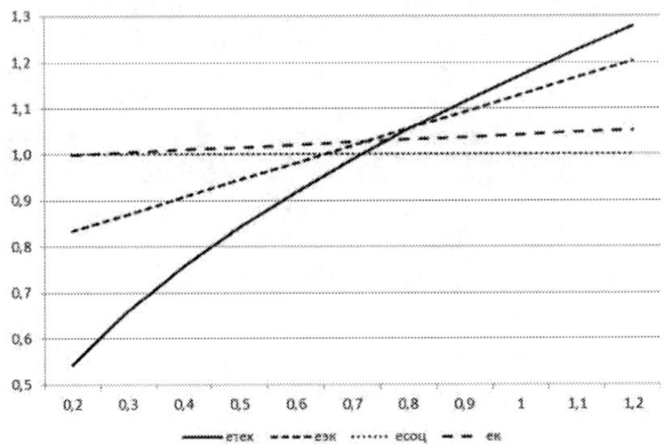

Fig. 4 Graphs of elasticity function values (11) depending on the train traffic intensity μ_ζ (with the fixed train composition μ_ψ =40 cars).

Analyzing the dependencies illustrated by figure 4 it can be identified that in the reviewed range of parameter μ_ζ for values $\mu_\zeta < 0,21$ the maximum value belongs to function $\varepsilon_{соц}$, in range $0,21 < \mu_\zeta < 0,72$ - function $\varepsilon_к$, in range $0,72 < \mu_\zeta < 0,81$ - function $\varepsilon_{эк}$ and function $\varepsilon_{тех}$ has the maximum value if $\mu_\zeta > 0,81$. Consequently, for the indicated example table 1 presents the most prioritized directions of sustainable development of a railway ground (international transport corridor) depending on the intensity of transportation process.

In table 1 the following reference codes are used: *T*- the improvement of technical and technological parameters, *E* – the reduction of harmful impact on the environment, *S* – the development of social programmes, *Q* – the increase of service quality.

Table 1. The most prioritized directions of resource expenditure for ensuring sustainable development for a railway ground (ITC)

μ_ζ	0,2	0,3	0,4	0,5	0,6	0,7	0,8	0,9	1,0	1,1	1,2
Direction	*S*	*Q*	*Q*	*Q*	*Q*	*Q*	*E*	*T*	*T*	*T*	*T*

5 Scientific novelty and practical relevance

The authors have offered a methodology of planning and estimation of actions, aimed at improvement of technical and technological parameters of railway transport system, first of all in international transport corridors, which is based on the concept of sustainable development of complex systems. The task of effective resource distribution in directions of sustainable development is formulated as an optimization task of non-linear programming, where variables represent functions of elasticity of resource expenditure on separate directions. To solve the task an economic and mathematical model of railway ground is designed which is used to carry out a series of tests for defining the nature of elasticity functions for one of the international transport corridors of Ukraine. The obtained dependencies make it possible to identify the most prioritized directions of resource distribution depending on the parameters of demand for transport services.

6 Conclusions

Within the context of integration of the railway transport of Ukraine into the European transport system, the issues of bringing up technical facilities and operation technology of Ukrainian railways to the requirements and standards of the EU are becoming particularly relevant. The EU requirements and standards stipulate not only technical and technological aspects of transportation process, but also ensuring ecological safety as well as quality increase of provided services. In connection with the above there arises a task of effective distribution of capital investments directed at improving transportation process on railway grounds, first of all in international transport corridors. The concept of sustainable development of complex systems makes it possible to take into account the multitude of directions of their operation in a comprehensive and systemic way. The authors offer approaches and mathematical models that are based on the principles of sustainable development of systems. These approaches and models enable to obtain efficiency evaluation and prioritize activities aimed at improving transportation process on railway grounds. Thus the obtained dependencies form the basis for the solution of the task of optimization of finance resources distribution in directions of sustainable development of railway grounds.

References

1. I. R. Yukhnovsky, *The transport complex of Ukraine. Rail: Problems and Prospects*, (FADA, 2006).
2. On approving the transport strategy of Ukraine for the period up to 2020. Electronic resource: http://www.strategy-center.ru/page.php?vrub=inf&vparid =1617&vid=2317

3. N. Ya. Kazhuro, *The concept of sustainable development as a new paradigm of social progress*. Sc. and Techn., **15**, 511-520, (2016).
4. A. M. Pasichnyk, O. M. Klen, S. V. Miroshnichenko, *Analysis and evaluation of the efficiency of using the transit potential of the Ukrainian transport system*. Electrom. compat. and safety in rail. transp., **12**, 88-97, (2016).
5. Official website of the State Statistics Service of Ukraine http://www.ukrstat.gov.ua.
6. V. N. Plakhotnik, Ju. V. Onyshchenko, L. A. Yaryshkina, *The environmental impacts of railway transportation in the Ukraine*. Transport. Research, P. D: Transp. and Env., **10 (3)**, 263-268, (2005).
7. G. M. Kirpa, *Integration of Ukrainian rail transport into European transport system: Monograph*, (2003).
8. Kh. N. Gizatullin, V. A. Troitsky, *The concept of sustainable development: a new socio-economic paradigm* . Soc. Scien. and the pres., **5**, 124-130, (1998).
9. Ya. K. Karyeva, E. A. Kamalov, *Organizational forms of interaction of different modes of transport in cargo transfer nodes*. Logistics: Theory and Practice, **1 (2)**, 56–60, (2012).
10. N. A. Tushin, *System integration in transport processes (theoretical foundations, organizational forms, optimization methods)*: Author. diss. doc. tech. sc., 05.22.08 - management of transportation processes, 43 (2012).
11. D. V. Lomotko, V. V. Kuleshov, A. V. Kuleshov, *Improvement of development of elements of transport systems of railway junctions on the basis of a unified system of management of the fleet of freight wagons for mixed transportation*. Coll. of sc. works of UkrDAZT, **122**, 22-26, (2011).
12. O. V. Friedrichson, *Formation of a transport and logistics container delivery system for products of a metallurgical enterprise*: Diss. cand. tech. sc., 05.22.01 - transport and transport-technological systems of the country, its regions and cities, organization of production in transport, UrGUPS, 190, (2012).
13. D. I. Kochneva, *Improving the functioning of the regional container transport and logistics system*: Diss. cand. tech. sc., 05.22.01 - transport and transport-technological systems of the country, its regions and cities, organization of production in transport, UrGUPS, 230, (2012).
14. A. M. Okorokov, *Improvement of management of transport cargo complexes in transport nodes*: Diss. cand. tech. sc., 05.22.01 - Transport systems, DNUZT, 210, (2013).
15. D. A. Timokha, *State regulation of railway transport*: Diss. cand. econom. sc., 25.00.02 - mechanisms of public administration, NADU, 217, (2007).
16. M. M. Chekhovska, *Organizational and economic mechanism of improvement of environmental activity on the railway transport of Ukraine*: Diss. cand. econom. sc., 08.08.01 - Economics of nature management and environmental protection, NAS of Ukraine, 220, (2003).
17. E. L. Kuzina, *Management of the development of the environmental management system in railway transport to ensure the environmental and economic safety of urban areas*: Diss. doct. econ. Sc., 08.00.05 - Economics and Management of the National Economy, Russian State Social University, 353, (2012).
18. Sustainable Development Strategy. Our vision and strategy: A railway fit for the future 2013-2024. Electronic resource: http://www.google.com.ua/url?sa=t&rct=j&q=&esrc=s&frm=1&source=web&cd=1&cad=rja&uact=8&ved=0ahUKEwiWweffu_zLAhUoIJoKHUl7BhQQFggpMAA&url=http%3A%2F%2Fwww.networkrail.co.uk%2Fbrowse%2520documents%2Fstrategicbusinessplan%2Fcp5%2Fsupporting%2520documents%2Ftransforming%2520network%2520rail%2Fsustainable%2520development%2520strategy.pdf&usg=AFQjCNGHC4NHwm4pVpCxmNRxwRdKspLhOA&sig2=4AY4i7dniZW7FDo8or4Pvg&bvm=bv.118817766,d.bGs
19. A. G. Kalchenko *Logistics*. (KNEU, 2004).
20. O. I. Kharchenko, *Determining the optimal distribution of capital investments for the sustainable development of railways*. Sc. and progr. of transp. Bull. of Dnipr. Nat. Un. of Rail. Trans., **2 (56)**, 52-59, (2015).
21. O. I. Kharchenko, *Assessment of parameters characterizing the sustainable development of railway transport units*. Proc. of Dnipr. nat. univ. of rail. transp. «Transp. syst. and techn. of transp.»: View of Dnipr. nat. univ. of rail. transp., **11**, 22-29, (2016).
22. V. I. Bobrovsky, D. M. Kozachenko, R. V. Vernigora, *Basic model of track development in simulation models of railway stations*. Coll. sc. works of UkrDAZT: Series «Improvement of Cargo and Commercial Work on the Railways of Ukraine», **62**, 20-25, (2004).
23. V. Bobrovskiy, D. Kozachenko, R. Vernigora *Functional simulation of railway stations on the basis of finite-state automata*. Transp. Probl., **9 (3)**, 57-66, (2014).
24. V. I. Bobrovsky, D. M. Kozachenko, R. V. Vernigora, V. V. Malashkin, *Functional modeling of railway stations: monograph*, (DNURT, 2015).
25. E. V. Alekseeva *Building mathematical models of integer linear programming. Examples and tasks*, (Novosib. state univ., 2012).
26. N. Draper, G. Smith, *Applied Regression Analysis. Book 1*. (Fin. and Stat., 1986).

Constructing a database structure in the problem of the assignment of air transport operators to commissioned tasks

Ilona Jacyna-Gołda[1], *Mariusz* Izdebski[2,*], *Piotr* Klimek[3], and *Rostislav* Vašek[4]

[1] Warsaw University of Technology, Faculty of Production Engineering, Poland,
[2] Warsaw University of Technology, Faculty of Transport, Poland,
[3] OLTIS Polska Sp. z o. o., Poland,
[4] CID International, Czech Republic, Ostrava

Abstract. The article defines the database structure used in the issue of the assignment of air transport operators to commissioned transport tasks. The purpose of the work is to develop a database structure that fulfills the function of recording information, presenting results and processing calculated data. Presenting the database in a modular manner is a new concept of data flow management in the problem of the assignment of air transport operators to tasks. The issue of the assignment of air service operators to tasks is a complex optimization problem. The complexity of the database architecture used in the assignment of operators to tasks depends on the number of modules responsible for individual functions implemented in the process of determining the assignment of operators to tasks. In addition, the paper presents a mathematical model of the assignment of operators to tasks, includes basic data entered into the model and collected in the database, describes the assignment restrictions and the criterion function. In order to solve the problem, an optimization algorithm concept has been proposed to solve the analyzed problem. The problem of the assignment has been presented in the multi-criteria aspect.

1 Introduction

In general, the problem of assigning performers, e.g. vehicles, employees, to tasks is a known issue in the literature and widely discussed [12], [16], [2], [5], [6], [15]. According to the literature, the classic issue of assignment consists in assigning available resources, e.g. vehicles, employees, to assigned tasks. The assumption of the assignment is that each task, if it is possible, is assigned to exactly one performer and each performer is assigned to only one task. The measures for assessing the correctly generated assignment are usually the minimum time of completing all tasks or the minimum cost of completing the tasks. The issue is often modified by introducing different combinations of the number of tasks assigned to resources, e.g. equal number of tasks and resources, more tasks than resources, less tasks than resources. In the literature, the assignment problem referring to air transport describes only the assignment of airplanes or crews to routes [14], [17], [13], [11], [10], [3]. The assignment problem of air transport operator to transport tasks is not analyzed in the literature therefore constructing the database structure for this problem as well as the problem itself broadens the topics of assignment problems in air transport.

In the issue of the assignment of air transport operators to commissioned tasks, the problem is to assign all tasks commissioned by individual clients to a specific operator operating on a given transport market. The task is defined as transporting cargo by air between two transhipment terminals. Each operator is able to perform a transport task at different times and according to its own transport rate depending on the amount of transported cargo. The main limitation is the time limit for completing tasks. The problem is analyzed in a multi-criteria aspect, where the duration of tasks and the total cost of transport play a decisive role. A graphic interpretation of the assignment issue in the context of air services is presented in Fig. 1.

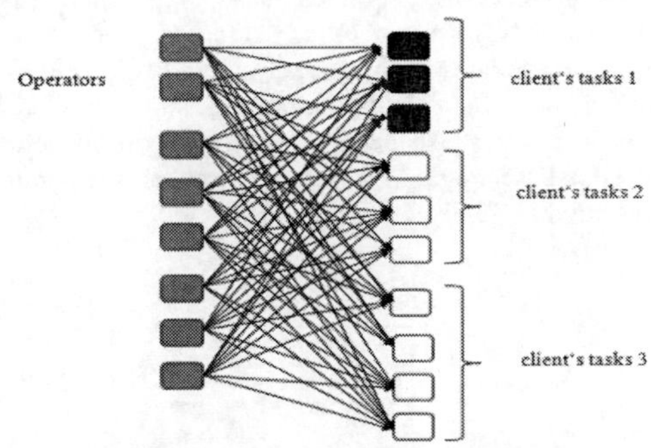

Fig. 1. The assignment of operators to tasks in air services

Connections between the operator and a given task mean that a given operator can perform a given task.

The final effect of the assignment is the selection of operators providing air transport services (service providers) for the service commissioned by the given client - the buyer of the transport service. The complexity of the discussed problem of the assignment of air transport operators to commissioned tasks requires the development of a database capable of processing

* Corresponding author: mizdeb@wt.pw.edu.pl

various types of information both entered by the user of this database and information resulting from complex computational operations. The purpose of this work is to propose a database structure adequate to the discussed issue of the allocation of operators to tasks. Presenting the database structure in a modular manner is a new concept of data flow management in the problem of the assignment of air transport operators to tasks. The complexity of the database architecture used in the assignment of operators to tasks depends on the number of modules responsible for individual functions implemented in the process of determining the assignment of operators to tasks.

2 The database structure in the problem of allocation of air transport operators to tasks

Each database record intended for information management in the problem of the allocation of air transport operators to tasks consists of the following elements: record name, type of input data, and source of input data. The names of the records identify both individual operators of air services existing on a given transport market as well as clients and tasks commissioned for implementation. In addition, the database stores information about the types of aircraft available to the operator, the cost of transporting cargo, the duration of tasks, the size of tasks to be realized. The type of input data entered determines the type of data stored in the record, e.g. text or number type.

The developed database is adapted to the mathematical model, which requires defined input data necessary to determine the allocation of operators to tasks. In addition, the modular structure allows database to combine with an optimization algorithm developed to determine the decision variables of the mathematical model.

The source of input data informs about the method of entering data into the database, e.g. manually or automatically by using other databases. The developed database operates on both data entered by the user and data generated in calculation processes or uploaded from other databases. An example of the record structure in the database regarding the air transport operator is presented in Table 1.

Table. 1. Air service operator database record

Record Name	Air Operator
The type of input data	Textual data
Source of data acquisition	Data entered manually

The database consists of the following modules:

- Module of fixed data – contains database tables about clients, operators, their locations, aircraft types, unit costs of transporting loads at individual operators, customer characteristics, customer preferences, transport tasks, transport costs,
- Module of dynamic data – is designed to handle data that can be added, modified and removed by users during the system's operation, e.g. characteristics of new customers, volume of transport tasks, in addition, the module contains data that is the result of computational operations,
- Archive module – is a part of the database intended for storing analysis reports, results of assigning tasks to operators,
- Module of entry data – is designed to support the process of direct data input by the user, additional data about customers, operators,
- Module of data presentation – mediates the process of sharing data collected in the database for authorized users,
- Optimization module – the module stores reports and data summaries regarding the assignment of operators to tasks, the module is based on a built-in optimization algorithm,
- Calibration module – saves the test settings of the optimization algorithm regarding its input parameters,
- Verification module – checking the correctness of the results generated by the optimization algorithm,
- Data exchange module – is the interface between applications based on different base systems.

The database structure is updated via the dynamic data entry module. The optimization module, calibration parameters of the optimization algorithm and verification of results operate independently of the other database modules. The correctness of database structure is measured by the speed of generated solution, saving data and modification. Graphic presentation of the database structure is shown in Fig. 2.

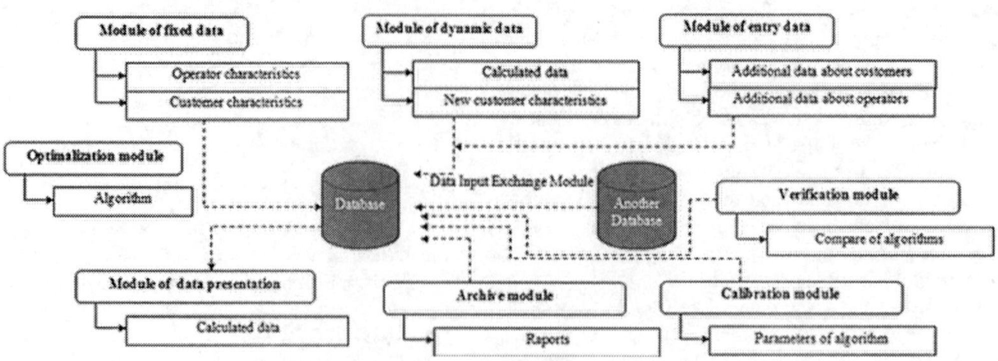

Fig. 2. Modular database structure for the assignment of air transport operators to clients' tasks

3 The mathematical model of assigning operators to tasks

The complexity of the mathematical model is dictated by a multi-criteria approach to determining the assignment of operators to tasks. The presented database includes both input data entered into the mathematical model in the form of sets, matrices and vectors, as well as stores calculated data related to the designated decision variables.

In order to develop a mathematical model, input data has been defined:
OL = {*ol*: *ol*=1, …, *OL*} – set of numbers of air transport service operators (transport service providers),
NU = {*nu*: *nu*=1, …, *NU*} – set of customer numbers,
Z = {*z*: *z* =1,…, *Z*} – set of tasks to be realized,
S = {*s*: *s* =1,…, *S*} – set of aircraft types,
LZ=[*lz*(*nu*): *lz*(*nu*)∈ **N**, *nu* ∈ **NU**] - the number of tasks commissioned by a given client for air transport;
LS=[*ls*(*s,ol*): *ls*(*s,ol*) ∈ **N**, *s*∈ **S**, *ol* ∈ **OL**] - number of aircraft of a given type owned by the operator;
T=[*t*(*ol,z,s*): *t*(*ol,z,s*) ∈ **R**, *ol* ∈ **OL**, *z*∈ **Z**, *s*∈ **S**] – the time of the task realization by the operator for a given type of aircraft
C=[*c*(*ol,z,s*): *c*(*ol,z,s*) ∈ **R**, *ol* ∈ **OL**, *z*∈ **Z**, *s*∈ **S**] – unit cost of the task realization by the operator for a given type of aircraft;
Q=[*q*(*z, nu*): *q*(*z*)∈ **R**, *z*∈ **Z**, *nu*∈ **NU**] – the size of the tasks commissioned by the client [tons];
TZ=[*tz*(*nu*): *tz*(*nu*)∈ **R**, *nu* ∈ **NU**] – the time required to complete all tasks required by the client;
Decision variables take the following form:
- **X**=[*x*(*ol, z, nu*): *x*(*ol, z, nu*)∈ {0,1}, *ol* ∈ **OL**, *z*∈ **Z**, *nu*∈ **NU**] – assigning a given operator to a task commissioned by a given client;
- **Y**=[*y*(*ol, s,z,nu*): *y*(*ol, s*)∈ {0,1}, *ol* ∈ **OL**, *s*∈ **S**, , *z*∈ **Z**, *nu*∈ **NU**] – assigning a given type of aircraft by the operator to a task commissioned by a given client. Restrictions take the form:
- total duration of all tasks ordered by the client:

$$\forall nu \in \mathbf{NU} \quad \sum_{ol\in OL}\sum_{z\in Z}\sum_{s\in S} x(ol,z,nu)\cdot y(ol,s,z,nu)\cdot t(ol,s,z) \le tz(nu) \tag{1}$$

- all client's tasks must be realized:

$$\forall nu \in \mathbf{NU} \quad \sum_{ol\in OL}\sum_{z\in Z}\sum_{s\in S} x(ol,z,nu)\cdot y(ol,s,z,nu) = lz(nu) \tag{2}$$

- the number of aircraft of a given type can not be exceeded:

$$\forall ol \in \mathbf{OL},\ \forall s \in \mathbf{S} \quad \sum_{z\in Z}\sum_{nu\in NU} x(ol,z,nu)\cdot y(ol,s,z,nu) \le ls(s,ol) \tag{3}$$

The criterion functions minimize the time (4) and the cost of implementing the tasks (5):

$$F1(X,Y) = \sum_{ol\in OL}\sum_{z\in Z}\sum_{s\in S}\sum_{nu\in NU} x(ol,z,nu)\cdot y(ol,s,z,nu)\cdot t(ol,s,z) \longrightarrow \min \tag{4}$$

$$F2(X,Y) = \sum_{ol\in OL}\sum_{z\in Z}\sum_{s\in S}\sum_{nu\in NU} x(ol,z,nu)\cdot y(ol,s,z,nu)\cdot c(ol,s,z)\cdot q(z,nu) \longrightarrow \min \tag{5}$$

4 The genetic algorithm for the assignment of the operator to the tasks

The database structure optimization module works based on genetic algorithm. Considering the complexity of the issue of determining the assignment of operators to tasks and the computational speed of generating the result, a genetic algorithm has been chosen that is able to solve the analyzed problem in a sufficiently short time.
Genetic algorithms are algorithms that base their operation on natural selection mechanisms and heredity. The results generated by a genetic algorithm are satisfactory in terms of the quality of the designated solution and the costs of obtaining it. They have found application as a practical optimization tool in complex decision problems determining the location of logistics facilities, the allocation of vehicles to tasks and shaping logistics networks [9], [7], [4], [8]. It should be emphasized that the genetic algorithm belongs to heuristic methods that give solutions close to optimal. Despite these drawbacks, genetic algorithms are successfully used in optimization problems. The genetic algorithm is an iterative algorithm, so the generated solution improves with each transition to the next iteration. Each individual in the population is assessed according to the adaptation function adopted for the problem. The selection process in the genetic algorithm consists in choosing the best individuals (chromosomes) from the initial population for the next generation. The crossing process involves the exchange of genetic material between two individuals. Depending on the structure, the crossing process takes different forms. The classic genetic algorithm is based on structures known as chromosomes. A classic representative of the chromosome are binary strings randomly generated. In the genetic algorithm, other representations of the chromosome structure are also possible, e.g. as a sequence of natural numbers, matrix structure. The mutation consists in randomly repositioning genes, changing their values. The genetic algorithm in the presented problem consists of the following stages:
- Step 1. Determining the structure of the input data.
- Stage 2. Determining the adaptation function.
- Stage 3. Determining the reproduction process.
- Stage 4. Determination of the crossing process.
- Stage 5. Determination of the mutation process.
The initial population was generated in randomly way. The stop condition is a certain number of iterations. The assessment of individuals is carried out on the basis of the adaptation function, which is developed based on the

criteria function (4), (5). Matrix structure has been used to represent the chromosome in the analyzed assignment problem. This structure of input data works well in complex decision problems, e.g. traveling salesman problem [1].

The algorithm of the reproduction process generates the population according to the principle of the roulette method. The reproduction algorithm requires the use of linear scaling to counteract premature convergence of the algorithm in the initial iterations [7]. The matrix structure has the interpretation of decision variables. The initial population consists of a specific number of matrix structures determined at the beginning of the algorithm.

The matrix structure consists of two parts. In the first part, the allocation of operators to tasks is specified, in the second, the type of aircraft is selected. The number of rows is the number of operators providing air transport services, the number of columns is the number of tasks to be carried out. The process of completing Part I concerning the allocation is a random process. However, it should be remembered that in each column there can be only one value "1", which means that the task is carried out by only one operator, and the sum of all "1" in the entire matrix equals the number of tasks to be carried out. After completing the entire matrix, the restriction for the duration of all tasks is checked (1). If the limit is not met, the process of completing the matrix should be repeated.

The number of structures in Part II depends on the number of aircraft types available to each operator. Aircraft types are assigned to tasks randomly. An example matrix structure for seven operators and six tasks and two types of aircraft included in the transport process is shown in Fig. 3.

Part I

	Z1	Z2	Z3	Z4	Z5	Z6
O1	1	0	0	1	0	1
O2	0	1	0	0	0	0
O3	0	0	0	0	0	0
O4	0	0	0	0	1	0
O5	0	0	0	0	0	0
O6	0	0	1	0	0	0
O7	0	0	0	0	0	0

Part II

First type

	Z1	Z2	Z3	Z4	Z5	Z6
O1	1	0	0	1	0	1
O2	0	0	0	0	0	0
O3	0	0	0	0	0	0
O4	0	0	0	0	0	0
O5	0	0	0	0	0	0
O6	0	0	0	0	0	0
O7	0	0	0	0	0	0

Second type

	Z1	Z2	Z3	Z4	Z5	Z6
O1	0	0	0	0	0	0
O2	0	1	0	0	0	0
O3	0	0	0	0	0	0
O4	0	0	0	0	1	0
O5	0	0	0	0	0	0
O6	0	0	1	0	0	0
O7	0	0	0	0	0	0

Fig. 3. Matrix structure of the genetic algorithm

Based on the adaptation function, the genetic algorithm determines the final solution. According to the principle of the genetic algorithm, it seeks the maximum solution. Considering various types of criteria functions in the analyzed assignment, the adaptation functions for the k-th matrix structure $\mathbf{M}(t, k)$ can be represented as follows ($\mathbf{K} = \{1, \ldots, k, \ldots, K\}$ - a set of structures in the population, t – an iteration):

$$F(k,t) = \frac{F1\min}{F1(k,t)} + \frac{F2\min}{F2(k,t)} \longrightarrow \max \qquad (6)$$

The problem of allocation has been shown in the context of a multi-criteria problem, so in order to be able to add values for various criteria functions, the adaptation functions were presented as the sum of quotients, where **F1**min defines the minimum structure value from the entire population in a given iteration for the criteria function determining the time of task completion, while F1(k,t) determines the values of individual structures in a given population according to this function of criteria. The designations **F2**min and F2(k,t) relate to the second function of the criterion (cost of carrying out tasks). The function F(k,t) will reach its maximum value when each function e.g. F1(k,t) reaches the value of **F1**min, or F2(k,t) approaches the value of **F2**min.

The crossing process consists in randomly selecting matrix structures and then randomly combining two structures in pairs. A sub matrix is randomly generated for each pair. Matrix substructures for two matrix structures are exchanged with each other. The mutation involves swapping the values of randomly selected genes. An example of the process of crossing the structure for part I is presented on Fig. 4, mutation and structure repair on Fig. 5. The processes taking place in Part II are consistent with the changes taking place in Part I of the structure.

The principle of operation of the genetic algorithm developed for the issue of the allocation of air transport operators can be presented in the following steps:

- The step 1. First, an initial population consisting of a fixed number of matrix structures is generated. These structures will be processed while the genetic algorithm is running.
- The step 2. An adaptation function is determined for each matrix structure.
- The step 3. Carrying out the crossing and mutation process. Crossover and mutation occur with a certain probability defined as input data. In the case of an incorrect matrix structure generated after the process of crossing and mutation that does not meet the limitations of the mathematical model, the repair process takes place.

The stages of the process of reproduction, crossing, mutation are repeated a certain number of iterations. After completing all iterations, a matrix structure with the minimum value of the adaptation function is selected from the entire population.

Before crossover process

	Z1	Z2	Z3	Z4	Z5	Z6
O1	1	0	0	1	0	1
O2	0	1	0	0	0	0
O3	0	0	0	0	0	0
O4	0	0	0	0	1	0
O5	0	0	0	0	0	0
O6	0	0	1	0	0	0
O7	0	0	0	0	0	0

	Z1	Z2	Z3	Z4	Z5	Z6
O1	0	0	0	0	0	1
O2	0	0	0	1	0	0
O3	0	1	0	0	0	0
O4	1	0	0	0	1	0
O5	0	0	0	0	0	0
O6	0	0	1	0	0	0
O7	0	0	0	0	0	0

After crossover process

	Z1	Z2	Z3	Z4	Z5	Z6
O1	1	0	0	1	0	1
O2	0	0	0	0	0	0
O3	0	1	0	0	0	0
O4	0	0	0	0	1	0
O5	0	0	0	0	0	0
O6	0	0	1	0	0	0
O7	0	0	0	0	0	0

	Z1	Z2	Z3	Z4	Z5	Z6
O1	0	0	0	0	0	1
O2	0	1	0	1	0	0
O3	0	0	0	0	0	0
O4	1	0	0	0	1	0
O5	0	0	0	0	0	0
O6	0	0	1	0	0	0
O7	0	0	0	0	0	0

Fig. 4. Crossing matrix structures

Before mutation

	Z1	Z2	Z3	Z4	Z5	Z6
O1	1	0	0	1	0	1
O2	0	1	0	0	0	0
O3	0	0	0	0	0	0
O4	0	0	0	0	1	0
O5	0	0	0	0	0	0
O6	0	0	1	0	0	0
O7	0	0	0	0	0	0

After mutation

	Z1	Z2	Z3	Z4	Z5	Z6
O1	1	0	0	1	0	1
O2	0	1	0	0	0	0
O3	0	0	0	0	0	0
O4	0	1	0	0	1	0
O5	0	0	0	0	0	0
O6	0	0	1	0	0	0
O7	0	0	0	0	0	0

Repairing process

	Z1	Z2	Z3	Z4	Z5	Z6
O1	1	0	0	1	0	1
O2	0	0	0	0	0	0
O3	0	0	0	0	0	0
O4	0	1	0	0	1	0
O5	0	0	0	0	0	0
O6	0	0	1	0	0	0
O7	0	0	0	0	0	0

Fig. 5. Gene mutation in the structure and repair process

5 Conclusions

The developed database architecture is an original proprietary concept presenting the database as an integrated modular structure taking into account various types of activities occurring during the process of assigning air transport operators to contracted tasks.

The application of the genetic algorithm in the optimization module will minimize the calculation time and accelerate the processes occurring in the database. The computational complexity of the implemented algorithm underlines the important role of developing an appropriate database for implemented computational processes.

In order to obtain the correct results generated by the genetic algorithm, first of all the analysis of the sensitivity of this algorithm should be performed. Sensitivity analysis of the proposed genetic algorithm is carried out in order to determine the optimal parameters for which the algorithm generates an optimal result. The parameters of the genetic algorithm to be determined are the probability of crossing, mutation as well as the size of the population and the number of iterations of the algorithm. The calibration module is the module in which the sensitivity analysis is carried out.

The correctness of the results generated by the genetic algorithm is confirmed in the verification process. The verification process is carried out in the results verification module. The algorithm verification process can be carried out based on a comparison of algorithm solutions with solutions obtained using a random algorithm. The number of comparisons is determined at the beginning of the verification. In the case where the genetic algorithm generated a better result in each comparison with the random algorithm, the verification process is completed. In the case where the random algorithm produces a better result than the genetic algorithm, the sensitivity analysis should be repeated. Further research will consist in implementing the database in a programming environment and checking its effectiveness.

References

1. Bräysy O., Gendreau M.: Vehicle Routing Problem with Time Windows, Part II: Metaheuristics, Transportation Science, Institute for Operations

https://doi.org/10.1051/matecconf/201929404004

Research and the Management Sciences ,Volume 39 Issue 1, pp. 119-139, USA 2005.

2. Burkard R., Dell'Amico M., Marttello S.: Assignment problems Society for Industrial and Applied Mathematics, Philadelphia 2009.
3. Etschmaier M. M., Mathaisel D. F. X.: Airline Scheduling: An Overview, Volume 19, Issue 2, page 127-138, Transportation Science (1985).
4. Harper P. R., Senna V., Vieria I. T., Shahani A. K.: A genetic algorithm for the project assignment problem, Computers Ops Res. 32, pp, 1225-1265, 2005.
5. Jacyna M. Cargo flow distribution on the transportation network of the national logistic system. International Journal of Logistics Systems and Management, 2013 pp. 197-218.
6. Jacyna M. The role of the cargo consolidation center in urban logistics system. International Journal of Sustainable Development and Planning, , 2013, pp.100-113
7. Jacyna M., Izdebski M., Szczepański E., Gołda P.: The task assignment of vehicles for a production company, w: Symmetry-Basel, vol. 11, nr 10, 2018, ss. 1-19, DOI:10.3390/sym10110551
8. Jacyna-Gołda I.: Inżynieria oceny efektywności sieci dostaw. Wydawnictwo Naukowe PWN, ISBN: 978-83-012-0526-3, Warszawa 2019.
9. Jacyna-Gołda Ilona, Izdebski Mariusz, Podviezko Askoldas: Assessment of the efficiency of assignment of vehicles to tasks in supply chains: A case-study of a municipul company, w: Transport, vol. 32, nr 3, 2017, ss. 243-251
10. Klincewicz J. G., Rosenwein M. B.: The Airline Exception Scheduling Problem, Volume 29, Issue 1, pp. 4-16, Transportation Science (1995).
11. Lohatepanont M., Barnhart C.: Airline Schedule Planning: Integrated Models and Algorithms for Schedule Design and Fleet Assignment, Volume 38, Issue 1, pp. 19-32, Transportation Science (2004).
12. Pentico D., W.: Assignment problems: A golden anniversary survey, European Journal of Operational Research, Volume 176, Issue 2, 2007, pp. 774-793.
13. Rexing B., Barnhart C., Kniker T., Jarrah A., Krishnamurthy N.: Airline Fleet Assignment with Time Windows, Volume 34, Issue 1, pp. 1-20, Transportation Science (2000).
14. Schaefer A. J., Johnson E. L., Kleywegt A. J., Nemhauser G. L.: Airline Crew Scheduling Under Uncertainty, Volume 39, Issue 3, pp. 340-348, Transportation Science (2005).
15. Wasiak M., Jacyna M., Lewczuk K., Szczepański E.: The method for evaluation of efficiency of the concept of centrally managed distribution in cities, Transport, vol. 32, is 4, 2017, pp.348-357;
16. Wilson J. M.: A genetic algorithm for the generalized assignment problem, Journal of the Operational Research Society, Vol: 48, pp. 804-809, 1997.
17. Yen J. W., Birge J. R.: A Stochastic Programming Approach to the Airline Crew Scheduling Problem, Volume 40, Issue 1, page 3-14, Transportation Science (2006).

Work completed within the framework of the EUREKA project project implemented on the basis of an agreement with the National Center for Research and Development No. EUREKA / EPLOS / 3/2017. on the European Logistics Services Portal (EPLOS)

Decision-making problems of designing database in the aspect of planning transport of cargoes by intermodal transport

Mariusz Izdebski[1], *Marianna* Jacyna[1], *Piotr* Klimek[3], *Ilona* Jacyna-Gołda[2], and *Rostislav* Vašek[4]

[1] Warsaw University of Technology, Faculty of Transport, Poland,
[2] Warsaw University of Technology, Faculty of Production Engineering, Poland,
[3] OLTIS Polska Sp. z o. o., Poland,
[4] CID International, Czech Republic, Ostrava

Abstract. The article presents issues related to database design for problems of planning transport of cargoes by intermodal transport using road and air transport. The main aim of this paper is to develop the structure of the database used in the problems of planning transport of cargoes by intermodal transport. The issue of intermodal transport is a complex decision problem which requires taking into account many technical, economic and qualitative aspects of various modes of transport. The multitude of these aspects contributes to the processing of a large amount of input data used by advanced calculation algorithms developed to determine the optimal routes of cargo transportation. Designing database architecture depends on the number of modules responsible for the final determination of transport plans. It is assumed that the database is responsible not only for collecting input data about air, road infrastructure, processing of information necessary for transport but also for saving results generated by calculation modules. In the article, to understand the problem of the issue, a mathematical model of transporting loads by intermodal transport was included, and for its solution an ant algorithm was proposed. The developed structure of the database with the presented modules is a newly approach in problems of planning transport of cargoes by intermodal transport.

1 Introduction

The problem of intermodal transport planning is one of the general problem of determining vehicle routes presented in the literature as *vehicle routing problem* [1], [2], [11], [8], [9], [13]. According to the definition, intermodal transport consists in the transport of loads with at least two types of transport means used [9], [10], [12]. Planning transport routes in intermodal transport is a complex optimization issue that requires taking into account many factors of a technical, qualitative and economic nature.

Technical factors characterize the logistics network and limit the flow of cargo streams between network facilities, e.g. the reloading capability of a given airport, the limitation of the road network resulting from permissible loads, distances between network facilities. Technical factors characterize the logistics network facilities and the relationships between them.

Economic factors include e.g. the unit costs of fuel consumption by vehicles performing transport, the costs of transport by air, the costs of cargo handling at the airport.

The quality factors include service time and transport time.

The transport task consists in transporting the load from the place of loading to the place of unloading. The logistics process within the carriages analyzed in this work can be defined as transporting cargo by road to the transshipment place, which is the airport, cargo transport between two airports and delivery by car to the final recipient. It is problematic to choose the first trans-shipment point, i.e. the airport. The determinants of choosing the right airports in the entire logistics network are not only the distance between the loading point (the place of loading) and the airport, the distance between two airports, the distance between the airport and the landing point, but also the cost and time of transhipment at a given point. The problem of intermodal transport is analyzed in a multi-criteria approach, which underlines the complexity of the analyzed issue. An exemplary route for moving a load from the loading point to the unloading point is shown in Fig. 1 (continuous lines define the selected route, dashed lines, potential routes selection options). The main aim of this paper is to develop the structure of the database used in the problems of planning transport of cargoes by intermodal transport. The developed structure of the database with the presented modules is a newly approach in problems of planning transport of cargoes by intermodal transport.

* Corresponding author: mizdeb@wt.pw.edu.pl

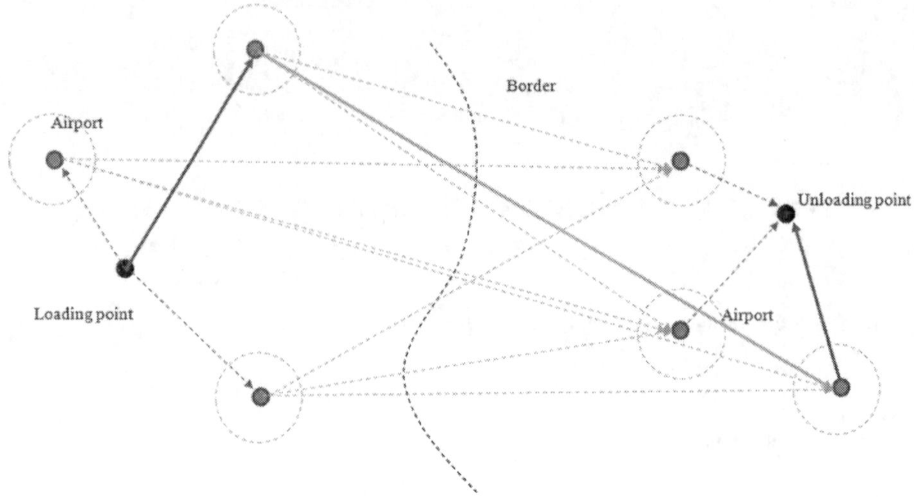

Fig. 1. The problem of intermodal transport including road and air transport

2 The concept of the structure of the intermodal transport infrastructure database

Due to the complexity of the problem of planning intermodal transport, it is necessary to process a large amount of input data used by advanced calculation algorithms developed to determine the optimal routes of cargo transportation. It is assumed that the database is responsible for collecting input data about air, road infrastructure, information processing and saving of results generated by calculation modules. The development of the database architecture for the issues of planning transport of goods by intermodal transport is determined by the number of modules responsible for determining the final transport plans. Data for the application defining the intermodal transport plan is entered from the database. The structure of the database is shown in Fig. 2.

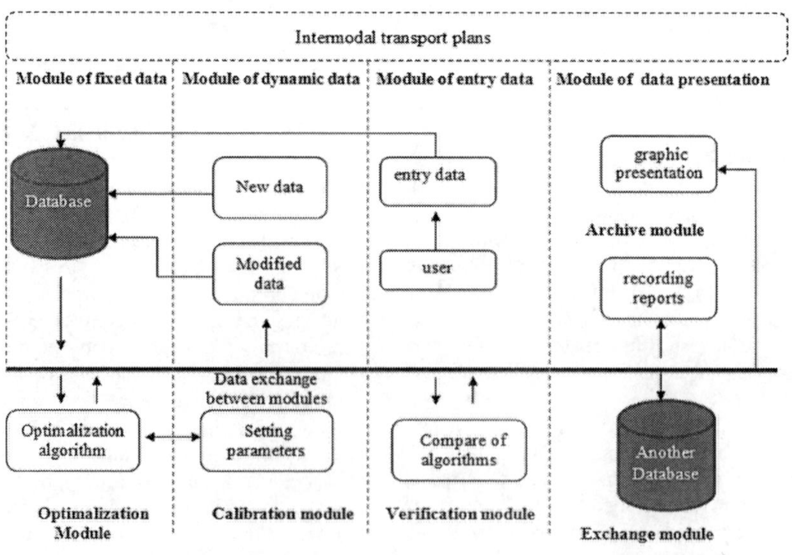

Fig. 2. Database architecture in intermodal transport

The database consists of the following modules:

- the module of fixed data – contains tables of the database on suppliers, recipients, airports, locations, types of aircraft and vehicles,
- the dynamic data module – designed to support the data that can be added, modified and removed by users during system operation, e.g. loading, unloading, cargo handling, transport costs,
- the archive module – is part of the database for storing analysis reports, results of intermodal transport plans,

- the data entry module – designed to support the process of direct input by the user,
- the data presentation module – mediates in the process of sharing data collected in the database for authorized users,
- the optimization and calculation module – the results of the module are reports and data collation on intermodal transport plans, this module is based on the built-in optimization algorithm,
- the calibration module for the optimization algorithm – sets and saves the parameters of the optimization algorithm,
- the verification module – validates the results generated by the optimization algorithm,
- the data exchange module – it is an interface between applications based on different base systems.

The database is updated via a dynamic data entry module. The optimization module, the calibration module of the optimization algorithm parameters and the results verification module operate independently of the other database modules. The correctness of the developed database is measured by the speed of solution generation, data storage and modification.

3 The mathematical model of intermodal transport

As mentioned in point 2, the designed database is responsible, among others, for saving and processing data generated in a mathematical model. In order to implement the database in the problem of intermodal transport, a mathematical model for planning these transports, including road and air transport, was developed. The necessary data for testing is:

- **PZ**={1,…,*pz*,…PZ} – the set of loading points;
- **A**={1,…,*a*,…A} – the set of airports;
- **S**={1,…,*s*,…S} – the set of types of aircraft used to carry loads,
- **P**={1,…,*p*,…P} – the set of types of vehicles used to carry loads,
- **PW**={1,…,*pwz*,…PW} – the set of unloading points,
- **ZAD**={1,…,*zad*,…ZAD} – a set of tasks,
- **D1**=[$d1(pz,a)$: $d1(pz,a) \in R^+$, $pz \in$ **PZ**, $a \in$ **A**] – the matrix of distances between the loading points and the airports,
- **D2**=[$d2(a,a')$: $d2(a,a') \in R^+$, $a \in$ **A**, $a' \in$ **A**] – the matrix of distances between airports,
- **D3**=[$d3(a,pw)$: $d3(a, pw) \in R^+$, $a \in$ **A**, $pw \in$ **PW**] – the matrix of distances between airports and the unloading points,
- **T1**=[$t1(pz,a)$: $t1(pz,a) \in R^+$, $pz \in$ **PZ**, $a \in$ **A**] – the matrix of transit times between the loading point and the airport,
- **T2**=[$t2(a,a')$: $t2(a,a') \in R^+$, $a \in$ **A**, $a' \in$ **A**] – the matrix of transit times between airports,
- **T3**=[$t3(a,pw)$: $t3(a, pw) \in R^+$, $a \in$ **A**, $pw \in$ **PW**] – the matrix of transit times between the airports and the unloading points,
- **TO**=[$to(a,s)$, $to(a,s) \in R^+$, $a \in$ **A**, $s \in$ **S**] – cargo handling time at a given airport depends on the type of aircraft chosen,
- **CL**=[$cl(s)$,$cl(s) \in R^+$, $s \in$ **S**] – the cost of transport by air,
- **CD**=[$cd(p)$,$cd(p) \in R^+$, $p \in$ **P**] – the cost of transport by road,
- tz – time of intermodal transport,
- **LS**=[$ls(s,a)$,$ls(s,a) \in N$,$s \in$ **S**, $a \in$ **A**] – the number of aircraft of a given type available at a given airport,
- **LP1**=[$lp1(p,pz)$, $lp1(p,pz) \in N$,$p \in$ **P**, $pz \in$ **PZ**] – number of vehicles of a given type available at a given loading point,
- **LP2**=[$lp2(p,pw)$, $lp2(p,pw) \in N$,$p \in$ **P**, $pw \in$ **PW**] – number of vehicles of a given type available at a given unloading point.

To identify the complexity of the problem, decision variables and the criterion function were also developed. It is assumed that the decision variables take the form of binary variables and define the connection between identified transport network nodes:

- **X1**=[$x1(pz,a,p,zad)$: $x1(pz,a,p, zad) \in \{0,1\}$, $pz \in$ **PZ**, $a \in$ **A**, $p \in$ **P**, $zad \in$ **ZAD**] – the matrix of connections between loading points and airports, if $x1(pz,a,p,zad)$=1 there is a connection between the given loading point and the airport carried out by a given type of the vehicle in the given task, otherwise 0.
- **X2**=[$x2(a,a',s,zad)$: $x2(a,a',s, zad) \in \{0,1\}$, $a \in$ **A**, $a' \in$ **A**, $s \in$ $s \in$ **S**, $zad \in$ **ZAD**] – the matrix of connections between airports, if $x2(a,a',s,zad)$=1 there is a connection between the given airports carried out by the a given type of the aircraft in a given task, otherwise 0,
- **X3**=[$x3(a,pw,p,zad)$: $x3(a, pw,p,zad) \in \{0,1\}$, $a \in$ **A**, $pw \in$ **PW**, $p \in$ **P** , $zad \in$ **ZAD**] – the matrix of connections between airports and unloading points, if $x3(a,pw,p,zad)$=1 there is a connection between the given airport and the unloading point carried out by a given type of the vehicle in a given task, otherwise 0.

The restrictions apply to, among others:

- time of completing tasks:

$$\forall pz \in \mathbf{PZ},\ \forall a \in \mathbf{A},\ \forall p \in \mathbf{P}, \forall a' \in \mathbf{A}, \forall s \in \mathbf{S},\ \forall pw \in \mathbf{PW}\ \forall zad \in \mathbf{ZAD}$$

$$\begin{aligned} &x1(pz,a,p,zad)\cdot[t1(pz,a)+to(a,s)]+\\ &x2(a,a',s,zad)\cdot[t2(a,a')+to(a',s)]+\\ &x3(a,pw,p,zad)\cdot t3(a,pw)\le tz \end{aligned} \quad (1)$$

- the number of possible allocations of vehicles and aircraft:

$$\forall p \in \mathbf{P}\ \ \forall pz \in \mathbf{PZ}$$

$$\sum_{a \in \mathbf{A}} \sum_{zad \in \mathbf{ZAD}} x1(pz,a,p,zad) \le lp1(p,pz) \quad (2)$$

$$\forall s \in \mathbf{S}\ \ \forall a \in \mathbf{A}$$

$$\sum_{a' \in \mathbf{A}} \sum_{zad \in \mathbf{ZAD}} x2(a,a',s,zad) \le ls(s,a) \quad (3)$$

$$\forall p \in \mathbf{P}\ \ \forall pw \in \mathbf{PW}$$

$$\sum_{a \in \mathbf{A}} \sum_{zad \in \mathbf{ZAD}} x3(a,pw,p,zad) \le lp2(p,pw) \quad (4)$$

and other restrictions important from the point of view of the problem being realized.

Defining the criterion function it is assumed that for service recipients the problem is to define a plan so that the execution time is as short as possible and the cost as low as possible. Therefore, the criterion functions take the form (5) of minimizing the implementation time of the intermodal transport plan and the cost of transport (6):

$$F(X1,X2,X3)= \sum_{pz\in PZ}\sum_{a\in A}\sum_{p\in P}\sum_{zad\in ZAD} x1(pz,a,p,zad)\cdot [t1(pz,a)+to(a,s)]+ \sum_{a\in A}\sum_{a'\in A}\sum_{s\in S}\sum_{zad\in ZAD} x2(a,a',s,zad) \cdot[t2(a,a')+to(a',s)]+ \sum_{a\in A}\sum_{pw\in PW}\sum_{p\in P}\sum_{zad\in ZAD} x3(a,pw,p,zad)\cdot t3(a,pw) \longrightarrow \min \quad (5)$$

$$F(X1,X2,X3)= \sum_{pz\in PZ}\sum_{a\in A}\sum_{p\in P}\sum_{zad\in ZAD} x1(pz,a,p,zad)\cdot cd(p)+ \sum_{a\in A}\sum_{a'\in A}\sum_{s\in S}\sum_{zad\in ZAD} x2(a,a',s,zad)\cdot cl(s)+ \sum_{a\in A}\sum_{pw\in PW}\sum_{p\in P}\sum_{zad\in ZAD} x3(a,pw,p,zad)\cdot cd(p) \longrightarrow \min \quad (6)$$

4 The ant algorithm for planning intermodal transport

The database structure optimization module works based on the ant algorithm. Ant algorithms are algorithms whose principle of operation consists in imitating the existence of ants in the natural environment [4], [5]. The form algorithm is an optimization algorithm with success used in various complex decision problems, e.g. the traveling salesman problem [3], [6], [7]. The deciding factor in choosing the route of each ant is the pheromone, chemical substance released by ants. Ants choose the path with the strongest concentration of pheromone. The selection of the appropriate type of formic algorithm is important in correctly generating results, i.e .:

- *ant-density* – leaving a constant amount of pheromone on each section of the ant route,
- *ant-quantity* – leaving the amount of pheromone inversely proportional to the length of the selected edge on each section of the ant route,
- *ant-cycle* – leaving the pheromone only when all the ants build the whole road, the trail is updated between successive iterations.

In the developed algorithm for intermodal transport, an *ant-cycle* update was applied. The advantage of the ant algorithm is speed of action and a simple implementation method. In databases that take into account the execution of complex calculations, the decisive factor is the speed of results generation. Taking into account the complexity of the problem of determining routes in intermodal transport and the time of generating the result, the ant algorithm was chosen as an algorithm solving the problem of transport in intermodal transport.

The developed algorithm determines the routes of cargo transport and selects the type of the vehicle and the aircraft to perform the task ordered. The ant algorithm has been developed to solve multi-criteria problems related to the determination of cargo transportation routes from suppliers to customers, including road and air transport. In order to solve optimization problems in the theory of ant algorithms, the concept of an artificial ant was introduced [4], [5]. Each ant in its route chooses the type of the vehicle and the type of the aircraft. The ant determines the minimum route of cargo transport in accordance with the adopted criteria. The route of a single ant consists of four layers that appear consecutively. The route of each ant consists of all completed tasks. During the construction of the route in the first layer, the type of the vehicle performing transport by road is selected, in the second layer the airport from which the cargo is sent is selected, the third layer shows the type of the aircraft performing air transport, while the fourth layer is the layer defining air route and road route transport between the airport and the unloading point. The starting and finishing point of the ant route is the loading and unloading point. These layers are repeated cyclically until all the tasks are completed by the ant. Ant route with highlighted layers is shown in Fig. 3. For the purpose of describing the components of the algorithm, a set of points of all layers of the route realized by a single ant was defined $\mathbf{WTP}$. The set of ants in the anthill $\mathbf{MR}$ was defined as:

$$\mathbf{WTP}=\{1,\ldots,y,\ldots,z,\ldots,\overline{\mathrm{WTP}}\} \quad (7)$$

where $\overline{\mathrm{WTP}}$ - the size of the set $\mathbf{WTP}$, $y \neq z$.

The starting point for the route of each ant is the loading point. The further route of the ant, and thus the selection of subsequent points, takes place with a certain probability:

$$PR^{mr}{}_{yz}(t)=\begin{cases}\dfrac{[\tau_{yz}(t)]^{\alpha}\cdot[\eta_{yz}(t)]^{\beta}}{\sum_{l\in\Omega^{mr}}[\tau_{yl}(t)]^{\alpha}\cdot[\eta_{yl}(t)]^{\beta}} & ,z\in\Omega^{mr}\\ 0 & ,z\notin\Omega^{mr}\end{cases} \quad (8)$$

where:

$\tau_{yz}(t)$ – the intensity of the pheromone trail between y-th and z-th punktem in t iteration,

$\eta_{yz}(t)$ – heuristic information, for each layer assumes a different value, e.g.

- for imaginary connections between the vehicle/the aircraft types and route points in each iteration of the algorithm is $\eta_{yz}(t)=0$.
- for real connections between route points:

$$n_{yz}(t) = \frac{1}{d(y,z)\cdot c + t(y,z) + to(z)} \tag{9}$$

where:

$d(y,z)$ – the route between the points of the route,

c – the transportation cost, $t(y,z)$ - driving time,

$to(z)$ – service time,

α, β – the impact of pheromones and heuristic data on ant behavior,

Ω^{mr} – the set of l points in a given layers.

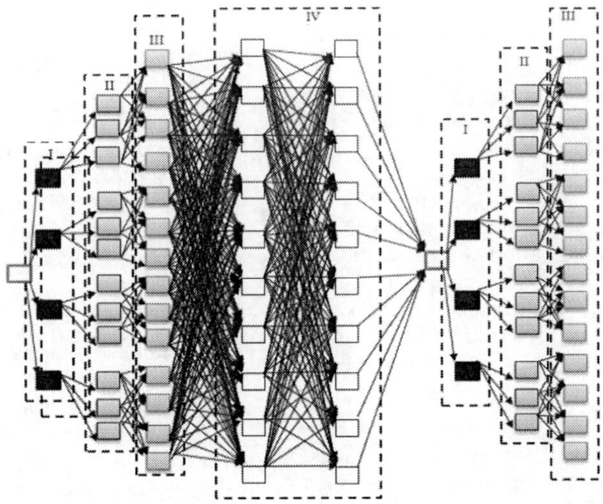

Fig. 3. Layers of the ant route

Random selection of the transition route between the point y and points l-th begins with calculating the probability of crossing to the points not yet visited by the ant according to the pattern (8). The next step is to calculate the distribution for each passage path and to draw a number r from the range [0,1]. The passage path tr about the value of the cumulative distribution q_{tr} which fulfills the dependence $q_{tr-1} < r \leq q_{tr}$ is selected, where tr is the route number between y-th point and l-th points defined in each layer.

After completing the construction of the route by all ants, the pheromone is updatedAt the beginning it is assumed that the trail on connections between points is equally strong. In subsequent iterations, the pheromone trail is calculated according to the formula:

$$\tau_{yz}(t+1) = (1-\rho)\tau_{yz}(t) + \sum_{mr=1}^{MR} \Delta\tau_{yz}^{\ mr}(t) \tag{10}$$

where:

mr – the another ant in the anthill $mr \in \mathbf{MR}$,

ρ – volatilization coefficient $(0 < \rho \leq 1)$,

$\tau_{yz}(t+1)$ – pheromone enhancement, for the first iteration, assumes a value at every link τ_0.

The first component of the formula (10) determines the speed of volatilization of the pheromone, while the second defines the enhancement of the pheromone and takes the value:

$$\Delta\tau_{yz}^{\ mr}(t) = \begin{cases} \frac{1}{L^{mr}(t)} \\ 0 \end{cases} \tag{11}$$

when the route (y,z) has been realized by the ant mr otherwise

where:

$L^{mr}(t)$ - the value of the route generated by the ant including the minimum task execution time and the cost, the value calculated for the entire ant route according to the denominator of the pattern (9).

The ant algorithm designing cargo transport plans in intermodal transport can be presented in the following steps:

- The step 0. Selection of the task (loading point) to be carried out at random way.
- The step 1. Selection of the type of vehicle performing transport by road. At the same time, checking the condition on the number of vehicles of a given type used to perform transport, limiting (2) the mathematical model. When there is a lack of the vehicle of a given type, the algorithm again selects a different type of the vehicle and checks the number of available vehicles. In the first iteration of the algorithm, the ant goes to layer I in a random manner, in subsequent iterations in accordance with the passage probability determined with patterns (8).
- The step 2. The passage of ants from the loading point to the airport in accordance with the probability determined from the formulas (8).
- The step 3. Selection of the type of aircraft at a given airport. The passage of ants from layer II to III in the first iteration in a random manner, in successive iterations in accordance with the passage probability determined with formulas (8).
- The step 4. Selection of the final airport in accordance with the passage probability determined with formulas (8). The choice of the final airport depends not only on

the distance between the ports, but also on the distance from the port to the unloading point. Step 4 completes the implementation of the selected task.

- The step 5. Checking the condition for the duration of the task, limiting (1) the model. If the restriction is not met, return to step 1.
- The step 6. Selecting the next task in a random way.
- The step 7. The ant carries out further tasks in accordance with steps 1-5.
- The step 8. After completing all the tasks by a single ant, the next ant from the anthill determines the cargo route for each task from the beginning in accordance with steps 0-8. Steps 0-8 are repeated until all ants have created routes consisting of all tasks.
- The step 9. After determining the routes by all ants, the pheromone is updated according to the pattern (10).

The steps of the 0-9 algorithm are repeated with the specified number of iterations until the stop condition of the algorithm is reached. The condition for stopping the algorithm is a certain amount of iteration determined at the beginning of the operation of the algorithm.

5 Conclusions

The developed database architecture for intermodal transport is an original proprietary concept presenting the database as an integrated modular structure that takes into account the various types of activities taking place during the planning of cargo transport in intermodal systems. The use of the ant algorithm in the optimization module will minimize the calculation time and speed up the processes taking place in the database.

In order to obtain the correct results generated by the ant algorithm, it is first necessary to analyze the sensitivity of this algorithm to change their input parameters. This analysis is carried out by a calibration module built into the database. The sensitivity analysis can be performed based on a set of fixed input parameters that are entered into the algorithms. The quality of generated solutions depends on correctly selected parameters. The maximum value generated by the ant algorithm in the sensitivity analysis indicates the parameters of this algorithm that determined this value. In the second place, the algorithm verification process should be carried out.

The verification process is carried out in the results verification module. The algorithm verification process can be carried out based on a comparison of algorithm solutions with solutions obtained using a random algorithm. The number of comparisons is determined at the beginning of the verification. In case when the random algorithm determines a better solution, the sensitivity stage should be performed for a different set of parameters of the algorithm. In the case where the ant algorithm generated a better result, the verification process is completed.

References

1. Chabrier A.: *Vehicle routing problem with elementary shortest path based column generation*, Computers & Operations Research, **33(10)**, pp. 2972-2990 (2006)
2. Chiang, W.C., Russell R.A.: *Simulated annealing metaheuristics for the vehicle routing problem with time windows*. European Journal of Operational Research, vol. **63** (1996)
3. Dorigo M., Blum Ch.: *Ant Colony Optimization Theory: a survey*, Theoretical Computer science, Vol.**344** pp. 243-278 (2005)
4. Dorigo M., Gambardela L. M.: *Ant Colonies for the Travelling Salesman Problem*, BioSystems, pp. 73-81, Vol.**43** (1997).
5. Dorigo M., Gambardela L. M.: *Ant Colony System: A cooperative learning approach to the traveling salesman problem*, IEEE Transactions on Evolutionary Computation, Vol.**1(1)**, pp. 53-66 (1997).
6. Dorigo, M., Stutzle, T.: *Ant Colony Optimization*, Bradford Books, USA, (2004).
7. Gołda Paweł, Izdebski Mariusz, Szczepański Emilian: *The Application of Ant Algorithm in the Assignment Problem of Aircrafts to Stops Points on the Apron*, w: Journal of KONES, vol. **25**, nr 4, ss. 479-488 (2018).
8. Gołda P.: *Selected decision problems in the implementation of airport operations*. Scientific Journal of Silesian University of Technology. Series Transport. 2018, 100.
9. Jacyna M., Izdebski M., Szczepański E., Gołda P.: *The task assignment of vehicles for a production company*, w: Symmetry-Basel, vol. **11**, nr 10, ss. 1-19 (2018) DOI:10.3390/sym10110551
10. Jacyna M.: *The role of the cargo consolidation center in urban logistics system*. International Journal of Sustainable Development and Planning, is.1, pp.100-113 (2013).
11. Jacyna-Gołda Ilona, Izdebski Mariusz, Podviezko Askoldas: *Assessment of the efficiency of assignment of vehicles to tasks in supply chains: A case-study of a municipul company*, w: Transport, vol. **32**, nr 3, ss. 243-251 (2017).
12. Jacyna-Gołda I.: *Evaluation of operational reliability of the supply chain in terms of the control and management of logistics processes*. W: Nowakowski T. [i in.] (red.), *Safety and Reliability: Methodology and Applications*, CRC Press Taylor & Francis Group, s. 549–558 (2015), ISBN 978-1-138-02681-0.
13. Jacyna-Gołda I.: *Inżynieria oceny efektywności sieci dostaw*. Wydawnictwo Naukowe PWN, ISBN: 978-83-012-0526-3, Warszawa (2019).

Work completed within the framework of the EUREKA project project implemented on the basis of an agreement with the National Center for Research and Development No. EUREKA / EPLOS / 3/2017. on the European Logistics Services Portal (EPLOS).

Designing of transshipment terminals in the aspect of selected intermodal transport systems

Dariusz Pyza*, and *Roland* Jachimowski

Warsaw University of Technology, Faculty of Transport, Koszykowa 75, 00-662, Warsaw, Poland

Abstract. The article deals with the problem of intermodal terminal designing for selected intermodal transport systems, determined by the type of intermodal transport unit. The purpose of the article is to present a comprehensive methodology for designing intermodal terminals, enabling the assessment of developed design solutions. Specificity of intermodal terminals due to the tasks carried out and their role in the integration of various modes of transport in freight transport was presented. In the following, the general principles and methodology of intermodal terminal design were characterized. Based on that, the conceptual project of the terminal was presented as a case study.

1 Introduction

The economic development of European Union (EU) countries determines the development of transport, which affects the improvement of economic systems and creates the possibility of moving from solutions based on the use of resources located near their production and consumption to solutions based to the greatest extent on specialization, i.e. the progressing international division of labor [15].

In this aspect, cargo shipments are more and more technologically advanced, so that supply networks are becoming more and more complex, covering more and more regions.

In supply networks, transport plays a crucial role in ensuring the most efficient and effective flow of raw materials, semi-finished products and finished products from the places of their shipment - sending them to the places of their collection - destination. The specificity of the supply network determined by the type of cargo and its transport susceptibility affects the organization of transport and the applied transport technology [9, 13, 14, 15].

The increase in the flow of cargo in economic systems implies the increase in the importance of sustainable transport. This approach requires not only the efficient use of various forms of transport capacity of, but also the search for alternative, environmentally friendly drive sources [5].

In the scope of freight transport, high expectations exist in relation to the development of high-speed railway transport, including intermodal transport. It should be noted, that in many cases these expectations are divergent, between various stakeholders of the intermodal transport market, which include: infrastructure managers and representatives of carriers, intermodal terminal operators, enterprises using transport services - shippers and consignees, as well as representatives of environmental entities, e.g. public administration entities or shippers at other transshipment terminals [2, 9, 15, 20].

An essential element of sustainable transport is forecasting future transport needs, that affect the development of transport networks using intermodality. In this sense, the efficiency of cargo transport is influenced by the multi-branch, integrated transport network, based on a modern and well-designed infrastructure [15, 16].

The dynamic development of transport affects both environmental and social costs. The negative impact of transport on the natural environment is caused by high external costs of transport, which include, among others: emissions of harmful exhaust compounds, increased noise levels and accidents.

Reducing the negative transport impact on the environment is one of the priorities of the transport policy of the European Union and the Member Countries. In this aspect, many actions are taken, including among others improving the competitiveness of the railway sector and increasing the share of rail in the distribution of transport tasks or the development of intermodal transport, including support for the creation of intermodal terminals [8].

In Poland, the volume of intermodal transport has been systematically increasing over recent years. In 2018, 17 million tons of cargo were transported by intermodal transport, which is an increase compared to 2017 by 15.6%. The share of intermodal transport in the entire railway transport market measured by the mass of transported cargo reached the level of 6.8% and was higher by 0.7% as compared to 2017. The transportation work performed by the intermodal transport of cargo was at the level of 6.2 billion tkm , which was an increase of 14.8% compared to 2017 [17].

* Corresponding author: dpz@wt.pw.edu.pl

The presented trends as well as the analysis of transport directions indicate the growing importance of Poland as a transit country connecting East and West.

In this aspect, intermodal terminals, including their modernization and construction of new ones, become important.

The purpose of the article is to present a comprehensive methodology for designing intermodal terminals, enabling the assessment of the developed design solutions.

2 Intermodal terminals and their specification

2.1. Transshipment terminal in the intermodal transport network

The implementation of transport processes, requires transfer points, where various types of activities related to cargo handling take place.

The transshipment terminals, which aim is to change the type and means of transport play a special role in intermodal transport networks [4].

Intermodal terminal is a spatial object connected with the infrastructure and proper organization, enabling efficient transshipment of intermodal transport units, between means of transport belonging to different modes of transport and performing operations on these units [10, 15].

Intermodal terminals are the points of concentration and distribution of cargo in the structures of transport and logistics systems. These points provide transshipment and warehousing services as well as other services related to the distribution of cargo [9].

There are transshipment terminals with various capacities, fulfilling various functions and tasks in the intermodal transport networks.

Because of that, it is possible to distinguishe sea, inland and land intermodal terminals. Land intermodal terminals are located on the railway network with access to road infrastructure and, in some cases, inland waterway infrastructure. Sea (marine) intermodal terminals are located in seaports, where they usually form a part of the sea port. Sea intermodal terminals, due to their functions in intermodal transport, have access to both, rail and road transport infrastructure, and in some cases to inland waterway infrastructure. Inland intermodal terminals are located in inland ports, usually with an access to railway and road infrastructure.

Classification of intermodal terminals can be carried out due to various criteria [6, 9, 10]:

- type of intermodal transport units,
- means of transport involved in the transport and the delivery of intermodal transport units (ITU) from and to the terminal,
- terminal capacity expressed by the annual turnover of intermodal transport units (TEU),
- functions performed by the terminal in the intermodal transport networks and
- others.

Detailed classification of intermodal terminals with division criteria is presented in Table 1.

Table 1. Classification of intermodal terminals

Classification criteria	Terminal type
Means of transport involved in the transport and the delivery of intermodal transport units from and to the terminal	Sea
	Land
	Inland
Type of intermodal transport units	Container
	Universal (handles all intermodal transport units)
	Selective (handle selected intermodal transport units)
Location on the railway network	Rail, rail-road
	Sea-rail-road
	Sea-rail-road-inland waterway
ITU handling system	Vertical transshipment lo-lo
	Horizontal transshipment ro-ro
	Horizontal transshipment ro-ro and vertical transshipment lo-lo

The sea terminals provide services of:

- oceanic container vessels: very large (1 million TEU), large (0.6 ÷ 1 million TEU), medium (less than 0.6 million TEU),
- container vessels and semi-containers vessels: medium (0.1 ÷ 0.6 million TEU), small (less than 0.1 million TEU).

365/5000

Land terminals are divided into: international (hubs) with over 70,000 TEU service, European terminals that support services for 30,000 ÷ 70,000 TEU and domestic terminals that support less than 30,000 TEU. Inland water terminals are usually medium-class container terminals servicing over 80,000 TEU and small container terminals with a service volume of less than 80,000 TEU.

Intermodal terminals are not only transshipment hubs, but also service centers for the transport of containerized cargoes that offers additional services [19].

2.2 Technical requirements for the intermodal terminals

The functional and spatial system, handling equipment and work organization of the intermodal terminals determine their capacity, i.e. the ability to receive and dispatch a certain amount of cargo per unit of time.

This involves costs, that are a function of the number of cargo handled as well as functional and spatial solutions and the type of equipment used. In this respect, the capacity of the intermodal terminals can be changed by modifying both the functional and spatial layout of the terminals as well as their handling equipment. However, it should be remembered that such activities entail expenditures and affect the total costs of the terminals' operation.

Integration of various forms of transport in the intermodal terminals is perceived as a way to increase the capacity of terminals. At the same time, the terminals' operation costs are reduced.

The choice of the functional and spatial solutions of the intermodal terminal and its handling equipment should take into account the specifics of the terminal and the number and types of ITU handled per unit of time.

In general, functional and spatial solutions include [9, 16]:

- rail and road system of the terminal,
- storage and handling yards for containers, swap bodies, semi-trailers, trucks, tractors with semi-trailers,
- parking for trucks with ITU waiting for service,
- loading fronts, buildings and technical facilities.

These solutions are determined by the type and the location of the intermodal terminal. Land terminals, marine (sea) and inland terminals have different functional and spatial systems as well as the handling equipment. Further research in this paper is focused on intermodal land terminals located on the railway network.

The operating capability of the intermodal terminal is conditioned by the handling capability of the transport means, storage capacity for intermodal transport units and the capacity of the ITU handling equipment [3]. The minimum requirements for intermodal terminals are regulated by the AGTC agreement. The AGTC agreement specifies the required parameters of the infrastructure of the railway lines and terminals. The guidelines in this respect are prepared by the International Union of Railways – UIC [9, 15].

Efficient ITU service in the terminal is provided by the rail tracks and roads system. Individual groups of tracks should have proper track connections, and the entire track system should be designed with appropriate guidelines and meet all conditions for safe work. The connection of the intermodal terminal with the public road network is provided by the road system, which also allows maneuvers of means of road transport at the terminal depending on the technology of their service. The intermodal terminal road system includes road lanes and maneuver roads [9, 15, 16].

The ITU storage and handling yards have a double function in the intermodal terminal. They enable both the storage of intermodal transport units and the sorting of intermodal transport units. The storage and handling yards should be designed taking into account technical and technological solutions of the terminal as well as the type and specificity of intermodal transport units. The efficiency of storage and handling yards (areas) is achieved by placing them in the range of the main loading device (handling equipment), thereby limiting the number of loading operations to the necessary minimum [1, 18].

Storage and handling yards are designed for ITU such as: semitrailers as well as tracks and tractors with semi-trailers, swap bodies and containers. These types of units can be transported in a Piggyback system, Rolling Highway system as well as in other rail-road systems.

Parking area is usually used as a place where the tractors with semitrailers, trucks with swap bodies as well as the handling equipment, can wait for the operations. Parking areas and storage yards should be designed in accordance with the requirements for internal roads and yards [9, 10, 12, 15].

Loading fronts enable handling of intermodal trains and other means of transport. Loading fronts and storage yards form a loading system, which is an important element of the terminal. The intermodal transport units handling technology at the terminal determines the manner of using individual elements of the loading system [9, 10, 11, 15, 16].

A schematic diagram of functional and spatial solutions of the intermodal terminal for the "rolling highway" system is presented in fig. 1.

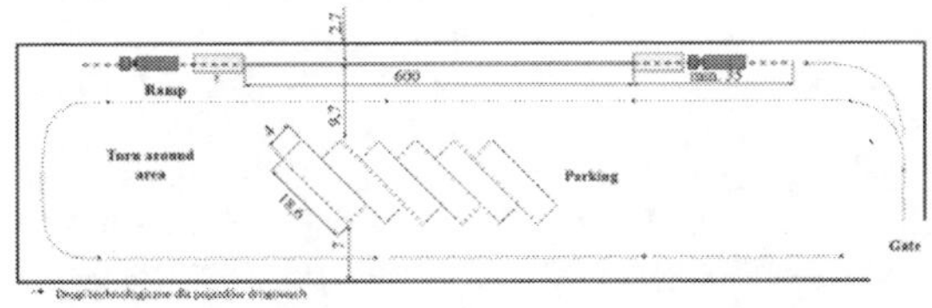

Fig. 1. The example of an intermodal terminal with the „rolling highway" system [9].

Functional and spatial configuration of the terminal is also determined by the handling equipment. The terminals are equipped with handling devices with various technical and operational characteristics as well as appropriate accessories enabling the handling of ITU [9].

3 Methodology for intermodal terminal designing

Intermodal terminal designing is an interdisciplinary problem, covering both technical and economic issues. It takes place in terms of process and object design.

Process design is shaping the layout of the terminal, requiring the consideration of the specificity of intermodal transport units and the processes of their flow in the terminal.

The aim of this part of the design is to determine the intermodal terminal layout and the connections between given areas of that layout in the technological aspect.

Object oriented design relies on dimensioning of the terminal, in the aspect of functional and spatial areas defined in the process design. At this stage of a design, parameters that characterize the layout (functional and spatial areas) of the terminal are obtained.

The design procedure for the intermodal terminal is carried out in four design stages (fig 2) [9]:

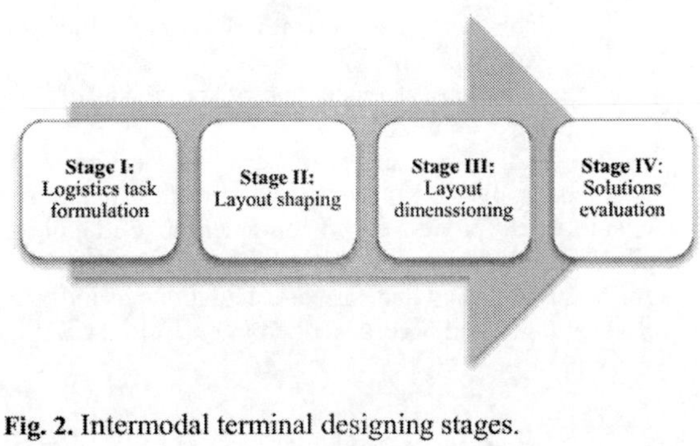

Fig. 2. Intermodal terminal designing stages.

A. **First stage:**
- definition of suppliers and recipients and determination of the intermodal transport units' (ITU) transition through by the terminal,
- estimation of the annual and average daily mass of cargo handled by the terminal and the structure of intermodal transport units,
- estimation of the cargo forecasts of the designed terminal and the structure of intermodal transport units in the aspect of the development of intermodal transport systems,
- estimation of the percentage share of ITUs handled in the terminal,
- determination of the average storage time of ITUs.

B. **Second stage:**
- description of the process of the intermodal transport units flow through the terminal,
- determination of the number of intermodal transport units moved through the terminal in different relations,
- determination of the layout (functional and spatial areas) of the proposed terminal,
- selection of the intermodal transport units storage method in the terminal,
- determination of the intermodal transport units flow cumulation factor.

C. **Third stage:**
- determination of transport cycles and calculation of transport cycle times for transport of intermodal transport units through the terminal,
- calculation of the daily labor consumption of the intermodal transport units flow through the terminal,
- calculation of the number of handling equipment and human resources necessary to service intermodal transport units, taking into account their type and a human work category,
- calculation of necessary handling equipment and human resources,
- calculation of necessary capacity of trains importing and exporting intermodal transport units,
- determination of the necessary capacity of storage yards,
- calculation of parameters of the loading fronts layout and parameters of communication routes,
- calculation of expenditures and operating costs of the planned intermodal terminal.

D. **Fourth stage:**
- calculation of assessment indicators for individual design solutions,
- comprehensive evaluation of intermodal terminal project solutions.

The individual stages of terminal design require detailed analyzes in the scope of rationalization of technological processes and equipment as well as layout configuration. Parameterization of input data necessary for terminal design and formal calculations of individual design stages had been described in detail in the authors' work [9].

4 Intermodal terminal project – case study

In accordance with the presented methodology of intermodal terminal design and using the formalization of individual design stages, presented in paper [9], a case study of the land intermodal terminal design was presented.

A. **Logistics task formulation**

Design assumptions:
- terminal services intermodal transport (loading) units such as: ISO containers (1A, 1B, 1C), semi-trailers (NS) and swap bodies (NW);
- terminal services different rail directions: import, export, transit;
- the intermodal transport units (ITU) service technology requires intermediate handling (temporary storage) and direct handling rail car-rail car and trailer-rail car;
- the average number of ITU serviced by the terminal per day is presented in table 2,
- transshipment operations are performer by rail mounted gantry.

Table 2. Physical ITU number

	n_N	n_P	$n_{P\text{-}N}$	$n_{S\text{-}S}$	$n_{T\text{-}BP}$	n_{PN}	n_{PP}
1A	25	17	17	12	10	0	8
1B	25	17	15	10	8	0	8
1C	25	17	10	8	3	0	8
NW	5	8	0	0	0	0	0
NS	8	10	0	0	0	0	0

where:
- n_N – number of ITU in the rail transport import
- n_P – number of ITU in the rail transport export
- $n_{P\text{-}N}$ – number of ITU in the rail transport transit
- $n_{S\text{-}S}$ – number of ITU in the road transport transit
- $n_{T\text{-}BP}$ – number of ITU in the rail transport transit without transshipment
- n_{PN} – number of empty ITU in the rail transport export
- n_{PP} – number of empty ITU in the rail transport import

In the intermodal terminal, ITUs may undergo several (max. three) transshipment operations.

The percentages of the number of operations for each type of ITU passing through the terminal are shown in table 3.

Table 3. Percentage number of operations

No	$u(ko^N)$	$u(ko^P)$	$u(ko^{P\text{-}N})$	$u(ko^{S\text{-}S})$	$u(ko^{PN})$	$u(ko^{PP})$
1	10	10	20	0	10	10
2	80	80	80	100	90	90
3	10	10	0	0	0	0

where:

- $u(ko^N)$ – percentage of the number of operations performed on ITU in rail exports;
- Average storage time: n_N – 24h; n_P – 12h; $n_{P\text{-}N}$ – 48h; $n_{S\text{-}S}$ – 24h;
- Intermodal transport units flow cumulation factor – 1,3;
- Rail car - Sdgmnss (434S) – 60' capacity.

B. Shaping and dimensioning of functional/spatial areas

Assuming that for a long period of time ITUs flow should be balanced, the capacity of trains for containers and swap bodies is:

$$LMz_{\_N} = LMz_{\_P} = 285 \text{ TEU}$$

Taking into consideration the possibility of operating two railway lines, the capacity of LM_{1kier} trains will be 143 TEUs in each direction. With the use of Sdgmnss (434S) cars (rail cars) of a capacity of 3TEU, the demand for cars in one direction (including rail cars utilization factor for semitrailers – 1 and 0.9 for containers and swap bodies) will be: **53** rail cars for semitrailers and swap bodies and **7** rail cars for semi-trailers.

Including the rail car length (18.34 m), the length of all cars for one rail direction will be 1104m. It means **two** trains a day for each direction with a length of 552 m (1104m: 2 = 552m) without a locomotive.

Then it was assumed that the length of one rail track on the terminal including a train, a locomotive and a certain clearance for inaccurate positioning is 582 m.

For ITU's subjected to intermediate handling, storage yard has to be provided. Taking into account the daily number of containers, swap bodies and semi-trailers moved through the terminal as well as their assumed storage time, the capacity of the storage yard should be :

- 583 TEU for containers
- 19 places for swap bodies
- 25 places for semi-trailers

Assuming that the containers are positioned along the loading tracks, the length of a single storage bay of containers should not exceed 582 m. Therefore, assuming that the length of a single TEU is 6.5 m, 582 / 6.5 = 89 TEU can be stored along one storage bay. Assuming the container stacking factor is k = 2, than 172 TEU can be stored along one bay. Therefore, the distribution of 582 TEU on the storage yard with a given stacking factor will require 582/172 = 4 storage bays.

Another element that has to be calculated is a number of cranes performing transshipment operations. In order to calculate that, it was necessary to calculate the number of operations (N_{ko}). In this case study, including the number of containers, swap bodies and semitrailers that has to be handled in a given direction (some of them are handled directly in the relation rail car – truck, some in the relations: rail car – storage – road truck), the necessary number of operations will be: N_{ko} = **644** operations.

Assuming that the average crane cycle time is 4 min/operation, daily working hours is 16h, as well as the working time usage factor of a crane is 0.8, then the necessary number of cranes n_u will be

$$n_u = \frac{N_{ko}}{q_u},\ q_u = \frac{T_o \cdot \beta_u}{t_{c\acute{s}r}} \tag{1}$$

where:
N_{ko} – number of operations
q_u – crane daily capacity
T_o – crane daily working time (2 shifts, 8hours every shift) $T_o = 2 \cdot 8 \cdot 60min = 960\ minutes$
β_u – working time usage factor of a crane $\beta_u = 0.8$
$t_{c\acute{s}r}$ –average crane cycle time – 4 min/operation
Then:

$$q_u = \frac{960\ minutes \cdot 0.8}{4\ min} = 192[\text{operations}/16\text{h}]$$

$$n_u = \frac{N_{ko}}{q_u} = \frac{644}{192} = 3.35 = 4\ cranes$$

With the use of 4 cranes, their transshipment potential will not be fully utilized, while 3 cranes will be insufficient to service the intermodal terminal at a higher ITUs cumulation than average. Therefore, 4 gantries were accepted.

C. Solution evaluation

In order to evaluate the solution it was necessary to calculate expenditures for setting up a terminal as well as its operational cost [7].

Expenditures in the intermodal terminal include those incurred for infrastructure, handling equipment and terminal management systems. In this study we focus on the infrastructure and handling equipment.

In the terminal, handling of intermodal trains requires two loading tracks. It was assumed that the sum of the length of all railway tracks L_{tk} is equal to 2 · 582 = 1164 [m], while the cost of 1 km c_{tk} track is equal to 6000 [PLN/m], and therefore the total expenditure on railway tracks is:

$$E_{tk} = L_{tk} \cdot c_{tk} = 1164 \cdot 6000 = 6\,984\,000 \text{ [PLN]}$$

In addition, rail tracks are connected by a single rail track crossover. It was assumed that the single crossover cost is $E_{rk} = c_{rk}$ = 380 000[PLN].

The expenditures for the E_{ps} storage yard are the product of the P_{ps} storage yard surface and the unit cost c_{ps} of the concrete pavement. For the purpose of calculations, it was assumed that c_{ps} = 240 [PLN / m2], while P_{ps} = 6402m2 was determined on the basis of the length and width of storage bays of containers, swap bodies and semi-trailers. Hence, expenditures on the storage yard are:

$$E_{ps} = P_{ps} \cdot c_{ps} = 6402 \cdot 240 = 1\,536\,480 \text{ [PLN]}$$

The cranes expenditures E_{UT} are the product of the unit cost c_{mt} for the crane with its assembly and the number n_{mt} of cranes;

$$E_{UT} = n_{mt} \cdot c_{mt} = 4 \cdot 4\,000\,000 = 16\,000\,000 \text{ [PLN]}$$

It was assumed that the annual costs of railway tracks maintenance K_{tk}^R and crossovers K_{rk}^R represent 1% of the expenditure incurred for their purchase (E_{tk} and E_{rk}), therefore:

$$K_{tk}^R + K_{rk}^R = (E_{tk} + E_{rk}) \cdot 0.01 = 73\ 640 \text{ [PLN/year]}$$

For the purpose of a research it was assumed that the cranes maintenance costs represent 10% of their purchase, therefore:

$$K_{UT}^R = 0.1 \cdot 16\ 000\ 000 = 1\ 600\ 000 \text{ [PLN/year]}$$

Handling equipment operations require the employment of qualified employees. For the purposes of calculations it was assumed that crane operator labor cost is 25 [PLN / h]. It was assumed that each employee works 250 days a year for 8 hours, which gives 2000 hours a year. Therefore, the annual labor costs is:

$$K_r^R = 25 \cdot 2000 \cdot 8 = 400\ 000 \text{ [PLN/year]}$$

The aim of this case study was to show, how to calculate the intermodal terminal and what expenditures and costs are the most important while making a decision on terminal layout, its equipment and operations technology. Other expenditures and costs related to terminal administration, terminal management system, as well as other services offered by the terminal were not included in these case study. The detailed analysis on all the element necessary for the terminal evolution can be found in [9].

Conclusion

Designing intermodal terminals is a multi-stage process that takes into account a number of technical and technological requirements resulting from normative documents.

The layout of the terminal and a technical equipment of the terminal is determined by the specificity of the ITUs and the technological processes. In addition, in many cases, terminal layout solutions are also conditioned by the availability of the land designated for investment.

Therefore, the determination of the size and capacity of terminal layout (functional and spatial areas) and the number of handling equipment necessary to service a given number of ITUs requires a series of calculations. In this aspect, designing should take into account variant solutions, which makes it possible to choose the optimal solution for the adopted assessment criteria.

The growing share of intermodal transport in Poland in the total number of transported cargo causes, that the existing intermodal terminals are very burdened, which means that in the near future they may reach their capacity limit.

Such a phenomenon determines decisions of intermodal terminal operators in the scope of modernization or construction of new terminals. The article proposes an author's approach to a comprehensive intermodal terminal design methodology, in which the area of process and object design is distinguished.

The methodology for designing intermodal terminals presented in the article may be the basis for its use in the design work of designers.

References

1. Alicke, K., *Modeling and optimization of the intermodal terminal Mega Hub.* OR Spectrum, **24**, pp. 1–17, (2002).
2. Beim M., Mazur B., Soczowka A., Zajdler R., *Transport intermodalny w Województwie Wielkopolskim w latach 2004-2014. Przemiany, stan obecny i perspektywy rozwoju.* Ekspertyza wykonana na zlecenie Wielkopolskiego Regionalnego Obserwatorium Terytorialnego. Poznań (2015).
3. Boysen, N. and Fliedner, M., *Determining crane areas in intermodal transshipment yards: The yard partition problem.* European Journal of Operational Research, **204**, pp. 336-342, (2010).
4. Carlo, J.H., Vis, F.A., Roodbergen, K.J., *Transport operations in container terminals: Literature overview, trends, research directions and classification scheme.* European Journal of Operational Research, **236**, pp.1-13, (2014).
5. Jachimowski R., *Ocena funkcjonowania terminali intermodalnych w aspekcie realizowanych procesów transportowo-przeładunkowych*, Oficyna Wydawnicza Politechniki Warszawskiej, (2018) ISBN 9788378147718, 150 s
6. Jachimowski R., *Review of transport decision problems in the marine intermodal terminal,* Archives of Transport, **44(4)**, pp. 35-45, 2017
7. Jachimowski R., *Selected aspects of costs shaping in the intermodal terminal*, Journal of KONES Powertrain and Transport, Institute of Aviation (Aeronautics) BK, Vol. **25**, No. 1, (2018)
8. Jachimowski R., Szczepański E., Kłodawski M. [i in.]: *Selection of a container storage strategy at the rail-road intermodal terminal as a function of minimization of the energy expenditure of transshipment devices and CO2 emission,* w: Annual Set The Environment Protection, vol. **20**, ss. 965-988 (2018)
9. Jacyna M., Pyza D., Jachimowski R., *Transport intermodalny. Projektowanie terminali przeładunkowych.* Wydawnictwo Naukowe PWN, (2017)
10. Jacyna M., Pyza D., *Rola intermodalnych terminali przeładunkowych w przewozach kolejowo-drogowych.* Problemy Kolejnictwa. Oficyna Wydawnicza Politechniki Warszawskiej, Tom 59 Zeszyt 169/2016. Warszawa (2016)
11. Jakubowski L., *Technologia prac ładunkowych,* Oficyna Wydawnicza Politechniki Warszawskiej, Warszawa (2003)

12. Poliński J., *Rola kolei w transporcie intermodalnym.* Oficyna Wydawnicza Politechniki Warszawskiej, Warszawa (2015)
13. Pyza D., Jachimowski R., *Modelling of Parcels Transport System.* Proceedings of 19th International Scientific Conference. Transport Means 2015. Kaunas University of Technology, Lithuania, (2015)
14. Pyza D., *Optimization of transport in distribution systems with restrictions on delivery times.* Archives of Transport, Polish Academy of Sciences Committee of Transport, vol. **21**, iss. 3-4, Warsaw (2009)
15. Pyza D., *Technologie przewozowe w transporcie intermodalnym.* Przegląd komunikacyjny, rocznik LXXIV, nr 4/2019. Stowarzyszenie Inżynierów i Techników Komunikacji Rzeczpospolitej Polskiej, (2019)
16. Pyza D., *Transport intermodalny – uwarunkowania techniczno-technologiczne, organizacyjne i funkcjonalne*, [w:] Wybrane zagadnienia logistyki stosowanej, tom IV, red. J. Feliks, Wydawnictwa AGH, Kraków (2016)
17. *Rok 2018 w przewozach intermodalnych.* Podsumowanie Prezesa UTK. Urząd Transportu Kolejowego, Warszawa (2019)
18. Steenken, D., Voß, S., and Stahlbock, R., *Container Terminal Operation and Operations Research - A Classification and Literature Review.* OR Spectrum, **26(1)**, pp. 3–49, (2004)
19. Szczepański E., Jachimowski R., Kłodawski M., [i in.]: *Containers storage strategy at the rail-road intermodal terminal,* w: Proceedings of 22nd International Scientific Conference Transport Means. Kaunas University of Technology (2018)
20. Wronka J., *Transport intermodalny/kombinowany w polityce transportowej Polski.* Przegląd komunikacyjny nr **8/2015**. Stowarzyszenie Inżynierów i Techników Komunikacji Rzeczpospolitej Polskiej, (2015).

Distribution of information flows in the advanced network of MPLS of railway transport by means of a neural model

Ihor Zhukovyts'kyy*, *Victoria* Pakhomova, *Halyna* Domanskay, and *Andrew* Nechaiev

Dnipro National University of Railway Transport named after Academician V. Lazaryan, Dep. «Electronic Computing Machines», Lazaryan Str., 2, 49010, Dnipro, Ukraine

Abstract. Ensuring interoperability of railway transport is possible only due to the developed information structure. Today, Ukraine uses the information-telecommunication system (ITS) of railway transport, which is based on a data communication network. The effectiveness of its work is largely determined by the routing system. The current algorithm for choosing the shortest route, which is used in the existing routing protocol (OSPF), does not always lead to an effective result. However, there is MPLS technology, which could improve the quality of the ITS network by creating virtual channels between its nodes. The authors proposed a scheme for selecting tunnels for the flows in the MPLS network, which is based on the neural model of a multilayer perceptron of configuration 18–3–3–10 with the activation function Softmax in hidden layers and a linear activation function in the input layer. To simulate the network operation, flow data is needed: class of service (CoS), sender and recipient identifiers, average flow rate vector and tunnel data (their initial load). The final load of the tunnels is taken as the resulting output of the neural network, on the basis of which the tunnel is selected for the flow of the *k*-th class of service.

Introduction

The interaction of individual elements of infrastructure, analysis of the railway transport work and management to date is impossible without support of the automated information systems. Ukraine introduced the information-telecommunication system (ITS) of railway transport, which is based on a data communication network [1, 2]. At the present stage, the railway transport ITS network routers use the OSPF (Open Shortest Path First) protocol, because it is a common standard supported by different hardware manufacturers and it avoids closed loops in the network development process. More often, the routing criterion in the ITS network is the data transmission time, which depends on many factors: channel bandwidth; traffic intensity that can change over time, routing delay, and others. The current algorithm for choosing the shortest route, which is widely used in the existing routing protocol, does not always lead to an effective result. Thus, there is a need to study the possibility of solving the routing problem (in particular, the distribution of traffic flows) in the ITS network of railway transport.

Review of performed studies

At present, there is MPLS technology that could significantly improve the Quality of Service (QoS), create easily the virtual channels between the nodes of the railway ITS network, encapsulate various link protocols for the transmission of various traffic, including IP packets, ATM cells, SONET/SDH frames and Ethernet family technology frames [3].

It is known that MPLS technology is based on the processing of MPLS header, which is appended to each data packet [4]. An MPLS header contains one or more labels that determine the packet route. Routers that are located on the MPLS network input or output are called LER (Label Edge Router). LER at the MPLS network input adds an MPLS label to the data packet, and LER at the MPLS network output removes the MPLS label from the data packet. Routers that perform routing of data packets that rely only on the specified labels are called LSR (Label Switching Router). Labels between LER and LSR are distributed using LDP (Label Distribution Protocol). In order to get a complete picture of the MPLS network, LSRs constantly exchange labels and information about each neighboring node using the standard procedure. Virtual channels (tunnels) called LSP (Label Switch Path) are set by providers to solve various tasks, for example, for the organization of virtual private networks or for the transmission of traffic over the MPLS network through the specified tunnel. In many respects, LSP is no different from PVC in ATM or Frame Relay networks, except that LSP does not depend on the features of data link layer technologies [3].

Many scholars around the world conduct the research and implementation of MPLS technology [5-11]. For example, [12] developed an analytical model of successive queues that describes the tunneling mechanism in the MPLS network. The developed model allowed researching the processes of burst chaining of

* Corresponding author: ivzhuk.@ua.fm

packets, chaining of bursts, and fragmentation of packet bursts in LSP tunnels. The influence of chaining and fragmentation processes on probabilistic-time characteristics of traffic in MPLS tunnels is shown. The comparative analysis in the MPLS network with tunnels and in the MPLS network without the use of a tunneling mechanism is conducted. On the example of the European Internet backbone, the comparative estimation of the effectiveness of IP routing and MPLS is performed [13]. The use of MPLS technology can provide an increase in traffic by 1.7 times on average.

The construction of analytical models of MPLS networks is complicated due to the comparatively high complexity of the technology; therefore, the simulation of MPLS is still a promising area of research. The colored Petri nets of CPN Tools modeling system are a combination of the Petri net graph and the CPN ML programming language used to describe the attributes of elements in [14]. However, the features of MPLS technology impose additional restrictions on the mathematical model of network management. Existing mathematical models of the MPLS network are implemented using a streaming approach and are often static. The main disadvantage of such models is non-traceability of data transmission losses and packet delay in routers queues. Elimination of these disadvantages is possible with the transition to quasi-flow models that operate not with the bandwidth requirements, but with the packet blocks of some length, peculiar for a particular type of traffic [6]. In [15], there is developed a model for distributing flows of different classes in the MPLS computer network subject to compliance with the specified values of the quality indicators: average delay and the share of lost packets.

In [17], based on the use of the method of Lagrange undetermined multipliers, there is developed an algorithm for optimizing the search for a switching path according to the LSP labels, a target function of costs, which includes bandwidth constraints for each class of service.

In [9] there is a study of dependence of the quality of solution to the MPLS TE traffic engineering problem on the flow assignment sequence for the simplified fragment of the ITS network of railway transport.

The final solution to the problem of distributing information flows in MPLS network has not been found yet. At the present stage, in order to improve the rail transport ITS, research is being carried out on the use of various network technologies and organization of traffic routing using artificial intelligence methods [1, 18, 19]. The [20] suggests an approach to identifying network flows and organizing Big Data traffic routes in a virtual data processing centre based on a neural network. For the integrated network of rail transport ITS, which in the long run should work on different technologies, the optimal route has already been determined using the software model "MLP34-2-410-34", the input of which is an array of network channels bandwidth [18].

Purpose

Based on the use of the neural model, to develop a methodology for rational distribution of traffic flows in the advanced network of MPLS in the rail transport ITS. In order to ensure the QoS parameters for different types of traffic, to provide a separate tunnel system for each class.

Methodology

Mathematical statement of the problem. As an example, we consider an advanced network of MPLS in the rail transport ITS, the fragment of which is shown in Fig. 1.

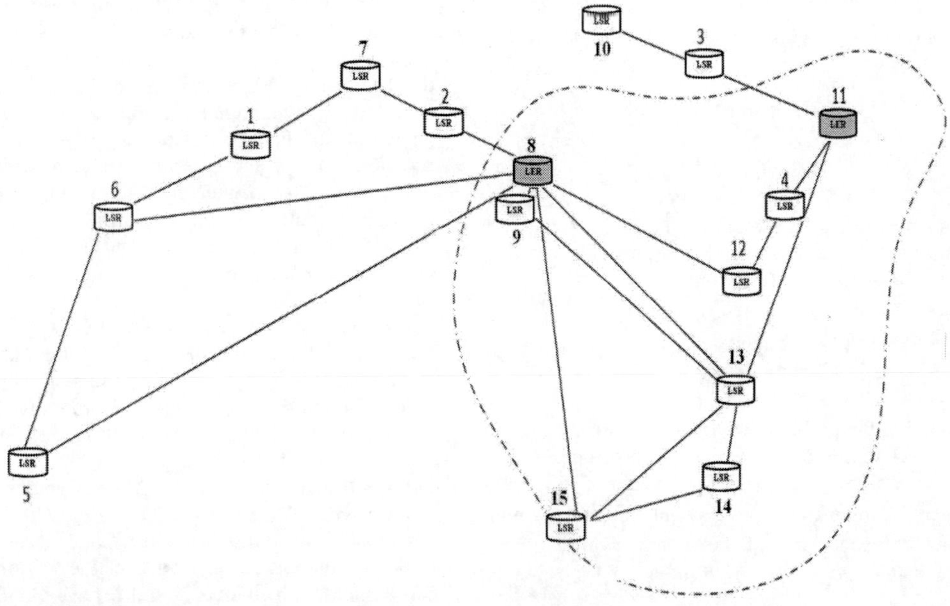

Fig. 1. Fragment of rail transport ITS

An advanced network of MPLS in the rail transport ITS can be represented as a graph:

$$G(X,E), \tag{1}$$

where $X=\{x_j\}$, $j=\overline{1,n}$ are the network nodes (MPLS switches) and $E=\{(r,s)\}$ – set of arcs (communication channels). Let the maximum number of tunnels is P_{max} and their current load $\{Pk_{load}\}$.

Let us introduce the Class of Service (CoS) of the traffic flows and specify the appropriate requirements matrix for the l-th class:

$$V=\left\|v_{i,j}(l)\right\|,\ l=\overline{0,2}, \tag{2}$$

where $v_{i,j}(l)$ is the average flow rate of the l-th class of service, which is transferred from node i to node j.

In addition, the Quality of Service (QoS) parameters for each class were introduced as an average delay constraint:

$$\bar{t}(l)<T(l), \tag{3}$$

with T(0)=50 ms, T(1)=75 ms, T(2)=100 ms.

It is necessary to select a tunnel in the MPLS network:

$$K(l)=\max_{0\le p\le P_{max}-1}\left\{Pk_{load}+\sum_{(i,j)\in p}\frac{v_{i,j}(l)}{C^p}\right\}, \tag{4}$$

where $K(l)$ is the maximum coefficient of use of the l-th class tunnel; C^p - the bandwidth of the p-th tunnel (in particular, $C^p=100$ Mbps using Fast Ethernet technology) and to find the relevant flow distribution:

$$F(l)=\left[f_{rs}(l)\right], \tag{5}$$

on the condition that the maximum possible load factor of the tunnel is:

$$K(l)\rightarrow \min, \tag{6}$$

with

$$K(l)\le 0.65. \tag{7}$$

Neural network as the main method of solving the problem. For the fragment of rail transport ITS (see fig. 1) $P_{max}=5$: $8\rightarrow 13\rightarrow 11$ (0-th tunnel);
$8\rightarrow 9\rightarrow 13\rightarrow 11$ (1-st tunnel);
$8\rightarrow 12\rightarrow 4\rightarrow 11$ (2-nd tunnel);
$8\rightarrow 15\rightarrow 13\rightarrow 11$ (3-rd tunnel);
$8\rightarrow 15\rightarrow 14\rightarrow 13\rightarrow 11$ (4-th tunnel).

For solving the problem, we took a multilayer neural network (NN), the structure of which is shown in Fig. 2.

The figure shows that the NN consists of five layers: input layer $\bar{X}$, three hidden layers containing 16, 12 and 8 neurons, respectively, and the output layer $\bar{Y}$. The structure of the vector $\bar{X}$ is as follows: X_1 – class of service (CoS0, CoS1, CoS2); X_2 – sending station (station i); X_3 – recipient station (station j); $X_4, X_5, ..., X_8$ – initial values of the p-th tunnel load, where $i=\overline{1,5}$; $X_9, X_{10}, ..., X_{18}$ are average values of flow rates v_{ij} (Mbps). The resulting vector $\bar{Y}$ shows the distribution of flows from the station i to station j (with given service class) through the MPLS network tunnels.

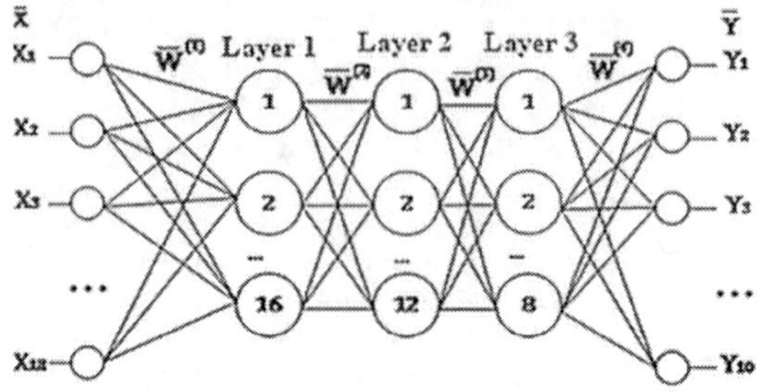

Fig. 2. Neural network of configuration 18-3-36-10

We proposed a scheme for selecting tunnels for the flows in the MPLS network (Fig. 3). The basis of the scheme is the software neuron model "CoSThDist". For simulation of work using NN, the following traffic flow data is needed: class of service (CoS); sender and recipient identifiers; average flow rate vector and tunnel data (their initial load). As the neural network output, we took the final load of the tunnels, on the basis of which the tunnel for the flow of the k-th class of service is selected. In the proposed scheme, the dashed arrow depicts the implementation of actions at the preparatory stage, which teaches the neural network and stores the prediction results. In real time, the system obtains data on traffic flows, tunnels and forms the output based on the final prediction of the neural model.

CoSThDist software model. For the distribution of traffic flows in the MPLS network, we created CoSThDist software model. The model has some constraints: there is chosen a simplified graph for displaying the MPLS network in the rail transport ITC (see Fig. 1); the communication channel load and the generation of average traffic rates is performed using a pseudo-random number generator; there are considered only those "sender-recipient" variants, which pass through the core of MPLS network; delays on data lines and routers of the rail transport ITS network are not taken into account; for a time no more than 10 flows are serviced.

During the simulation, five tunnels are considered (bidirectional). The tunnels laid in the direction, for example, from 8 to 11, will be called forward, and the tunnels laid in the direction from 11 to 8 - backward. The forward tunnels have indexes from zero to four, and the backward tunnels - from five to nine. Tunnels with indexes 0 and 1 have the CoS0 class (highest priority), tunnels with indexes 2 and 3 have the CoS1 class, and the tunnel with index 4 has the CoS2 class. Classes of backward tunnels correspond to classes of forward ones. So, tunnels with indexes 5 and 6 have CoS0 class and correspond to tunnels with indexes 0 and 1, tunnels with indexes 7 and 8 (class CoS1) – to tunnels with indexes 2 and 3, and the tunnel with index 9 (class CoS2) corresponds to the tunnel with index 4.

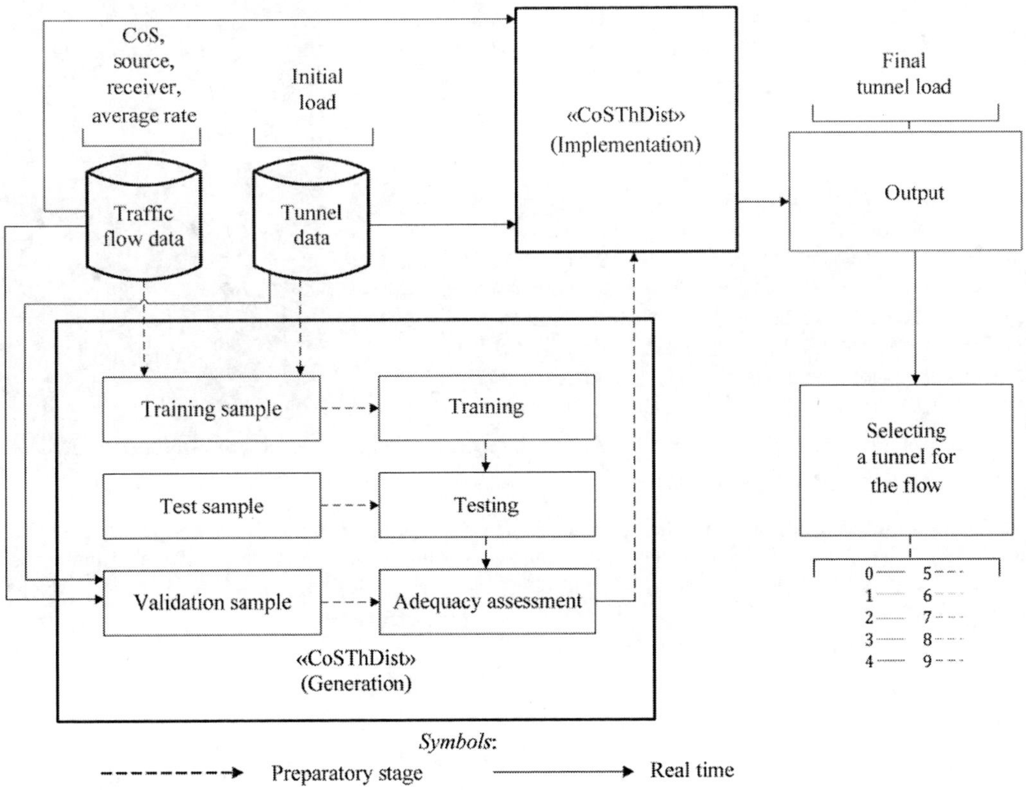

Fig. 3. Scheme for organizing tunnel selection in the MPLS network

If the load factor of the tunnel exceeds 0.65, then the subsequent flows will no longer be directed through this tunnel.

If there are no tunnels for the flow of the specified class, this flow should be directed to the tunnel with a less-priority CoS.

The sample is generated for each class of traffic and each possible route passing through the MPLS tunnel system. The size of the total sample can be calculated by the formula: $S_{sam} = k_1 \cdot k_2 \cdot k_3$, where k_1 – sample size for each class, k_2 – number of CoS, k_3 – number of possible "sender-recipient" combinations.

The input and output layers of the created NN use the Relu activation function, which has the following formula $f(x) = \max(0, x)$ and implements a simple threshold transition to zero. The Relu activation function graph is shown in Fig. 4.

NN training is carried out in 520 epochs. The total sample is divided into training and test ones. The test sample is allocated with 20% of the size of the total sample, that is, 10240 examples.

The research on the model showed that the accuracy graph on the test sample varies strongly from 0 to 0.8, which suggests that the algorithm for selecting the best tunnels for distributing traffic flows is rather complicated for the multilayer perceptron model, however, the accuracy of the trained NN was 74.5%, that is, such NN is capable of correctly distributing traffic flows in MPLS network in 74 out of 100 cases. For the MPLS network of the rail transport ITS fragment, this accuracy should be sufficient, but for a larger number of flows, a more complicated neural model may be needed.

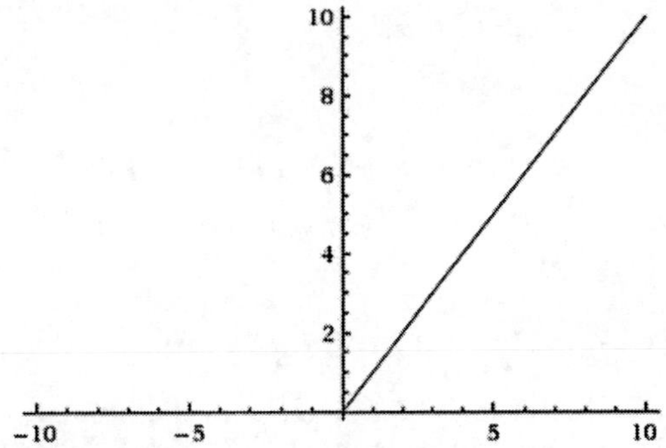

Fig. 4. Relu activation function graph

Findings

The control sample fragment and the NN prediction result are presented in Fig. 5

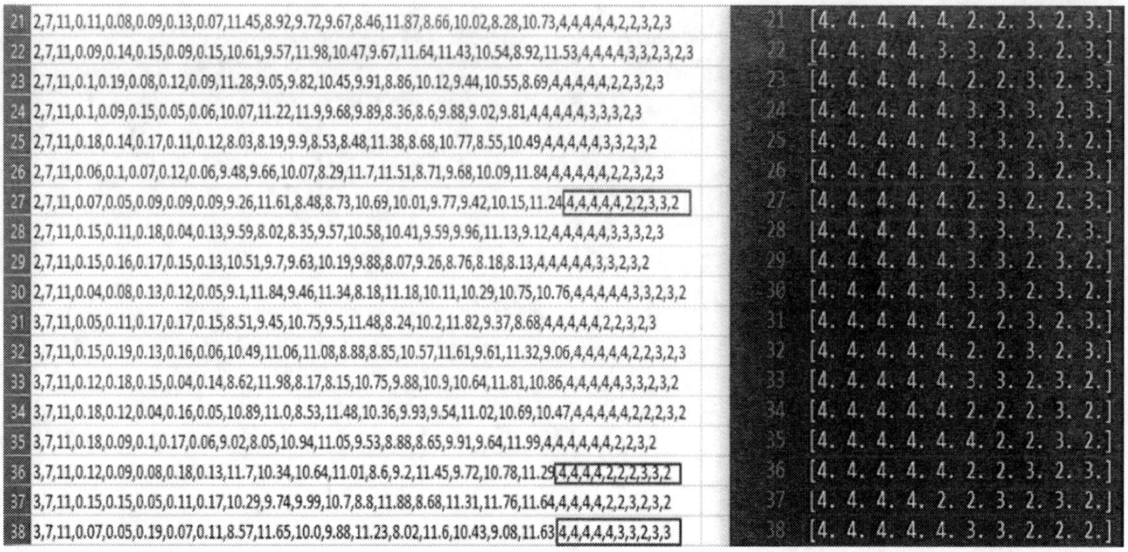

Fig. 5. Comparison of the control sample fragment with the NN prediction

The figure shows that in some lines the NN made an error. Line 21 (see Fig. 5) shows that for input parameters: l=2 (CoS); i=7 (sender); j=11 (recipient); at current load of tunnels k_{load} ={0.11, 0.08, 0.09, 0.13, 0.07} and at the average flow rates $v_{ij}(l)$ = {11.45, 8.92, 9.72, 9.67, 8.46, 11.87, 8.66, 10.02, 8.28, 10.73} we obtain the following traffic flow distribution $F(l)$ = {4, 4, 4, 4, 4, 2, 2, 3, 2, 3}, which is demonstrated in Fig. 6.

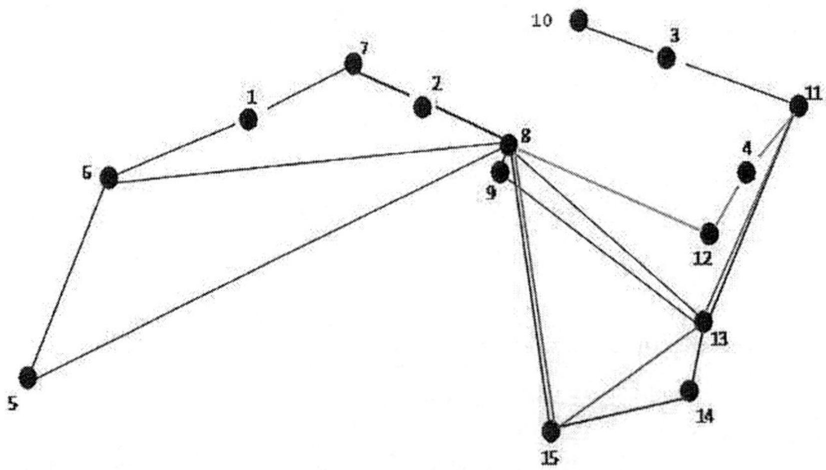

Symbols: use of tunnel with index 2 – bright green colour; with index 3 - blue, with index 4 - purple

Fig. 6. Distribution of traffic flows by tunnels

Originality and practical value

Research of the effect of the activation function on the NN prediction accuracy. The Keras framework, used for building the NN, contains many different activation functions, among which the most popular are: Relu; Linear; Softmax; Tanh; Sigmoid. Training of the NN of configuration 18-3-16-12-8-10 was performed under the following conditions: sample size – 6 400 examples; optimizer - Adam; learning rate - 0.0001; the number of epochs - 520. The research results are summarized in Table 1.

Table 1. NN prediction accuracy by the type of activation function

Parameter	Activation function of hidden layers			
	Relu	Softmax	Tanh	Sigmoid
Accuracy, %	37.23	74.5	37.69	0.59

According to the research results, it was found that the highest accuracy of the NN prediction occurs when using the activation function Softmax (74.5%). This is approximately 2 times higher than when using the Tanh activation function, with an accuracy of 37.69%. Other activation functions also provide less accuracy.

Research of the effect of the sample size on the NN prediction accuracy. Training of the NN of configuration 18-3-16-12-8-10 was performed under the following conditions: activation function - Softmax; optimizer - Adam; learning rate - 0.0001; the number of epochs - 520. The research results are summarized in Table 2.

Table 2. Accuracy of NN prediction by total sample size.

Parameter	Total sample size			
	2 560	6 400	25 600	128 000
Accuracy, %	44.34	74.5	76.25	43.42

From the table it is clear that when using the NN of configuration 18-3-16-12-8-10 it is enough to have a sample of 6400 examples. With the increase in the total sample size, the accuracy of the NN prediction rises insignificantly, and herewith one may observe over-training of the NN.

Conclusions

1. In order to improve the quality of network operation in rail transport ITS during interaction of different railways, it is expedient to use MPLS technology and organize the selection of a tunnel by means of a multilayer neural network based on data on traffic flows and communication channels that constitute the corresponding tunnels.
2. Using the Keras framework, the program model in Python is created for traffic flow distribution in the advanced MPLS network of the analysed rail transport ITS fragment "CoTShDist". The model generates a sample for the neural network and implements this network. The input of the neural network is the flow parameters (service class, sender and recipient identifiers; average flow rate), and the tunnel load, the output vector of the neural network is distribution of flows through tunnels in the MPLS network.
3. On the created CoSThDist software model, we researched the neural network parameters. It is determined that the optimal option is the NN of configuration 18-3-36-10 (with Softplus activation function in hidden layers), which, when using the Adam Optimizer, gives 74.5% accuracy in 520 epochs for a total sample of 6 400 examples.

References

1. V. M. Pakhomova *Research on the information-telecommunication system of railway transport using artificial intelligence*: monograph, *Dnipro: PF "Standard-Service" Publishing House*, 220, (2018)
2. I. Zhukovyts'kyy, V. Pakhomova, *Research of Token Ring network options in automation system of marshalling yard,* Transport Problems, **13**, Iss. 2, 145–154 (2018), doi: 10.20858/tp.2018.13.2.14
3. Vivek Alwayn, *Advanced MPLS Design and Implementation, - M.: Williams*, 474 (2004)
4. V.G.Olifer, N.A.Olifer, *Computer networks. Principles, technologies, protocols, St. Petersburg: Piter*, 944 (2010)
5. O.I. Romanov, S.P. Pasko, *Estimation of time delay in IP networks and MPLS in servicing messages in complex multitransit communication lines,* Scientific reports of the National Technical University of Ukraine "KPI", **5**, 11-20 (2011)
6. K. M. Rukkas, K. A. Ovchinnikov, N. A. Korolyuk, *Comparative analysis of modern control systems for MPLS networks,* Systems of Information Processing, **8**, 233-237 (2011)
7. O. Akinsipe & F. Goodarzi & M.Li, *Comparison of IP, MPLS and MPLS RSVP-TE Networks using OPNET* [electronic recourse], International Journal of Computer Applications, (2012), *Access mode: https://pdfs.semanticscholar.org/bef6fe5e.pdf*
8. A. Ghanwani et al, *Traffic Engineering Standards in IP Networks Using MPLS* [electronic resource] Communication Mag, (1999), IOS Press (55-76), Access mode: http://citeseerx.ist.psu.eduoi=10.1
9. R. S. Naoum, M. Maswady, *Performance evaluation for VOIP over IP and MPLS* [electronic resource], World of Computer Science and Information Technology Journal (WCSIT), **2**, 110-114 (2012), Access mode: http://file.scirp.org/Html/5-6101256_25011.htm
10. M. A. Rahman et al, *Performance Analysis of MPLS Protocols over conventional Network* [electronic resource] Microwave Conf., China-Japan Joint, Shanghai, Sept. (2008), Access mode: http://www.academia.edu/381853/Pel_Network
11. K. Herguner, R. S. Kalan, C. Cetinkaya, M. Sayit, *Towards QoS-aware routing for DASH utilizing MPTCP over SDN,* IEEE Conference on Network Function Virtualization and Software Defined Networks (NFV-SDN), (6–8 Nov. 2017). – Berlin, Germany, 1–6 (2017), doi: 10.1109/nfv-sdn.2017.8169844
12. A.B. Goldstein *Tunneling control model in MPLS network*, Informatization and Communication, **1**, 10-14 (2015)
13. O.Yu. Zaichenko, S.O.Kuhariev, O.V. Kuharieva, *Simulation of networks with MPLS technology for the purpose of managing various types of traffic, Bulletin of the National Technical University of Ukraine "KPI". - Ser: Computer Science, Management and Computing*, **47**, 154-161 (2007)
14. D. A. Zaitsev, A. L. Sakun, *Research on MPLS efficiency using colored Petri nets,* Zv'yazok, **5**, 49-55 (2006)
15. Ahmed A. M. Sharadka, *Method for optimizing flow distribution in computer networks using MPLS*

technology, Author's abstract ... cand. of Tech. Sciences: 05.13.13 / *NTU "Kyiv Polytechnic Institute"- Kyiv,* 20 (2007)

16. N.V. Budyldina D.S. Tribunsky, V.P. Shuvalov. *Optimization of networks with multiprotocol label switching, – M.: Goryachaya Liniya-Telecom*, 144 (2010)
17. V.M Pakhomova, *Research on traffic engineering in the computer network of UZ using MPLS TE technology, Science and Transport Progress. Bulletin of the Dnepropetrovsk National University of Railway Transport*, **1 (55)**, 154-157 (2015)
18. V. M. Pakhomova, I. D. Tsykalo, *Optimal route definition in the network based on the multilayer neural model*, Science and Transport Progress, **6 (78)**, 126-142 (2018), doi: 10.15802/stp2018/154443
19. Wu-Hsiao Hsua, Yuh-Pyng Shieha, Jenhui Chen, *Multiple path selection algorithm for Diff Servaware MPLS traffic engineering*, Computer Communications. – Access mode: https://pdfs.semanticscholar.org/5346/9a8d456c.pdf
20. I. P. Bolodurina, D. I. Parfenov, *Dynamic routing algorithms and methods for controlling traffic flows of cloud applications and services,* Bulletin of the South Ural State Univ. Series: Computational Mathematics and Software Engineering, **2**, **Vol. 6**, 84–98 (2017)

Estimation of the interaction level between urban passenger transport and city train

Olga Ozerova[1], *Petro* Yanovsky[2], *Viktoriia* Yanovska[3], *Sergiy* Lytvynenko[4,*], *Larysa* Lytvynenko[5], and *Serhii* Martseniuk[2]

[1] Mykolayv College of Transport Infrastructure of the Dnipropetrovsk National University of Railway Transport, Department of Transport Technologies, 54020, Mykolayv, Ukraine
[2] National Aviation University, Military Training Department, 03058 Kyiv, Ukraine
[3] State University of Infrastructure and Technology, Department of Economics, Marketing and Business Administration 04073 Kyiv, Ukraine
[4] National Aviation University, International Economy Department, 03058 Kyiv, Ukraine
[5] National Aviation University, Management of Foreign Economic Activity of Enterprises Department, 03058 Kyiv, Ukraine

Abstract. In the article the estimation of the interaction level between urban passenger transport and city train was made using the systems approach through application of modern methods of developing adequate easy-to-use mathematical models. Applying the systems approach, the transport node was considered as a comprehensive object, which is a single entity. It was identified that the transport node efficiency depends on the interaction level of the structure and its technology with the passenger traffic that requires designing a rational structure of the node and providing the technological interaction setting. The flow chart of determination the indicators of urban passenger transport modes' operation was proposed in order to improve the passenger service quality by increasing the level of interaction of urban passenger transport modes with each other, also the system of efficient use of urban transport means was developed. The system of criteria for assessing the level of interaction between urban passenger transport and city train was implemented, which consists of criteria for the individual assessment of urban passenger transport and city train and criteria for their system interoperability, including consideration of the system quality from the passenger's point of view.

1 Introduction

Because of great complexity of transport systems (TS), they are difficult to calculations to determine their perspective parameters. But transport science and practice have accumulated specific experience of using certain methodological approaches and models for the implementation of appropriate calculations over the past. As an example, it should be noted that complex transport nodes, where the biggest share of the TS' operation on the redistribution of passenger flows by directions and routes is concentrated, was viewed as mass service systems (MSS). Intrinsically, this is not a fairly accurate consideration of transport nodes (TN) since management plays a significant role in nodes, which is practically not taken into account in the MSS. Therefore, complex structures of TN should be investigated using a systems approach.

The systems approach involves considering such a complex object as a transport node, as an organic whole. In this case, the output parameters of the TN are affected by the interaction of its elements and subsystems, as well as their structures and technologies. Especially the interaction of passenger flows and technological structure should be highlighted. These aspects are practically not subject to formal description. Therefore, when evaluating the investment efficiency in the development of TN, and particularly in improving the passenger service quality in TN, a systems approach should be implemented using modern methods of developing adequate easy-to-use mathematical models.

2 Literature review and defining the problem

The papers [1-9] suggest approaches to solve the problem of train routing based on genetic algorithms. The authors in [1] proposed a genetic algorithm for routing and scheduling trains to achieve efficient and reliable travel itinerary and schedule. The algorithm can change the start time of trains, as well as some parts of the train trajectory. Computer modelling has shown that the developed algorithm allows reducing train delays.

The results of scientific research given in [2] reveal the ways of solving the issue of finding robust and stable solutions for the flexible job shop scheduling problem considering random machine breakdowns using a two-stage Hybrid Genetic Algorithm to generate the predictive schedule. The author determined that different measures have different sufficient implications on the relative performance of the offered rescheduling method in comparison with other existing methods.

The scientific work [3] is devoted to the high-value issue of the train scheduling improvement through the implementation of hybrid job shop approach. The author developed a constructive algorithm on the base of a disjunctive graph model of train operations and adapted

* Corresponding author: sllitvinenko1982@gmail.com

Simulated Annealing and Local Search meta-heuristic improvement algorithms allowing more flexible train scheduling which can be realized in real conditions.

The study [4] considering train scheduling issues aimed at determining a periodic timetable for a set of trains based on track capacities and given operational constraints. It is focused on solving the problem of one-way track linking two stations having a number of intermediate stations. It was made with the use of a model on the base of directed multigraph in which the variables in the relaxed constraints are associated only with nodes. The results of applying the proposed model confirm a sufficient relaxation solution speed-up.

The genetic algorithm and the hybrid genetic algorithm proposed by the authors in [9] were used for optimal planning of public transport vehicles based on their actual operating environment. The results of the study provided an opportunity for a smart planning of public transport vehicles based on the proposed algorithms, which eliminated the disadvantages of traditional planning methods.

3 Research aim and objectives

The research is aimed at the comprehensive estimation of the interaction level between urban passenger transport and city train using the systems approach and applying modern methods of developing adequate easy-to-use mathematical models.

Main objectives of the research are as follows: to consider the transport node as a comprehensive object and a single whole applying the systems approach; to develop the flow chart of determination the indicators of urban passenger transport modes' operation and the system of efficient use of urban transport means; to design the system of criteria for assessing the level of interaction between urban passenger transport and city train taking into account the system quality from the passenger's point of view.

4 Overview key findings

4.1 The flow chart of a transport passenger node of a large city

Primarily, the issue of transport nodes' development is closely linked with the beginning of the railway infrastructure development and the emergence of crossing points in the network structure, mergers, branching of several lines and transport modes.

In this regard, the question arises: have transport nodes kept their weight in the society in the result of the transition of the state from the planned to the market economy?

It should be emphasized that the structural reform in transport is aimed at improving the transport service quality of the clientele in its broadest sense: by volumes and duration of passenger delivery, by regularity and transportation costs, compliance with safety requirements. Therefore, it is necessary for the transport system, an important component of which are transport nodes, to ensure a high speed of passengers' delivery with full and effective use of fixed assets and rolling stock and absolute meeting of the growing demand of users for the provision of high-quality transport service. Based on this, a comprehensive and systems approaches to improving the functioning of transport nodes in large cities provide an opportunity to identify real and economically feasible ways to improve the efficiency of transport operations with passengers and achieve high rank by railways among other transport modes in the state and at the international level.

Consequently, the systems analysis most closely corresponds to the nature of the tasks for improving the passenger service system in transport nodes. So, their study should be based on the following principles [10]:

- the development of transport nodes is carried out within a single transport system, taking into account the state and prospects of the national economy;
- the development of transport nodes establishes economic, social, political and defence requirements;
- two main subsystems – flows of rolling stock and permanent means – technologically interact in transport nodes;
- capacities of transport nodes via indicators of carrying, throughput and processing capacity influenced by the state and system of technical facilities' operation and organization level are quantitatively assessed;
- the sequence of improving the organization of transport nodes is regular in the phased implementation of the formation of new structures in their composition and the introduction of modern technologies;
- emergence of new qualities and properties, different from the past are important features in the improvement of transport nodes.

In the transport node, the incoming flows of passengers are transformed in outgoing flows, which uses its infrastructure. It is a terminal with a complex structure of various transport modes for the processing and smoothing of uneven passenger flows. In order to determine the capabilities of subsystems and elements of the transport node, their processing and throughput capacities are determined.

In addition, the operation of all components is regulated by scientifically grounded technological processes. The structure of the transport node implies not only the spatial location of its elements, but also performing specific functions by them.

The role of specific elements is determined by the scheme and technology. Therefore, the technology needs to be developed and implemented on a specific scheme, providing an organized passage of flows in the structure of the node in compliance with technological standards (Fig. 1).

Efficiency of the transport node depends on the level of interaction between the structure and its technology with the flows of passengers. For this purpose it is important to build a rational structure of the node and to set up technology on it.

The rational structure of the transport node provides the required throughput capacity and volume of its elements at the lowest operation cost. The transport node's rational structure development involves an extensive network of its elements for flow passage, as well as the ability of smoothing the unevenness of the incoming and outgoing flows. Herewith, it is important to consider the interaction of elements, subsystems and management functions and to reflect interconnection and the influence of the node structure parameters.

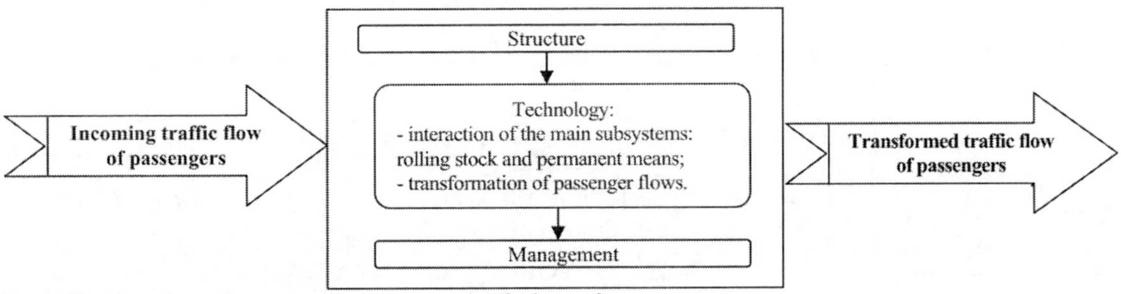

Fig. 1. The flow chart of a transport passenger node of a large city

The passage and transformation of flows should be carried out with minimal losses, and the node was able to adapt to changes in variable operating conditions, requiring reasonable reserves to ensure a high degree of economic stability and technological flexibility.

The transport node is a complex organizational system with nonlinear dependencies between its subsystems and elements. Its technological parameters (carrying, throughput and processing capacity, etc.) are determined not only by the parameters of its individual elements, but also by the quality of its structure, operation technology and management system.

The quality of management plays an important role in its operation, which provides flexibility, manageability of the implemented technology, and as a result of managerial actions – high economic result of the whole node operation. In addition, the fact that the management creates the availability of reserves should to be taken into account. It will provide passenger service cost savings of railways.

In the transport node of a large city, passenger transfer points (PTP) are important elements where the intersection of the transport routes of various modes of urban passenger transport (UPT) and the distribution of passenger flows between them (transfer of passengers) are carried out. The duration of each passenger's trip depends on the level of interaction between the modes of urban passenger transport in PTP.

4.2 Determining performance indicators of urban passenger transport modes in the node

Based on studying the practice of passenger transport service in large cities and the results of conducted researches on the rationalization of UPT operation [10], a flow chart of determining performance indicators of UPT modes was developed for the purpose of improving passenger service quality by increasing interaction level of UPT modes between each other (Fig. 2).

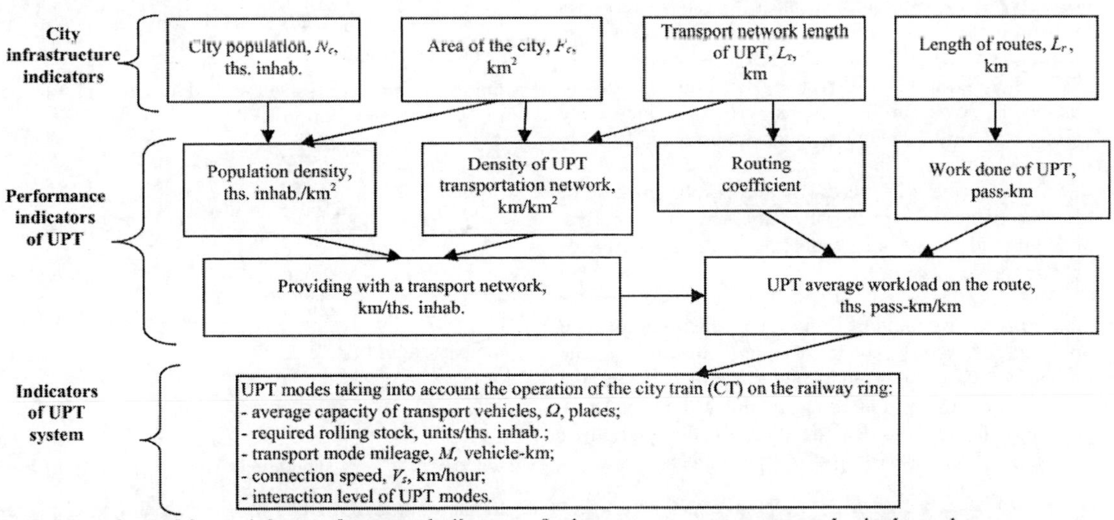

Fig. 2. Flow chart of determining performance indicators of urban passenger transport modes in the node

The interaction of UPT modes, primarily, will shorten the waiting times for the transport vehicles by passengers in transfer places, and as a result, the connection speed V_s will increase and the duration of passengers' delivery from places of residence to the places of destination (work, study, etc.) will decrease.

In addition, it should be noted that the implementation of certain measures to improve the functioning of UPT modes due to ensuring their interoperability will lead to increase in average capacity of transport vehicles, reduction of the need for rolling stock, reduction of transport mileage, and as a result – reduced costs for the UPT operation. The system of effective use of urban transport modes is presented in Fig. 3.

When solving transport problems of the city, it is necessary to study the time costs associated with the movement of the population from the place of residence to the place of work (study), in order to determine the impact degree of passenger transportation delays. The duration of the displacement of people from their place of residence to the place of work (study) consists of a separate moving by different UPT modes and walking. Separately, the time lost by passengers due to waiting UPT vehicles and city train (CT), as well as the number of transfers are taken into account.

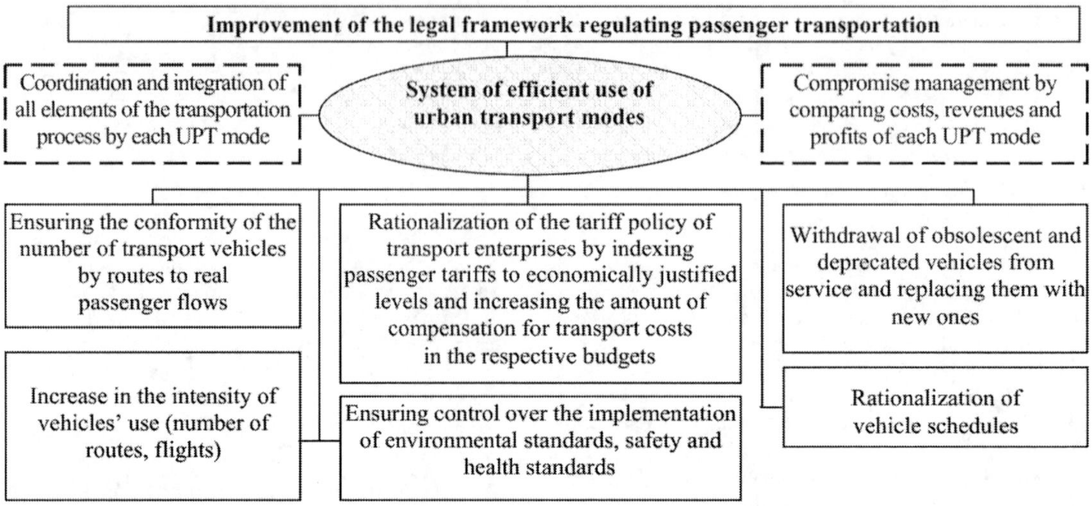

Fig. 3. System of effective use of urban transport modes

4.3 Modelling time costs of urban passengers in the system "Urban Passenger Transport – City Train" considering waiting times for the transport vehicles

To reduce the time spent on all types of trips, it is necessary to solve a complex set of the following issues in the city combined with suburban area due to the close interaction of the UPT with suburban passenger transport (SPT) modes, which is driven by the commonality of passenger flows: rolling stock renewal; increasing culture and service of passengers; improvement of the transport network in conjunction with intermodal interfaces; increase in speed of vehicles; organization of direct connection between the city and the suburban area; provide necessary intervals and regularity of vehicle traffic.

The time spent on the moving of passengers for labour purposes should be defined as the amount of time on transportation, vehicle waiting considering the number of passengers who are at the UPT stop, the duration of pick-up and drop-off, the vehicle capacity and load factor, and the duration of the pedestrian passage, and pick-ups (drop-offs) in a vehicle:

$$t_{mov} = t_{ped}^{dep} + t_{dep}^{wt} + t_{pup} + t_{tra}^{UPT} + t_{stp} + \\ + t_{tr}^{PTP} + t_{p}^{CT} + t_{tra}^{CT} + t_{d}^{CT} + t_{ppg}^{dst} \quad (1)$$

For passengers using the system "Urban Passenger Transport – City Train" it is important that the following condition is met:

$$t_{ats} \leq t_{sts}, \quad (2)$$

where t_{ats} – actual duration of the passenger transportation by UPT and CT considering real delays (waiting periods) in PTP; t_{sts} – scheduled transportation duration of the passenger by UPT and CT considering real transfer duration in PTP.

$$t_{ats} = t_{ats}^{UPT} + t_{awt}^{PTP} + t_{ats}^{CT}, \quad (3)$$

where t_{ats}^{UPT} – actual duration of passenger transportation by UPT; t_{awt}^{PTP} – actual waiting periods (transfers) of the passenger in the PTP; t_{ats}^{CT} – actual transportation time of the passenger by CT.

$$t_{sts} = t_{sts}^{UPT} + t_{tr}^{PTP} + t_{sts}^{CT}, \quad (4)$$

where t_{sts}^{UPT} – scheduled duration of passenger transportation by UPT; t_{tr}^{PTP} – duration of passenger transfer in PTP; t_{sts}^{CT} – scheduled duration of passenger transportation by CT.

Passenger costs when transporting by the "UPT – CT" system:

$$\Delta t_{cst}^{pss} = t_{ats} - t_{sts} \text{ at } T = const, \quad (5)$$

where T – tariff for transportation of one passenger, UAH.

Effectiveness criteria of the "UPT – CT" system operation (from city authorities' point of view):

$$K_{mun} = \frac{D_b}{K_b} \geq 1, \quad (6)$$

where D_b – receipt of funds to the municipal budget in the form of taxes paid by transport enterprises, UAH; K_b – funds sent from the municipal budget for commercial payments to carriers (subventions for subsidized transport of certain population categories, UAH).

Transport service coefficient (in terms of UPT):

$$K_{tra} = \frac{P_e}{B} \geq 1, \quad (7)$$

where P_e – profit of the transport enterprise from activity, UAH; C – transportation costs of the transport enterprise, UAH.

Quality criteria by the transportation duration in providing transport services by the "UPT – CT" system (from passenger's point of view):

$$K_q^{tra} = \frac{t_{sts}}{t_{ats}} \leq 1 . \quad (8)$$

The level of providing transport services by the "UPT – CT" system in terms of the passenger taking into account can be estimated as follows:

$$K_{pss} = \frac{(t_{sts} - t_{ats}) \cdot (B_{tra}^{UPT} + B_{tra}^{CT})}{T^{CT} + T^{UPT}}, \quad (9)$$

where $B_{tra}^{UPT} + B_{tra}^{CT}$ – cost of a 1-time passenger transportation by UPT and CT respectively, UAH; $T^{CT} + T^{UPT}$ – transportation tariff of one passenger by UPT and CT respectively, UAH.

Integrated efficiency criterion of the "UPT – CT" system on the basis of is determined as follows:

$$K_i = K_{mun} B_{mun} + K_{tra} B_{tra} + K_{pss} B_{pss}, \quad (10)$$

where W_{mun}, W_{tra}, W_{pss} – weight coefficients of efficiency criteria components.

The value of weight coefficients is determined taking into account the conditions of UPT and CT operation in the city using real statistics (it can be set by the value of determination coefficient R^2).

Also the functioning of UPT and CT should take into account the criteria for assessing their level of interaction, as well as providing quality service for the passenger. Further, a system of criteria for assessing the interaction level between UPT and CT should be proposed.

The railway and UPT interaction option should estimate the time spent by passengers on the trip. Moreover, it is necessary to take into account additional capital investment and operating costs caused by the development of a specific transfer station on the railway ring at the node, which will be carried out in further scientific researches.

5 Conclusions

The results of observations on the carriage of passengers in large cities testify to the following:

- there is a regularity of the influence of the city size on costs of citizens for their transportation (the greater the city, the more time spent on travel to the city and from the suburban area to the city;
- assessment of the population transport service level should be carried out according to the criterion of total transportation time, which varies significantly in different cities due to differences in the industrial structure of these cities;
- passengers use all available UPT modes in intra-city transportation of large cities;
- passengers in most cases use two transport modes (railway or bus and one UPT mode) to travel from a suburban area to the city;
- cultural and domestic movements within the territory of a large city are very complex, which depends on numerous factors and, above all, on the location and purpose of cultural and community establishments.

The above mentioned factors affecting the passenger transportation quality in large cities and necessary priority tasks for its increase influence on the economic justification of volumes and direction of investment in the urban passenger transport development.

References

1. O. Gholami, Y.N. Sotskov, *IFAC Proceedings Volumes*, **45**, 6: 158-163 (2012)
2. N. Al-Hinai, T.Y. ElMekkawy, *International Journal of Production Economics*, 132: 279–291 (2011)
3. R.I. Burdett, E. Kozan, *A sequencing approach for train timetabling* (School of Mathematical Sciences, Queensland University of Technology, 2007)
4. A. Caprara, M. Fischetti, P. Toth, *Operations Research*, **50**: 851-861 (2002)
5. A.E. Eiben, J.E. Smith, *Introduction to Evolutionary Computing* (Springer, 2003)
6. C.J. Hinde, M.S. Withall, I.W. Phillips, T. Jackson, S. Brown, R. Watson, *Proceedings of the International Conference on Evolutionary Computation*: 170-175 (2010)
7. S.Q. Liu, E. Kozan, *Transportation Science* **45**, 2: 175-198 (2011)
8. M.M. Nasiri, F. Kianfar, *Computers & Industrial Engineering*, **61**:161-170 (2011)
9. F. Zhang, X. Cao, D. Yang, *Tsinghua Science and Technology*, 13: 625-631 (2008)
10. E.A. Safronov, *Transport systems of cities and regions* (ASV Publishing House, 2007)

Implementation of road charging in the city of Žilina – a tool for reducing negative externalities

Alica Kalašová[1,*], *Veronika* Harantová[1], *Simona* Kubíková[1], and *Jerzy* Mikulski[2]

[1]Department of road and urban transport, Faculty of operation and economics of transport and communications, University of Žilina, Univerzitná 1, 010 26, Slovakia
[2]University of Economics, 50, 1-Maja str, 40-287, Katowice, Poland

Abstract. It has been proven on several implementations worldwide, that introducing of road charging scheme in urban area is very simple and efficient way to reduce traffic volumes, lower congestions and shorten travel time. Herewith, there is a possibility of utilization of charged money, e.g. for financing of public transportation. In our paper, based on analysis of traffic intensities and results of social survey, we describe a proposal of charging schema for the central part of the city of Žilina in Slovakia.

1 Introduction

Toll collection is one of the ways that users of road network can pay for its utilization. Type of the vehicle, the distance run off in charging zone or the time period spent there are usually main factors to determine the rate of charging. There are three main reasons, why toll collection is introduced to:
- obtain finance that can be further used for construction of new road infrastructure or improvement of public transportation,
- to reduce emissions, noise and other negative externalities produced by transportation,
- to reduce congestions on selected roads (areas) or time periods in order to increase capacity of road network.
Nowadays, there is tend to build up and operate systems of electronic toll collection.

In table 1, there you can see examples of toll rate dividing into two parts – toll rate outside rush hour and in the rush hour. The toll rates are set per day or per entry to city centre. In the last column of table 1, there is traffic reduction which is a result of toll collection implementation. A reduction of traffic volume in the city centres has an important positive effect on citizens, environment, as well as on better quality of life in cities.

Table 1 shows some examples of electronic toll systems in the chosen European cities [2]. These systems are based on DSRC and ANPR technologies.

Dedicated Short Range Communications (DSRC) is an open-source protocol for wireless communication, similar in some respects to WiFi. While WiFi is used mainly for wireless Local Area Networks, DSRC is intended for highly secure, high-speed wireless communication between vehicles and the infrastructure.

Table 1. Toll collection systems in European cities [10]

City	Start	Technology	Toll rate		Traffic reduction
			outside rush hour	in the rush hour	
Stockholm	2006	DSRC/ANPR	11-15 SEK/entry (1-1,4 €/entry)	25-35 SEK/entry (2,4-3,3 €/entry)	20%
London	2003	ANPR	10-14 £/day (11,2-15,7 €/entry)	10-14 £/day (11,2-15,7 €/entry)	30%
Bergen	1986	DSRC	10-30 NOK/entry (1-3,1 €/entry)	20-56 NOK €/entry (2,1-5,8 €/entry)	7%
Oslo	1990	DSRC	4-25 NOK/entry (0,4-2,6 €/entry)	5-31 NOK/entry (0,5-3,2 €/entry)	5%
Trondheim	1991	DSRC	11-15 NOK / entry (1,15-1,55 €/entry)	14-30 NOK / entry (1,45-3,1 €/entry)	10%
DSRC = Dedicated Short Range Communication, ANPR = Automated Number Plate Recognition					

* Corresponding author: simona.kubikova@fpedas.uniza.sk

Automatic number-plate recognition (ANPR) is a technology that uses optical character recognition on images to read vehicle registration plates to create vehicle location data. It can use existing closed-circuit television, road-rule enforcement cameras, or cameras specifically designed for the task. ANPR is used by police forces around the world for law enforcement purposes, including checking if a vehicle is registered or licensed. It is also used for electronic toll collection on pay-per-use roads and as a method of cataloguing the movements of traffic, for example by highways agencies. [10,11]

The key advantage, comparing to traditional systems with toll plazas, is that there is no need to stop vehicles when executing the process of charging. It means that such system is not a source of congestions and does not affect the fluency of traffic flow in a negative way. Recently, in many European countries electronic toll collection was introduced on highways and main roads.

However, the same principles can be utilized in urban areas as well.

2 Traffic characteristics of the city of Žilina

Žilina is the largest city situated on the Váh River, located 49°15′ of northern latitude and 18°45′ of eastern longitude. With population of 82 924 inhabitants (on 1st June 2019) and the area of 80,03 square kilometres it creates an important centre of northwest Slovakia.

As in other European cities, the traffic situation in Žilina is continually getting worse. Fast grow of individual transportation, descent of public transport together with development of new business and commercial activities in the city (manufacturers KIA MOTORS, MOBIS, and others) – all these processes make an important impact on the city communication system. There are about 68 792 registered cars (on 31th May 2019) in the Žilina county, the level of automobilization is about 1 car per 2,5 inhabitants [5].

The most important part of the city communication system is a 4-lane city ring, which surrounds the central part of the city and ensures that no transit traffic will pass through it (Figure 1.).

Fig. 1. Four – lane city ring of Žilina [5]

Although there can be found some bottlenecks on the city ring and in adjacent urban settlements, these problems are to be solved by better local organization of traffic or redesign of critical points. Also an intensive decrease of transit traffic is expected after completing of the highway next to the city.

The problem we deal with lies in the central part of the city where congestions and problems with parking irritate more and more people. Every year, Faculty of Operation and Economics of Transportation and Communications realizes a survey of traffic intensities on selected crossroads in Žilina. Results of these surveys, together with data gathered from several bachelor and diploma theses related to organization of traffic in the city, give us an opportunity to have a relatively detailed overview of traffic situation in Žilina within last 10 years. Based on the gathered data, we can see an unanimous increase of traffic intensities in covered area. The growth between years 2008 and 2018 reached up to 19 % on one of the monitored crossroads.

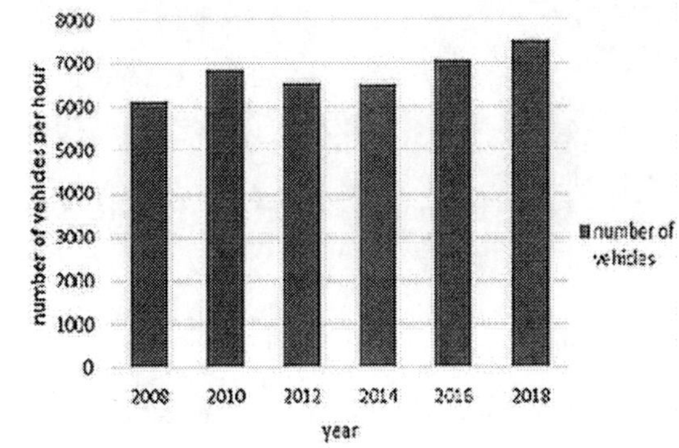

Fig. 2. Development of traffic intensities in covered area [6]

The growing traffic brings another related problem. As you can see in figure 3, the number of traffic accidents in Žilina County, is rising up [1].

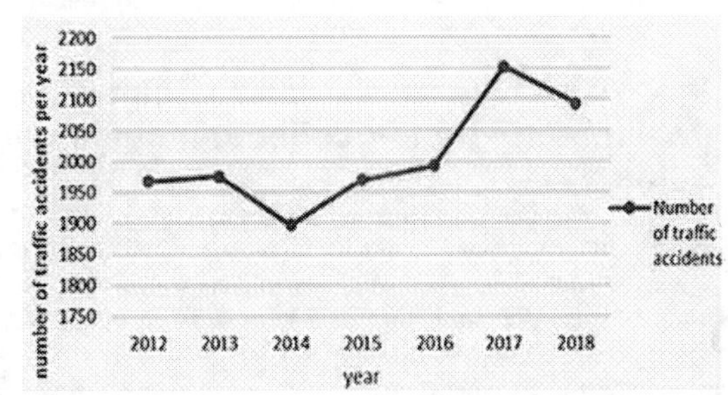

Fig. 3. The number of traffic accidents in Žilina [7]

On the other side, the utilization of public transport has dropped rapidly. As the statistics provided by Transport Company of the City of Žilina shows, there is a 21 % dropdown but since 2015, 12 % growth in passengers transported has been recorded. [4]. There are several reasons of this problem, inter alia insufficient financing of public transport and thus reduction of lines, disastrous organization of trolleybus network and at last but not least congestions that affect public transportation and make it unreliable [9,12].

Table 2 Utilization of public transport [4]

Year	Number of passengers
2008	13 505 000
2009	11 838 671
2010	11 312 297
2011	11 120 339
2012	10 890 930
2013	10 630 000
2014	10 809 000
2015	11 015 000
2016	11 411 000
2017	11 650 000
2018	12 053 000

Let have a look on parking situation in Žilina [3]. In 2002, Žilina-parking company was founded. 75 % of the company share has a private investor and 25% belongs to the city of Žilina. New parking concept was introduced since 2003 which includes parking places for residents and visitors. There are three parking zones for visitors in the central part of the city with following prices:

- Zone A, historical city centre and area surrounding area for pedestrians – 1,5 €/first hour, each additional hour 2 €,
- Zone B, around the historical centre – 0,8 €/first hour, each additional hour 1,6 €,
- Zone C, areas next to the city centre – 0,5 €/first hour, each additional hour 1 €.

Besides, above mentioned parking places, there are several parking places operated by other private companies, where the price can differ. After the first year of operation, when the number of cars in central city zone dropped partially, the drivers get used on the parking fee and the system is no more successful as a regulation of traffic, it was just an example of short - term regulation. Even the year-by-year increase of parking vehicles is 35% and the average fee paid by one car parking is approximately 0,6 €. Because there are not enough parking places, the cars are often parking on pavement and this is very disturbing for pedestrians.

3 The reasons and aims of implementing a toll collection system in Žilina

As all the previous regulations, especially the parking system does not lead to reduction of traffic and congestions. We think that it is necessary to come with more sophisticated solution of the problem. There is no place for increasing capacity of the city communication system, contrariwise we have to eliminate amount of the traffic and toll collection is the way that we can achieve it. The concrete aims of toll collection are:

- regulation of vehicles entering central part of the city,
- better environment and conditions of living for residents,
- larger utilization of public transport,
- source of financing for public transport and road infrastructure.

It is necessary to take into account that toll systems bring together some risks and negative impacts as well. For successful realization, it is necessary to focus on technological aspects and limitations, viability and financing of the system as well as public and political acceptance of the system. These factors will be discussed in the following paragraphs.

3.1 An acceptance survey of a toll collection system in Žilina

To know the level acceptance on toll system, an anonymous questionnaire survey in central city zone of Žilina was made in 2017. This survey included 15 questions prepared by Department of Road and Urban Transport, where it was also evaluated. In total, 383 questionnaires were collected.

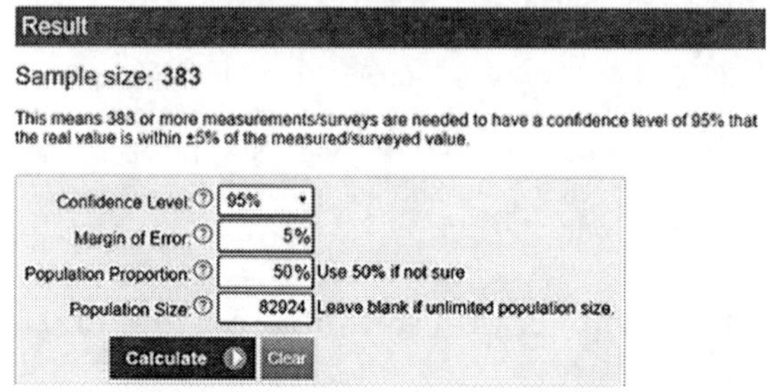

Fig. 4. The required number of respondents using Sample size calculator [8]

At first some basic facts about survey respondents: All the respondents were drivers parking in the central zone of the city - 62% men and 38% women. According to age structure, the largest part of them (45%) was between 23 and 45 years old; 81 % of respondents use private cars.

The survey gave us also the answer on how often the respondents visit the central zone. It is very interesting, that about 18 % of respondents enter the central zone more than 4 times a day and 21 % enter 3 times a day. Together with other respondents entering once (41%) or twice (20 %) a day, we can see that 65% of questioned drivers were regular visitors.

It is also very important to know reasons of respondents' visits in the central zone. The survey showed, that about 40% of trips are related with work and business. Nearly the same part of visits is related with shopping (24%) and private matters (25%), the rest remains for free time activities (11%). An alarming fact is that 65% of respondents do not use public transport for travelling to the central zone. Even other question showed that 44% of questioned people use the public transport less than once a day in average.

Other parts of questionnaires were related to satisfaction with traffic situation and the awareness of toll payment. Nearly 65 % consider the situation to be critical and see the increase of cars, 31 % are relatively satisfied, and only 4 % think that there is no problem. However, that the majority of respondents (87%) agree with the idea that there should be some regulation processes before the traffic system becomes uncontrollable.

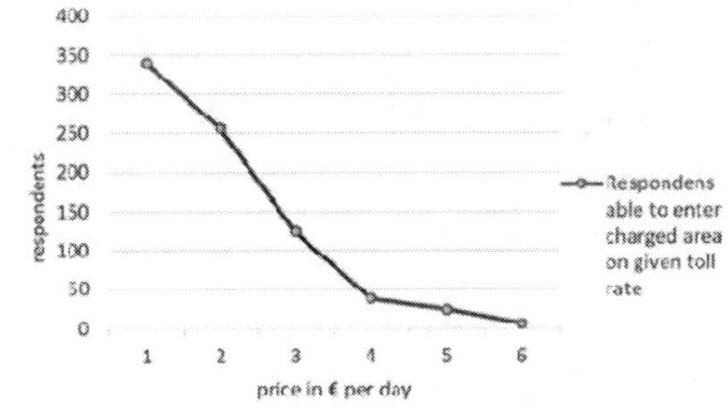

Fig. 5. Respondents able to enter charged area on given toll rate [own study]

Regarding to the awareness of the fee for entering central zone of the city, figure 5 shows how many respondents would be still able to enter the zone by car on given fee levels. The results of question „What is the fee level, which would certainly influence your decision what kind of transport mean to use to travel to the central zone?" are showed in figure 6.

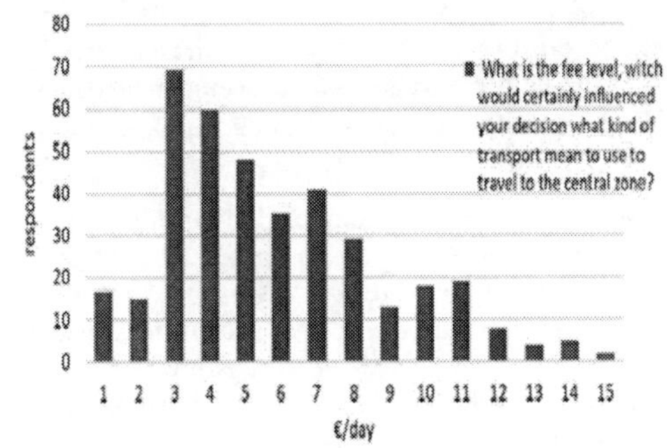

Fig. 6. The impact of given toll rate on respondents [own study]

3.2 The chosen area covered by charging

As was mentioned before, the main traffic related problems in Žilina are congestions and parking organisation in the central part of the city. We have prepared two possible variants, in both of them cordon method of charging where the payment is done on the entrance to the zone seems to be the best solution.

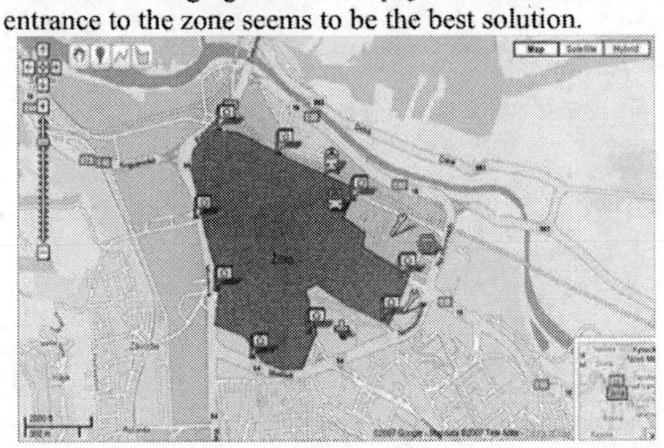

Fig. 7. The chosen area for charging with entrance gates [own study]

First, smaller variant assumed that only the interior part of city central part is affected by road charging. In that case, installation of 12 entrance gates would be necessary. However, with knowledge of traffic relations in solved area we suppose that this variant would not solve the problem. Probably it would make even increase problems with parking in surrounding areas and in this case, increase in utilization of public transport cannot be expected.

The second variant assumes to cover entire central part of the city with some exceptions (hospital, industrial park and shopping centre in the east). On the boundary of this area are opportunities to build parking places and provide the interchange to public transport. Another advantage is that in this configuration, only 9 entrance gates are needed (see figure 7).

4 A proposal of a toll collection system in Žilina

Other important step is to set the operation time of the system. That would have impact on the volume of traffic within traffic peaks and can help to decrease its oscillation during the day. Here we can utilize the knowledge earned in other cities where toll systems are in operation. The most common time of the operation is from 7.00 to 18.30 on working days.

4.1 Proposed technology for toll collection

When looking for the suitable technology, we are considering two factors:
- cost of the system – infrastructure and operation costs,
- cost of the equipment in vehicle

As can be seen in table 1, there are three main technologies for electronic toll collection. We assume that utilization of GPS system is not suitable for area of this size. Technology of DSRC is suitable, but it requires each car to be equipped with OBU. However, the system based on number plate recognition (ANPR) does not require any additional car equipment. When we consider that in all cases, there should be some kind of enforcement, this technology seems to be ideal for our solution and its reliability was proven in London. [3]

4.2 Toll rates

Setting the rate for entering solved area is the key step for successful realization of the project. When the price will be too high, there will be a negative impact on the traffic in the area. On the other side, when the fee will be too low, the toll collection will not solve anything, because it will not fulfil the aims, for which it was introduced. It is necessary to say that the fee would be probably paid once a day, no matter how often the car will enter the area. We also suppose that after introduction of toll there will be no other fees for parking in solved area. When determining the fee rate we took into account:
- average loan (it is 20 595 € in county of Žilina),
- current fees for parking (1-2 €),
- results of the acceptance survey.

Based on the mentioned criteria, we can consider 3 - 4 € to be an ideal fee rate. This rate is higher than the price for parking, so it is expected, that it will have the regulation effect, on the other side, it is relatively acceptable comparing to average loan. This level seems to be a good compromise according to results of acceptance survey.

4.3 Discounts, exceptions of payment

When thinking about who should have exception from payment, we have to go out from the aims of the toll system. As we remind that one of the aims is to reach higher utilization of public transport, it is clear that public transport vehicles should be free of charge. There can probably be some kind of discount for taxi vehicles. As we want to provide better environment conditions for residents, their vehicles should have some kind of discount as well. Of course there should be no fee for ecological vehicles (electrical, hybrid) and vehicles of police and rescue services.

4.4 Methods of payment

Together with realization of road charging, the system of payment must be designed. Several applications have showed that payment with utilization of internet and mobile phones is easy and simple. Of course, more alternatives how to pay help to better acceptance of the systems, so there should be other possibilities how to pay, for example on petrol stations, post offices or other selected places, a registered card payment can also be used.

5 Conclusion

There are many other important steps, which have to be taken prior to successful realization of road charging system. It must be clear, who will be the owner and operator of the system and who can provide the money for its installation. Not less important is the question how the profit from the operation will be used. According to our opinion, the profit should be used for improvement of the traffic situation in charged area, because its users will be affected by the system. However, it does not mean that we will build new roads and parking places at the expense of disruption green areas - we can rather focus on improvement of public transport. In Slovakia, there is not yet convenient situation for introduction of toll systems in cities. However, the capital Bratislava introduces low-emission zones in the city centre. Together with some legislative steps, it is necessary to inform the public about possibilities we have. By informing the public, we can reach our target – to increase the level of acceptance on road charging systems in cities and to show the benefits these systems can bring.

References

1. V. Konečný, Analýza dopadov dopravy na životné prostredie (Analysis of transport impacts on environment), *In proceedings of From Horse-drawn Railway to High-speed Transportation System*, ČVUT Praha (2007), ISBN 978-80-01-03699-0.
2. P. Urbánková, Úvod do mýtneho systému, Projekt Zavádění mýta ve městech v podmínkách ČR, available on : http://www.telematix.cz/projekty
3. Parking company Žilina, http://www.parkovaniezilina.sk
4. Transport Enterprise of the city of Žilina, http://www.dpmz.skz
5. Basic information about the city of Žilina and Žilina region, http://www.zilina.sk/
6. Traffic accidents in the Slovak republic https://www.ssc.sk/files/documents/dopravne-inzinierstvo
7. Slovak statistics in transport, https://www.minv.sk/?kompletna-statistika
8. Sample size calculator online, https://www.calculator.net/sample-size calculator.html?type=1&cl=95&ci=5&pp=50&ps=82924&x=72&y=35
9. R. Berežný, V. Konečný: The impact of the quality of transport services on passenger demand in the suburban bus transport, *Procedia Engineering,* Vol. **192,** p. 40-45 (2017)
10. A. Kalašová, S. Kubíková, J. Kupčuljaková: Traffic Time Delay Modelling on the Intersection in the City of Martin, Using Software Aimsun, *Recent advances in traffic engineering for transport networks and systems Book Series: Lecture Notes in Networks and Systems* **Volume: 21** Pages: 203-212 (2018)
11. A. Kalašová, K. Čulík, S. Kubíková: Smart City – Model of Sustainable Development of Cities, *Conference: 11th International Scientific and Technical Conference on Automotive Safety – proceedings paper* (2018)
12. A. Kalašová, J. Kupčuljaková: The future in the telematics applications as support for increased safety, *Transport problems,* **Vol. 7**, pages 103-109 (2012)

This contribution is made within project VEGA 1/0436/18 Externalities in road transport, an origin, causes and economic impacts of transport measures.

This contribution is the result of the project implementation: Centre of excellence for systems and services of intelligent transport II.,
ITMS 26220120050 supported by the Research & Development Operational Programme funded by the ERDF.

"Podporujeme výskumné aktivity na Slovensku/Projekt je spolufinancovaný zo zdrojov EÚ"

Improvement of international rail transportation of benzene between Ukraine and the European Union countries

Vasyl Gaba[1], *Tetiana* Hrushevska[1,*], *Oleg* Strelko[2], *Anna* Kyrychenko[2], *Marina* Rudyuk[1], and *Ievgen* Gusar[1]

[1]SUIT, Department of Management of Railway Commercial Activities, 29/18, Kotelnikov Street, Kyiv 03115 Ukraine
[2]SUIT, Department of Transportation Technologies and Process Control Traffic, 29/18, Kotelnikov Street, Kyiv 03115 Ukraine

Abstract. It is considered the possibility of improvement of the international railway transportation of benzene and other related cargoes between Ukraine and the European Union countries for the carriage of these cargoes in tank containers. The volume of chemical cargo transportation to the EU countries to the rolling stock types for 2018 has been analyzed. The research of the effectiveness of organization of international transportation according to the criterion of the total hours of downtime by wagons under accumulation. The mathematical model of chemical cargoes transportation by wagons, wagon groups and container trains with the accumulation to the technical standards on the cargo diversions was elaborated. The models take into account the unevenness of the wagon arrivals to the accumulation points and observance of the number of wagons to the established norm. A comparison of the effectiveness of the two models of the organization of the chemical cargo delivery and it was defined a more effective version of transportation. It has been established that for more efficient using of the rolling stock and for the rolling stock circulation decreasing is the cargo transportation by trains which consist of cisterns or/and tank container carriages. The influence of the components of the rolling stock circulation and its value are analyzed. The research of the cost of benzene transportation in cisterns and tank containers for the individual diversions was studied. According to the research results the cost of benzene transportation by platforms of the railways ownership is less than twice in comparison to private cisterns and tank-containers of the Ukrainian Railways fleet. These recommendations will improve the process of transportation and increase of volumes of international cargo by railroads between Ukraine and the EU countries.

1. Introduction

In today's conditions of globalization and the world economy integration Ukraine can not stay out of the world. Ukraine's entry into the world economy community gives access to international markets, expansion of trade and economic ties and improvement of transport and telecommunication systems.

Expansion of export of Ukrainian goods to the European market is one of the main factors for boosting the growth of Ukraine's gross domestic product, its economic prosperity and the population welfare.

In a scientific environment, specialists in transport technologies and organization of transportation by railroads have been constantly discussing the improving of international cargo delivery and different ways to organize trains along railroad tracks [1]. Two main variants are considered: the chemical cargo dispatch by wagons or wagon groups and by the composition of the container train to the standards during accumulation.

It is considered the cargo dispatch by wagons or wagon groups increases the rolling stock circulation due to downtime for wagons or the group wagons at the technical stations, waiting for accumulating and departure.

During the organization of the container train loading the transportation of chemical cargoes volume is increasing and the rolling stock circulation is decreasing.

It is difficult to study this problem without calculations due to the complexity and scale of the process, especially due to the uneven wagon arrivals to accumulation points. Therefore, the solution of the above task remains relevant nowadays.

2 Literature review and problem statement

An analysis of modern researches in the field of improving international rail transport shows that special attention is paid to the simulation of railway technological processes, which allows to take into account the technological carriage traffics characteristics and the development of mathematical models of the transport systems functioning. The development of a mathematical model for the bulk cargoes delivery by rail transport routes, which takes into account the interaction of production, transport and consumption, is proposed in the paper [2]. This model allows to solve a wide range of tasks that are faced by suppliers and consumers in the choice of railway transport, and is associated with the planning and operation of the stations. However, the

* Corresponding author: gtm1520@gmail.com

proposed mathematical model for the cargo delivery by rail routes does not take into account the cargo transportation by the container train and the influence of the main technological parameters which affect to the rolling stock circulation.

The development of a mathematical model for optimization of the sorting process is presented in [3]. This model allows temporarily to store freight wagons on a specially designed mixed rail, but does not include different variants of train departure.

International practice shows that in recent years, more than 60% of internationally transported cargoes were handled door-to-door. This gives the opportunity to optimize the transportation time, reduce the storage and transportation costs. International experience in improving the organization of intermodal freight transportation is revealed in [4]. The issue of developing approaches to determining the mode of transport involved in the transport process is addressed in [5].

The analysis of the influence of constituent elements on the value of the turnover of the freight wagon is analyzed in [6]. The parameters of the laws of their distribution are determined and the response function is obtained on the basis of the results of a complete factor experiment. However, this does not make it possible to determine the efficiency of sending rail freight by single wagons, groups of wagons and routes or container trains.

In [7], to improve the organization of transportation between different participants in the transport process, the authors propose to create an LCL export platform (LEP), using the blockchain concept to optimize LCL operations for international trade, by integrating and exchanging information between freight forwarding companies and their customers.

Development and active use of modern logistics concepts of multi-modal, inter-modal and combined transport are aimed at enhancement of the overall efficiency of cargo delivery processes, improvement of the package of services, provided in the process of motion from a supplier to a consumer. Comparison of external cost, assessment of economic benefits when using different modes of transport in the multi-modal chain was performed in [8].

A comprehensive business process and the model of the multi-modal transport system of container transportation of main based on the Multi-Agent technology for optimization and management of transport operations were presented in [9]. Optimal management in terms of using various kinds of transport in the container multi-modal system according to the criteria of cost, time and quality of cargo transport was considered on the method of dynamic programming [10].

For the plan for the formation of trains with the help of a mathematical model with fuzzy costs presented in [11] optimization of the process of forming (dismantling) warehouses at technical stations. The train formation plan determines the route, the frequency of trains, defines the requirements for trains, but does not take into account the departure of trains by accumulating to the norm of the receipt of wagons at points of accumulation.

The paper [12] analyzes the impact of systemic differences in rail transport between North America and Europe, provides a brief overview of relevant optimization models. Confirmed by the results that North American companies have lower transportation costs than European ones, due to the fact that North American models are mainly aimed at finding short distances for each wagon, and European ones – to a high level of accumulation to the standards of standards.

However, it should be noted that the above studies do not take into account the stochastic nature of the arrival of wagons in points of accumulation. There are no comprehensive models for assessing the efficiency of transportation by group or freight forwarding and freight transportation within the container train.

3. The aim and methods of the study

The aim of the work is to improve international rail transport and to conduct research on the most efficient rail freight shipments between Ukraine and EU countries.

The purpose of the article is to analyze and investigate the definition of methods of transportation of chemical cargo by rail between Ukraine and the countries of the European Union.

This goal is achieved by such commonly methods as analysis, synthesis, analogy, comparison, systems approach were used.

Scientific novelty. On the basis of the analysis and research, theoretical provisions and substantiated practical recommendations for increasing the volume of chemical cargo transportation to EU countries are formulated.

4. Presentation of the main research materials with substantiation of scientific results

During 2018, the European Union remained a key trading partner of Ukraine, with a share of trade in goods and services. The total volume of trade in goods and services between Ukraine and the EU in 2018 amounted to 49317 million dollars. The United States, and the volume of 41.1% of the total trade in goods and services between Ukrainian exports to the EU states was $ 23032 million. USA, which is 14.3% more than last year.

The main importers of Ukrainian products during 2018 were: Poland (14.9%), Italy (11.9%), Germany (11.3%), the Netherlands (7.9%), Hungary (7.4%), Spain (6.2%), Great Britain (4.9%).

The main goods of Ukrainian exports to the EU countries include: ferrous metals, ores, slag, grain crops, seeds and fruits of oilseeds, petroleum products, products of chemical related industries industry.

One of the products manufactured by industrial enterprises of Ukraine and exported to the EU countries is benzene – a colorless, volatile liquid with a characteristic smell. Density is 0.88 g/cm^3. Boiling point – 80.1 °C. Melting point – 5.5 °C. Benzene is a dangerous bulk load that according to the current classification of dangerous goods belongs to the third class (flammable liquids with the assignment of a special UN number 1114).

Today there are several fundamentally different methods of benzene producing:

1. Coking coal. This process was historically the first and served as the main source of benzene before the Second World War. Recently, the share of benzene obtained in this way is less than 10%. It should be added that benzene, obtained from coal tar, contains a significant amount of thiophene, which makes such a benzene unsuitable for certain processes.

2. Catalytic reforming (aromisation) of gasoline fractions of oil. This process is the main source of benzene in the US. In Western Europe, Russia and Japan, in this way they receive 40-60% of the total amount of the compound. In this process, in addition to benzene, toluene and xylenes are formed. Considering that toluene is formed in quantities exceeding the demand for it, it is also partially processed into: benzene - by the method of hydrodealkylation; a mixture of benzene and xylenes - the method of disproportionation.

3. Pyrolysis of gasoline and heavier petroleum fractions. Up to 50% of benzene is produced by this method. Along with benzene, toluene and xylenes are formed. In some cases, all this fraction is sent to the dealkylation stage, where both toluene and xylenes are converted to benzene.

The production of benzene in Ukraine is carried out in two ways:

– at coking coal (method 1);

– in the processing of crude oil (methods 2, 3).

Benzene is one of the ten most important chemicals in the chemical industry and has a high demand for enterprises in the EU. It is widely used in industry, is the raw material for the production of polymers (plastics, rubber), drugs, dyes, pesticides, detergents, explosives, as a solvent for varnishes, and the like.

The main areas of benzene use are the production of ethylbenzene, cumene and cyclohexane. The share of these products accounts for about 70% of the world's consumption of benzene.

Delivery of benzene to destinations is carried out both by road and by rail. Transportation of benzene by road is carried out in special tank-vehicles. This transport is equipped with the necessary equipment, which ensures the safe transportation of cargo. This kind of transportation has its advantages in mobility, safety, door-to-door delivery, and can be effectively used in domestic traffic and in small quantities for short distances.

Considering the small volume of international traffic by road, we will not determine the cost of such transportation. Comparison of costs for delivery of goods by road and tank-containers of the complete door-to-door supply chain between the sender and the recipient is specified in the contract and it is confidential information. Therefore, the calculations are made only by rail from the departure station to the destination station.

Transportation of large quantities of benzene over considerable distances is usually carried out by rail transport with subsequent unloading or overloading to other modes of transport.

Since benzene is an extremely flammable liquid with a flash point of 11.630 C in accordance with DSTU 4500-3: 2008 "Cargo is dangerous. Classification ", it is classified as dangerous goods of the 3rd grade of danger, and the average degree of danger (packing group II) has been assigned to indicators of flash temperature and boiling temperature.

Transportation of benzene by rail is carried out in railway cisterns model 15-1443-07, 15-5103-6B, 15-150-03B, 15-1527. In internal traffic, benzene is transported in accordance with the Rules for the carriage of dangerous goods, the Rules for the carriage of bulk cargoes, and in international traffic also in accordance with the "Regulations for the Transport of Dangerous Goods" (Annex 2 to SMGS). Analysis of the transportation of benzene in all types of connections (domestic, transit, exports to the EU and the Russian Federation) is presented in Table 1.

Table 1. Benzene transportation to EU countries in 2018

Country of destination	Export	Internal connection	Transit	Total
Poland	31 900		6 276	38 176
Slovakia	46 280		156	46 436
Netherlands	40 194			40 194
Italy	19 627			19 627
Czech Rep.	8 034			8 034
Germany	4 498			4 499
Ukraine		25 637		25 637
Russian Federation	2 325			2 326
Total	152 859	25 637	6 431	184 927

Diagram of imported petrol to the EU countries for 2018 (export) is shown in Figure 1.

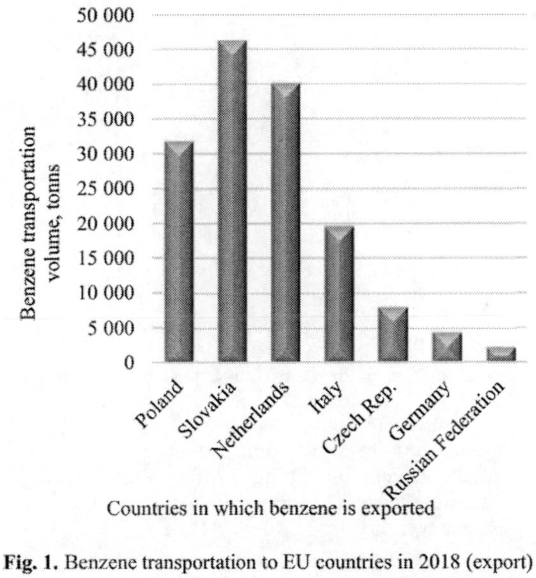

Fig. 1. Benzene transportation to EU countries in 2018 (export)

As it can be seen from the table and diagram, the benzene export from Ukraine is 83% or 152 859 tons and is advantageous for it. Comparing five months of 2019 to the same period in 2018, benzene export increased by 5917 tons and amounted to 7,357 tons. The main dispatchers of benzene for export are PJSC "Zaporizhkoks", PJSC "AKXZ", PJSC "ArcelorMittal", PJSC "DKHZ", PJSC "MK Azovstal" and others. An analysis of the use of rolling stock for the carriage of benzene in 2018 and the five months of 2019 is given in Table 2.

Table 2. Analysis of the use of rolling stock for the transportation of benzene

		2018				**5 months, 2018**				**5 months, 2019**			
Cargo type	**Rolling stock**	**inter-nal**	**export**	**transit**	**total**	**inter-nal**	**export**	**transit**	**total**	**internal**	**export**	**transit**	**total**
Benzene	Indefinite		35		35								
	Platforms			1250	1250			96	96			886	886
	Tank-containers		31505	5182	3668		13222	2566	1578		14876	1258	16134
	Cisterns	18711	62866		8157	11234	30217		4145	8941	31895		40837
	Total	18711	94406	6431	11954	11234	43440	2662	5733	8941	46772	2143	57856
Coal benzol	Indefinite		31		31								
	Tank-containers	114	20557		2067		9846		9846		6143		6143
	Cisterns	6812	37865		4467	1860	14367		1622	2579	20655		23234
	Total	6926	58454		6537	1860	24213		2607	2579	26798		29377
Total		25637	152859	6431	1849	13094	67653	2662	8340	11520	73570	2143	87233

The cargo is delivered to the border transmission station, where the permutation of 1520 mm rail wagons is carried out on the 1435 mm railroad tracks, or when transported by tank-containers, they are trans-loaded from the platform of 1520 mm to the platform of 1435 mm. When transporting benzene to Poland via the border crossing of Izov-ex Khrubushov-eks, the cargo, both in cisterns and in tank-containers, follows a 1520 mm track deep into Poland almost 400 km to the station 'Slavkiv LHS.

The technology of the transfer of cargoes at border transmission stations with the EU countries is determined by the protocols of meetings on harmonization of the estimated volumes and conditions of transportation of foreign trade by rail by rail between the railways administrations.

Figure 2 shows the use of rolling stock by type during 2018.

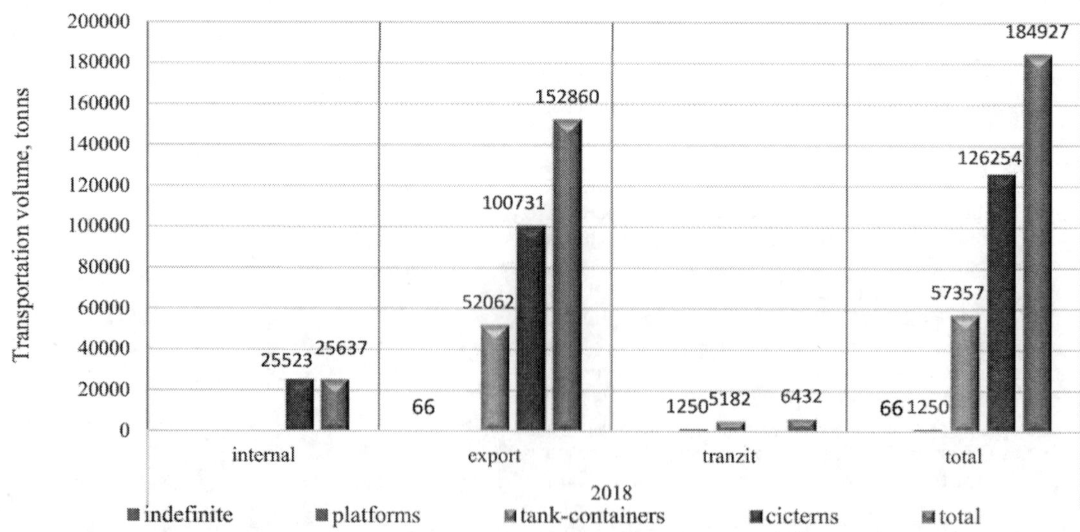

Fig. 2. Analysis of the use of rolling stock by type of rolling stock during 2018

From the schedule it is possible to conclude that 66% of benzene is exported from Ukraine and only 34% in tank containers on fitting platforms in Ukraine. In domestic transport, cisterns are used for transporting benzene by rail.

The shortage of rail cisterns due to the deadline for their operation and the lack of funds for the production of new, high tariffs for transportation in cisterns leads to the search for alternatives for the transportation of benzene. The modern kind of transportation of benzene is the tank-containers of such models EMT5, EMT6, EMKD. The characteristic difference between such containers is that they have a standard size of 30 feet in length, but the frame is standard for a 20 foot container. In some sources, their name is called "SWAP-container". To transport such containers it is necessary to involve the corresponding platforms in the length of 60-80 feet: railway fitting platforms of such models -13-4117, 13-7024, 13-1796-04. Load-carrying capacity of platforms – 70–72 tons, length of axles of carriages – 19500–25690 mm.

For the buyer of cargo, one of the main factors is the assessment of the cost of delivery of cargo [12]. Costs are considered for each delivery scheme depending on the established baseline conditions. Total costs to be borne by the seller or the buyer include expense groups:

$$P_1P_2 = P_{pv} + P_l + P_t + P_c + P_r + P_f + P_s, \quad (1)$$

where P_{pv} – expenses for preparation of cargo dispatch, UAH;

P_l – complex of expenses for loading, UAH;

P_t – payment for transportation taking into account the cost of additional services for cargo handling, UAH;

P_c – expenses for customs clearance of cargo at departure of cargo for export, UAH;

P_r – expenses for rental of rolling stock (cisterns, tank-containers and fitting platforms);

P_f – expenses for forwarding services, UAH;

P_s – other shipping costs.

The costs of the seller and the buyer are determined depending on the basic conditions when entering into a foreign trade contract. It is they that influence the formation of the composition of the costs of organizations, and, accordingly, the financial result of export-import operations.

Each expense metric consists of sub-groups of individual expenditures. Expences for preparing the cargo departure are calculated on the basis of the cost of one ton of cargo. For the most part, this is done by the sender of the goods and may include these costs in the cost of the goods.

Complex of expenses on load is determined by the formula:

$$P_l = P_l^m + P_{ss}, \quad (2)$$

where P_l^m – the costs of the main (leased or private) rolling stock loading; P_{ss} – survey services.

The cost of customs clearance during the cargo dispatch for export is determined by the formula:

$$P_c = P_{cp} + P_{cd} + P_d, \quad (3)$$

where P_{cp} – customs payments, UAH;

P_{cd} – customs duties, UAH;

P_d – registration of the customs declaration, UAH.

Costs of cargo delivery are calculated according the requirements of tariff policies and additional documents of the state administrations of the country of the cargo transporter.

The cost of renting cisterns or fitting platforms is made with the rolling stock owners.

Expeditor expenses for transportation of goods for export are determined by the interest rate on the cost of delivery of goods, which is indicated in the annex to the contract for provision of services in freight forwarding.

During transportation, there may be certain costs not provided for by the terms of transportation (expenses from improperly executed documents, technical problems of rolling stock, etc.). Such costs are considered on the basis of specific situations and, accordingly, completed documents, and sometimes in court.

The study of the cost of transportation of benzene in cisterns and tank-containers took place in one of the directions of Avdeyivka-Izov-Avdeyivka:

1. A private cistern of 60 tons load capacity – the cost of transportation of the loaded and empty cistern for infusion amounted to 88 950 UAH. for a wagon. The price does not include the fee for using a private wagon.

2. Cistern of the UZ fleet with a carrying capacity of 60 tons - the cost of transportation on the same route was 96 920 UAH. for a wagon. The cost includes the fare for transportation and the payment for the use of the wagon.

3. Private tank-containers with their transportation on the platforms of "UTSTS Liski" (UZ) - the loading of two tank-containers of total weight of 60 tons makes for the specified route 36 880 UAH per wagon. The cost includes the payment for the transportation and using of the platform without paying for using of tank-containers.

The analysis of the calculations shows the efficiency of benzene transportation in tank-containers and such transportation is cost-effective for enterprises receiving benzene. Transportation of benzene in cisterns, taking into account the constant increase of the cost of transportation for most of the recipient companies, is unprofitable.

Raising tariffs for rail freight transportation within Ukraine by 15% from October 31, 2017 has had a very negative impact on the financial situation of chemical enterprises, as these costs affect the cost of finished goods and raw materials supplies.

The advantage of transporting benzene in tank-containers is obvious, but the constraining factor is the limited number of railway fitting platforms of the above-mentioned brands.

The enterprises of Poland and partly Slovakia for the transportation of benzene from Ukraine use container containers in the amount of 90 units and there is an opportunity to increase them to 110, but the lack of long-standing fitting platforms of the Ukrainian railways park does not allow to increase the volumes of transportation.

The appointment of Ukrzaliznytsya JSC from 1 April 2019 was a positive factor in increasing the efficiency and, thus, increasing container traffic in the Ukraine-Poland interconnection, by heading the container train along the route Nizhnedneprovsk-Vuzol of the Pridneprovska railroad-Izov-export of the Lviv railway. The formation of a container train is carried out from the groups of wagons by departure from the stations

Mariupol-sorting, Avdeyivka and Sartana of the Donetsk railway and Zaporozhye-Live and Pravda Prydniprovska railway. Changing this container train actually reduces the time spent fitting platforms on routes and thus increases their circulation.

An important factor in increasing the volumes of transportation of benzene and other chemical cargo from Ukraine to the EU countries is the efficient use of rolling stock due to the reduction of the circulation of wagons and the shortening of its delivery time.

Intensification of the rolling stock circulation allows to increase the volume of transportation without increasing the wagon fleet, thereby reducing capital costs on the rolling stock. According to the statistics of recent years, the largest part of the circulation is the total wagon downtime under load operations at loading / unloading points - about 49%, the wagon downtime at technical and intermediate stations is about 37% and in the railway carriage is 14%. In fig. 3 shows the distribution of factors affecting the rolling stock circulation.

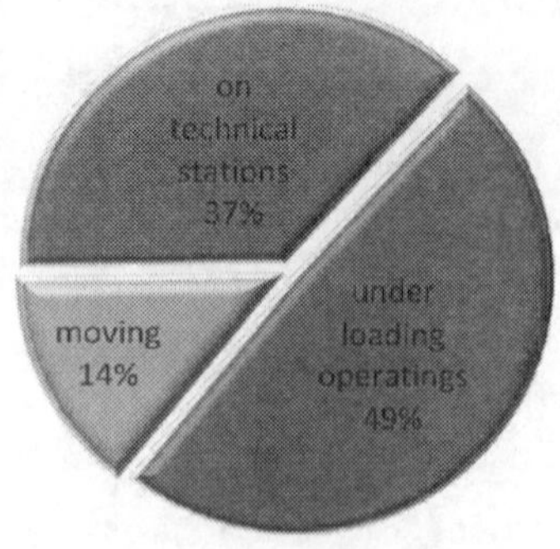

Fig. 3. Distribution of influencing factors on the rolling stock circulation

Transportation of benzene and other chemical cargo from Donetsk and Pridneprovska railways stations can be carried out by single wagons, groups of wagons and routes or container trains, the model of which is presented in Fig. 4, 5 respectively.

Model of chemical cargo transportation by carriage and carriage groups dispatch

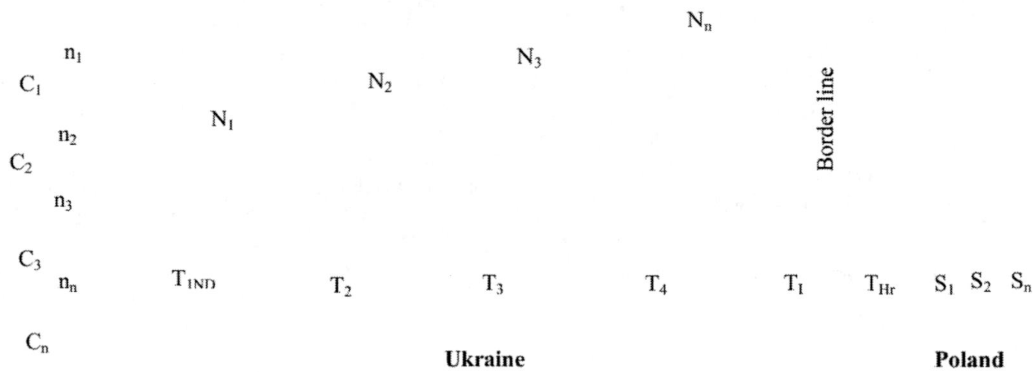

Fig. 4. Transportation of chemical goods by Donetsk and Prydniprovska railway by carriage and carriage groups dispatch

Model of the chemical cargo transportation by the container train

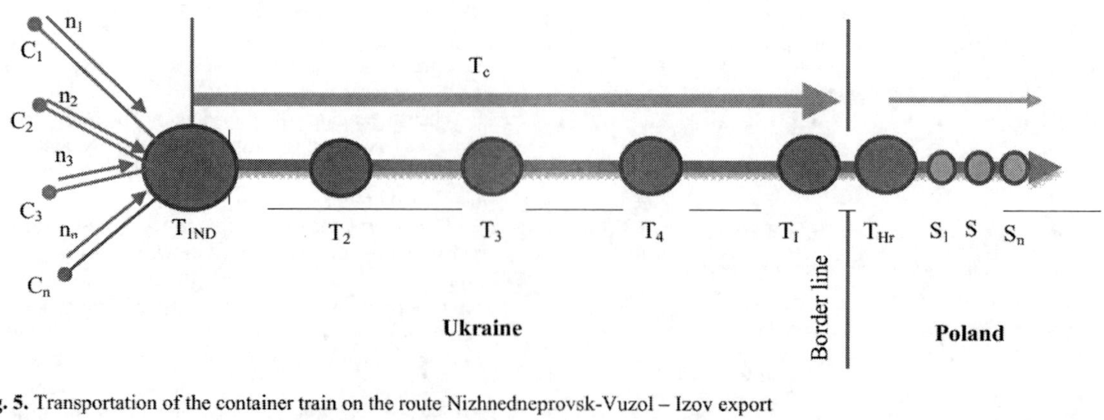

Fig. 5. Transportation of the container train on the route Nizhnedneprovsk-Vuzol – Izov export

where C_1, C_2, C_3, C_n – loading and unloading stations;
T_{IND}, T_2, T_3, T_n, T_I, T_{Hr} – technical stations;
S_1, S_2, S_n – stations located in Poland;
n_1, n_2, n_3, n_n – transportation of goods by group or single carriages;
T_c – transportation of goods in the container train.

To calculate the travel time of a wagon, apply a three-member formula:

$$\theta = \frac{1}{24} \cdot \left(\frac{(1+\alpha)l_v}{V_d} + \frac{(1+\alpha)l_v}{L_t} \cdot t_t + K_m \cdot t_v \right), \qquad (4)$$

where l_v – the trip of the wagon in the loaded condition, km;

α – coefficient of empty running;

V_d – average vehicle speed, km / h;

t_t – average wagon downtime at the technical stations during dispatching of wagons or wagon groups and container train, h;

L_t – "arm carriage" – the average distance between the technical stations, where trains are not subject to proof or subject to reorganization, km;

K_m – coefficient of local work;

t_v – average wagon downtime under a loading operation, h.

When considering each model of the magnitude K_m and α assume equal to 1.

Formula (4) allows to calculate the comparison of spent time spent on delivery of chemical cargoes by carriage by rail and container trains accumulated to the norm of the compositions by the direction of carriage by rail.

The conducted researches revealed that during carriage by group and single carriages the carriage turnover is 15 days, that is, two flights per month, and for transportations in a container train for up to 10 days, or three flights. Container trains are most effective because the wagons do not idle at the technical stations in anticipation of accumulation and shipment. Moving container trains gives an opportunity to increase the volume of transportation of chemical goods, including benzene to the EU countries in 1,5 times.

An important role in the organization of international transportation, in particular, and benzene, is given to freight forwarding or carrier companies, which can organize such transportation more efficiently at the least cost. One such company is "GalTrans", Lviv, which has been operating in the transport services market for 20 years and is one of the leaders in the organization of cargo transportation by railways in the western direction. The company provides services to companies in Poland, Slovakia, Germany in the organization of international cargo transportation, including benzene. The experience of the company proves its efficiency in the market of international forwarding services. "GalTrans" is a reliable partner between the dispatcher, the carrier and the consignee.

In our opinion, one of the restraining factors of the increase of the circulation of existing transportations in tank-containers and thus an increase in the volumes of transportation for the export of benzene is the poor security of goods on the Donetsk and Dnieper railways of Ukraine, which leads to frequent theft of spare parts both platforms and tank-containers. As a result of these thefts, wagons are idle in anticipation of replacement of spare parts for several weeks. Equally important factor affecting the turnover of wagons, and thus reducing the volumes of transportation of benzene in tank-containers, is the lack of work of the appropriate station personnel of the UZ who do not show any damage to the rolling stock during transportation or at the arrival stations, and the breakage The wagons are already on the access roads of the recipient. Thus, a wagon with a detected malfunction can no longer return to the arrival station for repair, and again, the idling time for repair during weeks.

5. Conclusion

To improve international transportation of benzene and related cargoes by rail between Ukraine and the countries of the European Union in the article:

- a mathematical model of transportation of chemical goods by wagon (group) shipments and transportation of a container train has been developed for accumulation to the norm of compositions in the direction of transportation. The most effective variant of transportation of chemical goods - within the container train;
- the research of the cost of transportation of benzene in cisterns and tank-containers in separate directions was carried out, as a result of which it is proposed to carry out a phased transition to the transport of benzene in tank containers, and for the more efficient use of rolling stock - to reduce the turnover of wagons by transporting wagons in container trains;
- the calculation of estimation of cost of delivery of loads by cost groups is resulted. The above suggestions will give an opportunity to increase the volumes of transportation of chemical goods in 1,5-2 times and reduce expenses for these transportation;
- it is proposed to increase the production of long-standing fitting platforms for tank-containers, attracting investors and the most professional forwarding / carrier companies working in the provision of transport services in the international market.

These recommendations will improve the transportation processes and increase of volumes of international cargo by railroads between Ukraine and the EU countries.

References

1. V. Matsiuk, V. Myronenko, V. Horoshko, A. Prokhorchenko, T. Hrushevska, R Shcherbyna, N. Matsiuk, J. Khokhlacheva, I. Biziuk, N. Tymchenko *Improvement of efficiency in the organization of transfer trains at developed railway nodes by implementing a «flexible model»* Eastern–European Journal of Enterprise Technologies: Control processes, **2/3(98)**, 32–39. (2019).
2. N. Shramenko, V. Shramenko *Mathematical model of the logistics chain for the delivery of bulk cargo by rail transport.* Naukovyi Visnyk Natsionalnoho Hirnychoho Universytetu, **5**, 136–141. (2018).

3. M. Bohlin, S. Gestrelius, F. Dahms, M. Mihalák, H. & Flier. *Optimization Methods for Multistage Freight Train Formation.* Transportation Science, **50(3),** 823-840. doi: 10.1287/trsc.2014.0580. (2016).
4. Bontekoning, Y. M., Priemus H. *Breakthrough innovations in intermodal freight transport.* Transportation Planning and Technology. **27 (5),** 335–345. (2004). doi: 10.1080/0308106042000273031
5. Konings R., Priemus H., Nijkamp P *The Future of Intermodal Freight Transport. Operations, Design and Policy. Transport economics, management and policy.* Series Editor: Kenneth Button,University Professor, School of Public Policy, George Mason University, USA, 135–151. (2008)
6. G. Peresta, T. Bolvanovskaya *Analiz vplyvu skladovykh elementiv na velychynu oborotu vantazhnoho vahona.* Zbirnyk naukovykh prats DNUZT im. akad. V.Lazariana. **Vyp. 1.** P. 75 – 77. (2011) [in Ukrainian].
7. Tan A. W. K., Zhao Y., Halliday T. A Blockchain *Model for Less Container Load Operations in China* International Journal of Information Systems and Supply Chain Management. **11(2),** 39–53. (2018). doi: https://doi.org/10.4018/ijisscm.2018040103
8. Kos S., Vukić L., Brčić D. *Comparison of External Costs in Multimodal Container Transport Chain* PROMET – Traffic&Transportation, **29 (2),** 243. (2017). doi: 10.7307/ptt.v29i2.2183
9. Liu T., Zheng G. Study *Logistics Architecture for Grain Container Multimodal Transport Based on Multi-Agent* Proceedings of The 7th International Conference on Computer Engineering and Networks – PoS(CENet2017). (2017). doi: 10.22323/1.299.0049
10. Hao C., Yue Y. *Optimization on Combination of Transport Routes and Modes on Dynamic Programming for a Container Multimodal Transport System* Procedia Engineering, **137**, 382–390. (2016). doi: 10.1016/j.proeng.2016.01.272
11. M. Yaghini, M. Momeni, M. & Sarmadi, *A hybrid solution method for fuzzy train formation planning.* Applied Soft Computing, **31,** 257-265. doi: 10.1016/j.asoc.2015.02.039. (2015).
12. Uwe Clausen, Robert Voll *A comparison of North American and European railway systems. Why Europe needs particular optimization models for railway freight traffic.* Eur. Transp. Res. Rev, **5,** 129–133. DOI 10.1007/s12544-013-0090-4. (2013).
13. Nagornyi Ye. *Transportno-ekspedytorska diialnist* Ye.V. Nahornyi, D.V. Lomotko, N.Iu. Shramenko ta in.: pidruchnyk. – Kh.: KhNADU, 2012. – 352 s. [in Ukrainian]

Improvement of the analysis methods of the working hours` distribution of locomotive crews by the types of train operation in freight traffic

Dmytro Kozachenko[1], *Mykola* Berezovyi[2], *Vyacheslav* Malashkin[2], *Tetiana* Bolvanovska[2], and *Yurii* Grimak[3]

[1] Dnipro National University of Railway Transport, Department of Operation, 41010 Dnipro, Ukraine
[2] Dnipro National University of Railway Transport, Department of Transport Junctions, 41010 Dnipro, Ukraine
[3] Lviv College of Transport Infrastructure of Dnipro National University of Railway Transport named after Academician V. Lazaryan, 79011 Lviv, Ukraine

Abstract. The work is aimed to increase the efficiency of cargo transportation by railway transport by improving the work organization of locomotive crews. The research is based on the methods of organization of operational work of railways and the methods of mathematical statistics. The problem of increasing the use efficiency of working hours of locomotive crews in freight traffic is considered in the article. Based on the statistical data processing of drivers' routes, the working hours` distribution of locomotive crews for different types of train operation in freight traffic for four locomotive depots was performed. During the research, a conditional division of the working time of locomotive crew by the type of work was performed, if the crew worked with trains of various categories. The algorithm for assigning the idle time and empty running of locomotive to certain types of train operation was also developed. The work originality lies in the fact that it improves the methods of analysis of the working hours` distribution of locomotive crews for certain types of train operation in freight traffic. The developed methods give an opportunity to evaluate the work specifics of individual locomotive depots and provide the initial data for planning the use of locomotives and the work and rest schedule of locomotive crews.

1 Introduction

A characteristic feature of the modern working conditions of railway transport in Ukraine is a critical level of traction rolling stock wear and an increasing misbalance between the number of locomotives and traffic volumes. In this regard, the task of improving the efficiency of using the existing fleet of locomotives and the staff of locomotive crews is of particular relevance. One of the ways to optimize and rationalize the transportation process is the development and implementation of a reliable system of operational planning of the work of locomotives and locomotive crews.

In Ukraine a system for traction train traffic, when the locomotive fleet and locomotive crews are distributed along the railway network and are assigned to the locomotive depot has been developed. The part of the railway network where the locomotives assigned to one locomotive depot turn around is called the locomotive operational area. Within long operational areas there may be intermediate points for locomotive turn-around, maintenance points, servicing terminals, locomotive crew change points, and the like. The section between the station of the main and turn-around depot is called the locomotive run. The locomotive operational areas may consist of one or several locomotive runs. The locomotive run may consist of one or more sections of the work of locomotive crews. In the process of its movement, the train can pass several sections of the work of locomotive crews and locomotive operational areas. The need to perform the regulated types of maintenance and servicing of locomotives, as well as compliance with the established norms of the working time of locomotive crews makes it necessary to stop the trains for their change.

When the standards for the operational work of Ukrzaliznytsia units are established for the planned month, calculations are made to determine the required fleet of locomotives for freight traffic and the attendant staff of locomotive crews, i.e., they are planned monthly. The planning provides for:

– calculation of the daily average attendant staff of locomotive crews for the mastering of the freight traffic volumes specified in the plan for the month;

* Corresponding author: dmytro.kozachenko@outlook.com

https://doi.org/10.1051/matecconf/201929404011

– the distribution of existing staff of locomotive crews in the assignment point over the work sections, types of movement and the types of work in order to bring its size in line with the upcoming workload;
– clarification of the annual holidays schedule for locomotive crews on the upcoming month.

Operational planning and management of the work of locomotives and locomotive crews at different levels is carried out by locomotive dispatchers of railway within the railway, locomotive dispatchers within the divisions of railway transportations and duty officers of locomotive depots. The development of methods for solving the problem of finding rational train servicing options by locomotives and crews has a rather long history and a significant number of scientific works are aimed at its solution. One of the first mathematical methods for solving this problem was proposed in the 60s of the 20th century. In the work [1] by Charnes and Miller used linear programming for the assignment of crew–engine pairings to a set of potential trips to provide each train in a given schedule with sufficient resources. Bartlett in [2] gave an algorithm for minimizing fleet size based on the idea that, for a fixed time horizon, this objective is tantamount to minimizing total idle time.

In the works [3] by Cordeau et al. and Piu and Speranza [6] two types of locomotive planning models were identified: *Single* and *Multiple*. Single locomotive planning models assume that there is only one type of locomotive available for assignment and train is drived by only one locomotive. Multiple locomotive planning models assume that one or several locomotives can service each train; herewith there are several types of locomotives. In the work [4] the problem of servicing the train movement of the Italian railways is considered and an algorithm for its solving is presented. In the work [5] Vaidyanathan et al. presented an algorithm for solving the task of planning the work of locomotives for USA railways. In the work [5] Teichmannet al. presented an algorithm for solving the task of planning the work of locomotives for Czech railway.

The studies aimed at solving the problems of train maintenance by locomotives and locomotive crews for the conditions of the Russian Federation are presented in the works [8-10]. It should be noted that the presented models and methods of planning the work of locomotives and locomotive crews are designed for cases when the train schedule is known. At the same time, the railway transport operation is largely influenced by random factors. In this regard, the train movement is characterized by significant monthly, daily and intra-day non-uniformity. At the same time, the occupation or non-occupation of train path by a freight train is a random event. Therefore, the problem of planning and operational planning of the work of locomotives and locomotive crews should be considered both in a dynamic and stochastic formulation. However, the lack of currently functioning models for constructing a graph of the predicted turn-around of locomotives and locomotive crews for a monthly period does not allow solving the problem by the method of transportation process simulation, therefore the index method remains the basic method; based on the formation of the working time expenditures of locomotive and locomotive crews' work according to statistical data for the past similar period. To construct simulation models and check their adequacy, it is necessary to evaluate the actual working conditions of railway transport and establish the distribution of the working time of locomotives and locomotive crews by the types of work when servicing trains of various categories.

2 Methodology

The document that makes it possible to establish the working time distribution of the locomotive crews when servicing trains of various categories is the driver's route. It is filled by the locomotive driver and processed programmatically by entering the main indicators using the software that provides functioning of the automated working station processing the driver's route (WS DR). The driver's route is the main document for the calculation and accounting of the operational work of locomotives, fuel and lubricants and electricity consumption, working time and wages of locomotive crews.

The main sections of the driver's route are as follows:
1. Data on the traction rolling stock (TRS) and the staff of locomotive crew.
2. Information on the working time of the TRS and the locomotive crew.
3. Data on the locomotives operating in different connections, such as pushing or double heading.
4. Electricity and fuel consumption data.
5. Data on the movement, weight of the train and the shunting operation performed.

Further, the locomotive and the locomotive crew during one travel are considered as a single whole.

Let us dwell in more detail on the sections containing information on the working time of the TRS and the locomotive crew, as well as on the use of locomotive with and without a train during the working hours of the locomotive crew.

The working time duration of the locomotive crew (travel time) in general can be determined by the expression [11]

$$t_{\mathrm{d}} = t_{\mathrm{a\text{-}ac}} + t_{\mathrm{ac\text{-}dep}} + t_{\mathrm{mov}} + t_{\mathrm{id}} + t_{\mathrm{arr\text{-}w.e}} \quad (1)$$

where$t_{\mathrm{a\text{-}ac}}$ – is the time interval between the moment of the locomotive crew appearance at the beginning of the working time and the moment of the locomotive acceptance, h;
$t_{\mathrm{ac\text{-}dep}}$ – is the time interval between the moment of locomotive acceptance and its departure with train, h;
t_{mov} – is the movement duration of locomotive with train or in single running at the service section, h;
t_{id} – is theidle time of locomotive with a train at intermediate stations and at the station, which is the boundary of the service section – at the point of turn-around, h;
$t_{\mathrm{arr\text{-}w.e}}$ – is the time interval from the moment of locomotive arrival with train or in a single running to the final station until the work end of the locomotive crew.

During one travel, depending on the characteristics of the service section, such as its length, the structure of train traffic volume, the fullness of the train movement schedule, and othersthe locomotive can be used both to service the trains of the same or different categories. The decision on how to use the locomotive is made by the dispatching staff, depending on the operational situation and the restrictions, taking into account the mode of operation of the locomotive crew and the locomotive mileage between maintenance and repair.

The sequence of technological operations performed by locomotive with locomotive crew during one travel can be formalized as a directed graph $G = (V, P)$, presented in Fig. 1. The vertices of the graph $v_n \in V$ are the technological operations corresponding to certain states of locomotive, and the arcs of the graph $p \in P$ – transitions from one state to another. Transitions are initiated by the person making the decision on the further use of locomotive.

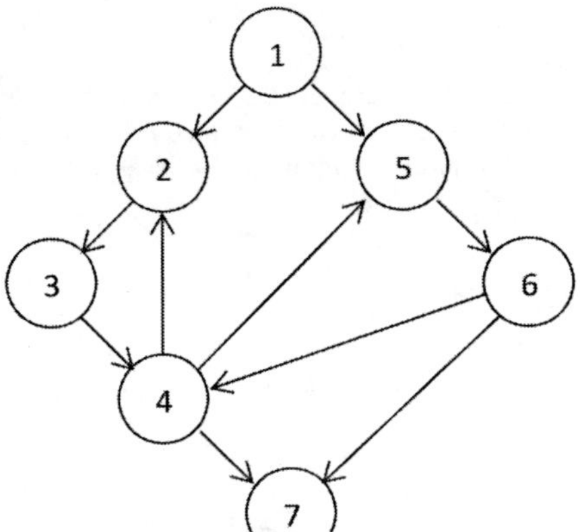

Fig. 1. Graphic representation of the use scheme of locomotive during one travel

Thus, the state «1» corresponds to the locomotive acceptance by the locomotive crew at the beginning of its work. After the crew's report on locomotive acceptance and readiness to carry out train work, a decision can be made to supply the locomotive to the freight train (state «2») or to dispatch the backup locomotive to another station to another train (state «5»). In state «2» the locomotive is waiting for the train, then itis supplied to the train, and thenthe operations are performed to test the brakes, receive a form of warnings and documents, waiting for the possibility of departure and departure, at the time of which the transition to state «3» occurs – the locomotive moves with the train through the section to the destination station. In state «5», a single locomotive waits for the possibility of departure and is departed, at this moment there is a transition to state «6» – backup running. When transiting to state «4» from state «6» one performs the operations with the locomotive upon arrival and supplies itto the freight train, i.e. transition to state «2». When transiting from state «3»to state «4»one also performs the stipulated technological operations with the locomotive upon arrival, and the further work plan may be as follows:

- supply to the freight train – the transition to state «2»;
- backup departure to another station to another freight train (state «5»);
- end of work and delivery of locomotive (state «7»).

The transition to state «7» is possible from state «6». This is possible if, after servicing one or several trains, the backup locomotive is dispatched to the station of the main or turn-around depot to further change of the locomotive crew. In this case, the locomotive crew at the locomotive delivers the locomotive and finishes the work.

Thus, this oriented graph can be used in a simulation model to simulate the sequence of individual operations with a locomotive during one travel.

The theory of locomotive operation includes three main ways of train servicing by locomotives – run, loop and ring, differing in the frequency of locomotive entry into the main depot, and three main ways of servicing locomotives by locomotive crews – calling system, job order system and using the personalized schedules.

A feature of the operation of locomotives and locomotive crews in the current conditions is that the main criterion for their use is the adherence to overhaul intervals. It should also be noted the lack of service sectionsassignment to the individual depots and the actual impersonality of the locomotive assigned depotsconcerning repairs. Maintenance and repair of locomotives can be performed not only in the assigned depots and the point of locomotives turn-around, but also in any other depot where such a need has arisen.

Another feature is that during one travel the locomotive can be used to service trains of various categories.

3 Results

The obtained statistical characteristics of the duration distribution of one travel and its main components for the four depots located at the site with different characteristics of the train traffic volume and the characteristics of its servicing were published in the article [11]. The average duration of a travel is fixed in the range from 8.8 hours for Znamenka depot to 11.30 hours for Odessa-Sortirovochnayadepot.

This is explained by the fact that Znamenka depot serves mainly through freight trains with and without processing with an insignificant share of district, clean-up and transfer trains. The Odessa-Sortirovochnaya depot is characterized by significant volumes of transfer trains at the junction, most of the locomotive crews working according to thepersonalized schedules with 12-hour duration of working time.

The average duration values of the use of locomotives in train work – the time spent by the locomotive with the train ranges from 5.5 hours for Znamenka depot (62.5% of the travel duration) to 7.44 hours for Nizhnedneprovsk-Uzel depot (71.5% of the travel duration).

The duration of locomotive movement with a train for Znamenka depot is 4.07 hours or 46.3% of the travel duration, for Nizhnedneprovsk-Uzel depot – 4.36 hours or 41.9% of the travel duration.

There is no the correlation relationship between the above indicators of the locomotives` use, and the correlation coefficient is r=0.053 and does not reach the minimum value of the loose couple according to the Chaddock scale (Fig. 2).

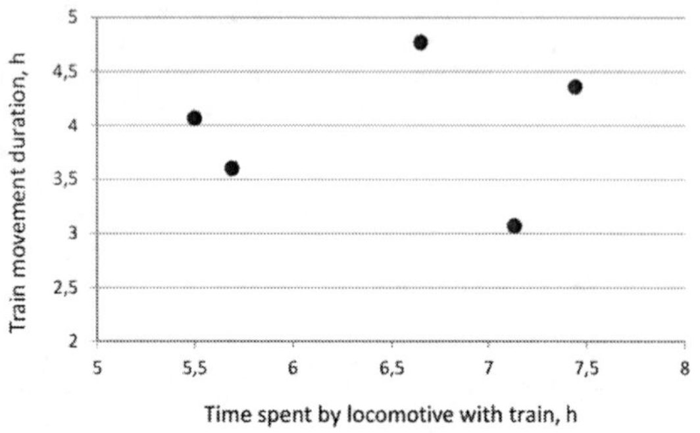

Fig.2. The dependence of the train movement duration on the time spent by locomotive with train.

Further studies have shown that the ratio of elements of train work – the time that a locomotive is with train and the time in motion with a train – have significant differences depending on the category of trains served by locomotives. Train driving on the extended runs of service cannot be compared, since such trains often move according to the schedule.

During one travel, the locomotive can serve trains of one category, while the calculation of statistical indicators has no differences from the calculation, the results of which are given above.

However, much of the travels is related to the driving of trains of various categories. In this regard, there is a need to develop a method of classifying the components of the working time of locomotives and locomotive crews as certain types of work when servicing trains of various categories.

The beginning of the work of locomotive crew is the moment of appearance T_a at work. After that, the crewgoes to locomotive and performs its acceptance in accordance with the rules and regulations at the momentT_{ac}. The presence of the specified period of time ($t_{a\text{-}ac} = T_{ac} - T_a$) is mandatory, and its duration must comply with the established standards t_{stand}. The time interval $t_{a\text{-}ac}$does not refer to a particular type of work.

The report on the locomotive readiness to perform work is the moment of the beginning of time period, which should be assigned to work with a train of a certain category. At the time of the locomotive crew`s statement on the readiness to work T_{ac}, the decision maker, as a rule, already knows further plan of work with the locomotive. The train category k_i in the process of research is determined by its number $n_i \in \mathrm{N}$ ($i = 1, 2, 3, \ldots$), and the vector of numbers Nand the order of their assignment to certain train categories is determined by the Order of direction of car traffic volumes and their organization into freight trains (Train formation plan).

The moment of the end of the time interval W_i, which is related to the work with the i-th train of a certain category k_i, is the moment $T_{end\,i} = T_{unc\,i} + t_{tech}$, where $T_{unc\,i}$ is the moment of the locomotive uncoupling from a train at the terminal station of the train running, t_{tech} is the technological interval associated with the assignment of the train and transfer of train documents to the employee of the station technological centre. Since the driver's route does not contain information about the duration of the interval t_{tech}, its duration in research is assumed to be 10 minutes.

It should be noted that the time moment $T_{end\,i}$ may be the starting point of the locomotive crew`s work with the next category of train $T_{start\,i+1}$ if the travel is not over. Obviously, for the first train, which the locomotive crew will work with at the start of shift $T_{start\,1} = T_{ac}$, where $i = 1$. Thus, the total work duration of locomotive crew$W_i$with a certain train category $k_i$is $W_i = T_{end\,i} - T_{start\,i}$.

The analysis of the drivers` routes showed the presence of works related to the backup locomotives running for the subsequent work with trains of certain categories. In the performed studies, the duration of backup locomotives running $t_{b\,i}$ is assigned to work with a train of the category whose servicing is the next technological element or state of the locomotive, i.e. $W_{i+1} = W_{i+1} + t_{b\,i}$. In the case of backup locomotive running to the station of locomotive crew's change and the end of work, the interval $t_{b\,i}$ is assigned to the type of work of a certain category of the previous train, i.e. $W_{i-1} = W_{i-1} + t_{b\,i}$.

The statistical processing of drivers` routes showed the following results.

For the four depots considered, Nizhnedneprovsk-Uzel, Znamenka, Lviv-Zapad and Odessa-Sortirovochnaya, one can distinguish the categories of trains with which locomotives work throughout the entire travel, i.e. the locomotive serves one or two trains of only one category. These include through trains, unit trains and district trains. At the same time for Odessa-Sortirovochnayadepot,the routes for servicing the sections with diesel-electric and electric traction were analyzed. Some of these travels, from the total number of the analyzed ones, are in the range from 19.25 to 52.8% with a predominance of travels in which trains of a certain category are serviced.

Thus, for Znamenka depot, the number of travels when the locomotives served only district trains accounted for 24.1% of the total; for Lviv-Zapad and Nizhnedneprovsk-Uzel depots, dominate the travelswhenthrough trains were serviced – 29.64% and 34.6%, respectively. Odessa-Sortirovochnaya depot in the electric traction is characterized by the minimum number of travels, when the locomotives serviced only through and unit trains – 3.87% of the total.

This confirms the need to take into account the specifics of the work of each locomotive depot.

The results of research of the duration of the main elements of the travel, when only trains of a certain category were serviced, are shown in Fig. 3.

Depot	Train category	Mathematical expectation, h		
		Travel duration, t_d	Time spent with train t_{train}	Movement duration with train t_{mov}
Nizhnedneprovsk-Uzel	Through	10.10	7.26	4.69
	Route	9.10	5.07	3.16
	District	8.60	5.59	3.74
Znamenka	Through	8.79	5.86	4.43
	Route	7.81	5.04	3.96
	District	7.25	4.34	3.35
Lviv-Zapad	Through	8.51	5.87	4.47
	Route	10.17	7.12	5.57
	District	10.20	6.96	4.19
Odessa-Sortirovochnaya, electric traction	Through	7.95	4.81	3.66
	Route	8.07	5.03	3.52
	District	7.82	4.97	3.04
Odessa-Sortirovochnaya, diesel-electric traction	Through	-	-	-
	Route	8.50	4.28	3.03
	District	11.16	6.88	4.27

Fig. 3. Duration of the main elements of the travel

The duration of time spent by locomotive with train t_{train} and the duration of movement with train t_{mov} correlates with the total durationof travel $t_d$for trains of all categories; according to the Chaddock scale they have a very significant relationship. As an example, Fig.4 shows the field of points and correlation coefficients.

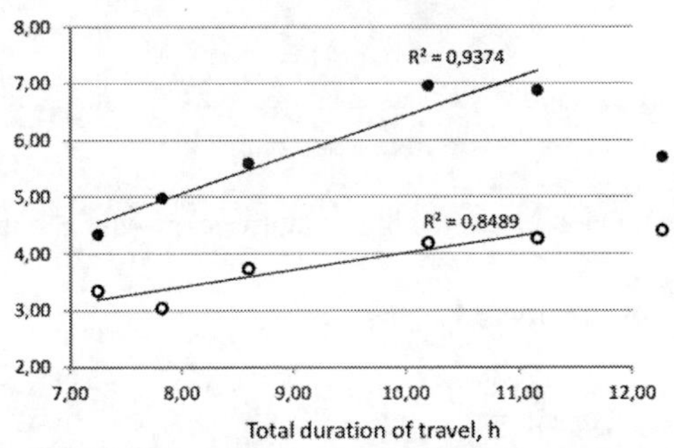

Fig. 4. Dependencies of time spent by locomotive with train $t_{train}$and the duration of movement with train t_{mov} on the travel duration t_d

Using the methods of regression analysis we obtained analytical dependencies of the type $t_d = f(t_{train})$ and $t_d = f(t_{mov})$, which are convenient to use when simulating the work of locomotive crews. Fields of points and analytical models are presented in Fig. 5.

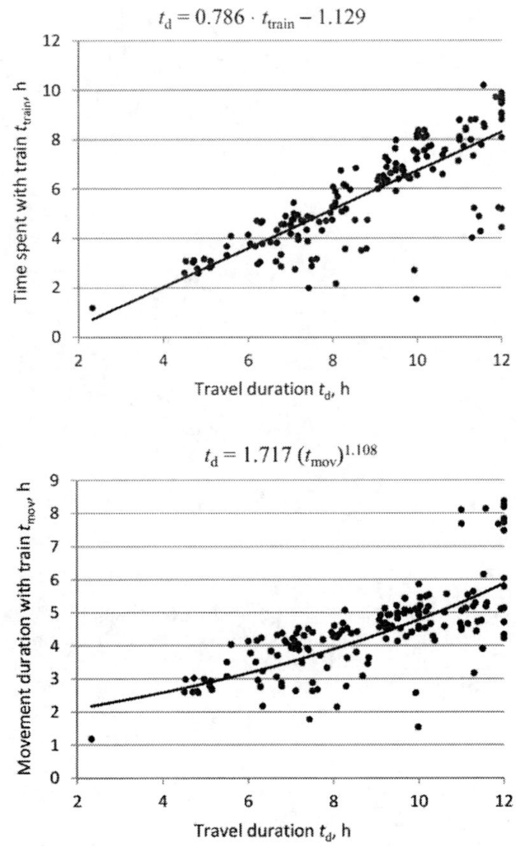

Fig. 5. Fields of points and corresponding analytical models

Analysis of the values of t_{train}, t_{mov} and t_d allowed us to obtain their statistical parameters, as well as the nature of distribution. The results of the analysis are given in Fig. 6, and histograms and distribution laws of the indicated values are shown in Fig. 7.

Statistical parameters	Time characteristic		
	Travel duration, t_d	Time spent with train t_{train}	Movement duration with train t_{mov}
t_{min}, h	2.33	1.17	1.17
t_{max}, h	12.00	10.18	8.37
Mathematical expectation m_{t_i}, h	8.789	5.855	4.430
Mean-square deviation σ_{t_i}, h	2.130	2.030	1.806
Distribution law	normal	normal	gamma
Chi-square test	2.42	8.77	9.23

Fig. 6. Statistical parameters of duration of the main elements of the travel

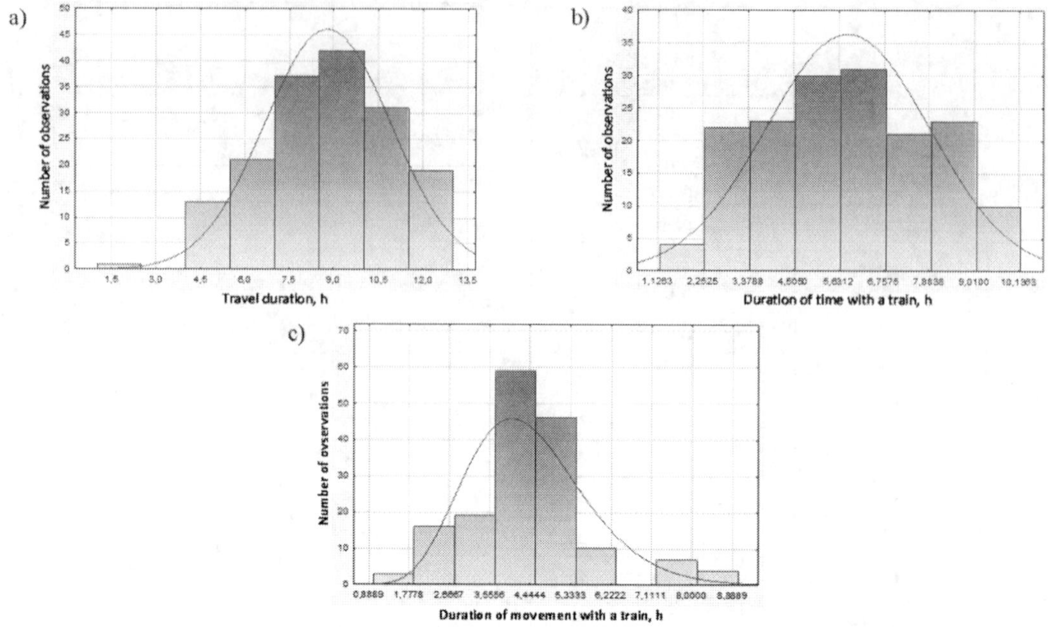

Fig. 7. Distribution histograms oftravel durationt_{d} (a), duration of stay of locomotive crew with a traint_{train} (b) and movement duration of locomotive crew with a traint_{mov} (c)

Similar dependencies were obtained for the travels, during which the locomotive serviced different categories of trains. In addition to the categories discussed above, they should include clean-up, transfer trains and for some depots an insignificant number of pick-up and pick-up-district trains. A feature of such travels is that an additional element appears – backup locomotive running. Of the total service time of a particular category of train during the travel, backup running ranges from 1.5% for through trains, to 28.6% for transfer trains, and up to 50% for pick-up trains.

Another feature of such travels is that the locomotive performance is generally lower than that when servicing only one category of trains during the travel.

Similar to the results of the studies presented above, we can state that there is correlation relationship between the servicing time of a train of certain category t_k and the duration of movement with a train of this category $t_{\text{mov}k}$; however, for various depots according to the Chaddock scale, this relationship is somewhat lower and significant.

As an example Fig. 8 shows the distribution histogram of the duration of locomotive crew`s stay with a train, when it served the trains of different categories during the travel.

4 Scientific novelty and practical significance of work

The work originality lies in the fact that it improves the methods of analysis of the working hours` distribution of locomotive crews for certain types of train operation in freight traffic. The developed methods give an opportunity to evaluate the work specifics of individual locomotive depots and provide the initial data for planning the use of locomotives and the work and rest schedule of locomotive crews.

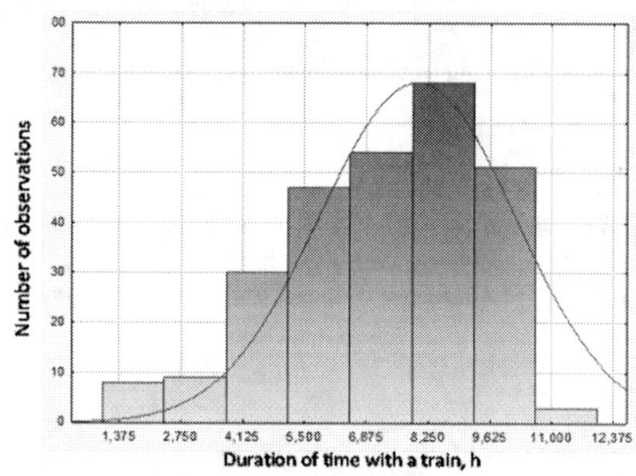

Fig. 8. Duration of stay of locomotive crew with a train

5 Conclusions

As a result of the research, the following conclusions were obtained.

1. The graphical formalization of the locomotive use scheme for one travel as a directed graph can be used in a simulation model to simulate the sequence of individual operations with the locomotive.

2. There are differences in travels, during which the locomotive serves trains of only one category and trains of different categories, in terms of the duration distribution of individual states of the locomotive, which requires a conditional division of such travels.

3. We proposed a method for assigning the travel time components of a locomotive crew to the trains of various categories that were serviced during the travel.

4. Using the methods of regression analysis, we obtained analytical dependencies, which are convenient to use when simulating the work of locomotives and locomotive crews.

5. Statistical parameters were obtained, as well as the distribution nature of the values that are components of the travel duration of locomotive crew.

References

1. A. Charnes, M. H. Miller *A model for the optimal programming of railway freight train movements,* Management Science, **Vol. 3**, 74-92 (1965).
2. T. E. Bartlett, *An Algorithm for the Minimum Number of Transport Units to Maintain a Fixed Schedule*, Naval Res. Logist. Quart, **Vol. 4**,139-149 (1957).
3. J. F. Cordeau, P. Toth, D. Vigo, *A survey of optimization models for train routing and scheduling*, Transportation Science, **Vol. 32**, 380-404 (1998).
4. A. Caprara, M. Fischetti, P. Toth, D. Vigo, P. L. Guida, *Algorithms for Railway Crew Management*, Math. Programming, **Vol. 79**, 125-141 (1997).
5. B. Vaidyanathan, R. K. Ahuja,J. B. Orlin, *The locomotive routing problem*, Transportation Science, **Vol. 42, Iss. 4**, 492-507 (2008). doi: 10.1287/trsc.1080.0244
6. F. Piu, M. G. Speranza, *The locomotive assignment problem: a survey on optimization models*, International Transactions in Operational Research, **Vol. 21**, 327-352 (2014). doi: 10.1111/itor.12062
7. D. Teichmann, M. Dorda, K. Golc, H. Binova, *Locomotive Assignment Problem with heterogeneous vehicle fleet and hiring external locomotives*, Mathematical Problems in Engineering, 1-7 (2015) doi: 10.1155/2015/583909
8. P. Kozlov, E. Timukhina, N. Tushin, *Coordination of locomotives turnover and servicing modes*, Transport problems, **Vol. 13**, **Iss.1**, 19-26 (2018). doi: 10.21307/tp.2018.13.1.2
9. P. A. Kozlov, S. P. Vakulenko, *Optimization of locomotive work modes with the use of system Labyrinth*, World of Transport and Transportation, **No. 4**, 10-14 (2016). [in Russian]
10. P. A. Kozlov, S. P. Vakulenko, *Optimization of locomotives turnover for given train flows*, Railway Transport, **No. 10**, 34-37 (2016). [in Russian]
11. Y. Grimak, *Analysis of the distribution of elements of working time of locomotive brigades within the trips in the freight traffic*, Transport systems and transportation technologies, **No. 17**, 4-8 (2019). [in Ukrainian]

Innovative concept of loading bimodal trailers with the organization of a loading terminal

Jan Strzemkowski[1,*], *Agnieszka* Merkisz – Guranowska[2], *Maciej* Andrzejewski[1], *Hanna* Stawecka[1], *Michalina* Kamińska[2], and *Mariusz* Far[1]

[1] Łukasiewicz Research Network – Rail Vehicles Institute „TABOR", ul. Warszawska 181, 61-055 Poznań, Poland,
[2] Poznan University of Technology, Pl. Marii Skłodowskiej-Curie 5, 60-965 Poznań, Poland,

Abstract. The article presents an innovative concept of horizontal transverse loading of bimodal semi-trailers by means of a transshipment terminal equipped with special the transshipment stands. An innovative switching station and a working principle improving the loading of bimodal trailers were described. The concept of the transshipment terminal equipped with several loading stands for bimodal semi-trailers is also presented. Various variants of organization of the terminal with horizontal transverse loading of bimodal semi-trailers are indicated. A preliminary computer simulation was carried out and the results are compared with the existing Polish system Tabor 1 and Tabor 2.

1 Introduction

Increasing traffic density is observed, due to the continuous development and growth of road transport. As a result, more and more pollutants, coming from exhaust fumes from combustion engines, are emitted [1,2]. To reduce environment pollution, the European Union's postulates presented in the White Paper encourage to look for new solutions that meet the "Vision of a competitive and sustainable transport system". One of the most important component of road transport is freight transport with the use of semi-trailers. Especially for this type of transportation an alternative solution will be presented. One of the possible solutions is the transport of semi-trailers by railway. It is of course possible to transport standard semi-trailers or containers on specially designed platforms. However, we must remember that the ratio of dead weight to the weight of transported goods, as well as the length of the train, does not put this type of transport at the top of the economic ladder.

And what about BIMODAL transport?

Fig. 1. Bimodal unit

Bimodal transport is a type of transport where an adapted semi-trailer, fixed with suitable adapters, is transported on railway bogies. Bimodal semi-trailer differs from the standard one and should be redesigned, according to the current road and railway requirements [3].

The main components of a bimodal semi-trailer are shown in Figure 2.

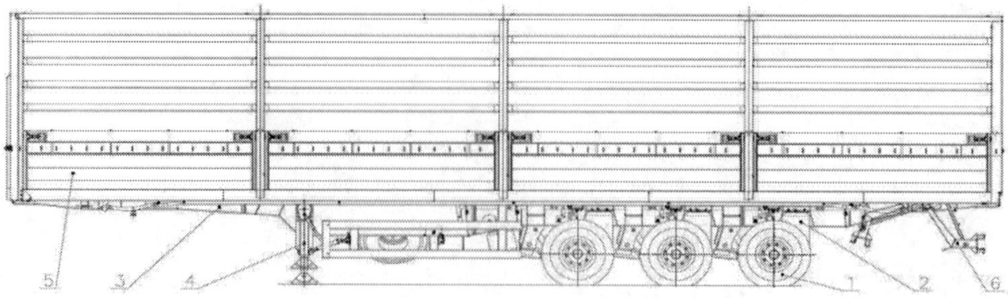

Fig. 2. Bimodal semi-trailer

* Corresponding author: jan.strzemkowski@tabor.com.pl

These are:
- axles with road wheels (1),
- load-bearing bellows (2),
- semi-trailer frame (3),
- supporting leg (4),
- built-in body (5),
- tilting buffer (6).

We had should remember, that in Poland there is a continuous increase of amount of semi-trailers, with important number of new semi-trailers. The increase in the number of semi-trailers is shown in Figure 3, referring to the data from the Central Register of Vehicles [4]. When designing new semi-trailers, we have the option of introducing structural changes that meet the requirements of rolling stock, which will allow us to gradually increase bimodal rolling stock.

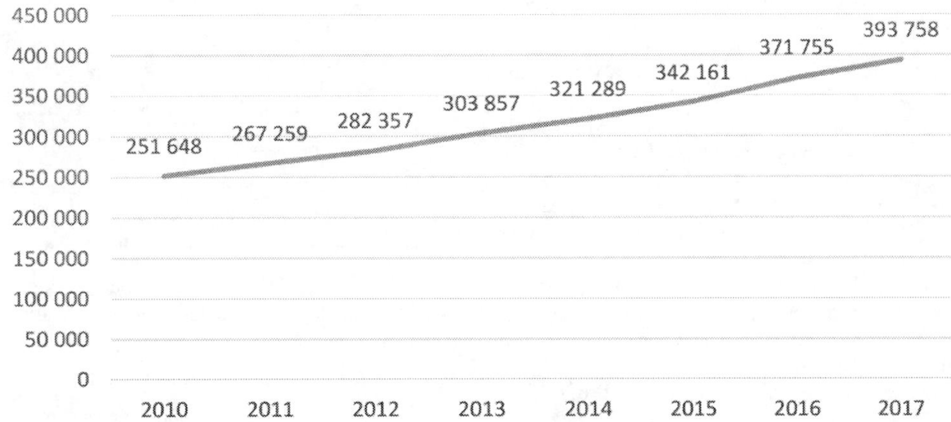

Fig. 3. The number of semi-trailers in road transport in Poland in the years 2010-2017

2 Bimodal fleet

The bimodal transport system is a qualitatively new technical solution for road and rail rolling stock, in which other design criteria and requirements as well as provisions on the basis of which road and rail rolling stock are designed and operated. The combination of these two systems into one bimodal system forces a correlation of requirements and restrictions in force separately for each transport system, e.g. in terms of permissible dimensions and masses, strength (road gauge - railway gauge, allowable road wheel load of the semi-trailer - allowable wheel set pressure on the track, higher strength requirements for a bimodal semi-trailer than for a road semi-trailer) [5].

Compared to other known modes of combined transport, the bimodal system has the following advantageous features [5]:

- low dead weight of rolling stock, i.e. the ratio expressed by the ratio of the weight of transported load to the weight of the rolling stock is the most favorable in the bimodal system
- the smallest distance between adjacent loading units
- the most favorable rolling stock gauge in the main zone, resulting from the position of the roof of the loading unit at a height limited to 4180 mm, measured from the rail head
- the smallest length of the bimodal train composition

Figure 4 shows a fragment of the bimodal train composition which consists of two end bogies, one middle bogie and two loading units [5]:
- bimodal bogies (1)
- end adapters (2)
- central adapters (3)
- bimodal semi-trailers (4)

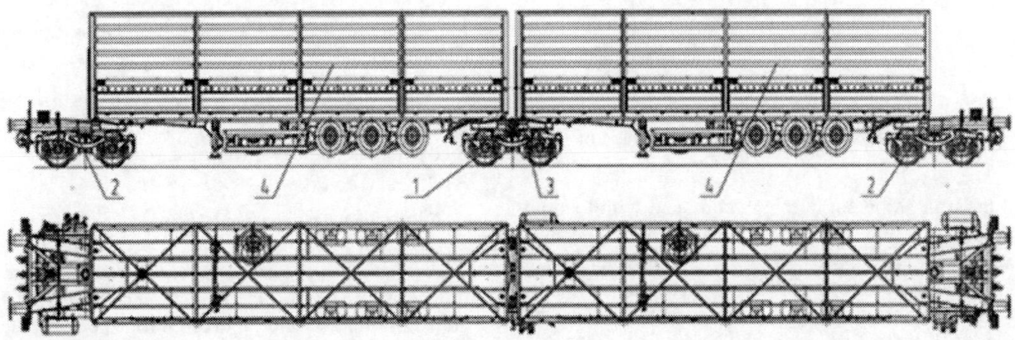

Fig. 4. Bimodal train unit

3 Innovative loading concept for semi-trailers

Loading of semi-trailers on the adapters is carried out in the following stages [3]:

1. Introduction of the semi-trailer on the loading stand,
2. Lifting of the semi-trailer to the maximum height with:
   – load-bearing bellows of tractor – front of the semi-trailer,
   – load-bearing bellows of semi-trailer – rear of the semi-trailer,
3. Lowering of the supporting leg of the semitrailer,
4. Take the tractor out under the semi-trailer,
5. Transverse movement of the central part of the stand with the semi-trailer to coincide of the longitudinal axle of the semi-trailer with the axles of bogies with adapters,
6. Lowering of the semi-trailer's body to support of the semi-trailer's frame on the adapters:
   – front of the semi-trailer using the supporting leg of the semi-trailer,
   – rear of the semi-trailer using the lifting bellows of the road axles,
7. Locking of the semi-trailers on the adapters,
8. Lifting of the road axles of the semi-trailer,
9. Locking of the road axles,
10. Returning of the central section of the transshipment stand to the starting position.

Transshipment stand for horizontal transverse loading of bimodal semi-trailers is shown in Figure 5 and consists of the following components:

- ramps (1):
- central movable segment with guide rollers (2)
- guide rails (3)

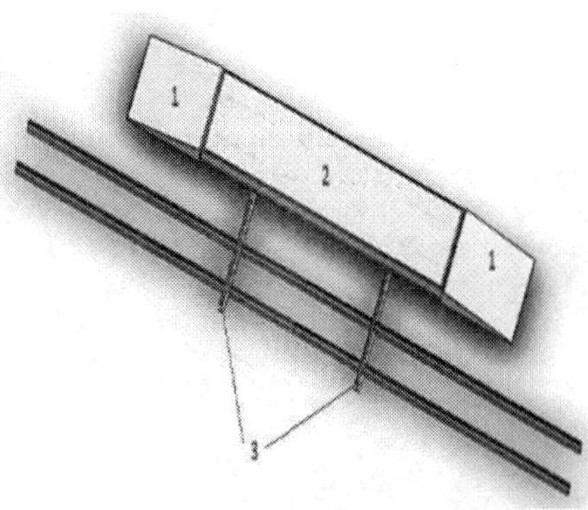

Fig. 5. Loading stand for horizontal transverse loading of bimodal semi-trailers

At the transshipment terminal, equipped with loading stands for horizontal transverse loading of bimodal semi-trailers, a tractor unit with a bimodal semi-trailer drives the semi-trailer into the middle segment of the semi-trailer loading stand using a loading ramp. When the trailer is in the right place on the middle segment, the trailer supports are lowered and the trailer and tractor are disconnected. The tractor unit is then leaving a stand by a ramp on the other side of the middle segment. The bimodal semi-trailer is transported to the central axis of the railway track, where railway bogies equipped with semi-trailer adapters are placed, with the appropriate distance, as illustrated in Figure 6.

Fig. 6. Transport of a semi-trailer to the railway track

Figure 7 shows a semi-trailer in the axis of the railway track before lowering onto the adaptors on the railway bogies.

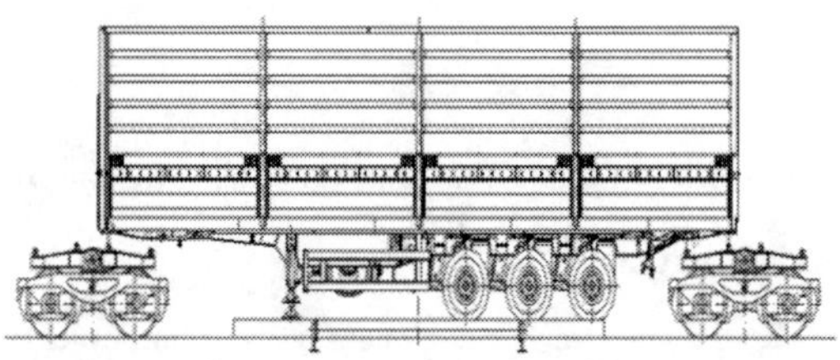

Fig. 7. Semi-trailer in the axis of the railway track

The next stage of unit forming is to lift on and block the road wheels and semi-trailer supporting legs. In result, semi-trailer is placed on the adaptors of railway bogies. The view of the trailer mounted on the adaptors is shown in Figure 8.

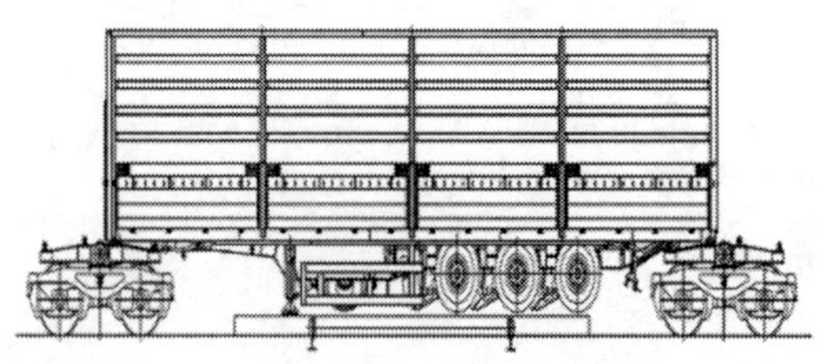

Fig. 8. Semi-trailer in the axis of the railway track mounted on adapters

The middle segment returns to its original position and, after locking the pins on the adapter, the trailer is ready for transport on rail track.

4 Terminal organization

In the design of the arrangement of stands for horizontal transverse loading of bimodal semi-trailers, the only constraint that should be taken into account is the size of the loading terminal: the area where the rail head is at the same level as the maneuvering place.

The first variant shown in Figure 9 is that the stands are located in series on one side of the track. With such an arrangement of stand, only small space of the terminal is occupied.

The second variant is shown in Figure 10. In this variant, loading stands are arranged alternately on both sides of the railway track. This variant facilitates the maneuverability of the tractor with semi-trailer.

Figure 11 shows the third variant, where the loading stands are arranged alternately on both sides of the railway track and offset by the width of the semi-trailer. This arrangement eliminates the need to maneuvering the tractor and trailer, which results in the easiest way to bring semi-trailers into loading stands.

As mentioned above, the only limitation of loading stand arrangement is the size of the transshipment terminal. So, several more examples may be presented of organization of loading stands and transshipment terminal can be organized to operate simultaneously on two railway tracks.

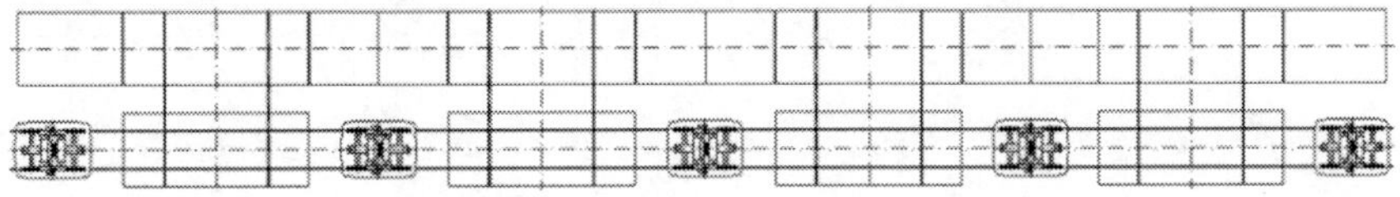

Fig. 9. Variant 1

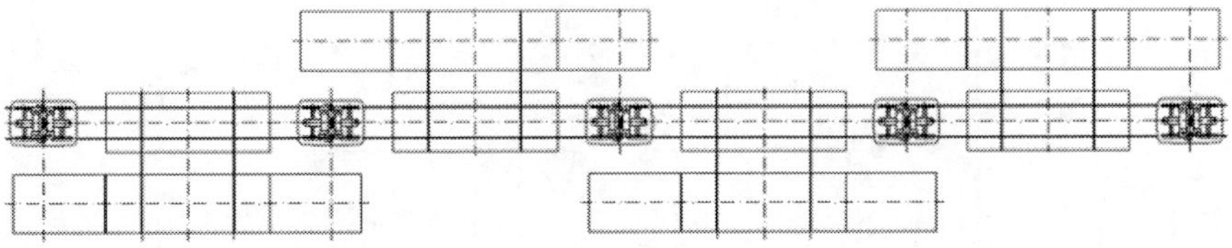

Fig 10. Variant 2

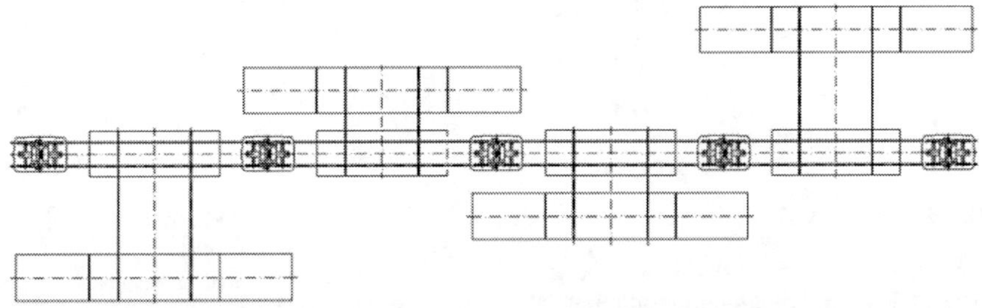

Fig. 11. Variant 3

The variants shown in Figs. 12, 13, 14 show two railway tracks with different arrangement of stations which are served by one central segment of the switching station for loading bimodal trailers. Each of the presented solutions for the arrangement of switching stations on the terminal eliminates various restrictions. Starting from the size of the terminal, through the ease of introducing the semi-trailer to the transshipment stand and ending with the skills of the operator of the truck tractor introducing the semi-trailer.

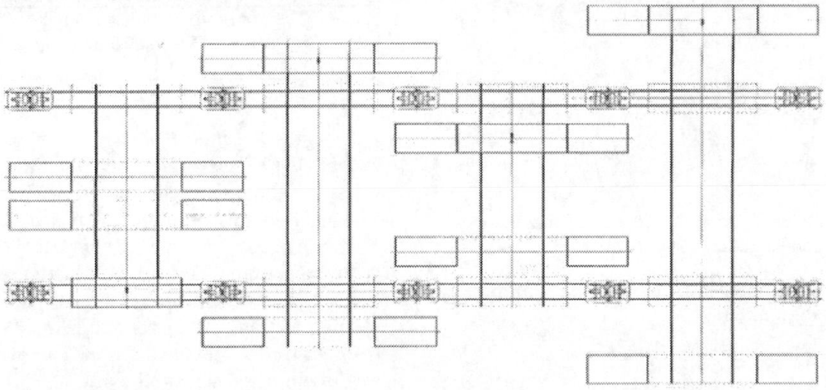

Fig. 12. Variant for 2 tracks operation

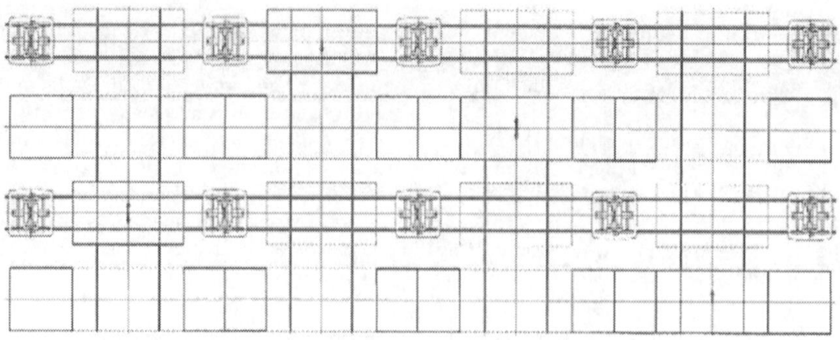

Fig. 13. Variant for 3 tracks operation

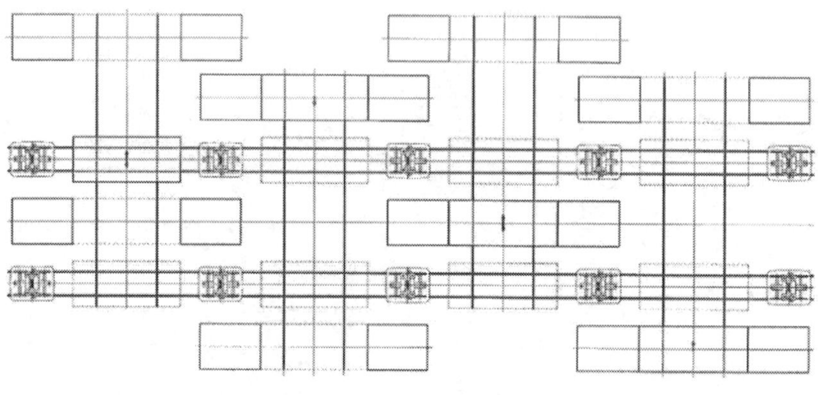

Fig. 14. Variant for 2 tracks operation

5 Computer simulation

A preliminary general computer simulation was carried out using the BUILDER C ++ program, introducing semi-trailers to switching stations and loading semi-trailers on railway bogies for three variants servicing one railway track.

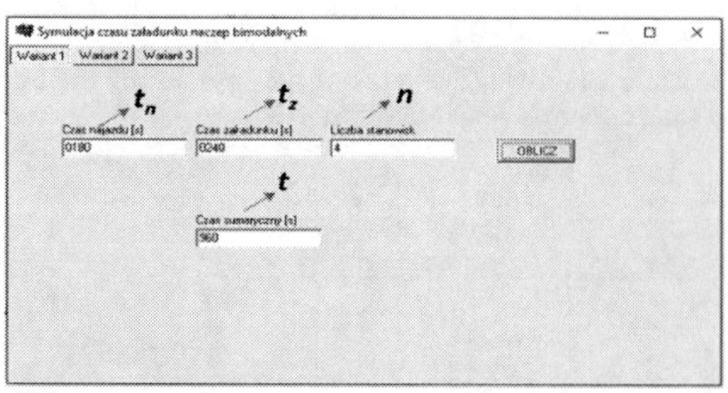

Fig. 15. View of the program window for simulating the loading time of bimodal trailers for Variant 1

After calculating, the estimated time for loading four semi-trailers in Variant 1 will be

$$t = t_n \cdot t_z + n$$

where:

t - total time
t_n – semi-trailer entry time
t_z - inroad time
n - number of positions

The calculation results for the three variants are summarized in tables 1 – 3.

Table 1. Variant 1

Lp.	List of activities	Summary time [s]
1.	Approach of the tractor with the semitrailer to the switching station	180
2.	The semi-trailer moves into the track axis and is loaded onto railway bogies	240
	Total Time	960

For the existing Tabor 1 and Tabor 2 systems in Poland [6], it was calculated that the parallel forming time of bimodal semi-trailers, whose equivalent is variant 1 in this article, is 850 seconds for one semi-trailer. To compare the Tabor 1 and Tabor 2 systems with the system equipped with loading stations, multiply the result by 4, because the proposed system has 4 loading stations. The obtained result of 3400 seconds is over 4 times longer than the innovative method of loading bimodal semi-trailers.

Table 2. Variant 2

Lp.	List of activities	Summary time [s]
1.	Approach of the tractor with the semitrailer to the switching station 1	90
2.	Approach of the tractor with the semitrailer to the switching station 2	280
3	The semi-trailer moves into the track axis and is loaded onto railway bogies	240
	Total Time	520

Table 3. Variant 3

Lp.	List of activities	Summary time [s]
1.	Approach of the tractor with the semitrailer to the switching station	90
2.	The semi-trailer moves into the track axis and is loaded onto railway bogies 1	240
3	The semi-trailer moves into the track axis and is loaded onto railway bogies 2	320
	Total Time	410

After comparing the three proposed variants, the obtained times clearly indicate that Variant 3, which was designed in such a way as to facilitate the operation of the truck tractor operator to introduce the semi-trailer to the switching station, obtained the shortest time to connect four semi-trailers. Detailed computational research with a breakdown into individual stages of bimodal formation will be carried out at a later stage of research.

6 Summary

In the 90s of the last century the bimodal system in Poland and in the world began to develop. In Poland the bimodal system called TABOR 1 and TABOR 2 was introduced. The Tabor 1 and Tabor 2 systems did not require additional loading devices at the loading terminal. The main disadvantage of this system was the time-consuming joining of the bimodal train, as well as the need to store semi-trailers in the parking lot. The long train preparation time resulted mainly from the complicated serial connection of the train unit. The article presents an innovative system for loading bimodal semi-trailers and the organization of a terminal equipped with positions for horizontal transverse loading of bimodal trailers. Train formation and de-formation with use of transverse semi-trailer loading stands should significantly shorten the time of connecting a bimodal train. A transshipment terminal with several loading stands enables simultaneous loading of several trailers at the same time. Equipping the terminal with loading stands allows also the direct loading of semi-trailers, without the need of storing them in the parking lot. The proposed method of loading semi-trailers makes possible total or partial automatization of semi-trailers loading process. Creating the train composition with the help of horizontal transverse loading of semi-trailers significantly reduces the time of connecting the bimodal train as shown by the results of the simulation. The loading terminal with several switching stations enables simultaneous loading of several trailers. Equipping the loading terminal with switching stations also allows for efficient direct loading without having to store the semi-trailers in the parking lot first. The proposed method of semi-trailers reloading allows for full or partial automation of the semi-trailers reloading process.

References

1. P. Daszkiewicz, M. Andrzejewski, MATEC Web of Conferences, Volume: **118**, Article Number: UNSP 00014, DOI: 10.1051/matecconf/201711800014, Published: (2017)
2. L. Rymaniak, MATEC Web of Conferences, Volume: **118**, Article Number: UNSP 00009, DOI: 10.1051/matecconf/201711800009, Published: (2017)
3. M. Medwid, A. Merkisz-Guranowska, J.Strzemkowski, J. Królikowski: RAIL VEHICLES, Volume: **3**, Pages: 1-10, Published: (2018)
4. J. Merkisz, L. Rymaniak, EKSPLOATACJA I NIEZAWODNOSC-MAINTENANCE AND RELIABILITY, Volume: **19**, Issue: 4, Pages: 522-529, DOI: 10.17531/ein.2017.4.5, Published: (2017)
5. M. Medwid, R. Cichy: ISBN 978-83-87350-87-1 Published: (2017)
6. K. Nowicki: Master thesis nr TR-M/S-11/2007 (2007)

Intellectualization of logistic supply chains on the basis of forecasting volumes of cargo transportation

Vitalii Gurnak, *Lyudmila* Volynets,* and *Ilona* Khalatska

NTU, Department of Transport Law and Logistics, 010101, Kyiv, 1, M. Omelianovycha-Pavlenka Str., Ukraine

Abstract. Research shows that the accelerated economic movement towards a more intelligent model of development and rapid dissemination of knowledge contributes to the emergence of new modern logistics technologies and approaches to managing global supply chains. The dominant feature of the sustainable development of transport services is the invariant concept of management, the basis of which is the socio-ecological and economic model. Indicators of the transport sector show a decline in volumes of transportation, but the positive dynamics of growth is demonstrated by the agrarian sector, which produces half of all export volumes and provides one third of the currency revenues of the state.Factors that inhibit the effective functioning of the logistics chain are revealed. In particular, the technological orientation of railways, mainly on route transportation, highlighted the problem of uneven development of ports at ports and railway ports, as well as inconsistencies in grain volumes in Stevedoring companies' deals with real port facilities. In fact, there is a schedule of seasonal fluctuations in the transportation of cereals, in particular, their absence for 3-4 months per year.The rational model of management and intellectualization of grain supply logistic chain, which takes into account the indicators of artificial neural network forecasting and seasonal grain transportation parameter, is proposed.

1. Introduction.

The typical features of the current state of the transport industry activity are as follows. The volumes of transportation are rapidly falling, the critical level reached the financial condition of the entire industry.

Long-term drop in rail freight transport objectively depends on changes in gross national product (GDP). The analysis showed that the decisive impact on the volumes of freight traffic is precisely the dynamics of GDP, that is, in fact, industrial production. If this indicator grows, rail transport is almost proportionally increasing, and vice versa.

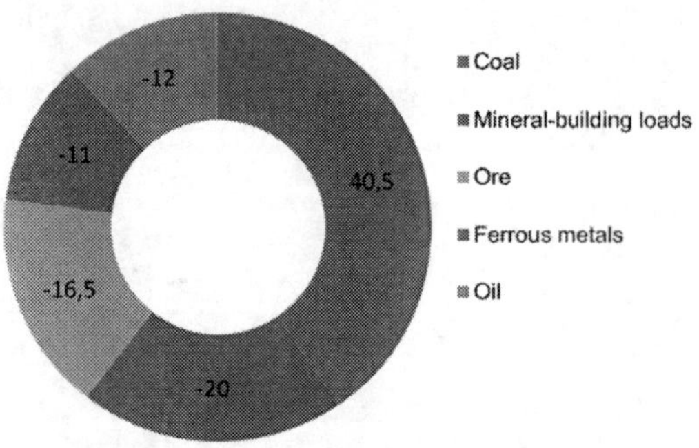

Fig.1. Falling volumes of rail transportation from 2013 to 2018, million tons

The main reasons are the temporary loss of the coal-mining capacities of the Donbas and the annexation of the Crimea; a significant drop in export volumes due to the loss of production capacities for the production of ferrous metals (from $ 17.6 to $ 11.6 billion); reduction of volumes of production and export of domestic machine-building products (from $ 3.5 to $ 0.7 bln); a significant drop in exports due to restrictions set by the Customs Union (43%); reorientation by the Customs Union of transit of oil and petroleum products from Ukrainian ports to others; the decline in domestic production and consumption of mineral-building cargoes (imports decreased from $ 21.2 to $ 13.4 billion).

Against this background, the dynamical development and growth of production volumes, especially of the grain and oilseed group, is demonstrated by the agrarian sector of the state.

If 28 years ago Ukraine exported grain to North Africa and Europe, today, 40% of exports go to Asian countries, where the population growth is steady and quite high. According to UN forecasts, in 2050, the world will have 9.7 billion inhabitants, with the largest population growth expected in India. That is, an increase in the volume of agricultural production in general will have a guaranteed market of sales. Therefore, problems of logistic nature are at the forefront.

Modern trends in the development of the world economy in the context of global civilization dynamics and the formation of an information society show the intellectualization of all aspects of society's vital activity. The vast majority of developed countries actively

* Corresponding author: volinec_3@ukr.net

develop a knowledge-based economic system that fundamentally changes the traditional approaches and models of logistics supply chain development. Logistics is not only a key function of the modern enterprise and business in general, but also, according to the figurative expression of German specialists, the "motor of the economy", that is, it affects the efficiency and competitiveness of the national economic system [1].

At the same time, it is important to note that the current stages of development of human society and strengthening its impact on the environment as a consequence of scientific and technological progress, increases the practical value of the environmental economy.

Problems of environmental substantiation of logistics chain intellectualization projects and management of investment processes taking into account the requirements of environmental protection are quite urgent. For today it is a priority because of the increasing importance of further implementation of sustainable development in the country's economic life. As you know, the concept of sustainable development includes three components: economic, social and environmental, which are inextricably linked. Therefore, the intellectualization of supply chain logistics should be based on the environmental concept of management based on the interconnection between ecosystems, social communities and economic systems, as well as conditions that ensure the steady state and progressive development of all three systems.

It should be noted that every year Ukrzaliznytsia transports grain more and more, but this dynamics lags behind the rate of productivity growth in the country. If the state carrier previously transported 80% of the grain collected in the country, three years ago - 68%, then today - 63%. At the same time, Ukraine has a unique, well-developed regionally-developed railway infrastructure. It makes it possible to reduce the cost of grain logistics. However, at the moment, we have a lot of low-impact stations and sections that have inherited from the socialist system of planned economy.

Based on the fact that the formation of a network of railways in different regions of Ukraine was conducted under the influence of specific factors. This necessitates the analysis of the current state of low-impact sections (DMDs) of railways and the decision of expediency of their further functioning, taking into account the peculiarities of economic development of certain geographical regions.

The results of the analysis of the current state of the MDS in the transport market show that on a significant number of them (up to 15.21% of the length of the network), the number of freight trains does not exceed 1-2 per day (Table 1).

Table 1. The number of low-impact sections in the railway transport

Name of regional offices	DN - 1		DN - 2		DN - 3		DN - 4		DN - 5		All		Specific weight MD%
	All	MD	All	MD	All	MD	All	МД	All	MD	All	MD	
Prydniprovska	20	1	18	1	15	2	13	0	0	0	66	4	6,06
Southern	5	2	22	1	5	1	16	0	7	0	55	4	7,27
South - West	19	3	23	8	24	3	14	2	19	9	99	25	25,25
Odesa	44	2	19	11	19	0	14	2	0	0	96	15	15,63
Lviv	32	6	15	1	22	5	31	6	18	0	118	18	15,25
PJSC "Ukrzaliznytsya"	120	14	97	22	85	11	88	10	44	9	434	66	15,21

* - the Donetsk railway was not taken into account in connection with the military events in the East of Ukraine and transportation in the territory of the annexed ARC.
DN is the directorate at each regional branch.

Special attention is paid to the work of low-impact sections and stations, which are 46% and 30% on the network of national carriers respectively. According to the criteria for defining few activities, those who carry out less than 10% of work are considered such as on average on the network of Ukrzaliznytsya. The negative result of getting such an almost unused infrastructure exceeds UAH 7 billion a year [2].

At the same time, the European transport directives, which Ukraine has in accordance with the Association Agreement with the EU to implement its legislation, contains a norm according to which none of the companies in this sector of the market should work at a loss. Such a norm is enshrined in the latest edition of the draft Law of Ukraine "On Rail Transport". It also applies to JSC "Ukrzaliznytsya", which, in accordance with European regulations, should be the operator of public utilities infrastructure, which remains in state ownership.

In order to assess the effectiveness of the railway infrastructure, a whole package of data on cargo operations at stations, operational cargo turnover, cargo tension in the polling stations and a number of financial indicators were analyzed. As a result, groups of the most disadvantaged sections of the infrastructure were identified. According to the findings of the analysis on operational and economic efficiency of Ukrzaliznytsia's work with static geo-information visualization, today almost half of the railway facilities generate billions of losses to the state carrier. Specifically: 45.7% of all contracting companies together perform only 2.1% tonkilometrovoy work. For the maintenance of these inefficient sections and their functioning, the company directed more than UAH 7.8 billion, and received a total of UAH 543 million. of income, that is, less than 1% of all revenues from freight transportation.

Moreover, inefficient districts are located throughout Ukraine. At the same time, in comparison with 2016 their number increased. In 2018, the worst financial result was obtained at Yahodyn-Kovel (66 km), Kovel-Sarny (142 km) and Uzhgorod-Syanki (90 km) (Lviv railway); T. Shevchenko - Pomichna (129 km), Tsvetkovy - Khrystynivka (148 km), Khrystynivka - Timanovka (113 km) and Kherson-Vadim (103 km) (Odesa railway); Surske - Apostolove (139 km) (Prydniprovska railroad), Ziatkivtsi - Vinnitsa (62 km) (Southwest railway) and Lyubotin - Smorodinov (112 km) (Southern railway).

The results of diagnostics of low-impact sections allowed to reveal a considerable reserve of carrier and throughput capacity. It is established that there is a significant under-utilization of production capacities. The main reason - in low-impact areas there is a need for special, simplified operation methods, implementation of which can significantly reduce the cost of transportation, the level of tariff, which, in turn, will allow attracting more consumers of rail transport products and approach the solution to the problem of loss-making low-impact areas and the possibility of realizing a forecast of their activities in the future.

In the prevailing volatile and difficult economic, political and social conditions, railways continue to be permanently working and attractive to most cargo owners by mode of transport [3].

The problem of developing the system of forecasting freight volumes in modern conditions has become particularly relevant, since analyzing and forecasting volumes of cargo transportation is an important tool for developing effective management decisions and will involve the creation of intelligent logistics supply chains and the planning of the need for material, labor and financial resources.

2. Analysis of literary data and problem statement

The volumes of grain production in Ukraine far exceed the needs of the domestic market. At present, most domestic grain is exported. At the same time, state strategic goals have been determined and continue to increase grain production to 80 million tons. However, the grain market infrastructure does not fully meet the needs of growing traders.

In particular, in the operation of the transport and logistics system, difficulties are observed with the delivery of export batches to ports, which affects the efficiency of the utilization of terminal capacities. First of all, this is due to the increasing load on the transport infrastructure. Thus, it became problematic for railway transport to submit wagons in all applications.

Today, rail transport occupy a dominant share (about 60%) in the total volumes of cereals transportation. Distribution and use of UZ rail cars takes place in a non-transparent manner.

Obviously, the rules and tariffs for using the railway infrastructure should be equally transparent for all participants in the logistics chain.

The logistics chain and its parts are the subject of research by many foreign and domestic scientists. Logistics aspect of supply chain management is devoted to work [3]. The paper [2] describes the need to change the priorities of investment in the development of transport infrastructure in modern conditions and increase the efficiency of the logistics supply chain. The main areas of logistics in the agricultural sector are reserves and transport. The decision of issues of optimization of export cargo flows and development of infrastructure of different types of transport for their effective interaction is considered in the paper [2,4].

Theoretical and methodological bases of research were the scientific works of scientists in the field of planning and forecasting of various spheres of activity. In [1,5] the necessity of forecasting of the future volumes of transportation by railroad during the formation of the level of tariff for transportation of goods is indicated. Empirical results [6] confirm that the artificial neural network is an effective tool for forecasting the exchange rate. This technique provides evidence that there is an opportunity to extract hidden information for reliable forecasting of the future.

The decision-making process when planning the required number of rolling stock is considered one of the most problematic issues for the transport industry. The paper [7] reveals a new approach to determining the size of the fleet in air travel with the help of Monte Carlo modeling is being developed. Applying this method, the airline receives maximum profit. Such a decision procedure can also be applied on rail transport to detect the required number of carriages for forecasting cargo transportation.

Forecasting flows is an important component of transport systems. The adoption of unwarranted decisions in the transport of goods leads to a rise in monetary expenditures in the railway industry. Transportation management should provide the choice and implementation of only optimal solutions that can comprehensively assess the performance of vehicles for different transportation options [8].

In order to accurately predict the volumes of freight transportation by railways of Ukraine and the possibility of extracting hidden information during the planning of passenger traffic, the above-mentioned current experience of forecasting was taken into account both in transport and in other sectors.

3. The purpose and tasks of the study

The aim of the work is to improve the management of logistic supply chains on the basis of forecast values of volumes of cargo transportation.

To achieve this goal, the following tasks were set:

- to analyze the work of the transport industry and the volume of transportation of grain;
- To develop a model of forecasting of volumes of cargo transportation and to intellectualize the logistic chain of supply of grain;
- to propose a mathematical model, increase the efficiency and intellectualization of the logistic supply chain.

4. Investigation of the forecasting method for determining the volume of cargo transportation

The task of forecasting was realized due to the structural model on the basis of artificial neural network (ANN). This method of forecasting has high properties of adaptation to input data variables through self-study. The main advantage of neural network models is non-linearity, that is, the ability to establish nonlinear relationships between future and actual values of processes. Other important advantages include: adaptability, mass-compliance (parallel structure of ANN accelerates computation) and uniformity of analysis and design. In this case, the disadvantages of ANN are the lack of transparency in modeling, the complexity of choosing architecture, high requirements for the consistency of the training sample, the complexity of the choice of learning algorithm and the resource-intensiveness of the learning process [9].

Equation of the mathematical structure of the neuron:

$$y = f(u), \quad (1)$$

$$U = \sum_{i=1}^{n} w_i x_i + w_0 \quad (2)$$

where w_i, x_i – respectively, the forces at the inputs in the neurons and the magnitudes of synaptic bonds;

w_0 – displacement coefficient of the neuron (weight of the additional input);

U – induced logical field;

$f(u)$ – transfer function or activation function.

The additional input and its corresponding weight are used to initialize the neuron. Under initialization, the shift in the activation function of the neuron along the horizontal axis, that is, the formation of the threshold of sensitivity of the neuron, is meant.

The method of communication of neurons defines the architecture of the neural network. According to the work [10], depending on the way neurons communicate, the networks are divided into:
- direct distribution networks;
- recurrent neural networks;
- radial basis functions;
- Self-organizing maps or Kohonen Networks

Probably, the architecture of multilayer neural networks is now used most often. The network is composed of many sensor elements (input nodes) that form the input layer of one or more hidden layers of computing neurons and one source layer of the neurons.

The multilayer neural network can simulate the function of practically any degree of complexity, and the number of layers and the number of elements in each layer determine the complexity of the function. Determining the number of intermediate layers and the number of elements in them is an important issue when constructing.

Neural networks are especially effective for systems that are characterized by a high degree of nonlinearity. In order to evaluate the ability of the neural network to effectively predict the volumes of freight transportation by rail, first of all, it is necessary to analyze the preliminary statistics of freight traffic.

Forecasting is realized with the help of the fully connected multilayer perceptron (MLP, Multilayer perceptron) with direct distribution, architecture 12-96-1. The signals propagate in one direction, starting from the input layer of the neurons, through the hidden layers to the output layer and on the output neurons the result of the processing of the signal is obtained.

The activation function of the input layer is identical, that is the input signals are not converted, but transmitted to the next layer. In this network, unlike its predecessor, perceptron Rosenblatt, uses nonlinear activation functions, namely sigmoid. It allows you to amplify weak signals and not overload with powerful signals. Such a nonlinear actuation function solves the dilemma of noise saturation. On the coiled layer, the logistic function (3) is used, and at the output - the hyperbolic tangent (4)

$$f(U) = \frac{1}{1+e^{-u}}, \quad (3)$$

$$f(U) = \frac{e^{u}-e^{-u}}{e^{u}+e^{-u}} \quad (4)$$

At the output of the neuron we obtain the forecast value of the volumes of cargo transportation by rail

4.1 Results of modeling and intellectualization of the grain supply logistics chain

As input data, statistical observations on volumes of transportation of grains and products of reminiscing from 2004 to 2018 were used. For the training sample 90% of the values were allocated, and for the control - 10%. As a test sample for cross-check, real volumes of grain shipments for the 8 months of 2018 were used. The results of forecasting are shown in Fig. 2.

The adequacy of the model can be judged by the diagram of the distribution of residues (Fig. 3), in the ideal case the graph resembles the normal distribution law and has a bell-shaped shape.

The relationship between incoming and outgoing data is presented in Fig. 4.

Judging by the figure, we can conclude that the meurezh catches the general tendency of a number.

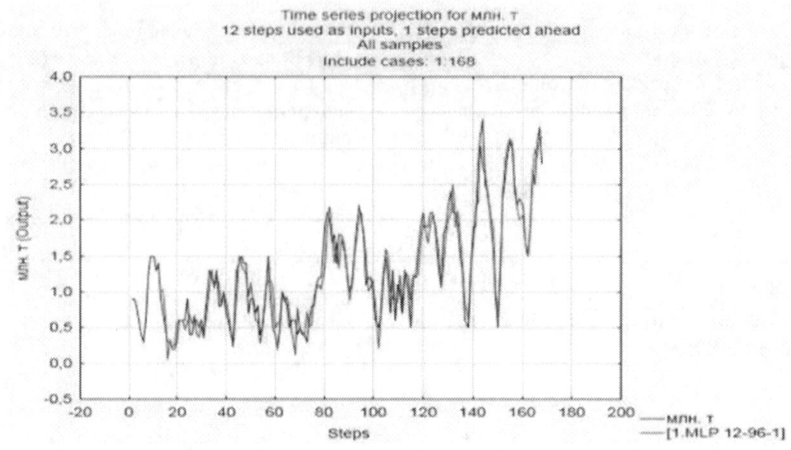

Fig. 2. Comparison of the predictive series with real values

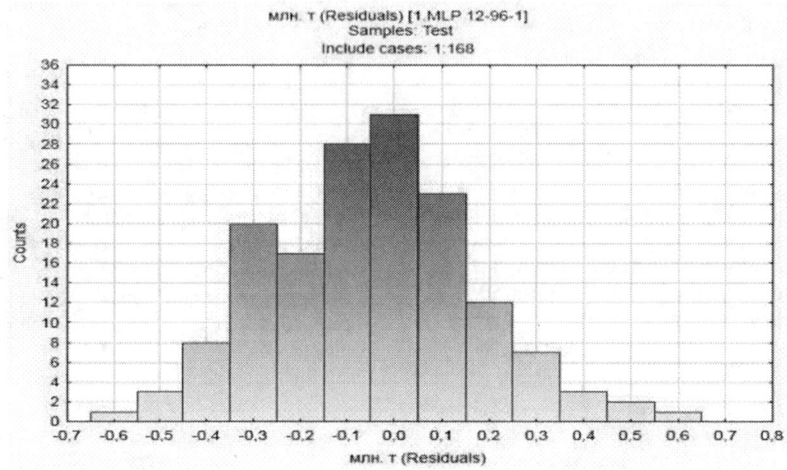

Fig. 3. Chart of residue distribution

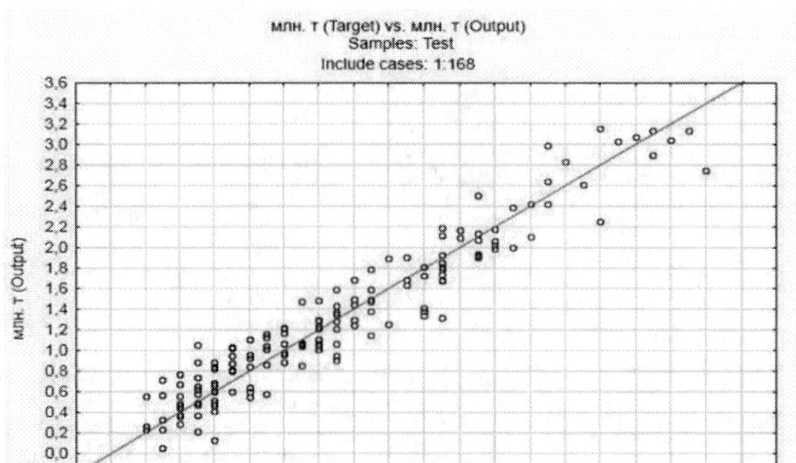

Fig. 4. Schedule of the relationship between incoming and outgoing signals

The results of the cross-check are shown in Fig. 5

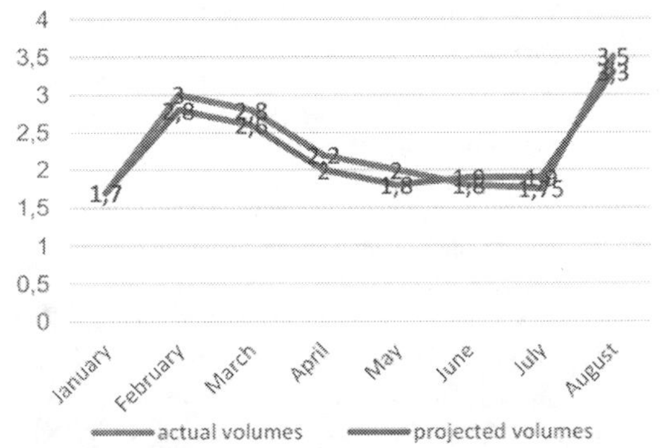

Fig. 5. Comparative schedule of forecast and actual volumes of transportation of grain and reminiscent products in 8 months 2018

The value of the accuracy of the forecasting of traffic is determined by the value of the average absolute percentage error (SAPS). The magnitude of the accuracy of pro-rigging is the concept of the exact opposite error of prediction. If the prediction error is high, then the accuracy is small, and vice versa, if the prediction error is small, then the accuracy is high.

Comparison of actual and forecast data of the investigated cargo transportation period is presented in Fig. 6

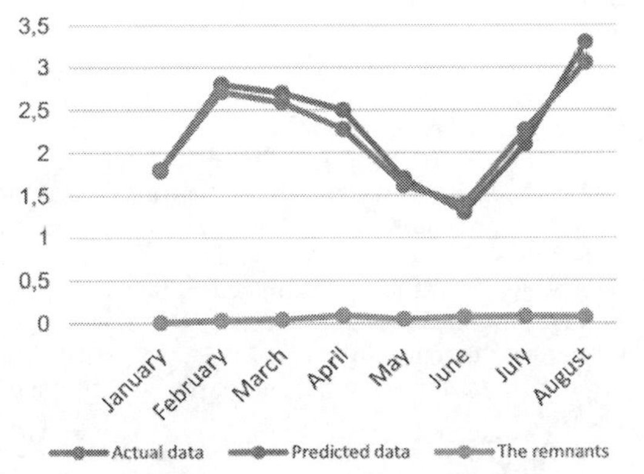

Fig. 6. Comparison of actual and forecast data on the volumes of transportation of grain and reminiscent products

Speaking of high accuracy, we always talk about a low forecast error. The forecast error is to a degree characterized by the accuracy of the applied forecasting technique. The experiment showed that the average absolute error rate of grains and maize products transmitted during the first eight months of 2018 amounted to 5.7%. Given that rail transport is a fairly inert system, the AMD indicator of 5.7% is sufficient to make managerial decisions.

Transport industry is quite inert system. It is difficult to make operational management decisions. Therefore, it is necessary to provide high-quality forecasting for taking tactical administrative decisions for some time before the event's time for effective work of the industry.

Knowing the predicted level of cargo transportation, it is possible to determine the required number of cars of the type of container that will meet the conditions for the transfer of this cargo, that is, to improve the logistics chain of grain supply. Forecast values of carriage rates for grain and reminiscent products are shown in Fig. 7

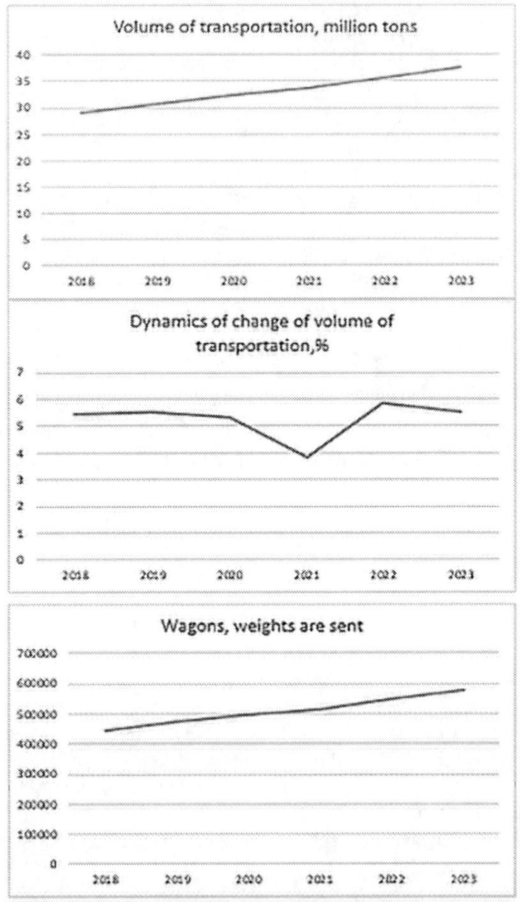

Fig. 7. Projected values of carriage rates for grain and maize products

Due to forecasting of volumes of transportation for the next period it is possible to improve the efficiency of management of supply chains and realize the agrarian potential, especially in peak periods. The foregoing values of freight transportation rates were highlighted in subsequent calculations.

4.2. Formalization of management of supply chains of grain

The presence of integrated systems of flow processes in logistics chains is considered an important condition for an industrial enterprise. Ensuring continuity and reducing aggregate costs in all branches of the logistics chain from manufacturer to consumer should be based on the rational use of rolling stock to carry out the planned volumes of transportation.

The drafting of the distribution plan for wagons depends mainly on the number of wagons that are not needed for carriage. Also, an important planning parameter is the location of wagons in a specific time period.

Under the optimal organization of logistic supply chains is minimization of operating costs for the carriage of goods. In calculating the plan for the formation and operational plan of vehicle management, the main criterion for choosing promotion is the minimum operating cost. But the same approach to calculating the transportation plan for different types of cargo is not optimal. Therefore, in order to find the optimal variant of the movement of material flow, it is suggested to take into account the parameter of uneven transport or seasonality. The above studies have shown that for different cargoes the coefficient of unevenness of transportation is uneven. The choice of a rational plan for the distribution of wagons on the railway network is presented in the form of a target function:

$$C(x_{ij}) = \sum_{i=1}\sum_{j=0} x_{ij} \times k_{н} \times \delta_{зав} \times d_{ij}^{зав} +$$
$$+ \sum_i \sum_j x_{ij} \times k_{н} \times \delta_{пор} \times d_{ij}^{пор} + {}^{\Delta D}/_{\Delta x_{ij}} \rightarrow min \quad (5)$$

where C – expenses for transportation of cargo, UAH;
x_{ij} – variable number of cars moving from station i to station j;
$k_{н}$ – coefficient of uneven cargo transportation;
$\delta_{зав}$, $\delta_{пор}$ – the proportion of loaded wagons and wagons correspondingly going from station i and stations j;
$d_{ij}^{зав}$, $d_{ij}^{пор}$ – expenses related to the transfer of loaded and empty wagons respectively;
ΔD – specific costs due to an increase in carload above average, UAH;
Δx_{ij} – a specific increase in the number of cars transported from stations i to the station j.

The following factors are taken into account as limitations:

– Limitations related to bandwidth of stations and stations:

$$\sum_{i=1} N_{наяв}^{ст/дільн} - \sum_{j=0} N_{потр}^{ст/дільн} \geq П;$$

– Limitation on the length of the the freight train:

$$\sum_{e=1}^{S} \mu_e(e) \leq L_e;$$

– Limit taking into account the time of delivery of the goods:

$$t_{нав} + t_{форм} + t_{ш.прям} \leq T$$

Provided that the required number of wagons is delivered to load points:

$$1 \leq x_{ij} \leq \frac{a_i}{q_{ст}} \quad (6)$$

where $\sum N_{наяв}^{ст/дільн}$ – available bandwidth of stations and sections respectively in trains;

$\sum N_{потр}^{ст/дільн}$ – need a bandwidth of stations and stations corresponding to what is necessary for carrying out a plan of transportation in trains;

П – the number of trains that can be involved in the distribution of wagons in directions for railways;

$\mu_e(e)$ – the intensity of the carriage at the e-th ironwork station;

S – the number of stations that the train enters;

L_e – the length of the receiving and sending tracks at the e-station;

$t_{нав}$ – load time xij of cars, h;

$t_{форм}$ – duration of train formation, h;

$t_{ш.прям}$ – time of finding a train on the route, hour;

T – time of delivery of cargo, hour;

a_i – quantity of cargo in i-th load point, t;

$q_{ст}$ – average static load of the car, t / weights.

Application of the result of the decision of the proposed model gives the opportunity to make rational management decisions and intellectualize the logistics supply chain. That is to plan and promptly execute the distribution depending on the demand for freight wagons. This model takes into account both annual fluctuations of volumes of transportations, and monthly. And also the traffic of vehicles and railway infrastructure.

Conclusions

1. The analysis of the management of supply chains in the structure of logistics services of domestic enterprises of the agrarian sector. Indicators of the transport sector indicate a decline in transportation volumes, but the positive dynamics of growth is demonstrated by the agrarian sector, which produces half of all export volumes and provides one-third of the currency revenues of the state.

2. Formed a neural network model for predicting traffic volumes. This method of forecasting has high properties of adaptation to input data variables through self-study. The magnitude of the accuracy of the forecasting of traffic is determined by the value of the average absolute percentage error, which was 5.56%. Given that the transport industry is a fairly inert system, such an indicator is sufficient for making managerial decisions.

3. In order to find rational management decisions and intellectualization of the grain supply logistic chain, it is suggested to take into account the parameter of non-uniformity of transportation or seasonality. For optimality we accept minimization of operating expenses for transportation of cargo.

On the basis of the received data on the volumes of transportation and using the optimization model, a solution was obtained for the optimal plan for distribution of wagons. The procedure for distributing wagons is quite universal and allows for a real-time adjustment of the train formation plan. The proposed model can be used to make rational management decisions while improving the efficiency and intellectualization of logistic supply chains.

References:

1. V.M. Gurnak, L.M. Volynets, M.V. Gurnak, PMSAL, **9**, 33-39 (2017)
2. V.M. Gurnak, L.M. Volynets, CSW SUIT, **41(1)**, 7-17 (2018)
3. T.A. Vorkut, O.E. Bilonog, A.M. Dmitrichenko, Yu.O. Tretijchenko, *Supply Chain Management: Logistic Aspect* (NTU, 2017)
4. L.M. Volynets, V.M. Gurnak, The Eighth ISPC, 247-250 (2018)
5. S. M. Proshchuk, G. E. Bogomazova, R. A. Puri, CSW UkrSURT, **164**, 161-169 (2016).
6. R. P. Pradhan, R. Kumar, JMR, **4**, 111–117 (2010)
7. Y. Wang, H. Sun, J. Zhu, B. Zhu, MPE, **2015**, 1–13 (2015)
8. S. Panchenko, O. Lavrukhin, O. Shapatina, EEJET, **3 (85)**, 39–45 (2017)
9. V. V. Kruglov, V. V. Borisov, *Artificial neural networks. Theory and practice* (Hotline – Telecom, 2002)
10. S. Khaikin, *Neural networks* (LLC I. D. Williams, 2006)

Methodical foundations of cargo owners' transport services logistics management

Petro Yanovsky[1], *Viktoriia* Yanovska[2], *Sergiy* Lytvynenko[3 *], *Larysa* Lytvynenko[4], and *Oleksandr* Vodchic[1]

[1] National Aviation University, Military Training Department, 03058 Kyiv, Ukraine
[2] State University of Infrastructure and Technology, Department of Economics, Marketing and Business Administration 04073 Kyiv, Ukraine
[3] National Aviation University, International Economy Department, 03058 Kyiv, Ukraine
[4] National Aviation University, Management of Foreign Economic Activity of Enterprises Department, 03058 Kyiv, Ukraine

Abstract. In the article methodical foundations of transport services logistics management of cargo owners were realized, on the basis of integrated estimation of economic efficiency of holistic progressive logistic chain operation of cargo delivery. In the making of the study, the authors have identified the provision of logistic management conditions by production and mainline rail transport in carrying out transportations. In the successful implementation of finished product loading and arriving cargo processing plans, clear interaction of all logistics chain elements is an important factor, achieved on the basis of introduction of a stable cargo train schedule at sections and the submission of local operation organization in nodes, at sections and stations. On the basis of modelling possible variants of different parameters ratios of the logistic interaction process of production with mainline rail transport, it was determined that at different parameters ratios results can be obtained that will vary by waiting time of carrying out loading in railway cars and waiting time of removal of finished products from manufacturer's warehouse, required car fleet and loading of material handling vehicles. In this case, the most economical option should be taken as the main condition for ensuring a rational interaction of industrial and mainline rail transport.

1 Introduction

Driven by the need to initiate the restoration of the state's economy and the need to bring up rapidly its functioning to the level desirable by the society, priority is given to solving problems on the improvement of relations between cargo owners and railways on mutually beneficial conditions, technical and technological re-equipment of mainline and industrial rail transport. The issue of searching for a complex of effective measures the implementation of which in the operation of railways will improve the use of limited resources and increase the quality of transport services, in current conditions of economic, financial and monetary relations' globalization among the states is of particular relevance.

The uneven distribution of producers and consumers of specific products in the state territory, and also the corresponding natural resources, leads to uneven concentration of works related to loading and unloading of cars on Ukrainian railways. Quite frequently, railway cars of one type arrive for unloading at the station, and for loading there is a need for railway cars of another type. The unevenness of territorial division of works on loading and unloading increase by seasonality of carrying specific cargoes, requiring the transfer a certain number of appropriate rolling stock to performance points in the short term. Notwithstanding such unevenness, any railway should have specific quantity of cargo resources of railway cars on a daily basis in order to provide stable and rhythmic production operation. Due to this full amount of scheduled and additional loading and departures for the purpose of finished products should be provided.

Transport production is the foundation of a market economy. Therefore, it acts as a subject of market relations, the quality of operation of which affects the effective functioning and development of all industries of national economy served by transport. The provision of rational interaction of railways, industrial structures and consumer organizations in establishing mutually beneficial economic-legal relations between transport and cargo owners is particularly important in economic development, i.e. through the implementation of a compromise management of the transport process on the equal participation of all its participants.

2 Literature review and defining the problem

Many Ukrainian and foreign scholars were engaged in studying the issues of logistics management in railway transport. The most significant contribution to solving scientific problems on this issue was made by E. Alyoshinsky, G. Bogomazova, T. Butko, H. Kyrychenko, O. Lavrukhin, D. Lomotko, S. Panchenko,

* Corresponding author: sllitvinenko1982@gmail.com

L. Parkhomenko, A. Prokhorchenko, O. Shander, O. Shapatina, G. Zambrybor, V. Zapara, Y. Zapara, and Y. Yue. Research papers of B. Abramovic, E. Nedeliakova, M. Panak, D. Sipus, J. Sabino, J. Leal, T. Stützle, M. Birattari, L. Kuttner are of particular interest.

The papers [1-6] suggest approaches to solve the problem of train routing. D.R. Martinelli and H. Teng in [1] explore neural networks for obtaining good solutions in short time periods for the train formation problem. G. de O. Ramos and A.L.C. Bazzan in [2] present the use of the GRASP metaheuristic, propose a novel performance evaluation function, which measures the number of vehicles having an incentive to deviate from the routes to which they were assigned. A novel performance metric was introduced, which measures the amount of vehicles assigned to non-least cost routes. This approach outperforms the well-known method of successive averages providing considerably fairer assignments. In paper [6] a genetic algorithm to compute an approximate solution that seeks to minimize the average travel time was applied, a non-trivial network was used departing from binary route choice scenarios.

L. Kuttner in [7] proposed an option to optimize designing networks in the cargo service of German railways, which is an interesting and original modern approach to solving the issue. Scholars J. Sabino, J. Leal, T. Stützle and M. Birattari in [8] offered an algorithm for optimizing the operational planning of the railway station performance. The proposed algorithm minimizes the multi-purpose function, which considers both fixed and variable transport costs associated with the movement of railway cars in the area of the railway station's yard. An attempt to monitor the synergy of logistic processes in railway transport was carried out by B. Abramovic, E. Nedeliakova, M. Panak and D. Sipus in [9]. The results of the study allowed estimating costs for quality logistics processes in railway transport.

3 Research aim and objectives

The aim of the research is to develop the methodical bases of cargo owners' transport service logistics management.

To achieve the purpose, the following tasks were identified:

- to carry out a comprehensive assessment of the economic efficiency of the functioning a holistic progressive logistic chain of cargo delivery (LCCD);
- to identify possible variants of different parameters ratios of the logistic interaction process of production with mainline rail transport;
- to evaluate the results of modelling the interaction between the finished products' accumulation process in the manufacturer's warehouse and the system of supplying empty car in the logistics chain "Production – Mainline Rail Transport".

4 Overview key findings

4.1 Providing the condition of production and mainline rail transport logistic management in performing carriage

Improvement of cargo owners' transport service should be carried out with a view to ensure competitiveness and profitability of railways at the stage of their reformation and integration into the EU. Due to limited resources, successful accomplishment of this complex task is provided by the implementation of logistic principles at all stages of the transport process, taking into account the interests of clients, intermediaries, operators and the carrier. As D. Lomotko in [10] and T. Yanovska in [11] rightly pointed out, in the context of economic market transformations the requirements of the clientele to the quality of transport services are increased while increasing competition from the side of railways. As indicated by the results of the researches carried out, a comprehensive solution to all problems with the cargo transportation can be achieved by designing logistic chains of cargo delivery (LCCD), minimizing the function of total costs of the operation of the logistics chain "Production – Mainline Rail Transport" (LCPMRT) presented by equation (1), which takes into account all costs of the carrier when sending a batch of finished products (Fig. 1).

$$\sum E_{tot} = 365 \cdot Q_d \cdot \frac{q_l^a}{q_l^p} \cdot [t_{del}^{st}\left(C_{del} + P_c + C_{str}^{st} + C_{pay}^{tc}\right) + \\ + t_{del}^{d} \cdot \left(\frac{C_{del}^{d}}{L_s} + C_{del}^{w}\right)] \geq C_{luv} \cdot E_n + C_{mn} \quad (1)$$

where Q_d – daily cargo traffic, t; q_l^a – actual hourly productivity of loading and unloading vehicle, t/h; q_l^p – planned hourly productivity of loading and unloading vehicle, t/h; t_{del}^{st} – delay of 1 tonne of cargo per day at the station; C_{del} – costs of delay 1 ton of cargo for 1 hour, UAH; P_c – price per 1 ton of cargo; C_{str}^{st} – amount of payment for the storage of 1 ton of cargo for 1 hour at the station by the cargo owner, UAH; C_{pay}^{tc} – amount of payment for the use of station tracks and cars by the cargo owner to the railway, UAH; t_{del}^{d} – duration of delay 1 ton of cargo on the driveway per day, h; C_{del}^{d} – the cost of delay for 1 hour 1 tonne of cargo on the driveway; L_s – static load of the railway car, t/car; C_{del}^{w} – costs for delays of 1 ton of finished products at manufacturer's warehouse for 1 hour, UAH; C_{luv} – cost of used loading and unloading vehicles and other technical means; E_n – normative coefficient of investment efficiency; C_{mn} – costs for maintenance of introduced modern loading and unloading vehicles and other technical means, UAH.

Result (1) considers the operation of the logistics chain "Production – Mainline Rail Transport" without taking into account its interaction with other parts of the entire transport and logistic system (TLS). Therefore, it must be evaluated as the previous one. To obtain the final result, it is necessary to carry out a comprehensive assessment of economic efficiency of the functioning of a progressive logistic chain of cargo delivery.

High-quality interaction of industrial and mainline transport involves the organization of the work of all basic elements of the logistics chain in a single technology providing rhythmic reception and dispatch of trains, coordinated shunting movements, even distribution of loading and unloading for 24 hours and by days of the week.

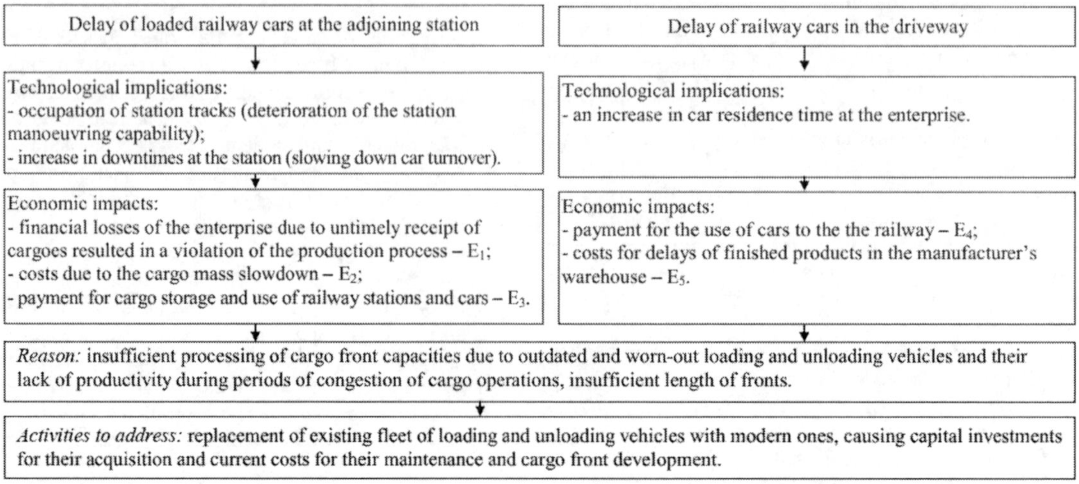

Fig. 1. Providing the condition of production and mainline rail transport logistic management in performing carriage

Herewith, the clear interaction of all elements of the logistics chain is an important factor in the successful implementation of the plans for loading finished products and processing of arriving cargo.

As evidenced by the best practices of railways and the results of scientific researches, such a clear interaction and harmonization can be achieved on the basis of the introduction of a stable schedule of cargo trains at sections and subordination of the local operation organization to it in nodes, at sections and stations.

To implement it in the practical conditions of servicing each cargo front, the supply-arrival of car groups should be subordinated to such a schedule. Given the rather high stability of production processes at large enterprises, the moments of departure of car groups with finished products should be established at the same time intervals, equal to the interval of accumulation of the shipment at manufacturer's warehouse $\overline{I_{ac}}$. Planned moments of the end of the accumulation of the shipment q need to be clearly aligned with the corresponding paths of a stable schedule of cargo trains at stations.

The scheme of accumulating the shipment of products at the manufacturer's station is shown in Fig. 2.

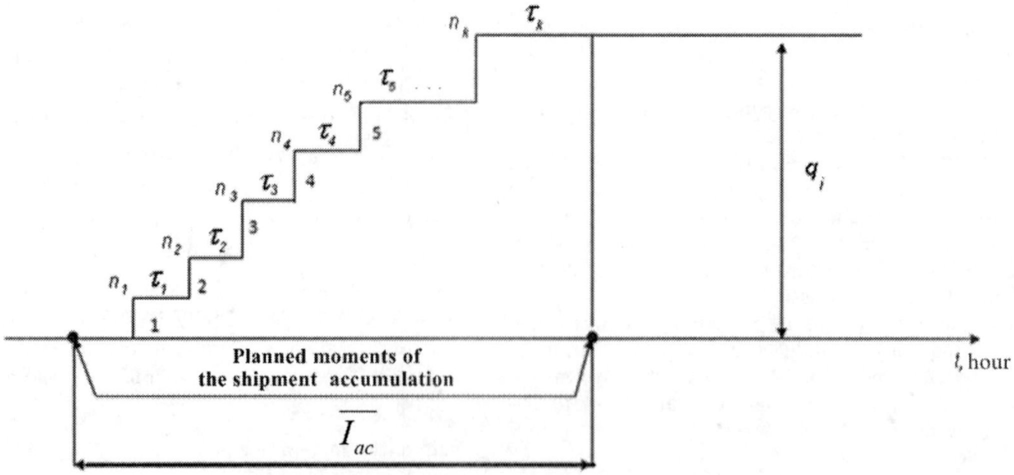

Fig. 2. The scheme of accumulating the shipment of products at the manufacturer's station

The cost of unit-hour accumulation of one shipment of *i* product type is determined by the formula:

$$O^{i}-g=n_1(\tau_1+\tau_2+\tau_3+...+\tau_i)+ , \\ +n_2(\tau_2+\tau_3+\tau_4+...+\tau_i)+...+n_i\tau_i \quad (2)$$

where $n, n_2, ... n_i$ – groups of products entering the warehouse of finished products and taking part in the accumulation q_i of the shipment; $\tau_1, \tau_2, \tau_3, ..., \tau_i$ – intervals between groups of product units arrived at the manufacture's warehouse.

Since n_i and τ_i are random variables (regardless sufficiently high regularity of the production process), therefore, $O^{i}-g$ is a random variable. In this case, the mathematical expectation is determined by the formula:

$$M[O^{i}-g]=M[n_1(\tau_1+\tau_2+...+\tau_k)]+ \\ +M[n_2(\tau_2+\tau_3+...+\tau_k)]+M[n_k\tau_k] \quad (3)$$

Using the property of average stability, the mathematical expectation of random variables with a sufficient degree of accuracy can be estimated from their average values. Having designated the number of intervals between groups of units of *i* product during the interval $\overline{I_{ac}}$ as *k*, unit-hour of accumulating the shipment can be written as:

$$O^{i}-g=\bar{n}\cdot\bar{\tau}\cdot k+\bar{n}\cdot\bar{\tau}(k-1)+\bar{n}\cdot\bar{\tau}(k-2)+...+\bar{n}\cdot\bar{\tau}$$

$$O^{i}-g=\bar{n}\cdot\bar{\tau}\cdot[k+(k-1)+(k-2)+...+1]$$

$$O^{i}-g=\bar{n}\cdot\bar{\tau}\cdot\frac{(1+k)k}{2}$$

$$\bar{n}\cdot\bar{\tau}\cdot\frac{(1+k)k}{2}=\frac{1}{2}I_{ac}\cdot q_i$$

$$\bar{n}\cdot\bar{\tau}\cdot(1+k)k=I_{ac}\cdot q_i.$$

From the last expression, the value of the shipment q_i is determined by the amount of its accumulation interval $\overline{I_{ac}}$, which should be used to select the train schedule path for 24 hours for sending specific shipment of its total consumption per day Q_i to specific consumer:

$$\overline{I_{ac}}=\frac{\bar{n}\cdot\bar{\tau}(1+k)k}{q_i}. \quad (4)$$

Equation (2) determines the value of the shipment q_i, inverse to the interval value $\overline{I_{ac}}$:

$$q_i=\frac{\bar{n}\cdot\bar{\tau}(1+k)k}{\overline{I_{ac}}}. \quad (5)$$

Taking into account the known relationship between the amount of the batch q_i, daily volume of products shipped to specific consumer Q_i and the length of the production cycle T_c on the output of *i* product by the enterprise $q_i=Q_i\cdot T_c$, we obtain the value of the production cycle:

$$T_c=\frac{\bar{n}\cdot\bar{\tau}(1+k)k}{\overline{I_u}\cdot O^{i}\text{-}g}. \quad (6)$$

Consequently, the linking of a stable schedule of dispatching finished products to the network with the production process of the enterprise is carried out through the process (duration) of the accumulating specific shipment. If, for convenience, intervals between the moments of the end of accumulating shipments indicate as T_{ac}^{pr}, the possible correlations between the value T_{ac}^{pr} and arrival interval of empty cars under load I_{ar} can be represented as follows: $I_{ar}=T_{ac}^{pr}\infty$; $I_{ar}<T_{ac}^{pr}$; $I_{ar}>T_{ac}^{pr}$.

In addition, there is a similar ratios between duration of the shipment loading t_l and shipment accumulation interval: $t_l=T_{ac}^{pr}$; $t_l<T_{ac}^{pr}$; $t_l>T_{ac}^{pr}$.

4.2 Accumulating finished products at manufacturer's warehouse and the supply of empty cars with equal arrival intervals of empty cars and period of finished products accumulation

Possible variants of different parameters ratios of the logistic interaction process of production with mainline rail transport are shown. On the basis of modelling possible variants of different parameters ratios of the logistic interaction process of production with mainline rail transport Table 1 was developed.

The proposed modelling is theoretical, based on a generalization of practical experience, in particular, JSC "Ukrainian Railway" and other railway operators.

The analysis of the results presented in Table 1 shows that different parameters ratios will give different results, which will vary depending on the duration of waiting for cargo operations by railway cars t_{wt}^{c} and waiting for dispatching finished products from manufacturer's warehouse t_{del}^{w}, required car fleet n_{req}^{c} and load of loading and unloading vehicles.

The most economical variant (variants №1 and №2 in Table 1) is necessary to accept as the basic condition for providing rational interaction of industrial and a mainline rail transport.

5 Conclusions

The transport factor occupies an important place in the development of all regions and the state as a whole, in international economy and foreign trade. Therefore, it is necessary to ensure the priority development of its components, taking into account the specific economic situation. Progress in the transport sector depends on the level of technical and technological development of all transport modes, which should be considered when using logistics economics for the development and implementation of logistic chain of cargo delivery (LCCD).

Table 1. Results of modelling the interaction of the process of accumulating finished products at manufacturer's warehouse and the system of supplying empty cars in the logistics chain "Production – Mainline Rail Transport"

№	Ratios of arrival intervals of empty cars I_{ar} and period of finished products accumulation T_{ac}^{pr}	Ratios of loading time t_l and period of finished products accumulation T_{ac}^{pr}	Car downtimes in waiting for cargo operations t_{wt}^{c}	Delays of finished products at the manufacturer's warehouse t_{del}^{w}	Level of the required car fleet to carry out transportation n_{req}^{c}	Consequences of the lack of harmonization in the accumulating finished products at the manufacturer's warehouse and supplying empty cars
1.	$I_{ar} = T_{ac}^{pr}$	$t_l = T_{ac}^{pr}$	$t_{wt}^{c} = 0$	$t_{del}^{w} = 0$	$n_{req}^{c} = optim$	Dispatch of finished products is rhythmic
2.	$I_{ar} = T_{ac}^{pr}$	$t_l < T_{ac}^{pr}$	$t_{wt}^{c} = 0$	$t_{del}^{w} = 0$	$n_{req}^{c} = optim$	Dispatch of finished products is rhythmic, breaks in the operation of loading and unloading vehicles
3.	$I_{ar} = T_{ac}^{pr}$	$t_l > T_{ac}^{pr}$	$t_{wt}^{c} > 0$	$t_{del}^{w} > 0$	$n_{req}^{c} = optim$	Dispatch of finished products is rhythmic, car downtimes and delays of finished products at manufacturer's warehouse
4.	$I_{ar} < T_{ac}^{pr}$	$t_l = T_{ac}^{pr}$	$t_{wt}^{c} >> 0$	$t_{del}^{w} = 0$	$n_{req}^{c} > optim$	Dispatch of finished products is rhythmic, car downtimes
5.	$I_{ar} < T_{ac}^{pr}$	$t_l < T_{ac}^{pr}$	$t_{wt}^{c} >> 0$	$t_{del}^{w} = 0$	$n_{req}^{c} > optim$	Dispatch of finished products is rhythmic, breaks in the operation of loading and unloading vehicles
6.	$I_{ar} < T_{ac}^{pr}$	$t_l > T_{ac}^{pr}$	$t_{wt}^{c} >> 0$	$t_{del}^{w} >> 0$	$n_{req}^{c} > optim$	Dispatch of finished products is non-rhythmic, car downtimes and delays of finished products at consignor's warehouse
7.	$I_{ar} > T_{ac}^{pr}$	$t_l = T_{ac}^{pr}$	$t_{wt}^{c} = 0$	$t_{del}^{w} >> 0$	$n_{req}^{c} < optim$	Dispatch of finished products is non-rhythmic, delays of finished products at consignor's warehouse, breaks in the operation of loading and unloading vehicles
8.	$I_{ar} > T_{ac}^{pr}$	$t_l < T_{ac}^{pr}$	$t_{wt}^{c} = 0$	$t_{del}^{w} >> 0$	$n_{req}^{c} < optim$	Dispatch of finished products is non-rhythmic, breaks in the operation of loading and unloading vehicles
9.	$I_{ar} > T_{ac}^{pr}$	$t_l > T_{ac}^{pr}$	$t_{wt}^{c} = 0$	$t_{del}^{w} >> 0$	$n_{req}^{c} < optim$	Dispatch of finished products is non-rhythmic, delays of finished products at consignor's warehouse

The efficient functioning of the logistics chain "Production – Transport – Consumption" can be provided under market conditions when building relationships between all its structures using the model of transport process compromise management. The principles of equal partnership in the transport services market between the railways and cargo owners, taking into account the specifics of their technological processes and diverse economic interests should be putted as the basis of such a model. To achieve this, a fair competitive environment in which all transport modes and private business should have equal rights and opportunities for operating and own development depending on the real position in the transport market should be formed. Moreover, it is important to provide market laws in the transport sector with the introduction of a fair tariff system, eliminating the current practice of subsidizing private companies in the sector at the expense of rail transport. Only in such a way it is possible to ensure the financial stability of railways, which will facilitate their rapid technical and technological upgrading and a significant increase in the transport services' quality of cargo owners.

References

1. D.R. Martinelli, H.Teng, *Transportation Research Part C: Emerging Technologies*, **4**, 1: 33-49 (1996)

2. G. de O. Ramos, A.L.C. Bazzan, *Proceedings of the 15th Annual Conference on Genetic and Evolutionary Computation*: 473-480 (2015)
3. H. Bar-Gera, *Origin-based algorithms for transportation network modeling* (National Institute of Statistical Sciences, 1999)
4. N. Haque, M.A. Hossain, T. Islam, J. Khalil, *A trip assignment software for auto trips.* Dhaka: Ahsanullah University of Science and Technology (2014)
5. A.L.C. Bazzan, *Proceedings of the 14th Annual Conference on Genetic and Evolutionary Computation*: 43-50 (2012)
6. A.L.C. Bazzan, D. Cagara, B. Scheuermann, *Proceedings of IEEE Symposium on Computational Intelligence in Vehicles and Transportation Systems*: 43-50 (2014)
7. L. Kuttner, *Tagungsband Multikonferenz Wirtschaftsinformatik*:1923-1934 (2018)
8. J.A. Sabino, J.E. Leal, T. Stützle, M. Birattari, *Pesquisa Operacional*, **30**, 2: 487-514 (2010)
9. B. Abramovic, E. Nedeliakova, M. Panak, D. Sipus, *Business Logistics in Modern Management*:17: 15-28 (2017)
10. D.V. Lomot'ko, *Formation transport process of railways Ukraine on the basis of logistical principles*. Kharkiv: Ukrainian State Academy of Railway Transport (2008)
11. P.O. Yanovsky, *Railway transport of Ukraine*, **1**: 41-49 (2014)

Multi-criteria method of selection the way of conducting railway traffic on the open line for modernized and revitalized railway lines

Piotr Gołębiowski[1], *Marianna* Jacyna[1], and *Jolanta* Żak[1,*]

[1] Warsaw University of Technology, Faculty of Transport, 00662 Warsaw, Poland

Abstract. Nowadays, thanks to funds coming from various sources - including the European Union, intensive modernization and revitalization works regarding railway transport infrastructure are carried out. These works include both the railway surface and the change in the way of conducting railway traffic on both open lines and operating control posts. During modernization and revitalization, efforts are being made to introduce more and more new solutions in both mentioned areas. However, there is no multi-criteria method that would allow to assess which method seems to be the most advantageous. The aim of this article is to present a multicriteria method of choosing the method of conducting railway traffic on the open line for modernized and revitalized railway lines. It should be noted that the basic group of criteria in the method applies to the organization of railway traffic. The article discusses the basic issues regarding the modernization and revitalization of railway lines, ways of conducting railway traffic on open lines and multi-criteria assessment methods. A multicriteria method for selecting the method of conducting trains was presented and it was verified for a selected railway line in Poland.

1 Introduction

The most important problem regarding railway transport is the need to maintain the highest level of safety. It is affected by the condition of tracks and turnouts, as well as control command and signalling equipment. The bad condition of the above-mentioned elements is the most common cause of accidents on the railways. Additional factors that affect many accidents are exceeding the speed limit by drivers [1-2], as well as poor state of rolling stock.

In order to ensure the level of safety [3-6] and improve the quality of provided services [7-9], the infrastructure manager constantly monitors any irregularities related to the condition of railway infrastructure. The activities related to the supervision of the railway line included [10]:
- control covering the entire length of tracks,
- control of sections under extraordinary circumstances (e.g. during floods, snowfall, rail ruptures),
- protection of dangerous places,
- viewing tours.

Visual inspection and assessment of the used surface during supervision are carried out using two techniques:
- surface inspection, i.e. visual assessment of the condition of the railway's elements and unit measurement of worn or damaged elements,
- measurement of the track made using appropriate measuring instruments, i.e. trolleys or measuring wagons.

According to the assessment of the technical condition of railway lines in Poland, about 15,000 km of railway tracks has significantly reduced operating parameters compared to the construction parameters [11]. Their technical condition requires repair related to the replacement of individual components or the entire track surface.

Lack of modernization of the existing infrastructure by the manager and failure to remove the problems [12] carries the consequences of introducing numerous speed limits [13-14] and the emergency mode of traffic control devices, e.g. by means of a substitute signal. The consequence of this is the slow removal of existing defects associated with the lack of financial resources and qualified personnel responsible for security. Therefore, the lack of a quick reaction of the manager in this respect results in an increase in the number of restrictions (i.e. speed reduction, closure of railway tracks) on the given lines, and consequently an increase in the travel time of trains and a reduction in the capacity of railway lines.

One of the decision problems in the modernization or revitalization of a railway line is the choice of the way of conducting railway traffic on adjacent routes. There are no methods that precisely determine the way of traffic on railway lines. The aim of the article is to develop a multicriteria method [15-19] of selection the way of conducting railway traffic on the open line for modernized and revitalized railway lines. The article presents the problems of revitalization and

* Corresponding author: j.zak@wt.pw.edu.pl

modernization of railway lines. The rules of running trains on the open lines are discussed and the multicriteria method for selecting the method is presented. An example of the method's operation based on the Ustronie Morskie station in the West Pomeranian Voivodeship is presented. The scientific novelty in the article is an attempt to use the selected method of multi-criteria assessment for choosing the best way of conducting railway traffic on the open lines for a selected operating control post. Another novelty is the use of the multi-criteria assessment method with interval assessments in rail transport.

2 Problems of revitalization and modernization of railway lines

The technical condition of many railway lines is not the best. Thanks to funds obtained from various sources, there is a chance to renew individual sections of the line so that the quality of provided services is on the high level. Renewal of the railway line can be done in two ways: revitalizing it or modernizing it.

Revitalization is a comprehensive and in-depth set of activities carried out to improve infrastructure conditions [1]. The purpose of the performed activities is to restore the value and significance of degraded areas and objects (in the event of losing their original functions) to improve the economic, social and environmental situation.

Modernization [20] includes the execution of construction works in order to modernize and improve the infrastructure, striving to increase its utility value. It should be noted that the commonly used concept of "modernization" does not legally appear in the Construction Law. The modernization process includes the following works [21]:

- renovation - execution of construction works in a given place or facility (not included in the set of maintenance operations) in order to restore its original condition using the same or other building materials (e.g. replacement of track surface elements without changing their technical parameters),
- reconstruction - construction works, as a result of which certain technical parameters change, not including surfaces and fixed dimensions, i.e. length, height and width (e.g. reconstruction of the track system, change of track geometry, replacement of track elements with another type),
- extension - execution of construction works as a result of which there are changes in the parameters of a given building, i.e.: building area, cubic volume, height, length and width (e.g. extension of the second railway track, extension of passenger stops to stations).

Therefore, revitalization, in contrast to modernization, which assumes improvement of the railway line parameters and increasing the permissible speed of trains, is only a reconstruction of the original condition of the railway line.

3 Rules for conducting traffic on railway open lines

3.1 General assumptions

During the modernization or revitalization of the railway line, the decision-maker is faced with the need to decide which way of conducting the railway traffic on the open lines is appropriate for the analysed case. A number of factors have to be taken into consideration, which influence the choice of the best solution [7, 22]. During making the decision, it is necessary to remember that when conducting of trains traffic along the open line, several rules are necessary to take into account:

- on the double track open line, the train traffic should be conduct on the right track, looking in the direction of the train; from this rule there are many exceptions - traffic can also be carried out on the left track when traffic control devices allow or require such a necessity,
- on each track of the open line or block section in one time can be only one train,
- the station inspector can send a train from a traffic post if the adjacent trail or distance is free; in addition, it must be ensured that the traffic officer in the neighbouring post does not train on the train track,
- about duty of the open line station inspector is informed on the basis of the line blocking devices state or an agreement with the station inspector of the neighbouring operating control post using radio communication devices (announcing of trains),
- permission on departure of the train station inspector may issue on the basis of receiving a permit from a neighbouring operating control post for the train's departure or other indications [1],
- when the devices do not allow for the train departure on the open line onto the permit signal, it can be sent to a substitute signal or a written order under certain conditions,
- when the line blocking devices do not work properly, carry out the traffic based on announcements by means of telephone communication.

3.2 Conducting of train traffic based on telephone announcements

The first way to conduct train traffic on the open lines is telephone train announcements. In this method, traffic is secured by telephone communication of the station inspectors and making appropriate notes in the Traffic Log. The rules for announcing trains should be defined for two situations.

The first one concerns the open line on which a single-track two-way traffic is conducted. This type of traffic can be carried out for two cases: on single-track open lines and double-track open lines, where one or two tracks allow on two-way traffic. For this situation, the rules for announcing trains are as follows:

- in order to departure a train on the open line, first you must call for a train request by phone (the station inspector from the train departure post is calling the station inspector from the neighbouring post),

- if there are no obstacles to driving, the station inspector of the neighbouring post gives permission to departure a train (during a telephone conversation in which the permit was requested),
- when the train leaves the departure post, it is necessary to announce the departure of the train to the station inspector of the neighbouring post by phone,
- when the train has arrived at the arrival post, it is necessary to confirm the arrival of it to the departure operating control post by phone.

The second situation concerns the open line on which double-track one-way traffic is conducted. This type of traffic can be carried out on double-track open lines along each track. Therefore, for each track a strictly defined train direction applies. With this kind of traffic, it is certain that no train was sent to the non-fundamental track. For this situation, the rules for announcing trains are as follows:
- due to the above-mentioned premises, there is no need to request a permit and give a permit,
- when the train leaves the departure post, it is necessary to announce the departure of the train to the station inspector of the neighbouring post by phone,
- when the train leaves the departure post, it is necessary to announce the departure of the train to the station inspector of the neighbouring post by phone.

As already mentioned, train announcements are made via short phone calls using pre-conceived telephone plans, having content strictly defined in [23]. Telephone announcements may also be carried out using radio communication devices.

3.3 Conducting of train traffic based on manual line block

The second way to conduct train traffic on the open lines is valid when the operating control posts and open line are equipped with devices of a manual block. In this method, the traffic is secured by telephone communication of station inspector, handling of manual block devices and making appropriate notes in the Traffic Log. The rules for announcing trains are also defined for two situations - as for telephone announcements.

When single-track two-way train movement is in force, the following line blocking devices are used - the device situate at the entrance end of the section to release the entry signal and device situate at the leaving end of the section to release the entry signal, in addition, a telephone request and give permission should be made and the permission mechanism should be handled. After the messages concerning the request and permission is exchanged, the station inspector giving the permission blocks the permission mechanism, which unblocks the permission mechanism of the station inspector requesting the permit. Then, after leaving the train from the leaving post, station inspector blocking device situate at the entrance end of the section to release the entry signal, which causes the device situate at the leaving end of the section to release the entry signal of the receiving unblock and has an interpretation of the time of departure. After the train entry to the receiving station, the station inspector blocks the device situate at the leaving end of the section to release the entry signal, which unblocks the device situate at the entrance end of the section to release the entry signal of the station inspector and is interpreted as the arrival announcement.

When double-track one-way train traffic applies, the following devices of a manual block are used - the device situate at the entrance end of the section to release the entry signal and device situate at the leaving end of the section to release the entry signal referring to a specific track. The station inspector has at his disposal the device situate at the entrance end of the section to release the entry signal, which blocking causes unblocking of the device situate at the leaving end of the section to release the entry signal in the receiving operating control post and has an interpretation of the announcement of the time of departure. The blocking of the device situates at the leaving end of the section to release the entry signal by the station inspector accepts the device situate at the entrance end of the section to release the entry signal of the station inspector and is interpreted as the arrival announcement.

It should be noted that manual block does not provide information about the train number that has been sent to the open line. Therefore, on the double-track open line, when one-way traffic is carried out, after each of them is sent, the number should be forwarded by phone. The information is given before blocking the device situate at the entrance end of the section to release the entry signal.

3.4 Conducting of train traffic based on automatic electric block system

The second way to conduct train traffic on the open lines is valid when the operating control posts and open line are equipped with devices of automatic electric block system. In this method, the traffic is secured by telephone communication of station inspector, handling of automatic electric block system devices and making appropriate notes in the Traffic Log.

The essence of the automatic electric block system is that it allows dividing the open line into many block sections. At each of the sections, theoretically, there can be only one train at one time. However, it should be noted that you cannot talk about traffic flow. In order to keep it, subsequent trains should be sent out of the operating control post in such a way that the indication of the block signal would allow you to drive at the maximum permitted speed for this and the next block signal (so-called green driving).

Block signal of automatic electric block system are not man operated. The signals on them appear automatically on the basis of the information about the occupancy of the block section, which is protected by the given block signal. The block system can only be turned on in one direction at one time. It can be set permanently (if the automatic electric block system function allows it) - then it is not necessary to re-engage it in the same direction after the train has passed, or only for passing one train (for an unprivileged direction). When a given track is adapted to conduct two-way traffic, the block signal in the second direction will be extinguished, except for the last block signal acting as a warning signal. On single track open lines equipped with devices of automatic electric block system, which are adapted to carry out two-way traffic, the automatic electric block system switches off after running off the open line of the last train. For such a case, re-setting the direction of automatic electric block system is possible only after obtaining permission from the station inspector. After the train is sent to the open line towards the neighbouring station inspector, he is obliged to provide information to the neighbouring station about the train number by phone.

In case the devices of automatic electric block system do not work properly, enter the train announcements by phone. Movement of trains then takes place at the interval of follow-on stations - only one train running in a given direction at a given moment can be on the route.

4 The algorithm of multi-criteria method of selection the way of conducting railway traffic on the open lines

4.1 The steps of algorithm

To choose the best solution use of the SAW interval method with objective weights obtained by interval entropy [23- 25]. This method is one of methods, which are used for solving multi-criteria optimization tasks. The SAW method was chosen for solving this problem because of the simplicity of the calculations and because of the possibility of using interval evaluations of variants against individual criteria.

SAW interval method consists of the following steps:

Step 1. Construction of the interval decision matrix (**V** – set of variants, **C** – set of criteria):

$$\mathbf{X}=\left[\left[\underline{x}_{vc},\overline{x}_{vc}\right]\right] \quad v\in\mathbf{V},\, c\in\mathbf{C} \tag{1}$$

Step 2. Normalization of the interval decision matrix

$$\left[\underline{n}_{vc},\overline{n}_{vc}\right]=\begin{cases}\left[\dfrac{\underline{x}_{vc}}{\sum\limits_{v\in\mathbf{V}}\left(\underline{x}_{vc}+\overline{x}_{vc}\right)},\dfrac{\overline{x}_{vc}}{\sum\limits_{v\in\mathbf{V}}\left(\underline{x}_{vc}+\overline{x}_{vc}\right)}\right] & \text{(stimulant)}\\ \left[\dfrac{\left(\overline{x}_{vc}\right)^{-1}}{\sum\limits_{v\in\mathbf{V}}\left(\left(\underline{x}_{vc}\right)^{-1}+\left(\overline{x}_{vc}\right)^{-1}\right)},\dfrac{\left(\underline{x}_{vc}\right)^{-1}}{\sum\limits_{v\in\mathbf{V}}\left(\left(\underline{x}_{vc}\right)^{-1}+\left(\overline{x}_{vc}\right)^{-1}\right)}\right] & \text{(destimulant)}\end{cases} \tag{2}$$

Step 3. Determination of objective weights

a) Determination of the entropy vector for each criterion:

$$\mathbf{e}=\left[\left[\underline{e}_c,\overline{e}_c\right]:c\in\mathbf{C}\right]$$

$$\underline{e}_c=\min\left\{-\left(\ln\overline{\overline{\mathbf{V}}}\right)^{-1}\sum_{v\in\mathbf{V}}\left(\underline{n}_{vc}\ln\underline{n}_{vc}\right),-\left(\ln\overline{\overline{\mathbf{V}}}\right)^{-1}\sum_{v\in\mathbf{V}}\left(\overline{n}_{vc}\ln\overline{n}_{vc}\right)\right\} \tag{3}$$

$$\overline{e}_c=\min\left\{-\left(\ln\overline{\overline{\mathbf{V}}}\right)^{-1}\sum_{v\in\mathbf{V}}\left(\underline{n}_{vc}\ln\underline{n}_{vc}\right),-\left(\ln\overline{\overline{\mathbf{V}}}\right)^{-1}\sum_{v\in\mathbf{V}}\left(\overline{n}_{vc}\ln\overline{n}_{vc}\right)\right\}$$

b) Determination of the volatility level vector for each criterion according to the formula:

$$\mathbf{d}=\left[\left[\underline{d}_c,\overline{d}_c\right]\equiv\left[1-\overline{e}_c,1-\underline{e}_c\right]:c\in\mathbf{C}\right] \tag{4}$$

c) Determination of the vector of objective weights

$$\mathbf{w}=\left[\left[\underline{w}_c,\overline{w}_c\right]\equiv\left[\frac{\underline{d}_{vc}}{\sum\limits_{c\in\mathbf{C}}\left(\underline{d}_{vc}+\overline{d}_{vc}\right)},\frac{\overline{d}_{vc}}{\sum\limits_{c\in\mathbf{C}}\left(\underline{d}_{vc}+\overline{d}_{vc}\right)}\right]:c\in\mathbf{C}\right] \tag{5}$$

Step 4. Designation for each variant of the linear combination of normalized assessments against criteria and weight vector.

Step 5. Line ordering of results and selection of the final variant.

4.2 The assumption of variants definition

In the multicriteria method of selection the way of conducting railway traffic on the open line for modernized and revitalized railway lines, many variants are defined and then evaluated on the basis of a series of partial criteria. Ten variants of railway traffic will be analysed in the developed method:

(1) telephone announcements of trains – single-track two-way train movement ($v = 1$),
(2) telephone announcements of trains – double-track one-way train movement ($v = 2$),
(3) manual line block – single-track two-way train movement ($v = 3$),
(4) manual line block – double-track one-way train movement ($v = 4$),
(5) two-state automatic electric block system – single-track two-way train movement ($v = 5$),
(6) two-state automatic electric block system – double-track one-way train movement ($v = 6$),
(7) three-state automatic electric block system – single-track two-way train movement ($v = 7$),
(8) three-state automatic electric block system – double-track one-way train movement ($v = 8$),
(9) four-state automatic electric block system – single-track two-way train movement ($v = 9$),
(10) four-state automatic electric block system – double-track one-way train movement ($v = 10$).

As criteria in the multicriteria method of selection the way of conducting railway traffic on the open line for modernized and revitalized railway lines, it is proposed to use:

- time of agreement to accept the train ($c = 1$) – it means the time which elapses from the moment when the neighbouring operating control post starts calling on the announcing links to the moment of agreement and termination of the conversation and possibly the time of blocking the permission mechanism and switching on the block direction,
- time of receiving information about the departure of the train ($c = 2$) – it means the time which elapses from the moment when the neighbouring post starts calling on the announcing links to the moment of receiving information

about the departure of the train and termination of the conversation and eventually the time of unblocking the device situate at the leaving end of the section to release the entry signal,
- time of preparing the train entry ($c = 3$) – it means the time which elapses from the moment of start of activities related to the entry of the train until the start of preparation of the path,
- preparation time of the path ($c = 4$) – it means the time it takes for the physical preparation of the path (taking into account, among others, cooperation with other persons involved in the conducting of trains traffic),
- driving surveillance ($c = 5$) – it means the observation time of the train entering,
- end time of the path ($c = 6$) – it means the time when the path of the train is physically finished,
- time of activities related to the end of the train entry ($c = 7$) – it means the time that elapses from the moment the path is completed to the end of the train entry documenting,
- time of announcing the arrival of the train ($c = 8$) – it means the time which elapses from the moment when the calling of the neighbouring post on the announcing links starts until the information on the departure of the train has been forwarded and the conversation has been terminated, and possibly the blocking time of the device situate at the leaving end of the section to release the entry signal,
- time of requesting permission to send a train ($c = 9$) – it means the time which elapses from the moment when calling the neighbouring post on the announcing links to the moment of agreement and termination of the conversation, and possibly the time of unblocking the permission mechanism and switching on the block direction,
- time of train departure preparation ($c = 10$) – it means the time that elapses from the start of activities related to the entry of the train until the start of the preparation of the path,
- preparation time of the path ($c = 11$) – it means the time it takes for the physical preparation of the path (taking into account, among others, cooperation with other persons involved in the conducting of trains traffic),
- driving surveillance ($c = 12$) – it means the observation time of the train entering,
- end time of the path ($c = 13$) – it means the time when the path of the train is physically finished,
- time of activities related to the end of the train departure ($c = 14$) – it means the time that elapses from the moment the path is completed to the end of the train departure documenting,
- time of giving information about the departure of the train ($c = 15$) – it means the time which elapses from the moment when the calling of the neighbouring post on the announcing links starts until the information on the departure of the train has been forwarded and the conversation has been terminated, and possibly the blocking time of the device situate at the entrance end of the section to release the entry signal,
- time of receiving information about the arrival of the train ($c = 16$) – it means the time which elapses from the moment when the neighbouring post starts calling on the announcing links to the moment of receiving information about the arrival of the train and termination of the conversation and eventually the time of unblocking the device situate at the entrance end of the section to release the entry signal,
- the sum of times from the criteria from $c = 1$ to $c = 16$ ($c = 17$) – it means the time of absorption of the station inspector associated with the passage of one train by the station; this is the time when the station inspector cannot do other activities.

The selection of criteria in the method allows to assess the time necessary to pass one train - that is activities related to the acceptance and dispatching of the train. On this basis, the capacity of the traffic post can be estimated. If we analyse an operating control post where the train ends or starts a run, certain activities will not be taken into account.

Data for calculating the value of individual criteria should be adopted in accordance with the local conditions of the analysed traffic station [23]. The key documents necessary to obtain the input data to choose the best solution is the schematic layout of the track system, the arrangement of devices in the control room, the type of devices and many more.

5 Case study

As part of the research, an example of operation of multicriteria method of selection the way of conducting railway traffic on the open line for modernized and revitalized railway lines was made. The analysis was carried out for the Ustronie Morskie station located on the single-track railway line No. 402 Koszalin - Goleniów. This station conducts traffic on Ustronie Morskie – Mścice open line (direction Koszalin) and Ustronie Morskie - Kołobrzeg open line (direction Goleniów).

The railway line No. 402 is currently single-track. The assumption was made for the research that the situation will not change, and it will be a single-track line. Therefore, all variants were considered where single-track two-way traffic is possible [23]. These are ($v \in \mathbf{V}$):
- telephone announcements of trains – single-track two-way train movement ($v = 1$),
- manual line block – single-track two-way train movement ($v = 3$),
- two-state automatic electric block system – single-track two-way train movement ($v = 5$),
- three-state automatic electric block system – single-track two-way train movement ($v = 7$),
- four-state automatic electric block system – single-track two-way train movement ($v = 9$).

Some partial criteria of the assessment were used to select the variant. These are ($c \in \mathbf{C}$):
- time of agreement to accept the train ($c = 1$),
- time of receiving information about the departure of the train ($c = 2$),
- time of preparing the train entry ($c = 3$),
- preparation time of the path ($c = 4$),

- driving surveillance ($c = 5$),
- end time of the path ($c = 6$),
- time of activities related to the end of the train entry ($c = 7$),
- time of announcing the arrival of the train ($c = 8$),
- time of requesting permission to send a train ($c = 9$),
- time of train departure preparation ($c = 10$),
- preparation time of the path ($c = 11$).

The values of evaluations of individual variants in relation to individual criteria are presented in Table 1. These values were determined based on consultations with the infrastructure manager PKP Polskie Linie Kolejowe expert, who works daily at the Ustronie Morskie station.

Table 1. Interval decision matrix (source: own work).

	$v = 1$	$v = 3$	$v = 5$	$v = 7$	$v = 9$
$c = 1$	[5;10]	[11;15]	[10;15]	[10;15]	[10;15]
$c = 2$	[20;40]	[20;40]	[10;30]	[10;30]	[10;30]
$c = 3$	[10;40]	[10;40]	[10;20]	[10;20]	[10;20]
$c = 4$	[5;10]	[11;15]	[10;15]	[10;15]	[10;15]
$c = 5$	[15;30]	[15;30]	[10;30]	[10;30]	[10;30]
$c = 6$	[10;40]	[10;40]	[10;20]	[10;20]	[10;20]
$c = 7$	[5;7]	[3;5]	[3;5]	[3;5]	[3;5]
$c = 8$	[5;7]	[3;5]	[1;3]	[1;3]	[1;3]
$c = 9$	[3;5]	[5;7]	[1;3]	[1;3]	[1;3]
$c = 10$	[3;5]	[5;7]	[1;3]	[1;3]	[1;3]
$c = 11$	[114;265]	[126;275]	[99;215]	[99;215]	[99;215]

The normalized interval decision matrix is shown in Table 2. On its basis, the entropy, the level of variation and the objective weights of the criteria were determined as presented in Table 3. The results obtained by the proposed SAW interval method with objective weights are shown in Table 4, where the results of linear combinations, elements of a normalized interval decision matrix (Table 2) and objective criteria weights (Table 3), and the ranking of decision variants are visible.

Table 2. Normalized interval decision matrix (source: own work).

	$v = 1$	$v = 3$	$v = 5$	$v = 7$	$v = 9$
$c = 1$	[0,04;0,09]	[0,09;0,13]	[0,09;0,13]	[0,09;0,13]	[0,09;0,13]
$c = 2$	[0,08;0,18]	[0,08;0,24]	[0,04;0,18]	[0,04;0,18]	[0,04;0,18]
$c = 3$	[0,05;0,21]	[0,05;0,21]	[0,05;0,11]	[0,05;0,11]	[0,05;0,11]
$c = 4$	[0,04;0,09]	[0,09;0,13]	[0,09;0,13]	[0,09;0,13]	[0,09;0,13]
$c = 5$	[0,07;0,14]	[0,07;0,14]	[0,05;0,14]	[0,05;0,14]	[0,05;0,14]
$c = 6$	[0,05;0,08]	[0,08;0,13]	[0,08;0,13]	[0,08;0,13]	[0,08;0,13]
$c = 7$	[0,06;0,09]	[0,09;0,13]	[0,09;0,13]	[0,09;0,13]	[0,09;0,13]
$c = 8$	[0,03;0,04]	[0,04;0,21]	[0,07;0,21]	[0,07;0,21]	[0,07;0,21]
$c = 9$	[0,04;0,07]	[0,03;0,04]	[0,07;0,21]	[0,07;0,21]	[0,07;0,21]
$c = 10$	[0,04;0,07]	[0,03;0,04]	[0,07;0,21]	[0,07;0,21]	[0,07;0,21]
$c = 11$	[0,06;0,13]	[0,05;0,12]	[0,07;0,15]	[0,07;0,15]	[0,07;0,15]

Table 3. Entropy (**e**), level of variation (**d**) and objective weights (**w**) (source: own work).

	e	d	w
$c = 1$	[0,62;0,79]	[0,21;0,38]	[0,36;0,64]
$c = 2$	[0,50;0,97]	[0,02;0,50]	[0,05;0,95]
$c = 3$	[0,48;0,85]	[0,15;0,52]	[0,23;0,77]
$c = 4$	[0,62;0,79]	[0,21;0,38]	[0,36;0,64]
$c = 5$	[0,50;0,86]	[0,14;0,50]	[0,22;0,78]
$c = 6$	[0,48;0,85]	[0,15;0,52]	[0,23;0,77]
$c = 7$	[0,61;0,80]	[0,20;0,40]	[0,34;0,66]
$c = 8$	[0,49;0,80]	[0,20;0,512]	[0,28;0,72]
$c = 9$	[0,49;0,80]	[0,20;0,51]	[0,28;0,72]
$c = 10$	[0,49;0,80]	[0,20;0,51]	[0,28;0,72]
$c = 11$	[0,54;0,85]	[0,15;0,46]	[0,25;0,75]

Table 3. Entropy (**e**), level of variation (**d**) and objective weights (**w**) (source: own work).

Table 4. Received results (source: own work).

Variant	SAW	Rank
$v = 1$	[0,138;1,00]	3
$v = 3$	[0,179;1,113]	2
$v = 5$	[0,199;1,252]	1
$v = 7$	[0,199;1,252]	1
$v = 9$	[0,199;1,252]	1

The calculations show that variant 5, 7 or 9 should be chosen - i.e. an automatic electric block system in two-state, three-state or four-state option.

Summary and conclusions

The aim of the article was to develop a multicriteria method of selection the way of conducting railway traffic on the open line for modernized and revitalized railway lines. It is very important problem because the more advanced the traffic system is, the more financial investment should be made.

The developed method takes into account many criteria that are expressed in units of time. These criteria describe the time of engagement of the person conducting the railway traffic (station inspector) in the performance of individual activities. Therefore, the assessment is made from the point of view of absorption of the station inspector. On the other hand, the developed method allows choosing the best solution from the point of view of the capacity of the lines and stations.

Summing individual times and multiplying by the number of trains that the station inspector must handle, we are able to estimate the capacity of the station, especially for peak hours. This allows us to obtain the number of trains that the station inspector is able to handle during the analysed time.

References

1. M. Andrzejewski, D. Gallas, P. Daszkiewicz, A. Merkisz-Guranowska, H. Stawecka, MATEC Web of Conferences, Volume: **118**, Article Number: UNSP 00015 (2017)
2. P. Daszkiewicz, M. Andrzejewski, MATEC Web of Conferences, Volume: **118**, Article Number: UNSP 00014 (2017)
3. S. Clarke, Work & stress, **12(3)**, 285-292 (1998)
4. J. Santos-Reyes, A. N. Beard, Civil Engineering and Environmental Systems, **20(1)**, 1-21 (2003)
5. J. Santos-Reyes, A. N. Beard, R. A. Smith, Proceedings of the Institution of Mechanical Engineers, Part F: Journal of Rail and Rapid Transit, **219(2)**, 47-65 (2005)

6. M. Sitarz, TTS Technika Transportu Szynowego, **19(5-6)**, 56-61 (2012)
7. M. Jacyna, P. Gołębiowski, D. Pyza, Archives of Transport System Telematics, **11(1)**, 27-33 (2018)
8. I. Jacyna-Gołda, J. Żak, P. Gołębiowski, Archives of Transport, **32(4)**, 17-28 (2014)
9. M. Bajerlein, L. Rymaniak, P. Swiatek, A. Ziolkowski, P. Daszkiewicz, M. Dobrzynski, Applied Mechanics and Materials, **518**, 108-113 (2014).
10. J. Sysak (red.), Drogi kolejowe, Warszawa (1982)
11. Urząd Transportu Kolejowego, Sprawozdanie ze stanu bezpieczeństwa ruchu kolejowego w 2016 r., Warszawa (2017)
12. M. Jabłoński, A. Jabłoński, Zeszyty Naukowo-Techniczne Stowarzyszenia Inżynierów i Techników Komunikacji w Krakowie. Seria: Materiały Konferencyjne, **2(104)**, 183-196 (2014)
13. M. Jacyna, P. Gołębiowski, M. Urbaniak, Communications in Computer and Information Science: Tools of Transport Telematics, 199-210 (2016)
14. J. Kwaśnikowski, G. Gramza, Analiza wybranych zakłóceń w ruchu kolejowym, Poznań (2007)
15. M. Jacyna, M. Izdebski, E. Szczepański, P. Gołda, Symmetry-Basel, **11(10)**, 1-19 (2018)
16. M. Jacyna, M. Wasiak, Communications in Computer and Information Science: Tools of Transport Telematics, 1-13 (2015)
17. M. Kiciński, K. Solecka, Archives of Transport, **46(2)**, 71-84 (2018)
18. T. Trzaskalik, Zeszyty Naukowe Politechniki Śląskiej. Seria: Organizacja i zarządzanie, **74**, 239-263 (2014)
19. J. Żak, I. Jacyna-Gołda, Archives of Transport, **25-26(1-2)**, 117-135 (2013)
20. R. Licow, F. Tomaszewski, Problemy Kolejnictwa, **171**, 51-56 (2016)
21. Ustawa z dnia 7 lipca 1994r. Prawo budowlane (Dz. U. z 2003r. Nr 207, poz. 2016 z późn. zm.)
22. A. Szeląg, Archives of Transport, **42(2)**, 75-83 (2017)
23. PKP Polskie Linie Kolejowe S.A., *Instrukcja o prowadzeniu ruchu pociągów Ir-1*, Warszawa (2017)
24. D. Kacprzak, Studia Ekonomiczne. Zeszyty Naukowe Uniwersytetu Ekonomicznego w Katowicach, **348**, 144-155 (2018)
25. F. Lotfi, R. Fallahnejad, Entropy, **12**, 53-62 (2010)

The Economic Guarantee of Logistic Connections Optimization at Railway Transport

Vasyl Kopytko[1], *Sergei* Dziuba[2], *Maxym* Mishchenko[3,*], and *Tetyana* Shargun[1]

[1] Lviv Branch of DNURT, Department of Humanities and Social-Economic Training, I. Blazhkevych St., 12a, Lviv, Ukraine, 79052;
[2] M.S. Polyakov Institute of Geotechnical Mechanics of NAS of Ukraine, Simferopolska St., 2a, Dnipro, Ukraine, 49005;
[3] DNURT, Department of Economics and Management, Lazariana St., 2, Dnipro, Ukraine, 49010;

Abstract. The efficiency increasing of railway transport functioning is a main factor in the economic growth of the national economy. Material reserves are the main factor that defines the strategy of the enterprise and influences on the economic optimization level of logistic connections in general. Waste on deficiency and expense of storing the taken out from the main turnover material reserves redouble the economic situation. In this research the sources of the arising an uncertainty by taking a decision about economic optimization of logistic connections and working out methods of need determination in conditions of a certain uncertainty that will be led to improvement of the economic effectiveness of activity are determined. The best methods that may be used for this purpose are analysis of temporary series, multiple regression and correlation analysis, pattern recognition. The paper presents the results of research that allow, based on the built neuro-fuzzy model, to evaluate the dependence of macroeconomic indicators of the national economy of Ukraine on the indicators characterizing its economic situation in order to improve the control efficiency of logistics systems. The built neuro-fuzzy model, unlike existing models, allows reproducing the complex nonlinear dependences.

1. Introduction

The economic situation of the state, its development and stability to effectively integrate into the world economic community depends on its resource potential and its effective use. The resource potential includes natural environment, raw materials, quantity and quality of manpower, intelligent and industrial potential created at the time of economic state assessing, as well as availability of infrastructure for its development.

The paper presents the results of research that allow, based on the built neuro-fuzzy model, to evaluate the dependence of macroeconomic indicators of the national economy of Ukraine on the indicators characterizing its economic situation in order to improve the control efficiency of logistics systems. The built neuro-fuzzy model, unlike existing models, allows reproducing the complex nonlinear dependences.

2. Development of mechanism for monitoring economic situation and development prosepcts of the country based on modelling of macroeconomic processes using intelligent technologies

The efficiency increasing of railway transport functioning is a main factor in the economic growth of the national economy. Material reserves are the main factor that defines the strategy of the enterprise and influences on the economic optimization level of logistic connections in general. Waste on deficiency and expense of storing the taken out from the main turnover material reserves redouble the economic situation. In this research the sources of the arising an uncertainty by taking a decision about economic optimization of logistic connections and working out methods of need determination in conditions of a certain uncertainty that will be led to improvement of the economic effectiveness of activity are determined. The best methods that may be used for this purpose are analysis of temporary series, multiple regression and correlation analysis, pattern recognition.

In international practice, there are several approaches to determining the socio-economic condition of the state and the prospects for its positive development. However, numerous theoretical studies do not provide a complete picture of the state development; while determining its prospects, they use, as a rule, the production functions such as [1]:

$$W = f(L, A) \quad (1)$$

where L – state labour potential;
A – productive assets owned by the state.

Indicators of a number of studies have different focus (economic, technical, technological, social, etc.), which complicates the task of determining the development vector. That is why the purpose of this work is to develop a mechanism for monitoring the economic situation and development prospects based on modelling of mac-

* Corresponding author: mmi@ua.fm

roeconomic processes using intelligent technologies. At present the existing approaches to determine the state of the country include the following ones. The position of each of the states that make up the International Monetary Fund (IMF) is assessed by the indicator system that includes more than 20 indicators of economic, environmental and social conditions [2]. Herewith the indicators are grouped according to such classification features as poverty and social sphere state (population, GNI per capita, poverty, urban population, life expectancy, infant mortality, child malnutrition, access to improved water, illiteracy, general primary education), level of key economic indicators (exports of goods and services, gross domestic investment, gross national savings, current account balance, interest payments, total debt, total debt service), long-term development trends (GNI growth rate per capita), economy structure (agriculture, industry, production, services, private consumption, state consumption, imports of goods and services), level of prices and public finances (consumer price, GDP deflator level, current revenues, current budget balance, total surplus/deficit), trade conditions (total exports, total imports, food, fuel, ore and metals, production), external payments position (balance of resources, net income, net operating transfers, financial instruments, changes in net reserves, reserves, exchange rate), foreign debt level (total debt to maturity, long-term external debt, short-term debt, total debt to maturity on the World Bank loans, total debt to maturity on the International Development Association loans, total debt service, official grants, private lenders, foreign direct investment, portfolio property, transactions, payments to maturity, principal debt repayment, net flows, interest payments, net transfers). Since the level of indicators listed above by country has a great difference, the IMF proposed a classification of countries. Namely, with the analytical purpose the IMF divides countries by groups of income, i.e. by gross national income per capita into the following groups: low-income countries (GNI per capita of less than 745 dollars), the countries with income below average (746 - 2975 dollars), countries with incomes above the average (2976 - 9205 dollars) and those with high income (9206 dollars or higher).

The study of interrelation between the growth rate of GDP per person and its level (Fig. 1) confirms the need for country division into income groups prior to analysing other indicators, such as economic growth. The relationship between the growth of GDP per capita and the level of GDP is confirmed by sufficiently high coefficient of multiple correlation R^2 (0.5278) [3]. Dependence is characterized by the following equation:

$$\Delta GDP = -2{,}3841 \cdot \ln(x) + 27{,}832 \qquad (2)$$

where ΔGDP – growth rate of GDP per capita (%); x – GDP per capita ($ / person).

That is, the higher the level of GDP per capita, the lower the growth rate. It follows that, to assess certain result, the growth performance analysis should be conducted within each classification group.

Furthermore IMF classifies countries by region (geographic location) and by level of foreign debt.

Classification of countries in terms of foreign debt involves the grouping of countries with low, average, moderate and high level of debt.

Herewith the average level of debt means that the ratio of the present value of debt to gross national income exceeds 80% or the ratio of the present value of debt to exports exceeds 220% [1].

This classification is proposed by UNO within Development Program [4]. According to this program such tasks include the following: elimination of poverty and famine; provision of basic primary education; stimulation of gender equality; reducing child mortality; improvement of maternal health; control of AIDS, malaria and other diseases; maintenance of ecological balance; global partnership for development; achieved common indicators (population, GNI, GNI per capita, literacy, final birth rate, life expectancy, aid, reducing foreign debt, increasing investment and trade in % to GDP).

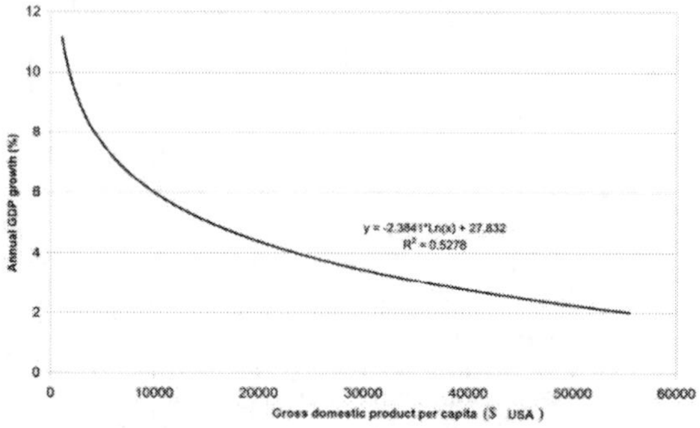

Fig. 1. Dependence of GDP growth rate on GDP per capita

One of the most influential international financial organizations is Multilateral Investment Guarantee Agency (MIGA), which insures foreign investment (direct and portfolio) and ensures their return in the event of insolvency of the country-borrower in the amount of 90% of the investment, giving such guarantees for 15-20 years [4]. Activities of the above Agency are associated with the risk of a particular country. In this context, Multilateral Investment Guarantee Agency analyses the country in terms of these commitments, determining the commercial, political and economic risks in each country, and guarantees the return of foreign investment [5]. Classification of factors that characterize the state of the country in terms of investment risk is presented in Fig. 2. Herewith the Multilateral Investment Guarantee Agency divides all countries in terms of political risk into seven groups (risk increases from the first to the seventh group), level of commercial risk – into three groups (A, B and C, where the most risky countries are those of group C), risk of war, expropriation and government action - into seven groups (the seventh one has the highest risk).

The process of creation of the European Union demanded determining the development indicators, which control an objective identity of any country in the group of countries that may enter the union without creating imbalance in the costs of countries-participants for stabilization of the EU economy as a whole.

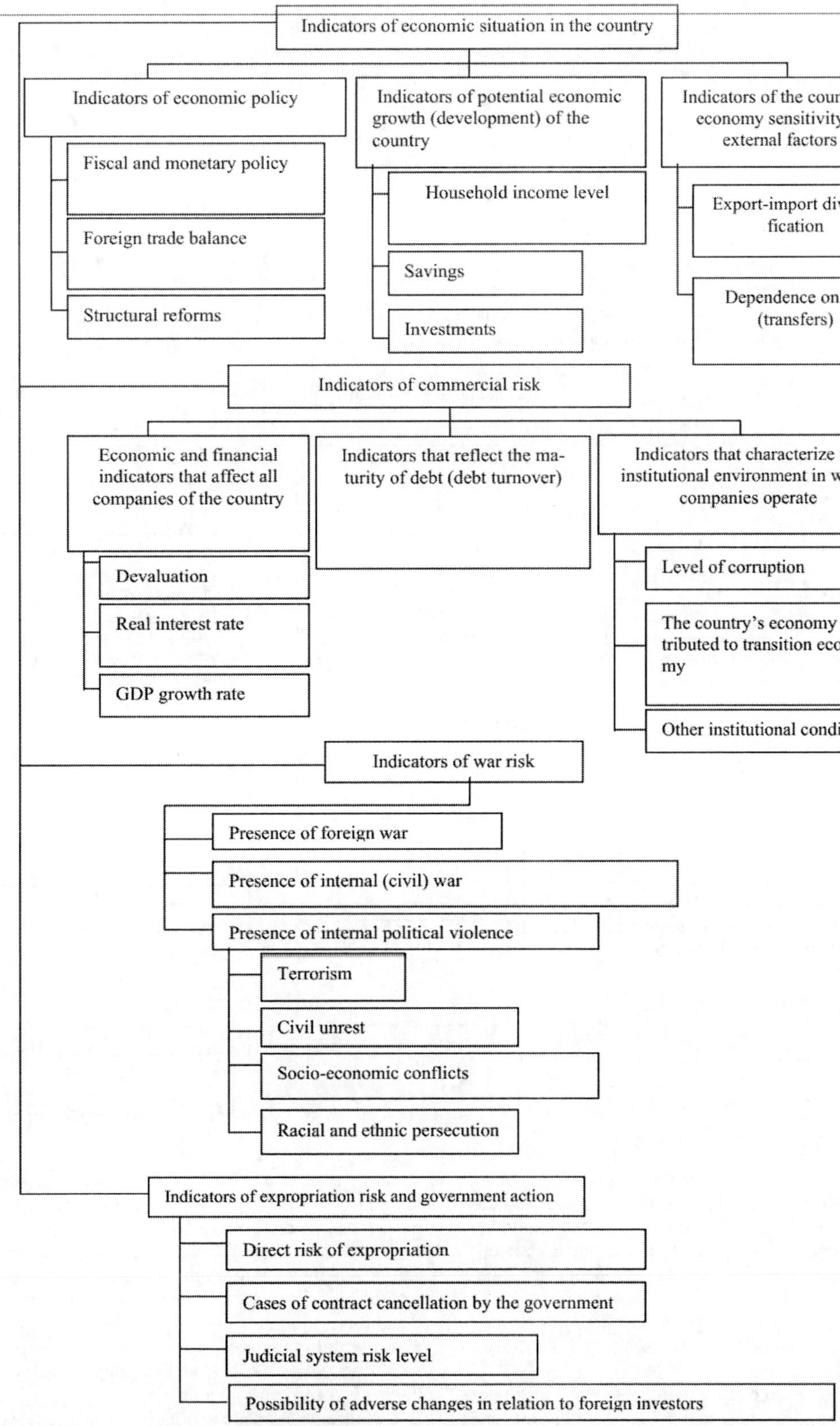

Fig. 2. Indicators of investment risk in the country economy

These indicators include the inflation rate (which must not exceed by 1.5% the average inflation rate of three EU countries characterized with highest price stability); the state budget deficit (which must not exceed 3% of GDP at the ratio of public debt to GDP of no more than 60%); the level of long-term interest rates should not be higher than 2% of the average level of the index in three EU countries that have achieved the highest price stability [6].

Continuing analysis of approaches to the assessment of country position and prospects of its development, it should be noted that, in order to monitor key growth indicators, each country develops a system of national accounts, which makes it possible to track a series of interrelated processes and phenomena in the economy (volume and structure of production, income, consumption, capital and finance formation) [7].

3. Classification of economic security factors

In the current research [8-10] the economic security factors (Figure 3) are divided into three groups: economic independence; economic stability and strength; ability for self-development and progress.

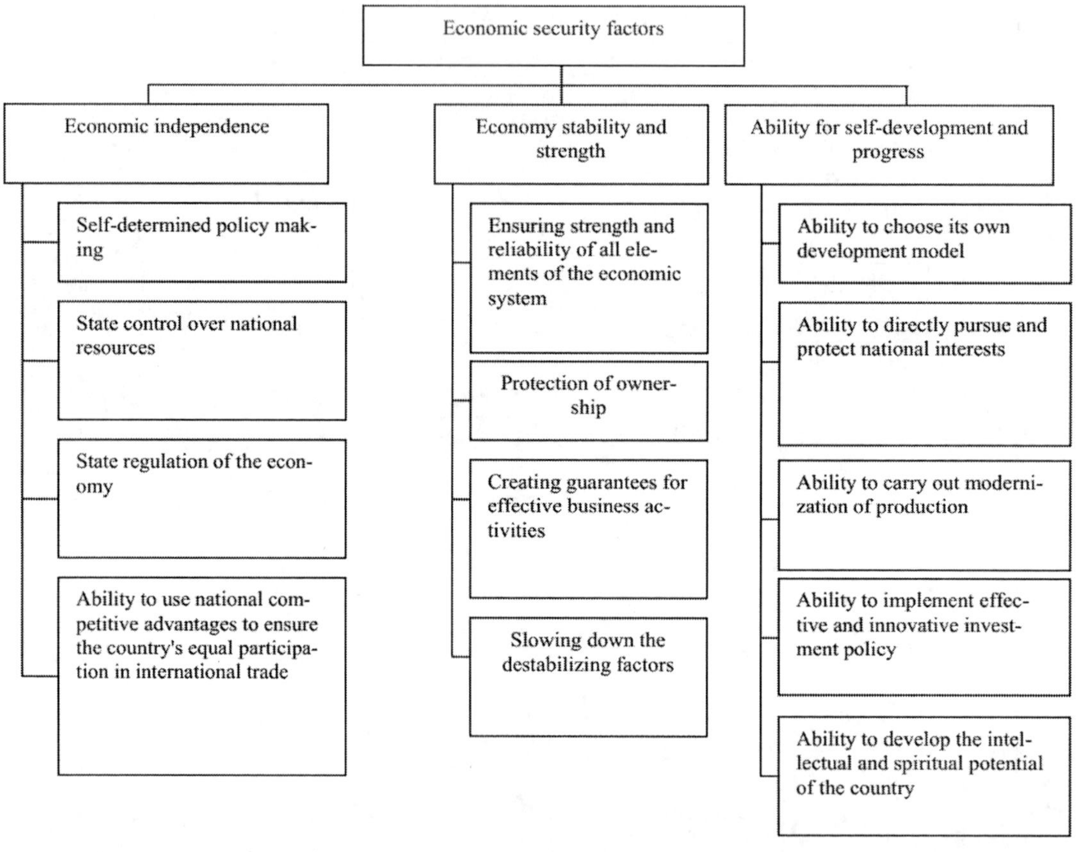

Fig. 3. Classification of economic security factors

The first group includes such factors as the country's ability to form its economic policies, the ability to control the national resources, the ability for state regulation of the economy and the ability to use national competitive advantages to ensure the country's equal participation in international trade.

Apparently, it is necessary to specify these factors and find the indicators that would characterize the level of these factors.

The second group of factors in contemporary studies includes those affecting the economy stability and strength. Namely, the ability of a country to ensure strength and reliability of all elements of the economic system; protection of ownership; country's guarantees for effective business activities; the ability of a country to slow down destabilizing factors.

The third group of factors included those providing self-development and progress. Namely, the opportunity and the ability to choose its own model of development; the ability to promote and protect its own national interests; the ability to modernize production, to pursue innovative and effective investment policy, to develop the intellectual and spiritual potential.

All these factors are qualitative in nature (ability, opportunity, etc.) and can be measured only using the expert method. However, indirectly the level of these factors can be identified with the help of indicators that answer the question whether the goal is reached, or whether there are conditions for its achievement. Table 1 shows the indicators, which, in our opinion, can indirectly determine the state's ability to develop.

Each of the above approaches to assess the economic situation of the country is complementary and designed for specific purposes (country's solvency assessment, its reliability for investors, stability of its economy). The indicators used for the specified purpose are the resultant, may be used as those for control of the country position and do not include factors subject to management on the level of current (operational) solutions. In other words, these are the indicators that characterize the result of average and long term development. To ensure operational control is necessary to determine the factors that affect the resulting indicators, level of influence of these factors on them and create operational tasks directed at positive current results in achieving the average and long-term goals.

Table 1. Indicators characterizing the level of economic security factors

		Factor	Factor level indicator
1	Economic independence		
	1.1.	Self-determined policy making	1.1.1. Capital share of the state and national investors in the economy (incl. by activities)
	1.2.	State control over national resources	1. 2.1. Share of resources controlled by the state and national investor in the economy (incl. by activities)
	1.3.	State regulation of economy	1. 3.1. Share of budget intended to finance the material production sectors (industry, agriculture, construction)
	1.4.	Ability to use national competitive advantages to ensure equal participation in international trade	1.4.1. Share of export of certain goods and services in world trade 1.4.2. Exports in % of GDP 1.4.3. Balance of visible trade of countries
2	Economy stability and strength		
	2.1.	Ensuring strength and reliability of all elements of the economic system	2.1.1. Financial stability of enterprises of the country's economy 2.1.2. Labor productivity in the economy
	2.2.	Protection of ownership	2.2.1. Economy structure by form of ownership
	2.3.	Creating guarantees for effective entrepreneurial activity	2.3.1. Labour productivity by form of ownership
	2.4.	Slowing down the destabilizing factors	2.4.1. Inflation rate in the country
3.	Ability for self-development and progress		
	3.1.	Ability to choose its own development model	3.1.1. Place of the country in the world economy by key indicators
	3.2.	Ability to directly pursue and protect national interests	3.2.1. Place of the country in the world economy by key indicators
	3.3.	Ability to carry out modernization of production	3.3.1. Gross national income per capita 3.3.2. Exports in % of GDP 3.3.3. Balance of visible trade 3.3.4. Export structure
	3.4.	Ability to implement effective and innovative investment policy	3.4.1.Gross national income per capita 3.4.2. Exports in % of GDP 3.4.3. Balance of visible trade 3.4.4. Export structure
	3.5.	Ability to develop the intellectual and spiritual potential of the country	3.5.1. Science expenditure level 3.5.2. Culture expenditure level 3.5.3. Education expenditure level

The state of socio-economic security is affected by the stability of the national currency, the level of balance of payments deficits, the level of internal and external debt. This, in turn, depends on the export potential of the country and the nature of exports (raw materials, products of processing industry, high technology, services). Export volumes enable the country to develop import without negative consequences for the national currency. The level of imports indicates the country's integration into the world economic community, the openness of its economy. The restraint of import growth indicates the inability of the country to develop its export potential and provide excess growth rate of exports of goods and services over their imports. Therefore, when describing the socio-economic development, it is advisable to analyse the above figures. The increased indicator of exports of goods and services as percentage of GDP indicates the innovation direction of the country's economy development, its increased competitiveness in the global market for goods and services, demand for the national currency of the country.

Analysis of economy of Ukraine conducted according to [7] shows the following. Dependency of GDP growth rates on the tax burden change (incl. and escl. subsidies) and tax revenue are characterized (with high impact coefficient: $R^2 \geq 0{,}8$) with polynomial dependencies, indicating the significant fluctuations in these indicators. The same can be said of the relationship between GDP and inflation, which can be approximated by a polynomial curve (R^2=0,5235). The most stable character among the set dependencies is shown by GDP dependency on net capital inflow, which is characterized with the third degree polynomial dependence at R^2 =0,77. This shows that the impact of factors on these indicators of the country's economy development as a complex system is characterized by a complex nonlinear dependencies of its outputs on inputs with fuzzy statistical information. Therefore, to build and analyse the model of impact of macroeconomic factors on these indicators it is advisable to use the neuro-fuzzy technology [12 - 14].

As the resultant indicators, we considered the gross national income (GNI) and the gross domestic product (GDP) per capita, as these indicators are the most generalizing when determining the economic situation of the country. The factors affecting the level of GDP per capita include the following: inflation level (X1), tax burden (X2), tax revenue share in the budget revenues (X3), real interest rate on deposits (X4), exports of goods and services as percentage of GDP (X5), import of goods and services as percentage of GDP (X6).

Inflation level (X1) characterizes the rationality of many processes in the economy. However, given that, according to classical theory [11], inflation is a manifestation of the cost crisis (increased cost of production of goods and services) or production volume crisis (production slump and the associated emergence of deficit and higher prices), the inflation indicator requires more in-depth study to identify factors that led to this level.

The tax burden (X2) is an indicator that should be optimized: positive impact of this factor on output has its upper limit, whereupon the share of shadow economy increases, tax revenue to state and local budgets reduces, net capital inflows to countries decreases. Establishing the optimal level of this indicator requires a detailed study of the country's tax system.

The tax revenue share in the budget revenues (X3) on the one hand characterizes the efficiency of the tax system of the state and on the other hand - the centralization of functions assumed by the state and local budgets (public administration, defence, health and environmental protection, development of science, education, culture, state participation in programs supporting the development of certain economic sectors, subsidies to loss-making industries). The study of these processes is the subject of special researches.

An important indicator of the country's position is the real interest rate on deposits (X4). As it is known, the deposit rate is the lower margin of investment efficiency. In addition, the real deposit rate is an indicator that determines the feasibility of forming the capital as investments into the banking system and hence the availability of credit resources in the country, which is an important component of successful activity of enterprises. Study if this indicator consists of two parts: study of factors affecting inflation, and study of efficiency of the country's banking system.

Exports of goods and services (X5) describes the export potential of the country and, affecting its economic state, in turn, depends on the availability of innovative infrastructure, competitiveness of domestic enterprises, development of foreign trade, the level of its liberalization enshrined in law.

Imports of goods and services (X6) show the openness of the economy and the competition level. Indices of exports and imports are consolidated in the index of balance of goods and services, which should be positive. The study of these indicators by the structure of exports and imports, as well as certain types of similar products (e.g., metal, grains, etc.) allows ascertaining the effectiveness of export and import activities, pricing policy of the companies, deficiencies in the regulation of foreign trade.

4. Development of gdp dependence model in relation to the selected macroeconomic factors

At the first stage of model building we analyse the relationship between inputs (X1, X2, X3, X4, X5, X6, X7, X8) and outputs (GDP), according to available annual data of the State Statistics Committee of Ukraine, the National Bank of Ukraine, for the period 10 years [7, 15], which are listed in Table 2.

Let us consider the GDP dependency model in relation to the selected macroeconomic factors X1, X2, X3, X4, X5, X6, X7, X8 as the following:

$$GDP = F(X1, X2, X3, X4, X5, X6, X7, X8),\ XI_{min} \le XI \le XI_{max},\ I = 1,...,8. \quad (3)$$

We start building the explicit model (3) by forming the fuzzy productive knowledge base, which is a set of fuzzy expert linguistic rules of IF-THEN type that bind together the linguistic evaluations of input variables (X1, X2, X3, X4, X5, X6, X7, X8) and output variable of GDP. Linguistic evaluations will be obtained on the basis of statistic data set out in Table 2.

To build the explicit model (3), we formalize the linguistic evaluations "Low", "Lower than average" and others using the membership functions. We set these functions in the following form [14]

$$\mu^T(x) = \frac{1}{1+(\frac{x-b}{c})^2}$$

where $\mu^T(x)$ is the number in the range [0,1], which describes the subjective measure of x value compliance with fuzzy term (*"low"*, *"lower than average"* et al.), b and c parameters, first selected by an expert and then configured to experimental data: b - coordinate of maximum of the function $\mu^T(x)$, herewith $\mu^T(b)=1$, c - factor of concentration - extension of the function $\mu^T(x)$.

Table 2. Annual data of key macroeconomic indicators

Years	GDP UAH/person	X1 % p.a.	X2	X3	X4 %	X5 mln. UAH	X6 mln. UAH	X7 mln. $	X8 mln. UAH
Year 1	1595.9	139.7	0.12	0.8	-5.4	60455.9	34740.4	896.9	4159.9
Year 2	1843.8	110.1	0.1	0.7	24.2	59552.5	35219.5	1438.2	6447.8
Year 3	2041.8	120.0	0.1	0.8	2.3	93601.7	55197.0	2063.6	2113.7
Year 4	2624.0	119.2	0.1	0.8	1.5	122205.9	67599.6	2810.7	1965.3
Year 5	3492.2	125.8	0.07	0.6	-16.6	106200.0	97643.0	3875.0	-969.3
Year 6	4236.3	106.1	0.05	0.7	4.7	113245.0	109924.0	4801.0	593.4
Year 7	4724.1	99.4	0.05	0.7	7.5	124392.0	114501.0	5924.0	-1635.4
Year 8	5640.2	108.2	0.05	0.7	-0.2	154394.0	147525.0	7566.0	506.7
Year 9	7327.2	112.3	0.04	0.7	-4.4	211248.0	185318.0	9606.0	11009.0
Year 10	9095.1	110.3	0.04	0.7	-2.3	227252.0	223555.0	17311.0	7806.3

As an example, Figure 4 shows the membership functions for X1 variable.

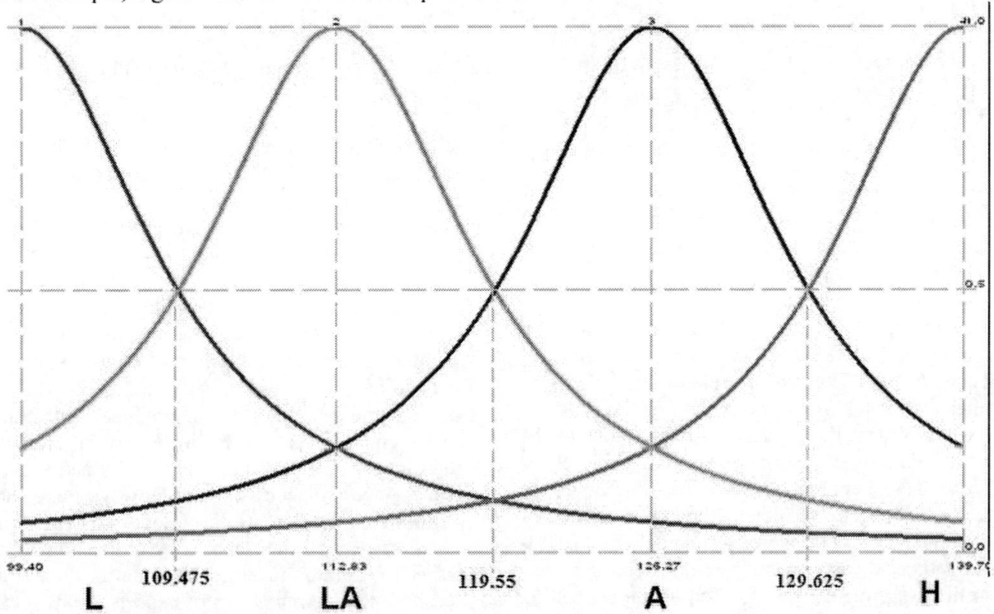

Fig. 4. Membership functions of X1variable values for corresponding fuzzy term:

L - low, LA – lower than average, A - average, H - high

Next, using the membership functions and the technology [12 - 14] we introduce, according to [12], the model (3) in an explicit form

$$GDP = \frac{\sum_{\kappa=1}^{5} d_k \mu(GDP_k)}{\sum_{\kappa=1}^{5} \mu(GDP_k)} \quad (4)$$

where GDP_k - value of GDP variable in the range D_k, $k = 1,...,5$ (model (5, 6, 7, 8, 9))

$$\mu(GDP_1)= \\ =\min\left\{\begin{array}{l}1, \\ w_1\mu_{1B}(X1)\cdot\mu_{2B}(X2)\cdot\mu_{3B}(X3)\cdot\mu_{4HC}(X4)\cdot\mu_{5H}(X5)\cdot\mu_{6H}(X6)\cdot\mu_{7H}(X7)\cdot\mu_{8HC}(X8)+ \\ +w_2\mu_{1HC}(X1)\cdot\mu_{2B}(X2)\cdot\mu_{3C}(X3)\cdot\mu_{4B}(X4)\cdot\mu_{5H}(X5)\cdot\mu_{6H}(X6)\cdot\mu_{7H}(X7)\cdot\mu_{8C}(X8)+ \\ +w_3\mu_{1C}(X1)\cdot\mu_{2B}(X2)\cdot\mu_{3B}(X3)\cdot\mu_{4HC}(X4)\cdot\mu_{5H}(X5)\cdot\mu_{6H}(X6)\cdot\mu_{7H}(X7)\cdot\mu_{8HC}(X8)+ \\ +w_4\mu_{1HC}(X1)\cdot\mu_{2B}(X2)\cdot\mu_{3B}(X3)\cdot\mu_{4HC}(X4)\cdot\mu_{5HC}(X5)\cdot\mu_{6H}(X6)\cdot\mu_{7H}(X7)\cdot\mu_{8HC}(X8)\end{array}\right\} \quad (5)$$

$$\mu(GDP_2)= \\ =\min\left\{\begin{array}{l}1, \\ w_5\mu_{1C}(X1)\cdot\mu_{2HC}(X2)\cdot\mu_{3H}(X3)\cdot\mu_{4H}(X4)\cdot\mu_{5HC}(X5)\cdot\mu_{6HC}(X6)\cdot\mu_{7H}(X7)\cdot\mu_{8H}(X8)+ \\ +w_6\mu_{1H}(X1)\cdot\mu_{2H}(X2)\cdot\mu_{3C}(X3)\cdot\mu_{4C}(X4)\cdot\mu_{5HC}(X5)\cdot\mu_{6HC}(X6)\cdot\mu_{7H}(X7)\cdot\mu_{8H}(X8)\end{array}\right\} \quad (6)$$

$$\mu(GDP_3)= \\ =\min\left\{\begin{array}{l}1, \\ w_7\mu_1(X1)\cdot\mu_{2H}(X2)\cdot\mu_{3C}(X3)\cdot\mu_{4C}(X4)\cdot\mu_{5HC}(X5)\cdot\mu_{6HC}(X6)\cdot\mu_{7HC}(X7)\cdot\mu_{8H}(X8)+ \\ +w_8\mu_{1H}(X1)\cdot\mu_{2H}(X2)\cdot\mu_{3C}(X3)\cdot\mu_{4HC}(X4)\cdot\mu_{5C}(X5)\cdot\mu_{6C}(X6)\cdot\mu_{7HC}(X7)\cdot\mu_{8H}(X8)\end{array}\right\} \quad (7)$$

$$\mu(GDP_4)=w_9\mu_{1HC}(X1)\cdot\mu_{2H}(X2)\cdot\mu_{3C}(X3)\cdot\mu_{4HC}(X4)\cdot\mu_{5B}(X5)\cdot\mu_{6B}(X6)\cdot\mu_{7C}(X7)\cdot\mu_{8B}(X8) \quad (8)$$

$$\mu(GDP_5)=w_{10}\mu_{1HC}(X1)\cdot\mu_{2H}(X2)\cdot\mu_{3C}(X3)\cdot\mu_{4HC}(X4)\cdot\mu_{5B}(X5)\cdot\mu_{6B}(X6)\cdot\mu_{7B}(X7)\cdot\mu_{8C}(X8) \quad (9)$$

Here w_i is the number in the range [0, 1], which describes a subjective measure of expert's confidence with respect to the statement with the number i from the knowledge base, that is the significance of the $i-$th rule.

It should be noted that use of the model (4) allows reliable assessment of the impact of changes in selected indicators on GDP. And the better the initial values for *b, c, w* parameters are selected, the closer to the experimental data the model (3) will be.

In order to obtain the model (4), rather close to the experimental data, one conducts its setting, i.e. selects such *b, c, w* parameters, which ensure minimal deviation of distance Δ between the model (theoretical ones, derived from model (4)) and the experimental data. Δ value, which is the standard deviation of the theoretical output from the experimental one, is used for model setting quality assessment.

To set the model, we used the gradient method modification in the case of undifferentiated functions, the so-called Shor's r-algorithm [12] with precision $\varepsilon = 0.001$.

Let us consider the impact of certain factors on the level of gross domestic product (GDP) per capita. As it is known, the inflation level (X1) causes a negative impact on GDP level. Herewith there are two ways of the negative impact of inflation on GDP. First, if during GDP calculation the inflation is not taken into account, the GDP indicator, as well as other cost indicators, increases without quality positive change in the country's economy, only due to price rise. On the other hand, high inflation is the indicator of inefficient functioning of the economy, which is manifested in relatively high cost of production of goods and services or their deficiency due to lower production volumes. The results of the analysis carried out by means of neuro-fuzzy technologies show that in Ukraine GDP per capita declines with rising inflation, which corresponds to the second of the above ways of inflation impact on GDP. Growth of export of goods and services (X5) has a positive effect on GDP per capita, as it indicates the rise in production of competitive goods and services, liberalization of foreign trade for enterprises, increased demand for the national currency in the global currency market. However, one should not be limited to the analysis of total exports. During detailed analysis of the impact of export on GDP it is necessary to examine its structure, identify the raw material and high-tech components of exports, focus on the level of services export growth. It should be noted that in Ukraine, according to the State Statistics Committee of Ukraine, as of 2005 the raw material component of export of goods makes 13.7%, 8.8% are for export of the plant products, 40.9% of exports are non-precious metals of products thereof [5]. Herewith the export of instruments and equipment is 0.4%, vehicles and road equipment 4.8% and mechanical equipment - 8.3% [5]. Regarding export of services, 32% of them are transport services of pipeline transportation [5]. In view of the fact that in recent years the structure of export has not changed, the export growth does not lead to qualitative changes in GDP (Fig. 5).

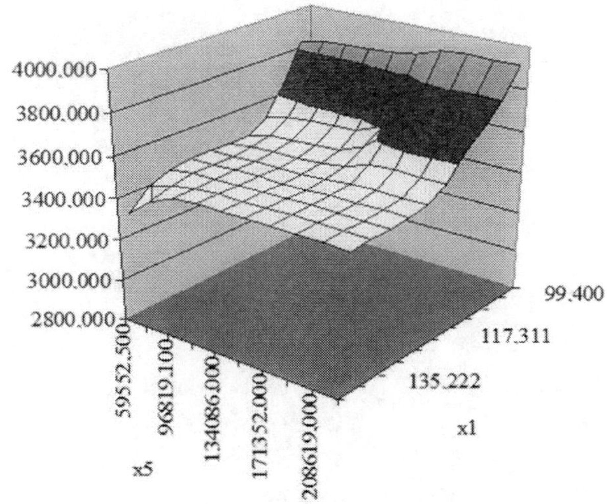

Fig. 5. GDP dependence on inflation and exports of goods and services

Therefore, at the moment, in order to increase GDP, it is necessary to restructure the export potential of the country. Growth of import of goods and services (X6) is the objective consequence of integration processes and globalization of the world economy. If the import volumes are accompanied by GDP growth, this shows the positive effects of the country's integration into the world economic community and that export is growing more rapidly than import. According to the analysis carried out using neuro-fuzzy technology, in Ukraine when import (X6) increases, GDP per capita remains virtually unchanged, indicating a lack of effective export potential formation policy. The growth of budget deficit has also the negative impact on GDP (X8) (Fig. 6).

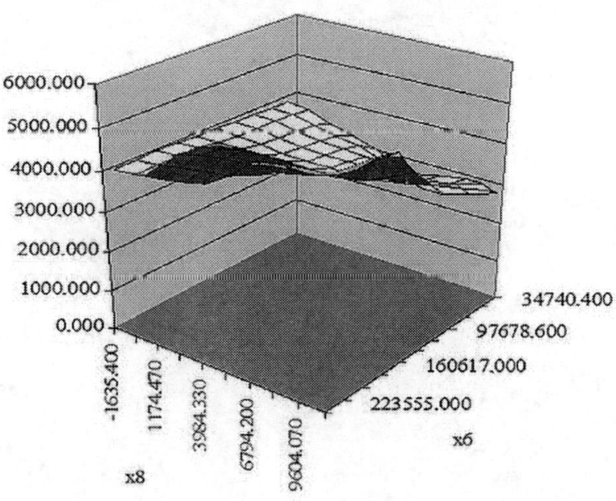

Fig. 6. GDP dependence on budget deficit and import of goods and services

The level of gross national income (GNI) per capita reflects the level of labour productivity in a country, the level of wages and income, falling to an average distribution per citizen through the existing distribution mechanisms in the country. Together with other indicators, defined as the average, it is characterized by a number of disadvantages associated with the lack of consideration of the distribution structure. However, it is generally recognized indicator that characterizes the welfare of the country as a whole. This is confirmed by comparative analysis of the socio-economic condition of the countries according to the World Bank data [2].

Gross national income per capita first of all is affected by the development of entrepreneurship in the country, which is directly related to tax burden (X2), real interest rate on deposits (X4) and direct investment in Ukraine (X7). The positive impact of foreign investments on GNI depends not only and not so much on their volumes, but on the effectiveness of their use. Therefore, Fig. 7 shows in these days a decline in efficiency of foreign investment coming into the country and not accompanied by the latest technology in the manufacturing sector.

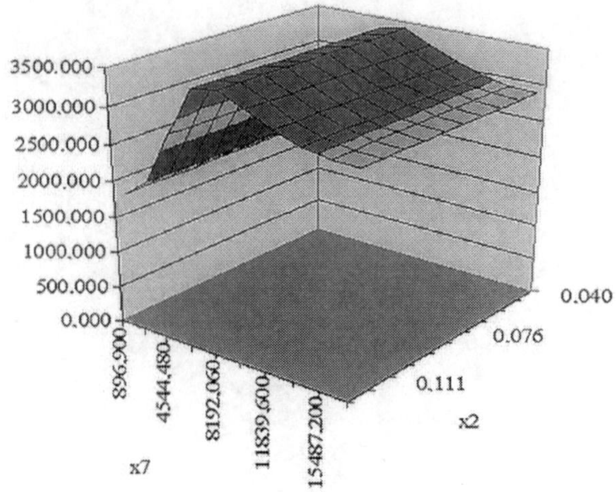

Fig. 7. GNI dependence on tax burden rate and amount of direct investment into the country

Real deposit rate according to the analysis (Figure 8) does not affect GNI, which is connected, first of all, with low capacity of the financial sector, which has been developing only since 1991 and is not directed at investment financing of the companies.

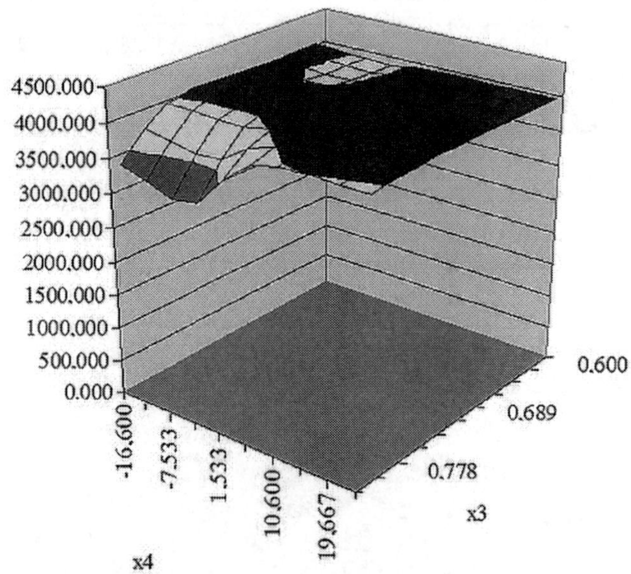

Fig. 8. GNI dependence on tax revenue share in budget revenues and real deposit interest rate

In addition, the present regulatory policy does not sufficiently encourage the banks to bring interest on deposits into conformity with inflation level, as evidenced by the high variability of this indicator. The tax revenue share in the budget revenues (X3) in turn depends on the tax burden (X2), but the growth of this indicator also shows the development of business and the reduction of the shadow economy. Objectively (X3) also depends on the degree of centralization of value added distribution in the country. In Ukraine, as shown in Fig. 8, X3 is not accompanied by GNI growth, indicating the shortcomings of distribution mechanism and lack of effectiveness of the tax system.

Thus the above research results allow, first, on the basis of the built neuro-fuzzy model, to evaluate the dependence of macroeconomic indicators, namely GDP per capita and gross national income per capita, on the indicators characterizing the economic situation of the country. Herewith this neuro-fuzzy model, unlike existing models, allows reproducing the complex nonlinear dependences.

Secondly, it is advisable to direct further research at determination of the hierarchy of factors and, on this basis, disclosure of economic events occurring in the country.

5. References

1. U. N. Kiselev. *Investigating a two-sector model of economic growth with the Cobb-Douglas production function*, **Vol. 34**, **issue 2**, pp 66–73 (2010)
2. http://web.wordbank.org/wsite/external/dataststastics/
3. A.K. Parida; R. Bisoi; P.K.Dash; S. Mishra. *Times Series Forecasting using Chebyshev Functions based Locally Recurrent neuro-Fuzzy Information System*. Int. J. Comput. Intell. Syst. **Vol. 10**, **issue 1**, pp 375 - 393 (2017).
4. V. Melnyk. *Development of scientific and methodical approach to process planning innovation development of railways* (Efektyvna ekonomika), **Ed 2** (2019) -. http://www.economy.nayka.com.ua/?n=2&y=2019
5. http://www.ondd.be/website.nsf/alllweb/
6. https://europa.eu/european-union/about-eu/euro_en
7. Statistical Yearbook of Ukraine, 575p. (2006)
8. M. Potapov. *European experience of tariff formation in the railway sector of railway transport* (Scientific Bulletin of the Odessa National Economic University). **ed. 2 (265)**, p. 157-172 (2019)
9. O.G. Bilorus. Globalization and Development Security: Monograph, 733 p. (2001)
10. A. Davity, S. Paklina. *Digital Manufacturing: New Challenges for Marketing and Business Models*. Russian Management Journal, **Vol. 15**, **issue 4**, pp 537 - 552 (2017)
11. McConnell R. Campbell, Bru Stanley L. Economics: Principles, problems and politics. 1028 p. (2015)
12. N.E. Boytsun, O.M. Kiselyova, O.M. Kiselev, O.M. Prytomanova. *Prediction of economic and financial processes based on neuron-fuzzy technologies*, **5**, P.79-93 (2005)
13. N.E. Boytsun, O.M. Kiselyova. *K-model of the impact of macroeconomic factors on exports of goods to Russia, Ukraine* (Economy: problems of theory and practice. Collection of sc. papers.) **Issue 104,** P.3-13 (2001)
14. A S. Sremac; I. Tanackov. *ANFIS model for determining the economic order quantity*. Decis. Mak. Appl. Manag. Eng. **Vol. 1, issue 2**, pp 81-92 (2018)
15. http://www.bank.gov.ua/statist/sndex.htm

The hybrid trains in international logistics transportation

Zoia Kaira[1], *Liudmila* Golovkova[2,*], *Ivan* Rekun[2], and *Yurii* Trubai[2]

[1] WSB University in Gdansk, Professor of management, 80-268 Gdansk, Grunwaldska 238A, Poland
[2] DNURT, Department of Finance and economic security, 49010 Dnipro, Lazaryan Street 2, Ukraine

Abstract. Analytical information for the market players concerning to the overall future hybrid train market and the subsegments is considered. The forecast of the volume railway transportation in Ukraine is represented in the paper. The aim of the paper is to examine the role of hybrid trains in logistics transportation segment under escalating importance of international logistics where transport segment is influenced in largely degree of political, economic, social, technological, environmental and legal changes. The paper is targeted the stakeholders to provide with information on key market drivers, restraints, challenges, and opportunities.

1. Background

Transportation logistics issues are of great importance for business, as customers location and resourcing opportunities are widely disperced. Neglect of logistics aspects brings nor only higher costs but eventual noncompetitiveness, which will result in diminished market share, more expensive supplies or lower profits.logistics problems can prevent exporters from valuably benefiting potentially profitable market overseas.

The aim of the paper is to examine the role of hybrid trains in logistics transportation segment under escalating importance of international logistics where transport segment is influenced in largely degree of political, economic, social, technological, environmental and legal changes. The paper devoted getting information the market players on the overall hybrid train market and the subsegments. This will help stakeholders understand the competitive landscape and gain more insights to better position their businesses and plan suitable go-to-market strategies.

2. Global trends in freight transport volumes

The increasing demand for reducing carbon emissions and developing more efficient trains are expected to impact the growth of the hybrid trains market, and regulations regarding environmental issues to accelerate the gowth of hydrogen trains in close future.

The transport and logistics sector driven by market trends is undergoing an important transformation together with new technological solutions. National logistics performance is linked to a number of factors including legislation, international agreements, shipper performance and technology. Inadequate infrastructure and poor logistics services increase transport costs and delivery times whereas competitiveness becomes increasingly dependent on cost efficiency.

Along with remoteness, they are major determinants of a country's ability to participate in the world economy such as connectivity [1]. The importance of connectivity is very high in today's globalised economy, where value chains are increasingly interconnected and spread out all over the world [2]. Well-functioning logistics systems facilitate trade by lowering the cost of access to international markets and by improving the connectivity and competitiveness of domestic firms [3]. Connectivity can be defined as the centrality of a country to its relevant networks [2]. Connectivity reflects not only geography and the global structure of transport networks, but also trade transaction costs – which may increase economic distances and hence reduce connectivity [4]. Transport connectivity is fundamental to increasing country or region competitiveness. Better connectivity promotes regional integration, reduces trade costs and increases trade volumes, thereby promoting economic growth, social integration and development [3, 4, 5] .

Freight transport volumes are growing and this is dependent on state of economy. In the European Union the removal of internal borders and other market barriers contribute to the growth. Efficiency improvements have resulted in a reduction of real transport costs. This allows companies to benefit from differences in labour costs and skills in different regions. Transport over long distances is becoming more and more profitable compared to alternatives like local production [7]. Volume of freight transportation by type of carrier such as railway and motor vehicles in Ukraine is illustrated in the Table 1.

Transport system has a lot constraints in terms of transportation time, price, quality and social as well as ecological impact consequently it is important the question of weights, dimensions and modes of transport

* Corresponding author: g.liudmila22@gmail.com

and IT support systems [7]. Taking into account globalisation impact, the growth of transport and the increasing environmental awareness of society, science and industry are seeking to identify solutions for sustainable freight transport. I n addition, policy makers in Europe are identifying a legislative framework to support a sustainable development for the anticipated growth of freight transport [6].

Table 1. Volume of freight transportation by type of carrier in Ukraine (thsd. tones)[1]

Year	Railway[1]		Motor vehicles
	shipment[1]	transportation[1]	
1995	360225,3		1816401,0
1996	296050,7	342558,1	1254540,2
1997	293523,5	341417,0	1249866,6
1998	286321,5	335052,5	1081326,2
1999	284244,3	334635,9	955329,1
2000	295921,0	357381,6	938916,1
2001	313089,0	370199,1	977268,8
2002	330188,3	392592,0	947263,8
2003	363364,7	445534,7	973283,0
2004	388295,0	462367,6	1027396,3
2005	378911,7	450277,3	1120715,3
2006	398148,3	478711,4	1167199,7
2007	415910,7	514192,9	1255225,3
2008	399679,7	498536,8	1266598,1
2009	322221,8	391523,4	1068857,9
2010	357969,1	432897,0	1168218,8
2011	388715,6	469308,1	1252390,3
2012	378102,3	457454,5	1259697,7
2013	377318,3	443601,5	1260767,5
2014	325171,0	386276,5	1131312,7
2015	294301,2	349994,8	1020604,0
2016	292104,7	343433,5	1085663,4
2017	277288,9	339550,5	1121673,6
2018	267639,1	322342,1	1205530,8

[1]According to the data of joint-stock company "Ukrzaliznytsia".

In the Table 2 the method weighted moving average was used to forecast D- the volume of railway transportation in Ukraine. The forecast for period t+1 is a continuation of the average demand observed through period t. Thus we have

$$F_{t+1} = A_t = W_1 D_t + W_2 D_{t-1} + W_3 D_{t-2},$$

with the condition

$$\sum_{i=1}^{3} W_i = 1,$$

where W_1=0,5; W_2=0,3; W_3=0,2.

The forecast error is also shown as the difference between actual meaning D_i and forecast F_i in the Table 2.

The forecast calculated for the volume of railway transportation in Ukraine illustrates Figure 1.ding to small growth.

Transport has several impacts on the environment. Emissions contribute to air pollution and climate change, noise causes nuisance and health risks and infrastructure has serious impacts on landscape and ecosystems. In addition to these impacts on the environment, transport has also other severe impacts on society [8]. The impacts of the transport sector as a whole are the sum of the impacts of the various transport modes. On the international freight transport market, rail transport is one of the most important modes, so it is needed to consider energy use of the transport sector.

Table 2. Moving-average forecasts of freight transportation in Ukraine, mln. t

Year	D_t	A_t (Three-Period Moving Average)	F_t (Three- Period Forecast)	Error (D_t-F_t)
1996	342,6			
1997	341,4			
1998	335,0	340,7		
1999	334,6	338,1	340,7	-6,1
2000	357,4	339,4	338,1	19,3
2001	370,2	348,6	339,4	30,8
2002	392,6	368,3	348,6	4,4
2003	445,5	340,0	368,3	77,2
2004	462,4	422,4	340,0	70,4
2005	450,3	451,5	422,4	27,9
2006	478,7	462,0	451,5	27,1
2007	514,2	471,6	462,0	52,2
2008	498,5	493,3	471,6	26,9
2009	391,5	485,0	493,3	-101,8
2010	432,9	453,3	485,0	-52,1
2011	469,3	419,6	453,3	16,2
2012	457,4	448,8	419,6	37,8
2013	443,6	460,7	448,8	-5,2
2014	386,3	439,1	460,7	-7,3
2015	350,0	407,7	439,1	-89,1
2016	343,4	366,8	407,7	-64,3
2017	340,4	346,0	366,8	-27,2
2018	322,3	338,0	346,0	-23,6
2019			338,0	

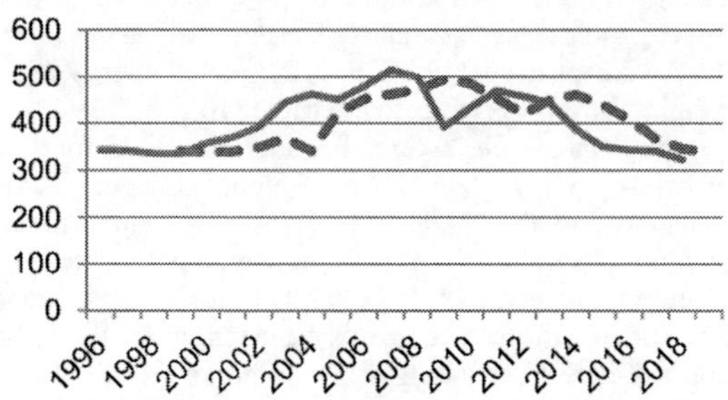

Fig.1. Forecast for the volume of freight transportation by railway carrier in Ukraine (1996-2018)
Source: authors' elaboration on the the data of joint-stock company "Ukrzaliznytsia".

2.1 Energy use of the transport sector

The trends in the energy use from transport over the last three decades of the previous century indicate the energy consumption of transport has almost doubled over this period [8]. The energy use of the transport sector is expected to grow much faster in the non-OECD countries than in the OECD countries. Where the non-OECD countries currently take about 36% of the worldwide transport related CO_2 emissions, their share is expected to equal that of the OECD countries somewhere around 2040. The growth in the non-OECD countries has been even higher: the energy use almost tripled in these 30 years time. Both for OECD and non-OECD countries, road transport had by far the largest share: about three quarters, and this share is steadily increasing [8].

As elaborated in [8], international transport has generally a minor share in road transport (8% in North America, 30% in Europe). In rail transport, the share of international transport varies a lot (5% in North America, 51% in Europe). Worldwide the share of road and rail transport are currently roughly the same [9]. Also within the OECD, the share of road and rail is comparable (OECD, 2007). Azar made an assessment of the growth in freight transport worldwide between 1990 and 2100 [9, 10]. The same study gives also estimates for the energy use in 2100.

The results of this assessment are illustrated in Table 3.

Table 3. Growth in global freight transport volumes

	Transport volume in Ttkm/year		Transport volume in Ttkm/year	
	11990	22100	11990	22100
Road	6,4	40	23	72
Rail	6,1	13	3,1	4,3
Domestic water	2,6	5	1,2	1,6
Ocean	29	126	5,8	16
Air	0,07	0,28	0,32	0,62
Total	44	184	33	95

Source: [9]

Table 3 reveals that despite of the growth in freight transport volume is expected to be strongest in road transport, the growth in rail transporty is estimated to be lower than the average. Despite of a probable improvement in fuel efficiency, the global energy use of freight transport is expected to triple [10].

This makes clear that the main driver behind the growing energy demand of freight transport is the transport growth [10]. Rail transport relies both on diesel and on electricity. The environmental performance of electric trains is generally better than that of diesel trains. The actual difference depends on electricity mix and the applied diesel technology. An important difference is that electric transport offers the possibility to use sustainably generated electricity. In that case the environmental performance of electric trains is much better than that of diesel trains. However, in an integrated electricity market, the marginal environmental impact from electric energy will be determined by the supplier of electricity. It is difficult to determine from which source any particular use of electricity stems [10].

Rail transport is responsible for 1.5% of the global transport energy use. Light duty vehicles (including passenger cars) have the highest share with 44%. The other main energy users within the transport sector are aviation (12%), maritime shipping (10%) and buses (6%). There are no worldwide statistics on the share of international road and rail freight transport in the energy use of total freight transport. However, data on the share of international freight transport in the transport volume can give a good indication.

The modal split differs a lot between countries. In Russia, the rail freight transport volume is several times larger than the road freight transport volume and also in China,the share of rail is much higher than that of road. In the USA, road and rail have rather comparable shares, while in Europe, rail transport has only a minor share in the inland freight market which is dominated by road transport [9]. As can be seen in Table 2, particularly the share of road transport is expected to increase strongly worldwide. The growth in road freight transport volumes is accompanied by a growth in both the number of vehicles and vehicle-kilometres. Figure 2 shows an estimate for the number of road freight vehicles worldwide [9].

Rail transport relies both on diesel and on electricity. The environmental performance of electric trains is generally better than that of diesel trains.
The actual difference depends on electricity mix and the applied diesel technology. An important difference is that electric transport offers the possibility to use sustainably generated electricity.

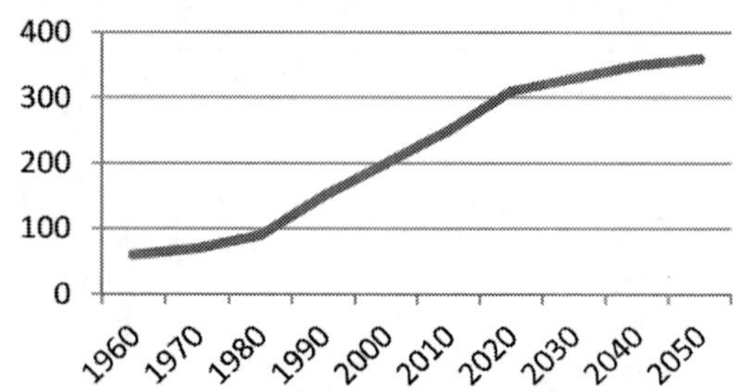

Fig. 2. Estimated trend of worldwide number of freight vehicles
Source:authors' elaboration on the base of [11]

In that case the environmental performance of electric trains is much better than that of diesel trains. However, in an integrated electricity market, the marginal environmental impact from electric energy will be determined by the "marginal" supplier of electricity. There is variation in logistical parameters, particularly load factors [9].

Differences exist in the energy mix of electricity used for electric trains. In specific markets, the differences between transport modes are generally small. Differences depend more on logistical characteristics and technology (e.g.emission standards) than on mode *per se* [8](Van Essen, et al., 2003). The current vehicle utilisation is a trade-off between the direct costs vehicle kilometres and the various cost of optimizing logistical chains. The second include costs related to time-losses, lower flexibility, storage, which might increase when vehicles are used in a more efficient way. Therefore, optimizing logistics is not just a task for the transport sector, but it is also related to governmental measures, in particular transport pricing [8].

2.2 Opportunities to improve the environmental performance of freight transport

Researches make clear the CO_2 emissions of transport an increasing trend [8]. This is contrasting with the CO_2 reduction targets discussed within the post-Kyoto climate policy and which have already been adopted by some countries (e.g. the European Union). In the short term, many developed countries will be able to meet their CO_2 reduction goals under the Kyoto protocol without drastic measures in the transport sector. However, for the long term CO_2 emission reductions of 40 to 60% compared to 1990 are expected to be necessary in order to limit the effects of global warming to acceptable levels [8]. Given the expected growth of the transport sector in the next decades and its strong reliance of fossil fuels, such long-term reduction goals cannot be met without significant contributions from the transport sector. Measures for CO_2 reduction in international rail freight transport aimed at volume reduction and modal shift may contribute to a reduction of pollutant emissions, e.g. a shift towards electric rail transport in combination with a shift to greening electricity production [8].

Diesel electric locomotives may have most of what they need for regenerative braking since they might already use dynamic braking. This uses the traction motors as generators to absorb much of the train's energy, but without a way to store the generated electricity it is simply thrown into the atmosphere as heat with large rooftop resistor banks and cooling fans [8]. Using a storage system means that a non-fully electric train can use regenerative (as opposed to merely dynamic) braking, and even shut down the main power source whilst idling or stationary. Reducing energy consumption provides environmental benefits and economic savings [12]. The most important environmental impacts from the transport sector are caused by emissions of air pollutants, CO2 and noise. The contribution to air pollution is decreasing in most parts of the world, mainly due to various vehicle emission standards that have been implemented all over the world and are periodically tightened [9].

Only in those parts of the world that have an extremely high growth in transport volumes,the overall emissions of air pollutants may not yet ecrease. Noise is an important environmental problem which, just like air pollution, has severe health impacts causing high numbers of deaths each year. There are various measures that could be taken to reduce the contribution of freight traffic to ambient noise levels. The most cost-effective measures are those addressing the noise at-source. A mix of measures, like improving fuel efficiency of vehicles, alternative fuels and logistical improvements, is needed. An effective policy for reducing the environmental impact of international rail transport should aim at improving the environmental performance of all modes of transport as well as ensuring a level playing field for the various modes. Regulation, infrastructure measures and pricing measures that take into account the environmental costs can contribute to this [9].

3 Transport and Logistics Trends

The transport and logistics sector is undergoing an important transformation as new technological solutions come into everyday use, driven by market trends [12]. This segment of the economy is influenced a large degree of political, economic, social, technological, environmental and legal changes. As complexity of modern transport and logistics grows, it is needed to realise what should be the short vs long term focus and what to invest in. An attempt to identify the key forces transforming the transport and logistics segment reveals that the hybrid train segment is expected to be the fastest growing market [12].

3.1 Rail trains evolution

It is worth to mention the idea of the railway started in about 600 BC during Greek and Roman times as ship dock ways, and the rails were simply grooves in limestone [13]. In the 16th century, railways appeared also in German and Austrian mines as they became the smartest way to transport large stone blocks. During the 18th century, before the steam industrial revolution, the wagons were pulled by ponies very often named "Captain" as they were leading the convoy of stone and ore-filled carts. The idea of the railway started in about 600 BC during Greek and Roman times as ship dock ways, and the rails were simply grooves in limestone. In the early 1800's, with steam gradually taking over as the means to supply large amounts of power, horses and ponies gave way to steam-powered locomotives. Some agreements were reached regarding railway gauges, and the first trains designed to carry goods and passengers started to appear [13].

In the 70s, the trend for non-electrically powered trains has been to reduce carbon dioxide emissions and to save energy [13]. The diesel-powered locomotives of today are already "hybrid" in a way that the diesel motor does not power the locomotive directly i.e. by means of a mechanical transmission. The high power delivery versus the load to be moved is so severe a contrast that it would rip apart most common mechanical transmission devices except for hydraulic converters. So, in order to deliver the power in a smooth manner, the diesel engine generates electricity on board the locomotive and powers a set of electric motors that gently transfer the power to the locomotive's wheels. Some of these engines also used a level of regenerative braking to save energy

The first truly hybrid locomotives appeared in the late 19th century and were equipped with lead-acid batteries. These were mainly used on tram lines or street railways. It is not known whether these early attempts really saved energy. Some smaller engines were equipped solely with lead-acid batteries to overhaul underground lines in London. The first near successful hybridization implementation attempt was initiated in the Czech Republic in 1986 but remained experimental [13].

In the 21st century, Japanese hybrid machines made real progress in term of usability and advancement and started to make sense in terms of efficiency.

3.2 Hybrid rail trains

A hybrid train can be defined as a locomotive, railcar or train that uses an onboard rechargeable energy storage system (RESS), placed between the power source (often a diesel engine prime mover) and the traction transmission system connected to the wheels. Since most diesel locomotives are diesel-electric, they have all the components of a series hybrid transmission except the storage battery, making this a relatively simple prospect. Surplus energy from the power source, or energy derived from regenerative braking, charges the storage system. During acceleration, stored energy is directed to the transmission system, boosting that available from the main power source. In existing designs, the storage system can be electric traction batteries, or a flywheel. The energy source is diesel, liquefied petroleum gas, or hydrogen (for fuel cells) and transmission is direct mechanical, electric or hydrostatic [14]

Hydrogen train which works on hydrogen fuel cells is said to be a true eco-friendly train. Hydrogen train does not require a dual fuel system, which makes it a completely pollution-free and noise-free in operation achieving the need of an emission-free train. North America market is expected to register the fastest growth during the forecast period. The increasing demand for reducing carbon emission and developing more efficient trains are expected to propel the growth. BNSF with GE Transportation has planned to launch a diesel-electric locomotive by the end of 2020 [14].

The increasing concerns of carbon emission due to conventional railways and efficiency in this mode of transport have led the European authorities to develop a hybrid train network that has propelled the growth of this market. Government initiatives have encouraged the key players to develop the infrastructure for the train network. These trains have provided an opportunity for railway manufacturers to increase their presence in the region.

While diesel still plays a major part on many routes in many countries, rail is experimenting with new forms of power, including hydrogen fuel cells and batteries, while making old, dirty locomotives more efficient.

Today, the best hybrid trains are much more efficient. The batteries on board are able to store more and more energy and with the significant progress made in increasing the level of energy density in battery technology, the amount of energy stored on board is increasing. Through regenerative braking and on-board energy management to enable stop/start in stations. It has shown up to 10 % fuel savings and 60% reduction in harmful emissions. However, batteries have a limited lifetime due to their limited number of cycles (c.3000) and they are still expensive to replace and difficult to recycle. The demand for the batteries is high during some of the charge-discharge cycles and this level of demand can damage them if it is not spread well enough across the battery pack. The bigger the pack, the higher the demand peak that can be dealt with. To some extent, any individual rolling stock could also move by itself and in that way ease the assembling of trains [13].

The transformation from diesel to hybrid-electric operation will allow a noise emissions reduction in stations by 75 percent, a 25 percent fuel consumption and CO_2 emissions reduction. NOx emissions can also be reduced by up to 70 percent and particulate emissions by around 90 percent.The total operating costs of rail vehicles with MTU hybrid drives are also significantly lower, and even brake pads and discs experience reduced wear due to energy recuperation during braking. As a result of the additional power of the electric machine, train operators also have the option to improve journey times through better acceleration.

The importance of connectivity is very high in today's globalised economy, where value chains are increasingly interconnected and spread out all over the world. This is particularly important for catching-up economies trying to reduce their productivity gaps with the advanced countries, because higher-tech value chains tend to be particularly internationalised. It is needed to evaluate the up and coming Hybrid Train Market and how it will contribute to developing more sustainable transport solutions.

3.3 Hybrid Train Market Trends

Development of infrastructure and reduction in carbon emissions are significant factors for the growth of the hybrid train market globally [15].

Hydrogen trains don't require a dual fuel system, which makes it a completely pollution-free and noise-free in operation achieving the need of an emission-free train. The railway industry in North America is inclined towards innovation, technology, and development of high-performance and energy efficient trains.The increasing demand for reducing carbon emissions and developing more efficient trains are expected to propel the growth. The World Market for Hybrid Trains, 2019 to 2030: Estimated to Grow from 5,429 Units in 2019 to 9,109 Units by 2030 [15].

The global hybrid train market is estimated to grow from 5,429 Units in 2019 to reach 9,109 Units by 2030, at a CAGR of 4.8% (Figure 3).

Increased emphasis on reduction in carbon emissions is expected to boost the market [16]. Strict regulations regarding environmental issues are probably to accelerate growth of hydrogen trains during the forecast period. The hydrogen train segment is the fastest growing market. Hydrogen train which works on hydrogen fuel cells is said to be a true eco-friendly train. Hydrogen train does not require a dual fuel system, which makes it a completely pollution-free and noise-free in operation achieving the need of an emission-free train. Alstom with Eversholt Rail has introduced the first hydrogen fuel cell powered train in Germany in September 2018, which has already entered into the commercial services [15].

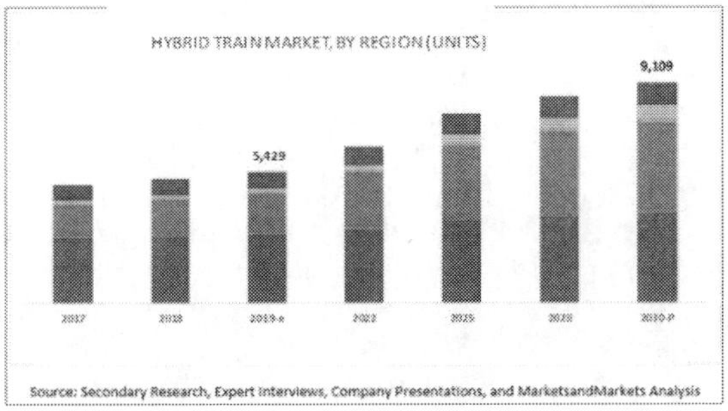

Fig. 3. Hybrid train market, by region (units)
Source:[16]

The North America market is expected to witness the fastest growth, followed by the Middle East & Africa. The US accounts for the largest share of the North American train market in 2018. The railway industry in the country is inclined towards innovation, technology, and development of high-performance and energy-efficient trains. The increasing demand for reducing carbon emission and developing more efficient trains are expected to propel the growth. BNSF, Bombardier, and GE are some of the key players in the US railway industry. BNSF with GE Transportation has planned to launch a diesel-electric locomotive by the end of 2020. The Asia Pacific is expected to be the largest market, followed by Europe. The increasing concerns of carbon emission due to conventional railways and efficiency in this mode of transport have led the European authorities to develop a hybrid train network that has propelled the growth of this market. Government initiatives have encouraged the key players to develop the infrastructure for the train network. These trains have provided an opportunity for railway manufacturers to increase their presence in the region [12].

While diesel still plays a major part on many routes in many countries, rail is experimenting with new forms of power, including hydrogen fuel cells and batteries, while making old, dirty locomotives more efficient [12].

Conclusions

The increasing concerns of carbon emission due to conventional railways and efficiency in this mode of transport have led the European authorities to develop a hybrid train network that has propelled the growth of this market. Government initiatives have encouraged the key players to develop the infrastructure for the train network. These trains have provided an opportunity for railway manufacturers to increase their presence in the region. While diesel still plays a major part on many routes in many countries, rail is experimenting with new forms of power, including hydrogen fuel cells and batteries, while making old, dirty locomotives more efficient.

Today, the best hybrid trains are much more efficient. The batteries on board are able to store more and more energy and with the significant progress made in increasing the level of energy density in battery technology, the amount of energy stored on board is increasing. A mix of measures, like improving fuel efficiency of vehicles, alternative fuels and logistical improvements, is needed. An effective policy for reducing the environmental impact of international rail transport should aim at improving the environmental performance of all modes of transport as well as ensuring a level playing field for the various modes. The up and coming Hybrid Train Market likely will contribute to developing more sustainable transport solutions.

References

1. OECD/ITF (2017), ITF Transport Outlook 2017, OECD Publishing, Paris. https://www.ttm.nl/wp-content/uploads/2017/01/itf_study.pdf
2. ITF (2019), "Enhancing Connectivity and Freight in Central Asia", International Transport Forum Policy Papers, No. 71, OECD Publishing, Paris.: https://www.itf-oecd.org/sites/default/files/docs/connectivity-freight-central-asia_2.pdf
3. J.-F. Arvis, , L. Ojala, B. Shepherd, D. Saslavsky, C. Busch, A. Raj (2014), *Connecting to Compete 2014*
4. The World Bank, 2014. Trade logistics in the global economy: the logistics performance index and its indicators; The World Bank, 2014.
5. ASEF (2016), *ASEF Outlook Report 2016/2017*, Asia-Europe Foundation, June 2016.
6. OECD (2018), *Enhancing Connectivity through Transport Infrastructure: The Role of Official Development Finance and Private Investment*, The Development Dimension, OECD Publishing, Paris, https://doi.org/10.1787/9789264304505-en
7. S. Jeschke. Global Trends in Transport Routes and Goods Transport: Influence on Future International Loading Units .https://www.acea.be/uploads/publications/SAG_16_Global_Trends_in_Transport_Routes__Goods_Transport.pdf
8. H. van Essen. The Environmental Impacts of Increased International Road and Rail Freight Transport Past Trends and Future Perspectives 10-12 November 2008 in Guadalajara, Mexico. CE Delft, Delft, the Netherlands. http://www.oecd.org/greengrowth/greening-transport/41380980.pdf
9. C. Azar, K. Lindgren and B. A. Andersson (2001). Hydrogen or methanol in the transportation sector? Goteburg, Goteburg University.
10. OECD. Globalisation, Transport and the Environment. OECD Publishing, 12 .01. 2010.– 276 p.

11. Metz (2005), World Wide Emission Trend 1950 to 2050 Road Transport and all Sources, Eurochamp Workshop in Andechs, Chemistry, Transport and Impacts of Atmospheric Pollutants With Focus on Fine Particulates, October 2005, www.eurochamp.org/datapool/page/28/Metz.pdf
12. Powering the trains of tomorrow. 26 January 2017. https://www.railway-technology.com/features/featurepowering-the-trains-of-tomorrow-5723499/
13. Olivier Chabilan, Hybrid Locomotives, Ultracapacitors for the Railway Vehicles and Infrastructure.2018 https://www.skeletontech.com/skeleton-blog/hybrid-locomotives-ultracapacitors-for-the-railway-vehicles-and-infrastructure
14. Olivier Chabilan, Hybrid Locomotives, Ultracapacitors for the Railway Vehicles and Infrastructure Posted by Olivier Chabilan on - 07 February 2018 https://www.skeletontech.com/skeleton-blog/hybrid-locomotives-ultracapacitors-for-the-railway-vehicles-and-infrastructure
15. UK, DVV Media. *"GE unveils hybrid locomotive"*. Railway Gazette. *Retrieved 2016-03-22.*
16. Research and Markets. Dublin, May 07, 2019 (GLOBE NEWSWIRE) -- The "Hybrid Train Market by Propulsion, Operating Speed, Application, and Region - Global Forecast to 2030"

Analysis of thermodynamic parameters in spark ignition VCR engine

Patryk Urbański[1,*], *Maciej* Bajerlein[2], *Jerzy* Merkisz[2], *Andrzej* Ziółkowski[2], and *Dawid* Gallas[1]

[1] Łukasiewicz Research Network - Rail Vehicles Institute "TABOR", Warszawska 181, 61-055 Poznan, Poland
[2] Poznan University of Technology, Institute of Combustion Engines and Transport, Piotrowo 3, 60-695 Poznan, Poland

Abstract. 3D models of Szymkowiak and conventional engines were created in the Solidworks program. During the motion analysis, the characteristics of the piston path were analyzed for the two considered engine units. The imported file with the generated piston routes was used in the AVL Fire program, which simulated combustion processes in the two engines with identical initial conditions. The configurations for two different compression ratios were taken into account. The basic thermodynamic parameters occurring during the combustion process in internal combustion engines were analyzed.

1. Introduction

Technological progress over the decades has enabled the engineers to move from the level of general efficiency of several percent in the case of steam engines up to over 45% in the most efficient internal combustion engines. Despite the huge success of such a large efficiency improvement, half of the fuel energy is still not converted into useful energy mostly due to thermal and mechanical losses.

One of the most important parameters of a reciprocating internal combustion engine, which affects the efficiency of the unit is the compression ratio ε. This parameter is the ratio of the maximum cylinder volume V_{max} when the piston is in the BDC to the volume of the combustion chamber when the piston is in TDC, so when the volume of the chamber is the smallest V_{min} [1, 2]. The internal combustion engine does not typically operate under stable conditions. This means that it often works in suboptimal conditions, where the constant value of the geometric compression ratio as a parameter is not the optimal solution, due to the resulting decrease in the overall efficiency of the system. Increasing ε brings with it a visible improvement in engine efficiency and power, by increasing the pressure inside the combustion chamber. Increasing the described parameter value is, however, only profitable up to a certain point due to the increased resistance during the compression stroke and the probability of knock during combustion, which may lead to engine damage.

The use of a variable compression ratio may prove to be a solution to the above problem [3, 4]. It is a type of a mechanism where it is possible to change the compression ratio during the engine operation, thus adjusting the engine operating parameters to the given operating conditions present at the time. The advantages of a system using this solution also include: the possibility of using alternative fuels and low-temperature combustion processes.

Ongoing research and development works seem to confirm the perceived potential of such a solution. The engine with this system will have a greater degree of complexity, however, the prospects of increased efficiency, reduced fuel consumption and reduction of exhaust emissions [5–8] make further research in this field necessary.

2. Test objects

Constructor Msc. Mirosław Szymkowiak designed a piston engine with an unprecedented crank and piston system (figure 1). The unit converts the reciprocating motion of the piston into the rotary motion of the crankshaft using a unique and innovative system. The energy coming from the piston motion is transferred to the swingarm by means of the first connecting rod. Then, through the swingarm and the second connecting rod, it is transmitted to the crankshaft. The swingarm is also mounted to pivot on the camshaft. Thanks to this last element, it becomes possible to adjust the compression ratio smoothly during operation simply by rotating it.

The engine has the shape of a cube with a side of 600 mm. The eight-cylinder engine was designed in a two-row system (4 cylinders in a row). The main advantages of the engine are its reduced height by placing the crankshaft between the rows of cylinders. The variable compression ratio also allows the engine to be adapted to run on various fuels with a wide range of characteristics. The crank-piston system used in the engine also caused a reduction of the normal force acting on the cylinder barrel, which translates into a reduction in mechanical losses and an increase in the service life of the ring-cylinder pair. Thanks to the described advantages, this

* Corresponding author: patryk.urbanski@tabor.com.pl

engine can be used in various types of vehicles [8] and industry.

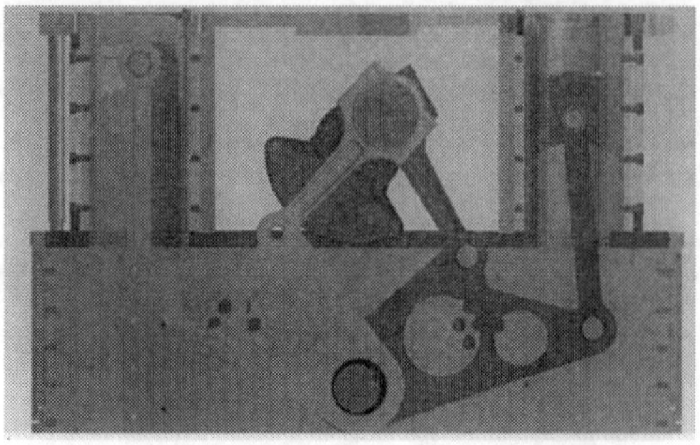

Fig. 1. Szymkowiak (or swingarm) engine design principle

For the purpose of simulating the combustion processes of the swingarm and conventional engines and their subsequent comparison, in 3D – Solidworks – models of both units were created that represent their real geometrical dimensions. Based on the created models in the – motion analysis - tab the piston motion paths were generated (Figure 2) for each of the engines. The obtained piston motion characteristics were saved in a .txt file for use in combustion simulation.

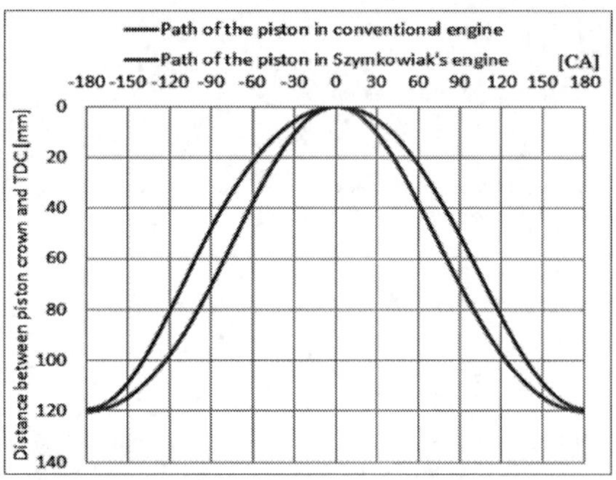

Fig. 2. Paths of motion for pistons in conventional and Szymkowiak's engines

3. Combustion process simulation

The AVL Fire program was used to simulate the combustion process. One of the program segments is the Workflow Manager tab. Using to this program it is possible to simulate processes of compression and spark ignition engines. This version of the program made it possible to simulate combustion processes in a spark-ignition engine for conventional and Szymkowiak engines.

Based on the available data, a movable mesh of the combustion chamber was created. The combustion chamber shape was adopted as a cylinder, which means that the piston used had a flat bottom. The height of the cylinder was selected in such a way that the compression ratio was equal to 8.6 and 11.6. With these configurations, the cylinder motion path of an actual swingarm engine maintains symmetry with respect to the line passing through the TDC point. An important assumption of the simulation is the lack of fuel exchange. The considered range of motion of the piston was determined at 540–800 °CA, which includes the compression stroke and the most important stage of the combustion stroke. The engine speed was set at 2000 rpm. The most important initial conditions for the simulation are presented in Table 1.

Table 1. Boundary conditions in simulations

Parameter	Value	Unit
Fresh air injected		
Pressure	97000	Pa
Air density	[-]	kg/m³
Temperature	340	K
Turbulent kinetic energy	11	m²/s²
Turbulent scale length	0.003	m
Initialization	Swirl/Tumble	[-]
Turbulence velocity	3000	rpm
Direction of the axis of turbulence rotation	X = 0 Y = 0 Z = 1	[-]
Exhaust recirculation		
Fuel type	Gasoline	[-]
EGR mass fraction	0.12	[-]
EGR composition	0.478	[-]
Laminar flame velocity	Default	[-]

The program has several combustion models. The Extended Coherent Flame Model (ECFM) with a spherical shape of the initial flame nucleus was chosen for the computational needs of the simulation. The ECFM model is the manufacturer's recommended AVL program for simulating combustion in a spark-ignition engine [10].

Due to the use of gasoline as a fuel, it was necessary to determine the parameters related to the injection and ignition of the fuel-air mixture (Table 2). The direct fuel injection model was chosen. The injector has been positioned along the cylinder axis. The nozzle tip has been set at a distance of 2 mm from the engine head (z axis). The injection was assumed to take place at 660 °CA. Due to the change in the compression ratio, the volume of the chamber varies depending on the setting of the camshaft and hence the air mass is also different. For this reason, in order to obtain a stoichiometric mixture, the mass of injected fuel will vary depending on the tested configuration. In the case of ε=8.6 the fuel weight for a single injection was 46 mg, and for ε=11.6 it was 43.5 mg.

The spark plug was set at a distance of 10 mm from the injector relative to the y axis and at a distance of 7 mm from the head (z axis). The injector and the spark plug lie on the common y-z plane. The ignition spark was determined to occur at 680 °CA.

Table 2. Ignition and injection parameters

Parameter	Value	Unit
Ignition parameters		
Spark timing	680	CA
Spark location	X = 0 Y = 10 Z = 7 (from the engine head)	mm
Injection parameters		
Injection timing	660	CA
Number of nozzle holes	1	[-]
Nozzle position	X = 0 Y = 0 Z = 2 (from the engine head)	mm
Mass of injected fuel	46; ε = 8.6 43.5; ε = 11.6	mg
Fuel temperature	330.15	K
Outer diameter	3.5	mm
Inner diameter	3.2	mm
Half outer cone	48	deg
Half inner cone	42	deg

The injection took place using a pintle nozzle type injector [11]. This is a single-hole injector designed for direct, high-pressure gasoline injection. Dimensions describing this element can be found in Table 2 and Figure 3.

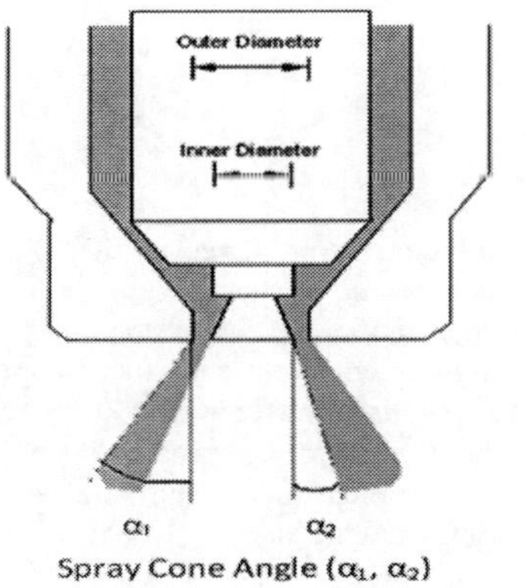

Fig. 3. Diameters of the injector used [11]

The injection process is described by the Injection Rate Diagram (Figure 4). The injection time in the graph is a reference to the actual injection duration, which in the simulation was assumed to last a total of 0.3 ms. Injection Rate has been selected to simplify the simulation and comparison with the real engine. At a later stage of research, it is planned to compare the simulation results with the actual performance of Szymkowiak's engine. For this reason, the simplification of the Injection Rate is a great help when controlling the injection in a real engine.

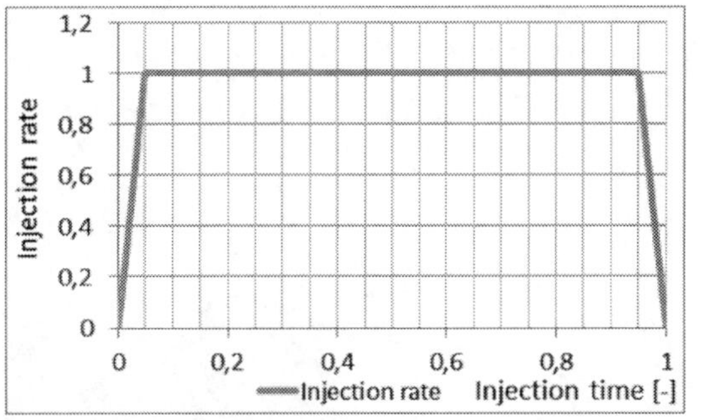

Fig. 4. Injection rate diagram

4. Simulation results

Simulations of engine processes were carried out for two Szymkowiak engine configurations. The first simulation was carried out for a compression ratio of 8.6, the second for a compression ratio of 11.6. At such camshaft settings, the piston motion path in the swingarm engine is characterized by symmetry with respect to the line passing the TDC. For comparative purposes, another combustion simulation was also carried out for a conventional engine with identical compression ratios. All initial conditions in the four performed simulations were identical. The exception was the mass of injected fuel, selected individually based on the value of the compression ratio. The subject of the analysis were: characteristics of the Equivalence Ratio distribution, pressure, temperature, rate of fuel evaporation and the heat recovery rate [12, 13].

4.1. Compression ratio = 8.6

The parameter λ describes the actual Air to Fuel Ratio (AFR) divided by the air-fuel stoichiometric ratio AFR_{stoich} (1). The equivalence ratio is in turn the inverse of the λ (2) parameter.

$$\lambda = \frac{AFR}{AFR_{stoich}} \quad (1)$$

$$EquivalenceRatio = \frac{1}{\lambda} \quad (2)$$

This parameter describes the flow behavior of the fuel stream in the AVL Fire program. The cross-section of the cylinder runs through its center and is visible in the y-z plane. The cross-sections shown (Table 3) give the Equivalence Ratio at points 10, 20 and 30 °CA after injection. For comparative purposes, cross-sections from simulation of conventional and Szymkowiak engines were compiled.

There is a difference in the chamber height in individual engines for the cross-sections shown. Due to the characteristics of the piston path, the volume in each analyzed cross section will be smaller in Szymkowiak's engine. This has an impact on the air/fuel mixture combustion process quality. In the last tested point in the conventional engine, the fuel stream does not touch any

 https://doi.org/10.1051/matecconf/201929405001

of the walls. On the other hand, in Szymkowiak's engine, the fuel comes into contact with the piston's bottom, which may cause fuel deposition on the surface.

Table 3. Equivalence ratio in cylinder for ε=8.6.
C – Conventional engine, S – Szymkowiak engine

°CA after injection	Type of engine	Equivalence Ratio [-]
10	C	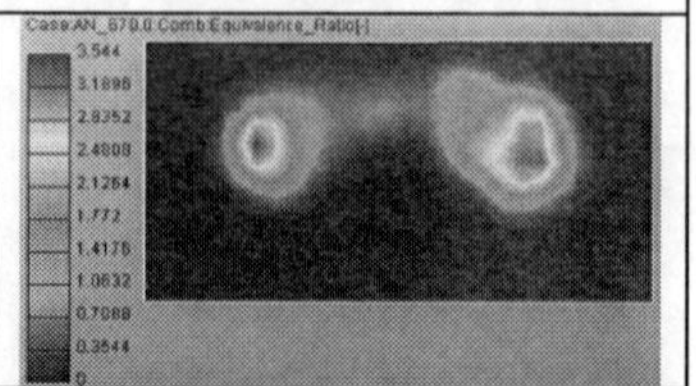
	S	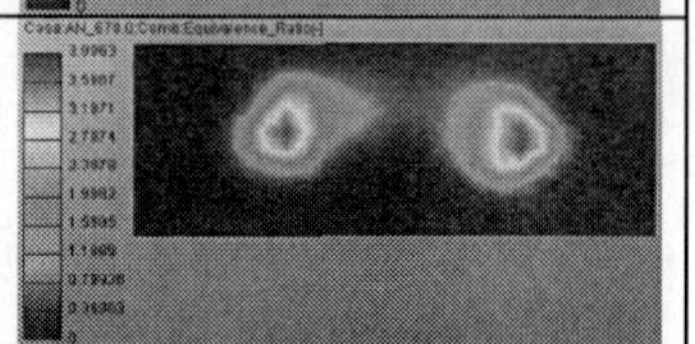
20	C	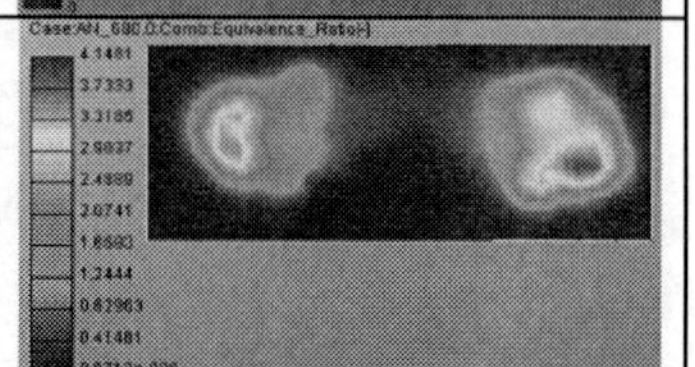
	S	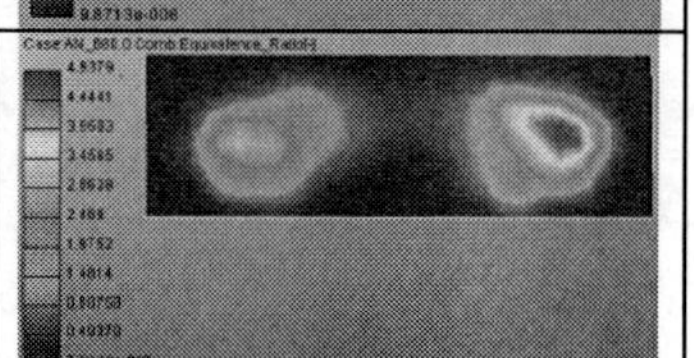
30	C	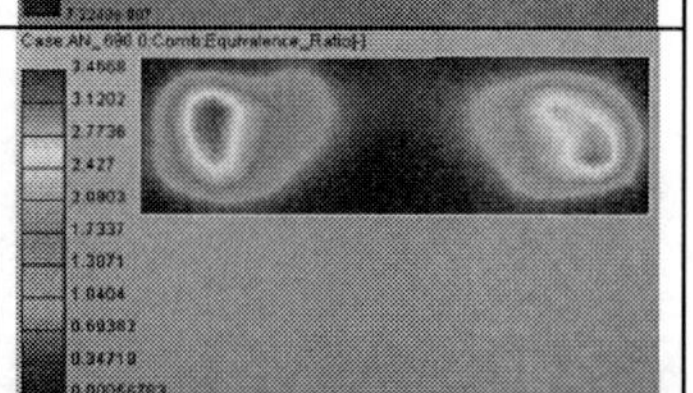
	S	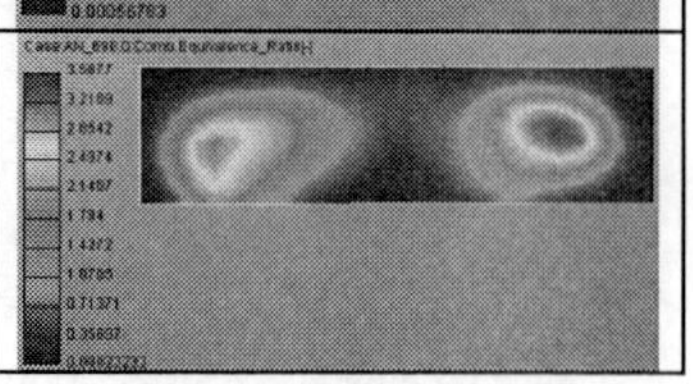

The distribution of mean cylinder pressure for simulations with a compression ratio 8.6 is shown in Figure 5. It is evident that the pressure in the swingarm engine has higher values in the wider CA range. The maximum value of the described parameter was 62.4 bar for Szymkowik's engine and was observed at 727.8 °CA. In the case of a conventional engine, the highest pressure value reached the level of 62.5 bar, which was found at 727.2 °CA. The points of reaching the maximum pressure are practically identical.

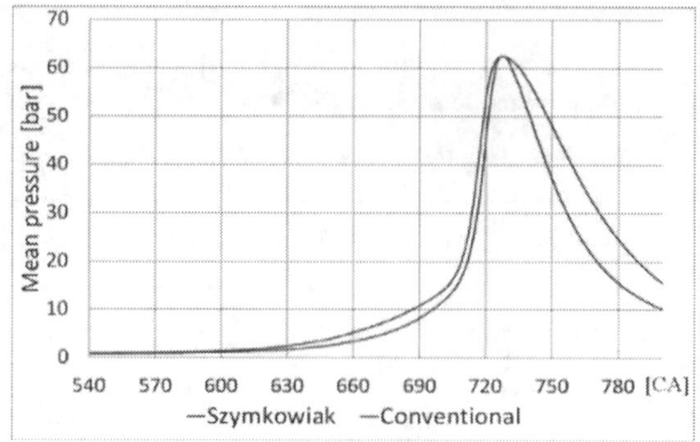

Fig. 5. Mean pressure in cylinder for ε=8.6

The temperature distribution characteristic is shown in Figure 6. In the compression stroke in Szymkowiak's engine, this parameter value is higher than for the conventional unit, which is due to the piston path characteristics, in which the compression element is located closer to the TDC at the same CA. The maximum values achieved in the simulation are 2374.19 K in the range of 736.6-736.9 °CA in the case of a swingarm engine and 2449.49 K in 730.6 °CA for a conventional engine.

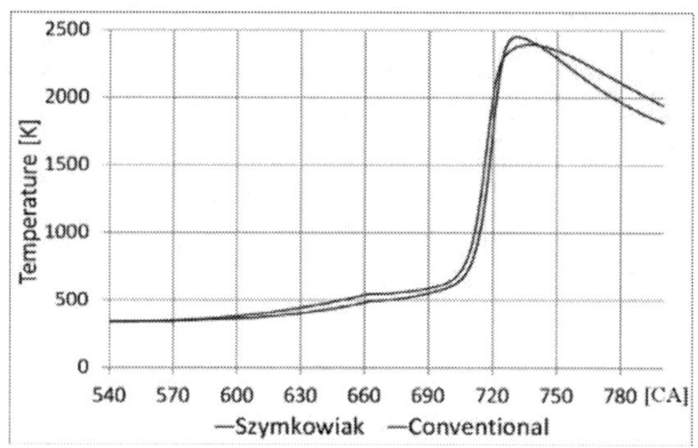

Fig. 6. Mean temperature in cylinder for ε=8.6

Due to the different characteristics of the piston path in both engines, the temperature value has a significant effect on the evaporation rate of the injected fuel (Figure 7). The higher temperature in the compression stroke causes the gasoline to evaporate faster in Szymkowiak's engine. The duration of the process from the moment of injection to complete fuel evaporation was 34.5 °CA in Szymkowiak's engine and 58.1 °CA in a conventional engine. The moment at which 50% of the fuel evaporated was at 672 °CA and 674 °CA in the swingarm and conventional engine.

The heat release rate (Figure 8) in Szymkowiak's engine reached the maximum value of 107.83 J/deg at 717.2 °CA. In the case of a conventional engine, this value was 127.69 J/deg and was reached at 719.8 °CA. This means that the value obtained in the conventional engine is 18.4% higher than for the swingarm engine. Heat dissipation is more distributed over time in the case of the swingarm engine. A measure of the fuel energy use is the Accumulated heat release (Figure 9), which is a sum of the heat produced. For Szymkowiak's engine, the result was 1615.76 J while for the conventional engine it was 1681.13. This indicates a better use of fuel energy by 4% in the conventional engine.

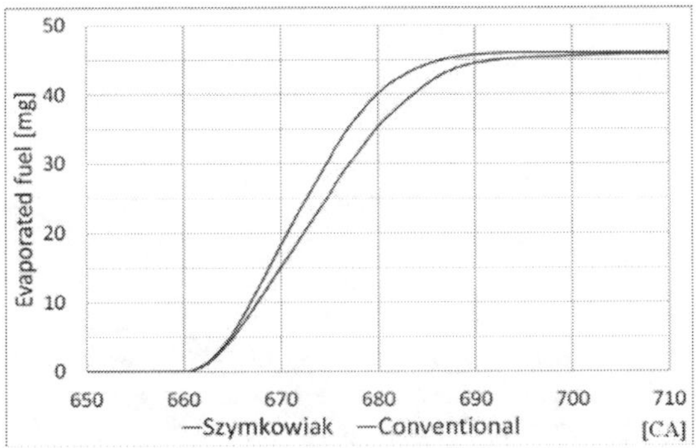

Fig. 7. Evaporation of injected fuel for ε=8.6

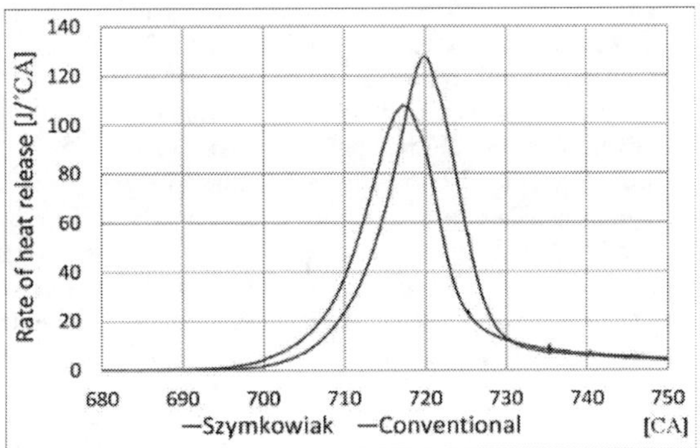

Fig. 8. Rate of heat release for ε=8.6

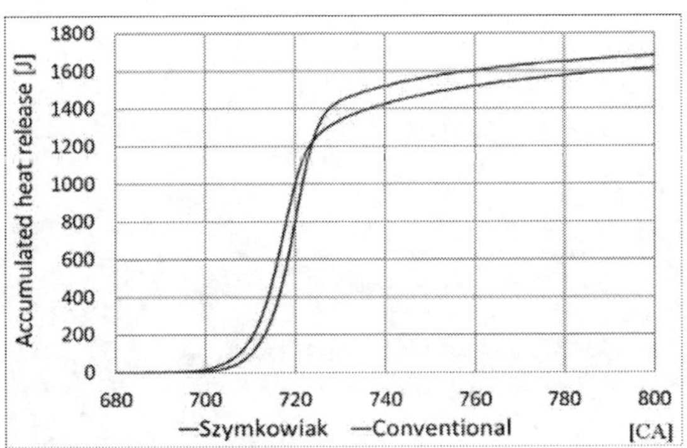

Fig. 9. Accumulated heat release for ε=8.6

4.2. Compression ratio = 11.6

The equivalence ratio for the compression ratio equal to 11.6 is shown in Table 4. It is clear that the shape and area of the Fuel Spray section is visibly different depending on the engine type. When increasing the compression ratio, the height of the chamber decreases at the same time, which translates into a smaller cylinder volume. In addition, in Szymkowiak's engine, the height will be smaller relative to the conventional engine due to the characteristics of the piston motion path. This factor contributes to the dispersion of the fuel mist in the cylinder. As can be seen at 30 °CA after injection in both engines there was a fuel contact with the surface of the combustion chamber. In Szymkowiak's engine, however, the contact surface is significantly larger, which indicates a lower combustion quality.

Table 4. Equivalence ratio in cylinder for ε=11.6. C – Conventional engine, S – Szymkowiak engine

°CA after injection	Type of engine	Equivalence Ratio [-]
10	C	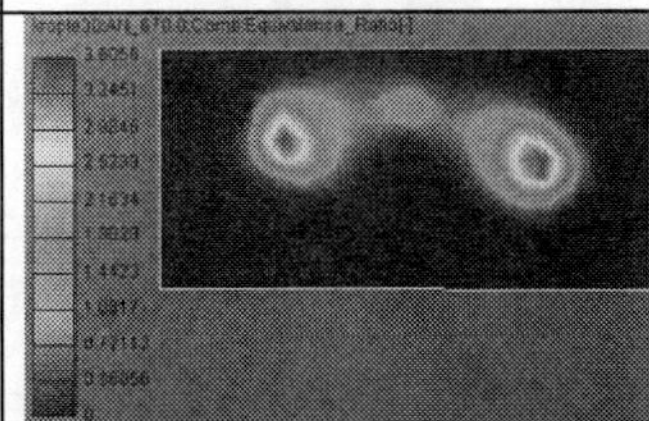
	S	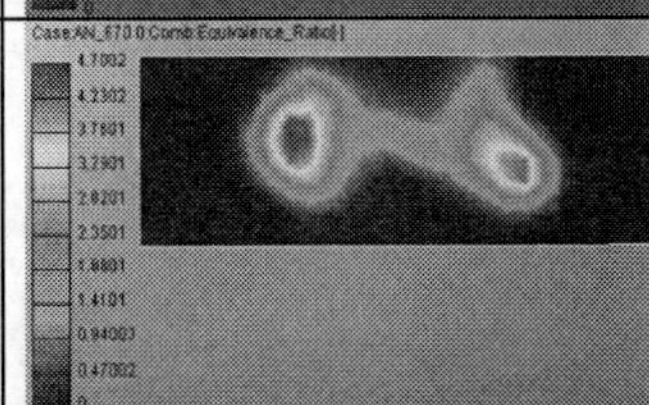
20	C	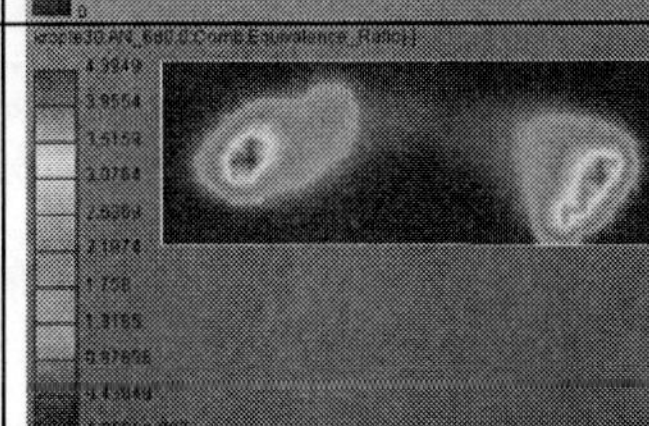
	S	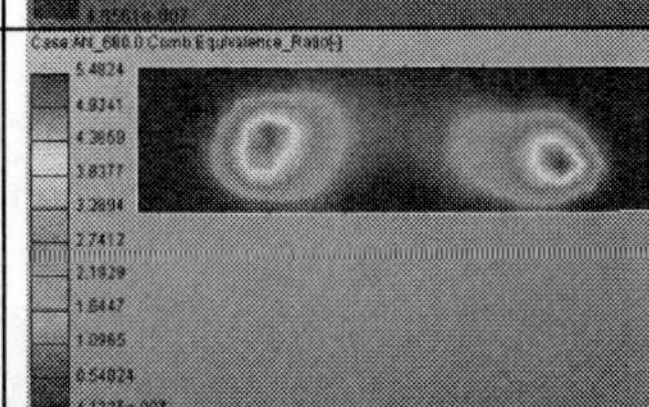
30	C	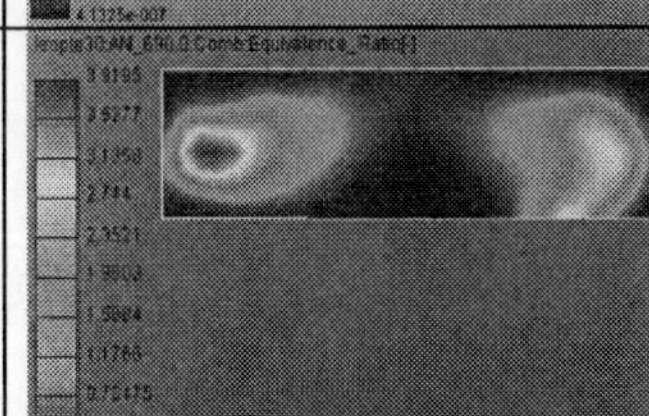
	S	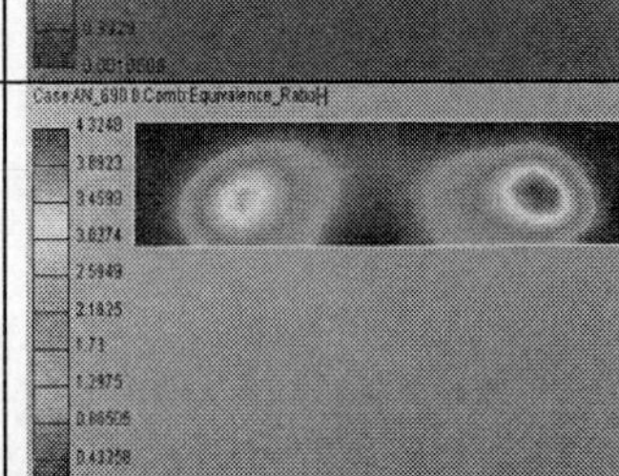

In simulations for the compression ratio of 11.6, the distribution of mean cylinder pressure is shown in Figure 10. In Szymkowiak's engine the pressure has a clearly wider range of high values which can have a positive effect on the torque distribution. The maximum of the described parameter is reached at 4.8 °CA after TDC in both engines. The values of the generated pressures, however, differ. In Szymkowiak's engine, the maximum pressure was 86.1 bar, while in the conventional engine as much as 90 bar.

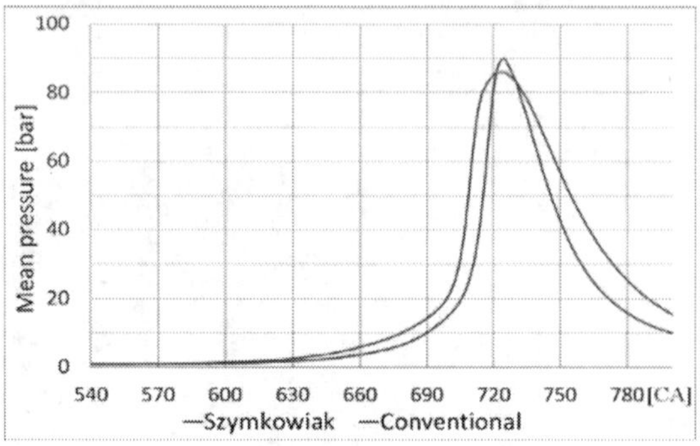

Fig. 10. Mean pressure in cylinder for ε=11.6

The average temperature prevailing in the cylinder (Figure 11) as in the case of ε=8.6 is higher in the compression stroke in the VCR engine. After exceeding 680 °CA, at which point the ignition occurred, the temperature in the initial phase of the combustion process is also higher in the swingarm engine. The reason is the accelerated heat release rate (Figure 13) which indicates a faster combustion process. The maximum temperature reached was 2349.6 K at 730 °CA in the case of Szymkowiak's engine and 2537.8 K at 726.3 °CA in a conventional engine. The difference was about 7.5% which means that the probability of nitrogen oxides formation in a conventional engine is much higher.

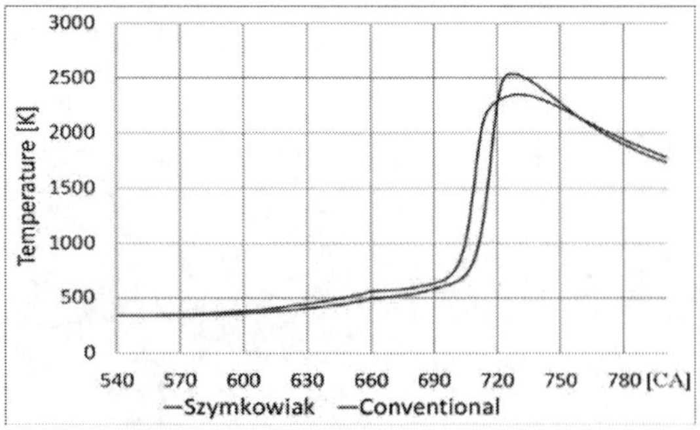

Fig. 11. Mean temperature in cylinder for ε=11.6

Evaporation of fuel for the compression ratio of 11.6 (Figure 13) also took less time in the case of Szymkowiak's engine – 32.5 °CA. The conventional engine process lasted 6.5 °CA longer. Increasing the compression ratio resulted in shortening the duration of fuel evaporation by 2 °CA in Szymkowiak's engine and by 19.1 °CA in the conventional one. The time required for evaporation of half the fuel mass was equal to 11 °CA in the VCR engine and 14 °CA in the conventional engine.

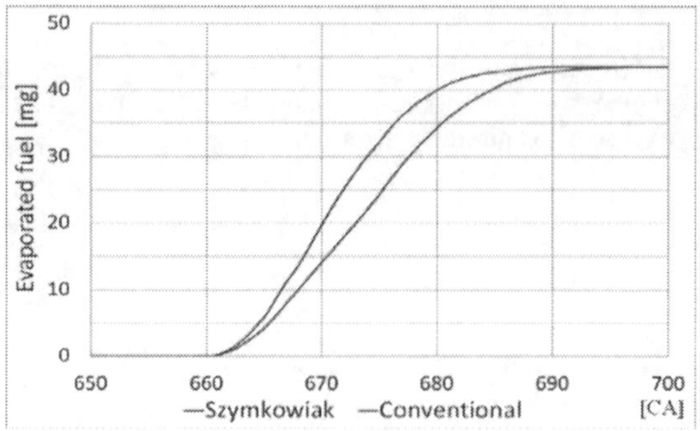

Fig. 12. Evaporation of injected fuel for ε=11.6

The heat release rate for ε=11.6 is shown in Figure 13. The graph shows that the combustion process took place faster in Szymkowiak's engine. The peak for the VCR unit was at 10 °CA before TDC and was 115.1 J/deg. In a conventional engine, the maximum achieved value is 24.3 units higher than for the swingarm engine and takes place at 2.6 °CA before TDC. The difference between the maximum values reaches 21%.

Accumulated Heat Release (fig. 14) for the conventional engine reached the final value of 1650.6 J. This result is 178.5 J greater than in the case of Szymkowiak's engine. The difference in the use of fuel energy at the level of 12% results in a significant reduction in the overall efficiency, remembering that this parameter is a product of several component efficiencies, including the thermal efficiency of the engine.

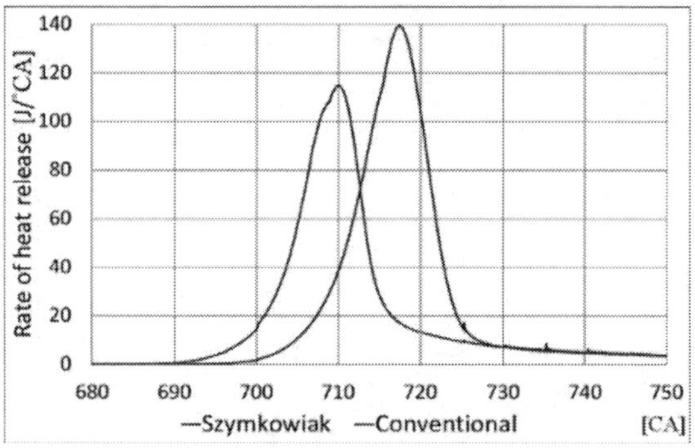

Fig. 13. Rate of heat release for ε=11.6

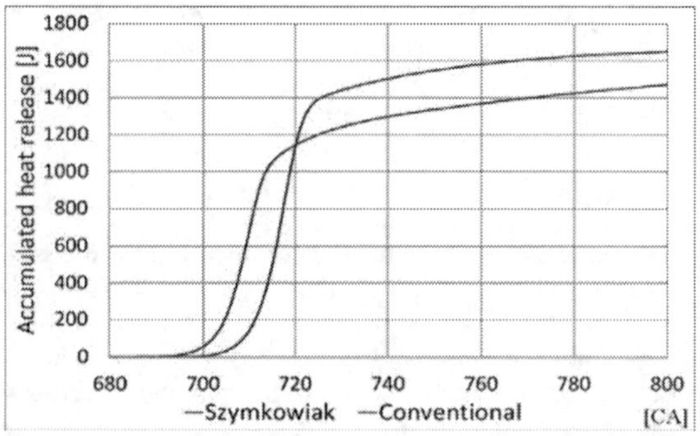

Fig. 14. Accumulated heat release for ε=11.6

5. Conclusions

The conducted simulations were aimed at the analysis and comparison of the thermodynamic processes taking place in Szymkowiak's and conventional engines. With the identical boundary conditions and the same moment of fuel injection for which the crankshaft position was the determinant, the most important conclusions drawn from the simulation are:

• The characteristics of the piston motion path has a significant impact on the Fuel Spray distribution in engines. In the Szymkowiak engine, due to the smaller volume of the combustion chamber, the probability of contact between the fuel spray and the elements forming the chamber is higher.

• The maximum pressure values obtained in simulations are quite similar to each other. The points at which these values were reached were also very similar. An interesting aspect is the cylinder pressure characteristics. In Szymkowiak's engine, the range of higher pressure values is wider. This means that despite the achievement of lower maximum values compared to a conventional engine, a wider range of higher pressure can translate into higher values of average useful pressure and, consequently, a better distribution of torque.

• Lower maximum temperatures in Szymkowiak's engine mean that the risk of nitrogen oxides being formed is significantly lower.

• The fuel evaporation rate for both tested compression ratio values is greater for Szymkowiak's engine. This means that the preparation of the fuel-air mixture will be faster in the swingarm engine.

• The maximum values and the location where heat is released were visibly different. The combustion process is faster in Szymkowiak's engine and it is more evenly spread over time. Accumulated heat release proves that in Szymkowiak's engine the use of fuel energy is slightly worse. This is the effect of the contact of the fuel stream with the walls of the combustion chamber.

• The assumed identical fuel injection timing for both engines, determined based on the crankshaft angle, is not optimal for Szymkowiak's engine due to the smaller volume of the combustion chamber and, as a consequence, worse conditions of the fuel stream propagation. For this reason, plans were made to perform simulations for fuel injection where the timing is determined based on the distance of the piston from TDC.

• Plans have also been made to compare the obtained simulation results to the results from a real engine. Such analysis will be aimed at verifying the accuracy and legitimacy of using a combustion processes simulation instead of real engine tests.

References

1. J. Heywood, *Internal Combustion Engine Fundamentals*. McGraw-Hill Education (2018)
2. T. Hoeltgebaum, R. Simoni, D. Martins, Elsevier, **96**, 308-322 (2016)
3. A. Shaik, N. Moorthi, R. Rudramoorthy, Professional Engineering Publishing LTD, **221**, 1159-1168 (2007)
4. S. Nagarajaa, K. Sooryaprakashb, R. Sudhakaran, Book Series: Procedia Earth and Planetary Science, **11**, 393-401 (2015)
5. P. Daszkiewicz, M. Andrzejewski, MATEC Web of Conferences, **118** (2017)
6. L. Rymaniak, MATEC Web of Conferences, **118**, (2017)
7. K. Andrzejczak, J. Selech, TRANSPORT PROBLEMS, **12**, 103-111 (2017)
8. M. Andrzejewski, D. Gallas, P. Daszkiewicz, A. Merkisz-Guranowska, H. Stawecka, MATEC Web of Conferences, **118** (2017)
9. L. Rymaniak, P. Daszkiewicz, J. Merkisz, Y.V. Bolzhelarskyi, AIP Conference Proceedings, **2078** (2019)
10. AVL Fire^{TM} Version 2017 – Combustion Module
11. AVL Fire^{TM} Version 2017 – Spray Module
12. I. Pielecha, **222**, 64-73, (2018)
13. I. Pielecha, W. Cieslik, JOURNAL OF THERMAL ANALYSIS AND CALORIMETRY, **126**, 815-827 (2016)

MATEC Web of Conferences **294**, 05002 (2019) https://doi.org/10.1051/matecconf/201929405002
EOT-2019

Contact arc time – important parameter of DC high-speed circuit-breakers

Artur Rojek*, and *Marek* Skrzyniarz

Railway Research Institute, Chlopickiego 50, 04-275 Warsaw, Poland

Abstract. The parameters of DC high-speed circuit-breakers (HSCB) and test methods are specified in the standards. These standards define requirements of the circuit breakers opening time and the total breaking time. During the direct current (DC) breaking after opening the contacts, the arc ignites, the voltage of which is so low that it does not limit the switched off current. The time in which this phenomenon occurs is defined as the contact arc time.
The contact arc time has a significant influence on the breaking process. It can take up to over 40% of arcing time. Therefore, tests were carried out, which were to answer the questions about what and to a what extent affects the contact arc time.
Since the standards for high-speed circuit-breakers do not refer to the contact arc time, definitions of this parameter are proposed in the article. Using this definition numerous tests of various factors have been carried out on contact arc time.
The conducted research shows that shortening the DC breaking time through high-speed breakers is possible by limiting the contact arc time, which can be obtained by modifying the circuit breaker's construction.

1. Introduction

High-speed circuit-breakers (HSCB) sorted in traction substations are subject to several requirements regarding their electrical and mechanical parameters as well as their durability and reliability. The malfunctioning of the circuit-breakers may cause railway infrastructure damage or failure, rail traffic disruptions and in extreme cases may even endanger human health or life.

Breaking time is one of basic HSCR parameters particularly affecting the protection of human life and health in case of short-circuits to the available elements.

Prior to HSCR placing in service, they are subject to numerous trials and laboratory tests. The basic reference documents relating to HSCR dedicated to working in traction substations and section cabins are EN 50123-1 [3] and EN 50123-2 [4] standards as well as EN 60077-3 [7] standard defining the requirements and tests for rolling stock breakers. The provisions of latter should be read together with EN 60077-1 [5] and EN 60077-2 [6].

Many publications have been devoted to issues connected with HSCR and short-circuit breaking, including [1, 2, 8, 9, 13, 14], and their tests [10, 12]. Only some of them deal with the breaking time problem and the correlation between that time and breaker's parameters, as well as circuit's parameters in which the breaker is working. The tests have shown that the parameters of the circuit and the breaking current have little influence on the contact arc time. The technical condition of the circuit breaker resulting from the number of breakings has an effect on the contact arc time, but much less than on the arcing time. The construction of the circuit breaker has the greatest influence on the tested parameter. In the case of some types of high-speed breakers, the value of the contact arc time causes that the breaking time and the limited short-circuit current have high values despite the minimum values of the circuit breaker's opening time. Tests were carried out on the contact arc time of such factors as the high-speed circuit-breakers' construction, the time constant of the circuit, the rate of rise of the breaking current and the expected peak value of the current. Tests have also been carried out examining the impact of the number of operating current breakings made by the circuit breaker on the contact arc time. Detailed results of these tests are presented below.

2. HSCR breaking time

EN 50123-2 [4] and EN 60077-3 [7] standards define breaking time t_b as the time from the moment when the current exceeds the trip setting level until the breaking current stops flowing. The breaking time t_b[1] is the sum of opening time t_i and arcing time t_a. The opening time t_i is defined as the time that passes from the time when the current exceeds the trip setting level until the contacts are opened by the circuit breaker. In turn, the arcing time t_a is the time counted from the moment the contacts are open until the current is turned off completely.

The opening time t_i depends largely on HSCB construction – the way it is tripped and the mechanism opening the contacts [12, 13, 15]. The arcing time t_a also depends on HSCB construction. Its duration is influenced by the construction of contacts, arc chute

[1] Symbols for parameters in compliance with EN 50123-1 [3]

*Corresponding author: arojek@ikolej.pl

parameters, the method of arc extinguishing and the method of magnetic blowout.

Due to breaking time and the capability to limit the current, EN 50123-1 [3] and EN 50123-2 [4] standards divide high-speed breakers designed for stationary equipment into 3 types, i.e. H, V and S. In accordance with provisions of the standard. HSCB of H type is supposed to prevent reaching the peak value by the short-circuit current, have opening time t_i not bigger than 5 ms, whereas the total break time t_b should not exceed 20 ms, when the value of sustained current is at least 7 times bigger than the trigger setting level and the initial steepness of the current rise has a minimum value of 5 kA/ms.

During arcing time t_a, the electric arc does not develop uniformly nor arc voltage u_a grows linearly. After the circuit breaker's contacts open, the arc is ignited between them. This arc is short and large in diameter which means that its voltage is low. This arc moves to the arc chute where it increases its diameter and the deionization takes place. From the time of the contacts opening to moment of the arc entering the arc chamber, the arc voltage is low and practically it does not cause any limitation of the short-circuit current. The author of the publication [14] defined the period of the low voltage arc appearance as the contact arc time t_s. EN 50123-1 [3], EN 50123-2 [4] and EN 60077-3 [7] standards neither characterize nor include this parameter, although it is an important factor affecting the speed of switching off the current and the degree of its limitation.

The duration of the contact arc time t_s and opening time t_i essentially affect the coordination of short-circuit protection in the traction vehicle – traction substation system [11].

For the purpose of testing and analyzing their results presented in the following sections, the contact arc time t_s is defined as the time from the moment of breaker's contacts opening (arc ignition between contacts) until the arc reaches voltage of $u_a = 0.1\ U_d$ (supply voltage of the circuit). Since in many cases the arc voltage u_a varies when the current is switched off, it rises and falls, the condition $u_a = 0.1\ U_d$ is deemed to be fulfilled if then there is no decrease of the voltage value below 0.1 U_d. The above definition of the contact arc time t_s is graphically presented in Fig.1.

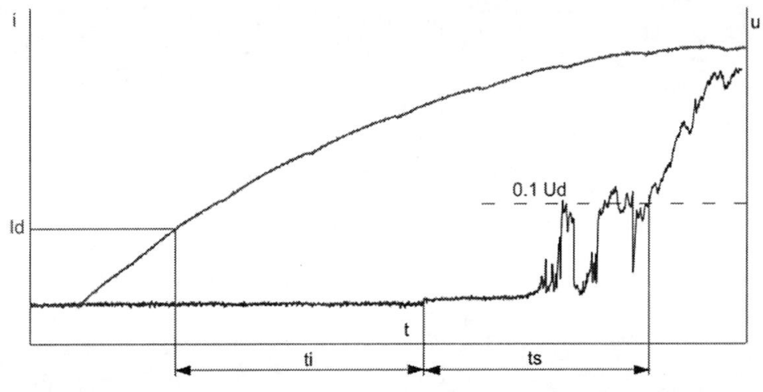

Fig. 1. Definition of contact arc time t_s.

3. Research of contact arc time

3.1. High-speed breaker's construction vs. contact arc time

Figures 2 – 4 present short-circuit current and voltage flows on three types of HSCB. All the flows were recorded during tests in the same short-circuit circuit and the tested breakers has identical setting levels. The breakers differed un main and arcing contacts construction, arc chute construction and magnetic blowout systems.

The analysis of flows presented in Figures 2 – 4 show that the current is still rising in the t_s time. Type 2 breaker, for which the flows are presented in Fig. 3, despite its shortest opening time is characterized by the longest t_s time. The consequence of this is the longest break time and the highest value of the limited (cut off) current. The breaker has a gap between the main and arcing contacts, the largest geometric dimensions of the arcing cones, and its arc chute does not contain conductive elements. This causes that the time required for the arc movement from the main contacts to the inside of the arc chute is the longest.

Paradoxically, the type 1 circuit breaker with the longest opening time t_i has the shortest t_s time (Fig. 2). This causes the limited short-circuit current $I_{cut\ off}$ to be the smallest. The breaker has a compact design of main and arcing contacts, short arcing cones and arc chute solutions allow for quick extension and extinguishing of the arc.

The parameters described in this section have a significant impact on the coordination of short-circuit protection in the vehicle-substation system. In order to ensure the highest level of coordination, the rolling stock breaker should limit the short-circuit current to the lowest level in the shortest possible time [11]. This means that in addition to the opening time, also the contact arc time should have possible small values. The higher the short-circuit current steepness, the higher the t_i and t_s time requirements are set for the rolling stock breakers.

Selecting circuit breakers only taking into consideration their opening time does not guarantee the coordination of protection. In the case of long t_s times, short-circuit current will rise to the value at which the breaker in the traction substation is tripped.

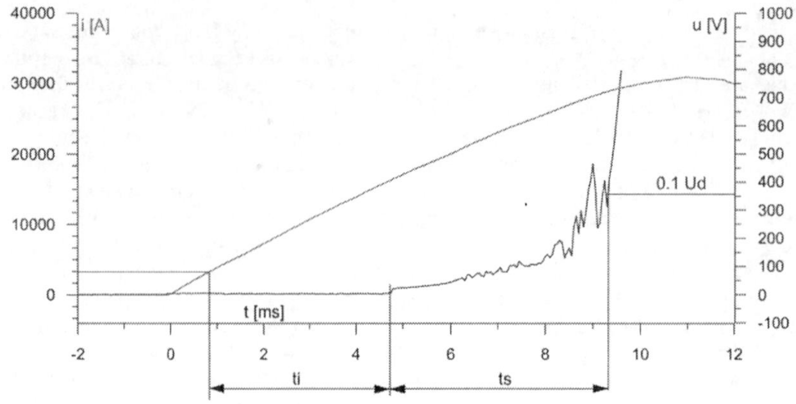

Fig. 2. Short-circuit current flows, voltage at HSCB type 1:t_i = 3.9 ms; t_s = 3.6 ms; $I_{\text{cut off}}$ = 30.9 kA, t_b = 23 ms.

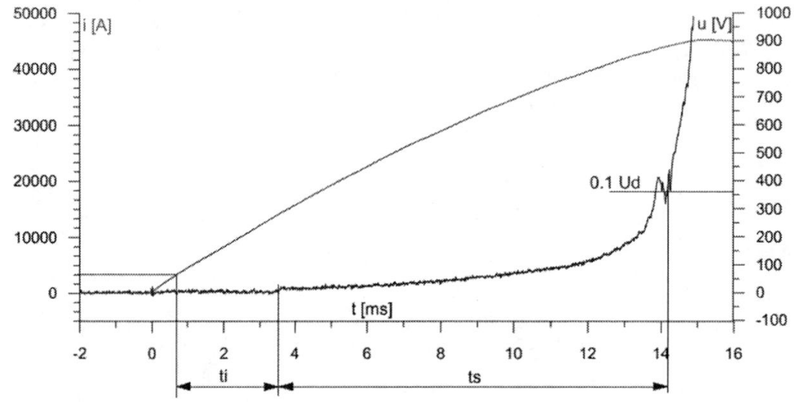

Fig. 3. Short-circuit current flows, voltage at HSCB type 2:t_i = 2.8 ms; t_s = 10.6 ms; $I_{\text{cut off}}$ = 45.3 kA, t_b = 39 ms.

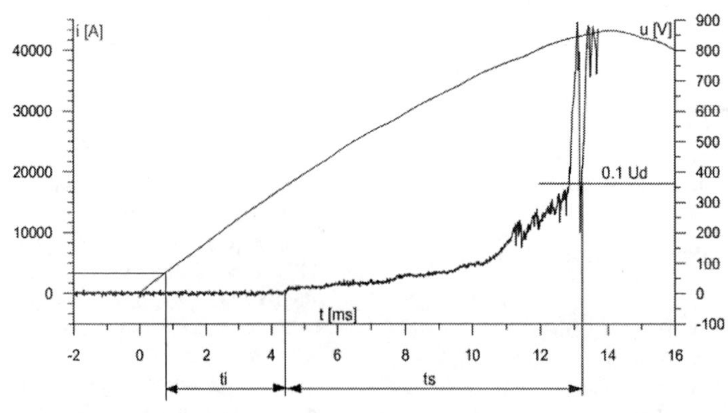

Fig. 4. Short-circuit current flows, voltage at HSCB type 3: t_i = 3.6 ms; t_s = 8.8 ms; $I_{\text{cut off}}$ = 43.2 kA, t_b = 26 ms.

3.2. The influence of the breaking current and circuit parameters on contact arc time

The tests were carried out on a system powered by a 6-pulse rectifier set. The values of resistance and inductance in the circuit were selected so that in each test a similar value of the time constant of the circuit would be obtained. This means that with the rise in the expected sustained short-circuit current I_{SS}, the steepness of the short-circuit current di/dt rose. The setting of the breaker in all tests was I_d = 1500 A.A HSCB blowout magnet was used for the tests with a rated voltage of 3.6 kV, fitted only with main contacts and a blowout coil.

Three series of tests were carried out: at a constant time $t_c \approx$ 16.4 ms (15.4 – 17.1 ms), $t_c \approx$ 9.2 ms (7.9 – 10.8 ms) and a system without additional glands (t_c from 0. 8 to 3.1 ms). On the basis of recorded voltage flows on the circuit-breaker and short-circuit current, for each test the contact arc time t_s and the arcing time t_a were determined. On the basis of the obtained results, correlations were determined:

- t_s = f(di/dt);
- *ts/ta = f(di/dt);*
- *ts = f(tc);*
- *ts/ta = f(tc);*
- *ts = f(Icut off);*
- *ts/ta = f(Icut off).*

The flows of the aforementioned correlations are presented in Figures 5 - 10. The results of measurements marked with points were approximated by a linear function. Examples of voltage flows on the breaker and the switched off current are shown in Figures 11 - 14.

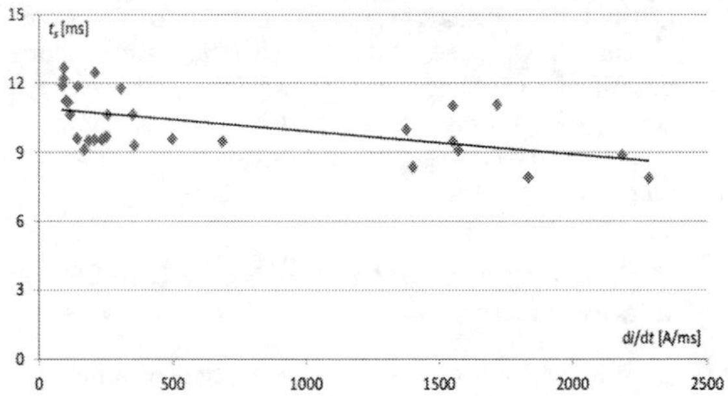

Fig. 5. Contact arc time t_s correlation with the speed of the current rise di/dt.

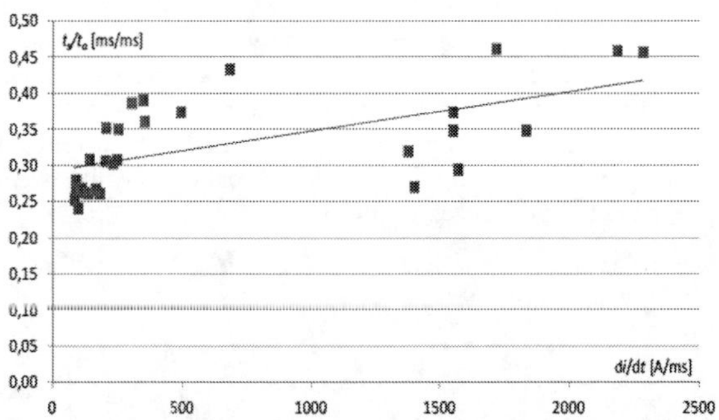

Fig. 6. Contact arc time t_s and arcing time t_a correlation with the speed of the current rise di/dt.

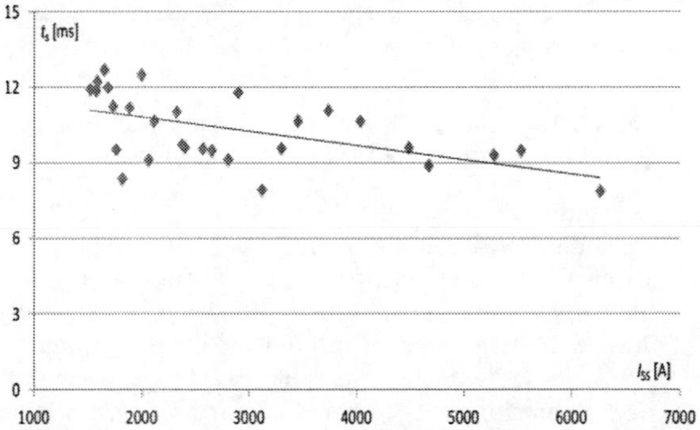

Fig. 7. Contact arc time t_s correlation with the prospective sustained short-circuit current I_{SS}.

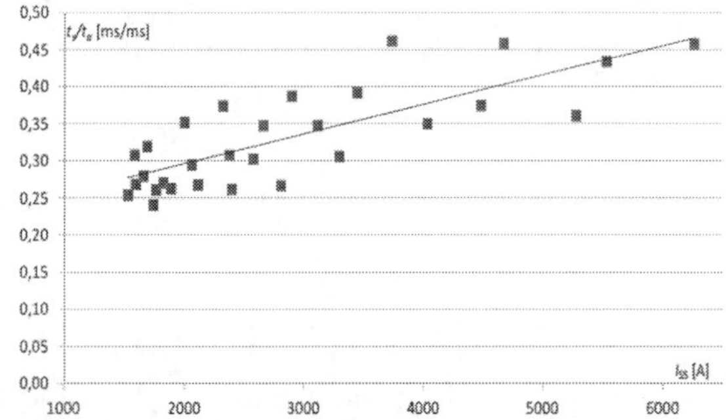

Fig. 8. Contact arc time t_s and arcing time t_a correlation with the prospective sustained short-circuit current I_{SS}.

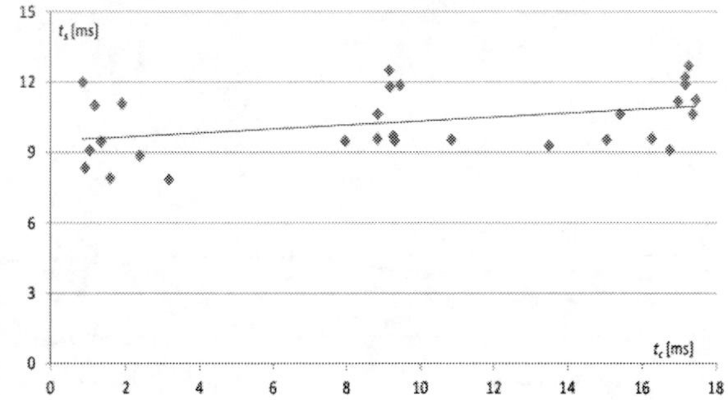

Fig. 9. Contact arc time t_s correlation with the circuit time constant t_c.

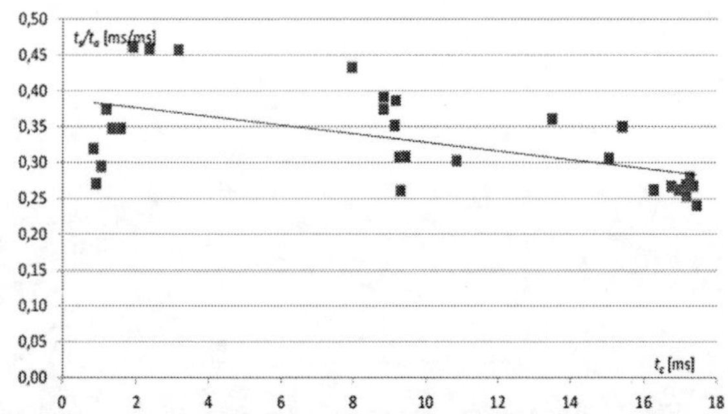

Fig. 10. Contact arc time t_s and arcing time t_a correlation with the circuit time constant t_c.

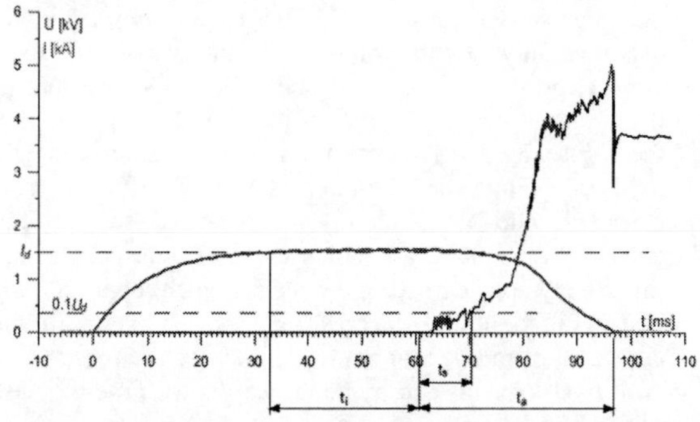

Fig. 11. The voltage flow on breaker's contacts and switch-off current at: di/dt = 143 A/ms, t_c = 9.4 ms [13].

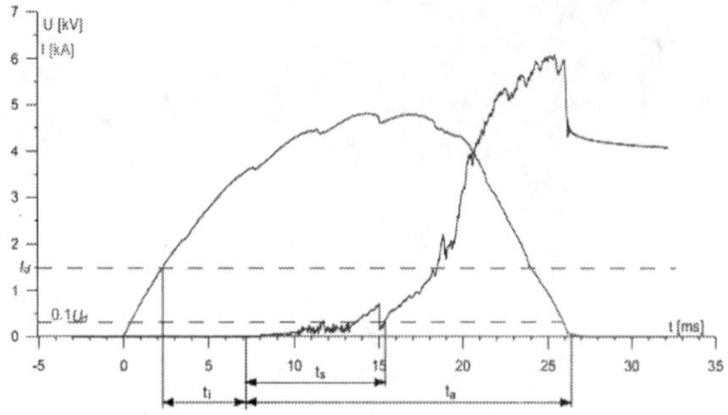

Fig. 12. The voltage flow on breaker's contacts and switch-off current at: di/dt = 353 A/ms, t_c = 13.5 ms [13].

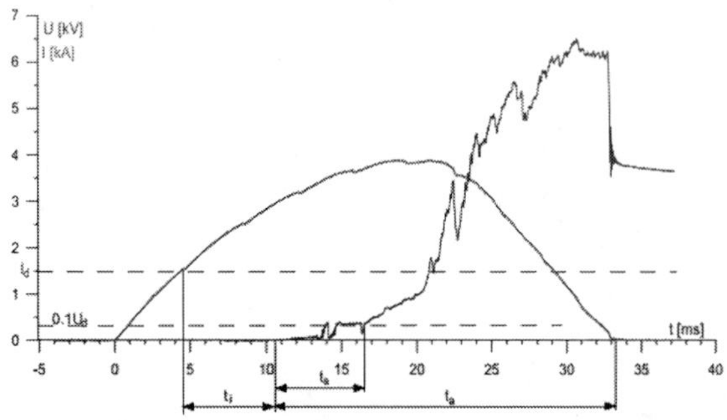

Fig. 13. The voltage flow on breaker's contacts and switch-off current at: di/dt = 686 A/ms, t_c = 7.9 ms [13].

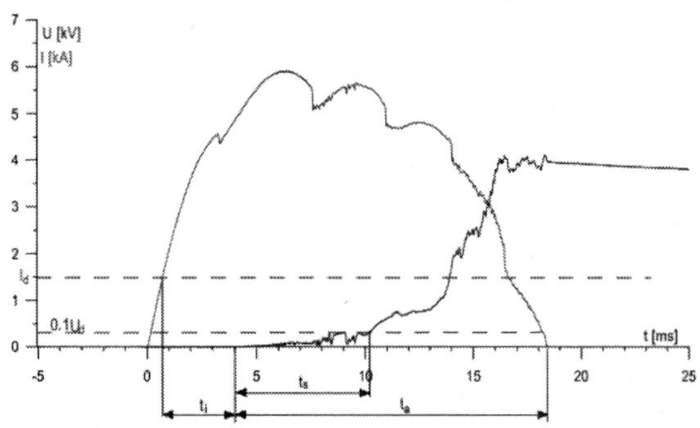

Fig. 14. The voltage flow on breaker's contacts and switch-off current at: di/dt = 2283 A/ms, t_c = 3.2 ms [13].

On the basis of Figures 5, 7 and 9 it can be concluded that the contact arc time t_s of the tested circuit breaker depends only slightly on the parameters of the short circuit and the current. The rise in the time constant of the t_c circuit causes the rise of the t_s. It is estimated that the t_c rise by 1 ms causes the t_s rise by approximately 0.08 ms. However, the increase in the di/dt current steepness and the value of the expected short-circuit current I_{SS} result in a small decrease in the contact arc time t_s. By approximating the measurement data with the linear function, the decrease can be estimated in the contact arc time t_s by about 1.1 ms with the increase in di/dt by 1 kA / ms and by about 0.5 ms with the increase of the I_{SS} by 1 kA.

The parameters of the circuit and short-circuit current affect the arcing time t_a, which value rises with the increase of the time constant of the t_c circuit and drops with the increase of di/dt and I_{SS} [13, 14]. The influence of these parameters on the arcing time value is much stronger than on the contact arc time t_s. Therefore, the t_s/t_a diagrams in the di/dt function (Fig. 6), I_{SS} (Fig. 8) and t_c (Fig. 10) are inverse in character than those shown in Figures 5, 7 and 9.

Analyzing the measurement results, the value of the current assumed as the end of the contact arc time was also determined, thus when the arc voltage u_a reaches 0.1 U_d (See: Fig. 1). In a series of measurements at arcing time $t_c \approx 16.4$ ms at $u_a = 0.1\ U_d$ had a value less than the maximum current of 0.6% to 3% (on average 1.5%). In the case of the time constant $t_c \approx 9.2$ ms the short-circuit current was lower by 0.3% to 3.3% (on average 1.4%).

The afore mentioned analysis for series of measurements for small values of time constant was not possible. This resulted from the character of short-circuit current flows (See: Fig.14). Due to strong current pulsations, the change in its value within 1 ms can reach almost 1.2 kA. This means that in case of t_s shorter or longer by 1ms, a significant difference in the short-circuit current momentary value can occur.

3.3. The influence of the number of breakings on contact arc time

The aim of subsequent tests was to determine if the number of breakings performed by the high-speed circuit breaker has an effect on the contact arc time of this circuit breaker. These tests were carried out using a new high-speed circuit breaker. 300 breakings of the operating current of approximately 1400 A were carried out in a circuit with a time constant tc = 15 ms. During the tests, no service and maintenance of the breaker were carried out. Examples of the switched off current and voltage flows at the circuit breaker terminals are presented in Figures 15 and 16.

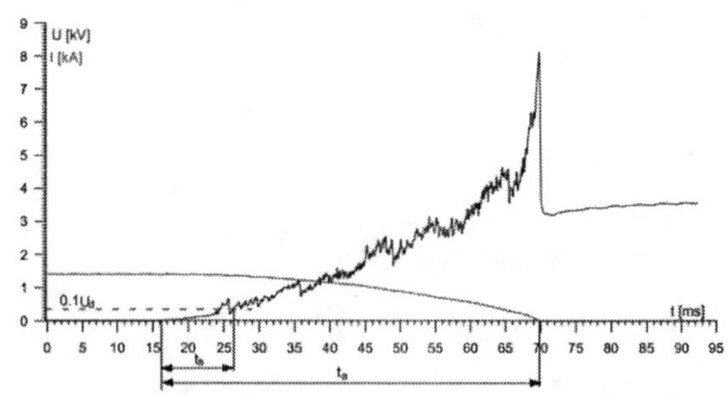

Fig. 15. The voltage flow on breaker's contacts and switch-off current – 1. breaking: t_s = 10.05 ms, t_a = 53.50 ms.

Figure 17 shows the changes of the arcing time t_a depending on the number of breakings n, whereas Fig. 18 presents the effect of the number of breakings n on the length of contact arc time t_s. Results of tests of every 20 breakings are shown in both pictures. According to predictions, the arcing time t_a increased together with the rise of breakings n. This is due to the degradation of contacts and arcing chute, which together with the growth of the wear rate has a smaller potential for extinguishing the arc.

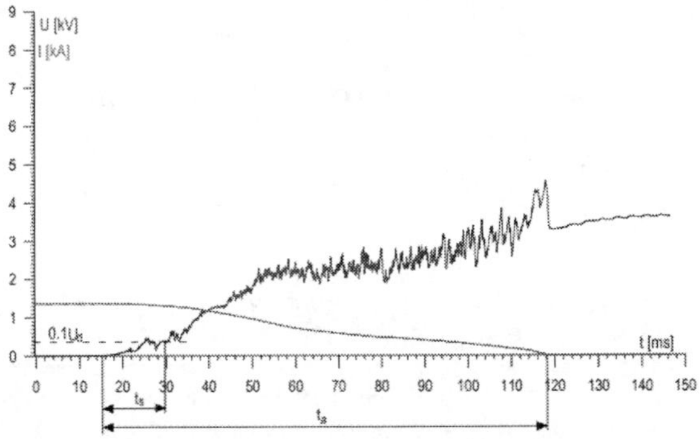

Fig. 16. The voltage flow on breaker's contacts and switch-off current – 300. breaking: t_s = 14.30 ms, t_a = 102.85 ms.

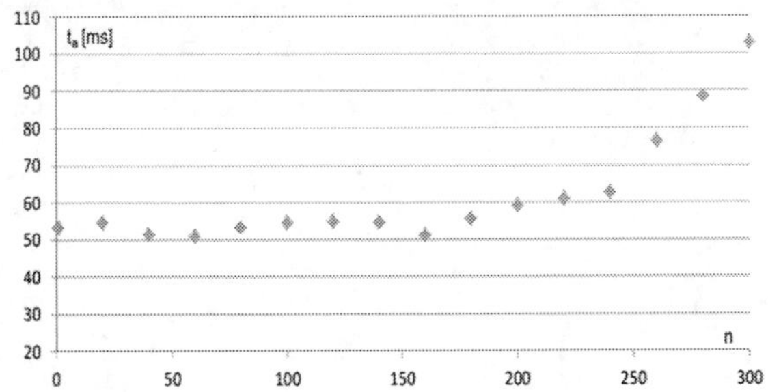

Fig. 17. Contact time t_a correlation with the number of breakings n.

Analyzing the measurement results in terms of contact arc time t_s , its rise was also noticed (Fig. 18) together with the growth of the number of breakings n. Comparing the contact arc time during the first and 300th breaking, t_s increased from 10.05 ms to 14.30 ms, i.e. by about 42 %, whereas the arcing time increased by about 92%. Approximating the measurement results of contact arcing time with the linear function, it can be estimated that t_s increases by about 1.4 ms for every 100 breakings. That means that together with the rise of the number of breakings n, the contact arc time t_s has a smaller share in the arcing time t_a.

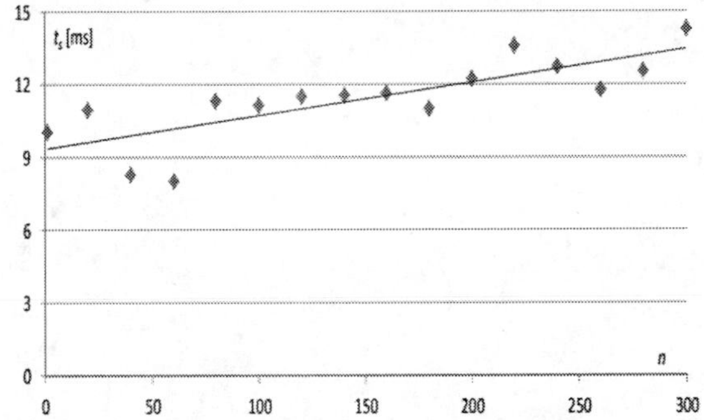

Fig. 18. Contact arc time t_s correlation with the number of breakings n.

The above correlations change when frequent high-speed breaker's service and maintenance activities are conducted.

Analyzing the results of the research on the influence of the number of breakings on the value of time ts, the value of the current at the time considered as the end of the contact arc time was also determined. In this series of measurements at short-circuit current at u_a = 0.1 U_d had a value lower than the maximum current by 0.3% to 4.3%, with the average short-circuit current limitation was 2.0% for the first 100 breakings, 2.7% for breakings from n = 101 to n = 200 and 3.3% for breakings from n = 201 to n = 300.

4. Summing up and conclusions

The article discusses the issue of time when the contact arc appears while DC breaking by a high-speed circuit breaker. This parameter is not defined in standards that deal with high-speed circuit breakers and their tests. This parameter is extremely important, among others, to achieve proper coordination of short–circuit protection in the traction substation–traction vehicle system. For this reason, a proposal is presented for the definition of contact arc time which is defined as the time from HSCB contacts opening until the arc reaches the voltage of 0.1 supply voltage. Using this definition, tests of this time and its correlation with other variables were carried out.

The test results presented in his article that the contact arc time t_s is a significant part of the arcing time t_a . The length of the arcing time t_a is influenced by many factors, including short-circuit parameters, switched off current and the breaker's construction.

The contact arc time t_s depends on the high-speed breaker's construction. This time can constitute even 33% of the total break time t_b and is longer than the breaker's opening time t_i. When the contact arc time t_s appears, the short circuit current is practically unlimited. For circuits of time constant t_c of over 7 ms the current limitation during contact arc does not exceed 4% of the current maximum value. For this reason the contact arc time strongly affects the potential of short-circuit current's limitation and the break time.

The contact arc time t_s depends in a very small degree on the time constant t_c, the speed of the current rise di/dt and expected value of short-circuit current I_{SS} switched off by a high-speed breaker. Basing on the conducted research it has been estimated that the rise of circuit time constant t_c of 1ms causes the extension of the contact arc time t_s by the time measured in a fraction of a millisecond for every 1 kA/ms or 1 kA.

The HSCB technical condition affects the contact arc time $t_{s,}$, however, it is much smaller than the influence on the arcing time t_a . The test results showed that performing 300 breakings of operating current caused the rise of t_s time in the degree more than half of the arcing time t_a .

Therefore it can be concluded that in order to significantly shorten the contact arc time $t_{s,}$, it is necessary to analyze the HSCB construction as well as elements responsible for the emergence of the arc in the arc chute.

References

1. Bartosik, M., Lasota, R., Wójcik,, F.: *New generation of ultra-high-speed current-limiting circuit – breakers for electric traction of direct current.* TTS Technika Transportu Szynowego no. **7-8**, 80 – 88 (2000).
2. Bartosik,M., Lasota, R., Wójcik, F.: *High speed circuits breakers type DCN for substation.* TTS Technika Transportu Szynowego no. **10**, 55 – 59 (2003).
3. EN 50123-1:2003. Railway applications. Fixed installations. D.C. switchgear. Part 1: General.
4. EN 50123-2:2003. Railway applications. Fixed installations. D.C. switchgear. Part 2: D.C. circuit breakers.
5. EN 60077-1:2002. Railway applications – Electric equipment for rolling stock. Part 1: General service condition and general rules.
6. EN 60077-2:2002. Railway applications – Electric equipment for rolling stock. Part 2: Electrotechnical components – General rules.
7. EN 60077-3:2002. Railway applications – Electric equipment for rolling stock. Part 3: Electrotechnical components – Rules for d.c. circuit-breakers.
8. Maciołek, T., Mierzejewski, L., Szeląg, A.: *Breaking the short-circuits in DC traction by high-speed circuits breakers and power-switch.* TTS Technika Transportu Szynowego no. **3**, 40 – 47 (2001).
9. Mierzejewski, L., Szeląg, A.: *Research of short circuits and overvoltages in DC electric traction circuits with single-stage transformations110/3 kV.* 4th International Conference MET'99, pp. 58 – 60, Warsaw (1999).
10. Rojek, A., Sidorowicz, M.: *Research and tests of high-speed circuit breakers for rolling stock and substations in 3 kV DC traction power system.* Railway Reports **159**, 7 – 26 (2013).
11. Rojek, A., Zbieć, A.: *Co-ordination of Short Circuit Protection Devices in Motive Power Unit - Sub-station System.* Railway Reports **159**, 27 – 46 (2012).
12. Rojek, A.: *Parameters of DC high-speed circuit-breakers.* MATEC Web of Conferences **180**, 06006 (2018).
13. Rojek, A.: *Switching-off time of electromagnetic blow-out high-speed DC circuit breaker.* Railway Reports **169**, 53 – 60 (2015).
14. Wójcik, F.: *Theoretical analysis of possibilities for improvement in the effective operation of DC magnetic blow-out circuit breakers.* Electrical Review R. 87 no. **5/2011**, 186 – 192 (2011).
15. Sechéron Homepage, www.secheron.com, last accessed 2019/03/18.

Identification of the railway reconstruction parameters at imposition of high speed traffic on the existing lines

Iryna Lebid[1,*], *Irina* Kravchenya[2], *Tatyana* Dubrovskaya[3], *Nataliia* Luzhanska[4], *Mykola* Berezovyi[5], and *Yevhen* Demchenko[5]

[1] National Transport University, Department of International Transportation and Customs Control, Mykhaila Omelianovycha-Pavlenka Str., 1, Kyiv, Ukraine, 01010
[2] BelSUT, Department «Road and Traffic Management», Kirov str., 34, Gomel, Belarus, 246653
[3] BelSUT, Department «Engineering, construction, and maintenance of transport facilities», Kirov str., 34, Gomel, Belarus, 246653
[4] National Transport University, Department of Transport Technologies, Mykhaila Omelianovycha-Pavlenka Str., 1, Kyiv, Ukraine, 01010
[5] Dnipro National University of Railway Transport named after Academician V. Lazaryan, Transport Junctions Department, Lazaryan St., 2, Dnipro, Ukraine, 49010

Abstract. The problems of railroad design technical reconstruction aimed at ensuring train protection in introduction of high-speed traffic on the rail lines have been addressed. The description of railroad technical system with certain specific features has been proposed. The pair of mutually dual problems regarding the optimal reconstruction of railway curves for raising train speeds at minimal capital investments has been addressed. The problem of determining the optimal speeds of trains of each category under optimum outer rail cant, increase of the unbalanced acceleration value, compliance with the condition of uniform load on rail tracks, comfortable carriage of passengers, in which reduced costs on increased speeds in curved tracks will be minimal has been formulated. To achieve the goals of the study, mathematical optimization techniques and algorithms of analysis of the rail lines of different nature are used. This, by finding and value engineering optimization, allows to minimize high-value and lengthy procedures of physical simulation. As a problem-solving technique, Lagrange multiplier method and its applicability have been proposed. Numeral solution to the problem of selection of certain stretches for increasing rolling stock speeds on operating lines without reconstruction or with minimum expenses on reconstruction of contour of track has been reviewed.

1 Introduction

One of the priorities of contemporary railway is raising train speeds on existing rail lines that is inseparably linked to assuring safety and comfortable carriage of passengers.

Economic and social development of the country is directly related to transport development. The economic growth has been accompanied by an increase in the demand for high-speed transport, especially railway one, as it provides citizens of the country with freedom of movement at minimal time expenses. The increase in train speeds produces considerable economic effect through increased line carrying capacity, lower operating costs, labour costs reduction (for locomotive and train crews), reduction of necessary car and locomotive fleet, and costs of fuels and electricity.

One of the major challenges of contemporary railway is reducing travel time of passengers on transregional lines to the maximum permissible, taken in accordance with the condition for ensuring minimal fatigue of organism. For the Republic of Belarus, this criterion shall be 3–5 hours per one way.

Despite the relevance of the problem of the high-speed operation introduction, it is quite difficult to implement it. This can be explained by the lack of the rolling stock, range territory for researches, and regulatory framework, which has just being established.

The works [1-5] are devoted to express and high-speed traffic in foreign countries. The works [6-13] are devoted to the possibility of implementation of high-sped operation on the existing lines.

Features of creation on the railway roads of Ukraine a network of high-speed railway lines are described in the work [6].

The calculation methodology of railway track when interacting with high-speed rolling stock is proposed by D. Kurgan in the work [7].

Theoretical foundations of designing express and high-speed railway lines without reconstruction or with minimal reorganization of existing lines are formulated by A. Gavrilenkov. In the work [9] he noticed that the improvement in design of track contour relates to an increase in the circular curve radius and rail cant.

Analysis of the impact of the unbalanced acceleration on the speed of passenger trains movement is considered

* Corresponding author: i.h.lebed@gmail.com

by P. Kovtun and T. Dubrovskaya. The work [10] proposes one of the options for increasing the speed of passing curves by increasing the permissible limit of the unbalanced acceleration.

The method of determining the technical parameters of the reconstruction of the line for the transition to high-speed passenger trains in mixed freight and passenger traffic, taking into account the uncertainty of the initial information is given in the work [11].

2 Train safety control on railroad transport

The railways must provide not only public demand in passenger transportation, but also passenger safety while travelling on railroad transport.

The specific instruction, which obliges responsible railway workers to pay particular attention to the train safety control, has been developed in order to ensure safe operation of high-speed passenger trains. Special aspects of ensuring safe operation of high-speed passenger trains have been reflected in the various sections, numerous instructions, directions, orders that is poorly adapted for use and awkward while technical training, instructing and trial work period.

Each railway worker connected to handling and service of passenger trains is obligated to fulfil particularly all the operations on preparing entrance, exit and pass routes of passengers trains.

Train safety control in introduction of high-speed operation is the most significant social and economic objective oriented to health care and human security.

Rail traffic safety means an ability of train operation to remain in non-hazardous condition during estimated time, when there is no any life and health threat to passengers, technical stuff, population, safety of cargo, business facilities and transport system technical means.

Increased locomotive capacity and speed used for train traffic lead to increased rate of negative consequences of traffic accidents. For example, the consequences of train or wagon derailment at speeds of 90 km/h and 160 km/h will differ significantly.

Increased magnitude of traffic flow on railroad transport increases the likelihood of negative consequences. Thereof, the more trains pass in unit time, the greater will be the total delay time of trains involuntarily idle due to the emergency stop of one train.

The difficulty of conducting rail traffic safety studies stems from the fact that tens of thousands of people of various specialties, whose activities are connected by unified technological process, take part in the process of transport organization and operation. And traffic safety depends both on those people actions and on technical condition of transport means under their authority.

During research, it is necessary to take into account the linkages and synergies of elements, which may be conditionally combined into four groups:

– human resources (psychophysiological status, professional qualifications, etc.);
– rolling stock (specifications, technical status, etc.);
– rail track (profile, technical status, etc.);
– traffic environment (service position, operation of signalling, communication and electrical supply devices, meteorological conditions, etc.).

Thus, the introduction of high-speed traffic requires comprehensive survey of all the different aspects operating within framework of this system: institutional, technical, psychological, social et. al. The present work reflects an attempt to apply such systematic approach to the problem of train safety control on railroad transport.

3 The relevance of computer optimization methods in introduction of high-speed traffic

Contemporary state of the theory in relation to railroad transport is characterized by the presence of developed system of mathematical models and algorithms for analysis of various characteristics of railroad lines.

An increase in train speeds on existing rail lines using computer engineering means (computer aided design) is one of the major tasks assigned to railroad transport for the foreseeable future.

The use of information technology makes it possible to provide the whole decision-making process at all stages of project development on railroad line reconstruction for high speeds with on-line informational support. Implementing such kind of support requires data source management, the choice of the best features of information model and an effective design management apparatus.

The need for modelling of reconstruction design process itself for express and high-speed railroad lines arises in order to provide a systems analyst with the description tools of design technology being developed.

Application of computer optimization methods in reconstruction of express and high-speed railroad lines will make it possible to search for the optimal solutions in the event of certain statement of the problem without significant material expenses that at this time are one of the most important criteria of each research.

4 The railroad as a simulation object

Being a simulation object, the railroad possesses the following features [9, 13]:

– unpredictability – behavior of subsystems of railroad system is not fully known to researcher. The knowledge of railroad system behavior in the past does not allow reliably predicted its future behavior. Therefore, any system research is carried out at high degree of uncertainty. Herewith, there is high probability of behavior multivalence of a single subsystem under similar conditions and external effects on railroad system;
– uniqueness – none of the existing rail lines has no parallel in respect of the parameters for the system itself and influence of external environmental on it. Hence the need for building of the model for each rail line separately;

– integrity, i.e. the need to consider railroad system as a whole object allowing for decomposition into subsystems during the modelling phase. There is certain difficulty in the multilevel railroad system decomposition connected with different scale of the processes in time ongoing on various hierarchy levels of component communication of the railroad system. Moreover, each component possesses its own "intellect" behavior in the form of complex decision-making and transferring information algorithm in accordance with hierarchy of control actions on components of lower performance level;
– non-interaction, i.e. the presence of autonomous space-time metric and autonomous principle of statefulness within the railroad system;
– modelability – railroad system may be presented by finite set of the models, each of which reflects a certain feature of its essence.

5 Problem statement

Increase in speeds, while insuring safety and maintaining maximum facilities for passengers, might be accessed by using new rolling stock, strengthening track superstructure, and improving a rail line design.

Suppose railroad as a technological system. The speed on the line is limited by capabilities of technical facilities. Speed limitations vary along the length of the line.

It is necessary to split the line into m consistent subsystems – stretches (railway hauls and interstations) with constant speed limit within them.

The stretch is characterized by parameter set of technical facilities (superstructure, including track switches, constructional works) determining speed limitations on the stretch [12].

After reconstruction activities and fixed equipment modernization of varying degrees (replacement of track switches, reinforcement of bridges, etc.) maximum limit speeds on the scratches must be increased up to 120, 140 or 160 km/h respectively.

The improvement in design of track contour relates primarily to an increase in the circular curve radius, rail cant, as well as the change of the unbalanced acceleration value.

Since, in reality, train flows are characterized by a great heterogeneity (high-speed, express, suburban, freight, long, heavy-tonnage, etc.), there is also a need for determination of optimal speeds of trains of each category within every line stretch.

The objective of this study is to find a common approach to solving two problems:
– the definition of optimal values of design radii increasing the speeds of trains on existing lines at minimal cost of reconstruction and maximum travel time reduction,
– the determination of optimal speeds of trains of each category with the introduction of high-speed traffic on existing railway lines with a change of the unbalanced acceleration value from 0,7 m/s^2 to 0,9 m/s^2 and minimal reduced expenditures.

The tasks should be solved with the unconditional safety and comfort of the passengers.

6 Determination of optimal railway curve radii in introduction of high-speed traffic

Travel time reduction ΔT serves as a quantity indicator of speed increase technical efficiency on the stretch, while capital investments K aimed at fixed equipment modernization is a quantitative indicator of economic efficiency.

Curve radii increase leads to higher train speeds and, as a consequence, travel time reduction ΔT. However, the more curve radius, the higher a need in capital investments K for line reconstruction. In reality, capital investments on reconstruction are limited $K \le K_0$. The restriction can also be imposed on travel time reduction $\Delta T \ge \Delta T_0$.

Let us consider several mutually dual problems regarding the optimal reconstruction of railway curves for raising train speeds at minimal expenditures.

6.1 The problem of optimal radii determination at the maximum travel time reduction

Assume, there is a railway stretch on which m independent (single-radius and compound) curves are located. Each i-curve $(i=\overline{1,m})$ has its:
– curvilinear distance l_i;
– speed limitation within this stretch v_i;
– angle of curvature α_i;
– capital investments K_i for line reconstruction per unit length of curve;
– parameter a dependent on rail cant and acceptable value of unbalanced acceleration.

The aim is to determinate the values of such design radii R_i restricting the speed on curves in which capital investments K will be equal to set K_0, and the travel time reduction ΔT will reach its maximum.

$$\Delta T = \sum_{i=1}^{m} l_i \left(\frac{1}{v_i} - \frac{1}{a\sqrt{R_i}} \right) \to \max \qquad (1)$$

when:

$$\sum_{i=1}^{m} K_i \alpha_i R_i^2 = K_0 \qquad (2)$$

To solve the problem of interest, there is a good reason to apply Lagrange multiplier method [14, 15]. First, compose Lagrange function:

$$L(R_i,\lambda) = \sum_{i=1}^{m} l_i \left(\frac{1}{v_i} - \frac{1}{a\sqrt{R_i}} \right) + \lambda \left(K_0 - \sum_{i=1}^{m} K_i \alpha_i R_i^2 \right), \quad (3)$$

where λ – Lagrange multiplier representing how much increase in the value of capital investments K_0 per unit will change the maximum reduction of travel time ΔT in optimal resolution while increasing capital investments K_0 per unit.

Let us identify partial derivative of Lagrange function for unknowns R_i $(i=\overline{1,m})$ and λ, and set them to zero. As a result, we obtain the following system of equations:

$$\begin{cases} \dfrac{\partial L(R_i,\lambda)}{\partial R_i} = \dfrac{l_i}{2aR_i^{3/2}} - 2\lambda K_i \alpha_i R_i = 0, \\ \dfrac{\partial L(R_i,\lambda)}{\partial \lambda} = K_0 - \sum_{i=1}^{m} K_i \alpha_i R_i^2 = 0. \end{cases} \quad (4)$$

After solving the system of equations (4) in unknowns R_i and λ for the set value of the capital investments K_0, optimal values of design radii R_i (5) and the maximum reduction of travel time ΔT (6):

$$R_i = \left(\frac{l_i}{K_i\alpha_i}\right)^{2/5} \frac{K_0^{1/2}}{\left(\sum_{i=1}^{m} l_i^{4/5}(K_i\alpha_i)^{1/5}\right)^{2/5}} \quad (5)$$

$$\Delta T = \sum_{i=1}^{m} l_i \left(\frac{1}{v_i} - \frac{1}{a}\left(\frac{K_i\alpha_i}{l_i}\right)^{1/5} \frac{\left(\sum_{i=1}^{m} l_i^{4/5}(K_i\alpha_i)^{1/5}\right)^{1/5}}{K_0^{1/4}} \right) \quad (6)$$

6.2 The problem of optimal radii determination at the maximum capital investments

The following setting of the dual problem of optimal reconstruction of railway curves could be considered:

It is necessary to define values of radii R_i $(i=\overline{1,m})$ restricting the speed on curves in which travel time reduction ΔT will be equal to set ΔT_0, and the capital investments K will be minimal:

$$K = \sum_{i=1}^{m} K_i \alpha_i R_i^2 \to \min \quad (7)$$

when:

$$\sum_{i=1}^{m} l_i \left(\frac{1}{v_i} - \frac{1}{a\sqrt{R_i}} \right) = \Delta T_0 \quad (8)$$

Let us compose Lagrange function:

$$L(R_i,\lambda) = \sum_{i=1}^{m} K_i \alpha_i R_i^2 + \\ + \lambda\left(\Delta T_0 - \sum_{i=1}^{m} l_i \left(\frac{1}{v_i} - \frac{1}{a\sqrt{R_i}} \right)\right), \quad (9)$$

where λ – Lagrange multiplier, representing how much the change of travel time reduction ΔT_0 per unit will decrease the value of capital investments K in optimal resolution.

As a result, we obtain the following system of equations:

$$\begin{cases} \dfrac{\partial L(R_i,\lambda)}{\partial R_i} = 2K_i\alpha_i R_i - \lambda \dfrac{l_i}{2aR_i^{3/2}} = 0, \\ \dfrac{\partial L(R_i,\lambda)}{\partial \lambda} = \Delta T_0 - \sum_{i=1}^{m} l_i \left(\dfrac{1}{v_i} - \dfrac{1}{a\sqrt{R_i}} \right) = 0. \end{cases} \quad (10)$$

After solving the system of equations (10) for the set reduction of travel time ΔT_0, the optimal values of design radii R_i (11) and minimal value of capital investments K (12) are obtained:

$$R_i = \left(\frac{l_i}{K_i\alpha_i}\right)^{2/5} \left[\frac{\sum_{i=1}^{m} l_i^{4/5}(K_i\alpha_i)^{1/5}}{a\left(\sum_{i=1}^{m} \frac{l_i}{v_i} - \Delta T_0\right)^{2/5}} \right]^2 \quad (11)$$

$$K = \sum_{i=1}^{m} K_i \alpha_i \left(\frac{l_i}{K_i\alpha_i}\right)^{4/5} \left[\frac{\sum_{i=1}^{m} l_i^{4/5}(K_i\alpha_i)^{1/5}}{a\left(\sum_{i=1}^{m} \frac{l_i}{v_i} - \Delta T_0\right)^{2/5}} \right]^4 \quad (12)$$

6.3 Numerical solution to the problem of the optimal radii determination

Quantitative solution to the problem of determining the optimal reconstructed radii R_i, using the Lagrangian approach, will be considered on the example of Belarusian Railways section on the Minsk – Krasnoye direction, with a length of 10 km (712 km – 722 km), involving nine independent curves with the radius $R < 2000$ m. Required for solving the set problem characteristics of these curves are presented in Table 1.

Table 1. Characteristics of curves.

№	l, m	v, m/s	α	K_i, у.е.	a, m/s²	ΔT, s
1	171,11	40,9	9,73	97777	0,7	3,5
2	226,07	33,3	20,73	129183		
3	436,38	31,3	40,72	249360		
4	196,12	31,4	18,37	112069		
5	245,22	32,1	22,68	140126		
6	174,41	32,3	15,60	99663		
7	183,70	30,8	15,10	104971		
8	522,27	31,8	48,13	298440		
9	305,66	36,8	21,10	174663		

Using computer algebra system MathCAD, the values of optimal radii R_i = (3595; 2656; 2028; 2788; 2563; 2976; 3015; 1897; 2638), are obtained (Figure 1).

$m := 8 \quad i := 0..m \quad \Delta T := 3.5 \quad a := 0.7$

$l := (171.11 \;\; 226.07 \;\; 436.38 \;\; 196.12 \;\; 245.22 \;\; 174.41 \;\; 183.70 \;\; 522.27 \;\; 305.66)^T$

$K := (97777 \;\; 129183 \;\; 249360 \;\; 112069 \;\; 140126 \;\; 99663 \;\; 104971 \;\; 298440 \;\; 174663)^T$

$\alpha := (9.73 \;\; 20.73 \;\; 40.72 \;\; 18.37 \;\; 22.68 \;\; 15.60 \;\; 15.10 \;\; 48.13 \;\; 21.10)^T$

$\upsilon := (40.9 \;\; 33.3 \;\; 31.3 \;\; 31.4 \;\; 32.1 \;\; 32.3 \;\; 30.8 \;\; 31.8 \;\; 36.8)^T$

$$R_i := \left(\frac{l_i}{K_i \cdot \alpha_i}\right)^{\frac{2}{5}} \cdot \left[\frac{\sum_{k=0}^{m} (l_k)^{\frac{4}{5}} \cdot (K_k \cdot \alpha_k)^{\frac{1}{5}}}{a \cdot \left(\sum_{k=0}^{m} \frac{l_k}{\upsilon_k} - \Delta T\right)}\right]^2$$

$R^T = (3.595 \times 10^3 \;\; 2.656 \times 10^3 \;\; 2.028 \times 10^3 \;\; 2.788 \times 10^3 \;\; 2.563 \times 10^3 \;\; 2.976 \times 10^3 \;\; 3.015 \times 10^3 \;\; 1.897 \times 10^3 \;\; 2.638 \times 10^3)$

Fig. 1. Quantitative solution to the problem using computer algebra system MathCAD

7 The problem of determining the optimal train speeds in curves

7.1 The impact of unbalanced acceleration in introduction of high-speed traffic

The train speed on curved tracks may also be enhanced through increasing of unbalanced acceleration in curves from its acceptable value of 0,7 m/s^2.

In this case, an acceptable value of unbalanced acceleration for rolling stock is determined by:
- human exposure;
- dynamic strength properties of the rolling stock;
- shear-resisting indicators of the track.

Functional state of a passenger is an integral quality indicator of organism properties that define directly or indirectly comfort level of passenger travelling in rolling stock.

Studies have shown that the majority of people can experience continuous and repeated exposure of unbalanced acceleration of up to 0.9 m/s^2 inclusive satisfactorily.

According to studies, experienced by passenger feeling of travel sickness, nausea and dizziness caused by unbalanced acceleration of 0.9 m/s^2 is increased by 0.05 c.u. and considered acceptable.

Unbalanced centrifugal acceleration of 1 m/s^2 at exposure of rare and short-term character can be experienced satisfactorily.

General functional state and working efficiency of locomotive crew at integrated effect of noise, vibration and unbalanced acceleration of 0.9 m/s^2 remains within the tolerances. Thus, increase in unbalanced acceleration to 0,9 m/s^2 will not have significant impact on organism of a passenger, but it can substantially reduce travel time owing to faster running through a curve [10].

The data obtained are enabled to make calculations at unbalanced acceleration of 0.9 m/s^2 and analyse the impact of the unbalanced acceleration increase on speeds of passenger train while rounding a curves of different radii.

Limitation of unbalanced acceleration acceptable rate in terms of dynamic and strength profile is established on the basis of provision on unacceptability of the violation of current requirements to the relevant safety indicators.

The acceptable values of unbalanced acceleration on indicators of impact on the track are determined, taking into account adjustments of individual indicators.

Limitation of train speed and unbalanced acceleration value in curve stretches of track are determined through the occurrence of the limitation in at least one of the foregoing indicators.

7.2 Setting of the problem of determining the optimal train speeds in curves

The problem of determining the optimal speeds of train of each category in curves may be formulated as follows:

Assume, there is a railway stretch with m independent (single-radius or compound) curves on it and running trains of j-category ($j = \overline{1,n}$).

Each curve has its:
- length of curve l (m);
- angle of curvature α;
- curve radius R (m).

It is necessary to determine speeds v_j of the j-category trains ($j = \overline{1,n}$) on a curve of certain radius at optimal rail cant h (mm), increased unbalanced acceleration a, compliance with the condition of uniform load on rail tracks, comfortable carriage of passengers in which reduced costs on increased speeds in curved tracks will be minimal:

$$E = f(v_1, v_2, ..., v_n, h, a) = l\sum_{j=1}^{n} c_j \frac{N_j}{v_j} \rightarrow \min \qquad (13)$$

in restricting:
- compliance with the condition of uniform load on rail tracks:

$$\sum_{j=1}^{n} \beta_j v_j^2 = 12{,}96\frac{ghR}{S}, \qquad (14)$$

– compliance with the condition of comfortable carriage of passengers:

$$h = 12{,}5\frac{v_{\max}^2}{R} - \frac{S}{g}a, \qquad (15)$$

– acceptable values of unbalanced acceleration:

$$a \geq 0{,}7, \qquad (16)$$

$$a \leq 0{,}9. \qquad (17)$$

where c_j – present value of a j-category train;
N_j – the number of trains of j-category;
β_j – weight factor of a j-category train;
g – gravity acceleration, m/s^2;
S – distance between the track axes, mm;
$v_{\max}$ – maximum speed in this curve, km/h;
a – unbalanced acceleration, m/s^2.

7.3 Solution to the problem of determining the optimal train speeds in curves using Lagrange multiplier method

In order to solve the problem of interest (13) – (17), it is appropriate to apply Lagrange multiplier method [14,15].

As the constraints within the problem take the form of equities (14) – (15) and inequalities (16) – (17), to find an absolute minimum of a function f, it is necessary to:

1. Solve the problem when all the constraints, taking the form of inequalities (16) – (17), are omitted, and then define f to each solution.
2. Add one more inequality to the initial constraint system, them another. It is necessary to consider only those solutions that meet all the constraints.

Thus, first we will be considering the problem (13) – (15) excluding the constraints (16) and (17).

Let us compose Lagrange function:

$$L(v_1, v_2, ..., v_n, h, \lambda_1, \lambda_2) = l\sum_{j=1}^{n} c_j \frac{N_j}{v_j} + \lambda_1\left(12{,}96\frac{ghR}{S} - \sum_{j=1}^{n}\beta_j v_j^2\right) + \lambda_2\left(12{,}5\frac{v_{\max}^2}{R} - 163a - h\right), \qquad (18)$$

Then identify partial derivative of Lagrange function for unknowns v_j $(j = \overline{1,n})$, h, λ_1, λ_2, and set them to zero. As a result, we obtain the following system of equations:

$$\frac{\partial L(v_1,...,v_n,h,a,\lambda_1,...,\lambda_4,x_1,x_2)}{\partial v_j} = -l c_j \frac{N_j}{v_j^2} + 2\lambda_1\beta_j v_j = 0, \qquad (19)$$

$$\frac{\partial L(v_1,...,v_n,h,a,\lambda_1,...,\lambda_4,x_1,x_2)}{\partial h} = \lambda_1 12{,}96\frac{gR}{S} - \lambda_2 = 0; \qquad (20)$$

$$\frac{\partial L(v_1,...,v_n,h,a,\lambda_1,...,\lambda_4,x_1,x_2)}{\partial \lambda_1} = 12{,}96\frac{ghR}{S} - \sum_{j=1}^{n}\beta_j v_j^2 = 0; \qquad (21)$$

$$\frac{\partial L(v_1,...,v_n,h,a,\lambda_1,...,\lambda_4,x_1,x_2)}{\partial \lambda_2} = 12{,}5\frac{v_{\max}^2}{R} - 163a - h = 0. \qquad (22)$$

As follows from the expression (19):

$$l c_j N_j = 2\lambda_1\beta_j v_j^3; \qquad (23)$$

$$v_j = \left(\frac{l c_j N_j}{2\lambda_1\beta_j}\right)^{\frac{1}{3}}. \qquad (24)$$

Substitute (24) into (21) and find $\frac{1}{2\lambda_1}$:

$$\sum_{j=1}^{n}\left(\frac{l c_j N_j}{2\lambda_1\beta_j}\right)^{\frac{2}{3}}\beta_j = 12{,}96\frac{ghR}{S}; \qquad (25)$$

$$\sum_{j=1}^{n}\left(\frac{l c_j N_j}{2\lambda_1}\right)^{\frac{2}{3}}\beta_j^{\frac{1}{3}} = 12{,}96\frac{ghR}{S}; \qquad (26)$$

$$\left(\frac{1}{2\lambda_1}\right)^{\frac{2}{3}} = \frac{12{,}96ghR}{S\sum_{j=1}^{n}(l c_j N_j)^{\frac{2}{3}}\beta_j^{\frac{1}{3}}}; \qquad (27)$$

$$\frac{1}{2\lambda_1} = \left(\frac{12{,}96ghR}{S\sum_{j=1}^{n}(l c_j N_j)^{\frac{2}{3}}\beta_j^{\frac{1}{3}}}\right)^{\frac{3}{2}}. \qquad (28)$$

Based on (22) and (28), we obtain:

$$v_j = \left(\frac{l c_j N_j}{\beta_j}\right)^{\frac{1}{3}}\left(\frac{12{,}96ghR}{S\sum_{j=1}^{n}(l c_j N_j)^{\frac{2}{3}}\beta_j^{\frac{1}{3}}}\right)^{\frac{1}{2}}. \qquad (29)$$

Optimal speeds of j-category trains in curves of certain radius at optimal rail cant, acceptable values of unbalanced acceleration, compliance with the condition of uniform load on rail tracks and comfortable carriage of passengers are the following:

$$v_j = \left(\frac{c_j N_j}{\beta_j}\right)^{\frac{1}{3}} \left(\frac{12{,}96gR\left(12{,}5\frac{v_{\max}^2}{R} - 163a\right)}{s\sum_{j=1}^{n}\left(c_j N_j\right)^{\frac{2}{3}}\beta_j^{\frac{1}{3}}}\right)^{\frac{1}{2}}. \quad (30)$$

Taking into account the optimal speeds of j-category trains in curve of certain radius, the amount of reduced expenditure will be minimal and of:

$$E = l\sum_{j=1}^{n} c_j N_j \left[\left(\frac{c_j N_j}{\beta_j}\right)^{\frac{1}{3}} \left(\frac{12{,}96gR\left(12{,}5\frac{v_{\max}^2}{R} - 163a\right)}{s\sum_{j=1}^{n}\left(c_j N_j\right)^{\frac{2}{3}}\beta_j^{\frac{1}{3}}}\right)^{\frac{1}{2}}\right]^{-1} \quad (31)$$

Varying the values of unbalanced acceleration from 0.7 to 0.9, it is possible to define the optimal speeds of the trains of each category at which reduced costs on increase of the speeds in curves will be minimal.

Conclusions

The problems reviewed above for each line stretch make it possible to determine:
– optimal values of design radii restricting the curve speeds at minimal capital investments and maximum travel time reduction;
– maximum allowed speed of the trains of different category in curves at optimal rail cant, increased unbalanced acceleration from its acceptable value, compliance with the condition of uniform load on rail tracks and comfortable carriage of passengers when reduced expenditures on increase of the speeds in curves will be minimal.

References

1. *Economic Analysis of High Speed Rail in Europe. Fundacion BBVA*, 132, (2009).
2. T. Limon, Y. Crozet Transportation Research Procedia, **25 C**, 2828 – 2846 (2017).
3. Y. Crozet, Round table report #155, ITF-OCDE, 73 –105 (2014)
4. F. Dobruszkes, C. Dehon, M. Givoni Transportation Research, Part A no 69, Elsevier, 461 – 475 (2014).
5. J. Birge, F. Louveaux Introduction to Stochastic Programming: Springer Series in Operations Research and Financial Engineering, Springer; **2nd ed.**, 485 (2011)
6. I. Zaitseva, The bulletin of transport and industry economics, **60**, 86 – 93 (2017)
7. D. Kurgan *Methodology of calculations of railway track when interacting with high-speed rolling stock*, 35 (2017)
8. P. Panchenko *Justification of standards for the maintenance of railway gauge in terms of high-speed traffic*, 20 (2014).
9. A. Gavrilenkov *Theoretical foundations of designing high-speed and high-speed railway lines*, 213 (2004).
10. P. Kovtun, T. Dubrovskaya Collection of scientific works, UkrSUT, **181**, 6 – 15 (2018)
11. I. Kravchenya, T. Dubrovskaya, Bulletin of MIIT Bryansk branch: Transport and transport logistics, **1(7)**, 24 – 29 (2015)
12. I. Kravchenya, T. Dubrovskaya, Current issues and prospects for the development of transport and construction complexes: materials of the 4th International applied research conference, **2**, 33 – 35 (2018)
13. I. Kravchenya, T. Dubrovskaya, Bulletin of MIIT Bryansk branch: Transport and transport logistics, **2(4)**, 15 – 17 (2013)
14. T. Hamdy, Operations research: an introduction, 815 (2007)
15. I. Kravchenya, E. Burduk, T. Alymova Mathematical modeling. Linear and nonlinear programming, network planning and control, 112 (2014)

Improving the performance parameters of railway wheels with the help of optimal design technologies of their electric pulse processing

Igor Vakulenko[1], *Nickolay* Kuzin[2,3,*], *Leonid* Vakulenko[4], *Sergey* Raksha[1], and *Svetlana* Proidak[1]

[1] DNURT, Department of Applied Mechanics and Materials Science,49010,Dnipro,Lazaryan St., 2, Ukraine
[2] DNURT, Lviv Branch, 79052, Lviv, I.Blazhkevych Str., 12-a, Ukraine
[3] Lviv Research Institute of Forensic Science, 79000 Lviv, Ukraine
[4] Head Department of the Pridniprovk Railway, 49600, Dnipro, D.Yavornizkogo Av., 108, Ukraine

Abstract. The processing by pulses of electric current (PEC) of a fragment of the rim of a railway wheel was carried out on the DS10D equipment. When the density of the electric current is from 3 to 17 A/mm^2, the processing cycle consisted of the action of a pulse with a duration of 2.5-3 s and a pause between pulses of 4 s. After 150 cycles of PEC, the hardness of cold-formed metal was reduced from 10 to 20%. Studies of the parameters of the metal structure of the wheels have determined that the processing of PEC leads to a decrease in the number of defects in the internal structure, which are accumulated as a result of cold deformation of the rim along the rolling surface of the railway wheel. It has been established that, according to the nature of the influence on the hardness distribution of carbon steel, the PEC treatment corresponds to changes during tempering in the average temperature range and allows to significantly increase the operating parameters of the wheels of railway equipment.

1. Introduction

Compared to the thermal softening technology of hardened metal, processes that are based on the use of other physical influences such as magnetic field processing [1] or electric current pulses [2] have received sufficient propagation. Combines such processing with the absence of significant thermal effects on the development of processes structural transformation.

Known effects of plasticization from the action of electric current pulses are associated with the influence on the motion cloud free electrons in the crystalline lattice of a metal material [3]. As the energy of moving electrons increases as a result of the electric current, their impulse is partially transmitted to the dislocations, resulting in a lowering of the activation energy for the start of displacement [4]. However, according to [5], the effect plasticization of metal materials is considered not be related with activation energy, but depends on the degree of divergence between the direction of action electric impulse and the propagation plastic deformation of the metal.

Taking into account that in the nodes of crystal lattices there are non-neutral metal atoms, and ions, a local violation in their arrangement, for example, in the formation of a dislocation, leads to the idea that the dislocation must also have a certain electric charge. Thus, the interaction of the dislocation with the purposeful movement of the cloud moving electrons should be based on the influence of electric pulse on effect plasticization of the metal material.

2. Status of the problem

The origin and distribution of the plastic flow in metals and alloys is accompanied by the movement of dislocations with different signs, which are characterized by a different reaction to the action of electric pulse [6]. Under these conditions, with a predetermined direction in relation to the active stress at the sample load, the summation of the pulses electric current will lead to acceleration of the dislocations one sign and complicate the motion of dislocations opposite sign. For dislocations, which is characterized by a relatively low resistance level by the obstacle, the pulse of electric energy will be sufficient to start their movement. As a result, a local acceleration of a number dislocations of such a sign will occur, which in turn will lead to decrease in the operating syress to ensure the conditions for the continuous propagation of plastic deformation [7]. For dislocations of the opposite sign, the separation from the blocking places can only be due to the occurrence of thermal fluctuations of a certain level. This is due to fact that the nature of effect pulse action

* Corresponding author: kuzin.nick81@gmail.com

electric current is an obstacle to the beginning of motion such dislocations. The ratio between the quantities of indicated dislocations and causes achievement the effect of plasticization metal on the action of pulses electric current. At the same time, the sensitivity of metal material to the direction of action of the pulses should be associated with the recombination of the dislocation structure, which will lead to an appropriate directional plastic deformation.

Taking into account that the first attempts to use the pulse action of electric current were aimed at increasing the plastic propertiesmonocrystalsof metals during plastic deformation, the electric current density in the pulse was equal to 10^3 - $2\cdot10^3$ A / mm². Approximately similar regimes were used in the treatment of cold drawn wire for increase its plastic properties [2-4]. The use of this technology was not limited only to the increased plasticity of cold drawn metal, it was found to be used with special surface finishes, coatings, etc. [8-10]. The high density of electric current in a pulse restricts the use of such technology, primarily the power of equipment. Indeed, according to [5,6], the intersection of single crystals and cold drawn wire basically did no more 1mm. The use of processing metal layers on the rim railway wheel, based on the action of electric current with such a high density in the pulse is limited to the equipment required power.

3.Material and methodology of research

The material for research was the fragment of rim railway wheel after operation with a carbon content of 0.61% and concentration of other chemical elements complied with the requirements of the Standard of Ukraine. The processing of electric current pulses of a fragment rim railway wheel was carried out on the equipment DS10D, using an alternating current.

The calculation of the pulse frequency was carried out under conditions where the thickness of the metal layer is limited to the appearance of the "skin effect" by the ratio:

$$f_k = (\pi \cdot \mu_0 \cdot \mu \cdot \gamma \cdot \delta^2)^{-1} \quad (1)$$

where f_k - the critical frequency (Hz) above which the "skin effect" occurs, γ - the specific electrical conductivity (Ohm / m), μ_0 , μ - the magnetic permeability of the vacuum and the test metal, respectively (Gn / m), δ - the thickness of the metal layer that is required process (m)

In developing the processing technology of electric current pulses fragment rim railway wheel determined the optimal shape of the pulse electric current, which looked like a symmetric trapezoid. The time interval when increasing and decreasing the amplitude value of the electric current was 0.5-1s. For the treatment of a layer metal in the thickness of 10-15mm there was enough current density to 17 A/mm², the processing cycle consisted of a pulse duration of 2.5-3s and a pause between them up to 4s.

The hardness used was measured using the Vickers method (Hv) as a strength characteristic. The evaluation of the parameters thin crystalline structure was carried out using X-ray structural analysis techniques. The microstructure of the metal was studied under a light microscope used methods of quantitative metallography.

4. Results and discussions

The metal on the rolling surface rim of the railway wheel during operation is subject to a very high degree of cold plastic deformation with considerable heterogeneity in its distribution. For ease of research, the rolling surface of the railway wheel was divided into three sections, with varying degrees of cold plastic deformation. The plot (I) - part of the rim near the crest, (II) – middle part of the rolling surface and (III) - near the lateral surface of the wheel. Measurement of hardness on the rolling surface confirmed the existence of a significant difference in the degree deformation of the metal rim railway wheel after operation. For regions (I) and (III), the hardness was in the range of 5 - 5,5 GPa, and for (II) it was 4 - 4,9 GPa. The lower level of hardness on the site (II) is due to the acceleration process of softening cold-deformed metal from heating during the braking stages of the rolling stock. Indeed, if we assume that the area (II) is characterized by a maximum wear metal, then the degree of plastic deformation should be higher compared with other sections of the rim. The results of studies [11] determined that in the case of heating to certain temperatures, the degree of softening metal will be proportional to the value of cold plastic deformation.

Given that the process of electric current pulses is accompanied by changes in the geometric dimensions of the sample (Fig.1) the nature of the influence on the hardness metal may be similar to that of a reversible load [4,5].

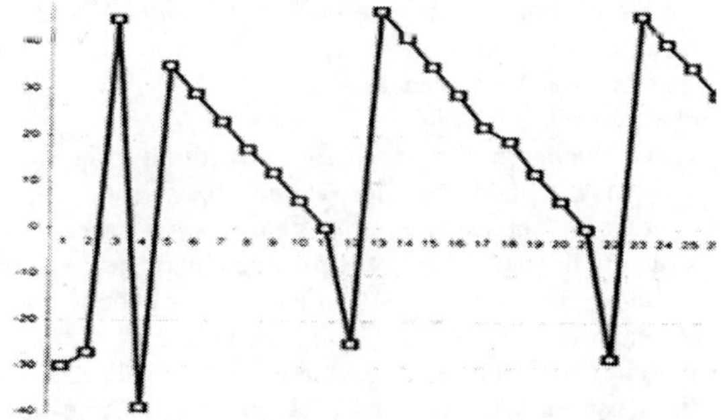

Fig. 1. The sample size change from the cycle number during the processing by pulses of electric current [12].

Indeed, regardless of the nature metal materials, the development of softening processes is observed for a wide range values of plastic deformation, in the event of divergence in the direction or form subsequent loading of the previous one. On the basis of this it can be assumed that in addition to the influence of electric current pulse on the acceleration of motion of cloud free electrons, one should expect the emergence of additional lever in the form of a mechanical component of the process. On the basis of the analysis obtained results, it was determined that as a result of the electric current pulses, a certain level of metal softening was achieved. Reduction of the hardness level for the region (I) was 20; for (II) 9 and for (III) 11% relative to the state after a cold plastic deformation [12]. Taking into account that in electric pulse processing the temperature of the metal did not exceed 50° C, the achieved level of softening can't be related to the thermal nature of the effect. More it is known that heating up to temperatures of $200\text{-}250^{\circ}$ C, due to the development of processes of deformation aging, should lead to an increase in the hardness of cold-deformed carbon steels [11]. The obtained of equal effect softening metal on rolling surface railway wheel after electric current pulses is confirmed by the appearance of qualitative changes in the internal structure of the metal.

Figure 2 shows the microstructure of metal volumes near the rolling surface railway wheel after operation. The structure corresponds to the metal after a high degree of cold plastic deformation.

In the background of high turbulence structure colonies of perlite (Fig. 2a,b), the presence sites of structurally free ferrites (Fig. 2c) can be regarded as evidence of heating metal near the rolling surface to temperatures higher than the beginning of phase transformations. On the other hand, the absence in this section signs formation of additional boundary sections, which are characteristic of the development processes of recrystallization [11], indicates a very short period of exposure metal at these temperatures [12,13].

After electric pulses treatment fragment of the rim railway wheel, reducing the level of hardness is accompanied by quite logical changes in the internal structure of the metal (Fig. 3). From the comparative analysis of the structural components, according to external features, the effect of electric pulses should be attributed to the thermal influence on metal after plastic deformation (Fig. 3a). Indeed, in comparison with the structural state metal wheels after operation (Fig. 2a,b), electric pulse treatment results not only in the coarsening of structural components (Fig. 3a,b), but also in some changes in their distribution.

Comparative analysis of sites structurally free ferrite up to (Fig.2c) and after electric pulse treatment (Fig. 3c), confirms the appearance of qualitative differences in the structure of steel. One of them is the emergence of more distinct additional boundaries in volumes of a structurally free ferrite (designation "a", Fig. 3c).The presence in volumes of structurally free ferrite distorted lines, similar to the lines slipping of dislocations, indicates a high degree of cold deformation with the obligatory change in the direction of loading.

The action of pulses of electric current is accompanied by the appearance of additional distortions, which significantly distinguishes them from the lines of slippage in the propagation of plastic flow in metallic materials [11].Thus, we can assume that degree of softening obtained from the action of pulses electric current in reality is accompanied by significant changes in the internal structure of the metal rim after cold plastic deformation.

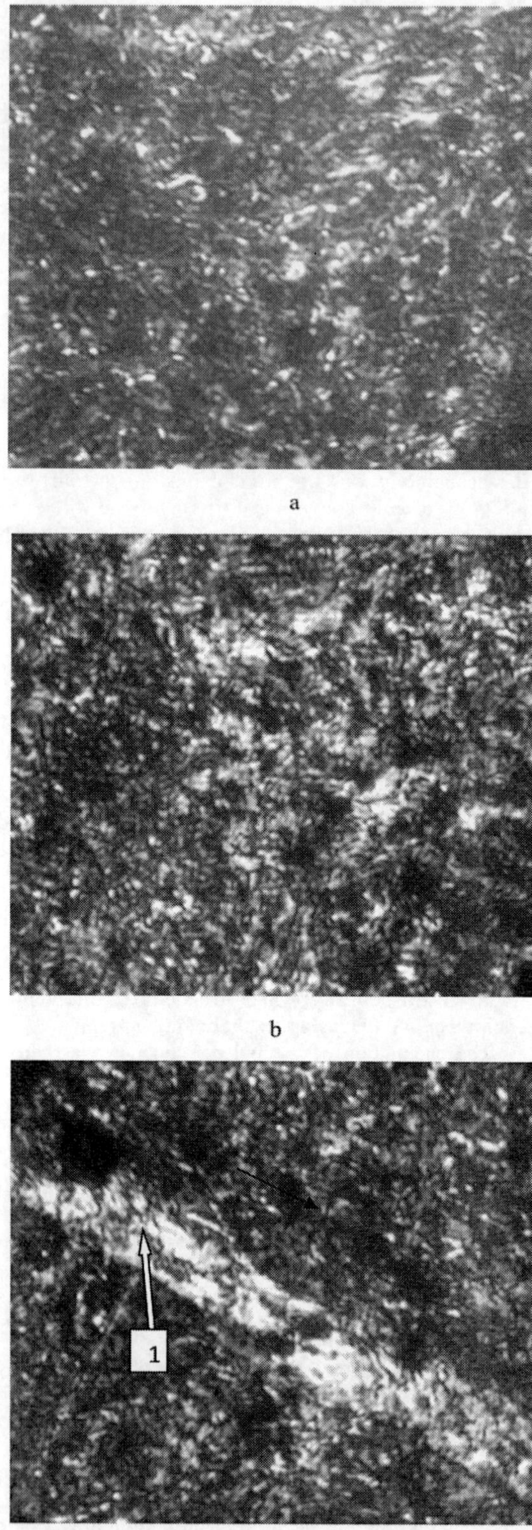

Fig.2. Microstructure of the steel near the rolling surface of the wheel after decommissioning. Pearlite colonies (a,b) and site of structurally free ferrite - (1) (c), magnification 250.

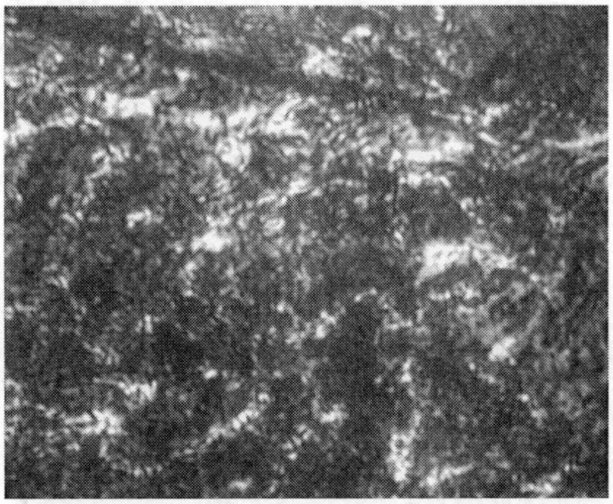

a

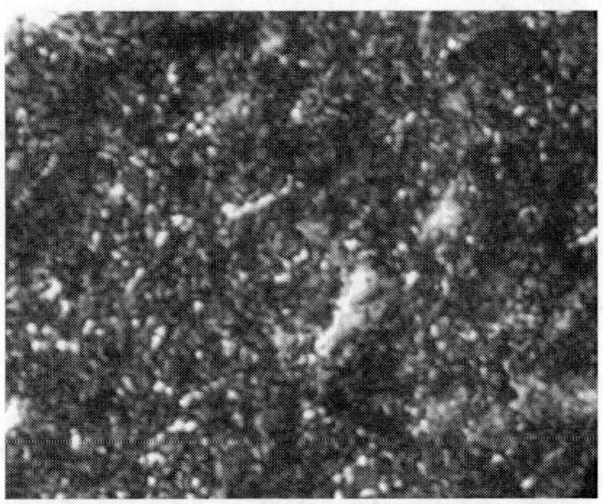

b

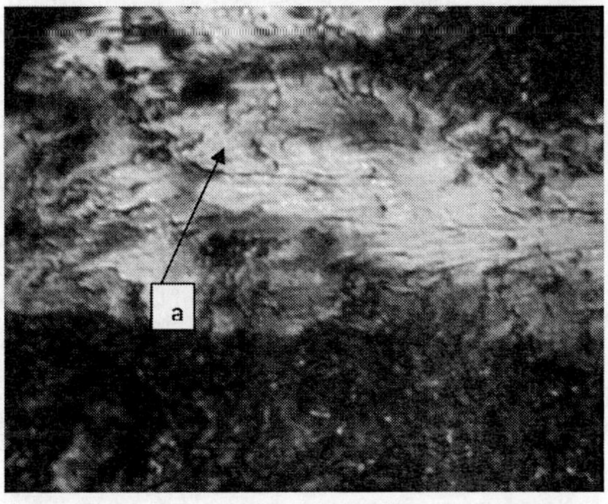

c

Fig.3. The microstructure of the steel near the rolling surface of the wheel after the treatment of the IES (a, b - perlite colonies, c - a section of structurally free ferrite), magnification 250.

Considering the mechanism of development process of softening in the cold deformed metal from the action of pulses electric current, the cyclic change in the size of the metal specimen undergoing processing becomes a definite value. Indeed, in the conditions of a cyclic loading, the micro volume of a metal is subject to the action of a definite total deformation. However, under the influence of the emerging internal stresses, the most soft structural components will first be deformed. As a result development of processes, the relaxation of internal stresses in volumes of a structurally free ferrite, the magnitude existing stresses may not be sufficient for the start deformation of pearlite colonies.

Given the very small change in the geometric dimensions of the specimen (Fig.1) with ferrite hardness about 2 times below the pearlite colony, a large part of the plastic deformation should be concentrated in the regions of a structurally free ferrite of steel. According to stoichiometry, in steel of railway wheels, the share of structurally free ferrite can reach up to 20-25%.

Assuming that the main effect of action pulses electric current is concentrated in ferrite, the additive contribution to the general effect of softening at the level of 10-15% is fully justified..

The results of the softening metal rim railway wheel on the rolling surface after deformation were confirmed by X-ray structural analysis (Table). The analysis of the results on the measurement of the internal stresses (σ), density of the dislocation by the interference (211) - (ρ_{211}), distortions of the second kind (μ) and the size of the coherent scattering regions (*D*), regardless studied areas of the wheel (I, II, III) development of very complicated by the nature of structural transformations in the metal under the influence of pulses electric current.

Table. Changing of parameters fine crystalline structure of the ferrite from pulses electric current.

Parameter	Number areas of rim railway wheel		
	I	II	III
σ, *GPa*	-0,064	-0,128	+0,210
$\rho_{211}\times10^{11}cm^{-2}$	3,6/2,8*	3,6/2,2*	3,9/3,0*
$\mu,\times10^{-3}$	1,09/0,84*	1,22/0,87*	1,18/0.97*
D, nm	28,9/32,7*	27,7/33,0*	27,7/31,6*

Given the very complex conditions of loading the railway wheel on the rolling surface (heterogeneous hardened from cold plastic deformation, rapid heating and cooling metal to different temperatures during braking of rolling stock, etc.), an attempt was made to assess the efficiency of pulses electric current treatment compared to thermal technology.

The figure 4 shows effect of temperature heating on the hardness steel of a railway wheel after cold plastic deformation with a different dispersion of perlite. At the conditions of equal degree of plastic deformation, the hardness of metal increases in proportion to as cending dispersion of the pearlite colony. At the same time, for equal degree of plastic deformation, the softening effect is to a large extent determined by the dispersion pearlite colony of the steel. This is due ability of cementite plate-like form to be plastically deformed [11]. An increase of the temperature heating will be accompanied by appropriate metal softening. From the comparative analysis it can be determined that the more finely the plate cementite, the degree of softening will be lower (Fig.4).

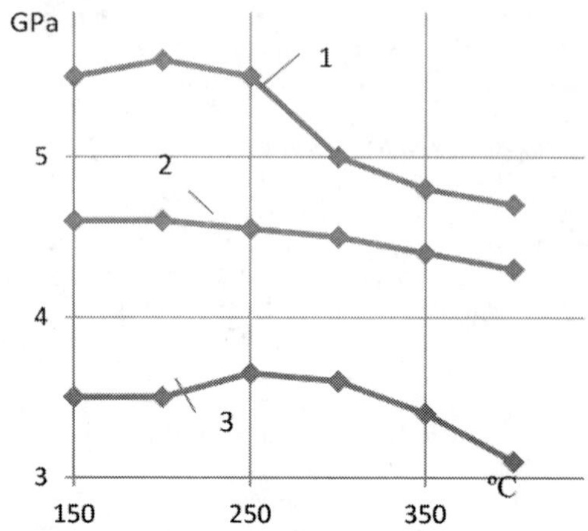

Fig.4. The influence temperature heating on hardness steel of railway wheel after cold deformation (1,3-75, 2-25%) at high (1,2) and low (3) disperse of pearlite colony.

The nature influence of the degree plastic deformation and dispersion of pearlite colony on the development of softening processes under heating is similar to the known data. So, for the carbon content close investigated steel after cold plastic deformation of 25%, softening at a level of 20% achieved by heating up to 400° C, with duration of exposure up to 1 hour. Reduction of hardness by 10% after cold plastic deformation by 25% achieved already after the heating of the metal to 300 - 350 ° C. Similar results are obtained and for higher degrees of plastic deformation. Thus, for a steel with a carbon content of 0.7%, deformed by 90% dragging, the effect of softening by 20% is achieved after heating to a temperature of 380°C, and a decrease in hardness by 10% - by heating up to 300°C [11].

Analysis of the experimental data (table) shows that, in general, there is a fairly good agreement both in character and in absolute values with known results [11 - 13], which should be considered as confirmation of the development of softening processes in carbon steel of wheel fragment after treatment with electric pulses current. Indeed, according results of the study, the process of softening cold-deformed metal is accompanied by decrease in the accumulated number of dislocations, distortions of the second kind and an increase in the size of the regions coherent dispersion (table). At the same time, it is certain interest to evaluate the rate of structural changes depending on the different levels of metal hardening after cold deformation. Taking only the magnitude of the decrease in hardness after treatment pulses electric current (ΔH_v) as an indicator of the degree metal softening, a formal construction of the degree of change in the characteristics of the fine-crystalline structure from (Fig. 5) was carried out. It follows from the presented relationships that, in proportion increase in the effect of metal softening after being treated with electric pulses, decrease in size rate of coherent scattering regions, dislocation density, and second-type distortions is observed.

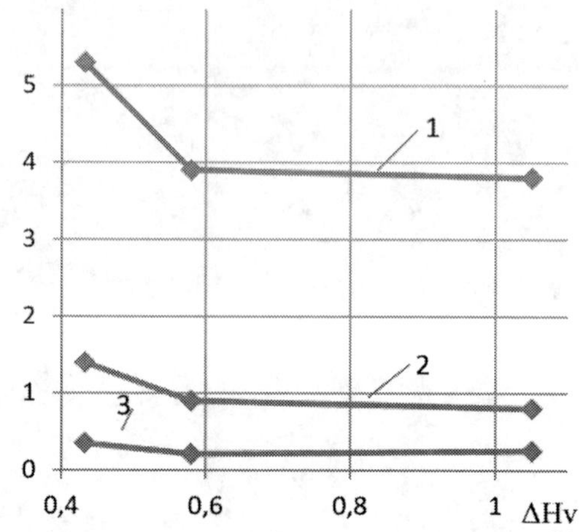

Fig.5. The dependence of change the size coherent scattering regions (ΔD , nm) - (1); dislocation density ($\Delta\rho_{211}$ $\times10^{11} cm^{-2}$) – (2) and distortions of the second kind ($\Delta\mu$ $\times10^{-3}$) - (3) of the degree softening (ΔH_v , GPa) steel after treatment with pulses of electric current.

Although, on the whole, the achievable softening effect from impulse treatment with electric current pulses is proportional to degree strain hardening of metal, the observed change in the sign of residual stresses indicates the development of quite complex structural changes. At the same time, cumulative effect of the non-uniformity of distribution plastic deformation and its directivity on the rolling surface wheel indicates the need to continue studying causes that led to a change in the sign of residual stresses after treatment with pulses electric current.

Thus, if we use the datas of the table and estimate the degree of metal railway wheel softening after treatment of pulses electric current according to the characteristics, compared with the thermal technology, the softening effect may correspond to the heating metal to temperatures of 400-450°C [13]. At the same time, the low temperatures of heating metal (up to 50°C) during the processing of pulses electric current can't be attributed development processes of diffusion mass transfer.

In order to determine mechanism of soften cold-deformed metal after processing of pulses electric current, an assessment was made of the stresses that caused movement of dislocations from various external influences. Low temperature heating of the metal at processing of pulses electric current immediately excludes the thermal nature of the influence. On other hand, due to high values of dislocation density accumulated in crystallographic sliding systems (table), one of the possible mechanisms explaining effect of softening at ambient temperatures may be non-conservative displacement of dislocations under the action of pulses of electric current.

From the balance of forces that determine resistance to displacement dislocation from one slip system to another [14] and facilitate the displacement fragment of the dislocation line into a parallel slip plane [15], the relation obtained:

$$\mathrm{Sin}\,\alpha \approx {}^{b}\!/_{h}, \qquad (2)$$

where α is the angle of disorientation between the line of dislocation and the Burgers vector (b), h is the distance between the parallel sliding planes. Taking into account that $\rho = h^{-2}$ [16], where ρ - the density of dislocations, the relation (2) takes the form:

$$\mathrm{Sin}\,\alpha \approx b\sqrt{\rho} \qquad (3)$$

After substituting (3) $b = 2{,}48 \cdot 10^{-7}$ mm [11], the experimental values of the dislocation density in accordance with areas rim of railway wheel under study (I, II, III), the calculation shows that as a result of pulses electric current treatment, the angle disorientation between the dislocation line and Biurges vector is reduced. So, for the regions I, II, and III cold-deformed metal, the value α is from $36'$ up to $48'$ respectively. As a result of the processing of pulses electric current according to the estimated calculations, there was a decrease of the angle of 48% for the III region, 28% for II and 17% for I. The obtained results can be regarded as evidence of a change in the ratio between the edge and screw components of the dislocation structure in the metal rim railway wheel in as a result of processing by pulses of electric current. Thus, one of the explanations softening effect of metal railway wheel as a result the development processes of recombination dislocation structure should be considered. Given that the edge component of the dislocation structure [14] is proportional to the angle, as a result of electrical impulse processing, the part of the screw component should be increased. As a result, the conditions for reducing the metal resistance, of processes creeping dislocations, which should contribute to the development of their annihilation. After processing of pulses electric current, the effect of reducing the hardness of cold-deformed metal on the rolling surface railway wheel practically adequate the heating in the middle temperature range.

4. Conclusions

1. Mechanism of softening of cold-deformed metal rim railway wheel after processing by pulses of electric current, based on recombination of dislocation structure.
2. Achieved the effect of softening cold-deformed metal in comparison with thermal technologies, adequate the heating in the middle temperature range.

5. Acknowledgements

The authors are grateful to Dr. V.G. Anofriev for providing material for research and Dr. V.A. Sokirko for carrying out the processing of fragment railway wheel by electrical pulses. The authors would like to express their gratitude to the staff of the Department of Applied Mechanics and Materials Science of the University for their participation in the preparation of samples for research.

6. Literature

1. M.L. Bernshteyjn., V.N. Pustovoyjt, *Termicheskaya obrabotka staljnihkh izdeliyj v magnitnom pole.* (Mashinostroenie, 1987).
2. V.I. Spicihn, O.A. Troickiyj, *Ehlektroplasticheskaya deformaciya metallov*. (Nauka. 1985).
3. A.F. Sprecher, S.L. Mannan, H. Conrad, Acta Met. **34**, 1145 (1986).
4. H. Conrad, Materials Science and Engineering **A287**, 227 (2000).
5. G. Tang, J. Zhang, M. Zheng, et al., Materials Science and Engineering,**A. 281**, 263 (2000).
6. K F. Yao, J. Wang, M. Zheng, et al., Scripta Materialia, **45**.,533 (2001).
7. F. Garofalo, Met. Trans.**2**, 2315 (1971).
8. K. Ait Aissa, A. Achour, J. Camus, et al.,Thin Solid Films**550**,264 (2014).
9. S. M. Razavian, B. Rezai, M. Irannajad, et al, Intern. J. of Mining Sci. and Tech., **25**.473 (2015).
10. W. L.Morgan,L. A. Rosocha, Chemical Physics of Low-Temperature Plasmas **398,** 255 (2012).
11. I. A. Vakulenko, V.I. Boljshakov *Morfologiya strukturih i deformacionnoe uprochnenie stali* (Makoveckiyj, 2008).
12. L.I.Vakulenko, *Problemi koroziyjno-mekhanichnogo ruyjnuvannya, inzheneriya poverkhni, diagnostichnie sistemi: materiali konferencij KMN – 2013*, (Fiziko-mekhanichnogoinstitutu, 2013).
13. L.Vakulenko, S.Myamlin, A.Kovalek. *New technologies and achievements in metallurgy, materials engineering and production engineering* (Czestochova,.**48**, 369, 2015).
14. J.Friedel,*Dislocations* (Pergamon press, 1964).
15. W.T. Read Jr. *Dislocations in crystals*. (McGraw Hill).
16. A. Guinier, *Theorie et Techniquedela Radiocristallographie* (Dunod, 1956).

Mathematical model of pantograph cooperation with two degrees of freedom with a catenary system

Marek Kaniewski[1,*], and *Michał* Cichoński[2]

[1] Railway Research Institute, Chłopickiego 50, 04-275 Warsaw, Poland
[2] Stemmann-Polska Sp. z o.o.. ul. Spółdzielcza 22, 55-080 Kąty Wrocławskie

Abstract. The authors present a method of modeling pantograph cooperation with a catenary system by means of a computer program based on a pantograph discrete model with two degrees of freedom and a catenary system model as a focused mass that in the dynamic state has variable stiffness along the span. After performing the appropriate tests, the validation was carried out according to the proposed method in the PN EN 50318:2019-02 standard required in the Technical Specifications for the interoperability of the subsystem "Energy" and "Rolling stock - locomotives and passenger rolling stock" of the trans-European rail system in Europe.

1. Introduction

In the scientific literature, many models are described that allow for more or less accurate description of the interaction of the pantograph with the catenary system. These are models based on solving equations or systems of second order differential equations using analogue computers [3] or digital [5] [7] [8]. To create programs calculating the parameters assessing the quality of cooperation, various computing environments are used, for example MES, MRS, MATLAB & Simulink [7], Delphi.

In studies [1] [2] [3] [4] [5], the authors describe as one of the possible models describing the cooperation of the pantograph with the overhead contact line, the pantograph model with two degrees of freedom and the overhead contact line represented as a one-dimensional model with clustered parameters at the contact point. The model of the overhead contact line is reduced to one concentrated mass with the stiffness specified in relationship (1) and (2). The pantograph consists of two concentrated masses connected by elastic-damping elements. For the model to be used, for example, in the process of designing the catenary system or checking a new type of pantograph on a new type of rail vehicle, it should be validated according to the requirements of PN-EN 50318:2019-02 [6]. The validation process is two-stage. First, the simulation result obtained using the mathematical model is compared with the reference results given in the standard [6] and then with the results obtained during the actual journeys on the catenary system. The simulations are carried out at the driving speed of 250 and 300 km/h and compared with the reference parameters given in the standard [6].

2. Description of the model used

To simulate the quality of the pantograph's cooperation with the catenary system, the model according to [3] was adopted (the model diagram is shown in Figure 1) with the following assumptions:

- the pantograph model has two degrees of freedom taking into account the division of the pantograph into two masses, m1 and m2,
- the mass of the catenary system represented by substitute mass is focused at the contact point, it is constant at the length of the span, this assumption is also given by the source [1],
- the frequency of natural vibrations of the overhead contact line is constant along the span,
- the stiffness of the overhead contact line changes sinusoidally along the span and has the minimum in the middle of the span and the maximum for the support structure,
- dry and sticky friction in the overhead contact line is omitted,
- the mass of the pantograph is reduced to the point of contact with the contact wire and fixed regardless of the height relative to the roof of the electric head,
- there is dry friction in the pantograph and shoe arms by a value of $-w_o$ or $+w_o$ in the opposite direction than the movement; changes in steps according to the signum function,
- sticky friction is taken into account by the proportionality factor to the speed of movement of the contact point and its value is reduced to the point of contact,
- the static force exerted by the pantograph drive system is constant for each electric head movement speed and the position of the pantograph,

* Corresponding author: mkaniewski@ikolej.pl

- the vertical component of the aerodynamic drag force is always positive (directed upwards) and proportional to the square of the speed of movement of the rail vehicle,
- the influence of the locomotive vibrations was omitted,
- the movement starts in the middle of the span and takes place at a constant speed.

The stiffness of the overhead contact line is described by dependence (1).

$$k_s = k_{s1} - k_{s2} * \cos\Omega(t) \quad (1),$$

where: k_{s1} – average value of overhead contact line stiffness in N/m,
k_{s2} – amplitude of the overhead contact line stiffness in N/m,
Ω - pantograph travel frequency under supporting structures in rad/s according to relationship (2),

$$\Omega = \frac{2\pi v}{L}, \quad (2)$$

where: v – displacement speed of the pantograph mounted on the rail vehicle

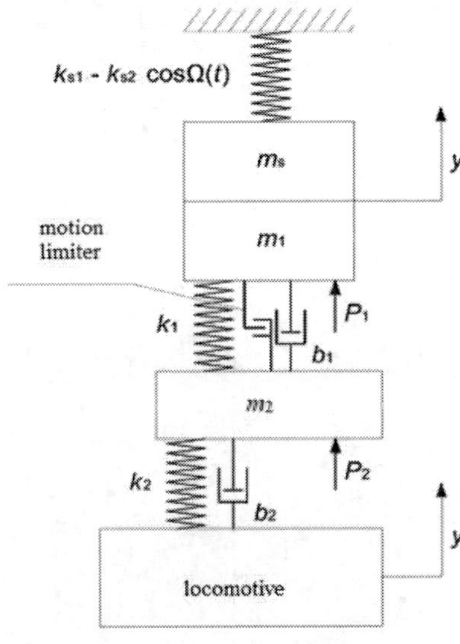

Fig. 1. The pantograph model of a two degrees of freedom. Source [3]

For the above-described assumptions, a system of equations describing the traffic equation according to [3] was written on the basis of the dynamic equilibrium of reduced masses to overhead contact line and pantograph in accordance with the principle of d'Alembert.

$$\begin{aligned} &(m_s + m_{s1})\ddot{y}_1 + b_1(\dot{y}_1 - \dot{y}_2) + \\ &+ k_1(y_1 - y_2) + (k_{s1} - k_{s2}\cos\Omega t)y_1 = P_1, \quad (3) \\ &m_2\ddot{y}_2 + b_1(\dot{y}_2 - \dot{y}_1) + k_1(y_2 - y_1) + b_2\dot{y} = P_2, \end{aligned}$$

where: m_1 - the mass of the pantograph head,

- m_2 - pantograph's shoulder mass reduced to the contact point,
- k_1 - stiffness of the electric head 's suspension,
- b_1 - damping coefficient of viscous displacement of the electric head,
- b_2 - damping coefficient of viscous pantograph arm displacement,
- y_1 - displacement of the contact point of the electric head with the contact wire,
- y_2 - vertical displacement of the contact of the upper pantograph arm node,
- P_1 - aerodynamic force acting on the electric head,
- P_2 - the sum of aerodynamic forces acting on the pantograph arms and the static force generated by the propulsion system.

For the validation process, the following value required by the standard [6] was adopted: P_1=0 N, P_2=120 N, m_1=7.2 kg, m_2=15 kg, k_1= 4200 N/m, k_2= 50 N/m, b_1= 10 Ns/m, b_2= 90 Ns/m; the overhead contact line, which has one contact wire with a unit mass of 1.07 kg/m and one supporting rope with a unit mass of 1.35 kg/m; consisting of 10 identical spans with a length of 60 m; the stagger of contact wire's is 0.2 m; the pulls are respectively the contact wire - 20000 N and the carrier rope 16000 N; aerodynamic force equal to 0. Adopted driving speed during 250 km/h simulation. The vibration damping for the contact wire is 0 Ns/m. The overhang of the contact wire is also 0. During the simulation, the overhead contact line cooperates with one pantograph. The parameters calculated for the spans 5 and 6 are analyzed – in the middle of the stress section where the lowest values of forces extending the contact wire and the supporting rope should be expected. The spacing of hangers was given in the standard [6].

3. Simulation results

The calculations of contact forces in the simulation process occurring between the reference pantograph and the reference overhead contact line were made using a computer program created using the Delphi 10.2 Professional package. The following parameters are loaded during the program startup: static pantograph force, as a mean constant value for the pantograph operating range, dry and viscous damping of the first and second pantograph spring, mass of the shoe and the mass of the articulated frame (arms), springing force of the electric head, springs' spacing, maximum deflection of the electric head's springs, length of the overhead contact line span, distance of the contact wire from the track axis, average value and amplitude of the overhead contact line rigidity, frequency of overhead contact line oscillation, reduced net mass in the half of the span and under the column, four factors concerning aerodynamic impact, vehicle movement speed the same as the pantograph, the vibration frequency of the rail vehicle, the amplitude of the locomotive's springing. Some values have assumed zero values (they are given in p.2). An analysis was carried out confirming the validation of the adopted method and additionally 5 calculation variants showing the possibilities of the program. The validation of the calculation method is shown in Figures 2 and 3. Fig. 2 shows the diagram of the contact force between the pantograph head and the contact wire of the overhead contact line obtained as a result of simulation for the reference model. The length of the simulated section corresponds to the length of 10 spans of the reference overhead contact line. The simulation discretization was 0.02 s. Therefore the frequency of the next calculated points is 50 Hz.

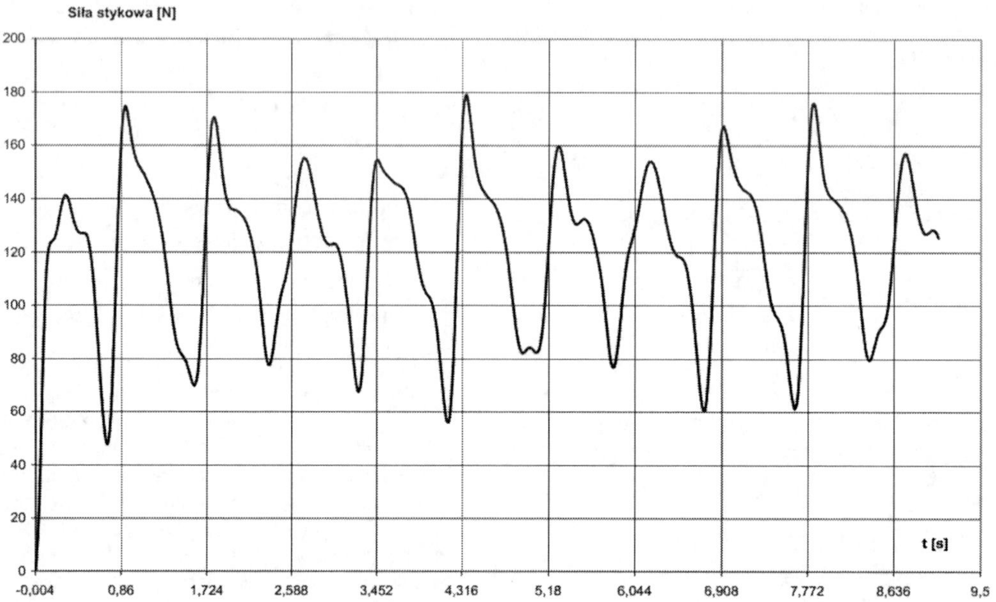

Fig. 2. The simulation result of the contact force for 10 spheres of the reference and reference pantograph, speed 250 km/h

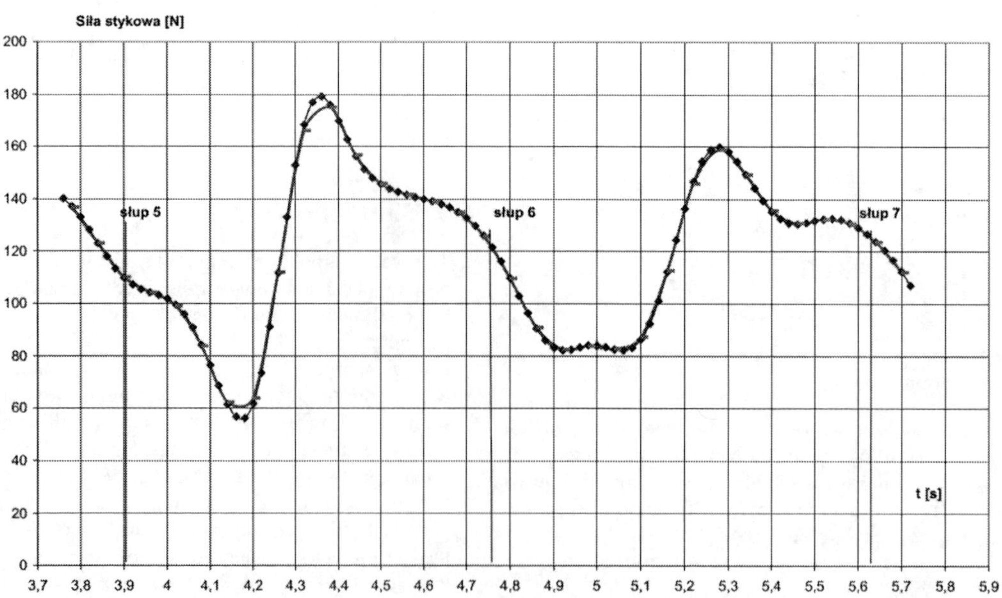

Fig. 3. Simulation result for 5 and 6 spans of contact force for the speed of eating 250 km/h (blue line frequency 50 Hz, red line 20 Hz)

Figure 3 shows the course of the contact force for the preferred by the standard [6] bays 5th and 6th. The thin blue line shows the course for the frequency of calculation 50 Hz and the thick red line for the frequency 17 Hz (frequency 20 Hz in this case can not be obtained). Point 3A of the standard [6] requires that the results be filtered with a low-pass filter from 0 to 20 Hz. Practically speaking, there are no fundamental differences between the course obtained in the simulation process and filtered in the range from 0 to 20 Hz.

Using the program, you can analyze the value of the contact force for the changed important parameters, for example the pantograph. Thus, changing the reduced mass of the electric head to 20 kg and the reduced mass of the pantograph arms to 30 kg, we get the course shown in Fig. 4. For this course, the mean value of the FM contact force has risen slightly and amounts to 121 N, the standard deviation is 69 N, so it has increased over 100% a minimum momentary contact force decreased to 2 N. Such parameters do not meet the criteria of proper cooperation of the pantograph with the overhead contact line. The contact force with a 2 N value will cause a spark between the pantograph electric head and the contact wire.

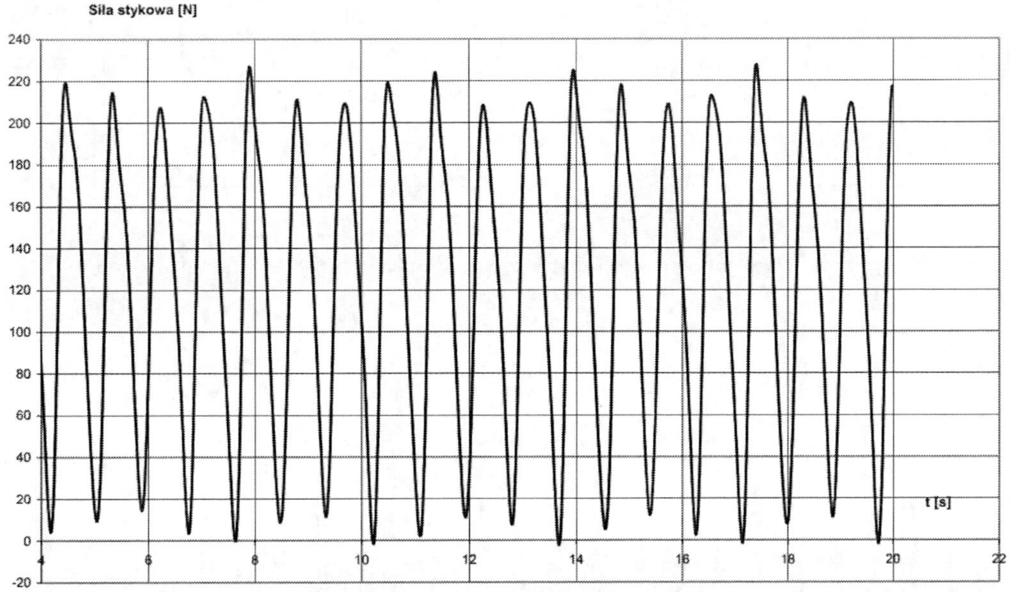

Fig. 4. Simulation result for the reference catenary system and pantograph with a reduced mass of 20 kg and a reduced pantograph arm mass of 30 kg, speed 250 km/h

The filtering procedure allows for the removal of accidental noise in the waveforms while emphasizing the most important characteristic frequencies occurring in the pantograph – catenary system. The basic frequency of system vibrations results from the length of the spans and the speed of travel. It also depends on the type of catenary system and ranges from 0.7 to 2 Hz. However, the vibration resulting from the distance between hangers at a speed of 250 km/h is up to 12 Hz with the distance between consecutive hangers 6 m.

The obtained results from stimulation were compared with the values given in the standard [6]. The speed during the simulation was 250 km/h.

In the next case under consideration (Fig. 5), the static force value changed to 90 N in the data. After the calculation, the average contact force was 90 N, the standard deviation of the contact force was 23 N, the maximum force value was 141 N and the minimum value of the contact force was 35 N. Such quality parameters of the pantograph's cooperation with the overhead contact line should be considered correct. There will be no sparking between the contact wire and the pantograph electric head. The contact wire will wear to a minimum.

We will also disagree with the case in which there was a significant increase in the dry friction force in the electric head's suspension (obliteration in the electric head or chafing caused by a collision with the catenary system elements). Let us assume then that the dry friction force of one side of the electric head is 190 N. After the calculation, the average contact force was 119 N, the standard deviation of the contact force is 48 N, the maximum value of the contact force 288 N and the minimum contact force 21 N. The calculated quality parameters of cooperation pantograph with overhead contact line proves that the contact between the contact wire and the electric head is incorrect.

The following phenomena will occur in the contact between the contact wire and the electric head cover: contact breaks, possibly overburning one conductor, faster contact wire wear. The course of contact forces in this case is shown in Figure 6.

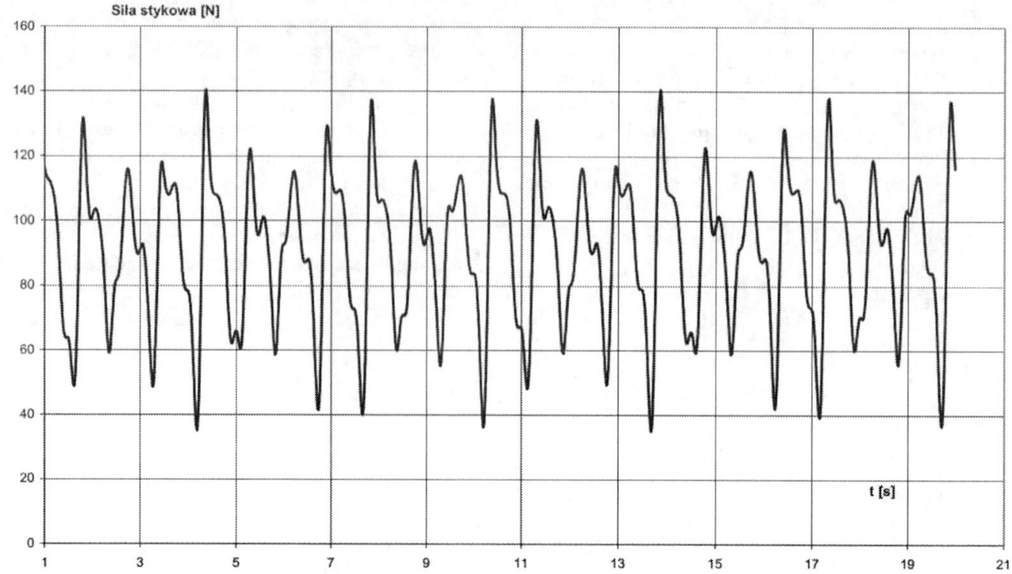

Fig. 5. Simulation result for reference catenary system and pantograph when static force F=90 N, speed 250 km/h

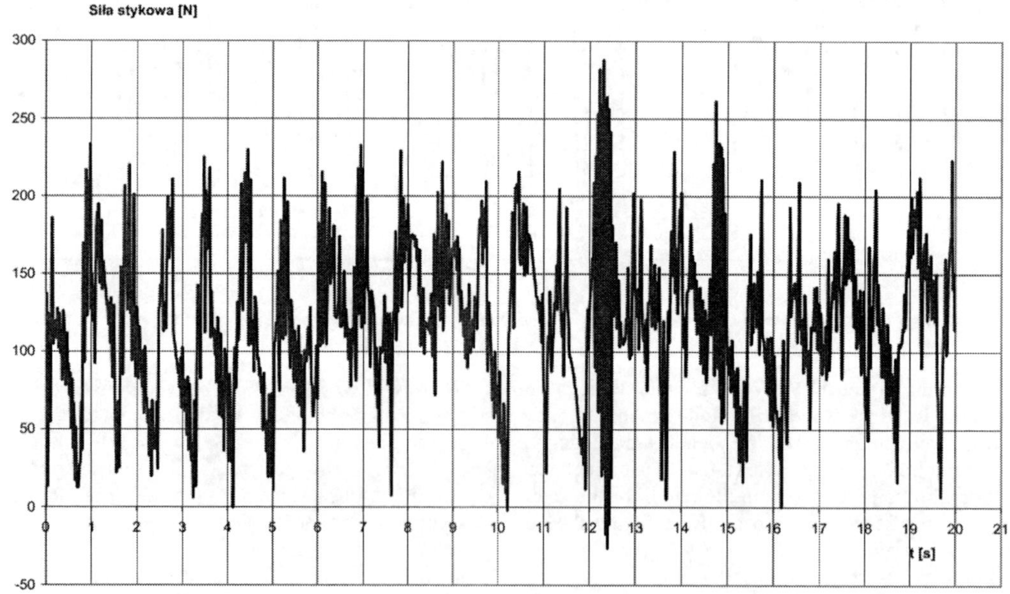

Fig. 6. Simulation result for reference catenary system and pantograph contact force F=90 N, travelling speed 250 km/h

In the next case under consideration, we assume that the pantograph's universal joint has become seized. The calculation of the force of such an event is shown in Figure 7. It is a simulation diagram of the reference catenary system and pantograph in which the stiffness of the electric head was changed 1.6 times compared to the reference pantograph. The following values of the rated parameters were obtained: the average value of the contact force was 120 N, the standard deviation of the contact force is 29 N, the maximum value of the force 173 N and the minimum contact force 57 N. The calculated parameters of the pantograph's cooperation with the catenary system show that between the contact wire and the electric head cooperation, despite increasing the rigidity of the electric head is correct

The next considered case in which there was a significant increase in the dry friction force is the increase of the friction force in the bearings of the articulated frame (seizing the bearings). Suppose that then the dry friction force is 10 times greater. After the calculation, the average contact force was 124 N, the standard deviation of the contact force is 40 N, the maximum value of the contact force is 200 N and the minimum contact force is 4 N. The calculated quality parameters of pantograph cooperation with the catenary system show that between the contact wire and the electric head the cooperation is incorrect. There will be cracks and contact gaps. The instantaneous value of the minimum contact force of 4 N and the minimum statistical value of 5 N is unacceptable from the point of view of correct cooperation of the pantograph with the catenary system. The course of contact forces in time for this case is shown in Figure 8.

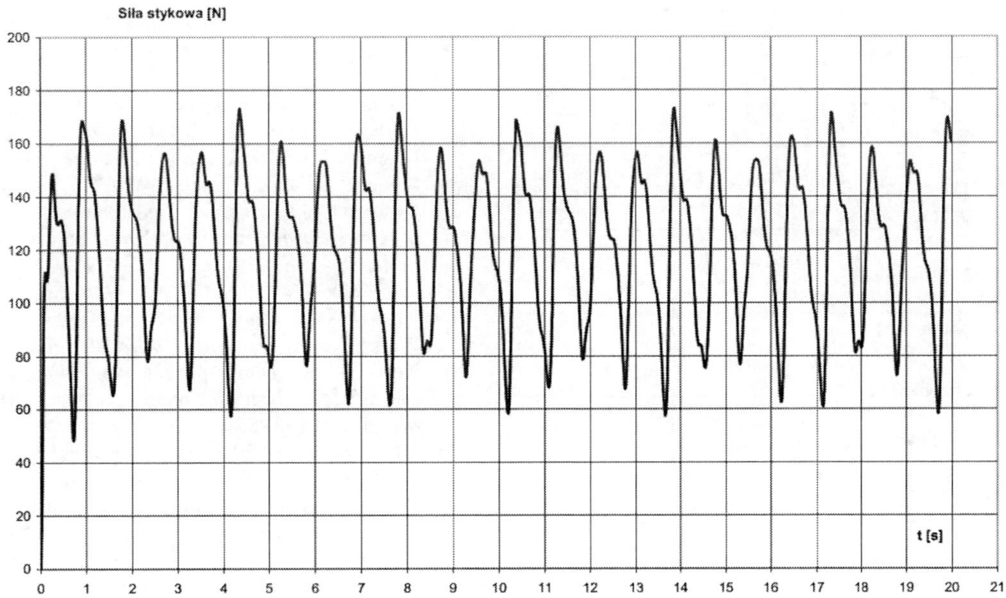

Fig. 7. Simulation result for reference catenary system and pantograph. The stiffness of the electric head has been increased 1.6 times, speed 250 km/h

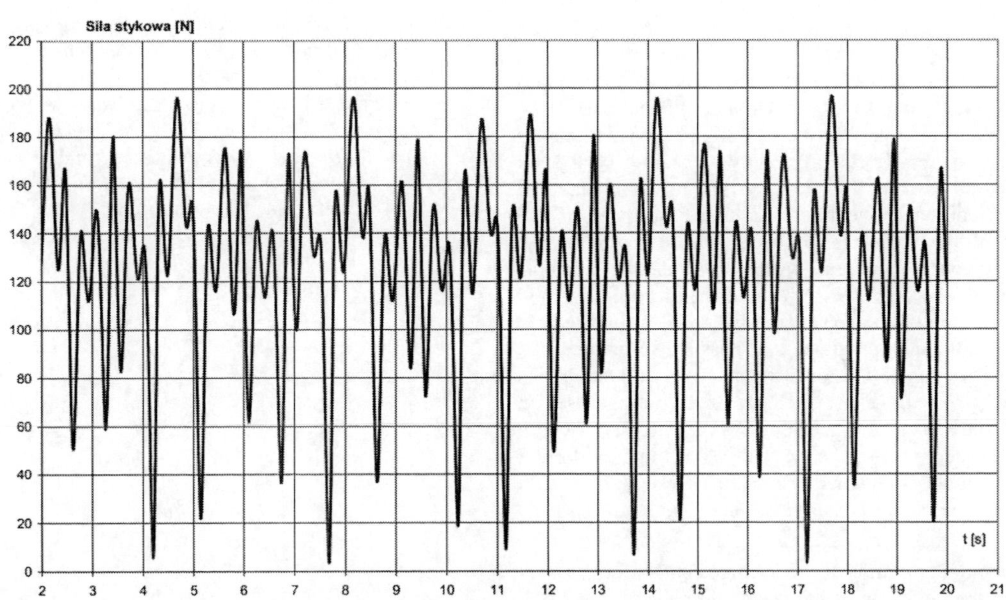

Fig. 8. Simulation result for reference catenary system and pantograph. In the pantograph, the rigidity of the frame has been increased by 10 times, speed 250 km/h

The comparison of simulation results is presented in the table 1 shows the correctness of the dynamics of the simulated system and thus the correctness of the simulation method used.

Table 1. Results of simulation analysis of pantograph cooperation - overhead contact line

Parameters	Required range	Simulation results
Speed [km/h]	250	
Obtained results	According to the standard [6]	According to the simulation
Average contact force F_M [N]	from 110 to 120	119
Standard deviation of the contact force σ [N]	from 26 to 31	31
Statistical maximum contact force $F_M+3\sigma$	from 190 to 210[1]	212
Statistical minimum contact force $F_M-3\sigma$	from 20 do 40	26
Actual maximum of the contact force [N]	from 175 to 210	175
Actual minimum of the contact force [N]	from 50 to 75	62
Maximum uplift at support [mm]	from 48 to 55	U_5 =48, U_6 =52, U_7=54
Percentage of losing contact [%]	0	0

4. Summary

The authors of the article presented a computer program (tool) with the help of which it is possible to make simulations allowing for the analysis of the pantograph's cooperation with the catenary system. Performed simulations with the data given by the standard confirmed that the obtained results for the speed of 250 km/h of the reference pantograph with the reference catenary system are in accordance with the requirements of PN-EN 50318:2019-02 [6] and therefore the method described should be considered validated. With the help of this computer program it is possible to analyze the quality of cooperation for different types of pantographs differing in parameters: stiffness of the springs used in the electric head suspension, dry or viscous friction, concentrated mass, static force setting, different aerodynamic conditions, travelling speed (up to 250 km/h). The analysis can also be subject to variable parameters of catenary system: that is, tensile forces, the traction cable, length of the overhead contact line span, distance of the contact wire, maximum and minimum stiffness of the overhead contact line, maximum and minimum mass of the reduced overhead contact line, aerodynamic forces, and speed of travel.

Using the model, you can analyze changes in contact force over time, displacement of the contact point, displacement, velocity and acceleration of the first and second stage of pantograph suspension.

[1] Accepting the limits of the average contact force 120 N and the deviation of the standard contact force 31 N, the maximum statistical contact force calculated from the relationship $FM + 3\sigma = 120 + (3 * 31) = 213$ N

References

1. Vologin V.A. Interaction between pantograph and contact line, VNIIZhT, Moscow (2006): (rus)
2. Grajnert J.: *Nowoczesne rozwiązania konstrukcyjne odbieraków prądu*. Trakcja i Wagony nr **5** (1978)
3. Grajnert J.; *Wpływ niektórych parametrów odbieraka prądu na siłę stykową,* Pojazdy szynowe nr **4** s. 7 – 13 (1978)
4. Kaniewski M.: *Model matematyczny odbieraka prądu i sieci jezdnej*. Prace Naukowe Politechniki Warszawskiej Transport z. **95**, s. 209-220 (2013).
5. Kobielski A.: *Dobór parametrów odbieraka prądu za pomocą systemu komputerowego*. Trakcja i Wagony nr **2** (1985)
6. PN-EN 50318:2019-02 - wersja angielska - *Zastosowania kolejowe. Systemy odbioru prądu. Walidacja symulacji oddziaływania dynamicznego pomiędzy pantografem a siecią jezdną górną*.
7. Zdziebki P., Martowicz D., Uhl T.: *Modelowanie hybrydowe do symulacji interakcji pantograf i sieć trakcyjnej*.
8. Thomas Reichmann, *Simulation des Systems Oberleitungskettenwerk und Stromabnehmer mit der Finite-Elemente-Methode*. Elektrische Bahnen **103**, 1-2. s. 69-75.
9. Rojek A.: *Short Circuit Current Breaking by High Seed Circuit Breakers in Traction Substatin*. Problemy Kolejnictwa, **177**, s. 51-59.
10. Głowacz M., Kaniewski M., Rojek A.: *Overhead contact line systems for high-speed rails. High-speed rail in Poland*. Advances and Perspectives. CRC Press/Taylor & Francis Group. ISBN 978-1-138-54469-7, str. 279-300 (2018)

Modelling of the electric locomotion DS3 working

Dmytro Bannikov[1,*], *Anatoliy* Radkevich[1], and *Antonina* Muntian[2]

[1] Dnipro National University of Railway Transport named after Academician V. Lazaryan, Department of Construction Production and Geodesy, 49010 Lazaryan St., 2, Dnipro, Ukraine
[2] Dnipro National University of Railway Transport named after Academician V. Lazaryan, Department of Foreign Languages, 49010 Lazaryan St., 2, Dnipro, Ukraine

Abstract. The purpose of the work is to evaluate the spatial distribution of the rigidity of the carrier frame and the body of the Ukrainian electric locomotive DC3 as a single system during operation and repair. The research was carried out on the basis of the finite element method with the application of design and computing complex SCAD for Windows. The numerical model of the electric locomotive was built, tested and then used to identify the causes and providing guidance on solving some specific operational questions. They are included, for example, the possibility of eliminating the body skew when jacking up on jacks and jamming the door as a result. The researches showed that the structure of the electric locomotive DS3 in general has a rather high spatial rigidity, both in transverse and longitudinal directions, and on torsion. However, for some practical repair tasks there is not enough for that rigidity. It was recommended to increase the thickness of the shell of its body up to 4 mm or the roof up to 8 mm, which leads to an increasing in the total mass of the machine by about 2,5 and 3,5 tons, respectively.

1 Problem statement

In the late 90's of the twentieth century, in Ukraine, at the premises of the Dnipro Electric Locomotive Building Plant (DELBP), the development of the constructional part of the new mainline cargo and passenger electric locomotive of alternative current – fig.1 was launched. 1. At the same time, Siemens Corporation partook in the constructional design of the electrical section of the car. The first 18 pilot electric locomotives were manufactured during 2002 – 2008 and put into operation at the Southwest Railway (place of registration – Kyiv-Passenger depot). It was planned that the car will be used on international routes as a new generation of modern electric locomotives additionally intended for the markets of China and India.

Structurally, the engine is a one-section 17 m long locomotive with a nominal weight of 90 t designated for a track of 1520 mm wide. Construction speed is 160 km/h.

Since 2008 the production of these cars has initially been slowed down and then completely suspended. According to various data, today there are only 3 engines, which are in satisfactory condition. Other engines are either decommissioned, mainly due to lack of spare parts, or under repair.

Along with the economic reasons associated with the global crisis of 2008 – 2010, some exploitation and maintenance issues of the engine in question have had no small share in this situation.

2 Analysis of recent research and publications

The constructional design of the electric locomotive DS3 was carried out in accordance with the norms [1], which at that time were, in fact, the only available norms that regulated the development of the supporting elements of electric locomotives. Concurrently, the constructional design was carried out on the basis of traditional approaches that were based on consideration of the electric locomotive as a system of solids [2], which together represented a mechanism. Without a doubt, such a simplified modeling of the engine required a great deal of ingenuity from constructional design engineers, since such models had far more assumptions than factual data. Therefore, to get a foresight about the actual work of an engine designed under such circumstances was possible only after its manufacturing on the basis of relevant testing.

Similar testing was carried out in 2003 by specialists of the branch research laboratory (BRL) of the dynamics and sustainability of the rolling stock of the Dnipro National University of Railway Transport named after academician V. Lazaryan. Their results were presented in a series of publications: static sustainability tests in the publication [3], dynamic sustainability tests – in the publication [4], and shock tests – in the publication [5].

* Corresponding author: bdo2 020@yahoo.com

Fig.1. Mainline electric locomotive DS3.

The main conclusion, based on the review of these works, made by the authors, is that, in general, the electric locomotive DS3 meets the requirements of the norms. However, there are individual places in its construction that require additional attention and should either be reinforced or refined, since the load in them is too big. Separately maximum speeds for the motion of the electric locomotive DS3 in the curves of different radii [6], in railroad switches [7] were outlined, and joints, which needed improving, in particular, brake transmission were defined [8].

On the basis of the given recommendations the modernization and improvement of the constructional design of the electric locomotive DS3 was carried out, which increased the engine's operational reliability. remain the only alternative.

2.1 Unresolved issues

Exploitation of part of the experimental batch of DS3 electric locomotives during 2004 – 2007 revealed the necessity to carry out the engine's lifting on jacks during the work on maintenance and repair of the car. In this case, there was often a jamming of the superstructure's door, which complicated the carrying out of repair work.

In this regard, the question about the supporting capacity of the superstructure of the engine in the event of its possible uneven supporting arose. Such calculation is not provided for by norms [1]. In addition, it needed a qualitatively different approach that would allow taking into account the dispensation of rigidity in the supporting frame and the superstructure of the electric locomotive, under the condition of changing the nature of the engine's supporting.

The rapid development of computer technology in the beginning of the twenty-first century gave impetus to the development of qualitatively different approaches to the constructional design of mechanical-engineering in general, and locomotives in particular. Thus, one of the methods of building mechanics – The Method of Finite Elements – got widespread development (MFE) [9], and on the basis of which there were developed many software products for its computer implementation. One of these products is the project-computational complex SCAD for Windows, which has become very popular on the domestic market since approximately 2005 – 2007. Despite its construction orientation, it has successfully been applied in the machine-building industry as well [10].

In international practice, more versatile software products under MFE have been developed, such as ANSYS [11], NASTRAN [12], CosmosWorks [13], or SolidWorks [14]. They allowed for machine-building constructions to perform not only linear static analysis, but also nonlinear analysis [15], dynamic analysis [16], and sometimes even investigating the processes of destruction of material [17]. However, up to now, their cost in the market is very high, so powerful domestic software development, such as SCAD for Windows [18] or Lira for Windows [19], remain the only alternative.

2.2 Purpose of the research

The assessment of the spatial distribution of the rigidity of the supporting frame and the superstructure of the DS3 electric locomotion as a single system with the use of software products for the MFE implementation was identified as the main goal of the research carried out by the authors.

Moreover a number of following tasks were subsequently formulated and solved. Firstly, it was necessary to construct a numerical model of the electric locomotive, which would include appropriate boundary conditions and load for different modes of its operation.

Secondly, the created model had to be checked by comparing the results of calculations with experimental data. Thirdly, on the basis of the tested model of the engine, to evaluate the nature of the distribution of the rigidity of the engine and provide recommendations for repair and maintenance work, in particular with the use of jacks.

3 Main material presentation

To construct a finite-element model (FE-model) of the DS3 electric locomotive and to conduct further research of the engine's work on its basis, SCAD for Windows versions 11.1-11.3 was selected. Due to its construction orientation during the simulation, it became necessary to adapt this product in a certain way for the purpose of modeling machine-building constructions. The main methods of calculation, developed and tested by the author in this regard, are highlighted in his publication [10]. As a result, it was possible to accurately illustrate in the model both necessary boundary conditions in the course of engine's functioning in different operating modes, and the nature and location of load of sources that created static and dynamic impacts on the structure.

Designed by the authors the FE-model of the electric locomotive DS3 is presented in Fig. 2. It is a plate-core system and consists of about 50,000 elements, mainly of isoparametric type. The total number of joints reaches 40,000, which results in about 250,000 degrees of freedom.

The carrier beam of the electric locomotive frame and the elements of the carcass and the superstructure of the engine were modeled by core finite elements, for which the required cross section was set. Coating of the superstructure, cab and deck of the electric locomotives, as well as the interior floor, were modeled with lamellar finite elements of required thickness. For the simulation of the location of internal equipment, special "rigids" the rods of considerable rigidity and at the same time insignificant mass were used. With their help, it was possible to accurately place on the model the available masses from the elements of the equipment.

Numerical analysis and all necessary calculations were carried out in a physically and geometrically linear formulation. Thus, the first of the formulated research tasks was solved.

For the solution of the second task of research, a comparison of the results of calculations based on the FE model of the electric locomotive DS3 with the results of its experimental studies, which, as already noted above, were performed in 2003 by the specialists of the GNDL of the dynamics and sustainability of the rolling stock of the Dnipro National University of Railway Transport named after academician V. Lazaryan.

The comparison was performed for the most loaded and therefore the most dangerous first mode of the engine's functioning, which assumed the effect of the compressive force of 2500 kN. The scheme of placing of experimental sensors on the elements of the electric locomotive, together with their original designations taken during the experiment, according to the functioning [3] is shown in Fig. 3. The stress values at these points, obtained after processing data measurements of strain indicators, according to the results of the summary report, are given in Table. 1. Additionally, there are the strains obtained from the FE modeling. The error in the last column is computed in relation to the experimental values.

In general, the comparison of the results showed a fairly good level of commonality between numerical and experimental data. The average error was about 15%. However, at the point YP1 (highlighted in darker colors) the computational stress was significantly different from the experimental data. Unfortunately, we could not determine the reason for this deviation. Most likely it is connected with the fault of the sensor itself or its poor-quality connection to the elements of the construction. On the cross-bearer beams of the superstructure the stress values were generally low in absolute magnitude and slight deviations from them already give a high error. Therefore, for some sensors with a stress level below 10 MPa (limit of instrumental error) this error was not calculated.

It should also be noted that the FE-model was originally constructed only for the engine frame with a part of its coating. However, in contrast to the experimental data, the discrepancy was very significant and reached 2-3 times. Therefore, it was decided to modulate also the superstructure, the roof and the entire internal system of load bearing partitions (the final model is shown in Fig.2. The results obtained after this are shown in Table 1.

Thus, one of the interim conclusions of the research was the conclusion on the significant impact of the superstructure on the overall stress-strain state (SSS) of the electric locomotive structure and the necessity of its mandatory consideration in conjunction with the carriage frame.

The tried and tested FE model of the electric locomotive DS3 was further used to assess its deformed state when using jacks for repair work. Fig. 4 shows the scheme and markings of the points in which the assessing of movement of superstructure and doorways of the electric locomotive DS3 in different directions during the change of the points of support was performed. In Tables 2 and 3 there are the results obtained for different variants of the location of the supports.

From the analysis of the obtained data it was found that approximately equally unfavorable are the cases of leaning of the superstructure of the electric locomotive DS3 against both two and three diagonally placed supports. Moreover, the largest vertical deflections occur in the buffer beams and reach values of approximately 13 mm, which practically does not affect the angle of twisting of the superstructure of the electric locomotive and can be considered as relatively acceptable. However, moving the doorways at the same time reaches the value of approximately 7 mm, which practically leads to the jamming the doors. It should also be noted that the strain in the elements of the supporting frame and the superstructure of the engine did not exceed 50 MPa, which is a totally acceptable level.

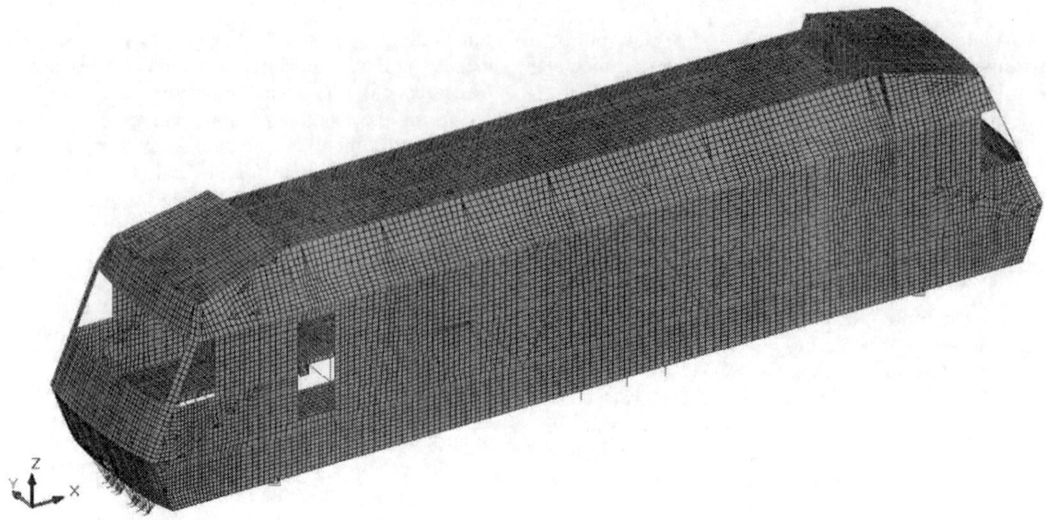

a)

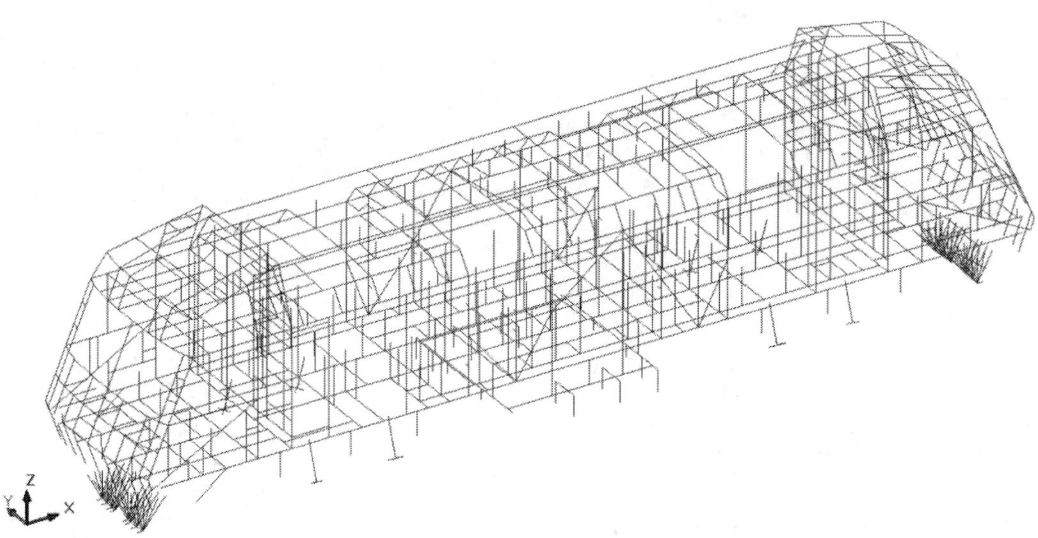

b)

Fig. 2. Numerical model of electric locomotive DS3: a – general view; b – rod frame of the model.

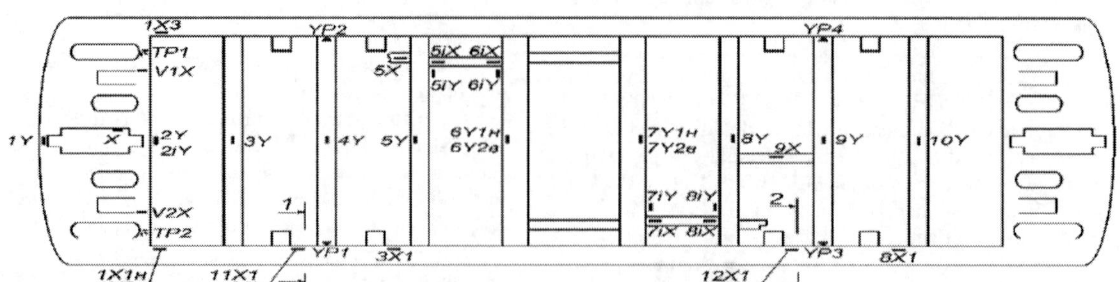

Fig. 3. Scheme of control sensors' placement.

After detecting this characteristic of the engine the developers offered several options for enhancement. It should be noted that all suggestions regarding the reinforcement of supporting structures of the electric locomotive DS3 encountered a constructive limitation of the weight of the engine in 90 tons, because the actual mass of the designed engine was already very close to it. There were even attempts to lower it due to the installation of hatchways in various elements of the engine. However, this did not add to the overall rigidity of the electric locomotive construction, but on the contrary, lowered it.

It should be noted that similar simulation of various variants of DS3 weight reduction was also carried out by the author of the publication using the project-computational complex SCAD for Windows.

In the result, as the main working decisions, two of the most effective options were selected – an increase of thickness of the superstructure's coating from 1.5 mm to 4 mm, or an increase of the thickness of the engine's roof's coating from 1.5 mm to 8 mm. According to numerical simulation, both options allow to increase the general rigidity of the electric locomotive superstructure and reduce deformation values by 15 – 25%. Under such conditions, the total weight of the engine increases in the first case by about 2,5 tons, and in the second by about 3,5 tons.

A corresponding change in the characteristics of the electric locomotive DS3 required additional special tests. However, due to the gradual curtailment of the production of this engine, such works were not carried out.

Table 1. Comparison of experimental and numerical data.

№	**Sensor**	**Stress values, MPa**		**Error (%)**
		Expe-riment	**FE-modeling**	
Main beams of the base frame of the carriage body				
1.	1X1	-185	-155 ÷ -230	-16 ÷ +24
2.	1X2	-77	-67	-13
3.	11X1	-116	-105 ÷ -125	-9 ÷ +8
4.	11X2	-85	-76	-11
5.	3X1	-157	-110 ÷ -120	-30 ÷ -24
6.	12X1	-108	-100 ÷ -120	-7 ÷ +11
7.	12X2	-69	-74	+7
8.	8X1	-123	-130 ÷ -135	+6 ÷ +10
9.	YП1	54	-25	-146
Cross-bearer beams of the base frame of the carriage body				
10.	3Y	0	+9	–
11.	4Y	-1	+16	–
12.	5Y	+3	+16	–
13.	5iY	+18	+16	-11
14.	5X	-9	-6	–
15.	5iX	-9	-7	–
16.	6Y1	+11	+13	+18
17.	6iY	-3	-6	–
18.	7iX	-13	-9	-31
19.	7iY	+10	+10	–
20.	8iX	-18	-7	-61
21.	8iY	+8	19	–
22.	9X	-48	-35	-27
Buffer beam				
23.	1Y	290	-225 ÷ -240	-22 ÷ -17
24.	2Y	103	+100 ÷ +115	-3 ÷ +12
25.	2iY	82	+100 ÷ +110	+22 ÷ +34
26.	B2X	14	+15	+7
27.	X	225	+150	-33
28.	Y	99	+100 ÷ +110	+1 ÷ +11

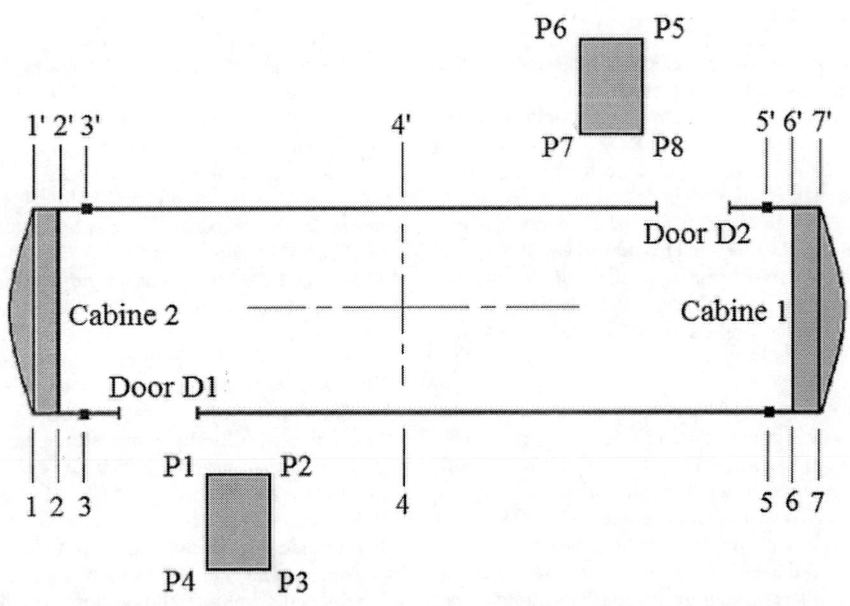

Fig. 4. Points for assessing the movement of the body and doorways of the electric locomotive DS3 while changing the points of support

Table 2. Deflection of the superstructure of the electric locomotive DS3.

Point	1 (1')	2 (2')	3 (3')	4 (4')	5 (5')	6 (6')	7 (7')
Coordinate	**0**	**1105**	**2080**	**7890**	**13700**	**14675**	**15780**
Supporting:							
4 points	-0,04 (-0,04)	-0,06 (-0,07)	-	-1,44 (-1,40)	-	-0,07 (-0,07)	-0,03 (-0,05)
3 points along the diagonal of the doors	-0,64 (-12,97)	-0,36 (-12,36)	0,00 (-11,66)	-1,58 (-7,59)	0,00 (0,00)	0,34 (0,46)	0,67 (1,10)
3 points diagonally without doors	0,62 (1,05)	0,31 (0,42)	0,00 (0,00)	-1,37 (-6,67)	0,00 (-10,98)	-0,36 (-11,64)	-0,59 (-12,22)
2 points along the diagonal of the doors	0,57 (-3,66)	0,20 (-3,70)	0,00 (-3,70)	-5,75 (-3,43)	-8,02 (0,00)	-8,43 (-0,11)	-8,75 (-0,10)
2 points diagonally without doors	-13,11 (-0,70)	-12,47 (-0,41)	-11,69 (0,00)	-7,07 (-1,03)	0,00 (0,64)	0,41 (0,99)	1,17 (1,34)

Table 3. Moving of the doorways of the electric locomotive DS3.

Supporting:	Moving	Door D1					Door D2				
		D1	D2	D3	D4	Δ	D1'	D2'	D3'	D4'	Δ
4 points	longitudinal	0,19	0,18	0,26	0,25	**0,08**	-0,19	-0,19	-0,27	-0,26	**0,08**
	transverse	0,15	0,18	0,12	0,13	**0,06**	1,61	1,65	1,68	1,63	**0,07**
	vertical	-0,33	-0,79	-0,66	-0,27	**0,52**	-0,33	-0,78	-0,64	-0,27	**0,51**
3 points along the diagonal of the doors	longitudinal	-0,55	-0,53	0,19	0,15	**0,70**	2,59	2,57	3,65	3,69	**1,12**
	transverse	9,59	8,76	3,02	3,53	**6,57**	-1,66	-0,74	-2,16	-2,71	**1,97**
	vertical	-0,44	-1,12	-0,84	-0,33	**0,79**	-0,97	-2,31	-2,03	-0,84	**1,47**
3 points diagonally without doors	longitudinal	0,87	0,84	0,33	0,34	**0,54**	-2,31	-2,29	-3,50	-3,51	**1,22**
	transverse	1,47	2,13	0,73	0,43	**1,70**	13,21	12,61	7,28	7,55	**5,93**
	vertical	-0,12	-0,40	-0,40	-0,13	**0,28**	-10,80	-10,42	-10,41	-10,81	**0,40**
2 points along the diagonal of the doors	longitudinal	1,34	1,36	0,84	0,80	**0,56**	4,17	4,15	3,99	4,03	**0,18**
	transverse	2,18	1,42	0,55	0,99	**1,63**	-7,95	-7,10	-3,61	-4,08	**4,34**
	vertical	-0,81	-1,96	-1,67	-0,69	**1,27**	-0,61	-1,49	-1,21	-0,49	**1,00**
2 points diagonally without doors	longitudinal	1,90	1,88	3,19	3,20	**1,32**	-0,59	-0,57	0,04	0,03	**0,56**
	transverse	-9,91	-9,40	-3,65	-3,79	**6,26**	-0,60	-1,05	0,71	0,82	**1,76**
	vertical	-11,55	-11,12	-11,12	-11,56	**0,44**	0,49	0,18	0,19	0,48	**0,31**

4 Conclusions

On the basis of the finite-element simulation of the mainline cargo and passenger electric locomotive of alternative current DS3 functioning, one should state the following:

1. The construction of the electric locomotive as a whole has a rather high spatial rigidity, both in transverse and longitudinal directions, and on torsion. This is confirmed by the conclusions made through a series of various tests by the specialists of the branch research laboratory (GNDL) of the dynamics and sustainability of the rolling stock of the Dnipro National University of Railway Transport named after academician V. Lazaryan.

2. In a number of situations related to the repair work of this engine, there is a need for lifting its superstructure on the jacks. Furthermore, even in the case of uneven lift on two diagonally placed supports, the spatial rigidity of the engine's body is sufficient to provide a suitable level of stress-strain state of the carrier frame and the superstructure of the electric locomotive. However, at the same time, there is a movement of the doorways, which is ample for superstructure doors jamming/

3. Elimination of this characteristic of the electric locomotive DS3 requires an increase in its superstructure's coating to 4 mm or the roof's coating to 8 mm, which leads to an increase in the total mass of the engine by about 2.5 and 3.5 tons, respectively. Taking into account the constructional limitation of the weight of the engine at 90 tons, as well as the need for special additional tests, such a reinforcement of the electric locomotive cannot be recommended as final.

4. Simulation with the help of FE-modeling, in particular on the basis of project-computational complex SCAD for Windows, can be used with high probability to simulate the operations of the electric stock structure of the railways s in general and electric locomotives in particular.

In concluding, the authors of the publication would like to express their sincere gratitude to the team of developers of project-computational complex SCAD for Windows, who kindly provided the authors at one time with the possibility of conducting scientific research and practical calculations – doctor of technical sciences. A.V. Perelmuter, candidate of technical sciences. A.M. Perelmuter, candidate of technical sciences V.S. Karpilovsky, Ye. Z. Kryksunov, O.M. Trofymchuk.

Special thanks to the authors would like to express the staff of the Dnipro Electric Locomotive Building Plant (DEBZ), in the person of the Chief Design Manager of the electric locomotive DS3, candidate of technical sciences. A.M. Hryvnak and the Head of the Design and Development Department G.O. Rozental for the sharing with the authors of the materials regarding the experimental tests of the electric locomotive DS3, its constructional part, as well as the available statistical data on the operation of this engine.

References

1. Norm for a calculation and estimation of durability of bearing elements, dynamic qualities and affecting way of carriage part of locomotives of railways of MPS Russia for tracks 1520 мм. Moscow, Russia. 145 (1998)
2. V.Ja. Panasenko, I.V. Klimenko. *Structural analysis of the suspension system of the electric locomotive DS3*. Visnik of Dnipropetrovsk National University of Railway Transport named after academician V. Lazarjan. 40, 37-39 (2012)
3. E.P. Blohin, S.A. Kostritsa, V.V. Chumak, N.P. Ostroverhov, A.V. Sultan, V.N. Datsenko, A.E. Krivchikov. *The results of the strength tests of the electric locomotive DS3*. Visnik of Dnipropetrovsk National University of Railway Transport named after academician V. Lazarjan. 5, 13-16 (2004)
4. E.P. Blohin, A.N. Pshinko, M.L. Korotenko, R.B. Granovsky, N. Ja. Garkavy, E.M. Dzichkovsky. *Dynamics of high-speed electric locomotive of type DS3*. Visnik of Dnipropetrovsk National University of Railway Transport named after academician V. Lazarjan. 5, 9-12 (2004)
5. E.P. Blohin, A.M. Bondarev, V.L. Gorobets, A.E. Krivchikov, V.V. Chumak, D.A. Jagoda. *The results of shock tests of the main electric locomotive DS-3*. Visnik of Dnipropetrovsk National University of Railway Transport named after academician V. Lazarjan. 8, 16-20 (2005)
6. V.V. Ribkin, A.P. Taturevitch, S.N. Sova S. N. *The results of studies to establish the conditions of circulation of the electric locomotive DS3-001 on the railway network*. Visnik of Dnipropetrovsk National University of Railway Transport named after academician V. Lazarjan. 8, 166-169 (2005)
7. V.V. Ribkin, M.I. Umanov, A.P. Taturevitch, V.V. Tsiganenko, N.V. Halipova, V.V. Kovalev, V. V. *The results of experimental studies on the impact on the path of the electric locomotive DS3-001*. Visnik of Dnipropetrovsk National University of Railway Transport named after academician V. Lazarjan. 5, 183-187 (2004)
8. V.Ja. Panasenko, I.V. Klimenko. *To the question of modernization of the brake lever transmission of the electric locomotive DS3*. Visnik of Dnipropetrovsk National University of Railway Transport named after academician V. Lazarjan. 8, 88-91 (2005)
9. G.R. Liu, S.S. Quek, S. *The Finite Element Method. A Practical Course*. Amsterdam, Elsevier LTD (2014)
10. D.O. Bannikov. *Use of building-oriented software SCAD for the analysis of the work of machine-building constructions*. Visnik DNURT "Science and Progress of Transport". 73, 98-111 (2018)
11. B. Wang, Zh. Yong, J.Ch. Rong. *Several Realizations of the Finite Element Software ANSYS*. Structural Analysis Applied Mechanics and Materials. 442, 507-510 (2013)
12. H.F. Gao. *The Development and Application of Parallel Computation for Structural Analysis Based on Nastran Software*. Applied Mechanics and Materials. 778, 41-45 (2015)
13. Ch. Khelifi, M. Ouali, F. Ferroudji, L. Adjilout L. *Modeling and Static Analysis of an Areogenerator Savonius Cracked by Using SolidWorks/CosmosWorks Software*. Applied Mechanics and Materials. 446-447. 744-750 (2015)
14. E.H. Toreh, M. Shahmohammadi, N. Khamseh N. *Kinematic and Kinetic Study of Rescue Robot by SolidWorks Software*. Research Journal of Applied Sciences, Engineering and Technology. 5-21, 5070-5076 (2013)
15. M. Kathrotiya, M. *Construction Stage Analysis of Flat Slab Structure with respect to Non Linear Time History Analysis using Software Aid*. International Journal for Research in Applied Science and Engineering Technology. 6-4, 4833-4839 (2018)
16. R. Vysoký, R. *Current Capabilities of Modal Analysis of Aircraft Propeller in ANSYS Mechanical Environment*. Advances in Military Technology. 12-1, 33-47 (2017)
17. W.M. Li, Ya.Ya. Yan, Yi. Xing. *The Numerical Simulation of Punching and the Fatigue Analysis of Punch Based on ANSYS/LS-DYNA and ANSYS-Fatigue Tool*. Advanced Materials Research. 904, 469-473 (2014)
18. V.S. Karpilovsky, E.Z. Kriksunov, A.A. Maljarenko, S.Yu. Fialko, A.V. Perelmuter, M.A. Perelmuter, M. *SCAD Office. Versiya 21. Vychislitelnyy kompleks SCAD++*. Moscow, SCAD SOFT. 848 (2015)
19. R.Yu. Vodopyanov, V.P. Titok, V. P., A.E. Artamonova. *Programmnyy kompleks Lira-SAPR 2014.Rukovodstvo polzovatelya. Obuchayushchie primery*. Moscow. 394 (2014)

Rationing of the number of signals and interlockings in the operational stock of railway stations

Iryna Lebid[1,*], *Dmitrii* Shevchenko[2], *Irina* Kravchenya[3], *Nataliia* Luzhanska[4], *Georgii* Prokudin[1], and *Myroslav* Oliskevych[1]

[1] National Transport University, Department of International Transportation and Customs Control, Mykhaila Omelianovycha-Pavlenka Str., 1, Kyiv, Ukraine, 01010
[2] BelSUT, department «Information management systems and technologies», Kirov str., 34, Gomel, Belarus, 246653
[3] BelSUT, department «Road and Traffic Management», Kirov str., 34, Gomel, Belarus, 246653
[4] National Transport University, Department of Transport Technologies, Mykhaila Omelianovycha-Pavlenka Str., 1, Kyiv, Ukraine, 01010

Abstract. The operational stock of signals and interlockings is necessary to ensure the uninterrupted operation of the electric interlocking systems in the event of their component failure. The existing standards for the composition and number of signals and interlockings available in the operational stock of railway stations are not mathematically based. It is considered the task of justifying the number of signals and interlockings in the operational stock of railway stations, and the factors affecting its quantity are indicated. It is proposed the mathematical model for rationing the amount of equipment necessary to ensure the uninterrupted operation of station signals and interlocking systems with a specified probability. The process of using and replenishment of reserve equipment is described using the the single-channel Markov queuing system model with a limited queue. The proposed probabilistic methodology approbation for the rationing of the amount of reserve equipment was carried out at the stations of the two Belarusian Railway divisions totaling to 58 stations equipped with relay and computer-based electric interlocking. It is shown the efficiency of the calculated amount of equipment in comparison with the existing values.

1 Problem statement

At railway stations, it is stored the stock of relay and other electronic equipment operated at a station (SPTA – spare parts, tools and accessories). This stock is needed to ensure uninterrupted operation of signals and interlocking systems in case of their component failure.

For stations equipped with relay electric interlocking (EL), there are no standards for the number of spare parts, and, in fact, their actual quantity was established historically and approved at a regulatory level. For stations equipped with computer-based interlocking (CBI), the number of SPTA, in accordance with the documents [1, 2], has to be at least 10% of the total amount of the used equipment of each type.

However, this standard incorrectly (only linearly) takes into account the amount of equipment in use and does not take into account:
– equipment failure rate;
– possibility and efficiency of stock replenishment.

For example, at a station there can be operated 10 impulse reed relays with a rectifier (plus 1 relay at stock) and 200 highly reliable neutral small-sized plug-in relays (plus 20 relays at stock).

Obviously, such a ratio of operated and spare relays is unjustified.

The most important thing is that the standard of the number of signals and interlockings given in [1, 2] and available in the operational stock of the railway stations of Belarusian Railway is not justified on the basis of the predetermined probability of SPTA absence.

2 Mathematical model of rationing of the number of signals and interlockings

2.1 Model description

Let us consider the process of SPTA usage and replenishment at a railway station.

Suppose at the station it is operated N items of signals and interlockings of the same type with a failure rate of λ, and there are X more devices at stock (Figure 1, *a*).

In case one of N devices fails (with the total rate of $N \cdot \lambda$) it is replaced with one of X spare devices and is given for repair to a maintenance area (MA). After some time T_s, a similar device will be returned to the station (see Figure 1, *b*).

If during the stock replenishment period T_s, X devices will additionally fail (a total of $X + 1$), then the last of them cannot be replaced, and the transportation process will be disrupted due to SPTA absence (see Figure 1, *c*).

* Corresponding author: i.h.lebed@gmail.com

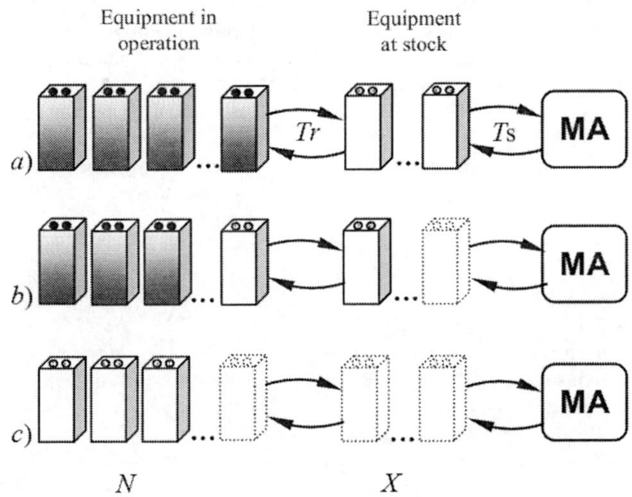

Fig. 1. Stock usage and replenishment chart [compiled by the authors]

The usage and replenishment process of a certain SPTA type can be described by a single-channel Markov queuing system (QS) model with a limited queue M / M / 1 / (X – 1) (Figure 2).

The incoming flow of requests is formed by the failure of X items of the same equipment type requiring replacement. The rack with N reserve equipment items is modeled by the servicing device (1 item) and the queue (N – 1 items).

At the same time, the servicing device also performs the functions of personnel engaged in SPTA replenishment (transferring faulty equipment to a maintenance area and returning similar operative parts to the station).

A repair request (a failed signaling arrangement) that faces N requests in the queuing system is rejected due to the absence of SPTA (all N spare signal and interlocking equipment is already in operation).

Thus, the task of SPTA rationing comes to determining (for each equipment type) the smallest value of N for which the actual probability of maintenance rejection Q_N would not exceed the standard probability P for given values of the equipment amount X and its failure rate λ [3].

$$Q_X = \rho^X \left(\sum_{i=0}^{X} \rho^i \right)^{-1} \quad (1)$$

where

$$\rho = \lambda \cdot N \cdot T_s,$$

T_s is stock replenishment average time.

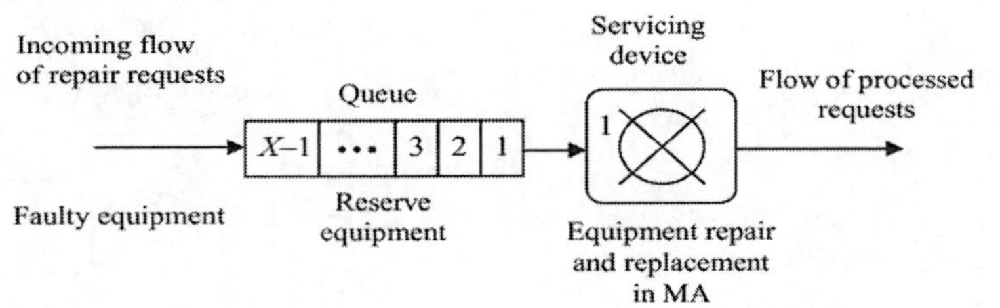

Fig. 2. Conceptual model of stock usage and replenishment
Source: compiled by the authors based on [3]

2.2 Mathematical model assumptions

Let us consider the basic assumptions of the proposed mathematical model.

The failure flow of all equipment types is assumed to be the simplest, i.e.

– stationary (the flow for which the mathematical expectation of the equipment failure rate per unit of time does not change in time);

- ordinary (meaning the practical impossibility of two or more simultaneous equipment failure manifestations);

- forgetful (means that the number of equipment failures in the past does not determine how many of them will occur in the future) [4].

This assumption is traditional when analyzing the reliability of electronic signal and interlocking systems. It is valid for systems operating in the period of normal use in which case component failures are practically independent [5-7].

It is assumed that stock replenishment time T_s is characterized by an exponential distribution.

Additional studies on the simulation model in the GPSS system [8-10] have shown a slight change in the probability of maintenance rejection when replacing the exponential distribution of service time with other distributions which, compared to the exponential law, have less variance with the same mean value.

It is assumed that SPTA replenishment is done gradually piece by piece (the Markov queuing system is single-channel).

In fact, several devices (and also of various types) can be delivered to / from a maintenance area at the same time. This assumption, obviously, overestimates the standard values of the number of SPTA.

The proposed model of stock usage and replenishment assumes the availability of at least one item at stock which is consistent with a number of Belarusian Railway's governing documents.

As a source of data on the equipment failure rate, in practice, it is used guaranteed estimations (by means of upper confidence regions) that exceed the true values.

2.3 Probability substantiation P of the absence of a certain SPTA type

One of the important criteria for rationing the number of SPTA is the probability P of the lack of stocks, a reduction in which leads to an increase in the number of spare parts. Until now, the probability of the lack of signal and interlocking stocks P has not been specified and normalized [11-13]. The following approach is proposed for P specification.

The average non-failure operating time of station signal and interlocking systems (for an average railway station), as a rule, does not exceed 2000 hours (about 2.8 months), and their average recovery time is at least 1 hour. Consequently, the signal and interlocking system (*SI*) stationary availability (*A*) [11, 14] at an average railway station usually does not exceed the following:

$$K_A^{SI} < \frac{5000}{5000+1} = 0{,}9998\,.$$

Note. In the standard [15] (p.18) it is indicated that the normative value of the availability factor of devices and railway automation systems should not be less than 0.98.

Consequently, the unavailability factor (*UN*) of the signal and interlocking system (*SI*) is:

$$K_{UN}^{SI} = 1 - K_A^{SI} > 5{,}00\cdot 10^{-4}\,. \quad (2)$$

There is no much sense in the lack of stock probability provision P which would be significantly less than the railway automation and telemechanics unavailability factor. Therefore, it is proposed to specify the absence probability P of each SPTA type at

$$P = 0{,}5\cdot K_{UN}^{SI} = 2{,}50\cdot 10^{-4}\,, \quad (3)$$

so that the influence of the «lack of SPTA» factor on the reliability of the signal and interlocking system is relatively small (at 50%) as compared to the influence of the «automation and telemechanics failure» factor.

On the other hand, for unreliable equipment, the requirements for the probability P of SPTA absence should be specified not at 50% of the signal and interlocking system unavailability factor, but of the unavailability factor of the very equipment of a particular type.

As a result, the absence probability P of i-type spare equipment should be chosen as the greatest out of two numbers

$$P = \max\{2{,}5\cdot 10^{-4};\, 0{,}5\cdot K_{UN}^{i}\}\,, \quad (4)$$

where K_{UN}^{i} is the unavailability factor of i-type equipment [14] defined by the expression:

$$K_{UN}^{i} = 1 - \frac{(N\lambda)^{-1}}{(N\lambda)^{-1} + T_r}\,, \quad (5)$$

where N is the amount of i-type equipment operated at a station; λ is the failure rate of i-type equipment, 1 / hour; T_r is the average time of equipment recovery, hours.

3 Methodology approbation

3.1 Obtained results

For some typical source data in Table 1, we give the proposed methodology recommendations and compare them with the existing computer-based interlocking standard [1].

Here X_1 is the quantity of spare equipment according to the existing standard;

T_s is the stock replenishment average time (delivery of similar signal and interlocking equipment from the maintenance area), hour;

X_2 is the amount of spare equipment according to the proposed mathematical model.

Table 1. Standards of SPTA quantity

№	N	X_1	λ	T_s	X_2
1	20	2	10^{-6}	24	2
2	20	2	10^{-6}	12	1
3	20	2	10^{-7}	24	1
4	500	50	10^{-7}	120	2
5	10	1	10^{-6}	48	2

Source: compiled by the authors based on [3]

Note. In the standard [16] it is indicated that the failure rate of neutral small-sized plug-in relays is $1{,}1\cdot 10^{-7}$ 1 / hour.

It can be seen that for the source data in the first line, the results of the existing and stochastic approach are identical ($X_1 = X_2$).

In case when average stock replenishment time T_s may be reduced from 24 to 12 hours (see line 2, table 1), the operational equipment stock could be narrowed down from 2 to 1 item.

Besides, the operational equipment stock can be reduced to 1 item for highly reliable equipment (with the failure rate of no more than 10^{-7} 1 / hour, see line 3, table 1).

The stochastic model provides the greatest benefits for massively operated highly reliable equipment (N = 500, $X_2 = 2 << X_1 = 50$, see line 4).

For small equipment of a low or medium level of reliability, the stochastic model can suggest a larger stock (see line 5) than it is stipulated by the existing standard [2].

The proposed stochastic methodology of the stock rationing has the following advantages over the existing approach:
– objectively takes into account (and nonlinearly) the amount of operated equipment;
– makes allowance for the equipment failure rate;
– takes into account the efficiency of SPTA replenishment;
– includes the requirements for the probability of SPTA absence.

The implementation of the proposed stochastic methodology of operational stock rationing is characterized by the following practical importance:
– a general increase in the reliability of railway automation and telemechanics systems during their operation due to an increase in SPTA number at stations where its quantity is insufficient;
– as an objective, reasonable tool for calculating the volume of SPTA. It is planned to introduce automation into the calculation of operational stock standards in automatic signal and interlocking control system (a subsystem of a maintenance area) which is a road-level information system;
– minimization of the operational stock of signals and interlockings by reducing the number of SPTA where it is redundant;
– minimization of the cost of introduced railway automation and telemechanics (including imported items) due to equipment operational stock optimization included in the delivery set (a manufacturer / supplier is interested in increasing the scope of delivery).

The dependence of the automation and telemechanics availability factor (indeed, the equipment of each individual type) on the SPTA number is of an exponential nature (Figure 3). SPTA reduction (where it is redundant) results in a slight decrease in the availability factor of equipment of the type under consideration.

Where the amount of spare equipment is not enough an increase in SPTA by one item results in a significant availability factor increment.

Thus, in Figure 3, the standard value of the availability factor requires two pieces of equipment of a certain type. An increase in the number of SPTA from one to two items entails a considerable availability factor increment.

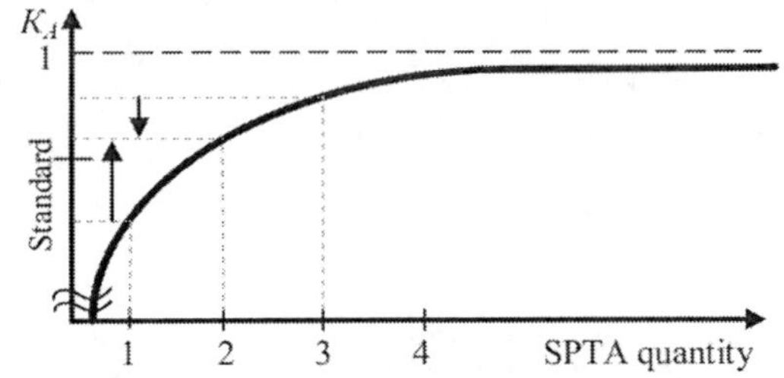

Fig. 3. Dependence of the equipment availability factor on the SPTA number [compiled by the authors based on [5, 14]

A reduction in the number of SPTA from three items to two implies an insignificant decrease in the availability factor.

As a result, the redistribution of SPTA equipment among signals and interlocking divisions is accompanied by a release of a certain amount of equipment.

This equipment can be used to eliminate the need for the purchase of a new signal and interlocking equipment.

Thus, in the course of the preliminary approbation of the stochastic methodology of SPTA rationing using the example of signals and interlocking division 1 (Minsk) of Belarusian railway, it has been established that the release of a certain SPTA amount can result in:
– more than 30 thousand rubles as a lump sum (on a one-off basis, due to the refusal to purchase similar equipment in the reporting period);
– more than 1.5 thousand rubles annually (due to cuts in expenditures on transportation and periodic inspection of SPTA in signal and interlocking maintenance areas).

3.2 Methodology realization

For engineering calculation of the number of SPTA equipment of each type at the station in accordance with the proposed methodology, the document [1] is proposed to be supplemented with an appendix where for typical values of failure rates λ (in columns) and the amount of operated equipment X (in rows), in the form of a two-dimensional table, it will be specified the required amount of spare parts. For the following typical values of average stock replenishment time T_s = 24, 48, 72, 120 hours and average recovery time T_r = 0.5, 1, 1.5, 2 hours, a separate table is filled in.

For equipment with known failure rate characteristics and stock replenishment time, as well as in case of non-standard values of these characteristics, an individual calculation of the number of SPTA is possible.

3.3 Specifying information on the equipment failure rate

The developed methodology involves the use of a known failure rate value of the equipment type under examination.

The most reliable statistical estimates of signals and interlocking reliability factors are those obtained during operation in real conditions on the basis of statistics on failures and total operating time.

The priorities in obtaining initial information on the reliability (intensity) of signals and interlockings are as follows:
1) statistical methods for estimating failure rates according to operating results;
2) manufacturer's data;
3) literary sources;
4) reliability manuals.

However, for a number of railway automation and telemechanics devices, the information on their failure rate may be missing. In this case, it is proposed the approach when the equipment failure rate is set indirectly based on the known maintenance interval of signal and interlocking equipment in a maintenance area.

Based on the comparison of information on maintenance intervals and known equipment failure rates, it has been found that the probability of failure during the period between inspections of various equipment types is mostly about 0.001-0.05. Moreover, for neutral small-sized plug-in relays 1, this probability does not exceed 0.0144. This value is selected as a reference one in determining the equipment failure rate based on the frequency of maintenance.

3.4 Approbation results

The approbation of the proposed stochastic methodology of SPTA rationing was carried out at a number of Belarusian Railway`s stations equipped with relay electric and computer-based interlocking. In the approbation process it was revealed the following:
– at stations equipped with electric interlocking, the actual number of SPTA is established by decrees of the division head. These values were formed "historically" in the process of more than 20 years of operating the equipment. At the same time, they practically coincide with the values proposed by the stochastic methodology;
– at stations equipped with computer-based interlocking, the number of spare equipment significantly exceeds the standard of 10%, which is largely due to the lack of objective information on the reliability of new equipment and the desire of staff to minimize the risks of SPTA shortage;
– the number of recorded failures of relay signals and interlockings is several times less than the real value, due to the existing approach of material incentives and punishment of division workers. It is almost impossible to hide computer-based interlocking equipment failures because of the automatic event documentation, and the fact that the repair of such an equipment type is usually performed by a manufacturer.

4 Conclusions

As compared with the existing deterministic approach, the proposed stochastic technique objectively takes into account the intensity of equipment failures and stock replenishment efficiency, as well as the likelihood of its absence (due to remote maintenance).

For the majority of signals and interlockings operated at stations (taking into account failure rates), the new approach justifies a reduction in the operational stock. It would make it possible to:
1) eliminate the need for the purchase of new equipment;
2) reduce the cost of SPTA storage.

For stations that are already in operation, the new standard would allow for clarifying and «legalizing» the existing number of SPTA.

References

1. Standard 19.292-2014 of the state association «Belarusian Railway». *Commissioning procedure of railway automation and telemechanics systems and devices* (2014)
2. Standard 09150.19.081-2008 of the state association «Belarusian Railway». *Railway automation and telemechanics equipment. Commissioning procedure, maintenance and repair of computer-based signals and interlocking* (2008)
3. D. Shevchenko *Probability Theory and Mathematical Statistics*, 318 (2006)
4. T. Hamdy, *Operations research: an introduction*, 815 (2007)
5. V. Sapozhnikov, Vl. Sapozhnikov, V. Shamanov, *Reliability of railway automatics, telemechanics and communication systems*, 263 (2003)
6. *ISA, «ISA-62443-1-1* – Models and Concepts,» in ISA-62443 – Security for Industrial Automation and Control Systems, **D6E2**, 3–115 (2016)
7. *Safety and Security requirements of Rail transport system in multi-stakeholder environments* (2017) http://www.cyrail.eu/IMG/pdf/cyrail-d21_-_safety_and_security_requirements_of_rail_transport_system.pdf
8. J. Thakur, N. Kumar, Int. J. Emerg, Technol. Adv. Eng., **vol. 1**, no. 2 (2011)
9. D. Shevchenko, I. Kravchenya, *Simulation modeling on GPSS*, 97 (2007)
10. GPSS World Reference Manual, Minuteman Software, USA (2009)
11. D.Shevchenko, I. Kravchenya, Autom., communic., comp. science, **10**, 10 – 12 (2017)
12. D.Shevchenko, I. Kravchenya, Safety problems in transport: materials of the 8th International applied research conference dedicated to the Year of Science, **1**, 206 – 207 (2017)
13. D.Shevchenko, I.Kravchenya, V. Golik, Bulletin of BSUT: Scie.&Trans, **2(33)**, 142 – 144 (2016)
14. D. Shevchenko *Fundamentals of the reliability theory*, 250 (2010)
15. Russian Federal Standard (GOST) 32668-2014. *Safety relays, relay units and racks. General specification* (2014)
16. Russian Federal Standard (GOST) 34012-2016. *Railway automatics and telemechanics equipment. General specification* (2016)

Selected aspects of proper integration between ERTMS/ETCS on-board and trackside devices.

Dominik Adamski*, *Andrzej* Białoń, and *Zwadka* Łukasz

Railway Research Institute, 04-275 Chłopickiego 50, Poland

Abstract. Achieving the interoperability of the European rail system requires many measures to unify technical solutions as well as regulations in each Member State. However, there is a possibility of some incompatibilities between individual subsystems despite generating them in accordance with unified applicable requirements. It is possible that the interoperable rolling stock will not be able to move freely over the interoperable railway line due to some incompatibilities and differences in the versions of the installed firmware in the ERTMS/ETCS system devices. In connection with the above the correct integration of rail vehicles with track-side equipment should be examined by means of tests under operating conditions. The article presents compliance tests of the correct integration of the on-board subsystem with the track-side subsystem which are carried out by the Railway Research Institute.

1 Introduciton

Over the years railway systems in particular Member States of the European Union have been developed independently using diversified technical solutions. The diversification of the technical rules, power supply, signaling, etc. has prevented the free movement of people and goods through the Member States. Therefore, it was decided to draw up the so-called White Paper - 'Roadmap to a single European transport area towards a competitive and resource efficient transport system' - where harmonization of technical requirements was considered to be the key issue in rail transport. The main goal was to lead to a situation in which trains will be able to cross the borders of European countries freely without stopping, on the contrary as it is today. In order to reach this goal it is necessary to achieve interoperability of the European rail system which will consequently lead to the creation of a single European railway area. This requires the elimination of a number of technical, legal and administrative obstacles. Thus, the railway system has been divided into structural subsystems [2]: infrastructure, energy, the control-command and signalling and rolling stock as well as functional subsystems: maintenance, operation and traffic management and telematics applications for passenger transport and freight transport. This division allowed drawing up specific Technical Specifications for Interoperability (TSIs) for particular subsystems, i.e. the requirements that must be met in order to achieve full technical harmonization of the railway. Dedicated to the relevant subsystems, the TSIs have the same structure, where the essential requirements, basic parameters, interfaces with other subsystems, the scope of necessary checks and inspections in order to obtain EC verification certificates, etc. are described.

The article focuses on issues related to the structural control-command and signalling subsystem implemented by the Commission Regulation (EU) No. 2016/919 of 27 May 2016 on the technical specification for interoperability within the control-command and signalling subsystems of the rail system in the European Union, hereinafter referred to as "TSI CCS "(from control-command and signalling) [1].

According to the provisions of Directive 2008/57/EC of June 17, 2008 command and signalling subsystem is described as: "All the equipment necessary to ensure safety and to command and control movements of trains authorised to travel on the network". Moreover [2] divides command and signalling subsystem into two separate subsystems i.e. " Control-Command and Signalling On-board Subsystem " and " Control-Command and Signalling Track-side Subsystem". It is clearly visible that in spite of the common denominator "control-command and signalling ", both subsystems are independent assessment subjects for their EC verification by notified bodies [12]. It should also be remembered that apart from the above mentioned control-command and signalling layers of the subsystem so-called basic layer is also distinguished. It consists of track vacancy detection devices, turnouts, station, line equipment etc. The requirements for the basic layer are defined by national regulations and this is conditioned by the diversity of traffic regulations and the applied technical solutions in individual Member States like it is in Poland in [7]. The confirmation of meeting the TSI requirements by the subsystem is to receive the EC certificate of verification which, after carrying out the necessary checks and inspections is issued by the notified body [9, 12].

* Corresponding author: dadamski@ikolej.pl

2 ERTMS System

2.1 Characteristics of the ERTMS system

Regardless of the case under consideration, the European Rail Traffic Management System (ERTMS) is installed on interoperable railway vehicles and interoperable railway lines. ERTMS is defined to meet the TSI control-command and signalling requirements despite where it is implemented. The ERTMS system is classified as Class A and is divided into the European Train Control System (ETCS) and the Global System for Mobile Communications-Railways (GSM-R). The operation of the ETCS system is based on calculating and controlling the braking curves [4], [5]. The aforementioned curves depend on many factors relating both to the vehicle and to the track. When devising the system, it was assumed that these factors can be separated into track-dependent and vehicle-dependent. Thus, vehicle data includes information such as vehicle weight, maximum axle load, maximum speed, brake system parameters, etc.

Track data, as opposed to vehicle data (which is provided once before the train's departure), are received by the vehicle during the entire driving time. They change both in time (depending on the traffic situation) and in space (depending on the location of the vehicle). These data include, above all, an movement authority, which consists of the maximum distance that the vehicle can cover and the maximum speed as a function of the distance from the reference point. Together with the movement authority vehicle also receives another information which is determined by track-dependent factors affecting the braking curves counted by the system. Such information includes, for example, the track profile (up and down gradients), distances to neighboring balises and information about other track – vehicle transmission channels.

ETCS systems in both on-board and track-side variants appear in three levels and depending on the requirements also several baselines. Baselines correspond to the version of the technical specifications designed and validated for particular project. TSI 2016/919 [1] indicates three sets of specifications that may be applied during EC verification. From 1 January 2019 the set of specifications only number 2 and 3 are valid for on-board subsystem. That means that since that date only vehicles equipped with baseline 3.4.0 and 3.6.0 are able to obtain EC certificate of verification excluding some specific cases with derogations. Differences between baselines consist in introducing additional functionalities clarifying the requirements as well as improving the stability of the system operation. Baseline 3.4.0 comparing to baseline 2.3.0d introduces new functions such as: Passive Shunting mode, Limited Supervision mode, level crossing not protected, new specification for the Non Leading mode, track conditions for the power supply, track conditions for sounding the horn, virtual balise cover etc. It is worth to mentioned that baselines backwards and forwards compatibility is defined in [6]. The rules for assigning the appropriate baseline for the GSM-R system are analogous to those applicable to ETCS and GSM-R system can be implemented in two versions: baseline 0 and baseline 1. It is important to mention that project baseline is not related to the levels installed. The article focuses on on-board systems so detailed information regarding trackside subsystem baseline version management will be intentionally omitted.

ETCS is based on digital track-vehicle transmission. The transmission can be carried out by balises, short, medium or long loops, a digital radio channel or specialized transmission modules. Data describing the track and the vehicle are used to calculate static and dynamic speed profiles. The calculated profile is constantly compared to the current speed in the position function. The necessary location function is based on clearly distinguishable (through a unique number) and accurately located devices for point transmission (balises or loops). The control and supervision functions always operate according to the same rules regardless of the channel in which information from the track was received.

2.2 System diversification

Depending on the specific project, both on-board and track-side equipment are provided by different manufacturers. In addition, individual hardware configurations may differ in software versions, versions of interoperability constituents or components used because a certain manufacturer may for example allow to use three types of modems or four types of sensors for odometry devices with their generic product [10]. Moreover, rolling stock manufacturers in consultation with manufacturers of on-board devices must jointly develop a dedicated ETCS/ERTMS system configuration specific to a particular vehicle type and the ETCS/ERTMS system itself must be properly implemented in order correctly communicate and interacts with train systems. ERTMS/ETCS trackside implementation is also subject to similar conditions. Challenges arise relating to the connection between the Radio Control Center (RBC) with relay interlockings (basic layer) or with neighbouring RBC, especially if it was built by another manufacturer.

Despite the introduction of standardization and unification of requirements relating to the control-command and signalling subsystem many issues still need to be clarified. Tests carried out in laboratories in simulated conditions as well as manufacturer's operational tests do not provide sufficient information on the correct integration of on-board equipment of the subsystem with track side equipment. This issue was already mentioned at the very introduction of [1] where it was stated: "Even a successful certification process cannot always exclude that, when an on-board CCS subsystem interacts with a trackside CCS subsystem, one of the subsystems repeatedly fails to function or perform as intended under certain conditions. This may be due to deficiencies in the specifications, different interpretations, design errors or equipment being installed incorrectly. A more coordinated way to perform compatibility tests should be introduced in order to help operators to take appropriate decisions."

Therefore table 6.2 of the [1] clearly states the checks than need to be carried out when verifying the onboard subsystem regardless of the chosen evaluation module. One of the aspects that has to be assess is "Integration with Control-Command and Signalling Trackside Subsystems and other subsystems: tests under operational conditions". These checks must be carried out on the line under as many different operational conditions as reasonably possible e.g. line gradient, train speed, traction power, weather conditions etc. The scope of these tests should not include tests already carried out at earlier stages for example tests carried out on the interoperability constituents or in the laboratory.

3 Tests of a proper integration of subsystems

Due to arisen situation the need to verify the correct integration of both subsystems in operational conditions was noticed and individual EU Member States were required by [1] to develop operational test scenarios. These tests are intended to check proper operation of the railway system in situations relevant to the ETCS and GSM-R and to demonstrate that the subsystems under examination are compatible with each other. Test scenarios created by each country's National Safety Authority (NSA) are to be sent to the European Railway Agency (ERA) in order to create a consolidated version of scenarios valid throughout the EU community in the future [11]. Currently scenarios elaborated by these entities are used in each Member State separately. In Poland the body that defined the test scenarios was the Office of Rail Transport (UTK). They were published in the form of tables describing individual test cases along with possible driver's actions, expected system reaction and finally the behavior of the tested vehicle.

Every vehicle subject to EC verification, irrespective of the chosen evaluation module based on [3], must pass the tests of correct integration with the track-side subsystem. Such tests must be carried out by an entity having appropriate competences and qualifications. Most often it is a Notified Body (NoBo) or the Designated Body (DeBo) cooperating with it. The Railway Research Institute conducts tests on the compliance of on-board subsystems with track-side as NoBo or as DeBo depending on the function performed in the given EC verification process. A similar situation takes place in the case of verification of track-side subsystems for which operational scenarios dedicated to specific implementations on specific lines are also provided by Infrastructure Managers. Issues related to the testing of trackside subsystems are not covered by scope of the article and therefore they will not be discussed.

3.1 Test results

The integration tests of a rolling stock with track-side infrastructure are carried out based on UTK operational scenarios and a set of additional checks developed by the Railway Research Institute. Train test runs are performed in fixed conditions in order to obtain a specific and expected results. Particular attention should be paid to the integration of the ERTMS/ETCS system with the Class B system, as it is the scope of a specific application for a given subsystem of a vehicle in a given country. During the tests, the correctness of the vehicle's transitions between levels and operating modes is checked, the response to changing the length of the Movement Authority (MA), speed control, reaction to the failure of individual components, etc. All indications on the Driver-Machine Interface (DMI) as well as the signals of semaphores are recorded using digital cameras. In addition, in the case of level 2 tests, it is coordinated from the level of RBC, where a representative of the Railway Research Institute is also present.

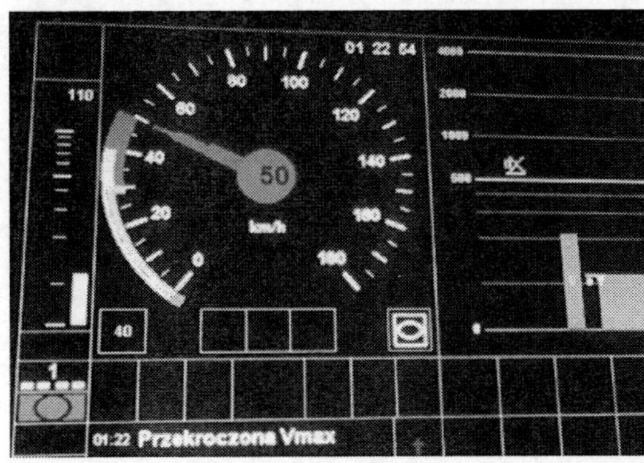

Fig. 1. DMI during test run [8].

The Juridical Recorder (JRU) is the source of all necessary data to evaluate the obtained results, in which all required train driving parameters are saved (see Fig.2). In addition, in order to verify the correctness of telegrams read from the balises and RBCs it is also possible to use the so-called Logs obtained from the European Vital Computer (EVC).

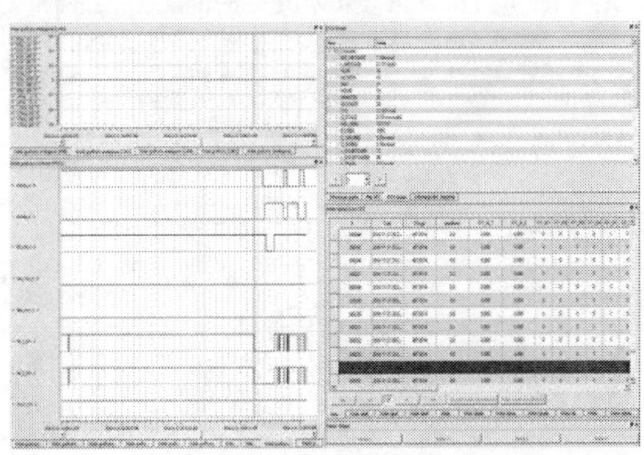

Fig. 2. Screen with raw data from JRU.

Data collected from the aforementioned sources, knowledge of the testing ground and visual observations made by qualified staff allow for an assessment of the correct integration of the on-board subsystem with trackside subsystem. In ambiguous cases, the assessment process is significantly extended, because it is necessary to analyze the error in detail in order to eliminate it. System errors that cause the TRIP - emergency vehicle stop , poor driving parameters (permitted speed reduced with no reason), problems with measuring the distance traveled, or measuring the distance from train antenna to the recently traveled balise are only a part of the difficulties encountered during discussed tests. In most cases, minor changes in the vehicle software are sufficient solution to the problems that arose but in extreme cases it is necessary to replace individual system components. Sometimes it is necessary to repeat whole test scope because necessary adjustments take time and solutions are not always obvious. For example, during the L2 tests, it turned out that the tested electric multiple traction unit that had two BTM components and EVC was incorrectly calculating distance between them. The result of this situation were problems with starting of a vehicle from known position after changing cab while RBC connection was still maintained. This issue was causing odometry errors and it was solved, after analyzing data from the vehicle and RBC telegrams, via software upgrade.

Observation of events and driving parameters in connection with the knowledge of individual subset requirements enables the analysis and subsequent assessment of the correct integration of on-board subsystem with trackside subsystem.

4 Conclusion

The issues of correct integration of the "Signalling On-board Subsystem" with the "Signalling Trackside Subsystem" discussed in the article are an important element in the approval of the rolling stock. This is the final verification of the vehicle before obtaining approval for the putting into service. Detected errors and irregularities must be eliminated by introducing changes to the vehicle before issuing the EC verification certificate. Each ERTMS/ETCS on-board equipment manufacturer has an individual approach to the manner in which he implements the requirements of subsets and TSIs what in the final result may require the implementation of additional measures to ensure full compatibility of subsystems. Therefore because of variance in national signalling equipment (e.g. interlocking's), operational rules, different interpretations, possible errors in design etc. tests mentioned above shall be performed in order to demonstrate the technical compatibility of the considered subsystems.

The last but not least It should be remembered that laboratory and commissioning tests, due to their nature, do not allow to formulate the conclusion that the vehicle will properly cooperate with the infrastructure dedicated to it. Therefore commissioning checks should be extended by the aforementioned tests before placing rolling stock in the service. Probably in the future ERA will formulate a unified and consolidated test plan that will allow for unambiguous confirmation of compliance of the tested subsystems.

References

1. Commission Regulation (EU) 2016/919 of 27 May 2016 on the technical specification for interoperability relating to the 'control-command and signalling' subsystems of the rail system in the European Union
2. Commission Directive 2011/18/EU of 1 March 2011 amending Annexes II, V and VI to Directive 2008/57/EC of the European Parliament and of the Council on the interoperability of the rail system within the Community
3. Commission Decision (EU) 2010/713 of 9 November 2010 on modules for the procedures for assessment of conformity, suitability for use and EC verification to be used in the technical specifications for interoperability adopted under Directive 2008/57/EC of the European Parliament and of the Council
4. System Requirements Specification UNISIG SUBSET-026 version 2.3.0d
5. System Requirements Specification UNISIG SUBSET-026 version 3.4.0
6. System Requirements Specification UNISIG SUBSET-104 version 3.3.0
7. The list of the President of the Office of Rail Transport on the relevant national technical specifications and standardization documents, the use of which allows meeting the essential requirements relating to the interoperability of the rail system of 19 January 2017.
8. Adamski D., Ortel K. *Problematyka współpracy pokładowego systemu ERTMS/ETCS z polską infrastrukturą,* Autobusy (6/2018)
9. Kycko M., *Metodyka certyfikacji podsystemu sterowanie,* Prace Naukowe Politechniki Warszawskiej, Warsaw (2016)
10. Pawlik M., *Europejski System Zarządzania Ruchem Kolejowym, przegląd funkcji i rozwiązań technicznych – od idei do wdrożeń i eksploatacji,* KOW, Warsaw (2015)
11. https://www.era.europa.eu/activities/european-rail-traffic-management-system-ertms_en.
12. Gradowski P.: *Upgrading the railway infrastructure technical parameters using the example of the Control-Command and Signalling subsystem with the EC verification certificate*, Railway Reports, issue **182** (March 2019), ISSN 0552-2145, Railway Research Institute, Warsaw 2019, pp. 131-146

Systems mutual loadings of traction asynchronous cars

Andriy Afanasov, *Oleksandr* Shapovalov*, and *Maryna* Voitenko

Dnipro National University of Railway Transport named after Academician V. Lazaryan, 49000 Dnipro, Ukraine

Abstract. The review of methods of post-repair tests of traction asynchronous motors under conditions of locomotive depot is carried out. The necessity of introducing special stands for post-repair tests is shown in order to improve their quality, reduce the number of failures of service equipment, improve safety of the railway transport. Let's consider a question of power supply of the research asynchronous engine in wide frequency range and also a possibility of creation of the universal stand for testing important asynchronous engines of the electrorolling stock. Survey of possible versions of the scheme mutual loadings of asynchronous electrical machines which can be used for testing traction asynchronous engines of the rolling stock is executed. Such systems can be constructed both with use of the static converter of frequency, and without it. The given short characteristic of each alternative, certain advantages and shortcomings of each of system options. It is shown that systems with use of static converters of frequency differ in high power efficiency, allow to carry out tests in the wide frequency range of power supply, however have higher prime cost. Schemes without converters of frequency differ in smaller power efficiency, impossibility to carry out tests in the wide range of frequency of power supply, high level of consumption of reactive power. The results of the carried-out analysis can be used when choosing rational option of a system mutual loadings of asynchronous electric motors which will differ in higher functionality and power efficiency. Use of such systems will allow to lower capital expenditure for creation new and modernization of traction electrical machines that exist at the station for test.

1 Introduction

According to DSTU GOST 2582:2017 [1] electrical machines of the traction rolling stock of the main and industrial transport at acceptance tests for one hour at current which gives excess of temperature of windings which corresponds to excess of temperature at the nominal mode.In fact this value of current, which corresponding to the sentry of the mode.

Requirements of the corresponding standards and rules of repair of the traction and motorvagonny rolling stock of the main and industrial transport provide carrying out acceptance tests of everyone newly the manufactured or repaired traction electrocar [1]. These tests represent an important part of technological process of production or repair of the electrical machine, material expenses on which enter prime cost of end products. About 15% of energy are spent for Postrepair tests which is necessary for repair.

The quality of technical control which is carried out at acceptance tests of traction electrical machines in a final case defines reliability and non-failure operation work of the locomotive, and respectively and cost efficiency of rail transportation on the main and industrial transport. Tests for heating, check of frequency of rotation and reversal requires mandatory loadings of traction electrical machines [1] at the testing bench.

The largest power efficiency of acceptance tests can be reached by use of a system mutual loading of traction electrical machines [2].

2 Analysis of literary data and statement of a problem

The choice of the scheme of test of electrical machines is defined by the electric equipment of the test station, but it is possible to describe all of them a number of quantitative and qualitative parameters which allow to define efficiency of this technological process.

One of quantitative parameters is the power of the additional equipment which is defined by frequency rate in relation to the power of the tested electrical machine. Also this power determines the necessary area of the test station:

$$K_к = \frac{P_{ae}}{P_{am}}, \quad (1)$$

where P_{ae} – total power of auxiliary equipment; P_{am} – the power of the engine examinee.

* Corresponding author: shapovalov93as@gmail.com

The second parameter is quantity of the consumed electric power at tests. As examinees engines differ on power, it is possible to express this parameter in relative units and to call coefficient of profitability of the scheme. It will be equal:

$$K_e = \frac{P_c - P_{am}}{P_{am}} \ , \tag{2}$$

where P_c – the power which is consumed by the stand; P_{am} – the power which is consumed by the examinee the engine.

From this point of view the most economic schemes of tests are schemes of tests of electrical machines on a method mutual loadings, when two electrical machines electrically and mechanically connected, so that one of them, working in the generator mode, gives all electric energy made by first to the second electrical machine, which works in the engine mode, and the last spends all mechanical energy for rotation of the first electric machine. Energy from the network is consumed only on a covering of losses in the scheme.

The most bulky and difficult there are schemes of tests of commutatorless traction electric motors of alternating current [3]. For schemes of direct test of loading it is frequency the operated asynchronous traction electric motors from a sinusoidal source of tension brought to [4-6] frequency rate of rated capacity makes from 3.5 to 7, the coefficient of profitability from 0.25 to 0.75, it means that the general power of the additional equipment exceeds the power of the examinee of the electric motor by 3.5-7 times, and electricity consumption by satellite machines which are included into the scheme makes 0.25-0.75 capacities of the examinee of the engine. In the provided schemes the examinee the traction engine receives food from sources sinusoidal tension. Schemes are provided have higher power efficiency in comparison with schemes direct loadings, however have smaller power efficiency in comparison with schemes with the static converter of frequency.

In work [7] is considered methods of increase in efficiency due to reduction of number of the service equipment, and recalculation of current of the engine. However the offered method complicates tests of cars with ventilation.

In the modern asynchronous electric drive are used static converters with pulse-duration regulation of tension. In works [8-10] schemes of test of asynchronous traction engines with a power supply of the examinee of the engine from the frequency converter are developed.

However uniform standards which would regulate work of a system the frequency converter - the asynchronous engine is not developed for today.There is an incomplete number of standards and technical requirements to separate components of a control system of the asynchronous drive, such as frequency converter, filters, asynchronous engines [11-13].

The purpose of the work is definition of possible alternative versions of the scheme mutual loadings of asynchronous traction electrical machines, definition of the main characteristics of such systems, and carrying out their comparative analysis.

3 Definitions of the conditions mutual loadings of asynchronous machines

As is known from [14, 15] for ensuring process mutual loadings of electrical machines from the power point of view necessary following conditions:
– ensuring transfer of electric power from the generator to the engine;
– ensuring transfer of mechanical power from the engine to the generator;
– coverings of all losses which arise by transfer of power.

Power indicators of an electric way of compensation of magnetic and mechanical losses at mutual loading in traction electric motors are given in [16].

For introduction of the asynchronous machine to the generating mode it is necessary to provide the frequency of rotation of a rotor of the machine higher then frequencies of rotation of magnetic field of the machine, that is to provide negative slidings. This condition imposes additional restrictions for carrying out tests as at test of two identical machines necessary different frequencies of rotation of their rotors, or different frequencies of power supply.

Let's define necessary conditions for introduction of the asynchronous engine to the generating mode, as slidings for the generating mode is in limits $-\infty < s < 0$, then it is possible to determine the frequency of power supply of the engine as follows:

$$n_{am} = n_{smr}\left(1 - s_e\right), \tag{3}$$

$$n_{ag} = n_{sgr}\left(1 - s_g\right), \tag{4}$$

where n_{am} – frequency of engine rotation; n_{ag} – frequency of generator rotation; s_e – size of sliding of the engine; s_g – size of sliding of the generator; n_{smr} – synchronous frequency of engine rotation; n_{sgr} – synchronous frequency of generator rotation.

It is possible to determine synchronous frequencies of rotation for each machinnes as follows:

$$n_{smr} = \frac{60 f_e}{\text{p}}, \tag{5}$$

$$n_{sgr} = \frac{60 f_g}{\text{p}}, \tag{6}$$

where f_e – frequency of power supply of the engine; f_g – frequency of power supply of the generator; p – number of couples of poles of the electrical machine.

As the unification of shafts of rotors of asynchronous machines are possible with both by means of the coupling, and by means of a reducer (variator), It is necessary to consider two options: $n_д = n_г$ and $n_д \neq n_г$.

Let's consider a condition of realization mutual loadings with an identical frequency of rotation of shaft of rotors and different frequency of supply voltages $n_\partial = n_г$:

$$\frac{60 f_e}{p}(1-s_e) = \frac{60 f_g}{p}(1-s_g). \quad (7)$$

From expression we will determine the frequency of power supply of the engine from the frequency converter:

$$f_e = f_g \frac{1-s_g}{1-s_e}. \quad (8)$$

In this dependence it should be noted that sliding of the generator is negative.

Let's consider a condition of realization mutual loadings with different frequencies of rotation of shaft of rotors $n_\partial \neq n_г$ in this form:

$$\frac{n_{am}}{n_{ag}} = \frac{1-s_e}{1-s_g}. \quad (9)$$

From the expression we will determine the necessary frequency of rotation of the generator in a form:

$$n_{ag} = n_{am} \frac{1-s_g}{1-s_e}. \quad (10)$$

Knowing frequencies of rotation of the engine and the generator it is possible to define the transfer relation of the gearing (variator):

$$\mu = n_{am} \frac{n_{ag}}{n_{am}}. \quad (11)$$

One more complexity of realization of a system mutual loadings of asynchronous is that the power of the asynchronous machines with an invariable frequency and supply voltage is sliding function, that is at increase in power consumption will increase slidings. It can be defined from this dependence:

$$s = \frac{n_1 - n_2}{n_1}. \quad (12)$$

where s – size of sliding of the asynchronous machine; n_1 – speed of rotation of magnetic field (synchronous speed); n_2 – speed of rotation of a rotor.

Let's express slidings s through angular speeds of rotation of magnetic field ω_1 and a rotor ω_2:

$$\omega_1 = \frac{2\pi n_1}{60}, \quad (13)$$

$$\omega_2 = \frac{2\pi n_2}{60}. \quad (14)$$

From where we will receive:

$$n_1 = \frac{60\omega_1}{2\pi}, \quad (15)$$

$$n_2 = \frac{60\omega_2}{2\pi}, \quad (16)$$

where ω_1 – angular speed of rotation of stator magnetic field; ω_2 – angular speed of rotation of a rotor.

We will substitute the received expressions in a sliding formula (12):

$$s = \frac{\omega_1 - \omega_2}{\omega_1}. \quad (17)$$

We find out the effect of sliding on the power that is developing by the engine:

$$P_1 = \frac{F_1 l}{t} = \frac{F_1 2\pi R}{T} = F_1 RF\omega_1, \quad (18)$$

$$P_2 = \frac{F_2 l}{t} = \frac{F_2 2\pi R}{T} = F_2 RF\omega_2, \quad (19)$$

where P_1 – the power consumed by the electric motor; P_2 – power on a rotor shaft; F_1 – the electromagnetic force which affects a rotor; F_2 – the mechanical force which affects a rotor; l – rotor circle length; R – rotor radius.

From formulas (18) and (19) let's express angular frequencies of rotation of magnetic field ω_1 and a rotor ω_2:

$$\omega_1 = \frac{P_1}{F_1 R}, \quad (20)$$

$$\omega_2 = \frac{P_2}{F_2 R}. \quad (21)$$

Taking the ratio $\omega_1 - \omega_2$ to ω_1 we get:

$$\frac{\omega_1 - \omega_2}{\omega_1} = \frac{1}{\omega_1}\left(\frac{P_1}{F_1 R} - \frac{P_2}{F_2 R}\right). \quad (22)$$

As in the settled mode $F_1 = F_2$ forces are balanced, we get such ratio:

$$\frac{\omega_1 - \omega_2}{\omega_1} = \frac{P_1 - P_2}{F_1 R\omega_1}. \quad (23)$$

Taking into account formulas (17) and (18) we get:

$$s = \frac{P_1 - P_2}{P_1}. \quad (24)$$

From where we have:

$$P_2 = P_1(1-s). \quad (25)$$

For today while carrying out tests of asynchronous machines widely use static converters of frequency. This is due to the fact that, according to [8] at tests the engine has to receive power supply from same power sources as well as during the work. However during the work the converter of frequency brings additional losses in the engine which are connected by it with the highest harmonicas of current. It is possible to reduce these losses when the frequency of switching of power keys of the converter having increased however in this case losses increase by switching in the converters. For this reason the frequency of pulse-duration modulation of modern converters of frequency is in limits 10-20 kHz. In general system development mutual loading with high power to indicators is rather difficult task which demands an integrated approach.

4 Possible variations of schemes mutual loadings

The scheme of tests of asynchronous engines with the mechanical converter of frequency is submitted in (Fig. 1)

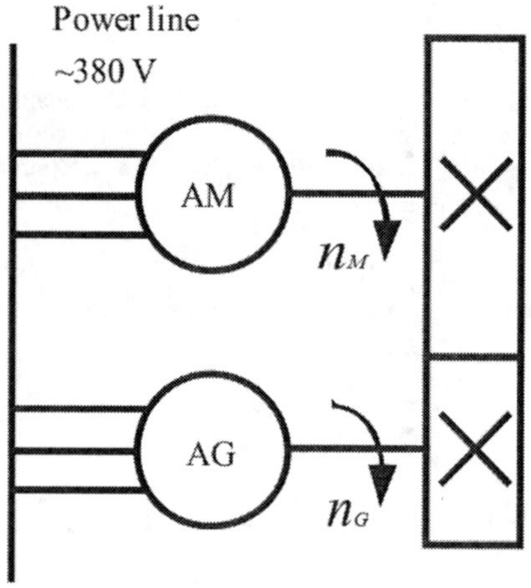

Fig. 1. The scheme mutual loadings of asynchronous engines with the mechanical converter of frequency

This scheme represents two same asynchronous machines, the asynchronous engine and the asynchronous generator, the shafts of the rotors are interconnected by a reducer (multiplier). In machines of low and average power instead of a reducer also possible use of belt variators, or hydraulic drives. The main feature of the scheme is that frequencies of rotation of shaft of rotors of the engine and the generator differ. At the same time there is possible a connection of both machines to the general network with power supply with an identical frequency of current. At test of engines of low and average power for smooth regulation of the trancemission ratio there will be expedient use of belt variators. It should be noted that when using the mechanical converter of frequency the scheme has rather high coefficient of power efficiency, as the scheme provides minimum number of possible service equipment. For carrying out tests of engines of high power can be used the scheme, which is submitted in (Fig.2).

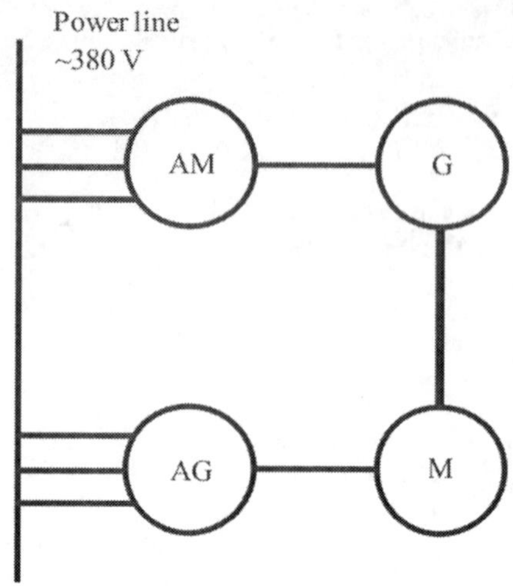

Fig. 2. The scheme mutual loadings the asynchronous engine - the asynchronous generator with a link of a direct current

This scheme consists of two machines of alternating current (the asynchronous engine and the asynchronous generator) and two machines of a direct current (the direct current motor and the direct current generator). The rotor of the asynchronous engine is connected to an anchor of the direct current generator, the rotor of the asynchronous generator is connected to an anchor of the direct current motor. In comparison with the previous option this scheme can be used at test of asynchronous traction engines of high power. However use of auxiliary machines in this scheme reduces profitability coefficient and also increases the area of the test station, necessary for placement of the equipment. Schemes with use of static converters of frequency differ in high power efficiency. The version of such scheme is given on (Fig. 3).

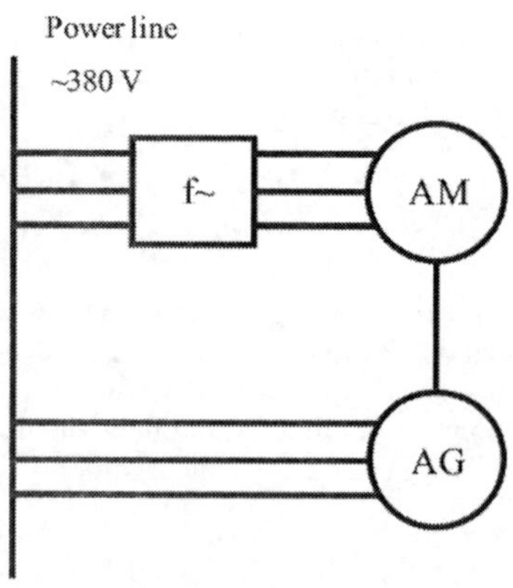

Fig. 3. The scheme mutual loadings of asynchronous engines with one converter of frequency

This scheme represents two same asynchronous machines, bring down rotors of which connected among themselves by means of the coupling. The asynchronous engine receives a power supply from the frequency converter. The generator works with excitation from network. As bring down rotors rotate with an identical frequency for introduction of the asynchronous engine to the mode of the generator it is necessary to increase the frequency of rotation of its rotor. By means of the converter of frequency it is possible to regulate smoothly the frequency of rotation of shaft of machines and consequently also power. Uses of small number of the service equipment provides high power efficiency of the scheme.

Lack of the schemes given above is the narrow range of examinees of frequencies, in fact it is possible to test engines with one certain frequency which is close to the frequency of a power line. For test of asynchronous machines in the wide range of the power supply frequencies the scheme is used with two converters of frequency provided on (Fig. 4).

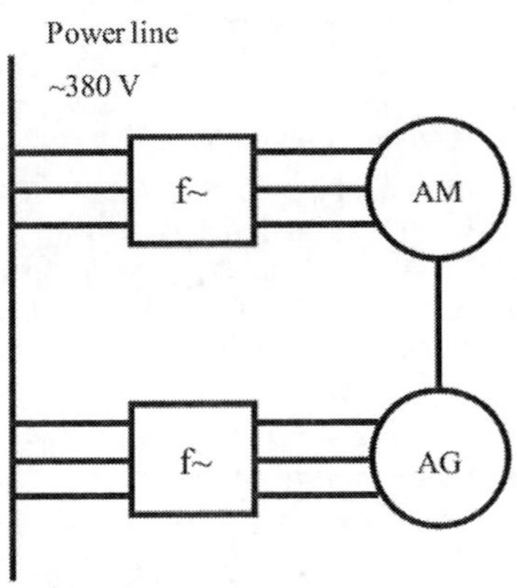

Fig. 4. The scheme mutual loadings of asynchronous engines with two converters of frequency

Feature of this scheme is use of two converters of frequency which allow to realize a system mutual loadings with any frequency of power supply of the engine and generator. At the same time the coefficient of power efficiency decreases and the number of the service equipment increases. The scheme with two converters of frequency and a link of a direct current allows to reduce losses during tests provided on (Fig. 5).

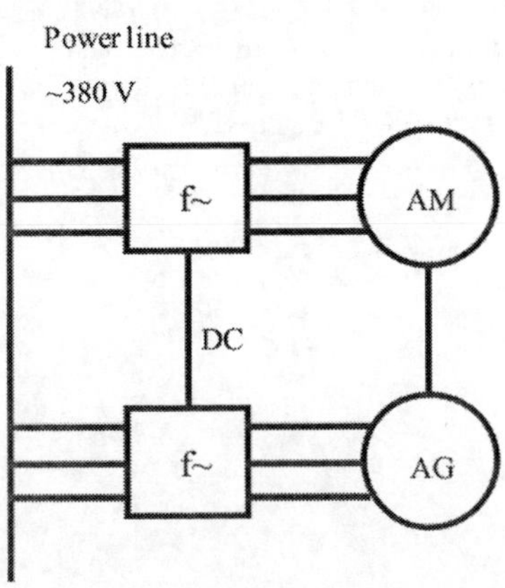

Fig. 5. The scheme mutual loadings of asynchronous engines with two converters of frequency and a link of a direct current

This scheme represents two same asynchronous machines, bring down rotors of which connected among themselves by means of the coupling. The asynchronous engine receives a power supply from the frequency converter. The induction generator also receives power from the frequency converter.

The unification of this scheme with the links of a direct current of converters of frequencies among themselves is feature of this scheme. This decision allows to increase coefficient of power efficiency in comparison with the scheme provided on (Fig. 4), due to reduction of number of transformations of energy. Electric energyIt which is developed by the generator gets to a link of a direct current of converters of frequency, and from it arrives to the engine. At the same time all advantages of the scheme remain provided on (Fig. 4).

The scheme of tests of asynchronous engines with one converter of frequency and a direct current generator provided on (Fig. 6).

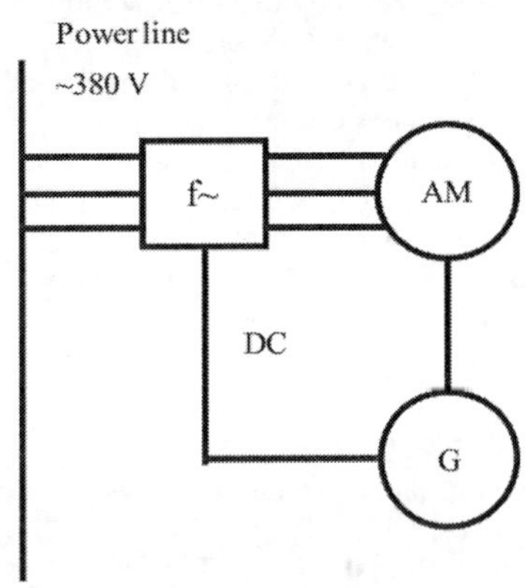

Fig. 6. The scheme mutual loadings the asynchronous engine - the direct current generator

This scheme represents frequency converter, asynchronous engine and a direct current generator. The rotor of induction motor is connected to the rotor direct current generator. The asynchronous engine receives a power supply from the frequency converter. The generator is used to load the induction motor. Generator energy is fed to the direct current link of the frequency converter.

The advantage of this scheme is the ability to test over a wide range of power frequencies using only one frequency converter. The scheme also allows you to test engines of different power.

Conclusions

It was analyzed systems of mutual loading of three-phase asynchronous engines, it is proved a possibility of realization mutual loadings both with the static converter of frequency, and without it. Each of the schemes mutual loadings has both the advantages and shortcomings. As a result of researches it was defined a condition of introduction of the asynchronous engine to the generator mode during the work in parallel with network. For common industrial electric motors of low power use of the scheme will be rational (Fig. 1), As possible use of a belt variator thanks to the profitability coefficient increases, and also it is possible to regulate the transfer relation smoothly. As the power of examinees of machines is low, during the work the scheme will consume low reactive power. The scheme (Fig. 1) is highlighted with the lowest cost as converters of frequency and also auxiliary machines are not its part.

For common industrial electric motors of an average and high power use of the scheme (Fig. 3) will be rational. Thanks to the converter of frequency it is possible to regulate smoothly the frequency of rotation of a rotor of the electric motor that provides input of the asynchronous engine in the generator mode and also to smoothly regulate electric motor power. For traction asynchronous electric motors of railway and industrial transport in which the frequency of power supply differs from common industrial there will be rational use of the scheme (Fig. 5). This scheme allows to realize any frequency of power supply of examinees of electrical machines. Thanks to the unification of links of a direct current of converters of frequency energy from the generator to the engine with the smallest number of transformations that allows to provide rather high effectiveness ratio, with the different frequencies of power supply of the electric motor other than the frequency of network.

References

1. GOST DSTU 2582:2017 *The rotating traction electrical machines. General specifications* (GOST 2582-2013, IDT; IEC 60349-1:2010, NEQ; IEC 60349-2:2010, NEQ), (2017)
2. H. K. Zherve, *Industrial tests of electrical machines*, publ. Energoatomizdat, 506 c., (1984)
3. V. H. Shcherbakov, *Traction electric motors of electric locomotives*, Nautilus Agency, 1998. - 672 c., (1998)
4. R. Y. Alkyn, E. A. Dolgosheyev, I. I. Tal'ya, I. L. Torgonskiy, T*he device for tests of the asynchronous traction engine with a short-circuited rotor*, A.c. the USSR № 1352424A1, **№42**, (1987)
5. I. I. Tal'ya, *The device for test of the asynchronous traction engine with a short-circuited rotor*, pat. RF№ 2023274 (13) C1, **№4**, (1994)
6. I. I. Tal'ya, *The device for tests of electrical machines of alternating current without collectors*, Pat.№ 2071608, **№1**, (1997)
7. P. O. Loza, *Improvement of power and other indicators of acceptance tests of traction engines of electric locomotives*, Messenger DnipropetrovskNational University of Railway Transport named after Academician V. Lazaryan, **27**, 81-83, (2008)
8. A. P. Kalinov, Є. Y. Danchenko, O. V. Korsun', V. V. Lotous, *Way of test of asynchronous engines under loading with use of converters of frequency*, Pat.№ 117504 , Base of patents of Ukraine (2017)
9. D. I. Popov, *The stand for test of asynchronous machines*, Pat.RU186188U1 (2018)
10. V. V. Kharlamov, D. I. Popov, *Snand of testing of asynchronous motors reciprocal load method*, Pat.RU178716U1 (2018)
11. DSTU IEC/TS 60034-17:2009. *The electrical machines rotating. Part 17. Asynchronous engines with a short-circuited rotor and power supply through the converters. Unit for application* (IEC/TS 60034-17:2006, IDT) (2009)
12. DSTU EN 61800-3:2015/ Change №1:2015 (EN 61800-3:2004/A1:2012,IDT). *Systems power to the electric drive with an adjustable speed. Part 3. Requirements to electromagnetic compatibility and special test methods* (2015)
13. DSTU IEC 61800-2:2008. *Adjustable speed electrical power drive systems. Part 2. General requirements and rated specifications of low-voltage, variable-frequency AC drives*. (IEC 61800-2:1998, IDT) (2008)
14. A. M. Afanasov, *Electromechanical principles of providing mutual loading of electrical machines of a direct current,* Messenger Dnipropetrovsk National University of Railway Transport named after Academician V. Lazaryan, **27,** 42-46 (2009)
15. A. M. Afanasov, *The systems of mutual loading of traction electrical machines of the constant and pulsing current, the monograph, Makovetsky,* 248 p (2012)
16. A. M. Afanasov, *Power indicators of an electric way of compensation of magnetic and mechanical losses in traction electric motors at their mutual loading,* Messenger DnipropetrovskNational University of Railway Transport named after Academician V. Lazaryan, **34**, 65-68 (2010)

The effect of ozone on wet atmospheric corrosion of aluminium of high-voltage lines

Galina Tatarchenko[1], *Nataliia* Beloshitskaya[1], *Viktor* Sychenko[2,*], and *Vitalii* Liashuk[2]

[1] SNU named after Dal
[2] DNURT, Ukraine

Annotation. One of the factors influencing traffic safety is the uninterrupted power supply of interlocking devices. As a rule transmission of electric energy is carried out using aluminum wires. In order to assess the reliability of power lines in the work, research was carried out on the influence of ozone on corrosion of aluminum in a humid atmosphere and acidic medium. It was found that the presence of oxygen dissolved in acetic acid leads to anode currents increase, while at crossover ozone, cathode currents increase by an order of magnitude and 2-3 times the anode decreases; the corrosion potential of an E_{cor} shifts to the positive side at 0.15V. The decrease in the density of anode currents in the presence of ozone proceeds due to the growth of the thickness of the aluminum oxide film. According to the Auger-spectra, the film is increased by 3-4 times compared with the nonzonized medium, due to which the corrosion resistance of aluminum K_m increases from 0.066 to 0.025 g/(m^2·h). The rate of formation of oxides depends on the concentration of oxidants in the solution, and on their oxidative activity. According to calculations of thermodynamic factors that determine the preferred route of oxidation, the probability of formation of oxides in the presence of ozone is significantly higher. Local destruction of the protective layer occurs mainly in defective places and places of local increase in pH of the electrode layer. Investigation of the cyclic polarization of a sample of aluminum, pre-cathode activated by a current of 1 mA/cm^2 density, showed the presence of hysteresis, which decreases in the next 2-3 cycles. In this case, the exterior of the aluminum shifts to (0.10 ÷ 0.15) V is more positive and the corrosion rate is reduced, which indicates the formation of an oxide film in which irreversible changes occur. The change in the inclination of the polarization curves indicates that the cathodic reaction is delayed and is associated with the change in the structure of the surface layer of the electrode and the diffusion restrictions of ozone recovery.

1 Introduction

Atmospheric corrosion is the most common type of corrosion of metals that occurs in a humid air environment: approximately 80% of metal structures, buildings, structures, bridges, cars, etc. exploited under atmospheric conditions. A distinctive feature of atmospheric corrosion is that it flows not in the volume of the electrolyte, but in thin films. In this case, the corrosion process proceeds according to the laws of electrochemical kinetics, but has its own specific features.

The main factors influencing the speed of atmospheric corrosion are:
- humidity of the atmosphere;
- chemical composition of the atmosphere;
- duration of humidifying and drying periods of moist films.

Wet atmospheric corrosion usually begins at relative humidity above 70%. At this moisture, called critical, there is capillary condensation of moisture and water begins to manifest the properties of the electrolyte. Capillary condensation can stimulate surface roughness, various inequalities, metal contamination with solid particles (dust), etc. The thickness of the moisture films formed on the metal ranges from 0.01 to 1 μm and under these conditions a very intense oxygen flow occurs to the surface of the metal, which leads to an acceleration of the corrosion process compared with the volume of the electrolyte.

At high voltage of the electric power line with a high degree of heterogeneity of the electric field near the surface there is a dissociation of the characteristic form: the region covered by the discharge, has a weak purple glow with non-stationary yellow thin channels of single discharges - streamers. In such a "crown case", the electric field strength is sufficient for the ionization and excitation of neutral gas particles. The appearance of corona in the air is accompanied by sound (noise), in the area of the cover is formed ozone, that is, the source of the active oxidizer. Loss of energy on the crown significantly reduces the economic performance and leads to active corrosion processes of all structures.

* Corresponding author: elpostz@i.ua

2 Problem statement

The deterioration of the power transmission lines in many countries is most often due to corrosion of aluminum wires. As noted in [1], one of the main possible reasons for changing the longitudinal resistance of the transmission lines - corrosion of wires under the influence of various climatic factors (including acid rain, humidity, high temperature, solar radiation). According to their calculations for the conditions of Ukraine, the rate of corrosion of aluminum wires should be taken in the range of 1.75 - 3.5 μm/year (for rural and urban areas). In this case, the longitudinal active resistance of phase aluminum wires consisting of wires of a minimum standard diameter (1.7 mm), during 30 years of operation can (at maximum corrosion rate) increase by 30%, and in 70 years - double (100%). For wires with a maximum wire diameter (4.1 mm) the degree of growth of the longitudinal resistance decreases and at the same rate of corrosion will be: for 30 years of operation 12%; over 70 years - 26%.

From the point of view of ensuring the reliability of power supply and extending the service life of conductors, new alloys with high corrosion resistance are often developed. For example, as suggested by the authors [2] conductors of aluminum alloy with manganese, which have high resistance to sea salt, their corrosion is 1.6-2 times higher than that of ordinary aluminum. This work is relevant because there is not much literature data on corrosive aluminum studies involving the active oxidant [3] in a humid atmosphere, especially with the participation of ozone formed during discharges.

3 Analysis of recent researches and publications

The corrosion resistance of aluminum and its alloys is mainly due to the presence of a protective oxide film $Al_2O_3{\cdot}3H_2O$ or $Al_2O_3{\cdot}H_2O$ [4]. Diluted acetic acid at temperatures up to 50 °C hardly affects aluminum, that is, in 5% of AcOH at a temperature of 200 °C, the metal collapses slowly [4]. The presence in the solution of oxygen-containing oxidizing agents (CrO_4^{2-}, VO^{3-}, MnO^{4-}, MoO_4^{2-}, WO_4^{2-}) changes the nature of the anode curves of aluminum dissolution - even small (0.01 N) additives $Cr_2O_7^{2-}$ slows down the overall process compared to the background solution [5]. However, with a further increase in the concentration of the oxidizer, the rate of corrosion is the same as in the background solution, and at concentrations greater than 0.5N $Cr_2\,O_7^{2-}$ activation of the anode reaction occurs. In acidic solutions, oxidizing agents increase the pH of the electrode layer on aluminum and promote the passivation of the electrode, contributing to a positive shift in the potential of the metal. The presence of protective layers on the surface of the metal (oxide, hydroxide, salt films) leads to a strong polarization of the electrode [6]. It should also be noted that the construction of the model of the mechanism of aluminum corrosion in environments containing oxoanions of oxide type, based on only the study of electrochemical characteristics of the metal is impossible. The identity of the polarization characteristics of aluminum, for example, in acidic and alkaline media [6-8], does not indicate the identity of corrosive properties. The electronegativity of the metal and the amphotericity of its oxides are required in electrochemical studies to account for internal self-dissolving currents, whose magnitude is not always a single-valued function of the concentration of the oxidizer, the properties of the products of restoration and stoichiometric ratios of the components involved in electrochemical reactions.

4 The purpose of the article

The purpose of the work is to study the influence of ozone on the corrosive behavior of aluminum in the conditions of the acidic environment of a humid atmosphere.

5 Materials and methods

The Electrochemical studies were carried out on samples of aluminum in solutions 5% acetic acid at a temperature of 20 ^{0}C. Polarization curves were made on electrodes in a three-electrode cell with separated space, an auxiliary electrode - platinum. All potentials are given in the n. Ozonation of solutions by passing the gas through the working solution, its concentration in the gas phase was 5×10^{-4} mol/l. To remove the traces of air-oxide films mechanically cleaned metal samples and steel were pre-exposed to cathode polarization for 3 minutes with a current density of 100 A/m^2. After the cathodic polarization was completed, cathode and anode curves were taken at a 1.44 V/h scan speed, and the anode curves captured in this way were used to estimate the dissolution rate. Extrapolation of curves was carried out to determine the teflon area.

The metallographic studies were carried out on 5 samples after 250 hours of their presence in 5% of AcOH with and without ozone.

6 Results and discussion

The according to the obtained polarization curves of an aluminum electrode (Fig.1, curve 3), when the nitrogen E_{cor} is bubbled through the working solution decreases to -0.3V, and the density of currents is reduced by an order of magnitude, that is, without the presence of oxygen in the medium of dissolution is carried out mainly at the expense of acetate ions to form a readily soluble aluminum acetate. The presence of oxygen dissolved in AcOH leads to the growth of anode currents, and when oxygen or ozone is bubbled, cathode currents increase by an order of magnitude, and the anode decreases 2-3 times, E_{cor} shifted to the positive side on 0,15V.

The presence of ozone in an acid solution increases the corrosion resistance of aluminum, compared with non-ozonized medium, the rate of corrosion of K_m

decreases from 0.066 to 0.025 g / (m^2· h) Considering that the pH of 5% solution of AcOH is equal to 2.5 the significant contribution to the process of anodic dissolution of aluminum is made by acetate ions, whose concentration in the working solution is ~ 5 · 10^{-4} g·ion/liter. The decrease in the density of the anode currents occurs due to the transfer of dissolved oxygen and ozone to the metal and the formation of the oxide layer. The value of the Tafelian coefficient b_c for the non-zoned medium, is 0.12V, corresponding to the stage of the discharge of protons on the metal, in the presence of ozone, the coefficient b_c almost twice as high.

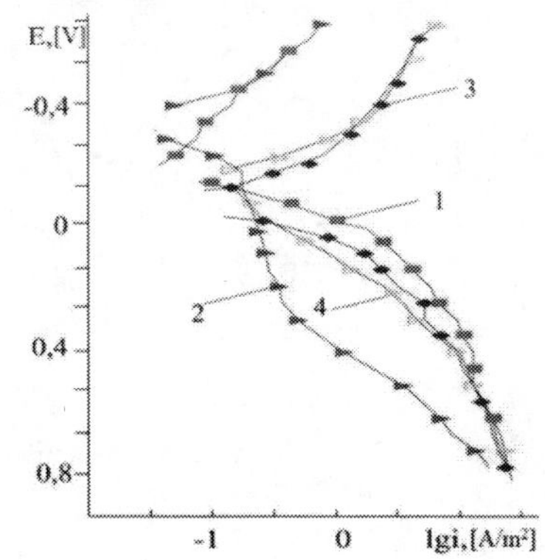

Fig. 1. Polarization curves of aluminum AD0 in 5% AcOH, 20 °C: 1 - natural aeration, 2 - nitriding, 3 - oxygen, 4 - ozonation of solutions.

However, the calculation of the rate of corrosion on the density of currents obtained from polarization curves shows that with natural aeration K_m is 0.025 g/(m^2·h), with bubbling of oxygen through a solution - 0.038 g/(m^2· year), and at bubbling the ozone-oxygen mixture - 0,216 g / (m^2· h) Consequently, the electrode eventually changes the oxide films and, accordingly, their conductivity. Such a result is possible in accordance with the proposed hypothesis of the restoration of ozone through its active forms formed in the electrode layer. As a result, it is possible a creation:

$$Al + 3OH^{-} \rightarrow Al(OH)_3 + 3e \qquad (1)$$

and further change

$$2Al(OH)3 + 3H+ + 3e = Al2O3 \cdot 3H2O \qquad (2)$$

As a result, at the initial moment of time, we obtain anode current growth, due to the strengthening of the cathode reaction of the reduction of ozone, and then its decrease, due to the formation of a thick hydrated oxide film.

As shown earlier, the oxygen and ozone concentrations in the solution can be compared, because the predominant route of oxidation is thermodynamic factors. According to the reference data [5], the most stable form of aluminum oxide when it is formed in water systems is trihydrate $Al_2O_3{\cdot}3H_2O$, the free energy of formation of which equals ΔG_{298}^0 = -2295 kJ/mol. In this regard, the probability of its formation is significantly higher than, for example, monohydrate (ΔG_{298}^0 = -1427 kJ/mol). We consider the thermodynamics of two possible ways of formation of aluminum oxides:

ozone $2Al + 3H2O + 3O3 \rightarrow Al2O3 \cdot 3H2O + 3O2$ (3)

and oxygen $2Al + 3H2O + 1.5O2 \rightarrow Al2O3 \cdot 3H2O$ (4)

As calculations have shown, free energy for the reaction

$$2Al + 6H2O \rightarrow Al2O3 \cdot 3H2O + 3H2, \qquad (5)$$

equals ΔG_{298}^0 = -336 kJ/mol for ozone oxidation of aluminum (3) is -1044 kJ /mol, and oxygen (4) - 791.5 kJ/mol. Given the concentration of ozone and oxygen in the solution, the contribution of ozone in the formation of oxide films and, consequently, in the passivation of aluminum is significant. The rate of formation of oxides depends on the concentration of oxidants in the solution, and on their oxidative activity.

Investigation of the cyclic polarization of the sample Al (Fig. 2), pre-cathode-activated current density 1 mA/cm^2, showed the presence of hysteresis, which decreases in the next 2-3 cycles. In this case, E_{cor} of aluminum is shifted by (0,10 ÷ 0,15) V is more positive and the rate of corrosion decreases, as evidenced on the formation of an oxide film in which irreversible changes occur, for example, the transition according to the reactions (1-2).

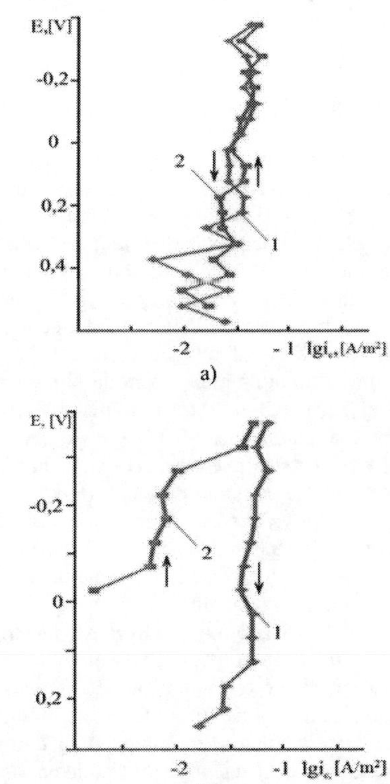

Fig. 2. Cathodic cyclic curves of Al electrode in ozonated acid solution, t = 20 °C: a) the first cycle, b) the sixth cycle, 1 - direct polarization, 2 - reverse polarization (→ - polarization direction indicator)

In the first cycle, in the cathode polarization of the aluminum electrode in the ozonized acid solution, the curves of the direct and reverse runs practically coincide (Fig. 2). At potentials close to the corrosion potential of aluminum in this medium, unstable values of current density (to a greater extent in direct polarization) are associated with the formation and dissolution of thin oxide layers. Repetitive cyclic polarization practically does not change the density of cathode currents in the forward direction and reduces one and a half or two orders of magnitude with the reverse (Fig.2b).

It is known [9-10] that a corrosion-electrochemical system is capable of generating resonant oscillations, the frequency of which depends on the corrosion-electrochemical properties of the metal. The frequency of oscillations of the corrosion-electrochemical circuit depends on the two-layer capacity of the metal and the polarization resistance. Changes in the sample in accordance with [9] occur primarily due to electrode reactions, which occur in parallel with similar velocities and affect the potential in a double electric layer. On the example of the aluminum sample, it should be noted that the second repetition of cyclic polarization leads to the displacement of the corrosion potential of E_{cor} in the direction of positive values. Over time, E_{cor} is set to a value of $\approx$ 0.2V. In cyclic trials, the anode-cathode transition was fixed at potentials close to 0.55V, both in direct polarization and in reverse. However, after five to six cycles of E_{cor} again set to values close to 0.2V (Fig. 2). Such changes in the magnitude of the potential for corrosion are possible as a result of the formation of soluble oxide layers on the surface of aluminum in acidic medium.

According to the obtained values of the density of cathode currents in direct and reverse polarization, depending on the number of the cycle, it has been found that in all experiments there is a tendency to increase the currents, the increase of the potential reduces their value and to a greater extent - with direct polarization. The change in the slope of the curves shows that the cathodic reaction is difficult to overcome, which is probably due to the change in the surface layer of the electrode and the diffusion restrictions of oxygen recovery. With anodic polarization of an aluminum electrode, the curves of the direct and reverse move form a negative hysteresis, reaching at a potential of 0.7V the difference in the density of currents to one order (Fig. 3). In the case of polarization in the cathodic direction, there is a "delay" in the dissolution of the metal, indicating the formation of a film that protects the surface of the metal well. By the nature of the curves, one can assume the formation of a film with anionic conductivity.

With cyclic polarization, the hysteresis decreases and even tends to form a positive loop and indicates a predisposition to the formation of pittens, which is one of the main factors preventing the use of aluminum as a structural material in the investigated medium. In this case, the angular coefficient of the anode curves is equal to 0.10 ÷ 0.12. This testifies to the kinetic complications of the anode process, apparently connected with the metal-electrolyte transformation with the participation of HO-ions. An increase in the size of pittenses, which may be the result of intercrystalline corrosion, is revealed. Changing the anode polarization of an aluminum electrode induces delayed changes in time in the state of its surface, as a result of which the stationary anode activity increases with increasing potential, and decreases with decreasing. Subsequently, the density of the anode currents in the forward direction is reduced to the values of the reverse, therefore, it can be assumed that there are changes in the structure of the oxide film.

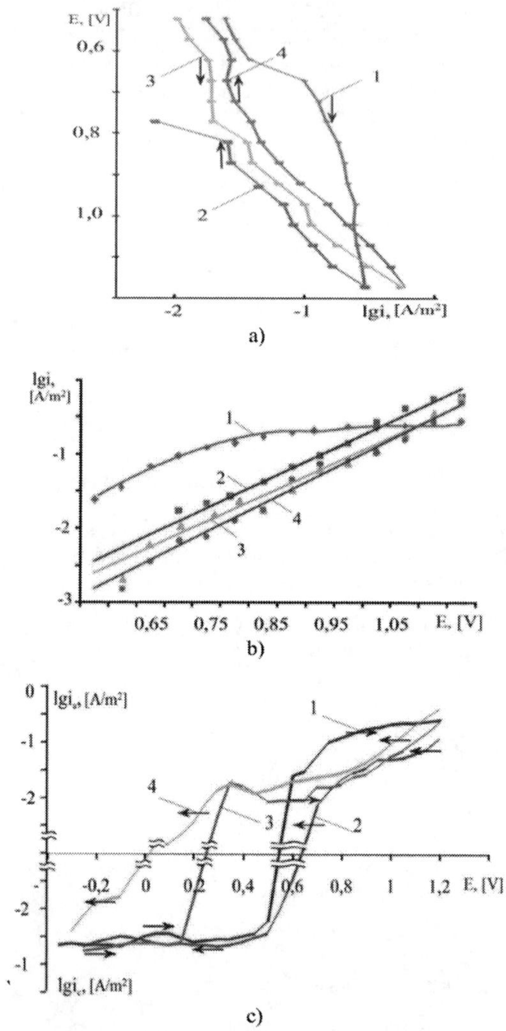

Fig. 3. Cyclic curves of Al electrode in ozonized acid solution, t = 20 ^{0}C anode (a, b) and full (c): a, c) 1, 3 - direct polarization; 2.4 - reverse. 1,2 - the first cycle; 3.4 - the sixth cycle; b) direct polarization; the numbers of the curves correspond to the number of the cycle.

Metallographic studies have shown that the aluminum sills were electroplated, resulting in strips of metal aluminum compounds with refractory impurities. Then, to detect the boundaries of grains, the silt was etched with 0.5% aqueous solution of hydrofluoric acid (Fig. 4a). It is determined that the boundaries of grains are thin without detecting impurities on them, the grain is large - 1 point on the scale of GOST 5639-82, corrosion on grain boundaries was not detected. However, samples after the tests in the ozone medium revealed ulcerous corrosion with a depth of 0.12 mm (Fig. 4b), the number of ulcers is small and they are chaotic, which can be associated with selective dissolution of microinclusions and heterogeneities in the crystalline structure of the material. The presence of ulcerous corrosion, suggests the existence in the film of pores, where there is an active dissolution of the metal.

Electron microscopic examination of aluminum samples after 24 hours in ozonized acetic acid also revealed a large number of slip lines on the surfaces resulting from preliminary mechanical preparation and proof of these samples. Any obviously pronounced zones or focal points of etching of the surface were absent.

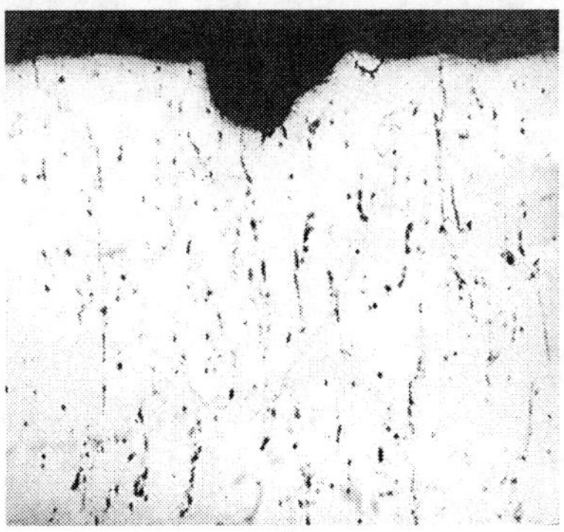

a)

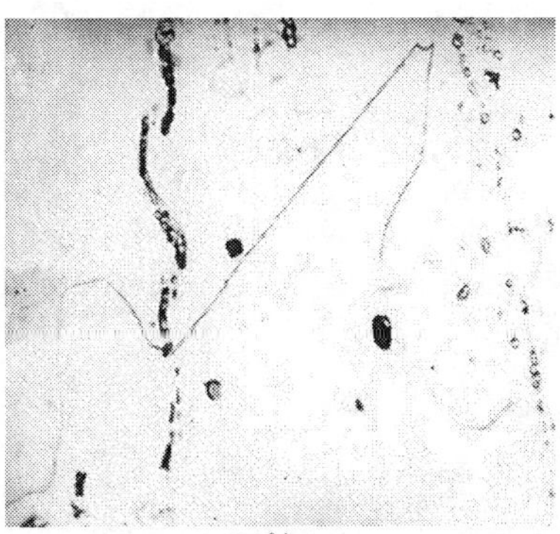

b)

Fig. 4. Microstructure of an aluminum sample after 24 a year stay in ozonated acid solution, 20 ^{0}C: a) x100; b) x500.

7 Conclusions

Based on the results obtained, we can say that the presence of ozone in an acidic environment:
- reduces the corrosion rate by 2-3 times, it is basically uniform, but in some places characterized by the presence of peptic corrosion;
- substantially changes the thickness of the aluminum oxide film and increases by 3-4 times in comparison with the nonzonized medium;
- leads to the local destruction of the protective layer, especially in defective places and places of local increase in pH of the electrode layer;
- the rate of formation of oxides depends on the concentration of oxidants in the solution, and on their oxidative activity, according to thermodynamic factors that determine the preferred route of oxidation, the probability of formation of oxides in the presence of ozone is significantly higher.

References

1. F. P Shkrabets, P. Yu. Krasovsky *Influence of aluminum corrosion on electrical parameters of power lines.* M. Electromechanics and Automatics: reas.-tech.coll, **Iss. 79** - P. 36-39 (2007)
2. M. Watabe, H. Ito, K. Nagano, T. Tsuji, N. Suga and Y. Akasofu. *High Corrosion Resistance Conductor for Overhead Transmission Lines* Sei Technical Review, **No. 84** - P. 47-52 (2017)
3. G. O. Tatarchenko *Restoration of ozone and oxygen on metals in the process of corrosion.* Bulletin of East-Ukrainian National University named after Volodymyr Dahl **№3 (220)** - P. 143-149 (2015)
4. A. M. Sukhotin, A. A. Pozdeeva, E. I. Antonovskaya *Passivity and corrosion of metals.* Proceedings of the GIPH. - L.: Chemistry, **Issue.67**. - P. 5-12 (1971)
5. V. S. Sinyavsky, V. D. Valkov, V. D. Kalinin. *Corrosion and Protection of Aluminum Alloys* - M .: Metallurgy. - P. 124-140 (1986)
6. A. M. Sukhotin, *Directory of Electrochemistry* - L.: Chemistry. – P. 237 (1981)
7. H. Kukhling, *Reference book on physics. Trans. from ger.* - M .: Mir, - P. 520 (1982)
8. M. S. Smidt, K. K. Baldridge [et. al] *General atomic and molecular electronic structure system.* J. Comput. Chem - **V.4, No. 11** - P.1347-1363 (1993)
9. Yu. N. Mikhailovsky, A. I. Marshakov. *Generation of harmonic oscillations by corrosion-electrochemical systems.* Protection of metals. - **V.38. - №1.** - P. 5-11 (2002)
10. G. O. Tatarchenko, K. V. Cherkasy *Fluctuations of the limiting current in metal dissolution reactions in ozonized sulfuric acid.* Fiz.chem material mechanics. - Special issue **№7.** - Vol.2. - P.911-915 (2008)

Estimation of Private Locomotives Usage Efficiency for Freight Transportation in Ukraine

Dmytro Kozachenko[1,*], *Bogdan* Gera[2], *Nazar* Sannytskyy[3], *Yuliia* Hermaniuk[2], and *Tetyana* Shargun[2]

[1] Dnipro National University of Railway Transport, Department of Operation, 41010 Dnipro, Ukraine
[2] Liviv Branch of Dnipro National University of Railway Transport, 79052 Lviv, Ukraine
[3] Lviv College of Transport Infrastructure of Dnipro National University of Railway Transport, 79011 Lviv, Ukraine

Abstract. Ukraine has undertaken the implementation of European Union Directives, which provide non-discriminatory admission to the railway infrastructure of independent carriers. A significant number of factors considerably affect working conditions of the carriers which are independent of Ukrzaliznytsia. One of the goals while performing transportation by independent carriers is the organization of private locomotives operation and their service by engine crews. The purpose of the article is to assess technical and economic indices of private locomotives use for goods transportation. The schemes of infrastructure objects arrangement for locomotives operation and engine crews work are presented. Performed calculations show the potential possibility of independent railway carriers to fulfil transportation to seaports with the use of its private locomotive infrastructure exclusively. The estimation of economically expedient volumes of cargo transportation by private locomotives in comparison with existing pricing of Ukrzaliznytsia is carried out. The nonlinear dependence of transportation volume depending on the distance is determined, at which the total cost of cargo transportation by Ukrzaliznytsia locomotives is equal to the cost of cargo transportation by private locomotives. The results of the study allow concerned companies to evaluate the investment efficiency into the development of locomotive fleet for rail transportation.

1 Introduction

Rail transport is the basis of the transport system of Ukraine and its share accounted 80% of cargo turnover (excluding pipeline transport) in 2018. At the same time, there was a negative trend in redistribution of freight volumes between railway and road transport in favor of the latter in recent years. This leads to the increase of expenditures for petroleum products for transportation and to deterioration of the country's energy balance. One of the main problems of modern railway transport in Ukraine that threatens both stability and safety of the industry is the crucial wear of its main means, locomotive fleet, in particular. Today, the electric locomotives deterioration is 93.0%, and the diesel locomotives wear is 99.8%. The reduction of rolling stock fleet leads to the reduction of transportation volume due to inability to master them, and to deterioration of infrastructure and car use and, consequently, to the loss of railway competitiveness and transition of freight traffic to competitive means of transport. Fig. 1 shows the dynamics of the number of Ukrzaliznytsia locomotives and the turnover of freight cars in relation to the indices of 2006, and Fig. 2 shows a field of points, which characterizes the relationship between these indicators. The analysis shows that between them there is a very strong inverse relationship with correlation coefficient -0.94. Therefore, the locomotive fleet deficiency leads to the decrease of investors' interest in railcar fleet upgrading.

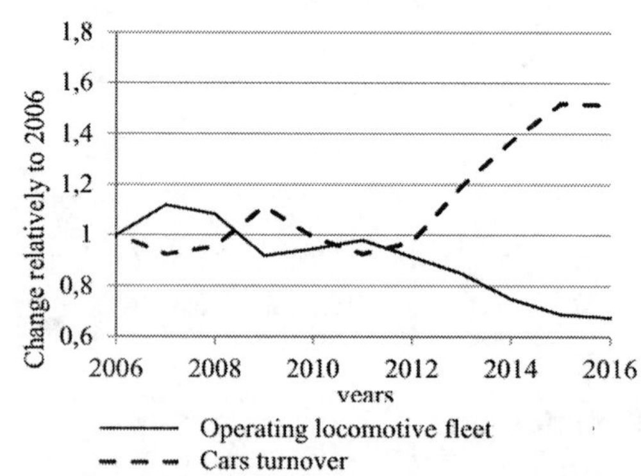

Fig. 1. Dynamics of operating locomotive fleet and cars turnover

Under these conditions, the issue of renewal of rolling stock fleet is topical and of crucial importance for the railway transport of Ukraine. It should be noted that in Ukraine there were attempts to renew the locomotive fleet massively. In particular, in the period from 2012 to 2016, the "Program of locomotive fleet updating for the railways of Ukraine" was functioning. The given

* Corresponding author: dmytro.kozachenko@outlook.com

program envisaged the purchase of 509 locomotives for 28.7 billion UAH, but it was not implemented. At present, according to the current program "Strategy of the Development of JSC "Ukrzaliznytsya"" acting until 2021, the state monopolist plans to buy only 120 freight locomotives. In fact, until February 2019, Ukrzaliznytsya received only 22 locomotives Evolution TE33A produced by General Electric on terms of financial leasing.

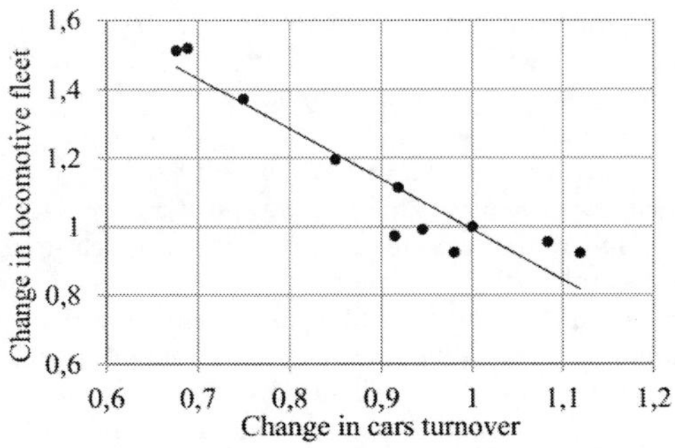

Fig. 2. Field of points that characterizes the correlation between locomotive fleet and cars turnover

In some cases, domestic enterprises carry out transportation of their cargoes by running between Ukrzaliznytsia stations using main locomotives. As an example, transportation of PJSC "Volyn-Cement" is executed between stations Kryvyn and Zdolbuniv (38 km), SE "Chervonogradvantazhtrans" carries out transportation between stations Chervonograd, Girnyk and Sosnivka (6-14 km), PJSC "Ukrgrafit" performs transportation between station named after Anatoliy Alimov and Zaporozhye Live (10 km). Particular attention deserves the work of PJSC "Ivano-Frankivsk Cement", which fleet of main locomotives includes a new locomotive TE33A, three locomotives 2M62 and one locomotive M62, which transports raw materials on the routes Dubivtsi-Yamnitsa (15 km) and Yamnitsa-Khodorov (70 km). The economic efficiency of locomotives in ownership of industrial enterprises which are used on short sections of the track is mainly due to possibility of significant reduction of irregularity in transportation, rolling stock productivity increase and reduction of freight mass on wheels. For instance, at the section of the track Dubivtsi-Yamnitsa, the locomotive of PJSC "Ivano-Frankivsk Cement" performs up to 4 turnovers per day when the normative term of cargoes delivery by Ukrzaliznytsya on the same track section is 2 days. However, all these transportations are carried out on low-activity lines and are experimental.

Crisis phenomena at railway markets, which are in a monopoly state, are typical for many countries. In the European Union, as a solution to this problem, the vertical division of the railway industry is realized by separating infrastructure of the railway transport from transportation activities. At the same time, four packages of Directives of the European Union were issued. One of the main objectives of these Directives is to ensure non-discriminatory access of independent carriers to railway infrastructure [1-3]. Studies of problems encountered by independent carriers are carried out in [4]. In particular, [4] studied the conditions of operation of the railways of Finland. As the main problems, railway carriers of Finland indicate the need for significant investments in rolling stock, discrimination at the stage of drawing up the timetable of trains traffic, the complexity of bureaucratic procedures while licensing. As for the conditions of Poland and Sweden, according to [5], the main barriers for entering the market are high cost of rolling stock traction and bureaucracy. To reduce the initial investment in locomotives, carriers prefer to exploit used locomotives. In [6], the analysis of the conditions for independent carriers' operation in Germany is presented, where the model of partial vertical division of the railway industry is implemented, when the infrastructure manager also has units that provide services for transportation of goods and passengers. It is noted that within the framework of regulatory norms in force in Germany, the infrastructure manager has the opportunity to create preferences for dependent on him carriers by varying the composition and quality of services access to the railway infrastructure. According to [7], the additional problems faced by independent carriers in Germany are that railway infrastructure in the country is primarily adapted for passenger transportation by DeutscheBahn group, so freight carriers are subject to increased technical standards; additional problems are created by the lack of full access to the information that characterizes infrastructure, etc.

In the Russian Federation, in accordance with the Federal Law "On Rail Transport in the Russian Federation" and the Federal Law "The Charter of Railway Transport of the Russian Federation", the access of carriers of different ownership to the main railway infrastructure is declared. However, since 2003, when these laws were passed, independent of Russian Railways carriers of goods have not appeared on the railways. At the same time, since 2003, private (rented) locomotives have been operating in the Russian Federation [8]. At the same time JSC Russian Railways remains the carrier. All private locomotives are mainly diesel locomotives and are owned by "Globaltrans", "Transgarant" and "Transoil". It is rather illustrative how private locomotives of OJSC "BaltTransServis" are exploited, which is part of the holding "Globaltrans". The main transportation by trains with private locomotives of OJSC "BaltTransServis" is carried out at Kuibyshev, North, South-Eastern, Moscow, Oktyabrskaya, Kaliningrad and North-Caucasian railways. The company "BaltTransServis" manages 60 main locomotives, but does not have its private depot or personnel of locomotive crews. Locomotives of the "BaltTransServis" fleet are used on closed routes, agreed with JSC Russian Railways. The shoulders of the locomotive turnover make up more than 1000 km, while locomotives are operated by Russian Railways crews, and their servicing is carried out in the locomotive depots of Russian Railways. Under these conditions OJSC "BaltTransServis" acts in relation to JSC Russian Railways as a locomotive traction operator.

At present, the fleet of private main-line locomotives makes up about 1% of main locomotives of the Russian Federation and accounts for about 3% of freight transportation. One of the main reasons for limiting the admission of private locomotives to the main railway network in the Russian Federation is the imperfection of internal regulations of JSC Russian Railways. The absence of real steps from the side of Russian Railways to change these documents gives the opportunity to hold the market of freight transportation on railways in the monopoly condition.

Thus, the main problems limiting the use of private locomotive traction in vertical segmentation of market models is the high cost of access to the market, high cost of rolling stock traction, as well as imperfection of legislative and regulatory framework, which is designed to maintain competitive modes in the market.

Ukraine, on the one hand, and the European Union, the European Atomic Energy Community and their member states, on the other hand, signed the Association Agreement, under which Ukraine undertook implementation of the norms of the European Union in its legislation, including the Directives of the European Union concerning organization of railway transportation market. Taking into account that one of the main principles of this market organization is the formation of a competitive sector in the field of railway transportation, Ukraine has external obligations to organize the admission of independent carriers to the main railway infrastructure. The specified factor creates external requirements for the admission of locomotives of independent carriers to the railway network. At the same time, the crucial wear of the locomotive fleet and the inability to develop scheduled volumes of transportation is an internal stimulus for introduction of private locomotive traction.

2 Methodology

The legislative and regulatory framework of Ukraine in force does not exclude operation of locomotives that do not belong to Ukrzaliznytsia on the main railway network. The activity of such locomotives on public railways is governed by clause 9.12 of the Rules of Technical Operation of the Railways of Ukraine. At the same time, "The collection of tariffs for freight transportation within Ukraine and related services" includes clause 20, according to which tariffing of cargo transportation on railways of general use by trains with private (rented) locomotives is carried out. In this case, payment for loaded or empty cars is determined according to tariff plans 29.1-29.4 for locomotives and cars separately. Running of private or rented locomotives "under its own power" on the general network of railways of Ukraine (except those associated with transferable operations) is paid by tariff plans 29.5 or 29.6.

Taking into account considerable length of legislative and regulatory framework development for creation of non-discriminatory conditions of independent carriers' access to the main railway infrastructure, it is important to determine the area of private locomotives effective use to ensure freight transportation on railways under current regulatory framework and tariffs.

When performing transportation by independent companies there is mutual financial responsibility between an owner and an infrastructure manager for infrastructure provision and its use as

– delays of trains of independent carriers caused by an infrastructure manager or other carriers' fault, may cause the breach in timing of their locomotive crews and significant secondary delays and unscheduled expenses due to the need for their delivery to rotation stations;

– delays of trains of independent carriers caused by an infrastructure manager or other carriers' fault may cause the breach in operation conditions of the companies they serve;

– delays of trains of independent carriers caused by their fault can provoke train delays of other carriers.

Under such conditions, the problem of ensuring the access of rolling stock to the main and receiving-entry tracks and elimination of conflict situations with train and shunting movements of rolling stock of Ukrzaliznytsya is solved by passing trains of independent carriers in accordance with the rigid schedule [8-11]. The organization of independent carriers' locomotives access to the locomotive infrastructure when transportation is carried out without compliance with freight trains schedule is a problematic issue connected with provision of a non-discriminatory access and the lack of tariffs for such services. Moreover, delays in locomotives maintenance will also cause trains demurrage on the main infrastructure; the tariffs for its use is also absent. In this regard, at the stage of transport market formation, the scheme will be effective, if locomotive service centers are located at points, where loading or unloading of train blocks are concentrated [12].

Performed research shows that diesel locomotives used on Ukrzaliznytsia can serve routes of up to 822 km, and electric locomotives can serve routes of up to 1000 km under condition of locomotive infrastructure concentration for performance of equipment and maintenance of locomotives at points of loading or unloading. Possible schemes for organization of private diesel locomotives and their locomotive crews' operation are given in Fig. 3.

General expenditures for transportation of 1 ton of cargo by Ukrzaliznytsia can be determined by the formula.

$$E_r = \frac{e_{il}(L,g) + e_{ie}(L) + \theta_r(L)e_c}{q} \quad (1)$$

where L – distance of transportation, km; g – type of cargo; $e_{il}(L,g), e_{ie}(L)$ – respectively, infrastructure services and locomotive traction of Ukrzaliznytsia costs during car transportation loaded with cargo g and an empty car depending on the distance of transportation L, UAH; $\theta_r(L)$ – car turnover when transporting by Ukrzaliznytsia locomotives depending on the distance of transportation L, days; q – load carrying capacity of the car, t.

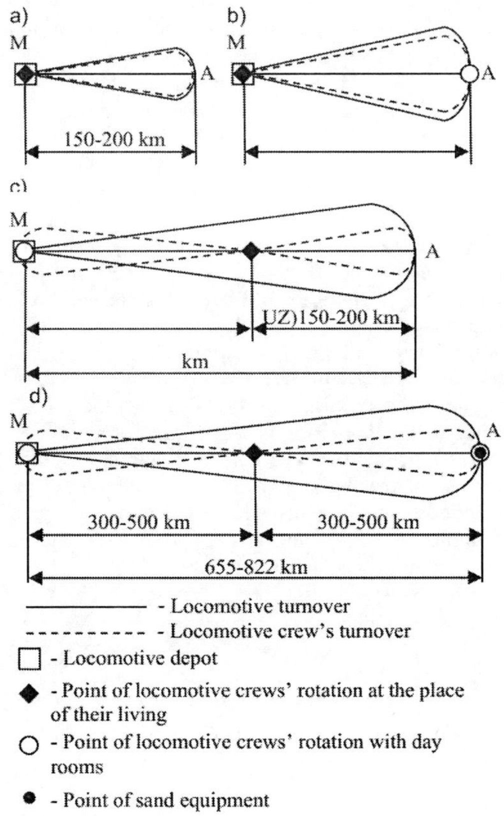

Fig. 3. Schemes of work organization of diesel locomotives and locomotive crews of independent carriers.

Total costs for cargo transportation by private (rented) locomotives can be determined by the formula.

$$E_p = \frac{e_{pl}(L,g)+e_{pe}(L)+E_f(L)+E_{cr}(L)+E_l(L)}{qm} + \frac{e_{cl}(L,g)+e_{ce}(L)+\theta_p(L)e_c}{q} \quad (2)$$

where $e_{pl}(L,g), e_{pe}(L)$ – respectively, the cost of Ukrzaliznytsia infrastructure services for accessing it with own (rented) locomotive during cars transportation with cargo g and empty cars for distance L, UAH; $e_{cl}(L,g)$, $e_{ce}(L)$ – respectively, the cost of infrastructure services of Ukrzaliznytsia for a car accessing it with load g and an empty car during transportation in the train formation with its own (rented) locomotive for distance L, UAH; $E_f(L), E_{cr}(L), E_l(L)$ – respectively, the costs connected with fuel (electricity) payment, the work of locomotive crews, as well as the purchase and maintenance of a locomotive for one trip at the length L, UAH; m – number of cars in train formation; $e_{cl}(L,g), e_{ce}(L)$ – respectively, the cost of Ukrzaliznytsia infrastructure services for a car accessing it with cargo g and an empty car in the train formation with own (rented) locomotive for distance L, UAH; $\theta_p(L)$ – car turnover during transportation in the train formation with its own (rented) locomotive, depending on the distance of transportation L, days.

3 Results

In the course of study, calculations were made of the costs connected with transportation for different distances of iron ore in gondolas (tariff scheme 1) and grain in grain cars (tariff scheme 2) by diesel locomotives and electric locomotives that are new and used. For example, Fig. 4 shows the distribution of costs connected with iron ore transportation for 500 km.

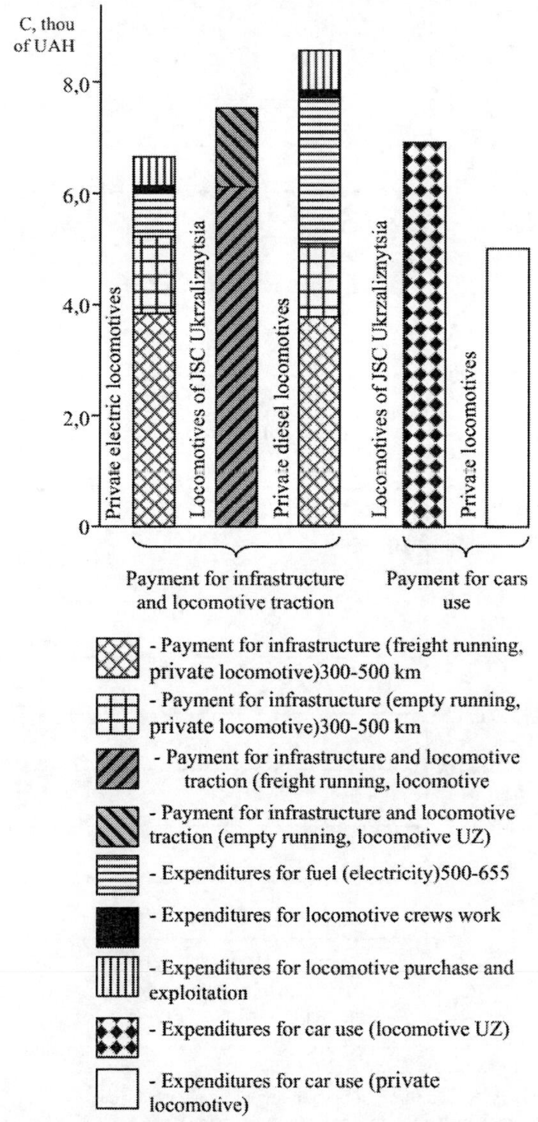

Fig. 4. Distribution of expenses for transportation of iron ore for distance of 500 km by locomotives of JSC Ukrzaliznytsya and own (rented) locomotives

 https://doi.org/10.1051/matecconf/201929406001

The analysis of Fig. 4 shows that the main effect of private traction implementation is achieved by increasing the efficiency of freight car use and reducing expenditures for its use. It should also be noted that Ukrzaliznytsia's revenues from provision of infrastructure for transportation by private diesel locomotives are close to Ukrzaliznytsia's revenues from transportation cargoes, excluding fuel costs and expenses for locomotive crews' labor costs.

Fig. 5, *a* and *b* represent the dependence of cost savings of consignors on the distance traveled by private locomotives.

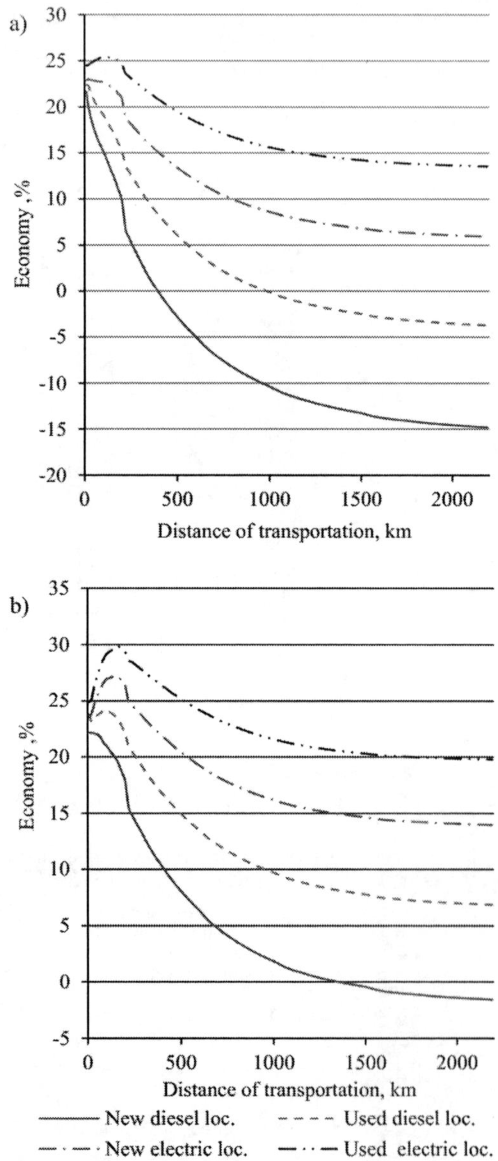

Fig. 5. Economy of private locomotives use during transportation of iron ore (a) and grain (b)

The maximum effect from implementation of private locomotive traction is achieved when transporting cargoes for distances up to 300 km, where transportation can be carried out by one locomotive crew during the locomotive turnover, after which the effect is reduced. Exploitation of private electric locomotives and used diesel locomotives has an economic benefit for all distances that are possible in Ukraine. The use of new diesel locomotives is inefficient for transporting iron ore and coal for distances of more than 400 km, and ineffective for grain transportation for distances of more than 1300 km.

To ensure the recoupment of capital investments into the development of rolling stock fleet, independent transport companies should carry out relevant volumes of transportation. As a result of the research, nonlinear dependencies of transportation volumes from the distance are established, where total expenses for cargoes transportation by Ukrzaliznytsya locomotive are equal to the costs for cargoes transportation by own (rented) locomotive. The corresponding curves allow to determine areas within which the exploitation of private locomotive traction provides an economic effect. An example of such areas for iron ore and grain transportation is shown in Fig. 6.

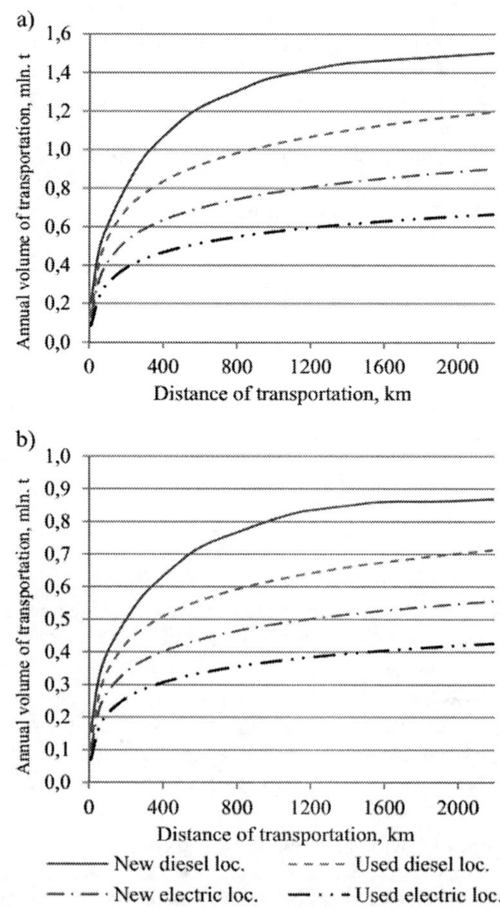

Fig. 6. Minimal volumes of transportation providing economic efficiency of private traction: a – iron ore; b – grain

4 Scientific novelty and practical significance of work

As a result of the performed researches, for the first time, dependencies are established that determine the areas of efficient use of private locomotives for railway transportation within Ukraine.

5 Conclusions

The admission of private locomotives to the main railway network of Ukraine is one of the methods that allows solving the problem of rolling stock shortage. The given measure meets the obligations undertaken by Ukraine under the Association Agreement between it and the European Union, the European Atomic Energy Community and their member states.

References

1. R. Pittman, Options for Restructuring the State-Owned Monopoly, Railway Research in Transportation Economics, **Vol. 20**, 179-198 (2007). https://doi.org/10.1016/S0739-8859(07)20007-1
2. A. Laurino, F. Ramellab, P. Beria, The economic regulation of railway networks: A worldwide survey, Transportation Research Part A: Policy and Practice, **Vol. 77**, 202–212 (2015). doi: 10.1016/j.tra.2015.04.011
3. R. Pittman, Reforming and restructuring Ukrzaliznytsia: a crucial task for ukrainian reformers, Science and Transport Progress, **No. 1 (67)**, 34-50 (2017). doi:10.15802/stp2017/92775
4. M. Makitalo, Market entry barriers in Finland's rail freight transport, World Review of Intermodal Transportation Research, **Vol. 3, No. 1/2**, 181-195 (2010). doi: 10.1504/WRITR.2010.031587
5. M. Laisi, M. Makitalo, O.-P. Hilmola, Stimulating competition in the liberalized railway freight market, *Baltic Journal of Management*, **Vol. 7, No. 1**, 68-85 (2012). doi 10.1108/17465261211195865
6. H. Link, Rail infrastructure charging and on-track competition in Germany, *International Journal of Transport Management*. **Vol. 2., Iss. 1**, 17–27 (2004). doi :10.1016/j.ijtm.2004.05.002
7. B. Slack, A. Vogt, Challenges confronting new traction providers of rail freight in Germany, Transport Policy, **Vol. 14, No. 5**, 399–409 (2007). doi:10.1016/j.tranpol.2007.04.003
8. D. Kozachenko, R. Vernigora, N. Berezovy, Perspectives of the Private Locomotives Use for Main Rail-freight Traffic in Ukraine, *Ukrainian Railways*, **No. 1**, 50-55 (2013). (In Russian)
9. D. Kozachenko, R. Vernigora, V. Balanov, N. Berezovy, L. Yelnikova, Y. Germanyuk, Evaluation of the transition to the organization of freight trains traffic by the schedule, Transport problems, **Vol. 11, Iss. 1**, P. 41-48 (2016). doi: 10.20858/tp.2016.11.1.4
10. D. Tsutskov, Selection of the optimal technological parameters of the organization of transportation process with solid thread chart, Proceedings VNIIAS, **No. 3**, 66-75 (2005). (In Russian)
11. N. Berezovy, Transportation of steel products circular route schedule with private locomotives, East European Journal of advanced technologies, **No. 2(3)**, 51-55 (2014). (In Russian). DOI: https://doi.org/10.15587/1729-4061.2014.23089
12. D. Kozachenko, R. Vernigora, V. Balanov, N. Sannytskyy, N. Berezovy, T. Bolvanovska, Improving the Methods of Estimation of the Unit Train Effectiveness, Transport problems, **Vol. 11, Iss. 3**, 91-101 (2016). doi: 10.20858/tp.2016.11.3.9

Improvement of the technique of calculating operational parameters using an automated system

Yurii Statyvka[1], *Hanna* Kyrychenko[2], *Oleh* Strelko[2], *Yuliia* Berdnychenko[2,*], *Vasyl* Gaba[2], and *Tetiana* Hrushevska[2]

[1] National Technical University of Ukraine "Igor Sikorsky Kyiv Polytechnic Institute", 37, Prosp. Peremohy, Kyiv, 03056, Ukraine
[2] State University of Infrastructure and Technologies, 9 Kyrylivska Street, Kyiv, 04071, Ukraine

Abstract. Impact of the existing operating conditions on terms of cargo delivery and compliance with the rolling stock circulation norms is studied. Planned and actual rolling stock circulation characteristics are analyzed. Deviations from the norms have been detected, which is indicative of non-compliance with technological processes. Nowadays the rolling stock circulation problem becomes more complex; for the service recipients, i.e. customers of the railway, it means delays in cargo deliveries, non-fulfillment of financial obligations before other participants in the railway transport process. A new approach to the methodology and principle of calculating anticipated key performance characteristics are proposed. The existing information technologies for railway management make it possible to monitor compliance with the rolling stock circulation norms and the terms of cargo deliveries and to further improve methods and techniques of performance parameters. This will enable establishing realistic rolling stock circulation norms and cargo delivery terms, thus ensuring compliance with technological requirements of the railway transport processes in order to provide good customer service.

1. Introduction

The main aspects of the strategy of development of the Ukrainian railways are compliance with the cargo delivery terms, intensification of the rolling stock circulation and improvement of the railway management system based upon information technologies. In our opinion, compliance with the cargo delivery terms and the rolling stock circulation norms is an indication of a perfect technological process upon which the railway operation is based. Unfortunately, currently and in recent years the rolling stock circulation is growing, as stated in the reports of the railway authorities and analysts. Nonconformity with the rolling stock circulation norms and cargo delivery terms is indicative of non-compliance with technological processes. Besides, there exists a significant difference between planned and actual rolling stock circulation. Therefore it is necessary to develop new methods and principles of calculating the key performance parameters.

The aim of this study is to establish principles for the formation of evaluation system for railway cargo transportation efficiency capable of adequate reflecting of the current state of cargo transportation system and improving efficiency of railway cargo delivery.

Subject of the research is the system of quality rating of railway cargo delivery.

The research focuses on the principles for the formation of evaluation system for railway cargo transportation efficiency and mechanism for its creation developed on the basis of circulation data in regard to rail cars which travel to ports' railway stations to be unloaded with subsequent loading on the territories of the corresponding railway divisions.

Scientific novelty of the study lies in development of principles, establishment of feasibility and mechanism of the formation of evaluation system for railway cargo transportation efficiency capable of adequate reflecting of the current state of railway services based upon the statistical data available.

The scientific novelty of the obtained results is in solving urgent scientific and practical problem of increasing the efficiency of using wagons fleet due to the new methodology and principle of calculating and forecasting key performance indicators. In particular, the following scientific results were obtained during the research:

1. Determine the freight cars turnover, based on the actual work, that is reflected in the database of ASFTO UZ – E.
2. It has been proposed to determine the freight cars turnover as an indicator of the service of a certain cargo flow.
3. When carrying out an incomplete wagon cycle on a railway, it has been proposed to calculate the share of each railway in turnover - a full cycle from the first loading to the next loading.
4. According to the turnover data, it has been proposed to calculate the required active fleet of freight cars for the transportation of a certain cargo in a specific rolling stock.

* Corresponding author: yb08@ukr.net

2 Literature review and defining problem

Today the expert community is analyzing different ways of solving the problem of adequate prediction of railway transport processes, taking into account the existing operating conditions [1-6], which will enable establishing realistic terms of cargo deliveries and the rolling stock circulation norms. Also, practicing specialists raise an issue of changing the norm setting principles used for planning rail shipments in order to comply with the cargo delivery terms, to intensify rolling stock circulation and to improve railway management system [7-8]. The issues of efficient use of rail transport and intensification of rolling stock circulation are also studied in publications [9-14] dealing with optimization of operation of industrial enterprises.

As known, average (relative) rail car circulation for a railway network is determined by dividing the number of the rail car fleet by the daily average load [15, p.24], and rail car circulation by circulation elements [15, p.50] is determined the following formula:

$$O_v = \frac{l}{24} \cdot \left(\frac{1}{V_d} + \frac{1}{L_{tex}} \cdot t_{tex} + k \cdot t_{van} \right), \qquad (1)$$

where l – complete trip of the rail car, km;

V_d – the average rail car service speed, km/h;

L_{tex} – average run between the train stations, km;

t_{tex} – average rail car detention at one train station, km;

k – cargo operations/ turn-round ratio;

t_{van} – average rail car detention at one train station including loading operations, h.

It is obvious that it's impossible to describe operation of the railway cargo transportation system adequately using average values of a limited number of parameters; that is why actual rolling stock circulation characteristics significantly exceed the values obtained by applying the formula (1).The existing formula do not take into account the rail car maintenance time, although technical condition of the rail car fleet has a significant effect on the rail car operation cycle (turnaround). This is one of the reasons for failures to achieve planned rolling stock circulation figures and its constant intensification [16].

Similar problems exist in the railway administrations of the neighboring states (Russian Railways (RZD), CIS and Baltic countries). In the formulas for calculation of the rolling stock circulation specialists use special coefficients determined by experts; forecasting techniques are also used [17].

Therefore, taking into account considerable difference between the rolling stock circulation characteristics determined on the basis of functional dependency on a limited set of parameters (see formula (1)) and the actual cargo transportation performance characteristics, it is necessary to create a railway cargo transportation efficiency evaluation system that would adequately reflect the railway cargo transportation processes. Such system should comply with the following requirements:

1. standard values must be relevant to the railway cargo transportation process in its current state;
2. if the standard values are exceeded, this is indicative of a problem in the technological process; it helps to find a problematic link in the chain;
3. the efficiency evaluation (norm setting) system should be flexible, i.e., it must be able to correct the standard values in a predictable way depending on the transport system's manageability in regard to reduction of handling time;
4. the efficiency evaluation (norm setting) system should be stable, i.e., not sensitive to accidental or manipulative deviations;
5. the efficiency evaluation (norm setting) system should be safe, i.e., the recommended standard values must not coincide with the boundary efficiency values, in order to avoid constant operation of the system at the limit of its capacity.

At the same time, the railway information system (Ukrzaliznytsia's Rail Freight Traffic Automatic Control System) is a gigantic database which processes information about the work performed, identifies the processes that have taken place [18] and stores information about all operations associated with each rail car. It means that there is a possibility of calculating circulation of each rail car or determining individual performance parameters for a group of rail cars. Similarly, the automated system calculates the time during which each rail car is loaded (to bill the customer), that is a part of the rail car circulation.

So, in order to determine if it is possible to create an adequate efficiency evaluation (norm setting) system based upon Ukrzaliznytsia's information system data, we perform a detailed analysis of data concerning transportation (including batch delivery) of ferrous metals from Kryvyi Rih to Odesa Port.

When developing a new methodology for determining performance characteristics of the railway, it is recommended to take into consideration the following:

The rolling stock circulation should not be calculated using outdated formulas; instead, a rail car circulation should be determined based upon actual operation, which is reflected in the database of Ukrzaliznytsia's Rail Freight Traffic Automatic Control System.

It is recommended to calculate the rolling stock circulation according to the type of the rolling stock used for transportation of a specific cargo, i.e., to determine the rail car circulation as an indicator associated with managing a specific cargo traffic.

If on the territory of a railway authority an incomplete rail car operation cycle (turnaround) takes place, the share of this railway in the circulation (full cycle from loading to the next loading) has to be calculated.

On the basis of circulation data and information about operations performed, it is necessary to determine the rail car fleet required to transport a specific cargo using specific rolling stock.

3 Methodology

In order to analyze the rolling stock circulation parameters of the rail transport processes, data pertaining to shipments (including group shipments) of ferrous metals from Kryvyi Rih to the port of Odessa along several different routes was processed. For analysis of the data a table of train traffic norms in accordance with route timetables has been developed, taking into account handling time at junction points and stations where the locomotive or traction system were changed.

Data concerning the travel time from the starting point of the route to the destination station carriage comprises a part of the parameter – the rolling stock circulation. The second part of the rail car operation cycle is the time when rail cars are on a spur track (feeder line), which is reflected in the database (LOADING or UNLOADING); duration of the operation is analyzed. And finally, the third part is the rail car travel to the next loading or unloading station.

On this route 2242 rail cars were sampled using the following statistical data: average circulation time — 10.14, standard deviation — 5.87, median — 8.77, minimal circulation time — 2.52 and maximum circulation time — 61.01. Here and hereinafter the time is given in days.

As can be seen from the histogram, see Fig. 1, the sampling includes information about 53 rail cars (2.4% of the number of the rail cars sampled) with circulation time of up to 4 days and about 123 rail cars (5.4% of the number of the rail cars sampled) with circulation time exceeding 20 days. The rest of the rail cars with circulation time between 4- and 20-days amount to 92% of the number of the rail cars sampled.

To help understand changes in operation of the transportation system on the basis of changes in statistical sampling data, Fig. 2 shows the calculated monthly minimal lower boundary values (Min), upper boundary values (Max), 1st and 3rd quartiles, the median and the average circulation time. The lower and upper boundaries were selected as the values which were distanced not more than 1.5 interquartile range (IQR) from the median. Data which go beyond the Min and Max boundaries is considered outliers and is not presented in the diagrams.

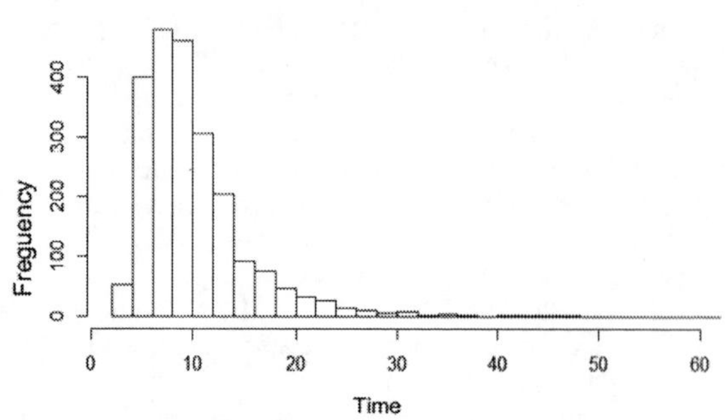

Fig. 1. Distribution of rail car travel time.

As can be seen, the most significant variations are observed for the upper boundary — from 24.30 in January to 12.53 in September, which amounts to 11.77 days or 48% of the maximum value. Much less significant absolute variations are observed for the median and the average value. Thus, the average value fluctuated in the range between 12.41 and 7.2, which amounts to 5.21 days or 42% of the maximum value, and the median — between 11.14 and 6.25, which amounts to 4.89 days or around 44% of the maximum value. The most insignificant absolute variations are observed for the lower boundary — from 4.45 to 2.52, that is 1.93 days or 43% of the maximum value. Similar changes are observed for the 1st and 3rd quartiles.

Hence, the above mentioned data demonstrates that there exist quite significant variations of all presented statistical characteristics of transportation of the specified cargoes. However, the data truthfully reflects operation of the railway transportation system and may be used as a basis for creation of the efficiency evaluation (norm setting) system.

Thus, the median, being a stable parameter as far as the outliers are concerned, unlike the average value, may be used as a current standard in regard to rail car circulation time. If the rail car circulation time exceeds the median, the loss function (penalty function) must be applied to those entities involved in the activities which caused exceedance of said standard value. In this sampling the median is 8.77 days, and during 7 out of 12 months the monthly sampling median is less than the total sampling median, which attests to the fact that such standard, or norm, is not too demanding. However, of course, it does not exclude selection of the loss function in specific cases, when the circulation time exceeds the median as the standard time value. Selection of the loss function is a separate cargo transportation management issue and requires a separate study.

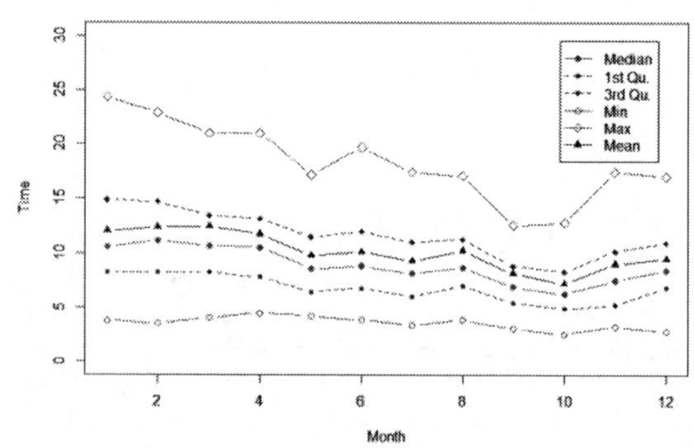

Fig. 2. Changes in rolling stock circulation time.

Based upon the data presented, it can be concluded that at the time of the sampling the state of the cargo transportation system on the specified route has the capacity limit of around 4 days, since such cases amount to only 2.4%, and the lower boundary, as was mentioned earlier, is in the range between 4.45 and 2.52 days.

Thus, the developed efficiency evaluation (norm setting) system complies with the requirements 1), 4) and 5), specified above, since: it is based upon actual conditions; the median is stable in regard to outliers; it

https://doi.org/10.1051/matecconf/201929406002

makes it possible to determine, and thus avoid, boundary values of efficiency — lower boundary values, although the median approaching the lower boundary is practically impossible for such a complex transportation system.

In order to analyses the behavior of the transportation system in more detail, let us consider statistical data pertaining to the sampling that will be built based upon the original sampling, separating only data concerning rail cars, the full circulation cycle of which began and ended on the territory of Prydniprovska Railway. With this, we exclude from the original sampling the data concerning rail cars which were subsequently loaded not on the territory of Prydniprovska Railway, but on the territory of Odesa Railway or any other railway division.

The sampling arranged in this way (let us call it S2) contains information about circulation of 1632 rail cars with the following statistical data: average circulation time — 10.35, standard deviation — 5.61, the median — 9.00, minimal circulation time — 3.16 and maximum circulation time — 53.28 (see Fig. 3).

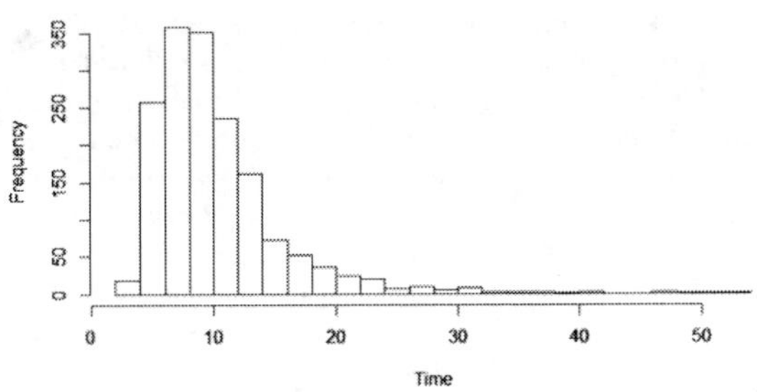

Fig. 3. Diagram of S2 sampling.

Comparison of time during which the rail cars travelled through the territory of Prydniprovska Railway, Fig. 4, and Odesa Railway, Fig. 5, illustrates performance of the transportation system operation, using the travel of the same rail cars as an example. In particular, diagrams of time series quantiles of the time during which the rail cars travelled through the territory of Prydniprovska Railway and Odesa Railway, Fig. 6, show that from among the rail cars with travel time less than 3 days around 30% of their total number travelled through territories of both Odesa and Prydniprovska Railway. However, at shorter time intervals Prydniprovska Railway showed higher efficiency – for 10% of rail cars with the shortest travel time on the territory of Prydniprovska Railway it did not exceed 1.8 days, while on the territory of Odesa Railway it was 2.58 days. On the other hand, at longer time intervals the rail cars travelled faster on the territory of Odesa Railway — with the median of 3.74 against 4.65 on the territory of Prydniprovska Railway, i.e., the median for Odesa Railway was less by almost one day (0.91 days). And the 90th percentile of travel time on the territory of Odesa Railway was less than the corresponding value for Prydniprovska Railway by 4.32 days.

The relationship mentioned above is defined by the entire set of factors which have an impact on the rail car circulation time — nomenclature and quantity of cargo handling operations, mileage, organizational and other features of specific railway divisions and can hardly be harmonized in a rational way. Therefore, in order to ensure compliance with the requirements in regard to relevance of the efficiency evaluation (norm setting) system to the actual state of the transportation system, we think it expedient to consider operation of each railway division a separate subsystem and to adopt the following standard median values — 4.65 days for Prydniprovska Railway and 3.74 days for Odesa Railway. Boundary efficiency values can be taken as 5% quantile — 1.52 and 2.44 for Prydniprovska Railway and Odesa Railway respectively.

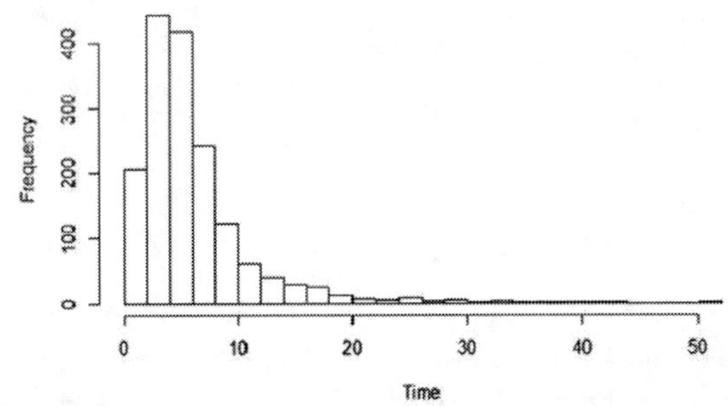

Fig. 4. Distribution of rail car travel time on the territory of Prydniprovska Railway.

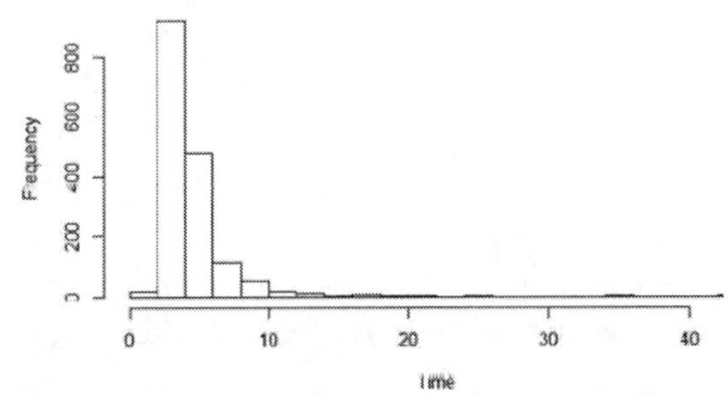

Fig. 5. Distribution of rail car travel time on the territory of Odesa Railway.

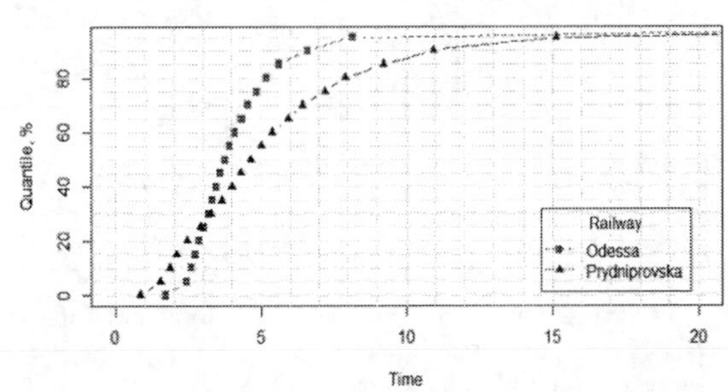

Fig. 6. Quantiles of rolling stock travel time distribution

In order to ensure compliance with the requirement 2) in regard to the efficiency evaluation (norm setting) system, let us determine distribution of S2 sampling rail cars according to the time during which they travel through the territory of each of the railway divisions compared to statistical data for the corresponding railway division.

In order to analyze the period of time during which rail cars were on the territory of different railway authorities (during operation cycle), let us make a logical substitution $b_1 = t^{Od} > bound^{Od}$, which adopts the value "true" if during the rail car travel on the territory of Odessa Railway t^{Od} is bigger than the value of the travel boundary parameter pertaining to Odessa Railway $bound^{Od}$. Similarly, let us make a logical substitution for Prydniprovska Railway $b_2 = t^{Pr} > bound^{Pr}$.

We will substitute $bound^{Od}$ and $bound^{Pr}$ with $bound^{R}$, where $R \in \{Od, Pr\}$. If $bound^{R}$ is the average value of time during which the rail cars travel through the territory of a specific railway $R \in \{Od, Pr\}$, we will substitute $bound^{R}$ with $mean^{R}$.

Table 1 presents data regarding sampling of rail cars according to the travel boundary $bound^{R}$ with values of the 25th percentile, median, mean value and the 75th percentile.

We will suppose that the rail car belongs to rail car type 1, if the period of time during which it is on the territory of Odessa Railway exceeds the value of $bound^{Od}$, specified in the table column, only for Odessa Railway, but not for Prydniprovska Railway (i.e., the same rail car's travel time on the territory of Prydniprovska Railway is less than the value of $bound^{Pr}$). Rail cars of type 2 include those rail cars the travel time of which exceeds the value of $bound^{Pr}$ only on the territory of Prydniprovska Railway, and on the territory of Odessa Railway the travel time is less than the value of $bound^{Od}$. Rail cars of type 3 include those rail cars the travel time of which exceeds the value of $bound^{R}$ on the territory of both railways. And, finally, type 4 means the rail cars, the travel time of which does not exceed the value of $bound^{R}$ on each of the railways.

Table 1. Sampling of rail cars according to the travel boundary.

Row No	$bound^R$	1, % $b_1 \wedge \overline{b_2}$	2, % $\overline{b_1} \wedge b_2$	3, % $b_1 \wedge b_2$	4,% $\overline{b_1} \wedge \overline{b_2}$
1	$t^R_{0.25}$	17,8	18	57	7,2
2	$t^R_{0.5}$	24,6	24,6	25,4	25,4
3	$mean^R$	21.93	21.93	13.23	42.89
4	$t^R_{0.75}$	18,8	18.3	6.7	56,7

Every sampled S2 rail car can belong only to one type, and combination of the four rail car types represents complete sampling S2.

As can be seen from Table 1, the numbers of rail cars of types 1 and 2 respectively are very close for different values of $bound^{R}$. Together they give an idea about the number of rail cars which overstepped the corresponding boundary $bound^{R}$ only on the territory of one of the railways.

Thus, in line No 2 of the table we see that almost 50 per cent of rail cars (49.2%) are among the half of the rolling stock with the longest travel time (exceeding the median for the corresponding Railways) only due to the problems occurring on the territory of one of the Railways. It is possible that it was due to common problems that 25.4% of rail cars had travel time which exceeded the median on the territory of both Railways at the same time – type 3. Travel time of the remaining 25.4% of rail cars is less that the median value for both Railways.

As far as the rail cars with $bound^R$=$mean^R$, see line No 3 of the table. It is easy to see that over 57% of rail cars have travel time which exceeds the average value for the corresponding Railways. With this, it should be noted that in 44% of the cases it was due to problems occurring on the territory of one of the Railways, and 13% of the rail cars exceeded the average travel time for both Railways.

Thus, percentage of rail cars which were delayed only on the territory of one railway division is from 35.8% to 49.2%.

Such analysis also makes it possible to establish the area of responsibility of each of the railway divisions within the boundaries of the entire system. In order to do that we will create sampling S3 based upon sampling S2, leaving only those rail cars the circulation time of which exceeds the circulation time median on a specified route in the original sampling — 8.77 days. The corresponding calculation results are given in Table 2.

It can be noted that the rail cars the delay of which was caused by only one railway division amount to 19.7% - 67.7%. With this, from among all rail cars the circulation time of which exceeds the median 8.77 days, 16.6% of rail cars were delayed exclusively with exceedance of their own median within the boundaries of Odesa Railway, 34.0% — exclusively with exceedance of their own median within the boundaries of Prydniprovska Railway, and 48.4% - their own median values were exceeded on the territories of both railway divisions.

Thus, we can state that the proposed approach to the formation of efficiency evaluation (norm setting) system complies with the requirement 2) in regard to localization of the problematic link.

Table 2. Calculations for rail cars the travel time of which exceeds the circulation time median value for the specified route.

Row No	$bound^R$	1, % $b_1 \wedge \overline{b_2}$	2, % $\overline{b_1} \wedge b_2$	3, % $b_1 \wedge b_2$	4,% $\overline{b_1} \wedge \overline{b_2}$
1	$t^R_{0.25}$	3,6	16,1	79,5	0,1
2	$t^R_{0.5}$	16,6	34,0	48,4	1,0
3	$mean^R$	26,7	41,0	25,2	7,1
4	$t^R_{0.75}$	27,8	34,7	12,8	24,5

As far as the requirement 3) is concerned (flexibility of the norm setting system), it is worth noting that flexibility (adaptability) is by all means is connected with selection of the loss function. However, it is clear that the loss function intended to reduce the number of rail cars the circulation time of which exceeds the median has to reduce the median itself, which eventually will be adopted as the standard value.

4 Conclusion

The study established the basic the formation of evaluation system for railway cargo transportation efficiency (norm setting) relevant to the actual state of the railway, which pertain to the requirements in regard to 1) relevance, 2) possibility to locate problematic links (sections), 3) flexibility (adaptability), 4) stability and 5) safety.

The possibility of creating such efficiency evaluation (norm setting) system has been demonstrated using Ukrzaliznytsia's information system data and circulation time of the rail cars which travel to ports' railway stations to be unloaded with subsequent loading on the territories of the corresponding railway divisions.

It was shown that relevance is ensured by using statistical data in regard to operation of the corresponding transportation subsystem, and stability and safety are ensured by using order statistics – for example, the median and the lower boundary (or the 5th percentile). The issue of the subsystem's size requires further research; however, the natural limit is the requirement to have the statistical data base sufficient to make well-grounded conclusions, i.e., it is defined by cargo handling efficiency.

It was proven that such system is capable of locating problematic sections and areas of responsibility of the subsystems, for example, by applying the method which is described here and which involves breaking the rail car fleet into separate categories (four in this particular case).

Flexibility (adaptability) of the system is ensured by linkage of the standard values to the order statistics that, provided the loss function is chosen properly, will be reduced due to reduction in the number of values which exceed the standard values. The problem of a rational, and moreover optimal, selection of the loss function requires a separate study.

Future research directions: the authors intend to study the issue of selecting the subsystem's boundary limits and evaluating flexibility (adaptability) depending on the loss function, using the statistical data available and mathematical modelling of railway cargo transportation processes.

References

1. E. Z. Zinder, SUBD i deystvitelno bolshie sistemyi, *Sistemyi upravleniya bazami dannyih: elektron. versiya zhur* (Aug. 2018). [Online]. Available: https://www.osp.ru/news/articles/1997/0801/13031555#part_1.
2. F. Y. Wang, Toward a revolution in transportation operations: AI for complex systems, *Intelligent Systems: IEEE*, **23** **(6)**, 8–13 (2008). doi: 10.1109/MIS.2008.112
3. H. I. Kyrychenko, O. H. Strelko, Yu. A. Berdnychenko, S. V. Hurinchuk, Automation of Work Processes at Ukrainian Sorting Stations, *International Journal of Engineering & Technology*, **7** **(2.23)**, 516–518 (2018). doi: 10.14419/ijet.v7i2.23.15346
4. N. A. Kuznetsov, F. F. Paschenko, N. G. Ryabyih, E. M. Zaharova, I. K. Minashina, Algoritmyi optimizatsii v zadachah planirovaniya na relsovom transporte, *Informatsionnyie protsessyi*, **14 (4)**, 307–318 (2014).
5. H. I. Kyrychenko, O. H. Strelko, Yu. A. Berdnychenko, O. V. Petrykovets, O. A. Kyrychenko, Modeliuvannia stsenariiv peremishchennia vantazhu u lantsiuhakh dostavky, *Transportni systemy ta tekhnolohii perevezen*, **12**, 32–37 (2016). doi: 10.15802/tstt2016/85882.
6. H. Kyrychenko, Yu. Statyvka, O. Strelko, Y. Berdnychenko and H. Nesterenko. Assessment of cargo delivery quality using fuzzy set apparatus, *International Journal of Engineering & Technology*, **7(4.3)**, 262-265 (2018).
7. Rozporiadzhennia kabinetu ministriv Ukrainy «Pro skhvalennia Natsionalnoi transportnoi stratehii Ukrainy na period do 2030 roku» vid 30 travnia 2018 roku № 430-r Kyiv. [Online]. Available: https://zakon.rada.gov.ua/laws/show/430-2018-%D1%80
8. *Stratehii rozvytku PAT "Ukrzaliznytsia" na 2017 – 2021 roky*, PAT «Ukrzaliznytsia» (2017).
9. N. Ueno, Y. Nakagawa, S. Morito and Y. Okuno, Steel product transportation and storage simulation: A combined simulation/optimization approach, *1988 Winter Simulation Conference Proceedings*, San Diego, CA, USA, 678-683 (1988).
10. A. O. Lyamzin. Efficiency of the transport system of the industrial region in the port logistics chain of grain supply: author's abstract *Theses for Cand. of Tech. Sci.*, 20 (2008).
11. A. Dávid. Innovation of handling systems in the world container ports and their terminals *Proceedings of the 17th International Conference Transport Means (Kaunas)* (Kaunas: University of Technology) 250-253 (*2013*).
12. O. L. Petrashevsky, A. M. Redzyuk and O. V. Alekseenko. Methodology of conceptual-logistic mapping and project modeling of the objectives of the traffic safety management system *NTU, Collection of Scientific Articles*, **6(10)**, 76-89 (2009).
13. N. I. Shapkin. Organization of rail transportation based on discrete methods of management and a strict schedule of trains traffic *Transport. Science, Technology, Management*, **8**, 2-8 (2008).

14. M. Marinov, T. Zunder, P. Mortimer, Rail Freight Services, *World Transport, Policy & Practice*, **18(2)**, 30-38 (2012).
15. M. V. Makarenko Kratkiy spravochnik pokazateley ekspluatatsionnoy rabotyi zheleznyih dorog Ukrainyi. *Uchebnoe posobie* (2001).
16. H. Ye. Bohomazova, Formuvannia avtomatyzovanoi tekhnolohii upravlinnia vahonopotokamy na zaliznychnykh napriamkakh, *Avtoreferat dysertatsii na zdobuttia naukovoho stupenia kandydata tekhnichnykh nauk*, Kharkiv (2019).
17. Rasporyazhenie OAO "RZhD" ot 18.08.2015 N 2075r "Ob utverzhdenii "Metodiki tehnicheskogo normirovaniya, ucheta rabochego parka i vremeni oborota gruzovogo vagona na infrastrukture obschego polzovaniya" i ustanovlenii statisticheskogo nablyudeniya za pokazatelyami ispolzovaniya gruzovyih vagonov na infrastrukture obschego polzovaniya", 14-17 (2015).
18. H. I. Kyrychenko, Metodyka stvorennia intelektualnoi avtomatyzovanoi systemy upravlinnia dostavkoiu vantazhiv na zaliznytsi, Nauka ta prohres transportu. *Visnyk Dnipropetrovskoho natsionalnoho universytetu zaliznychnoho transportu*, **2(68)**, 46-55 (2017). doi: 10.15802/stp2017/99950.

Investigation of possibility of hopper cars unloading on the car dumper VRS–134M

Stepan Dovhaniuk[1], *Volodymyr* Kalashnyk[1], *Alexei* Reidemeister[1,*], and *Oleksandr* Shykunov[1]

[1] Dnipro National University of Railway Transport named after Academician V. Lazaryan, 49000 Dnipro, Ukraine

Abstract. It was developed a computational model to evaluate the body strength of short base hopper models 19-758-01, 20-471, and 20-4015 when unloading them on the car dumper VRS–134M. Modeling of the stress-strain state of the car design was made using the finite element method. It was established that in terms of strength, bodies of the hopper models 19-758-01, 20-471, and 20-4015 are suitable for unloading on car dumper. The results of experimental studies showed sufficiently high accuracy of the chosen computational model for estimating the stress-strain state of design and confirmed the possibility of hopper car unloading on the car dumper VRS–134M.

1 Introduction

Open-type hopper cars of the models 19-758-01, 20-471, 20-4015 [1–2] have a typical design for this class of rolling stock. Their bodies consist of a frame (center sill, body bolster and end beams), lateral, inclined end walls and unloading bins closed by hatches from the outside. To the side wall frame formed by the top and bottom chords, posts and trusses, plating is attached. The fastening can be rigid, like that in a car for cement transportation, model 19–758–01, or movable, like that in a car for pellet transportation, models 20-471, 20-4015. The end walls are supported by the car frame through the inclined or vertical posts.

General view of such a car using the example of wagon for pellets transportation is shown in the (Fig. 1).

Fig. 1. The hopper car for transportation of pellets model 20-471

Owners of this type of rolling stock are interested in their maximal use, i.e. expansion of the range of cargoes and, accordingly, reduction of empty running and acceleration of car turnaround. A weighty argument for the enterprises using such cars is the possibility of their unloading on the car dumper.

A stationary rotary-type car dumper VRS–134M [3] used in the research unloads gondola cars and other types of open wagons by turning them over. The car dumper works in complex with electric trolley-pusher and shunting device. (Fig. 2).

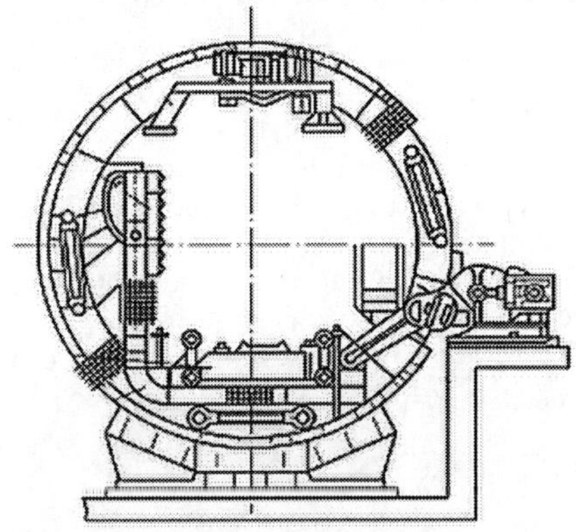

Fig. 2. The stationary rotary-type car dumper VRS–134M

The loaded cars are supplied by the car pusher and positioned on the platform inside the rotor of the car dumper. In the initial period of rotation of the rotor, the bridge with a car is shifted towards side wall of the car dumper until the car contacts it. With further rotation of the rotor, the cradle directed by the rollers in curved slots and crank rods, under the action of gravity moves with the car to the vibrator stops. In the overturned position, the car completely rests on the vibrators and the side wall of the car dumper. When the vibrators are turned on, the remaining of cargo is dumped out from the sides and bottom of the car.

The scheme of the car interaction with the parts of the car dumper is essential for subsequent analysis. The side wall of the car, without bending, rests on the side wall of the car dumper, and the upper top chord is fixed by three stops of the vibrators - the middle and the two outer ones. When the car is turned over, they serve as supports for the car body.

* Corresponding author: reidemeister.a@gmail.com

Design feature of the open-type hopper cars is their short base and, correspondingly, a shorter body. At the present time there is no method of estimating the influence of external factors on the elements of such cars when unloading on car dumper. In order to do this, it is necessary to know all the actual operational loads, however, there are no regulatory recommendations for their evaluation in expanded form. Thus, it is necessary to develop the corresponding methodology for both placing hopper cars in the car dumper, as well as for determining the actual operational loads.

2 Problem definition

The article is devoted to the research of the adaptation of short-base hopper cars to the conditions of unloading the bulk cargoes using the method of overturning and the development of methodology for evaluating the design integrity of the open-type hopper cars when unloading them on the car dumper. This will solve practical problems: reducing downtime, turnover of hopper cars and, ultimately, reducing the operating costs and increasing the economic efficiency of their use.

3 Analysis of research and publications

General requirements to the types of loads acting on the car when unloading on the car dumper are given in the «Norms ...» [4] and DSTU GOST 33211:2017 [5]. According to this, the side walls of gondola cars and open hoppers are calculated for the simultaneous action of vertical and horizontal forces evenly distributed over the chord width at a length of 0.8 m in the middle part of the chord between the neighboring posts.

The calculated vertical force per span q_v, MN, is calculated using the following formula:

$$q_v = \frac{Q_{gr}}{n} \qquad (1)$$

where Q_{gr} – is the gross gravity force of car, MN;

n – is the number of stops of car dumper.

The horizontal transverse force is taken to equal to $0.25q_v$.

The requirements for cars during strength tests in the process of loading and unloading operations are given in the GOST 33788-2016 [6], and the general requirements for ensuring the safety of freight cars are regulated by the DSTU GOST 22235-2015 [7].

Scientists in Russia [8–9] and in Ukraine [10] have dealt with the issues related to the improvement of methods for evaluating the strength of gondola car bodies.

Thus, in the work [8], calculation schemes for distribution of inertial loading from bulk cargo have been developed and a formula for determining the distribution of dynamic pressure along the height has been obtained. The calculation schemes were obtained by analyzing the interaction of the car dumper and gondola car with obtaining oscillograms of stresses in the structural elements.

The work [9] studies unloading process with determining dynamic forces and accelerations acting on the body elements when unloading bulk cargo using the car dumper. The mathematical model of movement of the car-dumper-gondola car body-bulk cargo system during unloading is specified, the influence of the system's rotation angle on the change in the loading value is investigated.

In the work [10] a new method of gondola cars unloading was developed and patented. It is based on the use of suspended car dumper, which practically eliminates all the drawbacks of the known machines and mechanisms designed to unload gondola cars in terms of power action on their bodies. Calculations of the strength of structural elements of the gondola car body when interacting with this car dumper are presented.

The proposed methodologies for calculating gondola car under the action of inertial loading of bulk cargo and from the action of forces during unloading on the car dumper make it possible to increase the reliability of calculating the strength of gondola car design.

4 Purpose and objectives of research

All previously discussed methodologies [8-10] consider the scheme of interaction of a gondola car with the parts of car dumper. Therefore, the task of development of the model and engineering methodology for calculating the bodies of short-based hopper cars is the relevant one. In the course of the work complex, it is supposed to evaluate design strength of hopper cars and correspondence of the strength characteristics of the elements of regulatory documentation based on theoretical and experimental studies.

The possibility of unloading on the car dumper of the cars transporting the following cargoes is considered: crushed stone (granite and limestone), limestone, clay, granulated slag, coal, sand. The cargo must freely spill out of the body, not freeze or stick to it. In the last case the excessive load is applied to the upper part of the body when the car is overturned.

Car bodies are made of various materials (steel grades), the least durable among them is the steel grade 09G2 of strength class 295. The yield strength for it is σt = 305 MPa [4, table 11], the permissible stress of the I design mode is [σ] = 0; 90σt = 274 MPa for the ridge and pivot beams and [σ] = 0; 95σt = 290 MPa for other elements of the body [4, table 10.1].

5 Main part of the research

Solution of the assigned tasks was carried out by combining analytical methods, numerical modeling and field experiments.

The calculation is carried out using finite element analysis. Geometric models of the body were automatically divided into flat finite elements of the «cover» type (resisting both deformations in the proper plane, as well as bending deformations) with a characteristic edge size 50 mm. The plating of side walls is excluded from geometric models, since, not being rigidly connected with the wall frame, it does not receive the loads acting on the car body.

The geometrical model and the loading scheme of the body elements are shown on the example of the car for transportation of pellets of the model 20-471 (Fig. 3).

It is assumed that the body rests on the middle and one of the outer stops. Loading groups: A, B – are hinged bearings of the pivoted joints, C – are dummy movable sliding bearings of side wall, D – is the gravity force of design, E – is the pressure on outer stop, F- is the pressure on the middle stop.

The stress state is estimated according to the values of the equivalent stresses Mises σ_{eq} acting on the external surfaces of the elements:

$$\sigma_{экв} = \frac{1}{\sqrt{2}}\sqrt{\begin{matrix}(\sigma_x - \sigma_y)^2 + (\sigma_y - \sigma_z)^2 + \\ +(\sigma_z - \sigma_x)^2 + 6(\tau_{xy}^2 + \tau_{yz}^2 + \tau_{zx}^2)\end{matrix}} \qquad (2)$$

where σ_x, σ_y, σ_z, τ_{xy}, τ_{yz}, τ_{zx} – are the components of stress tensor.

The pattern of distributions of stress arising in the elements of car bodies when unloading on the car dumper is shown in the (Fig. 5-7).

Fig. 3. Loads and supports for the car body 20-471 at unloading on the car dumper

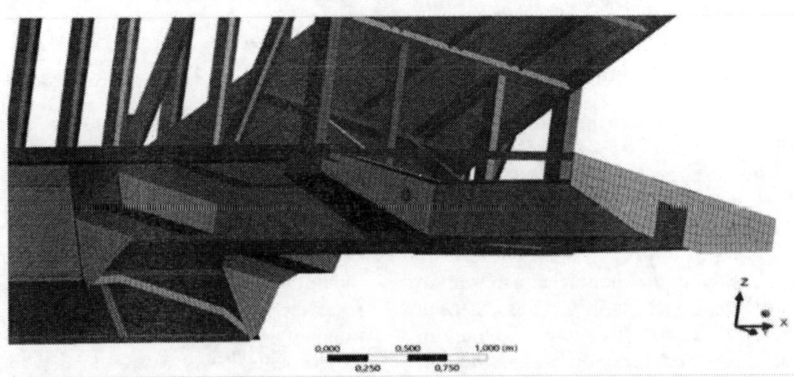

Fig. 4. The fragment of the finite element model of the console part of the car body 20-471.The console part of the body

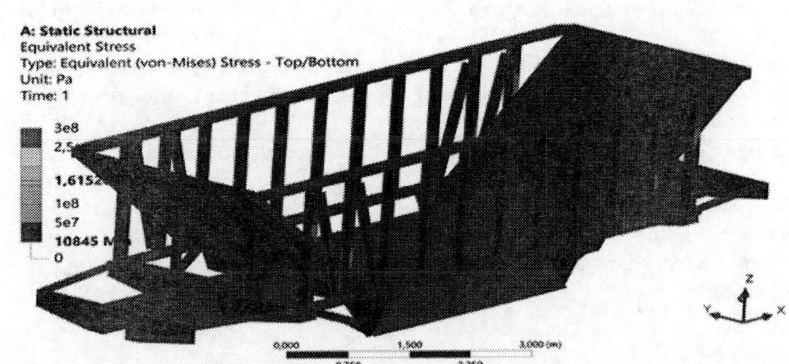

Fig. 5. Stresses in the car body 20-471 when unloading on the car dumper

 https://doi.org/10.1051/matecconf/201929406003

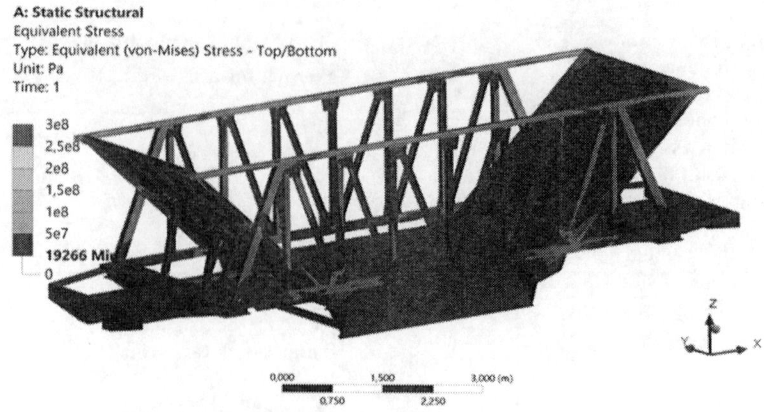

Fig. 6. Stresses in the car body 20-4015 when unloading on the car dumper

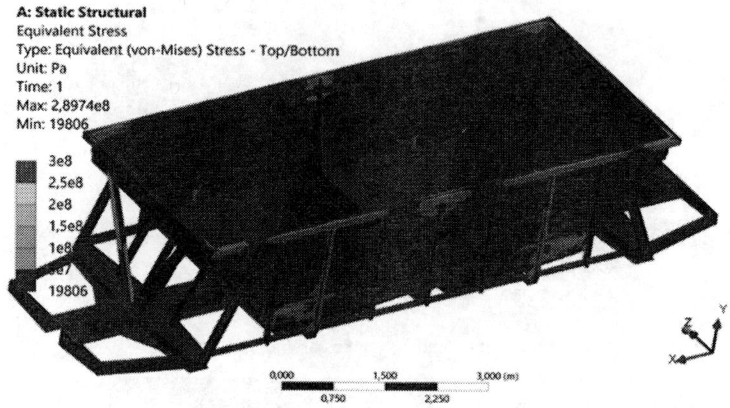

Fig. 7. Stresses in the car body 19-758-01 when unloading on the car dumper

Strength of the car body for cement transportation of the model 19–758–01 in the general case meets the requirements [4]. The stresses do not exceed 160 MPa (when the admissible ones [σ] = 290 MPa), except for the concentration area on the side wall post, where they are equal to 224 MPa.

Strength of the car body for pellets transportation 20-471 meets the requirements [4]. The stresses reach the values of 250 ... 272 MPa (the top and bottom chords, posts and trusses of the side walls, the bin jumper), which is less than the permissible [σ] = 290 MPa. The dimensions of the areas of construction with stresses greater than 250 MPa are relatively small, less than 100 mm.

The stresses in the side wall truss and at the base of the cantilever portion of the top chord in the car for transportation of pellets of model 20-4015, reaches the yield point. The strength of the other body elements meets the requirements [4]. This car can be unloaded on the car dumper, if it is excluded the possibility of non-dumping out of cargo from the body (for example, unloading the cars on the car dumper in the warm season, avoiding adfreezing).

According to the results of calculations for the body strength of all the above-mentioned cars, it is allowed to unload them on the car dumper VRS-134M (provided that the body is supported by the middle and one of the outer stops along the entire length of the latter).

For the purpose of approbation of the proposed methodology of theoretical studies, experimental research with unloading of hopper cars on the car dumper was carried out. When unloading the experimental cars were loaded with limestone for technological needs of up to 66 tons and equipped with sensors for recording the stresses. After that, the loaded cars one by one were supplied to the car dumper VRS-134M and positioned under the outer and middle stops. Each car was turned over twice in a row (the second time – in the empty state), after which the cars were inspected to identify possible damages and loss of pivots. During the unloading process, the continuous stress recording was performed. An example of the stress recording in the body elements of the car for transportation of pellets, model 20-471 when unloading is shown in the (Fig. 6). Here the oscillogram is gridded: stresses in MPa – the time of turning the car in sec.

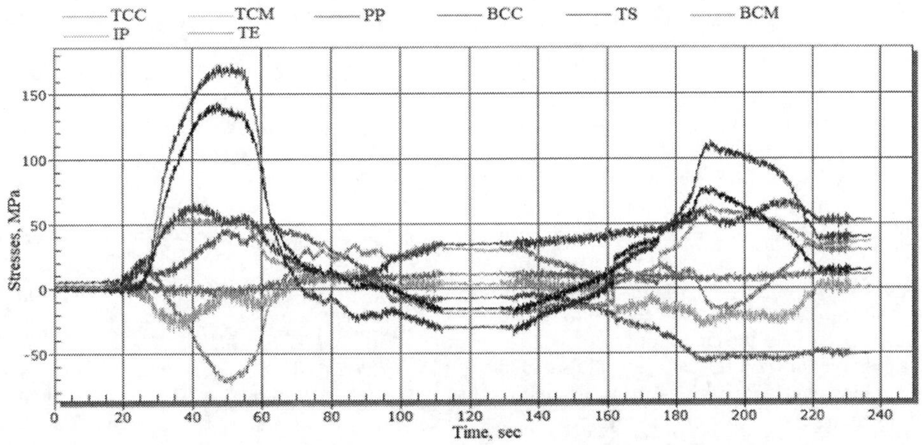

Fig. 8. The oscillogram of the stresses in the car body 20-471 when unloading on the car dumper

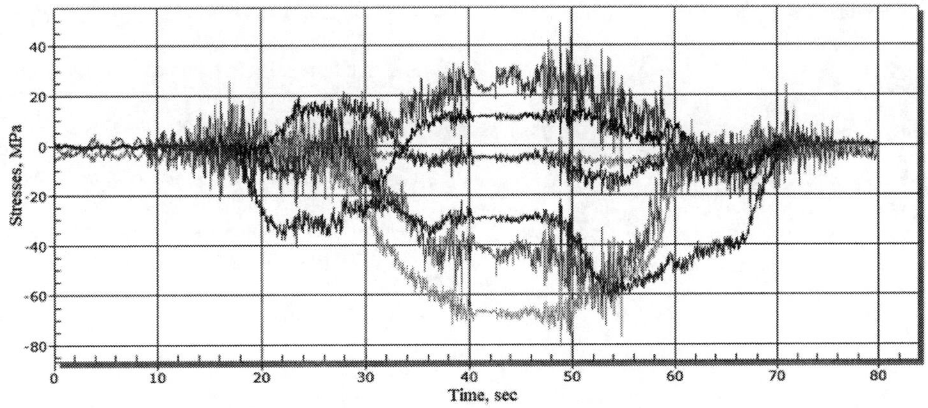

Fig. 9. Stresses in the car body 20-471, JSC DVM during overturning in an empty state

In the (Fig. 8,9) the following acronyms are accepted: TCC – is the top chord in the cantilever portion of the car, TCM – is the top chord in the middle part of the car, PP – is the pivot post, BCC – is the bottom chord in the cantilever portion of the car, TS – is the truss of the car side wall, BCM – is the bottom chord in the middle part of the car, IP – is the intermediate post, TE – is the truss of end wall.

On the oscillogram: 20-60 seconds – is the turn of the loaded car and return to its initial position when unloading, 160-220 seconds – is the turn of the empty car and return to its initial position.

After processing the oscillograms according to the experiments and recalculating the results for full load capacity, the body strength of the all cars was evaluated.

When unloading on the car dumper, the maximum total stresses recalculated for the full load capacity of the car were as follows:

- for the model 19-758-01: 174.5 MPa (bin edge), -138.5 MPa (side wall post), -118.8 MPa (side wall truss);
- for the model 19-758-01: 174.5 MPa (bin edge), -138.5 MPa (side wall post), -118.8 MPa (side wall truss);
- for the model 20-4015: 211.4 MPa (bottom chord), -205.6 MPa (outer truss of the side wall post), -117.1 MPa (side wall post);
- for the model 20-471: 225.1 MPa (bottom chord), -176.0 MPa (side wall post).

It can be seen that for all elements of cars, the values of total stresses obtained are much lower than those allowed.

Thus, the stress values obtained in the course of experimental studies according to the developed methodology confirmed the results of earlier analytical calculations. And on the whole, the results of the research have shown the possibility of unloading the short base hopper cars on this type of car dumper.

Conclusions

As a result of the studies, it was developed the model and engineering methodology for calculating the bodies of short-based hopper cars for strength when unloading on the car dumper VRS-134M. In the process of performing the work, a scheme for interaction of the short-based hopper car with the parts of the car dumper is found out. It makes it possible to represent the computational model more accurately. Reliability of the obtained results is confirmed by a comparison of the results of the calculated and experimental data. Justification of the developed recommendations is confirmed by the positive test results.

The stresses in car bodies did not exceed the permissible 280 MPa (steel 09G2S), i.e., the strength of the cars of the models 19-758-01, 20-4015, 20-471 is sufficient for unloading on the car dumper VRS-134M provided that the cargo is free to dump out and the car is positioned under the outer and middle stops of the car dumper.

The proposed method can be used to calculate the bodies of the short-based hopper cars for strength when unloading on other types of the car dumper.

The research on the hopper cars adaptation to the conditions of unloading bulk cargoes using the method of overturning and strength estimation of their bodies made it possible to reliably estimate a stress condition of the car bodies and to justify the possibility of unloading these cars on the car dumper. The latter is essential for reducing the downtime and turnaround of hopper cars.

References

1. *Freight railroad cars. Reference catalogue. Part 5. Hoppers.*
2. *Freight cars of gauge of 1520 mm of USSR railways (reference catalogue)*, Moscow, Transport, 177 p (1989)
3. N.P. Zhuravlev, O.B. Malikov *Transport-Cargo Systems*, Moscow, Marshrut, 367 p (2006)
4. *Norms for the calculation and design of railroad cars of the Ministry of Railways of the gauge of 1520mm (non-self-propelled)*, Moscow, GosNIIV-VNIIZhT, 319 p (1996)
5. *DSTU GOST 33211:2017. Freight cars. Requirements for structural strength and dynamic qualities. (GOST 33211-2014, IDT)*, Kyiv, SE UkrNDNC, 58 p (2017)
6. *GOST 33788-2016. Freight and passenger railcars. Methods of testing structural strength and dynamic performance*, Moscow, Standardinform, 41 p (2016)
7. *DSTU GOST 22235-2015. Freight cars fot 1520 mm gauge main railways. General requirements for safety in loading-unloading and shunting operations (GOST 22235-2010, IDT)*, Kyiv, SE UkrNDNC, 22 p (2015)
8. A.V. Dodonov, I.A. Hilov, A.A. Bityuk *Improvement of the method of calculating the strength of the gondola body structure of the gondola during unloading on the wagon breaker. Approval of new methods of freight wagons designing: a collection of scientific works.* Engineering center of car building, St. Petersburg, **11** (2012)
9. G.I. Petrov, I.V. Chepurnenko, O.I. Panachev *Adaptation of the mathematical model of the gondola to the conditions of body unloading by overturning*, Transport business of Russia, Moscow, **1** (2017)
10. R.I. Visnyak *Determination of the strength of the body of the gondola during unloading of bulk cargoes*, Author's abstract of candidate's thesis (Engineering), Kharkiv, UkrSURT, 20 p (2003)

The impact of asymmetry on equilibrium of logistics systems

Irina Savelieva[1,*], *Serhij* Melnikov[1], and *Alexandra* Orlovska[2]

[1] ONMU, Chair of Entrepreneurship & Tourism, 65029 Odessa, Mechnikov Str., 34, Ukraine
[2] DNURT, Lviv branch, 79052 Lviv, I. Blazhkevich Str., 12a, Ukraine

Abstract. The article examines one of the actual problems of the theory and practice of the logistics systems functioning related to the asymmetry of information in the transport services market. Assessing the level of asymmetry in logistics systems is of fundamental importance to increase the competitiveness of relevant systems. New approach formulated herein estimates the logistics systems asymmetry by taking into account information asymmetry about the quality of transport products, as well as information asymmetry about the quality of products within the boundaries of the corresponding supply chain. This approach allows to take into account its potential capabilities and increase functional stability in the process of logistics systems design.

1 Introduction

The basis of a classical economic theory is the assumption of a completeness and accuracy of information held by the economic agents. Based on this assumption, a conclusion is made about the principal possibility of rational behavior of economic agents and achieving the Pareto efficiency. However, this assumption does not correspond to economic reality and a lot of research in the twentieth century was devoted to the impact of informational and other types of asymmetry in market processes. The result of this research was the first in this century, the Nobel Prize in economics that was awarded Akerlof, Spence, and Stiglitz for the development of the theory of markets with asymmetric information.

Note that until now there is no clear agreement on the role of asymmetric information in market processes. For Akerlof, the asymmetric information plays a negative role, prevents economic agents from making rational decisions, restricts the effective allocation of resources, and therefore needs state regulation. Another viewpoint on the role of the asymmetric information was expressed by Hayek [1] – the Nobel Prize winner of 1974. They believe that it is the asymmetric information that is the prerequisite for the existence of markets and profit for entrepreneurs.

Asymmetry is one of the problems that hinders the sustainable development of logistics systems. An important step in solving this problem is to develop a theoretical basis for determining the extent to which the asymmetry is assessed.

The functionality of logistics systems is affected by the following types of information asymmetry related to quality:

1) information asymmetry of the quality of goods between producers and consumers within the relevant supply chains;
2) information asymmetry regarding the quality of transport products between the participants of the market of transport services: state authorities, cargo owners, carriers, logistics centers, intermediaries;
3) both types of asymmetry combined.

Indicators of information asymmetry in the market of logistics services should include:
- regularity and timeliness of transportation;
- delivery time;
- quality and conditions of cargo storage;
- vehicles traffic safety;
- schedule reliability and speed of passenger transportation;
- passengers comfort and service level throughout the route;
- guarantees of carrying out of the reloading process in the specified time in the logistic centers;
- possibility to receive in the real time information about cargo operations,
- presence of specialized equipment with the required technical characteristics in the logistics centers,
- availability and quality of additional operations and services.

All these indicators have a direct impact on the financial performance of economic agents of the logistics system.

In this paper, we consider a duopoly model of supply chains, which compete under conditions of the impact of real and information asymmetries. The equilibrium states of Cournot and Stackelberg are determined and the influence of all asymmetries is explicitly shown.

Akerlof was the first one to describe the impact of asymmetric information on market equilibrium. In his seminal work [2], Akerlof formalized the adverse

* Corresponding author: savirina@gmail.com

selection that occurs in the used car market with the asymmetry of information about car quality between supply chain and buyer.

Vives [3] considers the impact of private information about an uncertain linear demand on Cournot and Bertrand equilibria in the duopoly model. It is shown that if the goods are substitutes (not) to share information is a dominant strategy for each firm in Bertrand (Cournot) competition. In her work, Gal-or [4] examines the impact of cost information asymmetry on Cournot and Bertrand equilibria in the duopoly model. The author investigates how incentives for two duopolists to honestly share information change depending upon the type of equilibrium. Roller and Sinclair-Desgagne [5] have analyzed the sources of persistent asymmetry between firms and between markets in a Cournot's duopoly model.

Zanchettin [6] analyzed the impact of cost asymmetry and demand asymmetry on Bertrand and Cournot equilibria in a differentiated duopoly. The paper has shown that both the efficient firm's and industry profits are higher under Bertrand competition when asymmetry is strong and/or products are weakly differentiated. Wang at el. [7] has proposed an agent-based model to study the impact of asymmetric information on market evolution. The model proposed is able to demonstrate how the asymmetry of information leads to the adverse selection effect. The model also explains the coexistence of low- and high-quality goods in a market with asymmetric information. Ledvina and Sircar [8] have investigated how costs asymmetry impact on entry/exit decisions of firms at Cournot and Bertrand equilibria. The paper shows that due of cost asymmetry the differentiated goods result in more active firms in equilibrium than homogeneous goods.

Nagurney at el. [9] have developed a spatial price equilibrium model with information asymmetry in quality in that the producers at the supply markets are aware of their product quality whereas consumers at the demand markets are only aware of the average quality of the products. In the paper provided qualitative analysis of conditions for existence and uniqueness of equilibria as well as stability analysis for the solutions. Wang [10] has examined a supply chain model consisting of a dominant supplier and a buyer, in which the latter possesses superior knowledge of his private cost information. The supplier's imperfect knowledge about the buyer's cost is described by a uniform distribution. In the paper derived the explicit expressions of the optimal equilibrium outcomes of two contract formats offered by the supplier, the simple price-only contract and sophisticated menu of contracts, respectively.

Brangewitz and Manegold [11] have analyzed the impact of quality asymmetry on Cournot and Bertrand equilibria in an intermediate goods market. Melnikov [12] has investigated Cournot and Stackelberg equilibria in the duopoly model under conditions of asymmetric information of quality. It has found that optimal for both duopolists is a Stackelberg equilibrium when the leader is a manufacturer of high-quality good. It is received that adverse selection effects all equilibrium states.

Based on the analysis of the literature, the following types of the asymmetries impact on market equilibrium can be distinguished.

The first type is the impact of real asymmetries. This is the asymmetry of the values of market indicators. This asymmetry type is natural and always present in the economy. The examples of indicators at the micro level: costs, quality level, prices, activity strategy, production capacities, volumes of activity, location, etc., at the macro level: gross national income, gross regional product, equilibrium state (stability), a degree of monopolization, etc.

The second type is the impact of information asymmetries. This is the asymmetry of completeness, reliability and an availability of information between economic agents about market indicators. For example, asymmetry of Akerlof (asymmetry of information about a quality of goods between a seller and a buyer) and Stackelberg (asymmetry of information about a competitor's strategy between sellers).

The third type is the simultaneous impact of real and information asymmetries. An example of this is, the classical model from [2], where there is a real asymmetry of quality and asymmetry of information about quality.

In [12], the authors found that the Stackelberg asymmetry does not help the high-quality seller overcome the Akerlof asymmetry and earn more than a competitor. The leadership of a high-quality seller paradoxically increases a profit of the low-quality seller under conditions of the Akerlof asymmetry.

It is of interest to develop the results obtained by Melnikov [12] to the case of a location asymmetry between supply chains. The aim of this paper is the analysis of the asymmetry impact on equilibrium states in the duopoly model of supply chains.

2 The model

Two supply chains sell similar goods in the same markets, $m = 2 \cdot n + 1$, $n \in N$, N – set of natural numbers. Markets are located along a line, the distance between any neighboring markets is l.

One of the supply chains sells low-quality goods (index 0), the other – high-quality goods (index 1). The relationship between goods quality levels is described by the quality asymmetry coefficient, $k > 1$. Assume that all unit costs of high-quality goods are higher in k times comparison with low-quality goods.

Transportation delivery costs per unit of the low-quality product per unit distance are equal t. Deliveries are made on a DDP agreement, goods stocks of supply chains are unlimited.

Preferences of consumers in each market are described by the Cobb-Douglas utility function. In these conditions, the demand for goods is expressed by isoelastic functions. Consumers have full information about quality and form a separate demand for low-quality goods: $p_0 = 1/q_0$, and high-quality goods: $p_1 = k/q_1$, where p_0, p_1 – market prices, q_0, q_1 –

quantity supplied. Consumers are willing to pay for the same volume of high-quality goods k times greater. In each market, prices and sales volumes are the same.

We introduce into the model the location asymmetry. Location asymmetry arises when one of the supply chains has a competitive advantage because of their location. In this model, location asymmetry is measured by the ratio of the total distance of supply chains' transportations and depends on their location and the number of markets.

Assume that the low-quality supply chain is located on the 1st market, and the high-quality supply chain is located at the center, of the market with index $(m+1)/2$ (Fig.1).

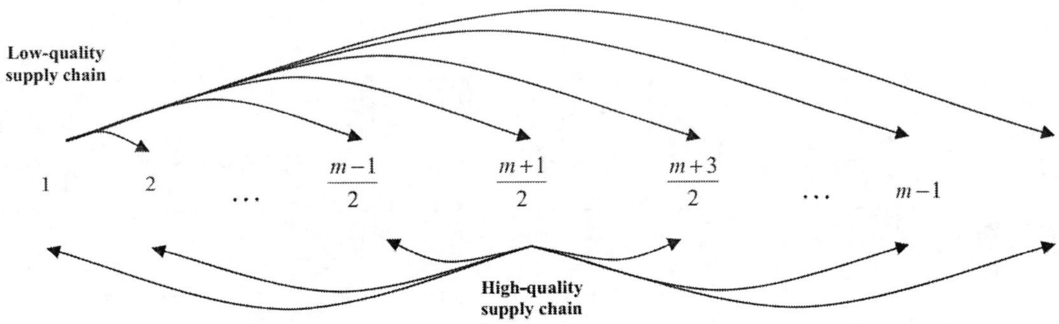

Fig. 1. Location of supply chains

Obviously, the low-quality supply chain has a maximum and the high-quality supply chain has a minimum delivery distance of goods. The total transportation distance of the high-quality supply chain is: $L=l\cdot(m^2-1)/4$, low-quality supply chain: $d\cdot L$, where $d=2\cdot m/(m+1)$ – location asymmetry coefficient, $d\in[1{,}5;2)$.

In conditions of complete information about goods quality the supply chains are monopolists. The profit functions that need maximizing over the quantities supplied for each supply chain are given by:

$$F_0^{\text{mono}}=q_0\cdot\left(m\cdot\frac{1}{q_0}-t\cdot d\cdot L\right)\to\max_{q_0},$$

$$F_1^{\text{mono}}=q_1\cdot\left(m\cdot\frac{k}{q_1}-t\cdot k\cdot L\right)\to\max_{q_1}\quad(1)$$

From (1) we see that for identical volumes of sales, $q_0=q_1$, a profit of the high-quality supply chain is always higher than the low-quality supply chain. The real asymmetries of quality and location "work" in favor of the high-quality supply chain: $\partial F_0/\partial d<0$, $\partial F_1/\partial k>0$.

3 The impact of Akerlof information asymmetry

Suppose that the low-quality supply chain began to advertise its product as a quality one. If consumers cannot distinguish a quality of goods, then the Akerlof information asymmetry of arises. In conditions of the Akerlof asymmetry, consumers form the demand for both goods already in the form of a single function: $p=(1-\alpha)\cdot p_0+\alpha\cdot p_1=(k+1)/(q_0+q_1)$, where $\alpha=q_1/(q_0+q_1)$ – market share of the high-quality goods, $(1-\alpha)$ – market share of the low-quality goods.

In conditions of incomplete information about goods quality the supply chains are duopolists. The profit functions:

$$F_0^{duo}=q_0\cdot\left(m\cdot\frac{k+1}{q_0+q_1}-t\cdot d\cdot L\right)\to\max_{q_0},$$

$$F_1^{duo}=q_1\cdot\left(m\cdot\frac{k+1}{q_0+q_1}-t\cdot k\cdot L\right)\to\max_{q_1}.$$

Let us find the Cournot equilibrium and analyze how the Akerlof asymmetry affects the profits of supply chains and impact of real asymmetries. Putting the first derivatives, $\partial F_0^{duo}/\partial q_0=0$ and $\partial F_1^{duo}/\partial q_1=0$, and solving for q_0,q_1 one obtains:

$$q_0^*=\sqrt{\frac{m\cdot(k+1)\cdot q_1}{t\cdot d\cdot L}}-q_1,\ q_1^*=\sqrt{\frac{m\cdot(k+1)\cdot q_0}{t\cdot k\cdot L}}-q_0,\quad(2)$$

which are the reaction functions.

The second derivatives are negative, $d^2F_0/dq_0^2<0$, $d^2F_1/dq_1^2<0$, what means the profit functions of reach a maximum.

Let us find the following indicators (Table 1):

- the range of permissible values of the quality asymmetry coefficient k;
- the sales volumes: q_0^e,q_1^e, $Q^e=q_0^e+q_1^e$;
- the profits: F_0^e,F_1^e;
- the prices under conditions of effect of the Akerlof asymmetry p^e and in its absence p_0^e,p_1^e;
- the market share of high-quality goods α^e;

• the Akerlof point k^e. The Akerlof point corresponds to the value of the quality asymmetry coefficient, in which the duopoly disappears ("market failure"). A sign of the Akerlof point will be considered the absence of sales or profit or loss of equilibrium stability;

• the market share of high-quality goods in the Akerlof point $\alpha^e\left(k^e\right)$.

Table 1. The Cournot equilibrium

Supply chain	k	q^C	Q^C	F^C
low quality	$1<k<d\cdot\text{P}$	$\frac{m\cdot k\cdot(k+1)}{t\cdot L\cdot(k+d)^2}$	$\frac{m\cdot(k+1)}{t\cdot L\cdot(k+d)}$	$\frac{m\cdot k^2\cdot(k+1)}{(k+d)^2}$
high quality		$\frac{m\cdot d\cdot(k+1)}{t\cdot L\cdot(k+d)^2}$		$\frac{m\cdot d^2\cdot(k+1)}{(k+d)^2}$

p_0^C	p^C	p_1^C	α^C	k^C	$\alpha^C\left(k^C\right)$
$\frac{t\cdot L\cdot(k+d)^2}{m\cdot k\cdot(k+1)}$	$\frac{t\cdot L\cdot(k+d)}{m}$	$\frac{t\cdot k\cdot L\cdot(k+d)^2}{m\cdot d\cdot(k+1)}$	$\frac{d}{k+d}$	$d\cdot\text{P}$	$\frac{1}{\text{P}+1}=14{,}6\%$

Let us analyze the results obtained. To analyze the impact of the Akerlof asymmetry on the profit of the low-quality supply chain, we calculate the coefficient: $A_0=\text{p}^C/p_0^C=k\cdot(k+1)/(k+d)$, where p_0^C – the selling price of the low-quality good in volume q_0^C in the absence of the Akerlof asymmetry. Then the profit of the low-quality supply chain can be represented in the form:

$$F_0^C=q_0^C\cdot\left(m\cdot A_0\cdot p_0^C-t\cdot d\cdot L\right). \tag{3}$$

The impact of the coefficient A_0 on profit (3) is determined by conditions: $\partial A_0/\partial k>0$, $\partial A_0/\partial d<0$, $\text{sign}(A_0-1)=\text{sign}\left(k-\sqrt{d}\right)$. Because of location asymmetry, the low-quality supply chain benefits from the Akerlof asymmetry only when $k>\sqrt{d}$. At $k=\sqrt{d}$, the Akerlof asymmetry does not affect profit: $F_0^{\text{mono}}\left(q_0^C\right)=F_0^C\left(q_0^C\right)$.

To analyze the impact of the Akerlof asymmetry on profit of the high-quality supply chain, we calculate the coefficient: $A_1=p^C/p_1^C=d\cdot(k+1)/(k\cdot(k+d))$, where p_1^C – the selling price of the high-quality good in volume q_1^C in the absence of the Akerlof asymmetry. Then the profit of the high-quality supply chain can be represented in the form:

$$F_1^C=q_1^C\cdot\left(m\cdot A_1\cdot \text{p}_1^C-t\cdot k\cdot L\right). \tag{4}$$

The impact of the coefficient A_1 on profit (4) is determined by conditions: $\partial A_1/\partial k<0$, $\partial A_1/\partial d>0$, $\text{sign}(1-A_1)=\text{sign}\left(k-\sqrt{d}\right)$. With the Akerlof asymmetry, the high-quality supply chain benefits from location asymmetry only up to $k<\sqrt{d}$. At $k=\sqrt{d}$, the Akerlof asymmetry does not affect profit: $F_1^{\text{mono}}\left(q_1^C\right)=F_1^C\left(q_1^C\right)$.

From Table 1 it follows that sale volumes and profits at the Cournot equilibrium are always positive. To find the Akerlof point, we will analyze the equilibrium stability. Let us consider the two-dimensional map:

$$q_0^*(t+1)=\sqrt{\frac{m\cdot(k+1)\cdot q_1(t)}{t\cdot d\cdot L}}-q_1(t),$$

$$q_1^*(t+1)=\sqrt{\frac{m\cdot(k+1)\cdot q_0(t)}{t\cdot k\cdot L}}-q_0(t). \tag{5}$$

The stability of the map fixed point (5), $\left(q_0^C,q_1^C\right)$, is defined by multiplicators μ_1,μ_2, which are eigenvalues of Jacobian matrix at the fixed point: $J=\begin{pmatrix}0 & \frac{k-d}{2\cdot d}\\ -\frac{k-d}{2\cdot k} & 0\end{pmatrix}$. The multiplicators are roots μ_1,μ_2 of the characteristic equation

$$\mu^2+|J|=\mu^2+\frac{(k-d)^2}{4\cdot k\cdot d}=0. \tag{6}$$

From (6) it follows that the eigenvalues are pure imaginary, since both k and d are positive. As is known, the boundary of the stability region of a two-dimensional map for pure imaginary multiplicators is found from condition $|J|=1$. Equating $|J|=1$, we find the value of the quality asymmetry coefficient at which the Cournot equilibrium loses stability: $k^C=d\cdot\text{P}$, where $\text{P}=3+2\cdot\sqrt{2}$ – the Puu point (bifurcation point in a duopoly model with one market [13].

Equating $\mu_1 = \mu_2 = 0$, we find the value of the quality asymmetry coefficient at which the Cournot equilibrium is a superstable: $k = d$.

The relationship between equilibrium sales volumes depends on the superstable point: $\operatorname{sign}\left(q_0^C - q_1^C\right) = \operatorname{sign}(k - \mathrm{d})$. Dynamics of equilibrium sales volumes, depending on the quality asymmetry coefficient, is shown in Fig.2 (a). Data: $t = 1$, $l = 0{,}1$, $m = 3$, $k \in (1; 9{,}375]$. The Akerlof point: $k^C = 8{,}74$, the superstable point: $k = d = 1{,}5$. An increase in the number of markets leads to a decrease in equilibrium sales volumes and an increase in the level of the Akerlof point. This is clearly seen in the bifurcation diagrams of the high-quality supply chain (Fig.2(b)). Data: $t = 1$, $l = 0{,}1$, $m = 3$, $m = 5$, $k \in [8; 10{,}415]$. The points of Akerlof: $k^C(m = 3) = 8{,}74$, $k^C(m = 5) = 9{,}71$.

From Fig. 2 we see that an increase of investment in quality paradoxically leads to the ousting of high-quality goods from the market. If consumers cannot distinguish the quality of goods, then the high-quality supply chain will either have to leave the market or switch to sale of the low-quality good. As a result, the duopoly is reduced to the market of low-quality goods. Thus, given model illustrates the adverse selection that results from the information asymmetry about quality [2].

4 The impact of Stackelberg information asymmetry

Now we introduce into the model the Stackelberg information asymmetry. Stackelberg information asymmetry arises when one of the supply chains (leader) knows the competitor's reaction curve, and competitor (follower) does not own such information. Assume that the low-quality supply chain is a leader, and the high-quality supply chain is a follower. The new profit function of the low-quality supply chain is: $F_0 = \sqrt{t \cdot k \cdot L \cdot m \cdot (k+1) \cdot q_0} - t \cdot d \cdot L \cdot q_0 \rightarrow \max_{q_0}$.

Using the standard procedure, we find the equilibrium indicators (Table 2).

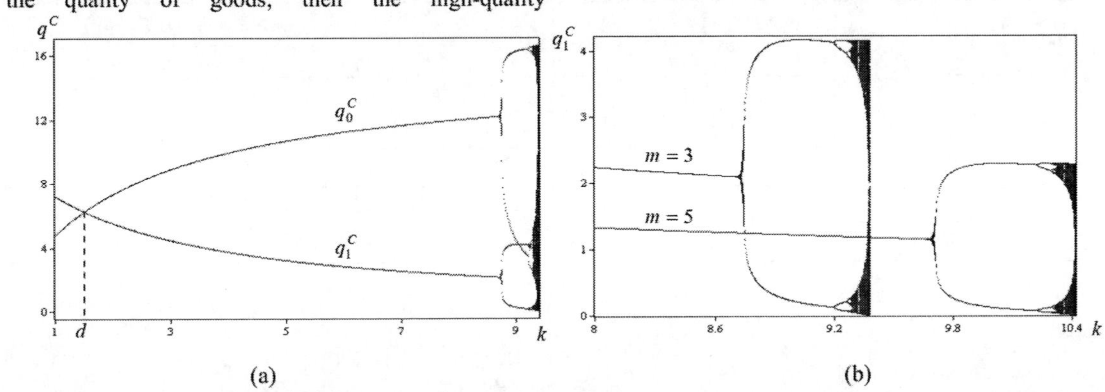

Fig.2. Bifurcation diagrams of supply chains

Table 2. The Stackelberg equilibrium, leader is the low-quality supply chain

Supply chain	k	q^{S_0}	Q^{S_0}	F^{S_0}
low quality	$1 < k < 2 \cdot d$	$\dfrac{m \cdot k \cdot (k+1)}{4 \cdot t \cdot L \cdot d^2}$	$\dfrac{m \cdot (k+1)}{2 \cdot t \cdot L \cdot d}$	$\dfrac{m \cdot k \cdot (k+1)}{4 \cdot d}$
high quality		$\dfrac{m \cdot (k+1) \cdot (2 \cdot d - k)}{4 \cdot t \cdot L \cdot d^2}$		$\dfrac{m \cdot (k+1) \cdot (2 \cdot d - k)^2}{4 \cdot d^2}$

$p_0^{S_0}$	p^{S_0}	$p_1^{S_0}$	α^{S_0}	k^{S_0}	$\alpha^{S_0}\left(k^{S_0}\right)$
$\dfrac{4 \cdot t \cdot L \cdot d^2}{m \cdot k \cdot (k+1)}$	$\dfrac{2 \cdot t \cdot L \cdot d}{\mathrm{m}}$	$\dfrac{4 \cdot k \cdot t \cdot L \cdot d^2}{m \cdot (k+1) \cdot (2 \cdot d - k)}$	$\dfrac{2 \cdot d - k}{2 \cdot d}$	$2 \cdot d$	0%

Let us analyze how a leadership of the low-quality supply chain affects the Cournot equilibrium. The impact on the equilibrium price we express in the form of a coefficient, which is defined as: $S_p^0 = \mathrm{p}^{S_0} / p^C = 2 \cdot d / (k + d)$, where the superscript indicates the leader, and the lower index indicates the influence indicator. The impact of real asymmetries on the equilibrium price is determined by conditions: $\partial S_p^0 / \partial k < 0$, $\partial S_p^0 / \partial d > 0$, $\operatorname{sign}\left(1 - S_p^0\right) = \operatorname{sign}(k - d)$.

The impact on the sales volume of the low-quality supply chain we express in the form of a coefficient, which is defined as: $S^0_{q_0} = q_0^{S_0}/q_0^C = (k+d)^2/(4\cdot d^2)$. Then the profit of the low-quality supply chain can be represented in the form:

$$F_0^{S_0} = S^0_{q_0}\cdot q_0^C\cdot\left(m\cdot S_p^0\cdot p^C - t\cdot d\cdot L\right). \quad (7)$$

The impact of the coefficient $S^0_{q_0}$ on profit (7) is determined by conditions: $\partial S^0_{q_0}/\partial k > 0$, $\partial S^0_{q_0}/\partial d < 0$, $\mathrm{sign}\left(S^0_{q_0} - 1\right) = \mathrm{sign}(k-d)$. At $k=d$, the Stackelberg asymmetry does not affect profit: $F_0^{S_0}\left(q_0^{S_0}\right) = F_0^C\left(q_0^C\right)$.

The impact on the sales volume of the high-quality supply chain we express in the form of a coefficient, which is defined as: $S^0_{q_1} = q_1^{S_0}/q_1^C = (2\cdot d - k)\cdot(k+d)^2/(4\cdot d^3)$. Then the profit of the high-quality supply chain can be represented in the form:

$$F_1^{S_0} = S^0_{q_1}\cdot q_1^C\cdot\left(m\cdot S_p^0\cdot p^C - t\cdot k\cdot L\right). \quad (8)$$

The impact of the coefficient $S^0_{q_1}$ on profit (8) is determined by conditions: $\mathrm{sign}\left(\partial S^0_{q_1}/\partial \mathrm{d}\right) = \mathrm{sign}(k-d)$, $\mathrm{sign}\left(\partial S^0_{q_1}/\partial k\right) = \mathrm{sign}(d-k)$, $S^0_{q_1} \le 1$. At $k=d$, the Stackelberg asymmetry does not affect profit: $F_1^{S_0}\left(q_1^{S_0}\right) = F_1^C\left(q_1^C\right)$.

Assume that the high-quality supply chain is a leader, and the low-quality supply chain is a follower. The new profit function of the high-quality supply chain is: $F_1 = \sqrt{t\cdot d\cdot L\cdot m\cdot(k+1)\cdot q_1} - t\cdot k\cdot L\cdot q_1 \to \max_{q_1}$.

Using the standard procedure, we find the equilibrium indicators (Table 3).

Table 3. The Stackelberg equilibrium, leader is the high-quality supply chain

Supply chain	k	q^{S_1}	Q^{S_1}	F^{S_1}
low quality	$k>1$	$\frac{m\cdot(k+1)\cdot(2\cdot k-d)}{4\cdot t\cdot L\cdot k^2}$	$\frac{m\cdot(k+1)}{2\cdot t\cdot L\cdot k}$	$\frac{m\cdot(k+1)\cdot(2\cdot k-d)^2}{4\cdot k^2}$
high quality		$\frac{m\cdot d\cdot(k+1)}{4\cdot t\cdot L\cdot k^2}$		$\frac{m\cdot d\cdot(k+1)}{4\cdot k}$

$p_0^{S_1}$	p^{S_1}	$p_1^{S_1}$	α^{S_1}
$\frac{4\cdot t\cdot L\cdot k^2}{m\cdot(k+1)\cdot(2\cdot k-d)}$	$\frac{2\cdot t\cdot L\cdot k}{\mathrm{m}}$	$\frac{4\cdot t\cdot L\cdot k^3}{m\cdot d\cdot(k+1)}$	$\frac{d}{2\cdot k}$

To determine the Akerlof point in the Stackelberg equilibrium $\left(q_0^{S_1}, q_1^{S_1}\right)$, it is necessary to specify the lower boundary of a market share of the high-quality good, at which the supply chain will leave the market.

Let us analyze how the leadership of the high-quality supply chain impacted the Cournot equilibrium. The impact on the equilibrium price we express in the form of a coefficient, which is defined as: $S_p^1 = p^{S_1}/p^C = 2\cdot k/(k+d)$. The impact of the real asymmetries on the equilibrium price is determined by conditions: $\partial S_p^1/\partial k > 0$, $\partial S_p^1/\partial d < 0$, $\mathrm{sign}\left(S_p^1 - 1\right) = \mathrm{sign}(k-d)$.

The impact on the sales volume of the low-quality supply chain we express in the form of a coefficient, which is defined as: $S^1_{q_0} = q_0^{S_1}/q_0^C = (2\cdot k - d)\cdot(k+d)^2/(4\cdot k^3)$. Then the profit of the low-quality supply chain can be represented in the form:

$$F_0^{S_1} = S^1_{q_0}\cdot q_0^C\cdot\left(m\cdot S_p^1\cdot p^C - t\cdot d\cdot L\right). \quad (9)$$

The impact of the coefficient $S^1_{q_0}$ on profit (9) is determined by conditions: $\mathrm{sign}\left(\partial S^1_{q_0}/\partial k\right) = \mathrm{sign}(d-k)$, $\mathrm{sign}\left(\partial S^1_{q_0}/\partial d\right) = \mathrm{sign}(k-d)$, $S^1_{q_0} \le 1$. At $k=d$, the Stackelberg asymmetry does not affect profit: $F_0^{S_1}\left(q_0^{S_1}\right) = F_0^C\left(q_0^C\right)$.

The impact on the sales volume of the high-quality supply chain we express in the form of a coefficient, which is defined as: $S^1_{q_1} = q_1^{S_1}/q_1^C = (k+d)^2/(4\cdot k^2)$. Then the profit of the high-quality supply chain can be represented in the form:

$$F_1^{S_1} = S^1_{q_1}\cdot q_1^C\cdot\left(m\cdot S_p^1\cdot \mathrm{p}^C - t\cdot k\cdot L\right). \quad (10)$$

The impact of the coefficient $S^1_{q_1}$ on profit (10) is determined by conditions: $\partial S^1_{q_1}/\partial k < 0$, $\partial S^1_{q_1}/\partial d > 0$, $\mathrm{sign}\left(1 - S^1_{q_1}\right) = \mathrm{sign}(k-d)$. At $k=d$, the Stackelberg asymmetry does not affect profit: $F_1^{S_1}\left(q_1^{S_1}\right) = F_1^C\left(q_1^C\right)$.

Thus, information asymmetries can change an action direction of real asymmetries, and real asymmetries can dwindle ("turn off") an action of information asymmetries.

5 The comparative analysis of equilibria

The relationship between sales volumes and profits in different equilibrium states depends significantly on the superstable point. Let us consider all cases.

1) $1<k<\mathrm{d}$. The supply chains' volumes sales: $q_1^{S_1} > q_1^C > q_1^{S_0} > q_0^C > \{ q_0^{S_0}; q_0^{S_1} \}$, $\operatorname{sign}(q_0^{S_0} - q_0^{S_1}) = \operatorname{sign}(d \cdot (\Phi - 1) - \mathrm{k})$, where $\Phi = (1+\sqrt{5})/2$ – the golden ratio.

The profits: $\mathrm{F}_1^{S_0} > \mathrm{F}_1^{S_1} > \mathrm{F}_1^C > \mathrm{F}_0^{S_0} > \mathrm{F}_0^C > \mathrm{F}_0^{S_1}$. The prices and total sales: $p^{S_0} > p^C > p^{S_1}$, $Q^{S_0} < \mathrm{Q}^C < Q^{S_1}$.

Thanks to the location asymmetry, the high-quality supply chain sells more and receives more profit. Note that the optimal for both supply chains is the equilibrium $(F_0^{S_0}, F_1^{S_0})$. Thus, the profit of the high-quality supply chain on the follower position turned out to be greater than on the leader position.

2) $k=d$. The supply chains' volumes sales: $q_0^C = q_0^{S_0} = q_0^{S_1} = q_1^C = q_1^{S_0} = q_1^{S_1} = m \cdot (k+1)/(4 \cdot t \cdot L \cdot k)$. The profits: $F_0^C = F_0^{S_0} = F_0^{S_1} = F_1^C = F_1^{S_0} = F_1^{S_1} = m \cdot (k+1)/4$. The prices and total sales: $p^C = p^{S_0} = p^{S_1}$, $\mathrm{Q}^C = Q^{S_0} = Q^{S_1}$.

It is obtained that in a state of the superstable Cournot equilibrium supply chains cannot increase their profits with the help of the Stackelberg asymmetry. In this equilibrium state, a transport costs of the supply chains are equalized, and the real asymmetries compensate each other.

3) $k>d$. The volumes sales: $q_0^{S_0} > q_0^C > q_0^{S_1} > q_1^C > \{ q_1^{S_0}, q_1^{S_1} \}$, $\operatorname{sign}(q_1^{S_1} - q_1^{S_0}) = \operatorname{sign}(k - d \cdot \Phi)$. The profits: $F_0^{S_1} > F_0^{S_0} > F_0^C > F_1^{S_1} > F_1^C > F_1^{S_0}$. The prices and total sales: $p^{S_0} < p^C < p^{S_1}$, $Q^{S_0} > \mathrm{Q}^C > Q^{S_1}$.

It is obtained that the high-quality supply chain cannot overcome the Akerlof information asymmetry with the help of the Stackelberg asymmetry. Note that the optimal for both supply chains is the equilibrium $(F_0^{S_1}, F_1^{S_1})$. Thus, the profit of the low-quality supply chain on the follower position turned out to be greater than on the leader position.

Let us illustrate the comparative analysis of the equilibria on a numerical example (Table 4). Data: $\mathrm{m}=9$, $t=1$, $\mathrm{l}=0{,}1$, $d=1{,}8$.

Table 4. Numerical example

Indicators	*States of equilibrium*								
	$1<\mathrm{k}<\mathrm{d}$, $\mathrm{k}=\mathrm{d}\cdot(\Phi-1)$			$k=d=1{,}8$			$\mathrm{k}>\mathrm{d}$, $\mathrm{k}=2{,}2$		
	C	S_0	S_1	*C*	S_0	S_1	*C*	S_0	S_1
q_0^e	1,247	0,816	0,816	1,750	1,750	1,750	1,980	2,444	1,934
q_1^e	2,017	1,825	3,457	1,750	1,750	1,750	1,620	1,556	1,339
Q^e	3,264	2,641	4,273	3,500	3,500	3,500	3,600	4,000	3,273
F_0^e	2,774	2,938	0,693	6,300	6,300	6,300	8,712	8,800	10,056
F_1^e	7,262	9,077	7,691	6,300	6,300	6,300	5,832	4,356	5,891
$p_0(q_0^e)$	0,802	1,226	1,226	0,571	0,571	0,571	0,505	0,409	0,517
p^e	0,647	0,800	0,494	0,800	0,800	0,800	0,889	0,800	0,978
$p_1(q_1^e)$	0,551	0,610	0,322	1,029	1,029	1,029	1,358	1,414	1,643
α	0,618	0,691	0,809	0,500	0,500	0,500	0,450	0,389	0,409
A_0 (A_1)	0,807 (1,174)			1,400 (0,778)			1,760 (0,655)		
S_p^0 (S_p^1)	1,236 (0,764)			1,000 (1,000)			0,900 (1,100)		
$S_{q_0}^0$ $(S_{q_0}^1)$	0,655 (0,655)			1,000 (1,000)			1,235 (0,977)		
$S_{q_1}^0$ $(S_{q_1}^1)$	0,905 (1,714)			1,000 (1,000)			0,960 (0,826)		

6 Conclusions

The article presents a method for assessing the level of informational asymmetry in logistics systems. A qualitative assessment of the level of asymmetry allows to increase the functional stability of the logistics systems as a whole and their individual agents. For this, it is proposed to use a mathematical tools that takes into account that informational asymmetry can change the direction of real asymmetry, and real asymmetry can reduce ("turn off") the effect of informational asymmetry. The article defines a superstable condition for logistics systems in which the decisions of the duopolists in the Cournot and Stackelberg equilibrias coincide.

References

1. F. Hayek, Economica, **4**, (1937)
2. G. Akerlof, Quarterly Journal of Economics, **84**, 3 (1970)
3. X. Vives, Journal of Economic Theory, **34** (1984)
4. E. Gal-or, The Review of Economic Studies, **53**, 1 (1986)
5. L. Roller, B. Sinclair-Desgagne, CIRANO: Scientific Series, https://cirano.qc.ca/pdf/publication/96s-23.pdf (1996)
6. P. Zanchettin, Journal of Economics & Management Strategy, **15** (2006)
7. Y. Wang, Y. Li, M. Liu, Physica A, **373** (2007)
8. A.F. Ledvina, R. Sircar, SSRN Electronic Journal, https://doi.org/10.2139/ssrn.1692957 (2011)
9. A. Nagurney, D. Li, L.S. Nagurney, International Journal of Production Economics, **158** (2014)
10. J.C. Wang, L. Yang, Y.Y. Wang, Z.H. Wang, International Transactions in Operational Research, doi: 10.1111/itor.12237 (2015).
11. S. Brangewitz, J. Manegold, Theoretical Economics Letters, **6** (2016)
12. S.V. Melnikov, Odesa National University Herald. Series Economy, **22** (2017)
13. T. Puu, Chaos, Solitons & Fractals, **1** (1991)

AUTHOR INDEX

AUTHOR INDEX

AUTHOR INDEX

EDP Sciences – Web of Conferences
17, Avenue du Hoggar
Parc d'Activité de Courtabouf
BP 112
F-91944 Les Ulis Cedex A
France

ISBN 978-1-5108-9775-5

CURRAN ASSOCIATES INC.
proceedings.com